STUDENT SOLUTIONS MANUAL

MARK McCOMBS
UNIVERSITY OF NORTH CAROLINA, CHAPEL HILL

PRECALCULUS

ENHANCED WITH GRAPHING UTILITIES

THIRD EDITION

MICHAEL SULLIVAN • MICHAEL SULLIVAN, III

Prentice
Hall

Upper Saddle River, NJ 07458

Editor in Chief: Sally Yagan
Senior Acquisitions Editor: Eric Frank
Associate Editor: Dawn Murrin
Supplement Editor: Aja Shevelew
Assistant Managing Editor: John Matthews
Production Editor: Donna Crilly
Supplement Cover Manager: Paul Gourhan
Supplement Cover Designer: Joanne Alexandris
Manufacturing Buyer: Ilene Kahn

© 2003 by Pearson Education, Inc.
Pearson Education, Inc.
Upper Saddle River, NJ 07458

The author and publisher of this book have used their best efforts in preparing this book. These efforts include the development, research, and testing of the theories and programs to determine their effectiveness. The author and publisher make no warranty of any kind, expressed or implied, with regard to these programs or the documentation contained in this book. The author and publisher shall not be liable in any event for incidental or consequential damages in connection with, or arising out of, the furnishing, performance, or use of these programs.

Printed in the United States of America

10 9 8 7 6 5 4 3 2 1
ISBN 0-13-099481-2

Pearson Education Ltd., *London*
Pearson Education Australia Pty. Ltd., *Sydney*
Pearson Education Singapore, Pte. Ltd.
Pearson Education North Asia Ltd., *Hong Kong*
Pearson Education Canada, Inc., *Toronto*
Pearson Educacíon de Mexico, S.A. de C.V.
Pearson Education—Japan, *Tokyo*
Pearson Education Malaysia, Pte. Ltd.
Pearson Education, *Upper Saddle River, New Jersey*

Contents

Chapter 5   Trigonometric Functions

Chapter 6   Analytic Trigonometry

Chapter 7   Applications of Trigonometric Functions

Chapter 8   Polar Coordinates; Vectors

Chapter 9   Analytic Geometry

Chapter 10   Systems of Equations and Inequalities

Chapter 11   Sequences; Induction; The Binomial Theorem

Chapter 12   Counting and Probability

Chapter 13   A Preview of Calculus: The Limit, Derivative and Integral of a Function

Appendix   Review

Preface

The <u>Student</u> <u>Solutions</u> <u>Manual</u> to accompany <u>Precalculus Enhanced with Graphing Utilities</u>, <u>3rd</u> <u>Edition</u> by Michael Sullivan and Michael Sullivan, III contains detailed solutions to all of the odd-numbered problems in the textbook. TI-83 graphing calculator screens have been included to demonstrate the use of the graphics calculator in solving and in checking solutions to the problems where requested. Every attempt has been made to make this manual as error free as possible. If you have suggestions, error corrections, or comments please feel free to write to me about them.

A number of people need to be recognized for their contributions in the preparation of this manual. Thanks go to Sally Yagan, Dawn Murrin and Audra Walsh at Prentice Hall. Thanks also to Warren Wegner for his meticulous error-checking of the solutions.

I especially wish to thank my mother, Sarah, and my brothers, Kirk and Doug, for their unwavering support and encouragement.

Finally, I am also greatly indebted to the Hank Williams and Muddy Waters for helping me endure the long hours of editing the manuscript.

Mark A. McCombs
Department of Mathematics
Campus Box 3250
University of North Carolina at Chapel Hill
Chapel Hill, NC 27599
mccombs@math.unc.edu

Chapter 1

Graphs

1.1   Rectangular Coordinates; Graphing Utilities

1.   (a)   Quadrant II
      (b)   Positive x-axis
      (c)   Quadrant III
      (d)   Quadrant I
      (e)   Negative y-axis
      (f)   Quadrant IV

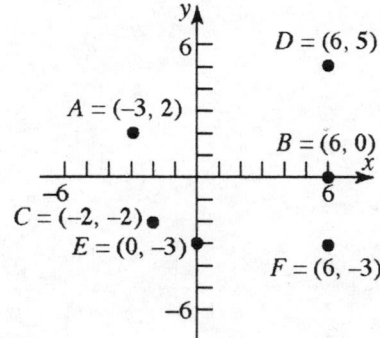

3.   The points will be on a vertical line that is two units to the right of the y-axis.

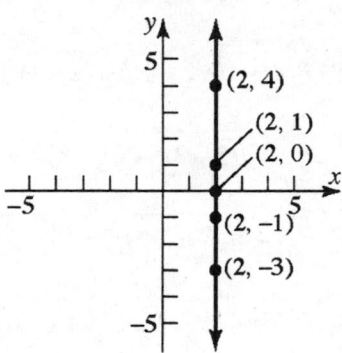

5. $(-1, 4)$; Quadrant II

7. $(3, 1)$; Quadrant I

9.
   $X \min = -11$
   $X \max = 5$
   $X \operatorname{scl} = 1$
   $Y \min = -3$
   $Y \max = 6$
   $Y \operatorname{scl} = 1$

11.
   $X \min = -30$
   $X \max = 50$
   $X \operatorname{scl} = 10$
   $Y \min = -90$
   $Y \max = 50$
   $Y \operatorname{scl} = 10$

13.
$X\min = -10$
$X\max = 110$
$X\operatorname{scl} = 10$
$Y\min = -10$
$Y\max = 160$
$Y\operatorname{scl} = 10$

15.
$X\min = -6$
$X\max = 6$
$X\operatorname{scl} = 2$
$Y\min = -4$
$Y\max = 4$
$Y\operatorname{scl} = 2$

17.
$X\min = -6$
$X\max = 6$
$X\operatorname{scl} = 2$
$Y\min = -1$
$Y\max = 3$
$Y\operatorname{scl} = 1$

19.
$X\min = 3$
$X\max = 9$
$X\operatorname{scl} = 1$
$Y\min = 2$
$Y\max = 10$
$Y\operatorname{scl} = 2$

21. $d(P_1,P_2) = \sqrt{(2-0)^2 + (1-0)^2} = \sqrt{4+1} = \sqrt{5}$

23. $d(P_1,P_2) = \sqrt{(-2-1)^2 + (2-1)^2} = \sqrt{9+1} = \sqrt{10}$

25. $d(P_1,P_2) = \sqrt{(5-3)^2 + (4-8)^2} = \sqrt{2^2 + (-4)^2} = \sqrt{4+16} = \sqrt{20} = 2\sqrt{5}$

27. $d(P_1,P_2) = \sqrt{(6-(-3))^2 + (0-2)^2} = \sqrt{9^2 + (-2)^2} = \sqrt{81+4} = \sqrt{85}$

29. $d(P_1,P_2) = \sqrt{(6-4)^2 + (4-(-3))^2} = \sqrt{2^2 + 7^2} = \sqrt{4+49} = \sqrt{53}$

31. $d(P_1,P_2) = \sqrt{(2.3-(-0.2))^2 + (1.1-0.3)^2} = \sqrt{(2.5)^2 + (0.8)^2}$
$= \sqrt{6.25+0.64} = \sqrt{6.89} \approx 2.625$

33. $d(P_1,P_2) = \sqrt{(0-a)^2 + (0-b)^2} = \sqrt{a^2 + b^2}$

35. $P_1 = (1,3); P_2 = (5,15)$
$d(P_1,P_2) = \sqrt{(5-1)^2 + (15-3)^2}$
$= \sqrt{(4)^2 + (12)^2}$
$= \sqrt{16+144}$
$= \sqrt{160} = 4\sqrt{10}$

37. $P_1 = (-4,6); P_2 = (4,-8)$
$d(P_1,P_2) = \sqrt{(4-(-4))^2 + (-8-6)^2}$
$= \sqrt{(8)^2 + (-14)^2}$
$= \sqrt{64+196}$
$= \sqrt{260} = 2\sqrt{65}$

39. $A = (-2,5)$, $B = (1,3)$, $C = (-1,0)$

$$d(A,B) = \sqrt{(1-(-2))^2 + (3-5)^2} = \sqrt{3^2 + (-2)^2}$$
$$= \sqrt{9+4} = \sqrt{13}$$
$$d(B,C) = \sqrt{(-1-1)^2 + (0-3)^2} = \sqrt{(-2)^2 + (-3)^2}$$
$$= \sqrt{4+9} = \sqrt{13}$$
$$d(A,C) = \sqrt{(-1-(-2))^2 + (0-5)^2} = \sqrt{1^2 + (-5)^2}$$
$$= \sqrt{1+25} = \sqrt{26}$$

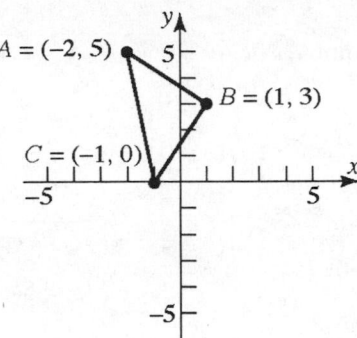

Verifying that $\triangle$ ABC is a right triangle by the Pythagorean Theorem:
$$[d(A,B)]^2 + [d(B,C)]^2 = [d(A,C)]^2$$
$$\left(\sqrt{13}\right)^2 + \left(\sqrt{13}\right)^2 = \left(\sqrt{26}\right)^2$$
$$13 + 13 = 26$$
$$26 = 26$$

The area of a triangle is $A = \frac{1}{2} \cdot bh$. In this problem,

$$A = \frac{1}{2} \cdot [d(A,B)] \cdot [d(B,C)] = \frac{1}{2} \cdot \sqrt{13} \cdot \sqrt{13} = \frac{1}{2} \cdot 13 = \frac{13}{2} \text{ square units}$$

41. $A = (-5,3)$, $B = (6,0)$, $C = (5,5)$

$$d(A,B) = \sqrt{(6-(-5))^2 + (0-3)^2} = \sqrt{11^2 + (-3)^2}$$
$$= \sqrt{121+9} = \sqrt{130}$$
$$d(B,C) = \sqrt{(5-6)^2 + (5-0)^2} = \sqrt{(-1)^2 + 5^2}$$
$$= \sqrt{1+25} = \sqrt{26}$$
$$d(A,C) = \sqrt{(5-(-5))^2 + (5-3)^2} = \sqrt{10^2 + 2^2}$$
$$= \sqrt{100+4} = \sqrt{104}$$

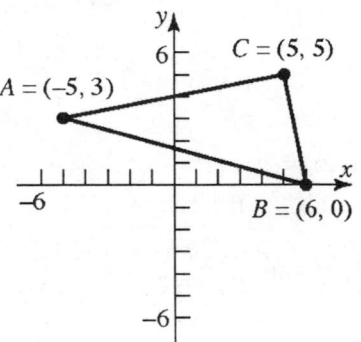

Verifying that $\triangle$ ABC is a right triangle by the Pythagorean Theorem:
$$[d(A,C)]^2 + [d(B,C)]^2 = [d(A,B)]^2$$
$$\left(\sqrt{104}\right)^2 + \left(\sqrt{26}\right)^2 = \left(\sqrt{130}\right)^2$$
$$104 + 26 = 130$$
$$130 = 130$$

The area of a triangle is $A = \frac{1}{2} \cdot bh$. In this problem,

$$A = \frac{1}{2} \cdot [d(A,C)] \cdot [d(B,C)] = \frac{1}{2} \cdot \sqrt{104} \cdot \sqrt{26} = \frac{1}{2} \cdot \sqrt{2704} = \frac{1}{2} \cdot 52 = 26 \text{ square units}$$

43. $A = (4,-3),\ B = (0,-3),\ C = (4,2)$

$$d(A,B) = \sqrt{(0-4)^2 + (-3-(-3))^2} = \sqrt{(-4)^2 + 0^2}$$
$$= \sqrt{16+0} = \sqrt{16} = 4$$
$$d(B,C) = \sqrt{(4-0)^2 + (2-(-3))^2} = \sqrt{4^2 + 5^2}$$
$$= \sqrt{16+25} = \sqrt{41}$$
$$d(A,C) = \sqrt{(4-4)^2 + (2-(-3))^2} = \sqrt{0^2 + 5^2}$$
$$= \sqrt{0+25} = \sqrt{25} = 5$$

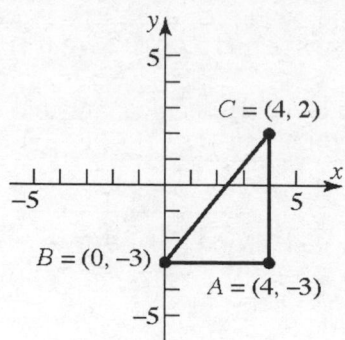

Verifying that $\Delta\ ABC$ is a right triangle by the Pythagorean Theorem:
$$[d(A,B)]^2 + [d(A,C)]^2 = [d(B,C)]^2$$

$$4^2 + 5^2 = \left(\sqrt{41}\right)^2 \rightarrow 16+25 = 41 \rightarrow 41 = 41$$

The area of a triangle is $A = \dfrac{1}{2} \cdot bh$. In this problem,

$$A = \frac{1}{2} \cdot [d(A,B)] \cdot [d(A,C)] = \frac{1}{2} \cdot 4 \cdot 5 = 10 \text{ square units}$$

45.   All points having an x-coordinate of 2 are of the form (2, y). Those which are 5 units from $(-2, -1)$ are:

$$\sqrt{(2-(-2))^2 + (y-(-1))^2} = 5 \rightarrow \sqrt{4^2 + (y+1)^2} = 5$$
$$\text{Squaring both sides}: \ 4^2 + (y+1)^2 = 25$$
$$16 + y^2 + 2y + 1 = 25$$
$$y^2 + 2y - 8 = 0$$
$$(y+4)(y-2) = 0$$
$$y = -4 \ \text{ or } \ y = 2$$

Therefore, the points are (2, –4) or (2, 2).

47.   All points on the x-axis are of the form (x, 0). Those which are 5 units from (4, –3) are:
$$\sqrt{(x-4)^2 + (0-(-3))^2} = 5 \rightarrow \sqrt{(x-4)^2 + 3^2} = 5$$
$$\text{Squaring both sides}: \ (x-4)^2 + 9 = 25$$
$$x^2 - 8x + 16 + 9 = 25 \rightarrow x^2 - 8x = 0$$
$$x(x-8) = 0 \rightarrow x = 0 \ \text{ or } \ x = 8$$

Therefore, the points are (0, 0) or (8, 0).

49.   The coordinates of the midpoint are:
$$(x,y) = \left(\frac{x_1 + x_2}{2}, \frac{y_1 + y_2}{2}\right) = \left(\frac{5+3}{2}, \frac{4+2}{2}\right) = \left(\frac{8}{2}, \frac{6}{2}\right) = (4,3)$$

51. The coordinates of the midpoint are:

$$(x, y) = \left(\frac{x_1 + x_2}{2}, \frac{y_1 + y_2}{2} \right) = \left(\frac{-3 + 6}{2}, \frac{2 + 0}{2} \right) = \left(\frac{3}{2}, \frac{2}{2} \right) = \left(\frac{3}{2}, 1 \right)$$

53. The coordinates of the midpoint are:

$$(x, y) = \left(\frac{x_1 + x_2}{2}, \frac{y_1 + y_2}{2} \right) = \left(\frac{4 + 6}{2}, \frac{-3 + 1}{2} \right) = \left(\frac{10}{2}, \frac{-2}{2} \right) = (5, -1)$$

55. The coordinates of the midpoint are:

$$(x, y) = \left(\frac{x_1 + x_2}{2}, \frac{y_1 + y_2}{2} \right) = \left(\frac{-0.2 + 2.3}{2}, \frac{0.3 + 1.1}{2} \right) = \left(\frac{2.1}{2}, \frac{1.4}{2} \right) = (1.05, 0.7)$$

57. The coordinates of the midpoint are:

$$(x, y) = \left(\frac{x_1 + x_2}{2}, \frac{y_1 + y_2}{2} \right) = \left(\frac{a + 0}{2}, \frac{b + 0}{2} \right) = \left(\frac{a}{2}, \frac{b}{2} \right)$$

59. The midpoint of AB is: $D = \left(\dfrac{0 + 6}{2}, \dfrac{0 + 0}{2} \right) = (3, 0)$

The midpoint of AC is: $E = \left(\dfrac{0 + 4}{2}, \dfrac{0 + 4}{2} \right) = (2, 2)$

The midpoint of BC is: $F = \left(\dfrac{6 + 4}{2}, \dfrac{0 + 4}{2} \right) = (5, 2)$

$$d(C, D) = \sqrt{(0 - 4)^2 + (3 - 4)^2} = \sqrt{(-4)^2 + (-1)^2} = \sqrt{16 + 1} = \sqrt{17}$$

$$d(B, E) = \sqrt{(2 - 6)^2 + (2 - 0)^2} = \sqrt{(-4)^2 + 2^2} = \sqrt{16 + 4} = \sqrt{20} = 2\sqrt{5}$$

$$d(A, F) = \sqrt{(2 - 0)^2 + (5 - 0)^2} = \sqrt{2^2 + 5^2} = \sqrt{4 + 25} = \sqrt{29}$$

61. $d(P_1, P_2) = \sqrt{(-4 - 2)^2 + (1 - 1)^2} = \sqrt{(-6)^2 + 0^2} = \sqrt{36} = 6$

$d(P_2, P_3) = \sqrt{(-4 - (-4))^2 + (-3 - 1)^2} = \sqrt{0^2 + (-4)^2} = \sqrt{16} = 4$

$d(P_1, P_3) = \sqrt{(-4 - 2)^2 + (-3 - 1)^2} = \sqrt{(-6)^2 + (-4)^2} = \sqrt{36 + 16} = \sqrt{52} = 2\sqrt{13}$

Since $\left[d(P_1, P_2) \right]^2 + \left[d(P_2, P_3) \right]^2 = \left[d(P_1, P_3) \right]^2$, the triangle is a right triangle.

63. $d(P_1, P_2) = \sqrt{(0 - (-2))^2 + (7 - (-1))^2} = \sqrt{2^2 + 8^2} = \sqrt{4 + 64} = \sqrt{68} = 2\sqrt{17}$

$d(P_2, P_3) = \sqrt{(3 - 0)^2 + (2 - 7)^2} = \sqrt{3^2 + (-5)^2} = \sqrt{9 + 25} = \sqrt{34}$

$d(P_1, P_3) = \sqrt{(3 - (-2))^2 + (2 - (-1))^2} = \sqrt{5^2 + 3^2} = \sqrt{25 + 9} = \sqrt{34}$

Since $d(P_2, P_3) = d(P_1, P_3)$, the triangle is isosceles.

Since $\left[d(P_1, P_3) \right]^2 + \left[d(P_2, P_3) \right]^2 = \left[d(P_1, P_2) \right]^2$, the triangle is also a right triangle.

Therefore, the triangle is an isosceles right triangle.

65. Using the Pythagorean Theorem:

$$90^2 + 90^2 = d^2$$
$$8100 + 8100 = d^2$$
$$16200 = d^2$$
$$d = \sqrt{16200} = 90\sqrt{2} \approx 127.28 \text{ feet}$$

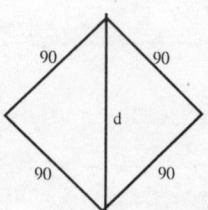

67. (a) First: (90, 0), Second: (90, 90)
         Third: (0, 90)

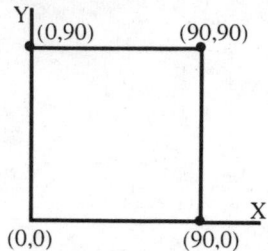

(b) Using the distance formula:
$$d = \sqrt{(310 - 90)^2 + (15 - 90)^2}$$
$$= \sqrt{220^2 + (-75)^2}$$
$$= \sqrt{54025} \approx 232.4 \text{ feet}$$

(c) Using the distance formula:
$$d = \sqrt{(300 - 0)^2 + (300 - 90)^2}$$
$$= \sqrt{300^2 + 210^2}$$
$$= \sqrt{134100} \approx 366.2 \text{ feet}$$

69. The Intrepid heading east moves a distance $30t$ after t hours. The truck heading south moves a distance $40t$ after t hours. Their distance apart after t hours is:

$$d = \sqrt{(30t)^2 + (40t)^2}$$
$$= \sqrt{900t^2 + 1600t^2}$$
$$= \sqrt{2500t^2}$$
$$= 50t$$

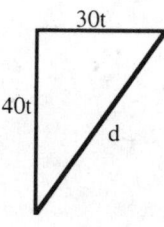

Graphs

1.2 Introduction to Graphing Equations

1. $y = x^4 - \sqrt{x}$

    $0 = 0^4 - \sqrt{0}$ $1 = 1^4 - \sqrt{1}$ $0 = (-1)^4 - \sqrt{-1}$

    $0 = 0$ $1 \neq 0$ $0 \neq 1 - \sqrt{-1}$

    $(0, 0)$ is on the graph of the equation.

3. $y^2 = x^2 + 9$

    $3^2 = 0^2 + 9$ $0^2 = 3^2 + 9$ $0^2 = (-3)^2 + 9$

    $9 = 9$ $0 \neq 18$ $0 \neq 18$

    $(0, 3)$ is on the graph of the equation.

5. $x^2 + y^2 = 4$

    $0^2 + 2^2 = 4$ $(-2)^2 + 2^2 = 4$ $\sqrt{2}^2 + \sqrt{2}^2 = 4$

    $4 = 4$ $8 \neq 4$ $4 = 4$

    $(0, 2)$ and $\left(\sqrt{2}, \sqrt{2}\right)$ are on the graph of the equation.

7. $(-1, 0), (1, 0)$ 9. $\left(-\dfrac{\pi}{2}, 0\right), \left(\dfrac{\pi}{2}, 0\right), (0, 1)$

11. $(0, 0)$ 13. $(-4, 0), (-1, 0), (4, 0), (0, -3)$

15. $(-1.5, 0), (1.5, 0), (0, -2)$ 17. none

19. $y = 5x + 4$

    $2 = 5a + 4$

    $-2 = 5a \Rightarrow a = -\dfrac{2}{5}$

21. $2x + 3y = 6$

    $2a + 3b = 6$

23. $y = x + 2$

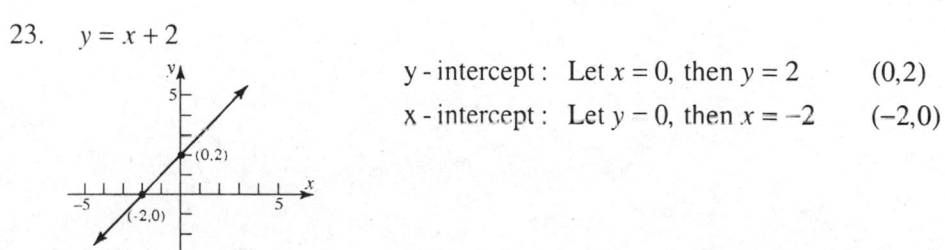

    y-intercept : Let $x = 0$, then $y = 2$ $(0, 2)$

    x-intercept : Let $y = 0$, then $x = -2$ $(-2, 0)$

25. $y = 2x + 8$

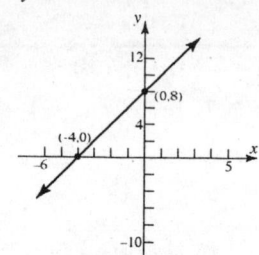

y - intercept :   Let $x = 0$, then $y = 8$ (0,8)

x - intercept :   Let $y = 0$, then $x = -4$ (−4,0)

27. $y = x^2 - 1$

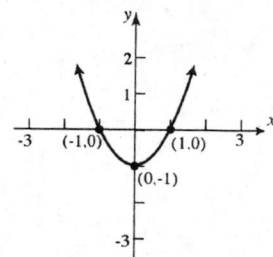

y - intercept :   Let $x = 0$, then $y = -1$ (0,−1)

x - intercept :   Let $y = 0$, then $x = \pm 1$ (−1,0);(1,0)

29. $y = -x^2 + 4$

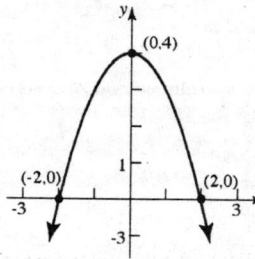

y - intercept :   Let $x = 0$, then $y = 4$ (0,4)

x - intercept :   Let $y = 0$, then $x = \pm 2$ (−2,0);(2,0)

31. $2x + 3y = 6$

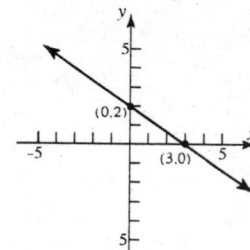

y - intercept :   Let $x = 0$, then $y = 2$ (0,2)

x - intercept :   Let $y = 0$, then $x = 3$ (3,0)

33. $9x^2 + 4y = 36$

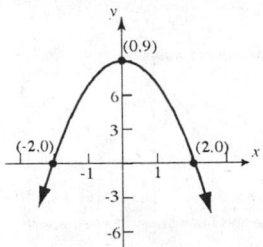

y-intercept :   Let $x = 0$, then $y = 9$ $(0,9)$

x-intercept :   Let $y = 0$, then $x = \pm 2$ $(-2,0); (2,0)$

35. $y = 2x - 13$
Use VALUE and ZERO (or ROOT) on the graph of $y_1 = 2x - 13$.

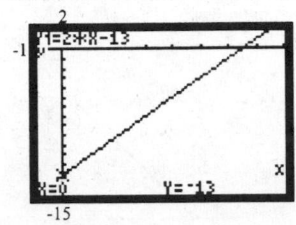
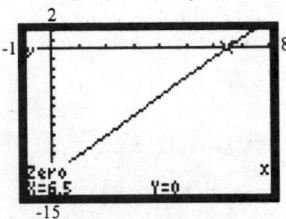

y-intercept: $(0,-13)$ x-intercept: $(6.5,0)$

37. $y = 2x^2 - 15$
Use VALUE and ZERO (or ROOT) on the graph of $y_1 = 2x^2 - 15$.

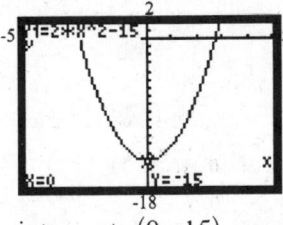
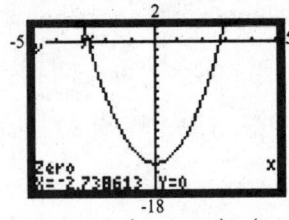
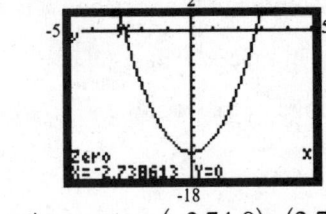

y-intercept: $(0,-15)$ x-intercepts: $(-2.74,0)$, $(2.74,0)$

39. $3x - 2y = 43$
Use VALUE and ZERO (or ROOT) on the graph of $y_1 = 1.5x - 43/2$.

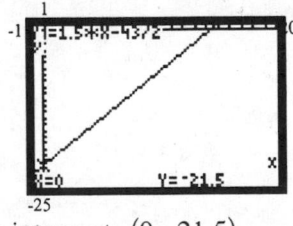
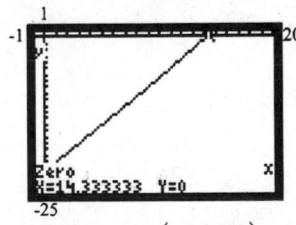

y-intercept: $(0,-21.5)$ x-intercept: $(14.33,0)$

9

41. $5x^2 + 3y = 37$

Use VALUE and ZERO (or ROOT) on the graph of $y_1 = (-5/3)x^2 + 37/3$.

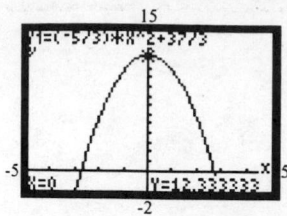 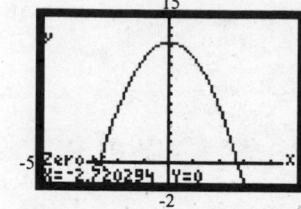 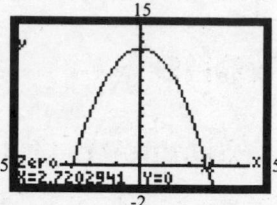

y-intercept: $(0, 12.33)$ x-intercepts: $(-2.72, 0), \ (2.72, 0)$

43.   (a)

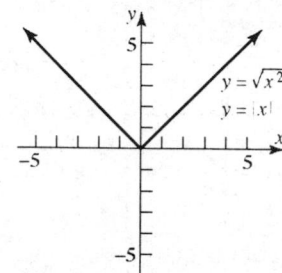 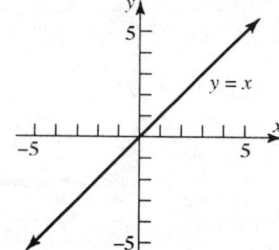 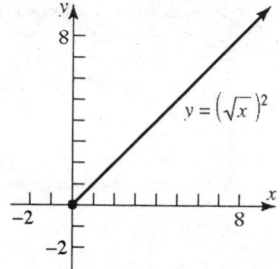

(b)   Since $\sqrt{x^2} = |x|$, then for all x, the graphs of $y = \sqrt{x^2}$ and $y = |x|$ are the same.

(c)   For $y = \left(\sqrt{x}\right)^2$, the domain of the variable x is $x \geq 0$; for $y = x$, the domain of the variable x is all real numbers. Thus, $\left(\sqrt{x}\right)^2 = x$ only for $x \geq 0$.

(d)   For $y = \sqrt{x^2}$, the range of the variable y is $y \geq 0$; for $y = x$, the range of the variable y is all real numbers. Also, $\sqrt{x^2} = x$ only if $x \geq 0$.

45.   Answers will vary

Graphs

1.3   Symmetry; Graphing Key Equations; Circles

1.

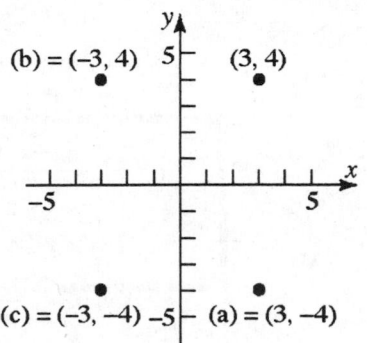

3.

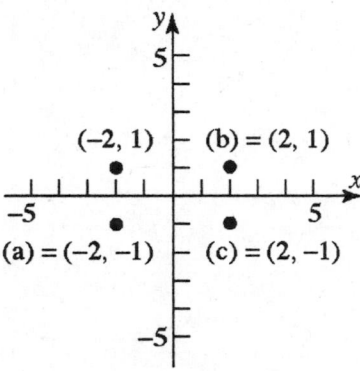

5.

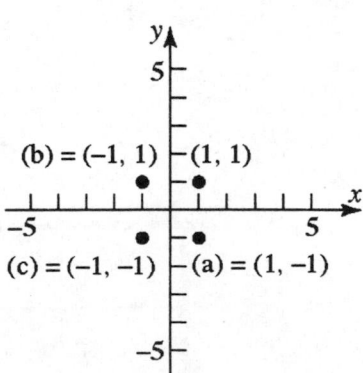

7.

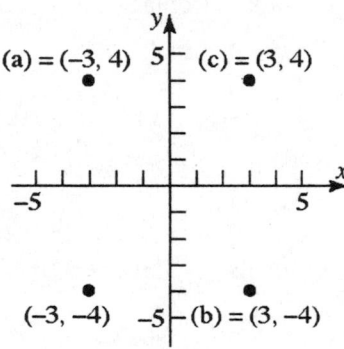

9.

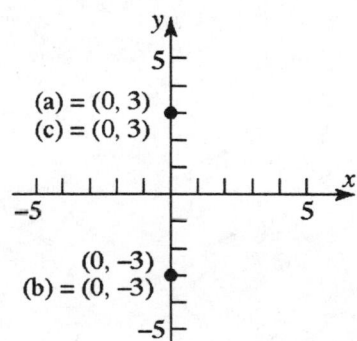

11.   symmetric to the x-axis, y-axis and origin

13.   symmetric to the y-axis

15.   symmetric to the x-axis

17.   not symmetric to x-axis, y-axis, or origin

19.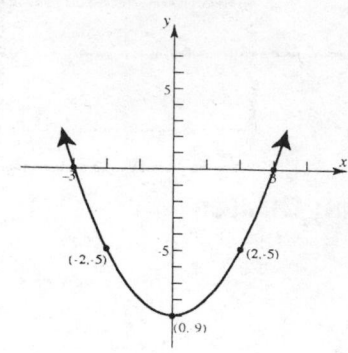

21.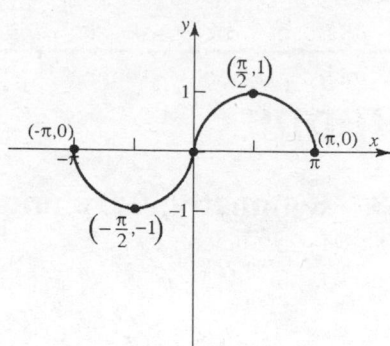

23. $x^2 = y + 5$

 y - intercept :   Let $x = 0$, then $y = -5$ $(0,-5)$

 x - intercept :   Let $y = 0$, then $x = \pm\sqrt{5}$ $\left(-\sqrt{5},0\right),\left(\sqrt{5},0\right)$

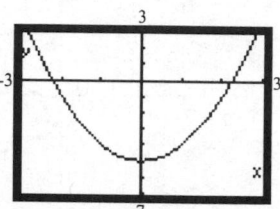

 Test for symmetry:

 x - axis :   Replace y by $-y$ so $x^2 = -y + 5$, which is not equivalent to $x^2 = y + 5$.

 y - axis :   Replace x by $-x$ so $(-x)^2 = y$ or $x^2 = y + 5$, which is equivalent to $x^2 = y + 5$.

 Origin :   Replace x by $-x$ and y by $-y$ so $(-x)^2 = -y + 5$ or $x^2 = -y + 5$,

            which is not equivalent to $x^2 = y + 5$.

 Therefore, the graph is symmetric with respect to the y - axis.

25. $y = 3x$

 y - intercept: Let $x = 0$, then $y = 0$ $(0,0)$

 x - intercept: Let $y = 0$, then $x = 0$ $(0,0)$

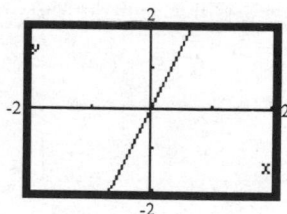

 Test for symmetry:

 x - axis :   Replace y by $-y$ so $-y = 3x$, which is not equivalent to $y = 3x$.

 y - axis :   Replace x by $-x$ so $y = 3(-x)$ or $y = -3x$, which is not equivalent to $y = 3x$.

 Origin :   Replace x by $-x$ and y by $-y$ so $-y = 3(-x) \Rightarrow y = 3x$,

            which is equivalent to $y = 3x$.

 Therefore, the graph is symmetric with respect to the origin.

27. $x^2 + y - 9 = 0$

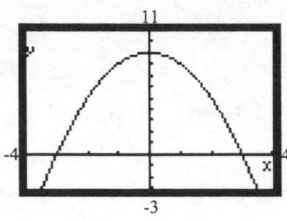

y - intercept : Let $x = 0$, then $y = 9$ $(0,9)$

x - intercept : Let $y = 0$, then $x = \pm 3$ $(-3,0),(3,0)$

Test for symmetry:

x - axis :   Replace y by $-y$ so $x^2 + (-y) - 9 = 0 \Rightarrow x^2 - y - 9 = 0$,

which is not equivalent to $x^2 + y - 9 = 0$.

y - axis :   Replace x by $-x$ so $(-x)^2 + y - 9 = 0 \Rightarrow x^2 + y - 9 = 0$,

which is equivalent to $x^2 + y - 9 = 0$.

Origin :   Replace x by $-x$ and y by $-y$ so $(-x)^2 + (-y) - 9 = 0 \Rightarrow x^2 - y - 9 = 0$,

which is not equivalent to $x^2 + y - 9 = 0$.

Therefore, the graph is symmetric with respect to the y-axis.

29. $y = x^3 - 27$

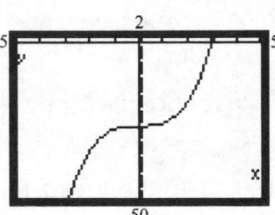

y - intercept : Let $x = 0$, then $y = 0^3 - 27$

$y = -27$ $(0,-27)$

x - intercept : Let $y = 0$, then $0 = x^3 - 27$

$x^3 = 27 \Rightarrow x = 3$ $(3,0)$

Test for symmetry:

x - axis :   Replace y by $-y$ so $-y = x^3 - 27$, which is not equivalent to $y = x^3 - 27$.

y - axis :   Replace x by $-x$ so $y = (-x)^3 - 27 \Rightarrow y = -x^3 - 27$,

which is not equivalent to $y = x^3 - 27$.

Origin :   Replace x by $-x$ and y by $-y$ so $-y = (-x)^3 - 27$

$\Rightarrow y = x^3 + 27$, which is not equivalent to $y = x^3 - 27$.

Therefore, the graph is not symmetric to the x-axis, the y-axis, or the origin.

31. $y = x^2 - 3x - 4$

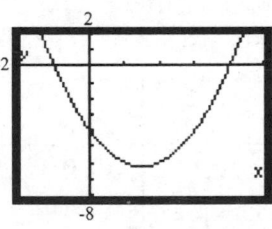

y - intercept: Let $x = 0$, then $y = 0^2 - 3(0) - 4$

$y = -4$ $(0,-4)$

x - intercept: Let $y = 0$, then $0 = x^2 - 3x - 4$

$(x - 4)(x + 1) = 0$

$x = 4$ $x = -1$ $(4,0), (-1,0)$

Test for symmetry:

x - axis: Replace y by $-y$ so $-y = x^2 - 3x - 4$,

which is not equivalent to $y = x^2 - 3x - 4$.

y-axis: Replace x by $-x$ so $y = (-x)^2 - 3(-x) - 4 \Rightarrow y = x^2 + 3x - 4,$

which is not equivalent to $y = x^2 - 3x - 4$.

Origin: Replace x by $-x$ and y by $-y$ so $-y = (-x)^2 - 3(-x) - 4$

$\Rightarrow y = -x^2 - 3x + 4$, which is not equivalent to $y = x^2 - 3x - 4$.

Therefore, the graph is not symmetric to the x-axis, the y-axis, or the origin.

33. $y = \dfrac{3x}{x^2 + 9}$

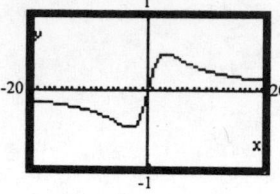

y-intercept : Let $x = 0$, then $y = \dfrac{0}{0 + 9}$

$$y = 0 \qquad (0, 0)$$

x-intercept : Let $y = 0$, then $0 = \dfrac{3x}{x^2 + 9}$

$$3x = 0 \Rightarrow x = 0 \quad (0, 0)$$

Test for symmetry:

x-axis: Replace y by $-y$ so $-y = \dfrac{3x}{x^2 + 9}$, which is not equivalent to $y = \dfrac{3x}{x^2 + 9}$.

y-axis: Replace x by $-x$ so $y = \dfrac{3(-x)}{(-x)^2 + 9} \Rightarrow y = \dfrac{-3x}{x^2 + 9},$

which is not equivalent to $y = \dfrac{3x}{x^2 + 9}$.

Origin: Replace x by $-x$ and y by $-y$ so $-y = \dfrac{-3x}{(-x)^2 + 9}$

$\Rightarrow y = \dfrac{3x}{x^2 + 9}$, which is equivalent to $y = \dfrac{3x}{x^2 + 9}$.

Therefore, the graph is symmetric with respect to the origin.

35. $y = x^3$ 37. $y = \sqrt{x}$

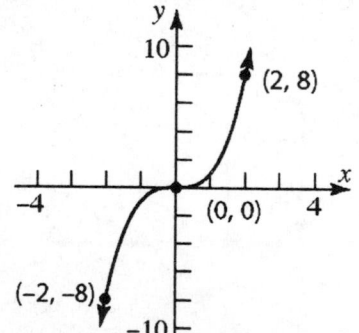   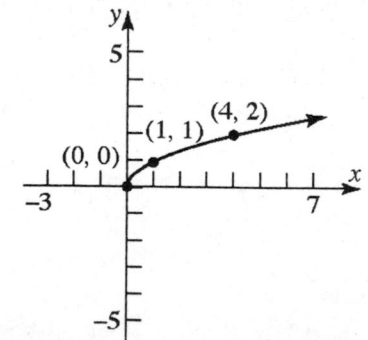

x-intercept : (0,0);   y-intercept : (0,0)   x-intercept : (0,0);   y-intercept : (0,0)

39. Center = (2, 1)

Radius = distance from (0,1) to (2,1)

$$= \sqrt{(2-0)^2 + (1-1)^2} = \sqrt{4} = 2$$

$$(x-2)^2 + (y-1)^2 = 4$$

41. Center = midpoint of (1,2) and (4,2)

$$= \left(\frac{1+4}{2}, \frac{2+2}{2}\right) = \left(\frac{5}{2}, 2\right)$$

Radius = distance from $\left(\frac{5}{2}, 2\right)$ to (4,2)

$$= \sqrt{\left(4 - \frac{5}{2}\right)^2 + (2-2)^2} = \sqrt{\frac{9}{4}} = \frac{3}{2}$$

$$\left(x - \frac{5}{2}\right)^2 + (y-2)^2 = \frac{9}{4}$$

43. $(x-h)^2 + (y-k)^2 = r^2$

$(x-0)^2 + (y-0)^2 = 2^2$

$x^2 + y^2 = 4$

General form:

$x^2 + y^2 - 4 = 0$

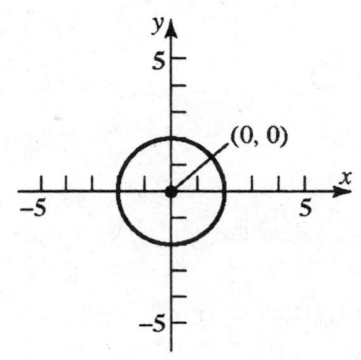

45. $(x-h)^2 + (y-k)^2 = r^2$

$(x-0)^2 + (y-2)^2 = 2^2$

$x^2 + (y-2)^2 = 4$

General form:

$x^2 + y^2 - 4y + 4 = 4$

$x^2 + y^2 - 4y = 0$

47. $(x-h)^2 + (y-k)^2 = r^2$

$(x-4)^2 + (y-(-3))^2 = 5^2$

$(x-4)^2 + (y+3)^2 = 25$

General form:

$x^2 - 8x + 16 + y^2 + 6y + 9 = 25$

$x^2 + y^2 - 8x + 6y = 0$

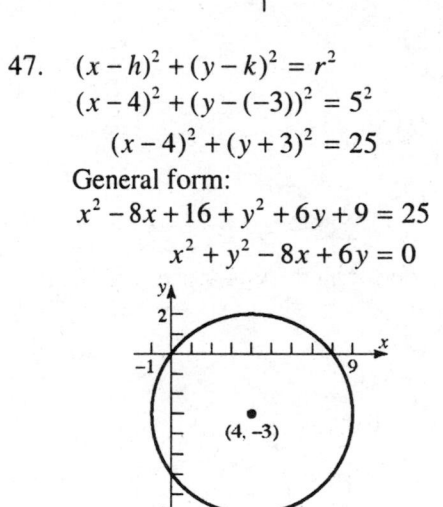

49. $(x-h)^2 + (y-k)^2 = r^2$

$(x-(-2))^2 + (y-1)^2 = 4^2$

$(x+2)^2 + (y-1)^2 = 16$

General form:

$x^2 + 4x + 4 + y^2 - 2y + 1 = 16$

$x^2 + y^2 + 4x - 2y - 11 = 0$

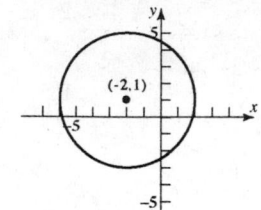

51. $(x-h)^2 + (y-k)^2 = r^2$

$\left(x-\dfrac{1}{2}\right)^2 + (y-0)^2 = \left(\dfrac{1}{2}\right)^2$

$\left(x-\dfrac{1}{2}\right)^2 + y^2 = \dfrac{1}{4}$

General form:

$x^2 - x + \dfrac{1}{4} + y^2 = \dfrac{1}{4}$

$x^2 + y^2 - x = 0$

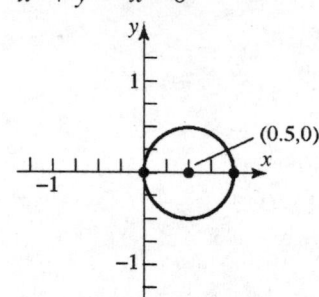

53. $x^2 + y^2 = 25$

$x^2 + y^2 = 5^2$

Center : (0,0)

Radius = 5

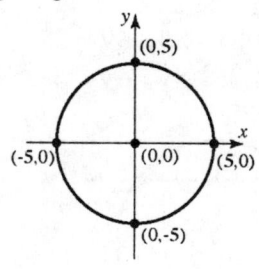

55. $(x-2)^2 + y^2 = 4$

$(x-2)^2 + y^2 = 2^2$

Center: (2,0)

Radius = 2

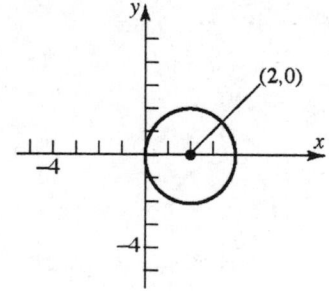

57. $x^2 + y^2 + 4x - 4y - 1 = 0$

$x^2 + 4x + y^2 - 4y = 1$

$(x^2 + 4x + 4) + (y^2 - 4y + 4) = 1 + 4 + 4$

$(x+2)^2 + (y-2)^2 = 3^2$

Center: (−2,2)

Radius = 3

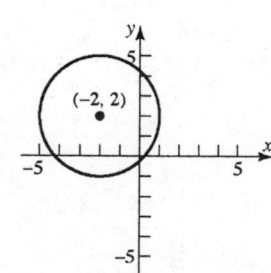

59. $x^2 + y^2 - x + 2y + 1 = 0$

$$x^2 - x + y^2 + 2y = -1$$

$$\left(x^2 - x + \frac{1}{4}\right) + (y^2 + 2y + 1) = -1 + \frac{1}{4} + 1$$

$$\left(x - \frac{1}{2}\right)^2 + (y + 1)^2 = \left(\frac{1}{2}\right)^2$$

Center: $\left(\frac{1}{2}, -1\right)$

Radius = $\frac{1}{2}$

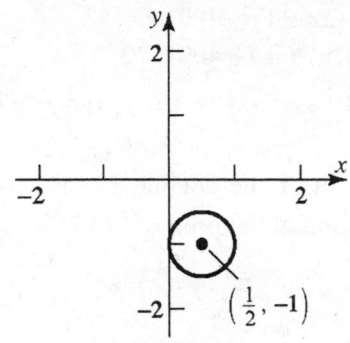

61. $2x^2 + 2y^2 - 12x + 8y - 24 = 0$

$$x^2 + y^2 - 6x + 4y = 12$$

$$x^2 - 6x + y^2 + 4y = 12$$

$$(x^2 - 6x + 9) + (y^2 + 4y + 4) = 12 + 9 + 4$$

$$(x - 3)^2 + (y + 2)^2 = 5^2$$

Center: $(3, -2)$
Radius = 5

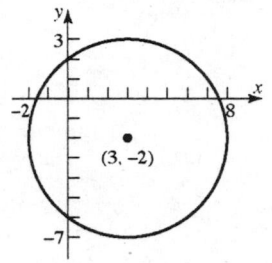

63. Center at $(0,0)$; containing point $(-2, 3)$.

$$r = \sqrt{(-2 - 0)^2 + (3 - 0)^2} = \sqrt{4 + 9} = \sqrt{13}$$

Equation: $(x - 0)^2 + (y - 0)^2 = \left(\sqrt{13}\right)^2$

$$x^2 + y^2 = 13 \Rightarrow x^2 + y^2 - 13 = 0$$

65. Center at $(2,3)$; tangent to the x-axis.

$r = 3$

Equation: $(x - 2)^2 + (y - 3)^2 = 3^2$

$$x^2 - 4x + 4 + y^2 - 6y + 9 = 9 \Rightarrow x^2 + y^2 - 4x - 6y + 4 = 0$$

67. Endpoints of a diameter are $(1,4)$ and $(-3,2)$.
The center is at the midpoint of that diameter:

Center: $\left(\dfrac{1 + (-3)}{2}, \dfrac{4 + 2}{2}\right) = (-1, 3)$

Radius: $r = \sqrt{(1 - (-1))^2 + (4 - 3)^2} = \sqrt{4 + 1} = \sqrt{5}$

Equation: $(x - (-1))^2 + (y - 3)^2 = \left(\sqrt{5}\right)^2$

$$x^2 + 2x + 1 + y^2 - 6y + 9 = 5 \Rightarrow x^2 + y^2 + 2x - 6y + 5 = 0$$

69. (c)   71.   (b)

73. $(x + 3)^2 + (y - 1)^2 = 16$ 75. $(x - 2)^2 + (y - 2)^2 = 9$

77. (b), (c), (e) and (g)

79. $x^2 + y^2 + 2x + 4y - 4091 = 0$
   $x^2 + 2x + y^2 + 4y - 4091 = 0$

   $x^2 + 2x + 1 + y^2 + 4y + 4 = 4091 + 5 \Rightarrow (x+1)^2 + (y+2)^2 = 4096$

   The circle representing Earth has center $(-1,-2)$ and radius $= \sqrt{4096} = 64$

   So the radius of the satellite's orbit is $64 + 0.6 = 64.6$ units.

   The equation of the orbit is $(x+1)^2 + (y+2)^2 = (64.6)^2$

   $$x^2 + y^2 + 2x + 4y - 4168.6 = 0$$

Graphs

1.4   Solving Equations Using a Graphing Utility

1. $x^3 - 4x + 2 = 0$; Use ZERO (or ROOT) on the graph of $y_1 = x^3 - 4x + 2$.

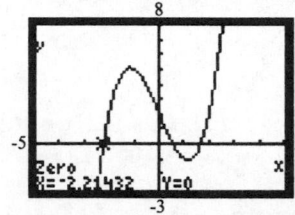 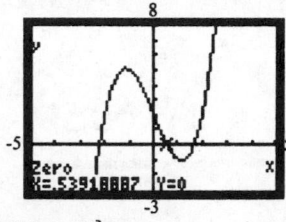 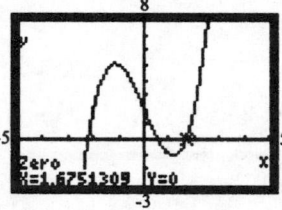

The solution set is $\{-2.21, 0.54, 1.68\}$.

3. $-2x^4 + 5 = 3x - 2$; Use ZERO (or ROOT) on the graph of $y_1 = -2x^4 - 3x + 7$.

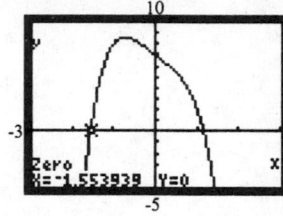 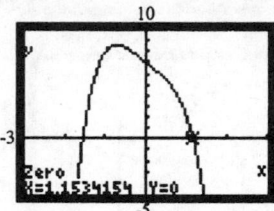

The solution set is $\{-1.55, 1.15\}$.

5. $x^4 - 2x^3 + 3x - 1 = 0$; Use ZERO (or ROOT) on the graph of $y_1 = x^4 - 2x^3 + 3x - 1$.

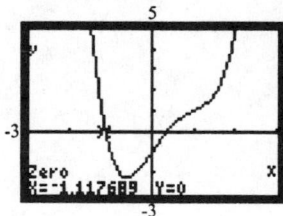 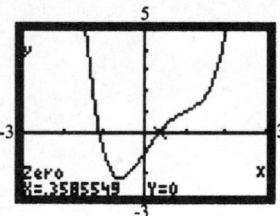

The solution set is $\{-1.12, 0.36\}$.

7. $-x^3 - \dfrac{5}{3}x^2 + \dfrac{7}{2}x + 2 = 0;$

Use ZERO (or ROOT) on the graph of $y_1 = -x^3 - (5/3)x^2 + (7/2)x + 2$.

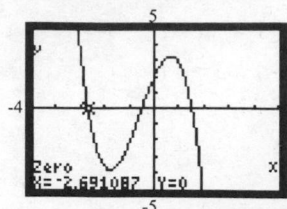 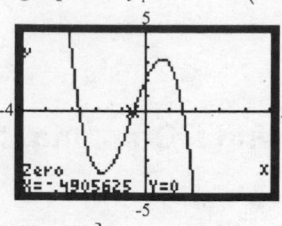 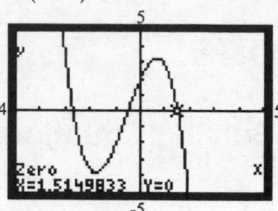

The solution set is $\{-2.69, -0.49, 1.51\}$.

9. $-\dfrac{2}{3}x^4 - 2x^3 + \dfrac{5}{2}x = -\dfrac{2}{3}x^2 + \dfrac{1}{2}$

Use ZERO (or ROOT) on the graph of $y_1 = -(2/3)x^4 - 2x^3 + (2/3)x^2 + (5/2)x - 1/2$.

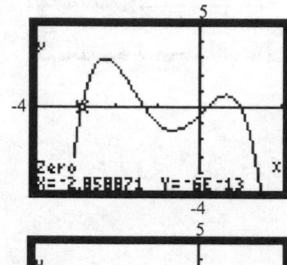 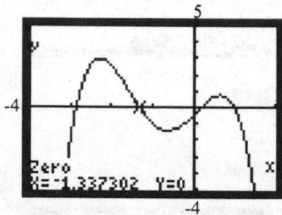

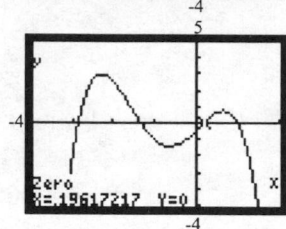 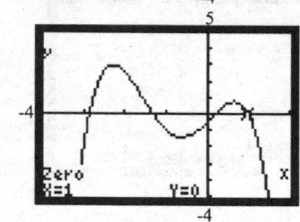

The solution set is $\{-2.86, -1.34, 0.20, 1.00\}$.

11. $x^4 - 5x^2 + 2x + 11 = 0;$ Use ZERO (or ROOT) on the graph of $y_1 = x^4 - 5x^2 + 2x + 11$.

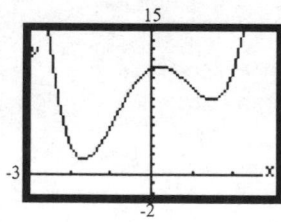

There are no real solutions.

13.
$$3x + 2 = 6$$
$$3x + 2 - 2 = 6 - 2$$
$$3x = 4$$
$$\frac{3x}{3} = \frac{4}{3} \Rightarrow x = \frac{4}{3}$$
The solution set is $\left\{\dfrac{4}{3}\right\}$.

15.
$$6 - x = 2x + 12$$
$$6 - x - 6 = 2x + 12 - 6$$
$$-x = 2x + 6$$
$$-x - 2x = 2x + 6 - 2x$$
$$-3x = 6 \Rightarrow \frac{-3x}{-3} = \frac{6}{-3} \Rightarrow x = -2$$
The solution set is $\{-2\}$.

17. $2(3 + 2x) = 3(x - 4)$
$6 + 4x = 3x - 12$
$6 + 4x - 6 = 3x - 12 - 6$
$4x = 3x - 18$
$4x - 3x = 3x - 18 - 3x$
$x = -18$
The solution set is $\{-18\}$.

19. $8x - (2x + 1) = 3x - 13$
$8x - 2x - 1 = 3x - 13$
$6x - 1 = 3x - 13$
$6x - 1 + 1 = 3x - 13 + 1$
$6x = 3x - 12$
$6x - 3x = 3x - 12 - 3x$
$3x = -12$
$\dfrac{3x}{3} = \dfrac{-12}{3} \Rightarrow x = -4$
The solution set is $\{-4\}$.

21. $\dfrac{2}{3}p = \dfrac{1}{2}p + \dfrac{1}{3}$
$6\left(\dfrac{2}{3}p\right) = 6\left(\dfrac{1}{2}p + \dfrac{1}{3}\right)$
$4p = 3p + 2$
$4p - 3p = 3p + 2 - 3p$
$p = 2$
The solution set is $\{2\}$.

23. $0.9t = 0.4 + 0.1t$
$0.9t - 0.1t = 0.4 + 0.1t - 0.1t$
$0.8t = 0.4$
$\dfrac{0.8t}{0.8} = \dfrac{0.4}{0.8}$
$t = 0.5$
The solution set is $\{0.5\}$.

25. $\dfrac{x+1}{3} + \dfrac{x+2}{7} = 5$
$(21)\left(\dfrac{x+1}{3} + \dfrac{x+2}{7}\right) = (5)(21)$
$(21)\left(\dfrac{x+1}{3}\right) + (21)\left(\dfrac{x+2}{7}\right) = 105$
$7(x+1) + (3)(x+2) = 105$
$7x + 7 + 3x + 6 = 105$
$10x + 13 = 105$
$10x + 13 - 13 = 105 - 13$
$10x = 92$
$\dfrac{10x}{10} = \dfrac{92}{10}$
$x = 9.2$
The solution set is $\{9.2\}$.

27. $\dfrac{5}{y} + \dfrac{4}{y} = 3$
$y\left(\dfrac{5}{y} + \dfrac{4}{y}\right) = y(3)$
$5 + 4 = 3y$
$9 = 3y \Rightarrow \dfrac{9}{3} = \dfrac{3y}{3} \Rightarrow y = 3$

and since y = 3 does not cause a denominator to equal zero, the solution set is {3}.

29. $$(x+7)(x-1)=(x+1)^2$$
$$x^2-x+7x-7=x^2+2x+1$$
$$x^2+6x-7=x^2+2x+1$$
$$x^2+6x-7-x^2=x^2+2x+1-x^2$$
$$6x-7=2x+1$$
$$6x-7-2x=2x+1-2x$$
$$4x-7=1$$
$$4x-7+7=1+7$$
$$4x=8$$
$$\frac{4x}{4}=\frac{8}{4}\Rightarrow x=2$$
The solution set is $\{2\}$.

31. $x^2-9x=0$
$$x(x-9)=0$$
$$x=0$$
or $x-9=0\Rightarrow x=9$
The solution set is $\{0,9\}$.

33. $$z^2+z-6=0$$
$$(z+3)(z-2)=0$$
$$z+3=0\Rightarrow z=-3$$
or $z-2=0\Rightarrow z=2$
The solution set is $\{-3,2\}$.

35. $$2x^2-5x-3=0$$
$$(2x+1)(x-3)=0$$
$$2x+1=0\Rightarrow x=-\frac{1}{2}$$
or $x-3=0\Rightarrow x=3$
The solution set is $\left\{-\frac{1}{2},3\right\}$

37. $$3t^2-48=0$$
$$3(t^2-16)=0$$
$$3(t+4)(t-4)=0$$
$$t+4=0\Rightarrow t=-4$$
or $t-4=0\Rightarrow t=4$
The solution set is $\{-4,4\}$.

39. $$x(x+8)+12=0$$
$$x^2+8x+12=0$$
$$(x+6)(x+2)=0\Rightarrow x=-6,x=-2$$
The solution set is $\{-6,-2\}$.

41. $$4x^2+9=12x$$
$$4x^2-12x+9=0$$
$$(2x-3)^2=0\Rightarrow x=\frac{3}{2}$$
The solution set is $\left\{\frac{3}{2}\right\}$.

43. $2x^2-x=15$
$$2x^2-x-15=0$$
$$(2x+5)(x-3)=0$$
$$2x+5=0\Rightarrow x=-\frac{5}{2}$$
or $x-3=0\Rightarrow x=3$
The solution set is $\left\{-\frac{5}{2},3\right\}$.

45.
$$\frac{4(x-2)}{x-3}+\frac{3}{x}=\frac{-3}{x(x-3)}$$

$$x(x-3)\left(\frac{4(x-2)}{x-3}+\frac{3}{x}\right)=\left(\frac{-3}{x(x-3)}\right)x(x-3)$$

$$x(x-3)\left(\frac{4(x-2)}{x-3}\right)+x(x-3)\left(\frac{3}{x}\right)=-3$$

$$x(4(x-2))+(x-3)(3)=-3$$

$$4x^2-8x+3x-9=-3$$

$$4x^2-5x-6=0$$

$$(4x+3)(x-2)=0$$

$$4x+3=0\Rightarrow x=-\frac{3}{4}$$

or $x-2\Rightarrow x=2$

Since neither of these values causes a denominator to equal zero, the solution set is $\left\{-\frac{3}{4},2\right\}$.

47. $x^2-4x+2=0$
$a=1,\quad b=-4,\quad c=2$

$$x=\frac{-(-4)\pm\sqrt{(-4)^2-4(1)(2)}}{2(1)}$$

$$=\frac{4\pm\sqrt{16-8}}{2}=\frac{4\pm\sqrt{8}}{2}$$

$$=\frac{4\pm2\sqrt{2}}{2}=2\pm\sqrt{2}$$

The solution set is $\left\{2-\sqrt{2},2+\sqrt{2}\right\}$.

49. $x^2-4x-1=0$
$a=1,\quad b=-4,\quad c=-1$

$$x=\frac{-(-4)\pm\sqrt{(-4)^2-4(1)(-1)}}{2(1)}$$

$$=\frac{4\pm\sqrt{16+4}}{2}=\frac{4\pm\sqrt{20}}{2}$$

$$=\frac{4\pm2\sqrt{5}}{2}=2\pm\sqrt{5}$$

The solution set is $\left\{2-\sqrt{5},2+\sqrt{5}\right\}$.

51. $2x^2-5x+3=0$
$a=2,\quad b=-5,\quad c=3$

$$x=\frac{-(-5)\pm\sqrt{(-5)^2-4(2)(3)}}{2(2)}$$

$$=\frac{5\pm\sqrt{25-24}}{4}=\frac{5\pm1}{4}$$

The solution set is $\left\{1,\ \frac{3}{2}\right\}$.

53. $4y^2-y+2=0$
$a=4,\quad b=-1,\quad c=2$

$$y=\frac{-(-1)\pm\sqrt{(-1)^2-4(4)(2)}}{2(4)}$$

$$=\frac{1\pm\sqrt{1-32}}{8}=\frac{1\pm\sqrt{-31}}{8}$$

No real solution.

55. $4x^2=1-2x$
$$4x^2+2x-1=0$$
$a=4,\quad b=2,\quad c=-1$

$$x=\frac{-2\pm\sqrt{2^2-4(4)(-1)}}{2(4)}=\frac{-2\pm\sqrt{4+16}}{8}=\frac{-2\pm\sqrt{20}}{8}$$

$$=\frac{-2\pm2\sqrt{5}}{8}=\frac{-1\pm\sqrt{5}}{4}$$

The solution set is $\left\{\frac{-1-\sqrt{5}}{4},\frac{-1+\sqrt{5}}{4}\right\}$.

57. $4x^2 = 9x + 2$
$4x^2 - 9x - 2 = 0$
$a = 4, \quad b = -9, \quad c = -2$

$$x = \frac{-(-9) \pm \sqrt{(-9)^2 - 4(4)(-2)}}{2(4)} = \frac{9 \pm \sqrt{81 + 32}}{8} = \frac{9 \pm \sqrt{113}}{8}$$

The solution set is $\left\{ \dfrac{9 - \sqrt{113}}{8}, \dfrac{9 + \sqrt{113}}{8} \right\}$.

59. $\dfrac{3}{4}x^2 - \dfrac{1}{4}x - \dfrac{1}{2} = 0$

$4\left(\dfrac{3}{4}x^2 - \dfrac{1}{4}x - \dfrac{1}{2} \right) = (0)(4)$

$3x^2 - x - 2 = 0$

$a = 3, \quad b = -1, \quad c = -2$

$x = \dfrac{-(-1) \pm \sqrt{(-1)^2 - 4(3)(-2)}}{2(3)}$

$= \dfrac{1 \pm \sqrt{1 + 24}}{6} = \dfrac{1 \pm \sqrt{25}}{6} = \dfrac{1 \pm 5}{6}$

$\Rightarrow x = \dfrac{1 + 5}{6}$ or $x = \dfrac{1 - 5}{6}$

$x = \dfrac{6}{6} = 1$ or $x = \dfrac{-4}{6} = -\dfrac{2}{3}$

The solution set is $\left\{ -\dfrac{2}{3}, 1 \right\}$.

61. $4 - \dfrac{1}{x} - \dfrac{2}{x^2} = 0$

$\left(x^2 \right)\left(4 - \dfrac{1}{x} - \dfrac{2}{x^2} \right) = (0)\left(x^2 \right)$

$4x^2 - x - 2 = 0$

$a = 4, \quad b = -1, \quad c = -2$

$x = \dfrac{-(-1) \pm \sqrt{(-1)^2 - 4(4)(-2)}}{2(4)}$

$= \dfrac{1 \pm \sqrt{1 + 32}}{8} = \dfrac{1 \pm \sqrt{33}}{8}$

Since neither of these values causes a denominator to equal zero, the solution set is

$\left\{ \dfrac{1 + \sqrt{33}}{8}, \dfrac{1 - \sqrt{33}}{8} \right\}$.

63. $\sqrt{y + 3} = 5$

$\left(\sqrt{y + 3} \right)^2 = 5^2$

$y + 3 = 25 \Rightarrow y = 22$

Check : $\sqrt{22 + 3} = \sqrt{25} = 5$

The solution is $y = 22$.

65. $\sqrt{2t - 1} = 1$

$\left(\sqrt{2t - 1} \right)^2 = 1^2$

$2t - 1 = 1 \Rightarrow 2t = 2 \Rightarrow t = 1$

Check: $\sqrt{2(1) - 1} = \sqrt{1} = 1$

The solution is $t = 1$.

67. $\sqrt{3t + 1} = -6$
Since the principal square root is always a non-negative number, this equation has no real solution.

69. $\sqrt[3]{1 - 2x} - 3 = 0$

$\sqrt[3]{1 - 2x} = 3$

$\left(\sqrt[3]{1 - 2x} \right)^3 = 3^3$

$1 - 2x = 27 \Rightarrow -2x = 26 \Rightarrow x = -13$

Check : $\sqrt[3]{1 - 2(-13)} - 3 = \sqrt[3]{27} - 3 = 0$

The solution is $x = -13$.

71. $\sqrt{15-2x} = x$

$\left(\sqrt{15-2x}\right)^2 = x^2$

$15 - 2x = x^2 \Rightarrow x^2 + 2x - 15 = 0$

$(x+5)(x-3) = 0 \Rightarrow x = -5$ or $x = 3$

Check $x = -5$: $\sqrt{15-2(-5)} = \sqrt{25}$

$= 5 \neq -5$

Check $x = 3$: $\sqrt{15-2(3)} = \sqrt{9} = 3 = 3$

The solution is $x = 3$.

73. $x = 2\sqrt{x-1}$

$x^2 = \left(2\sqrt{x-1}\right)^2$

$x^2 = 4(x-1) \Rightarrow x^2 = 4x - 4$

$x^2 - 4x + 4 = 0 \Rightarrow (x-2)^2 = 0 \Rightarrow x = 2$

Check: $2 = 2\sqrt{2-1} \Rightarrow 2 = 2$

The solution is $x = 2$.

75. $\sqrt{x^2 - x - 4} = x + 2$

$\left(\sqrt{x^2-x-4}\right)^2 = (x+2)^2$

$x^2 - x - 4 = x^2 + 4x + 4$

$-8 = 5x \Rightarrow -\dfrac{8}{5} = x$

Check

$x = -\dfrac{8}{5} : \sqrt{\left(-\dfrac{8}{5}\right)^2 - \left(-\dfrac{8}{5}\right) - 4} = \left(-\dfrac{8}{5}\right) + 2$

$\sqrt{\dfrac{64}{25} + \dfrac{8}{5} - 4} = \dfrac{2}{5} \Rightarrow \sqrt{\dfrac{64 + 40 - 100}{25}} = \dfrac{2}{5}$

$\sqrt{\dfrac{4}{25}} = \dfrac{2}{5} \Rightarrow \dfrac{2}{5} = \dfrac{2}{5}$, The solution is $x = -\dfrac{8}{5}$.

77. $3 + \sqrt{3x+1} = x$

$\sqrt{3x+1} = x - 3$

$\left(\sqrt{3x+1}\right)^2 = (x-3)^2$

$3x + 1 = x^2 - 6x + 9$

$0 = x^2 - 9x + 8$

$(x-1)(x-8) = 0$

$x = 1$ or $x = 8$

Check $x = 1$: $3 + \sqrt{3(1)+1}$

$= 3 + \sqrt{4} = 5 \neq 1$

Check $x = 8$: $3 + \sqrt{3(8)+1}$

$= 3 + \sqrt{25} = 8 = 8$

The solution is $x = 8$.

79. $\sqrt{2x+3} - \sqrt{x+1} = 1$

$\sqrt{2x+3} = 1 + \sqrt{x+1}$

$\left(\sqrt{2x+3}\right)^2 = \left(1 + \sqrt{x+1}\right)^2$

$2x + 3 = 1 + 2\sqrt{x+1} + x + 1$

$x + 1 = 2\sqrt{x+1}$

$(x+1)^2 = \left(2\sqrt{x+1}\right)^2$

$x^2 + 2x + 1 = 4(x+1)$

$x^2 + 2x + 1 = 4x + 4$

$x^2 - 2x - 3 = 0$

$(x+1)(x-3) = 0 \Rightarrow x = -1$ or $x = 3$

Check $x = -1$: $\sqrt{2(-1)+3} - \sqrt{-1+1}$

$= \sqrt{1} - \sqrt{0} = 1 - 0 = 1 = 1$

Check $x = 3$: $\sqrt{2(3)+3} - \sqrt{3+1}$

$= \sqrt{9} - \sqrt{4} = 3 - 2 = 1 = 1$

The solution set is $\{-1, 3\}$.

81. $\sqrt{3x+1} - \sqrt{x-1} = 2$

$$\sqrt{3x+1} = 2 + \sqrt{x-1}$$

$$\left(\sqrt{3x+1}\right)^2 = \left(2 + \sqrt{x-1}\right)^2$$

$$3x+1 = 4 + 4\sqrt{x-1} + x - 1$$

$$2x-2 = 4\sqrt{x-1}$$

$$(2x-2)^2 = \left(4\sqrt{x-1}\right)^2$$

$$4x^2 - 8x + 4 = 16(x-1)$$

$$x^2 - 2x + 1 = 4x - 4$$

$$x^2 - 6x + 5 = 0$$

$$(x-1)(x-5) = 0 \Rightarrow x = 1 \ \text{ or } \ x = 5$$

Check $x = 1$: $\sqrt{3(1)+1} - \sqrt{1-1}$

$$= \sqrt{4} - \sqrt{0} = 2 - 0 = 2 = 2$$

Check $x = 5$: $\sqrt{3(5)+1} - \sqrt{5-1}$

$$= \sqrt{16} - \sqrt{4} = 4 - 2 = 2 = 2$$

The solution set is $\{1, 5\}$.

83. $(3x+1)^{1/2} = 4$

$$\left((3x+1)^{1/2}\right)^2 = (4)^2$$

$$3x+1 = 16 \Rightarrow 3x = 15 \Rightarrow x = 5$$

Check

$x = 5$: $(3(5)+1)^{1/2} = 4$

$$16^{1/2} = 4 \Rightarrow 4 = 4$$

The solution is $x = 5$.

85. $(5x-2)^{1/3} = 2$

$$\left((5x-2)^{1/3}\right)^3 = (2)^3$$

$$5x-2 = 8 \Rightarrow 5x = 10 \Rightarrow x = 2$$

Check

$x = 2$: $(5(2)-2)^{1/3} = 2$

$$8^{1/3} = 2 \Rightarrow 2 = 2$$

The solution is $x = 2$.

87. $(x^2+9)^{1/2} = 5$

$$\left((x^2+9)^{1/2}\right)^2 = (5)^2$$

$$x^2 + 9 = 25 \Rightarrow x^2 = 16$$

$$x = -4 \ \text{ or } \ x = 4$$

Check

$x = -4$: $\left((-4)^2 + 9\right)^{1/2} = 5$

$$25^{1/2} = 5 \Rightarrow 5 = 5$$

$x = 4$: $\left((4)^2 + 9\right)^{1/2} = 5$

$$25^{1/2} = 5 \Rightarrow 5 = 5$$

The solution set is $\{-4, 4\}$.

89. $|2x+3| = 5$

$$2x+3 = 5 \ \text{ or } \ 2x+3 = -5$$

$$2x = 2 \ \text{ or } \quad 2x = -8$$

$$x = 1 \ \text{ or } \quad x = -4$$

The solution set is $\{-4, 1\}$.

91. $|1-4t| + 8 = 13 \Rightarrow |1-4t| = 5$

$$1-4t = 5 \ \text{ or } \ 1-4t = -5$$

$$-4t = 4 \ \text{ or } \quad -4t = -6$$

$$t = -1 \ \text{ or } \quad t = \frac{3}{2}$$

The solution set is $\left\{-1, \dfrac{3}{2}\right\}$.

93. $|-2x| = 8$
$\quad -2x = 8 \quad$ or $\quad -2x = -8$
$\quad\quad x = -4 \quad$ or $\quad\quad x = 4$
The solution set is $\{-4, 4\}$.

95. $\dfrac{2}{3}|x| = 9$

$|x| = \dfrac{27}{2} \Rightarrow x = \dfrac{27}{2}$ or $x = -\dfrac{27}{2}$

The solution set is $\left\{-\dfrac{27}{2}, \dfrac{27}{2}\right\}$.

97. $\left|\dfrac{x}{3} + \dfrac{2}{5}\right| = 2$

$\quad \dfrac{x}{3} + \dfrac{2}{5} = 2 \quad$ or $\quad \dfrac{x}{3} + \dfrac{2}{5} = -2$

$\quad 5x + 6 = 30 \quad$ or $\quad 5x + 6 = -30$

$\quad\quad 5x = 24 \quad$ or $\quad\quad 5x = -36$

$\quad\quad x = \dfrac{24}{5} \quad$ or $\quad\quad x = -\dfrac{36}{5}$

The solution set is $\left\{-\dfrac{36}{5}, \dfrac{24}{5}\right\}$.

99. $|u - 2| = -\dfrac{1}{2}$

impossible, since absolute value always yields a non-negative number.

101. $|x^2 - 9| = 0$

$\quad x^2 - 9 = 0$

$\quad x^2 = 9 \Rightarrow x = \pm 3$
The solution set is $\{-3, 3\}$.

103. $|x^2 - 2x| = 3$

$\quad x^2 - 2x = 3 \quad$ or $\quad x^2 - 2x = -3$

$\quad x^2 - 2x - 3 = 0 \quad$ or $\quad x^2 - 2x + 3 = 0$

$(x - 3)(x + 1) = 0 \quad$ or $\quad x = \dfrac{2 \pm \sqrt{4 - 12}}{2} = \dfrac{2 \pm \sqrt{-8}}{2} \Rightarrow$ no real solution

$x = 3$ or $x = -1 \quad$ The solution set is $\{-1, 3\}$.

105. $|x^2 + x - 1| = 1$

$\quad x^2 + x - 1 = 1 \quad$ or $\quad x^2 + x - 1 = -1$

$\quad x^2 + x - 2 = 0 \quad$ or $\quad x^2 + x = 0$

$(x - 1)(x + 2) = 0 \quad$ or $\quad x(x + 1) = 0$

$\quad x = 1, x = -2 \quad$ or $\quad x = 0, x = -1$

The solution set is $\{-2, -1, 0, 1\}$.

107. $\quad ax - b = c, \quad a \neq 0$
$ax - b + b = c + b$
$\quad\quad ax = c + b$

$\quad\quad \dfrac{ax}{a} = \dfrac{c + b}{a} \Rightarrow x = \dfrac{c + b}{a}$

109. $\quad \dfrac{x}{a} + \dfrac{x}{b} = c, \quad a \neq 0, b \neq 0, a \neq -b$

$ab\left(\dfrac{x}{a} + \dfrac{x}{b}\right) = ab \cdot c$

$\quad\quad bx + ax = abc$

$\quad\quad x(a + b) = abc$

$\quad\quad \dfrac{x(a + b)}{a + b} = \dfrac{abc}{a + b} \Rightarrow x = \dfrac{abc}{a + b}$

110. $\quad \dfrac{a}{x} + \dfrac{b}{x} = c, \quad c \neq 0$

$x\left(\dfrac{a}{x} + \dfrac{b}{x}\right) = x \cdot c$

$\quad\quad a + b = cx$

$\quad\quad \dfrac{a + b}{c} = \dfrac{cx}{c} \Rightarrow x = \dfrac{a + b}{c}$

$\quad\quad$ such that $a \neq -b$

111.   Solving for R:

$$\frac{1}{R} = \frac{1}{R_1} + \frac{1}{R_2}$$

$$RR_1R_2\left(\frac{1}{R}\right) = RR_1R_2\left(\frac{1}{R_1} + \frac{1}{R_2}\right)$$

$$R_1R_2 = RR_2 + RR_1$$

$$R_1R_2 = R(R_2 + R_1)$$

$$\frac{R_1R_2}{R_2 + R_1} = \frac{R(R_2 + R_1)}{R_2 + R_1}$$

$$\frac{R_1R_2}{R_2 + R_1} = R$$

113.   Solving for R:

$$F = \frac{mv^2}{R}$$

$$RF = R\left(\frac{mv^2}{R}\right)$$

$$RF = mv^2 \Rightarrow \frac{RF}{F} = \frac{mv^2}{F} \Rightarrow R = \frac{mv^2}{F}$$

115.   Graph the equations $y_1 = \sqrt{x}\,/4 + x\,/1100$ and $y_2 = 4$; then use INTERSECT to find the x-coordinate of the points of intersection:

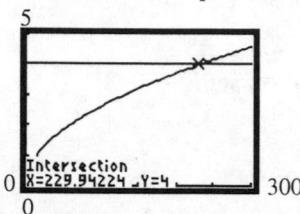

The distance to the water's surface is approximately 229.94 feet.

117. $x^2 = 9$ is not equivalent to $x = 3$ since $x^2 = 9$ also has $x = -3$ as a solution.

$x = \sqrt{9}$ is equivalent to $x = 3$ since the equations have equivalent solutions

$(x-1)(x-2) = (x-1)^2$ is not equivalent to $x - 2 = x - 1$ since the first equation has solution set $\{1\}$, but the second equation has no solution.

119 - 121.   Answers will vary.

Graphs

1.5 Solving Inequalities

1. $[0, 2]$ $0 \leq x \leq 2$

3. $(-1, 2)$ $-1 < x < 2$

5. $[0, 3)$ $0 \leq x < 3$

7. (a) $6 < 8$
   (b) $-2 < 0$
   (c) $9 < 15$
   (d) $-6 > -10$

9. (a) $7 > 0$
   (b) $-1 > -8$
   (c) $12 > -9$
   (d) $-8 < 6$

11. (a) $2x + 4 < 5$
    (b) $2x - 4 < -3$
    (c) $6x + 3 < 6$
    (d) $-4x - 2 > -4$

13. $[0, 4]$

15. $[4, 6)$

17. $[4, \infty)$

19. $(-\infty, -4)$

21. $2 \leq x \leq 5$

23. $-3 < x < -2$

25. $x \geq 4$

27. $x < -3$

29. If $x < 5$, then $x - 5 < 0$.

31. If $x > -4$, then $x + 4 > 0$.

33. If $x \geq -4$, then $3x \geq -12$.

35. If $x > 6$, then $-2x < -12$.

37. If $x \leq 5$, then $-4x \geq -20$.

39. If $2x < 6$, then $x < 3$.

41.   If $-\dfrac{1}{2}x \le 3$, then $x \ge -6$.

43. $x + 1 < 5$

$x + 1 - 1 < 5 - 1 \Rightarrow x < 4$

$\{x \mid x < 4\}$ or $(-\infty, 4)$

45. $1 - 2x \le 3$

$-2x \le 2 \Rightarrow x \ge -1$

$\{x \mid x \ge -1\}$ or $[-1, +\infty)$

47. $3x - 7 > 2$

$3x > 9 \Rightarrow x > 3$

$\{x \mid x > 3\}$ or $(3, +\infty)$

49. $3x - 1 \ge 3 + x$

$2x \ge 4 \Rightarrow x \ge 2$

$\{x \mid x \ge 2\}$ or $[2, +\infty)$

51. $-2(x + 3) < 8$

$-2x - 6 < 8$

$-2x < 14 \Rightarrow x > -7$

$\{x \mid x > -7\}$ or $(-7, +\infty)$

53. $4 - 3(1 - x) \le 3$

$4 - 3 + 3x \le 3$

$3x + 1 \le 3$

$3x \le 2 \Rightarrow x \le \dfrac{2}{3}$

$\left\{x \mid x \le \dfrac{2}{3}\right\}$ or $\left(-\infty, \dfrac{2}{3}\right]$

55. $\dfrac{1}{2}(x - 4) > x + 8$

$\dfrac{1}{2}x - 2 > x + 8$

$-\dfrac{1}{2}x > 10 \Rightarrow x < -20$

$\{x \mid x < -20\}$ or $(-\infty, -20)$

57. $\dfrac{x}{2} \ge 1 - \dfrac{x}{4}$

$2x \ge 4 - x$

$3x \ge 4 \Rightarrow x \ge \dfrac{4}{3}$

$\left\{x \mid x \ge \dfrac{4}{3}\right\}$ or $\left[\dfrac{4}{3}, +\infty\right)$

59. $0 \le 2x - 6 \le 4$
$6 \le 2x \le 10$

$3 \le x \le 5$
$\{x \mid 3 \le x \le 5\}$ or $[3, 5]$

61. $-5 \le 4 - 3x \le 2$
$-9 \le -3x \le -2$

$3 \ge x \ge \dfrac{2}{3}$

$\left\{x \mid \dfrac{2}{3} \le x \le 3\right\}$ or $\left[\dfrac{2}{3}, 3\right]$

63. $-3 < \dfrac{2x - 1}{4} < 0$
$-12 < 2x - 1 < 0$

$-11 < 2x < 1 \Rightarrow -\dfrac{11}{2} < x < \dfrac{1}{2}$

$\left\{x \mid -\dfrac{11}{2} < x < \dfrac{1}{2}\right\}$ or $\left(-\dfrac{11}{2}, \dfrac{1}{2}\right)$

65. $1 < 1 - \dfrac{1}{2}x < 4$

$0 < -\dfrac{1}{2}x < 3$

$0 > x > -6 \Rightarrow -6 < x < 0$
$\{x \mid -6 < x < 0\}$ or $(-6, 0)$

67. $(x + 2)(x - 3) > (x - 1)(x + 1)$
$x^2 - x - 6 > x^2 - 1$

$-x - 6 > -1$

$-x > 5 \Rightarrow x < -5$
$\{x \mid x < -5\}$ or $(-\infty, -5)$

69. $x(4x + 3) \le (2x + 1)^2$
$4x^2 + 3x \le 4x^2 + 4x + 1$

$3x \le 4x + 1$

$-x \le 1 \Rightarrow x \ge -1$
$\{x \mid x \ge -1\}$ or $[-1, +\infty)$

71. $\dfrac{1}{2} \le \dfrac{x + 1}{3} < \dfrac{3}{4}$
$6 \le 4x + 4 < 9$

$2 \le 4x < 5$

$\dfrac{1}{2} \le x < \dfrac{5}{4}$

$\left\{x \mid \dfrac{1}{2} \le x < \dfrac{5}{4}\right\}$ or $\left[\dfrac{1}{2}, \dfrac{5}{4}\right)$

73. $|x| < 6$
$-6 < x < 6$
$\{x \mid -6 < x < 6\}$ or $(-6, 6)$

75. $|x| > 4$
$x < -4$ or $x > 4$
$\{x \mid x < -4 \text{ or } x > 4\}$ or

$(-\infty, -4) \cup (4, +\infty)$

77. $|2x| < 8$
$-8 < 2x < 8$

$-4 < x < 4$
$\{x | -4 < x < 4\}$ or $(-4, 4)$

79. $|3x| > 12$
$3x < -12$ or $3x > 12$

$x < -4$ or $x > 4$
$\{x | x < -4 \text{ or } x > 4\}$ or $(-\infty, -4) \cup (4, \infty)$

81. $|x - 2| + 2 < 3 \Rightarrow |x - 2| < 1$
$-1 < x - 2 < 1$

$1 < x < 3$
$\{x | 1 < x < 3\}$ or $(1, 3)$

83. $|3t - 2| \le 4$
$-4 \le 3t - 2 \le 4$

$-2 \le 3t \le 6 \Rightarrow -\dfrac{2}{3} \le t \le 2$

$\left\{t \left| -\dfrac{2}{3} \le t \le 2\right.\right\}$ or $\left[-\dfrac{2}{3}, 2\right]$

85. $|x - 3| \ge 2$
$x - 3 \le -2$ or $x - 3 \ge 2$

$x \le 1$ or $x \ge 5$
$\{x | x \le 1 \text{ or } x \ge 5\}$ or

$(-\infty, 1] \cup [5, +\infty)$

87. $|1 - 4x| - 7 < -2 \Rightarrow |1 - 4x| < 5$
$-5 < 1 - 4x < 5$

$-6 < -4x < 4$

$\dfrac{-6}{-4} > x > \dfrac{4}{-4}$

$\dfrac{3}{2} > x > -1 \Rightarrow -1 < x < \dfrac{3}{2}$

$\{x | -1 < x < 1.5\}$ or $(-1, 1.5)$

89. $|1 - 2x| > |-3| \Rightarrow |1 - 2x| > 3$
$1 - 2x < -3$ or $1 - 2x > 3$

$-2x < -4$ or $-2x > 2$

$x > 2$ or $x < -1$
$\{x | x < -1 \text{ or } x > 2\}$ or $(-\infty, -1) \cup (2, \infty)$

91. $|2x + 1| < -1$
No solution since absolute value is always non-negative.

93. $|x - 2| < 0.5$
$-0.5 < x - 2 < 0.5$

$-0.5 + 2 < x < 0.5 + 2$

$1.5 < x < 2.5$
Solution set: $\{x | 1.5 < x < 2.5\}$

95. $|x - (-3)| > 2$
$x - (-3) < -2$ or $x - (-3) > 2$

$x + 3 < -2$ or $x + 3 > 2$

$x < -5$ or $x > -1$
Solution set: $\{x | x < -5 \text{ or } x > -1\}$

97.   21 < young adult's age < 30

99.   A temperature x that differs from 98.6°F by at least 1.5°
$$|x - 98.6°| \geq 1.5°$$
$$x - 98.6° \leq -1.5° \quad \text{or} \quad x - 98.6° \geq 1.5°$$
$$x \leq 97.1° \quad \text{or} \quad \quad x \geq 100.1°$$
The temperatures that are considered unhealthy are those that are less than 97.1°F or greater than 100.1°F, inclusive.

101. (a) An average 25-year-old male can expect to live at least 48.4 more years. $25 + 48.4 = 73.4$. Therefore, the average age of a 25-year-old male will be ≥ 73.4.
      (b) An average 25-year-old female can expect to live at least 54.7 more years. $25 + 54.7 = 79.7$. Therefore, the average age of a 25-year-old female will be ≥ 79.7.
      (c) By the given information, a female can expect to live 6.3 years longer.

103. Let P represent the selling price and C represent the commission.
Calculating the commission:
$$C = 45,000 + 0.25(P - 900,000) = 45,000 + 0.25P - 225,000 = 0.25P - 180,000$$
Calculate the commission range, given the price range:
$$900,000 \leq \quad\quad P \quad\quad \leq 1,100,000$$
$$0.25(900,000) \leq \quad 0.25P \quad\quad \leq 0.25(1,100,000)$$
$$225,000 \leq \quad 0.25P \quad\quad \leq 275,000$$
$$225,000 - 180,000 \leq 0.25P - 180,000 \leq 275,000 - 180,000$$
$$45,000 \leq \quad\quad C \quad\quad \leq 95,000$$
The agent's commission ranges from \$45,000 to \$95,000, inclusive.
$$\frac{45,000}{900,000} = 0.05 = 5\% \quad \text{to} \quad \frac{95,000}{1,100,000} = 0.086 = 8.6\%, \text{ inclusive.}$$
As a percent of selling price, the commission ranges from 5% to 8.6%.

105. Let W represent the weekly wage and T represent the withholding tax.
Calculating the tax:
$$T = 69.90 + 0.28(W - 517) = 69.90 + 0.28W - 144.76 = 0.28W - 74.86$$
Calculating the withholding tax range, given the range of weekly wages:
$$525 \leq \quad\quad W \quad \leq 600$$
$$0.28(525) \leq \quad 0.28W \quad \leq 0.28(600)$$
$$147 \leq \quad 0.28W \quad \leq 168$$
$$147 - 74.86 \leq 0.28W - 74.86 \leq 168 - 74.86$$
$$72.14 \leq \quad\quad T \quad\quad \leq 93.14$$
The amount of withholding tax ranges from \$72.14 to \$93.14, inclusive.

107. Let K represent the monthly usage in kilowatt-hours.
Let C represent the monthly customer bill.
Calculating the bill:
$$C = 0.10494K + 9.36$$

Calculating the range of kilowatt-hours, given the range of bills:

$$80.24 \leq \qquad C \qquad \leq 271.80$$
$$80.24 \leq 0.10494K + 9.36 \leq 271.80$$
$$70.88 \leq \qquad 0.10494K \qquad \leq 262.44$$
$$675.43 \leq \qquad K \qquad \leq 2500.86$$

The range of usage in kilowatt-hours varied from 675.43 to 2500.86.

109. Let C represent the dealer's cost and M represent the markup over dealer's cost.
If the price is \$8800, then $8800 = C + MC = C(1 + M)$

Solving for C: $\quad C = \dfrac{8800}{1 + M}$

Calculating the range of dealer costs, given the range of markups:

$$0.12 \leq \quad M \quad \leq 0.18$$
$$1.12 \leq 1 + M \leq 1.18$$
$$\frac{1}{1.12} \geq \frac{1}{1 + M} \geq \frac{1}{1.18}$$
$$\frac{8800}{1.12} \geq \frac{8800}{1 + M} \geq \frac{8800}{1.18}$$
$$7857.14 \geq \quad C \quad \geq 7457.63$$

The dealer's cost ranged from \$7457.63 to \$7857.14, inclusive.

111. Let T represent the score on the last test and G represent the course grade.
Calculating the course grade and solving for the last test:

$$G = \frac{68 + 82 + 87 + 89 + T}{5} = \frac{326 + T}{5} \rightarrow T = 5G - 326$$

Calculating the range of scores on the last test, given the grade range:

$$80 \leq \quad G \quad < 90$$
$$400 \leq \quad 5G \quad < 450$$
$$74 \leq 5G - 326 < 124$$
$$74 \leq \quad T \quad < 124$$

The fifth test must be greater than or equal to 74.

113. Let g represent the number of gallons of gasoline in the gas tank.
Since the car averages 25 miles per gallon, a trip of at least 300 miles will require

at least $\dfrac{300}{25} = 12$ gallons of gas.

Therefore the range of the amount of gasoline is $12 \leq g \leq 20$.

115. Since $a < b$

$$\frac{a}{2} < \frac{b}{2} \qquad\qquad\qquad \frac{a}{2} < \frac{b}{2}$$
$$\frac{a}{2} + \frac{a}{2} < \frac{a}{2} + \frac{b}{2} \qquad\qquad \frac{a}{2} + \frac{b}{2} < \frac{b}{2} + \frac{b}{2} \qquad \text{Thus, } a < \frac{a+b}{2} < b$$
$$a < \frac{a+b}{2} \qquad\qquad\qquad \frac{a+b}{2} < b$$

117.   If $0 < a < b$, then $0 < a^2 < ab$ and $0 < ab < b^2$

$$ab - a^2 > 0 \qquad\qquad\qquad b^2 - ab > 0$$
$$ab > a^2 > 0 \qquad\qquad\qquad b^2 > ab > 0$$
$$\left(\sqrt{ab}\right)^2 > a^2 \qquad\qquad\qquad b^2 > \left(\sqrt{ab}\right)^2$$
$$\sqrt{ab} > a \qquad\qquad\qquad\qquad b > \sqrt{ab}$$

Thus, $a < \sqrt{ab} < b$

119.   For $0 < a < b$, $\dfrac{1}{h} = \dfrac{1}{2}\left(\dfrac{1}{a} + \dfrac{1}{b}\right)$

$$h \cdot \frac{1}{h} = \frac{1}{2}\left(\frac{b+a}{ab}\right) \cdot h \;\Rightarrow\; 1 = \frac{1}{2}\left(\frac{b+a}{ab}\right) \cdot h \;\Rightarrow\; \frac{2ab}{a+b} = h$$

$$h - a = \frac{2ab}{a+b} - a \qquad\qquad\qquad\qquad b - h = b - \frac{2ab}{a+b}$$

$$= \frac{2ab - a(a+b)}{a+b} \qquad\qquad\qquad\qquad = \frac{b(a+b) - 2ab}{a+b}$$

$$= \frac{2ab - a^2 - ab}{a+b} = \frac{ab - a^2}{a+b} \qquad\qquad = \frac{ab + b^2 - 2ab}{a+b} = \frac{b^2 - ab}{a+b}$$

$$= \frac{a(b-a)}{a+b} > 0 \qquad\qquad\qquad\qquad = \frac{b(b-a)}{a+b} > 0$$

Therefore, $h > a$.   Therefore, $h < b$.   Thus, $a < h < b$.

121. Answers will vary.

Chapter 1

Graphs

1.6 Lines

1. (a) Slope $= \dfrac{1-0}{2-0} = \dfrac{1}{2}$

    (b) If x increases by 2 units, y will increase by 1 unit.

3. (a) Slope $= \dfrac{1-2}{1-(-2)} = -\dfrac{1}{3}$

    (b) If x increases by 3 units, y will decrease by 1 unit.

5. $(x_1, y_1) \quad (x_2, y_2)$
    $(2, 3) \quad (4, 0)$

    Slope $= \dfrac{y_2 - y_1}{x_2 - x_1} = \dfrac{0-3}{4-2} = \dfrac{-3}{2}$

    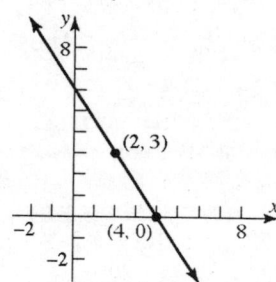

7. $(x_1, y_1) \quad (x_2, y_2)$
    $(-2, 3) \quad (2, 1)$

    Slope $= \dfrac{y_2 - y_1}{x_2 - x_1} = \dfrac{1-3}{2-(-2)} = \dfrac{-2}{4} = -\dfrac{1}{2}$

    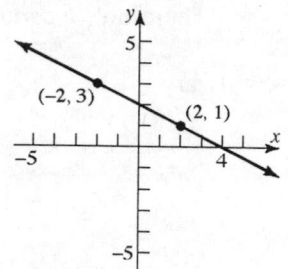

9. $(x_1, y_1) \quad (x_2, y_2)$
    $(-3, -1) \quad (2, -1)$

    Slope $= \dfrac{y_2 - y_1}{x_2 - x_1} = \dfrac{-1-(-1)}{2-(-3)} = \dfrac{0}{5} = 0$

    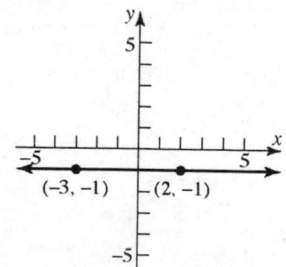

11. $(x_1, y_1) \quad (x_2, y_2)$
    $(-1, 2) \quad (-1, -2)$

    Slope $= \dfrac{y_2 - y_1}{x_2 - x_1} = \dfrac{-2-2}{-1-(-1)} = \dfrac{-4}{0}$

    Slope is undefined.

    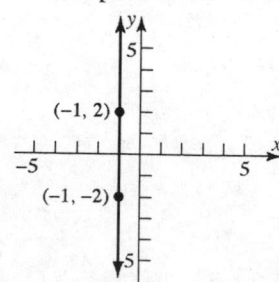

13.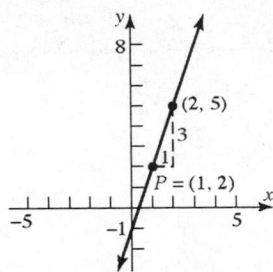

15.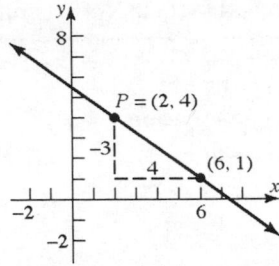

17.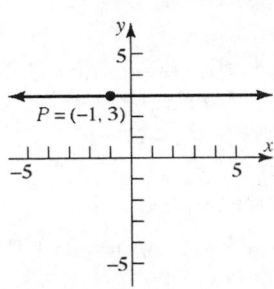

19.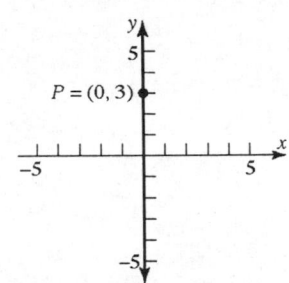

21. Slope $= 4 \rightarrow$ if x increases by 1, y increases by 4

original point $(1,2)$

Answers will vary. Three possible points are:
$x = 1 + 1 = 2$ and $y = 2 + 4 = 6$

$(2,6)$

$x = 2 + 1 = 3$ and $y = 6 + 4 = 10$

$(3,10)$

$x = 3 + 1 = 4$ and $y = 10 + 4 = 14$

$(4,14)$

23. Slope $= -\dfrac{3}{2} \rightarrow$ if x increases by 2, y decreases by 3

original point $(2,-4)$

Answers will vary. Three possible points are:
$x = 2 + 2 = 4$ and $y = -4 - 3 = -7$

$(4,-7)$

$x = 4 + 2 = 6$ and $y = -7 - 3 = -10$

$(6,-10)$

$x = 6 + 2 = 8$ and $y = -10 - 3 = -13$

$(8,-13)$

25. Slope $= -2 \rightarrow$ if x increases by 1, y decreases by 2

original point $(-2,-3)$

Answers will vary. Three possible points are:
$x = -2 + 1 = -1$ and $y = -3 - 2 = -5$

$(-1,-5)$

$x = -1 + 1 = 0$ and $y = -5 - 2 = -7$

$(0,-7)$

$x = 0 + 1 = 1$ and $y = -7 - 2 = -9$

$(1,-9)$

27. $(0,0)$ and $(2,1)$ are points on the line.

Slope $= \dfrac{1-0}{2-0} = \dfrac{1}{2}$

y-intercept is 0; using $y = mx + b$:

$y = \dfrac{1}{2}x + 0$

$2y = x$

$0 = x - 2y$

$x - 2y = 0$ or $y = \dfrac{1}{2}x$

29. $(-1,3)$ and $(1,1)$ are points on the line.

Slope $= \dfrac{1-3}{1-(-1)} = \dfrac{-2}{2} = -1$

Using $y - y_1 = m(x - x_1)$

$y - 1 = -1(x - 1)$

$y - 1 = -x + 1$

$\quad y = -x + 2$

$x + y = 2$ or $y = -x + 2$

31. $y - y_1 = m(x - x_1),\ \ m = 2$

$y - 3 = 2(x - 3)$

$y - 3 = 2x - 6$

$\quad y = 2x - 3$

$2x - y = 3$ or $y = 2x - 3$

33. $y - y_1 = m(x - x_1),\ \ m = -\dfrac{1}{2}$

$y - 2 = -\dfrac{1}{2}(x - 1)$

$y - 2 = -\dfrac{1}{2}x + \dfrac{1}{2}$

$\quad y = -\dfrac{1}{2}x + \dfrac{5}{2}$

$x + 2y = 5$ or $y = -\dfrac{1}{2}x + \dfrac{5}{2}$

35. Slope $= 3$; containing $(-2,3)$

$y - y_1 = m(x - x_1)$

$y - 3 = 3(x - (-2))$

$y - 3 = 3x + 6$

$\quad y = 3x + 9$

$3x - y = -9$ or $y = 3x + 9$

37. Slope $= -\dfrac{2}{3}$; containing $(1,-1)$

$y - y_1 = m(x - x_1)$

$y - (-1) = -\dfrac{2}{3}(x - 1)$

$y + 1 = -\dfrac{2}{3}x + \dfrac{2}{3}$

$\quad y = -\dfrac{2}{3}x - \dfrac{1}{3}$

$2x + 3y = -1$ or $y = -\dfrac{2}{3}x - \dfrac{1}{3}$

39. Containing $(1,3)$ and $(-1,2)$

$m = \dfrac{2-3}{-1-1} = \dfrac{-1}{-2} = \dfrac{1}{2}$

$y - y_1 = m(x - x_1)$

$y - 3 = \dfrac{1}{2}(x - 1)$

$y - 3 = \dfrac{1}{2}x - \dfrac{1}{2}$

$\quad y = \dfrac{1}{2}x + \dfrac{5}{2}$

$x - 2y = -5$ or $y = \dfrac{1}{2}x + \dfrac{5}{2}$

41. Slope $= -3$; y-intercept $= 3$

$y = mx + b$

$y = -3x + 3$

$3x + y = 3$ or $y = -3x + 3$

43. x-intercept $= 2$; y-intercept $= -1$

Points are $(2,0)$ and $(0,-1)$

$m = \dfrac{-1-0}{0-2} = \dfrac{-1}{-2} = \dfrac{1}{2}$

$y = mx + b$

$y = \dfrac{1}{2}x - 1$

$x - 2y = 2$ or $y = \dfrac{1}{2}x - 1$

45. Slope undefined; passing through (2,4)
    This is a vertical line.
    $x = 2$
        No slope intercept form.

47. Parallel to $y = 2x$; Slope $= 2$
    Containing $(-1,2)$
    $y - y_1 = m(x - x_1)$
    $y - 2 = 2(x - (-1))$
    $y - 2 = 2x + 2 \rightarrow y = 2x + 4$
    $2x - y = -4$ or $y = 2x + 4$

49. Parallel to $2x - y = -2$; Slope $= 2$
    Containing the point $(0,0)$
    $y - y_1 = m(x - x_1)$
    $y - 0 = 2(x - 0)$
    $y = 2x$
    $2x - y = 0$ or $y = 2x$

51. Parallel to $x = 5$;
    Containing $(4,2)$
    This is a vertical line.
    $x = 4$
        No slope intercept form.

53. Perpendicular to $y = \dfrac{1}{2}x + 4$;
    Slope of perpendicular $= -2$
    Containing $(1,-2)$
    $y - y_1 = m(x - x_1)$
    $y - (-2) = -2(x - 1)$
    $y + 2 = -2x + 2 \rightarrow y = -2x$
    $2x + y = 0$ or $y = -2x$

55. Perpendicular to $2x + y = 2$;
    Containing $(-3,0)$
    Slope of perpendicular $= \dfrac{1}{2}$
    $y - y_1 = m(x - x_1)$
    $y - 0 = \dfrac{1}{2}(x - (-3)) \rightarrow y = \dfrac{1}{2}x + \dfrac{3}{2}$
    $x - 2y = -3$ or $y = \dfrac{1}{2}x + \dfrac{3}{2}$

57. Perpendicular to $x = 8$;
    Slope of perpendicular $= 0$
    Containing $(3,4)$
    $y - y_1 = m(x - x_1)$
    $y - 4 = 0(x - 3) \rightarrow y - 4 = 0 \rightarrow y = 4$
    $y = 4$ or $y = 0x + 4$

59. $y = 2x + 3$
    Slope $= 2$
    y-intercept $= 3$

61. $\dfrac{1}{2}y = x - 1$
    $y = 2x - 2$
    Slope $= 2$
    y-intercept $= -2$

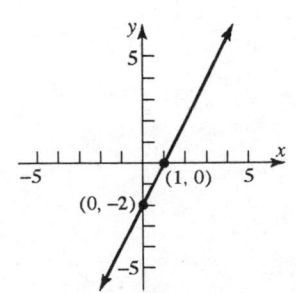

63. $y = \dfrac{1}{2}x + 2$

Slope $= \dfrac{1}{2}$

y-intercept $= 2$

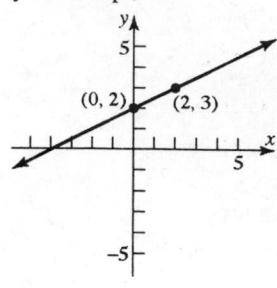

65. $x + 2y = 4$

$2y = -x + 4 \rightarrow y = -\dfrac{1}{2}x + 2$

Slope $= -\dfrac{1}{2}$

y-intercept $= 2$

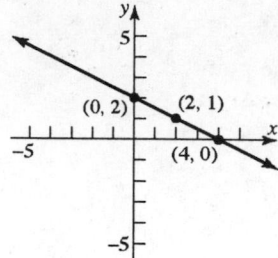

67. $2x - 3y = 6$

$-3y = -2x + 6 \rightarrow y = \dfrac{2}{3}x - 2$

Slope $= \dfrac{2}{3}$

y-intercept $= -2$

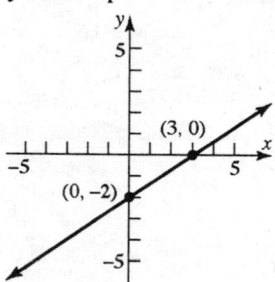

69. $x + y = 1$

$y = -x + 1$

Slope $= -1$

y-intercept $= 1$

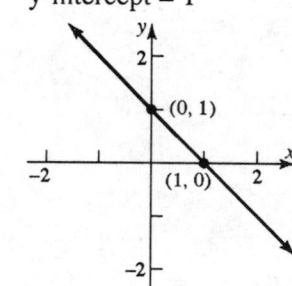

71. $x = -4$

Slope is undefined

y-intercept - none

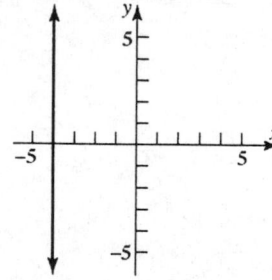

73. $y = 5$

Slope $= 0$

y-intercept $= 5$

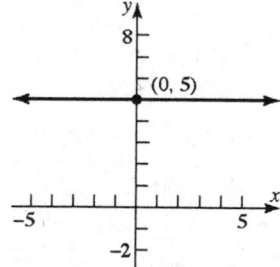

75. $y - x = 0$
    $y = x;$ Slope = 1
    y-intercept = 0

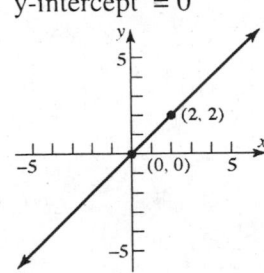

77. $2y - 3x = 0$
    $2y = 3x \rightarrow y = \dfrac{3}{2}x$

    Slope = $\dfrac{3}{2}$; y-intercept = 0

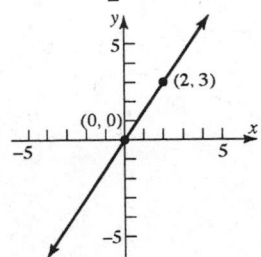

79. The equation of the x-axis is $y = 0$. (The slope is 0 and the y-intercept is 0.)

81. (b)

83. (d)

85. Slope = 1; y-intercept = 2

    $y = x + 2$ or $x - y = -2$

87. Slope = $-\dfrac{1}{3}$; y-intercept = 1

    $y = -\dfrac{1}{3}x + 1$ or $x + 3y = 3$

89. Let x = number of miles driven, and let C = cost in dollars.
    Total cost = (cost per mile)(number of miles) + fixed cost
    $C = 0.07x + 29$
    When $x = 110,$ $C = (0.07)(110) + 29 = \$36.70.$
    When $x = 230,$ $C = (0.07)(230) + 29 = \$45.10.$

91. $(^{\circ}C, ^{\circ}F) = (0, 32);$ $(^{\circ}C, ^{\circ}F) = (100, 212)$

    slope $= \dfrac{212 - 32}{100 - 0} = \dfrac{180}{100} = \dfrac{9}{5}$

    $^{\circ}F - 32 = \dfrac{9}{5}(^{\circ}C - 0)$

    $^{\circ}F - 32 = \dfrac{9}{5}(^{\circ}C)$

    $^{\circ}C = \dfrac{5}{9}(^{\circ}F - 32)$

    If $^{\circ}F = 70,$ then
    $^{\circ}C = \dfrac{5}{9}(70 - 32) = \dfrac{5}{9}(38)$
    $^{\circ}C \approx 21^{\circ}$

93. (a) Since there is only a profit of
        $0.50 per copy and the expense
        of $100 must be deducted, the
        profit is:
        $P = 0.50x - 100$
    (b) $P = 0.50(1000) - 100$
        $= 500 - 100 = \$400$
    (c) $P = 0.50(5000) - 100$
        $= 2500 - 100 = \$2400$

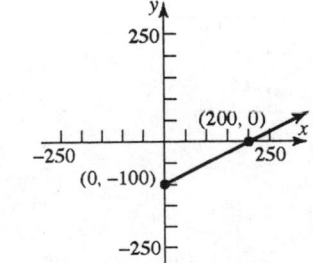

95. $C = 0.06543x + 5.65$
     For 300 kWh,
         $C = 0.06543(300) + 5.65 = \$25.28$
     For 750 kWh,
         $C = 0.06543(750) + 5.65 = \$54.72$

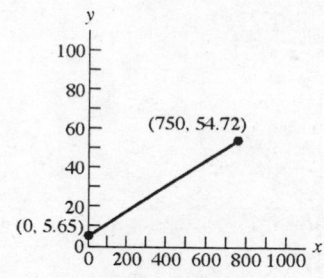

97. $2x - y = C$
     Graph the lines:
         $2x - y = -4$
         $2x - y = 0$
         $2x - y = 2$
     All the lines have the same slope, 2.
     The lines are parallel.

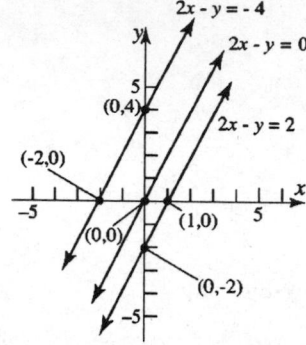

99. (a) $x^2 + (mx + b)^2 = r^2$

           $x^2 + m^2x^2 + 2bmx + b^2 = r^2 \Rightarrow (1 + m^2)x^2 + 2bmx + b^2 - r^2 = 0$
           There is one solution if and only if the discriminant is zero.
               $(2bm)^2 - 4(1 + m^2)(b^2 - r^2) = 0$

           $4b^2m^2 - 4b^2 + 4r^2 - 4b^2m^2 + 4m^2r^2 = 0 \Rightarrow -4b^2 + 4r^2 + 4m^2r^2 = 0$

                       $-b^2 + r^2 + m^2r^2 = 0 \Rightarrow r^2(1 + m^2) = b^2$

     (b)   Using the quadratic formula, knowing that the discriminant is zero:

           $$x = \frac{-2bm}{2(1 + m^2)} = \frac{-bm}{\left(\dfrac{b^2}{r^2}\right)} = \frac{-bmr^2}{b^2} = \frac{-mr^2}{b}$$

           $$y = m\left(\frac{-mr^2}{b}\right) + b = \frac{-m^2r^2}{b} + b = \frac{-m^2r^2 + b^2}{b} = \frac{r^2}{b}$$

     (c)   The slope of the tangent line is m.
           The slope of the line joining the point of tangency and the center is:

           $$\frac{\left(\dfrac{r^2}{b} - 0\right)}{\left(\dfrac{-mr^2}{b} - 0\right)} = \frac{r^2}{b} \cdot \frac{b}{-mr^2} = -\frac{1}{m}$$

101. $x^2 + y^2 - 4x + 6y + 4 = 0$

$(x^2 - 4x + 4) + (y^2 + 6y + 9) = -4 + 4 + 9$

$(x-2)^2 + (y+3)^2 = 9$

Center: $(2, -3)$

Slope from center to $\left(3, 2\sqrt{2} - 3\right)$ is $\dfrac{2\sqrt{2} - 3 - (-3)}{3 - 2} = \dfrac{2\sqrt{2}}{1} = 2\sqrt{2}$

Slope of the tangent line is: $\dfrac{-1}{2\sqrt{2}} = -\dfrac{\sqrt{2}}{4}$

Equation of the tangent line:

$$y - \left(2\sqrt{2} - 3\right) = -\frac{\sqrt{2}}{4}(x - 3) \Rightarrow y - 2\sqrt{2} + 3 = -\frac{\sqrt{2}}{4}x + \frac{3\sqrt{2}}{4}$$

$$4y - 8\sqrt{2} + 12 = -\sqrt{2}x + 3\sqrt{2} \Rightarrow \sqrt{2}x + 4y = 11\sqrt{2} - 12$$

103. Find the centers of the two circles:

$$x^2 + y^2 - 4x + 6y + 4 = 0$$

$$(x^2 - 4x + 4) + (y^2 + 6y + 9) = -4 + 4 + 9$$

$$(x-2)^2 + (y+3)^2 = 9 \qquad \text{Center:}\ (2, -3)$$

$$x^2 + y^2 + 6x + 4y + 9 = 0$$

$$(x^2 + 6x + 9) + (y^2 + 4y + 4) = -9 + 9 + 4$$

$$(x+3)^2 + (y+2)^2 = 4 \qquad \text{Center:}\ (-3, -2)$$

Find the slope of the line containing the centers:

$$m = \frac{-2 - (-3)}{-3 - 2} = -\frac{1}{5}$$

Find the equation of the line containing the centers:

$$y + 3 = -\frac{1}{5}(x - 2) \Rightarrow 5y + 15 = -x + 2 \Rightarrow x + 5y = -13$$

105. (b), (c), (e) and (g)   107. (c)

109 – 111. Answers will vary.

43

Graphs

1.R   Chapter Review

1. $2 - \dfrac{x}{3} = 8$

$6 - x = 24 \Rightarrow x = -18$
The solution set is $\{-18\}$.

3. $-2(5 - 3x) + 8 = 4 + 5x$
$-10 + 6x + 8 = 4 + 5x$
$6x - 2 = 4 + 5x$
$x = 6$
The solution set is $\{6\}$.

5. $\dfrac{3x}{4} - \dfrac{x}{3} = \dfrac{1}{12}$

$9x - 4x = 1 \Rightarrow 5x = 1 \Rightarrow x = \dfrac{1}{5}$

The solution set is $\left\{\dfrac{1}{5}\right\}$.

7. $\dfrac{x}{x-1} = \dfrac{6}{5}$

$5x = 6x - 6$

$6 = x$
and since $x = 6$ does not cause a denominator
to equal zero, the solution set is $\{6\}$.

9. $x(1 - x) = 6$
$x - x^2 = 6$

$0 = x^2 - x + 6$

$b^2 - 4ac = (-1)^2 - 4(1)(6) = 1 - 24 = -23 < 0$
No real solution.

11. $\dfrac{1}{2}\left(x - \dfrac{1}{3}\right) = \dfrac{3}{4} - \dfrac{x}{6}$

$\dfrac{x}{2} - \dfrac{1}{6} = \dfrac{3}{4} - \dfrac{x}{6}$

$6x - 2 = 9 - 2x$

$8x = 11$

$x = \dfrac{11}{8}$

13. $(x - 1)(2x + 3) = 3$
$2x^2 + x - 3 = 3$
$2x^2 + x - 6 = 0$
$(2x - 3)(x + 2) = 0$

$x = \dfrac{3}{2}$ or $x = -2$

15. $2x + 3 = 4x^2$
$0 = 4x^2 - 2x - 3$

$x = \dfrac{2 \pm \sqrt{4 + 48}}{8} = \dfrac{2 \pm \sqrt{52}}{8}$

$= \dfrac{2 \pm 2\sqrt{13}}{8} = \dfrac{1 \pm \sqrt{13}}{4}$

17. $\sqrt[3]{x^2 - 1} = 2$

$\left(\sqrt[3]{x^2 - 1}\right)^3 = (2)^3$

$x^2 - 1 = 8$

$x^2 = 9$

$x = \pm 3$

Check:

$x = -3$	$x = 3$
$\sqrt[3]{(-3)^2 - 1} = 2$	$\sqrt[3]{(3)^2 - 1} = 2$
$\sqrt[3]{9 - 1} = 2$	$\sqrt[3]{9 - 1} = 2$
$\sqrt[3]{8} = 2$	$\sqrt[3]{8} = 2$
$2 = 2$	$2 = 2$

so the solution set is $\{-3, 3\}$.

19. $x(x + 1) + 2 = 0$

$x^2 + x + 2 = 0$

$x = \dfrac{-1 \pm \sqrt{1 - 8}}{2} = \dfrac{-1 \pm \sqrt{-7}}{2}$

no real solutions

21. $\sqrt{2x - 3} + x = 3$

$\sqrt{2x - 3} = 3 - x$

$2x - 3 = 9 - 6x + x^2$

$x^2 - 8x + 12 = 0$

$(x - 2)(x - 6) = 0$

$x = 2$ or $x = 6$

Check $x = 2$: $\sqrt{2(2) - 3} + 2 = \sqrt{1} + 2 = 3$

Check $x = 6$: $\sqrt{2(6) - 3} + 6 = \sqrt{9} + 6$

$= 9 \neq 3$

The solution is $x = 2$.

23. $x^{3/2} + 5x^{1/2} = 0$

$x^{3/2} = -5x^{1/2}$

$\left(x^{3/2}\right)^2 = \left(-5x^{1/2}\right)^2$

$x^3 = 25x$

$x^3 - 25x = 0$

$x\left(x^2 - 25\right) = 0$

$x(x - 5)(x + 5) = 0$

$x = 0$

$x - 5 = 0 \Rightarrow x = 5$

$x + 5 = 0 \Rightarrow x = -5$

Check $x = 0$:

$0^{3/2} + 5(0)^{1/2} = 0$

$0 + 0 = 0$

$0 = 0$

Check $x = 5$:

$5^{3/2} + 5(5)^{1/2} = 0$

$5^{3/2} + 5^{3/2} = 0$

$2\left(5^{3/2}\right) \neq 0$

Check $x = -5$:

$(-5)^{3/2} + 5(-5)^{1/2} = 0$

$\left(\sqrt{-5}\right)^3 + 5\sqrt{-5} = 0$

but $\sqrt{-5}$ is undefined.
The solution set is $\{0\}$.

25. $\sqrt{x+1}+\sqrt{x-1}=\sqrt{2x+1}$

$\left(\sqrt{x+1}+\sqrt{x-1}\right)^2=\left(\sqrt{2x+1}\right)^2$

$x+1+2\sqrt{x+1}\sqrt{x-1}+x-1=2x+1$

$2x+2\sqrt{x+1}\sqrt{x-1}=2x+1$

$2\sqrt{x+1}\sqrt{x-1}=1$

$\left(2\sqrt{x+1}\sqrt{x-1}\right)^2=(1)^2$

$4(x+1)(x-1)=1\rightarrow4x^2-4=1$

$4x^2=5\rightarrow x^2=\dfrac{5}{4}\rightarrow x=\pm\dfrac{\sqrt{5}}{2}$

Check:

$x=\dfrac{\sqrt{5}}{2}\Rightarrow\sqrt{\dfrac{\sqrt{5}}{2}+1}+\sqrt{\dfrac{\sqrt{5}}{2}-1}=\sqrt{2\left(\dfrac{\sqrt{5}}{2}\right)+1}$

$1.79890743995=1.79890743995$

$x=-\dfrac{\sqrt{5}}{2}\Rightarrow\sqrt{-\dfrac{\sqrt{5}}{2}+1}+\sqrt{-\dfrac{\sqrt{5}}{2}-1}=\sqrt{2\left(-\dfrac{\sqrt{5}}{2}\right)+1}$

impossible since $-\dfrac{\sqrt{5}}{2}-1<0$

The solution set is $\left\{\dfrac{\sqrt{5}}{2}\right\}$.

27. $2\sqrt[3]{x^2}-\sqrt[3]{x}=1$

$2\sqrt[3]{x^2}-\sqrt[3]{x}-1=0\Rightarrow2x^{2/3}-x^{1/3}-1=0$

$p=x^{1/3}\Rightarrow p^2=x^{2/3}$

$2p^2-p-1=0\Rightarrow(2p+1)(p-1)=0$

$p=-\dfrac{1}{2}$ or $p=1$

$p=-\dfrac{1}{2}\Rightarrow x^{1/3}=-\dfrac{1}{2}$ $p=1\Rightarrow x^{1/3}=1$

$\Rightarrow\left(x^{1/3}\right)^3=\left(-\dfrac{1}{2}\right)^3\Rightarrow x=-\dfrac{1}{8}$ $\Rightarrow\left(x^{1/3}\right)^3=(1)^3\Rightarrow x=1$

Check

$x=-\dfrac{1}{8}:2\sqrt[3]{\left(-\dfrac{1}{8}\right)^2}-\sqrt[3]{-\dfrac{1}{8}}-1=0$

$2\left(\dfrac{1}{4}\right)-\left(-\dfrac{1}{2}\right)-1=0\Rightarrow\dfrac{1}{2}+\dfrac{1}{2}-1=0\Rightarrow0=0$

$x=1:2\sqrt[3]{1^2}-\sqrt[3]{1}-1=0$

$2-1-1=0\Rightarrow2-2=0\Rightarrow0=0$

the solution set is $\left\{-\dfrac{1}{8},1\right\}$.

29. $\sqrt{x^2+3x+7}-\sqrt{x^2-3x+9}+2=0$

$\sqrt{x^2+3x+7}=\sqrt{x^2-3x+9}-2\Rightarrow\left(\sqrt{x^2+3x+7}\right)^2=\left(\sqrt{x^2-3x+9}-2\right)^2$

$x^2+3x+7=x^2-3x+9-4\sqrt{x^2-3x+9}+4$

$6x-6=-4\sqrt{x^2-3x+9}$

$$\left(6(x-1)\right)^2 = \left(-4\sqrt{x^2-3x+9}\right)^2 \Rightarrow 36\left(x^2-2x+1\right)=16\left(x^2-3x+9\right)$$

$$36x^2-72x+36=16x^2-48x+144 \Rightarrow 20x^2-24x-108=0 \Rightarrow 5x^2-6x-27=0$$

$$(5x+9)(x-3)=0 \Rightarrow x=-\frac{9}{5}\ \text{ or }\ x=3$$

Check $\ x=-\dfrac{9}{5}$:

$$\sqrt{\left(-\frac{9}{5}\right)^2+3\left(-\frac{9}{5}\right)+7}-\sqrt{\left(-\frac{9}{5}\right)^2-3\left(-\frac{9}{5}\right)+9}+2=0$$

$$\sqrt{\frac{81}{25}-\frac{27}{5}+7}-\sqrt{\frac{81}{25}+\frac{27}{5}+9}+2=0 \Rightarrow \sqrt{\frac{81-135+175}{25}}-\sqrt{\frac{81+135+225}{25}}+2=0$$

$$\sqrt{\frac{121}{25}}-\sqrt{\frac{441}{25}}+2=0 \Rightarrow \frac{11}{5}-\frac{21}{5}+2=0 \Rightarrow 0=0$$

Check:

$$x=3:\sqrt{(3)^2+3(3)+7}-\sqrt{(3)^2-3(3)+9}+2=0$$

$$\sqrt{9+9+7}-\sqrt{9-9+9}+2=0 \Rightarrow \sqrt{25}-\sqrt{9}+2=0 \Rightarrow 2+2=0 \Rightarrow 4\neq 0$$

the solution set is $\left\{-\dfrac{9}{5}\right\}$.

31. $\ |2x+3|=7$
    $2x+3=7$ or $2x+3=-7$
    $\quad 2x=4$ or $\quad\ \ 2x=-10$
    $\quad\ \ x=2$ or $\quad\quad\ \ x=-5$
    The solution set is $\{-5, 2\}$.

33. $\ |2-3x|+2=9 \Rightarrow |2-3x|=7$
    $2-3x=7$ or $2-3x=-7$
    $\quad -3x=-5$ or $\quad\ -3x=9$
    $\quad\ \ x=-\dfrac{5}{3}$ or $\quad\quad x=3$
    The solution set is $\left\{-\dfrac{5}{3},3\right\}$.

35. $x^3-5x+3=0$; Use ZERO (or ROOT) on the graph of $y_1=x^3-5x+3$.

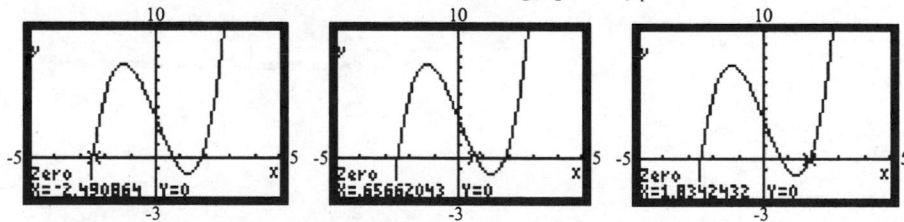

The solution set is $\{-2.49, 0.66, 1.83\}$.

37. $x^4 - 3 = 2x + 1$; Use ZERO (or ROOT) on the graph of $y_1 = x^4 - 2x - 4$.

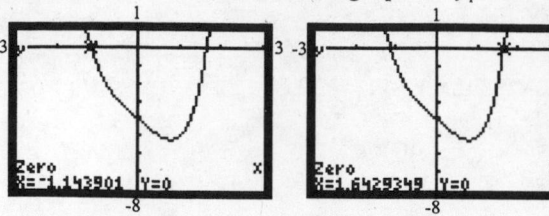

The solution set is $\{-1.14, 1.64\}$.

39. $\dfrac{2x-3}{5} + 2 \leq \dfrac{x}{2}$

$2(2x-3) + 10(2) \leq 5x$

$4x - 6 + 20 \leq 5x$

$14 \leq x \Rightarrow x \geq 14$

$\{x \mid x \geq 14\}$ or $[14, +\infty)$

41. $-9 \leq \dfrac{2x+3}{-4} \leq 7$

$36 \geq 2x + 3 \geq -28$

$33 \geq \ 2x \ \geq -31$

$\dfrac{33}{2} \geq \ x \ \geq -\dfrac{31}{2}$

$-\dfrac{31}{2} \leq \ x \ \leq \dfrac{33}{2}$

$\{x \mid -15.5 \leq x \leq 16.5\}$ or $[-15.5, 16.5]$

43. $6 > \dfrac{3-3x}{12} > 2$

$72 > 3 - 3x > 24$

$69 > \ -3x \ > 21$

$-23 < \ x \ < -7$

$\{x \mid -23 < x < -7\}$ or $(-23, -7)$

45. $|3x + 4| < \dfrac{1}{2}$

$\dfrac{-1}{2} < 3x + 4 < \dfrac{1}{2}$

$-\dfrac{9}{2} < \ 3x \ < -\dfrac{7}{2}$

$-\dfrac{3}{2} < \ x \ < -\dfrac{7}{6}$

$\left\{x \mid -\dfrac{3}{2} < x < -\dfrac{7}{6}\right\}$ or $\left(-\dfrac{3}{2}, -\dfrac{7}{6}\right)$

47. $|2x - 5| \geq 9$

$2x - 5 \leq -9$ or $2x - 5 \geq 9$

$2x \leq -4$ or $\quad 2x \geq 14$

$x \leq -2$ or $\quad x \geq 7$

$\{x \mid x \leq -2 \text{ or } x \geq 7\}$

or $(-\infty, -2] \cup [7, +\infty)$

49. $2+|2-3x| \le 4$

$|2-3x| \le 2$

$-2 \le 2-3x \le 2$

$-4 \le -3x \le 0 \rightarrow \dfrac{4}{3} \ge x \ge 0$

$0 \le x \le \dfrac{4}{3}$

$\left\{ x \Big| 0 \le x \le \dfrac{4}{3} \right\}$ or $\left[0, \dfrac{4}{3}\right]$

51. $1-|2-3x| < -4$

$-|2-3x| < -5 \rightarrow |2-3x| > 5$

$2-3x < -5$ or $2-3x > 5$

$7 < 3x$ or $-3 > 3x$

$\dfrac{7}{3} < x$ or $-1 > x$

$\Rightarrow x < -1$ or $x > \dfrac{7}{3}$

$\left\{ x \Big| x < -1 \text{ or } x > \dfrac{7}{3} \right\}$

or $(-\infty, -1) \cup \left(\dfrac{7}{3}, +\infty \right)$

53. $2x - 3y = 6 \Rightarrow y = \dfrac{2}{3}x - 2$

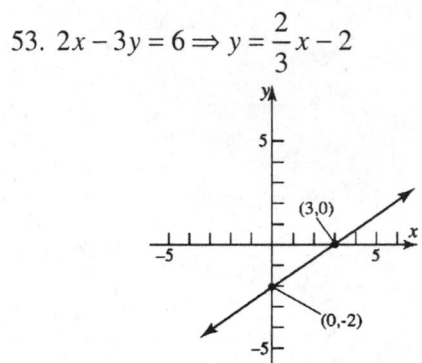

55. $y = x^2 - 9$

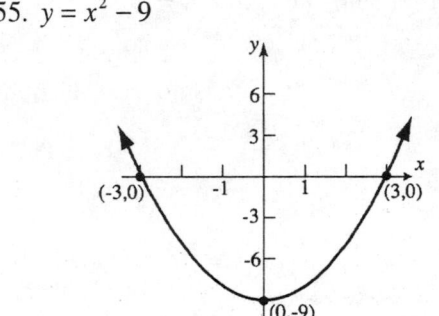

57. $x^2 + 2y = 16 \Rightarrow y = -\dfrac{1}{2}x^2 + 8$

59. Test for symmetry: $2x = 3y^2$

x - axis: Replace y by $-y$ so $2x = 3(-y)^2 \Rightarrow 2x = 3y^2$, which is

equivalent to $2x = 3y^2$.

y - axis: Replace x by $-x$ so $2(-x) = 3y^2 \Rightarrow -2x = 3y^2$,

which is not equivalent to $2x = 3y^2$.

Origin: Replace x by $-x$ and y by $-y$ so $2(-x) = 3(-y)^2$
$$\Rightarrow -2x = 3y^2, \text{ which is not equivalent to } 2x = 3y^2.$$
Therefore, the graph is symmetric with respect to the x-axis.

61. Test for symmetry: $x^2 + 4y^2 = 16$

x-axis: Replace y by $-y$ so $x^2 + 4(-y)^2 = 16 \Rightarrow x^2 + 4y^2 = 16$,

which is equivalent to $x^2 + 4y^2 = 16$.

y-axis: Replace x by $-x$ so $(-x)^2 + 4y^2 = 16 \Rightarrow x^2 + 4y^2 = 16$,

which is equivalent to $x^2 + 4y^2 = 16$.

Origin: Replace x by $-x$ and y by $-y$ so $(-x)^2 + 4(-y)^2 = 16 \Rightarrow x^2 + 4y^2 = 16$,

which is equivalent to $x^2 + 4y^2 = 16$.

Therefore, the graph is symmetric with respect to the x-axis, the y-axis and the origin.

63. Test for symmetry: $y = x^4 + 2x^2 + 1$

x-axis: Replace y by $-y$ so $-y = x^4 + 2x^2 + 1$,

which is not equivalent to $y = x^4 + 2x^2 + 1$.

y-axis: Replace x by $-x$ so $y = (-x)^4 + 2(-x)^2 + 1 \Rightarrow y = x^4 + 2x^2 + 1$,

which is equivalent to $y = x^4 + 2x^2 + 1$.

Origin: Replace x by $-x$ and y by $-y$ so $-y = (-x)^4 + 2(-x)^2 + 1 \Rightarrow -y = x^4 + 2x^2 + 1$,

which is not equivalent to $y = x^4 + 2x^2 + 1$.

Therefore, the graph is symmetric with respect to the y-axis.

65. Test for symmetry: $x^2 + x + y^2 + 2y = 0$

x-axis: Replace y by $-y$ so $x^2 + x + (-y)^2 + 2(-y) = 0 \Rightarrow x^2 + x + y^2 - 2y = 0$,

which is not equivalent to $x^2 + x + y^2 + 2y = 0$.

y-axis: Replace x by $-x$ so $(-x)^2 + (-x) + y^2 + 2y = 0 \Rightarrow x^2 - x + y^2 + 2y = 0$,

which is not equivalent to $x^2 + x + y^2 + 2y = 0$.

Origin: Replace x by $-x$ and y by $-y$ so $(-x)^2 + (-x) + (-y)^2 + 2(-y) = 0$
$$\Rightarrow x^2 - x + y^2 - 2y = 0, \text{ which is not equivalent to } x^2 + x + y^2 + 2y = 0.$$
Therefore, the graph is not symmetric to the x-axis, the y-axis, or the origin.

67. $y = x^3$

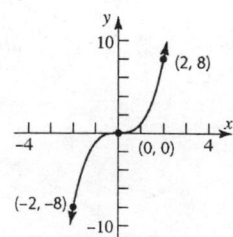

69. $x^2 + y^2 - 2x + 4y - 4 = 0$
$x^2 - 2x + y^2 + 4y = 4$
$(x^2 - 2x + 1) + (y^2 + 4y + 4) = 4 + 1 + 4$
$(x - 1)^2 + (y + 2)^2 = 9$
$(x - 1)^2 + (y + 2)^2 = 3^2$
Center: (1,–2)   Radius = 3

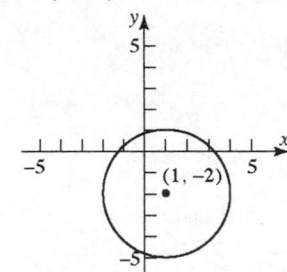

71. $3x^2 + 3y^2 - 6x + 12y = 0$
$x^2 + y^2 - 2x + 4y = 0$
$x^2 - 2x + y^2 + 4y = 0$
$(x^2 - 2x + 1) + (y^2 + 4y + 4) = 1 + 4$
$(x - 1)^2 + (y + 2)^2 = 5$
$(x - 1)^2 + (y + 2)^2 = \left(\sqrt{5}\right)^2$
Center: (1,–2)   Radius = $\sqrt{5}$

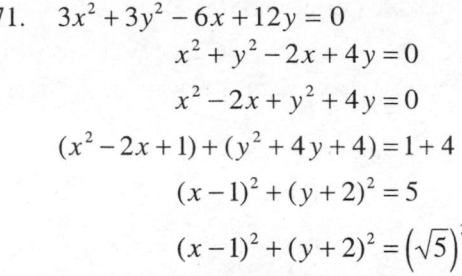

73. Center: (–1,3)   Radius = 5
$\left(x - (-1)\right)^2 + \left(y - 3\right)^2 = 5^2$
$\left(x + 1\right)^2 + \left(y - 3\right)^2 = 25$

75. Slope = –2; containing (3,–1)
$y - y_1 = m(x - x_1)$
$y - (-1) = -2(x - 3)$
$y + 1 = -2x + 6 \Rightarrow y = -2x + 5$
$2x + y = 5$ or $y = -2x + 5$

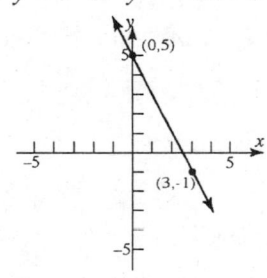

77. Slope undefined; containing (–3,4)
This is a vertical line.
$x = -3$; No slope intercept form.

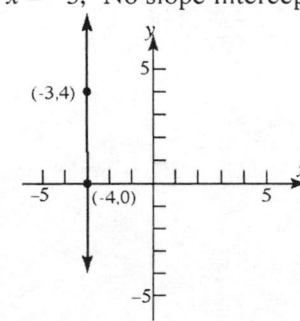

79. y-intercept = –2; containing (5,–3)
    Points are (5,–3) and (0,–2)
    $$m = \frac{-2-(-3)}{0-5} = \frac{1}{-5} = -\frac{1}{5}$$
    $$y = mx + b$$
    $$y = -\frac{1}{5}x - 2$$
    $$x + 5y = -10 \text{ or } y = -\frac{1}{5}x - 2$$

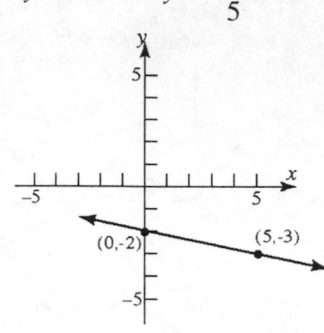

81. Parallel to $2x - 3y = -4$;
    Slope = $\frac{2}{3}$; containing (–5,3)
    $$y - y_1 = m(x - x_1)$$
    $$y - 3 = \frac{2}{3}(x - (-5)) \rightarrow y - 3 = \frac{2}{3}x + \frac{10}{3}$$
    $$y = \frac{2}{3}x + \frac{19}{3}$$
    $$2x - 3y = -19 \text{ or } y = \frac{2}{3}x + \frac{19}{3}$$

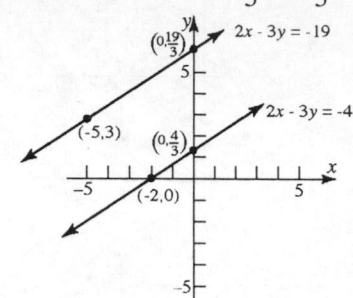

83. Perpendicular to $x + y = 2$;
    Containing (4,–3)
    Slope of perpendicular = 1
    $$y - y_1 = m(x - x_1)$$
    $$y - (-3) = 1(x - 4)$$
    $$y + 3 = x - 4$$
    $$y = x - 7$$
    $$x - y = 7 \text{ or } y = x - 7$$

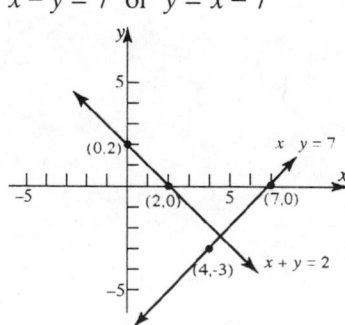

85. $2x + 5y = 10$
    $5y = -2x + 10$
    $$y = -\frac{2}{5}x + 2$$
    slope = $-\frac{2}{5}$; y-intercept = 2

87. Given the points (7,4) and (–3,2).
    Slope: $m = \frac{2-4}{-3-7} = \frac{-2}{-10} = \frac{1}{5}$; If x
    increases by 5 units, y will
    increase by 1 unit.
    Distance:
    $$d = \sqrt{(-3-7)^2 + (2-4)^2} = \sqrt{100+4}$$
    $$= \sqrt{104} = 2\sqrt{26}$$
    Midpoint: $\left(\frac{7+(-3)}{2}, \frac{4+2}{2}\right) = (2,3)$

89. Find the distance between each pair of points.
    $$d(A,B) = \sqrt{(1-3)^2 + (1-4)^2} = \sqrt{4+9} = \sqrt{13}$$
    $$d(B,C) = \sqrt{(-2-1)^2 + (3-1)^2} = \sqrt{9+4} = \sqrt{13}$$
    $$d(A,C) = \sqrt{(-2-3)^2 + (3-4)^2} = \sqrt{25+1} = \sqrt{26}$$
    Since $AB = BC$, triangle ABC is isosceles.

91. slope of $\overline{AB} = \dfrac{1-5}{6-2} = -1;$ slope of $\overline{AC} = \dfrac{-1-5}{8-2} = -1;$ slope of $\overline{BC} = \dfrac{-1-1}{8-6} = -1$

Therefore the points are collinear.

93. Endpoints of the diameter are $(-3, 2)$ and $(5, -6)$.
The center is at the midpoint of the diameter:

Center: $\left(\dfrac{-3+5}{2}, \dfrac{2+(-6)}{2} \right) = (1, -2)$

Radius: $r = \sqrt{(1-(-3))^2 + (-2-2)^2} = \sqrt{16+16} = \sqrt{32} = 4\sqrt{2}$

Equation:

$$(x-1)^2 + (y+2)^2 = \left(4\sqrt{2}\right)^2$$
$$x^2 - 2x + 1 + y^2 + 4y + 4 = 32$$
$$x^2 + y^2 - 2x + 4y - 27 = 0$$

95.   Answers will vary.

97. In each problem, we need to use the Least Common Multiple of the expressions
$x-2$ and x^2-4, namely x^2-4.

(a) We use x^2-4 as the LCD in order to combine the given expressions.

(b) We multiply each side of the equation by x^2-4 in order to clear out the denominators
before solving the equation.

(c) We use x^2-4 as the LCD in order to combine the terms on the left hand side before
solving the inequality.

Chapter 2

Linear and Quadratic Functions

2.1 Functions

1. Function
   Domain: {Dad, Colleen, Kaleigh,
               Marissa}
   Range: {Jan. 8, Mar. 15, Sept. 17}

3. Not a function

5. Not a function.

7. Function
   Domain: {1, 2, 3, 4}
   Range: {3}

9. Not a function

11. Function
   Domain: {−2, −1, 0, 1}
   Range: {4, 1, 0}

13. $f(x) = 2x + 5$
   (a) $f(0) = 2(0) + 5 = 5$
   (b) $f(1) = 2(1) + 5 = 7$
   (c) $f(-1) = 2(-1) + 5 = 3$
   (d) $f(-x) = 2(-x) + 5 = -2x + 5$
   (e) $-f(x) = -(2x + 5) = -2x - 5$
   (f) $f(x + 1) = 2(x + 1) + 5 = 2x + 2 + 5 = 2x + 7$
   (g) $f(2x) = 2(2x) + 5 = 4x + 5$
   (h) $f(x + h) = 2(x + h) + 5 = 2x + 2h + 5$

15. $f(x) = 3x^2 + 2x - 4$
   (a) $f(0) = 3(0)^2 + 2(0) - 4 = -4$
   (b) $f(1) = 3(1)^2 + 2(1) - 4 = 3 + 2 - 4 = 1$
   (c) $f(-1) = 3(-1)^2 + 2(-1) - 4 = 3 - 2 - 4 = -3$
   (d) $f(-x) = 3(-x)^2 + 2(-x) - 4 = 3x^2 - 2x - 4$
   (e) $-f(x) = -(3x^2 + 2x - 4) = -3x^2 - 2x + 4$
   (f) $f(x + 1) = 3(x + 1)^2 + 2(x + 1) - 4 = 3(x^2 + 2x + 1) + 2x + 2 - 4$
   $\qquad\qquad = 3x^2 + 6x + 3 + 2x + 2 - 4 = 3x^2 + 8x + 1$
   (g) $f(2x) = 3(2x)^2 + 2(2x) - 4 = 12x^2 + 4x - 4$
   (h) $f(x + h) = 3(x + h)^2 + 2(x + h) - 4 = 3(x^2 + 2xh + h^2) + 2x + 2h - 4$
   $\qquad = 3x^2 + 6xh + 3h^2 + 2x + 2h - 4$

17. $f(x) = \dfrac{x}{x^2 + 1}$

    (a) $f(0) = \dfrac{0}{0^2 + 1} = \dfrac{0}{1} = 0$

    (b) $f(1) = \dfrac{1}{1^2 + 1} = \dfrac{1}{2}$

    (c) $f(-1) = \dfrac{-1}{(-1)^2 + 1} = \dfrac{-1}{1+1} = -\dfrac{1}{2}$

    (d) $f(-x) = \dfrac{-x}{(-x)^2 + 1} = \dfrac{-x}{x^2 + 1}$

    (e) $-f(x) = -\dfrac{x}{x^2 + 1} = \dfrac{-x}{x^2 + 1}$

    (f) $f(x + 1) = \dfrac{x+1}{(x+1)^2 + 1} = \dfrac{x+1}{x^2 + 2x + 1 + 1} = \dfrac{x+1}{x^2 + 2x + 2}$

    (g) $f(2x) = \dfrac{2x}{(2x)^2 + 1} = \dfrac{2x}{4x^2 + 1}$

    (h) $f(x + h) = \dfrac{x+h}{(x+h)^2 + 1} = \dfrac{x+h}{x^2 + 2xh + h^2 + 1}$

19. $f(x) = |x| + 4$

    (a) $f(0) = |0| + 4 = 0 + 4 = 4$

    (b) $f(1) = |1| + 4 = 1 + 4 = 5$

    (c) $f(-1) = |-1| + 4 = 1 + 4 = 5$

    (d) $f(-x) = |-x| + 4 = |x| + 4$

    (e) $-f(x) = -(|x| + 4) = -|x| - 4$

    (f) $f(x + 1) = |x + 1| + 4$

    (g) $f(2x) = |2x| + 4 = 2|x| + 4$

    (h) $f(x + h) = |x + h| + 4$

21. Graph $y = x^2$. The graph passes the vertical line test. Thus, the equation represents a function.

23. Graph $y = \dfrac{1}{x}$. The graph passes the vertical line test. Thus, the equation represents a function.

25. $y^2 = 4 - x^2$

Solve for y: $y = \pm\sqrt{4 - x^2}$

For $x = 0$, $y = \pm 2$. Thus, $(0,2)$ and $(0,-2)$ are on the graph. This is not a function, since a distinct x corresponds to two different y's.

27. $x = y^2$

Solve for y: $y = \pm\sqrt{x}$

For $x = 1$, $y = \pm 1$. Thus, $(1,1)$ and $(1,-1)$ are on the graph. This is not a function, since a distinct x corresponds to two different y's.

29. Graph $y = 2x^2 - 3x + 4$. The graph passes the vertical line test. Thus, the equation represents a function.

31. $2x^2 + 3y^2 = 1$
Solve for y:

$$2x^2 + 3y^2 = 1 \rightarrow 3y^2 = 1 - 2x^2 \rightarrow y^2 = \frac{1 - 2x^2}{3} \rightarrow y = \pm\sqrt{\frac{1 - 2x^2}{3}}$$

For $x = 0$, $y = \pm\sqrt{\frac{1}{3}}$. Thus, $\left(0, \sqrt{\frac{1}{3}}\right)$ and $\left(0, -\sqrt{\frac{1}{3}}\right)$ are on the graph. This is not a function,

since a distinct x corresponds to two different y's.

33. $f(x) = -5x + 4$
Domain: {Real Numbers}

35. $f(x) = \frac{x}{x^2 + 1}$
Domain: {Real Numbers}

37. $g(x) = \frac{x}{x^2 - 16}$

$x^2 - 16 \neq 0$

$x^2 \neq 16 \Rightarrow x \neq \pm 4$
Domain: $\left\{x \mid x \neq -4, x \neq 4\right\}$

39. $F(x) = \frac{x - 2}{x^3 + x}$

$x^3 + x \neq 0$

$x(x^2 + 1) \neq 0$

$x \neq 0, \quad x^2 \neq -1$
Domain: $\left\{x \mid x \neq 0\right\}$

41. $h(x) = \sqrt{3x - 12}$

$3x - 12 \geq 0$

$3x \geq 12$

$x \geq 4$
Domain: $\left\{x \mid x \geq 4\right\}$

43. $f(x) = \frac{4}{\sqrt{x - 9}}$

$x - 9 > 0$

$x > 9$
Domain: $\left\{x \mid x > 9\right\}$

45. $p(x) = \sqrt{\frac{2}{x - 1}}$

$\frac{2}{x - 1} \geq 0 \Rightarrow x - 1 > 0 \Rightarrow x > 1$
Domain: $\left\{x \mid x > 1\right\}$

47. (a) $f(0) = 3$ since $(0, 3)$ is on the graph.
$f(-6) = -3$ since $(-6, -3)$ is on the graph.
(b) $f(6) = 0$ since $(6, 0)$ is on the graph.
$f(11) = 1$ since $(11, 1)$ is on the graph.
(c) $f(3)$ is positive since $f(3) \approx 3.7$.
(d) $f(-4)$ is negative since $f(-4) = -1$.
(e) $f(x) = 0$ when $x = -3$, $x = 6$, and $x = 10$.
(f) $f(x) > 0$ when $-3 < x < 6$, and $10 < x \leq 11$.
(g) The domain of f is $\left\{x \mid -6 \leq x \leq 11\right\}$ or $[-6, 11]$

(h)   The range of f is $\{y\,|-3 \le y \le 4\}$ or $[-3, 4]$
(i)   The x-intercepts are $(-3, 0)$, $(6, 0)$, and $(11, 0)$.
(j)   The y-intercept is $(0, 3)$
(k)   The line $y = \dfrac{1}{2}$ intersect the graph 3 times.
(l)   The line $x = 5$ intersects the graph 1 time
(m) $f(x) = 3$ when $x = 0$ and $x = 4$.
(n) $f(x) = -2$ when $x = -5$ and $x = 8$.

49.   Not a function since vertical lines will intersect the graph in more than one point.

51.   Function   (a)   Domain: $\{x\,|-\pi \le x \le \pi\}$;   Range: $\{y\,|-1 \le y \le 1\}$
                    (b) $\left(-\dfrac{\pi}{2},0\right)$, $\left(\dfrac{\pi}{2},0\right)$, $(0,1)$

53.   Not a function since vertical lines will intersect the graph in more than one point.

55.   Function   (a)   Domain: $\{x\,|\,x > 0\}$;   Range: $\{y\,|\,y \in \text{Real Numbers}\}$
                    (b) $(1, 0)$

57.   Function   (a)   Domain: $\{x\,|\,x \in \text{Real Numbers}\}$;   Range: $\{y\,|\,y \le 2\}$
                    (b) $(-3,0)$, $(3,0)$, $(0,2)$

59.   Function   (a)   Domain: $\{x\,|\,x \in \text{Real Numbers}\}$;   Range: $\{y\,|\,y \ge -3\}$
                    (b) $(1,0)$, $(3,0)$, $(0,9)$

61. $f(x) = 2x^2 - x - 1$
(a) $f(-1) = 2(-1)^2 - (-1) - 1 = 2$ $(-1,2)$ is on the graph of f.
(b) $f(-2) = 2(-2)^2 - (-2) - 1 = 9$ $(-2,9)$ is on the graph of f.
(c)   Solve for x:
$$-1 = 2x^2 - x - 1 \rightarrow 0 = 2x^2 - x$$
$$0 = x(2x - 1) \rightarrow x = 0, x = \frac{1}{2}$$
$(0,-1)$ and $\left(\dfrac{1}{2},-1\right)$ are points on the graph of f.
(d)   The domain of f is: $\{x\,|\,x \text{ is any real number}\}$.
(e)   x-intercepts:
$$f(x) = 0 \Rightarrow 2x^2 - x - 1 = 0$$
$$(2x + 1)(x - 1) = 0 \Rightarrow x = -\frac{1}{2}, x = 1$$
$\left(-\dfrac{1}{2},0\right)$ and $(1,0)$
(f)   y-intercept: $f(0) = 2(0)^2 - 0 - 1 = -1 \rightarrow (0,-1)$

63. $f(x) = \dfrac{x+2}{x-6}$

    (a) $f(3) = \dfrac{3+2}{3-6} = -\dfrac{5}{3} \neq 14$ $(3,14)$ is not on the graph of f.

    (b) $f(4) = \dfrac{4+2}{4-6} = \dfrac{6}{-2} = -3$ $(4,-3)$ is the point on the graph of f.

    (c)   Solve for x:

$$2 = \dfrac{x+2}{x-6}$$
$$2x - 12 = x + 2 \qquad\qquad (14, 2) \text{ is a point on the graph of } f.$$
$$x = 14$$

    (d)   The domain of f is: $\{x \mid x \neq 6\}$.

    (e)   x-intercepts:

$$f(x) = 0 \Rightarrow \dfrac{x+2}{x-6} = 0$$
$$x + 2 = 0 \Rightarrow x = -2$$
$$(-2, 0)$$

    (f)   y-intercept: $f(0) = \dfrac{0+2}{0-6} = -\dfrac{1}{3} \rightarrow \left(0, -\dfrac{1}{3}\right)$

65. $f(x) = \dfrac{2x^2}{x^4 + 1}$

    (a) $f(-1) = \dfrac{2(-1)^2}{(-1)^4 + 1} = \dfrac{2}{2} = 1$ $(-1,1)$ is a point on the graph of f.

    (b) $f(2) = \dfrac{2(2)^2}{(2)^4 + 1} = \dfrac{8}{17}$ $\left(2, \dfrac{8}{17}\right)$ is a point on the graph of f.

    (c)   Solve for x:

$$1 = \dfrac{2x^2}{x^4 + 1}$$
$$x^4 + 1 = 2x^2$$
$$x^4 - 2x^2 + 1 = 0 \qquad\qquad (1,1) \text{ and } (-1,1) \text{ are points on the graph of } f.$$
$$(x^2 - 1)^2 = 0$$

$$x^2 - 1 = 0 \Rightarrow x = \pm 1$$

    (d)   The domain of f is: $\{\text{Real Numbers}\}$.

    (e)   x-intercepts:

$$f(x) = 0 \Rightarrow \dfrac{2x^2}{x^4 + 1} = 0$$
$$2x^2 = 0 \Rightarrow x = 0$$
$$(0, 0)$$

    (f)   y-intercept: $f(0) = \dfrac{2x^2}{x^4 + 1} = \dfrac{0}{0+1} = 0 \Rightarrow (0,0)$

67. Solving for C:
$$f(x) = 2x^3 - 4x^2 + 4x + C \text{ and } f(2) = 5$$
$$f(2) = 2(2)^3 - 4(2)^2 + 4(2) + C$$
$$5 = 16 - 16 + 8 + C$$
$$-3 = C$$

69. Solving for A:
$$f(x) = \frac{3x + 8}{2x - A} \text{ and } f(0) = 2$$
$$f(0) = \frac{3(0) + 8}{2(0) - A}$$
$$2 = \frac{8}{-A} \Rightarrow -2A = 8 \Rightarrow A = -4$$

71. Solving for A:
$$f(x) = \frac{2x - A}{x - 3} \text{ and } f(4) = 0$$
$$f(4) = \frac{2(4) - A}{4 - 3}$$
$$0 = \frac{8 - A}{1}$$
$$0 = 8 - A$$
$$A = 8$$
f is undefined when $x = 3$.

73. $$f(x) = 4x + 3$$
$$\frac{f(x + h) - f(x)}{h} = \frac{4(x + h) + 3 - 4x - 3}{h} = \frac{4x + 4h + 3 - 4x - 3}{h} = \frac{4h}{h} = 4$$

75. $$f(x) = x^2 - x + 4$$
$$\frac{f(x + h) - f(x)}{h} = \frac{(x + h)^2 - (x + h) + 4 - (x^2 - x + 4)}{h}$$
$$= \frac{x^2 + 2xh + h^2 - x - h + 4 - x^2 + x - 4}{h} =$$
$$= \frac{2xh + h^2 - h}{h} = 2x + h - 1$$

77. $$f(x) = x^3 - 2$$
$$\frac{f(x + h) - f(x)}{h} = \frac{(x + h)^3 - 2 - (x^3 - 2)}{h}$$
$$= \frac{x^3 + 3x^2h + 3xh^2 + h^3 - 2 - x^3 + 2}{h}$$
$$= \frac{3x^2h + 3xh^2 + h^3}{h} = 3x^2 + 3xh + h^2$$

79. (a) III   (b) IV   (c) I   (d) V   (e) II

81.

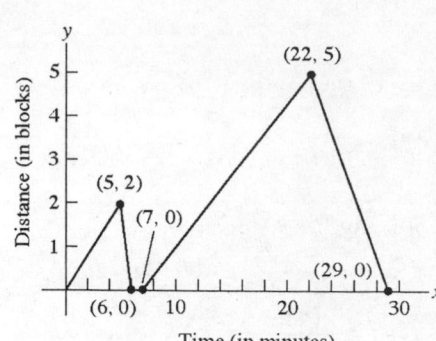

83. (a) 2 hours elapsed; $0 \le d \le 3$ miles.
    (b) 0.5 hours elapsed; $d = 3$ miles.
    (c) 0.3 hours elapsed; $0 \le d \le 3$ miles.
    (d) 0.2 hours elapsed; $d = 0$ miles.
    (e) 0.9 hours elapsed; $0 \le d \le 2.8$ miles.
    (f) 0.3 hours elapsed; $d = 2.8$ miles.
    (g) 1.1 hours elapsed; $0 \le d \le 2.8$ miles.
    (h) The furthest distance Kevin is from home is 3 miles.
    (i) 2 times.

85. (a) Graphing: $H(x) = 20 - 4.9x^2$

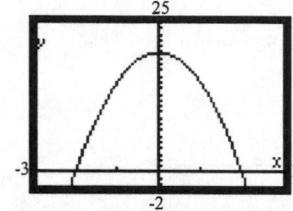

(b) $H(1) = 20 - 4.9(1)^2 = 20 - 4.9 = 15.1$ meters
    $H(1.1) = 20 - 4.9(1.1)^2 = 20 - 4.9(1.21) = 20 - 5.929 = 14.071$ meters
    $H(1.2) = 20 - 4.9(1.2)^2 = 20 - 4.9(1.44) = 20 - 7.056 = 12.944$ meters
    $H(1.3) = 20 - 4.9(1.3)^2 = 20 - 4.9(1.69) = 20 - 8.281 = 11.719$ meters

(c)
$$H(x) = 15$$
$$15 = 20 - 4.9x^2$$
$$-5 = -4.9x^2$$
$$x^2 \approx 1.0204$$
$$x \approx 1.01 \text{ seconds}$$

$$H(x) = 10$$
$$10 = 20 - 4.9x^2$$
$$-10 = -4.9x^2$$
$$x^2 \approx 2.0408$$
$$x \approx 1.43 \text{ seconds}$$

$$H(x) = 5$$
$$5 = 20 - 4.9x^2$$
$$-15 = -4.9x^2$$
$$x^2 \approx 3.0612$$
$$x \approx 1.75 \text{ seconds}$$

(d)
$$H(x) = 0$$
$$0 = 20 - 4.9x^2$$
$$-20 = -4.9x^2$$
$$x^2 \approx 4.0816$$
$$x \approx 2.02 \text{ seconds}$$

87. Let x represent the length of the rectangle.

Then $\dfrac{x}{2}$ represents the width of the rectangle, since the length is twice the width.

The function for the area is: $A(x) = x \cdot \dfrac{x}{2} = \dfrac{x^2}{2} = \dfrac{1}{2}x^2$

89. Let x represent the number of hours worked.
The function for the gross salary is: $G(x) = 10x$

91. $h(x) = \dfrac{-32x^2}{130^2} + x$

(a) $h(100) = \dfrac{-32(100)^2}{130^2} + 100 = \dfrac{-320000}{16900} + 100 \approx -18.93 + 100 = 81.07$ feet

(b) $h(300) = \dfrac{-32(300)^2}{130^2} + 300 = \dfrac{-2880000}{16900} + 300 \approx -170.41 + 300 = 129.59$ feet

(c) $h(500) = \dfrac{-32(500)^2}{130^2} + 500 = \dfrac{-8000000}{16900} + 500 \approx -473.37 + 500 = 26.63$ feet

(d)   Graphing $h(x) = \dfrac{-32x^2}{130^2} + x$

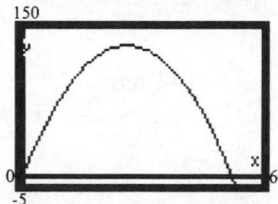

(e)   Solve $h(x) = \dfrac{-32x^2}{130^2} + x = 90$

$$\left(-\dfrac{32}{130^2}\right)x^2 + x - 90 = 0$$

$$x = \dfrac{-1 \pm \sqrt{1^2 - 4\left(-\dfrac{32}{130^2}\right)(-90)}}{2\left(-\dfrac{32}{130^2}\right)} \approx \dfrac{-1 \pm \sqrt{1 - 0.68166}}{-0.00379} \approx \dfrac{-1 \pm 0.5642}{-0.00379}$$

$x = \dfrac{-1 + 0.5642}{-0.00379} \approx 115$ feet   or $x = \dfrac{-1 - 0.5642}{-0.00379} \approx 413$ feet

Therefore, the ball reaches a height of 90 feet twice. The first time is when the ball has traveled 115 feet, and the second time is when the ball has traveled 413 feet.

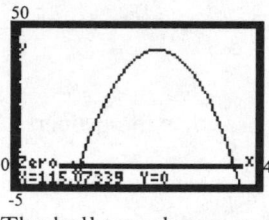
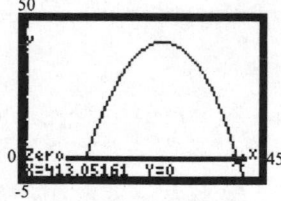

(f)   The ball travels approximately 275 feet before it reaches its maximum height of approximately 131.8 feet.

X	Y1	
0	0	
25	23.817	
50	45.266	
75	64.349	
100	81.065	
125	95.414	
150	107.4	
X=0		

X	Y1	
200	124.26	
225	129.14	
250	131.66	
275	131.8	
300	129.59	
325	125	
350	118.05	
X=350		

(g)   The ball travels approximately 265.5 feet before it reaches its maximum height of approximately 132.03 feet.

X	Y1	
263.5	132.03	
264	132.03	
264.5	132.03	
265	132.03	
265.5	132.03	
266	132.02	
266.5	132.02	

X=265.5

(h)   Solving $h(x) = \dfrac{-32x^2}{130^2} + x = 0$

$$\dfrac{-32x^2}{130^2} + x = 0 \Rightarrow x\left(\dfrac{-32x}{130^2} + 1\right) = 0 \Rightarrow x = 0 \ \text{ or } \ \dfrac{-32x}{130^2} + 1 = 0$$

$$\dfrac{-32x}{130^2} + 1 = 0 \Rightarrow 1 = \dfrac{32x}{130^2} \Rightarrow 130^2 = 32x \Rightarrow x = \dfrac{130^2}{32} \approx 528.125 \text{ feet}$$

Therefore, the domain of h is $\{x \mid 0 \le x \le 528.125\}$.

93. $C(x) = 100 + \dfrac{x}{10} + \dfrac{36000}{x}$

(a) $C(500) = 100 + \dfrac{500}{10} + \dfrac{36000}{500} = 100 + 50 + 72 = \$222$

(b) $C(450) = 100 + \dfrac{450}{10} + \dfrac{36000}{450} = 100 + 45 + 80 = \$225$

(c) $C(600) = 100 + \dfrac{600}{10} + \dfrac{36000}{600} = 100 + 60 + 60 = \$220$

(d) $C(400) = 100 + \dfrac{400}{10} + \dfrac{36000}{400} = 100 + 40 + 90 = \$230$

(e)   Graphing:

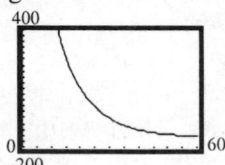

(f)   As x varies from 400 to 600 mph, the cost decreases from $230 to $220.

95.   (a) $h(x) = 2x$

$h(a+b) = 2(a+b) = 2a + 2b = h(a) + h(b);$ $h(x) = 2x$ has the property.

(b) $g(x) = x^2$

$g(a+b) = (a+b)^2 = a^2 + 2ab + b^2 \ne a^2 + b^2 = h(a) + h(b)$

$g(x) = x^2$ does not have the property.

(c) $F(x) = 5x - 2$

$F(a+b) = 5(a+b) - 2 = 5a + 5b - 2 \ne 5a - 2 + 5b - 2 = h(a) + h(b)$

$F(x) = 5x - 2$ does not have the property.

(d) $G(x) = \dfrac{1}{x}$

$G(a+b) = \dfrac{1}{a+b} \ne \dfrac{1}{a} + \dfrac{1}{b} = h(a) + h(b)$

$G(x) = \dfrac{1}{x}$ does not have the property.

97.   No, $x = -1$ is not in the domain of g, but it is in the domain of f.

Chapter 2

Linear and Quadratic Functions

2.2   Linear Functions and Models

1. $f(x) = 2x + 3$
      Slope = 2; y-intercept = 3

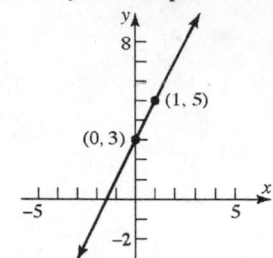

3. $h(x) = -3x + 4$
      Slope = -3; y-intercept = 4

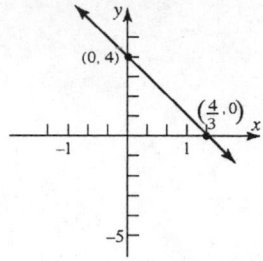

5. $f(x) = \dfrac{1}{4}x - 3$

      Slope = $\dfrac{1}{4}$; y-intercept = -3

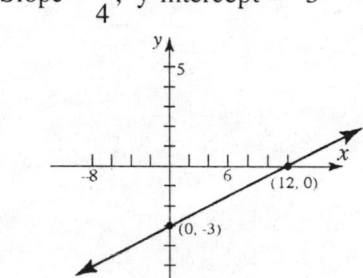

7. $F(x) = 4$
      Slope = 0; y-intercept = 4

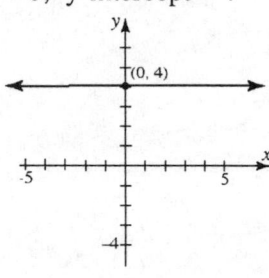

9. (a) Consider the data points (x, y), where x = the age in years of the computer and y = the value in dollars of the computer. So we have the points $(0, 3000)$ and $(3, 0)$.
   The slope formula yields:
   $$\text{slope} = \frac{\Delta y}{\Delta x} = \frac{0 - 3000}{3 - 0} = \frac{-3000}{3} = -1000 = m$$
   $(0, 3000)$ is the y-intercept, so $b = 3000$
   Therefore, the linear function is $f(x) = mx + b = -1000x + 3000$.

(b)   The graph of $f(x) = -1000x + 3000$

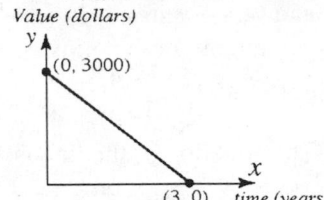

(c) The computer's value after 2 years is given by

$$f(2) = -1000(2) + 3000$$
$$= -2000 + 3000 = \$1000$$

11.(a)   Let x = the number of bicycles manufactured. We can use the cost function
$C(x) = mx + b$, with $m = 90$ and $b = 1800$.
Therefore $C(x) = 90x + 1800$

(b)   The graph of $C(x) = 90x + 1800$

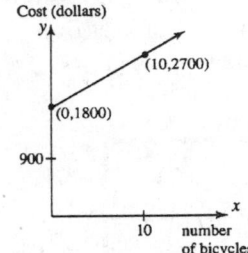

(c)   The cost of manufacturing 14 bicycles is given by $C(14) = 90(14) + 1800 = \$3060$.

13.   Linear, $m > 0$ 15.   Linear, $m < 0$ 17.   Nonlinear

19.   (a)

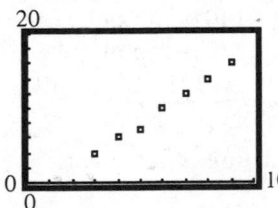

(b)   Answers will vary. We select (3,4) and (9,16). The slope of the line containing these points is:
$$m = \frac{16 - 4}{9 - 3} = \frac{12}{6} = 2$$
The equation of the line is:
$$y - y_1 = m(x - x_1)$$
$$y - 4 = 2(x - 3)$$
$$y - 4 = 2x - 6$$
$$y = 2x - 2$$

(c)

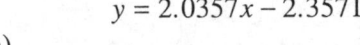

(d)   Using the LINear REGresssion program, the line of best fit is:
$$y = 2.0357x - 2.3571$$

(e)

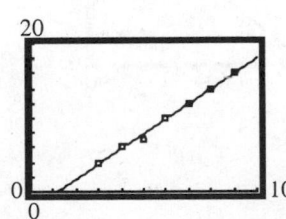

21. (a)

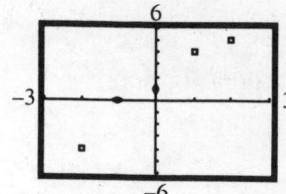

(b) Answers will vary. We select (–2,–4) and (2,5). The slope of the line containing these points is:
$$m = \frac{5-(-4)}{2-(-2)} = \frac{9}{4} = 2.25$$
The equation of the line is:
$$y - y_1 = m(x - x_1)$$
$$y - (-4) = 2.25(x - (-2))$$
$$y + 4 = 2.25x + 4.5$$
$$y = 2.25x + 0.5$$

(c)

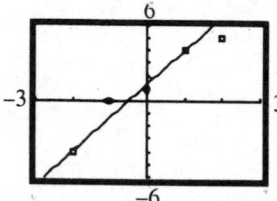

(d) Using the LINear REGresssion program, the line of best fit is:
$$y = 2.2x + 1.2$$

(e)

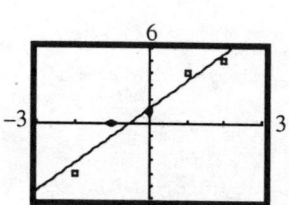

23. (a)

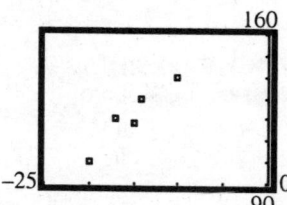

(b) Answers will vary. We select (–20,100) and (–10,140). The slope of the line containing these points is:
$$m = \frac{140-100}{-10-(-20)} = \frac{40}{10} = 4$$
The equation of the line is:
$$y - y_1 = m(x - x_1)$$
$$y - 100 = 4(x - (-20))$$
$$y - 100 = 4x + 80$$
$$y = 4x + 180$$

(c)

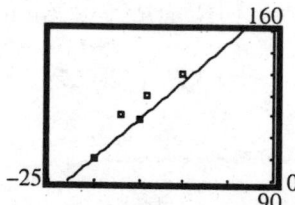

(d) Using the LINear REGresssion program, the line of best fit is:
$$y = 3.8613x + 180.2920$$

(e)

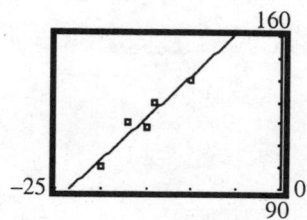

25. (a)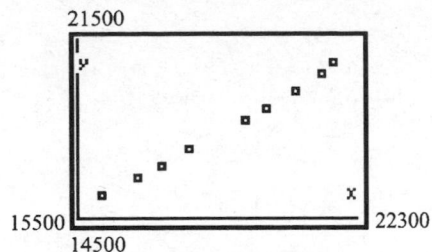

(b) Using the LINear REGression program, the line of best fit is:
$$C = 1.038I - 1897.507$$

(c) For each \$1 increase in disposable income, the per capita consumption increases by \$1.04

(d) $C = 1.038(21500) - 1897.507$
$$\approx \$20419.49$$

27. (a)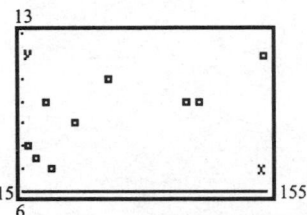

(b) Using the LINear REGression program, the line of best fit is:
$$L = 0.026G + 7.874$$

(c) For each 1 day increase in Gestation period, the life expectancy increases by 0.026 years.

(d) $L = 0.026(89) + 7.874$
$$\approx 10.188 \text{ years}$$

29. (a) The relation is not a function because 23 is paired with both 56 and 53.

(e) $D(p) = -1.3355p + 86.1974$

(b)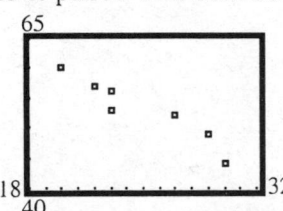

(f) Domain: $\{p \mid p > 0\}$

(c) Using the LINear REGression program, the line of best fit is:
$$D = -1.3355p + 86.1974$$

(g) $D(28) = -1.3355(28) + 86.1974$
$$\approx 48.8034$$
Demand is about 49 pairs.

(d) As the price of the jeans increases by \$1, the demand for the jeans decreases by 1.3355.

31. Let $p =$ the monthly payment and $B =$ the amount borrowed.
Consider the ordered pair (B, p).
We can use the points $(0,0)$ and $(1000, 6.49)$.
Now compute the slope:
$$\text{slope} = \frac{\Delta y}{\Delta x} = \frac{6.49 - 0}{1000 - 0} = \frac{6.49}{1000} = 0.00649$$

Therefore we have the linear function $p(B) = 0.00649B + 0 = 0.00649B$.
If $B = 145000$, then $p = (0.00649)(145000) = \$941.05$.

33. Let R = the revenue and g = the number of gallons of gasoline sold.
Consider the ordered pair (g, R).
We can use the points $(0, 0)$ and $(12, 15.84)$.
Now compute the slope:
$$\text{slope} = \frac{\Delta y}{\Delta x} = \frac{15.84 - 0}{12 - 0} = \frac{15.84}{12} = 1.32$$
Therefore we have the linear function $R(g) = 1.32g + 0 = 1.32g$.
If $g = 10.5$, then $R = (1.32)(10.5) \approx \$13.86$.

35. $v = kt$
$$64 = k(2) \Rightarrow k = 32$$

in 3 seconds
$$v = (32)(3) = 96 \text{ feet per second}$$

Chapter 2

Functions and Their Graphs

2.3   Properties of Functions

1.   Yes

3.   No. It only increases on (5, 10).

5. f is increasing on the intervals: $(-8, -2)$, $(0, 2)$, $(5, 10)$.

7.   Yes. The local maximum at $x = 2$ is 10.

9. f has local maxima at $x = -2$ and $x = 2$. The local maxima are 6 and 10, respectively.

11.   (a)   Intercepts: $(-2,0)$, $(2,0)$, and $(0,3)$.
        (b)   Domain: $\{x \mid -4 \le x \le 4\}$; Range: $\{y \mid 0 \le y \le 3\}$.
        (c)   Interval notation: Increasing: $(-2, 0)$ and $(2, 4)$; Decreasing: $(-4, -2)$ and $(0, 2)$.
        Inequality notation: Increasing: $-2 < x < 0$ and $2 < x < 4$
                            Decreasing: $-4 < x < -2$ and $0 < x < 2$
        (d)   Since the graph is symmetric to the y-axis, the function is <u>even</u>.

13.   (a)   Intercepts: $(0,1)$.
        (b)   Domain: { Real Numbers }; Range: $\{y \mid y > 0\}$.
        (c)   Interval notation: Increasing: $(-\infty, +\infty)$; Decreasing: never.
        Inequality notation: Increasing: $-\infty < x < +\infty$
                            Decreasing: never
        (d)   Since the graph is not symmetric to the y-axis or the origin, the function is <u>neither</u> even nor odd.

15.   (a)   Intercepts: $(-\pi,0)$, $(0,0)$, and $(\pi,0)$.
        (b)   Domain: $\{x \mid -\pi \le x \le \pi\}$; Range: $\{y \mid -1 \le y \le 1\}$.
        (c)   Interval notation: Increasing: $\left(-\dfrac{\pi}{2}, \dfrac{\pi}{2}\right)$; Decreasing: $\left(-\pi, -\dfrac{\pi}{2}\right)$ and $\left(\dfrac{\pi}{2}, \pi\right)$.

        Inequality notation: Increasing: $-\dfrac{\pi}{2} < x < \dfrac{\pi}{2}$

                            Decreasing: $-\pi < x < -\dfrac{\pi}{2}$ and $\dfrac{\pi}{2} < x < \pi$

        (d)   Since the graph is symmetric to the origin, the function is <u>odd</u>.

17. (a) Intercepts: $\left(0, \dfrac{1}{2}\right), \left(\dfrac{1}{2}, 0\right),$ and $\left(\dfrac{5}{2}, 0\right)$.

     (b) Domain: $\{x \mid -3 \le x \le 3\}$; Range: $\{y \mid -1 \le y \le 2\}$.

     (c) Interval notation: Increasing: $(2, 3)$; Decreasing: $(-1, 1)$;

                                Constant: $(-3, -1)$ and $(1, 2)$.

           Inequality notation: Increasing: $2 < x < 3$; Decreasing: $-1 < x < 1$;

                                 Constant: $-3 < x < -1$ and $1 < x < 2$.

     (d) Since the graph is not symmetric to the y-axis or the origin, the function is <u>neither</u> even nor odd.

19. (a) Intercepts: $(0, 2), (-2, 0),$ and $(2, 0)$.

     (b) Domain: $\{x \mid -4 \le x \le 4\}$; Range: $\{y \mid 0 \le y \le 2\}$.

     (c) Interval notation: Increasing: $(-2, 0)$ and $(2, 4)$;

                             Decreasing: $(-4, -2)$ and $(0, 2)$.

           Inequality notation: Increasing: $-2 < x < 0$ and $2 < x < 4$;

                             Decreasing: $-4 < x < -2$ and $0 < x < 2$.

     (d) Since the graph is symmetric to the y-axis, the function is <u>even</u>.

21. (a) f has a local maximum of 3 at $x = 0$.

     (b) f has a local minimum of 0 at both $x = -2$ and $x = 2$.

23. (a) f has a local maximum of 1 at $x = \dfrac{\pi}{2}$.

     (b) f has a local minimum of -1 at $x = -\dfrac{\pi}{2}$.

25. $f(x) = 5x$

(a) $\dfrac{f(x) - f(1)}{x - 1} = \dfrac{5x - 5}{x - 1} = \dfrac{5(x - 1)}{x - 1} = 5$

(b) $\dfrac{f(2) - f(1)}{2 - 1} = \dfrac{10 - 5}{2 - 1} = \dfrac{5}{1} = 5$

(c) Slope = 5; Containing $(1, 5)$:

$y - 5 = 5(x - 1)$

$y - 5 = 5x - 5 \Rightarrow y = 5x$

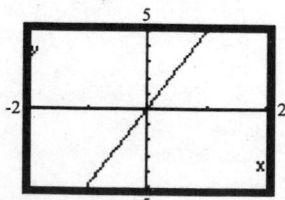

27. $f(x) = 1 - 3x$

(a) $\dfrac{f(x) - f(1)}{x - 1} = \dfrac{1 - 3x - (-2)}{x - 1}$

$= \dfrac{-3x + 3}{x - 1} = \dfrac{-3(x - 1)}{x - 1} = -3$

(b) $\dfrac{f(2) - f(1)}{2 - 1} = \dfrac{1 - 3(2) - (-2)}{2 - 1} = \dfrac{-3}{1} = -3$

(c) Slope = -3; Containing $(1, -2)$:

$y - (-2) = -3(x - 1)$

$y + 2 = -3x + 3 \Rightarrow y = -3x + 1$

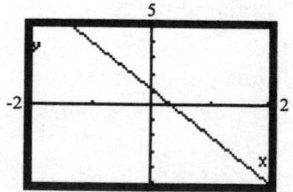

29. $f(x) = x^2 - 2x$

(a) $\dfrac{f(x) - f(1)}{x-1} = \dfrac{x^2 - 2x - (-1)}{x-1}$

    $= \dfrac{x^2 - 2x + 1}{x-1} = \dfrac{(x-1)^2}{x-1} = x - 1$

(b) $\dfrac{f(2) - f(1)}{2-1} = \dfrac{2^2 - 2(2) - (-1)}{2-1} = \dfrac{1}{1} = 1$

(c)   Slope = 1; Containing $(1, -1)$:

    $y - (-1) = 1(x-1)$

      $y + 1 = 1x - 1 \Rightarrow y = x - 2$

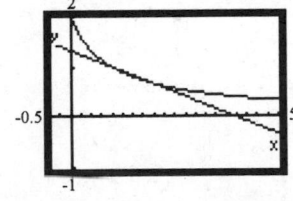

31. $f(x) = x^3 - x$

(a) $\dfrac{f(x) - f(1)}{x-1} = \dfrac{x^3 - x - 0}{x-1} = \dfrac{x^3 - x}{x-1}$

    $= \dfrac{x(x-1)(x+1)}{x-1} = x^2 + x$

(b) $\dfrac{f(2) - f(1)}{2-1} = \dfrac{2^3 - 2 - 0}{2-1} = \dfrac{6}{1} = 6$

(c)   Slope = 6; Containing $(1, 0)$:

    $y - 0 = 6(x-1) \Rightarrow y = 6x - 6$

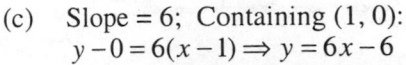
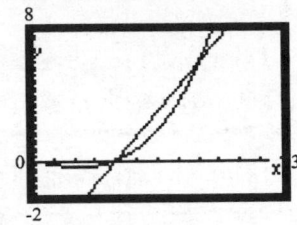

33. $f(x) = \dfrac{2}{x+1}$

(a)

$\dfrac{f(x) - f(1)}{x-1} = \dfrac{\left(\dfrac{2}{x+1} - 1\right)}{x-1} = \dfrac{\left(\dfrac{2 - x - 1}{x+1}\right)}{x-1}$

    $= \dfrac{1-x}{(x-1)(x+1)} = \dfrac{-1}{x+1}$

(b)

$\dfrac{f(2) - f(1)}{2-1} = \dfrac{\left(\dfrac{2}{2+1} - 1\right)}{2-1} = \dfrac{\left(-\dfrac{1}{3}\right)}{1} = -\dfrac{1}{3}$

(c)

Slope $= -\dfrac{1}{3}$; Containing $(1, 1)$:

$y - 1 = -\dfrac{1}{3}(x-1)$

$y - 1 = -\dfrac{1}{3}x + \dfrac{1}{3} \Rightarrow y = -\dfrac{1}{3}x + \dfrac{4}{3}$

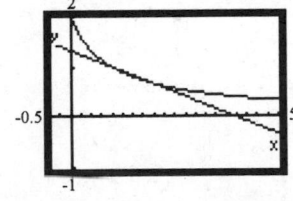

35. $f(x) = \sqrt{x}$

(a) $\dfrac{f(x) - f(1)}{x-1} = \dfrac{\sqrt{x} - 1}{x-1}$

(b) $\dfrac{f(2) - f(1)}{2-1} = \dfrac{\sqrt{2} - 1}{1} = \sqrt{2} - 1$

(c)   Slope $= \sqrt{2} - 1$; Containing $(1, 1)$:

    $y - 1 = \left(\sqrt{2} - 1\right)(x-1)$

    $y - 1 = \left(\sqrt{2} - 1\right)x - \left(\sqrt{2} - 1\right)$

      $y = \left(\sqrt{2} - 1\right)x - \sqrt{2} + 2$

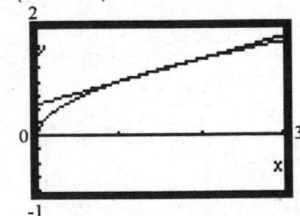

37. $f(x) = 4x^3$

$\quad\quad f(-x) = 4(-x)^3 = -4x^3$

$\quad$ f is odd.

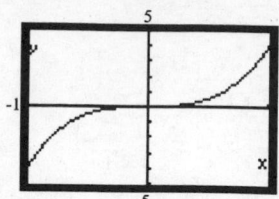

39. $g(x) = -3x^2 - 5$

$\quad\quad g(-x) = -3(-x)^2 - 5 = -3x^2 - 5$

$\quad$ g is even.

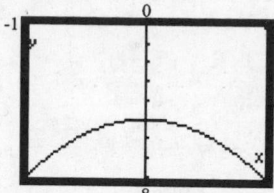

41. $F(x) = \sqrt[3]{x}$

$\quad\quad F(-x) = \sqrt[3]{-x} = -\sqrt[3]{x}$

$\quad$ F is odd.

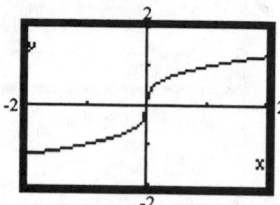

43. $f(x) = x + |x|$

$\quad\quad f(-x) = -x + |-x| = -x + |x|$

$\quad$ f is neither even nor odd.

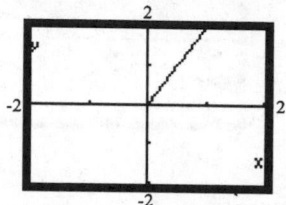

45. $g(x) = \dfrac{1}{x^2}$

$\quad\quad g(-x) = \dfrac{1}{(-x)^2} = \dfrac{1}{x^2}$

$\quad$ g is even.

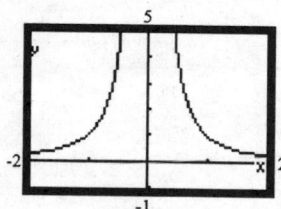

47. $h(x) = \dfrac{-x^3}{3x^2 - 9}$

$\quad\quad h(-x) = \dfrac{-(-x)^3}{3(-x)^2 - 9} = \dfrac{x^3}{3x^2 - 9}$

$\quad$ h is odd.

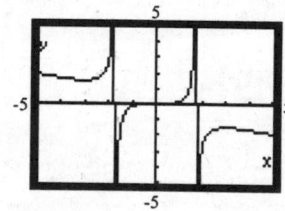

49. One at most because if f is increasing it could only cross the x-axis at most one time. It could not "turn" and cross it again or it would start to decrease.

51. $f(x) = x^3 - 3x + 2$ on the interval $(-2,2)$

Use MAXIMUM and MINIMUM on the graph of $y_1 = x^3 - 3x + 2$.

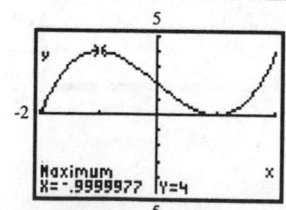
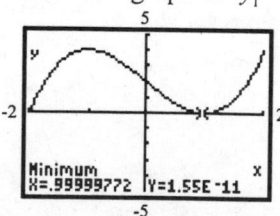

local maximum at: $(-1,4)$;   local minimum at: $(1,0)$

f is increasing on: $(-2,-1) \cup (1,2)$; f is decreasing on: $(-1,1)$

53. $f(x) = x^5 - x^3$ on the interval $(-2,2)$

Use MAXIMUM and MINIMUM on the graph of $y_1 = x^5 - x^3$.

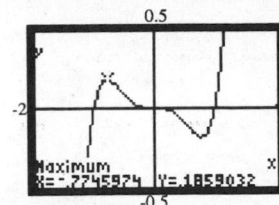 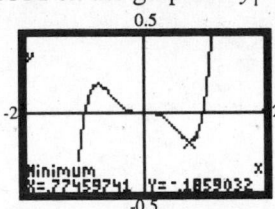

local maximum at: $(-0.78, 0.19)$;   local minimum at: $(0.78, -0.19)$

f is increasing on: $(-2, -0.78) \cup (0.78, 2)$; f is decreasing on: $(-0.78, 0.78)$

55. $f(x) = -0.2x^3 - 0.6x^2 + 4x - 6$ on the interval $(-6,4)$

Use MAXIMUM and MINIMUM on the graph of $y_1 = -0.2x^3 - 0.6x^2 + 4x - 6$.

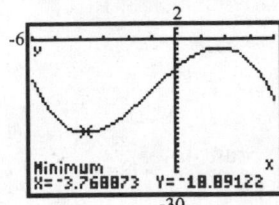 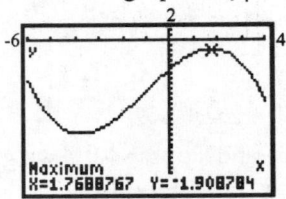

local maximum at: $(1.77, -1.91)$;   local minimum at: $(-3.77, -18.89)$

f is increasing on: $(-3.77, 1.77)$; f is decreasing on: $(-6, -3.77) \cup (1.77, 4)$

57. $f(x) = 0.25x^4 + 0.3x^3 - 0.9x^2 + 3$ on the interval $(-3,2)$

Use MAXIMUM and MINIMUM on the graph of $y_1 = 0.25x^4 + 0.3x^3 - 0.9x^2 + 3$.

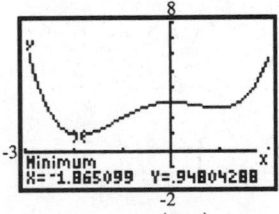 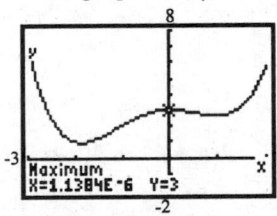 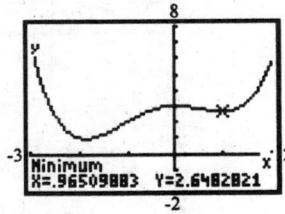

local maximum at: $(0,3)$; local minimum at: $(-1.86, 0.95)$, $(0.96, 2.65)$

f is increasing on: $(-1.86, 0) \cup (0.96, 2)$; f is decreasing on: $(-3, -1.86) \cup (0, 0.96)$

59. $f(x) = 2x + 5$

(a) $m_{sec} = \dfrac{f(x+h) - f(x)}{h} = \dfrac{2(x+h) + 5 - 2x - 5}{h} = \dfrac{2h}{h} = 2$

(b)   When $x = 1$,

$h = 0.5 \Rightarrow m_{sec} = 2$

$h = 0.1 \Rightarrow m_{sec} = 2$

$h = 0.01 \Rightarrow m_{sec} = 2$

as $h \to 0$, $m_{sec} \to 2$

(c)   Using point $(1, f(1)) = (1, 7)$ and slope = 2, we get the secant line:

$y - 7 = 2(x - 1) \Rightarrow y - 7 = 2x - 2 \Rightarrow y = 2x + 5$

(d) Graphing:

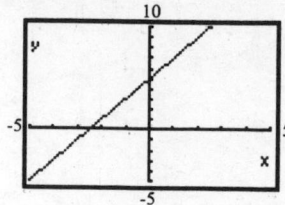

61. $f(x) = x^2 + 2x$

(a) $m_{sec} = \dfrac{f(x+h) - f(x)}{h} = \dfrac{(x+h)^2 + 2(x+h) - (x^2 + 2x)}{h}$

$= \dfrac{x^2 + 2xh + h^2 + 2x + 2h - x^2 - 2x}{h} = \dfrac{2xh + h^2 + 2h}{h} = 2x + h + 2$

(b) When $x = 1$,

$h = 0.5 \Rightarrow m_{sec} = 2 \cdot 1 + 0.5 + 2 = 4.5$

$h = 0.1 \Rightarrow m_{sec} = 2 \cdot 1 + 0.1 + 2 = 4.1$

$h = 0.01 \Rightarrow m_{sec} = 2 \cdot 1 + 0.01 + 2 = 4.01$

as $h \to 0$, $m_{sec} \to 2 \cdot 1 + 0 + 2 = 4$

(c) Using point $(1, f(1)) = (1, 3)$ and slope $= 4.01$, we get the secant line:

$y - 3 = 4.01(x - 1) \Rightarrow y - 3 = 4.01x - 4.01 \Rightarrow y = 4.01x - 1.01$

(d) Graphing:

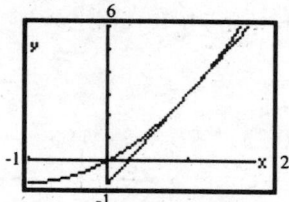

63. $f(x) = 2x^2 - 3x + 1$

(a) $m_{sec} = \dfrac{f(x+h) - f(x)}{h} = \dfrac{2(x+h)^2 - 3(x+h) + 1 - (2x^2 - 3x + 1)}{h}$

$= \dfrac{2(x^2 + 2xh + h^2) - 3x - 3h + 1 - 2x^2 + 3x - 1}{h}$

$= \dfrac{2x^2 + 4xh + 2h^2 - 3x - 3h + 1 - 2x^2 + 3x - 1}{h}$

$= \dfrac{4xh + 2h^2 - 3h}{h} = 4x + 2h - 3$

(b) When $x = 1$,

$h = 0.5 \Rightarrow m_{sec} = 4 \cdot 1 + 2(0.5) - 3 = 2$

$h = 0.1 \Rightarrow m_{sec} = 4 \cdot 1 + 2(0.1) - 3 = 1.2$

$h = 0.01 \Rightarrow m_{sec} = 4 \cdot 1 + 2(0.01) - 3 = 1.02$

as $h \to 0$, $m_{sec} \to 4 \cdot 1 + 2(0) - 3 = 1$

(c) Using point $(1, f(1)) = (1, 0)$ and slope $= 1.02$, we get the secant line:

$y - 0 = 1.02(x - 1) \Rightarrow y = 1.02x - 1.02$

(d)   Graphing:

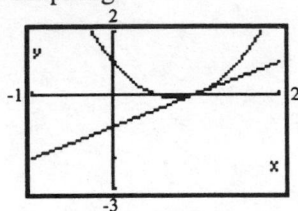

65. $f(x) = \dfrac{1}{x}$

(a)

$$m_{\text{sec}} = \frac{f(x+h) - f(x)}{h} = \frac{\left(\dfrac{1}{x+h} - \dfrac{1}{x}\right)}{h}$$

$$= \frac{\left(\dfrac{x - (x+h)}{(x+h)x}\right)}{h} = \left(\frac{x - x - h}{(x+h)x}\right)\left(\frac{1}{h}\right) = \left(\frac{-h}{(x+h)x}\right)\left(\frac{1}{h}\right) = -\frac{1}{(x+h)x}$$

(b)   When $x = 1$,

$$h = 0.5 \Rightarrow m_{\text{sec}} = -\frac{1}{(1+0.5)(1)} = -\frac{1}{1.5} \approx -0.667$$

$$h = 0.1 \Rightarrow m_{\text{sec}} = -\frac{1}{(1+0.1)(1)} = -\frac{1}{1.1} \approx -0.909$$

$$h = 0.01 \Rightarrow m_{\text{sec}} = -\frac{1}{(1+0.01)(1)} = -\frac{1}{1.01} \approx -0.990$$

$$\text{as } h \to 0, \ m_{\text{sec}} \to -\frac{1}{(1+0)(1)} = -\frac{1}{1} = -1$$

(c)   Using point $(1, f(1)) = (1,1)$ and slope $= -0.990$, we get the secant line:
$$y - 1 = -0.99(x - 1) \Rightarrow y - 1 = -0.99x + 0.99 \Rightarrow y = -0.99x + 1.99$$

(d)   Graphing:

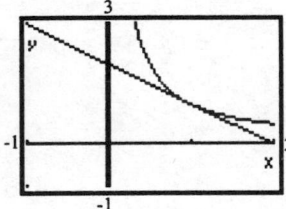

67.   Graphing: $V(x) = x(24 - 2x^2)$

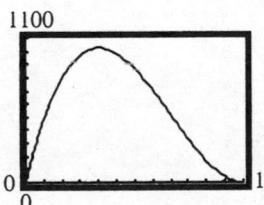   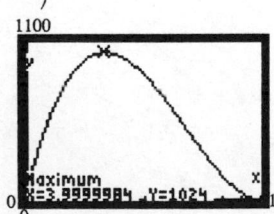

Use MAXIMUM. The volume is largest when $x = 4$ inches.

69. $s(t) = -16t^2 + 80t + 6$

(a) Graphing:

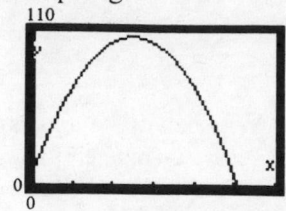

(b) The graph of $s(t) = -16t^2 + 80t + 6$ is a parabola that opens up since $a = -16 < 0$. The maximum value for s occurs at the vertex. That is when

$$t = \frac{-b}{2a} = \frac{-80}{(2)(-16)} = 2.5 \text{ seconds.}$$

(c) The maximum height is $s(2.5) = -16(2.5)^2 + (80)(2.5) + 6 = -100 + 200 + 6 = 106$ feet.

71. (a), (b), (e)

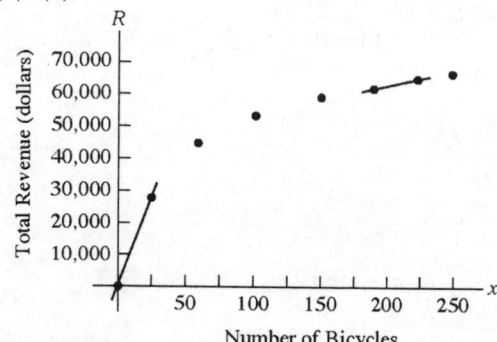

(c) Average rate of change $= \dfrac{28000 - 0}{25 - 0} = \dfrac{28000}{25} = 1120$

(d) For each additional bicycle sold between 0 and 25, the total revenue increases by $1120.

(f) Average rate of change $= \dfrac{64835 - 62360}{223 - 190} = \dfrac{2475}{33} = 75$

(g) For each additional bicycle sold between 190 and 223, the total revenue increases by $75.

73. (a), (b), (e)

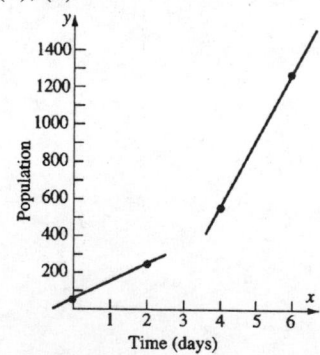

(c)   Average rate of change $= \dfrac{234-50}{2-0} = \dfrac{184}{2} = 92$

(d)   The population is increasing at a rate of 92 per day between day 0 and day 2.

(f)   Average rate of change $= \dfrac{1280-547}{6-4} = \dfrac{733}{2} = 366.5$

(g)   The population is increasing at a rate of 366.5 per day between day 4 and day 6.

(h)   As time passes, the average rate of change of the population is increasing.

Functions and Their Graphs

2.4 Library of Functions; Piecewise-Defined Functions

1. C   3. E   5. B   7. F

9.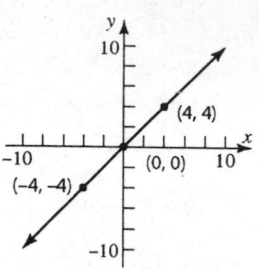

11.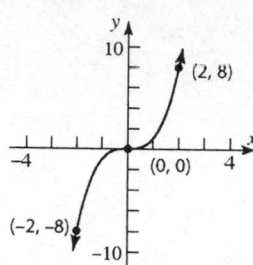

13.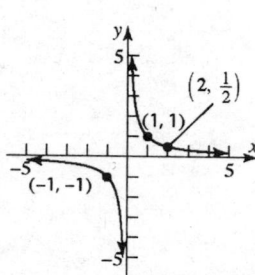

15. (a) $f(-2) = (-2)^2 = 4$
    (b) $f(0) = 2$
    (c) $f(2) = 2(2) + 1 = 5$

17. (a) $f(1.2) = \text{int}(2(1.2)) = \text{int}(2.4) = 2$
    (b) $f(1.6) = \text{int}(2(1.6)) = \text{int}(3.2) = 3$
    (c) $f(-1.8) = \text{int}(2(-1.8)) = \text{int}(-3.6) = -4$

19. $f(x) = \begin{cases} 2x & \text{if } x \neq 0 \\ 1 & \text{if } x = 0 \end{cases}$

    (a) Domain: {Real Numbers}
    (b) x-intercept: none
        y-intercept: (0,1)

    (c)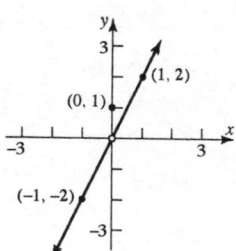

    (d) Range: $\left\{ y \mid y \neq 0 \right\}$
    (e) graphing utility:

    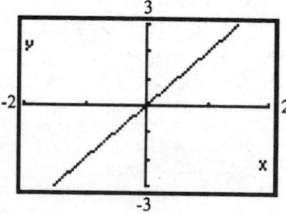

21. $f(x) = \begin{cases} -2x+3 & \text{if } x < 1 \\ 3x-2 & \text{if } x \geq 1 \end{cases}$

(a) Domain: {Real Numbers}

(b) x-intercept: none
y-intercept: (0,3)

(c)

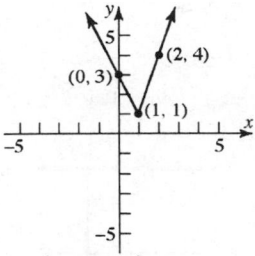

(d) Range: $\{y \mid y \geq 1\}$

(e) graphing utility:

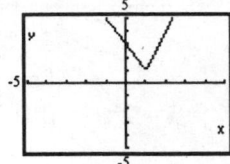

23. $f(x) = \begin{cases} x+3 & \text{if } -2 \leq x < 1 \\ 5 & \text{if } x = 1 \\ -x+2 & \text{if } x > 1 \end{cases}$

(a) Domain: $\{x \mid x \geq -2\}$

(b) x-intercept: (2, 0)
y-intercept: (0, 3)

(c)

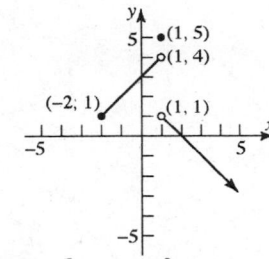

(d) Range: $\{y \mid y < 4\} \cup \{5\}$

(e) graphing utility:

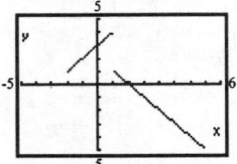

25. $f(x) = \begin{cases} 1+x & \text{if } x < 0 \\ x^2 & \text{if } x \geq 0 \end{cases}$

(a) Domain: {Real Numbers}

(b) x-intercept: (−1,0), (0,0)
y-intercept: (0,0)

(c)

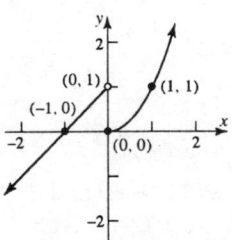

(d) Range: {Real Numbers}

(e) graphing utility:

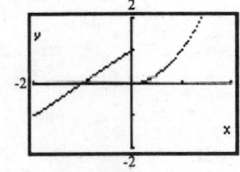

27. $f(x) = \begin{cases} |x| & \text{if } -2 \leq x < 0 \\ 1 & \text{if } x = 0 \\ x^3 & \text{if } x > 0 \end{cases}$

(a) Domain: $\{x \mid x \geq -2\}$

(b) x-intercept: none
y-intercept: (0, 1)

(c)

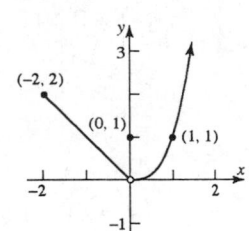

(d) Range: $\{y \mid y > 0\}$

(e) graphing utility:

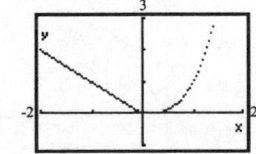

79

29. $h(x) = 2\,\text{int}(x)$

   (a) Domain: {Real Numbers}

   (b) x-intercept: all ordered pairs

        $(x,0)$ when $0 \le x < 1$.

       y-intercept: (0,0)

(c)

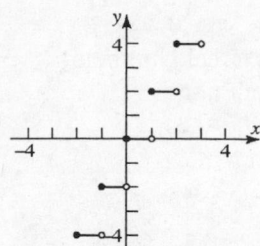

(d) Range: {Even Integers}

(e) graphing utility:

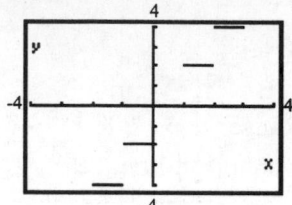

31. $f(x) = \begin{cases} -x & \text{if } -1 \le x \le 0 \\ \dfrac{1}{2}x & \text{if } 0 < x \le 2 \end{cases}$

33. $f(x) = \begin{cases} -x & \text{if } x \le 0 \\ -x+2 & \text{if } 0 < x \le 2 \end{cases}$

35. (a) Let x represent the number of miles and C be the cost of transportation.

$$C(x) = \begin{cases} 0.50x & \text{if } 0 \le x \le 100 \\ 0.50(100) + 0.40(x-100) & \text{if } 100 < x \le 400 \\ 0.50(100) + 0.40(300) + 0.25(x-400) & \text{if } 400 < x \le 800 \\ 0.50(100) + 0.40(300) + 0.25(400) + 0(x-800) & \text{if } 800 < x \le 960 \end{cases}$$

$$C(x) = \begin{cases} 0.50x & \text{if } 0 \le x \le 100 \\ 10 + 0.40x & \text{if } 100 < x \le 400 \\ 70 + 0.25x & \text{if } 400 < x \le 800 \\ 270 & \text{if } 800 < x \le 960 \end{cases}$$

Graphing:

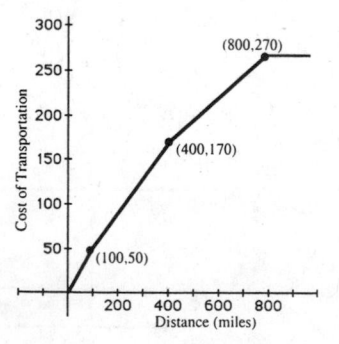

(b) For hauls between 100 and 400 miles the cost is: $C(x) = 10 + 0.40x$.

(c) For hauls between 400 and 800 miles the cost is: $C(x) = 70 + 0.25x$.

37.   Let x = the amount of the bill in dollars. The minimum payment due is given by

$$f(x) = \begin{cases} x & \text{if } x < 10 \\ 10 & \text{if } 10 \le x < 500 \\ 30 & \text{if } 500 \le x < 1000 \\ 50 & \text{if } 1000 \le x < 1500 \\ 70 & \text{if } 1500 \le x \end{cases}$$

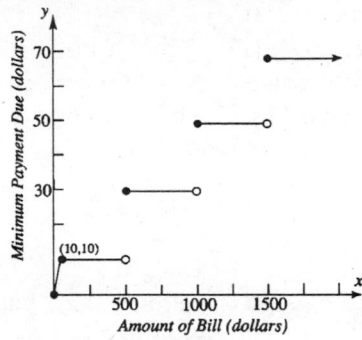

39.   For schedule X:

$$f(x) = \begin{cases} 0.15x & \text{if } 0 < x \le 27{,}050 \\ 4057.50 + 0.28(x - 27{,}050) & \text{if } 27{,}050 < x \le 65{,}550 \\ 14{,}837.50 + 0.31(x - 65{,}550) & \text{if } 65{,}550 < x \le 136{,}750 \\ 36{,}909.50 + 0.36(x - 136{,}750) & \text{if } 136{,}750 < x \le 297{,}300 \\ 94{,}707.50 + 0.396(x - 297{,}300) & \text{if } x > 297{,}300 \end{cases}$$

For Schedule Y-1:

$$f(x) = \begin{cases} 0.15x & \text{if } 0 < x \le 45{,}200 \\ 6{,}780.00 + 0.28(x - 45{,}200) & \text{if } 45{,}200 < x \le 109{,}250 \\ 24{,}714.00 + 0.31(x - 109{,}250) & \text{if } 109{,}250 < x \le 166{,}450 \\ 42{,}446.00 + 0.36(x - 166{,}450) & \text{if } 166{,}450 < x \le 297{,}300 \\ 89{,}552.00 + 0.396(x - 297{,}300) & \text{if } x > 297{,}300 \end{cases}$$

41.   (a)   Charge for 40 therms:
$C = 6.45 + 0.2012(20) + 0.1117(20) + 0.3209(40) = \$25.54$

(b)   Charge for 202 therms:
$C = 6.45 + 0.2012(20) + 0.1117(30) + 0.0374(152) + 0.3209(202) = \$84.33$

(c)   The monthly charge function:

$$C = \begin{cases} 6.45 + 0.2012x + 0.3209x & \text{for } 0 \le x \le 20 \\ 6.45 + 0.2012(20) + 0.1117(x - 20) + 0.3209x & \text{for } 20 < x \le 50 \\ 6.45 + 0.2012(20) + 0.1117(30) + 0.0374(x - 50) + 0.3209x & \text{for } x > 50 \end{cases}$$

$$= \begin{cases} 6.45 + 0.5221x & \text{for } 0 \le x \le 20 \\ 6.45 + 4.024 + 0.1117x - 2.234 + 0.3209x & \text{for } 20 < x \le 50 \\ 6.45 + 4.024 + 3.351 + 0.0374x - 1.87 + 0.3209x & \text{for } x > 50 \end{cases}$$

$$= \begin{cases} 6.45 + 0.5221x & \text{for } 0 \le x \le 20 \\ 8.24 + 0.4326x & \text{for } 20 < x \le 50 \\ 11.955 + 0.3583x & \text{for } x > 50 \end{cases}$$

(d)

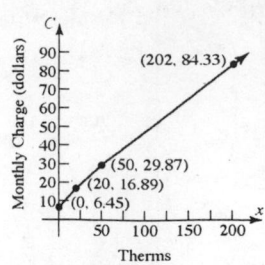

43.   Each graph is that of $y = x^2$, but shifted vertically.

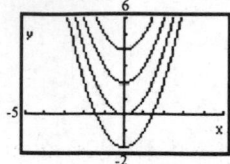

If $y = x^2 + k$, $k > 0$, the shift is up k units; if $y = x^2 + k$, $k < 0$, the shift is down $|k|$ units. The graph of $y = x^2 - 4$ is the same as the graph of $y = x^2$, but shifted down 4 units. The graph of $y = x^2 + 5$ is the graph of $y = x^2$, but shifted up 5 units.

45.   Each graph is that of $y = |x|$, but either compressed or stretched vertically.

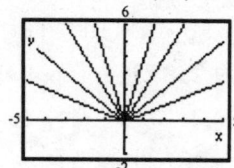

If $y = k|x|$ and $k > 1$, the graph is stretched; if $y = k|x|$ and $0 < k < 1$, the graph is compressed. The graph of $y = \frac{1}{4}|x|$ is the same as the graph of $y = |x|$, but compressed. The graph of $y = 5|x|$ is the same as the graph of $y = |x|$, but stretched.

47.   The graph of $y = \sqrt{-x}$ is the reflection about the y-axis of the graph of $y = \sqrt{x}$.

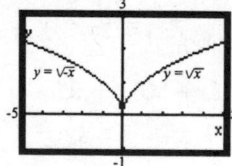

The same type of reflection occurs when graphing $y = 2x + 1$ and $y = 2(-x) + 1$.

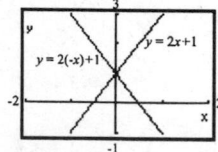

The conclusion is that the graph of $y = f(-x)$ is the reflection about the y-axis of the graph of $y = f(x)$.

49.   For the graph of $y = x^n$, n a positive even integer, as n increases, the graph of the function is narrower for $|x| > 1$ and flatter for $|x| < 1$.

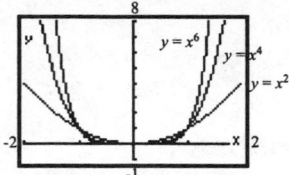

51. $f(x) = \begin{cases} 1 & \text{if } x \text{ is rational} \\ 0 & \text{if } x \text{ is irrational} \end{cases}$ Domain = { all real numbers}   Range = {0,1}

y-intercept: $x = 0 \Rightarrow x$ is rational $\Rightarrow y = 1$, so the y-intercept is (0, 1).

x-intercept: $y = 0 \Rightarrow x$ is irrational, so the graph has infinitely many x-intercepts, namely, there is an x-intercept at each irrational value for x.

$f(-x) = 1 = f(x)$ when x is rational; $f(-x) = 0 = f(x)$ when x is irrational $\therefore f$ is even.

The graph of f consists of 2 infinite clusters of distinct points, extending horizontally in both directions.

One cluster is located 1 unit above the x-axis, and the other is located along the x-axis.

Chapter **2**

Functions and Their Graphs

2.5   Graphing Techniques; Transformations

1. B   3. H   5. I   7. L

9. F   11. G   13. C   15. B

17. $y = (x-4)^3$ 19. $y = x^3 + 4$ 21. $y = -x^3$ 23. $y = 4x^3$

25. (1) $y = \sqrt{x} + 2$
     (2) $y = -\left(\sqrt{x} + 2\right)$
     (3) $y = -\left(\sqrt{-x} + 2\right)$

27. (1) $y = -\sqrt{x}$
     (2) $y = -\sqrt{x} + 2$
     (3) $y = -\sqrt{x+3} + 2$

29. c

31. c

33. $f(x) = x^2 - 1$
     Using the graph of $y = x^2$, vertically shift downward 1 unit.

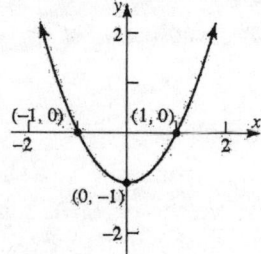

35. $g(x) = x^3 + 1$
     Using the graph of $y = x^3$, vertically shift upward 1 unit.

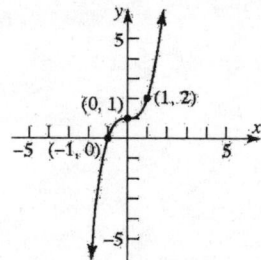

37. $h(x) = \sqrt{x-2}$
     Using the graph of $y = \sqrt{x}$, horizontally shift to the right 2 units.

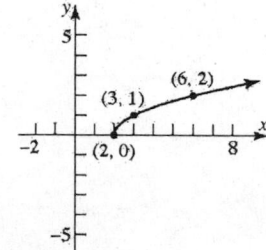

39. $f(x) = (x-1)^3 + 2$
     Using the graph of $y = x^3$, horizontally shift to the right 1 unit, then vertically shift up 2 units.

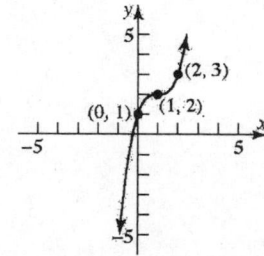

41. $g(x) = 4\sqrt{x}$

Using the graph of $y = \sqrt{x}$, vertically stretch by a factor of 4.

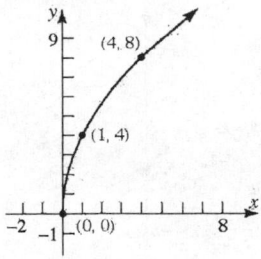

43. $h(x) = \dfrac{1}{2x} = \left(\dfrac{1}{2}\right)\left(\dfrac{1}{x}\right)$

Using the graph of $y = \dfrac{1}{x}$, vertically compress by a factor of $\dfrac{1}{2}$.

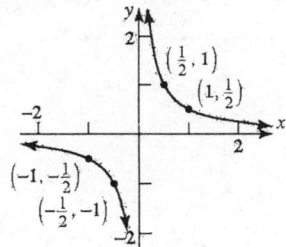

45. $f(x) = -\dfrac{1}{x}$

Reflect the graph of $y = \dfrac{1}{x}$, about the x-axis.

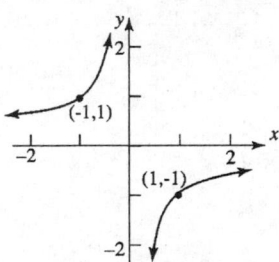

47. $g(x) = \left|-x\right|$

Reflect the graph of $y = \left|x\right|$, about the y-axis.

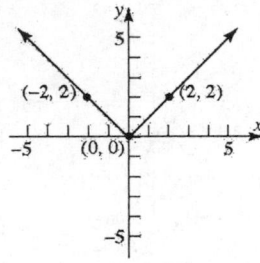

49. $h(x) = -x^3 + 2$

Reflect the graph of $y = x^3$ on the x-axis, vertically shift upward 2 units.

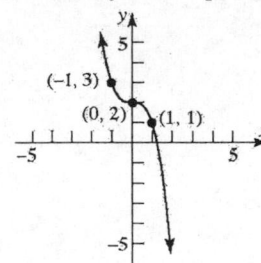

51. $f(x) = 2(x+1)^2 - 3$

Using the graph of $y = x^2$, horizontally shift to the left 1 unit, vertically stretch by a factor of 2, and vertically shift downward 3 units.

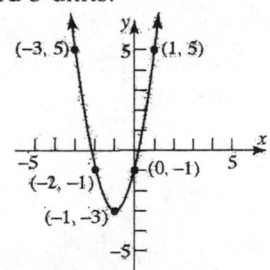

53. $g(x) = \sqrt{x-2} + 1$

Using the graph of $y = \sqrt{x}$, horizontally shift to the right 2 units and vertically shift upward 1 unit.

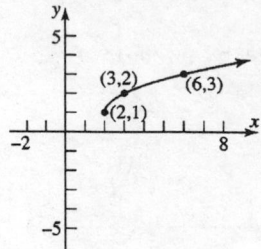

55. $h(x) = \sqrt{-x} - 2$

Reflect the graph of $y = \sqrt{x}$, about the y-axis and vertically shift downward 2 units.

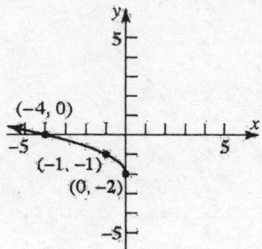

57. $f(x) = -(x+1)^3 - 1$

Using the graph of $y = x^3$, horizontally shift to the left 1 units, reflect the graph on the x-axis, and vertically shift downward 1 unit.

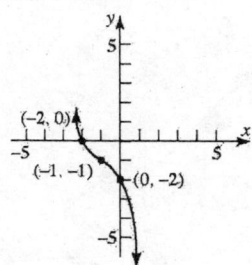

59. $g(x) = 2|1-x| = 2|-(-1+x)| = 2|x-1|$

Using the graph of $y = |x|$, horizontally shift to the right 1 unit, and vertically stretch by a factor or 2.

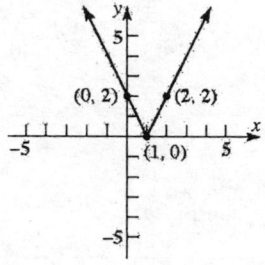

61. $h(x) = 2\operatorname{int}(x-1)$

Using the graph of $y = \operatorname{int}(x)$, horizontally shift to the right 1 unit, and vertically stretch by a factor of 2.

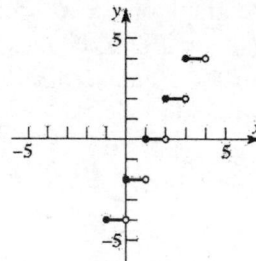

63. (a) $F(x) = f(x) + 3$
          Shift up 3 units.

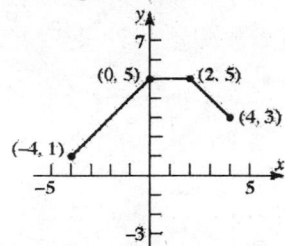

(b) $G(x) = f(x + 2)$
      Shift left 2 units.

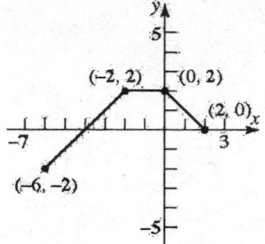

(c) $P(x) = -f(x)$
      Reflect about the x-axis.

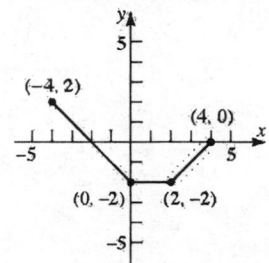

(d) $H(x) = f(x + 1) - 2$
      Shift left 1 unit and shift down 2 units.

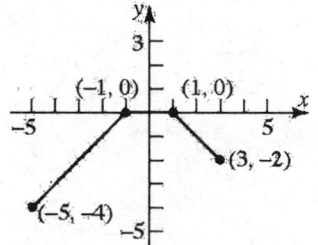

(e) $Q(x) = \dfrac{1}{2} f(x)$

Compress vertically by a factor of $\dfrac{1}{2}$.

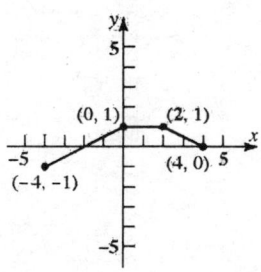

(f) $g(x) = f(-x)$
      Reflect about y-axis.

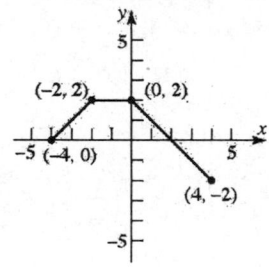

(g) $h(x) = f(2x)$

Compress horizontally by a factor of $\dfrac{1}{2}$.

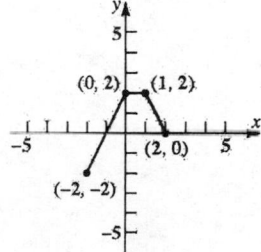

65.   (a) $F(x) = f(x) + 3$
            Shift up 3 units.

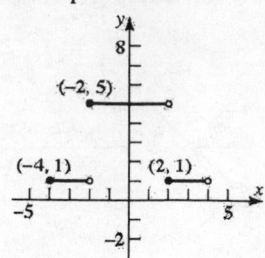

      (b) $G(x) = f(x + 2)$
            Shift left 2 units.

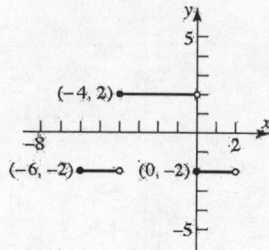

      (c) $P(x) = -f(x)$
            Reflect about the x-axis.

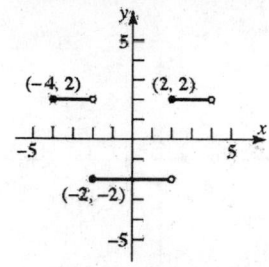

      (d) $H(x) = f(x + 1) - 2$
            Shift left 1 unit and shift down 2
            units.

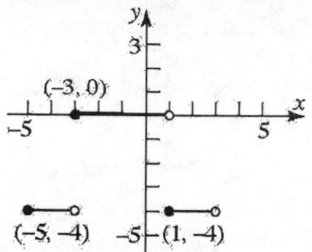

      (e) $Q(x) = \dfrac{1}{2} f(x)$

            Compress vertically by a factor of $\dfrac{1}{2}$.

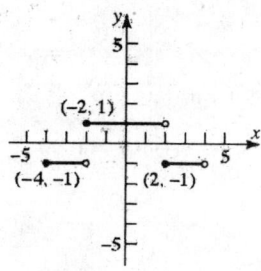

      (f) $g(x) = f(-x)$
            Reflect about y-axis.

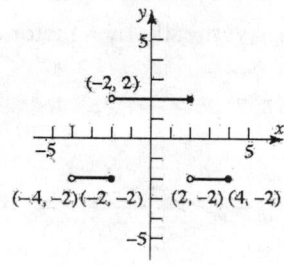

      (g) $h(x) = f(2x)$

            Compress horizontally by a factor of $\dfrac{1}{2}$.

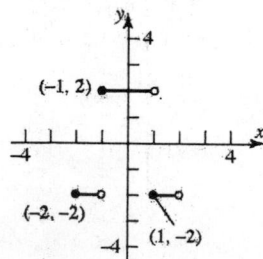

67. (a) $F(x) = f(x) + 3$
Shift up 3 units.

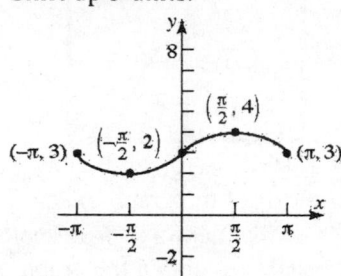

(b) $G(x) = f(x + 2)$
Shift left 2 units.

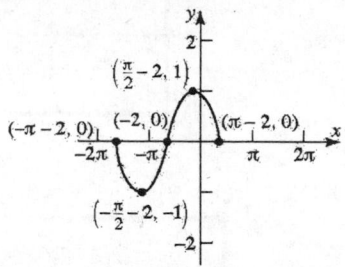

(c) $P(x) = -f(x)$
Reflect about the x-axis.

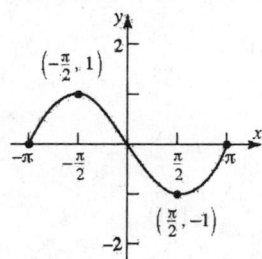

(d) $H(x) = f(x + 1) - 2$
Shift left 1 unit and shift down 2 units.

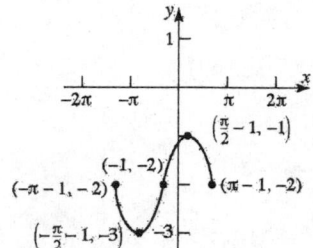

(e) $Q(x) = \dfrac{1}{2} f(x)$

Compress vertically by a factor of $\dfrac{1}{2}$.

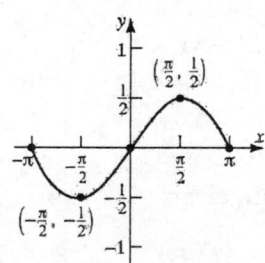

(f) $g(x) = f(-x)$
Reflect about y-axis.

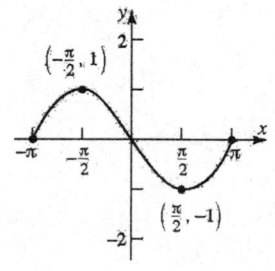

(g) $h(x) = f(2x)$

Compress horizontally by a factor of $\dfrac{1}{2}$.

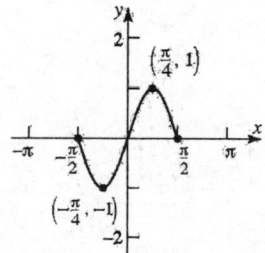

69.   (a) $y = |x + 1|$

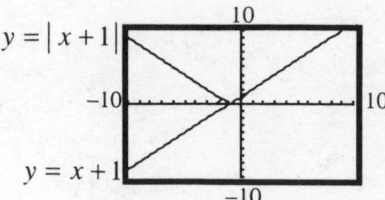

$y = x + 1$

(b) $y = |4 - x^2|$

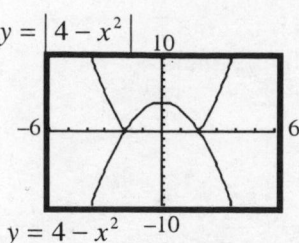

$y = 4 - x^2$

(c) $y = |x^3 + x|$

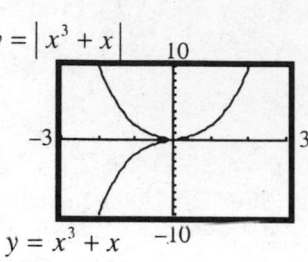

$y = x^3 + x$

(d)   Any part of the graph of $y = f(x)$ that lies below the x-axis is reflected about the x-axis to obtain the graph of $y = |f(x)|$.

71.   (a) $y = |f(x)|$

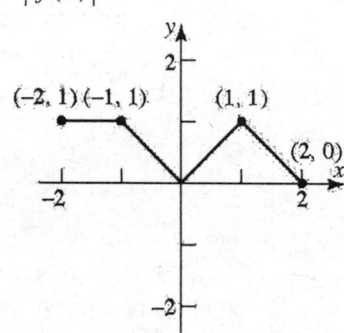

(b) $y = f(|x|)$

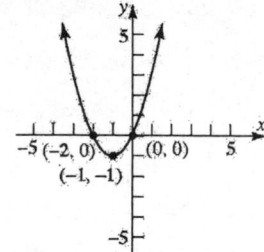

73. $f(x) = x^2 + 2x$

$f(x) = (x^2 + 2x + 1) - 1$

$f(x) = (x + 1)^2 - 1$

Using $f(x) = x^2$, shift left 1 unit and shift down 1 unit.

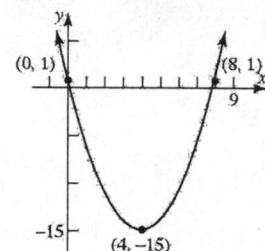

75. $f(x) = x^2 - 8x + 1$

$f(x) = (x^2 - 8x + 16) + 1 - 16$

$f(x) = (x - 4)^2 - 15$

Using $f(x) = x^2$, shift right 4 units and shift down 15 units.

77. $f(x) = x^2 + x + 1$

$$f(x) = \left(x^2 + x + \frac{1}{4} \right) + 1 - \frac{1}{4}$$

$$f(x) = \left(x + \frac{1}{2} \right)^2 + \frac{3}{4}$$

Using $f(x) = x^2$, shift left $\dfrac{1}{2}$ unit and

shift up $\dfrac{3}{4}$ unit.

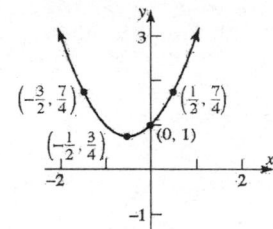

79. $y = (x - c)^2$

If $c = 0$, $y = x^2$.

If $c = 3$, $y = (x - 3)^2$; shift right 3 units.

If $c = -2$, $y = (x + 2)^2$; shift left 2 units.

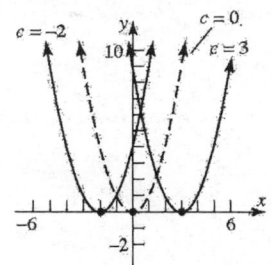

81. $F = \dfrac{9}{5}C + 32$

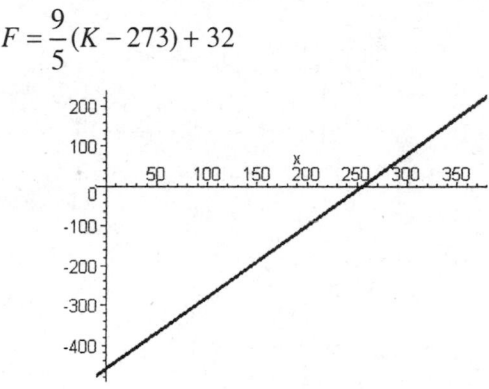

$$F = \frac{9}{5}(K - 273) + 32$$

Shift the graph 273 units to the right.

83.   (a)

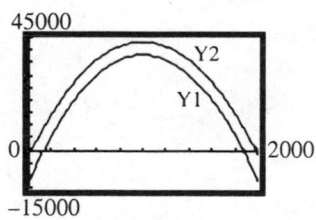

(b)   Select the 10% tax since the profits are higher.

(c)   The graph of Y1 is obtained by shifting the graph of $p(x)$ vertically down 10,000. The graph of Y2 is obtained by multiplying the y-coordinate of the graph of $p(x)$ by 0.9. Thus, Y2 is the graph of $p(x)$ vertically compressed by a factor of 0.9.

(d)   Select the 10% tax since the graph of $Y1 = 0.9p(x) \geq Y2 = -0.05x^2 + 100x - 6800$ for all x in the domain.

Chapter 2

Functions and Their Graphs

2.6   Operations on Functions; Composite Functions

1. $f(x) = 3x + 4$ $g(x) = 2x - 3$

   (a) $(f + g)(x) = 3x + 4 + 2x - 3 = 5x + 1$ The domain is all real numbers.

   (b) $(f - g)(x) = (3x + 4) - (2x - 3) = 3x + 4 - 2x + 3 = x + 7$
      The domain is all real numbers.

   (c) $(f \cdot g)(x) = (3x + 4)(2x - 3) = 6x^2 - 9x + 8x - 12 = 6x^2 - x - 12$
      The domain is all real numbers.

   (d) $\left(\dfrac{f}{g}\right)(x) = \dfrac{3x + 4}{2x - 3}$ The domain is all real numbers except $\dfrac{3}{2}$.

3. $f(x) = x - 1$ $g(x) = 2x^2$

   (a) $(f + g)(x) = x - 1 + 2x^2 = 2x^2 + x - 1$ The domain is all real numbers.

   (b) $(f - g)(x) = (x - 1) - (2x^2) = x - 1 - 2x^2 = -2x^2 + x - 1$
      The domain is all real numbers.

   (c) $(f \cdot g)(x) = (x - 1)(2x^2) = 2x^3 - 2x^2$ The domain is all real numbers.

   (d) $\left(\dfrac{f}{g}\right)(x) = \dfrac{x - 1}{2x^2}$ The domain is all real numbers except 0.

5. $f(x) = \sqrt{x}$ $g(x) = 3x - 5$

   (a) $(f + g)(x) = \sqrt{x} + 3x - 5$ The domain is $\left\{x \mid x \geq 0\right\}$.

   (b) $(f - g)(x) = \sqrt{x} - (3x - 5) = \sqrt{x} - 3x + 5$ The domain is $\left\{x \mid x \geq 0\right\}$.

   (c) $(f \cdot g)(x) = \sqrt{x}(3x - 5) = 3x\sqrt{x} - 5\sqrt{x}$ The domain is $\left\{x \mid x \geq 0\right\}$.

   (d) $\left(\dfrac{f}{g}\right)(x) = \dfrac{\sqrt{x}}{3x - 5}$ The domain is $\left\{x \mid x \geq 0 \text{ and } x \neq \dfrac{5}{3}\right\}$.

7. $f(x) = 1 + \dfrac{1}{x}$ $g(x) = \dfrac{1}{x}$

   (a) $(f + g)(x) = 1 + \dfrac{1}{x} + \dfrac{1}{x} = 1 + \dfrac{2}{x}$ The domain is $\left\{x \mid x \neq 0\right\}$.

   (b) $(f - g)(x) = 1 + \dfrac{1}{x} - \dfrac{1}{x} = 1$ The domain is $\left\{x \mid x \neq 0\right\}$.

   (c) $(f \cdot g)(x) = \left(1 + \dfrac{1}{x}\right)\dfrac{1}{x} = \dfrac{1}{x} + \dfrac{1}{x^2}$ The domain is $\left\{x \mid x \neq 0\right\}$.

   (d) $\left(\dfrac{f}{g}\right)(x) = \dfrac{\left(1 + \dfrac{1}{x}\right)}{\left(\dfrac{1}{x}\right)} = \dfrac{\left(\dfrac{x + 1}{x}\right)}{\left(\dfrac{1}{x}\right)} = \dfrac{x + 1}{x} \cdot \dfrac{x}{1} = x + 1$ The domain is $\left\{x \mid x \neq 0\right\}$.

9. $f(x) = \dfrac{2x+3}{3x-2}$ $g(x) = \dfrac{4x}{3x-2}$

   (a) $(f+g)(x) = \dfrac{2x+3}{3x-2} + \dfrac{4x}{3x-2} = \dfrac{2x+3+4x}{3x-2} = \dfrac{6x+3}{3x-2}$

   The domain is $\left\{x \,\middle|\, x \neq \dfrac{2}{3}\right\}$.

   (b) $(f-g)(x) = \dfrac{2x+3}{3x-2} - \dfrac{4x}{3x-2} = \dfrac{2x+3-4x}{3x-2} = \dfrac{-2x+3}{3x-2}$

   The domain is $\left\{x \,\middle|\, x \neq \dfrac{2}{3}\right\}$.

   (c) $(f \cdot g)(x) = \left(\dfrac{2x+3}{3x-2}\right)\left(\dfrac{4x}{3x-2}\right) = \dfrac{8x^2+12x}{(3x-2)^2}$

   The domain is $\left\{x \,\middle|\, x \neq \dfrac{2}{3}\right\}$.

   (d) $\left(\dfrac{f}{g}\right)(x) = \dfrac{\left(\dfrac{2x+3}{3x-2}\right)}{\left(\dfrac{4x}{3x-2}\right)} = \dfrac{2x+3}{3x-2} \cdot \dfrac{3x-2}{4x} = \dfrac{2x+3}{4x}$

   The domain is $\left\{x \,\middle|\, x \neq \dfrac{2}{3} \text{ and } x \neq 0\right\}$.

11. $f(x) = 3x+1$ $(f+g)(x) = 6 - \dfrac{1}{2}x$

   $6 - \dfrac{1}{2}x = 3x+1+g(x) \Rightarrow 5 - \dfrac{7}{2}x = g(x) \Rightarrow g(x) = 5 - \dfrac{7}{2}x$

13. $f(x) = 2x$ $g(x) = 3x^2+1$

   (a) $(f \circ g)(4) = f(g(4)) = f\left(3(4)^2+1\right) = f(49) = 2(49) = 98$

   (b) $(g \circ f)(2) = g(f(2)) = g(2 \cdot 2) = g(4) = 3(4)^2+1 = 48+1 = 49$

   (c) $(f \circ f)(1) = f(f(1)) = f(2(1)) = f(2) = 2(2) = 4$

   (d) $(g \circ g)(0) = g(g(0)) = g\left(3(0)^2+1\right) = g(1) = 3(1)^2+1 = 4$

15. $f(x) = 4x^2-3$ $g(x) = 3 - \dfrac{1}{2}x^2$

   (a) $(f \circ g)(4) = f(g(4)) = f\left(3 - \dfrac{1}{2}(4)^2\right) = f(-5) = 4(-5)^2-3 = 97$

   (b) $(g \circ f)(2) = g(f(2)) = g(4(2)^2-3) = g(13) = 3 - \dfrac{1}{2}(13)^2 = 3 - \dfrac{169}{2} = -\dfrac{163}{2}$

   (c) $(f \circ f)(1) = f(f(1)) = f(4(1)^2-3) = f(1) = 4(1)^2-3 = 1$

   (d) $(g \circ g)(0) = g(g(0)) = g\left(3 - \dfrac{1}{2}(0)^2\right) = g(3) = 3 - \dfrac{1}{2}(3)^2 = 3 - \dfrac{9}{2} = -\dfrac{3}{2}$

17. $f(x) = \sqrt{x}$ $g(x) = 2x$

     (a) $(f \circ g)(4) = f(g(4)) = f(2(4)) = f(8) = \sqrt{8} = 2\sqrt{2}$

     (b) $(g \circ f)(2) = g(f(2)) = g\left(\sqrt{2}\right) = 2\sqrt{2}$

     (c) $(f \circ f)(1) = f(f(1)) = f\left(\sqrt{1}\right) = f(1) = \sqrt{1} = 1$

     (d) $(g \circ g)(0) = g(g(0)) = g(2(0)) = g(0) = 2(0) = 0$

19. $f(x) = |x|$ $g(x) = \dfrac{1}{x^2 + 1}$

     (a) $(f \circ g)(4) = f(g(4)) = f\left(\dfrac{1}{4^2 + 1}\right) = f\left(\dfrac{1}{17}\right) = \left|\dfrac{1}{17}\right| = \dfrac{1}{17}$

     (b) $(g \circ f)(2) = g(f(2)) = g(|2|) = g(2) = \dfrac{1}{2^2 + 1} = \dfrac{1}{5}$

     (c) $(f \circ f)(1) = f(f(1)) = f(|1|) = f(1) = |1| = 1$

     (d) $(g \circ g)(0) = g(g(0)) = g\left(\dfrac{1}{0^2 + 1}\right) = g(1) = \dfrac{1}{1^2 + 1} = \dfrac{1}{2}$

21. $f(x) = \dfrac{3}{x + 1}$ $g(x) = \sqrt[3]{x}$

     (a) $(f \circ g)(4) = f(g(4)) = f\left(\sqrt[3]{4}\right) = \dfrac{3}{\sqrt[3]{4} + 1}$

     (b) $(g \circ f)(2) = g(f(2)) = g\left(\dfrac{3}{2 + 1}\right) = g\left(\dfrac{3}{3}\right) = \sqrt[3]{1} = 1$

     (c) $(f \circ f)(1) = f(f(1)) = f\left(\dfrac{3}{1 + 1}\right) = f\left(\dfrac{3}{2}\right) = \dfrac{3}{\left(\dfrac{3}{2}\right) + 1} = \dfrac{3}{\left(\dfrac{5}{2}\right)} = \dfrac{6}{5}$

     (d) $(g \circ g)(0) = g(g(0)) = g\left(\sqrt{0}\right) = g(0) = \sqrt{0} = 0$

23.   The domain of g is $\{x \mid x \neq 0\}$. The domain of f is $\{x \mid x \neq 1\}$.

     Thus, $g(x) \neq 1$, so we solve:

        $g(x) = 1$

        $\dfrac{2}{x} = 1$ Thus, $x \neq 2$; so the domain of $f \circ g$ is $\{x \mid x \neq 0, x \neq 2\}$.

        $x = 2$

25.   The domain of g is $\{x \mid x \neq 0\}$. The domain of f is $\{x \mid x \neq 1\}$.

     Thus, $g(x) \neq 1$, so we solve:

        $g(x) = 1$

        $-\dfrac{4}{x} = 1$ Thus, $x \neq -4$; so the domain of $f \circ g$ is $\{x \mid x \neq -4, x \neq 0\}$.

        $x = -4$

27.   The domain of g is {Real Numbers}. The domain of f is $\{x \mid x \geq 0\}$.
      Thus, $g(x) \geq 0$, so we solve:

$$g(x) \geq 0$$
$$2x + 3 \geq 0$$
$$x \geq -\frac{3}{2}$$

Thus, the domain of $f \circ g$ is $\left\{x \mid x \geq -\frac{3}{2}\right\}$.

29.   The domain of g is $\{x \mid x \geq 1\}$. The domain of f is {Real Numbers}.
      Thus, the domain of $f \circ g$ is $\{x \mid x \geq 1\}$.

31. $f(x) = 2x + 3 \qquad g(x) = 3x$
      The domain of f is all real numbers. The domain of g is all real numbers.
      (a) $(f \circ g)(x) = f(g(x)) = f(3x) = 2(3x) + 3 = 6x + 3$ Domain: All real numbers.
      (b) $(g \circ f)(x) = g(f(x)) = g(2x + 3) = 3(2x + 3) = 6x + 9$
                  Domain: All real numbers.
      (c) $(f \circ f)(x) = f(f(x)) = f(2x + 3) = 2(2x + 3) + 3 = 4x + 6 + 3 = 4x + 9$
                  Domain: All real numbers.
      (d) $(g \circ g)(x) = g(g(x)) = g(3x) = 3(3x) = 9x$ Domain: All real numbers.

33. $f(x) = 3x + 1 \qquad g(x) = x^2$
      The domain of f is all real numbers. The domain of g is all real numbers.
      (a) $(f \circ g)(x) = f(g(x)) = f\left(x^2\right) = 3x^2 + 1$ Domain: All real numbers.
      (b) $(g \circ f)(x) = g(f(x)) = g(3x + 1) = (3x + 1)^2 = 9x^2 + 6x + 1$
                  Domain: All real numbers.
      (c) $(f \circ f)(x) = f(f(x)) = f(3x + 1) = 3(3x + 1) + 1 = 9x + 3 + 1 = 9x + 4$
                  Domain: All real numbers.
      (d) $(g \circ g)(x) = g(g(x)) = g\left(x^2\right) = \left(x^2\right)^2 = x^4$ Domain: All real numbers.

35. $f(x) = x^2 \qquad g(x) = x^2 + 4$
      The domain of f is all real numbers. The domain of g is all real numbers.
      (a) $(f \circ g)(x) = f(g(x)) = f\left(x^2 + 4\right) = \left(x^2 + 4\right)^2 = x^4 + 8x^2 + 16$
                  Domain: All real numbers.
      (b) $(g \circ f)(x) = g(f(x)) = g\left(x^2\right) = \left(x^2\right)^2 + 4 = x^4 + 4$ Domain: All real numbers.
      (c) $(f \circ f)(x) = f(f(x)) = f\left(x^2\right) = \left(x^2\right)^2 = x^4$ Domain: All real numbers.
      (d) $(g \circ g)(x) = g(g(x)) = g\left(x^2 + 4\right) = \left(x^2 + 4\right)^2 + 4 = x^4 + 8x^2 + 16 + 4$
                  $= x^4 + 8x^2 + 20$ Domain: All real numbers.

37. $f(x) = \dfrac{3}{x-1}$ $g(x) = \dfrac{2}{x}$ The domain of f is $\{x \mid x \neq 1\}$.
The domain of g is $\{x \mid x \neq 0\}$.

(a) $(f \circ g)(x) = f(g(x)) = f\left(\dfrac{2}{x}\right) = \dfrac{3}{\left(\dfrac{2}{x} - 1\right)} = \dfrac{3}{\left(\dfrac{2-x}{x}\right)} = \dfrac{3x}{2-x}$

Domain of $f \circ g$ is $\{x \mid x \neq 0, \ x \neq 2\}$.

(b) $(g \circ f)(x) = g(f(x)) = g\left(\dfrac{3}{x-1}\right) = \dfrac{2}{\left(\dfrac{3}{x-1}\right)} = \dfrac{2(x-1)}{3}$

Domain of $g \circ f$ is $\{x \mid x \neq 1\}$

(c) $(f \circ f)(x) = f(f(x)) = f\left(\dfrac{3}{x-1}\right) = \dfrac{3}{\left(\dfrac{3}{x-1} - 1\right)} = \dfrac{3}{\left(\dfrac{3-(x-1)}{x-1}\right)} = \dfrac{3(x-1)}{4-x}$

Domain of $f \circ f$ is $\{x \mid x \neq 1, \ x \neq 4\}$.

(d) $(g \circ g)(x) = g(g(x)) = g\left(\dfrac{2}{x}\right) = \dfrac{2}{\left(\dfrac{2}{x}\right)} = \dfrac{2x}{2} = x$

Domain of $g \circ g$ is $\{x \mid x \neq 0\}$.

39. $f(x) = \dfrac{x}{x-1}$ $g(x) = -\dfrac{4}{x}$
The domain of f is $\{x \mid x \neq 1\}$. The domain of g is $\{x \mid x \neq 0\}$.

(a) $(f \circ g)(x) = f(g(x)) = f\left(-\dfrac{4}{x}\right) = \dfrac{\left(-\dfrac{4}{x}\right)}{\left(-\dfrac{4}{x} - 1\right)} = \dfrac{\left(-\dfrac{4}{x}\right)}{\left(\dfrac{-4-x}{x}\right)} = \dfrac{-4}{-4-x} = \dfrac{4}{4+x}$

Domain of $f \circ g$ is $\{x \mid x \neq -4, \ x \neq 0\}$.

(b) $(g \circ f)(x) = g(f(x)) = g\left(\dfrac{x}{x-1}\right) = -\dfrac{4}{\left(\dfrac{x}{x-1}\right)} = \dfrac{-4(x-1)}{x}$

Domain of $g \circ f$ is $\{x \mid x \neq 0, \ x \neq 1\}$.

(c) $(f \circ f)(x) = f(f(x)) = f\left(\dfrac{x}{x-1}\right) = \dfrac{\left(\dfrac{x}{x-1}\right)}{\left(\dfrac{x}{x-1} - 1\right)} = \dfrac{\left(\dfrac{x}{x-1}\right)}{\left(\dfrac{x-(x-1)}{x-1}\right)} = \dfrac{x}{1} = x$

Domain of $f \circ f$ is $\{x \mid x \neq 1\}$.

(d) $(g \circ g)(x) = g(g(x)) = g\left(\dfrac{-4}{x}\right) = -\dfrac{4}{\left(-\dfrac{4}{x}\right)} = \dfrac{-4x}{-4} = x$

Domain of $g \circ g$ is $\{x \mid x \neq 0\}$.

41. $f(x) = \sqrt{x}$ $g(x) = 2x + 3$
The domain of f is $\{x \mid x \geq 0\}$. The domain of g is {Real Numbers}.

(a) $(f \circ g)(x) = f(g(x)) = f(2x + 3) = \sqrt{2x + 3}$ Domain of $f \circ g$ is $\left\{x \mid x \geq -\dfrac{3}{2}\right\}$.

(b) $(g \circ f)(x) = g(f(x)) = g\left(\sqrt{x}\right) = 2\sqrt{x} + 3$ Domain of $g \circ f$ is $\{x \mid x \geq 0\}$.

(c) $(f \circ f)(x) = f(f(x)) = f\left(\sqrt{x}\right) = \sqrt{\sqrt{x}} = x^{1/4} = \sqrt[4]{x}$
Domain of $f \circ f$ is $\{x \mid x \geq 0\}$.

(d) $(g \circ g)(x) = g(g(x)) = g(2x + 3) = 2(2x + 3) + 3 = 4x + 6 + 3 = 4x + 9$
Domain of $g \circ g$ is {Real Numbers}.

43. $f(x) = x^2 + 1$ $g(x) = \sqrt{x - 1}$
The domain of f is {Real Numbers}. The domain of g is $\{x \mid x \geq 1\}$.

(a) $(f \circ g)(x) = f(g(x)) = f\left(\sqrt{x - 1}\right) = \left(\sqrt{x - 1}\right)^2 + 1 = x - 1 + 1 = x$
Domain of $f \circ g$ is $\{x \mid x \geq 1\}$.

(b) $(g \circ f)(x) = g(f(x)) = g\left(x^2 + 1\right) = \sqrt{x^2 + 1 - 1} = \sqrt{x^2} = |x|$
Domain of $g \circ f$ {Real Numbers}.

(c) $(f \circ f)(x) = f(f(x)) = f\left(x^2 + 1\right) = \left(x^2 + 1\right)^2 + 1 = x^4 + 2x^2 + 1 + 1 = x^4 + 2x^2 + 2$
Domain of $f \circ f$ is {Real Numbers}.

(d) $(g \circ g)(x) = g(g(x)) = g\left(\sqrt{x - 1}\right) = \sqrt{\sqrt{x - 1} - 1}$ Domain of $g \circ g$ is $\{x \mid x \geq 2\}$.

45. $f(x) = ax + b$ $g(x) = cx + d$ The domain of f is {Real Numbers}.
The domain of g is {Real Numbers}.

(a) $(f \circ g)(x) = f(g(x)) = f(cx + d) = a(cx + d) + b = acx + ad + b$
Domain of $f \circ g$ is {Real Numbers}.

(b) $(g \circ f)(x) = g(f(x)) = g(ax + b) = c(ax + b) + d = acx + bc + d$
Domain of $g \circ f$ is {Real Numbers}.

(c) $(f \circ f)(x) = f(f(x)) = f(ax + b) = a(ax + b) + b = a^2 x + ab + b$
Domain of $f \circ f$ is {Real Numbers}.

(d) $(g \circ g)(x) = g(g(x)) = g(cx + d) = c(cx + d) + d = c^2 x + cd + d$
Domain of $g \circ g$ is {Real Numbers}.

47. $(f \circ g)(x) = f(g(x)) = f\left(\dfrac{1}{2}x\right) = 2\left(\dfrac{1}{2}x\right) = x$

$(g \circ f)(x) = g(f(x)) = g(2x) = \dfrac{1}{2}(2x) = x$

49. $(f \circ g)(x) = f(g(x)) = f\left(\sqrt[3]{x}\right) = \left(\sqrt[3]{x}\right)^3 = x$

$(g \circ f)(x) = g(f(x)) = g\left(x^3\right) = \sqrt[3]{x^3} = x$

51.　$(f \circ g)(x) = f(g(x)) = f\left(\frac{1}{2}(x+6)\right) = 2\left(\frac{1}{2}(x+6)\right) - 6 = x + 6 - 6 = x$

　　$(g \circ f)(x) = g(f(x)) = g(2x-6) = \frac{1}{2}((2x-6)+6) = \frac{1}{2}(2x) = x$

53.　$(f \circ g)(x) = f(g(x)) = f\left(\frac{1}{a}(x-b)\right) = a\left(\frac{1}{a}(x-b)\right) + b = x - b + b = x$

　　$(g \circ f)(x) = g(f(x)) = g(ax+b) = \frac{1}{a}((ax+b)-b) = \frac{1}{a}(ax) = x$

55.　$H(x) = (2x+3)^4$　　　　$f(x) = x^4, \quad g(x) = 2x+3$

57.　$H(x) = \sqrt{x^2+1}$　　　　$f(x) = \sqrt{x}, \quad g(x) = x^2+1$

59.　$H(x) = |2x+1|$　　　　$f(x) = |x|, \quad g(x) = 2x+1$

61.　$f(x) = 2x^3 - 3x^2 + 4x - 1$　　$g(x) = 2$
　　$(f \circ g)(x) = f(g(x)) = f(2) = 2(2)^3 - 3(2)^2 + 4(2) - 1 = 16 - 12 + 8 - 1 = 11$
　　$(g \circ f)(x) = g(f(x)) = g\left(2x^3 - 3x^2 + 4x - 1\right) = 2$

63.　$f(x) = 2x^2 + 5$　　$g(x) = 3x + a$
　　$(f \circ g)(x) = f(g(x)) = f(3x+a) = 2(3x+a)^2 + 5$
　　When $x = 0$, $(f \circ g)(0) = 23$
　　Solving:
$$2(3 \cdot 0 + a)^2 + 5 = 23 \rightarrow 2a^2 + 5 = 23 \rightarrow 2a^2 = 18 \rightarrow a^2 = 9 \rightarrow a = -3 \text{ or } 3$$

65.　$S(r) = 4\pi r^2$　　$r(t) = \frac{2}{3}t^3, \ t \geq 0$

　　$S(r(t)) = S\left(\frac{2}{3}t^3\right) = 4\pi\left(\frac{2}{3}t^3\right)^2 = 4\pi\left(\frac{4}{9}t^6\right) = \frac{16}{9}\pi t^6$

67.　$N(t) = 100t - 5t^2, \ 0 \leq t \leq 10$　　$C(N) = 15000 + 8000N$
　　$C(N(t)) = C\left(100t - 5t^2\right) = 15000 + 8000\left(100t - 5t^2\right)$
　　　　　$= 15,000 + 800,000t - 40,000t^2$

69.　$p = -\frac{1}{4}x + 100$　　$0 \leq x \leq 400$

　　$\frac{1}{4}x = 100 - p \rightarrow x = 4(100 - p)$

　　$C = \frac{\sqrt{x}}{25} + 600 = \frac{\sqrt{4(100-p)}}{25} + 600 = \frac{2\sqrt{100-p}}{25} + 600, \quad 0 \leq p \leq 100$

71. $V = \pi r^2 h \qquad h = 2r \qquad \rightarrow V(r) = \pi r^2(2r) = 2\pi r^3$

73. $R(x) = \left(\dfrac{L}{P}\right)(x) = \dfrac{L(x)}{P(x)}$

75. $H(x) = (P \cdot I)(x) = P(x) \cdot I(x)$

77. $f(x) =$ number of Euros bought for x dollars; $g(x) =$ number of Yen bought for x Euros
   (a) $f(x) = 1.136235x$
   (b) $g(x) = 109.846x$
   (c) $g(f(x)) = 109.846(1.136235x) = 124.8108698x$
   (d) $g(f(1000)) = 124.8108698(1000) = 124810.8698$ Yen

79.   Given that f is odd and g is even, we know that
   $f(-x) = -f(x)$ and $g(-x) = g(x)$ for all x in the domain of f and g respectively.
   The composite function $f \circ g = f(g(x))$ has the following property:
   $f(g(-x)) = f(g(x))$ since g is even, $\therefore f \circ g$ is even
   The composite function $g \circ f = g(f(x))$ has the following property:
   $g(f(-x)) = g(-f(x))$ since f is odd
   $= g(f(x))$ since g is even, $\therefore g \circ f$ is even

Functions and Their Graphs

2.7 Mathematical Models; Constructing Functions

1. (a) The distance d from P to the origin is $d = \sqrt{x^2 + y^2}$. Since P is a point on the graph of $y = x^2 - 8$, we have:
$$d(x) = \sqrt{x^2 + (x^2 - 8)^2} = \sqrt{x^4 - 15x^2 + 64}$$

   (b) $d(0) = \sqrt{0^4 - 15(0)^2 + 64} = \sqrt{64} = 8$

   (c) $d(1) = \sqrt{(1)^4 - 15(1)^2 + 64} = \sqrt{1 - 15 + 64} = \sqrt{50} = 5\sqrt{2} \approx 7.07$

   (d) Graphing:

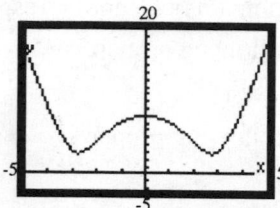

   (e) d is smallest when $x \approx -2.74$ and when $x \approx 2.74$.

3. (a) The distance d from P to the point $(1, 0)$ is $d = \sqrt{(x-1)^2 + y^2}$. Since P is a point on the graph of $y = \sqrt{x}$, we have:
$$d(x) = \sqrt{(x-1)^2 + \left(\sqrt{x}\right)^2} = \sqrt{x^2 - x + 1}$$

   (b) Graphing: (c) d is smallest when x is 0.50.

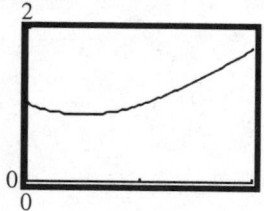

5. By definition, a triangle has area $A = \dfrac{1}{2}bh, b =$ base, $h =$ height. Because a vertex of the triangle is at the origin, we know that $b = x$ and $h = y$. Expressing the area of the triangle as a function of x, we have: $A(x) = \dfrac{1}{2}xy = \dfrac{1}{2}x\left(x^3\right) = \dfrac{1}{2}x^4$.

7. (a) $A(x) = xy = x(16 - x^2) = -x^3 + 16x$

(b) Domain: $\{x \mid 0 < x < 4\}$

(c) Graphing: The area is largest when x is approximately 2.31.

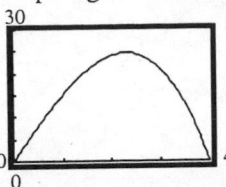

9. (a) $A(x) = (2x)(2y) = 4x(4 - x^2)^{1/2}$

(b) $p(x) = 2(2x) + 2(2y) = 4x + 4(4 - x^2)^{1/2}$

(c) Graphing:

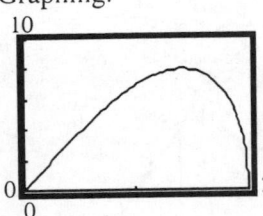

The area is largest when x is approximately 1.41.

(d) Graphing:

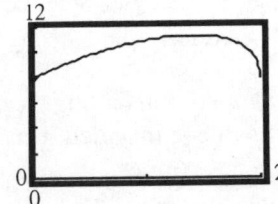

The perimeter is largest when x is approximately 1.41.

11. (a) $C =$ circumference, $A =$ area, $r =$ radius, $x =$ side of square

$$C = 2\pi r = 10 - 4x \quad \rightarrow \quad r = \frac{5 - 2x}{\pi}$$

$$A(x) = x^2 + \pi r^2 = x^2 + \pi \left(\frac{5 - 2x}{\pi}\right)^2 = x^2 + \frac{25 - 20x + 4x^2}{\pi}$$

(b) Since the lengths must be positive, we have:
$$10 - 4x > 0 \quad \text{and } x > 0$$
$$-4x > -10 \rightarrow x < 2.5 \text{ and } x > 0$$
Domain: $\{x \mid 0 < x < 2.5\}$

(c) Graphing: The area is smallest when x is approximately 1.40 meters.

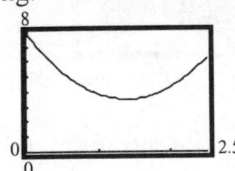

13. (a) Since the wire of length x is bent into a circle, the circumference is x.
Therefore, $C(x) = x$.

(b) Since $C = x = 2\pi r$, $r = \frac{x}{2\pi}$.

$$A(x) = \pi r^2 = \pi \left(\frac{x}{2\pi}\right)^2 = \frac{x^2}{4\pi}.$$

15. (a) A = area, r = radius; diameter = $2r$ (b) p = perimeter

$A(r) = (2r)(r) = 2r^2$ $p(r) = 2(2r) + 2r = 6r$

17. Area of the equilateral triangle $= \dfrac{1}{2}x \cdot \dfrac{\sqrt{3}}{2}x = \dfrac{\sqrt{3}}{4}x^2$

Area of $\dfrac{1}{3}$ of the equilateral triangle $= \dfrac{1}{2}x\sqrt{r^2 - \left(\dfrac{x}{2}\right)^2} = \dfrac{1}{2}x\sqrt{r^2 - \dfrac{x^2}{4}} = \dfrac{1}{3} \cdot \dfrac{\sqrt{3}}{4}x^2$

Solving for r^2:

$$\dfrac{1}{2}x\sqrt{r^2 - \dfrac{x^2}{4}} = \dfrac{1}{3} \cdot \dfrac{\sqrt{3}}{4}x^2 \Rightarrow \sqrt{r^2 - \dfrac{x^2}{4}} = \dfrac{2}{x} \cdot \dfrac{\sqrt{3}}{12}x^2$$

$$\sqrt{r^2 - \dfrac{x^2}{4}} = \dfrac{\sqrt{3}}{6}x \Rightarrow r^2 - \dfrac{x^2}{4} = \dfrac{3}{36}x^2 \Rightarrow r^2 = \dfrac{x^2}{3}$$

Area inside the circle, but outside the triangle:

$$A(x) = \pi r^2 - \dfrac{\sqrt{3}}{4}x^2 = \pi \dfrac{x^2}{3} - \dfrac{\sqrt{3}}{4}x^2 = \left(\dfrac{\pi}{3} - \dfrac{\sqrt{3}}{4}\right)x^2$$

19. (a)   The total cost of installing the cable along the road is $10x$. If cable is installed x miles along the road, there are $5 - x$ miles left from the road to the house and where the cable ends.

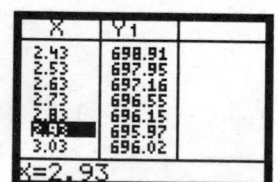

$$d = \sqrt{(5-x)^2 + 2^2} = \sqrt{25 - 10x + x^2 + 4}$$

$$= \sqrt{x^2 - 10x + 29}$$

The total cost of installing the cable is:

$$C(x) = 100x + 140\sqrt{x^2 - 10x + 29}$$

Domain: $\{x \mid 0 < x < 5\}$

(b) $C(1) = 100(1) + 140\sqrt{1^2 - 10(1) + 29} = 100 + 140\sqrt{20} \approx 100 + 626.10 = \$726.10$

(c) $C(3) = 100(3) + 140\sqrt{3^2 - 10(3) + 29} = 300 + 140\sqrt{8} \approx 300 + 396.00 = \$696.00$

(d)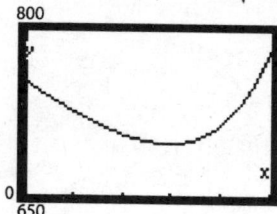

(e)

The table indicates that $x = 2.93$ results in the least cost.

(f)   Using MINIMUM, the graph indicates that $x = 2.96$ results in the least cost.

21.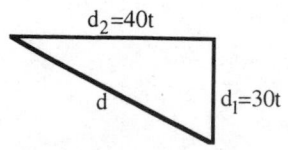

$$d^2 = d_1^2 + d_2^2$$

$$d^2 = (30t)^2 + (40t)^2$$

$$d(t) = \sqrt{900t^2 + 1600t^2}$$

$$d(t) = \sqrt{2500t^2} = 50t$$

23. (a)   length $= 24 - 2x$; width $= 24 - 2x$; height $= x$
$V(x) = x(24 - 2x)(24 - 2x) = x(24 - 2x)^2$

(b) $V(3) = 3(24 - 2(3))^2 = 3(18)^2 = 3(324) = 972$ cu. in.

(c) $V(10) = 10(24 - 2(10))^2 = 10(4)^2 = 10(16) = 160$ cu. in.

(d)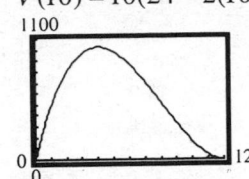

The volume is largest when $x = 4$ inches.

25. $r =$ radius of cylinder, $h =$ height of cylinder, $V =$ volume of cylinder

$$r^2 + \left(\frac{h}{2}\right)^2 = R^2 \Rightarrow r^2 + \frac{h^2}{4} = R^2$$

$$r^2 = R^2 - \frac{h^2}{4} \Rightarrow r^2 = \frac{4R^2 - h^2}{4}$$

$$V = \pi r^2 h \Rightarrow V(h) = \pi\left(\frac{4R^2 - h^2}{4}\right)h = \frac{\pi}{4}\left(4R^2 h - h^3\right)$$

Chapter 2

Functions and Their Graphs

2.R   Chapter Review

1. $f(4) = -5$ gives the ordered pair $(4,-5)$. $f(0) = 3$ gives $(0,3)$.

Finding the slope: $m = \dfrac{3-(-5)}{0-4} = \dfrac{8}{-4} = -2$

Using slope-intercept form: $f(x) = -2x + 3$

3. $f(x) = \dfrac{Ax+5}{6x-2}$ and $f(1) = 4$

Solving:

$$\frac{A(1)+5}{6(1)-2} = 4 \Rightarrow \frac{A+5}{4} = 4 \Rightarrow A+5 = 16 \Rightarrow A = 11$$

5. (b), (c), and (d) pass the vertical line test and therefore are functions.

7. $f(x) = \dfrac{3x}{x^2-4}$

(a) $f(-x) = \dfrac{3(-x)}{(-x)^2-4} = \dfrac{-3x}{x^2-4}$

(b) $-f(x) = -\left(\dfrac{3x}{x^2-4}\right) = \dfrac{-3x}{x^2-4}$

(c) $f(x+2) = \dfrac{3(x+2)}{(x+2)^2-4} = \dfrac{3x+6}{x^2+4x+4-4} = \dfrac{3x+6}{x^2+4x}$

(d) $f(x-2) = \dfrac{3(x-2)}{(x-2)^2-4} = \dfrac{3x-6}{x^2-4x+4-4} = \dfrac{3x-6}{x^2-4x}$

9. $f(x) = \sqrt{x^2-4}$

(a) $f(-x) = \sqrt{(-x)^2-4} = \sqrt{x^2-4}$

(b) $-f(x) = -\sqrt{x^2-4}$

(c) $f(x+2) = \sqrt{(x+2)^2-4} = \sqrt{x^2+4x+4-4} = \sqrt{x^2+4x}$

(d) $f(x-2) = \sqrt{(x-2)^2-4} = \sqrt{x^2-4x+4-4} = \sqrt{x^2-4x}$

11. $f(x) = \dfrac{x^2 - 4}{x^2}$

   (a) $f(-x) = \dfrac{(-x)^2 - 4}{(-x)^2} = \dfrac{x^2 - 4}{x^2}$

   (b) $-f(x) = -\left(\dfrac{x^2 - 4}{x^2}\right) = \dfrac{4 - x^2}{x^2}$

   (c) $f(x+2) = \dfrac{(x+2)^2 - 4}{(x+2)^2} = \dfrac{x^2 + 4x + 4 - 4}{x^2 + 4x + 4} = \dfrac{x^2 + 4x}{x^2 + 4x + 4}$

   (d) $f(x-2) = \dfrac{(x-2)^2 - 4}{(x-2)^2} = \dfrac{x^2 - 4x + 4 - 4}{x^2 - 4x + 4} = \dfrac{x^2 - 4x}{x^2 - 4x + 4}$

13. $f(x) = \dfrac{x}{x^2 - 9}$

   The denominator cannot be zero:
   $$x^2 - 9 \neq 0$$
   $$(x+3)(x-3) \neq 0$$
   $$x \neq -3 \text{ or } 3$$
   Domain: $\{x \mid x \neq -3, x \neq 3\}$

15. $f(x) = \sqrt{2 - x}$

   The radicand must be positive:
   $$2 - x \geq 0$$
   $$x \leq 2$$
   Domain: $\{x \mid x \leq 2\}$ or $(-\infty, 2]$

17. $f(x) = \dfrac{\sqrt{x}}{|x|}$

   The radicand must be positive and the denominator cannot be zero: $x > 0$
   Domain: $\{x \mid x > 0\}$ or $(0, +\infty)$

19. $f(x) = \dfrac{x}{x^2 + 2x - 3}$

   The denominator cannot be zero:
   $$x^2 + 2x - 3 \neq 0$$
   $$(x+3)(x-1) \neq 0$$
   $$x \neq -3 \text{ or } 1$$
   Domain : $\{x \mid x \neq -3, x \neq 1\}$

21. $f(x) = 2x - 5$

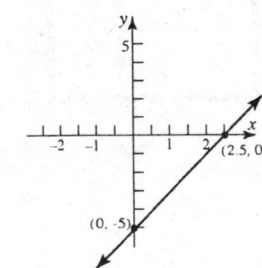

23. $h(x) = \dfrac{4}{5}x - 6$

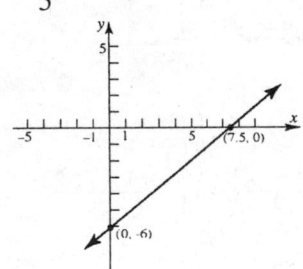

25. (a) Domain: $\{x \mid -4 \leq x \leq 4\}$ Range: $\{y \mid -3 \leq y \leq 1\}$

   (b) Increasing: $(-4, -1) \cup (3, 4)$; Decreasing: $(-1, 3)$; Constant: $(-5, -1)$

   (c) Local minimum $(3, -3)$; Local maximum $(-1, 1)$

   (d) The graph is not symmetric to the x-axis, the y-axis or the origin.

   (e) The function is neither even nor odd.

   (f) x-intercepts: $-2, 0, 4$;   y-intercept: 0

27. $y = x^3$

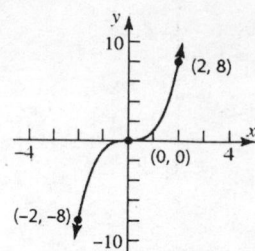

29. $f(x) = \begin{cases} 3x & \text{if } -2 < x \le 1 \\ x+1 & \text{if } x > 1 \end{cases}$

(a) Domain: $\{x \mid x > -2\}$

(b) x-intercept: $(0,0)$

    y-intercept: $(0,0)$

(c)

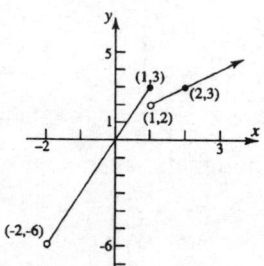

(d) Range: $\{y > -6\}$

(e) Graphing utility:

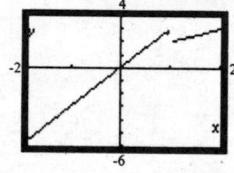

31. $f(x) = \begin{cases} x & \text{if } -4 \le x < 0 \\ 1 & \text{if } x = 0 \\ 3x & \text{if } x > 0 \end{cases}$

(a) Domain: $\{x \mid x \ge -4\}$

(b) x-intercept: none

    y-intercept: $(0, 1)$

(c)

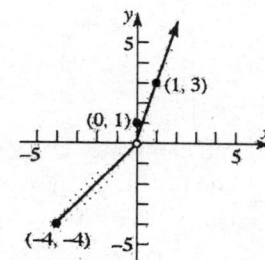

(d) Range: $\{y \mid y \ge -4, y \ne 0\}$

(e) Graphing utility:

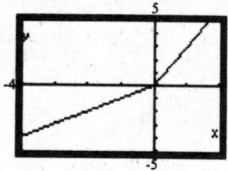

33. $f(x) = 2 - 5x$

$$\frac{f(x) - f(2)}{x - 2} = \frac{2 - 5x - (-8)}{x - 2} = \frac{-5x + 10}{x - 2} = \frac{-5(x - 2)}{x - 2} = -5$$

35. $f(x) = 3x - 4x^2$

$$\frac{f(x) - f(2)}{x - 2} = \frac{3x - 4x^2 - (-10)}{x - 2} = \frac{-4x^2 + 3x + 10}{x - 2}$$

$$= \frac{-(4x^2 - 3x - 10)}{x - 2} = \frac{-(4x + 5)(x - 2)}{x - 2} = -4x - 5$$

37. $f(x) = x^3 - 4x$

$$f(-x) = (-x)^3 - 4(-x) = -x^3 + 4x = -(x^3 - 4x) = -f(x) \qquad f \text{ is odd.}$$

39. $h(x) = \dfrac{1}{x^4} + \dfrac{1}{x^2} + 1$

$h(-x) = \dfrac{1}{(-x)^4} + \dfrac{1}{(-x)^2} + 1 = \dfrac{1}{x^4} + \dfrac{1}{x^2} + 1 = h(x)$ h is even.

41. $G(x) = 1 - x + x^3$

$G(-x) = 1 - (-x) + (-x)^3 = 1 + x - x^3 \neq -G(x) \neq G(x)$

G is neither even nor odd.

43. $f(x) = \dfrac{x}{1+x^2}$

$f(-x) = \dfrac{-x}{1+(-x)^2} = \dfrac{-x}{1+x^2} = -f(x)$ f is odd.

45. $F(x) = |x| - 4$
Using the graph of $y = |x|$, vertically
shift the graph downward 4 units.

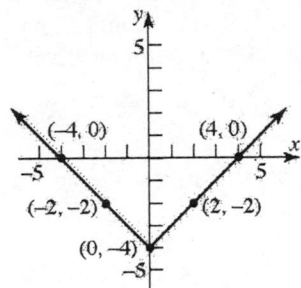

Intercepts: $(-4,0), (4,0), (0,-4)$
Domain: {Real Numbers}
Range: $\{y \mid y \geq -4\}$

47. $g(x) = -2|x|$
Reflect the graph of $y = |x|$ about the x-
axis and vertically stretch the graph by a
factor of 2.

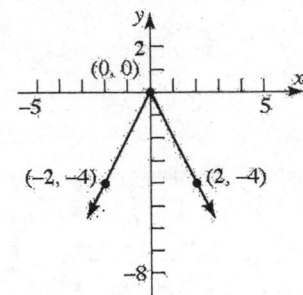

Intercepts: $(0,0)$
Domain: {Real Numbers}
Range: $\{y \mid y \leq 0\}$

49. $h(x) = \sqrt{x-1}$
Using the graph of $y = \sqrt{x}$, horizontally
shift the graph to the right 1 unit.
Intercepts: $(1,0)$
Domain: $\{x \mid x \geq 1\}$
Range: $\{y \mid y \geq 0\}$

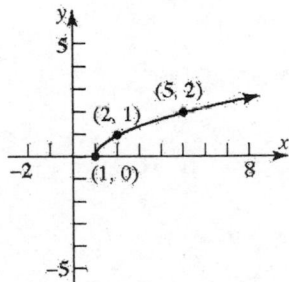

51. $f(x) = \sqrt{1-x} = \sqrt{-1(x-1)}$
Reflect the graph of $y = \sqrt{x}$ about the
y-axis and horizontally shift the graph to
the right 1 unit..

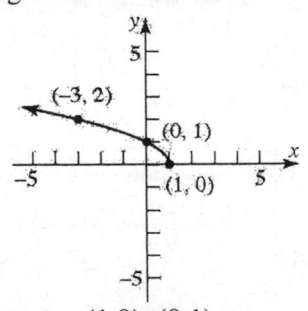

Intercepts: $(1,0), (0,1)$
Domain: $\{x \mid x \leq 1\}$
Range: $\{y \mid y \geq 0\}$

53. $h(x) = (x-1)^2 + 2$

Using the graph of $y = x^2$, horizontally shift the graph to the right 1 unit and vertically shift the graph up 2 units.

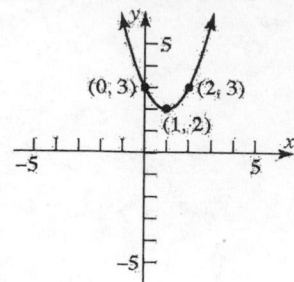

Intercepts: $(0,3)$
Domain: $\{$Real Numbers$\}$
Range: $\{y \mid y \geq 2\}$

55. $g(x) = 3(x-1)^3 + 1$

Using the graph of $y = x^3$, horizontally shift the graph to the right 1 unit, vertically stretch the graph by a factor of 3, and vertically shift the graph up 1 unit.

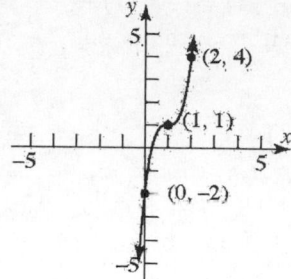

Intercepts: $(0,-2), \left(1 - \dfrac{\sqrt[3]{3}}{3}, 0\right)$

Domain: $\{$Real Numbers$\}$
Range: $\{$Real Numbers$\}$

57. $f(x) = 2x^3 - 5x + 1$ on the interval $(-3,3)$

Use MAXIMUM and MINIMUM on the graph of $y_1 = 2x^3 - 5x + 1$.

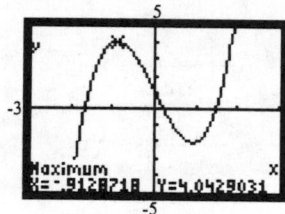 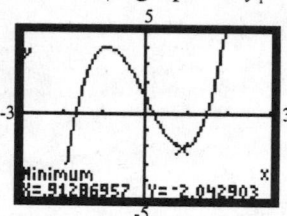

local maximum at: $(-0.91, 4.04)$;   local minimum at: $(0.91, -2.04)$

f is increasing on: $(-3, -0.91) \cup (0.91, 3)$; f is decreasing on: $(-0.91, 0.91)$

59. $f(x) = 2x^4 - 5x^3 + 2x + 1$ on the interval $(-2,3)$

Use MAXIMUM and MINIMUM on the graph of $y_1 = 2x^4 - 5x^3 + 2x + 1$.

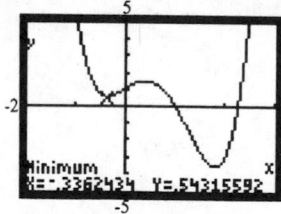 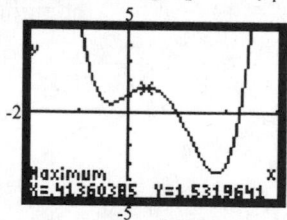 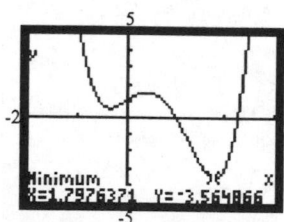

local maximum at: $(0.41, 1.53)$; local minimum at: $(-0.34, 0.54)$, $(1.80, -3.56)$

f is increasing on: $(-0.34, 0.41) \cup (1.80, 3)$; f is decreasing on: $(-2, -0.34) \cup (0.41, 1.80)$

61. $f(x) = 2 - x$ $g(x) = 3x + 1$

(a) $(f + g)(x) = f(x) + g(x) = 2 - x + 3x + 1 = 2x + 3$, Domain: {Real Numbers}

(b) $(f - g)(x) = f(x) - g(x) = 2 - x - (3x + 1) = 2 - x - 3x - 1 = -4x + 1$
       Domain: {Real Numbers}

(c) $(f \cdot g)(x) = f(x) \cdot g(x) = (2 - x)(3x + 1) = 6x + 2 - 3x^2 - x = -3x^2 + 5x + 2$
       Domain: {Real Numbers}

(d) $\left(\dfrac{f}{g}\right)(x) = \dfrac{f(x)}{g(x)} = \dfrac{2 - x}{3x + 1}$, Domain: $\left\{x \,\middle|\, x \ne -\dfrac{1}{3}\right\}$

63. $f(x) = 3x^2 + x + 1$ $g(x) = 3x$

(a) $(f + g)(x) = f(x) + g(x) = 3x^2 + x + 1 + 3x = 3x^2 + 4x + 1$
       Domain: {Real Numbers}

(b) $(f - g)(x) = f(x) - g(x) = 3x^2 + x + 1 - 3x = 3x^2 - 2x + 1$
       Domain: {Real Numbers}

(c) $(f \cdot g)(x) = f(x) \cdot g(x) = (3x^2 + x + 1)(3x) = 9x^3 + 3x^2 + 3x$
       Domain: {Real Numbers}

(d) $\left(\dfrac{f}{g}\right)(x) = \dfrac{f(x)}{g(x)} = \dfrac{3x^2 + x + 1}{3x}$, Domain: $\left\{x \,\middle|\, x \ne 0\right\}$

65. $f(x) = \dfrac{x + 1}{x - 1}$ $g(x) = \dfrac{1}{x}$

(a) $(f + g)(x) = f(x) + g(x) = \dfrac{x + 1}{x - 1} + \dfrac{1}{x} = \dfrac{x(x + 1) + 1(x - 1)}{x(x - 1)} = \dfrac{x^2 + x + x - 1}{x(x - 1)}$

$$= \dfrac{x^2 + 2x - 1}{x(x - 1)}$$

Domain: $\left\{x \,\middle|\, x \ne 0, x \ne 1\right\}$

(b) $(f - g)(x) = f(x) - g(x) = \dfrac{x + 1}{x - 1} - \dfrac{1}{x} = \dfrac{x(x + 1) - 1(x - 1)}{x(x - 1)} = \dfrac{x^2 + x - x + 1}{x(x - 1)}$

$$= \dfrac{x^2 + 1}{x(x - 1)}$$

Domain: $\left\{x \,\middle|\, x \ne 0, x \ne 1\right\}$

(c) $(f \cdot g)(x) = f(x) \cdot g(x) = \left(\dfrac{x + 1}{x - 1}\right)\left(\dfrac{1}{x}\right) = \dfrac{x + 1}{x(x - 1)}$

Domain: $\left\{x \,\middle|\, x \ne 0, x \ne 1\right\}$

(d) $\left(\dfrac{f}{g}\right)(x) = \dfrac{f(x)}{g(x)} = \dfrac{\left(\dfrac{x + 1}{x - 1}\right)}{\left(\dfrac{1}{x}\right)} = \left(\dfrac{x + 1}{x - 1}\right)\left(\dfrac{x}{1}\right) = \dfrac{x^2 + x}{x - 1}$

Domain: $\left\{x \,\middle|\, x \ne 0, x \ne 1\right\}$

67. $f(x) = 3x - 5$ $g(x) = 1 - 2x^2$
   (a) $(f \circ g)(2) = f(g(2)) = f(1 - 2(2)^2) = f(-7) = 3(-7) - 5 = -26$
   (b) $(g \circ f)(-2) = g(f(-2)) = g(3(-2) - 5) = g(-11) = 1 - 2(-11)^2 = -241$
   (c) $(f \circ f)(4) = f(f(4)) = f(3(4) - 5) = f(7) = 3(7) - 5 = 16$
   (d) $(g \circ g)(-1) = g(g(-1)) = g(1 - 2(-1)^2) = g(-1) = 1 - 2(-1)^2 = -1$

69. $f(x) = \sqrt{x + 2}$ $g(x) = 2x^2 + 1$
   (a) $(f \circ g)(2) = f(g(2)) = f(2(2)^2 + 1) = f(9) = \sqrt{9 + 2} = \sqrt{11}$
   (b) $(g \circ f)(-2) = g(f(-2)) = g(\sqrt{-2 + 2}) = g(0) = 2(0)^2 + 1 = 1$
   (c) $(f \circ f)(4) = f(f(4)) = f(\sqrt{4 + 2}) = f(\sqrt{6}) = \sqrt{\sqrt{6} + 2}$
   (d) $(g \circ g)(-1) = g(g(-1)) = g(2(-1)^2 + 1) = g(3) = 2(3)^2 + 1 = 19$

71. $f(x) = \dfrac{1}{x^2 + 4}$ $g(x) = 3x - 2$
   (a) $(f \circ g)(2) = f(g(2)) = f(3(2) - 2) = f(4) = \dfrac{1}{4^2 + 4} = \dfrac{1}{20}$
   (b) $(g \circ f)(-2) = g(f(-2)) = g\left(\dfrac{1}{(-2)^2 + 4}\right) = g\left(\dfrac{1}{8}\right) = 3\left(\dfrac{1}{8}\right) - 2 = \dfrac{-13}{8}$
   (c) $(f \circ f)(4) = f(f(4)) = f\left(\dfrac{1}{4^2 + 4}\right) = f\left(\dfrac{1}{20}\right) = \dfrac{1}{\left(\dfrac{1}{20}\right)^2 + 4} = \dfrac{1}{\left(\dfrac{1601}{400}\right)} = \dfrac{400}{1601}$
   (d) $(g \circ g)(-1) = g(g(-1)) = g(3(-1) - 2) = g(-5) = 3(-5) - 2 = -17$

73. $f(x) = 2 - x$ $g(x) = 3x + 1$
   The domain of f is all real numbers. The domain of g is all real numbers.
   (a) $(f \circ g)(x) = f(g(x)) = f(3x + 1) = 2 - (3x + 1) = 2 - 3x - 1 = 1 - 3x$
        Domain: All real numbers.
   (b) $(g \circ f)(x) = g(f(x)) = g(2 - x) = 3(2 - x) + 1 = 6 - 3x + 1 = 7 - 3x$
        Domain: All real numbers.
   (c) $(f \circ f)(x) = f(f(x)) = f(2 - x) = 2 - (2 - x) = 2 - 2 + x = x$
        Domain: All real numbers.
   (d) $(g \circ g)(x) = g(g(x)) = g(3x + 1) = 3(3x + 1) + 1 = 9x + 3 + 1 = 9x + 4$
        Domain: All real numbers.

75. $f(x) = 3x^2 + x + 1$ $g(x) = |3x|$
   The domain of f is all real numbers. The domain of g is all real numbers.
   (a) $(f \circ g)(x) = f(g(x)) = f(|3x|) = 3(|3x|)^2 + (|3x|) + 1 = 27x^2 + 3|x| + 1$
        Domain: All real numbers.
   (b) $(g \circ f)(x) = g(f(x)) = g(3x^2 + x + 1) = |3(3x^2 + x + 1)| = |9x^2 + 3x + 3|$
        Domain: All real numbers.

(c) $(f \circ f)(x) = f(f(x)) = f(3x^2 + x + 1) = 3(3x^2 + x + 1)^2 + (3x^2 + x + 1) + 1$
$$= 3(9x^4 + 6x^3 + 7x^2 + 2x + 1) + 3x^2 + x + 1 + 1$$
$$= 27x^4 + 18x^3 + 24x^2 + 7x + 5$$
Domain: All real numbers.

(d) $(g \circ g)(x) = g(g(x)) = g(|3x|) = |3|3x|| = 9|x|$ Domain: All real numbers.

77. $f(x) = \dfrac{x+1}{x-1}$ $g(x) = \dfrac{1}{x}$

The domain of f is $\{x \mid x \neq 1\}$. The domain of g is $\{x \mid x \neq 0\}$.

(a) $(f \circ g)(x) = f(g(x)) = f\left(\dfrac{1}{x}\right) = \dfrac{\left(\dfrac{1}{x}+1\right)}{\left(\dfrac{1}{x}-1\right)} = \dfrac{\left(\dfrac{1+x}{x}\right)}{\left(\dfrac{1-x}{x}\right)} = \left(\dfrac{1+x}{x}\right)\left(\dfrac{x}{1-x}\right) = \left(\dfrac{1+x}{1-x}\right)$

Domain of $f \circ g$ is $\{x \mid x \neq 0,\ x \neq 1\}$.

(b) $(g \circ f)(x) = g(f(x)) = g\left(\dfrac{x+1}{x-1}\right) = \dfrac{1}{\left(\dfrac{x+1}{x-1}\right)} = \left(\dfrac{x-1}{x+1}\right)$

Domain of $g \circ f$ is $\{x \mid x \neq -1,\ x \neq 1\}$.

(c) $(f \circ f)(x) = f(f(x)) = f\left(\dfrac{x+1}{x-1}\right) = \dfrac{\left(\dfrac{x+1}{x-1}+1\right)}{\left(\dfrac{x+1}{x-1}-1\right)} = \dfrac{\left(\dfrac{x+1+1(x-1)}{x-1}\right)}{\left(\dfrac{x+1-1(x-1)}{x-1}\right)}$

$$= \dfrac{\left(\dfrac{x+1+x-1}{x-1}\right)}{\left(\dfrac{x+1-x+1}{x-1}\right)} = \dfrac{\left(\dfrac{2x}{x-1}\right)}{\left(\dfrac{2}{x-1}\right)} = \left(\dfrac{2x}{x-1}\right)\left(\dfrac{x-1}{2}\right) = x$$

Domain of $f \circ f$ is $\{x \mid x \neq 1\}$.

(d) $(g \circ g)(x) = g(g(x)) = g\left(\dfrac{1}{x}\right) = \dfrac{1}{\left(\dfrac{1}{x}\right)} = x$, Domain of $g \circ g$ is $\{x \mid x \neq 0\}$.

79.   (a) $y = f(-x)$
Reflect about the y-axis.

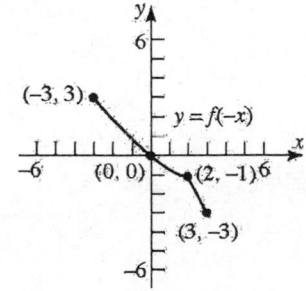

(b) $y = -f(x)$
Reflect about the x-axis.

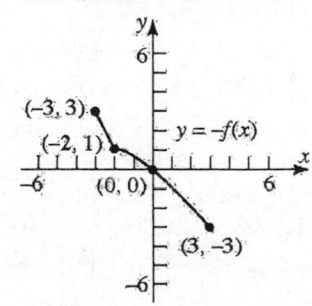

(c) $y = f(x+2)$
       Horizontally shift left 2 units.

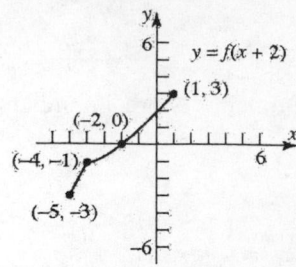

(d) $y = f(x) + 2$
       Vertically shift up 2 units.

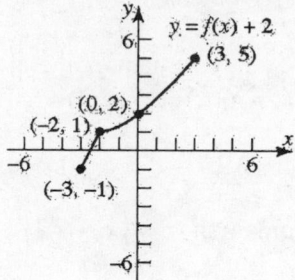

(e) $y = 2f(x)$
       Vertical stretch by a factor of 2.

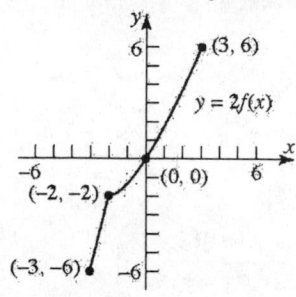

(f) $y = f(3x)$

       Horizontal compression by $\frac{1}{3}$.

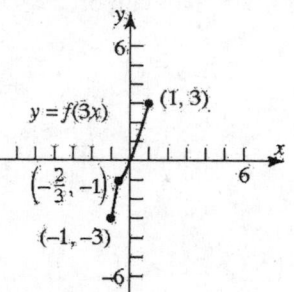

81. (a)   The relation is a function. Each HS GPA value is paired with exactly one College GPA value.

    (b)

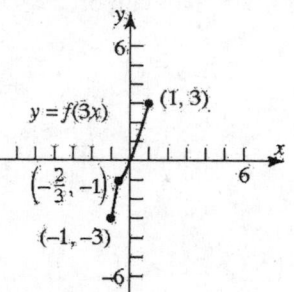

    (c)   Using the LINear REGression program, the line of best fit is:
           $G = 0.964x + 0.072$
    (d)   As the high school GPA increases by 1 point, the college GPA increases by 0.964 point.
    (e) $G(x) = 0.964x + 0.072$
    (f)   Domain: $\{x \mid 0 \le x \le 4\}$
    (g) $G(3.23) = (0.964)(3.23) + 0.072 \approx 3.186$
           The college GPA is approximately 3.19.

83.   Let R = the revenue in dollars, and g = the number of gallons of gasoline sold.
       Consider the ordered pair (g, R).
       We can use the points $(0,0)$ and $(13.5, 15.93)$.

Now compute the slope:

$$\text{slope} = \frac{\Delta y}{\Delta x} = \frac{15.93 - 0}{13.5 - 0} = \frac{15.93}{13.5} \approx 1.18$$

Therefore we have the linear function $R(g) = 1.18g + 0 = 1.18g$.
If $g = 11.2$, then $R = (1.18)(11.2) \approx \$13.22$.

85.

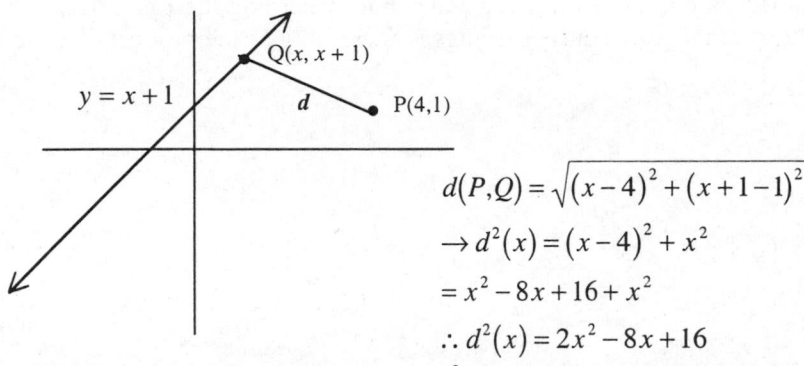

$$d(P,Q) = \sqrt{(x-4)^2 + (x+1-1)^2}$$
$$\rightarrow d^2(x) = (x-4)^2 + x^2$$
$$= x^2 - 8x + 16 + x^2$$
$$\therefore d^2(x) = 2x^2 - 8x + 16$$

Use MINIMUM on the graph of $y_1 = 2x^2 - 8x + 16$ to determine that the minimum occurs when $x = 2$. Therefore the point Q on the line $y = x + 1$ will be closest to the point $P = (4,1)$ when $Q = (2,3)$.

87. (a) $x^2 h = 10 \implies h = \dfrac{10}{x^2}$ $A(x) = 2x^2 + 4xh = 2x^2 + 4x\left(\dfrac{10}{x^2}\right) = 2x^2 + \dfrac{40}{x}$

(b) $A(1) = 2 \cdot 1^2 + \dfrac{40}{1} = 2 + 40 = 42 \text{ ft}^2$

(c) $A(2) = 2 \cdot 2^2 + \dfrac{40}{2} = 8 + 20 = 28 \text{ ft}^2$

(d)   Graphing:

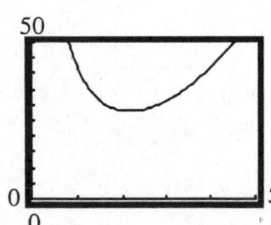

The area is smallest when $x \approx 2.15$.

89. (a), (b), (e)

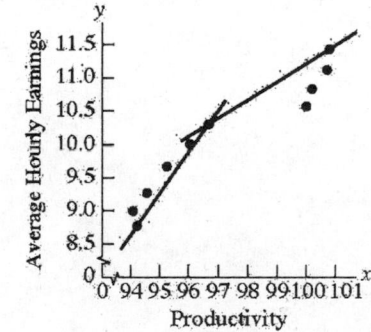

(c)   average rate of change $= \dfrac{10.32 - 8.76}{96.7 - 94.2} = \dfrac{1.56}{2.5} = 0.624$ dollars per unit of productivity

(d)   for each 1 unit increase in productivity, the avg. hourly earning increases by 0.624 dollars.

(f)   average rate of change $= \dfrac{11.44 - 10.32}{100.8 - 96.7} = \dfrac{1.12}{4.1} = 0.273$ dollars per unit of productivity

(g) for each 1 unit increase in productivity, the avg. hourly earning increases by 0.273 dollars

(h) the average rate of change of hourly earnings is decreasing as the productivity increases

Functions and Their Graphs

2.CR Cumulative Review

1. $x^3 - 6x^2 + 8x = 0$
    $x(x^2 - 6x + 8) = 0$
    $x(x-4)(x-2) = 0$
    $x = 0$ or $x = 4$ or $x = 2$

3. $x^2 + 4x + y^2 - 2y - 4 = 0$
    $(x^2 + 4x + 4) + (y^2 - 2y + 1) = 4 + 4 + 1$
    $\qquad\qquad (x+2)^2 + (y-1)^2 = 9$
    Center: $(-2,1)$
    Radius $= 3$

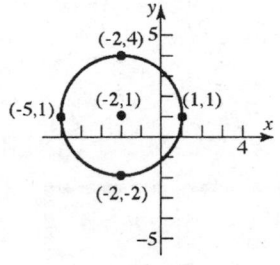

5. Perpendicular to $3x - 2y = 7 \Rightarrow y = \dfrac{3}{2}x - \dfrac{7}{2}$;

    Slope of perpendicular $= -\dfrac{2}{3}$

    Containing $(1,5)$

    $y - y_1 = m(x - x_1) \Rightarrow y - 5 = -\dfrac{2}{3}(x-1)$

    $y - 5 = -\dfrac{2}{3}x + \dfrac{2}{3} \Rightarrow y = -\dfrac{2}{3}x + \dfrac{17}{3}$

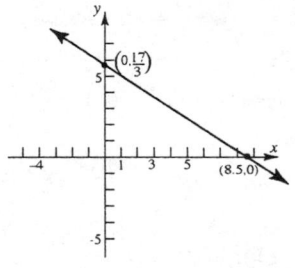

7. $f(x) = x^2 + 5x - 2$
    (a) $f(3) = 3^2 + 5 \cdot 3 - 2 = 9 + 15 - 2 = 22$
    (b) $f(-x) = (-x)^2 + 5(-x) - 2 = x^2 - 5x - 2$
    (c) $-f(x) = -(x^2 + 5x - 2) = -x^2 - 5x + 2$
    (d) $f(3x) = (3x)^2 + 5(3x) - 2 = 9x^2 + 15x - 2$
    (e) $\dfrac{f(x+h) - f(x)}{h} = \dfrac{(x+h)^2 + 5(x+h) - 2 - (x^2 + 5x - 2)}{h}$

    $\qquad = \dfrac{x^2 + 2xh + h^2 + 5x + 5h - 2 - x^2 - 5x + 2}{h} = \dfrac{2xh + h^2 + 5h}{h} = 2x + h + 5$

9. $f(x) = -3x + 7$

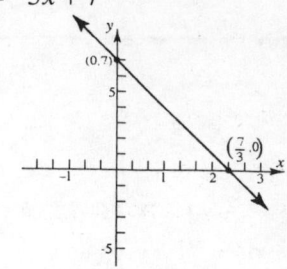

11. $f(x) = x^2 + 3x + 1$

$$\begin{array}{l} \text{avg rate of change of } f \\ \text{from 1 to } x \end{array} = \frac{f(x) - f(1)}{x-1} = \frac{x^2 + 3x + 1 - \left(1^2 + 3 \cdot 1 + 1\right)}{x-1}$$

$$= \frac{x^2 + 3x + 1 - 5}{x-1} = \frac{x^2 + 3x - 4}{x-1} = \frac{(x+4)(x-1)}{x-1} = x + 4 = m_{\text{sec}}$$

when $x = 2$, $m_{\text{sec}} = 2 + 4 = 6$.

13. $f(x) = -x^3 + 7x - 2$, for $-3 < x < 3$

Use MAXIMUM and MINIMUM on the graph of $y_1 = -x^3 + 7x - 2$.

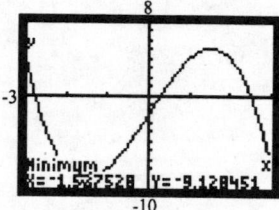

local maximum at: $(1.53, 5.13)$;   local minimum at: $(-1.53, -9.13)$

f is increasing on: $(-1.53, 1.53)$ f is decreasing on: $(-3, 1.53) \cup (1.53, 3)$

15. $f(x) = \begin{cases} 2x + 1 & \text{if } -3 < x < 2 \\ -3x + 4 & \text{if } x \geq 2 \end{cases}$

    (d)   Range: $\{ y < 5 \}$

    (e) Graphing Utility:

(a)   Domain: $(-3, \infty)$

(b)   x-intercept: $\left(-\frac{1}{2}, 0 \right)$

     y-intercept: $(0, 1)$

(c)

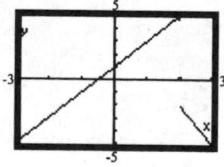

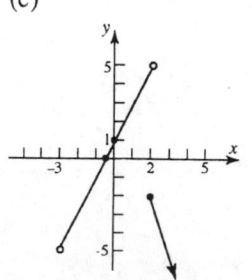

17. $f(x) = x^2 - 5x + 1$ $g(x) = -4x - 7$

   (a) $(f + g)(x) = x^2 - 5x + 1 + (-4x - 7) = x^2 - 9x - 6$

       The domain is all real numbers.

   (b) $\left(\dfrac{f}{g}\right)(x) = \dfrac{f(x)}{g(x)} = \dfrac{x^2 - 5x + 1}{-4x - 7}$ The domain is all real numbers except $-\dfrac{7}{4}$.

   (c) $(f \circ g)(2) = f(g(2)) = f(-4(2) - 7) = f(-15) = (-15)^2 - 5(-15) + 1$

$$= 225 + 75 + 1 = 301$$

   (d) $(f \circ g)(x) = f(g(x)) = f(-4x - 7) = (-4x - 7)^2 - 5(-4x - 7) + 1$

$$= 16x^2 + 56x + 49 + 20x + 35 + 1 = 16x^2 + 76x + 85$$

       Domain: All real numbers.

19.   (a) $R(x) = x \cdot p = x\left(-\dfrac{1}{10}x + 150\right) = -\dfrac{1}{10}x^2 + 150x$

   (b) $R(100) = -\dfrac{1}{10}(100)^2 + 150(100) = -1000 + 15000 = \$14000$

   (c) Using the MAXIMUM function on the graph of $y_1 = -x^2 / 10 + 150x$
       shows that $x = 750$ maximizes revenue.

$$R(750) = -\dfrac{1}{10}(750)^2 + 150(750) = -56250 + 112500 = \$56250 = \text{maximum revenue}$$

   (d) $p = -\dfrac{1}{10}(750) + 150 = -75 + 150 = \$75$ maximizes revenue

Chapter 3

Polynomial and Rational Functions

3.1   Quadratic Functions and Models

1.   C   3.   F   5.   G   7.   H

9. $f(x) = \dfrac{1}{4}x^2$

Using the graph of $y = x^2$, compress vertically

by a factor of $\dfrac{1}{4}$.

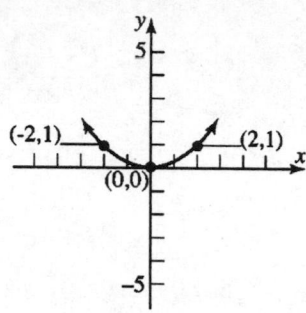

11. $f(x) = \dfrac{1}{4}x^2 - 2$

Using the graph of $y = x^2$, compress vertically

by a factor of $\dfrac{1}{4}$, then shift down 2 units.

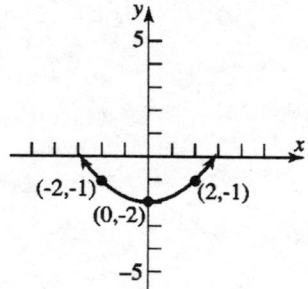

13. $f(x) = \dfrac{1}{4}x^2 + 2$

Using the graph of $y = x^2$, compress vertically

by a factor of $\dfrac{1}{4}$, then shift up 2 units.

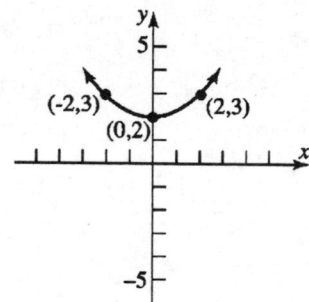

15. $f(x) = \dfrac{1}{4}x^2 + 1$

Using the graph of $y = x^2$, compress vertically

by a factor of $\dfrac{1}{4}$, then shift up 1 unit.

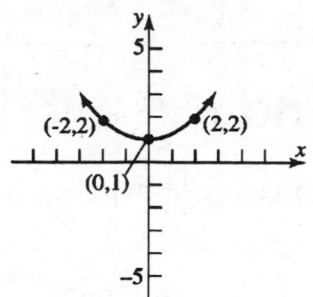

17. $f(x) = x^2 + 4x + 2$

$= \left(x^2 + 4x + 4\right) + 2 - 4 = \left(x + 2\right)^2 - 2$

Using the graph of $y = x^2$, shift left 2 units,
then shift down 2 units.

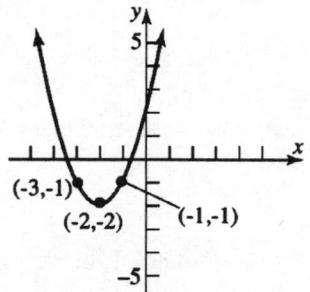

19. $f(x) = 2x^2 - 4x + 1$

$= 2\left(x^2 - 2x + 1\right) + 1 - 2 = 2\left(x - 1\right)^2 - 1$

Using the graph of $y = x^2$, shift right 1 unit,
stretch vertically by a factor of 2, then shift
down 1 unit.

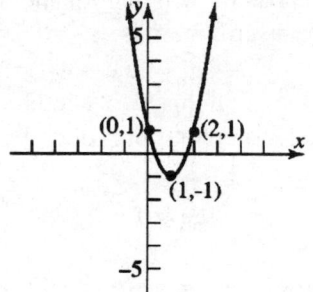

21. $f(x) = -x^2 - 2x$

$= -\left(x^2 + 2x + 1\right) + 1 = -\left(x + 1\right)^2 + 1$

Using the graph of $y = x^2$, shift left 1 unit,
reflect across the x-axis, then shift
up 1 unit.

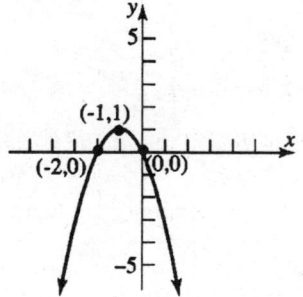

23. $f(x) = \dfrac{1}{2}x^2 + x - 1$

$= \dfrac{1}{2}(x^2 + 2x + 1) - 1 - \dfrac{1}{2}$

$= \dfrac{1}{2}(x+1)^2 - \dfrac{3}{2}$

Using the graph of $y = x^2$, shift left 1 unit,

compress vertically by a factor of $\dfrac{1}{2}$, then shift

down $\dfrac{3}{2}$ units.

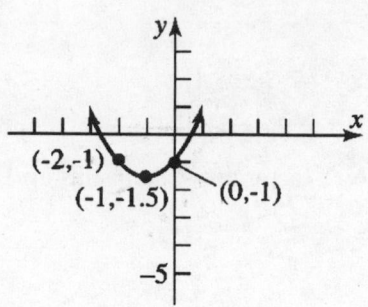

25. $f(x) = x^2 + 2x$
$a = 1, b = 2, c = 0$. Since $a = 1 > 0$, the graph opens up.

The x-coordinate of the vertex is $x = \dfrac{-b}{2a} = \dfrac{-(2)}{2(1)} = \dfrac{-2}{2} = -1$.

The y-coordinate of the vertex is $f\left(\dfrac{-b}{2a}\right) = f(-1) = (-1)^2 + 2(-1) = 1 - 2 = -1$.

Thus, the vertex is $(-1, -1)$.
The axis of symmetry is the line $x = -1$.
The discriminant is:

$b^2 - 4ac = (2)^2 - 4(1)(0) = 4 > 0$,

so the graph has two x-intercepts.
The x-intercepts are found by solving:

$x^2 + 6x = 0$

$x(x+2) = 0$

$x = 0$ or $x = -2$
The x-intercepts are –2 and 0.
The y-intercept is $f(0) = 0$.

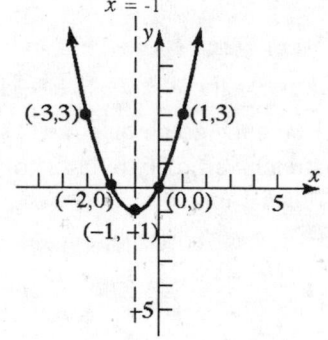

27. $f(x) = -x^2 - 6x$
$a = -1, b = -6, c = 0$. Since $a = -1 < 0$, the graph opens down.

The x-coordinate of the vertex is $x = \dfrac{-b}{2a} = \dfrac{-(-6)}{2(-1)} = \dfrac{6}{-2} = -3$.

The y-coordinate of the vertex is $f\left(\dfrac{-b}{2a}\right) = f(-3) = -(-3)^2 - 6(-3) = -9 + 18 = 9$.

Thus, the vertex is $(-3, 9)$.
The axis of symmetry is the line $x = -3$.
The discriminant is:

$b^2 - 4ac = (-6)^2 - 4(-1)(0) = 36 > 0$,

so the graph has two x-intercepts.

The x-intercepts are found by solving:

$$-x^2 - 6x = 0$$
$$-x(x + 6) = 0$$
$$x = 0 \ \text{ or } \ x = -6$$

The x-intercepts are –6 and 0.
The y-intercept is $f(0) = 0$.

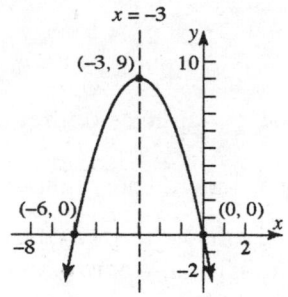

29. $f(x) = 2x^2 - 8x$

$a = 2, b = -8, c = 0$. Since $a = 2 > 0$, the graph opens up.

The x-coordinate of the vertex is $x = \dfrac{-b}{2a} = \dfrac{-(-8)}{2(2)} = \dfrac{8}{4} = 2$.

The y-coordinate of the vertex is $f\left(\dfrac{-b}{2a}\right) = f(2) = 2(2)^2 - 8(2) = 8 - 16 = -8$.

Thus, the vertex is $(2, -8)$.
The axis of symmetry is the line $x = 2$.
The discriminant is:

$$b^2 - 4ac = (-8)^2 - 4(2)(0) = 64 > 0,$$

so the graph has two x-intercepts.
The x-intercepts are found by solving:

$$2x^2 - 8x = 0$$
$$2x(x - 4) = 0$$
$$x = 0 \ \text{ or } \ x = 4$$

The x-intercepts are 0 and 4.
The y-intercept is $f(0) = 0$.

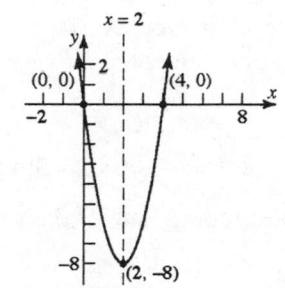

31. $f(x) = x^2 + 2x - 8$

$a = 1, b = 2, c = -8$. Since $a = 1 > 0$, the graph opens up.

The x-coordinate of the vertex is $x = \dfrac{-b}{2a} = \dfrac{-2}{2(1)} = \dfrac{-2}{2} = -1$.

The y-coordinate of the vertex is $f\left(\dfrac{-b}{2a}\right) = f(-1) = (-1)^2 + 2(-1) - 8 = 1 - 2 - 8 = -9$.

Thus, the vertex is $(-1, -9)$.
The axis of symmetry is the line $x = -1$.
The discriminant is:

$$b^2 - 4ac = 2^2 - 4(1)(-8) = 4 + 32 = 36 > 0,$$

so the graph has two x-intercepts.
The x-intercepts are found by solving:

$$x^2 + 2x - 8 = 0$$
$$(x + 4)(x - 2) = 0$$
$$x = -4 \ \text{ or } \ x = 2$$

The x-intercepts are –4 and 2.
The y-intercept is $f(0) = -8$.

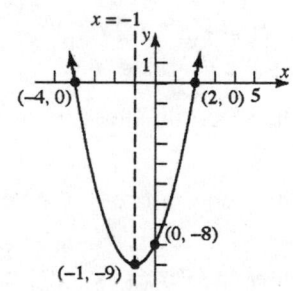

33. $f(x) = x^2 + 2x + 1$

$a = 1, b = 2, c = 1.$ Since $a = 1 > 0$, the graph opens up.

The x-coordinate of the vertex is $x = \dfrac{-b}{2a} = \dfrac{-2}{2(1)} = \dfrac{-2}{2} = -1.$

The y-coordinate of the vertex is $f\left(\dfrac{-b}{2a}\right) = f(-1) = (-1)^2 + 2(-1) + 1 = 1 - 2 + 1 = 0.$

Thus, the vertex is $(-1, 0)$.
The axis of symmetry is the line $x = -1$.
The discriminant is:

$\qquad b^2 - 4ac = 2^2 - 4(1)(1) = 4 - 4 = 0,$

so the graph has one x-intercept.
The x-intercept is found by solving:

$\qquad x^2 + 2x + 1 = 0$

$\qquad\quad (x+1)^2 = 0$

$\qquad\qquad\quad x = -1$

The x-intercept is -1.
The y-intercept is $f(0) = 1$.

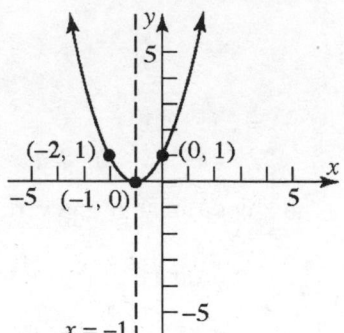

35. $f(x) = 2x^2 - x + 2$

$a = 2, b = -1, c = 2.$ Since $a = 2 > 0$, the graph opens up.

The x-coordinate of the vertex is $x = \dfrac{-b}{2a} = \dfrac{-(-1)}{2(2)} = \dfrac{1}{4}.$

The y-coordinate of the vertex is $f\left(\dfrac{-b}{2a}\right) = f\left(\dfrac{1}{4}\right) = 2\left(\dfrac{1}{4}\right)^2 - \dfrac{1}{4} + 2 = \dfrac{1}{8} - \dfrac{1}{4} + 2 = \dfrac{15}{8}.$

Thus, the vertex is $\left(\dfrac{1}{4}, \dfrac{15}{8}\right)$.

The axis of symmetry is the line $x = \dfrac{1}{4}$.

The discriminant is:

$\qquad b^2 - 4ac = (-1)^2 - 4(2)(2) = 1 - 16 = -15,$

so the graph has no x-intercepts.
The y-intercept is $f(0) = 2$.

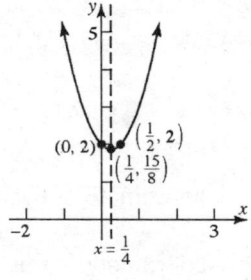

37. $f(x) = -2x^2 + 2x - 3$

$a = -2, b = 2, c = -3.$ Since $a = -2 < 0$, the graph opens down.

The x-coordinate of the vertex is $x = \dfrac{-b}{2a} = \dfrac{-(2)}{2(-2)} = \dfrac{-2}{-4} = \dfrac{1}{2}.$

The y-coordinate of the vertex is $f\left(\dfrac{-b}{2a}\right) = f\left(\dfrac{1}{2}\right) = -2\left(\dfrac{1}{2}\right)^2 + 2\left(\dfrac{1}{2}\right) - 3 = -\dfrac{1}{2} + 1 - 3 = -\dfrac{5}{2}.$

Thus, the vertex is $\left(\dfrac{1}{2}, -\dfrac{5}{2}\right)$.

The axis of symmetry is the line $x = \dfrac{1}{2}$.

The discriminant is:
$$b^2 - 4ac = 2^2 - 4(-2)(-3) = 4 - 24 = -20,$$
so the graph has no x-intercepts.
The y-intercept is $f(0) = -3$.

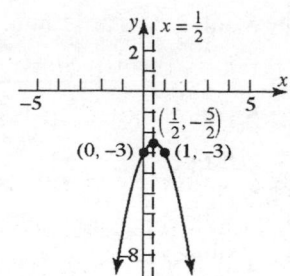

39. $f(x) = 3x^2 + 6x + 2$
$a = 3, b = 6, c = 2$. Since $a = 3 > 0$, the graph opens up.

The x-coordinate of the vertex is $x = \dfrac{-b}{2a} = \dfrac{-6}{2(3)} = \dfrac{-6}{6} = -1$.

The y-coordinate of the vertex is $f\left(\dfrac{-b}{2a}\right) = f(-1) = 3(-1)^2 + 6(-1) + 2 = 3 - 6 + 2 = -1.$

Thus, the vertex is $(-1, -1)$.
The axis of symmetry is the line $x = -1$.
The discriminant is:
$$b^2 - 4ac = 6^2 - 4(3)(2) = 36 - 24 = 12,$$
so the graph has two x-intercepts.
The x-intercepts are found by solving:

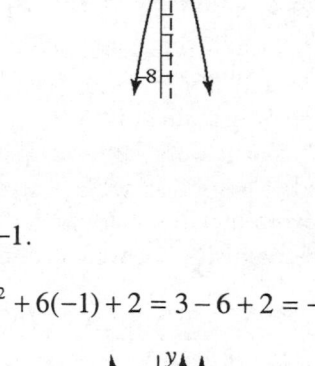

$$x = \frac{-b \pm \sqrt{b^2 - 4ac}}{2a} = \frac{-6 \pm \sqrt{12}}{2(3)}$$
$$= \frac{-6 \pm 2\sqrt{3}}{6} = \frac{-3 \pm \sqrt{3}}{3} \approx \frac{-3 \pm 1.732}{3}$$

The x-intercepts are approximately -0.42 and -1.58.
The y-intercept is $f(0) = 2$.

41. $f(x) = -4x^2 - 6x + 2$
$a = -4, b = -6, c = 2$. Since $a = -4 < 0$, the graph opens down.

The x-coordinate of the vertex is $x = \dfrac{-b}{2a} = \dfrac{-(-6)}{2(-4)} = \dfrac{6}{-8} = -\dfrac{3}{4}.$

The y-coordinate of the vertex is
$$f\left(\frac{-b}{2a}\right) = f\left(-\frac{3}{4}\right) = -4\left(-\frac{3}{4}\right)^2 - 6\left(-\frac{3}{4}\right) + 2 = -\frac{9}{4} + \frac{9}{2} + 2 = \frac{17}{4}.$$

Thus, the vertex is $\left(-\dfrac{3}{4}, \dfrac{17}{4}\right)$.

The axis of symmetry is the line $x = -\dfrac{3}{4}$.

The discriminant is:
$$b^2 - 4ac = (-6)^2 - 4(-4)(2) = 36 + 32 = 68,$$
so the graph has two x-intercepts.

The x-intercepts are found by solving:

$$x = \frac{-b \pm \sqrt{b^2 - 4ac}}{2a} = \frac{-(-6) \pm \sqrt{68}}{2(-4)}$$

$$= \frac{6 \pm 2\sqrt{17}}{-8} = \frac{-3 \pm \sqrt{17}}{4} \approx \frac{-3 \pm 4.123}{4}$$

The x-intercepts are approximately -1.78 and 0.28.
The y-intercept is $f(0) = 2$.

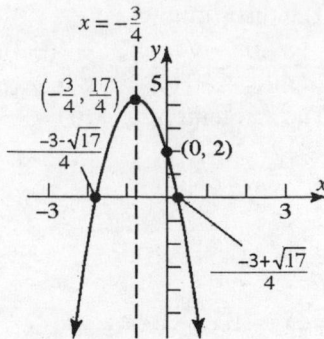

43.   Given that the graph of $f(x) = ax^2 + bx + c$ has vertex $(-1,-2)$ and passes through the point $(0,-1)$, we can conclude

$$\frac{-b}{2a} = -1, \qquad f(-1) = -2, \quad \text{and} \qquad f(0) = -1$$

Notice that $f(0) = -1 \Rightarrow a(0)^2 + b(0) + c = -1 \Rightarrow c = -1$

Therefore $f(x) = ax^2 + bx + c = ax^2 + bx - 1$.

Furthermore, $\dfrac{-b}{2a} = -1 \Rightarrow b = 2a$

     and $f(-1) = -2 \Rightarrow a(-1)^2 + b(-1) - 1 = -2 \Rightarrow a - b - 1 = -2$

$$\Rightarrow a - b = -1$$

Replacing b with $2a$ in this equation yields

     $a - 2a = -1 \Rightarrow -a = -1 \Rightarrow a = 1$.

     So $b = 2a = 2(1) = 2$.

Therefore, we have the function $f(x) = x^2 + 2x - 1$.

45.   Given that the graph of $f(x) = ax^2 + bx + c$ has vertex $(-3,5)$ and passes through the point $(0,-4)$, we can conclude

$$\frac{-b}{2a} = -3, \qquad f(-3) = 5, \quad \text{and} \qquad f(0) = -4$$

Notice that $f(0) = -4 \Rightarrow a(0)^2 + b(0) + c = -4 \Rightarrow c = -4$

Therefore $f(x) = ax^2 + bx + c = ax^2 + bx - 4$.

Furthermore, $\dfrac{-b}{2a} = -3 \Rightarrow b = 6a$

     and $f(-3) = 5 \Rightarrow a(-3)^2 + b(-3) - 4 = 5 \Rightarrow 9a - 3b - 4 = 5 \Rightarrow 9a - 3b = 9$

$$\Rightarrow 3a - b = 3$$

Replacing b with $6a$ in this equation yields

     $3a - 6a = 3 \Rightarrow -3a = 3 \Rightarrow a = -1$.

     So $b = 6a = 6(-1) = -6$.

Therefore, we have the function $f(x) = -x^2 - 6x - 4$.

47.   Given that the graph of $f(x) = ax^2 + bx + c$ has vertex $(1,-3)$ and passes through the point $(3,5)$, we can conclude

$$\frac{-b}{2a} = 1, \qquad f(1) = -3, \quad \text{and} \qquad f(3) = 5$$

Notice that $f(3) = 5 \Rightarrow a(3)^2 + b(3) + c = 5 \Rightarrow 9a + 3b + c = 5$

Furthermore, $\dfrac{-b}{2a} = 1 \Rightarrow b = -2a$

and $f(1) = -3 \Rightarrow a(1)^2 + b(1) + c = -3 \Rightarrow a + b + c = -3$

Replacing b with $-2a$ in each of these equations yields

$$9a + 3b + c = 5 \rightarrow 9a + 3(-2a) + c = 5 \rightarrow 9a - 6a + c = 5$$

$$\Rightarrow 3a + c = 5 \Rightarrow c = 5 - 3a$$

$$a + (-2a) + c = -3 \rightarrow a - 2a + c = -3$$

$$\Rightarrow -a + c = -3 \Rightarrow c = -3 + a$$

Finally, since $c = 5 - 3a$ and $c = -3 + a$, setting these two expressions equal and solving for a yields

$$5 - 3a = -3 + a$$

$$8 = 4a$$

$$2 = a$$

So $c = 5 - 3a = 5 - 3 \cdot 2 = 5 - 6 = -1$.

And $b = -2a = (-2)(2) = -4$.

Therefore, we have the function $f(x) = 2x^2 - 4x - 1$.

49. (a) $f(x) = 1(x - (-3))(x - 1) = 1(x + 3)(x - 1) = 1(x^2 + 2x - 3) = x^2 + 2x - 3$

$f(x) = 2(x - (-3))(x - 1) = 2(x + 3)(x - 1) = 2(x^2 + 2x - 3) = 2x^2 + 4x - 6$

$f(x) = -2(x - (-3))(x - 1) = -2(x + 3)(x - 1) = -2(x^2 + 2x - 3) = -2x^2 - 4x + 6$

$f(x) = 5(x - (-3))(x - 1) = 5(x + 3)(x - 1) = 5(x^2 + 2x - 3) = 5x^2 + 10x - 15$

(b) The value of a multiplies the value of the y-intercept by the value of a. The values of the x-intercepts are not changed.

(c) The axis of symmetry is unaffected by the value of a.

(d) The y-coordinate of the vertex is multiplied by the value of a.

(e) The x-coordinate of the vertex is the midpoint of the x-intercepts.

51. Given that the graph of $f(x) = ax^2 + bx + c$ has vertex $(0, 2)$ and passes through the point $(1, 8)$, we can conclude

$$\dfrac{-b}{2a} = 0, \qquad f(0) = 2, \qquad \text{and} \qquad f(1) = 8$$

Notice that $\dfrac{-b}{2a} = 0 \Rightarrow b = 0$

and $f(0) = 2 \Rightarrow a(0)^2 + b(0) + c = 2 \Rightarrow c = 2$

Therefore $f(x) = ax^2 + bx + c = ax^2 + 2$.

And since $f(1) = 8$, we have

$$f(1) = a(1)^2 + 2 = 8 \Rightarrow a + 2 = 8 \rightarrow a = 6$$

So we have the function $f(x) = 6x^2 + 2$.

53. $f(x) = x^2 + 2$, $a = 1, b = 0, c = 2$. Since $a = 1 > 0$, the graph opens up, so

the vertex is a minimum point. The minimum occurs at $x = \dfrac{-b}{2a} = \dfrac{-(0)}{2(1)} = \dfrac{0}{2} = 0$.

The minimum value is $f\left(\dfrac{-b}{2a}\right) = f(0) = (0)^2 + 2 = 0 + 2 = 2$.

X	Y1	
-3	11	
-2	6	
-1	3	
0	**2**	
1	3	
2	6	
3	11	

X=0

55. $f(x) = -x^2 - 4x$, $a = -1, b = -4, c = 0$. Since $a = -1 < 0$, the graph opens down, so the vertex is a maximum point. The maximum occurs at $x = \dfrac{-b}{2a} = \dfrac{-(-4)}{2(-1)} = \dfrac{4}{-2} = -2$.

The maximum value is $f\left(\dfrac{-b}{2a}\right) = f(-2) = -(-2)^2 - 4(-2) = -4 + 8 = 4$.

X	Y1	
-5	-5	
-4	0	
-3	3	
-2	**4**	
-1	3	
0	0	
1	-5	

X=-2

57. $f(x) = 2x^2 + 12x$, $a = 2, b = 12, c = 0$. Since $a = 2 > 0$, the graph opens up, so the vertex is a minimum point. The minimum occurs at $x = \dfrac{-b}{2a} = \dfrac{-12}{2(2)} = \dfrac{-12}{4} = -3$.

The minimum value is $f\left(\dfrac{-b}{2a}\right) = f(-3) = 2(-3)^2 + 12(-3) = 18 - 36 = -18$.

X	Y1	
-5	-10	
-4	-16	
-3	**-18**	
-2	-16	
-1	-10	
0	0	
1	14	

X=-3

59. $f(x) = 2x^2 + 12x - 3$, $a = 2, b = 12, c = -3$. Since $a = 2 > 0$, the graph opens up, so the vertex is a minimum point. The minimum occurs at $x = \dfrac{-b}{2a} = \dfrac{-12}{2(2)} = \dfrac{-12}{4} = -3$.

The minimum value is $f\left(\dfrac{-b}{2a}\right) = f(-3) = 2(-3)^2 + 12(-3) - 3 = 18 - 36 - 3 = -21$.

X	Y1	
-6	-3	
-5	-13	
-4	-19	
-3	**-21**	
-2	-19	
-1	-13	
0	-3	

X=-3

61. $f(x) = -x^2 + 10x - 4$

$a = -1, b = 10, c = -4$. Since $a = -1 < 0$, the graph opens down, so the vertex is a maximum point. The maximum occurs at $x = \dfrac{-b}{2a} = \dfrac{-10}{2(-1)} = \dfrac{-10}{-2} = 5$.

The maximum value is $f\left(\dfrac{-b}{2a}\right) = f(5) = -(5)^2 + 10(5) - 4 = -25 + 50 - 4 = 21$.

63. $f(x) = -3x^2 + 12x + 1$

$a = -3,\ b = 12,\ c = 1$. Since $a = -3 < 0$, the graph opens down, so the vertex is a maximum point. The maximum occurs at $x = \dfrac{-b}{2a} = \dfrac{-12}{2(-3)} = \dfrac{-12}{-6} = 2$.

The maximum value is $f\left(\dfrac{-b}{2a}\right) = f(2) = -3(2)^2 + 12(2) + 1 = -12 + 24 + 1 = 13$.

65. $R(p) = -4p^2 + 4000p,\ a = -4,\ b = 4000,\ c = 0$. Since $a = -4 < 0$, the graph is a parabola that opens down, so the vertex is a maximum point. The maximum occurs at

$$p = \dfrac{-b}{2a} = \dfrac{-4000}{2(-4)} = 500.$$

Thus, the unit price should be \$500 for maximum revenue.
The maximum revenue is

$$R(500) = -4(500)^2 + 4000(500) = -1000000 + 2000000 = \$1{,}000{,}000$$

67. $C(x) = x^2 - 80x + 2000,\ a = 1, b = -80, c = 2000$. Since $a = 1 > 0$, the graph opens up, so the vertex is a minimum point.

The minimum cost occurs at $x = \dfrac{-b}{2a} = \dfrac{-(-80)}{2(1)} = \dfrac{80}{2} = 40$ televisions produced.

The minimum cost is $f\left(\dfrac{-b}{2a}\right) = f(40) = (40)^2 - 80(40) + 2000 = 1600 - 3200 + 2000 = \$400$.

69.   (a) $R(x) = x\left(-\dfrac{1}{6}x + 100\right) = -\dfrac{1}{6}x^2 + 100x$

(b) $R(200) = -\dfrac{1}{6}(200)^2 + 100(200) = \dfrac{-20000}{3} + 20000 = \dfrac{40000}{3} \approx \$13{,}333$

(c) $x = \dfrac{-b}{2a} = \dfrac{-100}{2\left(-\dfrac{1}{6}\right)} = \dfrac{-100}{\left(-\dfrac{1}{3}\right)} = \dfrac{300}{1} = 300$ maximizes revenue

$$R(300) = -\dfrac{1}{6}(300)^2 + 100(300) = -15000 + 30000 = \$15{,}000 = \text{maximum revenue}$$

(d) $p = -\dfrac{1}{6}(300) + 100 = -50 + 100 = \$50$ maximizes revenue

71. (a)   If $x = -5p + 100$, then $p = \dfrac{100 - x}{5}$. $R(x) = x\left(\dfrac{100 - x}{5}\right) = -\dfrac{1}{5}x^2 + 20x$

    (b) $R(15) = -\dfrac{1}{5}(15)^2 + 20(15) = -45 + 300 = \$255$

    (c) $x = \dfrac{-b}{2a} = \dfrac{-20}{2\left(-\dfrac{1}{5}\right)} = \dfrac{-20}{\left(-\dfrac{2}{5}\right)} = \dfrac{100}{2} = 50$ maximizes revenue

        $R(50) = -\dfrac{1}{5}(50)^2 + 20(50) = -500 + 1000 = \$500$ = maximum revenue

    (d) $p = \dfrac{100 - 50}{5} = \dfrac{50}{5} = \$10$ maximizes revenue

73. (a)   Let x = width and y = length of the rectangular area.
$$P = 2x + 2y = 400 \Rightarrow y = \frac{400 - 2x}{2} = 200 - x$$
       Then $A(x) = (200 - x)x = 200x - x^2 = -x^2 + 200x$

    (b) $x = \dfrac{-b}{2a} = \dfrac{-200}{2(-1)} = \dfrac{-200}{-2} = 100$ yards maximizes area

    (c) $A(100) = -100^2 + 200(100) = -10000 + 20000 = 10,000$ sq. yds. = maximum area

75. Let x = width and y = length of the rectangular area.
$$2x + y = 4000 \quad \Rightarrow \quad y = 4000 - 2x$$
Then $A(x) = (4000 - 2x)x = 4000x - 2x^2 = -2x^2 + 4000x$
$$x = \frac{-b}{2a} = \frac{-4000}{2(-2)} = \frac{-4000}{-4} = 1000 \text{ maximizes area}$$
$$A(1000) = -2(1000)^2 + 4000(1000) = -2000000 + 4000000 = 2,000,000$$
The largest area that can be enclosed is 2,000,000 square meters.

77. (a) $a = -\dfrac{32}{2500}, b = 1, c = 200.$ The maximum height occurs when
$$x = \frac{-b}{2a} = \frac{-1}{2\left(-\dfrac{32}{2500}\right)} = \frac{2500}{64} = 39.0625 \text{ feet from base of the cliff.}$$

    (b)   The maximum height is
$$h(39.0625) = \frac{-32(39.0625)^2}{2500} + 39.0625 + 200 = 219.53 \text{ feet.}$$

    (c)   Solving when $h(x) = 0$:
$$-\frac{32}{2500}x^2 + x + 200 = 0$$

$$x = \frac{-1 \pm \sqrt{1^2 - 4\left(-\dfrac{32}{2500}\right)(200)}}{2\left(-\dfrac{32}{2500}\right)} = \frac{-1 \pm \sqrt{11.24}}{-0.0256} \Rightarrow x \approx -91.90 \text{ or } x \approx 170.02$$

        Since the distance cannot be negative, the projectile strikes the water 170.02 feet from the base of the cliff.

(d) Graphing:

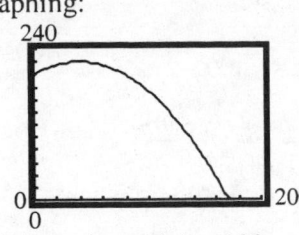

(e) Using the MAXIMUM function

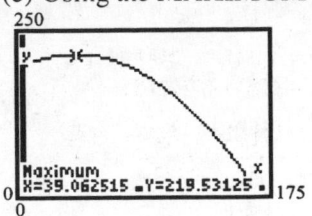

(f)   Solving when $h(x) = 100$:

$$-\frac{32}{2500}x^2 + x + 200 = 100 \Rightarrow -\frac{32}{2500}x^2 + x + 100 = 0$$

$$x = \frac{-1 \pm \sqrt{1^2 - 4\left(-\frac{32}{2500}\right)(100)}}{2\left(-\frac{32}{2500}\right)} = \frac{-1 \pm \sqrt{6.12}}{-0.0256}; \quad x \approx -57.57 \text{ or } x \approx 135.70$$

Since the distance cannot be negative, the projectile is 100 feet above the water 135.70 feet from the base of the cliff.

79. Locate the origin at the point where the cable touches the road. Then the equation of the parabola is of the form: $y = ax^2$, where $a > 0$. Since the point (200, 75) is on the parabola, we can find the constant a:

$$75 = a(200)^2 \quad \Rightarrow \quad a = \frac{75}{200^2} = 0.001875$$

When $x = 100$, we have:
$$y = 0.001875(100)^2 = 18.75 \text{ meters}.$$

81. Let x = the depth of the gutter and y = the width of the gutter.
Then $A = xy$ is the cross-sectional area of the gutter.
Since the aluminum sheets for the gutter are 12 inches wide, we have
$$2x + y = 12 \Rightarrow y = 12 - 2x.$$
The area is to be maximized, so: $A = xy = x(12 - 2x) = -2x^2 + 12x$.
This equation is a parabola opening down; thus, it has a maximum when
$$x = \frac{-b}{2a} = \frac{-12}{2(-2)} = \frac{-12}{-4} = 3.$$
Thus, a depth of 3 inches produces a maximum cross-sectional area.

83. Let x = the width of the rectangle or the diameter of the semicircle.
Let y = the length of the rectangle.

The perimeter of each semicircle is $\dfrac{\pi x}{2}$.

The perimeter of the track is given by: $\dfrac{\pi x}{2} + \dfrac{\pi x}{2} + y + y = 400$.

Chapter 3   Polynomial and Rational Functions

Solving for x:

$$\frac{\pi x}{2} + \frac{\pi x}{2} + y + y = 1500 \to \pi x + 2y = 400 \Rightarrow \pi x = 400 - 2y \to x = \frac{400 - 2y}{\pi}$$

The area of the rectangle is: $A = xy = \left(\frac{400 - 2y}{\pi}\right) y = \frac{-2}{\pi} y^2 + \frac{400}{\pi} y$

This equation is a parabola opening down; thus, it has a maximum when

$$y = \frac{-b}{2a} = \frac{\frac{-400}{\pi}}{2\left(\frac{-2}{\pi}\right)} = \frac{-400}{-4} = 100. \quad \text{Thus, } x = \frac{400 - 2(100)}{\pi} = \frac{200}{\pi} \approx 63.66.$$

The dimensions for the rectangle with maximum area are $\dfrac{200}{\pi} \approx 63.66$ meters by 100 meters.

Problems 85 – 87. The equations for the curves that are graphed on the screens use many more decimal places in order to get the desired accuracy.

85. (a)   Graphing: The data appear to be quadratic with $a < 0$.

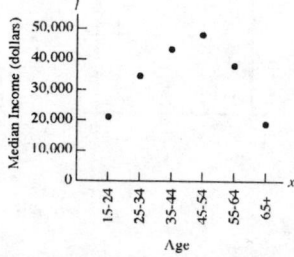

(b)   Using the QUADratic REGression program, the quadratic function of best fit is:
$$I(x) = -47.71x^2 + 4262.66x - 42777.73$$

(c) $x = \dfrac{-b}{2a} = \dfrac{-4262.66}{2(-47.71)} \approx 44.673$

An individual will earn the most income at an age of 44.7 years.

(d)   The maximum income will be:
$$I(44.7) = -47.71(44.7)^2 + 4262.66(44.7) - 42777.73 = \$52,434.30$$

(e)   Graphing the quadratic function of best fit:

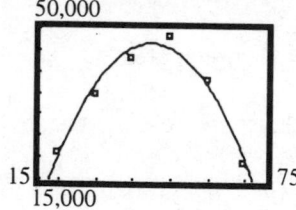

87. (a) Graphing: The data appears to be quadratic with $a < 0$.

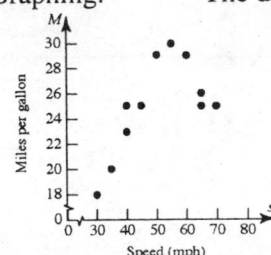

(b) Using the QUADratic REGression program, the quadratic function of best fit is:

$M(s) = -0.0175s^2 + 1.93s - 25.34$

(c) $s = \dfrac{-b}{2a} = \dfrac{-1.93}{2(-0.0175)} = 55.14$ The speed that maximizes miles per gallon is about 55 miles per hour.

(d) The predicted miles per gallon when the speed is 63 miles per hour is:

$M(63) = -0.0175(63)^2 + 1.93(63) - 25.34 \approx 26.79$ miles per gallon.

(e) Graphing the quadratic function of best fit:

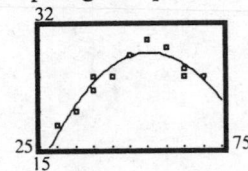

89. We are given: $V(x) = kx(a - x) = -kx^2 + akx$

The reaction rate is a maximum when: $x = \dfrac{-b}{2a} = \dfrac{-ak}{2(-k)} = \dfrac{ak}{2k} = \dfrac{a}{2}$

91. $f(x) = -5x^2 + 8$ $h = 1$:

$\text{Area} = \dfrac{h}{3}\left(2ah^2 + 6c\right) = \dfrac{1}{3}\left(2(-5)(1)^2 + 6(8)\right) = \dfrac{1}{3}(-10 + 48) = \dfrac{38}{3} \approx 12.67$ sq. units

93. $f(x) = x^2 + 3x + 5$, $h = 4$

$\text{Area} = \dfrac{h}{3}\left(2ah^2 + 6c\right) = \dfrac{4}{3}\left(2(1)(4)^2 + 6(5)\right) = \dfrac{4}{3}(32 + 30) = \dfrac{248}{3} \approx 82.67$ sq. units.

95. Answers will vary.

97. $y = x^2 - 4x + 1$; $y = x^2 + 1$; $y = x^2 + 4x + 1$

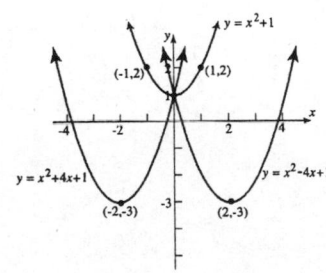

Each member of this family will be a parabola with the following characteristics:

(i) opens upwards since a > 0
(ii) y-intercept occurs at (0, 1)

Chapter 3

Polynomial and Rational Functions

3.2   Power Functions and Models

1. $f(x) = (x+1)^4$

Using the graph of $y = x^4$, shift the graph horizontally, 1 unit to the left.

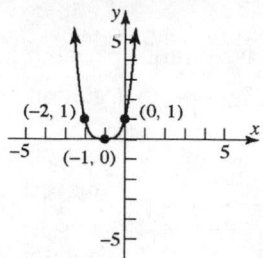

3. $f(x) = x^5 - 3$

Using the graph of $y = x^5$, shift the graph vertically, 3 units down.

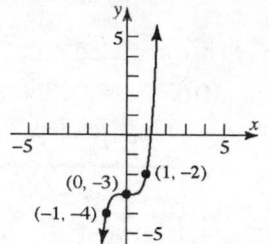

5. $f(x) = \dfrac{1}{2}x^4$

Using the graph of $y = x^4$, compress the graph vertically by a factor of $\dfrac{1}{2}$.

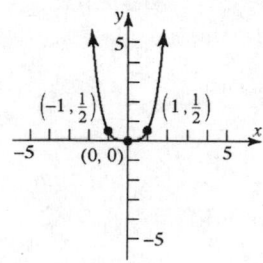

7. $f(x) = -x^5$

Using the graph of $y = x^5$, reflect the graph about the x-axis.

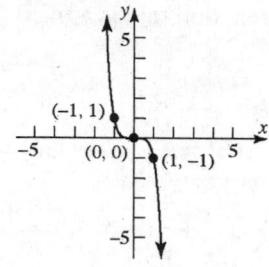

9. $f(x) = (x-1)^5 + 2$

Using the graph of $y = x^5$, shift the graph horizontally, 1 unit to the right, and shift vertically 2 units up.

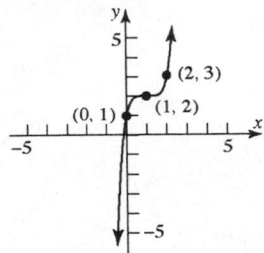

11. $f(x) = 2(x+1)^4 + 1$

Using the graph of $y = x^4$, shift the graph horizontally, 1 unit to the left, stretch vertically by a factor of 2, and shift vertically 1 unit up.

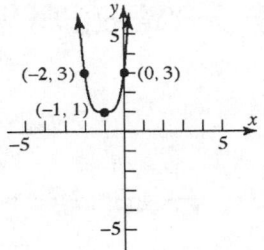

13. $f(x) = 4 - (x-2)^5 = -(x-2)^5 + 4$

Using the graph of $y = x^5$, shift the graph horizontally, 2 units to the right, reflect about the x-axis, and shift vertically 4 units up.

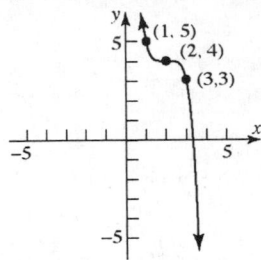

15. $f(x) = -\dfrac{1}{2}(x-2)^4 - 1$

Using the graph of $y = x^4$, shift the graph horizontally, 2 units to the right, reflect about the x-axis, compress vertically by a factor of $\dfrac{1}{2}$ and shift vertically 1 unit down.

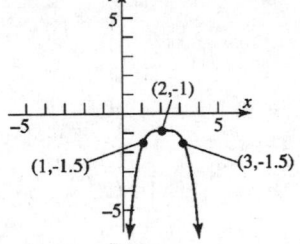

17.   (a)

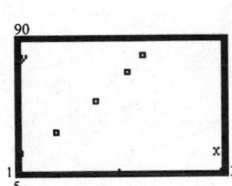

(b) $s = 15.973t^{2.002}$
(c)

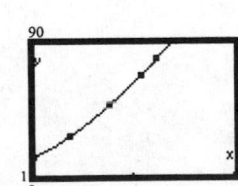

(d)   Solve $s = 15.973t^{2.002} = 100$ for t.

$15.973t^{2.002} = 100$

$t^{2.002} = \dfrac{100}{15.973} \approx 6.261$

$t \approx (6.261)^{1/2.002} \approx 2.4998$ seconds

(e) $s = 15.973t^2 = \dfrac{1}{2}gt^2$

$15.973 = \dfrac{1}{2}g$

$g = 31.946$

$s = \left(\dfrac{1}{2}\right)(31.946)t^2$

Chapter 3

Polynomial and Rational Functions

3.3 Polynomial Functions and Models

1. $f(x) = 4x + x^3$ is a polynomial function of degree 3.

3. $g(x) = \dfrac{1-x^2}{2} = \dfrac{1}{2} - \dfrac{1}{2}x^2$ is a polynomial function of degree 2.

5. $f(x) = 1 - \dfrac{1}{x} = 1 - x^{-1}$ is not a polynomial function because it contains a negative exponent.

7. $g(x) = x^{3/2} - x^2 + 2$ is not a polynomial function because it contains a fractional exponent.

9. $F(x) = 5x^4 - \pi x^3 + \dfrac{1}{2}$ is a polynomial function of degree 4.

11. $f(x) = a(x - (-1))(x - 1)(x - 3)$

   For $a = 1$: $f(x) = (x+1)(x-1)(x-3) = \left(x^2 - 1\right)(x-3) = x^3 - 3x^2 - x + 3$

13. $f(x) = a(x - (-3))(x - 0)(x - 4)$

   For $a = 1$: $f(x) = (x+3)(x)(x-4)$

   $\qquad f(x) = \left(x^2 + 3x\right)(x-4) = x^3 - 4x^2 + 3x^2 - 12x = x^3 - x^2 - 12x$

15. $f(x) = a(x - (-4))(x - (-1))(x - 2)(x - 3)$

   For $a = 1$: $f(x) = (x+4)(x+1)(x-2)(x-3)$

   $\qquad f(x) = \left(x^2 + 5x + 4\right)\left(x^2 - 5x + 6\right)$

   $\qquad f(x) = x^4 - 5x^3 + 6x^2 + 5x^3 - 25x^2 + 30x + 4x^2 - 20x + 24$

   $\qquad f(x) = x^4 - 15x^2 + 10x + 24$

17. $f(x) = a(x - (-1))(x - 3)^2$

   For $a = 1$: $f(x) = (x+1)(x-3)^2$

   $\qquad f(x) = (x+1)\left(x^2 - 6x + 9\right)$

   $\qquad f(x) = x^3 - 6x^2 + 9x + x^2 - 6x + 9 = x^3 - 5x^2 + 3x + 9$

19.   The real zeros of $f(x) = 3(x - 7)(x + 3)^2$ are: 7, with multiplicity one; and -3, with multiplicity two. The graph crosses the x-axis at 7 and touches it at -3.
The function resembles $y = 3x^3$ for large values of $|x|$.

21.   The real zeros of $f(x) = 4(x^2 + 1)(x - 2)^3$ is: 2, with multiplicity three. $x^2 + 1 = 0$ has no real solution. The graph crosses the x-axis at 2.
The function resembles $y = 4x^5$ for large values of $|x|$.

23.   The real zeros of $f(x) = -2\left(x + \dfrac{1}{2}\right)^2 (x^2 + 4)^2$ is: $-\dfrac{1}{2}$, with multiplicity two. $x^2 + 4 = 0$ has no real solution. The graph touches the x-axis at $-\dfrac{1}{2}$.
The function resembles $y = -2x^6$ for large values of $|x|$.

25.   The real zeros of $f(x) = (x - 5)^3 (x + 4)^2$ are: 5, with multiplicity three; and -4, with multiplicity two. The graph crosses the x-axis at 5 and touches it at -4.
The function resembles $y = x^5$ for large values of $|x|$.

27. $f(x) = 3(x^2 + 8)(x^2 + 9)^2$ has no real zeros. $x^2 + 8 = 0$ and $x^2 + 9 = 0$ have no real solutions. The graph neither touches nor crosses the x-axis. The function resembles $y = 3x^6$ for large values of $|x|$.

29.   The real zeros of $f(x) = -2x^2(x^2 - 2)$ are: $-\sqrt{2}$ and $\sqrt{2}$ with multiplicity one; and 0, with multiplicity two. The graph touches the x-axis at 0 and crosses the x-axis at $-\sqrt{2}$ and $\sqrt{2}$. The function resembles $y = -2x^4$ for large values of $|x|$.

31. $f(x) = (x - 1)^2$
(a)   x-intercept: 1; y-intercept: 1
(b)   touches x-axis at x = 1
(c) $y = x^2$
(d)   graphing utility:

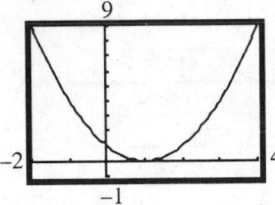

(e)   1 turning point; local minimum $(1, 0)$

(f)   graphing by hand

Interval	$(-\infty, 1)$	$(1, \infty)$
Number Chosen	-1	2
Value of f	$f(-1) = 4$	$f(2) = 1$
Location of Graph	Above x-axis	Above x-axis
Point on Graph	(-1, 4)	(2, 1)

f is above the x-axis for $(-\infty, 1) \cup (1, \infty)$

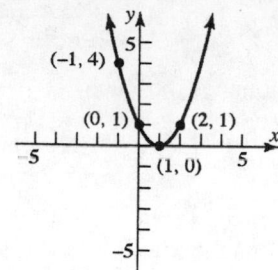

33. $f(x) = x^2(x - 3)$

(a)   x-intercepts: 0, 3; y-intercept: 0
(b)   touches x-axis at $x = 0$; crosses x-axis at $x = 3$
(c) $y = x^3$
(d)   graphing utility:

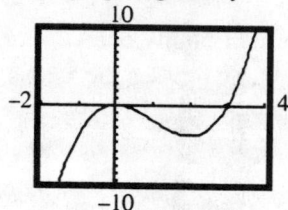

(e)   2 turning points; local maximum: (0, 0); local minimum: (2, –4)
(f)   graphing by hand

Interval	$(-\infty, 0)$	$(0, 3)$	$(3, \infty)$
Number Chosen	-1	2	4
Value of f	$f(-1) = -4$	$f(2) = -4$	$f(4) = 16$
Location of Graph	Below x-axis	Below x-axis	Above x-axis
Point on Graph	(-1, -4)	(2, -4)	(4, 16)

f is below the x-axis for $(-\infty, 0) \cup (0, 3)$; f is above the x-axis for $(3, \infty)$

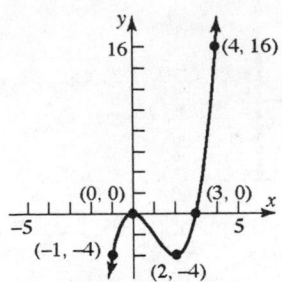

35. $f(x) = 6x^3(x + 4)$

(a)   x-intercepts: –4, 0; y-intercept: 0   (b)   crosses x-axis at x = –4 and x = 0

(c) $y = 6x^4$

(d)   graphing utility:

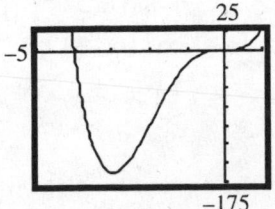

(e)   1 turning point; local minimum: (–3, –162)

(f)   graphing by hand

Interval	$(-\infty, -4)$	$(-4, 0)$	$(0, \infty)$
Number Chosen	-5	-2	1
Value of f	$f(-5) = 750$	$f(-2) = -96$	$f(1) = 30$
Location of Graph	Above x-axis	Below x-axis	Above x-axis
Point on Graph	(-5, 750)	(-2, -96)	(1, 30)

f is above the x-axis for $(-\infty, -4) \cup (0, \infty)$; f is below the x-axis for $(-4, 0)$

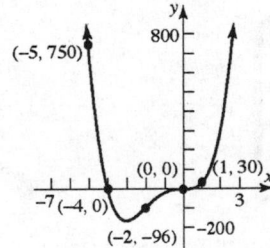

37. $f(x) = -4x^2(x + 2)$

(a)   x-intercepts: 0, –2; y-intercept: 0

(b)   crosses x-axis at $x = -2$; touches x-axis at $x = 0$

(c) $y = -4x^3$

(d)   graphing utility:

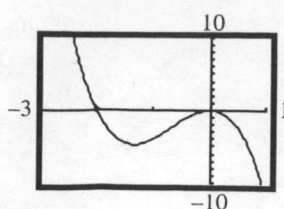

(e)　2 turning points;　local maximum: (0, 0); local minimum: (–1.33, –4.74)

(f)　graphing by hand

	-2		0

Interval	$(-\infty, -2)$	$(-2, 0)$	$(0, \infty)$
Number Chosen	-3	-1	1
Value of f	$f(-3) = 36$	$f(-1) = -4$	$f(1) = -12$
Location of Graph	Above x-axis	Below x-axis	Below x-axis
Point on Graph	(-3, 36)	(-1, -4)	(1, -12)

f is above the x-axis for $(-\infty,-2)$;　f　is below the x-axis for $(-2,0)\cup(0,\infty)$

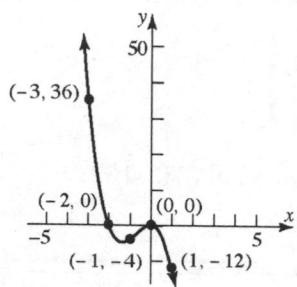

39.　　　$f(x) = (x+1)(x-2)(x+4)$

(a)　x-intercepts: $-1, 2, -4$; y-intercept: -8

(b)　crosses x-axis at $x = -1, 2, -4$

(c)　$y = x^3$

(d)　graphing utility:

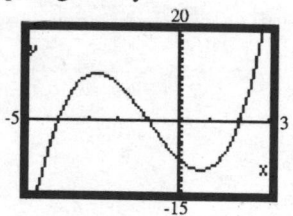

(e)　2 turning points; local maximum: (–2.73, 10.39);　local minimum: (0.73, –10.39)

(f)　graphing by hand

	-4	-1		2

Interval	$(-\infty, -4)$	$(-4, -1)$	$(-1, 2)$	$(2, \infty)$
Number Chosen	-5	-2	0	3
Value of f	$f(-5) = -28$	$f(-2) = 8$	$f(0) = -8$	$f(3) = 28$
Location of Graph	Below x-axis	Above x-axis	Below x-axis	Above x-axis
Point on Graph	(-5, -28)	(-2, 8)	(0, -8)	(3, 28)

f is above the x-axis for $(-4,-1)\cup(2,\infty)$;　f is below the x-axis for $(-\infty,-4)\cup(-1,2)$

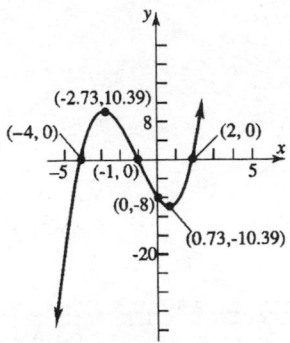

41. $f(x) = 4x - x^3 = x(4 - x^2) = x(2 + x)(2 - x)$

(a)   x-intercepts: $0, -2, 2$; y-intercept: 0

(b)   crosses x-axis at $x = 0, x = -2, x = 2$

(c) $y = -x^3$

(d)   graphing utility:

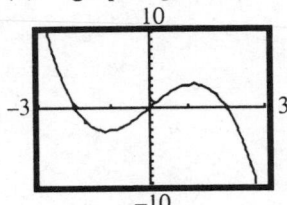

(e)   2 turning points; local maximum: $(1.15, 3.08)$;
              local minimum: $(-1.15, -3.08)$

(f)   graphing by hand

Interval	$(-\infty, -2)$	$(-2, 0)$	$(0, 2)$	$(2, \infty)$
Number Chosen	-3	-1	1	3
Value of f	$f(-3) = 15$	$f(-1) = -3$	$f(1) = 3$	$f(3) = -15$
Location of Graph	Above x-axis	Below x-axis	Above x-axis	Below x-axis
Point on Graph	(-3, 15)	(-1, -3)	(1, 3)	(3, -15)

f is above the x-axis for $(-\infty, -2) \cup (0, 2)$; f is below the x-axis for $(-2, 0) \cup (2, \infty)$

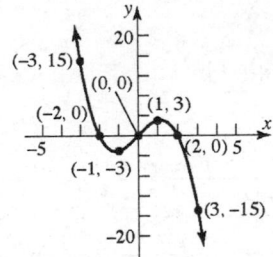

43. $f(x) = x^2(x - 2)(x + 2)$

(a)   x-intercepts: $0, 2, -2$; y-intercept: 0

(b)   crosses x-axis at $x = 2, x = -2$; touches x-axis at $x = 0$

(c) $y = x^4$

(d)   graphing utility:

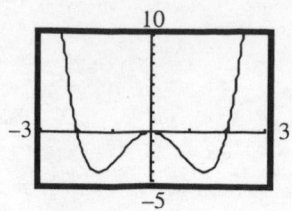

(e)   3 turning points; local maximum: $(0, 0)$; local minima: $(-1.41, -4)$, $(1.41, -4)$

(f)   graphing by hand

Interval	$(-\infty, -2)$	$(-2, 0)$	$(0, 2)$	$(2, \infty)$
Number Chosen	-3	-1	1	3
Value of f	$f(-3) = 45$	$f(-1) = -3$	$f(1) = -3$	$f(3) = 45$
Location of Graph	Above x-axis	Below x-axis	Below x-axis	Above x-axis
Point on Graph	$(-3, 45)$	$(-1, -3)$	$(1, -3)$	$(3, 45)$

f is above the x-axis for $\left(-\infty,-2\right)\cup\left(2,\infty\right)$; f is below the x-axis for $\left(-2,0\right)\cup\left(0,2\right)$

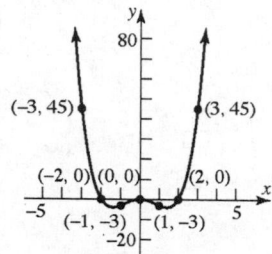

45. $f(x) = (x+1)^2(x-2)^2$

(a)   x-intercepts: $-1, 2$; y-intercept: 4   (b)   touches x-axis at $x = -1, x = 2$

(c) $y = x^4$

(d)   graphing utility:

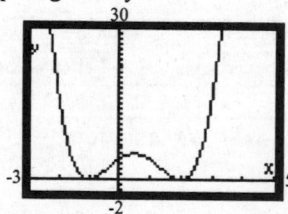

(e)   3 turning points; local maximum: $(0.5, 5.06)$; local minima: $(-1, 0), (2, 0)$

(f)   graphing by hand

Interval	$(-\infty, -1)$	$(-1, 2)$	$(2, \infty)$
Number Chosen	-2	0	3
Value of f	$f(-2) = 16$	$f(0) = 4$	$f(3) = 16$
Location of Graph	Above x-axis	Above x-axis	Above x-axis
Point on Graph	$(-2, 16)$	$(0, 4)$	$(3, 16)$

f is above the x-axis for $\left(-\infty,-1\right)\cup\left(-1,2\right)\cup\left(2,\infty\right)$

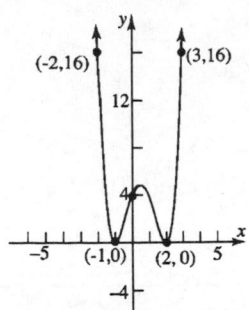

47. $f(x) = x^2(x-3)(x+1)$

(a)   x-intercepts: $0, 3, -1$; y-intercept: 0

(b)   crosses x-axis at $x = -1, x = 3$; touches x-axis at x = 0

(c) $y = x^4$

(d)   graphing utility:

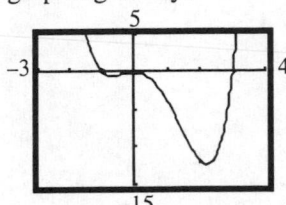

(e)   3 turning points; local maximum: (0, 0); local minima: (–0.69, -0.54), (2.19, –12.39)

(f)   graphing by hand

Interval	$(-\infty, -1)$	$(-1, 0)$	$(0, 3)$	$(3, \infty)$
Number Chosen	-2	$-\frac{1}{2}$	2	4
Value of f	$f(-2) = 20$	$f(-\frac{1}{2}) = -\frac{7}{16}$	$f(2) = -12$	$f(4) = 80$
Location of Graph	Above x-axis	Below x-axis	Below x-axis	Above x-axis
Point on Graph	(-2, 20)	$(-\frac{1}{2}, -\frac{7}{16})$	(2, -12)	(4, 80)

f is above the x-axis for $(-\infty, -1) \cup (3, \infty)$; f is below the x-axis for $(-1, 0) \cup (0, 3)$

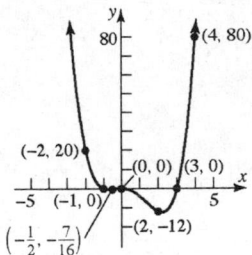

49. $f(x) = (x+2)^2(x-4)^2$

(a)   x-intercepts: $-2, 4$; y-intercept: 64

(b)   touches x-axis at $x = -2, x = 4$

(c) $y = x^4$

(d)   graphing utility:

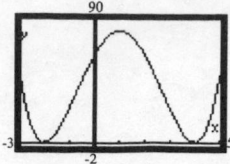

(e)   3 turning points; local maximum $(1,81)$; local minima $(-2,0)$ and $(4,0)$

(f)   graphing by hand

Interval	$(-\infty, -2)$	$(-2, 4)$	$(4, \infty)$
Number Chosen	-3	0	5
Value of f	$f(-3) = 49$	$f(0) = 64$	$f(5) = 49$
Location of Graph	Above x-axis	Above x-axis	Above x-axis
Point on Graph	(-3, 49)	(0, 64)	(5, 49)

f is above the x-axis for $(-\infty, -2) \cup (-2, 4) \cup (4, \infty)$

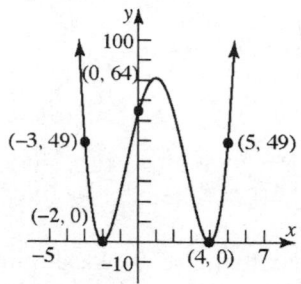

51. $f(x) = x^2(x-2)(x^2 + 3)$

(a)   x-intercepts: 0, 2; y-intercept: 0

(b)   crosses x-axis at $x = 2$; touches x-axis at x = 0

(c) $y = x^5$

(d)   graphing utility:

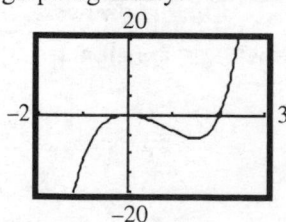

(e)   2 turning points; local maximum: (0, 0); local minimum: (1.48, -5.91)

(f)   graphing by hand

Interval	$(-\infty, 0)$	$(0, 2)$	$(2, \infty)$
Number Chosen	-1	1	3
Value of f	$f(-1) = -12$	$f(1) = -4$	$f(3) = 108$
Location of Graph	Below x-axis	Below x-axis	Above x-axis
Point on Graph	(-1, -12)	(1, -4)	(3, 108)

f is above the x-axis for $(2, \infty)$; f is below the x-axis for $(-\infty, 0) \cup (0, 2)$

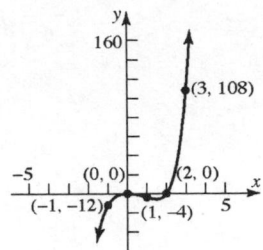

53. $f(x) = -x^2(x^2 - 1)(x + 1) = -x^2(x - 1)(x + 1)(x + 1) = -x^2(x - 1)(x + 1)^2$

(a)   x-intercepts: $0, 1, -1$; y-intercept: 0

(b)   crosses x-axis at $x = 1$; touches x-axis at $x = 0$ and $x = -1$

(c) $y = -x^5$

(d)   graphing utility:

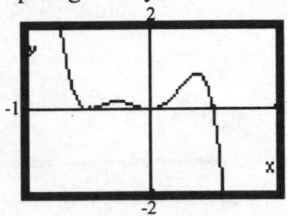

(e)   4 turning points; local maxima: $(-0.52, 0.10), (0.74, 0.43)$; local minima: $(-1, 0), (0, 0)$

(f)   graphing by hand

Interval	$(-\infty, -1)$	$(-1, 0)$	$(0, 1)$	$(1, \infty)$
Number Chosen	-2	$-\frac{1}{2}$	$\frac{1}{2}$	2
Value of f	$f(-2) = 12$	$f\left(-\frac{1}{2}\right) = \frac{3}{32}$	$f\left(\frac{1}{2}\right) = \frac{9}{32}$	$f(2) = -36$
Location of Graph	Above x-axis	Above x-axis	Above x-axis	Below x-axis
Point on Graph	$(-2, 12)$	$\left(-\frac{1}{2}, \frac{3}{32}\right)$	$\left(\frac{1}{2}, \frac{9}{32}\right)$	$(2, -36)$

f is above the x-axis for $(-\infty, -1) \cup (-1, 0) \cup (0, 1)$; f is below the x-axis for $(1, \infty)$

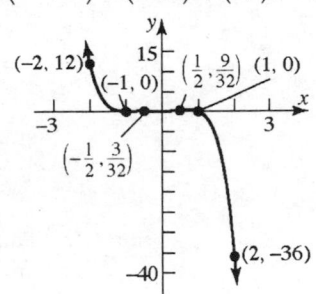

55. $f(x) = x^3 + 0.2x^2 - 1.5876x - 0.31752$

    (a) graphing utility:

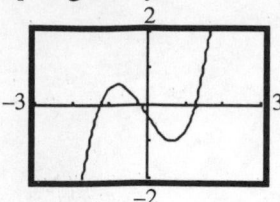

    (b)   x-intercepts: $-1.26, -0.2, 1.26$;
         y-intercept: -0.31752

    (c) $y = x^3$

    (d)   2 turning points;
         local maximum: $(-0.80, 0.57)$
         local minimum: $(0.66, -0.99)$

57. $f(x) = x^3 + 2.56x^2 - 3.31x + 0.89$

    (a) graphing utility

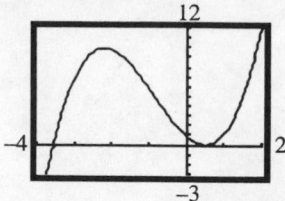

    (b)   x-intercepts: $-3.56, 0.50$;
         y-intercept: 0.89

    (c) $y = x^3$

    (d)   2 turning points;
         local maximum: $(-2.21, 9.91)$
         local minimum: $(0.50, 0)$

59. $f(x) = x^4 - 2.5x^2 + 0.5625$

    (a)   graphing utility:

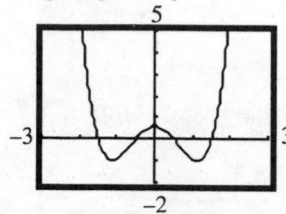

    (b)   x-intercepts: $-1.50, -0.50, 0.50, 1.50$;
         y-intercept: 0.5625

    (c) $y = x^4$

    (d)   3 turning points:
         local maximum: $(0, 0.5625)$
         local minima: $(-1.12, -1), (1.12, -1)$

61. $f(x) = 2x^4 - \pi x^3 + \sqrt{5}x - 4$

    (a) graphing utility:

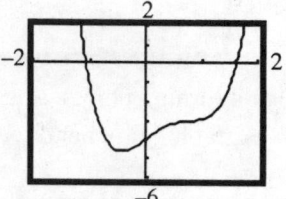

    (b)   x-intercepts: $-1.07, 1.62$;
         y-intercept: -4

    (c) $y = 2x^4$

    (c)   1 turning point;
         local minimum: $(-0.42, -4.64)$

63. $f(x) = -2x^5 - \sqrt{2}x^2 - x - \sqrt{2}$

    (a)   graphing utility:

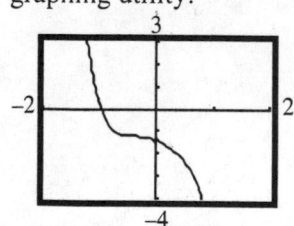

    (b)   x-intercept: -0.98;
         y-intercept: -1.41

    (c) $y = -2x^5$

    (d)   No turning points
         No local extrema

65.   Answers will vary. One possible
      answer is $f(x) = x(x-1)(x-2)$,
      since the graph crosses the x-axis at
      $x = 0, 1$ and 2.

67.   Answers will vary. One possible
      answer is

$$f(x) = -\frac{1}{2}(x+1)(x-1)(x-2), \text{ since}$$

      the graph crosses the x-axis at
      $x = -1, 1$ and 2 and has a y-intercept
      at -1.

69. (a)   Graphing, we see that the graph may be a cubic relation.

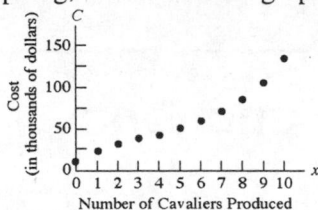

   (b)   Average rate of change $= \dfrac{50-43}{5-4} = \dfrac{7}{1} = 7$ thousand dollars per car

   (c)   Average rate of change $= \dfrac{105-85}{9-8} = \dfrac{20}{1} = 20$ thousand dollars per car

   (d) $C(x) = 0.2x^3 - 2.3x^2 + 14.3x + 10.2$
          $C(11) = 0.2(11)^3 - 2.3(11)^2 + 14.3(11) + 10.2 \approx 155.4$
          The cost of manufacturing 11 Cavaliers in 1 hour would be approximately \$155,400.

   (e) and (f)   Graphing the cubic function of best fit:

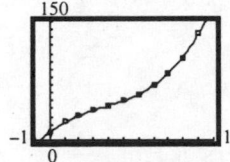

   (g)   The y-intercept would indicate the fixed costs before any cars are made.

71. (a)   Graphing, we see that the graph may be a cubic relation..

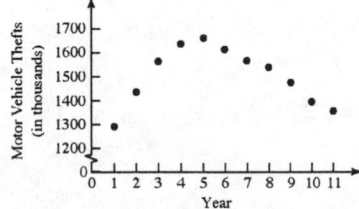

   (b) $T(x) = 1.52x^3 - 39.81x^2 + 282.29x + 1035.5$

   (c)   Graphing the cubic function of best fit:

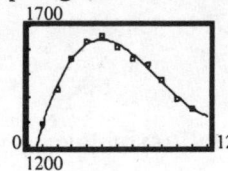

   (d) $T(12) = 1.52(12)^3 - 39.81(12)^2 + 282.29(12) + 1035.5 = 1316.9$
          According to the function there would be approximately 1,316,900 motor vehicle thefts in 1998.

   (e) answers will vary

73. Answers will vary

75. Answers will vary , $f(x) = (x+2)(x-1)^2$ and $g(x) = (x+2)^3(x-1)^2$ are two such polynomials.

Chapter 3

Polynomial and Rational Functions

3.4 Rational Functions I

1. In $R(x) = \dfrac{4x}{x-3}$, the denominator, $q(x) = x - 3$, has a zero at 3. Thus, the domain of $R(x)$ is all real numbers except 3.

3. In $H(x) = \dfrac{-4x^2}{(x-2)(x+4)}$, the denominator, $q(x) = (x-2)(x+4)$, has zeros at 2 and –4. Thus, the domain of $H(x)$ is all real numbers except 2 and –4.

5. In $F(x) = \dfrac{3x(x-1)}{2x^2 - 5x - 3}$, the denominator, $q(x) = 2x^2 - 5x - 3 = (2x+1)(x-3)$, has zeros at $-\dfrac{1}{2}$ and 3. Thus, the domain of $F(x)$ is all real numbers except $-\dfrac{1}{2}$ and 3.

7. In $R(x) = \dfrac{x}{x^3 - 8}$, the denominator, $q(x) = x^3 - 8 = (x-2)(x^2 + 2x + 4)$, has a zero at 2. ($x^2 + 2x + 4$ has no real zeros.) Thus, the domain of $R(x)$ is all real numbers except 2.

9. In $H(x) = \dfrac{3x^2 + x}{x^2 + 4}$, the denominator, $q(x) = x^2 + 4$, has no real zeros. Thus, the domain of $H(x)$ is all real numbers.

11. In $R(x) = \dfrac{3(x^2 - x - 6)}{4(x^2 - 9)}$, the denominator, $q(x) = 4(x^2 - 9) = 4(x-3)(x+3)$, has zeros at 3 and –3. Thus, the domain of $R(x)$ is all real numbers except 3 and –3.

13. (a) Domain: $\{x \mid x \neq 2\}$; Range: $\{y \mid y \neq 1\}$
    (b) Intercept: $(0, 0)$
    (c) Horizontal Asymptote: $y = 1$
    (d) Vertical Asymptote: $x = 2$
    (e) Oblique Asymptote: none

15. (a) Domain: $\{x \mid x \neq 0\}$; Range: all real numbers
    (b) Intercepts: $(-1, 0), (1, 0)$
    (c) Horizontal Asymptote: none
    (d) Vertical Asymptote: $x = 0$
    (e) Oblique Asymptote: $y = 2x$

17. (a) Domain: $\{x \mid x \neq -2, x \neq 2\}$; Range: $\{y \mid y \leq 0 \text{ or } y > 1\}$
    (b) Intercept: $(0, 0)$ (c) Horizontal Asymptote: $y = 1$
    (d) Vertical Asymptotes: $x = -2, x = 2$ (e) Oblique Asymptote: none

19. (a) Domain: $\{x \mid x \neq -1\}$; Range: $\{y \mid y \neq 2\}$
    (b) Intercepts: $(-1.5, 0); (0, 3)$ (c) Horizontal Asymptote: $y = 2$
    (d) Vertical Asymptote: $x = -1$ (e) Oblique Asymptote: none

21. (a) Domain: $\{x \mid x \neq -4, x \neq 3\}$; Range: all real numbers
    (b) Intercept: $(0, 0)$ (c) Horizontal Asymptote: $y = 0$
    (d) Vertical Asymptotes: $x = -4, x = 3$ (e) Oblique Asymptote: none

23. $R(x) = \dfrac{1}{(x-1)^2}$

Using the function, $y = \dfrac{1}{x^2}$, shift the graph horizontally 1 unit to the right.

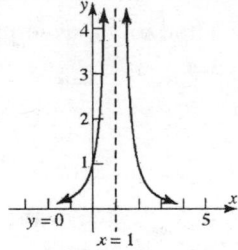

25. $H(x) = \dfrac{-2}{x+1} = -2\left(\dfrac{1}{x+1}\right)$

Using the function $y = \dfrac{1}{x}$, shift the graph horizontally 1 unit to the left, reflect about the x-axis, and stretch vertically by a factor of 2.

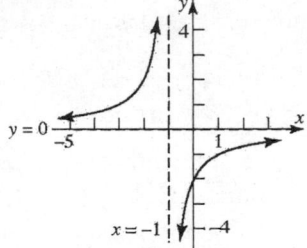

27. $R(x) = \dfrac{1}{x^2 + 4x + 4} = \dfrac{1}{(x+2)^2}$

$\qquad = \dfrac{1}{(x+2)^2}$

Using the function $y = \dfrac{1}{x^2}$, shift the graph horizontally 2 units to the left.

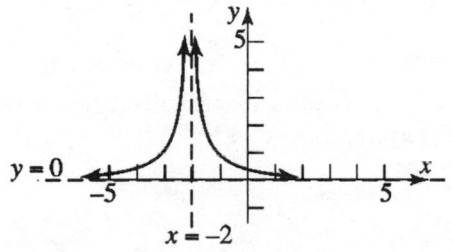

29. $F(x) = 1 - \dfrac{1}{x} = -\dfrac{1}{x} + 1$

Using the function, $y = \dfrac{1}{x}$, reflect the graph across the x-axis and then shift the graph vertically 1 unit up.

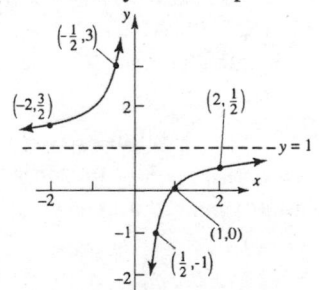

31. $R(x) = \dfrac{x^2 - 4}{x^2} = 1 - \dfrac{4}{x^2} = -4\left(\dfrac{1}{x^2}\right) + 1$

Using the function $y = \dfrac{1}{x^2}$, reflect about
the x-axis, stretch vertically by a factor of
4 and shift vertically 1 unit up.

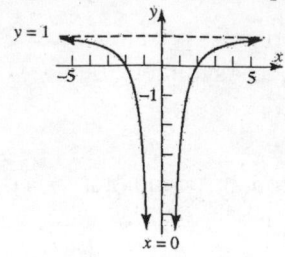

33. $G(x) = 1 + \dfrac{2}{(x-3)^2} = \dfrac{2}{(x-3)^2} + 1$

$= 2\left(\dfrac{1}{(x-3)^2}\right) + 1$

Using the function $y = \dfrac{1}{x^2}$, shift the

graph 3 units right, stretch vertically by a
factor of 2, and shift vertically 1 unit up.

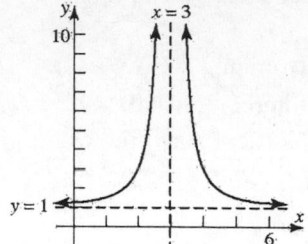

35. $R(x) = \dfrac{3x}{x+4}$

The degree of the numerator, $p(x) = 3x$, is $n = 1$. The degree of the denominator,
$q(x) = x + 4$, is $m = 1$. Since $n = m$, the line $y = \dfrac{3}{1} = 3$ is a horizontal asymptote. The
denominator is zero at $x = -4$, so $x = -4$ is a vertical asymptote.

37. $H(x) = \dfrac{x^4 + 2x^2 + 1}{x^2 - x + 1}$

The degree of the numerator, $p(x) = x^4 + 2x^2 + 1$, is $n = 4$. The degree of the
denominator, $q(x) = x^2 - x + 1$, is $m = 2$. Since $n > m + 1$, there is no horizontal
asymptote or oblique asymptote. The denominator has no real zeros, so there is no vertical
asymptote.

39. $T(x) = \dfrac{x^3}{x^4 - 1}$

The degree of the numerator, $p(x) = x^3$, is $n = 3$. The degree of the denominator,
$q(x) = x^4 - 1$ is $m = 4$. Since $n < m$, the line $y = 0$ is a horizontal asymptote. The
denominator is zero at $x = -1$ and $x = 1$, so $x = -1$ and $x = 1$ are vertical asymptotes.

41. $Q(x) = \dfrac{5 - x^2}{3x^4}$

The degree of the numerator, $p(x) = 5 - x^2$, is $n = 2$. The degree of the denominator,
$q(x) = 3x^4$ is $m = 4$. Since $n < m$, the line $y = 0$ is a horizontal asymptote. The
denominator is zero at $x = 0$, so $x = 0$ is a vertical asymptote.

43. $R(x) = \dfrac{3x^4 - 4}{x^3 + 3x}$

The degree of the numerator, $p(x) = 3x^4 - 4$, is $n = 4$. The degree of the denominator, $q(x) = x^3 + 3x$ is $m = 3$. Since $n = m + 1$, there is an oblique asymptote.

Dividing:

$$
\begin{array}{r}
3x \\
x^3 + 3x \overline{)\,3x^4 + 0x^3 + 0x^2 + 0x - 4} \\
\underline{3x^4 + 9x^2 } \\
-9x^2 + 0x - 4
\end{array}
\qquad R(x) = 3x + \dfrac{-9x^2 - 4}{x^3 + 3x}
$$

Thus, the oblique asymptote is $y = 3x$.

The denominator is zero at $x = 0$, so $x = 0$ is a vertical asymptote.

45. $G(x) = \dfrac{x^3 - 1}{x - x^2}, \; x \neq 1$

The degree of the numerator, $p(x) = x^3 - 1$, is $n = 3$. The degree of the denominator, $q(x) = x - x^2$ is $m = 2$. Since $n = m + 1$, there is an oblique asymptote.

Dividing:

$$
\begin{array}{r}
-x - 1 \\
-x^2 + x \overline{)\,x^3 + 0x^2 + 0x - 1} \\
\underline{x^3 - x^2 } \\
x^2 + 0x \\
\underline{x^2 - x } \\
x - 1
\end{array}
\qquad
\begin{array}{l}
G(x) = -x - 1 + \dfrac{x - 1}{x - x^2} = -x - 1 - \dfrac{1}{x}, \; x \neq 1 \\[2mm]
\text{Thus, the oblique asymptote is } y = -x - 1.
\end{array}
$$

$G(x)$ must be in lowest terms to find the vertical asymptote:

$$G(x) = \dfrac{x^3 - 1}{x - x^2} = \dfrac{(x - 1)(x^2 + x + 1)}{-x(x - 1)} = \dfrac{x^2 + x + 1}{-x}$$

The denominator is zero at $x = 0$, so $x = 0$ is a vertical asymptote.

47.   A rational function $R(x) = \dfrac{p(x)}{q(x)}$ has a vertical asymptote at x = c in each of these cases:

Case 1: $R(c) = \dfrac{nonzero}{zero}$

That is, whenever x = c yields a zero in the denominator of the function formula. And the denominator will equal zero only if it contains the factor $(x - c)^n$, for some $n > 0$.

Case 2: $R(c) = \dfrac{(x - c)^m}{(x - c)^n}$, where $n > 0$, $m > 0$ and $n > m$.

That is, whenever x = c yields a zero in the numerator and denominator of the function formula such that the multiplicity is greater in the denominator.

49. We need $R(x) = \dfrac{p(x)}{q(x)} = 2x + 1 + \dfrac{h(x)}{q(x)}$, with $h(x) \neq 0$ for some x.

Letting $q(x) = x + 1$, we have

$$2x + 1 + \frac{h(x)}{q(x)} = 2x + 1 + \frac{h(x)}{x+1} = \frac{(2x+1)(x+1) + h(x)}{x+1} = \frac{2x^2 + 3x + 1 + h(x)}{x+1}.$$

So choosing $h(x) = 1$, we get $R(x) = \dfrac{2x^2 + 3x + 1 + 1}{x+1} = \dfrac{2x^2 + 3x + 2}{x+1}$.

Polynomial and Rational Functions

3.5 Rational Functions II: Analyzing Graphs

In problems 1–37, we will use the terminology: $R(x) = \dfrac{p(x)}{q(x)}$, *where the degree of* $p(x) = n$ *and the degree of* $q(x) = m$.

1. $R(x) = \dfrac{x+1}{x(x+4)}$ $p(x) = x+1$; $q(x) = x(x+4) = x^2 + 4x$; $n = 1$; $m = 2$

    Step 1: Domain: $\{x \mid x \neq -4,\ x \neq 0\}$

    Step 2: (a) The x-intercept is the zero of $p(x)$: -1

    (b) There is no y-intercept; $R(0)$ is not defined, since $q(0) = 0$.

    Step 3: $R(-x) = \dfrac{-x+1}{-x(-x+4)} = \dfrac{-x+1}{x^2 - 4x}$; this is neither $R(x)$ nor $-R(x)$, so there is no symmetry.

    Step 4: The vertical asymptotes are the zeros of $q(x)$: $x = -4$ and $x = 0$

    Step 5: Since $n < m$, the line $y = 0$ is the horizontal asymptote.

    $R(x)$ intersects $y = 0$ at $(-1, 0)$.

    Step 6:

Interval	$(-\infty, -4)$	$(-4, -1)$	$(-1, 0)$	$(0, \infty)$
Number Chosen	-5	-2	$-\frac{1}{2}$	1
Value of R	$R(-5) = -\frac{4}{5}$	$R(-2) = \frac{1}{4}$	$R(-\frac{1}{2}) = -\frac{2}{7}$	$R(1) = \frac{2}{5}$
Location of Graph	Below x-axis	Above x-axis	Below x-axis	Above x-axis
Point on Graph	$(-5, -\frac{4}{5})$	$(-2, \frac{1}{4})$	$(-\frac{1}{2}, -\frac{2}{7})$	$(1, \frac{2}{5})$

Step 7: Graphing Utility:

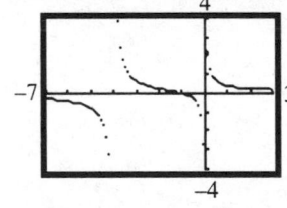

Step 8: Graphing by hand:

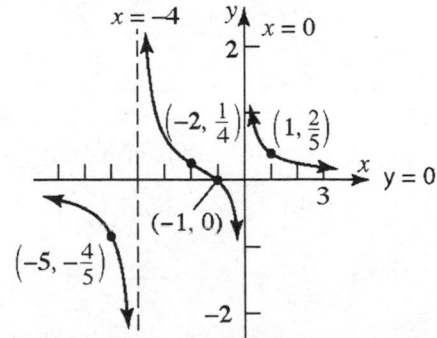

3. $R(x) = \dfrac{3x+3}{2x+4}$ $p(x) = 3x + 3;$ $q(x) = 2x + 4;$ $n = 1;$ $m = 1$

Step 1:   Domain: $\{x \mid x \neq -2\}$

Step 2:   (a)   The x-intercept is the zero of $p(x)$: -1

          (b)   The y-intercept is $R(0) = \dfrac{3(0)+3}{2(0)+4} = \dfrac{3}{4}$.

Step 3: $R(-x) = \dfrac{3(-x)+3}{2(-x)+4} = \dfrac{-3x+3}{-2x+4} = \dfrac{3x-3}{2x-4}$; this is neither $R(x)$ nor $-R(x)$, so there is no symmetry.

Step 4:   The vertical asymptote is the zero of $q(x)$: $x = -2$

Step 5:   Since $n = m$, the line $y = \dfrac{3}{2}$ is the horizontal asymptote.

        $R(x)$ does not intersect $y = \dfrac{3}{2}$.

Step 6:

Interval	$(-\infty, -2)$	$(-2, -1)$	$(-1, \infty)$
Number Chosen	-3	$-\frac{3}{2}$	0
Value of R	$R(-3) = 3$	$R(-\frac{3}{2}) = -\frac{3}{2}$	$R(0) = \frac{3}{4}$
Location of Graph	Above x-axis	Below x-axis	Above x-axis
Point on Graph	$(-3, 3)$	$(-\frac{3}{2}, -\frac{3}{2})$	$(0, \frac{3}{4})$

Step 7:   Graphing Utility:

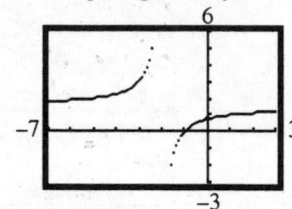

Step 8:   Graphing by hand:

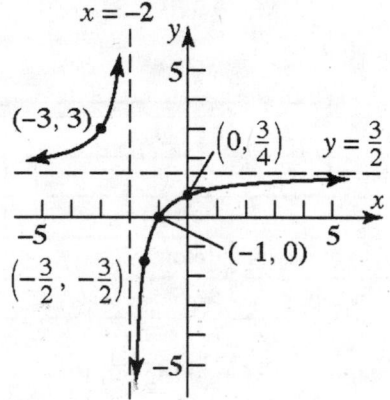

5. $R(x) = \dfrac{3}{x^2 - 4}$ $p(x) = 3;$ $q(x) = x^2 - 4;$ $n = 0;$ $m = 2$

Step 1:   Domain: $\{x \mid x \neq -2, \ x \neq 2\}$

Step 2:   (a)   There is no x-intercept.

          (b)   The y-intercept is $R(0) = \dfrac{3}{0^2 - 4} = \dfrac{3}{-4} = -\dfrac{3}{4}$.

Step 3: $R(-x) = \dfrac{3}{(-x)^2 - 4} = \dfrac{3}{x^2 - 4} = R(x)$; $R(x)$ is symmetric to the y-axis.

Step 4:   The vertical asymptotes are the zeros of $q(x)$: $x = -2$ and $x = 2$

Step 5:   Since $n < m$, the line $y = 0$ is the horizontal asymptote.

152

$R(x)$ does not intersect $y = 0$.

Step 6:

-2   2

Interval	$(-\infty, -2)$	$(-2, 2)$	$(2, \infty)$
Number Chosen	-3	0	3
Value of R	$R(-3) = \frac{3}{5}$	$R(0) = -\frac{3}{4}$	$R(3) = \frac{3}{5}$
Location of Graph	Above x-axis	Below x-axis	Above x-axis
Point on Graph	$(-3, \frac{3}{5})$	$(0, -\frac{3}{4})$	$(3, \frac{3}{5})$

Step 7:   Graphing Utility:

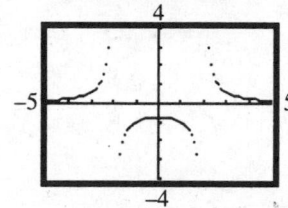

Step 8:   Graphing by hand:

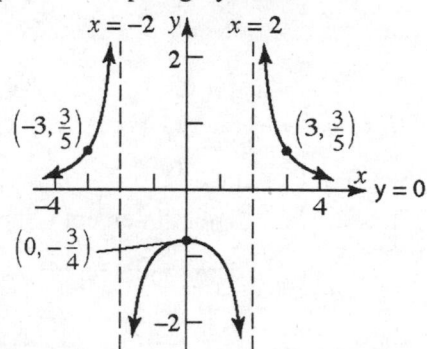

7. $P(x) = \dfrac{x^4 + x^2 + 1}{x^2 - 1}$ $p(x) = x^4 + x^2 + 1$; $q(x) = x^2 - 1$; $n = 4$; $m = 2$

Step 1:   Domain: $\{x \mid x \neq -1, x \neq 1\}$

Step 2:   (a)   There is no x-intercept.

(b)   The y-intercept is $P(0) = \dfrac{0^4 + 0^2 + 1}{0^2 - 1} = \dfrac{1}{-1} = -1$.

Step 3: $P(-x) = \dfrac{(-x)^4 + (-x)^2 + 1}{(-x)^2 - 1} = \dfrac{x^4 + x^2 + 1}{x^2 - 1} = P(x)$; $P(x)$ is symmetric to the y - axis.

Step 4:   The vertical asymptotes are the zeros of $q(x)$: $x = -1$ and $x = 1$

Step 5:   Since $n > m + 1$, there is no horizontal asymptote and no oblique asymptote.

Step 6:

-1   1

Interval	$(-\infty, -1)$	$(-1, 1)$	$(1, \infty)$
Number Chosen	-2	0	2
Value of P	$P(-2) = 7$	$P(0) = -1$	$P(2) = 7$
Location of Graph	Above x-axis	Below x-axis	Above x-axis
Point on Graph	(-2, 7)	(0, -1)	(2, 7)

Step 7:   Graphing Utility:

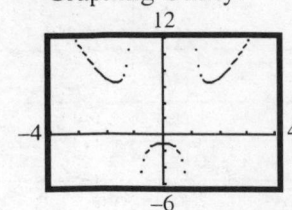

Step 8:   Graphing by hand:

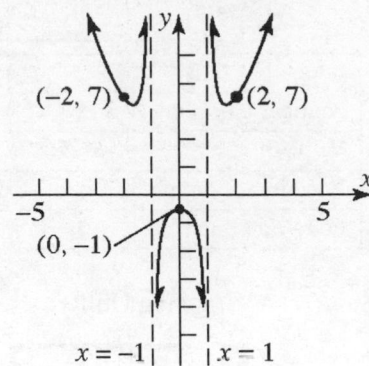

9. $H(x) = \dfrac{x^3 - 1}{x^2 - 9}$ $p(x) = x^3 - 1;\ q(x) = x^2 - 9;\ n = 3;\ m = 2$

Step 1:   Domain: $\{x \mid x \neq -3,\ x \neq 3\}$

Step 2:   (a)   The x-intercept is the zero of $p(x)$: 1.

(b)   The y-intercept is $H(0) = \dfrac{0^3 - 1}{0^2 - 9} = \dfrac{-1}{-9} = \dfrac{1}{9}$.

Step 3: $H(-x) = \dfrac{(-x)^3 - 1}{(-x)^2 - 9} = \dfrac{-x^3 - 1}{x^2 - 9}$; this is neither $H(x)$ nor $-H(x)$, so there is no symmetry.

Step 4:   The vertical asymptotes are the zeros of $q(x)$: $x = -3$ and $x = 3$

Step 5:   Since $n = m + 1$, there is an oblique asymptote. Dividing:

$$x^2 - 9 \overline{\smash{)}\,x^3 + 0x^2 + 0x - 1} \qquad H(x) = x + \dfrac{9x - 1}{x^2 - 9}$$

$$\underline{x^3 \qquad\ -9x}$$

$$9x - 1$$

The oblique asymptote is $y = x$.

Solve to find intersection points:

$$\dfrac{x^3 - 1}{x^2 - 9} = x \rightarrow x^3 - 1 = x^3 - 9x$$

$$-1 = -9x \rightarrow x = \dfrac{1}{9}$$

The oblique asymptote intersects $H(x)$ at $\left(\dfrac{1}{9}, \dfrac{1}{9}\right)$.

Step 6:

Interval	$(-\infty, -3)$	$(23, 1)$	$(1, 3)$	$(3, \infty)$
Number Chosen	-4	0	2	4
Value of H	$H(-4) \approx -9.3$	$H(0) = \frac{1}{9}$	$H(2) = -1.4$	$H(4) = 9$
Location of Graph	Below x-axis	Above x-axis	Below x-axis	Above x-axis
Point on Graph	(-4, -9.3)	$(0, \frac{1}{9})$	(2, -1.4)	(4, 9)

Step 7:   Graphing Utility:

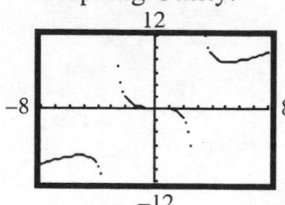

Step 8:   Graphing by hand:

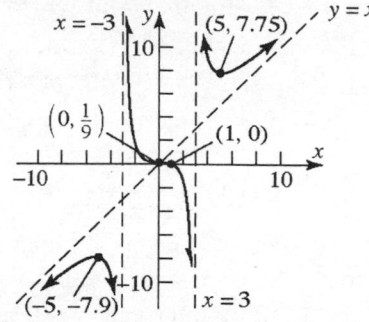

11. $R(x) = \dfrac{x^2}{x^2 + x - 6} = \dfrac{x^2}{(x+3)(x-2)}$ $p(x) = x^2$; $q(x) = x^2 + x - 6$; $n = 2$; $m = 2$

Step 1:   Domain: $\{x \mid x \neq -3, x \neq 2\}$

Step 2:   (a)   The x-intercept is the zero of $p(x)$: 0

        (b)   The y-intercept is $R(0) = \dfrac{0^2}{0^2 + 0 - 6} = \dfrac{0}{-6} = 0$.

Step 3: $R(-x) = \dfrac{(-x)^2}{(-x)^2 + (-x) - 6} = \dfrac{x^2}{x^2 - x - 6}$; this is neither $R(x)$ nor $-R(x)$, so there is no symmetry.

Step 4:   The vertical asymptotes are the zeros of $q(x)$: $x = -3$ and $x = 2$

Step 5:   Since $n = m$, the line $y = 1$ is the horizontal asymptote.

        $R(x)$ intersects $y = 1$ at (6, 1), since:

$$\frac{x^2}{x^2 + x - 6} = 1 \rightarrow x^2 = x^2 + x - 6 \rightarrow 0 = x - 6 \rightarrow x = 6$$

Step 6:

Interval	$(-\infty, -3)$	$(-3, 0)$	$(0, 2)$	$(2, \infty)$
Number Chosen	-6	-1	1	3
Value of R	$R(-6) = 1.5$	$R(-1) = -\frac{1}{6}$	$R(1) = -0.25$	$R(3) = 1.5$
Location of Graph	Above x-axis	Below x-axis	Below x-axis	Above x-axis
Point on Graph	(-6, 1.5)	$(-1, -\frac{1}{6})$	(1, -0.25)	(3, 1.5)

Step 7:   Graphing Utility:

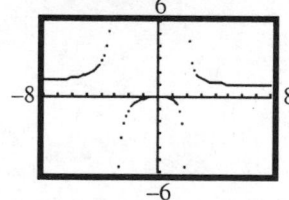

Step 8:   Graphing by hand:

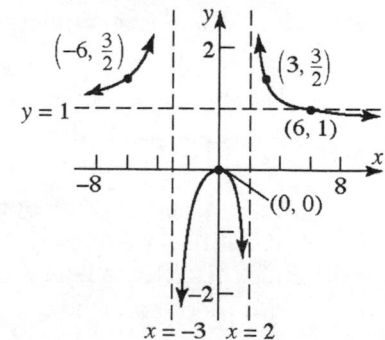

13. $G(x) = \dfrac{x}{x^2 - 4} = \dfrac{x}{(x+2)(x-2)}$　$p(x) = x;\ q(x) = x^2 - 4;\ n = 1;\ m = 2$

　　Step 1:　Domain: $\{x \mid x \neq -2,\ x \neq 2\}$

　　Step 2:　(a)　The x-intercept is the zero of $p(x)$: 0

　　　　　　(b)　The y-intercept is $G(0) = \dfrac{0}{0^2 - 4} = \dfrac{0}{-4} = 0$.

　　Step 3:　$G(-x) = \dfrac{-x}{(-x)^2 - 4} = \dfrac{-x}{x^2 - 4} = -G(x);\ G(x)$ is symmetric to the origin.

　　Step 4:　The vertical asymptotes are the zeros of $q(x)$: $x = -2$ and $x = 2$

　　Step 5:　Since $n < m$, the line $y = 0$ is the horizontal asymptote.

　　　　　　$G(x)$ intersects $y = 0$ at $(0, 0)$.

　　Step 6:

Interval	$(-\infty, -2)$	$(-2, 0)$	$(0, 2)$	$(2, \infty)$
Number Chosen	-3	-1	1	3
Value of G	$G(-3) = -\frac{3}{5}$	$G(-1) = \frac{1}{3}$	$G(1) = -\frac{1}{3}$	$G(3) = \frac{3}{5}$
Location of Graph	Below x-axis	Above x-axis	Below x-axis	Above x-axis
Point on Graph	$(-3, -\frac{3}{5})$	$(-1, \frac{1}{3})$	$(1, -\frac{1}{3})$	$(3, \frac{3}{5})$

　　Step 7:　Graphing Utility:　　　　　　Step 8:　Graphing by hand:

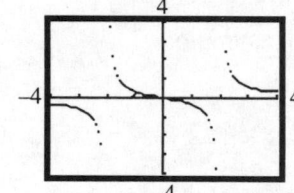
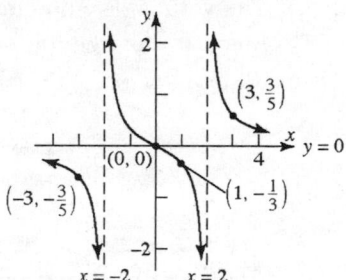

15. $R(x) = \dfrac{3}{(x-1)(x^2 - 4)} = \dfrac{3}{(x-1)(x+2)(x-2)}$　$p(x) = 3;\ q(x) = (x-1)(x^2 - 4);$

　　　　　　　　　　　　　　　　　　　　$n = 0;\ m = 3$

　　Step 1:　Domain: $\{x \mid x \neq -2,\ x \neq 1,\ x \neq 2\}$

　　Step 2:　(a)　There is no x-intercept.

　　　　　　(b)　The y-intercept is $R(0) = \dfrac{3}{(0-1)(0^2 - 4)} = \dfrac{3}{4}$.

　　Step 3:　$R(-x) = \dfrac{3}{(-x-1)((-x)^2 - 4)} = \dfrac{3}{(-x-1)(x^2 - 4)}$; this is neither $R(x)$ nor $-R(x)$,

　　　　　　so there is no symmetry.

　　Step 4:　The vertical asymptotes are the zeros of $q(x)$: $x = -2,\ x = 1,$ and $x = 2$

　　Step 5:　Since $n < m$, the line $y = 0$ is the horizontal asymptote.

　　　　　　$R(x)$ does not intersect $y = 0$.

Step 6:

Interval	$(-\infty, -2)$	$(-2, 1)$	$(1, 2)$	$(2, \infty)$
Number Chosen	-3	0	1.5	3
Value of R	$R(-3) = -\frac{3}{20}$	$R(0) = \frac{3}{4}$	$R(1.5) = -\frac{24}{7}$	$R(3) = \frac{3}{10}$
Location of Graph	Below x-axis	Above x-axis	Below x-axis	Above x-axis
Point on Graph	$(-3, -\frac{3}{20})$	$(0, \frac{3}{4})$	$(1.5, -\frac{24}{7})$	$(3, \frac{3}{10})$

Step 7:   Graphing Utility:

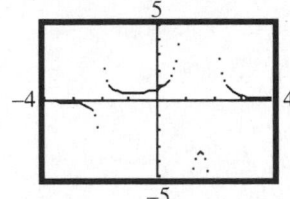

Step 8:   Graphing by hand:

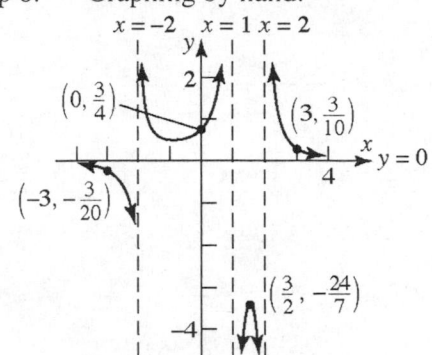

17. $H(x) = \dfrac{4(x^2 - 1)}{x^4 - 16} = \dfrac{4(x-1)(x+1)}{(x^2+4)(x+2)(x-2)}$ $p(x) = 4(x^2 - 1);\ \ q(x) = x^4 - 16;$

$n = 2;\ m = 4$

Step 1:   Domain: $\{x \mid x \neq -2,\ x \neq 2\}$

Step 2:   (a)   The x-intercepts are the zeros of $p(x)$: -1 and 1

(b)   The y-intercept is $H(0) = \dfrac{4(0^2 - 1)}{0^4 - 16} = \dfrac{-4}{-16} = \dfrac{1}{4}$.

Step 3: $H(-x) = \dfrac{4((-x)^2 - 1)}{(-x)^4 - 16} = \dfrac{4(x^2 - 1)}{x^4 - 16} = H(x);\ H(x)$ is symmetric to the y-axis.

Step 4:   The vertical asymptotes are the zeros of $q(x)$: $x = -2$, and $x = 2$

Step 5:   Since $n < m$, the line $y = 0$ is the horizontal asymptote.

$H(x)$ intersects $y = 0$ at $(-1, 0)$ and $(1, 0)$.

Step 6:

Interval	$(-\infty, -2)$	$(-2, -1)$	$(-1, 1)$	$(1, 2)$	$(2, \infty)$
Number Chosen	-3	-1.5	0	1.5	3
Value of H	$H(-3) \approx 0.49$	$H(-1.5) \approx -0.46$	$H(0) = \frac{1}{4}$	$H(1.5) \approx -0.46$	$H(3) \approx 0.49$
Location of Graph	Above x-axis	Below x-axis	Above x-axis	Below x-axis	Above x-axis
Point on Graph	$(-3, 0.49)$	$(-1.5, -0.46)$	$(0, \frac{1}{4})$	$(1.5, -0.46)$	$(3, 0.49)$

Step 7:   Graphing Utility:

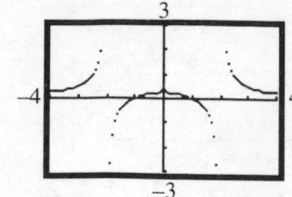

Step 8:   Graphing by hand:

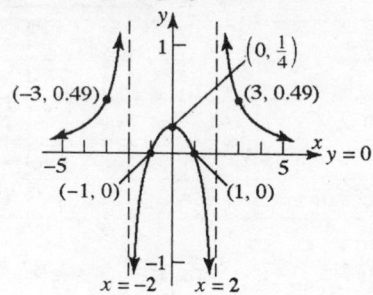

19. $F(x) = \dfrac{x^2 - 3x - 4}{x + 2} = \dfrac{(x+1)(x-4)}{x+2}$ $p(x) = x^2 - 3x - 4; \ q(x) = x + 2; \ n = 2; \ m = 1$

Step 1:   Domain: $\{x \mid x \neq -2\}$

Step 2:   (a)   The x-intercepts are the zeros of $p(x)$: -1 and 4.

   (b)   The y-intercept is $F(0) = \dfrac{0^2 - 3(0) - 4}{0 + 2} = \dfrac{-4}{2} = -2$.

Step 3: $F(-x) = \dfrac{(-x)^2 - 3(-x) - 4}{-x + 2} = \dfrac{x^2 + 3x - 4}{-x + 2}$; this is neither $F(x)$ nor $-F(x)$, so there is no symmetry.

Step 4:   The vertical asymptote is the zero of $q(x)$: $x = -2$

Step 5:   Since $n = m + 1$, there is an oblique asymptote. Dividing:

$$\begin{array}{r} x - 5 \\ x + 2 \overline{) x^2 - 3x - 4} \\ \underline{x^2 + 2x} \\ -5x - 4 \\ \underline{-5x - 10} \\ 6 \end{array} \qquad F(x) = x - 5 + \dfrac{6}{x + 2}$$

The oblique asymptote is $y = x - 5$.

Solve to find intersection points:

$$\dfrac{x^2 - 3x - 4}{x + 2} = x - 5 \rightarrow x^2 - 3x - 4 = x^2 - 3x - 10 \rightarrow -4 = -10$$

Since there is no solution, the oblique asymptote does not intersect $F(x)$.

Step 6:

Interval	$(-\infty, -2)$	$(-2, -1)$	$(-1, 4)$	$(4, \infty)$
Number Chosen	-3	-1.5	0	5
Value of F	$F(-3) = -14$	$F(-1.5) = 5.5$	$F(0) = -2$	$F(5) \approx 0.86$
Location of Graph	Below x-axis	Above x-axis	Below x-axis	Above x-axis
Point on Graph	(-3, -14)	(-1.5, 5.5)	(0, -2)	(5, 0.86)

Step 7:   Graphing Utility: Step 8:   Graphing by hand:

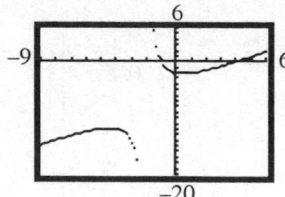

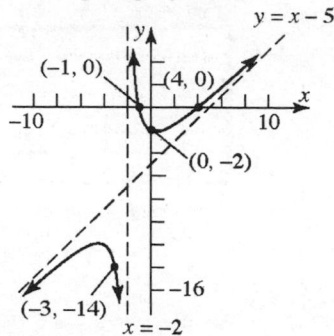

21. $R(x) = \dfrac{x^2 + x - 12}{x - 4} = \dfrac{(x + 4)(x - 3)}{x - 4}$ $p(x) = x^2 + x - 12;\ q(x) = x - 4;\ n = 2;\ m = 1$

Step 1:   Domain: $\{x \mid x \neq 4\}$

Step 2:   (a)   The x-intercepts are the zeros of $p(x)$: -4 and 3.

        (b)   The y-intercept is $R(0) = \dfrac{0^2 + 0 - 12}{0 - 4} = \dfrac{-12}{-4} = 3$.

Step 3: $R(-x) = \dfrac{(-x)^2 + (-x) - 12}{-x - 4} = \dfrac{x^2 - x - 12}{-x - 4}$; this is neither $R(x)$ nor $-R(x)$, so
there is no symmetry.

Step 4:   The vertical asymptote is the zero of $q(x)$: $x = 4$

Step 5:   Since $n = m + 1$, there is an oblique asymptote. Dividing:

$$\begin{array}{r} x + 5 \\ x - 4 \overline{)\, x^2 + x - 12} \\ \underline{x^2 - 4x} \\ 5x - 12 \\ \underline{5x - 20} \\ 8 \end{array} \qquad R(x) = x + 5 + \dfrac{8}{x - 4}$$

The oblique asymptote is $y = x + 5$.
Solve to find intersection points:

$$\dfrac{x^2 + x - 12}{x - 4} = x + 5$$
$$x^2 + x - 12 = x^2 + x - 20$$
$$-12 = -20$$

Since there is no solution, the oblique asymptote does not intersect $R(x)$.

Step 6:

Interval	$(-\infty, -4)$	$(-4, 3)$	$(3, 4)$	$(4, \infty)$
Number Chosen	-5	0	3.5	5
Value of R	$R(-5) = -\frac{8}{9}$	$R(0) = 3$	$R(3.5) = -7.5$	$R(5) = 18$
Location of Graph	Below x-axis	Above x-axis	Below x-axis	Above x-axis
Point on Graph	$(-5, -\frac{8}{9})$	$(0, 3)$	$(3.5, -7.5)$	$(5, 18)$

Step 7:   Graphing Utility: Step 8:   Graphing by hand:

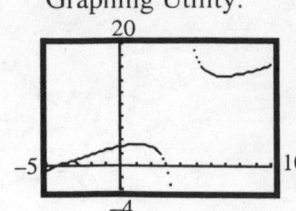

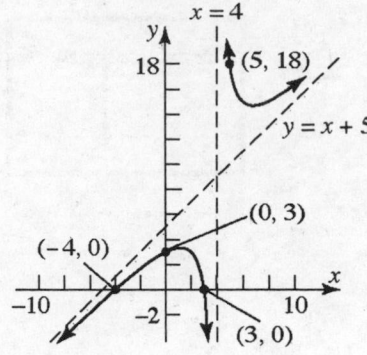

23. $F(x) = \dfrac{x^2 + x - 12}{x + 2} = \dfrac{(x+4)(x-3)}{x+2}$ $p(x) = x^2 + x - 12; \quad q(x) = x + 2; \quad n = 2; \quad m = 1$

Step 1:   Domain: $\{x \,|\, x \neq -2\}$

Step 2:   (a)   The x-intercepts are the zeros of $p(x)$: -4 and 3.

            (b)   The y-intercept is $F(0) = \dfrac{0^2 + 0 - 12}{0 + 2} = \dfrac{-12}{2} = -6$.

Step 3: $F(-x) = \dfrac{(-x)^2 + (-x) - 12}{-x + 2} = \dfrac{x^2 - x - 12}{-x + 2}$; this is neither $F(x)$ nor $-F(x)$, so there is no symmetry.

Step 4:   The vertical asymptote is the zero of $q(x)$: $x = -2$

Step 5:   Since $n = m + 1$, there is an oblique asymptote. Dividing:

$$\begin{array}{r} x - 1 \\ x+2 \overline{)\, x^2 + \; x - 12} \\ \underline{x^2 + 2x} \\ -x - 12 \\ \underline{-x - 2} \\ -10 \end{array} \qquad F(x) = x - 1 + \dfrac{-10}{x+2}$$

The oblique asymptote is $y = x - 1$.

Solve to find intersection points:

$$\dfrac{x^2 + x - 12}{x + 2} = x - 1 \rightarrow x^2 + x - 12 = x^2 + x - 2 \rightarrow -12 = -2$$

Since there is no solution, the oblique asymptote does not intersect $F(x)$.

Step 6:

Interval	$(-\infty, -4)$	$(-4, -2)$	$(-2, 3)$	$(3, \infty)$
Number Chosen	-5	-3	0	4
Value of F	$F(-5) = -\frac{8}{3}$	$F(-3) = 6$	$F(0) = -6$	$F(4) = \frac{4}{3}$
Location of Graph	Below x-axis	Above x-axis	Below x-axis	Above x-axis
Point on Graph	$(-5, -\frac{8}{3})$	$(-3, 6)$	$(0, -6)$	$(4, \frac{4}{3})$

160

Step 7:   Graphing Utility:

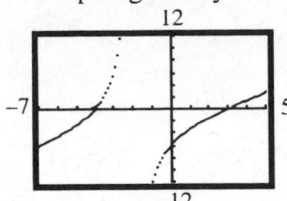

Step 8:   Graphing by hand:

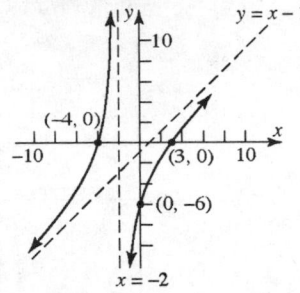

25. $R(x) = \dfrac{x(x-1)^2}{(x+3)^3}$ $p(x) = x(x-1)^2$; $q(x) = (x+3)^3$; $n = 3$; $m = 3$

Step 1:   Domain: $\{x \mid x \neq -3\}$

Step 2:   (a)   The x-intercepts are the zeros of $p(x)$: 0 and 1

   (b)   The y-intercept is $R(0) = \dfrac{0(0-1)^2}{(0+3)^3} = \dfrac{0}{27} = 0$.

Step 3: $R(-x) = \dfrac{-x(-x-1)^2}{(-x+3)^3}$; this is neither $R(x)$ nor $-R(x)$, so there is no symmetry.

Step 4:   The vertical asymptote is the zero of $q(x)$: $x = -3$

Step 5:   Since $n = m$, the line $y = 1$ is the horizontal asymptote.

   Solve to find intersection points:

$$\dfrac{x(x-1)^2}{(x+3)^3} = 1 \Rightarrow x^3 - 2x^2 + x = x^3 + 9x^2 + 27x + 27$$

$$0 = 11x^2 + 26x + 27$$

Since there is no real solution, $R(x)$ does not intersect $y = 1$.

Step 6:

	-3	0	1

Interval	$(-\infty, -3)$	$(-3, 0)$	$(0, 1)$	$(1, \infty)$
Number Chosen	-4	-1	$\frac{1}{2}$	2
Value of R	$R(-4) = 100$	$R(-1) = -0.5$	$R(\frac{1}{2}) \approx 0.003$	$R(2) = 0.016$
Location of Graph	Above x-axis	Below x-axis	Above x-axis	Above x-axis
Point on Graph	(-4, 100)	(-1, -0.5)	$(\frac{1}{2}, 0.003)$	(2, 0.016)

Step 7:   Graphing Utility:

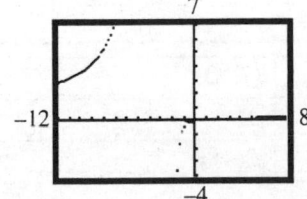

Step 8:   Graphing by hand:

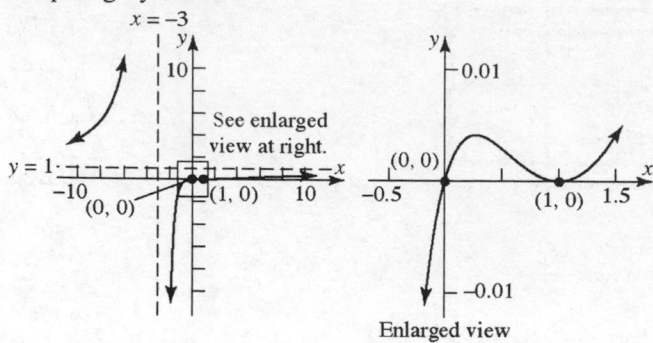

27. $R(x) = \dfrac{x^2 + x - 12}{x^2 - x - 6} = \dfrac{(x+4)(x-3)}{(x-3)(x+2)} = \dfrac{x+4}{x+2}$ $p(x) = x^2 + x - 12;$ $q(x) = x^2 - x - 6;$
$$n = 2; \ m = 2$$

Step 1:   Domain: $\{x \mid x \neq -2, \ x \neq 3\}$

Step 2:   (a)   The x-intercept is the zero of $p(x)$: -4 ; 3 is not a zero because reduced
         form must be used to find the zeros.

         (b)   The y-intercept is $R(0) = \dfrac{0^2 + 0 - 12}{0^2 - 0 - 6} = \dfrac{-12}{-6} = 2$.

Step 3: $R(-x) = \dfrac{(-x)^2 + (-x) - 12}{(-x)^2 - (-x) - 6} = \dfrac{x^2 - x - 12}{x^2 + x - 6}$; this is neither $R(x)$ nor $-R(x)$, so
         there is no symmetry.

Step 4:   The vertical asymptote is the zero of $q(x)$: $x = -2$; $x = 3$ is not a vertical
         asymptote because reduced form must be used to find the them. The graph
         has a hole at $\left(3, \dfrac{7}{5}\right)$.

Step 5:   Since $n = m$, the line $y = 1$ is the horizontal asymptote.
         $R(x)$ does not intersect $y = 1$ because $R(x)$ is not defined at $x = 3$.
$$\frac{x^2 + x - 12}{x^2 - x - 6} = 1$$
$$x^2 + x - 12 = x^2 - x - 6$$
$$2x = 6$$
$$x = 3$$

Step 6:

Interval	$(-\infty, -4)$	$(-4, -2)$	$(-2, 3)$	$(3, \infty)$
Number Chosen	-5	-3	0	4
Value of R	$R(-5) = \frac{1}{3}$	$R(-3) = -1$	$R(0) = 2$	$R(4) = \frac{4}{3}$
Location of Graph	Above x-axis	Below x-axis	Above x-axis	Above x-axis
Point on Graph	$(-5, \frac{1}{3})$	$(-3, -1)$	$(0, 2)$	$(4, \frac{4}{3})$

Step 7:   Graphing Utility:

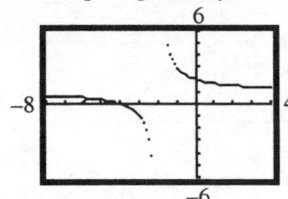

Step 8:   Graphing by hand:

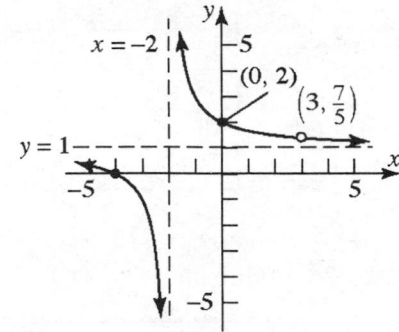

29. $R(x) = \dfrac{6x^2 - 7x - 3}{2x^2 - 7x + 6} = \dfrac{(3x+1)(2x-3)}{(2x-3)(x-2)} = \dfrac{3x+1}{x-2}$ $p(x) = 6x^2 - 7x - 3$;

$q(x) = 2x^2 - 7x + 6$; $n = 2$; $m = 2$

Step 1:   Domain: $\left\{ x \,\middle|\, x \neq \dfrac{3}{2},\, x \neq 2 \right\}$

Step 2:   (a)   The x-intercept is the zero of $p(x)$: $-\dfrac{1}{3}$; $x = \dfrac{3}{2}$ is not a zero because

reduced form must be used to find the zeros.

(b)   The y-intercept is $R(0) = \dfrac{6(0)^2 - 7(0) - 3}{2(0)^2 - 7(0) + 6} = \dfrac{-3}{6} = -\dfrac{1}{2}$.

Step 3: $R(-x) = \dfrac{6(-x)^2 - 7(-x) - 3}{2(-x)^2 - 7(-x) + 6} = \dfrac{6x^2 + 7x - 3}{2x^2 + 7x + 6}$; this is neither $R(x)$ nor $-R(x)$, so

there is no symmetry.

Step 4:   The vertical asymptote is the zero of $q(x)$: $x = 2$; $x = \dfrac{3}{2}$ is not a vertical

asymptote because reduced form must be used to find the them. The graph

has a hole at $\left(\dfrac{3}{2}, -11 \right)$.

Step 5:   Since $n = m$, the line $y = 3$ is the horizontal asymptote.

$R(x)$ does not intersect $y = 3$ because $R(x)$ is not defined at $x = \dfrac{3}{2}$.

$$\dfrac{6x^2 - 7x - 3}{2x^2 - 7x + 6} = 3$$

$$6x^2 - 7x - 3 = 6x^2 - 21x + 18$$

$$14x = 21$$

$$x = \dfrac{3}{2}$$

Step 6:

$-\frac{1}{3}$ $\frac{3}{2}$ 2

Interval	$(-\infty, -\frac{1}{3})$	$(-\frac{1}{3}, \frac{3}{2})$	$(\frac{3}{2}, 2)$	$(2, \infty)$
Number Chosen	-1	0	1.7	6
Value of R	$R(-1) = \frac{2}{3}$	$R(0) = -\frac{1}{2}$	$R(1.7) \approx -20.3$	$R(6) = 4.75$
Location of Graph	Above x-axis	Below x-axis	Below x-axis	Above x-axis
Point on Graph	$(-1, \frac{2}{3})$	$(0, -\frac{1}{2})$	$(1.7, -20.3)$	$(6, 4.75)$

Step 7:   Graphing Utility:

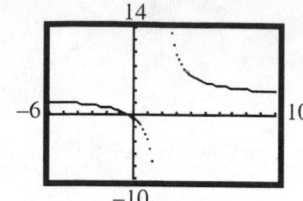

Step 8:   Graphing by hand:

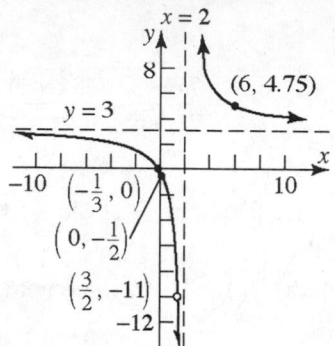

31. $R(x) = \dfrac{x^2 + 5x + 6}{x + 3} = \dfrac{(x+2)(x+3)}{x+3} = x + 2$ $p(x) = x^2 + 5x + 6$; $q(x) = x + 3$;
     $n = 2$; $m = 1$

Step 1:   Domain: $\{x \mid x \neq -3\}$

Step 2:   (a)   The x-intercept is the zero of $p(x)$: -2 ; -3 is not a zero because reduced
          form must be used to find the zeros.

          (b)   The y-intercept is $R(0) = \dfrac{0^2 + 5(0) + 6}{0 + 3} = \dfrac{6}{3} = 2$.

Step 3: $R(-x) = \dfrac{(-x)^2 + 5(-x) + 6}{-x + 3} = \dfrac{x^2 - 5x + 6}{-x + 3}$; this is neither $R(x)$ nor $-R(x)$, so
          there is no symmetry.

Step 4:   There are no vertical asymptotes. $x = -3$ is not a vertical asymptote because
          reduced form must be used to find the them. The graph has a hole at $(-3, -1)$.

Step 5:   Since $n = m + 1$ there is a oblique asymptote. The line $y = x + 2$ is the oblique
          asymptote. The oblique asymptote intersects $R(x)$ at every point of the
          form $(x, x + 2)$ except $(-3, -1)$.

Step 6:

                              -3   -2

Interval	$(-\infty, -3)$	$(-3, -2)$	$(-2, \infty)$
Number Chosen	-4	-2.5	0
Value of R	$R(-4) = -2$	$R(-2.5) = -\frac{1}{2}$	$R(0) = 2$
Location of Graph	Below x-axis	Below x-axis	Above x-axis
Point on Graph	$(-4, -2)$	$(-2.5, -\frac{1}{2})$	$(0, 2)$

Step 7:   Graphing Utility:

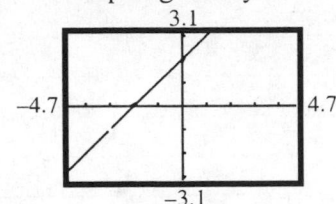

Step 8:   Graphing by hand:

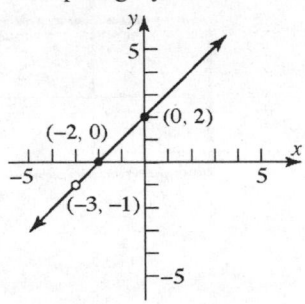

33. $f(x) = x + \dfrac{1}{x} = \dfrac{x^2 + 1}{x}$ $p(x) = x^2 + 1;\ \ q(x) = x;\ \ n = 2;\ \ m = 1$

Step 1:   Domain: $\left\{x \mid x \neq 0\right\}$

Step 2:   (a)   There are no x-intercepts.

         (b)   There is no y-intercept because 0 is not in the domain.

Step 3: $f(-x) = \dfrac{(-x)^2 + 1}{-x} = \dfrac{x^2 + 1}{-x} = -f(x)$; The graph of $f(x)$ is symmetric to the origin.

Step 4:   The vertical asymptote is the zero of $q(x)$: $x = 0$

Step 5:   Since $n = m + 1$, there is an oblique asymptote. Dividing:

$$\begin{array}{r} x \\ x\overline{)x^2 + 1} \\ \underline{x^2 } \\ 1 \end{array} \qquad f(x) = x + \dfrac{1}{x}$$

The oblique asymptote is $y = x$.

Solve to find intersection points:

$$\dfrac{x^2 + 1}{x} = x \rightarrow x^2 + 1 = x^2 \rightarrow 1 = 0$$

Since there is no solution, the oblique asymptote does not intersect $f(x)$.

Step 6:

Interval	$(-\infty, 0)$	$(0, \infty)$
Number Chosen	-1	1
Value of f	$f(-1) = -2$	$f(1) = 2$
Location of Graph	Below x-axis	Above x-axis
Point on Graph	(-1, -2)	(1, 2)

Step 7:   Graphing Utility:

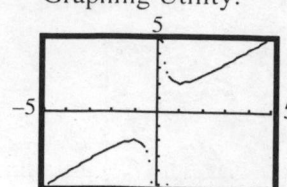

Step 8:   Graphing by hand:

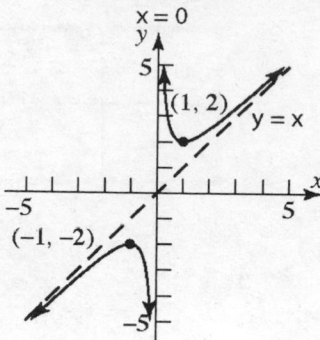

35. $f(x) = x^2 + \dfrac{1}{x} = \dfrac{x^3 + 1}{x}$ $p(x) = x^3 + 1;\ q(x) = x;\ n = 3;\ m = 1$

Step 1:   Domain: $\{x \mid x \neq 0\}$

Step 2:   (a)   The x-intercept is the zero of $p(x)$: -1

(b)   There is no y-intercept because 0 is not in the domain.

Step 3: $f(-x) = \dfrac{(-x)^3 + 1}{-x} = \dfrac{-x^3 + 1}{-x}$; this is neither $f(x)$ nor $-f(x)$, so there is no symmetry.

Step 4:   The vertical asymptote is the zero of $q(x)$: $x = 0$

Step 5:   Since $n > m + 1$, there is no horizontal or oblique asymptote.

Step 6:

Interval	$(-\infty, -1)$	$(-1, 0)$	$(0, \infty)$
Number Chosen	-2	$-\frac{1}{2}$	1
Value of f	$f(-2) = 3.5$	$f(-\frac{1}{2}) = -1.75$	$f(1) = 2$
Location of Graph	Above x-axis	Below x-axis	Above x-axis
Point on Graph	$(-2, 3.5)$	$(-\frac{1}{2}, -1.75)$	$(1, 2)$

Step 7:   Graphing Utility:

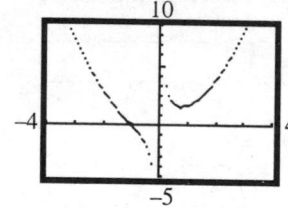

Step 8:   Graphing by hand:

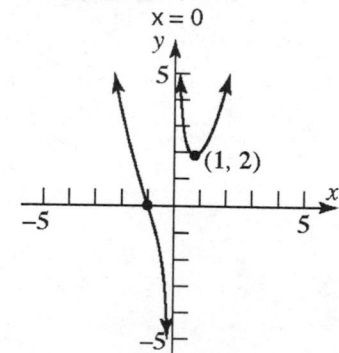

37. $f(x) = x + \dfrac{1}{x^3} = \dfrac{x^4 + 1}{x^3}$ $p(x) = x^4 + 1;\ \ q(x) = x^3;\ \ n = 4;\ \ m = 3$

Step 1:   Domain: $\{x \mid x \neq 0\}$

Step 2:   (a)   There are no x-intercepts.
          (b)   There is no y-intercept because 0 is not in the domain.

Step 3: $f(-x) = \dfrac{(-x)^4 + 1}{(-x)^3} = \dfrac{x^4 + 1}{-x^3} = -f(x)$; the graph of $f(x)$ is symmetric with respect to

the origin.

Step 4:   The vertical asymptote is the zero of $q(x)$: $x = 0$

Step 5:   Since $n = m + 1$, there is an oblique asymptote. Dividing:

$$f(x) = x + \dfrac{1}{x^3}$$ The oblique asymptote is $y = x$.

Solve to find intersection points: $\dfrac{x^4 + 1}{x^3} = x \Rightarrow x^4 + 1 = x^4 \Rightarrow 1 = 0$

Since there is no solution, the oblique asymptote does not intersect $f(x)$.

Step 6:

Interval	$(-\infty, 0)$	$(0, \infty)$
Number Chosen	-1	1
Value of f	$f(-1) = -2$	$f(1) = 2$
Location of Graph	Below x-axis	Above x-axis
Point on Graph	(-1, -2)	(1, 2)

Step 7:   Graphing Utility: Step 8:   Graphing by hand:

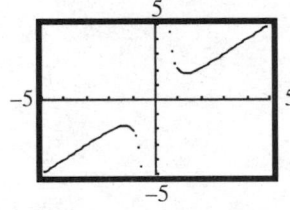   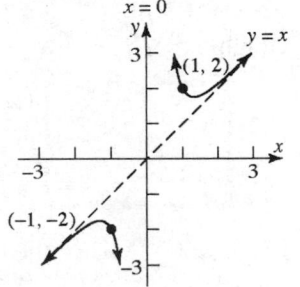

39. $f(x) = \dfrac{x^2}{x^2 - 4}$ 41. $f(x) = \dfrac{(x-1)^3(x-3)}{(x+1)^2(x-2)^2}$

43.   Answers will vary, one example is $R(x) = \dfrac{3(x-2)(x+1)^2}{(x+5)(x-6)^2}$

45. $g(h) = \dfrac{3.99 \times 10^{14}}{\left(6.374 \times 10^6 + h\right)^2}$

(a) $g(0) = \dfrac{3.99 \times 10^{14}}{\left(6.374 \times 10^6 + 0\right)^2} \approx 9.821 \ m/s^2$

(b) $g(443) = \dfrac{3.99 \times 10^{14}}{\left(6.374 \times 10^6 + 443\right)^2} \approx 9.8195 \ m/s^2$

(c) $g(8848) = \dfrac{3.99 \times 10^{14}}{\left(6.374 \times 10^6 + 8848\right)^2} \approx 9.794 \ m/s^2$

(d) $g(h) = \dfrac{3.99 \times 10^{14}}{\left(6.374 \times 10^6 + h\right)^2} \approx \dfrac{3.99 \times 10^{14}}{h^2} \to 0 \ \text{as} \ h \to \infty$

$\therefore y = 0$ is the horizontal asymptote.

(e)

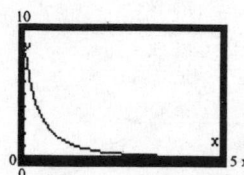

(f) $g(h) = \dfrac{3.99 \times 10^{14}}{\left(6.374 \times 10^6 + h\right)^2} = 0$, to solve this equation would require that

$3.99 \times 10^{14} = 0$, which is impossible. Therefore, there is no height above sea level at which $g = 0$. In other words, there is no point in the entire universe that is unaffected by the Earth's gravity!

47.   (a)   Graphing:

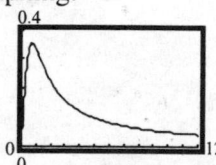

(b)   Using MAXIMUM, the concentration is highest when $t = 0.71$ hours.

(c) $C(t) = \dfrac{t}{2t^2 + 1} \approx \dfrac{t}{2t^2} = \dfrac{1}{2t} \to 0 \ \text{as} \ t \to \pm\infty$

therefore the horizontal asymptote is $y = 0$.

The concentration of the drug decreases to 0 as time increases.

49.   (a)   The average cost function is: $\overline{C}(x) = \dfrac{0.2x^3 - 2.3x^2 + 14.3x + 10.2}{x}$

(b) $\overline{C}(6) = \dfrac{0.2(6)^3 - 2.3(6)^2 + 14.3(6) + 10.2}{6} = \dfrac{56.4}{6} = 9.4$

The average cost of producing 6 Cavaliers per hour is $9400 per car.

(c) $\overline{C}(9) = \dfrac{0.2(9)^3 - 2.3(9)^2 + 14.3(9) + 10.2}{9} = \dfrac{98.4}{9} = 10.933$

The average cost of producing 9 Cavaliers per hour is $10,933 per car.

(d) Graphing:

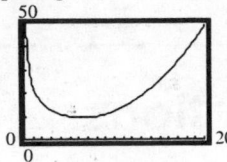

(e) Using MINIMUM, the number of Cavaliers that should be produced per hour to minimize cost is 6.38 cars

(f) The minimum average cost is

$$\overline{C}(6.38) = \frac{0.2(6.38)^3 - 2.3(6.38)^2 + 14.3(6.38) + 10.2}{6.38} \approx \$9366 \text{ per car.}$$

51. (a) The surface area is the sum of the areas of the six sides.

$$S = xy + xy + xy + xy + x^2 + x^2 = 4xy + 2x^2$$

The volume is $x \cdot x \cdot y = x^2 y = 10,000 \implies y = \dfrac{10000}{x^2}$

Thus, $S(x) = 4x\left(\dfrac{10000}{x^2}\right) + 2x^2 = 2x^2 + \dfrac{40000}{x} = \dfrac{2x^3 + 40000}{x}$

(b) Graphing:

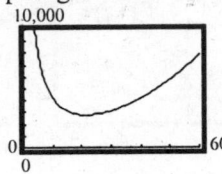

(c) Using MINIMUM, the minimum surface area (amount of cardboard) is 2,785 square inches.

(d) The surface area is a minimum when $x = 21.544$.

$$y = \frac{10000}{21.544^2} \approx 21.545$$

The dimensions of the box are: 21.544 in. by 21.544 in. by 21.545 in.

53. (a) $500 = \pi r^2 h \implies h = \dfrac{500}{\pi r^2}$

$$C(r) = 6(2\pi r^2) + 4(2\pi rh) = 12\pi r^2 + 8\pi r\left(\frac{500}{\pi r^2}\right) = 12\pi r^2 + \frac{4000}{r}$$

(b) Graphing:

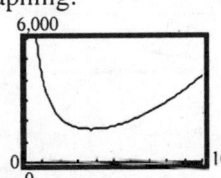

Using MINIMUM, the cost is least for $r \approx 3.76$ cm.

55. Answers will vary.

Chapter 3

Polynomial and Rational Functions

3.6 Polynomial and Rational Inequalities

1. $(x-5)(x+2) < 0$ $\qquad$ $f(x) = (x-5)(x+2)$

$x = 5$, $x = -2$ are the zeros.

Interval	Test Number	$f(x)$	Positive/Negative
$-\infty < x < -2$	-3	8	Positive
$-2 < x < 5$	0	-10	Negative
$5 < x < \infty$	6	8	Positive

The solution set is $\left\{ x \mid -2 < x < 5 \right\}$.

Graph $f(x) = (x-5)(x+2)$.

The x-intercepts are $x = -2$ and $x = 5$. The graph of f is below the x-axis for $-2 < x < 5$. Thus, the solution set is $\left\{ x \mid -2 < x < 5 \right\}$.

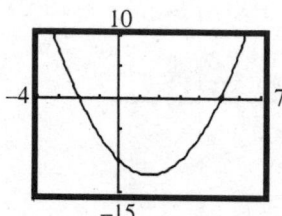

3. $x^2 - 4x > 0$ $\qquad$ $f(x) = x^2 - 4x$

$x(x-4) > 0$

$x = 0$, $x = 4$ are the zeros.

Interval	Test Number	$f(x)$	Positive/Negative
$-\infty < x < 0$	-1	5	Positive
$0 < x < 4$	1	-3	Negative
$4 < x < \infty$	5	5	Positive

The solution set is $\left\{ x \mid x < 0 \ \text{or} \ x > 4 \right\}$.

Graph $f(x) = x^2 - 4x$.

The x-intercepts are $x = 0$ and $x = 4$. The graph of f is above the x-axis for $x < 0$ or $x > 4$. Thus, the solution set is $\left\{ x \mid x < 0 \ \text{or} \ x > 4 \right\}$.

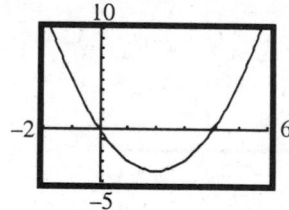

170

5. $x^2 - 9 < 0$ $f(x) = x^2 - 9$

   $(x + 3)(x - 3) < 0$

   $x = -3$, $x = 3$ are the zeros

Interval	Test Number	$f(x)$	Positive/Negative
$-\infty < x < -3$	-4	7	Positive
$-3 < x < 3$	0	-9	Negative
$3 < x < \infty$	4	7	Positive

The solution set is $\left\{ x \mid -3 < x < 3 \right\}$.

Graph $f(x) = x^2 - 9$.

The x-intercepts are $x = -3$ and $x = 3$. The graph of f is below the x-axis for $-3 < x < 3$. Thus, the solution set is $\left\{ x \mid -3 < x < 3 \right\}$.

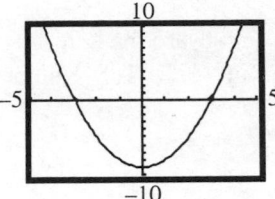

7. $x^2 + x > 12$ $f(x) = x^2 + x - 12$

   $x^2 + x - 12 > 0 \Rightarrow (x + 4)(x - 3) > 0$

   $x = -4$, $x = 3$ are the zeros.

Interval	Test Number	$f(x)$	Positive/Negative
$-\infty < x < -4$	-5	8	Positive
$-4 < x < 3$	0	-12	Negative
$3 < x < \infty$	4	8	Positive

The solution set is $\left\{ x \mid x < -4 \text{ or } x > 3 \right\}$.

Graph $y_1 = x^2 + x$, $y_2 = 12$.

y_1 intersects y_2 at $x = -4$ and $x = 3$. $y_1 > y_2$ for $x < -4$ or $x > 3$. Thus, the solution set is $\left\{ x \mid x < -4 \text{ or } x > 3 \right\}$.

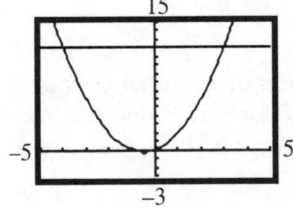

9. $2x^2 < 5x + 3$ $f(x) = 2x^2 - 5x - 3$

   $2x^2 - 5x - 3 < 0 \Rightarrow (2x + 1)(x - 3) < 0$

   $x = -\dfrac{1}{2}$, $x = 3$ are the zeros.

Interval	Test Number	$f(x)$	Positive/Negative
$-\infty < x < -1/2$	-1	4	Positive
$-1/2 < x < 3$	0	-3	Negative
$3 < x < \infty$	4	9	Positive

The solution set is $\left\{ x \mid -\dfrac{1}{2} < x < 3 \right\}$.

Graph $y_1 = 2x^2$, $y_2 = 5x + 3$.
y_1 intersects y_2 at $x = -0.50$ and $x = 3$.
$y_1 < y_2$ for $-0.50 < x < 3$. Thus, the solution set
is $\{ x \mid -0.50 < x < 3 \}$.

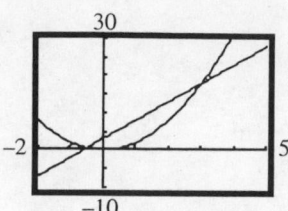

11. $x(x-7) > 8$ $f(x) = x^2 - 7x - 8$

$x^2 - 7x > 8 \Rightarrow x^2 - 7x - 8 > 0 \Rightarrow (x+1)(x-8) > 0$

$x = -1$, $x = 8$ are the zeros.

Interval	Test Number	$f(x)$	Positive/Negative
$-\infty < x < -1$	-2	10	Positive
$-1 < x < 8$	0	-8	Negative
$8 < x < \infty$	9	10	Positive

The solution set is $\{ x \mid x < -1 \text{ or } x > 8 \}$.

Graph $y_1 = x(x-7)$, $y_2 = 8$.
y_1 intersects y_2 at $x = -1$ and $x = 8$.
$y_1 > y_2$ for $x < -1$ or $x > 8$. Thus, the solution set
is $\{ x \mid x < -1 \text{ or } x > 8 \}$.

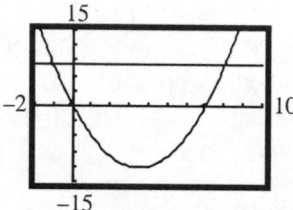

13. $4x^2 + 9 < 6x$ $f(x) = 4x^2 - 6x + 9$

$4x^2 - 6x + 9 < 0$

$b^2 - 4ac = (-6)^2 - 4(4)(9) = 36 - 144 = -108$

Since the discriminant is negative, there are no real zeros.
There is only one interval, the entire number line; choose any value and test.
For $x = 0$, $4x^2 - 6x + 9 = 9 > 0$. Thus, there is no solution.

Graph $y_1 = 4x^2 + 9$, $y_2 = 6x$.
y_1 is always greater than y_2. Thus, there is no
solution.

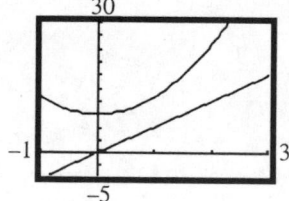

15. $6(x^2 - 1) > 5x$ $f(x) = 6x^2 - 5x - 6$

$6x^2 - 6 > 5x$

$6x^2 - 5x - 6 >$

$(3x + 2)(2x - 3) > 0$

$x = -\dfrac{2}{3}$, $x = \dfrac{3}{2}$ are the zeros.

Interval	Test Number	$f(x)$	Positive/Negative
$-\infty < x < -2/3$	-1	5	Positive
$-2/3 < x < 3/2$	0	-6	Negative
$3/2 < x < \infty$	2	8	Positive

The solution set is $\left\{ x \middle| x < -\dfrac{2}{3} \text{ or } x > \dfrac{3}{2} \right\}$.

Graph $y_1 = 6(x^2 - 1)$, $y_2 = 5x$.

y_1 intersects y_2 at $x = -0.67$ and $x = 1.5$. $y_1 > y_2$ for $x < -0.67$ or $x > 1.5$. Thus, the solution set is $\left\{ x \middle| x < -0.67 \text{ or } x > 1.5 \right\}$.

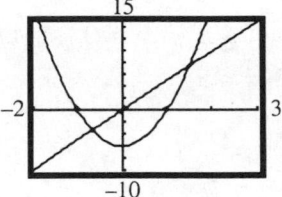

17. $(x - 1)(x^2 + x + 1) > 0$ $f(x) = (x - 1)(x^2 + x + 1)$

$x = 1$ is the zero. $x^2 + x + 1 = 0$ has no real zeros.

Interval	Test Number	$f(x)$	Positive/Negative
$-\infty < x < 1$	0	-1	Negative
$1 < x < \infty$	2	7	Positive

The solution set is $\left\{ x \middle| x > 1 \right\}$.

Graph $f(x) = (x - 1)(x^2 + x + 1)$.

The x-intercept is $x = 1$. The graph of f is above the x-axis for $x > 1$. Thus, the solution set is $\left\{ x \middle| x > 1 \right\}$.

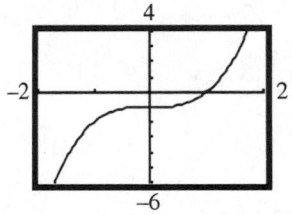

19. $(x - 1)(x - 2)(x - 3) < 0$ $f(x) = (x - 1)(x - 2)(x - 3)$

$x = 1$, $x = 2$, $x = 3$ are the zeros.

Interval	Test Number	$f(x)$	Positive/Negative
$-\infty < x < 1$	0	-6	Negative
$1 < x < 2$	1.5	0.375	Positive
$2 < x < 3$	2.5	-0.375	Negative
$3 < x < \infty$	4	6	Positive

The solution set is $\left\{ x \middle| x < 1 \text{ or } 2 < x < 3 \right\}$.

Graph $f(x) = (x - 1)(x - 2)(x - 3)$.

The x-intercepts are and $x = 3$. The graph of f is below the x-axis for $x < 1$ or $2 < x < 3$. Thus, the solution set is $\left\{ x \middle| x < 1 \text{ or } 2 < x < 3 \right\}$.

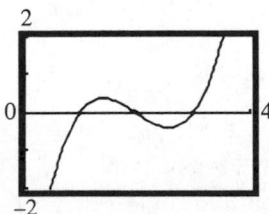

21. $x^3 - 2x^2 - 3x > 0$ $f(x) = x^3 - 2x^2 - 3x$

$x(x^2 - 2x - 3) > 0 \Rightarrow x(x + 1)(x - 3) > 0$

$x = -1$, $x = 0$, $x = 3$ are the zeros.

Interval	Test Number	$f(x)$	Positive/Negative
$-\infty < x < -1$	-2	-10	Negative
$-1 < x < 0$	-0.5	0.875	Positive
$0 < x < 3$	1	-4	Negative
$3 < x < \infty$	4	20	Positive

The solution set is $\left\{ x \mid -1 < x < 0 \text{ or } x > 3 \right\}$.

Graph $f(x) = x^3 - 2x^2 - 3x$.
The x-intercepts are $x = -1$, $x = 0$, and $x = 3$.
The graph of f is above the x-axis for
$-1 < x < 0$ or $x > 3$. Thus, the solution set is
$\left\{ x \mid -1 < x < 0 \text{ or } x > 3 \right\}$.

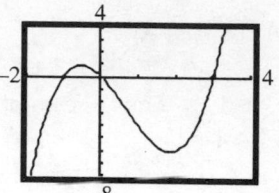

23. $x^4 > x^2$ $f(x) = x^4 - x^2$

$x^4 - x^2 > 0 \Rightarrow x^2\left(x^2 - 1\right) > 0 \Rightarrow x^2(x+1)(x-1) > 0$

$x = -1$, $x = 0$, $x = 1$ are the zeros.

Interval	Test Number	$f(x)$	Positive/Negative
$-\infty < x < -1$	-2	12	Positive
$-1 < x < 0$	-0.5	-0.1875	Negative
$0 < x < 1$	0.5	-0.1875	Negative
$1 < x < \infty$	2	12	Positive

The solution set is $\left\{ x \mid x < -1 \text{ or } x > 1 \right\}$.

Graph $y_1 = x^4$, $y_2 = x^2$.
y_1 intersects y_2 at $x = -1$, $x = 0$, and $x = 1$.
$y_1 > y_2$ for $x < -1$ or $x > 1$. Thus, the solution set
is $\left\{ x \mid x < -1 \text{ or } x > 1 \right\}$.

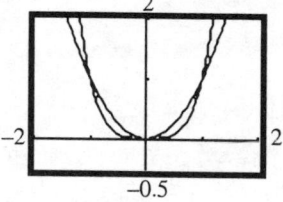

25. $x^3 > x^2$ $f(x) = x^3 - x^2$

$x^3 - x^2 > 0 \Rightarrow x^2(x-1) > 0$

$x = 0$, $x = 1$ are the zeros.

Interval	Test Number	$f(x)$	Positive/Negative
$-\infty < x < 0$	-1	-2	Negative
$0 < x < 1$	0.5	-0.125	Negative
$1 < x < \infty$	2	4	Positive

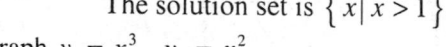

The solution set is $\left\{ x \mid x > 1 \right\}$.

Graph $y_1 = x^3$, $y_2 = x^2$.
y_1 intersects y_2 at $x = 0$ and $x = 1$. $y_1 > y_2$ for
$x > 1$. Thus, the solution set is $\left\{ x \mid x > 1 \right\}$.

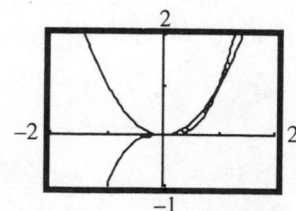

27. $x^4 > 1$ $f(x) = x^4 - 1$

$x^4 - 1 > 0 \Rightarrow (x^2 + 1)(x^2 - 1) > 0 \Rightarrow (x^2 + 1)(x + 1)(x - 1) > 0$

$x = -1$, $x = 1$ are the zeros.

Interval	Test Number	$f(x)$	Positive/Negative
$-\infty < x < -1$	-2	15	Positive
$-1 < x < 1$	0	-1	Negative
$1 < x < \infty$	2	15	Positive

The solution set is $\{ x \mid x < -1 \text{ or } x > 1 \}$.

Graph $y_1 = x^4$, $y_2 = 1$.
y_1 intersects y_2 at $x = -1$ and $x = 1$. $y_1 > y_2$ for
$x < -1$ or $x > 1$. Thus, the solution set is
$\{ x \mid x < -1 \text{ or } x > 1 \}$.

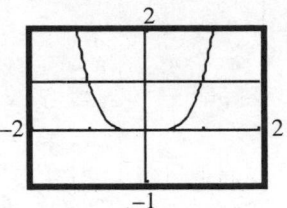

29. $x^2 - 7x - 8 < 0$ $f(x) = x^2 - 7x - 8$

$(x + 1)(x - 8) < 0$

$x = -1$, $x = 8$ are the zeros.

Interval	Test Number	$f(x)$	Positive/Negative
$-\infty < x < -1$	-2	10	Positive
$-1 < x < 8$	0	-8	Negative
$8 < x < \infty$	9	10	Positive

The solution set is $\{ x \mid -1 < x < 8 \}$.

Graph $f(x) = x^2 - 7x - 8$.
The x-intercepts are $x = -1$ and $x = 8$. The graph
of f is below the x-axis for $-1 < x < 8$. Thus, the
solution set is $\{ x \mid -1 < x < 8 \}$.

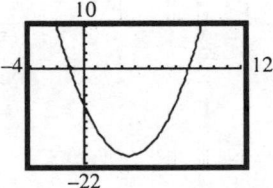

31.   Graph $f(x) = x^3 + x - 12$ and use ZERO to find the zero of the function.

$x = 2.14$ is the zero (rounded to two decimal places).

Interval	Test Number	$f(x)$	Positive/Negative
$-\infty < x < 2.14$	0	-12	Negative
$2.14 < x < \infty$	3	18	Positive

The solution set is $\{ x \mid x \geq 2.14 \}$.

Graph $f(x) = x^3 + x - 12$.
The x-intercept is $x = 2.14$. The graph of f is
above the x-axis for $x > 2.14$. Thus, the solution
set is $\{ x \mid x \geq 2.14 \}$.

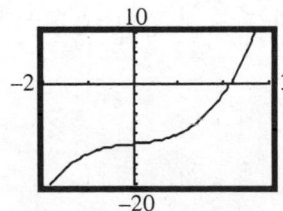

175

33. $x^4 - 3x^2 - 4 > 0$ $f(x) = x^4 - 3x^2 - 4$

$\left(x^2 - 4\right)\left(x^2 + 1\right) > 0 \Rightarrow (x+2)(x-2)\left(x^2+1\right) > 0$

$x = -2, x = 2$ are the zeros.

Interval	Test Number	$f(x)$	Positive/Negative
$-\infty < x < -2$	-3	50	Positive
$-2 < x < 2$	0	-4	Negative
$2 < x < \infty$	3	50	Positive

The solution set is $\left\{ x \mid x < -2 \ \text{or} \ x > 2 \right\}$.

Graph $f(x) = x^4 - 3x^2 - 4$.

The x-intercepts are $x = -2$, and $x = 2$. The graph
of f is above the x-axis for $x < -2$ or $x > 2$.

Thus, the solution set is $\left\{ x \mid x < -2 \ \text{or} \ x > 2 \right\}$.

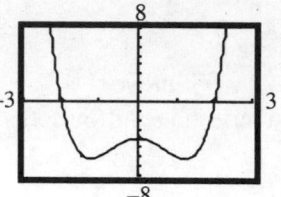

35. $x^3 - 4 \geq 3x^2 + 5x - 3$ $f(x) = x^3 - 3x^2 - 5x - 1$

$x^3 - 3x^2 - 5x - 1 \geq 0$

Do synthetic division by -1:

$$
\begin{array}{r|rrrr}
-1 & 1 & -3 & -5 & -1 \\
   &   & -1 & 4 & 1 \\
\hline
   & 1 & -4 & -1 & 0
\end{array}
$$

$x = -1$ is a solution. Solving the remaining factor $x^2 - 4x - 1 = 0$:

$$x = \frac{-(-4) \pm \sqrt{(-4)^2 - 4(1)(-1)}}{2(1)} = \frac{4 \pm \sqrt{20}}{2} = \frac{4 \pm 2\sqrt{5}}{2} = 2 \pm \sqrt{5}$$

$x = -1, x = 2 - \sqrt{5}, x = 2 + \sqrt{5}$ or $x = -1, x = -0.24, x = 4.24$ are the zeros.

Interval	Test Number	$f(x)$	Positive/Negative
$-\infty < x < -1$	-2	-11	Negative
$-1 < x < 2 - \sqrt{5}$	-0.5	0.625	Positive
$2 - \sqrt{5} < x < 2 + \sqrt{5}$	0	-1	Negative
$2 + \sqrt{5} < x < \infty$	5	24	Positive

The solution set is $\left\{ x \mid -1 \leq x \leq 2 - \sqrt{5} \ \text{or} \ x \geq 2 + \sqrt{5} \right\}$.

Graph $y_1 = x^3 - 4$, $y_2 = 3x^2 + 5x - 3$.

y_1 intersects y_2 at $x = -1$, $x = -0.24$,
and $x = 4.24$. $y_1 \geq y_2$ for
$-1 \leq x \leq -0.24$ or $x \geq 4.24$. Thus, the solution set
is $\left\{ x \mid -1 \leq x \leq -0.24 \ \text{or} \ x \geq 4.24 \right\}$.

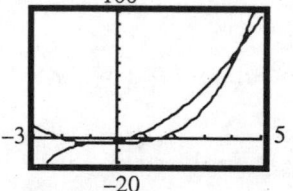

176

37. $\dfrac{x+1}{x-1} > 0$ $\qquad\qquad\qquad$ $f(x) = \dfrac{x+1}{x-1}$

The zeros and values where the expression is undefined are $x = -1$, and $x = 1$.

Interval	Test Number	$f(x)$	Positive/Negative
$-\infty < x < -1$	-2	$1/3$	Positive
$-1 < x < 1$	0	-1	Negative
$1 < x < \infty$	2	3	Positive

The solution set is $\left\{ x \mid x < -1 \text{ or } x > 1 \right\}$.

Graph $f(x) = \dfrac{x+1}{x-1}$.

The x-intercept is $x = -1$. The expression is undefined at $x = 1$. The graph of f is above the x-axis for $x < -1$ or $x > 1$. Thus, the solution set is $\left\{ x \mid x < -1 \text{ or } x > 1 \right\}$.

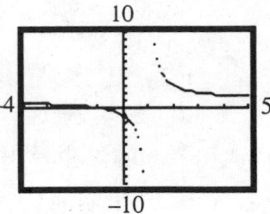

39. $\dfrac{(x-1)(x+1)}{x} < 0$ $\qquad\qquad$ $f(x) = \dfrac{(x-1)(x+1)}{x}$

The zeros and values where the expression is undefined are $x = -1$, $x = 0$, and $x = 1$.

Interval	Test Number	$f(x)$	Positive/Negative
$-\infty < x < -1$	-2	-1.5	Negative
$-1 < x < 0$	-0.5	1.5	Positive
$0 < x < 1$	0.5	-1.5	Negative
$1 < x < \infty$	2	1.5	Positive

The solution set is $\left\{ x \mid x < -1 \text{ or } 0 < x < 1 \right\}$.

Graph $f(x) = \dfrac{(x-1)(x+1)}{x}$.

The x-intercepts are $x = -1$ and $x = 1$. The expression is undefined at $x = 0$. The graph of f is below the x-axis for $x < -1$ or $0 < x < 1$. Thus, the solution set is $\left\{ x \mid x < -1 \text{ or } 0 < x < 1 \right\}$.

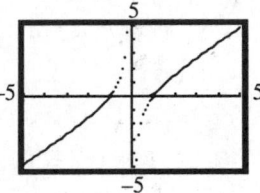

41. $\dfrac{(x-2)^2}{x^2-1} \geq 0$ $\qquad\qquad$ $f(x) = \dfrac{(x-2)^2}{x^2-1}$

$\dfrac{(x-2)^2}{(x+1)(x-1)} \geq 0$

The zeros and values where the expression is undefined are $x = -1$, $x = 1$, and $x = 2$.

Interval	Test Number	$f(x)$	Positive/Negative
$-\infty < x < -1$	-2	$16/3$	Positive
$-1 < x < 1$	0	-4	Negative
$1 < x < 2$	1.5	0.2	Positive
$2 < x < \infty$	3	0.125	Positive

The solution set is $\left\{ x \mid x < -1 \text{ or } x > 1 \right\}$.

Graph $f(x) = \dfrac{(x-2)^2}{x^2-1}$.

The x-intercept is $x = 2$. The expression is undefined at $x = -1$ and $x = 1$. The graph of f is above the x-axis for $x < -1$ or $x > 1$. Thus, the solution set is $\left\{ x \mid x < -1 \text{ or } x > 1 \right\}$.

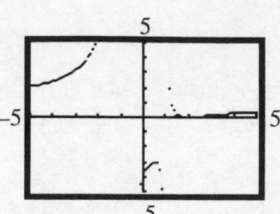

43. $6x - 5 < \dfrac{6}{x}$ $f(x) = 6x - 5 - \dfrac{6}{x}$

$$6x - 5 - \frac{6}{x} < 0 \Rightarrow \frac{6x^2 - 5x - 6}{x} < 0 \Rightarrow \frac{(2x-3)(3x+2)}{x} < 0$$

The zeros and values where the expression is undefined are $x = -\dfrac{2}{3}, x = 0,$ and $x = \dfrac{3}{2}$.

Interval	Test Number	$f(x)$	Positive/Negative
$-\infty < x < -2/3$	-1	-5	Negative
$-2/3 < x < 0$	-0.5	4	Positive
$0 < x < 3/2$	1	-5	Negative
$3/2 < x < \infty$	2	4	Positive

The solution set is $\left\{ x \mid x < -\dfrac{2}{3} \text{ or } 0 < x < \dfrac{3}{2} \right\}$.

Graph $y_1 = 6x - 5, \ y_2 = \dfrac{6}{x}$.

y_1 intersects y_2 at $x = -0.67$ and $x = 1.5$.
y_2 is undefined at $x = 0$. $y_1 < y_2$ for
$x < -0.67$ or $0 < x < 1.5$. Thus, the solution set is
$\left\{ x \mid x < -0.67 \text{ or } 0 < x < 1.5 \right\}$.

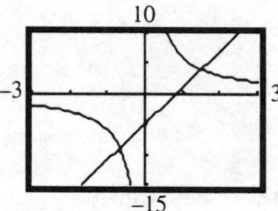

45. $\dfrac{x+4}{x-2} \le 1$ $f(x) = \dfrac{x+4}{x-2} - 1$

$$\frac{x+4}{x-2} - 1 \le 0 \Rightarrow \frac{x+4-(x-2)}{x-2} \le 0 \Rightarrow \frac{6}{x-2} \le 0$$

The value where the expression is undefined is $x = 2$.

Interval	Test Number	$f(x)$	Positive/Negative
$-\infty < x < 2$	0	-3	Negative
$2 < x < \infty$	3	6	Positive

The solution set is $\left\{ x \mid x < 2 \right\}$.

Graph $y_1 = \dfrac{x+4}{x-2}, \ y_2 = 1$.

y_1 is not defined at $x = 2$. $y_1 < y_2$ for $x < 2$. Thus, the solution set is $\left\{ x \mid x < 2 \right\}$.

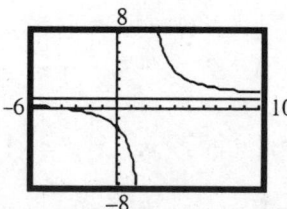

47. $$\frac{3x-5}{x+2} \le 2 \qquad f(x) = \frac{3x-5}{x+2} - 2$$

$$\frac{3x-5}{x+2} - 2 \le 0 \Rightarrow \frac{3x-5-2(x+2)}{x+2} \le 0 \Rightarrow \frac{x-9}{x+2} \le 0$$

The zeros and values where the expression is undefined are $x = -2$, and $x = 9$.

Interval	Test Number	$f(x)$	Positive/Negative
$-\infty < x < -2$	-3	12	Positive
$-2 < x < 9$	0	-4.5	Negative
$9 < x < \infty$	10	$1/12$	Positive

The solution set is $\left\{ x \,\middle|\, -2 < x \le 9 \right\}$.

Graph $y_1 = \dfrac{3x-5}{x+2}$, $y_2 = 2$.

y_1 is not defined at $x = -2$. y_1 intersects y_2 at $x = 9$. $y_1 < y_2$ for $-2 < x < 9$. Thus, the solution set is $\left\{ x \,\middle|\, -2 < x \le 9 \right\}$.

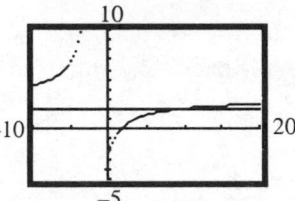

49. $$\frac{1}{x-2} < \frac{2}{3x-9} \qquad\qquad f(x) = \frac{1}{x+2} - \frac{2}{3x-9}$$

$$\frac{1}{x-2} - \frac{2}{3x-9} < 0 \Rightarrow \frac{3x-9-2(x-2)}{(x-2)(3x-9)} < 0 \Rightarrow \frac{x-5}{(x-2)(3x-9)} < 0$$

The zeros and values where the expression is undefined are $x = 2$, $x = 3$, and $x = 5$.

Interval	Test Number	$f(x)$	Positive/Negative
$-\infty < x < 2$	0	$-5/18$	Negative
$2 < x < 3$	2.5	$10/3$	Positive
$3 < x < 5$	4	$-1/6$	Negative
$5 < x < \infty$	6	$1/36$	Positive

The solution set is $\left\{ x \,\middle|\, x < 2 \text{ or } 3 < x < 5 \right\}$.

Graph $y_1 = \dfrac{1}{x-2}$, $y_2 = \dfrac{2}{3x-9}$.

The undefined values are at $x = 2$, $x = 3$.

y_1 intersects y_2 at $x = 5$. $y_1 < y_2$ for $x < 2$ or $3 < x < 5$. Thus, the solution set is $\left\{ x \,\middle|\, x < 2 \text{ or } 3 < x < 5 \right\}$.

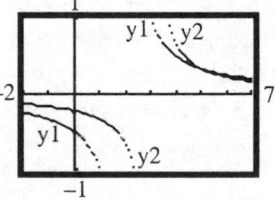

51. $$\frac{2x+5}{x+1} > \frac{x+1}{x-1} \qquad\qquad f(x) = \frac{2x+5}{x+1} - \frac{x+1}{x-1}$$

$$\frac{2x+5}{x+1} - \frac{x+1}{x-1} > 0 \Rightarrow \frac{(2x+5)(x-1)-(x+1)(x+1)}{(x+1)(x-1)} > 0$$

$$\frac{2x^2+3x-5-\left(x^2+2x+1\right)}{(x+1)(x-1)} > 0 \Rightarrow \frac{x^2+x-6}{(x+1)(x-1)} > 0 \Rightarrow \frac{(x+3)(x-2)}{(x+1)(x-1)} > 0$$

The zeros and values where the expression is undefined are
$x = -3$, $x = -1$, $x = 1$, and $x = 2$.

Interval	Test Number	$f(x)$	Positive/Negative
$-\infty < x < -3$	-4	2/5	Positive
$-3 < x < -1$	-2	$-4/3$	Negative
$-1 < x < 1$	0	6	Positive
$1 < x < 2$	1.5	$-9/5$	Negative
$2 < x < \infty$	3	3/4	Positive

The solution set is $\left\{ x \mid x < -3, \ -1 < x < 1, \ x > 2 \right\}$.

Graph $y_1 = \dfrac{2x+5}{x+1}$, $y_2 = \dfrac{x+1}{x-1}$.

The undefined values are at $x = -1$, $x = 1$.

y_1 intersects y_2 at $x = -3$ and $x = 2$. $y_1 > y_2$ for $x < -3$, $-1 < x < 1$, or $x > 2$. Thus, the solution set is $\left\{ x \mid x < -3, \ -1 < x < 1, \ x > 2 \right\}$.

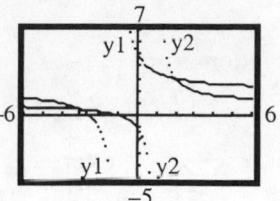

53. $\dfrac{x^2(3+x)(x+4)}{(x+5)(x-1)} > 0$ $f(x) = \dfrac{x^2(3+x)(x+4)}{(x+5)(x-1)}$

The zeros and values where the expression is undefined are

$x = -5$, $x = -4$, $x = -3$, $x = 0$ and $x = 1$.

Interval	Test Number	$f(x)$	Positive/Negative
$-\infty < x < -5$	-6	216/7	Positive
$-5 < x < -4$	-4.5	$-243/44$	Negative
$-4 < x < -3$	-3.5	49/108	Positive
$-3 < x < 0$	-1	$-3/4$	Negative
$0 < x < 1$	0.5	$-63/44$	Negative
$1 < x < \infty$	2	120/7	Positive

The solution set is $\left\{ x \mid x < -5, \ -4 < x < -3, \ x > 1 \right\}$.

Graph $f(x) = \dfrac{x^2(3+x)(x+4)}{(x+5)(x-1)}$.

The x-intercepts are $x = -4$, $x = -3$, and $x = 0$.
The expression is undefined at $x = -5$ and $x = 1$.
The graph of f is above the x-axis for $x < -5$, $-4 < x < -3$, or $x > 1$. Thus, the solution set is $\left\{ x \mid x < -5, \ -4 < x < -3, \ x > 1 \right\}$.

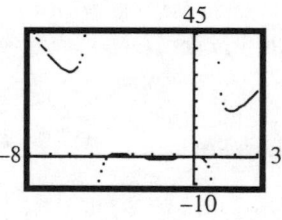

55. $\dfrac{2x^2 - x - 1}{x - 4} \le 0$ $f(x) = \dfrac{2x^2 - x - 1}{x - 4}$

$\dfrac{(2x+1)(x-1)}{x-4} \le 0$

The zeros and values where the expression is undefined are $x = -\dfrac{1}{2}$, $x = 1$, and $x = 4$

Interval	Test Number	$f(x)$	Positive/Negative
$-\infty < x < -1/2$	-1	$-2/5$	Negative
$-1/2 < x < 1$	0	1/4	Positive
$1 < x < 4$	2	$-5/2$	Negative
$4 < x < \infty$	5	44	Positive

The solution set is $\left\{ x \mid x \le -\dfrac{1}{2}, \text{or } 1 \le x < 4 \right\}$.

Graph $f(x) = \dfrac{2x^2 - x - 1}{x - 4}$.

The x-intercepts are $x = -0.5$, and $x = 1$. The expression is undefined at $x = 4$. The graph of f is below the x-axis for $x < -0.5$, or $1 < x < 4$. Thus, the solution set is $\left\{ x \mid x \le -0.5, \text{or } 1 \le x < 4 \right\}$.

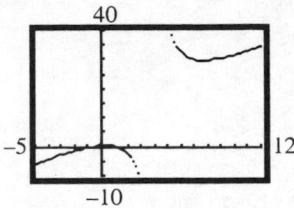

57. $\dfrac{x^2 + 3x - 1}{x + 3} > 0$ $\qquad f(x) = \dfrac{x^2 + 3x - 1}{x + 3}$

Solving when the numerator $= 0$: $x = \dfrac{-3 \pm \sqrt{3^2 - 4(1)(-1)}}{2(1)} = \dfrac{-3 \pm \sqrt{13}}{2}$

The zeros and values where the expression is undefined are

$x = \dfrac{-3 - \sqrt{13}}{2}$, $x = \dfrac{-3 + \sqrt{13}}{2}$, and $x = -3$ or $x = -3.30$, $x = 0.30$, $x = -3$.

Interval	Test Number	$f(x)$	Positive/Negative
$-\infty < x < \dfrac{-3 - \sqrt{13}}{2}$	-4	-3	Negative
$\dfrac{-3 - \sqrt{13}}{2} < x < -3$	-3.2	$9/5$	Positive
$-3 < x < \dfrac{-3 + \sqrt{13}}{2}$	0	$-1/3$	Negative
$\dfrac{-3 + \sqrt{13}}{2} < x < \infty$	1	$3/4$	Positive

The solution set is $\left\{ x \mid \dfrac{-3 - \sqrt{13}}{2} < x < -3 \text{ or } x > \dfrac{-3 + \sqrt{13}}{2} \right\}$.

Graph $f(x) = \dfrac{x^2 + 3x - 1}{x + 3}$.

The x-intercepts are $x = -3.30$ and $x = 0.30$. The expression is undefined at $x = -3$. The graph of f is above the x-axis for $-3.30 < x < -3$, or $x > 0.30$. Thus, the solution set is $\left\{ x \mid -3.30 < x < -3, \text{or } x > 0.30 \right\}$.

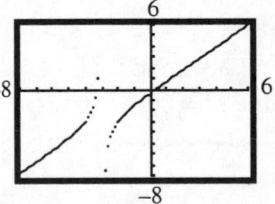

59. Let x be the positive number. Then

$$x^3 > 4x^2 \Rightarrow x^3 - 4x^2 > 0 \Rightarrow x^2(x - 4) > 0$$

The zeros are $x = 0$ and $x = 4$. $\qquad f(x) = x^3 - 4x^2$

Interval	Test Number	$f(x)$	Positive/Negative
$-\infty < x < 0$	-1	-5	Negative
$0 < x < 4$	1	-3	Negative
$4 < x < \infty$	5	25	Positive

The solution set is $\left\{ x \mid x > 4 \right\}$. All real number larger than 4 satisfy the condition.

61. The domain of the expression includes all values for which
$$x^2 - 16 \geq 0 \Rightarrow (x+4)(x-4) \geq 0$$
The zeros are $x = -4$ and $x = 4$. $f(x) = x^2 - 16$

Interval	Test Number	$f(x)$	Positive/Negative
$-\infty < x < -4$	-5	9	Positive
$-4 < x < 4$	0	-16	Negative
$4 < x < \infty$	5	9	Positive

The solution or domain is $\{ x \mid x \leq -4 \text{ or } x \geq 4 \}$.

63. The domain of the expression includes all values for which
$$\frac{x-2}{x+4} \geq 0$$
The zeros and values where the expression is undefined are $x = -4$ and $x = 2$

$$f(x) = \frac{x-2}{x+4}$$

Interval	Test Number	$f(x)$	Positive/Negative
$-\infty < x < -4$	-5	7	Positive
$-4 < x < 2$	0	$-1/2$	Negative
$2 < x < \infty$	3	$1/7$	Positive

The solution or domain is $\{ x \mid x < -4 \text{ or } x \geq 2 \}$.

65. (a) Find the values of t for which
$$80t - 16t^2 > 96 \Rightarrow -16t^2 + 80t - 96 > 0$$
$$16t^2 - 80t + 96 < 0 \Rightarrow 16(t^2 - 5t + 6) < 0 \Rightarrow 16(t-2)(t-3) < 0$$
The zeros are $t = 2$ and $t = 3$. $s(t) = 16t^2 - 80t + 96$

Interval	Test Number	$s(t)$	Positive/Negative
$-\infty < t < 2$	1	32	Positive
$2 < t < 3$	2.5	-4	Negative
$3 < t < \infty$	4	32	Positive

The solution set is $\{ t \mid 2 < t < 3 \}$. The ball is more than 96 feet above the ground for times between 2 and 3 seconds.

(b) Graphing: $s = 80t - 16t^2$

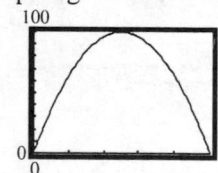

(c) Using MAXIMUM, the maximum height is 100 feet.

(d) The maximum height occurs at 2.5 seconds.

67. (a) Profit = Revenue − Cost
$$x(40 - 0.2x) - 32x \geq 50 \Rightarrow 40x - 0.2x^2 - 32x \geq 50$$
$$-0.2x^2 + 8x - 50 \geq 0 \Rightarrow 2x^2 - 80x + 500 \leq 0 \Rightarrow x^2 - 40x + 250 \leq 0$$
The zeros are approximately $x = 7.75$ and $x = 32.25$.

$$f(x) = x^2 - 40x + 250$$

Interval	Test Number	$f(x)$	Positive/Negative
$0 < x < 7.75$	7	19	Positive
$7.75 < x < 32.25$	10	-50	Negative
$32.25 < x < \infty$	40	250	Positive

The profit is at least $50 when at least 8 and no more than 32 watches are sold.

(b) Graphing the revenue function: $\left(R(x) = x(40 - 0.2x)\right)$

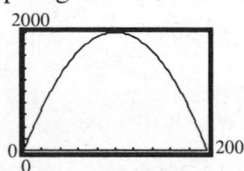

(c) Using MAXIMUM, the maximum revenue is $2,000.

(d) Using MAXIMUM, the company should sell 100 wristwatches to maximize revenue.

(e) Graphing the profit function: $P(x) = x(40 - 0.2x) - 32x$

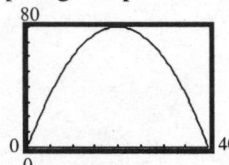

(f) Using MAXIMUM, the maximum profit is $80.

(g) Using MAXIMUM, the company should sell 20 watches for maximum profit.

69. The cost of manufacturing x Chevy Cavaliers in a day was found to be:

$$C(x) = 0.216x^3 - 2.347x^2 + 14.328x + 10.224$$

Since the budget constraints require that the cost must be less than or equal to $97,000, we need to solve: $C(x) \le 97$ or

$$0.216x^3 - 2.347x^2 + 14.328x + 10.224 \le 97$$

Graphing $y_1 = 0.216x^3 - 2.347x^2 + 14.328x + 10.224$ and $y_2 = 97$:

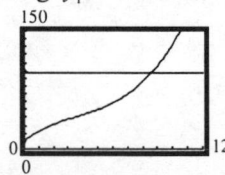

y_1 intersects y_2 at $x = 8.59$. $y_1 < y_2$ when $x < 8.59$. Chevy can produce at most eight Cavaliers in a day, assuming cars cannot be partially completed.

71. Prove that if a, b are real numbers and $a \ge 0$, $b \ge 0$, then $a \le b$ is equivalent to $\sqrt{a} \le \sqrt{b}$.

First note that if $a = 0$, then $a \le b \Rightarrow b = 0$.

Similarly, if $\sqrt{a} = 0$ then $a = 0$ and $\sqrt{a} \le \sqrt{b} \Rightarrow \sqrt{b} = 0 \Rightarrow b = 0$.

Therefore, $a \le b$ is equivalent to $\sqrt{a} \le \sqrt{b}$.

Now suppose $a \ne 0$.

We must show that $a \le b \Rightarrow \sqrt{a} \le \sqrt{b}$ and that $\sqrt{a} \le \sqrt{b} \Rightarrow a \le b$.

Assume $a \leq b$, this means that $b - a \geq 0$. We can factor the left hand side by treating it as a difference of two perfect squares:

$$\left(\sqrt{b} - \sqrt{a}\right)\left(\sqrt{b} + \sqrt{a}\right) \geq 0$$

Case 1: $\sqrt{b} - \sqrt{a} \geq 0$ and $\sqrt{b} + \sqrt{a} \geq 0$

$$\Rightarrow \sqrt{b} \geq \sqrt{a} \text{ and } \sqrt{b} \geq -\sqrt{a}$$

$$\Rightarrow \sqrt{a} \leq \sqrt{b} \text{ and } -\sqrt{a} \leq \sqrt{b}$$

or

Case 2: $\sqrt{b} \leq \sqrt{a}$ and $\sqrt{b} \leq -\sqrt{a}$

this is impossible since $\sqrt{a} \geq 0$ and $\sqrt{b} \geq 0$ which means

$$\sqrt{b} \leq -\sqrt{a} \text{ is impossible since } a \neq 0.$$

Since Case 2 is impossible, Case 1 must be true. Therefore, $a \leq b \Rightarrow \sqrt{a} \leq \sqrt{b}$. Assume $\sqrt{a} \leq \sqrt{b}$. This means that $\sqrt{a} - \sqrt{b} \leq 0$. Since $\sqrt{a} \geq 0$ and $\sqrt{b} \geq 0$, we can conclude that $\sqrt{a} + \sqrt{b} \geq 0$. Therefore,

$$\left(\sqrt{a} - \sqrt{b}\right)\left(\sqrt{a} + \sqrt{b}\right) \leq 0\left(\sqrt{a} + \sqrt{b}\right)$$

$$a - b \leq 0$$

$$a \leq b$$

Thus $\sqrt{a} \leq \sqrt{b} \Rightarrow a \leq b$.

Chapter 3

Polynomial and Rational Functions

3.7 The Real Zeros of a Polynomial Function

1. $f(x) = 4x^3 - 3x^2 - 8x + 4;\ \ c = 3$
   $f(3) = 4(3)^3 - 3(3)^2 - 8(3) + 4 = 108 - 27 - 24 + 4 = 61 \neq 0$
   Thus, 3 is not a zero of $f \therefore x - 3$ is not a factor of f.

3. $f(x) = 3x^4 - 6x^3 - 5x + 10;\ \ c = 1$
   $f(1) = 3(1)^4 - 6(1)^3 - 5(1) + 10 = 3 - 6 - 5 + 10 = 2 \neq 0$
   Thus, 1 is not a zero of $f \therefore x - 1$ is not a factor of f.

5. $f(x) = 3x^6 + 2x^3 - 176;\ \ c = -2$
   $f(-2) = 3(-2)^6 + 2(-2)^3 - 176 = 192 - 16 - 176 = 0$
   Thus, -2 is a zero of $f \therefore x + 2$ is a factor of f. Use synthetic division to find the factors.

   $$
   \begin{array}{r|rrrrrrr}
   -2 & 3 & 0 & 0 & 2 & 0 & 0 & -176 \\
      &   & -6 & 12 & -24 & 44 & -88 & 176 \\
   \hline
      & 3 & -6 & 12 & -22 & 44 & -88 & 0
   \end{array}
   $$

   The factored form is: $f(x) = (x + 2)(3x^5 - 6x^4 + 12x^3 - 22x^2 + 44x - 88)$.

7. $f(x) = 4x^6 - 64x^4 + x^2 - 16;\ \ c = 4$
   $f(4) = 4(4)^6 - 64(4)^4 + (4)^2 - 16 = 16384 - 16384 + 16 - 16 = 0$
   Thus, 4 is a zero of $f \therefore x - 4$ is a factor of f. Use synthetic division to find the factors.

   $$
   \begin{array}{r|rrrrrr}
   4 & 4 & 0 & -64 & 0 & 1 & 0 & -16 \\
     &   & 16 & 64 & 0 & 0 & 4 & 16 \\
   \hline
     & 4 & 16 & 0 & 0 & 1 & 4 & 0
   \end{array}
   $$

   The factored form is: $f(x) = (x - 4)(4x^5 + 16x^4 + x + 4)$.
   We can use grouping to factor $4x^5 + 16x^4 + x + 4 = 4x^4(x + 4) + x + 4 = (x + 4)(4x^4 + 1)$
   So $f(x) = (x - 4)(x + 4)(4x^4 + 1)$.

9. $f(x) = 2x^4 - x^3 + 2x - 1;\ \ c = -\dfrac{1}{2}$

   $f\left(-\dfrac{1}{2}\right) = 2\left(-\dfrac{1}{2}\right)^4 - \left(-\dfrac{1}{2}\right)^3 + 2\left(-\dfrac{1}{2}\right) - 1 = \dfrac{1}{8} + \dfrac{1}{8} - 1 - 1 = -\dfrac{7}{4} \neq 0$

   Thus, $-\dfrac{1}{2}$ is not a zero of $f \therefore x + \dfrac{1}{2}$ is not a factor of f.

185

11. $f(x) = 3x^4 - 3x^3 + x^2 - x + 1$
The maximum number of zeros is the degree of the polynomial which is 4.
p must be a factor of 1: $p = \pm 1$
q must be a factor of 3: $q = \pm 1, \pm 3$

The possible rational zeros are: $\dfrac{p}{q} = \pm 1, \pm \dfrac{1}{3}$

13. $f(x) = x^5 - 6x^2 + 9x - 3$
The maximum number of zeros is the degree of the polynomial which is 5.
p must be a factor of –3: $p = \pm 1, \pm 3$
q must be a factor of 1: $q = \pm 1$

The possible rational zeros are: $\dfrac{p}{q} = \pm 1, \pm 3$

15. $f(x) = -4x^3 - x^2 + x + 2$
The maximum number of zeros is the degree of the polynomial which is 3.
p must be a factor of 2: $p = \pm 1, \pm 2$
q must be a factor of –4: $q = \pm 1, \pm 2, \pm 4$

The possible rational zeros are: $\dfrac{p}{q} = \pm 1, \pm 2, \pm \dfrac{1}{2}, \pm \dfrac{1}{4}$

17. $f(x) = 3x^4 - x^2 + 2$
The maximum number of zeros is the degree of the polynomial which is 4.
p must be a factor of 2: $p = \pm 1, \pm 2$
q must be a factor of 3: $q = \pm 1, \pm 3$

The possible rational zeros are: $\dfrac{p}{q} = \pm 1, \pm \dfrac{1}{3}, \pm 2, \pm \dfrac{2}{3}$

19. $f(x) = 2x^5 - x^3 + 2x^2 + 4$
The maximum number of zeros is the degree of the polynomial which is 5.
p must be a factor of 4: $p = \pm 1, \pm 2, \pm 4$
q must be a factor of 2: $q = \pm 1, \pm 2$

The possible rational zeros are: $\dfrac{p}{q} = \pm 1, \pm \dfrac{1}{2}, \pm 2, \pm 4$

21. $f(x) = 6x^4 + 2x^3 - x^2 + 2$
The maximum number of zeros is the degree of the polynomial which is 4.
p must be a factor of 2: $p = \pm 1, \pm 2$
q must be a factor of 6: $q = \pm 1, \pm 2, \pm 3, \pm 6$

The possible rational zeros are: $\dfrac{p}{q} = \pm 1, \pm \dfrac{1}{2}, \pm \dfrac{1}{3}, \pm \dfrac{1}{6}, \pm 2, \pm \dfrac{2}{3}$

23. $f(x) = 2x^3 + x^2 - 1 = 2\left(x^3 + \frac{1}{2}x^2 - \frac{1}{2}\right)$

Note: The leading coefficient must be 1.

$a_2 = \frac{1}{2}, a_1 = 0, a_0 = -\frac{1}{2}$

$Max\left\{1, \left|-\frac{1}{2}\right| + |0| + \left|\frac{1}{2}\right|\right\} = Max\left\{1, \frac{1}{2} + 0 + \frac{1}{2}\right\} = Max\{1,1\} = 1$

$1 + Max\left\{\left|-\frac{1}{2}\right|, |0|, \left|\frac{1}{2}\right|\right\} = 1 + Max\left\{\frac{1}{2}, 0, \frac{1}{2}\right\} = 1 + \frac{1}{2} = 1.5$

The smaller of the two numbers is 1. Thus, every zero of f lies between -1 and 1.
Graphing using the bounds and ZOOM-FIT:

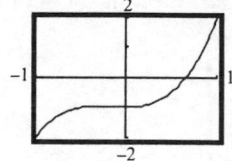

25. $f(x) = x^3 - 5x^2 - 11x + 11$
$a_2 = -5, a_1 = -11, a_0 = 11$

$Max\{1, |-5| + |-11| + |11|\} = Max\{1, 5 + 11 + 11\} = Max\{1, 27\} = 27$

$1 + Max\{|-5|, |-11|, |11|\} = 1 + Max\{1, 5, 11, 11\} = 1 + 11 = 12$

The smaller of the two numbers is 12. Thus, every zero of f lies between -12 and 12.
Graphing using the bounds and ZOOM-FIT: (Second graph has a better window.)

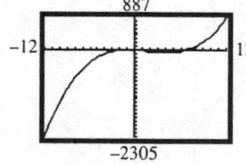   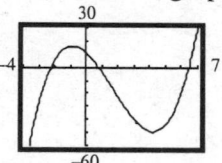

27. $f(x) = x^4 + 3x^3 - 5x^2 + 9$
$a_3 = 3, a_2 = -5, a_1 = 0, a_0 = 9$

$Max\{1, |9| + |0| + |-5| + |3|\} = Max\{1, 9 + 0 + 5 + 3\} = Max\{1, 17\} = 17$

$1 + Max\{|9|, |0|, |-5|, |3|\} = 1 + Max\{9, 0, 5, 3\} = 1 + 9 = 10$

The smaller of the two numbers is 10. Thus, every zero of f lies between -10 and 10.
Graphing using the bounds and ZOOM-FIT: (Second graph has a better window.)

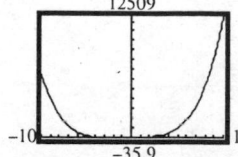   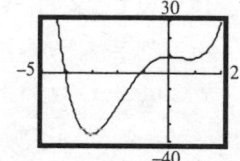

Chapter 3   Polynomial and Rational Functions

29. $f(x) = x^3 + 2x^2 - 5x - 6$

Step 1: $f(x)$ has at most 3 real zeros.

Step 2:   Possible rational zeros:

$$p = \pm 1, \pm 2, \pm 3, \pm 6; \quad q = \pm 1; \quad \frac{p}{q} = \pm 1, \pm 2, \pm 3, \pm 6$$

Step 3:   Using the Bounds on Zeros Theorem:

$$a_2 = 2, \quad a_1 = -5, \quad a_0 = -6$$

$$\text{Max}\left\{1, |-6| + |-5| + |2|\right\} = \text{Max}\left\{1, 13\right\} = 13$$

$$1 + \text{Max}\left\{|-6|, |-5|, |2|\right\} = 1 + 6 = 7$$

The smaller of the two numbers is 7. Thus, every zero of f lies between –7 and 7.

Graphing using the bounds and ZOOM-FIT: (Second graph has a better window.)

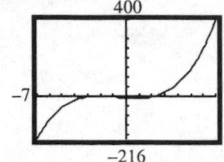 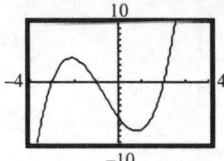

Step 4:   (a)   From the graph it appears that there are x-intercepts at –3, –1, and 2.

(b)   Using synthetic division:

$$-3\overline{)\begin{array}{cccc} 1 & 2 & -5 & -6 \\ & -3 & 3 & 6 \\ \hline 1 & -1 & -2 & 0 \end{array}}$$

Since the remainder is 0, $x - (-3) = x + 3$ is a factor. The other factor is the quotient: $x^2 - x - 2$.

(c)   Thus, $f(x) = (x + 3)\left(x^2 - x - 2\right) = (x + 3)(x + 1)(x - 2)$.

The zeros are –3, –1, and 2.

31. $f(x) = 2x^3 - 13x^2 + 24x - 9$

Step 1: $f(x)$ has at most 3 real zeros.

Step 2:   Possible rational zeros:

$$p = \pm 1, \pm 3, \pm 9; \quad q = \pm 1, \pm 2; \quad \frac{p}{q} = \pm 1, \pm 3, \pm 9, \pm \frac{1}{2}, \pm \frac{3}{2}, \pm \frac{9}{2}$$

Step 3:   Using the Bounds on Zeros Theorem:

$$f(x) = 2\left(x^3 - 6.5x^2 + 12x - 4.5\right)$$

$$a_2 = -6.5, \quad a_1 = 12, \quad a_0 = -4.5$$

$$\text{Max}\left\{1, |-4.5| + |12| + |-6.5|\right\} = \text{Max}\left\{1, 23\right\} = 23$$

$$1 + \text{Max}\left\{|-4.5|, |12|, |-6.5|\right\} = 1 + 12 = 13$$

The smaller of the two numbers is 13. Thus, every zero of f lies between –13 and 13.

Graphing using the bounds and ZOOM-FIT: (Second graph has a better window.)

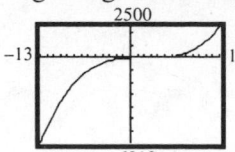 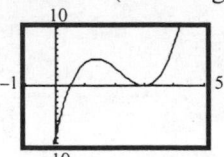

Step 4: (a)   From the graph it appears that there are x-intercepts at 0.5 and 3.

(b)   Using synthetic division:

$$3{\overline{\smash{\big)}\,2\quad -13\quad 24\quad -9}}$$
$$\quad\quad\ \ 6\quad -21\quad\ \ 9$$
$$\overline{\,2\quad\ -7\quad\ \ 3\quad\ \ 0}$$

Since the remainder is 0, $x-3$ is a factor. The other factor is the quotient: $2x^2 - 7x + 3$.

(c)   Thus, $f(x) = (x-3)\left(2x^2 - 7x + 3\right) = (x-3)(2x-1)(x-3)$.

The zeros are 0.5 and 3 (multiplicity 2).

33. $f(x) = 3x^3 + 4x^2 + 4x + 1$

Step 1: $f(x)$ has at most 3 real zeros..

Step 2:   Possible rational zeros:

$$p = \pm 1; \quad q = \pm 1, \pm 3; \quad \frac{p}{q} = \pm 1, \pm \frac{1}{3}$$

Step 3:   Using the Bounds on Zeros Theorem:

$$f(x) = 3\left(x^3 + \frac{4}{3}x^2 + \frac{4}{3}x + \frac{1}{3}\right)$$

$$a_2 = \frac{4}{3}, \quad a_1 = \frac{4}{3}, \quad a_0 = \frac{1}{3}$$

$$\text{Max}\left\{1, \left|\frac{1}{3}\right| + \left|\frac{4}{3}\right| + \left|\frac{4}{3}\right|\right\} = \text{Max}\{1, 3\} = 3$$

$$1 + \text{Max}\left\{\left|\frac{1}{3}\right|, \left|\frac{4}{3}\right|, \left|\frac{4}{3}\right|\right\} = 1 + \frac{4}{3} = \frac{7}{3}$$

The smaller of the two numbers is $\frac{7}{3}$. Thus, every zero of f lies

between $-\frac{7}{3}$ and $\frac{7}{3}$.

Graphing using the bounds and ZOOM-FIT: (Second graph has a better window.)

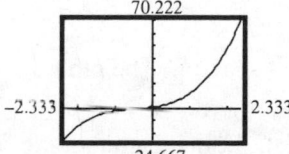 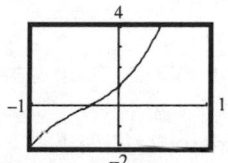

Step 4: (a)   From the graph it appears that there is an x-intercepts at $-\frac{1}{3}$.

(b)   Using synthetic division:

$$-\tfrac{1}{3}\overline{)3 \quad 4 \quad 4 \quad 1}$$

$$\begin{array}{rrrr} & -1 & -1 & -1 \\ \hline 3 & 3 & 3 & 0 \end{array}$$

Since the remainder is 0, $x-\left(-\dfrac{1}{3}\right)=x+\dfrac{1}{3}$ is a factor. The other factor is

the quotient: $3x^2+3x+3$.

(c)   Thus, $f(x)=\left(x+\dfrac{1}{3}\right)\left(3x^2+3x+3\right)=3\left(x+\dfrac{1}{3}\right)\left(x^2+x+1\right).$

$x^2+x+1=0$ has no real solution.

The zero is $-\dfrac{1}{3}$.

35. $f(x)=x^3-8x^2+17x-6$

Step 1: $f(x)$ has at most 3 real zeros.

Step 2:   Possible rational zeros:

$$p=\pm1,\,\pm2,\,\pm3,\,\pm6; \quad q=\pm1; \quad \dfrac{p}{q}=\pm1,\,\pm2,\,\pm3,\,\pm6$$

Step 3:   Using the Bounds on Zeros Theorem:

$a_2=-8,\quad a_1=17,\quad a_0=-6$

$\text{Max}\left\{1,\,|-6|+|17|+|-8|\right\}=\text{Max}\left\{1,\,31\right\}=31$

$1+\text{Max}\left\{|-6|,\,|17|,\,|-8|\right\}=1+17=18$

The smaller of the two numbers is 18. Thus, every zero of f lies
between -18 and 18.

Graphing using the bounds and ZOOM-FIT: (Second graph has a better window.)

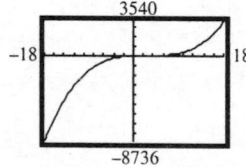
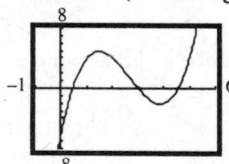

Step 4:   (a)   From the graph it appears that there are x-intercepts at 0.5, 3,
and 4.5.

(b)   Using synthetic division:

$$3\overline{)1 \quad -8 \quad 17 \quad -6}$$

$$\begin{array}{rrrr} & 3 & -15 & 6 \\ \hline 1 & -5 & 2 & 0 \end{array}$$

Since the remainder is 0, $x-3$ is a factor. The other factor is the quotient:
x^2-5x+2.

(c)   Thus, $f(x)=(x-3)\left(x^2-5x+2\right)$. Using the quadratic formula to find the
solutions of the depressed equation $x^2-5x+2=0$:

$$x=\dfrac{-(-5)\pm\sqrt{(-5)^2-4(1)(2)}}{2(1)}=\dfrac{5\pm\sqrt{17}}{2}$$

Thus, $f(x) = (x-3)\left(x - \left(\dfrac{5+\sqrt{17}}{2}\right)\right)\left(x - \left(\dfrac{5-\sqrt{17}}{2}\right)\right)$.

The zeros are 3, $\dfrac{5+\sqrt{17}}{2}$, and $\dfrac{5-\sqrt{17}}{2}$ or 3, 4.56, and 0.44.

37. $f(x) = x^4 + x^3 - 3x^2 - x + 2$
Step 1: $f(x)$ has at most 4 real zeros.
Step 2:   Possible rational zeros:

$$p = \pm 1, \pm 2; \quad q = \pm 1; \quad \frac{p}{q} = \pm 1, \pm 2$$

Step 3:   Using the Bounds on Zeros Theorem:
$$a_3 = 1, \quad a_2 = -3, \quad a_1 = -1, \quad a_0 = 2$$
$$\text{Max}\left\{1, |2| + |-1| + |-3| + |1|\right\} = \text{Max}\left\{1, 7\right\} = 7$$
$$1 + \text{Max}\left\{|2|, |-1|, |-3|, |1|\right\} = 1 + 3 = 4$$

The smaller of the two numbers is 4. Thus, every zero of f lies between -4 and 4.

Graphing using the bounds and ZOOM-FIT: (Second graph has a better window.)

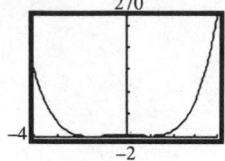   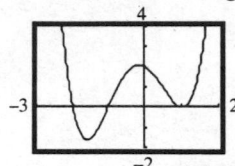

Step 4:   (a)   From the graph it appears that there are x-intercepts at -2, -1, and 1.
　　　　　 (b)   Using synthetic division:

$$
\begin{array}{r|rrrrr}
-2 & 1 & 1 & -3 & -1 & 2 \\
   &   & -2 & 2 & 2 & -2 \\
\hline
   & 1 & -1 & -1 & 1 & 0
\end{array}
\qquad
\begin{array}{r|rrrr}
-1 & 1 & -1 & -1 & 1 \\
   &   & -1 & 2 & -1 \\
\hline
   & 1 & -2 & 1 & 0
\end{array}
$$

Since the remainder is 0, $x + 2$ and $x + 1$ are factors. The other factor is the quotient: $x^2 - 2x + 1$.
　　　　　 (c)   Thus, $f(x) = (x+2)(x+1)(x-1)^2$.
The zeros are -2, -1, and 1 (multiplicity 2).

39. $f(x) = 2x^4 + 17x^3 + 35x^2 - 9x - 45$
Step 1: $f(x)$ has at most 4 real zeros.
Step 2:   Possible rational zeros:

$$p = \pm 1, \pm 3, \pm 5, \pm 9, \pm 15, \pm 45; \quad q = \pm 1, \pm 2;$$
$$\frac{p}{q} = \pm 1, \pm 3, \pm 5, \pm 9, \pm 15, \pm 45, \pm \frac{1}{2}, \pm \frac{3}{2}, \pm \frac{5}{2}, \pm \frac{9}{2}, \pm \frac{15}{2}, \pm \frac{45}{2}$$

Step 3:   Using the Bounds on Zeros Theorem:
$$f(x) = 2\left(x^4 + 8.5x^3 + 17.5x^2 - 4.5x - 22.5\right)$$
$$a_3 = 8.5, \quad a_2 = 17.5, \quad a_1 = -4.5, \quad a_0 = -22.5$$
$$\text{Max}\left\{1, |-22.5| + |-4.5| + |17.5| + |8.5|\right\} = \text{Max}\left\{1, 53\right\} = 53$$
$$1 + \text{Max}\left\{|-22.5|, |-4.5|, |17.5|, |8.5|\right\} = 1 + 22.5 = 23.5$$

The smaller of the two numbers is 23.5. Thus, every zero of f lies between -23.5 and 23.5.

Graphing using the bounds and ZOOM-FIT: (Second graph has a better window.)

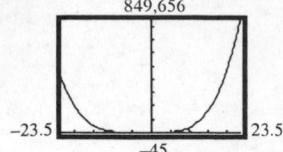 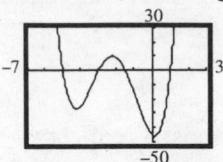

Step 4: (a) From the graph it appears that there are x-intercepts at $-5, -3, -1.5$, and 1.

(b) Using synthetic division:

$$-5\overline{)2\quad 17\quad 35\quad -9\quad -45}$$
$$\underline{-10\quad -35\quad 0\quad 45}$$
$$2\quad 7\quad 0\quad -9\quad 0$$

$$-3\overline{)2\quad 7\quad 0\quad -9}$$
$$\underline{-6\quad -3\quad 9}$$
$$2\quad 1\quad -3\quad 0$$

Since the remainder is 0, $x + 5$ and $x + 3$ are factors. The other factor is the quotient: $2x^2 + x - 3$.

(c) Thus, $f(x) = (x + 5)(x + 3)(2x + 3)(x - 1)$.
The zeros are $-5, -3, -1.5$ and 1.

41. $f(x) = 2x^4 - 3x^3 - 21x^2 - 2x + 24$

Step 1: $f(x)$ has at most 4 real zeros.

Step 2: Possible rational zeros:
$$p = \pm 1, \pm 2, \pm 3, \pm 4, \pm 6, \pm 8, \pm 12, \pm 24; \quad q = \pm 1, \pm 2;$$
$$\frac{p}{q} = \pm 1, \pm 2, \pm 3, \pm 4, \pm 6, \pm 8, \pm 12, \pm 24, \pm \frac{1}{2}, \pm \frac{3}{2}$$

Step 3: Using the Bounds on Zeros Theorem:
$$f(x) = 2\left(x^4 - 1.5x^3 - 10.5x^2 - x + 12\right)$$
$$a_3 = -1.5, \quad a_2 = -10.5, \quad a_1 = -1, \quad a_0 = 12$$
$$\text{Max}\left\{1, |12| + |-1| + |-10.5| + |-1.5|\right\} = \text{Max}\left\{1, 25\right\} = 25$$
$$1 + \text{Max}\left\{|12|, |-1|, |-10.5|, |-1.5|\right\} = 1 + 12 = 13$$

The smaller of the two numbers is 13. Thus, every zero of f lies between -13 and 13.

Graphing using the bounds and ZOOM-FIT: (Second graph has a better window.)

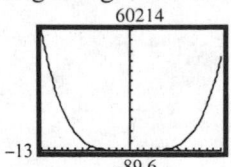 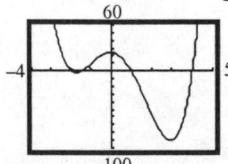

Step 4: (a) From the graph it appears that there are x-intercepts at $-2, -1.5, 1$, and 4.

(b)   Using synthetic division:

$$-2\overline{)\,2 \quad -3 \quad -21 \quad -2 \quad 24\,}$$
$$ -4 \quad 14 \quad 14 \quad -24$$
$$2 \quad -7 \quad -7 \quad 12 \quad 0$$

$$4\overline{)\,2 \quad -7 \quad -7 \quad 12\,}$$
$$ 8 \quad 4 \quad -12$$
$$2 \quad 1 \quad -3 \quad 0$$

Since the remainder is 0, $x + 2$ and $x - 4$ are factors. The other factor is the quotient: $2x^2 + x - 3$.

(c)   Thus, $f(x) = (x + 2)(2x + 3)(x - 1)(x - 4)$.
The zeros are $-2, -1.5, 1$ and 4.

43. $f(x) = 4x^4 + 7x^2 - 2$
Step 1: $f(x)$ has at most 4 real zeros.
Step 2:   Possible rational zeros:

$$p = \pm 1, \pm 2; \quad q = \pm 1, \pm 2, \pm 4; \quad \frac{p}{q} = \pm 1, \pm 2, \pm \frac{1}{2}, \pm \frac{1}{4}$$

Step 3:   Using the Bounds on Zeros Theorem:

$$f(x) = 4\left(x^4 + 1.75x^2 - 0.5\right)$$

$$a_3 = 0, \quad a_2 = 1.75, \quad a_1 = 0, \quad a_0 = -0.5$$

$$\text{Max } \left\{1, |-0.5| + |0| + |1.75| + |0|\right\} = \text{Max } \left\{1, 2.25\right\} = 2.25$$

$$1 + \text{Max } \left\{|-0.5|, |0|, |1.75|, |0|\right\} = 1 + 1.75 = 2.75$$

The smaller of the two numbers is 2.25. Thus, every zero of f lies between -2.25 and 2.25.
Graphing using the bounds and ZOOM-FIT: (Second graph has a better window.)

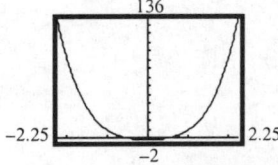   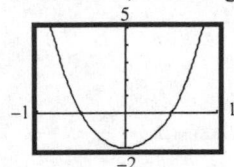

Step 4:   (a)   From the graph it appears that there are x-intercepts at -0.5, and 0.5.
            (b)   Using synthetic division:

$$-0.5\overline{)\,4 \quad 0 \quad 7 \quad 0 \quad -2\,}$$
$$ -2 \quad 1 \quad -4 \quad 2$$
$$4 \quad -2 \quad 8 \quad -4 \quad 0$$

$$0.5\overline{)\,4 \quad -2 \quad 8 \quad -4\,}$$
$$ 2 \quad 0 \quad 4$$
$$4 \quad 0 \quad 8 \quad 0$$

Since the remainder is 0, $x + 0.5$ and $x - 0.5$ are factors. The other factor is the quotient: $4x^2 + 8$.

(c)   Thus, $f(x) = 4(x + 0.5)(x - 0.5)\left(x^2 + 2\right) = (2x + 1)(2x - 1)\left(x^2 + 2\right)$.
The depressed equation has no real zeros.
The zeros are -0.5, and 0.5.

45. $f(x) = 4x^5 - 8x^4 - x + 2$
Step 1: $f(x)$ has at most 5 real zeros.
Step 2:   Possible rational zeros:

$$p = \pm 1, \pm 2; \quad q = \pm 1, \pm 2, \pm 4; \quad \frac{p}{q} = \pm 1, \pm 2, \pm \frac{1}{2}, \pm \frac{1}{4}$$

Step 3:   Using the Bounds on Zeros Theorem:

$$f(x) = 4\left(x^5 - 2x^4 - 0.25x + 0.5\right)$$

$$a_4 = -2, \ a_3 = 0, \ a_2 = 0, \ a_1 = -0.25, \ a_0 = 0.5$$

$$\text{Max}\left\{1, |0.5| + |-0.25| + |0| + |0| + |-2|\right\} = \text{Max}\left\{1, 2.75\right\} = 2.75$$

$$1 + \text{Max}\left\{|0.5|, |-0.25|, |0|, |0|, |-2|\right\} = 1 + 2 = 3$$

The smaller of the two numbers is 2.75. Thus, every zero of f lies between -2.75 and 2.75.

Graphing using the bounds and ZOOM-FIT: (Second graph has a better window.)

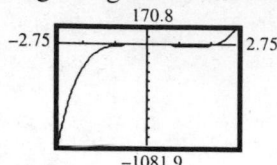 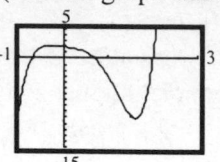

Step 4:   (a)   From the graph it appears that there are x-intercepts at -0.7, 0.7 and 2.

(b)   Using synthetic division:

$$2\overline{)\begin{array}{cccccc} 4 & -8 & 0 & 0 & -1 & 2 \\ & 8 & 0 & 0 & 0 & -2 \\ \hline 4 & 0 & 0 & 0 & -1 & 0 \end{array}}$$

Since the remainder is 0, $x - 2$ is a factor. The other factor is the quotient: $4x^4 - 1$.

(c)   Factoring,

$$f(x) = (x - 2)\left(4x^4 - 1\right) = (x - 2)(2x^2 - 1)(2x^2 + 1)$$

$$= (x - 2)\left(\sqrt{2}x - 1\right)\left(\sqrt{2}x + 1\right)\left(2x^2 + 1\right)$$

The zeros are $-\dfrac{\sqrt{2}}{2}, \dfrac{\sqrt{2}}{2}$, and 2 or $-0.71, -0.71$, and 2.

47. $f(x) = x^3 + 3.2x^2 - 16.83x - 5.31$

$f(x)$ has at most 3 real zeros. Solving by graphing (using ZERO):

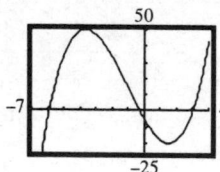

The zeros are approximately -5.9, -0.3, and 3.

49. $f(x) = x^4 - 1.4x^3 - 33.71x^2 + 23.94x + 292.41$

$f(x)$ has at most 4 real zeros. Solving by graphing (using ZERO):

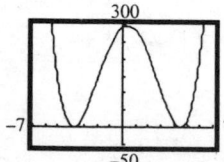

The zeros are approximately -3.80 and 4.50. These zeros are each of multiplicity 2.

51. $f(x) = x^3 + 19.5x^2 - 1021x + 1000.5$
       $f(x)$ has at most 3 real zeros. Solving by graphing (using ZERO):

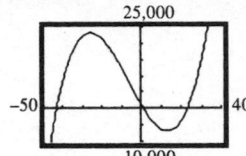

The zeros are approximately –43.5, 1, and 23.

53. $x^4 - x^3 + 2x^2 - 4x - 8 = 0$
       The solutions of the equation are the zeros of $f(x) = x^4 - x^3 + 2x^2 - 4x - 8$.
       Step 1: $f(x)$ has at most 4 real zeros.
       Step 2:   Possible rational zeros:

$$p = \pm 1, \pm 2, \pm 4, \pm 8; \quad q = \pm 1; \quad \frac{p}{q} = \pm 1, \pm 2, \pm 4, \pm 8$$

       Step 3:   Using the Bounds on Zeros Theorem:
$$a_3 = -1, \quad a_2 = 2, \quad a_1 = -4, \quad a_0 = -8$$
$$\text{Max}\left\{1, |-8| + |-4| + |2| + |-1|\right\} = \text{Max}\left\{1, 15\right\} = 15$$
$$1 + \text{Max}\left\{|-8|, |-4|, |2|, |-1|\right\} = 1 + 8 = 9$$

The smaller of the two numbers is 9. Thus, every zero of f lies
between –9 and 9.
Graphing using the bounds and ZOOM-FIT: (Second graph has a better window.)

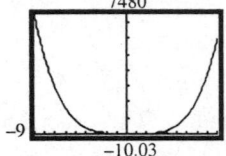   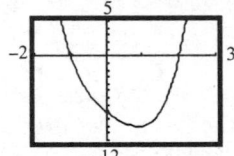

       Step 4:   (a)   From the graph it appears that there are x-intercepts at –1 and 2.
                  (b)   Using synthetic division:

$$\begin{array}{r|rrrr} -1) & 1 & -1 & 2 & -4 & -8 \\ & & -1 & 2 & -4 & 8 \\ \hline & 1 & -2 & 4 & -8 & 0 \end{array} \qquad \begin{array}{r|rrrr} 2) & 1 & -2 & 4 & -8 \\ & & & 2 & 0 & 8 \\ \hline & 1 & 0 & 4 & 0 \end{array}$$

Since the remainder is 0, $x + 1$ and $x - 2$ are factors. The other factor is the
quotient: $x^2 + 4$.
                  (c)   The zeros are –1 and 2. ($x^2 + 4 = 0$ has no real solutions.)

55. $3x^3 + 4x^2 - 7x + 2 = 0$
       The solutions of the equation are the zeros of $f(x) = 3x^3 + 4x^2 - 7x + 2$.
       Step 1: $f(x)$ has at most 3 real zeros.
       Step 2:   Possible rational zeros:

$$p = \pm 1, \pm 2; \quad q = \pm 1, \pm 3; \quad \frac{p}{q} = \pm 1, \pm 2, \pm \frac{1}{3}, \pm \frac{2}{3}$$

Step 3:   Using the Bounds on Zeros Theorem:

$$f(x) = 3\left(x^3 + \frac{4}{3}x^2 - \frac{7}{3}x + \frac{2}{3}\right)$$

$$a_2 = \frac{4}{3}, \quad a_1 = -\frac{7}{3}, \quad a_0 = \frac{2}{3}$$

$$\text{Max}\left\{1, \left|\frac{2}{3}\right| + \left|-\frac{7}{3}\right| + \left|\frac{4}{3}\right|\right\} = \text{Max}\left\{1, \frac{13}{3}\right\} = \frac{13}{3} \approx 4.333$$

$$1 + \text{Max}\left\{\left|\frac{2}{3}\right|, \left|-\frac{7}{3}\right|, \left|\frac{4}{3}\right|\right\} = 1 + \frac{7}{3} = \frac{10}{3} \approx 3.333$$

The smaller of the two numbers is 3.33. Thus, every zero of f lies between -3.33 and 3.33.

Graphing using the bounds and ZOOM FIT: (Second graph has a better window.)

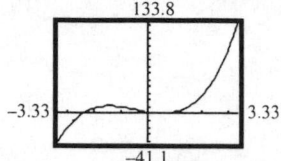 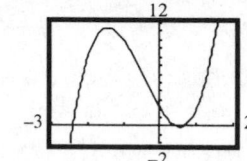

Step 4:   (a)   From the graph it appears that there are x-intercepts at $\frac{1}{3}, \frac{2}{3}$, and -2.4.

(b)   Using synthetic division:

$$\frac{2}{3}\overline{\smash{\big)}3 \quad 4 \quad -7 \quad 2}$$
$$\phantom{\frac{2}{3}\big)}\underline{2 \quad 4 \quad -2}$$
$$\phantom{\frac{2}{3}\big)}3 \quad 6 \quad -3 \quad 0$$

Since the remainder is 0, $x - \frac{2}{3}$ is a factor. The other factor is the quotient: $3x^2 + 6x - 3$.

$$f(x) = \left(x - \frac{2}{3}\right)(3x^2 + 6x - 3) = 3\left(x - \frac{2}{3}\right)(x^2 + 2x - 1)$$

Using the quadratic formula to solve $x^2 + 2x - 1 = 0$:

$$x = \frac{-2 \pm \sqrt{4 - 4(1)(-1)}}{2(1)} = \frac{-2 \pm \sqrt{8}}{2} = \frac{-2 \pm 2\sqrt{2}}{2} = -1 \pm \sqrt{2}$$

(c)   The zeros are $\frac{2}{3}, -1 + \sqrt{2}$, and $-1 - \sqrt{2}$ or $0.67, 0.41$, and -2.41.

57. $3x^3 - x^2 - 15x + 5 = 0$

Solving by factoring: $x^2(3x - 1) - 5(3x - 1) = 0 \Rightarrow (3x - 1)(x^2 - 5) = 0$

$$(3x - 1)(x - \sqrt{5})(x + \sqrt{5}) = 0$$

The solutions of the equation are $\frac{1}{3}, \sqrt{5}$, and $-\sqrt{5}$ or $0.33, 2.24$, and -2.24.

59. $x^4 + 4x^3 + 2x^2 - x + 6 = 0$

The solutions of the equation are the zeros of $f(x) = x^4 + 4x^3 + 2x^2 - x + 6$.

Step 1: $f(x)$ has at most 4 real zeros.

Step 2:   Possible rational zeros:

$$p = \pm 1, \pm 2, \pm 3, \pm 6; \quad q = \pm 1; \quad \frac{p}{q} = \pm 1, \pm 2, \pm 3, \pm 6$$

Step 3:   Using the Bounds on Zeros Theorem:

$$a_3 = 4, \quad a_2 = 2, \quad a_1 = -1, \quad a_0 = 6$$

$$\text{Max} \left\{ 1, |6| + |-1| + |2| + |4| \right\} = \text{Max} \left\{ 1, 13 \right\} = 13$$

$$1 + \text{Max} \left\{ |6|, |-1|, |2|, |4| \right\} = 1 + 6 = 7$$

The smaller of the two numbers is 7. Thus, every zero of f lies between -7 and 7.

Graphing using the bounds and ZOOM-FIT: (Second graph has a better window.)

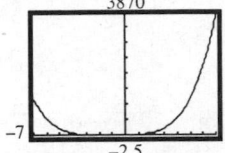   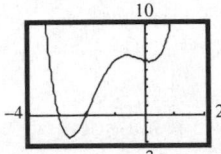

Step 4:   (a)   From the graph it appears that there are x-intercepts at -3 and -2.

(b)   Using synthetic division:

$$\begin{array}{r|rrrrr} -3) & 1 & 4 & 2 & -1 & 6 \\ & & -3 & -3 & 3 & -6 \\ \hline & 1 & 1 & -1 & 2 & 0 \end{array} \qquad \begin{array}{r|rrrr} -2) & 1 & 1 & -1 & 2 \\ & & -2 & 2 & -2 \\ \hline & 1 & -1 & 1 & 0 \end{array}$$

Since the remainder is 0, $x + 3$ and $x + 2$ are factors. The other factor is the quotient: $x^2 - x + 1$.

(c)   The zeros are -3 and -2. ($x^2 - x + 1 = 0$ has no real solutions.)

61. $x^3 - \frac{2}{3}x^2 + \frac{8}{3}x + 1 = 0$

The solutions of the equation are the zeros of $f(x) = x^3 - \frac{2}{3}x^2 + \frac{8}{3}x + 1 = 0$.

Step 1: $f(x)$ has at most 3 real zeros.

Step 2:   Use the equivalent equation $3x^3 - 2x^2 + 8x + 3 = 0$ to find the possible rational zeros:

$$p = \pm 1, \pm 3; \quad q = \pm 1, \pm 3; \quad \frac{p}{q} = \pm 1, \pm 3, \pm \frac{1}{3}$$

Step 3:   Using the Bounds on Zeros Theorem:

$$a_2 = -\frac{2}{3}, \quad a_1 = \frac{8}{3}, \quad a_0 = 1$$

$$\text{Max} \left\{ 1, |1| + \left| \frac{8}{3} \right| + \left| \frac{2}{3} \right| \right\} = \text{Max} \left\{ 1, \frac{13}{3} \right\} = \frac{13}{3} \approx 4.333$$

$$1 + \text{Max} \left\{ |1|, \left| \frac{8}{3} \right|, \left| -\frac{2}{3} \right| \right\} = 1 + \frac{8}{3} = \frac{11}{3} \approx 3.667$$

The smaller of the two numbers is 3.67. Thus, every zero of f lies between -3.67 and 3.67.

Graphing using the bounds and ZOOM-FIT: (Second graph has a better window.)

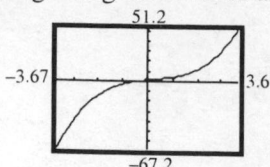 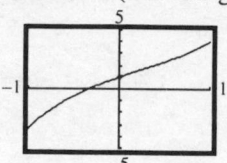

Step 4:   (a)   From the graph it appears that there is an x-intercepts at $-\dfrac{1}{3}$.

(b)   Using synthetic division:

$$-\dfrac{1}{2}\bigg)\begin{array}{cccc} 1 & -\dfrac{2}{3} & \dfrac{8}{2} & 1 \\[2mm] & -\dfrac{1}{3} & \dfrac{1}{3} & -1 \\ \hline 1 & -1 & 3 & 0 \end{array}$$

Since the remainder is 0, $x+\dfrac{1}{3}$ is a factor. The other factor is the quotient: $x^2 - x + 3$.

(c)   The real zero is $-\dfrac{1}{3}$. ($x^2 - x + 3 = 0$ has no real solutions.)

63.   Using the TABLE feature to show that there is a zero in the interval:

$f(x) = 8x^4 - 2x^2 + 5x - 1;\quad [0,1]$

X	Y1
-1	0
0	-1
1	10
2	129
3	644
4	2035
5	4974

X=-1

$f(0) = -1 < 0$ and $f(1) = 10 > 0$
Since one is positive and one is negative, there is a zero in the interval.

Using the TABLE feature to approximate the zero to two decimal places:

X	Y1
.213	-.0093
.214	-.0048
.215	-4E-4
.216	.0041
.217	.00856
.218	.01302
.219	.01748

X=.219

The zero is approximately 0.22.

65.   Using the TABLE feature to show that there is a zero in the interval:

$f(x) = 2x^3 + 6x^2 - 8x + 2;\quad [-5, -4]$

X	Y1
-8	-574
-7	-334
-6	-166
-5	-58
-4	2
-3	26
-2	26

Y1=2X^3+6X²−8X+2

$f(-5) = -58 < 0$ and $f(-4) = 2 > 0$
Since one is positive and one is negative, there is a zero in the interval.

Using the TABLE feature to approximate the zero to two decimal places:

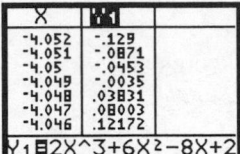 The zero is approximately –4.05.

67.   Using the TABLE feature to show that there is a zero in the interval:

$$f(x) = x^5 - x^4 + 7x^3 - 7x^2 - 18x + 18; \quad [1.4, 1.5]$$

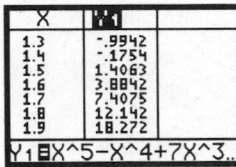 $f(1.4) = -0.1754 < 0$ and $f(1.5) = 1.4063 > 0$
Since one is positive and one is negative,
there is a zero in the interval.

Using the TABLE feature to approximate the zero to two decimal places:

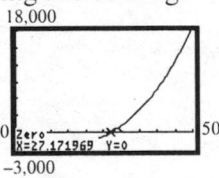   The zero is approximately 1.41.

69. $C(x) = 3000$

$$0.216x^3 - 2.347x^2 + 14.328x + 10.224 = 3000$$

$$0.216x^3 - 2.347x^2 + 14.328x - 2989.776 = 0$$

There are at most 3 solutions.

Graphing and solving:

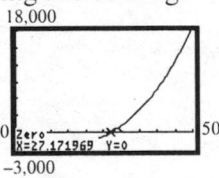

About 27 Cavaliers can be manufactured.

71. $x - 2$ is a factor of $f(x) = x^3 - kx^2 + kx + 2$ only if the remainder that results when $f(x)$ is divided by $x - 2$ is 0. Dividing, we have:

$$
\begin{array}{r|rrrr}
2 & 1 & -k & k & 2 \\
  &   & 2 & -2k+4 & -2k+8 \\
\hline
  & 1 & -k+2 & -k+4 & -2k+10 \\
\end{array}
$$

Since we want the remainder to equal 0, set the remainder equal to zero and solve:

$$-2k + 10 = 0 \Rightarrow -2k = -10 \Rightarrow k = 5$$

73.   By the Remainder Theorem we know that the remainder from synthetic division by c is equal to $f(c)$. Thus the easiest way to find the remainder is to evaluate:

$$f(1) = 2(1)^{20} - 8(1)^{10} + 1 - 2 = 2 - 8 + 1 - 2 = -7$$

The remainder is –7.

75. $f(x) = 2x^6 - 5x^4 + x^3 - x + 1$

By the Rational Zero Theorem, the only possible rational zeros are:

$$\frac{p}{q} = \pm 1, \pm \frac{1}{2}$$

Since $\frac{3}{5}$ is not in the list of possible rational zeros, it is not a zero of $f(x)$.

77. Let x be the length of a side of the original cube.
After removing the 1 inch slice, one dimension will be $x - 1$.
The volume of the new solid will be:

$$(x-1) \cdot x \cdot x = 294 \Rightarrow x^3 - x^2 = 294 \Rightarrow x^3 - x^2 - 294 = 0$$

The possible rational zeros are:

$p = \pm 1, \pm 2, \pm 3, \pm 6, \pm 7, \pm 14, \pm 21, \pm 42, \pm 49, \pm 98, \pm 147, \pm 294; \quad q = \pm 1$

The rational zeros are the same as the values for p.
Using synthetic division:

$$
\begin{array}{r|rrrr}
7 & 1 & -1 & 0 & -294 \\
  &   & 7 & 42 & 294 \\
\hline
  & 1 & 6 & 42 & 0 \\
\end{array}
$$

7 is a zero, so the length of the original edge of the cube was 7 inches.

Polynomial and Rational Functions

3.8 Complex Zeros; Fundamental Theorem of Algebra

1. Since complex zeros appear in conjugate pairs, $4+i$, the conjugate of $4-i$, is the remaining zero of f.

3. Since complex zeros appear in conjugate pairs, $-i$, the conjugate of i, and $1-i$, the conjugate of $1+i$, are the remaining zeros of f.

5. Since complex zeros appear in conjugate pairs, $-i$, the conjugate of i, and $-2i$, the conjugate of $2i$, are the remaining zeros of f.

7. Since complex zeros appear in conjugate pairs, $-i$, the conjugate of i, is the remaining zero of f.

9. Since complex zeros appear in conjugate pairs, $2-i$, the conjugate of $2+i$, and $-3+i$, the conjugate of $-3-i$, are the remaining zeros of f.

11. Since $3+2i$ is a zero, its conjugate $3-2i$ is also a zero of f. Finding the function:
$$f(x) = (x-4)(x-4)(x-(3+2i))(x-(3-2i)) = \left(x^2 - 8x + 16\right)((x-3)-2i)((x-3)+2i)$$
$$= \left(x^2 - 8x + 16\right)\left(x^2 - 6x + 9 - 4i^2\right) = \left(x^2 - 8x + 16\right)\left(x^2 - 6x + 13\right)$$
$$= x^4 - 6x^3 + 13x^2 - 8x^3 + 48x^2 - 104x + 16x^2 - 96x + 208$$
$$= x^4 - 14x^3 + 77x^2 - 200x + 208$$

13. Since $-i$ is a zero, its conjugate i is also a zero, and since $1+i$ is a zero, its conjugate $1-i$ is also a zero of f. Finding the function:
$$f(x) = (x-2)(x+i)(x-i)(x-(1+i))(x-(1-i)) = (x-2)\left(x^2 - i^2\right)((x-1)-i)((x-1)+i)$$
$$= (x-2)\left(x^2 + 1\right)\left(x^2 - 2x + 1 - i^2\right) = \left(x^3 - 2x^2 + x - 2\right)\left(x^2 - 2x + 2\right)$$
$$= x^5 - 2x^4 + 2x^3 - 2x^4 + 4x^3 - 4x^2 + x^3 - 2x^2 + 2x - 2x^2 + 4x - 4$$
$$= x^5 - 4x^4 + 7x^3 - 8x^2 + 6x - 4$$

15. Since $-i$ is a zero, its conjugate i is also a zero of f. Finding the function:
$$f(x) = (x-3)(x-3)(x+i)(x-i) = \left(x^2 - 6x + 9\right)\left(x^2 - i^2\right)$$
$$= \left(x^2 - 6x + 9\right)\left(x^2 + 1\right) = x^4 + x^2 - 6x^3 - 6x + 9x^2 + 9$$
$$= x^4 - 6x^3 + 10x^2 - 6x + 9$$

17.   Since $2i$ is a zero, its conjugate $-2i$ is also a zero of f. $x - 2i$ and $x + 2i$ are factors of f.
Thus, $(x - 2i)(x + 2i) = x^2 + 4$ is a factor of f. Using division to find the other factor:

$$
\begin{array}{r}
x - 4 \\
x^2 + 4 \overline{\smash{)}\, x^3 - 4x^2 + 4x - 16} \\
\underline{x^3 \qquad\quad + 4x} \\
-4x^2 \qquad\quad - 16 \\
\underline{-4x^2 \qquad\quad - 16}
\end{array}
$$

$x - 4$ is a factor and the remaining zero is 4. The zeros of f are $4, 2i, -2i$.

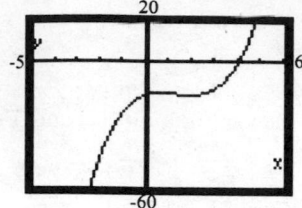

19.   Since $-2i$ is a zero, its conjugate $2i$ is also a zero of f. $x - 2i$ and $x + 2i$ are factors of f.
Thus, $(x - 2i)(x + 2i) = x^2 + 4$ is a factor of f. Using division to find the other factor:

$$
\begin{array}{r}
2x^2 + 5x - 3 \\
x^2 + 4 \overline{\smash{)}\, 2x^4 + 5x^3 + 5x^2 + 20x - 12} \\
\underline{2x^4 \qquad\quad + 8x^2} \\
5x^3 - 3x^2 + 20x \\
\underline{5x^3 \qquad\quad + 20x} \\
-3x^2 \qquad\quad - 12 \\
\underline{-3x^2 \qquad\quad - 12}
\end{array}
$$

$2x^2 + 5x - 3 = (2x - 1)(x + 3)$ are factors and the remaining zeros are $\dfrac{1}{2}$ and -3. The

zeros of f are $2i, -2i, -3, \dfrac{1}{2}$.

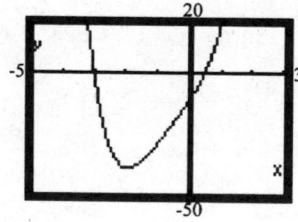

21.   Since $3 - 2i$ is a zero, its conjugate $3 + 2i$ is also a zero of h. $x - (3 - 2i)$ and $x - (3 + 2i)$
are factors of h. Thus,
$(x - (3 - 2i))(x - (3 + 2i)) = ((x - 3) + 2i)((x - 3) - 2i) = x^2 - 6x + 9 - 4i^2 = x^2 - 6x + 13$ is
a factor of h.

Using division to find the other factor:

$$
\begin{array}{r}
x^2 - 3x - 10 \\
x^2 - 6x + 13\overline{)x^4 - 9x^3 + 21x^2 + 21x - 130} \\
\underline{x^4 - 6x^3 + 13x^2} \\
-3x^3 + 8x^2 + 21x \\
\underline{-3x^3 + 18x^2 - 39x} \\
-10x^2 + 60x - 130 \\
\underline{-10x^2 + 60x - 130}
\end{array}
$$

$x^2 - 3x - 10 = (x + 2)(x - 5)$ are factors and the remaining zeros are –2 and 5. The zeros of h are $3 - 2i, 3 + 2i, -2, 5$.

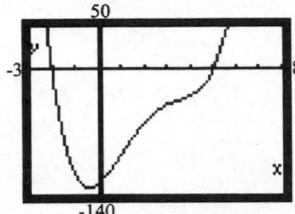

23. Since $-4i$ is a zero, its conjugate $4i$ is also a zero of h. $x - 4i$ and $x + 4i$ are factors of h.
Thus, $(x - 4i)(x + 4i) = x^2 + 16$ is a factor of h. Using division to find the other factor:

$$
\begin{array}{r}
3x^3 + 2x^2 - 33x - 22 \\
x^2 + 16\overline{)3x^5 + 2x^4 + 15x^3 + 10x^2 - 528x - 352} \\
\underline{3x^5 \qquad\quad + 48x^3} \\
2x^4 - 33x^3 + 10x^2 \\
\underline{2x^4 \qquad\quad + 32x^2} \\
-33x^3 - 22x^2 - 528x \\
\underline{-33x^3 \qquad\quad - 528x} \\
-22x^2 \qquad\quad - 352 \\
\underline{-22x^2 \qquad\quad - 352}
\end{array}
$$

$3x^3 + 2x^2 - 33x - 22 = x^2(3x + 2) - 11(3x + 2) = (3x + 2)(x^2 - 11)$

$= (3x + 2)\left(x - \sqrt{11}\right)\left(x + \sqrt{11}\right)$ are factors and the remaining zeros are $-\dfrac{2}{3}, \sqrt{11}$, and $-\sqrt{11}$.

The zeros of h are $4i, -4i, -\sqrt{11}, \sqrt{11}, -\dfrac{2}{3}$.

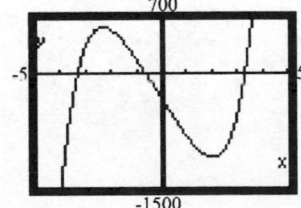

25. $f(x) = x^3 - 1 = (x - 1)(x^2 + x + 1)$ The zeros of $x^2 + x + 1 = 0$ are:

$$x = \frac{-1 \pm \sqrt{1^2 - 4(1)(1)}}{2(1)} = \frac{-1 \pm \sqrt{-3}}{2} = -\frac{1}{2} + \frac{\sqrt{3}}{2}i \text{ or } -\frac{1}{2} - \frac{\sqrt{3}}{2}i$$

The zeros are: $1, -\frac{1}{2} + \frac{\sqrt{3}}{2}i, -\frac{1}{2} - \frac{\sqrt{3}}{2}i$.

Evaluating f at each zero:

27. $f(x) = x^3 - 8x^2 + 25x - 26$

Step 1: $f(x)$ has 3 complex zeros.

Step 2: Possible rational zeros:

$$p = \pm 1, \pm 2, \pm 13, \pm 26; \quad q = \pm 1; \quad \frac{p}{q} = \pm 1, \pm 2, \pm 13, \pm 26$$

Step 3: Using synthetic division:

$$
\begin{array}{r|rrrr}
2 & 1 & -8 & 25 & -26 \\
  &   & 2 & -12 & 26 \\
\hline
  & 1 & -6 & 13 & 0 \\
\end{array}
$$

Since the remainder is 0, $x - 2$ is a factor. The other factor is the quotient: $x^2 - 6x + 13$.

Using the quadratic formula to find the zeros of $x^2 - 6x + 13 = 0$:

$$x = \frac{-(-6) \pm \sqrt{(-6)^2 - 4(1)(13)}}{2(1)} = \frac{6 \pm \sqrt{-16}}{2} = \frac{6 \pm 4i}{2} = 3 \pm 2i.$$

The complex zeros are $2, 3 - 2i, 3 + 2i$.

Evaluating f at each zero:

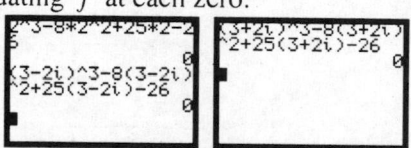

29. $f(x) = x^4 + 5x^2 + 4 = (x^2 + 4)(x^2 + 1) = (x + 2i)(x - 2i)(x + i)(x - i)$

The zeros are: $-2i, -i, i, 2i$.

Evaluating f at each zero:

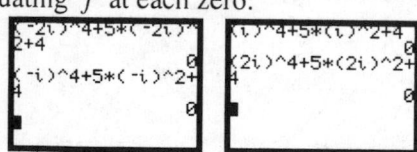

31. $f(x) = x^4 + 2x^3 + 22x^2 + 50x - 75$

Step 1: $f(x)$ has 4 complex zeros.

Step 2:   Possible rational zeros:

$$p = \pm 1, \pm 3, \pm 5, \pm 15, \pm 25, \pm 75; \quad q = \pm 1;$$

$$\frac{p}{q} = \pm 1, \pm 3, \pm 5, \pm 15, \pm 25, \pm 75$$

Step 3:   Using synthetic division:

$$
\begin{array}{r|rrrrr}
-3 & 1 & 2 & 22 & 50 & -75 \\
   &   & -3 & 3 & -75 & 75 \\
\hline
   & 1 & -1 & 25 & -25 & 0
\end{array}
$$

Since the remainder is 0, $x + 3$ is a factor. The other factor is the quotient:

$$x^3 - x^2 + 25x - 25 = x^2(x - 1) + 25(x - 1) = (x - 1)(x^2 + 25)$$

$$= (x - 1)(x + 5i)(x - 5i)$$

The complex zeros are $-3,\ 1,\ -5i,\ 5i$.

Evaluating f at each zero:

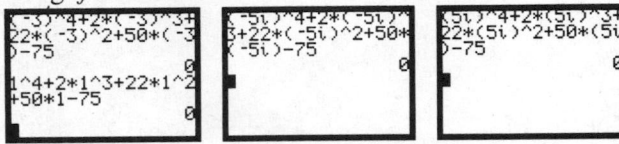

33. $f(x) = 3x^4 - x^3 - 9x^2 + 159x - 52$

Step 1: $f(x)$ has 4 complex zeros.

Step 2:   Possible rational zeros:

$$p = \pm 1, \pm 2, \pm 4, \pm 13, \pm 26, \pm 52; \quad q = \pm 1, \pm 3;$$

$$\frac{p}{q} = \pm 1, \pm 2, \pm 4, \pm 13, \pm 26, \pm 52, \pm \frac{1}{3}, \pm \frac{2}{3}, \pm \frac{4}{3}, \pm \frac{13}{3}, \pm \frac{26}{3}, \pm \frac{52}{3}$$

Step 3:   Using synthetic division:

$$
\begin{array}{r|rrrrr}
-4 & 3 & -1 & -9 & 159 & -52 \\
   &   & -12 & 52 & -172 & 52 \\
\hline
   & 3 & -13 & 43 & -13 & 0
\end{array}
\qquad
\begin{array}{r|rrrr}
\frac{1}{3} & 3 & -13 & 43 & -13 \\
   &   & 1 & -4 & 13 \\
\hline
   & 3 & -12 & 39 & 0
\end{array}
$$

Since the remainder is 0, $x + 4$ and $x - \dfrac{1}{3}$ are factors. The other factor is the

quotient: $3x^2 - 12x + 39 = 3(x^2 - 4x + 13)$.

Using the quadratic formula to find the zeros of $x^2 - 4x + 13 = 0$:

$$x = \frac{-(-4) \pm \sqrt{(-4)^2 - 4(1)(13)}}{2(1)} = \frac{4 \pm \sqrt{-36}}{2} = \frac{4 \pm 6i}{2} = 2 \pm 3i.$$

The complex zeros are $-4,\ \dfrac{1}{3},\ 2 - 3i,\ 2 + 3i$.

Evaluating f at each zero:

35. If the coefficients are real numbers and $2 + i$ is a zero, then $2 - i$ would also be a zero. This would then require a polynomial of degree 4.

Polynomial and Rational Functions

3.R   Chapter Review

1. $f(x) = \dfrac{1}{4}x^2 - 16$, $a = \dfrac{1}{4}, b = 0, c = -16$. Since $a = \dfrac{1}{4} > 0$, the graph opens up.

The x‑coordinate of the vertex is $x = \dfrac{-b}{2a} = \dfrac{-0}{2\left(\dfrac{1}{4}\right)} = \dfrac{0}{\left(\dfrac{1}{2}\right)} = 0$.

The y-coordinate of the vertex is $f\left(\dfrac{-b}{2a}\right) = f(0) = \dfrac{1}{4}(0)^2 - 16 = -16$.

Thus, the vertex is $(0, -16)$.
The axis of symmetry is the line $x = 0$.
The discriminant is:

$$b^2 - 4ac = (0)^2 - 4\left(\dfrac{1}{4}\right)(-16) = 16 > 0,$$

so the graph has two x-intercepts.
The x-intercepts are found by solving:

$$\dfrac{1}{4}x^2 - 16 = 0$$

$$x^2 - 64 = 0 \Rightarrow x^2 = 64 \Rightarrow x = 8 \ \text{ or } \ x = -8$$

The x-intercepts are -8 and 8.
The y-intercept is $f(0) = -16$.

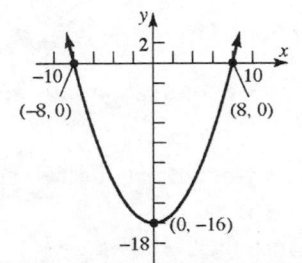

Graphing Utility:

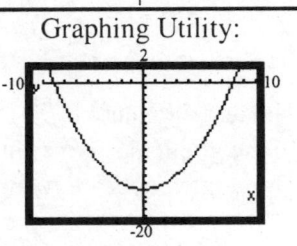

3. $f(x) = -4x^2 + 4x$, $a = -4, b = 4, c = 0$. Since $a = -4 < 0$, the graph opens down.

The x-coordinate of the vertex is $x = \dfrac{-b}{2a} = \dfrac{-4}{2(-4)} = \dfrac{-4}{-8} = \dfrac{1}{2}$.

The y-coordinate of the vertex is $f\left(\dfrac{-b}{2a}\right) = f\left(\dfrac{1}{2}\right) = -4\left(\dfrac{1}{2}\right)^2 + 4\left(\dfrac{1}{2}\right) = -1 + 2 = 1$.

Thus, the vertex is $\left(\dfrac{1}{2}, 1\right)$.

The axis of symmetry is the line $x = \dfrac{1}{2}$.

The discriminant is:
$$b^2 - 4ac = 4^2 - 4(-4)(0) = 16 > 0,$$
so the graph has two x-intercepts.

The x-intercepts are found by solving:

$-4x^2 + 4x = 0 \Rightarrow -4x(x-1) = 0 \Rightarrow x = 0$ or $x = 1$

The x-intercepts are 0 and 1.

The y-intercept is $f(0) = -4(0)^2 + 4(0) = 0$.

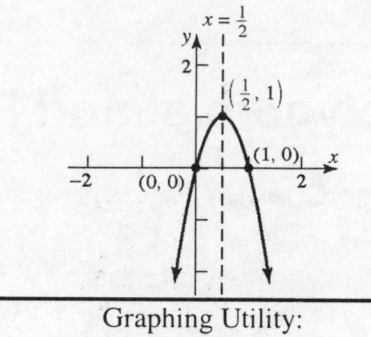

Graphing Utility:

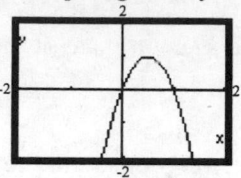

5. $f(x) = 3x^2 + 4x - 1$, $a = 3, b = 4, c = -1$. Since $a = 3 > 0$, the graph opens up.

The x-coordinate of the vertex is $x = \dfrac{-b}{2a} = \dfrac{-4}{2(3)} = \dfrac{-4}{6} = -\dfrac{2}{3}$.

The y-coordinate of the vertex is $f\left(\dfrac{-b}{2a}\right) = f\left(-\dfrac{2}{3}\right) = 3\left(-\dfrac{2}{3}\right)^2 + 4\left(-\dfrac{2}{3}\right) - 1 = \dfrac{4}{3} - \dfrac{8}{3} - 1 = -\dfrac{7}{3}$.

Thus, the vertex is $\left(-\dfrac{2}{3}, -\dfrac{7}{3}\right)$.

The axis of symmetry is the line $x = -\dfrac{2}{3}$.

The discriminant is: $b^2 - 4ac = (4)^2 - 4(3)(-1) = 16 + 12 = 28 > 0$,

so the graph has two x-intercepts.

The x-intercepts are found by solving: $3x^2 + 4x - 1 = 0$

$$x = \frac{-b \pm \sqrt{b^2 - 4ac}}{2a} = \frac{-4 \pm \sqrt{28}}{2(3)} = \frac{-4 \pm 2\sqrt{7}}{6} = \frac{-2 \pm \sqrt{7}}{3} \approx \frac{-2 \pm 2.646}{3}$$

The x-intercepts are approximately 0.22 and -1.55.

The y-intercept is $f(0) = 3(0)^2 + 4(0) - 1 = -1$.

Graphing Utility:

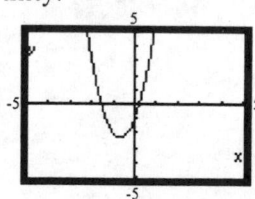

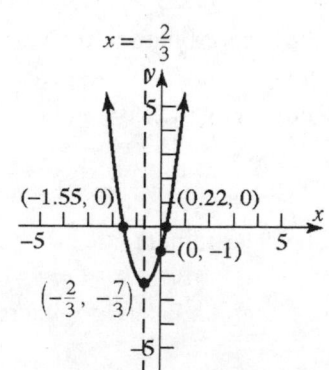

7. $f(x) = x^2 - 4x + 6$, $a = 1, b = -4, c = 6$. Since $a = 1 > 0$, the graph opens up.

The x-coordinate of the vertex is $x = \dfrac{-b}{2a} = \dfrac{-(-4)}{2(1)} = \dfrac{4}{2} = 2$.

The y-coordinate of the vertex is $f\left(\dfrac{-b}{2a}\right) = f(2) = (2)^2 - 4(2) + 6 = 4 - 8 + 6 = 2$.

Thus, the vertex is $(2, 2)$.

The axis of symmetry is the line $x = 2$.

The discriminant is:

$b^2 - 4ac = (-4)^2 - 4(1)(6) = 16 - 24 = -8 < 0$,

so the graph has no x-intercepts.

The y-intercept is $f(0) = (0)^2 - 4(0) + 6 = 6$.

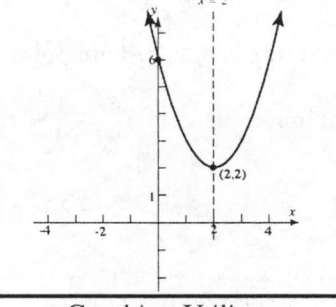

Graphing Utility:

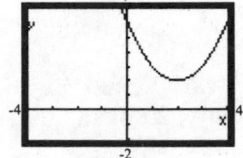

9. $f(x) = 2x^2 + 1$

Using the graph of $y = x^2$, stretch vertically
by a factor of 2, then shift up 1 unit.

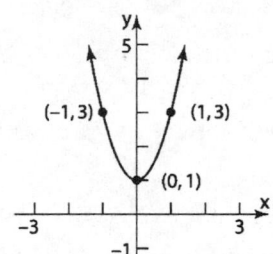

11. $f(x) = \dfrac{1}{2}(x + 1)^2 - 3$

Using the graph of $y = x^2$, shift left 1 unit,

compress vertically by a factor of $\dfrac{1}{2}$, then shift

down 3 units.

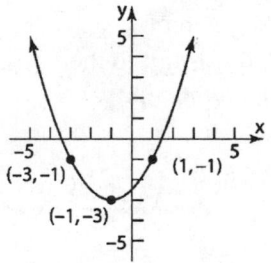

13. $f(x) = 3x^2 - 6x + 4$, $a = 3, b = -6, c = 4$. Since $a = 3 > 0$, the graph opens up, so the

vertex is a minimum point. The minimum occurs at $x = \dfrac{-b}{2a} = \dfrac{-(-6)}{2(3)} = \dfrac{6}{6} = 1$. The

minimum value is $f\left(\dfrac{-b}{2a}\right) = f(1) = 3(1)^2 - 6(1) + 4 = 3 - 6 + 4 = 1$.

15. $f(x) = -x^2 + 8x - 4$, $a = -1$, $b = 8$, $c = -4$. Since $a = -1 < 0$, the graph opens down, so the vertex is a maximum point. The maximum occurs at $x = \dfrac{-b}{2a} = \dfrac{-8}{2(-1)} = \dfrac{-8}{-2} = 4$. The maximum value is $f\left(\dfrac{-b}{2a}\right) = f(4) = -(4)^2 + 8(4) - 4 = -16 + 32 - 4 = 12$.

17. $f(x) = -3x^2 + 12x + 4$, $a = -3$, $b = 12$, $c = 4$. Since $a = -3 < 0$, the graph opens down, so the vertex is a maximum point. The maximum occurs at $x = \dfrac{-b}{2a} = \dfrac{-12}{2(-3)} = \dfrac{-12}{-6} = 2$. The maximum value is $f\left(\dfrac{-b}{2a}\right) = f(2) = -3(2)^2 + 12(2) + 4 = -12 + 24 + 4 = 16$.

19. $f(x) = 4x^5 - 3x^2 + 5x - 2$ is a polynomial function of degree 5.

21. $f(x) = 3x^2 + 5x^{1/2} - 1$ is not a polynomial function because there is a fractional exponent on the variable.

23. $f(x) = (x + 2)^3$
     Using the graph of $y = x^3$, shift the graph horizontally, 2 units to the left.

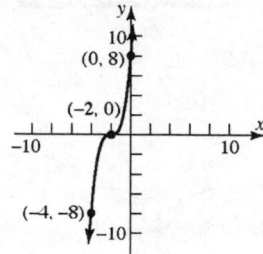

25. $f(x) = -(x - 1)^4$
     Using the graph of $y = x^4$, shift the graph horizontally, 1 unit right, and reflect about the x-axis.

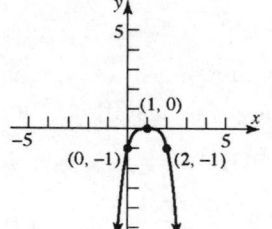

27. $f(x) = (x - 1)^4 + 2$
     Using the graph of $y = x^4$, shift the graph horizontally, 1 unit to the right, and shift vertically 2 units up.

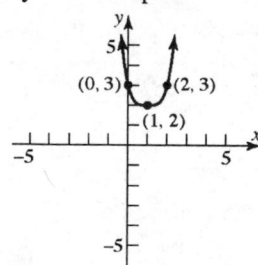

29. $f(x) = x(x + 2)(x + 4)$
     (a) x-intercepts: $-4, -2, 0$;   y-intercept: 0   (b) crosses x axis at $x = -4, -2, 0$
     (c) $y = x^3$

(d)   graphing utility

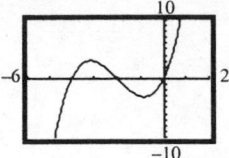

(e)   2 turning points; Local maximum: (–3.15, 3.08)
Local minimum: (–0.85, –3.08)

(f)   graphing by hand

		–4	–2	0
Interval	(–∞, –4)	(–4, –2)	(–2, 0)	(0, ∞)
Number Chosen	–5	–3	–1	1
Value of f	$f(-5) = -15$	$f(-3) = 3$	$f(-1) = -3$	$f(1) = 15$
Location of Graph	Below x-axis	Above x-axis	Below x-axis	Above x-axis
Point on Graph	(-5, -15)	(-3, 3)	(-1, -3)	(1, 15)

Graph of f is above the x-axis for $(-4, -2) \cup (0, \infty)$
Graph of f is below the x-axis for $(-\infty, -4) \cup (-2, 0)$

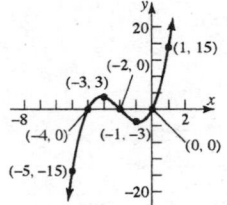

31. $f(x) = (x - 2)^2 (x + 4)$

(a)   x-intercepts: –4, 2;   y-intercept: 16
(b)   crosses x axis at x = –4 and touches the x axis at x = 2
(c) $y = x^3$ (d)   graphing utility

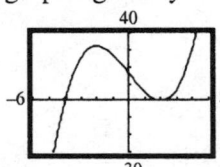

(e)   2 turning points; Local maximum: (–2, 32)
Local minimum: (2, 0)

(f)   graphing by hand

	–4	2	
Interval	(–∞, –4)	(–4, 2)	(2, ∞)
Number Chosen	–5	–2	3
Value of f	$f(-5) = -49$	$f(-2) = 32$	$f(3) = 7$
Location of Graph	Below x-axis	Above x-axis	Above x-axis
Point on Graph	(-5, -49)	(-2, 32)	(3, 7)

Graph of f is above the x-axis for $(-4, 2) \cup (2, \infty)$;
Graph of f is below the x-axis for $(-\infty, -4)$

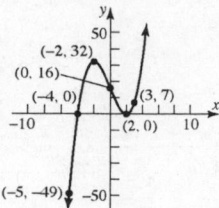

33. $f(x) = x^3 - 4x^2 = x^2(x-4)$
   (a)   x-intercepts: 0, 4;   y-intercept: 0
   (b)   crosses x axis at x = 4 and touches the x axis at x = 0
   (c) $y = x^3$ (d)   graphing utility

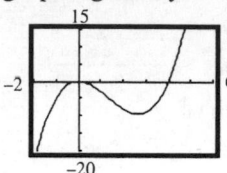

   (e)   2 turning points; Local maximum: (0, 0)
          Local minimum: (2.67, –9.48)
   (f)   graphing by hand

Interval	$(-\infty, 0)$	$(0, 4)$	$(4, \infty)$
Number Chosen	-1	1	5
Value of f	$f(-1) = -5$	$f(1) = -3$	$f(5) = 25$
Location of Graph	Below x-axis	Below x-axis	Above x-axis
Point on Graph	(-1, -5)	(1, -3)	(5, 25)

Graph of f is above the x-axis for $(4, \infty)$; Graph of f is below the x-axis for $(-\infty, 0) \cup (0, 4)$

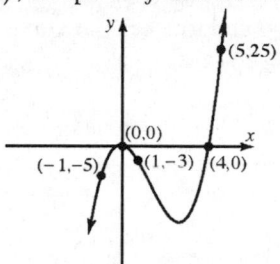

35. $f(x) = (x-1)^2(x+3)(x+1)$
   (a)   x-intercepts: –3, –1, 1;   y-intercept: 3
   (b)   crosses x axis at x = –3, –1 and touches x-axis at x = 1
   (c) $y = x^4$ (d)   graphing utility

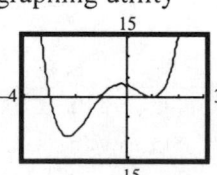

   (e)   3 turning points; Local maximum: (–0.22, 3.23)
          Local minima: (–2.28, –9.91), (1,0)

(f)   graphing by hand

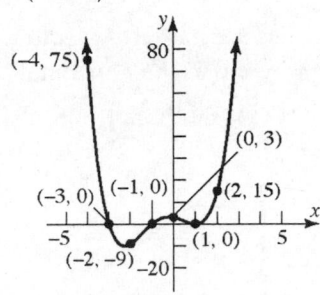

Interval	$(-\infty, -3)$	$(-3, -1)$	$(-1, 1)$	$(1, \infty)$
Number Chosen	-4	-2	0	2
Value of f	$f(-4) = 75$	$f(-2) = -9$	$f(0) = 3$	$f(2) = 15$
Location of Graph	Above x-axis	Below x-axis	Above x-axis	Above x-axis
Point on Graph	$(-4, 75)$	$(-2, -9)$	$(0, 3)$	$(2, 15)$

Graph of f is above the x-axis for $(-\infty, -3) \cup (-1, 1) \cup (1, \infty)$

Graph of f is below the x-axis for $(-3, -1)$

37. $R(x) = \dfrac{2x - 6}{x}$ $p(x) = 2x - 6;\ q(x) = x;\ n = 1;\ m = 1$

Step 1:   Domain: $\{x \mid x \neq 0\}$

Step 2: R is in lowest terms.

Step 3:   (a)   The x-intercept is the zero of $p(x)$: 3

          (b)   There is no y-intercept because 0 is not in the domain.

Step 4: $R(-x) = \dfrac{2(-x) - 6}{-x} = \dfrac{-2x - 6}{-x} = \dfrac{2x + 6}{x}$; this is neither $R(x)$ nor $-R(x)$, so there is

          no symmetry.

Step 5:   The vertical asymptote is the zero of $q(x)$: $x = 0$

Step 6:   Since $n = m$, the line $y = 2$ is the horizontal asymptote.

          $R(x)$ does not intersect $y = 2$.

Step 7:   Graphing utility:

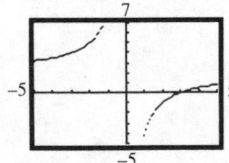

Step 8: Graphing by hand

Interval	$(-\infty, 0)$	$(0, 3)$	$(3, \infty)$
Number Chosen	-2	1	4
Value of R	$R(-2) = 5$	$R(1) = -4$	$R(4) = \frac{1}{2}$
Location of Graph	Above x-axis	Below x-axis	Above x-axis
Point on Graph	$(-2, 5)$	$(1, -4)$	$(4, \frac{1}{2})$

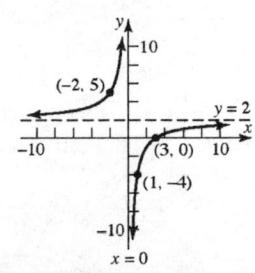

39. $H(x) = \dfrac{x+2}{x(x-2)}$ $p(x) = x+2;$ $q(x) = x(x-2) = x^2 - 2x;$ $n = 1;$ $m = 2$

Step 1:   Domain: $\{x \mid x \neq 0,\ x \neq 2\}$

Step 2: H is in lowest terms.

Step 3:   (a)   The x-intercept is the zero of $p(x)$: -2

(b)   There is no y-intercept because 0 is not in the domain.

Step 4: $H(-x) = \dfrac{-x+2}{-x(-x-2)} = \dfrac{-x+2}{x^2+2x}$; this is neither $H(x)$ nor $-H(x)$, so there is no symmetry.

Step 5:   The vertical asymptotes are the zeros of $q(x)$: $x = 0$ and $x = 2$

Step 6:   Since $n < m$, the line $y = 0$ is the horizontal asymptote.

$R(x)$ intersects $y = 0$ at $(-2, 0)$.

Step 7:   Graphing utility:

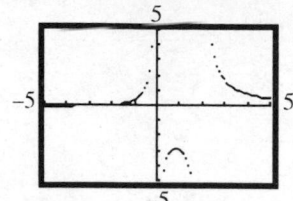

Step 8:   Graphing by hand:

Interval	$(-\infty, -2)$	$(-2, 0)$	$(0, 2)$	$(2, \infty)$
Number Chosen	-3	-1	1	3
Value of H	$H(-3) = -\frac{1}{15}$	$H(-1) = \frac{1}{3}$	$H(1) = -3$	$H(3) = \frac{5}{3}$
Location of Graph	Below x-axis	Above x-axis	Below x-axis	Above x-axis
Point on Graph	$(-3, -\frac{1}{15})$	$(-1, \frac{1}{3})$	$(1, -3)$	$(3, \frac{5}{3})$

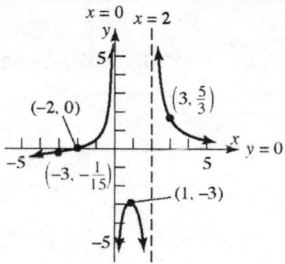

41. $R(x) = \dfrac{x^2+x-6}{x^2-x-6} = \dfrac{(x+3)(x-2)}{(x-3)(x+2)}$ $p(x) = x^2+x-6;$ $q(x) = x^2-x-6;$

$n = 2;$ $m = 2$

Step 1:   Domain: $\{x \mid x \neq -2,\ x \neq 3\}$

Step 2: R is in lowest terms.

Step 3:   (a)   The x-intercepts are the zeros of $p(x)$: -3 and 2

(b)   The y-intercept is $R(0) = \dfrac{0^2 + 0 - 6}{0^2 - 0 - 6} = \dfrac{-6}{-6} = 1.$

Step 4: $R(-x) = \dfrac{(-x)^2 + (-x) - 6}{(-x)^2 - (-x) - 6} = \dfrac{x^2 - x - 6}{x^2 + x - 6}$; this is neither $R(x)$ nor $-R(x)$, so there is no symmetry.

Step 5:   The vertical asymptotes are the zeros of $q(x)$: $x = -2$ and $x = 3$

Step 6:   Since $n = m$, the line $y = 1$ is the horizontal asymptote.

$R(x)$ intersects $y = 1$ at $(0, 1)$, since:

$$\frac{x^2 + x - 6}{x^2 - x - 6} = 1 \Rightarrow x^2 + x - 6 = x^2 - x - 6$$

$$2x = 0 \Rightarrow x = 0$$

Step 7:   Graphing utility:

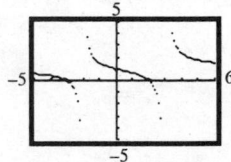

Step 8:   Graphing by hand:

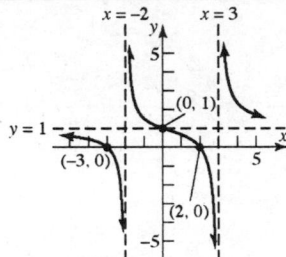

Interval	$(-\infty, -3)$	$(-3, -2)$	$(-2, 2)$	$(2, 3)$	$(3, \infty)$
Number Chosen	-4	-2.5	0	2.5	4
Value of R	$R(-4) \approx 0.43$	$R(-2.5) \approx -0.82$	$R(0) = 1$	$R(2.5) \approx -1.22$	$R(4) = \frac{7}{3}$
Location of Graph	Above x-axis	Below x-axis	Above x-axis	Below x-axis	Above x-axis
Point on Graph	$(-4, 0.43)$	$(-2.5, -0.82)$	$(0, 1)$	$(2.5, -1.22)$	$(4, \frac{7}{3})$

43. $F(x) = \dfrac{x^3}{x^2 - 4}$ $p(x) = x^3$; $q(x) = x^2 - 4$; $n = 3$; $m = 2$

Step 1:   Domain: $\{x \mid x \neq -2, x \neq 2\}$

Step 2: F is in lowest terms.

Step 3:   (a)   The x-intercept is the zero of $p(x)$: 0.

(b)   The y-intercept is $F(0) = \dfrac{0^3}{0^2 - 4} = \dfrac{0}{-4} = 0$.

Step 4: $F(-x) = \dfrac{(-x)^3}{(-x)^2 - 4} = \dfrac{-x^3}{x^2 - 4} = -F(x)$; $F(x)$ is symmetric to the origin.

Step 5:   The vertical asymptotes are the zeros of $q(x)$: $x = -2$ and $x = 2$

Step 6:   Since $n = m + 1$, there is an oblique asymptote. Dividing:

$$
\begin{array}{r}
x \phantom{{}+0x^2+0x+0} \\
x^2 - 4 \overline{\smash{)}x^3 + 0x^2 + 0x + 0} \\
\underline{x^3 - 4x } \\
4x
\end{array}
\qquad F(x) = x + \frac{4x}{x^2 - 4}
$$

The oblique asymptote is $y = x$.

Solve to find intersection points:

$$\frac{x^3}{x^2-4} = x \Rightarrow x^3 = x^3 - 4x \Rightarrow 0 = -4x \Rightarrow x = 0$$

The oblique asymptote intersects $F(x)$ at $(0, 0)$.

Step 7: Graphing utility:

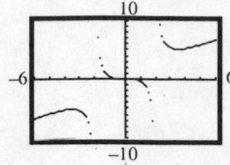

Step 8: Graphing by hand:

Interval	$(-\infty, -2)$	$(-2, 0)$	$(0, 2)$	$(2, \infty)$
Number Chosen	-3	-1	1	3
Value of F	$F(-3) = -\frac{27}{5}$	$F(-1) = \frac{1}{3}$	$F(1) = -\frac{1}{3}$	$F(3) = \frac{27}{5}$
Location of Graph	Below x-axis	Above x-axis	Below x-axis	Above x-axis
Point on Graph	$(-3, -\frac{27}{5})$	$(-1, \frac{1}{3})$	$(1, -\frac{1}{3})$	$(3, \frac{27}{5})$

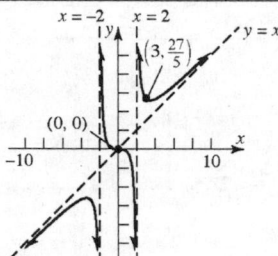

45. $R(x) = \dfrac{2x^4}{(x-1)^2}$ $p(x) = 2x^4;$ $q(x) = (x-1)^2;$ $n = 4;$ $m = 2$

Step 1: Domain: $\{x \mid x \neq 1\}$

Step 2: R is in lowest terms.

Step 3: (a)   The x-intercept is the zero of $p(x)$: 0

         (b)   The y-intercept is $R(0) = \dfrac{2(0)^4}{(0-1)^2} = \dfrac{0}{1} = 0.$

Step 4: $R(-x) = \dfrac{2(-x)^4}{(-x-1)^2} = \dfrac{2x^4}{(x+1)^2}$; this is neither $R(x)$ nor $-R(x)$, so there is no symmetry.

Step 5: The vertical asymptote is the zero of $q(x)$: $x = 1$

Step 6: Since $n > m + 1$, there is no horizontal asymptote and no oblique asymptote.

Step 7: Graphing utility:

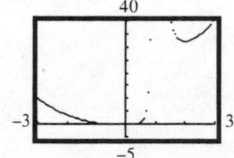

Step 8: Graphing by hand:

Interval	$(-\infty, 0)$	$(0, 1)$	$(1, \infty)$
Number Chosen	-2	$\frac{1}{2}$	2
Value of R	$R(-2) \approx \frac{32}{9}$	$R(\frac{1}{2}) = \frac{1}{2}$	$R(2) = 32$
Location of Graph	Above x-axis	Above x-axis	Above x-axis
Point on Graph	$(-2, \frac{32}{9})$	$(\frac{1}{2}, \frac{1}{2})$	$(2, 32)$

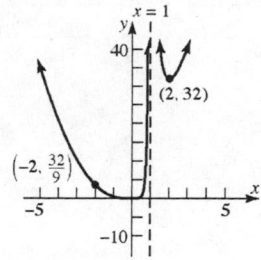

47. $G(x) = \dfrac{x^2 - 4}{x^2 - x - 2}$ $p(x) = x^2 - 4;\ q(x) = x^2 - x - 2;$

$$n = 2;\ m = 2$$

Step 1: Domain: $\{x \mid x \neq -1,\ x \neq 2\}$

Step 2: Write G in lowest terms: $G(x) = \dfrac{x^2 - 4}{x^2 - x - 2} = \dfrac{(x + 2)(x - 2)}{(x - 2)(x + 1)} = \dfrac{x + 2}{x + 1}$

Step 3: (a) The x-intercept is the zero of $p(x)$: -2 ; 2 is not a zero because reduced form must be used to find the zeros.

(b) The y-intercept is $G(0) = \dfrac{0^2 - 4}{0^2 - 0 - 2} = \dfrac{-4}{-2} = 2$.

Step 4: $G(-x) = \dfrac{(-x)^2 - 4}{(-x)^2 - (-x) - 2} = \dfrac{x^2 - 4}{x^2 + x - 2}$; this is neither $G(x)$ nor $-G(x)$, so there is no symmetry.

Step 5: The vertical asymptote is the zero of $q(x)$: $x = -1$; $x = 2$ is not a vertical asymptote because reduced form must be used to find the them. The graph has a hole at $\left(2, \dfrac{4}{3}\right)$.

Step 6: Since $n = m$, the line $y = 1$ is the horizontal asymptote.

$G(x)$ does not intersect $y = 1$ because $G(x)$ is not defined at $x = 2$.

$$\dfrac{x^2 - 4}{x^2 - x - 2} = 1 \Rightarrow x^2 - 4 = x^2 - x - 2$$

$$-2 = -x \Rightarrow x = 2$$

Step 7: Graphing utility:

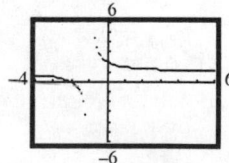

Step 8: Graphing by hand

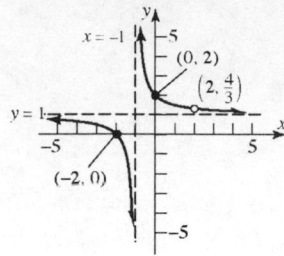

Interval	$(-\infty, -2)$	$(-2, -1)$	$(-1, 2)$	$(2, \infty)$
Number Chosen	-3	-1.5	0	3
Value of G	$G(-3) = \frac{1}{2}$	$G(-1.5) = -1$	$G(0) = 2$	$G(3) = 1.25$
Location of Graph	Above x-axis	Below x-axis	Above x-axis	Above x-axis
Point on Graph	$(-3, \frac{1}{2})$	$(-1.5, -1)$	$(0, 2)$	$(3, 1.25)$

49. (a) $2x^2 + 5x - 12 < 0$ $f(x) = 2x^2 + 5x - 12$

$(x + 4)(2x - 3) < 0$

$x = -4, \; x = \dfrac{3}{2}$ are the zeros.

Interval	Test Number	$f(x)$	Positive/Negative
$-\infty < x < -4$	-5	13	Positive
$-4 < x < 3/2$	0	-12	Negative
$3/2 < x < \infty$	2	6	Positive

The solution set is $\left\{ x \middle| -4 < x < \dfrac{3}{2} \right\}$.

(b) $2x^2 + 5x - 12 < 0$

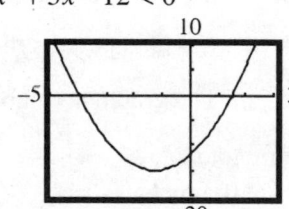

Graph $f(x) = 2x^2 + 5x - 12$.

The x-intercepts are $x = -4$ and $x = 1.5$. The graph of f is below the x-axis for $-4 < x < 1.5$. Thus, the solution set is $\left\{ x \middle| -4 < x < 1.5 \right\}$.

51. (a) $\dfrac{6}{x + 3} \geq 1$ $f(x) = \dfrac{6}{x + 3} - 1$

$\dfrac{6}{x + 3} - 1 \geq 0 \Rightarrow \dfrac{6 - 1(x + 3)}{x + 3} \geq 0 \Rightarrow \dfrac{-x + 3}{x + 3} \geq 0$

The zeros and values where the expression is undefined are $x = -3$, and $x = 3$.

Interval	Test Number	$f(x)$	Positive/Negative
$-\infty < x < -3$	-4	-7	Negative
$-3 < x < 3$	0	1	Positive
$3 < x < \infty$	4	$-1/7$	Negative

The solution set is $\left\{ x \middle| -3 < x \leq 3 \right\}$.

(b) $\dfrac{6}{x+3} \ge 1$

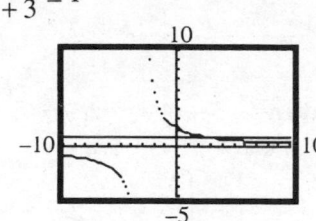

Graph $y_1 = \dfrac{6}{x+3}$, $y_2 = 1$.

y_1 is undefined at $x = -3$. y_1 intersects y_2 at $x = 3$. $y_1 > y_2$ for $-3 < x < 3$. Thus, the solution set is $\left\{ x \mid -3 < x \le 3 \right\}$.

53. (a) $\dfrac{2x-6}{1-x} < 2$ $f(x) = \dfrac{2x-6}{1-x} - 2$

$\dfrac{2x-6}{1-x} - 2 < 0 \Rightarrow \dfrac{2x-6-2(1-x)}{1-x} < 0 \Rightarrow \dfrac{4x-8}{1-x} < 0$

The zeros and values where the expression is undefined are $x = 1$, and $x = 2$.

Interval	Test Number	$f(x)$	Positive/Negative
$-\infty < x < 1$	0	-8	Negative
$1 < x < 2$	1.5	4	Positive
$2 < x < \infty$	3	-2	Negative

The solution set is $\left\{ x \mid x < 1 \text{ or } x > 2 \right\}$.

(b) $\dfrac{2x-6}{1-x} < 2$

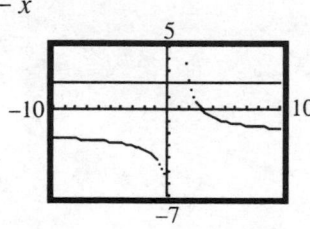

Graph $y_1 = \dfrac{2x-6}{1-x}$, $y_2 = 2$.

y_1 is undefined at $x = 1$. y_1 intersects y_2 at $x = 2$. $y_1 < y_2$ for $x < 1$ or $x > 2$. Thus, the solution set is $\left\{ x \mid x < 1 \text{ or } x > 2 \right\}$.

55. (a) $\dfrac{(x-2)(x-1)}{x-3} > 0$ $f(x) = \dfrac{(x-2)(x-1)}{x-3}$

The zeros and values where the expression is undefined are $x = 1$, $x = 2$, and $x = 3$.

Interval	Test Number	$f(x)$	Positive/Negative
$-\infty < x < 1$	0	$-2/3$	Negative
$1 < x < 2$	1.5	$1/6$	Positive
$2 < x < 3$	2.5	$-3/2$	Negative
$3 < x < \infty$	4	6	Positive

The solution set is $\left\{ x \mid 1 < x < 2 \text{ or } x > 3 \right\}$.

(b) $\dfrac{(x-2)(x-1)}{x-3} > 0$

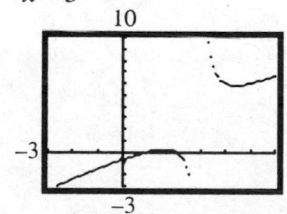

Graph $f(x) = \dfrac{(x-2)(x-1)}{x-3}$.

The x-intercepts are $x = 2$ and $x = 1$. The expression is undefined at $x = 3$. The graph of f is above the x-axis for $1 < x < 2$ or $x > 3$. Thus, the solution set is $\left\{ x \mid 1 < x < 2 \text{ or } x > 3 \right\}$.

57. (a) $\dfrac{x^2 - 8x + 12}{x^2 - 16} > 0$ $f(x) = \dfrac{x^2 - 8x + 12}{x^2 - 16}$

$\dfrac{(x-2)(x-6)}{(x+4)(x-4)} > 0$

The zeros and values where the expression is undefined are $x = -4$, $x = 2$, $x = 4$, and $x = 6$.

Interval	Test Number	$f(x)$	Positive/Negative
$-\infty < x < -4$	-5	77/9	Positive
$-4 < x < 2$	0	$-3/4$	Negative
$2 < x < 4$	3	3/7	Positive
$4 < x < 6$	5	$-1/3$	Negative
$6 < x < \infty$	7	5/33	Positive

The solution set is $\left\{ x \mid x < -4,\, 2 < x < 4,\, x > 6 \right\}$.

(b) $\dfrac{x^2 - 8x + 12}{x^2 - 16} > 0$

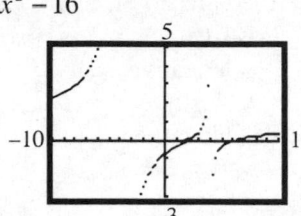

Graph $f(x) = \dfrac{x^2 - 8x + 12}{x^2 - 16}$.

The x-intercepts are $x = 2$, and $x = 6$. The expression is undefined at $x = -4$ and $x = 4$. The graph of f is above the x-axis for $x < -4$, $2 < x < 4$, or $x > 6$. Thus, the solution set is $\left\{ x \mid x < -4,\, 2 < x < 4,\, x > 6 \right\}$.

59. Use synthetic division: Quotient: $8x^2 + 5x + 6$ Remainder: $10 \neq 0$

$$1\overline{)8 \quad -3 \quad 1 \quad 4}$$
$$\underline{\quad\quad 8 \quad 5 \quad 6}$$
$$8 \quad 5 \quad 6 \quad 10$$

Therefore, g is not a factor of f.

61. Use synthetic division: Quotient: $x^3 - 4x^2 + 8x - 1$ Remainder: 0

$$-2\overline{)1 \quad -2 \quad 0 \quad 15 \quad -2}$$
$$\underline{\quad\quad -2 \quad 8 \quad -16 \quad 2}$$
$$1 \quad -4 \quad 8 \quad -1 \quad 0$$

Therefore, g is a factor of f.

63. $f(x) = 2x^8 - x^7 + 8x^4 - 2x^3 + x + 3$

Step 1: $f(x)$ has at most 8 real zeros.

Step 2:   Possible rational zeros:

p must be a factor of 3: $p = \pm 1, \pm 3$

q must be a factor of 2: $q = \pm 1, \pm 2$

The possible rational zeros are: $\dfrac{p}{q} = \pm 1, \pm \dfrac{1}{2}, \pm 3, \pm \dfrac{3}{2}$

65. $f(x) = x^3 - 3x^2 - 6x + 8$
Step 1: $f(x)$ has at most 3 real zeros.
Step 2:   Possible rational zeros:

$$p = \pm 1, \pm 2, \pm 4, \pm 8; \quad q = \pm 1; \quad \frac{p}{q} = \pm 1, \pm 2, \pm 4, \pm 8$$

Step 3:   Using the Bounds on Zeros Theorem:

$a_2 = -3, \quad a_1 = -6, \quad a_0 = 8$

Max $\{1, |8| + |-6| + |-3|\}$ = Max $\{1, 17\}$ = 17

$1 + $ Max $\{|8|, |-6|, |-3|\}$ = $1 + 8 = 9$

The smaller of the two numbers is 9. Thus, every zero of f lies between –9 and 9.

Graphing using the bounds and ZOOM-FIT: (Second graph has a better window.)

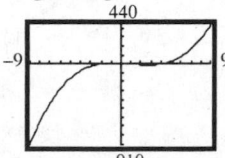 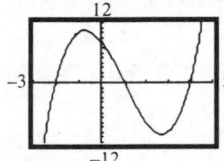

Step 4:   (a)   From the graph it appears that there are x-intercepts at –2, 1, and 4.
(b)   Using synthetic division:

$$
\begin{array}{r}
-2 \overline{)\begin{array}{rrrr} 1 & -3 & -6 & 8 \\ & -2 & 10 & -8 \end{array}} \\
\hline
\begin{array}{rrrr} 1 & -5 & 4 & 0 \end{array}
\end{array}
$$

Since the remainder is 0, $x - (-2) = x + 2$ is a factor. The other factor is the quotient: $x^2 - 5x + 4$.

(c)   Thus, $f(x) = (x + 2)(x^2 - 5x + 4) = (x + 2)(x - 1)(x - 4)$.

The zeros are –2, 1, and 4.

67. $f(x) = 4x^3 + 4x^2 - 7x + 2$
Step 1: $f(x)$ has at most 3 real zeros.
Step 2:   Possible rational zeros:

$$p = \pm 1, \pm 2; \quad q = \pm 1, \pm 2, \pm 4; \quad \frac{p}{q} = \pm 1, \pm 2, \pm \frac{1}{2}, \pm \frac{1}{4}$$

Step 3:   Using the Bounds on Zeros Theorem:

$f(x) = 4\left(x^3 + x^2 - 1.75x + 0.5\right)$

$a_2 = 1, \quad a_1 = -1.75, \quad a_0 = 0.5$

Max $\{1, |0.5| + |-1.75| + |1|\}$ = Max $\{1, 3.25\}$ = 3.25

$1 + $ Max $\{|0.5|, |-1.75|, |1|\}$ = $1 + 1.75 = 2.75$

The smaller of the two numbers is 2.75. Thus, every zero of f lies between –2.75 and 2.75.

Graphing using the bounds and ZOOM-FIT: (Second graph has a better window.)

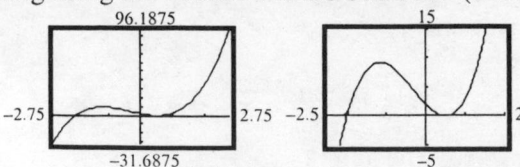

Step 4:   (a)   From the graph it appears that there are x-intercepts at –2 and 0.5.

(b)   Using synthetic division:

$$-2\overline{)\begin{array}{rrrr} 4 & 4 & -7 & 2 \\ & -8 & 8 & -2 \\ \hline 4 & -4 & 1 & 0 \end{array}}$$

Since the remainder is 0, $x-(-2)=x+2$ is a factor. The other factor is the quotient: $4x^2-4x+1$.

(c)   Thus, $f(x)=(x+2)\left(4x^2-4x+1\right)=(x+2)(2x-1)(2x-1)$.

The zeros are –2 and 0.5 (multiplicity 2).

69. $f(x)=x^4-4x^3+9x^2-20x+20$

Step 1: $f(x)$ has at most 4 real zeros.

Step 2:   Possible rational zeros:

$p=\pm1,\pm2,\pm4,\pm5,\pm10,\pm20; \quad q=\pm1;$

$\dfrac{p}{q}=\pm1,\pm2,\pm4,\pm5,\pm10,\pm20$

Step 3:   Using the Bounds on Zeros Theorem:

$a_3=-4,\ a_2=9,\ a_1=-20,\ a_0=20$

$\text{Max}\left\{1,|20|+|-20|+|9|+|-4|\right\}=\text{Max}\left\{1,53\right\}=53$

$1+\text{Max}\left\{|20|,|-20|,|9|,|-4|\right\}=1+20=21$

The smaller of the two numbers is 21. Thus, every zero of f lies between –21 and 21.

Graphing using the bounds and ZOOM-FIT: (Second graph has a better window.)

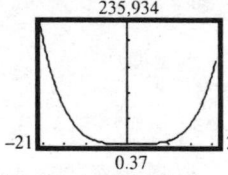 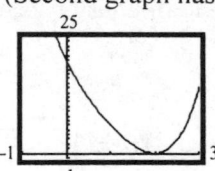

Step 4:   (a)   From the graph it appears that there is an x-intercept at 2.

(b)   Using synthetic division:

$$2\overline{)\begin{array}{rrrrr} 1 & -4 & 9 & -20 & 20 \\ & 2 & -4 & 10 & -20 \\ \hline 1 & -2 & 5 & -10 & 0 \end{array}} \qquad 2\overline{)\begin{array}{rrrr} 1 & -2 & 5 & -10 \\ & 2 & 0 & 10 \\ \hline 1 & 0 & 5 & 0 \end{array}}$$

Since the remainder is 0, $x-2$ is a factor twice. The other factor is the quotient: x^2+5.

(c)   Thus, $f(x) = (x-2)(x-2)(x^2 + 5) = (x-2)^2(x^2 + 5)$.

The zero is 2 (multiplicity 2). ($x^2 + 5 = 0$ has no real solutions.)

73. $g(x) = 15x^4 - 21.5x^3 - 1718.3x^2 + 5308x + 3796.8$

$g(x)$ has at most 4 real zeros.

Solving by graphing (using ZERO):

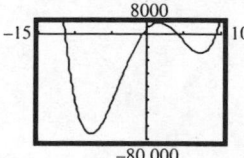

The zeros are approximately –11.3, –0.6, 4, and 9.33.

75. $2x^4 + 2x^3 - 11x^2 + x - 6 = 0$

The solutions of the equation are the zeros of $f(x) = 2x^4 + 2x^3 - 11x^2 + x - 6$.

Step 1: $f(x)$ has at most 4 real zeros.

Step 2:   Possible rational zeros:

$$p = \pm 1, \pm 2, \pm 3, \pm 6; \quad q = \pm 1, \pm 2; \quad \frac{p}{q} = \pm 1, \pm 2, \pm 3, \pm 6, \pm \frac{1}{2}, \pm \frac{3}{2}$$

Step 3:   Using the Bounds on Zeros Theorem:

$$f(x) = 2\left(x^4 + x^3 - 5.5x^2 + 0.5x - 3\right)$$

$$a_3 = 1, \ a_2 = -5.5, \ a_1 = 0.5, \ a_0 = -3$$

$$\text{Max } \left\{1, |-3| + |0.5| + |-5.5| + |1|\right\} = \text{Max } \{1, 10\} = 10$$

$$1 + \text{Max } \left\{|-3|, |0.5|, |-5.5|, |1|\right\} = 1 + 5.5 = 6.5$$

The smaller of the two numbers is 6.5. Thus, every zero of f lies between –6.5 and 6.5.

Graphing using the bounds and ZOOM-FIT: (Second graph has a better window.)

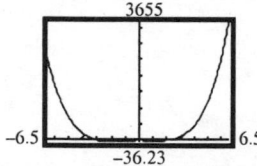   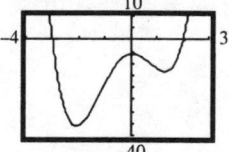

Step 4:   (a)   From the graph it appears that there are x-intercepts at –3 and 2.

(b)   Using synthetic division:

$$
\begin{array}{r|rrrr}
-3 & 2 & 2 & -11 & 1 & -6 \\
   &   & -6 & 12 & -3 & 6 \\
\hline
   & 2 & -4 & 1 & -2 & 0
\end{array}
\qquad
\begin{array}{r|rrrr}
2 & 2 & -4 & 1 & -2 \\
  &   & 4 & 0 & 2 \\
\hline
  & 2 & 0 & 1 & 0
\end{array}
$$

Since the remainder is 0, $x + 3$ and $x - 2$ are factors. The other factor is the quotient: $2x^2 + 1$.

(c)   The zeros are –3 and 2. ($2x^2 + 1 = 0$ has no real solutions.)

77. $2x^4 + 7x^3 + x^2 - 7x - 3 = 0$

The solutions of the equation are the zeros of $f(x) = 2x^4 + 7x^3 + x^2 - 7x - 3$.

Step 1: $f(x)$ has at most 4 real zeros.

Step 2: Possible rational zeros:

$$p = \pm 1, \pm 3; \quad q = \pm 1, \pm 2; \quad \frac{p}{q} = \pm 1, \pm 3, \pm \frac{1}{2}, \pm \frac{3}{2}$$

Step 3: Using the Bounds on Zeros Theorem:

$$f(x) = 2\left(x^4 + 3.5x^3 + 0.5x^2 - 3.5x - 1.5\right)$$

$$a_3 = 3.5, \ a_2 = 0.5, \ a_1 = -3.5, \ a_0 = -1.5$$

$$\text{Max}\left\{1, |-1.5| + |-3.5| + |0.5| + |3.5|\right\} = \text{Max}\{1, 9\} = 9$$

$$1 + \text{Max}\left\{|-1.5|, |-3.5|, |0.5|, |3.5|\right\} = 1 + 3.5 = 4.5$$

The smaller of the two numbers is 4.5. Thus, every zero of f lies between –4.5 and 4.5.

Graphing using the bounds and ZOOM-FIT: (Second graph has a better window.)

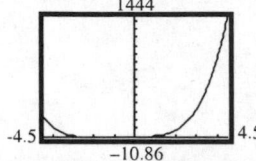   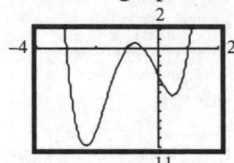

Step 4: (a)   From the graph it appears that there are x-intercepts at –3, –1, –0.5, and 1.

(b)   Using synthetic division:

$$-3\overline{)\begin{array}{ccccc} 2 & 7 & 1 & -7 & -3 \\ & -6 & -3 & 6 & 3 \\ \hline 2 & 1 & -2 & -1 & 0 \end{array}} \qquad -1\overline{)\begin{array}{cccc} 2 & 1 & -2 & -1 \\ & -2 & 1 & 1 \\ \hline 2 & -1 & -1 & 0 \end{array}}$$

Since the remainder is 0, $x + 3$ and $x + 1$ are factors. The other factor is the quotient: $2x^2 - x - 1$.

(c)   Thus, $f(x) = (x + 3)(x + 1)\left(2x^2 - x - 1\right) = (x + 3)(x + 1)(2x + 1)(x - 1)$.

The zeros are –3, –1, –0.5, and 1.

79. $f(x) = x^3 - x^2 - 4x + 2$

$$a_2 = -1, \ a_1 = -4, \ a_0 = 2$$

$$\text{Max}\left\{1, |2| + |-4| + |-1|\right\} = \text{Max}\{1, 7\} = 7$$

$$1 + \text{Max}\left\{|2|, |-4|, |-1|\right\} = 1 + 4 = 5$$

The smaller of the two numbers is 5. Thus, every zero of f lies between –5 and 5.

Graphing using the bounds and ZOOM-FIT: (Second graph has a better window.)

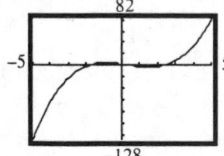   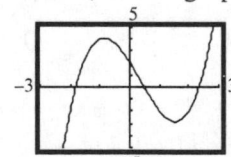

81. $f(x) = 2x^3 - 7x^2 - 10x + 35 = 2\left(x^3 - 3.5x^2 - 5x + 17.5\right)$

$a_2 = -3.5, \quad a_1 = -5, \quad a_0 = 17.5$

$\text{Max}\left\{1, |17.5| + |-5| + |-3.5|\right\} = \text{Max}\left\{1, 26\right\} = 26$

$1 + \text{Max}\left\{|17.5|, |-5|, |-3.5|\right\} = 1 + 17.5 = 18.5$

The smaller of the two numbers is 18.5. Thus, every zero of f lies between -18.5 and 18.5.

Graphing using the bounds and ZOOM-FIT: (Second graph has a better window.)

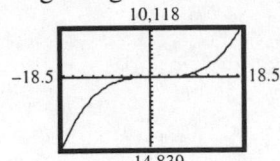   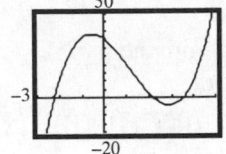

83. Using the TABLE feature to show that there is a zero in the interval:

$f(x) = 3x^3 - x - 1; \quad [0, 1]$

X	Y1
-2	-23
-1	-3
0	-1
1	1
2	21
3	77
4	187

Y1 = 3X^3 - X - 1

$f(0) = -1 < 0$ and $f(1) = 1 > 0$

Since one is positive and one is negative, there is a zero in the interval.

Using the TABLE feature to approximate the zero to two decimal places:

X	Y1
.82	-.1659
.83	-.1146
.84	-.0619
.85	-.0076
.86	.04817
.87	.10551
.88	.16442

Y1 = 3X^3 - X - 1

The zero is approximately 0.85.

85. Using the TABLE feature to show that there is a zero in the interval:

$f(x) = 8x^4 - 4x^3 - 2x - 1; \quad [0, 1]$

X	Y1
-2	163
-1	13
0	-1
1	1
2	91
3	533
4	1783

Y1 = 8X^4 - 4X^3 - 2X...

$f(0) = -1 < 0$ and $f(1) = 1 > 0$

Since one is positive and one is negative, there is a zero in the interval.

Using the TABLE feature to approximate the zero to two decimal places:

X	Y1
.9	-.4672
.91	-.3483
.92	-.2236
.93	-.093
.94	.04366
.95	.18655
.96	.33583

Y1 = 8X^4 - 4X^3 - 2X... The zero is approximately 0.94.

87.   Since complex zeros appear in conjugate pairs, $4 - i$, the conjugate of $4 + i$, is the remaining zero of f.

$$f(x) = (x-6)(x-(4-i))(x-(4+i)) = (x-6)(x-4+i)(x-4-i)$$
$$= (x-6)\left((x-4)^2 - i^2\right) = (x-6)\left((x-4)^2 + 1\right)$$
$$= (x-6)(x^2 - 8x + 16 + 1) = (x-6)(x^2 - 8x + 17)$$
$$= x^3 - 8x^2 + 17x - 6x^2 + 48x - 102$$

$$f(x) = x^3 - 14x^2 + 65x - 102$$

89.   Since complex zeros appear in conjugate pairs, $-i$, the conjugate of i, and $1 - i$, the conjugate of $1 + i$, are the remaining zeros of f.

$$f(x) = (x-i)(x+i)(x-(1-i))(x-(1+i)) = (x^2 - i^2)(x-1+i)(x-1-i)$$
$$= (x^2 + 1)\left((x-1)^2 - i^2\right) = (x^2 + 1)\left((x-1)^2 + 1\right)$$
$$= (x^2 + 1)(x^2 - 2x + 1 + 1) = (x^2 + 1)(x^2 - 2x + 2)$$
$$= x^4 - 2x^3 + 2x^2 + x^2 - 2x + 2$$

$$f(x) = x^3 - 2x^3 + 3x^2 - 2x + 2$$

91. $x^4 + 2x^2 - 8 = 0$
$(x^2 + 4)(x^2 - 2) = 0$
$x^2 + 4 = 0$ or $x^2 - 2 = 0$
$x^2 = -4$ or $x^2 = 2$
$x = \pm 2i$ or $x = \pm\sqrt{2}$
The solution set is $\left\{-2i,\ 2i,\ -\sqrt{2},\ \sqrt{2}\right\}$.

93. $x^3 - x^2 - 8x + 12 = 0$
The solutions of the equation are the zeros of the function $f(x) = x^3 - x^2 - 8x + 12$.
Step 1: $f(x)$ has 3 complex zeros.
Step 2:   Possible rational zeros:

$$p = \pm 1, \pm 2, \pm 3, \pm 4, \pm 6, \pm 12; \quad q = \pm 1; \quad \frac{p}{q} = \pm 1, \pm 2, \pm 3, \pm 4, \pm 6, \pm 12$$

Step 3:   Graphing the function:

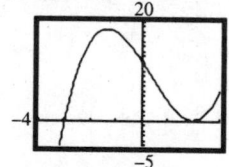

From the graph it appears that there are x-intercepts at -3 and 2.
Step 4:   Using synthetic division:

$$\begin{array}{r|rrrr} 2 & 1 & -1 & -8 & 12 \\ & & 2 & 2 & -12 \\ \hline & 1 & 1 & -6 & 0 \end{array}$$

Since the remainder is 0, $x - 2$ is a factor. The other factor is the quotient:
$$x^2 + x - 6 = (x + 3)(x - 2).$$
The complex zeros are –3, 2 (multiplicity 2).

95. $3x^4 - 4x^3 + 4x^2 - 4x + 1 = 0$
The solutions of the equation are the zeros of the function
$$f(x) = 3x^4 - 4x^3 + 4x^2 - 4x + 1$$
Step 1: $f(x)$ has 4 complex zeros.
Step 2:   Possible rational zeros:
$$p = \pm 1; \quad q = \pm 1, \pm 3; \quad \frac{p}{q} = \pm 1, \pm \frac{1}{3}$$
Step 3:   Graphing the function:

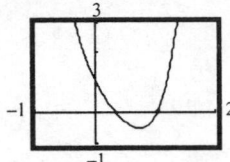

From the graph it appears that there are x-intercepts at $\frac{1}{3}$ and 1.

Step 4:   Using synthetic division:

$$1\overline{)\begin{array}{ccccc} 3 & -4 & 4 & -4 & 1 \\ & 3 & -1 & 3 & -1 \\ \hline 3 & -1 & 3 & -1 & 0 \end{array}} \qquad \frac{1}{3}\overline{)\begin{array}{cccc} 3 & -1 & 3 & -1 \\ & 1 & 0 & 1 \\ \hline 3 & 0 & 3 & 0 \end{array}}$$

Since the remainder is 0, $x - 1$ and $x - \frac{1}{3}$ are factors. The other factor is the

quotient: $3x^2 + 3 = 3(x^2 + 1)$.
Solving $x^2 + 1 = 0$:
$$x^2 = -1 \Rightarrow x = \pm i$$
The complex zeros are 1, $\frac{1}{3}$, $-i$, i.

97.   Let x represent the length and y represent the width of the rectangle.
$2x + 2y = 20 \Rightarrow y = 10 - x$.
$x \cdot y = 16 \Rightarrow x(10 - x) = 16$.
Solving the area equation:
$$10x - x^2 = 16 \Rightarrow x^2 - 10x + 16 = 0$$
$$(x - 8)(x - 2) = 0 \Rightarrow x = 8 \text{ or } x = 2$$
The length and width of the rectangle are 8 feet by 2 feet.

99. $C(x) = 4.9x^2 - 617.40x + 19,600$; $a = 4.9, b = -617.40, c = 19,600$. Since $a = 4.9 > 0$, the graph opens up, so the vertex is a minimum point.

(a)   The minimum marginal cost occurs at $x = \dfrac{-b}{2a} = \dfrac{-(-617.40)}{2(4.9)} = \dfrac{617.40}{9.8} = 63$.

(b)   The minimum marginal cost is

$$C\left(\frac{-b}{2a}\right) = C(63) = 4.9(63)^2 - (617.40)(63) + 19600 = \$151.90$$

101. (a)   Graphing, the data appear to be quadratic.

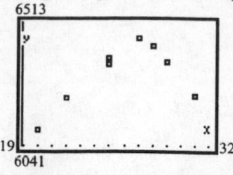

(b)   Using the QUADratic REGression program, the quadratic function of best fit is:
$R(A) = -7.76A^2 + 411.88A + 942.72$

(c)   The maximum revenue occurs at $A = \dfrac{-b}{2a} = \dfrac{-(411.88)}{2(-7.76)} = \dfrac{-411.88}{-15.52}$

≈ 26.53865979 thousand dollars $\approx \$26,538.66$

(d)   The maximum revenue is

$$R\left(\frac{-b}{2a}\right) = R(26.53866) = -7.76(26.53866)^2 + (411.88)(26.53866) + 942.72$$

$$\approx 6408.091683 \text{ thousand dollars } \approx \$6,408,091.68$$

(e)   graphing:

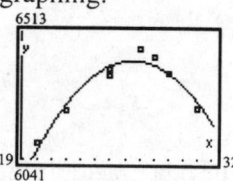

103. (a) $250 = \pi r^2 h \;\Rightarrow\; h = \dfrac{250}{\pi r^2}; \; A(r) = 2\pi r^2 + 2\pi rh = 2\pi r^2 + 2\pi r\left(\dfrac{250}{\pi r^2}\right) = 2\pi r^2 + \dfrac{500}{r}$

(b) $A(3) = 2\pi \cdot 3^2 + \dfrac{500}{3} = 18\pi + \dfrac{500}{3} \approx 223.22$ square cm

(c) $A(5) = 2\pi \cdot 5^2 + \dfrac{500}{5} = 50\pi + 100 \approx 257.08$ square cm

(d)   Use MINIMUM on the graph of

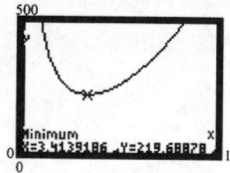

The area is smallest when the radius is approximately 3.41 cm.

105. Answers will vary.

Polynomial and Rational Functions

3.CR Cumulative Review

1. $4x^2 + 2x = 5 \Rightarrow 4x^2 + 2x - 5 = 0$
$a = 4, b = 2, c = -5$

$$x = \frac{-b \pm \sqrt{b^2 - 4ac}}{2a} = \frac{-2 \pm \sqrt{2^2 - 4(4)(-5)}}{2(4)} = \frac{-2 \pm \sqrt{4 + 80}}{8}$$

$$= \frac{-2 \pm \sqrt{84}}{8} = \frac{-2 \pm 2\sqrt{21}}{8} = \frac{-1 \pm \sqrt{21}}{4}$$

The solution set is $\left\{ \dfrac{-1 - \sqrt{21}}{4}, \dfrac{-1 + \sqrt{21}}{4} \right\}$.

3. $3x - 4y = 5 \Rightarrow y = \dfrac{3}{4}x - \dfrac{5}{4} \Rightarrow$ slope $= \dfrac{3}{4}$

$4x + 3y = 10 \Rightarrow y = -\dfrac{4}{3}x + \dfrac{10}{3} \Rightarrow$ slope $= -\dfrac{4}{3}$

Therefore, the lines are perpendicular since their slopes are negative reciprocals.

5. $g(t) = -16t^2 + 100t + 100 \Rightarrow g(4) = -16(4)^2 + 100(4) + 10 = -256 + 400 + 10 = 154$

7. $f(x) = -2(x + 3)^2 + 5$

Using the graph of $y = x^2$, shift the graph horizontally, 3 units to the left, reflect about the x-axis, stretch vertically by a factor of 2 and shift vertically 3 units up.

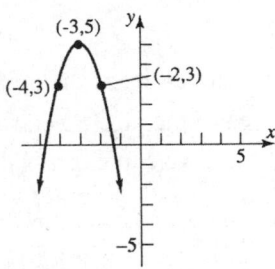

9. $f(x) = 0.2x^3 + x^2 + 1$

Use MAXIMUM and MINIMUM on the graph of $y_1 = 0.2x^3 + x^2 + 1$.

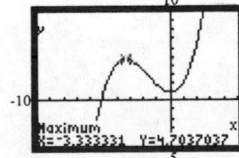 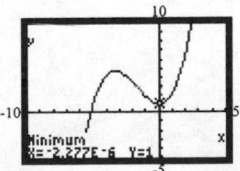

local maximum at: $(-3.33, 4.70)$;   local minimum at: $(0, 1)$

f is increasing on: $(-\infty, -3.33) \cup (0, \infty)$; f is decreasing on: $(-3.33, 0)$

11. $f(x) = (x+3)(x-4)^2$

    (a)   x-intercepts: -3, 4;   y-intercept: 48

    (b)   crosses x axis at $x = -3$, touches x-axis at $x = 4$

    (c) $y = x^3$

    (d)   graphing utility

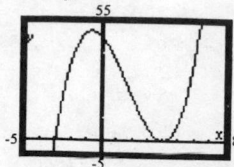

    (e)   2 turning points; Local maximum: $(-0.67, 50.81)$; Local minimum: $(4, 0)$

    (f)   graphing by hand

Interval	$(-\infty, -3)$	$(-3, 4)$	$(4, \infty)$
Number Chosen	-4	3.5	5
Value of f	$f(-4) = -64$	$f(3.5) = 1.625$	$f(5) = 8$
Location of Graph	Below x-axis	Above x-axis	Above x-axis
Point on Graph	$(-4, -64)$	$(3.5, 1.625)$	$(5, 8)$

Graph of f is above the x-axis for $(-3,4) \cup (4,\infty)$

Graph of f is below the x-axis for $(-\infty,-3)$

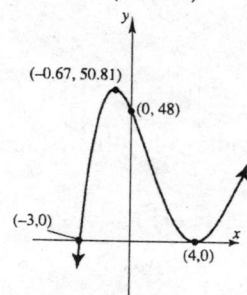

13.   Let $r =$ the radius of the circle after t seconds. Then $r(t) = 2t$.

The area of the circle as a function of the radius is given by $A(r) = \pi \cdot r^2$.

Therefore, we can express the area of the circle as a function of time:

$$(A \circ r)(t) = A(r(t)) = \pi \cdot (2t)^2 = 4\pi \cdot t^2$$

Chapter 4

Exponential and Logarithmic Functions

4.1 One-to-One Functions; Inverse Functions

1. (a)

Domain	Range
$200	20 hours
$300	25 hours
$350	30 hours
$425	40 hours

   (b) Inverse is a function.

3. (a)

Domain	Range
	20 hours
$200	25 hours
$350	30 hours
$425	40 hours

   (b) Inverse is not a function since $200 corresponds to two elements in the range.

5. (a) $\{(6,2), (6,-3), (9,4), (10,1)\}$

   (b) Inverse is not a function since 6 corresponds to 2 and -3.

7. (a) $\{(0,0), (1,1), (16,2), (81,3)\}$

   (b) Inverse is a function.

9. Every horizontal line intersects the graph of f at exactly one point. One-to-One.

11. There are horizontal lines that intersect the graph of f at more than one point. Not One-to-One.

13. Every horizontal line intersects the graph of f at exactly one point. One-to-One.

15. Graphing the inverse:

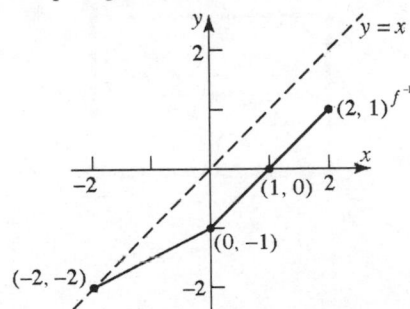

17. Graphing the inverse:

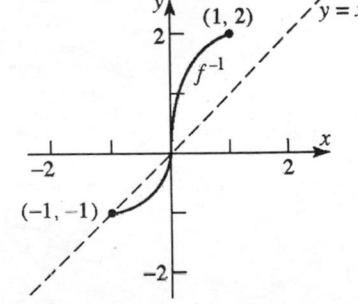

19. Graphing the inverse:

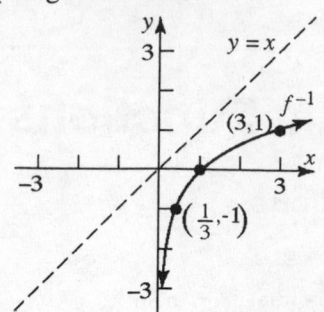

21. $f(x) = 3x + 4, \qquad g(x) = \dfrac{1}{3}(x - 4)$

$$f(g(x)) = f\left(\dfrac{1}{3}(x-4)\right) = 3\left(\dfrac{1}{3}(x-4)\right) + 4 = (x-4)+4 = x$$

$$g(f(x)) = g(3x+4) = \dfrac{1}{3}\big((3x+4)-4\big) = \dfrac{1}{3}(3x) = x$$

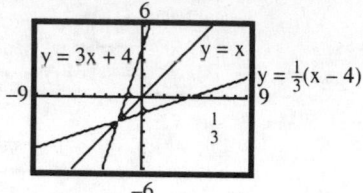

23. $f(x) = 4x - 8, \qquad g(x) = \dfrac{x}{4} + 2$

$$f(g(x)) = f\left(\dfrac{x}{4}+2\right) = 4\left(\dfrac{x}{4}+2\right) - 8 = (x+8)-8 = x$$

$$g(f(x)) = g(4x-8) = \dfrac{4x-8}{4} + 2 = x - 2 + 2 = x$$

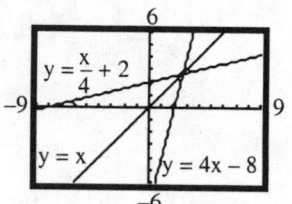

25. $f(x) = x^3 - 8, \qquad g(x) = \sqrt[3]{x+8}$

$$f(g(x)) = f\left(\sqrt[3]{x+8}\right) = \left(\sqrt[3]{x+8}\right)^3 - 8 = (x+8)-8 = x$$

$$g(f(x)) = g(x^3 - 8) = \sqrt[3]{(x^3 - 8)+8} = \sqrt[3]{x^3} = x$$

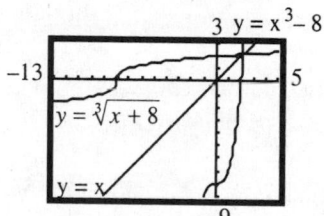

27. $f(x) = \dfrac{1}{x}, \qquad g(x) = \dfrac{1}{x}$

$$f(g(x)) = f\left(\dfrac{1}{x}\right) = \dfrac{1}{\left(\dfrac{1}{x}\right)} = x$$

$$g(f(x)) = g\left(\dfrac{1}{x}\right) = \dfrac{1}{\left(\dfrac{1}{x}\right)} = x$$

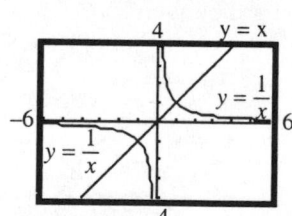

29. $f(x) = \dfrac{2x+3}{x+4}, \qquad g(x) = \dfrac{4x-3}{2-x}$

$$f(g(x)) = f\left(\dfrac{4x-3}{2-x}\right) = \dfrac{2\left(\dfrac{4x-3}{2-x}\right) + 3}{\left(\dfrac{4x-3}{2-x}\right) + 4} = \dfrac{\left(\dfrac{8x-6+6-3x}{2-x}\right)}{\left(\dfrac{4x-3+8-4x}{2-x}\right)} = \dfrac{\left(\dfrac{5x}{2-x}\right)}{\left(\dfrac{5}{2-x}\right)}$$

$$= \dfrac{5x}{2-x} \cdot \dfrac{2-x}{5} = x$$

$$g(f(x)) = g\left(\dfrac{2x+3}{x+4}\right) = \dfrac{4\left(\dfrac{2x+3}{x+4}\right) - 3}{2 - \left(\dfrac{2x+3}{x+4}\right)} = \dfrac{\left(\dfrac{8x+12-3x-12}{x+4}\right)}{\left(\dfrac{2x+8-2x-3}{x+4}\right)} = \dfrac{\left(\dfrac{5x}{x+4}\right)}{\left(\dfrac{5}{x+4}\right)}$$

$$= \dfrac{5x}{x+4} \cdot \dfrac{x+4}{5} = x$$

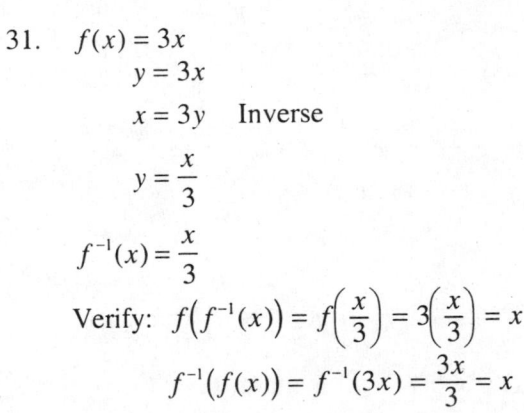

31. $f(x) = 3x$

$y = 3x$

$x = 3y$ Inverse

$y = \dfrac{x}{3}$

$f^{-1}(x) = \dfrac{x}{3}$

Verify: $f\left(f^{-1}(x)\right) = f\left(\dfrac{x}{3}\right) = 3\left(\dfrac{x}{3}\right) = x$

$\qquad\quad f^{-1}\left(f(x)\right) = f^{-1}(3x) = \dfrac{3x}{3} = x$

Domain of f = range of $f^{-1} = (-\infty, \infty)$

Range of f = domain of $f^{-1} = (-\infty, \infty)$

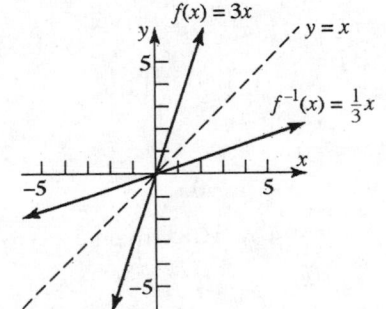

33. $f(x) = 4x + 2$

$\quad\quad y = 4x + 2$

$\quad\quad x = 4y + 2$ Inverse

$\quad\quad 4y = x - 2$

$\quad\quad y = \dfrac{x-2}{4}$

$\quad f^{-1}(x) = \dfrac{x-2}{4}$

Domain of f = range of $f^{-1} = (-\infty, \infty)$

Range of f = domain of $f^{-1} = (-\infty, \infty)$

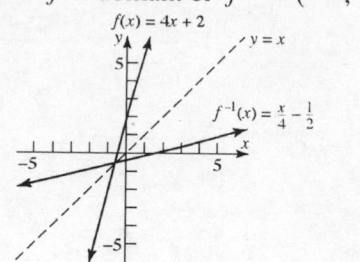

Verify: $f\left(f^{-1}(x)\right) = f\left(\dfrac{x-2}{4}\right) = 4\left(\dfrac{x-2}{4}\right) + 2 = x - 2 + 2 = x$

$\quad\quad\quad f^{-1}\left(f(x)\right) = f^{-1}(4x+2) = \dfrac{(4x+2)-2}{4} = \dfrac{4x}{4} = x$

35. $f(x) = x^3 - 1$

$\quad\quad y = x^3 - 1$

$\quad\quad x = y^3 - 1$ Inverse

$\quad\quad y^3 = x + 1$

$\quad\quad y = \sqrt[3]{x+1}$

$\quad f^{-1}(x) = \sqrt[3]{x+1}$

Domain of f = range of $f^{-1} = (-\infty, \infty)$

Range of f = domain of $f^{-1} = (-\infty, \infty)$

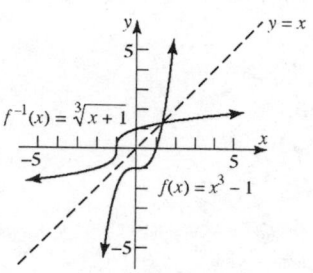

Verify: $f\left(f^{-1}(x)\right) = f\left(\sqrt[3]{x+1}\right) = \left(\sqrt[3]{x+1}\right)^3 - 1 = x + 1 - 1 = x$

$\quad\quad\quad f^{-1}\left(f(x)\right) = f^{-1}\left(x^3 - 1\right) = \sqrt[3]{\left(x^3-1\right)+1} = \sqrt[3]{x^3} = x$

37. $f(x) = x^2 + 4, \ x \ge 0$

$\quad\quad y = x^2 + 4 \ \ x \ge 0$

$\quad\quad x = y^2 + 4 \ \ y \ge 0$ Inverse

$\quad\quad y^2 = x - 4 \ \ y \ge 0$

$\quad\quad y = \sqrt{x-4} \Rightarrow f^{-1}(x) = \sqrt{x-4}$

Verify:

$f\left(f^{-1}(x)\right) = f\left(\sqrt{x-4}\right) =$

$\left(\sqrt{x-4}\right)^2 + 4 = x - 4 + 4 = x$

$\quad f^{-1}\left(f(x)\right) = f^{-1}\left(x^2 + 4\right) = \sqrt{\left(x^2+4\right)-4}$

$\quad = \sqrt{x^2} = |x| = x, \ x \ge 0$

Domain of f = range of $f^{-1} = [0, \infty)$

Range of f = domain of $f^{-1} = [4, \infty)$

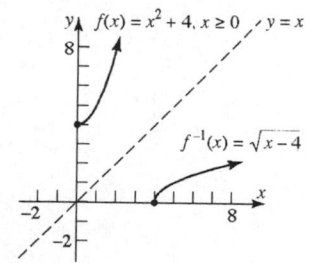

39. $f(x) = \dfrac{4}{x}$

$y = \dfrac{4}{x}$

$x = \dfrac{4}{y}$ Inverse

$xy = 4$

$y = \dfrac{4}{x}$

$f^{-1}(x) = \dfrac{4}{x}$

Domain of f = range of f^{-1}
= all real numbers except 0

Range of f = domain of f^{-1}
= all real numbers except 0

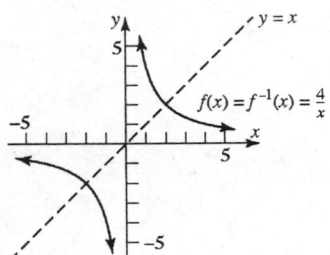

Verify: $f\left(f^{-1}(x)\right) = f\left(\dfrac{4}{x}\right) = \dfrac{4}{\left(\dfrac{4}{x}\right)} = 4 \cdot \left(\dfrac{x}{4}\right) = x$

$f^{-1}\left(f(x)\right) = f^{-1}\left(\dfrac{4}{x}\right) = \dfrac{4}{\left(\dfrac{4}{x}\right)} = 4 \cdot \left(\dfrac{x}{4}\right) = x$

41. $f(x) = \dfrac{1}{x-2}$

$y = \dfrac{1}{x-2}$

$x = \dfrac{1}{y-2}$ Inverse

$x(y-2) = 1$

$xy - 2x = 1$

$xy = 2x + 1$

$y = \dfrac{2x+1}{x}$

$f^{-1}(x) = \dfrac{2x+1}{x}$

Domain of f = range of f^{-1}
= all real numbers except 2

Range of f = domain of f^{-1}
= all real numbers except 0

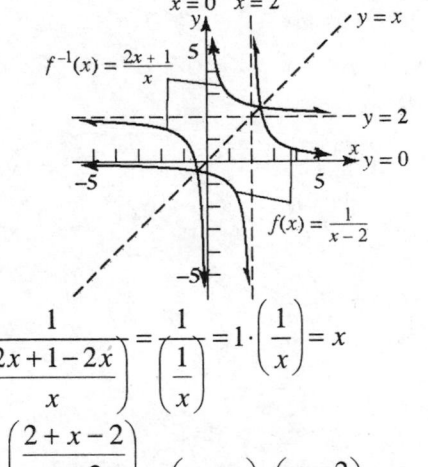

Verify: $f\left(f^{-1}(x)\right) = f\left(\dfrac{2x+1}{x}\right) = \dfrac{1}{\left(\dfrac{2x+1}{x} - 2\right)} = \dfrac{1}{\left(\dfrac{2x+1-2x}{x}\right)} = \dfrac{1}{\left(\dfrac{1}{x}\right)} = 1 \cdot \left(\dfrac{1}{x}\right) = x$

$f^{-1}\left(f(x)\right) = f^{-1}\left(\dfrac{1}{x-2}\right) = \dfrac{2\left(\dfrac{1}{x-2}\right) + 1}{\left(\dfrac{1}{x-2}\right)} = \dfrac{\left(\dfrac{2+x-2}{x-2}\right)}{\left(\dfrac{1}{x-2}\right)} = \left(\dfrac{x}{x-2}\right) \cdot \left(\dfrac{x-2}{1}\right) = x$

43. $f(x) = \dfrac{2}{3+x}$

$y = \dfrac{2}{3+x}$

$x = \dfrac{2}{3+y}$ Inverse

$x(3+y) = 2$

$3x + xy = 2$

$xy = 2 - 3x$

$y = \dfrac{2-3x}{x}$

$f^{-1}(x) = \dfrac{2-3x}{x}$

Domain of f =
range of f^{-1} = all real numbers except -3

Range of f =
domain of f^{-1} = all real numbers except 0

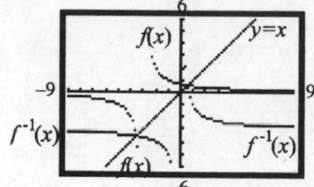

Verify: $f\left(f^{-1}(x)\right) = f\left(\dfrac{2-3x}{x}\right) = \dfrac{2}{3+\left(\dfrac{2-3x}{x}\right)} = \dfrac{2}{\left(\dfrac{3x+2-3x}{x}\right)} = \dfrac{2}{\left(\dfrac{2}{x}\right)} = 2\cdot\left(\dfrac{x}{2}\right) = x$

$f^{-1}\left(f(x)\right) = f^{-1}\left(\dfrac{2}{3+x}\right) = \dfrac{2-3\left(\dfrac{2}{3+x}\right)}{\left(\dfrac{2}{3+x}\right)} = \dfrac{\left(\dfrac{6+2x-6}{3+x}\right)}{\left(\dfrac{2}{3+x}\right)} = \left(\dfrac{2x}{3+x}\right)\cdot\left(\dfrac{3+x}{2}\right) = x$

45. $f(x) = (x+2)^2,\ \ x \geq -2$

$y = (x+2)^2 \ \ x \geq -2$

$x = (y+2)^2 \ \ y \geq -2$ Inverse

$\sqrt{x} = y+2, \ \ \ x \geq 0$

$y = \sqrt{x} - 2, \ \ x \geq 0$

$f^{-1}(x) = \sqrt{x} - 2, \ \ x \geq 0$

Domain of f = range of f^{-1} = $[-2, \infty)$
Range of f = domain of f^{-1} = $[0, \infty)$

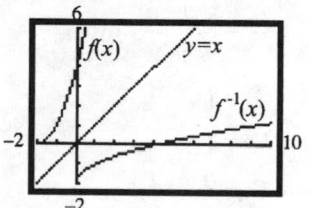

Verify: $f\left(f^{-1}(x)\right) = f\left(\sqrt{x}-2\right) = \left(\sqrt{x}-2+2\right)^2 = \left(\sqrt{x}\right)^2 = x$

$\ \ \ \ \ \ \ \ \ \ \ f^{-1}\left(f(x)\right) = f^{-1}\left((x+2)^2\right) = \sqrt{(x+2)^2} - 2 = x+2-2 = x, \ \ x \geq -2$

47. $f(x) = \dfrac{2x}{x-1}$

$y = \dfrac{2x}{x-1}$

$x = \dfrac{2y}{y-1}$ Inverse

$x(y-1) = 2y \Rightarrow xy - x = 2y$

$xy - 2y = x \Rightarrow y(x-2) = x$

$y = \dfrac{x}{x-2} \Rightarrow f^{-1}(x) = \dfrac{x}{x-2}$

Domain of f = range of f^{-1}
= all real numbers except 1
Range of f = domain of f^{-1}
= all real numbers except 2

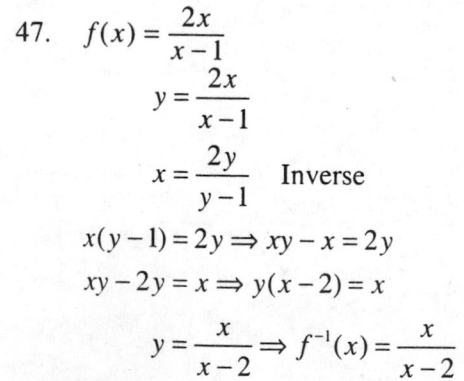

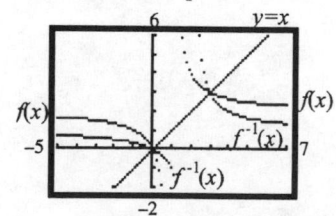

Verify: $f\left(f^{-1}(x)\right)=f\left(\dfrac{x}{x-2}\right)=\dfrac{2\left(\dfrac{x}{x-2}\right)}{\left(\dfrac{x}{x-2}\right)-1}=\dfrac{\left(\dfrac{2x}{x-2}\right)}{\left(\dfrac{x-x+2}{x-2}\right)}=\dfrac{\left(\dfrac{2x}{x-2}\right)}{\left(\dfrac{2}{x-2}\right)}=\left(\dfrac{2x}{x-2}\right)\cdot\left(\dfrac{x-2}{2}\right)=x$

$f^{-1}\left(f(x)\right)=f^{-1}\left(\dfrac{2x}{x-1}\right)=\dfrac{\left(\dfrac{2x}{x-1}\right)}{\left(\dfrac{2x}{x-1}\right)-2}=\dfrac{\left(\dfrac{2x}{x-1}\right)}{\left(\dfrac{2x-2x+2}{x-1}\right)}=\left(\dfrac{2x}{x-1}\right)\cdot\left(\dfrac{x-1}{2}\right)=x$

49.　$f(x)=\dfrac{3x+4}{2x-3}$

$y=\dfrac{3x+4}{2x-3}$

$x=\dfrac{3y+4}{2y-3}$　Inverse

$x(2y-3)=3y+4$

$2xy-3x=3y+4$

$2xy-3y=3x+4$

$y(2x-3)=3x+4$

$y=\dfrac{3x+4}{2x-3}\Rightarrow f^{-1}(x)=\dfrac{3x+4}{2x-3}$

Domain of f = range of f^{-1} = all real numbers except $\dfrac{3}{2}$

Range of f = domain of f^{-1} = all real numbers except $\dfrac{3}{2}$

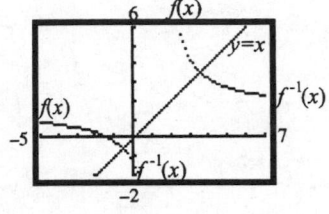

Verify:

$f\left(f^{-1}(x)\right)=f\left(\dfrac{3x+4}{2x-3}\right)=\dfrac{3\left(\dfrac{3x+4}{2x-3}\right)+4}{2\left(\dfrac{3x+4}{2x-3}\right)-3}=\dfrac{\left(\dfrac{9x+12+8x-12}{2x-3}\right)}{\left(\dfrac{6x+8-6x+9}{2x-3}\right)}=\dfrac{\left(\dfrac{17x}{2x-3}\right)}{\left(\dfrac{17}{2x-3}\right)}$

$=\dfrac{17x}{2x-3}\cdot\dfrac{2x-3}{17}=x$

$f^{-1}\left(f(x)\right)=f^{-1}\left(\dfrac{3x+4}{2x-3}\right)=\dfrac{3\left(\dfrac{3x+4}{2x-3}\right)+4}{2\left(\dfrac{3x+4}{2x-3}\right)-3}=\dfrac{\left(\dfrac{9x+12+8x-12}{2x-3}\right)}{\left(\dfrac{6x+8-6x+9}{2x-3}\right)}=\dfrac{\left(\dfrac{17x}{2x-3}\right)}{\left(\dfrac{17}{2x-3}\right)}$

$=\dfrac{17x}{2x-3}\cdot\dfrac{2x-3}{17}=x$

51. $f(x) = \dfrac{2x+3}{x+2}$

$y = \dfrac{2x+3}{x+2}$

$x = \dfrac{2y+3}{y+2}$ Inverse

$x(y+2) = 2y+3$

$xy + 2x = 2y + 3$

$xy - 2y = -2x + 3$

$y(x-2) = -2x + 3$

$y = \dfrac{-2x+3}{x-2}$

$f^{-1}(x) = \dfrac{-2x+3}{x-2}$

Domain of $f =$
range of f^{-1} = all real numbers except -2

Range of $f =$
domain of f^{-1} = all real numbers except 2

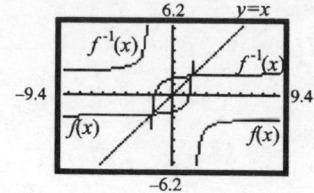

Verify:

$$f\left(f^{-1}(x)\right) = f\left(\dfrac{-2x+3}{x-2}\right) = \dfrac{2\left(\dfrac{-2x+3}{x-2}\right)+3}{\left(\dfrac{-2x+3}{x-2}\right)+2} = \dfrac{\left(\dfrac{-4x+6+3x-6}{x-2}\right)}{\left(\dfrac{-2x+3+2x-4}{x-2}\right)} = \dfrac{\left(\dfrac{-x}{x-2}\right)}{\left(\dfrac{-1}{x-2}\right)}$$

$$= \left(\dfrac{-x}{x-2}\right) \cdot \left(\dfrac{x-2}{-1}\right) = x$$

$$f^{-1}\left(f(x)\right) = f^{-1}\left(\dfrac{2x+3}{x+2}\right) = \dfrac{-2\left(\dfrac{2x+3}{x+2}\right)+3}{\left(\dfrac{2x+3}{x+2}\right)-2} = \dfrac{\left(\dfrac{-4x-6+3x+6}{x+2}\right)}{\left(\dfrac{2x+3-2x-4}{x+2}\right)} = \dfrac{\left(\dfrac{-x}{x+2}\right)}{\left(\dfrac{-1}{x+2}\right)}$$

$$= \left(\dfrac{-x}{x+2}\right) \cdot \left(\dfrac{x+2}{-1}\right) = x$$

53. $f(x) = 2\sqrt[3]{x}$

$y = 2\sqrt[3]{x}$

$x = 2\sqrt[3]{y}$ Inverse

$x^3 = 8y$

$y = \dfrac{x^3}{8}$

$f^{-1}(x) = \dfrac{x^3}{8}$

Domain of $f =$ range of $f^{-1} = (-\infty, \infty)$

Range of $f =$ domain of $f^{-1} = (-\infty, \infty)$

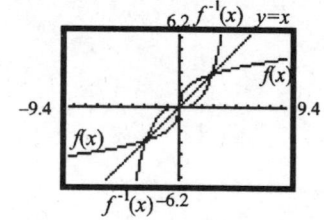

Verify: $f\left(f^{-1}(x)\right) = f\left(\dfrac{x^3}{8}\right) = 2\left(\sqrt[3]{\dfrac{x^3}{8}}\right) = 2\cdot\left(\dfrac{x}{2}\right) = x$

$$f^{-1}\left(f(x)\right) = f^{-1}\left(2\sqrt[3]{x}\right) = \dfrac{\left(2\sqrt[3]{x}\right)^3}{8} = \dfrac{8x}{8} = x$$

55. $f(x) = mx + b, \quad m \neq 0$

$\quad\quad y = mx + b$

$\quad\quad x = my + b \quad$ Inverse

$\quad x - b = my$

$\quad\quad\quad y = \dfrac{x - b}{m}$

$\quad f^{-1}(x) = \dfrac{x - b}{m}, \quad m \neq 0$

57. f^{-1} lies in quadrant I. Whenever (a,b) is on f, then (b,a) is on f^{-1}. Since both coordinates of (a,b) are positive, both coordinates of (b,a) are positive and it is in quadrant I.

59. $f(x) = |x|, x \geq 0$ is one-to-one. Thus, $f(x) = x, x \geq 0$ and $f^{-1}(x) = x, x \geq 0$.

61. $f(x) = \dfrac{9}{5}x + 32 \quad\quad g(x) = \dfrac{5}{9}(x - 32)$

$\quad f(g(x)) = f\left(\dfrac{5}{9}(x - 32)\right) = \dfrac{9}{5}\left(\dfrac{5}{9}(x - 32)\right) + 32 = x - 32 + 32 = x$

$\quad g(f(x)) = g\left(\dfrac{9}{5}x + 32\right) = \dfrac{5}{9}\left(\dfrac{9}{5}x + 32 - 32\right) = \dfrac{5}{9}\left(\dfrac{9}{5}x\right) = x$

63. $T(l) = 2\pi\sqrt{\dfrac{l}{g}}, \quad g \approx 32.2$

$\quad\quad T = 2\pi\sqrt{\dfrac{l}{g}} \quad \Rightarrow \quad \dfrac{T}{2\pi} = \sqrt{\dfrac{l}{g}}$

$\quad\quad\quad \Rightarrow \quad \dfrac{T^2}{4\pi^2} = \dfrac{l}{g} \quad \Rightarrow \quad l = \dfrac{gT^2}{4\pi^2}$

$\quad l(T) = \dfrac{gT^2}{4\pi^2}$

65. Yes, consider the function $f(x) = \dfrac{1}{x}$. In general the graph of f must have symmetry across the line $y = x$.

67. $f(x) = \begin{cases} \dfrac{1}{x}, & \text{if } x < 0 \\ x, & \text{if } x \geq 0 \end{cases}$

Exponential and Logarithmic Functions

4.2 Exponential Functions

1. (a) $3^{2.2} \approx 11.212$ (b) $3^{2.23} \approx 11.587$ (c) $3^{2.236} \approx 11.664$ (d) $3^{\sqrt{5}} \approx 11.665$

3. (a) $2^{3.14} \approx 8.815$ (b) $2^{3.141} \approx 8.821$ (c) $2^{3.1415} \approx 8.824$ (d) $2^{\pi} \approx 8.825$

5. (a) $3.1^{2.7} \approx 21.217$ (b) $3.14^{2.71} \approx 22.217$
   (c) $3.141^{2.718} \approx 22.440$ (d) $\pi^{e} \approx 22.459$

7. $e^{1.2} \approx 3.320$

9. $e^{-0.85} \approx 0.427$

11. B   13. D   15. A   17. E

19. $f(x) = 2^{x} + 1$
    Using the graph of $y = 2^{x}$, shift the graph up 1 unit.
    Domain: $(-\infty, \infty)$
    Range: $(1, \infty)$
    Horizontal Asymptote: $y = 1$

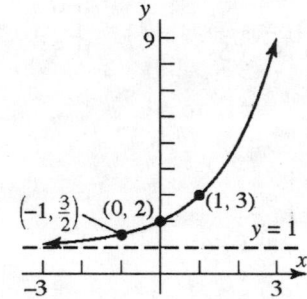

21. $f(x) = 3^{-x} - 2$
    Using the graph of $y = 3^{x}$, reflect the graph about the y-axis, and shift down 2 units.
    Domain: $(-\infty, \infty)$
    Range: $(-2, \infty)$
    Horizontal Asymptote: $y = -2$

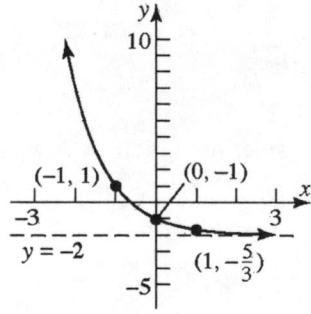

23. $f(x) = 3(4^x)$

Using the graph of $y = 4^x$, stretch the
graph vertically by a factor of 3.
Domain: $(-\infty, \infty)$
Range: $(0, \infty)$
Horizontal Asymptote: $y = 0$

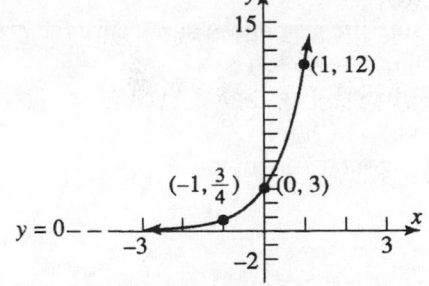

25. $f(x) = 3^{x/2}$

Using the graph of $y = 3^x$, stretch the
graph horizontally by a factor of 2.
Domain: $(-\infty, \infty)$
Range: $(0, \infty)$
Horizontal Asymptote: $y = 0$

27. $f(x) = 5 - 2(3^{(x+1)}) = -2(3^{(x+1)}) + 5$

Using the graph of $y = 3^x$, shift the graph
one unit to the left, stretch vertically by a
factor of 2, reflect on the x-axis, and shift
up 5 units.
Domain: $(-\infty, \infty)$
Range: $(-\infty, 5)$
Horizontal Asymptote: $y = 5$

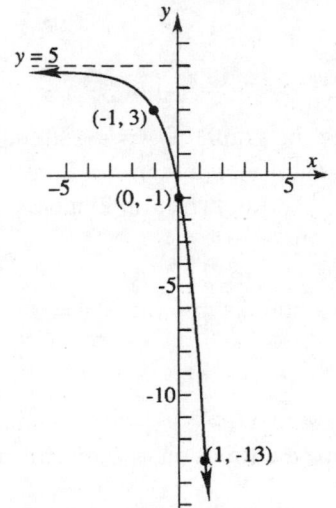

29. $f(x) = e^{-x}$

Using the graph of $y = e^x$, reflect the
graph about the y-axis.
Domain: $(-\infty, \infty)$
Range: $(0, \infty)$
Horizontal Asymptote: $y = 0$

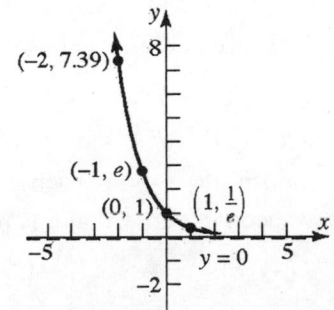

31. $f(x) = e^{x+2}$
Using the graph of $y = e^x$, shift the graph
2 units to the left.
Domain: $(-\infty, \infty)$
Range: $(0, \infty)$
Horizontal Asymptote: $y = 0$

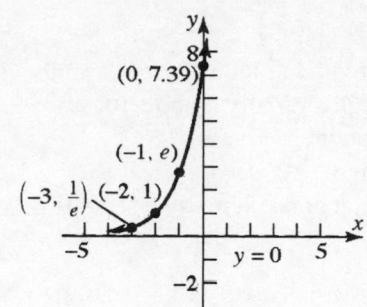

33. $f(x) = 5 + e^{-x} = e^{-x} + 5$
Using the graph of $y = e^x$, reflect the
graph about the y-axis and shift up 5
units.
Domain: $(-\infty, \infty)$
Range: $(5, \infty)$
Horizontal Asymptote: $y = 5$

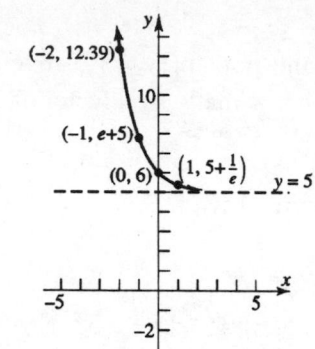

35. $f(x) = 2 - e^{x/2} = -e^{x/2} + 2$
Using the graph of $y = e^x$, stretch
horizontally by a factor of 2, reflect about
the x-axis, and shift up 2 units.
Domain: $(-\infty, \infty)$
Range: $(-\infty, 2)$
Horizontal Asymptote: $y = 2$

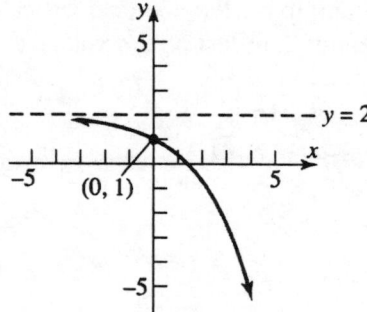

37. We need a function of the form $f(x) = k \cdot a^{p \cdot x}$, with $a > 0, a \neq 1$.

The graph contains the points $\left(-1, \dfrac{1}{3}\right), (0,1), (1,3)$ and $(2,9)$.

In other words, $f(-1) = \dfrac{1}{3}, f(0) = 1, f(1) = 3$ and $f(2) = 9$.

Therefore, $f(0) = k \cdot a^{p \cdot (0)} = k \cdot a^0 = k \cdot 1 = k \Rightarrow k = 1$.

and $f(1) = a^{p \cdot (1)} = a^p \Rightarrow a^p = 3$.

Let's choose $a = 3, p = 1$. Then $f(x) = 3^x$.

Now we need to verify that this function yields the other known points on the graph.

$$f(-1) = 3^{-1} = \frac{1}{3}; \qquad f(2) = 3^2 = 9$$

So we have the function $f(x) = 3^x$.

39. We need a function of the form $f(x) = k \cdot a^{p \cdot x}$, with $a > 0, a \neq 1$.

The graph contains the points $\left(-1, \frac{1}{2}\right), (0,2), (1,8)$ and $(2,32)$.

In other words, $f(-1) = \frac{1}{2}, f(0) = 2, f(1) = 8$ and $f(2) = 32$.

Therefore, $f(0) = k \cdot a^{p \cdot (0)} = k \cdot a^0 = k \cdot 1 = k \Rightarrow k = 2$.

and $f(1) = 2a^{p \cdot (1)} = 2a^p \Rightarrow 2a^p = 8 \Rightarrow a^p = 4$.

Let's choose $a = 4, p = 1$. Then $f(x) = 2 \cdot 4^x$.

Now we need to verify that this function yields the other known points on the graph.

$$f(-1) = 2 \cdot 4^{-1} = 2 \cdot \frac{1}{4} = \frac{1}{2}; \quad f(2) = 2 \cdot 4^2 = 2 \cdot 16 = 32$$

So we have the function $f(x) = 2 \cdot 4^x$.

41. We need a function of the form $f(x) = k \cdot a^{p \cdot x}$, with $a > 0, a \neq 1$.

The graph contains the points $\left(-1, -\frac{1}{6}\right), (0,-1), (1,-6)$ and $(2,-36)$.

In other words, $f(-1) = -\frac{1}{6}, f(0) = -1, f(1) = -6$ and $f(2) = -36$.

Therefore, $f(0) = k \cdot a^{p \cdot (0)} = k \cdot a^0 = k \cdot 1 = k \Rightarrow k = -1$.

and $f(1) = -a^{p \cdot (1)} = -a^p \Rightarrow -a^p = -6 \Rightarrow a^p = 6$.

Let's choose $a = 6, p = 1$. Then $f(x) = -6^x$.

Now we need to verify that this function yields the other known points on the graph.

$$f(-1) = -6^{-1} = -\frac{1}{6}; \quad f(2) = -6^2 = -36$$

So we have the function $f(x) = -6^x$.

43. $2^{2x+1} = 4$
$2^{2x+1} = 2^2$

$2x + 1 = 2 \Rightarrow 2x = 1 \Rightarrow x = \frac{1}{2}$

The solution set is $\left\{\frac{1}{2}\right\}$.

45. $3^{x^3} = 9^x$
$3^{x^3} = \left(3^2\right)^x \Rightarrow 3^{x^3} = 3^{2x}$

$x^3 = 2x \Rightarrow x^3 - 2x = 0 \Rightarrow x(x^2 - 2) = 0$

$x = 0$ or $x^2 = 2 \Rightarrow x = 0$ or $x = \pm\sqrt{2}$

The solution set is $\left\{-\sqrt{2}, \ 0, \ \sqrt{2}\right\}$.

47. $8^{x^2 - 2x} = \frac{1}{2}$

$\left(2^3\right)^{x^2 - 2x} = 2^{-1} \Rightarrow 2^{3x^2 - 6x} = 2^{-1}$

$3x^2 - 6x = -1 \Rightarrow 3x^2 - 6x + 1 = 0$

$$x = \frac{-(-6) \pm \sqrt{(-6)^2 - 4(3)(1)}}{2(3)} = \frac{6 \pm \sqrt{24}}{6} = \frac{6 \pm 2\sqrt{6}}{6} = \frac{3 \pm \sqrt{6}}{3}$$

The solution set is $\left\{\frac{3 - \sqrt{6}}{3}, \ \frac{3 + \sqrt{6}}{3}\right\}$.

49. $2^x \cdot 8^{-x} = 4^x$
$$2^x \cdot \left(2^3\right)^{-x} = \left(2^2\right)^x$$
$$2^x \cdot 2^{-3x} = 2^{2x}$$
$$2^{-2x} = 2^{2x}$$
$$-2x = 2x \rightarrow -4x = 0 \rightarrow x = 0$$
The solution set is $\{0\}$.

51. $\left(\dfrac{1}{5}\right)^{2-x} = 25$
$$\left(5^{-1}\right)^{2-x} = 5^2$$
$$5^{x-2} = 5^2$$
$$x - 2 = 2 \rightarrow x = 4$$
The solution set is $\{4\}$.

53. $4^x = 8$
$$\left(2^2\right)^x = 2^3$$
$$2^{2x} = 2^3 \rightarrow 2x = 3 \rightarrow x = \dfrac{3}{2}$$
The solution set is $\left\{\dfrac{3}{2}\right\}$.

55. $e^{x^2} = e^{3x} \cdot \dfrac{1}{e^2}$
$$e^{x^2} = e^{3x-2}$$
$$x^2 = 3x - 2$$
$$x^2 - 3x + 2 = 0$$
$$(x-1)(x-2) = 0 \rightarrow x = 1 \text{ or } x = 2$$
The solution set is $\{1, 2\}$.

57. $4^x = 7$
$$\left(4^x\right)^{-2} = 7^{-2} \quad \rightarrow \quad 4^{-2x} = \dfrac{1}{7^2} = \dfrac{1}{49}$$

59. $3^{-x} = 2$
$$(3^{-x})^{-2} = 2^{-2} \quad \rightarrow \quad 3^{2x} = \dfrac{1}{2^2} = \dfrac{1}{4}$$

61. $p = 100e^{-0.03n}$
  (a) $p = 100e^{-0.03(10)} = 100e^{-0.3} \approx 100(0.741) = 74.1\%$ of light
  (b) $p = 100e^{-0.03(25)} = 100e^{-0.75} \approx 100(0.472) = 47.2\%$ of light

63. $w(d) = 50e^{-0.004d}$
  (a) $w(30) = 50e^{-0.004(30)} = 50e^{-0.12} \approx 50(0.887) = 44.35$ watts
  (b) $w(365) = 50e^{-0.004(365)} = 50e^{-1.46} \approx 50(0.232) = 11.61$ watts

65. $D(h) = 5e^{-0.4h}$
$$D(1) = 5e^{-0.4(1)} = 5e^{-0.4} \approx 5(0.670) = 3.35 \text{ milligrams}$$
$$D(6) = 5e^{-0.4(6)} = 5e^{-2.4} \approx 5(0.091) = 0.45 \text{ milligrams}$$

67. $F(t) = 1 - e^{-0.1t}$
  (a) $F(10) = 1 - e^{-0.1(10)} = 1 - e^{-1} \approx 1 - 0.368 = 0.632 = 63.2\%$
  (b) $F(40) = 1 - e^{-0.1(40)} = 1 - e^{-4} \approx 1 - 0.018 = 0.982 = 98.2\%$
  (c) as $t \rightarrow +\infty$, $F(t) = 1 - e^{-0.1t} \rightarrow 1 - 0 = 1$
  (d) Graphing the function:

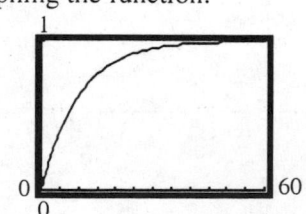

69. $P(x) = \dfrac{20^x e^{-20}}{x!}$

(a) $P(15) = \dfrac{20^{15} e^{-20}}{15!} \approx 0.0516 = 5.16\%$ The probability that 15 cars will arrive between 5:00 p.m. and 6:00 p.m. is 5.16%.

(b) $P(20) = \dfrac{20^{20} e^{-20}}{20!} \approx 0.0888 = 8.88\%$ The probability that 20 cars will arrive between 5:00 p.m. and 6:00 p.m. is 8.88%.

71. $R = 10^{\left(\frac{2345}{T} - \frac{2345}{D} + 2\right)}$

(a) $R = 10^{\left(\frac{2345}{283} - \frac{2345}{278} + 2\right)} \approx 10^{1.851} \approx 70.96\%$

(b) $R = 10^{\left(\frac{2345}{293} - \frac{2345}{288} + 2\right)} \approx 10^{1.861} \approx 72.61\%$

(c) $R = 10^{\left(\frac{2345}{x} - \frac{2345}{x} + 2\right)} = 10^2 = 100\%$

73. $I = \dfrac{E}{R}\left[1 - e^{-\left(\frac{R}{L}\right)t}\right]$

(a) $I = \dfrac{120}{10}\left[1 - e^{-\left(\frac{10}{5}\right)0.3}\right] = 12\left[1 - e^{-0.6}\right] \approx 5.414$ amperes after 0.3 second

$I = \dfrac{120}{10}\left[1 - e^{-\left(\frac{10}{5}\right)0.5}\right] = 12\left[1 - e^{-1}\right] \approx 7.585$ amperes after 0.5 second

$I = \dfrac{120}{10}\left[1 - e^{-\left(\frac{10}{5}\right)1}\right] = 12\left[1 - e^{-2}\right] \approx 10.376$ amperes after 1 second

(b) As $t \to \infty$, $e^{-\left(\frac{10}{5}\right)t} \to 0$. Therefore, the maximum current is 12 amperes.

(c), (f) Graphing the function:

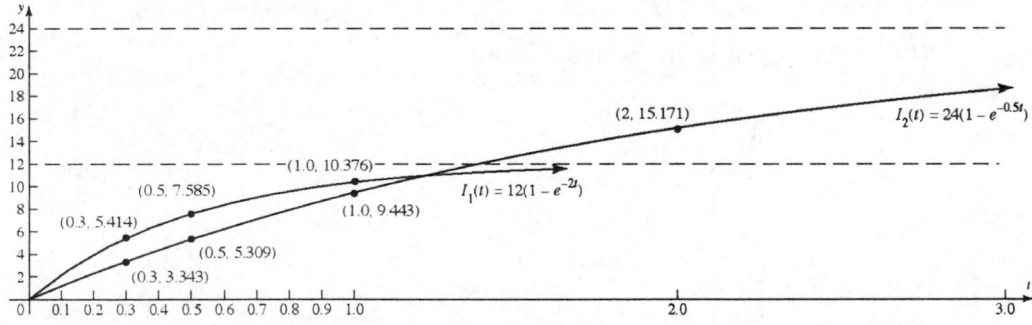

(d) $I = \dfrac{120}{5}\left[1 - e^{-\left(\frac{5}{10}\right)0.3}\right] = 24\left[1 - e^{-0.15}\right] \approx 3.343$ amperes after 0.3 second

$$I = \frac{120}{5}\left[1 - e^{-\left(\frac{5}{10}\right)^{0.5}}\right] = 24\left[1 - e^{-0.25}\right] \approx 5.309 \text{ amperes after 0.5 second}$$

$$I = \frac{120}{5}\left[1 - e^{-\left(\frac{5}{10}\right)^{1}}\right] = 24\left[1 - e^{-0.5}\right] \approx 9.443 \text{ amperes after 1 second}$$

(e)　　As $t \to \infty$, $e^{-\left(\frac{5}{10}\right)^{t}} \to 0$. Therefore, the maximum current is 24 amperes.

75.　$2 + \dfrac{1}{2!} + \dfrac{1}{3!} + \dfrac{1}{4!} + \dots + \dfrac{1}{n!}$

$n = 4$;　$2 + \dfrac{1}{2!} + \dfrac{1}{3!} + \dfrac{1}{4!} = 2.7083$

$n = 6$;　$2 + \dfrac{1}{2!} + \dfrac{1}{3!} + \dfrac{1}{4!} + \dfrac{1}{5!} + \dfrac{1}{6!} = 2.7181$

$n = 8$;　$2 + \dfrac{1}{2!} + \dfrac{1}{3!} + \dfrac{1}{4!} + \dfrac{1}{5!} + \dfrac{1}{6!} + \dfrac{1}{7!} + \dfrac{1}{8!} = 2.7182788$

$n = 10$;　$2 + \dfrac{1}{2!} + \dfrac{1}{3!} + \dfrac{1}{4!} + \dfrac{1}{5!} + \dfrac{1}{6!} + \dfrac{1}{7!} + \dfrac{1}{8!} + \dfrac{1}{9!} + \dfrac{1}{10!} = 2.7182818$

$e \approx 2.718281828$

77.　$f(x) = a^x$

$$\frac{f(x+h) - f(x)}{h} = \frac{a^{x+h} - a^x}{h} = \frac{a^x a^h - a^x}{h} = \frac{a^x\left(a^h - 1\right)}{h} = a^x\left(\frac{a^h - 1}{h}\right)$$

79.　$f(x) = a^x$

$$f(-x) = a^{-x} = \frac{1}{a^x} = \frac{1}{f(x)}$$

81.　$\sinh x = \dfrac{1}{2}\left(e^x - e^{-x}\right)$

(a)　$f(-x) = \sinh(-x) = \dfrac{1}{2}\left(e^{-x} - e^x\right) = -\dfrac{1}{2}\left(e^x - e^{-x}\right) = -\sinh x = -f(x)$

Therefore, $f(x) = \sinh x$ is an odd function.

(b)　Graphing:

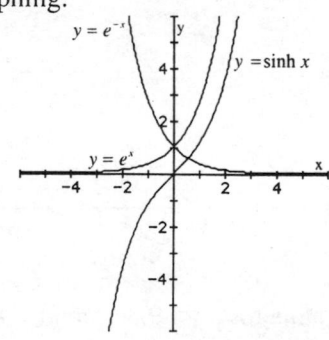

83. $f(x) = 2^{(2^x)} + 1$

$f(1) = 2^{(2^1)} + 1 = 2^2 + 1 = 4 + 1 = 5$

$f(2) = 2^{(2^2)} + 1 = 2^4 + 1 = 16 + 1 = 17$

$f(3) = 2^{(2^3)} + 1 = 2^8 + 1 = 256 + 1 = 257$

$f(4) = 2^{(2^4)} + 1 = 2^{16} + 1 = 65536 + 1 = 65537$

$f(5) = 2^{(2^5)} + 1 = 2^{32} + 1 = 4,294,967,296 + 1 = 4,294,967,297$

$4,294,967,297 = 641 \times 6,700,417$

85. Answers will vary.

Chapter 4

Exponential and Logarithmic Functions

4.3 Logarithmic Functions

1. $9 = 3^2$ is equivalent to $2 = \log_3 9$

3. $a^2 = 1.6$ is equivalent to $2 = \log_a 1.6$

5. $1.1^2 = M$ is equivalent to $2 = \log_{1.1} M$

7. $2^x = 7.2$ is equivalent to $x = \log_2 7.2$

9. $x^{\sqrt{2}} = \pi$ is equivalent to $\sqrt{2} = \log_x \pi$

11. $e^x = 8$ is equivalent to $x = \ln 8$

13. $\log_2 8 = 3$ is equivalent to $2^3 = 8$

15. $\log_a 3 = 6$ is equivalent to $a^6 = 3$

17. $\log_3 2 = x$ is equivalent to $3^x = 2$

18. $\log_2 6 = x$ is equivalent to $2^x = 6$

19. $\log_2 M = 1.3$ is equivalent to $2^{1.3} = M$

21. $\log_{\sqrt{2}} \pi = x$ is equivalent to $\left(\sqrt{2}\right)^x = \pi$

23. $\ln 4 = x$ is equivalent to $e^x = 4$

25. $\log_2 1 = 0$ since $2^0 = 1$

27. $\log_5 25 = 2$ since $5^2 = 25$

29. $\log_{\frac{1}{2}} 16 = -4$ since $\left(\dfrac{1}{2}\right)^{-4} = 2^4 = 16$

31. $\log_{10} \sqrt{10} = \dfrac{1}{2}$ since $10^{1/2} = \sqrt{10}$

33. $\log_{\sqrt{2}} 4 = 4$ since $\left(\sqrt{2}\right)^4 = 4$

35. $\ln \sqrt{e} = \dfrac{1}{2}$ since $e^{1/2} = \sqrt{e}$

37. The domain of $f(x) = \ln(x - 3)$ is:
$$x - 3 > 0 \Rightarrow x > 3$$
$$\{x \mid x > 3\}$$

39. The domain of $F(x) = \log_2 x^2$ is:
$$x^2 > 0$$
$$\{x \mid x \neq 0\}$$

41. The domain of $h(x) = \log_{\frac{1}{2}}\left(x^2 - 2x + 1\right)$ is:
$$x^2 - 2x + 1 > 0 \Rightarrow (x - 1)^2 > 0$$
$$\{x \mid x \neq 1\}$$

43. The domain of $f(x) = \ln\left(\dfrac{1}{x+1}\right)$ is:
$$\frac{1}{x+1} > 0 \Rightarrow x + 1 > 0$$
$$x > -1$$
$$\{x \mid x > -1\}$$

45. The domain of $g(x) = \log_5\left(\dfrac{x+1}{x}\right)$ requires that $\dfrac{x+1}{x} > 0$.

The expression is zero or undefined when $x = -1$ or $x = 0$.

Interval	Test Number	$f(x) = \dfrac{x+1}{x}$	Positive/Negative
$-\infty < x < -1$	-2	0.5	Positive
$-1 < x < 0$	-0.5	-1	Negative
$0 < x < \infty$	1	2	Positive

The domain is $\{x \mid x < -1 \text{ or } x > 0\}$

47. $\ln\left(\dfrac{5}{3}\right) \approx 0.511$

49. $\dfrac{\ln(10/3)}{0.04} \approx 30.099$

51. For $f(x) = \log_a x$, find a so that $f(2) = \log_a 2 = 2$ or $a^2 = 2$ or $a = \sqrt{2}$.
(The base a must be positive by definition.)

53. B   55. D   57. A   59. E

61. $f(x) = \ln(x+4)$
Using the graph of $y = \ln x$, shift the
graph 4 units to the left.
Domain: $(-4, \infty)$
Range: $(-\infty, \infty)$
Vertical Asymptote: $x = -4$

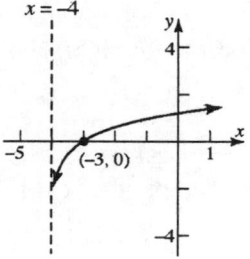

63. $f(x) = 2 + \ln(x) = \ln(x) + 2$
Using the graph of $y = \ln x$, shift up 2
units.
Domain: $(0, \infty)$
Range: $(-\infty, \infty)$
Vertical Asymptote: $x = 0$

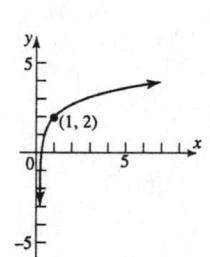

65. $g(x) = \ln(2x)$
Using the graph of $y = \ln x$, compress the
graph horizontally by a factor of $\dfrac{1}{2}$.
Domain: $(0, \infty)$
Range: $(-\infty, \infty)$
Vertical Asymptote: $x = 0$

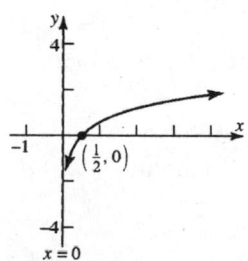

67. $f(x) = 3\ln x$
     Using the graph of $y = \ln x$, stretch the
     graph vertically by a factor of 3.
     Domain: $(0, \infty)$
     Range: $(-\infty, \infty)$
     Vertical Asymptote: $x = 0$

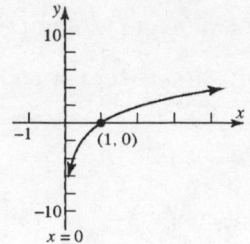

69. $g(x) = \ln(3 - x) = \ln(-(x - 3))$
     Using the graph of $y = \ln x$, reflect the
     graph about the y-axis, and shift 3 units to
     the right.
     Domain: $(-\infty, 3)$
     Range: $(-\infty, \infty)$
     Vertical Asymptote: $x = 3$

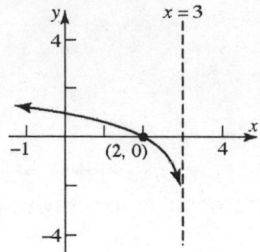

71. $f(x) = -\ln(x - 1)$
     Using the graph of $y = \ln x$, shift the
     graph 1 unit to the right, and reflect about
     the x-axis.
     Domain: $(1, \infty)$
     Range: $(-\infty, \infty)$
     Vertical Asymptote: $x = 1$

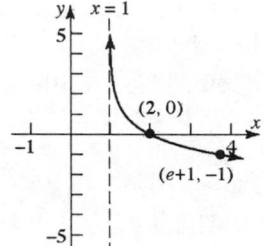

73. $f(x) = \log(x - 4)$
     Using the graph of $y = \log x$, shift 4 units
     to the right.
     Domain: $(4, \infty)$
     Range: $(-\infty, \infty)$
     Vertical Asymptote: $x = 4$

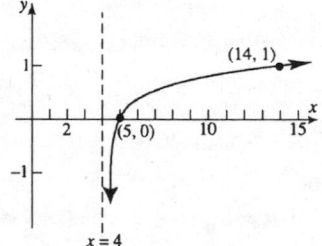

75. $h(x) = 4\log x$
     Using the graph of $y = \log x$, stretch the
     graph vertically by a factor of 4.
     Domain: $(0, \infty)$
     Range: $(-\infty, \infty)$
     Vertical Asymptote: $x = 0$

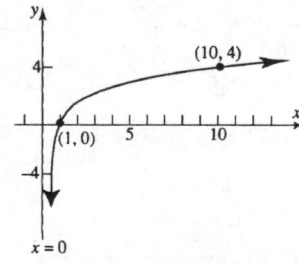

77. $f(x) = \log(2x)$
Using the graph of $y = \log x$, compress
the graph horizontally by a factor of $\dfrac{1}{2}$.

Domain: $(0, \infty)$
Range: $(-\infty, \infty)$
Vertical Asymptote: $x = 0$

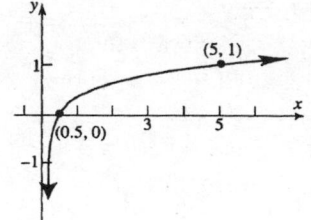

79. $f(x) = 2\log(x + 3)$
Using the graph of $y = \log x$, shift 3 units
to the left and stretch vertically by a
factor of 2.
Domain: $(-3, \infty)$
Range: $(-\infty, \infty)$
Vertical Asymptote: $x = -3$

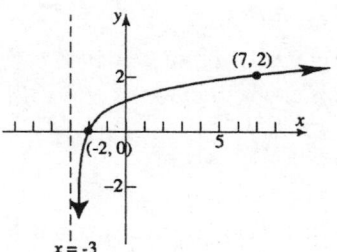

81. $f(x) = 3 + \log(x + 2) = \log(x + 2) + 3$
Using the graph of $y = \log x$, shift 2 units
to the left, and shift up 3 units.
Domain: $(-2, \infty)$
Range: $(-\infty, \infty)$
Vertical Asymptote: $x = -2$

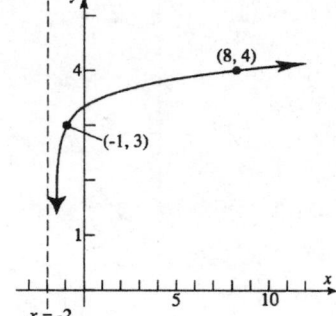

83. $f(x) = \log(2 - x) = \log(-(x - 2))$
Using the graph of $y = \log x$, reflect the
graph about the y-axis, and shift 2 units to
the right.
Domain: $(-\infty, 2)$
Range: $(-\infty, \infty)$
Vertical Asymptote: $x = 2$

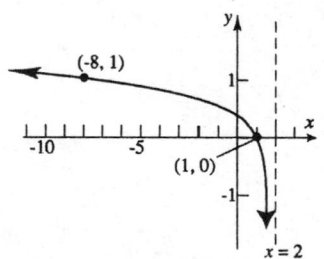

85. $\log_3 x = 2$
$x = 3^2 \Rightarrow x = 9$

87. $\log_2(2x + 1) = 3$
$2x + 1 = 2^3 \Rightarrow 2x + 1 = 8$
$2x = 7 \Rightarrow x = \dfrac{7}{2}$

89. $\log_x 4 = 2$
$x^2 = 4$
$x = 2$ $(x \neq -2,$ base is positive$)$

91. $\ln e^x = 5$
$e^x = e^5 \Rightarrow x = 5$

93. $\log_4 64 = x$

$$4^x = 64 \Rightarrow 4^x = 4^3 \Rightarrow x = 3$$

95. $\log_3 243 = 2x + 1$

$$3^{2x+1} = 243$$
$$3^{2x+1} = 3^5$$
$$2x + 1 = 5 \Rightarrow 2x = 4 \Rightarrow x = 2$$

97. $e^{3x} = 10$

$$3x = \ln(10) \Rightarrow x = \frac{\ln(10)}{3}$$

99. $e^{2x+5} = 8$

$$2x + 5 = \ln(8)$$
$$2x = -5 + \ln(8) \Rightarrow x = \frac{-5 + \ln(8)}{2}$$

101. $\log_3\left(x^2 + 1\right) = 2$

$$x^2 + 1 = 3^2$$
$$x^2 + 1 = 9 \Rightarrow x^2 = 8$$
$$x = \pm\sqrt{8} = \pm 2\sqrt{2}$$

103. $\log_2 8^x = -3$

$$8^x = 2^{-3}$$
$$8^x = \frac{1}{8} \Rightarrow 8^x = 8^{-1} \Rightarrow x = -1$$

105. (a)   Graphing $f(x) = 2^x$:

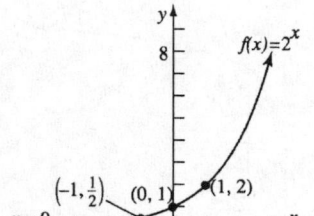

Domain: $(-\infty, \infty)$
Range: $(0, \infty)$
Horizontal asymptote: $y = 0$

(b)   Finding the inverse:
$$f(x) = 2^x$$
$$y = 2^x$$
$$x = 2^y \quad \text{Inverse}$$
$$y = \log_2 x$$
$$f^{-1}(x) = \log_2 x$$

Domain: $(0, \infty)$
Range: $(-\infty, \infty)$
Vertical asymptote: $x = 0$

(c)   Graphing the inverse:

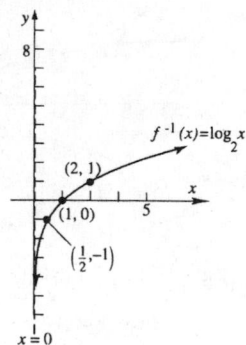

107. (a)   Graphing $f(x) = 2^{x+3}$:

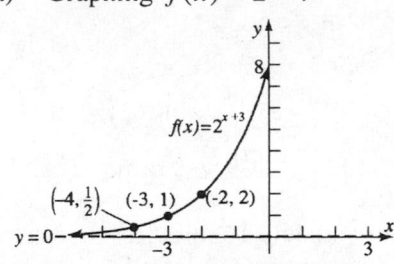

Domain: $(-\infty, \infty)$
Range: $(0, \infty)$
Horizontal asymptote: $y = 0$

(b)   Finding the inverse:
$$f(x) = 2^{x+3}$$
$$y = 2^{x+3}$$
$$x = 2^{y+3} \qquad \text{Inverse}$$
$$y + 3 = \log_2 x$$
$$y = -3 + \log_2 x$$
$$f^{-1}(x) = -3 + \log_2 x$$

(c)   Graphing the inverse:

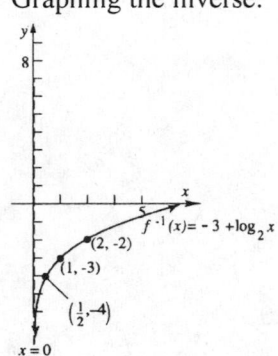

Domain: $(0, \infty)$
Range: $(-\infty, \infty)$
Vertical asymptote: $x = 0$

109. $P = 100e^{-0.1n}$

(a) $50 = 100e^{-0.1n}$
          $0.5 = e^{-0.1n}$

$\ln(0.5) = -0.1n$

$n = \dfrac{\ln(0.5)}{-0.1} \Rightarrow n \approx 6.93$

7 panes of glass are needed.

(b) $25 = 100e^{-0.1n}$
          $0.25 = e^{-0.1n}$

$\ln(0.25) = -0.1n$

$n = \dfrac{\ln(0.25)}{-0.1} \Rightarrow n \approx 13.86$

14 panes of glass are needed.

111. $w = 50e^{-0.004d}$

(a) $30 = 50e^{-0.004d}$
          $0.6 = e^{-0.004d}$

$\ln(0.6) = -0.004d$

$d = \dfrac{\ln(0.6)}{-0.004} \approx 127.7$

Approximately 128 days.

(b) $5 = 50e^{-0.004d}$
          $0.1 = e^{-0.004d}$

$\ln(0.1) = -0.004d$

$d = \dfrac{\ln(0.1)}{-0.004} \approx 575.6$

Approximately 576 days.

253

113. $F(t) = 1 - e^{-0.1t}$

    (a) $0.5 = 1 - e^{-0.1t}$

            $-0.5 = -e^{-0.1t}$

            $0.5 = e^{-0.1t} \Rightarrow \ln(0.5) = -0.1t$

            $t = \dfrac{\ln(0.5)}{-0.1} \approx 6.93$

    Approximately 7 minutes.

    (b) $0.8 = 1 - e^{-0.1t}$

            $-0.2 = -e^{-0.1t}$

            $0.2 = e^{-0.1t} \Rightarrow \ln(0.2) = -0.1t$

            $t = \dfrac{\ln(0.2)}{-0.1} \approx 16.09$

    Approximately 16 minutes.

    (c)   It is impossible for the probability to reach 100% because $e^{-0.1t}$ will never equal zero.

115. $D = 5e^{-0.4h}$

         $2 = 5e^{-0.4h}$

         $0.4 = e^{-0.4h}$

    $\ln(0.4) = -0.4h$

         $h = \dfrac{\ln(0.4)}{-0.4} \approx 2.29$ hours

117. $I = \dfrac{E}{R}\left[1 - e^{-\left(\frac{R}{L}\right)t} \right]$

    0.5 ampere:

     $0.5 = \dfrac{12}{10}\left[1 - e^{-\left(\frac{10}{5}\right)t} \right]$

     $0.4167 = 1 - e^{-2t}$

     $e^{-2t} = 0.5833$

     $-2t = \ln(0.5833)$

     $t = \dfrac{\ln(0.5833)}{-2} \approx 0.2695$ seconds

    1.0 ampere:

     $1.0 = \dfrac{12}{10}\left[1 - e^{-\left(\frac{10}{5}\right)t} \right]$

     $0.8333 = 1 - e^{-2t}$

     $e^{-2t} = 0.1667$

     $-2t = \ln(0.1667)$

     $t = \dfrac{\ln(0.1667)}{-2} \approx 0.8958$ seconds

119. $L(10^{-7}) = 10 \log\left(\dfrac{10^{-7}}{10^{-12}}\right) = 10 \log(10^5) = 10 \cdot 5 = 50$ decibels

121. $L(10^{-1}) = 10 \log\left(\dfrac{10^{-1}}{10^{-12}}\right) = 10 \log(10^{11}) = 10 \cdot 11 = 110$ decibels

123. $M(125{,}892) = \log\left(\dfrac{125{,}892}{10^{-3}}\right) = 8.1$

125. $R = 3e^{kx}$

(a) $$10 = 3e^{k(0.06)}$$
$$3.3333 = e^{0.06k}$$
$$\ln(3.3333) = 0.06k$$
$$k = \frac{\ln(3.3333)}{0.06}$$
$$k \approx 20.066$$

(b) $$R = 3e^{20.066(0.17)}$$
$$R = 3e^{3.41122}$$
$$R = 90.9\%$$

(c) $$100 = 3e^{20.066x}$$
$$33.3333 = e^{20.066x}$$
$$\ln(33.3333) = 20.066x$$
$$x = \frac{\ln(33.3333)}{20.066}$$
$$x \approx 0.175$$

(d) $$15 = 3e^{20.066x}$$
$$5 = e^{20.066x}$$
$$\ln(5) = 20.066x$$
$$x = \frac{\ln(5)}{20.066}$$
$$x \approx 0.08$$

Chapter 4

Exponential and Logarithmic Functions

4.4 Properties of Logarithms

1. $\log_3 3^{71} = 71$ 3. $\ln e^{-4} = -4$ 5. $2^{\log_2 7} = 7$

7. $\log_8 2 + \log_8 4 = \log_8(4 \cdot 2) = \log_8(8) = 1$

9. $\log_6 18 - \log_6 3 = \log_6\left(\dfrac{18}{3}\right) = \log_6(6) = 1$

11. $\log_2 6 \cdot \log_6 4$

$= \log_6\left(4^{\log_2 6}\right) = \log_6\left(\left(2^2\right)^{\log_2 6}\right)$

$= \log_6\left((2)^{2\log_2 6}\right) = \log_6\left((2)^{\log_2 6^2}\right) = \log_6\left(6^2\right) = 2$

13. $3^{\log_3 5 - \log_3 4} = 3^{\log_3\left(\frac{5}{4}\right)} = \dfrac{5}{4}$

15. $e^{\log_{e^2} 16}$

Simplify the exponent:

$$\text{Let} \qquad a = \log_{e^2} 16$$

$$\left(e^2\right)^a = 16$$

$$e^{2a} = 16 = 4^2$$

$$e^a = 4 \Rightarrow a = \ln 4$$

$$\text{Thus, } e^{\log_{e^2} 16} = e^{\ln 4} = 4$$

17. $\ln 6 = \ln(3 \cdot 2) = \ln 3 + \ln 2 = b + a$ 19. $\ln(1.5) = \ln\left(\dfrac{3}{2}\right) = \ln 3 - \ln 2 = b - a$

21. $\ln 8 = \ln 2^3 = 3 \cdot \ln 2 = 3a$

256

23. $\ln\left(\sqrt[5]{6}\right) = \ln\left(6^{1/5}\right) = \frac{1}{5} \cdot \ln 6 = \frac{1}{5} \cdot \ln(2 \cdot 3) = \frac{1}{5} \cdot (\ln 2 + \ln 3) = \frac{1}{5} \cdot (a + b)$

25. $\log_5(25x) = \log_5(25) + \log_5(x) = 2 + \log_5(x)$

27. $\log_2\left(z^3\right) = 3\log_2(z)$ 29. $\ln(ex) = \ln e + \ln x = 1 + \ln x$

31. $\ln\left(xe^x\right) = \ln x + \ln e^x = \ln x + x$

33. $\log_a\left(u^2 v^3\right) = \log_a u^2 + \log_a v^3 = 2\log_a u + 3\log_a v$

35. $\ln\left(x^2\sqrt{1-x}\right) = \ln x^2 + \ln\sqrt{1-x} = \ln x^2 + \ln(1-x)^{1/2} = 2\ln x + \frac{1}{2}\ln(1-x)$

37. $\log_2\left(\dfrac{x^3}{x-3}\right) = \log_2 x^3 - \log_2(x-3) = 3\log_2 x - \log_2(x-3)$

39. $\log\left[\dfrac{x(x+2)}{(x+3)^2}\right] = \log(x(x+2)) - \log(x+3)^2 = \log x + \log(x+2) - 2\log(x+3)$

41. $\ln\left[\dfrac{x^2-x-2}{(x+4)^2}\right]^{1/3} = \frac{1}{3}\ln\left[\dfrac{(x-2)(x+1)}{(x+4)^2}\right] = \frac{1}{3}\left[\ln(x-2)(x+1) - \ln(x+4)^2\right]$

$= \frac{1}{3}\left[\ln(x-2) + \ln(x+1) - 2\ln(x+4)\right] = \frac{1}{3}\ln(x-2) + \frac{1}{3}\ln(x+1) - \frac{2}{3}\ln(x+4)$

43. $\ln\left(\dfrac{5x\sqrt{1-3x}}{(x-4)^3}\right) = \ln\left(5x\sqrt{1-3x}\right) - \ln(x-4)^3 = \ln 5 + \ln x + \ln\sqrt{1-3x} - 3\ln(x-4)$

$= \ln 5 + \ln x + \ln(1-3x)^{1/2} - 3\ln(x-4) = \ln 5 + \ln x + \frac{1}{2}\ln(1-3x) - 3\ln(x-4)$

45. $3\log_5 u + 4\log_5 v = \log_5 u^3 + \log_5 v^4 = \log_5(u^3 v^4)$

47. $\log_3\sqrt{x} - \log_3 x^3 = \log_3\left(\dfrac{\sqrt{x}}{x^3}\right) = \log_3\left(\dfrac{x^{1/2}}{x^3}\right) = \log_3\left(x^{-5/2}\right) = -\frac{5}{2}\log_3 x$

49. $\log_4\left(x^2-1\right) - 5\log_4(x+1) = \log_4\left(x^2-1\right) - \log_4(x+1)^5 = \log_4\left(\dfrac{x^2-1}{(x+1)^5}\right)$

$= \log_4\left(\dfrac{(x+1)(x-1)}{(x+1)^5}\right) = \log_4\left(\dfrac{x-1}{(x+1)^4}\right)$

51. $\ln\left(\dfrac{x}{x-1}\right)+\ln\left(\dfrac{x+1}{x}\right)-\ln\left(x^2-1\right)=\ln\left|\dfrac{x}{x-1}\cdot\dfrac{x+1}{x}\right|-\ln\left(x^2-1\right)=\ln\left|\dfrac{x+1}{x-1}\div\left(x^2-1\right)\right|$

$$=\ln\left|\dfrac{x+1}{(x-1)(x-1)(x+1)}\right|=\ln\left(\dfrac{1}{(x-1)^2}\right)=\ln(x-1)^{-2}=-2\ln(x-1)$$

53. $8\log_2\sqrt{3x-2}-\log_2\left(\dfrac{4}{x}\right)+\log_2 4=\log_2\left(\sqrt{3x-2}\right)^8-\left(\log_2 4-\log_2 x\right)+\log_2 4$

$$=\log_2(3x-2)^4-\log_2 4+\log_2 x+\log_2 4=\log_2\left[x(3x-2)^4\right]$$

55. $2\log_a\left(5x^3\right)-\dfrac{1}{2}\log_a(2x+3)=\log_a\left(5x^3\right)^2-\log_a(2x-3)^{1/2}=\log_a\left[\dfrac{25x^6}{(2x-3)^{1/2}}\right]$

57. $2\log_2(x+1)-\log_2(x+3)-\log_2(x-1)=\log_2(x+1)^2-\log_2(x+3)-\log_2(x-1)$

$$=\log_2\left(\dfrac{(x+1)^2}{(x+3)}\right)-\log_2(x-1)=\log_2\left(\dfrac{\dfrac{(x+1)^2}{(x+3)}}{(x-1)}\right)=\log_2\left(\dfrac{(x+1)^2}{(x+3)(x-1)}\right)$$

59. $y=ab^x$

$\log(y)=\log\left(ab^x\right)=\log(a)+\log\left(b^x\right)=\log(a)+x\log(b)$

61. $\log_3 21=\dfrac{\log 21}{\log 3}\approx\dfrac{1.32222}{0.47712}\approx 2.771$

63. $\log_{\frac{1}{3}}71=\dfrac{\log 71}{\log\left(\dfrac{1}{3}\right)}=\dfrac{\log 71}{-\log 3}\approx\dfrac{1.85126}{-0.47712}\approx -3.880$

65. $\log_{\sqrt{2}}7=\dfrac{\log 7}{\log\sqrt{2}}=\dfrac{\log 7}{\log 2^{1/2}}=\dfrac{\log 7}{\left(\dfrac{1}{2}\log 2\right)}\approx\dfrac{0.84510}{0.5(0.30103)}\approx 5.615$

67. $\log_\pi e=\dfrac{\ln e}{\ln\pi}\approx\dfrac{1}{1.14473}\approx 0.874$ 69. $y=\log_4 x=\dfrac{\ln x}{\ln 4}$ or $y=\dfrac{\log x}{\log 4}$

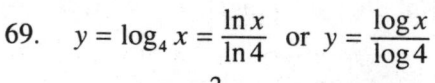

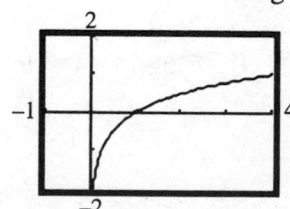

71. $y = \log_2(x + 2) = \dfrac{\ln(x + 2)}{\ln 2}$

or $y = \dfrac{\log(x + 2)}{\log 2}$

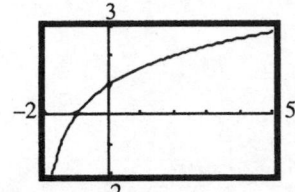

73. $y = \log_{x-1}(x + 1) = \dfrac{\ln(x + 1)}{\ln(x - 1)}$

or $y = \dfrac{\log(x + 1)}{\log(x - 1)}$

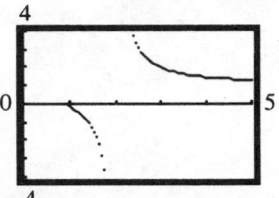

75. $\ln y = \ln x + \ln C$

$\ln y = \ln(xC)$

$y = Cx$

77. $\ln y = \ln x + \ln(x + 1) + \ln C$

$\ln y = \ln\big(x(x + 1)C\big)$

$y = Cx(x + 1)$

79. $\ln y = 3x + \ln C$

$\ln y = \ln e^{3x} + \ln C$

$\ln y = \ln(Ce^{3x})$

$y = Ce^{3x}$

81. $\ln(y - 3) = -4x + \ln C$

$\ln(y - 3) = \ln e^{-4x} + \ln C$

$\ln(y - 3) = \ln\big(Ce^{-4x}\big)$

$y - 3 = Ce^{-4x} \Rightarrow y = Ce^{-4x} + 3$

83. $3\ln y = \dfrac{1}{2}\ln(2x + 1) - \dfrac{1}{3}\ln(x + 4) + \ln C$

$\ln y^3 = \ln(2x + 1)^{1/2} - \ln(x + 4)^{1/3} + \ln C$

$\ln y^3 = \ln\left[\dfrac{C(2x + 1)^{1/2}}{(x + 4)^{1/3}}\right] \Rightarrow y^3 = \dfrac{C(2x + 1)^{1/2}}{(x + 4)^{1/3}}$

$y = \left[\dfrac{C(2x + 1)^{1/2}}{(x + 4)^{\frac{1}{3}}}\right]^{1/3} = \dfrac{\sqrt[3]{C}(2x + 1)^{1/6}}{(x + 4)^{1/9}}$

85. $\log_2 3 \cdot \log_3 4 \cdot \log_4 5 \cdot \log_5 6 \cdot \log_6 7 \cdot \log_7 8$

$= \dfrac{\log 3}{\log 2} \cdot \dfrac{\log 4}{\log 3} \cdot \dfrac{\log 5}{\log 4} \cdot \dfrac{\log 6}{\log 5} \cdot \dfrac{\log 7}{\log 6} \cdot \dfrac{\log 8}{\log 7} = \dfrac{\log 8}{\log 2} = \dfrac{\log 2^3}{\log 2} = \dfrac{3\log 2}{\log 2} = 3$

87. $\log_2 3 \cdot \log_3 4 \cdot \ldots \cdot \log_n(n + 1) \cdot \log_{n+1} 2$

$= \dfrac{\log 3}{\log 2} \cdot \dfrac{\log 4}{\log 3} \cdot \ldots \cdot \dfrac{\log(n + 1)}{\log n} \cdot \dfrac{\log 2}{\log(n + 1)} = \dfrac{\log 2}{\log 2} = 1$

89. Verifying:

$$\log_a\!\left(x + \sqrt{x^2 - 1}\right) + \log_a\!\left(x - \sqrt{x^2 - 1}\right) = \log_a\!\left[\left(x + \sqrt{x^2 - 1}\right)\!\left(x - \sqrt{x^2 - 1}\right)\right]$$

$$= \log_a\!\left[x^2 - \left(x^2 - 1\right)\right] = \log_a\!\left[x^2 - x^2 + 1\right] = \log_a 1 = 0$$

91. Verifying:

$$2x + \ln\!\left(1 + e^{-2x}\right) = \ln e^{2x} + \ln\!\left(1 + e^{-2x}\right) = \ln\!\left(e^{2x}\!\left(1 + e^{-2x}\right)\right) = \ln\!\left(e^{2x} + e^0\right) = \ln\!\left(e^{2x} + 1\right)$$

93. $f(x) = \log_a x$

$$x = a^{f(x)} \implies x^{-1} = a^{-f(x)} = \left(a^{-1}\right)^{f(x)} = \left(\frac{1}{a}\right)^{f(x)}$$

$$\log_{\frac{1}{a}} x^{-1} = f(x) \implies -\log_{\frac{1}{a}} x = f(x) \implies -f(x) = \log_{\frac{1}{a}} x$$

95. $f(x) = \log_a x$

$$a^{f(x)} = x \implies \frac{1}{a^{f(x)}} = \frac{1}{x} \implies a^{-f(x)} = \frac{1}{x} \implies -f(x) = \log_a \frac{1}{x} = f\left(\frac{1}{x}\right)$$

97.   If $A = \log_a M$ and $B = \log_a N$, then $a^A = M$ and $a^B = N$.

$$\log_a\left(\frac{M}{N}\right) = \log_a\left(\frac{a^A}{a^B}\right) = \log_a a^{A-B} = A - B = \log_a M - \log_a N$$

Chapter 4

Exponential and Logarithmic Functions

4.5 Logarithmic and Exponential Equations

1. $\log_4(x+2) = \log_4 8$

$x + 2 = 8 \Rightarrow x = 6$

3. $\dfrac{1}{2}\log_3 x = 2\log_3 2$

$\log_3 x^{1/2} = \log_3 2^2$

$x^{1/2} = 4 \Rightarrow x = 16$

5. $2\log_5 x = 3\log_5 4$

$\log_5 x^2 = \log_5 4^3$

$x^2 = 64$

$x = \pm 8$

Since $\log_5(-8)$ is undefined, the only solution is $x = 8$.

7. $3\log_2(x-1) + \log_2 4 = 5$

$\log_2(x-1)^3 + \log_2 4 = 5$

$\log_2\!\left(4(x-1)^3\right) = 5$

$4(x-1)^3 = 2^5$

$(x-1)^3 = \dfrac{32}{4}$

$(x-1)^3 = 8$

$x - 1 = 2$

$x = 3$

9. $\log x + \log(x+15) = 2$

$\log\!\left(x(x+15)\right) = 2$

$x(x+15) = 10^2$

$x^2 + 15x - 100 = 0 \Rightarrow (x+20)(x-5) = 0$

$x = -20$ or $x = 5$

Since $\log(-20)$ is undefined, the only solution is $x = 5$.

11. $\ln x + \ln(x+2) = 4$

$\ln\!\left(x(x+2)\right) = 4$

$x(x+2) = e^4 \Rightarrow x^2 + 2x - e^4 = 0$

$x = \dfrac{-2 \pm \sqrt{2^2 - 4(1)(-e^4)}}{2(1)} = \dfrac{-2 \pm \sqrt{4 + 4e^4}}{2} = \dfrac{-2 \pm 2\sqrt{1+e^4}}{2} = -1 \pm \sqrt{1+e^4}$

Since $\ln\!\left(-1 - \sqrt{1+e^4}\right)$ is undefined, the only solution is $x = -1 + \sqrt{1+e^4} \approx 6.456$.

13. $2^{2x} + 2^x - 12 = 0$

$\left(2^x\right)^2 + 2^x - 12 = 0 \Rightarrow \left(2^x - 3\right)\left(2^x + 4\right) = 0$

$2^x - 3 = 0 \qquad$ or $\; 2^x + 4 = 0$

$2^x = 3 \qquad$ or $\qquad 2^x = -4$

$x = \log_2 3 \qquad\qquad$ No solution

$x \approx 1.585$

The only solution is $x = \log_2 3 \approx 1.585$.

15. $3^{2x} + 3^{x+1} - 4 = 0$

$\left(3^x\right)^2 + 3 \cdot 3^x - 4 = 0 \Rightarrow \left(3^x - 1\right)\left(3^x + 4\right) = 0$

$3^x - 1 = 0 \quad$ or $\qquad 3^x + 4 = 0$

$3^x = 1 \quad$ or $\qquad 3^x = -4$

$x = 0 \qquad\qquad$ No solution

The only solution is $x = 0$.

17. $2^x = 10$

$\log\left(2^x\right) = \log 10$

$x \log 2 = 1$

$x = \dfrac{1}{\log 2} \approx 3.322$

19. $8^{-x} = 1.2$

$\log\left(8^{-x}\right) = \log(1.2)$

$-x \log 8 = \log(1.2)$

$x = \dfrac{\log(1.2)}{-\log 8} \approx -0.088$

21. $3^{1-2x} = 4^x$

$\log\left(3^{1-2x}\right) = \log\left(4^x\right) \Rightarrow (1 - 2x)\log 3 = x \log 4$

$\log 3 - 2x \log 3 = x \log 4 \Rightarrow \log 3 = x \log 4 + 2x \log 3 \Rightarrow \log 3 = x(\log 4 + 2 \log 3)$

$x = \dfrac{\log 3}{\log 4 + 2 \log 3} \approx 0.307$

23. $\left(\dfrac{3}{5}\right)^x = 7^{1-x}$

$\log\left(\left(\dfrac{3}{5}\right)^x\right) = \log\left(7^{1-x}\right)$

$x \log\left(\dfrac{3}{5}\right) = (1 - x)\log 7 \Rightarrow x(\log 3 - \log 5) = \log 7 - x \log 7$

$x \log 3 - x \log 5 + x \log 7 = \log 7 \Rightarrow x(\log 3 - \log 5 + \log 7) = \log 7$

$x = \dfrac{\log 7}{\log 3 - \log 5 + \log 7} \approx 1.356$

25. $1.2^x = (0.5)^{-x}$

$$\log 1.2^x = \log(0.5)^{-x}$$
$$x\log(1.2) = -x\log(0.5)$$
$$x\log(1.2) + x\log(0.5) = 0$$
$$x(\log(1.2) + \log(0.5)) = 0$$
$$x = 0$$

27. $\pi^{1-x} = e^x$

$$\ln \pi^{1-x} = \ln e^x$$
$$(1-x)\ln \pi = x$$
$$\ln \pi - x\ln \pi = x$$
$$\ln \pi = x + x\ln \pi$$
$$\ln \pi = x(1 + \ln \pi)$$
$$x = \frac{\ln \pi}{1 + \ln \pi} \approx 0.534$$

29. $5\left(2^{3x}\right) = 8$

$$2^{3x} = \frac{8}{5}$$

$$\log 2^{3x} = \log\left(\frac{8}{5}\right)$$

$$3x\log 2 = \log 8 - \log 5$$

$$x = \frac{\log 8 - \log 5}{3\log 2} \approx 0.226$$

31. $\log_a(x-1) - \log_a(x+6) = \log_a(x-2) - \log_a(x+3)$

$$\log_a\left(\frac{x-1}{x+6}\right) = \log_a\left(\frac{x-2}{x+3}\right) \Rightarrow a^{\left(\log_a\left(\frac{x-1}{x+6}\right)\right)} = a^{\left(\log_a\left(\frac{x-2}{x+3}\right)\right)}$$

so

$$\frac{x-1}{x+6} = \frac{x-2}{x+3} \Rightarrow (x-1)(x+3) = (x-2)(x+6)$$

$$x^2 + 2x - 3 = x^2 + 4x - 12 \Rightarrow 2x - 3 = 4x - 12 \Rightarrow 9 = 2x \to x = \frac{9}{2}$$

Since each of the original logarithms is defined for $x = \frac{9}{2}$, the solution is $x = \frac{9}{2}$.

33. $\log_{\frac{1}{3}}(x^2 + x) - \log_{\frac{1}{3}}(x^2 - x) = -1$

$$\log_{\frac{1}{3}}\left(\frac{x^2 + x}{x^2 - x}\right) = -1$$

$$\frac{x^2 + x}{x^2 - x} = \left(\frac{1}{3}\right)^{-1}$$

$$\frac{x(x+1)}{x(x-1)} = 3 \Rightarrow x + 1 = 3(x-1)$$

$$x + 1 = 3x - 3 \Rightarrow -2x = -4 \Rightarrow x = 2$$

35. $\log_2(x+1) - \log_4 x = 1$

$\log_2(x+1) - \dfrac{\log_2 x}{\log_2 4} = 1$

$\log_2(x+1) - \dfrac{\log_2 x}{2} = 1$

$2\log_2(x+1) - \log_2 x = 2$

$\log_2(x+1)^2 - \log_2 x = 2$

$\log_2\left(\dfrac{(x+1)^2}{x}\right) = 2$

$\dfrac{(x+1)^2}{x} = 2^2$

$x^2 + 2x + 1 = 4x$

$x^2 - 2x + 1 = 0$

$(x-1)^2 = 0$

$x - 1 = 0$

$x = 1$

37. $\log_{16} x + \log_4 x + \log_2 x = 7$

$\dfrac{\log_2 x}{\log_2 16} + \dfrac{\log_2 x}{\log_2 4} + \log_2 x = 7$

$\dfrac{\log_2 x}{4} + \dfrac{\log_2 x}{2} + \log_2 x = 7$

$\log_2 x + 2\log_2 x + 4\log_2 x = 28$

$7\log_2 x = 28$

$\log_2 x = 4$

$x = 2^4 = 16$

39. $\left(\sqrt[3]{2}\right)^{2-x} = 2^{x^2}$

$\left(2^{1/3}\right)^{2-x} = 2^{x^2}$

$2^{\frac{1}{3}(2-x)} = 2^{x^2}$

$\dfrac{1}{3}(2-x) = x^2 \Rightarrow 2 - x = 3x^2$

$3x^2 + x - 2 = 0 \Rightarrow (3x-2)(x+1) = 0$

$x = \dfrac{2}{3}$　or　$x = -1$

41. $\dfrac{e^x + e^{-x}}{2} = 1$

$e^x + e^{-x} = 2$

$e^x\left(e^x + e^{-x}\right) = 2e^x$

$e^{2x} + 1 = 2e^x$

$(e^x)^2 - 2e^x + 1 = 0$

$\left(e^x - 1\right)^2 = 0$

$e^x - 1 = 0$

$e^x = 1$

$x = 0$

43. $\dfrac{e^x - e^{-x}}{2} = 2$

$e^x - e^{-x} = 4$

$e^x\left(e^x - e^{-x}\right) = 4e^x$

$e^{2x} - 1 = 4e^x$

$(e^x)^2 - 4e^x - 1 = 0$

$e^x = \dfrac{-(-4) \pm \sqrt{(-4)^2 - 4(1)(-1)}}{2(1)}$

$= \dfrac{4 \pm \sqrt{20}}{2} = \dfrac{4 \pm 2\sqrt{5}}{2} = 2 \pm \sqrt{5}$

$x = \ln\left(2 + \sqrt{5}\right)$

Since $\ln\left(2 - \sqrt{5}\right)$ is undefined; it is not a solution.

45. Using INTERSECT to solve:
$$y_1 = \ln(x) / \ln(5) + \ln(x) / \ln(3)$$
$$y_2 = 1$$

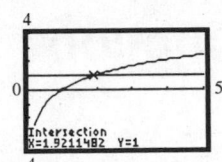

The solution is 1.92.

47. Using INTERSECT to solve:
$$y_1 = \ln(x+1) / \ln(5) - \ln(x-2) / \ln(4)$$
$$y_2 = 1$$

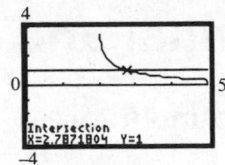

The solution is 2.79.

49. Using INTERSECT to solve:
$$y_1 = e^x; \quad y_2 = -x$$

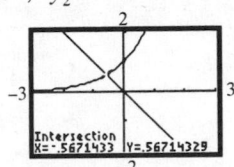

The solution is -0.57.

51. Using INTERSECT to solve:
$$y_1 = e^x; \quad y_2 = x^2$$

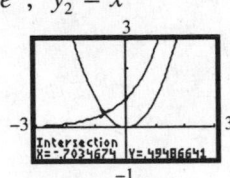

The solution is -0.70.

53. Using INTERSECT to solve:
$$y_1 = \ln x; \quad y_2 = -x$$

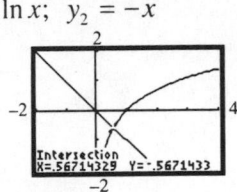

The solution is 0.57.

55. Using INTERSECT to solve:
$$y_1 = \ln x; \quad y_2 = x^3 - 1$$

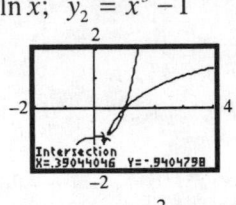

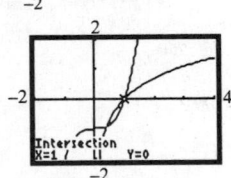

The solutions are 0.39, 1.00.

57. Using INTERSECT to solve:
$$y_1 = e^x + \ln x; \quad y_2 = 4$$

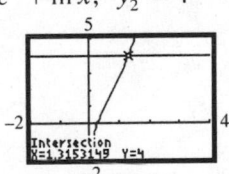

The solution is 1.32.

59. Using INTERSECT to solve:
$$y_1 = e^{-x}; \quad y_2 = \ln x$$

The solution is 1.31.

Exponential and Logarithmic Functions

4.6 Compound Interest

1. $P = \$100,\ r = 0.04,\ n = 4,\ t = 2$
$$A = P\left(1 + \frac{r}{n}\right)^{nt} = 100\left(1 + \frac{0.04}{4}\right)^{(4)(2)} \approx \$108.29$$

3. $P = \$500,\ r = 0.08,\ n = 4,\ t = 2.5$
$$A = P\left(1 + \frac{r}{n}\right)^{nt} = 500\left(1 + \frac{0.08}{4}\right)^{(4)(2.5)} \approx \$609.50$$

5. $P = \$600,\ r = 0.05,\ n = 365,\ t = 3$
$$A = P\left(1 + \frac{r}{n}\right)^{nt} = 600\left(1 + \frac{0.05}{365}\right)^{(365)(3)} \approx \$697.09$$

7. $P = \$10,\ r = 0.11,\ t = 2$
$$A = Pe^{rt} = 10e^{(0.11)(2)} \approx \$12.46$$

9. $P = \$100,\ r = 0.10,\ t = 2.25$
$$A = Pe^{rt} = 100e^{(0.10)(2.25)} \approx \$125.23$$

11. $A = \$100,\ r = 0.06,\ n = 12,\ t = 2$
$$P = A\left(1 + \frac{r}{n}\right)^{-nt} = 100\left(1 + \frac{0.06}{12}\right)^{(-12)(2)} \approx \$88.72$$

13. $A = \$1000,\ r = 0.06,\ n = 365,\ t = 2.5$
$$P = A\left(1 + \frac{r}{n}\right)^{-nt} = 1000\left(1 + \frac{0.06}{365}\right)^{(-365)(2.5)} \approx \$860.72$$

15. $A = \$600,\ r = 0.04,\ n = 4,\ t = 2$
$$P = A\left(1 + \frac{r}{n}\right)^{-nt} = 600\left(1 + \frac{0.04}{4}\right)^{(-4)(2)} \approx \$554.09$$

17. $A = \$80,\ r = 0.09,\ t = 3.25$
$$P = Ae^{-rt} = 80e^{(-0.09)(3.25)} \approx \$59.71$$

19. $A = \$400,\ r = 0.10,\ t = 1$
$$P = Ae^{-rt} = 400e^{(-0.10)(1)} \approx \$361.93$$

21. $r_e = \left(1 + \dfrac{r}{n}\right)^n - 1 = \left(1 + \dfrac{0.0525}{4}\right)^4 - 1 \approx 1.0535 - 1 = 0.0535 = 5.35\%$

23. $2P = P(1 + r)^3$

    $2 = (1 + r)^3$

    $\sqrt[3]{2} = 1 + r$

    $r = \sqrt[3]{2} - 1 \approx 1.26 - 1 = 0.26 = 26\%$

25. 6% compounded quarterly:

    $$A = 10,000\left(1 + \dfrac{0.06}{4}\right)^{(4)(1)} \approx \$10,613.64$$

    $6\frac{1}{4}\%$ compounded annually:

    $$A = 10,000(1 + 0.0625)^1 \approx \$10,625$$

    $6\frac{1}{4}\%$ compounded annually yields the larger amount.

27. 9% compounded monthly:

    $$A = 10,000\left(1 + \dfrac{0.09}{12}\right)^{(12)(1)} \approx \$10,938.07$$

    8.8% compounded daily:

    $$A = 10,000\left(1 + \dfrac{0.088}{365}\right)^{365} \approx \$10,919.77$$

    9% compounded monthly yields the larger amount.

29. Compounded monthly:

    $$2P = P\left(1 + \dfrac{0.08}{12}\right)^{12t}$$

    $2 \approx (1.00667)^{12t}$

    $\ln 2 \approx 12t \ln(1.00667)$

    $t \approx \dfrac{\ln 2}{12 \ln(1.00667)} \approx 8.69 \text{ years}$

    Compounded continuously:

    $2P = Pe^{0.08t}$

    $2 = e^{0.08t}$

    $\ln 2 = 0.08t$

    $t = \dfrac{\ln 2}{0.08} \approx 8.66 \text{ years}$

31. Compounded monthly:

    $$150 = 100\left(1 + \dfrac{0.08}{12}\right)^{12t}$$

    $1.5 \approx (1.00667)^{12t}$

    $\ln(1.5) \approx 12t \ln(1.00667)$

    $t \approx \dfrac{\ln(1.5)}{12 \ln(1.00667)} \approx 5.083 \text{ years}$

    Compounded continuously:

    $150 = 100e^{0.08t}$

    $1.5 = e^{0.08t}$

    $\ln(1.5) = 0.08t$

    $t = \dfrac{\ln(1.5)}{0.08} \approx 5.068 \text{ years}$

33.　$25,000 = 10,000e^{0.06t}$

　　　$2.5 = e^{0.06t}$

　　$\ln(2.5) = 0.06t$

　　　　$t = \dfrac{\ln(2.5)}{0.06} \approx 15.27$ years

35.　$A = 90,000(1 + 0.03)^5 \approx \$104,335$

37.　$P = 15,000e^{(-0.05)(3)} \approx \$12,910.62$

39.　$A = 1500(1 + 0.15)^5 = 1500(1.15)^5 \approx \$3017$

41.　$850,000 = 650,000(1 + r)^3$

　　$\dfrac{85}{65} = (1 + r)^3 \Rightarrow \sqrt[3]{\dfrac{85}{65}} = 1 + r \Rightarrow r \approx \sqrt[3]{1.3077} - 1 \approx 0.0935 = 9.35\%$

43.　5.6% compounded continuously:

　　　　$A = 1000e^{(0.056)(1)} \approx \$1057.60$

　　Jim does not have enough money to buy the computer.

　　5.9% compounded monthly:

　　　　$A = 1000\left(1 + \dfrac{0.059}{12}\right)^{12} \approx \$1060.62$

　　The second bank offers the better deal.

45.　Will - 9% compounded semiannually:

　　　　$A = 2000\left(1 + \dfrac{0.09}{2}\right)^{(2)(20)} \approx \$11,632.73$

　　Henry - 8.5% compounded continuously:

　　　　$A = 2000e^{(0.085)(20)} \approx \$10,947.89$

　　Will has more money after 20 years.

47.　$P = 50,000; \ t = 5$

　　(a)　Simple interest at 12% per annum:

　　　　　$A = 50,000 + 50,000(0.12)(5) = \$80,000$

　　(b)　11.5% compounded monthly:

　　　　　$A = 50,000\left(1 + \dfrac{0.115}{12}\right)^{(12)(5)} \approx \$88,613.59$

　　(c)　11.25% compounded continuously:

　　　　　$A = 50,000e^{(0.1125)(5)} \approx \$87,752.73$

　　　　Subtract $50,000 from each to get the amount of interest:

　　　　　(a)　$30,000　　(b)　$38,613.59　　　(c)　$37.752.73

　　Option (a) results in the least interest.

49.　(a)　$A = \$10,000, \ r = 0.10, \ n = 12, \ t = 20$ (compounded monthly)

　　　　　$P = 10,000\left(1 + \dfrac{0.10}{12}\right)^{(-12)(20)} \approx \$1364.62$

　　(b)　$A = \$10,000, \ r = 0.10, \ t = 20$ (compounded continuously)

　　　　　$P = 10,000e^{(-0.10)(20)} \approx \$1353.35$

51. $A = \$10,000$, $r = 0.08$, $n = 1$, $t = 10$ (compounded annually)

$$P = 10,000\left(1 + \frac{0.08}{1}\right)^{(-1)(10)} \approx \$4631.93$$

53.   Answers will vary.

55.   (a) $y = \dfrac{\ln(2)}{1 \cdot \ln\left(1 + \dfrac{0.12}{1}\right)} = \dfrac{\ln(2)}{\ln(1.12)} \approx 6.12$ years

       (b) $y = \dfrac{\ln(3)}{4 \cdot \ln\left(1 + \dfrac{0.06}{4}\right)} = \dfrac{\ln(3)}{4\ln(1.015)} \approx 18.45$ years

       (c) $mP = P\left(1 + \dfrac{r}{n}\right)^{nt}$

              $m = \left(1 + \dfrac{r}{n}\right)^{nt} \implies \ln(m) = nt \cdot \ln\left(1 + \dfrac{r}{n}\right) \implies t = \dfrac{\ln(m)}{n \cdot \ln\left(1 + \dfrac{r}{n}\right)}$

57.   Answers will vary.

Chapter **4**

Exponential and Logarithmic Functions

4.7 Growth and Decay

1. $P(t) = 500e^{0.02t}$

(a) $P(0) = 500e^{(0.02)\cdot(0)} = 500$ flies

(b) growth rate = 2 %

(c) graphing:

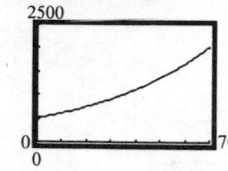

(d) $P(10) = 500e^{(0.02)\cdot(10)} \approx 611$ flies

(e) Find t when $P = 800$:

$$800 = 500e^{0.02t}$$

$$1.6 = e^{0.02t}$$

$$\ln(1.6) = 0.02t$$

$$t = \frac{\ln(1.6)}{0.02} \approx 23.5 \text{ days}$$

(f) Find t when $P = 1000$:

$$1000 = 500e^{0.02t}$$

$$2 = e^{0.02t}$$

$$\ln(2) = 0.02t$$

$$t = \frac{\ln(2)}{0.02} \approx 34.7 \text{ days}$$

3. $A(t) = A_0 e^{-0.0244t} = 500e^{-0.0244t}$

(a) decay rate = 2.44 %

(b) graphing

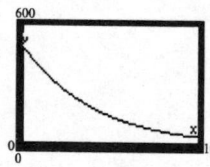

(c) $A(10) = 500e^{(-0.0244)(10)} \approx 391.74$ grams

(d) Find t when $A = 400$:

$$400 = 500e^{-0.0244t}$$

$$0.8 = e^{-0.0244t}$$

$$\ln(0.8) = -0.0244t$$

$$t = \frac{\ln(0.8)}{-0.0244} \approx 9.15 \text{ years}$$

(e) Find t when $A = 250$:

$$250 = 500e^{-0.0244t}$$

$$0.5 = e^{-0.0244t}$$

$$\ln(0.5) = -0.0244t$$

$$t = \frac{\ln(0.5)}{-0.0244} \approx 28.4 \text{ years}$$

5.   Use $N(t) = N_0 e^{kt}$ and solve for k:

$$1800 = 1000 e^{k(1)}$$

$$1.8 = e^k$$

$$k = \ln(1.8) \approx 0.5878$$

When $t = 3$:

$$N(3) = 1000 e^{0.5878\,(3)} = 5832 \text{ mosquitos}$$

Find t when $N(t) = 10,000$:

$$10,000 = 1000 e^{0.5878\,t}$$

$$10 = e^{0.5878\,t}$$

$$\ln(10) = 0.5878t \Rightarrow t = \frac{\ln(10)}{0.5878} \approx 3.9 \text{ days}$$

7.   Use $P(t) = P_0 e^{kt}$ and solve for k:

$$2P_0 = P_0 e^{k(1.5)}$$

$$2 = e^{1.5k}$$

$$\ln(2) = 1.5k$$

$$k = \frac{\ln(2)}{1.5} \approx 0.4621$$

When $t = 2$:

$$P(2) = 10,000 e^{0.4621\,(2)} = 25,199 \text{ is the population 2 years from now.}$$

9.   Use $A = A_0 e^{kt}$ and solve for k:

$$0.5 A_0 = A_0 e^{k(1690)}$$

$$0.5 = e^{1690\,k}$$

$$\ln(0.5) = 1690k \Rightarrow k = \frac{\ln(0.5)}{1690} \approx -0.00041$$

When $A_0 = 10$ and $t = 50$: $A = 10 e^{-0.00041\,(50)} \approx 9.797 \text{ grams}$

11.   Use $A = A_0 e^{kt}$ and solve for k:

half-life $= 5600$ years $\Rightarrow 0.5 A_0 = A_0 e^{k(5600)}$

$$0.5 = e^{5600\,k} \Rightarrow \ln(0.5) = 5600k \Rightarrow k = \frac{\ln(0.5)}{5600} \approx -0.000124$$

(a) Solve for t when $A = 0.3 A_0$:

$$0.3 A_0 = A_0 e^{(-0.000124)t}$$

$$0.3 = e^{(-0.000124)t}$$

$$\ln(0.3) = -0.000124t$$

$$t = \frac{\ln(0.3)}{-0.000124} \approx 9709 \text{ years ago}$$

(b) graphing $y_1 = e^{(-0.000124)x}$

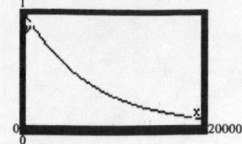

(c) Using INTERSECT with
$y_1 = e^{(-0.000124)x}$ and $y_2 = 0.5$

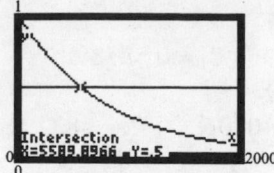

$t \approx 5589.90$ years

(d) Using INTERSECT with
$y_1 = e^{(-0.000124)x}$ and $y_2 = 0.3$

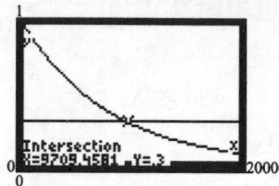

$x \approx 9708.46$ years

13. (a)   Using $u = T + (u_0 - T)e^{kt}$ where $t = 5$,
$T = 70$, $u_0 = 450$, $u = 300$:

$$300 = 70 + (450 - 70)e^{k(5)}$$
$$230 = 380e^{5k}$$
$$0.6053 = e^{5k}$$
$$5k = \ln(0.6053)$$
$$k = \frac{\ln(0.6053)}{5} \approx -0.1004$$

$T = 70$, $u_0 = 450$, $u = 135$:

$$135 = 70 + (450 - 70)e^{-0.1004\,t}$$
$$65 = 380e^{-0.1004\,t}$$
$$0.17105 = e^{-0.1004\,t}$$
$$-0.1004\,t = \ln(0.17105)$$
$$t = \frac{\ln(0.17105)}{-0.1004} \approx 17.6 \text{ minutes}$$

The pizza will be cool enough to eat at 5:18 p.m.

(b) graphing: $y_1 = 70 + 380e^{(-0.1004)x}$

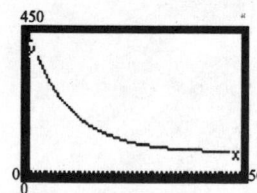

(c) Using INTERSECT with $y_1 = 70 + 380e^{(-0.1004)x}$ and $y_2 = 160$

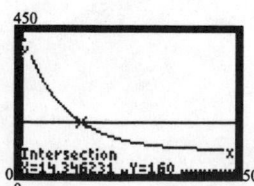

The pizza will be 160°F after about 14.35 minutes.

(d)   Use TRACE with $y_1 = 70 + 380e^{(-0.1004)x}$

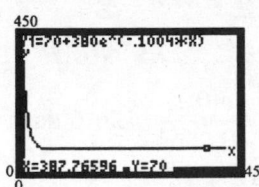

As time passes the temperature gets closer to 70°F.

15.   (a)   Using $u = T + (u_0 - T)e^{kt}$ where $t = 3$,
$T = 35$, $u_0 = 8$, $u = 15$:

$$15 = 35 + (8 - 35)e^{k(3)} \Rightarrow -20 = -27e^{3k} \Rightarrow 0.74074 = e^{3k}$$

$$3k = \ln(0.74074) \Rightarrow k = \frac{\ln(0.74074)}{3} \approx -0.100035$$

At $t = 5$:
$$u = 35 + (8 - 35)e^{-0.100035\,(5)} = 18.63°C$$
At $t = 10$:
$$u = 35 + (8 - 35)e^{-0.100035\,(10)} = 25.1°C$$

(b)   graphing $y_1 = 35 - 27e^{(-0.100035)x}$

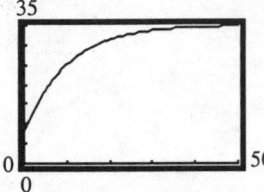

Use TRACE on the graph of $y_1 = 35 - 27e^{(-0.100035)x}$:

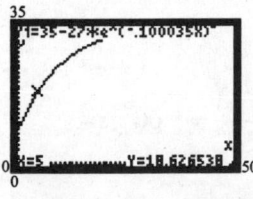   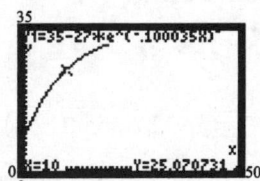

17.   Use $A = A_0 e^{kt}$ and solve for k:
$$15 = 25e^{k(10)}$$

$$0.6 = e^{10k} \Rightarrow \ln(0.6) = 10k \Rightarrow k = \frac{\ln(0.6)}{10} \approx -0.0511$$

When $A_0 = 25$ and $t = 24$:
$$A = 25e^{-0.0511(24)} \approx 7.33 \text{ kilograms}$$

Find t when $A = 0.5A_0$:
$$0.5 = 25e^{-0.0511\,t}$$

$$0.02 = e^{-0.0511\,t} \Rightarrow \ln(0.02) = -0.0511t \Rightarrow t = \frac{\ln(0.02)}{-0.0511} \approx 76.6 \text{ hours}$$

19.   Use $A = A_0 e^{kt}$ and solve for k: Find t when $A = 0.1A_0$:

$$0.5A_0 = A_0 e^{k(8)} \qquad\qquad 0.1A_0 = A_0 e^{-0.0866\,t}$$

$$0.5 = e^{8k} \Rightarrow \ln(0.5) = 8k \qquad\qquad 0.1 = e^{-0.0866\,t} \Rightarrow \ln(0.1) = -0.0866\,t$$

$$k = \frac{\ln(0.5)}{8} \approx -0.0866 \qquad\qquad t = \frac{\ln(0.1)}{-0.0866} \approx 26.6 \text{ days}$$

The farmers need to wait about 27 days before using the hay.

21.   (a)   The maximum proportion is the carrying capacity, $0.9 = 90\%$.

     (b) $P(0) = \dfrac{0.9}{1 + 6e^{-0.32(0)}} = \dfrac{0.9}{1 + 6 \cdot 1} = \dfrac{0.9}{7} = 0.1286$

     (c)   graphing:

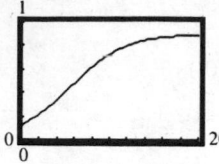

     (d) $P(15) = \dfrac{0.9}{1 + 6e^{-0.32(15)}} \approx 0.8577$, so 85.77% owned a VCR in 1999.

     (e) $0.8 = \dfrac{0.9}{1 + 6e^{-0.32t}}$

$$0.8\left(1 + 6e^{-0.32t}\right) = 0.9 \Rightarrow 1 + 6e^{-0.32t} = 1.125 \Rightarrow 6e^{-0.32t} = 0.125$$

$$e^{-0.32t} = 0.020833 \Rightarrow -0.32t = \ln(0.020833)$$

$$t = \frac{\ln(0.020833)}{-0.32} \approx 12.1$$

80% of households will own VCR's in 1996 (t = 12).

     (f) $0.45 = \dfrac{0.9}{1 + 6e^{-0.32t}}$

$$0.45\left(1 + 6e^{-0.32t}\right) = 0.9 \Rightarrow 1 + 6e^{-0.32t} = 2 \Rightarrow 6e^{-0.32t} = 1$$

$$e^{-0.32t} = \frac{1}{6} \Rightarrow -0.32t = \ln(1/6) \Rightarrow t = \frac{\ln(1/6)}{-0.32} \approx 5.60 \text{ years}$$

80% of households will own VCR's in 1996 (t = 12).

23.   (a)   As $t \to \infty$, $e^{-0.439t} \to 0$. Thus, $P(t) \to 1000$. The carrying capacity is 1000g.
            Growth rate = 43.9%.

     (b) $P(0) = \dfrac{1000}{1 + 32.33e^{-0.439(0)}} = \dfrac{1000}{33.33} = 30$ g

     (c)   graphing:

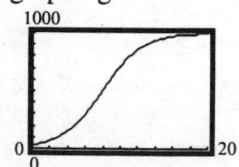

(d) $P(9) = \dfrac{1000}{1 + 32.33e^{-0.439(9)}} \approx 616.6 \text{ g}$

(e) $700 = \dfrac{1000}{1 + 32.33e^{-0.439t}}$

$700\left(1 + 32.33e^{-0.439t}\right) = 1000 \Rightarrow 1 + 32.33e^{-0.439t} = \dfrac{1000}{700} = \dfrac{10}{7}$

$32.33e^{-0.439t} = \dfrac{3}{7} \Rightarrow e^{-0.439t} = \dfrac{(3/7)}{32.33}$

$-0.439t = \ln\left(\dfrac{(3/7)}{32.33}\right) \Rightarrow t = \dfrac{\ln\left(\dfrac{(3/7)}{32.33}\right)}{-0.439} \approx 9.85 \text{ hours}$

(f)   To reach half-carrying capacity, we need

$500 = \dfrac{1000}{1 + 32.33e^{-0.439t}}$

$500\left(1 + 32.33e^{-0.439t}\right) = 1000 \Rightarrow 1 + 32.33e^{-0.439t} = 2$

$32.33e^{-0.439t} = 1 \Rightarrow e^{-0.439t} = \dfrac{1}{32.33}$

$-0.439t = \ln\left(\dfrac{1}{32.33}\right) \Rightarrow t = \dfrac{\ln\left(\dfrac{1}{32.33}\right)}{-0.439} \approx 7.92 \text{ hours}$

25.   (a) $y = \dfrac{6}{1 + e^{-(5.085 - 0.1156(100))}} \approx 0.0092 \text{ O-rings}$

(b) $y = \dfrac{6}{1 + e^{-(5.085 - 0.1156(60))}} \approx 0.8145 \text{ O-rings}$

(c) $y = \dfrac{6}{1 + e^{-(5.085 - 0.1156(30))}} \approx 5.0063 \text{ O-rings}$

(d)   Graphing:

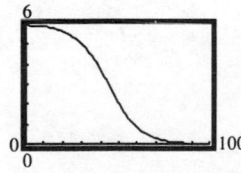

At 58°F, there would be 1 leaky O-ring.
At 44°F, there would be 3 leaky O-rings.
At 30°F, there would be 5 leaky O-rings.

Chapter 4

Exponential and Logarithmic Functions

4.8 Exponential, Logarithmic and Logistic Curve Fitting

1. (a) scatter diagram

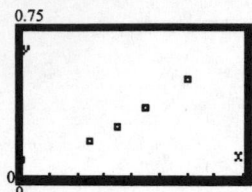

(b) Using EXPonential REGression on the data yields: $y = (0.0903)(1.3384)^x$

(c) $y = (0.0903)(1.3384)^x = (0.0903)\left(e^{\ln(1.3384)}\right)^x = (0.0903)\left(e^{\ln(1.3384)x}\right)$

$N(t) = (0.0903)\left(e^{\ln(1.3384)t}\right) = (0.0903)\left(e^{0.2915t}\right)$

(d) graphing: $y_1 = (0.0903)\left(e^{0.2915x}\right)$

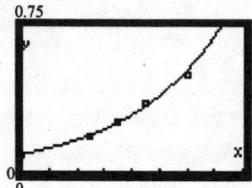

(e) $N(7) = (0.0903)\left(e^{(0.2915)\cdot 7}\right) \approx 0.695$ bacteria

(f) Find t when $N(t) = 0.75$

$(0.0903)\left(e^{(0.2915)\cdot t}\right) = 0.75$

$\left(e^{(0.2915)\cdot t}\right) = \dfrac{0.75}{0.0903} \approx 8.306$

$(0.2915)\cdot t = \ln(8.306)$

$t \approx \dfrac{\ln(8.306)}{0.2915} \approx 7.26$ hours

3. (a)   scatter diagram

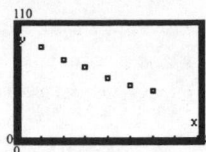

(b)   Using EXPonential REGression on the data yields: $y = (100.3262)(0.8769)^x$

(c) $y = (100.3262)(0.8769)^x = (100.3262)\left(e^{\ln(0.8769)}\right)^x = (100.3262)\left(e^{\ln(0.8769)x}\right)$

$A(t) = (100.3262)\left(e^{(-0.1314)t}\right)$

(d)   graphing: $y_1 = (100.3262)\left(e^{(-0.1314)x}\right)$

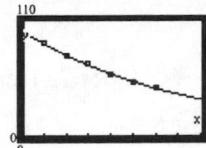

(e)   Find t when $A(t) = 0.5 \cdot A_0$

$$(100.3262)\left(e^{(-0.1314)t}\right) = (0.5)(100.3262)$$

$$\left(e^{(-0.1314)t}\right) = 0.5$$

$$(-0.1314)t = \ln(0.5)$$

$$t = \frac{\ln(0.5)}{-0.1314} \approx 5.28 \text{ weeks}$$

(f) $A(50) = (100.3262)\left(e^{(-0.1314)\cdot 50}\right) \approx 0.141$ grams

(g)   Find t when $A(t) = 20$

$$(100.3262)\left(e^{(-0.1314)t}\right) = 20$$

$$\left(e^{(-0.1314)t}\right) = \frac{20}{100.3262} \approx 0.1993$$

$$(-0.1314)t = \ln(0.1993)$$

$$t = \frac{\ln(0.1993)}{-0.1314} \approx 12.28 \text{ weeks}$$

5. (a)   Let $x = 1$ correspond to 1991, $x = 2$ correspond to 1992, etc.

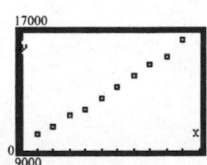

(b)   Using EXPonential REGression on the data yields: $y = (9478.4453)(1.0566)^x$

(c)   The average annual rate of return over the past 10 years is given by $0.0566 = 5.66\%$.

(d)   In the year 2021, $x = 21$, so $y = (9478.4453)(1.0566)^{21} \approx \$30,120.34$.

(e)   Find x when $y = 80000$

$$(9478.4453)(1.0566)^x = 80000$$

$$(1.0566)^x = \frac{80000}{9478.4453} \approx 8.84402$$

$$x\ln(1.0566) = \ln(8.84402) \Rightarrow x = \frac{\ln(8.84402)}{\ln(1.0566)} \approx 38.74 \text{ years}$$

7.   (a)   scatter diagram

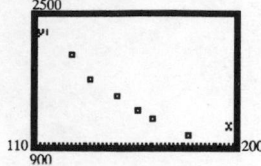

(b)   Using LnREGression on the data yields: $y = 32741.02369 - 6070.956754\ln(x)$

(c)   graphing $y_1 = 32741.02369 - 6070.956754\ln(x)$

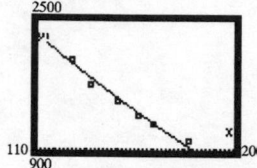

(d)   find x when $y = 1650$:

$$1650 = 32741.02369 - 6070.956754\ln(x)$$

$$-31091.02369 = -6070.956754\ln(x)$$

$$\frac{-31091.02369}{-6070.956754} = \ln(x)$$

$$5.1213 \approx \ln(x) \Rightarrow e^{5.1213} \approx x \Rightarrow x \approx 168 \text{ computers}$$

9.   (a)   Let $x = 0$ correspond to 1900, $x = 1$ correspond to 1910, etc.

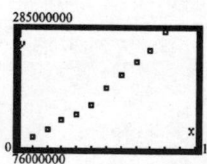

(b)   Using LOGISTIC REGression on the data yields:

$$y = \frac{799475916.5}{1 + 9.1968e^{-0.1603x}}$$

(c)   graphing $y_1 = \dfrac{799475916.5}{1 + 9.1968e^{-0.1603x}}$:

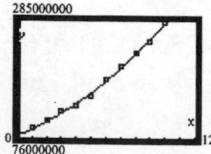

(d)   as $x \to \infty$, $y = \dfrac{799475916.5}{1+9.1968e^{-0.1603x}} \to \dfrac{799475916.5}{1+0} = 799475916.5$

Therefore, the carrying capacity of the United States is approximately 799,475,916 people.

(e)   in the year 2001, $x = 10.1$, so $y = \dfrac{799475916.5}{1+9.1968e^{-0.1603(10.1)}} \approx 283,321,149$ people

(f)   Find x when $y = 300,000,000$

$\dfrac{799475916.5}{1+9.1968e^{-0.1603x}} = 300000000$

$799475916.5 = 300000000\left(1+9.1968e^{-0.1603x}\right)$

$\dfrac{799475916.5}{300000000} = 1+9.1968e^{-0.1603x}$

$\dfrac{799475916.5}{300000000} - 1 = 9.1968e^{-0.1603x}$

$1.6649 \approx 9.1968e^{-0.1603x} \Rightarrow \dfrac{1.6649}{9.1968} = e^{-0.1603x}$

$0.1810 \approx e^{-0.1603x} \Rightarrow \ln(0.1810) \approx -0.1603x$

$\dfrac{\ln(0.1810)}{-0.1603} = x \Rightarrow x \approx 10.7$

Therefore, the United States population will be 300,000,000 in the year 2007.

11.   (a)   Let $x = 0$ correspond to 1900, $x = 1$ correspond to 1910, etc.

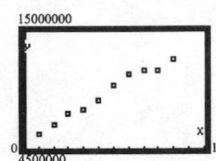

(b)   Using LOGISTIC REGression on the data yields:

$$y = \frac{14471245.24}{1+2.01534e^{-0.2458x}}$$

(c)   graphing $y_1 = \dfrac{14471245.24}{1+2.01534e^{-0.2458x}}$:

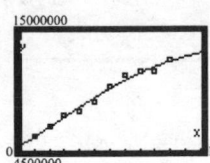

(d)   as $x \to \infty$, $y = \dfrac{14471245.24}{1+2.01534e^{-0.2458x}} \to \dfrac{14471245.24}{1+0} = 14471245.24$

Therefore, the carrying capacity of Illinois is approximately 14,471,245 people.

(e)   in the year 2010, $x = 11$, so $y = \dfrac{14471245.24}{1+2.01534e^{-0.2458(11)}} \approx 12,750,802$ people

Exponential and Logarithmic Functions

4.R Chapter Review

1. inverse $\{(2,1),(5,3),(8,5),(10,6)\}$ is a function

3.

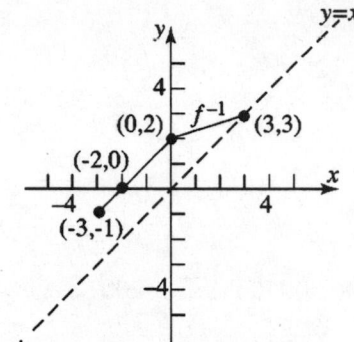

5. $f(x) = \dfrac{2x+3}{5x-2}$

$y = \dfrac{2x+3}{5x-2}$

$x = \dfrac{2y+3}{5y-2}$ Inverse

$x(5y-2) = 2y+3$

$5xy-2x = 2y+3$

$5xy-2y = 2x+3$

$y(5x-2) = 2x+3$

$y = \dfrac{2x+3}{5x-2}$

$f^{-1}(x) = \dfrac{2x+3}{5x-2}$

Domain of $f =$

range of f^{-1} = all real numbers except $\dfrac{2}{5}$.

Range of $f =$

domain of f^{-1} = all real numbers except $\dfrac{2}{5}$.

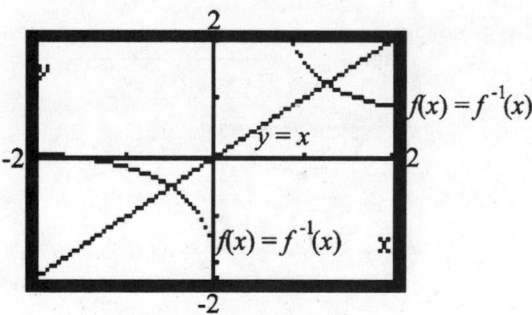

7. $f(x) = \dfrac{1}{x-1}$

$y = \dfrac{1}{x-1}$

$x = \dfrac{1}{y-1}$ Inverse

$x(y-1) = 1$

$xy - x = 1$

$xy = x + 1$

$y = \dfrac{x+1}{x}$

$f^{-1}(x) = \dfrac{x+1}{x}$

Domain of f =
range of f^{-1} = all real numbers except 1
Range of f =
domain of f^{-1} = all real numbers except 0

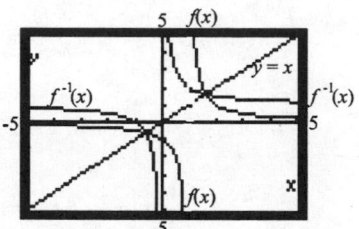

9. $f(x) = \dfrac{3}{x^{1/3}}$

$y = \dfrac{3}{x^{1/3}}$

$x = \dfrac{3}{y^{1/3}}$ Inverse

$xy^{1/3} = 3$

$y^{1/3} = \dfrac{3}{x}$

$y = \dfrac{27}{x^3}$

$f^{-1}(x) = \dfrac{27}{x^3}$

Domain of f =
range of f^{-1} = all real numbers except 0
Range of f =
domain of f^{-1} = all real numbers except 0

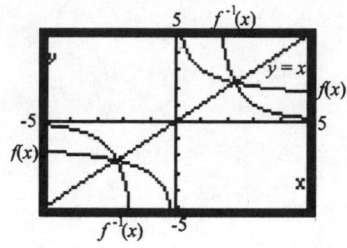

11.   (a) $f(4) = 3^4 = 81$
      (b) $g(9) = \log_3(9) = \log_3(3^2) = 2$

      (c) $f(-2) = 3^{-2} = \dfrac{1}{9}$
      (d) $g\left(\dfrac{1}{27}\right) = \log_3\left(\dfrac{1}{27}\right) = \log_3(3^{-3}) = -3$

13. $5^2 = z$ is equivalent to $2 = \log_5 z$
15. $\log_5 u = 13$ is equivalent to $5^{13} = u$

17.   The domain of $y = \log(3x-2)$ is:

$3x - 2 > 0 \Rightarrow x > \dfrac{2}{3}$

$\left\{x \,\middle|\, x > \dfrac{2}{3}\right\}$

19.   The domain of $y = \log_2(x^2 - 3x + 2)$ is:

$x^2 - 3x + 2 > 0$

$(x-2)(x-1) > 0$

$x > 2 \text{ or } x < 1$
$\{x \mid x < 1 \text{ or } x > 2\}$

21. $\log_2\left(\dfrac{1}{8}\right) = \log_2 2^{-3} = -3\log_2 2 = -3$

23. $\ln e^{\sqrt{2}} = \sqrt{2}$

25.　$2^{\log_2 0.4} = 0.4$

27.　$\log_3\left(\dfrac{uv^2}{w}\right) = \log_3 uv^2 - \log_3 w = \log_3 u + \log_3 v^2 - \log_3 w = \log_3 u + 2\log_3 v - \log_3 w$

29.　$\log\left(x^2\sqrt{x^3+1}\right) = \log x^2 + \log\left(x^3+1\right)^{1/2} = 2\log x + \tfrac{1}{2}\log\left(x^3+1\right)$

31.　$\ln\left(\dfrac{x\sqrt[3]{x^2+1}}{x-3}\right) = \ln\left(x\sqrt[3]{x^2+1}\right) - \ln(x-3) = \ln x + \ln\left(x^2+1\right)^{1/3} - \ln(x-3)$

$$= \ln x + \frac{1}{3}\ln\left(x^2+1\right) - \ln(x-3)$$

33.　$3\log_4 x^2 + \dfrac{1}{2}\log_4\sqrt{x} = \log_4\left(x^2\right)^3 + \log_4\left(x^{1/2}\right)^{1/2} = \log_4 x^6 + \log_4 x^{1/4} = \log_4 x^6 \cdot x^{1/4}$

$$= \log_4 x^{25/4} = \frac{25}{4}\log_4 x$$

35.　$\ln\left(\dfrac{x-1}{x}\right) + \ln\left(\dfrac{x}{x+1}\right) - \ln\left(x^2-1\right) = \ln\left(\dfrac{x-1}{x}\cdot\dfrac{x}{x+1}\right) - \ln\left(x^2-1\right) = \ln\left[\dfrac{\left|\dfrac{x-1}{x+1}\right|}{x^2-1}\right]$

$$= \ln\left(\frac{x-1}{x+1}\cdot\frac{1}{(x-1)(x+1)}\right) = \ln\frac{1}{(x+1)^2} = \ln(x+1)^{-2} = -2\ln(x+1)$$

37.　$2\log 2 + 3\log x - \dfrac{1}{2}\left[\log(x+3) + \log(x-2)\right] = \log 2^2 + \log x^3 - \dfrac{1}{2}\log\left[(x+3)(x-2)\right]$

$$= \log 4x^3 - \log\left((x+3)(x-2)\right)^{1/2} = \log\left[\frac{4x^3}{\left((x+3)(x-2)\right)^{1/2}}\right]$$

39.　$\log_4 19 = \dfrac{\log 19}{\log 4} \approx 2.124$

41.　$y = \log_3(x) = \dfrac{\ln(x)}{\ln(3)}$

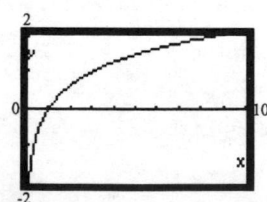

43. $f(x) = 2^{x-3}$
Using the graph of $y = 2^x$, shift the graph 3 units to the right.
Domain: $(-\infty, \infty)$
Range: $(0, \infty)$
Horizontal Asymptote: $y = 0$

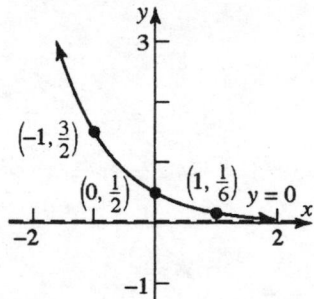

45. $f(x) = \dfrac{1}{2} \cdot 3^{-x}$
Using the graph of $y = 3^x$, reflect the graph about the y-axis, and shrink vertically by a factor of $\dfrac{1}{2}$.
Domain: $(-\infty, \infty)$
Range: $(0, \infty)$
Horizontal Asymptote: $y = 0$

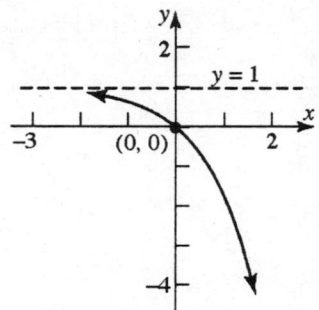

47. $f(x) = 1 - e^x$
Using the graph of $y = e^x$, reflect about the x-axis, and shift up 1 unit.
Domain: $(-\infty, \infty)$
Range: $(-\infty, 1)$
Horizontal Asymptote: $y = 1$

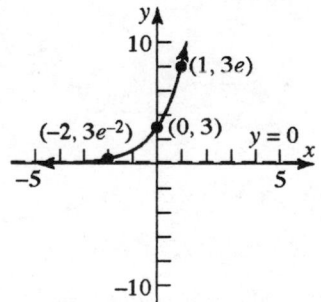

49. $f(x) = 3e^x$
Using the graph of $y = e^x$, stretch vertically by a factor of 3.
Domain: $(-\infty, \infty)$
Range: $(0, \infty)$
Horizontal Asymptote: $y = 0$

51. $f(x) = 3 - e^{-x}$
Using the graph of $y = e^x$, reflect the
graph about the y-axis, reflect about the
x-axis, and shift up 3 units.
Domain: $(-\infty, \infty)$
Range: $(-\infty, 3)$
Horizontal Asymptote: $y = 3$

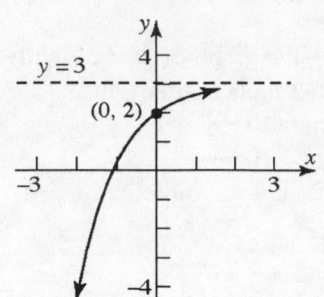

53. $4^{1-2x} = 2$
$$\left(2^2\right)^{1-2x} = 2$$
$$2^{2-4x} = 2^1$$
$$2 - 4x = 1 \Rightarrow -4x = -1 \Rightarrow x = \frac{1}{4}$$

55. $3^{x^2+x} = \sqrt{3}$
$$3^{x^2+x} = 3^{1/2}$$
$$x^2 + x = \frac{1}{2} \Rightarrow 2x^2 + 2x - 1 = 0 \Rightarrow x = \frac{-2 \pm \sqrt{4 - 4(2)(-1)}}{2(2)} = \frac{-2 \pm \sqrt{12}}{4} = \frac{-2 \pm 2\sqrt{3}}{4} = \frac{-1 \pm \sqrt{3}}{2}$$
$$x = \frac{-1 - \sqrt{3}}{2} \quad \text{or} \quad x = \frac{-1 + \sqrt{3}}{2}$$

57. $\log_x 64 = -3$
$$x^{-3} = 64$$
$$\left(x^{-3}\right)^{-1/3} = 64^{-1/3} \Rightarrow x = \frac{1}{\sqrt[3]{64}} = \frac{1}{4}$$

59. $5^x = 3^{x+2}$
$$\log\left(5^x\right) = \log\left(3^{x+2}\right)$$
$$x \log 5 = (x + 2) \log 3$$
$$x \log 5 = x \log 3 + 2 \log 3$$
$$x \log 5 - x \log 3 = 2 \log 3$$
$$x(\log 5 - \log 3) = 2 \log 3$$
$$x = \frac{2 \log 3}{\log 5 - \log 3}$$
$$x \approx 4.301$$

61. $9^{2x} = 27^{3x-4}$
$$\left(3^2\right)^{2x} = \left(3^3\right)^{3x-4}$$
$$3^{4x} = 3^{9x-12}$$
$$4x = 9x - 12 \Rightarrow -5x = -12$$
$$x = \frac{12}{5}$$

63. $\log_3 \sqrt{x-2} = 2$
$$\sqrt{x-2} = 3^2$$
$$x - 2 = 9^2 \Rightarrow x - 2 = 81 \Rightarrow x = 83$$

65. $8 = 4^{x^2} \cdot 2^{5x}$

$2^3 = \left(2^2\right)^{x^2} \cdot 2^{5x}$

$2^3 = 2^{2x^2 + 5x}$

$3 = 2x^2 + 5x \Rightarrow 0 = 2x^2 + 5x - 3$

$0 = (2x-1)(x+3) \Rightarrow x = \dfrac{1}{2}$ or $x = -3$

67. $\log_6(x+3) + \log_6(x+4) = 1$

$\log_6(x+3)(x+4) = 1$

$(x+3)(x+4) = 6^1$

$x^2 + 7x + 12 = 6$

$x^2 + 7x + 6 = 0$

$(x+6)(x+1) = 0$

$x = -6$ or $x = -1$

The logarithms are undefined when $x = -6$, so $x = -1$ is the only solution.

69. $e^{1-x} = 5$

$1 - x = \ln 5$

$-x = -1 + \ln 5$

$x = 1 - \ln 5 \approx -0.609$

71. $2^{3x} = 3^{2x+1}$

$\ln 2^{3x} = \ln 3^{2x+1}$

$3x \ln 2 = (2x+1)\ln 3$

$3x \ln 2 = 2x \ln 3 + \ln 3$

$3x \ln 2 - 2x \ln 3 = \ln 3$

$x(3\ln 2 - 2\ln 3) = \ln 3$

$x = \dfrac{\ln 3}{3\ln 2 - 2\ln 3}$

$x \approx -9.327$

73. $h(300) = \left(30(0) + 8000\right)\log\left(\dfrac{760}{300}\right) \approx 8000\log(2.53333) \approx 3229.5$ meters

75. $P = 25e^{0.1d}$

(a) $P = 25e^{0.1(4)}$

$= 25e^{0.4}$

≈ 37.3 watts

(b) $50 = 25e^{0.1d}$

$2 = e^{0.1d}$

$\ln(2) = (0.1)d$

$d = \dfrac{\ln(2)}{0.1} = 6.9$ decibels

77. (a) $n = \dfrac{\log(10000) - \log(90000)}{\log(1 - 0.20)} \approx 9.85$ years

(b) $n = \dfrac{\log(0.5i) - \log(i)}{\log(1 - 0.15)} = \dfrac{\log\left(\dfrac{0.5i}{i}\right)}{\log(0.85)} = \dfrac{\log(0.5)}{\log(0.85)} \approx 4.27$ years

79. $P = A\left(1 + \dfrac{r}{n}\right)^{-nt} = 85000\left(1 + \dfrac{0.04}{2}\right)^{-2(18)} \approx \$41,668.97$

81. $A = A_0 e^{kt}$

$0.5A_0 = A_0 e^{k(5600)} \Rightarrow 0.5 = e^{5600\,k}$

$\ln(0.5) = 5600k \Rightarrow k = \dfrac{\ln(0.5)}{5600} \approx -0.000124$

$$0.05A_0 = A_0 e^{-0.000124\,t}$$

$$0.05 = e^{-0.000124\,t} \Rightarrow \ln(0.05) = -0.000124\,t \Rightarrow t = \frac{\ln(0.05)}{-0.000124} \approx 24{,}159 \text{ years ago}$$

83. $P = P_0 e^{k\,t} = 5{,}840{,}445{,}216 e^{0.0133\,(3)} \approx 6{,}078{,}190{,}457$

85. (a) $P(0) = \dfrac{0.8}{1 + 1.67 e^{-0.16\,(0)}} = \dfrac{0.8}{1 + 1.67} = 0.2996$

(b) 0.8

(c) Graphing: (d) Using INTERSECT we have:

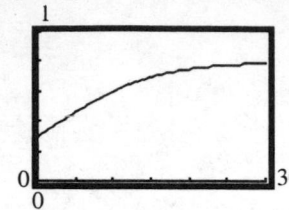   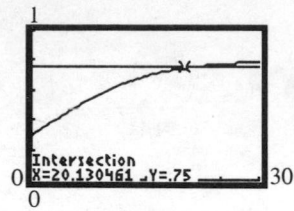

75% use Windows 98 in 2018.

87. (a) scatter diagram

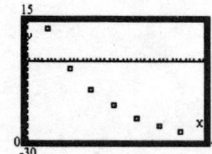

(b) Using LnREGression on the data yields: $y = 44.1978 - 20.3314 \ln(x)$

(c) graphing $y_1 = 44.1978 - 20.3314 \ln(x)$

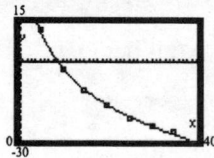

(d) if $x = 23,\ y = 44.1978 - 20.3314 \ln(23) \approx -19.55^\circ F$.

Chapter 4

Exponential and Logarithmic Functions

4.CR   Cumulative Review

1.   The graph represents a function since it passes the Vertical Line Test. The function is not a one-to-one function since the graph fails the Horizontal Line Test.

3. $x^2 + y^2 = 1$

   (a) $\left(\dfrac{1}{2}\right)^2 + \left(\dfrac{1}{2}\right)^2 = \dfrac{1}{4} + \dfrac{1}{4} = \dfrac{1}{2} \neq 1;$ $\left(\dfrac{1}{2}, \dfrac{1}{2}\right)$ is not on the graph.

   (b) $\left(\dfrac{1}{2}\right)^2 + \left(\dfrac{\sqrt{3}}{2}\right)^2 = \dfrac{1}{4} + \dfrac{3}{4} = 1;$ $\left(\dfrac{1}{2}, \dfrac{\sqrt{3}}{2}\right)$ is on the graph.

5. $2x - 4y = 16$

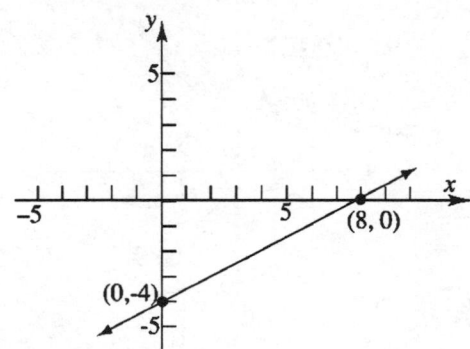

7.   Given that the graph of $f(x) = ax^2 + bx + c$ has vertex $(4, -8)$ and passes through the point $(0, 24)$, we can conclude

$$\dfrac{-b}{2a} = 4, \qquad\qquad f(4) = -8, \quad\text{and}\quad\qquad\qquad f(0) = 24$$

Notice that $f(0) = 24 \Rightarrow a(0)^2 + b(0) + c = 24 \Rightarrow c = 24$

Therefore $f(x) = ax^2 + bx + c = ax^2 + bx + 24$.

Furthermore, $\dfrac{-b}{2a} = 4 \Rightarrow b = -8a$

and $f(4) = -8 \Rightarrow a(4)^2 + b(4) + 24 = -8 \Rightarrow 16a + 4b + 24 = -8 \Rightarrow 16a + 4b = -32$

$$\Rightarrow 4a + b = -8$$

Replacing b with $-8a$ in this equation yields

$4a - 8a = -8 \Rightarrow -4a = -8 \Rightarrow a = 2$.

So $b = -8a = -8(2) = -16$.

Therefore, we have the function $f(x) = 2x^2 - 16x + 24$.

9. $f(x) = x^2 + 2$ $g(x) = \dfrac{2}{x-3}$

$(f \circ g)(x) = f(g(x)) = f\left(\dfrac{2}{x-3}\right) = \left(\dfrac{2}{x-3}\right)^2 + 2 = \dfrac{4}{(x-3)^2} + 2$

The domain of g is $\{x \mid x \neq 3\}$. The domain of f is {Real Numbers}

So the domain of $f \circ g$ is $\{x \mid x \neq 3\}$.

11.   (a) $g(x) = 3^x + 2$

Using the graph of $y = 3^x$, shift vertically
3 units up.
Domain: $(-\infty, \infty)$
Range: $(2, \infty)$
Horizontal Asymptote: $y = 2$

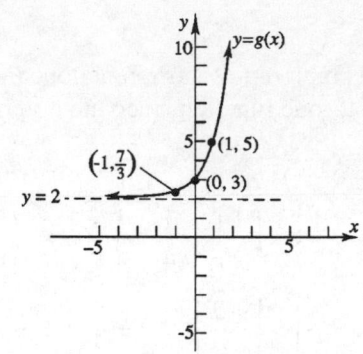

(b) $g(x) = 3^x + 2$

$y = 3^x + 2$

$x = 3^y + 2$ Inverse

$x - 2 = 3^y$

$\ln(x - 2) = \ln(3^y)$

$\ln(x - 2) = y \cdot \ln(3)$

$\dfrac{\ln(x - 2)}{\ln(3)} = y$

(c)

$g^{-1}(x) = \dfrac{\ln(x - 2)}{\ln(3)} = \log_3(x - 2)$

Domain: $(2, \infty)$
Range: $(-\infty, \infty)$
Vertical Asymptote: $x = 2$

13. $\log_3(x + 1) + \log_3(2x - 3) = 1$

$\log_3(x + 1)(2x - 3) = 1$

$(x + 1)(2x - 3) = 3^1 \Rightarrow 2x^2 - x - 3 = 3$

$2x^2 - x - 6 = 0 \Rightarrow (2x + 3)(x - 2) = 0$

$x = -\dfrac{3}{2}$ or $x = 2$

The logarithms are undefined when $x = -\dfrac{3}{2}$, so $x = 2$ is the only solution.

Trigonometric Functions

5.1   Angles and Their Measure

1.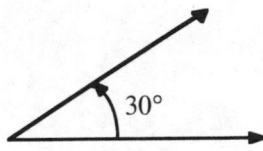

3.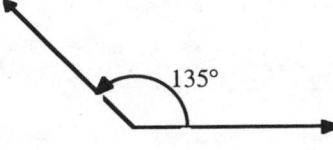

5.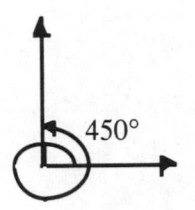

7.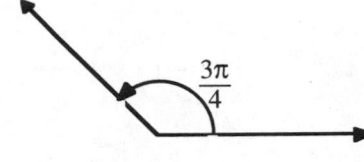

9.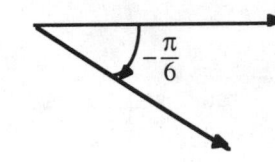

11.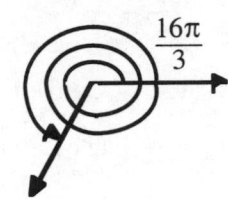

13. $30° = 30 \cdot \dfrac{\pi}{180}$ radian $= \dfrac{\pi}{6}$ radians

15. $240° = 240 \cdot \dfrac{\pi}{180}$ radian $= \dfrac{4\pi}{3}$ radians

17. $-60° = -60 \cdot \dfrac{\pi}{180}$ radian $= -\dfrac{\pi}{3}$ radians

19. $180° = 180 \cdot \dfrac{\pi}{180}$ radian $= \pi$ radians

21. $-135° = -135 \cdot \dfrac{\pi}{180}$ radian $= -\dfrac{3\pi}{4}$ radians

23. $-90° = -90 \cdot \dfrac{\pi}{180}$ radian $= -\dfrac{\pi}{2}$ radians

25. $\dfrac{\pi}{3} = \dfrac{\pi}{3} \cdot \dfrac{180}{\pi}$ degrees $= 60°$

27. $-\dfrac{5\pi}{4} = -\dfrac{5\pi}{4} \cdot \dfrac{180}{\pi}$ degrees $= -225°$

29. $\dfrac{\pi}{2} = \dfrac{\pi}{2} \cdot \dfrac{180}{\pi}$ degrees $= 90°$

31. $\dfrac{\pi}{12} = \dfrac{\pi}{12} \cdot \dfrac{180}{\pi}$ degrees $= 15°$

33. $-\dfrac{\pi}{2} = -\dfrac{\pi}{2} \cdot \dfrac{180}{\pi}$ degrees $= -90°$

35. $-\dfrac{\pi}{6} = -\dfrac{\pi}{6} \cdot \dfrac{180}{\pi}$ degrees $= -30°$

37. $r = 10$ meters; $\theta = \dfrac{1}{2}$ radian; $s = r\theta = 10 \cdot \dfrac{1}{2} = 5$ meters

39. $\theta = \dfrac{1}{3}$ radian; $s = 2$ feet; $s = r\theta \ \rightarrow \ r = \dfrac{s}{\theta} = \dfrac{2}{\left(\dfrac{1}{3}\right)} = 6$ feet

41. $r = 5$ miles; $s = 3$ miles; $s = r\theta$ or $\theta = \dfrac{s}{r} = \dfrac{3}{5} = 0.6$ radians

43. $r = 2$ inches; $\theta = 30°$; Convert to radians: $30° = 30 \cdot \dfrac{\pi}{180} = \dfrac{\pi}{6}$ radians

$s = r\theta = 2 \cdot \dfrac{\pi}{6} = \dfrac{\pi}{3}$ inches

45. $r = 10$ meters; $\theta = \dfrac{1}{2}$ radian

$A = \dfrac{1}{2}r^2\theta = \dfrac{1}{2}(10)^2\left(\dfrac{1}{2}\right) = \dfrac{100}{4} = 25$ square meters

47. $\theta = \dfrac{1}{3}$ radian; $A = 2$ square feet

$A = \dfrac{1}{2}r^2\theta \Rightarrow 2 = \dfrac{1}{2}r^2\left(\dfrac{1}{3}\right) = \dfrac{1}{6}r^2$

$2 = \dfrac{1}{6}r^2 \Rightarrow 12 = r^2 \Rightarrow r = \sqrt{12} \approx 3.464$ feet

49. $r = 5$ miles; $A = 3$ square miles

$A = \dfrac{1}{2}r^2\theta \Rightarrow 3 = \dfrac{1}{2}(5)^2\theta = \dfrac{25}{2}\theta$

$3 = \dfrac{25}{2}\theta \Rightarrow \dfrac{6}{25} = \theta \Rightarrow \theta \approx 0.24$ radians

51. $r = 2$ inches; $\theta = 30°$;   Convert to radians: $30° = 30 \cdot \dfrac{\pi}{180} = \dfrac{\pi}{6}$ radians

$$A = \frac{1}{2}r^2\theta = \frac{1}{2}(2)^2\left(\frac{\pi}{6}\right) = \frac{1}{2} \cdot 4\left(\frac{\pi}{6}\right) = \frac{\pi}{3} \approx 1.047 \text{ square inches}$$

53. $r = 2$ feet; $\theta = \dfrac{\pi}{3}$ radians

$$s = r\theta = 2 \cdot \frac{\pi}{3} = \frac{2\pi}{3} \approx 2.094 \text{ feet}$$

$$A = \frac{1}{2}r^2\theta = \frac{1}{2}(2)^2\left(\frac{\pi}{3}\right) = \frac{1}{2} \cdot 4\left(\frac{\pi}{3}\right) = \frac{2\pi}{3} \approx 2.094 \text{ square feet}$$

55. $r = 12$ yards; $\theta = 70°$;   Convert to radians: $70° = 70 \cdot \dfrac{\pi}{180} = \dfrac{7\pi}{18}$ radians

$$s = r\theta = 12 \cdot \frac{7\pi}{18} \approx 14.661 \text{ yards}$$

$$A = \frac{1}{2}r^2\theta = \frac{1}{2}(12)^2\left(\frac{7\pi}{18}\right) = \frac{1}{2} \cdot 144\left(\frac{7\pi}{18}\right) = 72\left(\frac{7\pi}{18}\right) \approx 87.965 \text{ square yards}$$

57. $17° = 17 \cdot \dfrac{\pi}{180}$ radian $= \dfrac{17\pi}{180}$ radians ≈ 0.30 radians

59. $-40° = -40 \cdot \dfrac{\pi}{180}$ radian $= -\dfrac{2\pi}{9}$ radians ≈ -0.70 radians

61. $125° = 125 \cdot \dfrac{\pi}{180}$ radian $= \dfrac{25\pi}{36}$ radians ≈ 2.18 radians

63. 3.14 radians $= 3.14 \cdot \dfrac{180}{\pi}$ degrees $\approx 179.91°$

65. 2 radians $= 2 \cdot \dfrac{180}{\pi}$ degrees $\approx 114.59°$

67. 6.32 radians $= 6.32 \cdot \dfrac{180}{\pi}$ degrees $\approx 362.11°$

69. $40°10'25'' = \left(40 + 10 \cdot \dfrac{1}{60} + 25 \cdot \dfrac{1}{60} \cdot \dfrac{1}{60}\right)° \approx (40 + 0.1667 + 0.00694)° \approx 40.17°$

71. $1°2'3'' = \left(1 + 2 \cdot \dfrac{1}{60} + 3 \cdot \dfrac{1}{60} \cdot \dfrac{1}{60}\right)° \approx (1 + 0.0333 + 0.00083)° \approx 1.03°$

73. $9°9'9'' = \left(9 + 9 \cdot \dfrac{1}{60} + 9 \cdot \dfrac{1}{60} \cdot \dfrac{1}{60}\right)° = (9 + 0.15 + 0.0025)° \approx 9.15°$

75. $40.32° = ?$

$0.32° = 0.32(1°) = 0.32(60') = 19.2'$

$0.2' = 0.2(1') = 0.2(60") = 12"$

$40.32° = 40° + 0.32° = 40° + 19.2' = 40° + 19' + 0.2' = 40° + 19' + 12" = 40°19'12"$

77. $18.255° = ?$

$0.255° = 0.255(1°) = 0.255(60') = 15.3'$

$0.3' = 0.3(1') = 0.3(60") = 18"$

$18.255° = 18° + 0.255° = 18° + 15.3' = 18° + 15' + 0.3' = 18° + 15' + 18" = 18°15'18"$

79. $19.99° = ?$

$0.99° = 0.99(1°) = 0.99(60') = 59.4'$

$0.4' = 0.4(1') = 0.4(60") = 24"$

$19.99° = 19° + 0.99° = 19° + 59.4' = 19° + 59' + 0.4' = 19° + 59' + 24" = 19°59'24"$

81. $r = 6$ inches; $\theta = 90° = \dfrac{\pi}{2}$ radians

$$s = r\theta = 6 \cdot \frac{\pi}{2} = 3\pi \text{ inches} \approx 9.42 \text{ inches}$$

$r = 6$ inches; $\theta = \dfrac{25}{60}$ rev $= \dfrac{5}{12} \cdot 360° = 150° = \dfrac{5\pi}{6}$ radians

$$s = r\theta = 6 \cdot \frac{5\pi}{6} = 5\pi \text{ inches} \approx 15.71 \text{ inches}$$

83. $r = 4$ m; $\theta = 45°$;   Convert to radians: $45° = 45 \cdot \dfrac{\pi}{180} = \dfrac{\pi}{4}$ radians

$$A = \frac{1}{2}r^2\theta = \frac{1}{2}(4)^2\left(\frac{\pi}{4}\right) = \frac{1}{2} \cdot 16\left(\frac{\pi}{4}\right) = 2\pi \approx 6.283 \text{ square meters}$$

85. $r = 30$ feet; $\theta = 135°$;   Convert to radians: $135° = 135 \cdot \dfrac{\pi}{180} = \dfrac{3\pi}{4}$ radians

$$A = \frac{1}{2}r^2\theta = \frac{1}{2}(30)^2\left(\frac{3\pi}{4}\right) = \frac{1}{2} \cdot (900)\left(\frac{3\pi}{4}\right) = \frac{2700\pi}{8} \approx 1060.29 \text{ square feet}$$

87. $r = 5$ cm.; $t = 20$ seconds; $\theta = \dfrac{1}{3}$ radian

$$\omega = \frac{\theta}{t} = \frac{\left(\dfrac{1}{3}\right)}{20} = \frac{1}{3} \cdot \frac{1}{20} = \frac{1}{60} \text{ radian/sec}$$

$$v = \frac{s}{t} = \frac{r\theta}{t} = \frac{5 \cdot \left(\dfrac{1}{3}\right)}{20} = \frac{5}{3} \cdot \frac{1}{20} = \frac{1}{12} \text{ cm/sec}$$

89. $d = 26$ inches; $r = 13$ inches; $v = 35$ mi / hr

$$v = \frac{35 \text{ mi}}{\text{hr}} \cdot \frac{5280 \text{ ft}}{\text{mi}} \cdot \frac{12 \text{ in}}{\text{ft}} \cdot \frac{1 \text{ hr}}{60 \text{ min}} = 36960 \text{ in/min}$$

$$\omega = \frac{v}{r} = \frac{36960 \text{ in/min}}{13 \text{ in}} = 2843.08 \text{ radians/min}$$

$$= \frac{2843.08 \text{ rad}}{\text{min}} \cdot \frac{1 \text{ rev}}{2\pi \text{ rad}} \approx 452.5 \text{ rev/min}$$

91. $r = 3960$ miles; $\theta = 35°9' - 29°57' = 5°12' = 5.2° = 5.2 \cdot \dfrac{\pi}{180} \approx 0.09076$ radian

$s = r\theta = 3960 \cdot 0.09076 \approx 359.4$ miles

93. $r = 3429.5$ miles; $\omega = 1$ rev / day $= 2\pi$ radians / day $= \dfrac{\pi}{12}$ radians / hr

$v = r\omega = 3429.5 \cdot \dfrac{\pi}{12} \approx 897.8$ miles/hr

95. $r = 2.39 \times 10^5$ miles;

$$\omega = 1 \text{ rev}/27.3 \text{ days} = 2\pi \text{ radians}/27.3 \text{ day} = \frac{\pi}{12 \cdot 27.3} \text{ radians/hr}$$

$$v = r\omega = \left(2.39 \times 10^5\right) \cdot \frac{\pi}{327.6} \approx 2292 \text{ miles/hr}$$

97. $r_1 = 2$ inches; $r_2 = 8$ inches; $\omega_1 = 3$ rev / min $= 6\pi$ radians / min
Find ω_2:

$v_1 = v_2$

$r_1\omega_1 = r_2\omega_2 \Rightarrow 2(6\pi) = 8\omega_2$

$\omega_2 = \dfrac{12\pi}{8} = 1.5\pi$ radians / min $= \dfrac{1.5\pi}{2\pi}$ rev / min $= \dfrac{3}{4}$ rev / min

99. $r = 4$ feet; $\omega = 10$ rev / min $= 20\pi$ radians / min

$v = r\omega = 4 \cdot 20\pi = 80\pi \dfrac{\text{ft}}{\text{min}} = \dfrac{80\pi \text{ ft}}{\text{min}} \cdot \dfrac{1 \text{ mi}}{5280 \text{ ft}} \cdot \dfrac{60 \text{ min}}{\text{hr}} \approx 2.86$ mi/hr

101. $d = 8.5$ feet; $r = 4.25$ feet; $v = 9.55$ mi/hr

$\omega = \dfrac{v}{r} = \dfrac{9.55 \text{ mi/hr}}{4.25 \text{ ft}} = \dfrac{9.55 \text{ mi}}{\text{hr}} \cdot \dfrac{1}{4.25 \text{ ft}} \cdot \dfrac{5280 \text{ ft}}{\text{mi}} \cdot \dfrac{1 \text{ hr}}{60 \text{ min}} \cdot \dfrac{1 \text{ rev}}{2\pi} \approx 31.47$ rev/min

103. The earth makes one full rotation in 24 hours. The distance traveled in 24 hours is the circumference of the earth. At the equator the circumference is $2\pi(3960)$ miles. Therefore, the linear velocity a person must travel to keep up with the sun is:

$v = \dfrac{s}{t} = \dfrac{2\pi(3960)}{24} \approx 1037$ miles / hr

105. r_1 rotates at ω_1 rev / min; r_2 rotates at ω_2 rev / min;
$v = r_1\omega_1 = r_2\omega_2$

So, $\dfrac{r_1}{r_2} = \dfrac{\omega_2}{\omega_1}$

107. Answers will vary.

Chapter 5

Trigonometric Functions

5.2   Trigonometric Functions: Unit Circle Approach

1. $P = \left(\dfrac{\sqrt{3}}{2}, \dfrac{1}{2} \right) \Rightarrow a = \dfrac{\sqrt{3}}{2}, b = \dfrac{1}{2}$

$\sin t = \dfrac{1}{2};$

$\csc t = \dfrac{1}{\left(\dfrac{1}{2} \right)} = 1 \cdot \left(\dfrac{2}{1} \right) = 2$

$\cos t = \dfrac{\sqrt{3}}{2};$

$\sec t = \dfrac{1}{\left(\dfrac{\sqrt{3}}{2} \right)} = 1 \cdot \left(\dfrac{2}{\sqrt{3}} \right) = \dfrac{2}{\sqrt{3}} \cdot \dfrac{\sqrt{3}}{\sqrt{3}} = \dfrac{2\sqrt{3}}{3}$

$\tan t = \dfrac{\left(\dfrac{1}{2} \right)}{\left(\dfrac{\sqrt{3}}{2} \right)} = \left(\dfrac{1}{2} \right) \cdot \left(\dfrac{2}{\sqrt{3}} \right)$

$\cot t = \dfrac{\left(\dfrac{\sqrt{3}}{2} \right)}{\left(\dfrac{1}{2} \right)} = \left(\dfrac{\sqrt{3}}{2} \right) \cdot \left(\dfrac{2}{1} \right) = \sqrt{3}$

$= \dfrac{1}{\sqrt{3}} \cdot \dfrac{\sqrt{3}}{\sqrt{3}} = \dfrac{\sqrt{3}}{3}$

3. $P = \left(-\dfrac{2}{5}, \dfrac{\sqrt{21}}{5} \right) \Rightarrow a = -\dfrac{2}{5}, b = \dfrac{\sqrt{21}}{5}$

$\sin t = \dfrac{\sqrt{21}}{5}$

$\csc t = \dfrac{1}{\left(\dfrac{\sqrt{21}}{5} \right)} = 1 \cdot \left(\dfrac{5}{\sqrt{21}} \right) \cdot = \dfrac{5}{\sqrt{21}} \cdot \dfrac{\sqrt{21}}{\sqrt{21}} = \dfrac{5\sqrt{21}}{21}$

$\cos t = -\dfrac{2}{5}$

$\sec t = \dfrac{1}{\left(-\dfrac{2}{5} \right)} = 1 \cdot \left(-\dfrac{5}{2} \right) \cdot = -\dfrac{5}{2}$

$\tan t = \dfrac{\left(\dfrac{\sqrt{21}}{5} \right)}{\left(-\dfrac{2}{5} \right)} = \left(\dfrac{\sqrt{21}}{5} \right) \left(-\dfrac{5}{2} \right) = -\dfrac{\sqrt{21}}{2}$

$\cot t = \dfrac{\left(-\dfrac{2}{5} \right)}{\left(\dfrac{\sqrt{21}}{5} \right)} = \left(-\dfrac{2}{5} \right) \left(\dfrac{5}{\sqrt{21}} \right)$

$= -\dfrac{2}{\sqrt{21}} \cdot \dfrac{\sqrt{21}}{\sqrt{21}} = -\dfrac{2\sqrt{21}}{21}$

5. $P = \left(-\dfrac{\sqrt{2}}{2}, \dfrac{\sqrt{2}}{2} \right) \Rightarrow a = -\dfrac{\sqrt{2}}{2}, b = \dfrac{\sqrt{2}}{2}$

$\sin t = \dfrac{\sqrt{2}}{2}$ $\qquad\qquad$ $\csc t = \dfrac{1}{\left(\dfrac{\sqrt{2}}{2} \right)} = 1 \cdot \left(\dfrac{2}{\sqrt{2}} \right) \cdot = \dfrac{2}{\sqrt{2}} \cdot \dfrac{\sqrt{2}}{\sqrt{2}} = \sqrt{2}$

$\cos t = -\dfrac{\sqrt{2}}{2}$ $\qquad\qquad$ $\sec t = \dfrac{1}{\left(-\dfrac{\sqrt{2}}{2} \right)} = 1 \cdot \left(-\dfrac{2}{\sqrt{2}} \right) \cdot = -\dfrac{2}{\sqrt{2}} \cdot \dfrac{\sqrt{2}}{\sqrt{2}} = -\sqrt{2}$

$\tan t = \dfrac{\left(\dfrac{\sqrt{2}}{2} \right)}{\left(-\dfrac{\sqrt{2}}{2} \right)} = -1$ $\qquad\qquad$ $\cot t = \dfrac{\left(-\dfrac{\sqrt{2}}{2} \right)}{\left(\dfrac{\sqrt{2}}{2} \right)} = -1$

7. $P = \left(\dfrac{2\sqrt{2}}{3}, -\dfrac{1}{3} \right) \Rightarrow a = \dfrac{2\sqrt{2}}{3}, b = -\dfrac{1}{3}$

$\sin t = -\dfrac{1}{3}$ $\qquad\qquad$ $\csc t = \dfrac{1}{\left(-\dfrac{1}{3} \right)} = 1 \cdot \left(-\dfrac{3}{1} \right) = -3$

$\cos t = \dfrac{2\sqrt{2}}{3}$ $\qquad\qquad$ $\sec t = \dfrac{1}{\left(\dfrac{2\sqrt{2}}{3} \right)} = 1 \cdot \left(\dfrac{3}{2\sqrt{2}} \right) = \dfrac{3}{2\sqrt{2}} \cdot \dfrac{\sqrt{2}}{\sqrt{2}} = \dfrac{3\sqrt{2}}{8}$

$\tan t = \dfrac{\left(-\dfrac{1}{3} \right)}{\left(\dfrac{2\sqrt{2}}{3} \right)} = \left(-\dfrac{1}{3} \right) \cdot \left(\dfrac{3}{2\sqrt{2}} \right)$ $\qquad\qquad$ $\cot t = \dfrac{\left(\dfrac{2\sqrt{2}}{3} \right)}{\left(-\dfrac{1}{3} \right)} = \left(\dfrac{2\sqrt{2}}{3} \right) \cdot \left(-\dfrac{3}{1} \right) = -2\sqrt{2}$

$\qquad\quad = -\dfrac{1}{2\sqrt{2}} \cdot \dfrac{\sqrt{2}}{\sqrt{2}} = -\dfrac{\sqrt{2}}{8}$

9. $\sin\left(\dfrac{11\pi}{2} \right) = \sin\left(\dfrac{3\pi}{2} + \dfrac{8\pi}{2} \right) = \sin\left(\dfrac{3\pi}{2} + 4\pi \right) = \sin\left(\dfrac{3\pi}{2} + 2 \cdot 2\pi \right) = \sin\left(\dfrac{3\pi}{2} \right) = -1$

11. $\tan(6\pi) = \tan(0 + 6\pi) = \tan(0) = 0$

13. $\csc\left(\dfrac{11\pi}{2} \right) = \csc\left(\dfrac{3\pi}{2} + \dfrac{8\pi}{2} \right) = \csc\left(\dfrac{3\pi}{2} + 4\pi \right) = \csc\left(\dfrac{3\pi}{2} + 2 \cdot 2\pi \right) = \csc\left(\dfrac{3\pi}{2} \right) = -1$

15. $\cos\left(-\dfrac{3\pi}{2} \right) = \cos\left(\dfrac{3\pi}{2} \right) = \cos\left(-\dfrac{\pi}{2} + \dfrac{4\pi}{2} \right) = \cos\left(-\dfrac{\pi}{2} + 2\pi \right) = \cos\left(-\dfrac{\pi}{2} \right) = \cos\left(\dfrac{\pi}{2} \right) = 0$

17. $\sec(-\pi) = \sec(\pi) = -1$

19. $\sin(45°) + \cos(60°) = \dfrac{\sqrt{2}}{2} + \dfrac{1}{2} = \dfrac{1+\sqrt{2}}{2}$

21. $\sin(90°) + \tan(45°) = 1 + 1 = 2$ 23. $\sin(45°)\cos(45°) = \dfrac{\sqrt{2}}{2} \cdot \dfrac{\sqrt{2}}{2} = \dfrac{2}{4} = \dfrac{1}{2}$

25. $\csc(45°)\tan(60°) = \sqrt{2} \cdot \sqrt{3} = \sqrt{6}$ 27. $4\sin(90°) - 3\tan(180°) = 4 \cdot 1 - 3 \cdot 0 = 4$

29. $2\sin\left(\dfrac{\pi}{3}\right) - 3\tan\left(\dfrac{\pi}{6}\right) = 2 \cdot \dfrac{\sqrt{3}}{2} - 3 \cdot \dfrac{\sqrt{3}}{3} = \sqrt{3} - \sqrt{3} = 0$

31. $\sin\left(\dfrac{\pi}{4}\right) - \cos\left(\dfrac{\pi}{4}\right) = \dfrac{\sqrt{2}}{2} - \dfrac{\sqrt{2}}{2} = 0$

33. $2\sec\left(\dfrac{\pi}{4}\right) + 4\cot\left(\dfrac{\pi}{3}\right) = 2 \cdot \sqrt{2} + 4 \cdot \dfrac{\sqrt{3}}{3} = 2\sqrt{2} + \dfrac{4\sqrt{3}}{3}$

35. $\tan(\pi) - \cos(0) = 0 - 1 = -1$ 37. $\csc\left(\dfrac{\pi}{2}\right) + \cot\left(\dfrac{\pi}{2}\right) = 1 + 0 = 1$

39. The point on the unit circle that corresponds to $\theta = \dfrac{2\pi}{3} = 120°$ is $\left(-\dfrac{1}{2}, \dfrac{\sqrt{3}}{2}\right)$.

$\sin\theta = \dfrac{\sqrt{3}}{2}$ $\csc\theta = \dfrac{1}{\left(\dfrac{\sqrt{3}}{2}\right)} = 1 \cdot \dfrac{2}{\sqrt{3}} \cdot \dfrac{\sqrt{3}}{\sqrt{3}} = \dfrac{2\sqrt{3}}{3}$

$\cos\theta = -\dfrac{1}{2}$ $\sec\theta = \dfrac{1}{\left(-\dfrac{1}{2}\right)} = 1 \cdot \left(-\dfrac{2}{1}\right) = -2$

$\tan\theta = \dfrac{\left(\dfrac{\sqrt{3}}{2}\right)}{\left(-\dfrac{1}{2}\right)} = \dfrac{\sqrt{3}}{2} \cdot \left(-\dfrac{2}{1}\right) = -\sqrt{3}$ $\cot\theta = \dfrac{\left(-\dfrac{1}{2}\right)}{\left(\dfrac{\sqrt{3}}{2}\right)} = -\dfrac{1}{2} \cdot \dfrac{2}{\sqrt{3}} \cdot \dfrac{\sqrt{3}}{\sqrt{3}} = -\dfrac{\sqrt{3}}{3}$

41. The point on the unit circle that corresponds to $\theta = 210° = \dfrac{7\pi}{6}$ is $\left(-\dfrac{\sqrt{3}}{2}, -\dfrac{1}{2}\right)$.

$\sin\theta = -\dfrac{1}{2}$ $\csc\theta = \dfrac{1}{\left(-\dfrac{1}{2}\right)} = 1 \cdot \left(-\dfrac{2}{1}\right) = -2$

$\cos\theta = -\dfrac{\sqrt{3}}{2}$ $\sec\theta = \dfrac{1}{\left(-\dfrac{\sqrt{3}}{2}\right)} = 1 \cdot \left(-\dfrac{2}{\sqrt{3}}\right) \cdot \dfrac{\sqrt{3}}{\sqrt{3}} = -\dfrac{2\sqrt{3}}{3}$

$$\tan\theta = \frac{\left(-\dfrac{1}{2}\right)}{\left(-\dfrac{\sqrt{3}}{2}\right)} = -\frac{1}{2}\cdot\left(-\frac{2}{\sqrt{3}}\right)\cdot\frac{\sqrt{3}}{\sqrt{3}} = \frac{\sqrt{3}}{3} \qquad \cot\theta = \frac{\left(-\dfrac{\sqrt{3}}{2}\right)}{\left(-\dfrac{1}{2}\right)} = -\frac{\sqrt{3}}{2}\cdot\left(-\frac{2}{1}\right) = \sqrt{3}$$

43.   The point on the unit circle that corresponds to $\theta = \dfrac{3\pi}{4} = 135°$ is $\left(-\dfrac{\sqrt{2}}{2}, \dfrac{\sqrt{2}}{2}\right)$.

$$\sin\theta = \frac{\sqrt{2}}{2} \qquad\qquad \csc\theta = \frac{1}{\left(\dfrac{\sqrt{2}}{2}\right)} = 1\cdot\frac{2}{\sqrt{2}}\cdot\frac{\sqrt{2}}{\sqrt{2}} = \frac{2\sqrt{2}}{2} = \sqrt{2}$$

$$\cos\theta = -\frac{\sqrt{2}}{2} \qquad\qquad \sec\theta = \frac{1}{\left(-\dfrac{\sqrt{2}}{2}\right)} = 1\cdot\left(-\frac{2}{\sqrt{2}}\right)\cdot\frac{\sqrt{2}}{\sqrt{2}} = -\frac{2\sqrt{2}}{2} = -\sqrt{2}$$

$$\tan\theta = \frac{\left(\dfrac{\sqrt{2}}{2}\right)}{\left(-\dfrac{\sqrt{2}}{2}\right)} = \frac{\sqrt{2}}{2}\cdot\left(-\frac{2}{\sqrt{2}}\right) = -1 \qquad \cot\theta = \frac{\left(-\dfrac{\sqrt{2}}{2}\right)}{\left(\dfrac{\sqrt{2}}{2}\right)} = -\frac{\sqrt{2}}{2}\cdot\frac{2}{\sqrt{2}} = -1$$

45.   The point on the unit circle that corresponds to $\theta = \dfrac{8\pi}{3} = 480°$ is $\left(-\dfrac{1}{2}, \dfrac{\sqrt{3}}{2}\right)$.

$$\sin\theta = \frac{\sqrt{3}}{2} \qquad\qquad \csc\theta = \frac{1}{\left(\dfrac{\sqrt{3}}{2}\right)} = 1\cdot\frac{2}{\sqrt{3}}\cdot\frac{\sqrt{3}}{\sqrt{3}} = \frac{2\sqrt{3}}{3}$$

$$\cos\theta = -\frac{1}{2} \qquad\qquad \sec\theta = \frac{1}{\left(-\dfrac{1}{2}\right)} = 1\cdot\left(-\frac{2}{1}\right) = -2$$

$$\tan\theta = \frac{\left(\dfrac{\sqrt{3}}{2}\right)}{\left(-\dfrac{1}{2}\right)} = \frac{\sqrt{3}}{2}\cdot\left(-\frac{2}{1}\right) = -\sqrt{3} \qquad \cot\theta = \frac{\left(-\dfrac{1}{2}\right)}{\left(\dfrac{\sqrt{3}}{2}\right)} = -\frac{1}{2}\cdot\frac{2}{\sqrt{3}}\cdot\frac{\sqrt{3}}{\sqrt{3}} = -\frac{\sqrt{3}}{3}$$

47.   The point on the unit circle that corresponds to $\theta = 405° = \dfrac{9\pi}{4}$ is $\left(\dfrac{\sqrt{2}}{2}, \dfrac{\sqrt{2}}{2}\right)$.

$$\sin\theta = \frac{\sqrt{2}}{2} \qquad\qquad \csc\theta = \frac{1}{\left(\dfrac{\sqrt{2}}{2}\right)} = 1\cdot\frac{2}{\sqrt{2}}\cdot\frac{\sqrt{2}}{\sqrt{2}} = \sqrt{2}$$

$$\cos\theta = \frac{\sqrt{2}}{2} \qquad\qquad \sec\theta = \frac{1}{\left(\dfrac{\sqrt{2}}{2}\right)} = 1 \cdot \frac{2}{\sqrt{2}} \cdot \frac{\sqrt{2}}{\sqrt{2}} = \sqrt{2}$$

$$\tan\theta = \frac{\left(\dfrac{\sqrt{2}}{2}\right)}{\left(\dfrac{\sqrt{2}}{2}\right)} = \frac{\sqrt{2}}{2} \cdot \frac{2}{\sqrt{2}} = 1 \qquad \cot\theta = \frac{\left(\dfrac{\sqrt{2}}{2}\right)}{\left(\dfrac{\sqrt{2}}{2}\right)} = 1$$

49. The point on the unit circle that corresponds to $\theta = -\dfrac{\pi}{6} = -30°$ is $\left(\dfrac{\sqrt{3}}{2}, -\dfrac{1}{2}\right)$.

$$\sin\theta = -\frac{1}{2} \qquad\qquad \csc\theta = \frac{1}{\left(-\dfrac{1}{2}\right)} - 2$$

$$\cos\theta = \frac{\sqrt{3}}{2} \qquad\qquad \sec\theta = \frac{1}{\left(\dfrac{\sqrt{3}}{2}\right)} = \frac{2}{\sqrt{3}} \cdot \frac{\sqrt{3}}{\sqrt{3}} = \frac{2\sqrt{3}}{3}$$

$$\tan\theta = \frac{\left(-\dfrac{1}{2}\right)}{\left(\dfrac{\sqrt{3}}{2}\right)} = -\frac{1}{2} \cdot \frac{2}{\sqrt{3}} \cdot \frac{\sqrt{3}}{\sqrt{3}} = -\frac{\sqrt{3}}{3} \qquad \cot\theta = \frac{\left(\dfrac{\sqrt{3}}{2}\right)}{\left(-\dfrac{1}{2}\right)} = \left(\dfrac{\sqrt{3}}{2}\right) \cdot \left(-\dfrac{2}{1}\right) = -\sqrt{3}$$

51. The point on the unit circle that corresponds to $\theta = -45° = -\dfrac{\pi}{4}$ is $\left(\dfrac{\sqrt{2}}{2}, -\dfrac{\sqrt{2}}{2}\right)$.

$$\sin\theta = -\frac{\sqrt{2}}{2} \qquad\qquad \csc\theta = \frac{1}{\left(-\dfrac{\sqrt{2}}{2}\right)} = 1 \cdot \left(-\dfrac{2}{\sqrt{2}}\right) \cdot \frac{\sqrt{2}}{\sqrt{2}} = -\sqrt{2}$$

$$\cos\theta = \frac{\sqrt{2}}{2} \qquad\qquad \sec\theta = \frac{1}{\left(\dfrac{\sqrt{2}}{2}\right)} = 1 \cdot \frac{2}{\sqrt{2}} \cdot \frac{\sqrt{2}}{\sqrt{2}} = \sqrt{2}$$

$$\tan\theta = \frac{\left(-\dfrac{\sqrt{2}}{2}\right)}{\left(\dfrac{\sqrt{2}}{2}\right)} = -\frac{\sqrt{2}}{2} \cdot \frac{2}{\sqrt{2}} = -1 \qquad \cot\theta = \frac{\left(\dfrac{\sqrt{2}}{2}\right)}{\left(-\dfrac{\sqrt{2}}{2}\right)} = -1$$

53. The point on the unit circle that corresponds to $\theta = \dfrac{5\pi}{2} = 450°$ is $(0,1)$.

$\sin\theta = 1$ $\csc\theta = \dfrac{1}{1} = 1$

$\cos\theta = 0$ $\sec\theta = \dfrac{1}{0} = \text{not defined}$

$\tan\theta = \dfrac{1}{0} = \text{not defined}$ $\cot\theta = \dfrac{0}{1} = 0$

55. The point on the unit circle that corresponds to $\theta = 720° = 4\pi$ is $(1,0)$.

$\sin\theta = 0$ $\csc\theta = \dfrac{1}{0} = \text{not defined}$

$\cos\theta = 1$ $\sec\theta = \dfrac{1}{1} = 1$

$\tan\theta = \dfrac{0}{1} = 0$ $\cot\theta = \dfrac{1}{0} = \text{not defined}$

57. Set the calculator to degree mode : $\sin(28°) \approx 0.47$

59. Set the calculator to degree mode : $\tan(21°) \approx 0.38$

61. Set the calculator to degree mode : $\sec(41°) = \dfrac{1}{\cos(41°)} \approx 1.33$

63. Set the calculator to radian mode : $\sin\left(\dfrac{\pi}{10}\right) \approx 0.31$

65. Set the calculator to radian mode : $\tan\left(\dfrac{5\pi}{12}\right) \approx 3.73$

67. Set the calculator to radian mode : $\sec\left(\dfrac{\pi}{12}\right) = \dfrac{1}{\cos\left(\dfrac{\pi}{12}\right)} \approx 1.04$

69. Set the calculator to radian mode: $\sin(1) \approx 0.84$.

71. Set the calculator to degree mode: $\sin(1°) \approx 0.02$.

73. For the point $(-3, 4)$, $x = -3$, $y = 4$, $r = \sqrt{x^2 + y^2} = \sqrt{9 + 16} = \sqrt{25} = 5$

$\sin\theta = \dfrac{4}{5}$ $\cos\theta = -\dfrac{3}{5}$ $\tan\theta = -\dfrac{4}{3}$

$\csc\theta = \dfrac{5}{4}$ $\sec\theta = -\dfrac{5}{3}$ $\cot\theta = -\dfrac{3}{4}$

75. For the point $(2, -3)$, $x = 2$, $y = -3$, $r = \sqrt{x^2 + y^2} = \sqrt{4+9} = \sqrt{13}$

$$\sin\theta = \frac{-3}{\sqrt{13}} \cdot \frac{\sqrt{13}}{\sqrt{13}} = -\frac{3\sqrt{13}}{13} \qquad \cos\theta = \frac{2}{\sqrt{13}} \cdot \frac{\sqrt{13}}{\sqrt{13}} = \frac{2\sqrt{13}}{13} \qquad \tan\theta = -\frac{3}{2}$$

$$\csc\theta = -\frac{\sqrt{13}}{3} \qquad\qquad \sec\theta = \frac{\sqrt{13}}{2} \qquad\qquad \cot\theta = -\frac{2}{3}$$

77. For the point $(-2, -2)$, $x = -2$, $y = -2$, $r = \sqrt{x^2 + y^2} = \sqrt{4+4} = \sqrt{8} = 2\sqrt{2}$

$$\sin\theta = \frac{-2}{2\sqrt{2}} \cdot \frac{\sqrt{2}}{\sqrt{2}} = -\frac{\sqrt{2}}{2} \qquad \cos\theta = \frac{-2}{2\sqrt{2}} \cdot \frac{\sqrt{2}}{\sqrt{2}} = -\frac{\sqrt{2}}{2} \qquad \tan\theta = \frac{-2}{-2} = 1$$

$$\csc\theta = \frac{2\sqrt{2}}{-2} = -\sqrt{2} \qquad \sec\theta = \frac{2\sqrt{2}}{-2} = -\sqrt{2} \qquad \cot\theta = \frac{-2}{-2} = 1$$

79. For the point $(-3, -2)$, $x = -3$, $y = -2$, $r = \sqrt{x^2 + y^2} = \sqrt{9+4} = \sqrt{13}$

$$\sin\theta = \frac{-2}{\sqrt{13}} \cdot \frac{\sqrt{13}}{\sqrt{13}} = -\frac{2\sqrt{13}}{13} \qquad \cos\theta = \frac{-3}{\sqrt{13}} \cdot \frac{\sqrt{13}}{\sqrt{13}} = -\frac{3\sqrt{13}}{13} \qquad \tan\theta = \frac{-2}{-3} = \frac{2}{3}$$

$$\csc\theta = \frac{\sqrt{13}}{-2} = -\frac{\sqrt{13}}{2} \qquad \sec\theta = \frac{\sqrt{13}}{-3} = -\frac{\sqrt{13}}{3} \qquad \cot\theta = \frac{-3}{-2} = \frac{3}{2}$$

81. For the point $\left(\frac{1}{3}, -\frac{1}{4}\right)$, $x = \frac{1}{3}$, $y = -\frac{1}{4}$, $r = \sqrt{x^2 + y^2} = \sqrt{\frac{1}{9} + \frac{1}{16}} = \sqrt{\frac{25}{144}} = \frac{5}{12}$

$$\sin\theta = \frac{\left(-\frac{1}{4}\right)}{\left(\frac{5}{12}\right)} = -\frac{1}{4} \cdot \frac{12}{5} = -\frac{3}{5} \qquad \csc\theta = \frac{\left(\frac{5}{12}\right)}{\left(-\frac{1}{4}\right)} = \frac{5}{12} \cdot -\frac{4}{1} = -\frac{5}{3}$$

$$\cos\theta = \frac{\left(\frac{1}{3}\right)}{\left(\frac{5}{12}\right)} = \frac{1}{3} \cdot \frac{12}{5} = \frac{4}{5} \qquad \sec\theta = \frac{\left(\frac{5}{12}\right)}{\left(\frac{1}{3}\right)} = \frac{5}{12} \cdot \frac{3}{1} = \frac{5}{4}$$

$$\tan\theta = \frac{\left(-\frac{1}{4}\right)}{\left(\frac{1}{3}\right)} = -\frac{1}{4} \cdot \frac{3}{1} = -\frac{3}{4} \qquad \cot\theta = \frac{\left(\frac{1}{3}\right)}{\left(-\frac{1}{4}\right)} = \frac{1}{3} \cdot -\frac{4}{1} = -\frac{4}{3}$$

83. $\sin(45°) + \sin(135°) + \sin(225°) + \sin(315°) = \frac{\sqrt{2}}{2} + \frac{\sqrt{2}}{2} - \frac{\sqrt{2}}{2} - \frac{\sqrt{2}}{2} = 0$

85. If $\sin\theta = 0.1$, then $\sin(\theta + \pi) = -0.1$

87. If $\tan\theta = 3$, then $\tan(\theta + \pi) = 3$

89.   If $\sin\theta = \dfrac{1}{5}$, then $\csc\theta = \dfrac{1}{\left(\dfrac{1}{5}\right)} = 1 \cdot \dfrac{5}{1} = 5$

91. $\sin(60°) = \dfrac{\sqrt{3}}{2}$

93. $\sin\left(\dfrac{60°}{2}\right) = \sin(30°) = \dfrac{1}{2}$

95. $(\sin 60°)^2 = \left(\dfrac{\sqrt{3}}{2}\right)^2 = \dfrac{3}{4}$

97. $\sin(2 \cdot 60°) = \sin(120°) = \dfrac{\sqrt{3}}{2}$

99. $2\sin(60°) = 2 \cdot \dfrac{\sqrt{3}}{2} = \sqrt{3}$

101. $\sin(-60°) = \sin(300°) = -\dfrac{\sqrt{3}}{2}$

103. Complete the table:

θ	0.5	0.4	0.2	0.1	0.01	0.001	0.0001	0.00001
$\sin\theta$	0.4794	0.3894	0.1987	0.0998	0.0100	0.0010	0.0001	0.00001
$\dfrac{\sin\theta}{\theta}$	0.9589	0.9735	0.9933	0.9983	1.0000	1.0000	1.0000	1.0000

As θ approaches 0, the ratio $\dfrac{\sin\theta}{\theta}$ approaches 1.

105. Use the formula $R = \dfrac{v_0^2 \sin(2\theta)}{g}$ with $g = 32.2 \text{ft} / \sec^2$; $\theta = 45°$; $v_0 = 100$ ft / sec :

$$R = \dfrac{100^2 \sin(2(45°))}{32.2} = \dfrac{10000 \sin(90°)}{32.2} = \dfrac{10000}{32.2} \approx 310.56 \text{ feet}$$

Use the formula $H = \dfrac{v_0^2 \sin^2\theta}{2g}$ with $g = 32.2 \text{ft} / \sec^2$; $\theta = 45°$; $v_0 = 100$ ft / sec :

$$H = \dfrac{100^2 \sin^2(45°)}{2(32.2)} \approx \dfrac{10000(0.7071)^2}{64.4} \approx 77.64 \text{ feet}$$

107. Use the formula $R = \dfrac{v_0^2 \sin(2\theta)}{g}$ with $g = 9.8 \text{ m} / \sec^2$; $\theta = 25°$; $v_0 = 500$ m / sec :

$$R = \dfrac{500^2 \sin(2(25°))}{9.8} = \dfrac{250,000 \sin 50°}{9.8} \approx 19,542 \text{ meters}$$

Use the formula $H = \dfrac{v_0^2 \sin^2\theta}{2g}$ with $g = 9.8 \text{ m} / \sec^2$; $\theta = 25°$; $v_0 = 500$ m / sec :

$$H = \dfrac{500^2 \sin^2(25°)}{2(9.8)} \approx \dfrac{250,000(0.4226)^2}{19.6} \approx 2278 \text{ meters}$$

109. Use the formula $t = \sqrt{\dfrac{2a}{g \sin\theta \cos\theta}}$ with $g = 32$ ft / sec^2 and $a = 10$ feet:

(a) $t = \sqrt{\dfrac{2(10)}{32\sin(30°)\cos(30°)}} = \sqrt{\dfrac{20}{\left(32 \cdot \dfrac{1}{2} \cdot \dfrac{\sqrt{3}}{2}\right)}} = \sqrt{\dfrac{20}{8\sqrt{3}}} = \sqrt{\dfrac{5}{2\sqrt{3}}} \approx 1.20$ seconds

(b) $t = \sqrt{\dfrac{2(10)}{32\sin(45°)\cos(45°)}} = \sqrt{\dfrac{20}{\left(32 \cdot \dfrac{\sqrt{2}}{2} \cdot \dfrac{\sqrt{2}}{2}\right)}} = \sqrt{\dfrac{20}{16}} = \sqrt{\dfrac{5}{4}} \approx 1.12$ seconds

(c) $t = \sqrt{\dfrac{2(10)}{32\sin(60°)\cos(60°)}} = \sqrt{\dfrac{20}{\left(32 \cdot \dfrac{\sqrt{3}}{2} \cdot \dfrac{1}{2}\right)}} = \sqrt{\dfrac{20}{8\sqrt{3}}} = \sqrt{\dfrac{5}{2\sqrt{3}}} \approx 1.20$ seconds

111. (a) $T(30°) = 1 + \dfrac{2}{3\sin(30°)} - \dfrac{1}{4\tan(30°)} = 1 + \dfrac{2}{\left(3 \cdot \dfrac{1}{2}\right)} - \dfrac{1}{\left(4 \cdot \dfrac{1}{\sqrt{3}}\right)} = 1 + \dfrac{4}{3} - \dfrac{\sqrt{3}}{4} \approx 1.9$ hrs

$\dfrac{1}{x} = \tan\theta \implies x = \dfrac{1}{\tan\theta}$

Distance traveled on road is: $8 - 2x = 8 - \dfrac{2}{\tan\theta}$

Time on road $= \dfrac{\text{distance on road}}{\text{rate on road}} = \dfrac{\left(8 - \dfrac{2}{\tan\theta}\right)}{8}$

Time on road $= \dfrac{\left(8 - \dfrac{2}{\tan(30°)}\right)}{8} \approx 0.57$ hours

(b) $T(45°) = 1 + \dfrac{2}{3\sin(45°)} - \dfrac{1}{4\tan(45°)} = 1 + \dfrac{2}{\left(3 \cdot \dfrac{1}{\sqrt{2}}\right)} - \dfrac{1}{4 \cdot 1} = 1 + \dfrac{2\sqrt{2}}{3} - \dfrac{1}{4} \approx 1.69$ hrs

Time on road $= \dfrac{\left(8 - \dfrac{2}{\tan(45°)}\right)}{8} \approx 0.75$ hours

(c) $T(60°) = 1 + \dfrac{2}{3\sin(60°)} - \dfrac{1}{4\tan(60°)} = 1 + \dfrac{2}{\left(3 \cdot \dfrac{\sqrt{3}}{2}\right)} - \dfrac{1}{4 \cdot \sqrt{3}}$

$= 1 + \dfrac{4}{3\sqrt{3}} - \dfrac{1}{4\sqrt{3}} \approx 1.63$ hrs

Time on road $= \dfrac{\left(8 - \dfrac{2}{\tan(60°)}\right)}{8} \approx 0.86$ hours

(d) $T(90°) = 1 + \dfrac{2}{3\sin(90°)} - \dfrac{1}{4\tan(90°)}$, but $\tan(90°)$ is undefined.

The distance would be 2 miles in the sand and 8 miles on the road. The total time would be: $\dfrac{2}{3} + 1 = \dfrac{5}{3} \approx 1.67$ hours.

113. (a) $R = \dfrac{(32^2)\sqrt{2}}{32} \cdot [\sin(2(60°)) - \cos(2(60°)) - 1] \approx 32\sqrt{2}(0.866 - (-0.5) - 1) \approx 16.6$ ft

(b)   Graph:

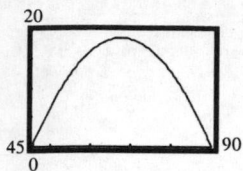

(c)   Using MAXIMUM, R is largest when $\theta \approx 67.5°$.

115. – 119.   Answers will vary.

Trigonometric Functions

5.3 Properties of the Trigonometric Functions

1. $\sin(405°) = \sin(360° + 45°) = \sin(45°) = \dfrac{\sqrt{2}}{2}$

3. $\tan(405°) = \tan(180° + 180° + 45°) = \tan(45°) = 1$

5. $\csc(450°) = \csc(360° + 90°) = \csc(90°) = 1$

7. $\cot(390°) = \cot(180° + 180° + 30°) = \cot(30°) = \sqrt{3}$

9. $\cos\left(\dfrac{33\pi}{4}\right) = \cos\left(\dfrac{\pi}{4} + \dfrac{32\pi}{4}\right) = \cos\left(\dfrac{\pi}{4} + 8\pi\right) = \cos\left(\dfrac{\pi}{4} + 4 \cdot 2\pi\right) = \cos\left(\dfrac{\pi}{4}\right) = \dfrac{\sqrt{2}}{2}$

11. $\tan(21\pi) = \tan(0 + 21\pi) = \tan(0) = 0$

13. $\sec\left(\dfrac{17\pi}{4}\right) = \sec\left(\dfrac{\pi}{4} + \dfrac{16\pi}{4}\right) = \sec\left(\dfrac{\pi}{4} + 4\pi\right) = \sec\left(\dfrac{\pi}{4} + 2 \cdot 2\pi\right) = \sec\left(\dfrac{\pi}{4}\right) = \sqrt{2}$

15. $\tan\left(\dfrac{19\pi}{6}\right) = \tan\left(\dfrac{\pi}{6} + \dfrac{18\pi}{6}\right) = \tan\left(\dfrac{\pi}{6} + 3\pi\right) = \tan\left(\dfrac{\pi}{6}\right) = \dfrac{\sqrt{3}}{3}$

17. Since $\sin\theta > 0$ for points in quadrants I and II, and $\cos\theta < 0$ for points in quadrants II and III, the angle θ lies in quadrant II.

19. Since $\sin\theta < 0$ for points in quadrants III and IV, and $\tan\theta < 0$ for points in quadrants II and IV, the angle θ lies in quadrant IV.

21. Since $\cos\theta > 0$ for points in quadrants I and IV, and $\tan\theta < 0$ for points in quadrants II and IV, the angle θ lies in quadrant IV.

23. Since $\sec\theta < 0$ for points in quadrants II and III, and $\sin\theta > 0$ for points in quadrants I and II, the angle θ lies in quadrant II.

25. $\sin\theta = -\dfrac{3}{5}, \quad \cos\theta = \dfrac{4}{5}$

$$\tan\theta = \frac{\sin\theta}{\cos\theta} = \frac{\left(-\dfrac{3}{5}\right)}{\left(\dfrac{4}{5}\right)} = -\frac{3}{5} \cdot \frac{5}{4} = -\frac{3}{4}$$

$$\sec\theta = \frac{1}{\cos\theta} = \frac{1}{\left(\dfrac{4}{5}\right)} = \frac{5}{4}$$

$$\csc\theta = \frac{1}{\sin\theta} = \frac{1}{\left(-\dfrac{3}{5}\right)} = -\frac{5}{3}$$

$$\cot\theta = \frac{1}{\tan\theta} = -\frac{4}{3}$$

27. $\sin\theta = \dfrac{2\sqrt{5}}{5}, \quad \cos\theta = \dfrac{\sqrt{5}}{5}$

$$\tan\theta = \frac{\sin\theta}{\cos\theta} = \frac{\left(\dfrac{2\sqrt{5}}{5}\right)}{\left(\dfrac{\sqrt{5}}{5}\right)} = \frac{2\sqrt{5}}{5} \cdot \frac{5}{\sqrt{5}} = 2$$

$$\sec\theta = \frac{1}{\cos\theta} = \frac{1}{\left(\dfrac{\sqrt{5}}{5}\right)}$$
$$= \frac{5}{\sqrt{5}} \cdot \frac{\sqrt{5}}{\sqrt{5}} = \sqrt{5}$$

$$\csc\theta = \frac{1}{\sin\theta} = \frac{1}{\left(\dfrac{2\sqrt{5}}{5}\right)} = 1 \cdot \frac{5}{2\sqrt{5}} \cdot \frac{\sqrt{5}}{\sqrt{5}} = \frac{\sqrt{5}}{2}$$

$$\cot\theta = \frac{1}{\tan\theta} = \frac{1}{2}$$

29. $\sin\theta = \dfrac{1}{2}, \quad \cos\theta = \dfrac{\sqrt{3}}{2}$

$$\tan\theta = \frac{\sin\theta}{\cos\theta} = \frac{\left(\dfrac{1}{2}\right)}{\left(\dfrac{\sqrt{3}}{2}\right)}$$
$$= \frac{1}{2} \cdot \frac{2}{\sqrt{3}} \cdot \frac{\sqrt{3}}{\sqrt{3}} = \frac{\sqrt{3}}{3}$$

$$\sec\theta = \frac{1}{\cos\theta} = \frac{1}{\left(\dfrac{\sqrt{3}}{2}\right)}$$
$$= \frac{2}{\sqrt{3}} \cdot \frac{\sqrt{3}}{\sqrt{3}} = \frac{2\sqrt{3}}{3}$$

$$\csc\theta = \frac{1}{\sin\theta} = \frac{1}{\left(\dfrac{1}{2}\right)}$$
$$= 1 \cdot \frac{2}{1} = 2$$

$$\cot\theta = \frac{1}{\tan\theta} = \frac{1}{\left(\dfrac{\sqrt{3}}{3}\right)}$$
$$= \frac{3}{\sqrt{3}} \cdot \frac{\sqrt{3}}{\sqrt{3}} = \sqrt{3}$$

31. $\sin\theta = -\dfrac{1}{3}, \quad \cos\theta = \dfrac{2\sqrt{2}}{3}$

$$\tan\theta = \frac{\sin\theta}{\cos\theta} = \frac{\left(-\dfrac{1}{3}\right)}{\left(\dfrac{2\sqrt{2}}{3}\right)} \qquad \sec\theta = \frac{1}{\cos\theta} = \frac{1}{\left(\dfrac{2\sqrt{2}}{3}\right)}$$

$$= -\frac{1}{3}\cdot\frac{3}{2\sqrt{2}}\cdot\frac{\sqrt{2}}{\sqrt{2}} = -\frac{\sqrt{2}}{4} \qquad = \frac{3}{2\sqrt{2}}\cdot\frac{\sqrt{2}}{\sqrt{2}} = \frac{3\sqrt{2}}{4}$$

$$\csc\theta = \frac{1}{\sin\theta} = \frac{1}{\left(-\dfrac{1}{3}\right)} \qquad \cot\theta = \frac{1}{\tan\theta} = \frac{1}{\left(-\dfrac{\sqrt{2}}{4}\right)}$$

$$= 1\cdot\left(-\frac{3}{1}\right) = -3 \qquad = -\frac{4}{\sqrt{2}}\cdot\frac{\sqrt{2}}{\sqrt{2}} = -2\sqrt{2}$$

33. $\sin\theta = \dfrac{12}{13}, \quad \theta$ in quadrant II

Solve for $\cos\theta$:
$$\sin^2\theta + \cos^2\theta = 1$$
$$\cos^2\theta = 1 - \sin^2\theta$$
$$\cos\theta = \pm\sqrt{1-\sin^2\theta}$$

Since θ is in quadrant II, $\cos\theta < 0$.

$$\cos\theta = -\sqrt{1-\sin^2\theta} = -\sqrt{1-\left(\frac{12}{13}\right)^2} = -\sqrt{1-\frac{144}{169}} = -\sqrt{\frac{25}{169}} = -\frac{5}{13}$$

$$\tan\theta = \frac{\sin\theta}{\cos\theta} = \frac{\left(\dfrac{12}{13}\right)}{\left(-\dfrac{5}{13}\right)} = \frac{12}{13}\cdot\left(-\frac{13}{5}\right) = -\frac{12}{5} \qquad \sec\theta = \frac{1}{\cos\theta} = \frac{1}{\left(-\dfrac{5}{13}\right)} = -\frac{13}{5}$$

$$\csc\theta = \frac{1}{\sin\theta} = \frac{1}{\left(\dfrac{12}{13}\right)} = \frac{13}{12} \qquad\qquad \cot\theta = \frac{1}{\tan\theta} = \frac{1}{\left(-\dfrac{12}{5}\right)} = -\frac{5}{12}$$

35. $\cos\theta = -\dfrac{4}{5}, \quad \theta$ in quadrant III

Solve for $\sin\theta$:
$$\sin^2\theta + \cos^2\theta = 1$$
$$\sin^2\theta = 1 - \cos^2\theta$$
$$\sin\theta = \pm\sqrt{1-\cos^2\theta}$$

Since θ is in quadrant III, $\sin\theta < 0$.

$$\sin\theta = -\sqrt{1-\cos^2\theta} = -\sqrt{1-\left(-\frac{4}{5}\right)^2} = -\sqrt{1-\frac{16}{25}} = -\sqrt{\frac{9}{25}} = -\frac{3}{5}$$

$$\tan\theta = \frac{\sin\theta}{\cos\theta} = \frac{\left(-\dfrac{3}{5}\right)}{\left(-\dfrac{4}{5}\right)} = -\frac{3}{5}\cdot\left(-\frac{5}{4}\right) = \frac{3}{4} \qquad \sec\theta = \frac{1}{\cos\theta} = \frac{1}{\left(-\dfrac{4}{5}\right)} = -\frac{5}{4}$$

$$\csc\theta = \frac{1}{\sin\theta} = \frac{1}{\left(-\dfrac{3}{5}\right)} = -\frac{5}{3} \qquad \cot\theta = \frac{1}{\tan\theta} = \frac{1}{\left(\dfrac{3}{4}\right)} = \frac{4}{3}$$

37. $\sin\theta = \dfrac{5}{13}, \quad 90° < \theta < 180°, \quad \theta$ in quadrant II

Solve for $\cos\theta$:
$$\sin^2\theta + \cos^2\theta = 1$$
$$\cos^2\theta = 1 - \sin^2\theta$$
$$\cos\theta = \pm\sqrt{1 - \sin^2\theta}$$
Since θ is in quadrant II, $\cos\theta < 0$.

$$\cos\theta = -\sqrt{1 - \sin^2\theta} = -\sqrt{1 - \left(\frac{5}{13}\right)^2} = -\sqrt{1 - \frac{25}{169}} = -\sqrt{\frac{144}{169}} = -\frac{12}{13}$$

$$\tan\theta = \frac{\sin\theta}{\cos\theta} = \frac{\left(\dfrac{5}{13}\right)}{\left(-\dfrac{12}{13}\right)} = \frac{5}{13}\cdot\left(-\frac{13}{12}\right) = -\frac{5}{12} \qquad \sec\theta = \frac{1}{\cos\theta} = \frac{1}{\left(-\dfrac{12}{13}\right)} = -\frac{13}{12}$$

$$\csc\theta = \frac{1}{\sin\theta} = \frac{1}{\left(\dfrac{5}{13}\right)} = \frac{13}{5} \qquad \cot\theta = \frac{1}{\tan\theta} = \frac{1}{\left(-\dfrac{5}{12}\right)} = -\frac{12}{5}$$

39. $\cos\theta = -\dfrac{1}{3}, \quad \dfrac{\pi}{2} < \theta < \pi, \quad \theta$ in quadrant II

Solve for $\sin\theta$:
$$\sin^2\theta + \cos^2\theta = 1$$
$$\sin^2\theta = 1 - \cos^2\theta$$
$$\sin\theta = \pm\sqrt{1 - \cos^2\theta}$$
Since θ is in quadrant II, $\sin\theta > 0$.

$$\sin\theta = \sqrt{1 - \cos^2\theta} = \sqrt{1 - \left(-\frac{1}{3}\right)^2} = \sqrt{1 - \frac{1}{9}} = \sqrt{\frac{8}{9}} = \frac{2\sqrt{2}}{3}$$

$$\tan\theta = \frac{\sin\theta}{\cos\theta} = \frac{\left(\dfrac{2\sqrt{2}}{3}\right)}{\left(-\dfrac{1}{3}\right)} = \frac{2\sqrt{2}}{3}\cdot\left(-\frac{3}{1}\right) = -2\sqrt{2}$$

$$\sec\theta = \frac{1}{\cos\theta} = \frac{1}{\left(-\dfrac{1}{3}\right)} = -3$$

$$\csc\theta = \frac{1}{\sin\theta} = \frac{1}{\left(\dfrac{2\sqrt{2}}{3}\right)} = \frac{3}{2\sqrt{2}}\cdot\frac{\sqrt{2}}{\sqrt{2}} = \frac{3\sqrt{2}}{4}$$

$$\cot\theta = \frac{1}{\tan\theta} = \frac{1}{-2\sqrt{2}}\cdot\frac{\sqrt{2}}{\sqrt{2}} = -\frac{\sqrt{2}}{4}$$

41. $\sin\theta = \dfrac{2}{3}$, $\tan\theta < 0$, $\Rightarrow$ θ in quadrant II

Solve for $\cos\theta$:

$$\sin^2\theta + \cos^2\theta = 1$$
$$\cos^2\theta = 1 - \sin^2\theta$$
$$\cos\theta = \pm\sqrt{1 - \sin^2\theta}$$

Since θ is in quadrant II, $\cos\theta < 0$.

$$\cos\theta = -\sqrt{1 - \sin^2\theta} = -\sqrt{1 - \left(\dfrac{2}{3}\right)^2} = -\sqrt{1 - \dfrac{4}{9}} = -\sqrt{\dfrac{5}{9}} = -\dfrac{\sqrt{5}}{3}$$

$$\tan\theta = \dfrac{\sin\theta}{\cos\theta} = \dfrac{\left(\dfrac{2}{3}\right)}{\left(-\dfrac{\sqrt{5}}{3}\right)} = \dfrac{2}{3}\cdot\left(-\dfrac{3}{\sqrt{5}}\right) = -\dfrac{2\sqrt{5}}{5} \qquad \sec\theta = \dfrac{1}{\cos\theta} = \dfrac{1}{\left(-\dfrac{\sqrt{5}}{3}\right)} = -\dfrac{3}{\sqrt{5}} = -\dfrac{3\sqrt{5}}{5}$$

$$\csc\theta = \dfrac{1}{\sin\theta} = \dfrac{1}{\left(\dfrac{2}{3}\right)} = \dfrac{3}{2} \qquad\qquad\qquad \cot\theta = \dfrac{1}{\tan\theta} = \dfrac{1}{\left(-\dfrac{2\sqrt{5}}{5}\right)} = -\dfrac{5}{2\sqrt{5}} = -\dfrac{\sqrt{5}}{2}$$

43. $\sec\theta = 2$, $\sin\theta < 0$, $\Rightarrow$ θ in quadrant IV

Solve for $\cos\theta$:

$$\cos\theta = \dfrac{1}{\sec\theta} = \dfrac{1}{2}$$

Solve for $\sin\theta$:

$$\sin^2\theta + \cos^2\theta = 1$$
$$\sin^2\theta = 1 - \cos^2\theta$$
$$\sin\theta = \pm\sqrt{1 - \cos^2\theta}$$

Since θ is in quadrant IV, $\sin\theta < 0$.

$$\sin\theta = -\sqrt{1 - \cos^2\theta} = -\sqrt{1 - \left(\dfrac{1}{2}\right)^2} = -\sqrt{1 - \dfrac{1}{4}} = -\sqrt{\dfrac{3}{4}} = -\dfrac{\sqrt{3}}{2}$$

$$\tan\theta = \dfrac{\sin\theta}{\cos\theta} = \dfrac{\left(-\dfrac{\sqrt{3}}{2}\right)}{\left(\dfrac{1}{2}\right)} = -\dfrac{\sqrt{3}}{2}\cdot\dfrac{2}{1} = -\sqrt{3}$$

$$\csc\theta = \dfrac{1}{\sin\theta} = \dfrac{1}{\left(-\dfrac{\sqrt{3}}{2}\right)} = -\dfrac{2}{\sqrt{3}} = -\dfrac{2\sqrt{3}}{3} \qquad \cot\theta = \dfrac{1}{\tan\theta} = \dfrac{1}{-\sqrt{3}} = -\dfrac{\sqrt{3}}{3}$$

45. $\tan\theta = \dfrac{3}{4}$, $\sin\theta < 0$, $\Rightarrow$ θ in quadrant III

Solve for $\sec\theta$:

$$\sec^2\theta = 1 + \tan^2\theta$$

$$\sec\theta = \pm\sqrt{1 + \tan^2\theta}$$

Since θ is in quadrant III, $\sec\theta < 0$.

$$\sec\theta = -\sqrt{1 + \tan^2\theta} = -\sqrt{1 + \left(\dfrac{3}{4}\right)^2} = -\sqrt{1 + \dfrac{9}{16}} = -\sqrt{\dfrac{25}{16}} = -\dfrac{5}{4}$$

$$\cos\theta = \dfrac{1}{\sec\theta} = -\dfrac{4}{5}$$

$$\sin\theta = -\sqrt{1 - \cos^2\theta} = -\sqrt{1 - \left(-\dfrac{4}{5}\right)^2} = -\sqrt{1 - \dfrac{16}{25}} = -\sqrt{\dfrac{9}{25}} = -\dfrac{3}{5}$$

$$\csc\theta = \dfrac{1}{\sin\theta} = \dfrac{1}{\left(-\dfrac{3}{5}\right)} = -\dfrac{5}{3} \qquad\qquad \cot\theta = \dfrac{1}{\tan\theta} = \dfrac{1}{\left(\dfrac{3}{4}\right)} = \dfrac{4}{3}$$

47. $\tan\theta = -\dfrac{1}{3}$, $\sin\theta > 0$, $\Rightarrow$ θ in quadrant II

Solve for $\sec\theta$:

$$\sec^2\theta = 1 + \tan^2\theta$$

$$\sec\theta = \pm\sqrt{1 + \tan^2\theta}$$

Since θ is in quadrant II, $\sec\theta < 0$.

$$\sec\theta = -\sqrt{1 + \tan^2\theta} = -\sqrt{1 + \left(-\dfrac{1}{3}\right)^2} = -\sqrt{1 + \dfrac{1}{9}} = -\sqrt{\dfrac{10}{9}} = -\dfrac{\sqrt{10}}{3}$$

$$\cos\theta = \dfrac{1}{\sec\theta} = \dfrac{1}{\left(-\dfrac{\sqrt{10}}{3}\right)} = -\dfrac{3}{\sqrt{10}} = -\dfrac{3\sqrt{10}}{10}$$

$$\sin\theta = \sqrt{1 - \cos^2\theta} = \sqrt{1 - \left(-\dfrac{3\sqrt{10}}{10}\right)^2} = \sqrt{1 - \dfrac{90}{100}} = \sqrt{\dfrac{10}{100}} = \dfrac{\sqrt{10}}{10}$$

$$\csc\theta = \dfrac{1}{\sin\theta} = \dfrac{1}{\left(\dfrac{\sqrt{10}}{10}\right)} = \sqrt{10} \qquad\qquad \cot\theta = \dfrac{1}{\tan\theta} = \dfrac{1}{\left(-\dfrac{1}{3}\right)} = -3$$

49. $\sin(-60°) = -\sin(60°) = -\dfrac{\sqrt{3}}{2}$

51. $\tan(-30°) = -\tan(30°) = -\dfrac{\sqrt{3}}{3}$

53. $\sec(-60°) = \sec(60°) = 2$

55. $\sin(-90°) = -\sin(90°) = -1$

57. $\tan\left(-\dfrac{\pi}{4}\right) = -\tan\left(\dfrac{\pi}{4}\right) = -1$

59. $\cos\left(-\dfrac{\pi}{4}\right) = \cos\left(\dfrac{\pi}{4}\right) = \dfrac{\sqrt{2}}{2}$

61. $\tan(-\pi) = -\tan(\pi) = 0$

63. $\csc\left(-\dfrac{\pi}{4}\right) = -\csc\left(\dfrac{\pi}{4}\right) = -\sqrt{2}$

65. $\sec\left(-\dfrac{\pi}{6}\right) = \sec\left(\dfrac{\pi}{6}\right) = \dfrac{2\sqrt{3}}{3}$

67. $\sin^2(40°) + \cos^2(40°) = 1$

69. $\sin(80°)\csc(80°) = \sin(80°) \cdot \dfrac{1}{\sin(80°)} = 1$

71. $\tan(40°) - \dfrac{\sin(40°)}{\cos(40°)} = \tan(40°) - \tan(40°) = 0$

73. $\cos(400°) \cdot \sec(40°) = \cos(40° + 360°) \cdot \sec(40°) = \cos(40°) \cdot \sec(40°)$
$$= \cos(40°) \cdot \dfrac{1}{\cos(40°)} = 1$$

75. $\sin\left(-\dfrac{\pi}{12}\right)\csc\left(\dfrac{25\pi}{12}\right) = -\sin\left(\dfrac{\pi}{12}\right)\csc\left(\dfrac{25\pi}{12}\right) = -\sin\left(\dfrac{\pi}{12}\right)\csc\left(\dfrac{\pi}{12} + \dfrac{24\pi}{12}\right) = -\sin\left(\dfrac{\pi}{12}\right)\csc\left(\dfrac{\pi}{12} + 2\pi\right)$
$$= -\sin\left(\dfrac{\pi}{12}\right)\csc\left(\dfrac{\pi}{12}\right) = -\sin\left(\dfrac{\pi}{12}\right) \cdot \dfrac{1}{\sin\left(\dfrac{\pi}{12}\right)} = -1$$

77. $\dfrac{\sin(-20°)}{\cos(380°)} + \tan(200°) = \dfrac{-\sin(20°)}{\cos(20° + 360°)} + \tan(20° + 180°)$
$$= \dfrac{-\sin(20°)}{\cos(20°)} + \tan(20°) = -\tan(20°) + \tan(20°) = 0$$

79. If $\sin\theta = 0.3$, then $\sin(\theta + \pi) = -0.3$

81. If $\tan\theta = 3$, then $\tan(\theta + \pi) = 3$

83. Find the value:
$$\sin(1°) + \sin(2°) + \sin(3°) + \ldots + \sin(357°) + \sin(358°) + \sin(359°)$$
$$= \sin(1°) + \sin(2°) + \sin(3°) + \ldots + \sin(360° - 3°) + \sin(360° - 2°) + \sin(360° - 1°)$$
$$= \sin(1°) + \sin(2°) + \sin(3°) + \ldots + \sin(-3°) + \sin(-2°) + \sin(-1°)$$
$$= \sin(1°) + \sin(2°) + \sin(3°) + \ldots - \sin(3°) - \sin(2°) - \sin(1°) = \sin(180°) = 0$$

85. The domain of the sine function is the set of all real numbers.

87. $f(\theta) = \tan\theta$ is not defined for numbers that are odd multiples of $\dfrac{\pi}{2}$.

89. $f(\theta) = \sec\theta$ is not defined for numbers that are odd multiples of $\frac{\pi}{2}$.

91. The range of the sine function is the set of all real numbers between -1 and 1, inclusive.

93. The range of the tangent function is the set of all real numbers.

95. The range of the secant function is the set of all real number greater than or equal to 1 and all real numbers less than or equal to -1.

97. The sine function is odd because $\sin(-\theta) = -\sin\theta$. Its graph is symmetric to the origin.

99. The tangent function is odd because $\tan(-\theta) = -\tan\theta$. Its graph is symmetric to the origin.

101. The secant function is even because $\sec(-\theta) = \sec\theta$. Its graph is symmetric to the y-axis.

103. (a) $f(-a) = -f(a) = -\frac{1}{3}$

(b) $f(a) + f(a+2\pi) + f(a+4\pi) = f(a) + f(a) + f(a) = \frac{1}{3} + \frac{1}{3} + \frac{1}{3} = 1$

105. (a) $f(-a) = -f(a) = -2$
(b) $f(a) + f(a+\pi) + f(a+2\pi) = f(a) + f(a) + f(a) = 2 + 2 + 2 = 6$

107. (a) $f(-a) = f(a) = -4$
(b) $f(a) + f(a+2\pi) + f(a+4\pi) = f(a) + f(a) + f(a) = (-4) + (-4) + (-4) = -12$

109. Since $\tan\theta = \frac{500}{1500} = \frac{1}{3}$, then $\sin\theta = \frac{1}{\sqrt{1+9}} = \frac{1}{\sqrt{10}}$.

$T = 5 - \dfrac{5}{\left(3\cdot\frac{1}{3}\right)} + \dfrac{5}{\left(\frac{1}{\sqrt{10}}\right)} = 5 - 5 + 5\sqrt{10} = 5\sqrt{10} \approx 15.8$ minutes

111. Let $P = (x, y)$ be the point on the unit circle that corresponds to an angle θ.
Consider the equation $\tan\theta = \frac{y}{x} = a$. Then $y = ax$. Now $x^2 + y^2 = 1$, so $x^2 + a^2x^2 = 1$.
Thus, $x = \pm\frac{1}{\sqrt{1+a^2}}$ and $y = \pm\frac{a}{\sqrt{1+a^2}}$; that is, for any real number a, there is a point
$P = (x, y)$ on the unit circle for which $\tan\theta = a$. In other words, $-\infty < \tan\theta < +\infty$, and the range of the tangent function is the set of all real numbers.

113. Suppose there is a number $p, 0 < p < 2\pi$, for which $\sin(\theta + p) = \sin\theta$ for all θ. If
$\theta = 0$, then $\sin(0 + p) = \sin p = \sin 0 = 0$; so that $p = \pi$. If $\theta = \frac{\pi}{2}$, then
$\sin\left(\frac{\pi}{2} + p\right) = \sin\left(\frac{\pi}{2}\right)$. But $p = \pi$. Thus, $\sin\left(\frac{3\pi}{2}\right) = -1 = \sin\left(\frac{\pi}{2}\right) = 1$, or $-1 = 1$.

This is impossible. The smallest positive number p for which $\sin(\theta + p) = \sin\theta$ for all θ is therefore $p = 2\pi$.

115. $\sec\theta = \dfrac{1}{\cos\theta}$: since $\cos\theta$ has period 2π, so does $\sec\theta$.

117. If $P = (a, b)$ is the point on the unit circle corresponding to θ, then $Q = (-a, -b)$ is the point on the unit circle corresponding to $\theta + \pi$.

Thus, $\tan(\theta + \pi) = \dfrac{-b}{-a} = \dfrac{b}{a} = \tan\theta$. If there exists a number p, $0 < p < \pi$, for which $\tan(\theta + p) = \tan\theta$ for all θ, then if $\theta = 0$, then $\tan(p) = \tan(0) = 0$.

But this means that p is a multiple of π. Since no multiple of π exists in the interval $(0, \pi)$, this is impossible. Therefore, the fundamental period of $f(\theta) = \tan\theta$ is π.

119. Let $P = (a, b)$ be the point on the unit circle corresponding to θ.

Then $\csc\theta = \dfrac{1}{b} = \dfrac{1}{\sin\theta}$; $\sec\theta = \dfrac{1}{a} = \dfrac{1}{\cos\theta}$; $\cot\theta = \dfrac{a}{b} = \dfrac{1}{\left(\dfrac{b}{a}\right)} = \dfrac{1}{\tan\theta}$

121. $(\sin\theta\cos\phi)^2 + (\sin\theta\sin\phi)^2 + \cos^2\theta = \sin^2\theta\cos^2\phi + \sin^2\theta\sin^2\phi + \cos^2\theta$
$$= \sin^2\theta(\cos^2\phi + \sin^2\phi) + \cos^2\theta = \sin^2\theta + \cos^2\theta = 1$$

123. Answers will vary.

Trigonometric Functions

5.4   Graphs of the Sine and Cosine Functions

1.   0

3.   The graph of $y = \sin x$ is increasing for $-\dfrac{\pi}{2} < x < \dfrac{\pi}{2}$.

5.   The greatest value of $y = \sin x$ is 1.

7. $\sin x = 0$ when $x = 0, \pi, 2\pi$

9. $\sin x = 1$ when $x = -\dfrac{3\pi}{2}, \dfrac{\pi}{2}$; $\sin x = -1$ when $x = -\dfrac{\pi}{2}, \dfrac{3\pi}{2}$

11.   B, C, F

13. $y = 3\sin x$; The graph of $y = \sin x$ is stretched vertically by a factor of 3.

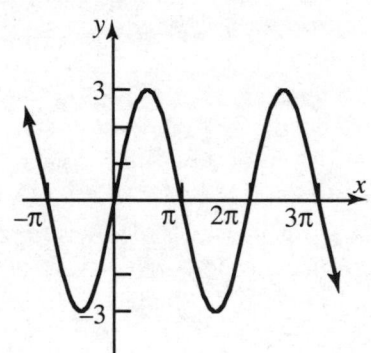

15. $y = \cos\left(x + \dfrac{\pi}{4}\right)$; The graph of $y = \cos x$ is shifted left $\dfrac{\pi}{4}$ units.

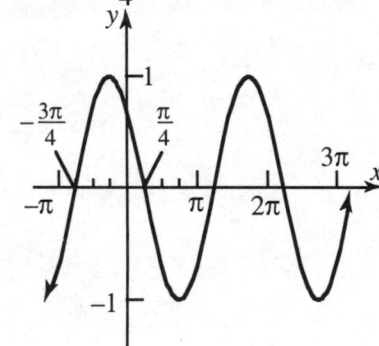

17. $y = \sin x - 1$; The graph of $y = \sin x$ is shifted down 1 unit.

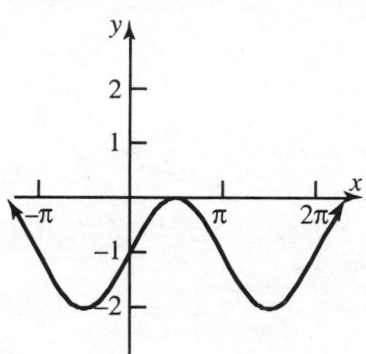

19. $y = -2\sin x$; The graph of $y = \sin x$ is stretched vertically by a factor of 2 and reflected across the x-axis.

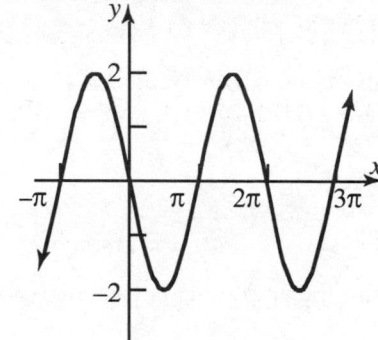

21. $y = \sin(\pi x)$; The graph of $y = \sin x$ is compressed horizontally by a factor of $\dfrac{1}{\pi}$.

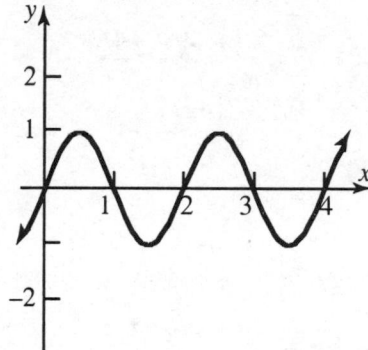

23. $y = 2\sin x + 2$; The graph of $y = \sin x$ is stretched vertically by a factor of 2 and shifted up 2 units.

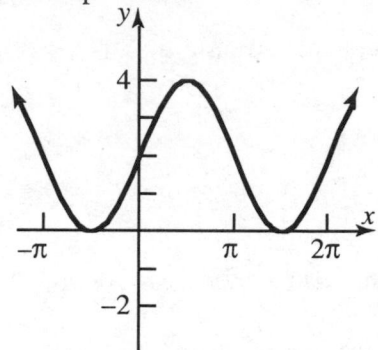

25. $y = -2\cos\left(x - \dfrac{\pi}{2}\right)$; The graph of $y = \cos x$ is shifted right $\dfrac{\pi}{2}$ units, stretched vertically by a factor of 2 and reflected across the x-axis.

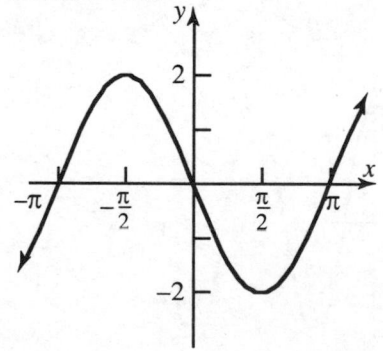

27. $y = 3\sin(\pi - x) = 3\sin(-(\pi + x))$
$\qquad\qquad = -3\sin(\pi + x)$

The graph of $y = \sin x$ is shifted left π units, stretched vertically by a factor of 3 and reflected across the x-axis.

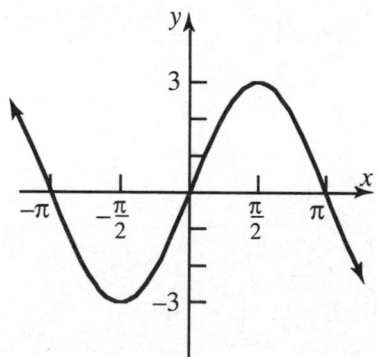

315

29. $y = 2\sin x$

This is in the form $y = A\sin(\omega x)$ where $A = 2$ and $\omega = 1$.

Thus, the amplitude is $|A| = |2| = 2$ and the period is $T = \dfrac{2\pi}{\omega} = \dfrac{2\pi}{1} = 2\pi$.

31. $y = -4\cos(2x)$

This is in the form $y = A\cos(\omega x)$ where $A = -4$ and $\omega = 2$.

Thus, the amplitude is $|A| = |-4| = 4$ and the period is $T = \dfrac{2\pi}{\omega} = \dfrac{2\pi}{2} = \pi$.

33. $y = 6\sin(\pi x)$

This is in the form $y = A\sin(\omega x)$ where $A = 6$ and $\omega = \pi$.

Thus, the amplitude is $|A| = |6| = 6$ and the period is $T = \dfrac{2\pi}{\omega} = \dfrac{2\pi}{\pi} = 2$.

35. $y = -\dfrac{1}{2}\cos\left(\dfrac{3}{2}x\right)$

This is in the form $y = A\cos(\omega x)$ where $A = -\dfrac{1}{2}$ and $\omega = \dfrac{3}{2}$.

Thus, the amplitude is $|A| = \left|-\dfrac{1}{2}\right| = \dfrac{1}{2}$ and the period is $T = \dfrac{2\pi}{\omega} = \dfrac{2\pi}{\left(\dfrac{3}{2}\right)} = \dfrac{4\pi}{3}$.

37. $y = \dfrac{5}{3}\sin\left(-\dfrac{2\pi}{3}x\right) = -\dfrac{5}{3}\sin\left(\dfrac{2\pi}{3}x\right)$

This is in the form $y = A\sin(\omega x)$ where $A = -\dfrac{5}{3}$ and $\omega = \dfrac{2\pi}{3}$.

Thus, the amplitude is $|A| = \left|-\dfrac{5}{3}\right| = \dfrac{5}{3}$ and the period is $T = \dfrac{2\pi}{\omega} = \dfrac{2\pi}{\left(\dfrac{2\pi}{3}\right)} = 3$.

39. F   41. A   43. H   45. C

47. J   49. A   51. B

53. $y = 5\sin(4x)$ $A = 5;\ \ T = \dfrac{\pi}{2}$ 55. $y = 5\cos(\pi x)$ $A = 5;\ \ T = 2$

57. $y = -2\cos(2\pi x)$ $A = -2$; $T = 1$ 59. $y = -4\sin\left(\dfrac{1}{2}x\right)$ $A = -4$; $T = 4\pi$

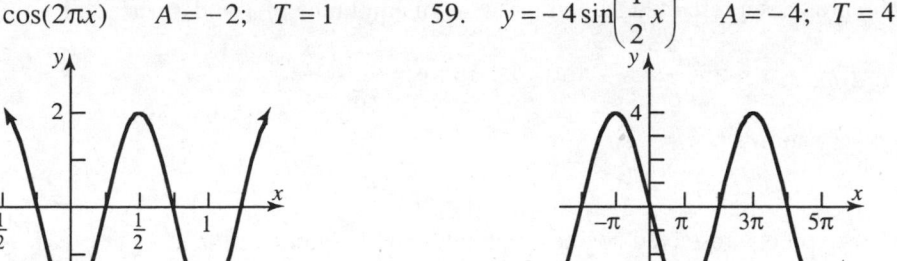

61. $y = \dfrac{3}{2}\sin\left(-\dfrac{2}{3}x\right) = -\dfrac{3}{2}\sin\left(\dfrac{2}{3}x\right)$

$$A = -\dfrac{3}{2};\quad T = 3\pi$$

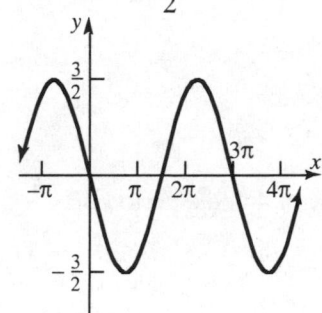

63. $|A| = 3$; $T = \pi$; $\omega = \dfrac{2\pi}{T} = \dfrac{2\pi}{\pi} = 2$; $y = \pm 3\sin(2x)$

65. $|A| = 3$; $T = 2$; $\omega = \dfrac{2\pi}{T} = \dfrac{2\pi}{2} = \pi$; $y = \pm 3\sin(\pi x)$

67. The graph is a reflected cosine graph with an amplitude of 3 and a period of 4π.

Find ω: $4\pi = \dfrac{2\pi}{\omega} \Rightarrow 4\pi\omega = 2\pi \Rightarrow \omega = \dfrac{2\pi}{4\pi} = \dfrac{1}{2}$

The equation is: $y = -3\cos\left(\dfrac{1}{2}x\right)$.

69. The graph is a sine graph with an amplitude of $\dfrac{3}{4}$ and a period of 1.

Find ω: $1 = \dfrac{2\pi}{\omega} \Rightarrow \omega = 2\pi$

The equation is: $y = \dfrac{3}{4}\sin(2\pi x)$.

71. The graph is a reflected sine graph with an amplitude of 1 and a period of $\dfrac{4\pi}{3}$.

Find ω: $\dfrac{4\pi}{3} = \dfrac{2\pi}{\omega} \implies 4\pi\omega = 6\pi \implies \omega = \dfrac{6\pi}{4\pi} = \dfrac{3}{2}$

The equation is: $y = -\sin\left(\dfrac{3}{2}x\right)$.

73. The graph is a reflected cosine graph with an amplitude of 2 and a period of $\dfrac{4}{3}$.

Find ω: $\dfrac{4}{3} = \dfrac{2\pi}{\omega} \implies 4\omega = 6\pi \implies \omega = \dfrac{6\pi}{4} = \dfrac{3\pi}{2}$

The equation is: $y = -2\cos\left(\dfrac{3\pi}{2}x\right)$.

75. The graph is a sine graph with an amplitude of 3 and a period of 4.

Find ω: $4 = \dfrac{2\pi}{\omega} \implies 4\omega = 2\pi \implies \omega = \dfrac{2\pi}{4} = \dfrac{\pi}{2}$

The equation is: $y = 3\sin\left(\dfrac{\pi}{2}x\right)$.

77. The graph is a reflected cosine graph with an amplitude of 4 and a period of $\dfrac{2\pi}{3}$.

Find ω: $\dfrac{2\pi}{3} = \dfrac{2\pi}{\omega} \implies 2\pi\omega = 6\pi \implies \omega = \dfrac{6\pi}{2\pi} = 3$

The equation is: $y = -4\cos(3x)$.

79. $I = 220\sin(60\pi t),\ t \geq 0$

Period: $T = \dfrac{2\pi}{\omega} = \dfrac{2\pi}{60\pi} = \dfrac{1}{30}$

Amplitude: $|A| = |220| = 220$

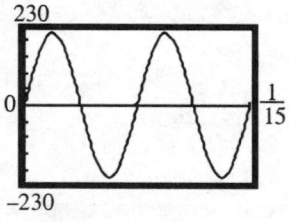

81. $V = 220\sin(120\pi t)$

  (a) Amplitude: $|A| = |220| = 220$

        Period: $T = \dfrac{2\pi}{\omega} = \dfrac{2\pi}{120\pi} = \dfrac{1}{60}$

  (b)

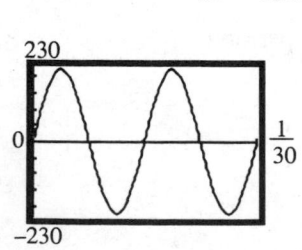

  (c) $V = IR$

        $220\sin(120\pi t) = 10I$

        $22\sin(120\pi t) = I$

  (d) Amplitude: $|A| = |22| = 22$

        Period: $T = \dfrac{2\pi}{\omega} = \dfrac{2\pi}{120\pi} = \dfrac{1}{60}$

  (e)

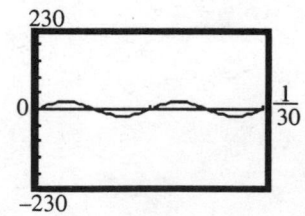

83. (a) $P = \dfrac{V^2}{R} = \dfrac{\left(V_0 \sin(2\pi f t)\right)^2}{R} = \dfrac{V_0^2 \sin^2(2\pi f t)}{R}$

     (b) The graph is the reflected cosine graph translated up a distance equivalent to the amplitude. The period is $\dfrac{1}{2f}$, so $\omega = 4\pi f$. The amplitude is $\dfrac{1}{2} \cdot \dfrac{V_0^2}{R} = \dfrac{V_0^2}{2R}$.

          The equation is: $P = -\dfrac{V_0^2}{2R} \cos(4\pi f t) + \dfrac{V_0^2}{2R} = \dfrac{V_0^2}{R} \cdot \dfrac{1}{2}\left(1 - \cos(4\pi f t)\right)$

     (c) Comparing the formulas:

          $$\sin^2(2\pi f t) = \frac{1}{2}\left(1 - \cos(4\pi f t)\right)$$

85. $y = \left|\cos x\right|, \quad -2\pi \le x \le 2\pi$

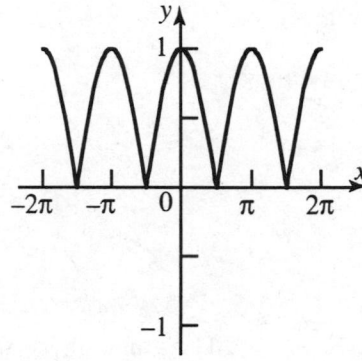

Trigonometric Functions

5.5 Graphs of the Tangent, Cotangent, Cosecant and Secant Functions

1. y-intercept: 0

3. y-intercept: 1

5. $\sec x = 1$ for $x = -2\pi, 0, 2\pi$; $\sec x = -1$ for $x = -\pi, \pi$

7. $y = \sec x$ has vertical asymptotes for $x = -\dfrac{3\pi}{2}, -\dfrac{\pi}{2}, \dfrac{\pi}{2}, \dfrac{3\pi}{2}$

9. $y = \tan x$ has vertical asymptotes for $x = -\dfrac{3\pi}{2}, -\dfrac{\pi}{2}, \dfrac{\pi}{2}, \dfrac{3\pi}{2}$

11. D

13. B

15. $y = -\sec x$; The graph of $y = \sec x$ is reflected across the x-axis vertically.

17. $y = \sec\left(x - \dfrac{\pi}{2}\right)$; The graph of $y = \sec x$ is shifted right $\dfrac{\pi}{2}$ units.

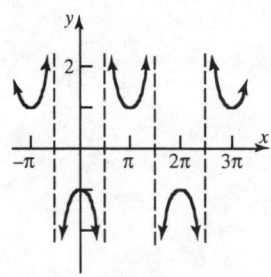

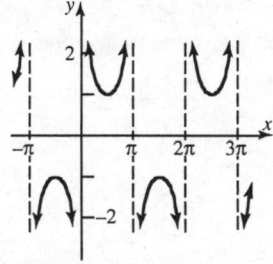

19. $y = \tan(x - \pi)$; The graph of $y = \tan x$ is shifted right π units.

21. $y = 3\tan(2x)$; The graph of $y = \tan x$ is compressed horizontally by a factor of $\dfrac{1}{2}$ and stretched vertically by a factor of 3.

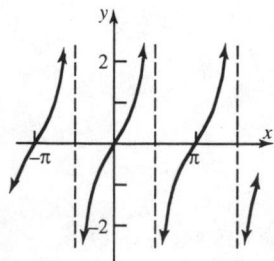

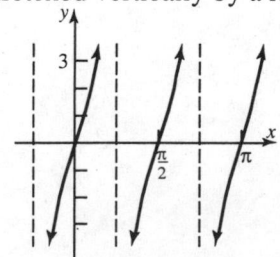

23. $y = \sec(2x)$; The graph of $y = \sec x$ is compressed horizontally by a factor of $\dfrac{1}{2}$.

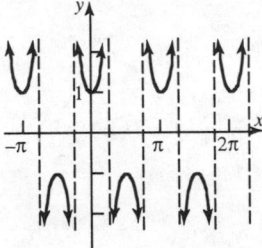

25. $y = \cot(\pi x)$; The graph of $y = \cot x$ is compressed horizontally by a factor of $\dfrac{1}{\pi}$.

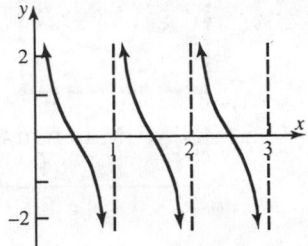

27. $y = -3\tan(4x)$; The graph of $y = \tan x$ is compressed horizontally by a factor of $\dfrac{1}{4}$, stretched vertically by a factor of 3 and reflected across the x-axis.

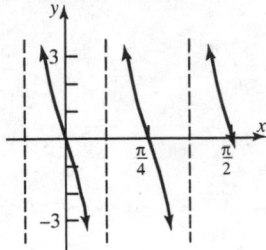

29. $y = 2\sec\left(\dfrac{1}{2}x\right)$; The graph of $y = \sec x$ is stretched horizontally by a factor of 2, stretched vertically by a factor of 2.

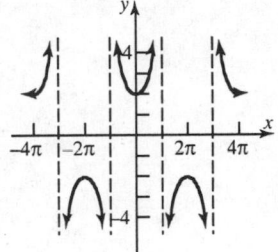

31. $y = -3\csc\left(x + \dfrac{\pi}{4}\right)$; The graph of $y = \csc x$ is shifted left $\dfrac{\pi}{4}$ units, stretched vertically by a factor of 3 and reflected across the x-axis.

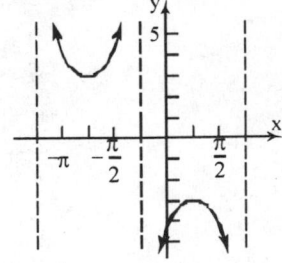

33. $y = \dfrac{1}{2}\cot\left(x + \dfrac{\pi}{4}\right)$; The graph of $y = \cot x$ is shifted left $\dfrac{\pi}{4}$ units and compressed vertically by a factor of $\dfrac{1}{2}$.

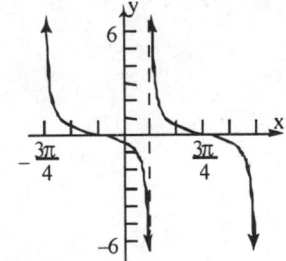

35.   (a) $L = \dfrac{3}{\cos\theta} + \dfrac{4}{\sin\theta} = 3\sec\theta + 4\csc\theta$

(b)   Graph:

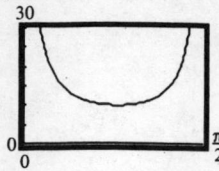

(c)   Use MINIMUM to find the least value: L is least when $\theta = 0.83$.

(d) $L = \dfrac{3}{\cos(0.83)} + \dfrac{4}{\sin(0.83)} \approx 9.86$ feet

Trigonometric Functions

5.6   Phase Shifts; Sinusoidal Curve Fitting

1. $y = 4\sin(2x - \pi)$

Amplitude: $|A| = |4| = 4$

Period: $T = \dfrac{2\pi}{\omega} = \dfrac{2\pi}{2} = \pi$

Phase Shift: $\dfrac{\phi}{\omega} = \dfrac{\pi}{2}$

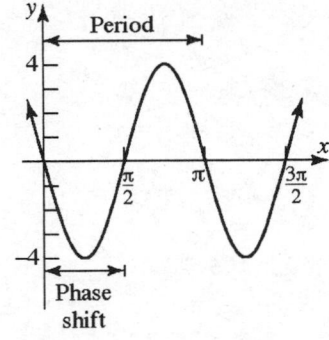

3. $y = 2\cos\left(3x + \dfrac{\pi}{2}\right)$

Amplitude : $|A| = |2| = 2$

Period : $T = \dfrac{2\pi}{\omega} = \dfrac{2\pi}{3}$

Phase Shift : $\dfrac{\phi}{\omega} = \dfrac{\left(-\dfrac{\pi}{2}\right)}{3} = -\dfrac{\pi}{6}$

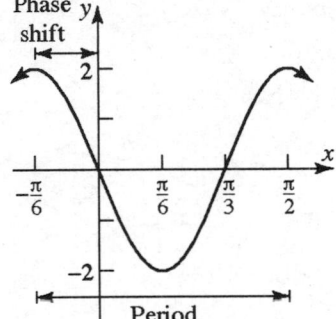

5. $y = -3\sin\left(2x + \dfrac{\pi}{2}\right)$

Amplitude : $|A| = |-3| = 3$

Period : $T = \dfrac{2\pi}{\omega} = \dfrac{2\pi}{2} = \pi$

Phase Shift : $\dfrac{\phi}{\omega} = \dfrac{\left(-\dfrac{\pi}{2}\right)}{2} = -\dfrac{\pi}{4}$

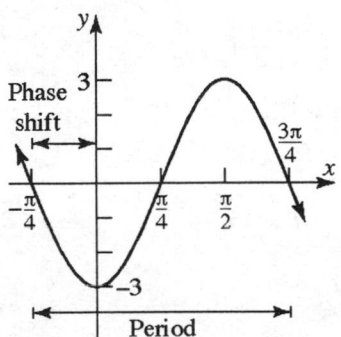

7. $y = 4\sin(\pi x + 2)$

Amplitude : $|A| = |4| = 4$

Period : $T = \dfrac{2\pi}{\omega} = \dfrac{2\pi}{\pi} = 2$

Phase Shift : $\dfrac{\phi}{\omega} = -\dfrac{2}{\pi}$

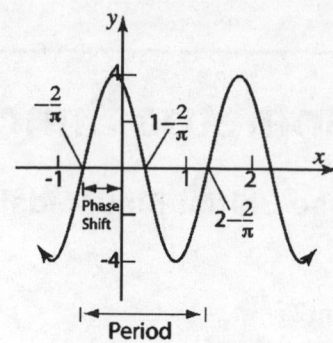

9. $y = 3\cos(\pi x - 2)$

Amplitude : $|A| = |3| = 3$

Period : $T = \dfrac{2\pi}{\omega} = \dfrac{2\pi}{\pi} = 2$

Phase Shift : $\dfrac{\phi}{\omega} = \dfrac{2}{\pi}$

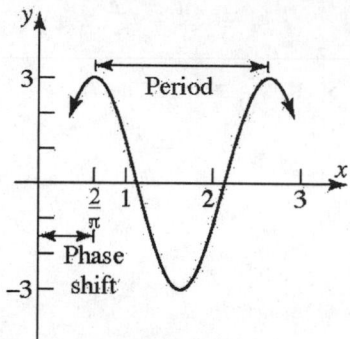

11. $y = 3\sin\left(-2x + \dfrac{\pi}{2}\right) = -3\sin\left(2x - \dfrac{\pi}{2}\right)$

Amplitude : $|A| = |-3| = 3$

Period : $T = \dfrac{2\pi}{\omega} = \dfrac{2\pi}{2} = \pi$

Phase Shift : $\dfrac{\phi}{\omega} = \dfrac{\left(\dfrac{\pi}{2}\right)}{2} = \dfrac{\pi}{4}$

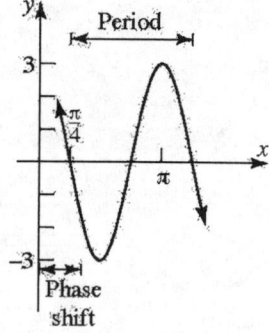

13. $|A| = 2;\quad T = \pi;\quad \dfrac{\phi}{\omega} = \dfrac{1}{2};\quad \omega = \dfrac{2\pi}{T} = \dfrac{2\pi}{\pi} = 2;\quad \dfrac{\phi}{\omega} = \dfrac{\phi}{2} = \dfrac{1}{2} \;\rightarrow\; \phi = 1$

$y = \pm 2\sin(2x - 1) = \pm 2\sin\left[2\left(x - \dfrac{1}{2}\right)\right]$

15. $|A| = 3;\quad T = 3\pi;\quad \dfrac{\phi}{\omega} = -\dfrac{1}{3};\quad \omega = \dfrac{2\pi}{T} = \dfrac{2\pi}{3\pi} = \dfrac{2}{3};$

$\dfrac{\phi}{\omega} = \dfrac{\phi}{\frac{2}{3}} = -\dfrac{1}{3} \;\rightarrow\; \phi = -\dfrac{1}{3} \cdot \dfrac{2}{3} = -\dfrac{2}{9} \qquad y = \pm 3\sin\left(\dfrac{2}{3}x + \dfrac{2}{9}\right) = \pm 3\sin\left[\dfrac{2}{3}\left(x + \dfrac{1}{3}\right)\right]$

17. $I = 120\sin\left(30\pi t - \dfrac{\pi}{3}\right), \quad t \geq 0$

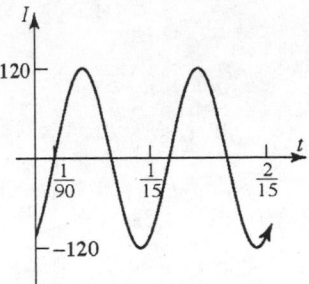

Period : $T = \dfrac{2\pi}{\omega} = \dfrac{2\pi}{30\pi} = \dfrac{1}{15}$

Amplitude : $|A| = |120| = 120$

Phase Shift : $\dfrac{\phi}{\omega} = \dfrac{\left(\dfrac{\pi}{3}\right)}{30\pi} = \dfrac{1}{90}$

19.   (a)   Draw a scatter diagram:

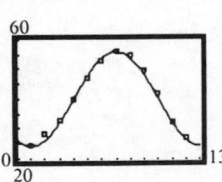

(b)   Amplitude: $A = \dfrac{56.0 - 24.2}{2} = \dfrac{31.8}{2} = 15.9$

Vertical Shift: $\dfrac{56.0 + 24.2}{2} = \dfrac{80.2}{2} = 40.1$

$\omega = \dfrac{2\pi}{12} = \dfrac{\pi}{6}$

Phase shift (use $y = 24.2$, $x = 1$):

$$24.2 = 15.9\sin\left(\dfrac{\pi}{6}\cdot 1 - \phi\right) + 40.1$$

$$-15.9 = 15.9\sin\left(\dfrac{\pi}{6} - \phi\right) \to -1 = \sin\left(\dfrac{\pi}{6} - \phi\right) \to -\dfrac{\pi}{2} = \dfrac{\pi}{6} - \phi$$

$$\phi = \dfrac{2\pi}{3}$$

Thus, $y = 15.9\sin\left(\dfrac{\pi}{6}x - \dfrac{2\pi}{3}\right) + 40.1$

(c)

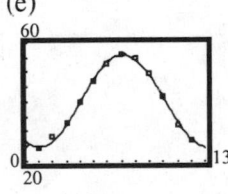

(e)

(d) $y = 15.62\sin(0.517x - 2.096) + 40.377$

21.   (a)   Draw a scatter diagram:

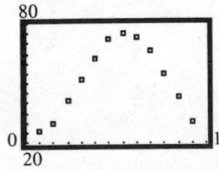

(b)   Amplitude: $A = \dfrac{75.4 - 25.5}{2} = \dfrac{49.9}{2} = 24.95$

Vertical Shift: $\dfrac{75.4 + 25.5}{2} = \dfrac{100.9}{2} = 50.45$

$\omega = \dfrac{2\pi}{12} = \dfrac{\pi}{6}$

Phase shift (use $y = 25.5$, $x = 1$):

$$25.5 = 24.95 \sin\left(\frac{\pi}{6} \cdot 1 - \phi\right) + 50.45$$

$$-24.95 = 24.95 \sin\left(\frac{\pi}{6} - \phi\right) \Rightarrow -1 = \sin\left(\frac{\pi}{6} - \phi\right) \Rightarrow -\frac{\pi}{2} = \frac{\pi}{6} - \phi \Rightarrow \phi = \frac{2\pi}{3}$$

Thus, $y = 24.95 \sin\left(\dfrac{\pi}{6}x - \dfrac{2\pi}{3}\right) + 50.45$

(c)

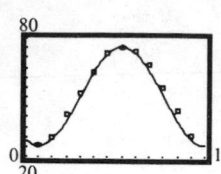

(e)

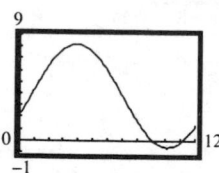

(d) $y = 25.693 \sin(0.476x - 1.814) + 49.854$

23. (a) $3.6333 + 12.5 = 16.1333$ hours which is at 4:08 p.m.

  (b)   Amplitude: $A = \dfrac{8.2 - (-0.6)}{2} = \dfrac{8.8}{2} = 4.4$

  Vertical Shift: $\dfrac{8.2 + (-0.6)}{2} = \dfrac{7.6}{2} = 3.8$

  $\omega = \dfrac{2\pi}{12.5} = \dfrac{\pi}{6.25}$

  Phase shift (use $y = -0.6$, $x = 10.1333$):

  $$-0.6 = 4.4 \sin\left(\frac{\pi}{6.25} \cdot 10.1333 - \phi\right) + 3.8$$

  $$-4.4 = 4.4 \sin\left(\frac{\pi}{6.25} \cdot 10.1333 - \phi\right) \rightarrow -1 = \sin\left(\frac{10.1333\pi}{6.25} - \phi\right)$$

  $$-\frac{\pi}{2} = \frac{10.1333\pi}{6.25} - \phi \rightarrow \phi = 6.6643$$

  Thus, $y = 4.4 \sin\left(\dfrac{\pi}{6.25}x - 6.6643\right) + 3.8$

  (c)

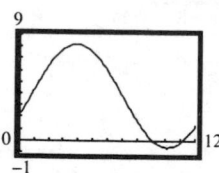

  (d) $y = 4.4 \sin\left(\dfrac{\pi}{6.25}(16.1333) - 6.6643\right) + 3.8 \approx 8.2$ feet

25.   (a)   Amplitude: $A = \dfrac{12.75 - 10.583}{2} = \dfrac{2.167}{2} = 1.0835$

Vertical Shift: $\dfrac{12.75 + 10.583}{2} = \dfrac{23.333}{2} = 11.6665$

$\omega = \dfrac{2\pi}{365}$

Phase shift (use $y = 10.583$, $x = 356$):

$$10.583 = 1.0835 \sin\left(\dfrac{2\pi}{365} \cdot 356 - \phi\right) + 11.6665$$

$$-1.0835 = 1.0835 \sin\left(\dfrac{2\pi}{365} \cdot 356 - \phi\right) \rightarrow -1 = \sin\left(\dfrac{712\pi}{365} - \phi\right)$$

$$-\dfrac{\pi}{2} = \dfrac{712\pi}{365} - \phi \rightarrow \phi = 7.6991$$

Thus, $y = 1.0835 \sin\left(\dfrac{2\pi}{365} x - 7.6991\right) + 11.6665$

(b)

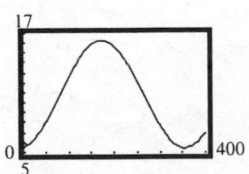

(c) $y = 1.0835 \sin\left(\dfrac{2\pi}{365}(92) - 7.6991\right) + 11.6665 \approx 11.85$ hours

27.   (a)   Amplitude: $A = \dfrac{16.233 - 5.45}{2} = \dfrac{10.783}{2} = 5.3915$

Vertical Shift: $\dfrac{16.233 + 5.45}{2} = \dfrac{21.683}{2} = 10.8415$

$\omega = \dfrac{2\pi}{365}$

Phase shift (use $y = 5.45$, $x = 356$):

$$5.45 = 5.3915 \sin\left(\dfrac{2\pi}{365} \cdot 356 - \phi\right) + 10.8415$$

$$-5.3915 = 5.3915 \sin\left(\dfrac{2\pi}{365} \cdot 356 - \phi\right) \rightarrow -1 = \sin\left(\dfrac{712\pi}{365} - \phi\right)$$

$$-\dfrac{\pi}{2} = \dfrac{712\pi}{365} - \phi \rightarrow \phi = 7.6991$$

Thus, $y = 5.3915 \sin\left(\dfrac{2\pi}{365} x - 7.6991\right) + 10.8415$

(b)

(c) $y = 5.3915 \sin\left(\dfrac{2\pi}{365}(92) - 7.6991\right) + 10.8415 \approx 11.74$ hours

Trigonometric Functions

5.R   Chapter Review

1. $135° = 135 \cdot \dfrac{\pi}{180}$ radian $= \dfrac{3\pi}{4}$ radians

3. $18° = 18 \cdot \dfrac{\pi}{180}$ radian $= \dfrac{\pi}{10}$ radians

5. $\dfrac{3\pi}{4} = \dfrac{3\pi}{4} \cdot \dfrac{180}{\pi}$ degrees $= 135°$ 7. $-\dfrac{5\pi}{2} = -\dfrac{5\pi}{2} \cdot \dfrac{180}{\pi}$ degrees $= -450°$

9. $\tan\left(\dfrac{\pi}{4}\right) - \sin\left(\dfrac{\pi}{6}\right) = 1 - \dfrac{1}{2} = \dfrac{1}{2}$

11. $3\sin(45°) - 4\tan\left(\dfrac{\pi}{6}\right) = 3 \cdot \dfrac{\sqrt{2}}{2} - 4 \cdot \dfrac{\sqrt{3}}{3} = \dfrac{3\sqrt{2}}{2} - \dfrac{4\sqrt{3}}{3}$

13. $6\cos\left(\dfrac{3\pi}{4}\right) + 2\tan\left(-\dfrac{\pi}{3}\right) = 6\left(-\dfrac{\sqrt{2}}{2}\right) + 2\left(-\sqrt{3}\right) = -3\sqrt{2} - 2\sqrt{3}$

15. $\sec\left(-\dfrac{\pi}{3}\right) - \cot\left(-\dfrac{5\pi}{4}\right) = \sec\left(\dfrac{\pi}{3}\right) + \cot\left(\dfrac{5\pi}{4}\right) = 2 + 1 = 3$

17. $\tan(\pi) + \sin(\pi) = 0 + 0 = 0$

19. $\cos(540°) - \tan(-45°) = -1 - (-1) = -1 + 1 = 0$

21. $\sin^2(20°) + \dfrac{1}{\sec^2(20°)} = \sin^2(20°) + \cos^2(20°) = 1$

23. $\sec(50°)\cos(50°) = \dfrac{1}{\cos(50°)} \cdot \cos(50°) = 1$

25. $\dfrac{\sin(50°)}{\cos(40°)} = \dfrac{\cos(40°)}{\cos(40°)} = 1$ 27. $\dfrac{\sin(-40°)}{\cos(50°)} = \dfrac{-\sin(40°)}{\cos(50°)} = \dfrac{-\cos(50°)}{\cos(50°)} = -1$

29. $\sin(400°)\sec(-50°) = \sin(40°+360°)\sec(50°) = \sin(40°)\csc(40°) = \sin(40°)\cdot\dfrac{1}{\sin(40°)} = 1$

31. $\sin\theta = \dfrac{4}{5},\quad \theta\text{ acute}\ \Rightarrow\ \theta$ in quadrant I

Solve for $\cos\theta$: $\sin^2\theta + \cos^2\theta = 1 \to \cos^2\theta = 1 - \sin^2\theta \to \cos\theta = \pm\sqrt{1-\sin^2\theta}$
Since θ is in quadrant I, $\cos\theta > 0$.

$$\cos\theta = \sqrt{1-\sin^2\theta} = \sqrt{1-\left(-\dfrac{4}{5}\right)^2} = \sqrt{1-\dfrac{16}{25}} = \sqrt{\dfrac{9}{25}} = \dfrac{3}{5}$$

$$\tan\theta = \dfrac{\sin\theta}{\cos\theta} = \dfrac{\left(\dfrac{4}{5}\right)}{\left(\dfrac{3}{5}\right)} = \dfrac{4}{5}\cdot\dfrac{5}{3} = \dfrac{4}{3}\qquad\qquad \sec\theta = \dfrac{1}{\cos\theta} = \dfrac{1}{\left(\dfrac{3}{5}\right)} = \dfrac{5}{3}$$

$$\csc\theta = \dfrac{1}{\sin\theta} = \dfrac{1}{\left(\dfrac{4}{5}\right)} = \dfrac{5}{4}\qquad\qquad \cot\theta = \dfrac{1}{\tan\theta} = \dfrac{1}{\left(\dfrac{4}{3}\right)} = \dfrac{3}{4}$$

33. $\tan\theta = \dfrac{12}{5},\quad \sin\theta < 0,\ \Rightarrow\ \theta$ in quadrant III

Solve for $\sec\theta$: $\sec^2\theta = 1 + \tan^2\theta \to \sec\theta = \pm\sqrt{1+\tan^2\theta}$
Since θ is in quadrant III, $\sec\theta < 0$.

$$\sec\theta = -\sqrt{1+\tan^2\theta} = -\sqrt{1+\left(\dfrac{12}{5}\right)^2} = -\sqrt{1+\dfrac{144}{25}} = -\sqrt{\dfrac{169}{25}} = -\dfrac{13}{5}$$

$$\cos\theta = -\dfrac{5}{13}$$

$$\sin\theta = -\sqrt{1-\cos^2\theta} = -\sqrt{1-\left(-\dfrac{5}{13}\right)^2} = -\sqrt{1-\dfrac{25}{169}} = -\sqrt{\dfrac{144}{169}} = -\dfrac{12}{13}$$

$$\csc\theta = \dfrac{1}{\sin\theta} = \dfrac{1}{\left(-\dfrac{12}{13}\right)} = -\dfrac{13}{12}\qquad\qquad \cot\theta = \dfrac{1}{\tan\theta} = \dfrac{1}{\left(\dfrac{12}{5}\right)} = \dfrac{5}{12}$$

35. $\sec\theta = -\dfrac{5}{4},\quad \tan\theta < 0,\ \Rightarrow\ \theta$ in quadrant II
Solve for $\cos\theta$:

$$\cos\theta = \dfrac{1}{\sec\theta} = \dfrac{1}{\left(-\dfrac{5}{4}\right)} = -\dfrac{4}{5}$$

Solve for $\sin\theta$: $\sin^2\theta + \cos^2\theta = 1 \to \sin^2\theta = 1 - \cos^2\theta \to \sin\theta = \pm\sqrt{1-\cos^2\theta}$
Since θ is in quadrant II, $\sin\theta > 0$.

$$\sin\theta = \sqrt{1-\cos^2\theta} = \sqrt{1-\left(-\dfrac{4}{5}\right)^2} = \sqrt{1-\dfrac{16}{25}} = \sqrt{\dfrac{9}{25}} = \dfrac{3}{5}$$

$$\tan\theta = \frac{\sin\theta}{\cos\theta} = \frac{\left(\dfrac{3}{5}\right)}{\left(-\dfrac{4}{5}\right)} = \frac{3}{5} \cdot -\frac{5}{4} = -\frac{3}{4}$$

$$\csc\theta = \frac{1}{\sin\theta} = \frac{1}{\left(\dfrac{3}{5}\right)} = \frac{5}{3} \qquad\qquad \cot\theta = \frac{1}{\tan\theta} = \frac{1}{\left(-\dfrac{3}{4}\right)} = -\frac{4}{3}$$

37. $\sin\theta = \dfrac{12}{13}, \quad \theta$ in quadrant II

Solve for $\cos\theta$: $\sin^2\theta + \cos^2\theta = 1 \rightarrow \cos^2\theta = 1 - \sin^2\theta \rightarrow \cos\theta = \pm\sqrt{1-\sin^2\theta}$
Since θ is in quadrant II, $\cos\theta < 0$.

$$\cos\theta = -\sqrt{1-\sin^2\theta} = -\sqrt{1-\left(\frac{12}{13}\right)^2} = -\sqrt{1-\frac{144}{169}} = -\sqrt{\frac{25}{169}} = -\frac{5}{13}$$

$$\tan\theta = \frac{\sin\theta}{\cos\theta} = \frac{\left(\dfrac{12}{13}\right)}{\left(-\dfrac{5}{13}\right)} = \frac{12}{13} \cdot -\frac{13}{5} = -\frac{12}{5} \qquad\qquad \sec\theta = \frac{1}{\cos\theta} = \frac{1}{\left(-\dfrac{5}{13}\right)} = -\frac{13}{5}$$

$$\csc\theta = \frac{1}{\sin\theta} = \frac{1}{\left(\dfrac{12}{13}\right)} = \frac{13}{12} \qquad\qquad \cot\theta = \frac{1}{\tan\theta} = \frac{1}{\left(-\dfrac{12}{5}\right)} = -\frac{5}{12}$$

39. $\sin\theta = -\dfrac{5}{13}, \quad \dfrac{3\pi}{2} < \theta < 2\pi, \Rightarrow \theta$ in quadrant IV

Solve for $\cos\theta$: $\sin^2\theta + \cos^2\theta = 1 \rightarrow \cos^2\theta = 1 - \sin^2\theta \rightarrow \cos\theta = \pm\sqrt{1-\sin^2\theta}$
Since θ is in quadrant IV, $\cos\theta > 0$.

$$\cos\theta = \sqrt{1-\sin^2\theta} = \sqrt{1-\left(-\frac{5}{13}\right)^2} = \sqrt{1-\frac{25}{169}} = \sqrt{\frac{144}{169}} = \frac{12}{13}$$

$$\tan\theta = \frac{\sin\theta}{\cos\theta} = \frac{\left(-\dfrac{5}{13}\right)}{\left(\dfrac{12}{13}\right)} = -\frac{5}{13} \cdot \frac{13}{12} = -\frac{5}{12} \qquad\qquad \sec\theta = \frac{1}{\cos\theta} = \frac{1}{\left(\dfrac{12}{13}\right)} = \frac{13}{12}$$

$$\csc\theta = \frac{1}{\sin\theta} = \frac{1}{\left(-\dfrac{5}{13}\right)} = -\frac{13}{5} \qquad\qquad \cot\theta = \frac{1}{\tan\theta} = \frac{1}{\left(-\dfrac{5}{12}\right)} = -\frac{12}{5}$$

41. $\tan\theta = \dfrac{1}{3}, \quad 180° < \theta < 270°, \Rightarrow \theta$ in quadrant III

Solve for $\sec\theta$: $\sec^2\theta = 1 + \tan^2\theta \rightarrow \sec\theta = \pm\sqrt{1+\tan^2\theta}$
Since θ is in quadrant III, $\sec\theta < 0$.

$$\sec\theta = -\sqrt{1+\tan^2\theta} = -\sqrt{1+\left(\frac{1}{3}\right)^2} = -\sqrt{1+\frac{1}{9}} = -\sqrt{\frac{10}{9}} = -\frac{\sqrt{10}}{3}$$

$$\cos\theta = \frac{1}{\sec\theta} = \frac{1}{\left(-\frac{\sqrt{10}}{3}\right)} = -\frac{3}{\sqrt{10}} \cdot \frac{\sqrt{10}}{\sqrt{10}} = -\frac{3\sqrt{10}}{10}$$

$$\sin\theta = -\sqrt{1-\cos^2\theta} = -\sqrt{1-\left(-\frac{3\sqrt{10}}{3}\right)^2} = -\sqrt{1-\frac{90}{100}} = -\sqrt{\frac{10}{100}} = -\frac{\sqrt{10}}{10}$$

$$\csc\theta = \frac{1}{\sin\theta} = \frac{1}{\left(-\frac{\sqrt{10}}{10}\right)} = -\frac{10}{\sqrt{10}} = -\sqrt{10} \qquad\qquad \cot\theta = \frac{1}{\tan\theta} = \frac{1}{\left(\frac{1}{3}\right)} = 3$$

43. $\sec\theta = 3, \;\; \frac{3\pi}{2} < \theta < 2\pi, \;\; \Rightarrow \;\; \theta$ in quadrant IV

$$\cos\theta = \frac{1}{\sec\theta} = \frac{1}{3}$$

Solve for $\sin\theta$: $\;\sin^2\theta + \cos^2\theta = 1 \;\rightarrow\; \sin^2\theta = 1-\cos^2\theta \;\rightarrow\; \sin\theta = \pm\sqrt{1-\cos^2\theta}$
Since θ is in quadrant IV, $\sin\theta < 0$.

$$\sin\theta = -\sqrt{1-\cos^2\theta} = -\sqrt{1-\left(\frac{1}{3}\right)^2} = -\sqrt{1-\frac{1}{9}} = -\sqrt{\frac{8}{9}} = -\frac{2\sqrt{2}}{3}$$

$$\tan\theta = \frac{\sin\theta}{\cos\theta} = \frac{\left(-\frac{2\sqrt{2}}{3}\right)}{\left(\frac{1}{3}\right)} = -\frac{2\sqrt{2}}{3} \cdot \frac{3}{1} = -2\sqrt{2}$$

$$\csc\theta = \frac{1}{\sin\theta} = \frac{1}{\left(-\frac{2\sqrt{2}}{3}\right)} = -\frac{3}{2\sqrt{2}} = -\frac{3\sqrt{2}}{4} \qquad\qquad \cot\theta = \frac{1}{\tan\theta} = \frac{1}{-2\sqrt{2}} = -\frac{\sqrt{2}}{4}$$

45. $\cot\theta = -2, \;\; \frac{\pi}{2} < \theta < \pi, \;\; \Rightarrow \;\; \theta$ in quadrant II

$$\tan\theta = \frac{1}{\cot\theta} = \frac{1}{-2} = -\frac{1}{2}$$

Solve for $\sec\theta$: $\;\sec^2\theta = 1+\tan^2\theta \rightarrow \sec\theta = \pm\sqrt{1+\tan^2\theta}$
Since θ is in quadrant II, $\sec\theta < 0$.

$$\sec\theta = -\sqrt{1+\tan^2\theta} = -\sqrt{1+\left(-\frac{1}{2}\right)^2} = -\sqrt{1+\frac{1}{4}} = -\sqrt{\frac{5}{4}} = -\frac{\sqrt{5}}{2}$$

$$\cos\theta = \frac{1}{\sec\theta} = \frac{1}{\left(-\frac{\sqrt{5}}{2}\right)} = -\frac{2}{\sqrt{5}} \cdot \frac{\sqrt{5}}{\sqrt{5}} = -\frac{2\sqrt{5}}{5}$$

$$\sin\theta = \sqrt{1-\cos^2\theta} = \sqrt{1-\left(-\frac{2\sqrt{5}}{2}\right)^2} = \sqrt{1-\frac{20}{25}} = \sqrt{\frac{5}{25}} = \frac{\sqrt{5}}{5}$$

$$\csc\theta = \frac{1}{\sin\theta} = \frac{1}{\left(\dfrac{\sqrt{5}}{5}\right)} = \frac{5}{\sqrt{5}} = \sqrt{5}$$

47. $y = 2\sin(4x)$

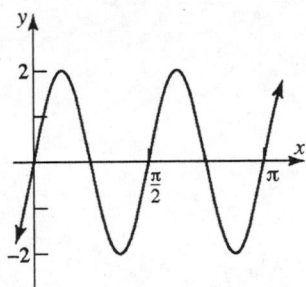

49. $y = -2\cos\left(x + \dfrac{\pi}{2}\right)$

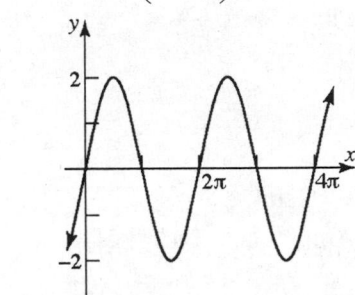

51. $y = \tan(x + \pi)$

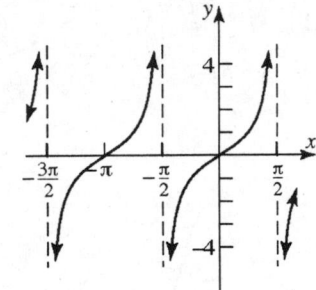

53. $y = -2\tan(3x)$

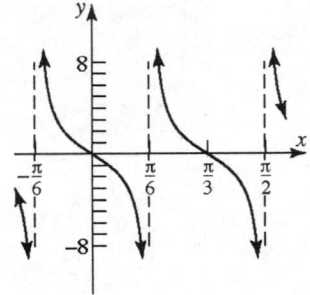

55. $y = \cot\left(x + \dfrac{\pi}{4}\right)$

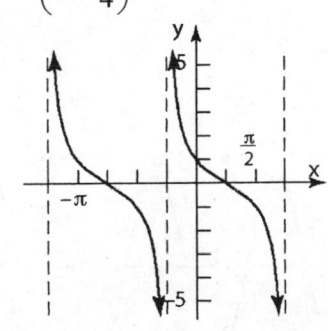

57. $y = \sec\left(x - \dfrac{\pi}{4}\right)$

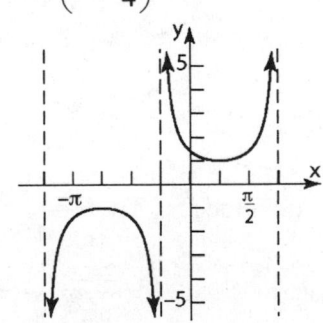

59. $y = 4\cos x$

Amplitude = 4
Period = 2π

61. $y = -8\sin\left(\dfrac{\pi}{2}x\right)$

Amplitude = 8
Period = 4

63. $y = 4\sin(3x)$
Amplitude: $|A| = |4| = 4$

Period: $T = \dfrac{2\pi}{\omega} = \dfrac{2\pi}{3}$

Phase Shift: $\dfrac{\phi}{\omega} = \dfrac{0}{3} = 0$

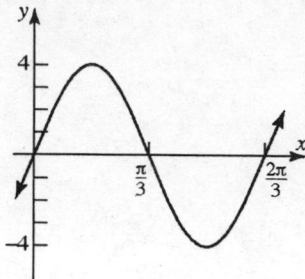

65. $y = 2\sin(2x - \pi)$
Amplitude: $|A| = |2| = 2$

Period: $T = \dfrac{2\pi}{\omega} = \dfrac{2\pi}{2} = \pi$

Phase Shift: $\dfrac{\phi}{\omega} = \dfrac{\pi}{2} = \dfrac{\pi}{2}$

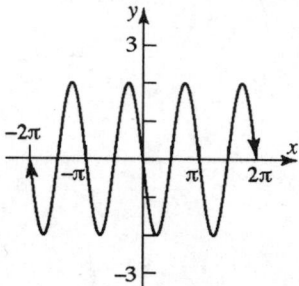

67. $y = \dfrac{1}{2}\sin\left(\dfrac{3}{2}x - \pi\right)$

Amplitude: $|A| = \left|\dfrac{1}{2}\right| = \dfrac{1}{2}$

Period: $T = \dfrac{2\pi}{\omega} = \dfrac{2\pi}{\left(\dfrac{3}{2}\right)} = \dfrac{4\pi}{3}$

Phase Shift: $\dfrac{\phi}{\omega} = \dfrac{\pi}{\left(\dfrac{3}{2}\right)} = \dfrac{2\pi}{3}$

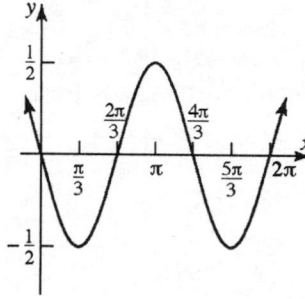

69. $y = -\dfrac{2}{3}\cos(\pi x - 6)$

Amplitude: $|A| = \left|-\dfrac{2}{3}\right| = \dfrac{2}{3}$

Period: $T = \dfrac{2\pi}{\omega} = \dfrac{2\pi}{\pi} = 2$

Phase Shift: $\dfrac{\phi}{\omega} = \dfrac{6}{\pi}$

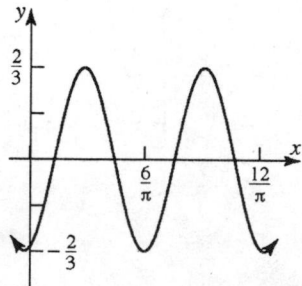

71. The graph is a cosine graph with an amplitude of 5 and a period of 8π. Find ω:
$$8\pi = \frac{2\pi}{\omega} \implies 8\pi\omega = 2\pi \implies \omega = \frac{2\pi}{8\pi} = \frac{1}{4}$$
The equation is: $y = 5\cos\left(\frac{1}{4}x\right)$.

73. The graph is a reflected cosine graph with an amplitude of 6 and a period of 8.
Find ω:
$$8 = \frac{2\pi}{\omega} \implies 8\omega = 2\pi \implies \omega = \frac{2\pi}{8} = \frac{\pi}{4}$$
The equation is: $y = -6\cos\left(\frac{\pi}{4}x\right)$.

75. Use calculator in radian mode: $\sin\left(\dfrac{\pi}{8}\right) \approx 0.38$

77. Terminal side of θ in Quadrant III implies

$\sin\theta < 0$ $\cos\theta < 0$

$\tan\theta > 0$ $\cot\theta > 0$

$\csc\theta < 0$ $\sec\theta < 0$

79. $P = \left(-\dfrac{1}{3}, \dfrac{2\sqrt{2}}{3}\right)$

$\cos t = -\dfrac{1}{3}$ $\sin t = \dfrac{2\sqrt{2}}{3}$

$\sec t = \dfrac{1}{\left(-\dfrac{1}{3}\right)} = -3$ $\csc t = \dfrac{1}{\left(\dfrac{2\sqrt{2}}{3}\right)} = \dfrac{3}{2\sqrt{2}} \cdot \dfrac{\sqrt{2}}{\sqrt{2}} = \dfrac{3\sqrt{2}}{8}$

$\tan t = \dfrac{\left(\dfrac{2\sqrt{2}}{3}\right)}{\left(-\dfrac{1}{3}\right)} = \left(\dfrac{2\sqrt{2}}{3}\right)\left(-\dfrac{3}{1}\right) = -2\sqrt{2}$

$\cot t = \dfrac{\left(-\dfrac{1}{3}\right)}{\left(\dfrac{2\sqrt{2}}{3}\right)} = \left(-\dfrac{1}{3}\right)\left(\dfrac{3}{2\sqrt{2}}\right) = -\dfrac{1}{2\sqrt{2}} \cdot \dfrac{\sqrt{2}}{\sqrt{2}} = -\dfrac{\sqrt{2}}{8}$

81. The domain of $y = \sec x$ is $\left\{x \mid -\infty < x < \infty, \text{ except odd multiples of } \dfrac{\pi}{2}\right\}$.
The range of $y = \sec x$ is $\left\{y \mid y < -1 \text{ or } y > 1\right\}$.

83. $r = 2$ feet, $\theta = 30°$ or $\theta = \dfrac{\pi}{6}$

$$s = r\theta = 2 \cdot \dfrac{\pi}{6} = \dfrac{\pi}{3} \text{ feet}$$

$$A = \dfrac{1}{2} \cdot r^2\theta = \dfrac{1}{2} \cdot (2)^2 \cdot \dfrac{\pi}{6} = \dfrac{\pi}{3} \text{ square feet}$$

85. $v = 180$ mi / hr, $d = \dfrac{1}{2}$ mile $\Rightarrow r = \dfrac{1}{4} = 0.25$ mile

$$\omega = \dfrac{v}{r} = \dfrac{180 \text{ mi/hr}}{0.25 \text{ mi}} = 720 \text{ rad/hr} = \dfrac{720 \text{ rad}}{\text{hr}} \cdot \dfrac{1 \text{ rev}}{2\pi \text{ rad}} = \dfrac{360 \text{ rev}}{\pi \text{ hr}} \approx 114.6 \text{ rev/hr}$$

87.   Since there are two lights on opposite sides and the light is seen every 5 seconds, the beacon makes 1 revolution every 10 seconds.

$$\omega = \dfrac{1 \text{ rev}}{10 \text{ sec}} \cdot \dfrac{2\pi}{1 \text{ rev}} = \dfrac{\pi}{5} \text{ radians / second}$$

89. $E(t) = 120\sin(120\pi t)$, $t \geq 0$

   (a)   The maximum value of E is the amplitude which is 120.

   (b)   Period $= \dfrac{2\pi}{120\pi} = \dfrac{1}{60}$ seconds

   (c)   Graphing:

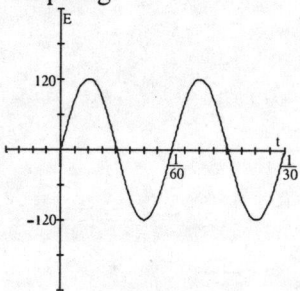

91.   (a)   Draw a scatter diagram:

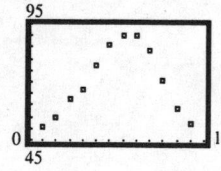

   (b)   Amplitude: $A = \dfrac{90 - 51}{2} = \dfrac{39}{2} = 19.5$

   Vertical Shift: $\dfrac{90 + 51}{2} = \dfrac{141}{2} = 70.5$

   $\omega = \dfrac{2\pi}{12} = \dfrac{\pi}{6}$

   Phase shift (use $y = 51$, $x = 1$):

$$51 = 19.5\sin\left(\frac{\pi}{6}\cdot 1 - \phi\right) + 70.5$$

$$-19.5 = 19.5\sin\left(\frac{\pi}{6} - \phi\right) \Rightarrow -1 = \sin\left(\frac{\pi}{6} - \phi\right) \Rightarrow -\frac{\pi}{2} = \frac{\pi}{6} - \phi \rightarrow \phi = \frac{2\pi}{3}$$

Thus, $y = 19.5\sin\left(\frac{\pi}{6}x - \frac{2\pi}{3}\right) + 70.5$

(c)

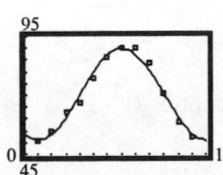

(e)

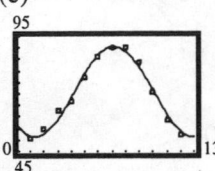

(d) $y = 19.518\sin(0.541x - 2.283) + 71.01$

93.   (a)   Amplitude: $A = \dfrac{13.367 - 9.667}{2} = \dfrac{3.7}{2} = 1.85$

Vertical Shift: $\dfrac{13.367 + 9.667}{2} = \dfrac{23.034}{2} = 11.517$

$\omega = \dfrac{2\pi}{365}$

Phase shift (use $y = 9.667$, $x = 355$):

$$9.667 = 1.85\sin\left(\frac{2\pi}{365}\cdot 355 - \phi\right) + 11.517$$

$$-1.85 = 1.85\sin\left(\frac{2\pi}{365}\cdot 355 - \phi\right) \rightarrow -1 = \sin\left(\frac{710\pi}{365} - \phi\right)$$

$$-\frac{\pi}{2} = \frac{710\pi}{365} - \phi \rightarrow \phi = 7.6818$$

Thus, $y = 1.85\sin\left(\dfrac{2\pi}{365}x - 7.6818\right) + 11.517$

(b)

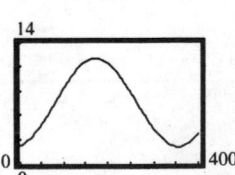

(c) $y = 1.85\sin\left(\dfrac{2\pi}{365}(91) - 7.6818\right) + 11.517 = 11.83$ hours

Trigonometric Functions

5.CR Cumulative Review

1. $2x^2 + x - 1 = 0$
$(2x - 1)(x + 1) = 0$

$x = \dfrac{1}{2}$ or $x = -1$

3. radius = 4, center $(0, -2)$
Using $(x - h)^2 + (y - k)^2 = r^2$
$(x - 0)^2 + (y - (-2))^2 = 4^2$
$x^2 + (y + 2)^2 = 16$
$x^2 + y^2 + 4y + 4 = 16$
$x^2 + y^2 + 4y - 12 = 0$

5. $x^2 + y^2 - 2x + 4y - 4 = 0$
$x^2 - 2x + 1 + y^2 + 4y + 4 = 4 + 1 + 4$

$(x - 1)^2 + (y + 2)^2 = 9$

$(x - 1)^2 + (y + 2)^2 = 3^2$
radius = 3, center $(1, -2)$

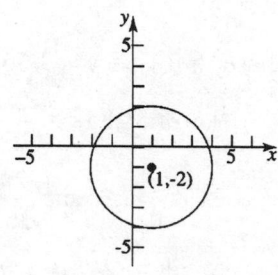

7. (a) $y = x^2$

(b) $y = x^3$

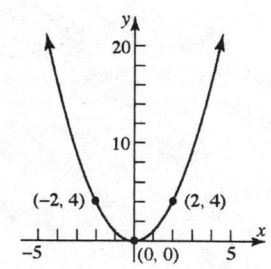

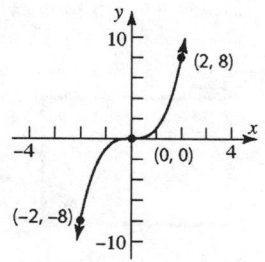

(c) $y = e^x$

(d) $y = \ln x$

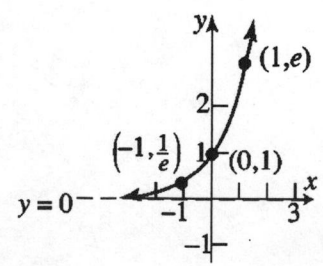

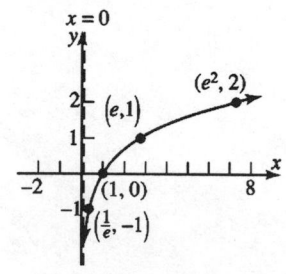

(e) $y = \sin x$

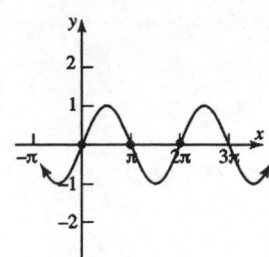

(f) $y = \tan x$

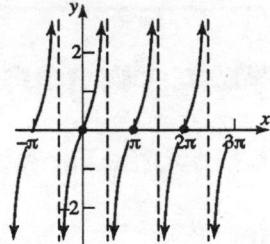

9. $\left(\sin\left(14°\right)\right)^2 + \left(\cos\left(14°\right)\right)^2 - 3 = 1 - 3 = -2$

11. $\tan\left(\dfrac{\pi}{4}\right) - 3\cos\left(\dfrac{\pi}{6}\right) + \csc\left(\dfrac{\pi}{6}\right) = 1 - 3\left(\dfrac{\sqrt{3}}{2}\right) + 2 = 3 - \dfrac{3\sqrt{3}}{2} = \dfrac{6 - 3\sqrt{3}}{2}$

13.   The graph is a cosine graph with an amplitude of 3 and a period of 12. Find ω:

$$12 = \frac{2\pi}{\omega} \implies 12\omega = 2\pi \implies \omega = \frac{2\pi}{12} = \frac{\pi}{6}$$

The equation is: $y = 3\cos\left(\dfrac{\pi}{6}x\right)$.

Chapter 6

Analytic Trigonometry

6.1 The Inverse Sine, Cosine and Tangent Functions

1. $\sin^{-1}(0)$

We are finding the angle θ, $-\dfrac{\pi}{2} \le \theta \le \dfrac{\pi}{2}$, whose sine equals 0.

$$\sin\theta = 0 \quad -\dfrac{\pi}{2} \le \theta \le \dfrac{\pi}{2}$$

$$\theta = 0 \Rightarrow \sin^{-1}(0) = 0$$

3. $\sin^{-1}(-1)$

We are finding the angle θ, $-\dfrac{\pi}{2} \le \theta \le \dfrac{\pi}{2}$, whose sine equals -1.

$$\sin\theta = -1 \quad -\dfrac{\pi}{2} \le \theta \le \dfrac{\pi}{2}$$

$$\theta = -\dfrac{\pi}{2} \Rightarrow \sin^{-1}(-1) = -\dfrac{\pi}{2}$$

5. $\tan^{-1}(0)$

We are finding the angle θ, $-\dfrac{\pi}{2} < \theta < \dfrac{\pi}{2}$, whose tangent equals 0.

$$\tan\theta = 0 \quad -\dfrac{\pi}{2} < \theta < \dfrac{\pi}{2}$$

$$\theta = 0 \Rightarrow \tan^{-1}(0) = 0$$

7. $\sin^{-1}\left(\dfrac{\sqrt{2}}{2}\right)$

We are finding the angle θ, $-\dfrac{\pi}{2} \le \theta \le \dfrac{\pi}{2}$, whose sine equals $\dfrac{\sqrt{2}}{2}$.

$$\sin\theta = \dfrac{\sqrt{2}}{2} \quad -\dfrac{\pi}{2} \le \theta \le \dfrac{\pi}{2}$$

$$\theta = \dfrac{\pi}{4} \Rightarrow \sin^{-1}\left(\dfrac{\sqrt{2}}{2}\right) = \dfrac{\pi}{4}$$

9. $\tan^{-1}\left(\sqrt{3}\right)$

We are finding the angle θ, $-\dfrac{\pi}{2} < \theta < \dfrac{\pi}{2}$, whose tangent equals $\sqrt{3}$.

$$\tan\theta = \sqrt{3} \qquad -\frac{\pi}{2} < \theta < \frac{\pi}{2}$$

$$\theta = \frac{\pi}{3} \Rightarrow \tan^{-1}\left(\sqrt{3}\right) = \frac{\pi}{3}$$

11. $\cos^{-1}\left(-\dfrac{\sqrt{3}}{2}\right)$

We are finding the angle θ, $0 \le \theta \le \pi$, whose cosine equals $-\dfrac{\sqrt{3}}{2}$.

$$\cos\theta = -\frac{\sqrt{3}}{2} \qquad 0 \le \theta \le \pi$$

$$\theta = \frac{5\pi}{6} \Rightarrow \cos^{-1}\left(-\frac{\sqrt{3}}{2}\right) = \frac{5\pi}{6}$$

13. $\sin^{-1}(0.1) \approx 0.10$

15. $\tan^{-1}(5) \approx 1.37$

17. $\cos^{-1}\left(\dfrac{7}{8}\right) \approx 0.51$

19. $\tan^{-1}(-0.4) \approx -0.38$

21. $\sin^{-1}(-0.12) \approx -0.12$

23. $\cos^{-1}\left(\dfrac{\sqrt{2}}{3}\right) \approx 1.08$

25. $\sin\left[\sin^{-1}(0.54)\right] = 0.54$

27. $\cos^{-1}\left[\cos\left(\dfrac{4\pi}{5}\right)\right] = \dfrac{4\pi}{5}$

29. $\tan\left[\tan^{-1}(-3.5)\right] = -3.5$

31. $\sin^{-1}\left[\sin\left(-\dfrac{3\pi}{7}\right)\right] = -\dfrac{3\pi}{7}$

33. yes, $\sin^{-1}\left[\sin\left(-\dfrac{\pi}{6}\right)\right] = -\dfrac{\pi}{6}$ since $\sin^{-1}\left[\sin(x)\right] = x$ where $-\dfrac{\pi}{2} \le x \le \dfrac{\pi}{2}$

and $-\dfrac{\pi}{6}$ is in the restricted domain of $f(x) = \sin(x)$.

35. no, $\sin\left[\sin^{-1}(2)\right] \ne 2$ since $\sin\left[\sin^{-1}(x)\right] = x$ where $-1 \le x \le 1$
and 2 is not in the domain of $f(x) = \sin^{-1}(x)$.

37. no, $\cos^{-1}\left[\cos\left(-\dfrac{\pi}{6}\right)\right] \neq -\dfrac{\pi}{6}$ since $\cos^{-1}[\cos(x)] = x$ where $0 \leq x \leq \pi$

and $-\dfrac{\pi}{6}$ is not in the restricted domain of $f(x) = \cos(x)$.

39. yes, $\cos\left[\cos^{-1}\left(-\dfrac{1}{2}\right)\right] = -\dfrac{1}{2}$ since $\cos[\cos^{-1}(x)] = x$ where $-1 \leq x \leq 1$

and $-\dfrac{1}{2}$ is in the domain of $f(x) = \cos^{-1}(x)$.

41. yes, $\tan^{-1}\left[\tan\left(-\dfrac{\pi}{3}\right)\right] = -\dfrac{\pi}{3}$ since $\tan^{-1}[\tan(x)] = x$ where $-\dfrac{\pi}{2} < x < \dfrac{\pi}{2}$

and $-\dfrac{\pi}{3}$ is in the restricted domain of $f(x) = \tan(x)$.

43. yes, $\tan\left[\tan^{-1}(2)\right] = 2$ since $\tan\left[\tan^{-1}(x)\right] = x$ where $-\infty < x < \infty$

45. (a) $D = 24 \cdot \left[1 - \dfrac{\cos^{-1}\left(\tan\left(23.5 \cdot \dfrac{\pi}{180}\right) \tan\left(29.75 \cdot \dfrac{\pi}{180}\right)\right)}{\pi} \right] \approx 13.92$ hours

(b) $D = 24 \cdot \left[1 - \dfrac{\cos^{-1}\left(\tan\left(0 \cdot \dfrac{\pi}{180}\right) \tan\left(29.75 \cdot \dfrac{\pi}{180}\right)\right)}{\pi} \right] \approx 12$ hours

(c) $D = 24 \cdot \left[1 - \dfrac{\cos^{-1}\left(\tan\left(22.8 \cdot \dfrac{\pi}{180}\right) \tan\left(29.75 \cdot \dfrac{\pi}{180}\right)\right)}{\pi} \right] \approx 13.85$ hours

47. (a) $D = 24 \cdot \left(1 - \dfrac{\cos^{-1}\left(\tan\left(23.5 \cdot \dfrac{\pi}{180}\right) \tan\left(21.3 \cdot \dfrac{\pi}{180}\right)\right)}{\pi} \right) \approx 13.30$ hours

(b) $D = 24 \cdot \left(1 - \dfrac{\cos^{-1}\left(\tan\left(0 \cdot \dfrac{\pi}{180}\right) \tan\left(21.3 \cdot \dfrac{\pi}{180}\right)\right)}{\pi} \right) \approx 12$ hours

(c) $D = 24 \cdot \left(1 - \dfrac{\cos^{-1}\left(\tan\left(22.8 \cdot \dfrac{\pi}{180} \right) \tan\left(21.3 \cdot \dfrac{\pi}{180} \right) \right)}{\pi} \right) \approx 13.26$ hours

49.   (a) $D = 24 \cdot \left(1 - \dfrac{\cos^{-1}\left(\tan\left(23.5 \cdot \dfrac{\pi}{180} \right) \tan\left(0 \cdot \dfrac{\pi}{180} \right) \right)}{\pi} \right) \approx 12$ hours

      (b) $D = 24 \cdot \left(1 - \dfrac{\cos^{-1}\left(\tan\left(0 \cdot \dfrac{\pi}{180} \right) \tan\left(0 \cdot \dfrac{\pi}{180} \right) \right)}{\pi} \right) \approx 12$ hours

      (c) $D = 24 \cdot \left(1 - \dfrac{\cos^{-1}\left(\tan\left(22.8 \cdot \dfrac{\pi}{180} \right) \tan\left(0 \cdot \dfrac{\pi}{180} \right) \right)}{\pi} \right) \approx 12$ hours

      (d)   The number of hours of daylight per day is approximately 12 hours
            throughout the year.

51.   At the latitude of Cadillac Mountain, the effective radius of the earth is 2710 miles.

$1530 \text{ ft} \cdot \dfrac{1 \text{ mile}}{5280 \text{ feet}} = 0.29$ mile

$\cos\theta = \dfrac{2710}{2710.29}$

$\theta \approx 0.01462$ radians

$s = r\theta = 3960(0.01462) = 39.62$ miles

$\dfrac{2\pi(2710)}{24} = \dfrac{39.62}{t}$

$t \approx 0.05583$ hour $= 3.35$ minutes

Analytic Trigonometry

6.2 The Inverse Trigonometric Functions (Continued)

1. $\cos\left(\sin^{-1}\left(\dfrac{\sqrt{2}}{2}\right)\right)$

Find the angle θ, $-\dfrac{\pi}{2} \leq \theta \leq \dfrac{\pi}{2}$, whose sine equals $\dfrac{\sqrt{2}}{2}$.

$$\sin\theta = \frac{\sqrt{2}}{2} \qquad -\frac{\pi}{2} \leq \theta \leq \frac{\pi}{2}$$

$$\theta = \frac{\pi}{4} \Rightarrow \cos\left(\sin^{-1}\left(\frac{\sqrt{2}}{2}\right)\right) = \cos\left(\frac{\pi}{4}\right) = \frac{\sqrt{2}}{2}$$

3. $\tan\left(\cos^{-1}\left(-\dfrac{\sqrt{3}}{2}\right)\right)$

Find the angle θ, $0 \leq \theta \leq \pi$, whose cosine equals $-\dfrac{\sqrt{3}}{2}$.

$$\cos\theta = -\frac{\sqrt{3}}{2} \qquad 0 \leq \theta \leq \pi$$

$$\theta = \frac{5\pi}{6} \Rightarrow \tan\left(\cos^{-1}\left(-\frac{\sqrt{3}}{2}\right)\right) = \tan\left(\frac{5\pi}{6}\right) = -\frac{\sqrt{3}}{3}$$

5. $\sec\left(\cos^{-1}\left(\dfrac{1}{2}\right)\right)$

Find the angle θ, $0 \leq \theta \leq \pi$, whose cosine equals $\dfrac{1}{2}$.

$$\cos\theta = \frac{1}{2} \qquad 0 \leq \theta \leq \pi$$

$$\theta = \frac{\pi}{3} \Rightarrow \sec\left(\cos^{-1}\left(\frac{1}{2}\right)\right) = \sec\left(\frac{\pi}{3}\right) = 2$$

7. $\csc\left(\tan^{-1}(1)\right)$

Find the angle θ, $-\dfrac{\pi}{2} < \theta < \dfrac{\pi}{2}$, whose tangent equals 1.

$$\tan\theta = 1 \qquad -\frac{\pi}{2} < \theta < \frac{\pi}{2}$$

$$\theta = \frac{\pi}{4} \Rightarrow \csc\left(\tan^{-1}(1)\right) = \csc\left(\frac{\pi}{4}\right) = \sqrt{2}$$

9. $\sin\left(\tan^{-1}(-1)\right)$

Find the angle θ, $-\dfrac{\pi}{2} < \theta < \dfrac{\pi}{2}$, whose tangent equals -1.

$$\tan\theta = -1 \qquad -\frac{\pi}{2} < \theta < \frac{\pi}{2}$$

$$\theta = -\frac{\pi}{4} \Rightarrow \sin\left(\tan^{-1}(-1)\right) = \sin\left(-\frac{\pi}{4}\right) = -\frac{\sqrt{2}}{2}$$

11. $\sec\left(\sin^{-1}\left(-\frac{1}{2}\right)\right)$

Find the angle θ, $-\dfrac{\pi}{2} \le \theta \le \dfrac{\pi}{2}$, whose sine equals $-\dfrac{1}{2}$.

$$\sin\theta = -\frac{1}{2} \qquad -\frac{\pi}{2} \le \theta \le \frac{\pi}{2}$$

$$\theta = -\frac{\pi}{6} \Rightarrow \sec\left(\sin^{-1}\left(-\frac{1}{2}\right)\right) = \sec\left(-\frac{\pi}{6}\right) = \frac{2\sqrt{3}}{3}$$

13. $\cos^{-1}\left(\cos\frac{5\pi}{4}\right) = \cos^{-1}\left(-\frac{\sqrt{2}}{2}\right)$

Find the angle θ, $0 \le \theta \le \pi$, whose cosine equals $-\dfrac{\sqrt{2}}{2}$.

$$\cos\theta = -\frac{\sqrt{2}}{2} \qquad 0 \le \theta \le \pi$$

$$\theta = \frac{3\pi}{4}$$

15. $\sin^{-1}\left(\sin\left(-\frac{7\pi}{6}\right)\right) = \sin^{-1}\left(\frac{1}{2}\right)$

Find the angle θ, $-\dfrac{\pi}{2} \le \theta \le \dfrac{\pi}{2}$, whose sine equals $\dfrac{1}{2}$.

$$\sin\theta = \frac{1}{2} \qquad -\frac{\pi}{2} \le \theta \le \frac{\pi}{2}$$

$$\theta = \frac{\pi}{6}$$

17. $\tan\left(\sin^{-1}\left(\dfrac{1}{3}\right)\right)$

Since $\sin\theta = \dfrac{1}{3}$, $-\dfrac{\pi}{2} \le \theta \le \dfrac{\pi}{2}$, let $y = 1$ and $r = 3$. Solve for x:

$$x^2 + 1 = 9 \Rightarrow x^2 = 8 \Rightarrow x = \pm\sqrt{8} = \pm 2\sqrt{2}$$

Since θ is in quadrant I, $x = 2\sqrt{2}$.

$$\tan\left(\sin^{-1}\left(\frac{1}{3}\right)\right) = \tan\theta = \frac{y}{x} = \frac{1}{2\sqrt{2}} \cdot \frac{\sqrt{2}}{\sqrt{2}} = \frac{\sqrt{2}}{4}$$

19. $\sec\left(\tan^{-1}\left(\dfrac{1}{2}\right)\right)$

Since $\tan\theta = \dfrac{1}{2}$, $-\dfrac{\pi}{2} < \theta < \dfrac{\pi}{2}$, let $x = 2$ and $y = 1$. Solve for r:

$$2^2 + 1 = r^2 \Rightarrow r^2 = 5 \Rightarrow r = \sqrt{5}$$

θ is in quadrant I.

$$\sec\left(\tan^{-1}\left(\frac{1}{2}\right)\right) = \sec\theta = \frac{r}{x} = \frac{\sqrt{5}}{2}$$

21. $\cot\left(\sin^{-1}\left(-\dfrac{\sqrt{2}}{3}\right)\right)$

Since $\sin\theta = -\dfrac{\sqrt{2}}{3}$, $-\dfrac{\pi}{2} \le \theta \le \dfrac{\pi}{2}$, let $y = -\sqrt{2}$ and $r = 3$. Solve for x:

$$x^2 + 2 = 9 \Rightarrow x^2 = 7 \Rightarrow x = \pm\sqrt{7}$$

Since θ is in quadrant IV, $x = \sqrt{7}$.

$$\cot\left(\sin^{-1}\left(-\frac{\sqrt{2}}{3}\right)\right) = \cot\theta = \frac{x}{y} = \frac{\sqrt{7}}{-\sqrt{2}} \cdot \frac{\sqrt{2}}{\sqrt{2}} = -\frac{\sqrt{14}}{2}$$

23. $\sin\left(\tan^{-1}(-3)\right)$

Since $\tan\theta = -3$, $-\dfrac{\pi}{2} < \theta < \dfrac{\pi}{2}$, let $x = 1$ and $y = -3$. Solve for r:

$$1 + 9 = r^2 \quad\Rightarrow\quad r^2 = 10 \quad\Rightarrow\quad r = \sqrt{10}; \ \theta \text{ is in quadrant IV.}$$

$$\sin\left(\tan^{-1}(-3)\right) = \sin\theta = \frac{y}{r} = \frac{-3}{\sqrt{10}} \cdot \frac{\sqrt{10}}{\sqrt{10}} = -\frac{3\sqrt{10}}{10}$$

25. $\sec\left(\sin^{-1}\left(\dfrac{2\sqrt{5}}{5}\right)\right)$

Since $\sin\theta = \dfrac{2\sqrt{5}}{5}$, $-\dfrac{\pi}{2} \le \theta \le \dfrac{\pi}{2}$, let $y = 2\sqrt{5}$ and $r = 5$. Solve for x:

$$x^2 + 20 = 25 \Rightarrow x^2 = 5 \Rightarrow x = \pm\sqrt{5}$$

Since θ is in quadrant I, $x = \sqrt{5}$.

$$\sec\left(\sin^{-1}\left(\frac{2\sqrt{5}}{5}\right)\right) = \sec\theta = \frac{r}{x} = \frac{5}{\sqrt{5}} \cdot \frac{\sqrt{5}}{\sqrt{5}} = \sqrt{5}$$

27. $\sin^{-1}\left(\cos\frac{3\pi}{4}\right) = \sin^{-1}\left(-\frac{\sqrt{2}}{2}\right) = -\frac{\pi}{4}$

29. $\cot^{-1}\left(\sqrt{3}\right)$

We are finding the angle θ, $0 < \theta < \pi$, whose cotangent equals $\sqrt{3}$.

$$\cot\theta = \sqrt{3} \quad 0 < \theta < \pi$$

$$\theta = \frac{\pi}{6} \Rightarrow \cot^{-1}\left(\sqrt{3}\right) = \frac{\pi}{6}$$

31. $\csc^{-1}(-1)$

We are finding the angle θ, $-\frac{\pi}{2} \le \theta \le \frac{\pi}{2}$, $\theta \ne 0$, whose cosecant equals -1.

$$\csc\theta = -1 \quad -\frac{\pi}{2} \le \theta \le \frac{\pi}{2}, \ \theta \ne 0$$

$$\theta = -\frac{\pi}{2} \Rightarrow \csc^{-1}(-1) = -\frac{\pi}{2}$$

33. $\sec^{-1}\left(\frac{2\sqrt{3}}{3}\right)$

We are finding the angle θ, $0 \le \theta \le \pi$, $\theta \ne \frac{\pi}{2}$, whose secant equals $\frac{2\sqrt{3}}{3}$.

$$\sec\theta = \frac{2\sqrt{3}}{3} \quad 0 \le \theta \le \pi, \ \theta \ne \frac{\pi}{2}$$

$$\theta = \frac{\pi}{6} \Rightarrow \sec^{-1}\left(\frac{2\sqrt{3}}{3}\right) = \frac{\pi}{6}$$

35. $\cot^{-1}\left(-\frac{\sqrt{3}}{3}\right)$

We are finding the angle θ, $0 < \theta < \pi$, whose cotangent equals $-\frac{\sqrt{3}}{3}$.

$$\cot\theta = -\frac{\sqrt{3}}{3} \quad 0 < \theta < \pi$$

$$\theta = \frac{2\pi}{3} \Rightarrow \cot^{-1}\left(-\frac{\sqrt{3}}{3}\right) = \frac{2\pi}{3}$$

37. $\sec^{-1}(4) = \cos^{-1}\left(\frac{1}{4}\right)$

We are finding the angle θ, $0 \le \theta \le \pi$, whose cosine equals $\frac{1}{4}$.

$$\cos\theta = \frac{1}{4} \Rightarrow \theta \text{ in quadrant I}$$

The calculator yields $\theta = \cos^{-1}\left(\frac{1}{4}\right) \approx 1.32$, which is an angle in quadrant I.

$$\therefore \quad \sec^{-1}(4) \approx 1.32$$

39. $\cot^{-1}(2) = \tan^{-1}\left(\frac{1}{2}\right)$

We are finding the angle θ, $0 \leq \theta \leq \pi$, whose tangent equals $\frac{1}{2}$.

$$\tan\theta = \frac{1}{2} \Rightarrow \theta \text{ in quadrant I}$$

The calculator yields $\theta = \tan^{-1}\left(\frac{1}{2}\right) \approx 0.46$, which is an angle in quadrant I.

$$\therefore \quad \cot^{-1}(2) \approx 0.46$$

41. $\csc^{-1}(-3) = \sin^{-1}\left(-\frac{1}{3}\right) \approx -0.34$

We are finding the angle θ, $-\frac{\pi}{2} \leq \theta \leq \frac{\pi}{2}$, whose sine equals $-\frac{1}{3}$.

$$\sin\theta = -\frac{1}{3} \Rightarrow \theta \text{ in quadrant IV}$$

The calculator yields $\theta = \sin^{-1}\left(-\frac{1}{3}\right) \approx -0.34$, which is an angle in quadrant IV.

$$\therefore \quad \csc^{-1}(-3) \approx -0.34$$

43. $\cot^{-1}\left(\sqrt{5}\right) = \tan^{-1}\left(\frac{1}{\sqrt{5}}\right)$

We are finding the angle θ, $0 \leq \theta \leq \pi$, whose tangent equals $\frac{1}{\sqrt{5}}$.

$$\tan\theta = \frac{1}{\sqrt{5}} \Rightarrow \theta \text{ in quadrant I}$$

The calculator yields $\tan^{-1}\left(\frac{1}{\sqrt{5}}\right) \approx 0.42$, which is an angle in quadrant I.

45. $\csc^{-1}\left(-\frac{3}{2}\right) = \sin^{-1}\left(-\frac{2}{3}\right)$

We are finding the angle θ, $-\frac{\pi}{2} \leq \theta \leq \frac{\pi}{2}$, whose sine equals $-\frac{2}{3}$.

$$\sin\theta = -\frac{2}{3} \Rightarrow \theta \text{ in quadrant IV}$$

The calculator yields $\sin^{-1}\left(-\frac{2}{3}\right) \approx -0.73$, which is an angle in quadrant IV.

$$\therefore \quad \csc^{-1}\left(-\frac{3}{2}\right) \approx -0.73$$

47. $\cot^{-1}\left(-\frac{3}{2}\right) = \tan^{-1}\left(-\frac{2}{3}\right)$

We are finding the angle θ, $0 \le \theta \le \pi$, whose tangent equals $-\frac{2}{3}$.

$$\tan\theta = -\frac{2}{3} \Rightarrow \theta \text{ in quadrant II}$$

The calculator yields $\tan^{-1}\left(-\frac{2}{3}\right) \approx -0.59$, which is an angle in quadrant IV.

Since θ is in quadrant II, $\theta \approx -0.59 + \pi \approx 2.55$.

$$\therefore \quad \cot^{-1}\left(-\frac{3}{2}\right) \approx 2.55$$

49. $y = \cot^{-1} x$

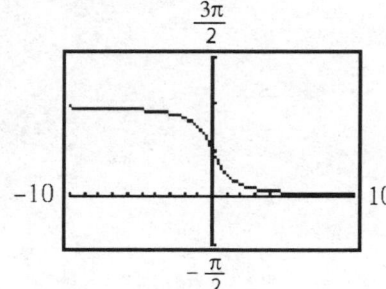

51. $y = \csc^{-1} x$

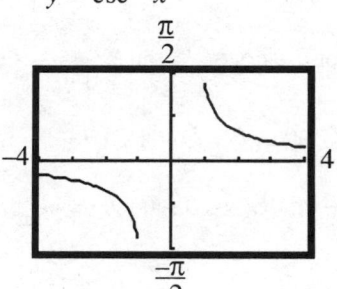

Analytic Trigonometry

6.3 Trigonometric Identities

1. $\csc\theta \cdot \cos\theta = \dfrac{1}{\sin\theta} \cdot \cos\theta = \dfrac{\cos\theta}{\sin\theta} = \cot\theta$

3. $1 + \tan^2(-\theta) = 1 + (-\tan\theta)^2 = 1 + \tan^2\theta = \sec^2\theta$

5. $\cos\theta(\tan\theta + \cot\theta) = \cos\theta\left(\dfrac{\sin\theta}{\cos\theta} + \dfrac{\cos\theta}{\sin\theta}\right) = \cos\theta\left(\dfrac{\sin^2\theta + \cos^2\theta}{\cos\theta\sin\theta}\right) = \dfrac{1}{\sin\theta} = \csc\theta$

7. $\tan\theta\cot\theta - \cos^2\theta = \tan\theta \cdot \dfrac{1}{\tan\theta} - \cos^2\theta = 1 - \cos^2\theta = \sin^2\theta$

9. $(\sec\theta - 1)(\sec\theta + 1) = \sec^2\theta - 1 = \tan^2\theta$

11. $(\sec\theta + \tan\theta)(\sec\theta - \tan\theta) = \sec^2\theta - \tan^2\theta = 1$

13. $\cos^2\theta(1 + \tan^2\theta) = \cos^2\theta \cdot \sec^2\theta = \cos^2\theta \cdot \dfrac{1}{\cos^2\theta} = 1$

15. $(\sin\theta + \cos\theta)^2 + (\sin\theta - \cos\theta)^2$
$$= \sin^2\theta + 2\sin\theta\cos\theta + \cos^2\theta + \sin^2\theta - 2\sin\theta\cos\theta + \cos^2\theta$$
$$= 2\sin^2\theta + 2\cos^2\theta = 2(\sin^2\theta + \cos^2\theta) = 2 \cdot 1 = 2$$

17. $\sec^4\theta - \sec^2\theta = \sec^2\theta(\sec^2\theta - 1) = (\tan^2\theta + 1)\tan^2\theta = \tan^4\theta + \tan^2\theta$

19. $\sec\theta - \tan\theta = \dfrac{1}{\cos\theta} - \dfrac{\sin\theta}{\cos\theta} = \left(\dfrac{1 - \sin\theta}{\cos\theta}\right) \cdot \left(\dfrac{1 + \sin\theta}{1 + \sin\theta}\right) = \dfrac{1 - \sin^2\theta}{\cos\theta(1 + \sin\theta)}$
$$= \dfrac{\cos^2\theta}{\cos\theta(1 + \sin\theta)} = \dfrac{\cos\theta}{1 + \sin\theta}$$

21. $3\sin^2\theta + 4\cos^2\theta = 3\sin^2\theta + 3\cos^2\theta + \cos^2\theta = 3(\sin^2\theta + \cos^2\theta) + \cos^2\theta$
$$= 3 \cdot 1 + \cos^2\theta = 3 + \cos^2\theta$$

23. $1 - \dfrac{\cos^2\theta}{1 + \sin\theta} = 1 - \dfrac{1 - \sin^2\theta}{1 + \sin\theta} = 1 - \dfrac{(1 - \sin\theta)(1 + \sin\theta)}{1 + \sin\theta} = 1 - 1 + \sin\theta = \sin\theta$

25. $$\frac{1+\tan\theta}{1-\tan\theta}=\frac{\left(1+\dfrac{1}{\cot\theta}\right)}{\left(1-\dfrac{1}{\cot\theta}\right)}=\frac{\left(\dfrac{\cot\theta+1}{\cot\theta}\right)}{\left(\dfrac{\cot\theta-1}{\cot\theta}\right)}=\frac{\cot\theta+1}{\cot\theta}\cdot\frac{\cot\theta}{\cot\theta-1}=\frac{\cot\theta+1}{\cot\theta-1}$$

27. $$\frac{\sec\theta}{\csc\theta}+\frac{\sin\theta}{\cos\theta}=\frac{\left(\dfrac{1}{\cos\theta}\right)}{\left(\dfrac{1}{\sin\theta}\right)}+\frac{\sin\theta}{\cos\theta}=\frac{\sin\theta}{\cos\theta}+\frac{\sin\theta}{\cos\theta}=\tan\theta+\tan\theta=2\tan\theta$$

29. $$\frac{1+\sin\theta}{1-\sin\theta}=\frac{\left(1+\dfrac{1}{\csc\theta}\right)}{\left(1-\dfrac{1}{\csc\theta}\right)}=\frac{\left(\dfrac{\csc\theta+1}{\csc\theta}\right)}{\left(\dfrac{\csc\theta-1}{\csc\theta}\right)}=\frac{\csc\theta+1}{\csc\theta}\cdot\frac{\csc\theta}{\csc\theta-1}=\frac{\csc\theta+1}{\csc\theta-1}$$

31. $$\frac{1-\sin\theta}{\cos\theta}+\frac{\cos\theta}{1-\sin\theta}=\frac{(1-\sin\theta)^2+\cos^2\theta}{\cos\theta(1-\sin\theta)}=\frac{1-2\sin\theta+\sin^2\theta+\cos^2\theta}{\cos\theta(1-\sin\theta)}$$
$$=\frac{1-2\sin\theta+1}{\cos\theta(1-\sin\theta)}=\frac{2-2\sin\theta}{\cos\theta(1-\sin\theta)}=\frac{2(1-\sin\theta)}{\cos\theta(1-\sin\theta)}=\frac{2}{\cos\theta}=2\sec\theta$$

33. $$\frac{\sin\theta}{\sin\theta-\cos\theta}=\frac{\sin\theta}{\sin\theta-\cos\theta}\cdot\frac{\left(\dfrac{1}{\sin\theta}\right)}{\left(\dfrac{1}{\sin\theta}\right)}=\frac{1}{\left(1-\dfrac{\cos\theta}{\sin\theta}\right)}=\frac{1}{1-\cot\theta}$$

35. $$(\sec\theta-\tan\theta)^2=\sec^2\theta-2\sec\theta\tan\theta+\tan^2\theta=\frac{1}{\cos^2\theta}-2\cdot\frac{1}{\cos\theta}\cdot\frac{\sin\theta}{\cos\theta}+\frac{\sin^2\theta}{\cos^2\theta}$$
$$=\frac{1-2\sin\theta+\sin^2\theta}{\cos^2\theta}=\frac{(1-\sin\theta)(1-\sin\theta)}{1-\sin^2\theta}=\frac{(1-\sin\theta)(1-\sin\theta)}{(1-\sin\theta)(1+\sin\theta)}=\frac{1-\sin\theta}{1+\sin\theta}$$

37. $$\frac{\cos\theta}{1-\tan\theta}+\frac{\sin\theta}{1-\cot\theta}=\frac{\cos\theta}{\left(1-\dfrac{\sin\theta}{\cos\theta}\right)}+\frac{\sin\theta}{\left(1-\dfrac{\cos\theta}{\sin\theta}\right)}=\frac{\cos\theta}{\left(\dfrac{\cos\theta-\sin\theta}{\cos\theta}\right)}+\frac{\sin\theta}{\left(\dfrac{\sin\theta-\cos\theta}{\sin\theta}\right)}$$
$$=\frac{\cos^2\theta}{\cos\theta-\sin\theta}+\frac{\sin^2\theta}{\sin\theta-\cos\theta}=\frac{\cos^2\theta-\sin^2\theta}{\cos\theta-\sin\theta}$$
$$=\frac{(\cos\theta-\sin\theta)(\cos\theta+\sin\theta)}{\cos\theta-\sin\theta}=\cos\theta+\sin\theta=\sin\theta+\cos\theta$$

39. $$\tan\theta+\frac{\cos\theta}{1+\sin\theta}=\frac{\sin\theta}{\cos\theta}+\frac{\cos\theta}{1+\sin\theta}=\frac{\sin\theta(1+\sin\theta)+\cos^2\theta}{\cos\theta(1+\sin\theta)}$$
$$=\frac{\sin\theta+\sin^2\theta+\cos^2\theta}{\cos\theta(1+\sin\theta)}=\frac{\sin\theta+1}{\cos\theta(1+\sin\theta)}=\frac{1}{\cos\theta}=\sec\theta$$

41. $\dfrac{\tan\theta + \sec\theta - 1}{\tan\theta - \sec\theta + 1} = \dfrac{\tan\theta + (\sec\theta - 1)}{\tan\theta - (\sec\theta - 1)} \cdot \dfrac{\tan\theta + (\sec\theta - 1)}{\tan\theta + (\sec\theta - 1)}$

$$= \frac{\tan^2\theta + 2\tan\theta(\sec\theta - 1) + \sec^2\theta - 2\sec\theta + 1}{\tan^2\theta - (\sec^2\theta - 2\sec\theta + 1)}$$

$$= \frac{\sec^2\theta - 1 + 2\tan\theta(\sec\theta - 1) + \sec^2\theta - 2\sec\theta + 1}{\sec^2\theta - 1 - \sec^2\theta + 2\sec\theta - 1}$$

$$= \frac{2\sec^2\theta - 2\sec\theta + 2\tan\theta(\sec\theta - 1)}{2\sec\theta - 2}$$

$$= \frac{2\sec\theta(\sec\theta - 1) + 2\tan\theta(\sec\theta - 1)}{2\sec\theta - 2}$$

$$= \frac{2(\sec\theta - 1)(\sec\theta + \tan\theta)}{2(\sec\theta - 1)} = \sec\theta + \tan\theta = \tan\theta + \sec\theta$$

43. $\dfrac{\tan\theta - \cot\theta}{\tan\theta + \cot\theta} = \dfrac{\left(\dfrac{\sin\theta}{\cos\theta} - \dfrac{\cos\theta}{\sin\theta}\right)}{\left(\dfrac{\sin\theta}{\cos\theta} + \dfrac{\cos\theta}{\sin\theta}\right)} = \dfrac{\left(\dfrac{\sin^2\theta - \cos^2\theta}{\cos\theta\sin\theta}\right)}{\left(\dfrac{\sin^2\theta + \cos^2\theta}{\cos\theta\sin\theta}\right)} = \dfrac{\sin^2\theta - \cos^2\theta}{1} = \sin^2\theta - \cos^2\theta$

45. $\dfrac{\tan\theta - \cot\theta}{\tan\theta + \cot\theta} + 1 = \dfrac{\left(\dfrac{\sin\theta}{\cos\theta} - \dfrac{\cos\theta}{\sin\theta}\right)}{\left(\dfrac{\sin\theta}{\cos\theta} + \dfrac{\cos\theta}{\sin\theta}\right)} + 1 = \dfrac{\left(\dfrac{\sin^2\theta - \cos^2\theta}{\cos\theta\sin\theta}\right)}{\left(\dfrac{\sin^2\theta + \cos^2\theta}{\cos\theta\sin\theta}\right)} + 1 = \dfrac{\sin^2\theta - \cos^2\theta}{1} + 1$

$$= \sin^2\theta - \cos^2\theta + 1 = \sin^2\theta + (1 - \cos^2\theta) = \sin^2\theta + \sin^2\theta = 2\sin^2\theta$$

47. $\dfrac{\sec\theta + \tan\theta}{\cot\theta + \cos\theta} = \dfrac{\left(\dfrac{1}{\cos\theta} + \dfrac{\sin\theta}{\cos\theta}\right)}{\left(\dfrac{\cos\theta}{\sin\theta} + \cos\theta\right)} = \dfrac{\left(\dfrac{1 + \sin\theta}{\cos\theta}\right)}{\left(\dfrac{\cos\theta + \cos\theta\sin\theta}{\sin\theta}\right)} = \dfrac{1 + \sin\theta}{\cos\theta} \cdot \dfrac{\sin\theta}{\cos\theta(1 + \sin\theta)}$

$$= \frac{\sin\theta}{\cos\theta} \cdot \frac{1}{\cos\theta} = \tan\theta\sec\theta$$

49. $\dfrac{1 - \tan^2\theta}{1 + \tan^2\theta} + 1 = \dfrac{1 - \tan^2\theta + 1 + \tan^2\theta}{1 + \tan^2\theta} = \dfrac{2}{\sec^2\theta} = 2 \cdot \dfrac{1}{\sec^2\theta} = 2\cos^2\theta$

51. $\dfrac{\sec\theta - \csc\theta}{\sec\theta\csc\theta} = \dfrac{\left(\dfrac{1}{\cos\theta} - \dfrac{1}{\sin\theta}\right)}{\left(\dfrac{1}{\cos\theta} \cdot \dfrac{1}{\sin\theta}\right)} = \dfrac{\left(\dfrac{\sin\theta - \cos\theta}{\cos\theta\sin\theta}\right)}{\left(\dfrac{1}{\cos\theta\sin\theta}\right)} = \sin\theta - \cos\theta$

53. $\sec\theta - \cos\theta - \sin\theta\tan\theta = \dfrac{1}{\cos\theta} - \cos\theta - \sin\theta \cdot \dfrac{\sin\theta}{\cos\theta} = \dfrac{1 - \cos^2\theta - \sin^2\theta}{\cos\theta}$

$$= \frac{\sin^2\theta - \sin^2\theta}{\cos\theta} = 0$$

55. $\dfrac{1}{1-\sin\theta}+\dfrac{1}{1+\sin\theta}=\dfrac{1+\sin\theta+1-\sin\theta}{(1-\sin\theta)(1+\sin\theta)}=\dfrac{2}{1-\sin^2\theta}=\dfrac{2}{\cos^2\theta}=2\sec^2\theta$

57. $\dfrac{\sec\theta}{1-\sin\theta}=\left(\dfrac{\sec\theta}{1-\sin\theta}\right)\cdot\left(\dfrac{1+\sin\theta}{1+\sin\theta}\right)=\dfrac{\sec\theta(1+\sin\theta)}{1-\sin^2\theta}=\dfrac{\sec\theta(1+\sin\theta)}{\cos^2\theta}$

$\qquad =\dfrac{1}{\cos\theta}\cdot\dfrac{1+\sin\theta}{\cos^2\theta}=\dfrac{1+\sin\theta}{\cos^3\theta}$

59. $\dfrac{(\sec\theta-\tan\theta)^2+1}{\csc\theta(\sec\theta-\tan\theta)}=\dfrac{\sec^2\theta-2\sec\theta\tan\theta+\tan^2\theta+1}{\csc\theta(\sec\theta-\tan\theta)}=\dfrac{2\sec^2\theta-2\sec\theta\tan\theta}{\csc\theta(\sec\theta-\tan\theta)}$

$\qquad =\dfrac{2\sec\theta(\sec\theta-\tan\theta)}{\csc\theta(\sec\theta-\tan\theta)}=\dfrac{2\sec\theta}{\csc\theta}=\dfrac{2\cdot\dfrac{1}{\cos\theta}}{\dfrac{1}{\sin\theta}}=2\cdot\dfrac{1}{\cos\theta}\cdot\dfrac{\sin\theta}{1}=2\tan\theta$

61. $\dfrac{\sin\theta+\cos\theta}{\cos\theta}-\dfrac{\sin\theta-\cos\theta}{\sin\theta}=\dfrac{\sin\theta}{\cos\theta}+\dfrac{\cos\theta}{\cos\theta}-\dfrac{\sin\theta}{\sin\theta}+\dfrac{\cos\theta}{\sin\theta}=\dfrac{\sin\theta}{\cos\theta}+1-1+\dfrac{\cos\theta}{\sin\theta}$

$\qquad =\dfrac{\sin^2\theta+\cos^2\theta}{\cos\theta\sin\theta}=\dfrac{1}{\cos\theta\sin\theta}=\sec\theta\csc\theta$

63. $\dfrac{\sin^3\theta+\cos^3\theta}{\sin\theta+\cos\theta}=\dfrac{(\sin\theta+\cos\theta)(\sin^2\theta-\sin\theta\cos\theta+\cos^2\theta)}{\sin\theta+\cos\theta}=1-\sin\theta\cos\theta$

65. $\dfrac{\cos^2\theta-\sin^2\theta}{1-\tan^2\theta}=\dfrac{\cos^2\theta-\sin^2\theta}{\left(1-\dfrac{\sin^2\theta}{\cos^2\theta}\right)}=\dfrac{\cos^2\theta-\sin^2\theta}{\left(\dfrac{\cos^2\theta-\sin^2\theta}{\cos^2\theta}\right)}=\cos^2\theta$

67. $\dfrac{(2\cos^2\theta-1)^2}{\cos^4\theta-\sin^4\theta}=\dfrac{\left[2\cos^2\theta-(\sin^2\theta+\cos^2\theta)\right]^2}{(\cos^2\theta-\sin^2\theta)(\cos^2\theta+\sin^2\theta)}$

$\qquad =\dfrac{(\cos^2\theta-\sin^2\theta)^2}{(\cos^2\theta-\sin^2\theta)(\cos^2\theta+\sin^2\theta)}=\dfrac{\cos^2\theta-\sin^2\theta}{\cos^2\theta+\sin^2\theta}$

$\qquad =\cos^2\theta-\sin^2\theta=1-\sin^2\theta-\sin^2\theta=1-2\sin^2\theta$

69. $\dfrac{1+\sin\theta+\cos\theta}{1+\sin\theta-\cos\theta}=\dfrac{(1+\sin\theta)+\cos\theta}{(1+\sin\theta)-\cos\theta}\cdot\dfrac{(1+\sin\theta)+\cos\theta}{(1+\sin\theta)+\cos\theta}$

$\qquad =\dfrac{1+2\sin\theta+\sin^2\theta+2\cos\theta(1+\sin\theta)+\cos^2\theta}{1+2\sin\theta+\sin^2\theta-\cos^2\theta}$

$\qquad =\dfrac{1+2\sin\theta+\sin^2\theta+2\cos\theta(1+\sin\theta)+(1-\sin^2\theta)}{1+2\sin\theta+\sin^2\theta-(1-\sin^2\theta)}$

$\qquad =\dfrac{2+2\sin\theta+2\cos\theta(1+\sin\theta)}{2\sin 0+2\sin^2\theta}=\dfrac{2(1+\sin\theta)+2\cos\theta(1+\sin\theta)}{2\sin\theta(1+\sin\theta)}$

$\qquad =\dfrac{2(1+\sin\theta)(1+\cos\theta)}{2\sin\theta(1+\sin\theta)}=\dfrac{1+\cos\theta}{\sin\theta}$

71. $(a\sin\theta + b\cos\theta)^2 + (a\cos\theta - b\sin\theta)^2$

$\qquad = a^2\sin^2\theta + 2ab\sin\theta\cos\theta + b^2\cos^2\theta + a^2\cos^2\theta - 2ab\sin\theta\cos\theta + b^2\sin^2\theta$

$\qquad = a^2(\sin^2\theta + \cos^2\theta) + b^2(\sin^2\theta + \cos^2\theta) = a^2 + b^2$

73. $\dfrac{\tan\alpha + \tan\beta}{\cot\alpha + \cot\beta} = \dfrac{\tan\alpha + \tan\beta}{\left(\dfrac{1}{\tan\alpha} + \dfrac{1}{\tan\beta}\right)} = \dfrac{\tan\alpha + \tan\beta}{\left(\dfrac{\tan\beta + \tan\alpha}{\tan\alpha\tan\beta}\right)}$

$\qquad\qquad = (\tan\alpha + \tan\beta) \cdot \left(\dfrac{\tan\alpha\tan\beta}{\tan\alpha + \tan\beta}\right) = \tan\alpha\tan\beta$

75. $(\sin\alpha + \cos\beta)^2 + (\cos\beta + \sin\alpha)(\cos\beta - \sin\alpha)$

$\qquad\qquad = \sin^2\alpha + 2\sin\alpha\cos\beta + \cos^2\beta + \cos^2\beta - \sin^2\alpha$

$\qquad\qquad = 2\sin\alpha\cos\beta + 2\cos^2\beta = 2\cos\beta(\sin\alpha + \cos\beta)$

77. $\ln\left|\sec\theta\right| = \ln\left|\dfrac{1}{\cos\theta}\right| = \ln\left|\cos\theta\right|^{-1} = -\ln\left|\cos\theta\right|$

79. $\ln\left|1 + \cos\theta\right| + \ln\left|1 - \cos\theta\right| = \ln\left(\left|1 + \cos\theta\right| \cdot \left|1 - \cos\theta\right|\right) = \ln\left|1 - \cos^2\theta\right|$

$\qquad\qquad = \ln\left|\sin^2\theta\right| = 2\ln\left|\sin\theta\right|$

81. Show that $\sec\left(\tan^{-1}v\right) = \sqrt{1 + v^2}$.

Let $\alpha = \tan^{-1}v$. Then $\tan\alpha = v$, $-\dfrac{\pi}{2} < \alpha < \dfrac{\pi}{2}$.

$\qquad \sec\left(\tan^{-1}v\right) = \sec\alpha = \sqrt{1 + \tan^2\alpha} = \sqrt{1 + v^2}$

83. Show that $\tan\left(\cos^{-1}v\right) = \dfrac{\sqrt{1 - v^2}}{v}$.

Let $\alpha = \cos^{-1}v$. Then $\cos\alpha = v$, $0 \le \alpha \le \pi$.

$\qquad \tan\left(\cos^{-1}v\right) = \tan\alpha = \dfrac{\sin\alpha}{\cos\alpha} = \dfrac{\sqrt{1 - \cos^2\alpha}}{\cos\alpha} = \dfrac{\sqrt{1 - v^2}}{v}$

85. Show that $\cos\left(\sin^{-1}v\right) = \sqrt{1 - v^2}$.

Let $\alpha = \sin^{-1}v$. Then $\sin\alpha = v$, $-\dfrac{\pi}{2} \le \alpha \le \dfrac{\pi}{2}$.

$\qquad \cos\left(\sin^{-1}v\right) = \cos\alpha = \sqrt{1 - \sin^2\alpha} = \sqrt{1 - v^2}$

Analytic Trigonometry

6.4   Sum and Difference Formulas

1. $\sin\left(\dfrac{5\pi}{12}\right) = \sin\left(\dfrac{3\pi}{12} + \dfrac{2\pi}{12}\right) = \sin\left(\dfrac{\pi}{4}\right)\cos\left(\dfrac{\pi}{6}\right) + \cos\left(\dfrac{\pi}{4}\right)\sin\left(\dfrac{\pi}{6}\right) = \dfrac{\sqrt{2}}{2}\cdot\dfrac{\sqrt{3}}{2} + \dfrac{\sqrt{2}}{2}\cdot\dfrac{1}{2}$

$\qquad = \dfrac{1}{4}\left(\sqrt{6} + \sqrt{2}\right)$

3. $\cos\left(\dfrac{7\pi}{12}\right) = \cos\left(\dfrac{4\pi}{12} + \dfrac{3\pi}{12}\right) = \cos\left(\dfrac{\pi}{3}\right)\cos\left(\dfrac{\pi}{4}\right) - \sin\left(\dfrac{\pi}{3}\right)\sin\left(\dfrac{\pi}{4}\right) = \dfrac{1}{2}\cdot\dfrac{\sqrt{2}}{2} - \dfrac{\sqrt{3}}{2}\cdot\dfrac{\sqrt{2}}{2}$

$\qquad = \dfrac{1}{4}\left(\sqrt{2} - \sqrt{6}\right)$

5. $\cos(165°) = \cos(120° + 45°) = \cos(120°)\cos(45°) - \sin(120°)\sin(45°)$

$\qquad = -\dfrac{1}{2}\cdot\dfrac{\sqrt{2}}{2} - \dfrac{\sqrt{3}}{2}\cdot\dfrac{\sqrt{2}}{2} = -\dfrac{1}{4}\left(\sqrt{2} + \sqrt{6}\right)$

7. $\tan(15°) = \tan(45° - 30°) = \dfrac{\tan(45°) - \tan(30°)}{1 + \tan(45°)\tan(30°)} = \dfrac{\left(1 - \dfrac{\sqrt{3}}{3}\right)}{\left(1 + 1\cdot\dfrac{\sqrt{3}}{3}\right)} = \dfrac{\left(\dfrac{3 - \sqrt{3}}{3}\right)}{\left(\dfrac{3 + \sqrt{3}}{3}\right)}$

$\qquad = \left(\dfrac{3 - \sqrt{3}}{3 + \sqrt{3}}\right)\cdot\left(\dfrac{3 - \sqrt{3}}{3 - \sqrt{3}}\right) = \dfrac{9 - 6\sqrt{3} + 3}{9 - 3} = \dfrac{12 - 6\sqrt{3}}{6} = \dfrac{6\left(2 - \sqrt{3}\right)}{6} = 2 - \sqrt{3}$

9. $\sin\left(\dfrac{17\pi}{12}\right) = \sin\left(\dfrac{15\pi}{12} + \dfrac{2\pi}{12}\right) = \sin\left(\dfrac{5\pi}{4}\right)\cos\left(\dfrac{\pi}{6}\right) + \cos\left(\dfrac{5\pi}{4}\right)\sin\left(\dfrac{\pi}{6}\right) = -\dfrac{\sqrt{2}}{2}\cdot\dfrac{\sqrt{3}}{2} + -\dfrac{\sqrt{2}}{2}\cdot\dfrac{1}{2}$

$\qquad = -\dfrac{1}{4}\left(\sqrt{6} + \sqrt{2}\right)$

11. $\sec\left(-\dfrac{\pi}{12}\right) = \dfrac{1}{\cos\left(-\dfrac{\pi}{12}\right)} = \dfrac{1}{\cos\left(\dfrac{3\pi}{12} - \dfrac{4\pi}{12}\right)} = \dfrac{1}{\cos\left(\dfrac{\pi}{4}\right)\cos\left(\dfrac{\pi}{3}\right) + \sin\left(\dfrac{\pi}{4}\right)\sin\left(\dfrac{\pi}{3}\right)}$

$$= \frac{1}{\left(\dfrac{\sqrt{2}}{2}\cdot\dfrac{1}{2}+\dfrac{\sqrt{2}}{2}\cdot\dfrac{\sqrt{3}}{2}\right)} = \frac{1}{\left(\dfrac{\sqrt{2}+\sqrt{6}}{4}\right)} = \left(\dfrac{4}{\sqrt{2}+\sqrt{6}}\right)\cdot\left(\dfrac{\sqrt{2}-\sqrt{6}}{\sqrt{2}-\sqrt{6}}\right)$$

$$= \frac{4\left(\sqrt{2}-\sqrt{6}\right)}{2-6} = \frac{4\left(\sqrt{2}-\sqrt{6}\right)}{-4} = -\left(\sqrt{2}-\sqrt{6}\right) = \sqrt{6}-\sqrt{2}$$

13. $\sin(20°)\cos(10°) + \cos(20°)\sin(10°) = \sin(20°+10°) = \sin(30°) = \dfrac{1}{2}$

15. $\cos(70°)\cos(20°) - \sin(70°)\sin(20°) = \cos(70°+20°) = \cos(90°) = 0$

17. $\dfrac{\tan(20°)+\tan(25°)}{1-\tan(20°)\tan(25°)} = \tan(20°+25°) = \tan(45°) = 1$

19. $\sin\left(\dfrac{\pi}{12}\right)\cos\left(\dfrac{7\pi}{12}\right) - \cos\left(\dfrac{\pi}{12}\right)\sin\left(\dfrac{7\pi}{12}\right) = \sin\left(\dfrac{\pi}{12}-\dfrac{7\pi}{12}\right) = \sin\left(-\dfrac{\pi}{2}\right) = -1$

21. $\cos\left(\dfrac{\pi}{12}\right)\cos\left(\dfrac{5\pi}{12}\right) + \sin\left(\dfrac{5\pi}{12}\right)\sin\left(\dfrac{\pi}{12}\right) = \cos\left(\dfrac{\pi}{12}-\dfrac{5\pi}{12}\right) = \cos\left(-\dfrac{\pi}{3}\right) = \cos\left(\dfrac{\pi}{3}\right) = \dfrac{1}{2}$

23. $\sin\alpha = \dfrac{3}{5},\ 0<\alpha<\dfrac{\pi}{2};$ $\qquad\qquad \cos\beta = \dfrac{2\sqrt{5}}{5},\ -\dfrac{\pi}{2}<\beta<0$

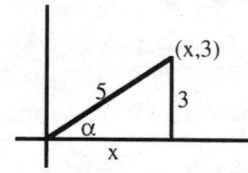   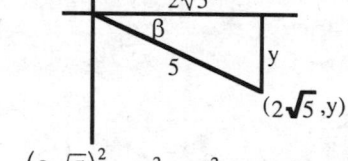

$x^2 + 3^2 = 5^2,\ x>0$ $\qquad\qquad\quad \left(2\sqrt{5}\right)^2 + y^2 = 5^2,\ y<0$

$\quad x^2 = 25-9 = 16,\ x>0$ $\qquad\qquad\quad y^2 = 25-20 = 5.\ y<0$

$\qquad\quad x = 4$ $\qquad\qquad\qquad\qquad\qquad y = -\sqrt{5}$

$\cos\alpha = \dfrac{4}{5},\ \tan\alpha = \dfrac{3}{4}$ $\qquad\qquad \sin\beta = -\dfrac{\sqrt{5}}{5},\ \tan\beta = \dfrac{-\sqrt{5}}{2\sqrt{5}} = -\dfrac{1}{2}$

(a) $\sin(\alpha+\beta) = \sin\alpha\cos\beta + \cos\alpha\sin\beta = \dfrac{3}{5}\cdot\dfrac{2\sqrt{5}}{5} + \dfrac{4}{5}\cdot-\dfrac{\sqrt{5}}{5} = \dfrac{6\sqrt{5}-4\sqrt{5}}{25} = \dfrac{2\sqrt{5}}{25}$

(b) $\cos(\alpha+\beta) = \cos\alpha\cos\beta - \sin\alpha\sin\beta = \dfrac{4}{5}\cdot\dfrac{2\sqrt{5}}{5} - \dfrac{3}{5}\cdot-\dfrac{\sqrt{5}}{5} = \dfrac{8\sqrt{5}+3\sqrt{5}}{25} = \dfrac{11\sqrt{5}}{25}$

(c) $\sin(\alpha-\beta) = \sin\alpha\cos\beta - \cos\alpha\sin\beta = \dfrac{3}{5}\cdot\dfrac{2\sqrt{5}}{5} - \dfrac{4}{5}\cdot-\dfrac{\sqrt{5}}{5} = \dfrac{6\sqrt{5}+4\sqrt{5}}{25}$

$\qquad = \dfrac{10\sqrt{5}}{25} = \dfrac{2\sqrt{5}}{5}$

(d) $\tan(\alpha - \beta) = \dfrac{\tan\alpha - \tan\beta}{1 + \tan\alpha\tan\beta} = \dfrac{\left(\dfrac{3}{4} - \left(-\dfrac{1}{2}\right)\right)}{\left(1 + \left(\dfrac{3}{4}\right)\cdot\left(-\dfrac{1}{2}\right)\right)} = \dfrac{\left(\dfrac{5}{4}\right)}{\left(\dfrac{5}{8}\right)} = 2$

25. $\tan\alpha = -\dfrac{4}{3}, \ \dfrac{\pi}{2} < \alpha < \pi;$ 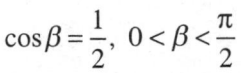 $\qquad \cos\beta = \dfrac{1}{2}, \ 0 < \beta < \dfrac{\pi}{2}$

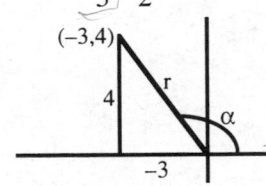
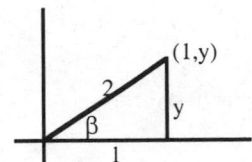

$r^2 = (-3)^2 + 4^2 = 25$

$r = 5$

$\sin\alpha = \dfrac{4}{5}, \ \cos\alpha = \dfrac{-3}{5} = -\dfrac{3}{5}$

$1^2 + y^2 = 2^2, \ \ y > 0$

$y^2 = 4 - 1 = 3. \ \ y > 0$

$y = \sqrt{3}$

$\sin\beta = \dfrac{\sqrt{3}}{2}, \quad \tan\beta = \dfrac{\sqrt{3}}{1} = \sqrt{3}$

(a) $\sin(\alpha + \beta) = \sin\alpha\cos\beta + \cos\alpha\sin\beta = \left(\dfrac{4}{5}\right)\cdot\left(\dfrac{1}{2}\right) + \left(-\dfrac{3}{5}\right)\cdot\left(\dfrac{\sqrt{3}}{2}\right) = \dfrac{4 - 3\sqrt{3}}{10}$

(b) $\cos(\alpha + \beta) = \cos\alpha\cos\beta - \sin\alpha\sin\beta = \left(-\dfrac{3}{5}\right)\cdot\left(\dfrac{1}{2}\right) - \left(\dfrac{4}{5}\right)\cdot\left(\dfrac{\sqrt{3}}{2}\right) = \dfrac{-3 - 4\sqrt{3}}{10}$

(c) $\sin(\alpha - \beta) = \sin\alpha\cos\beta - \cos\alpha\sin\beta = \left(\dfrac{4}{5}\right)\cdot\left(\dfrac{1}{2}\right) - \left(-\dfrac{3}{5}\right)\cdot\left(\dfrac{\sqrt{3}}{2}\right) = \dfrac{4 + 3\sqrt{3}}{10}$

(d) $\tan(\alpha - \beta) = \dfrac{\tan\alpha - \tan\beta}{1 + \tan\alpha\tan\beta} = \dfrac{\left(-\dfrac{4}{3} - \sqrt{3}\right)}{\left(1 + \left(-\dfrac{4}{3}\right)\cdot\sqrt{3}\right)} = \dfrac{\left(\dfrac{-4 - 3\sqrt{3}}{3}\right)}{\left(\dfrac{3 - 4\sqrt{3}}{3}\right)} = \left(\dfrac{-4 - 3\sqrt{3}}{3 - 4\sqrt{3}}\right)\cdot\left(\dfrac{3 + 4\sqrt{3}}{3 + 4\sqrt{3}}\right)$

$= \dfrac{-48 - 25\sqrt{3}}{-39} = \dfrac{48 + 25\sqrt{3}}{39}$

27. $\sin\alpha = \dfrac{5}{13}, \ -\dfrac{3\pi}{2} < \alpha < -\pi; \qquad \tan\beta = -\sqrt{3}, \ \dfrac{\pi}{2} < \beta < \pi$

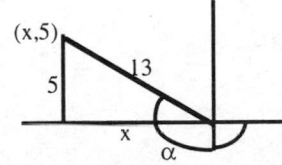
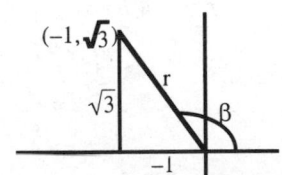

$$x^2 + 5^2 = 13^2, \ x < 0 \qquad\qquad r^2 = (-1)^2 + \sqrt{3}^2 = 4$$

$$x^2 = 169 - 25 = 144, \ x < 0 \qquad\qquad r = 2$$

$$x = -12$$

$$\cos\alpha = \frac{-12}{13} = -\frac{12}{13}, \ \tan\alpha = -\frac{5}{12} \qquad\qquad \sin\beta = \frac{\sqrt{3}}{2}, \ \cos\beta = \frac{-1}{2} = -\frac{1}{2}$$

(a) $\sin(\alpha + \beta) = \sin\alpha\cos\beta + \cos\alpha\sin\beta = \left(\dfrac{5}{13}\right)\cdot\left(-\dfrac{1}{2}\right) + \left(-\dfrac{12}{13}\right)\cdot\left(\dfrac{\sqrt{3}}{2}\right) = \dfrac{-5 - 12\sqrt{3}}{26}$

(b) $\cos(\alpha + \beta) = \cos\alpha\cos\beta - \sin\alpha\sin\beta = \left(-\dfrac{12}{13}\right)\cdot\left(-\dfrac{1}{2}\right) - \left(\dfrac{5}{13}\right)\cdot\left(\dfrac{\sqrt{3}}{2}\right) = \dfrac{12 - 5\sqrt{3}}{26}$

(c) $\sin(\alpha - \beta) = \sin\alpha\cos\beta - \cos\alpha\sin\beta = \left(\dfrac{5}{13}\right)\cdot\left(-\dfrac{1}{2}\right) - \left(-\dfrac{12}{13}\right)\cdot\left(\dfrac{\sqrt{3}}{2}\right) = \dfrac{-5 + 12\sqrt{3}}{26}$

(d) $\tan(\alpha - \beta) = \dfrac{\tan\alpha - \tan\beta}{1 + \tan\alpha\tan\beta} = \dfrac{\left(-\dfrac{5}{12} - (-\sqrt{3})\right)}{\left(1 + \left(-\dfrac{5}{12}\right)\cdot(-\sqrt{3})\right)} = \dfrac{\left(\dfrac{-5 + 12\sqrt{3}}{12}\right)}{\left(\dfrac{12 + 5\sqrt{3}}{12}\right)}$

$$= \left(\dfrac{-5 + 12\sqrt{3}}{12 + 5\sqrt{3}}\right)\left(\dfrac{12 - 5\sqrt{3}}{12 - 5\sqrt{3}}\right) = \dfrac{-240 + 169\sqrt{3}}{69}$$

29. $\sin\theta = \dfrac{1}{3}, \quad \theta$ in quadrant II

(a) $\cos\theta = -\sqrt{1 - \sin^2\theta} = -\sqrt{1 - \left(\dfrac{1}{3}\right)^2} = -\sqrt{1 - \dfrac{1}{9}} = -\sqrt{\dfrac{8}{9}} = -\dfrac{2\sqrt{2}}{3}$

(b) $\sin\left(\theta + \dfrac{\pi}{6}\right) = \sin\theta\cos\left(\dfrac{\pi}{6}\right) + \cos\theta\sin\left(\dfrac{\pi}{6}\right) = \left(\dfrac{1}{3}\right)\cdot\left(\dfrac{\sqrt{3}}{2}\right) + \left(-\dfrac{2\sqrt{2}}{3}\right)\cdot\left(\dfrac{1}{2}\right) = \dfrac{\sqrt{3} - 2\sqrt{2}}{6}$

(c) $\cos\left(\theta - \dfrac{\pi}{3}\right) = \cos\theta\cos\left(\dfrac{\pi}{3}\right) + \sin\theta\sin\left(\dfrac{\pi}{3}\right) = \left(-\dfrac{2\sqrt{2}}{3}\right)\cdot\left(\dfrac{1}{2}\right) + \left(\dfrac{1}{3}\right)\cdot\left(\dfrac{\sqrt{3}}{2}\right) = \dfrac{-2\sqrt{2} + \sqrt{3}}{6}$

(d) $\tan\left(\theta + \dfrac{\pi}{4}\right) = \dfrac{\tan\theta + \tan\left(\dfrac{\pi}{4}\right)}{1 - \tan\theta\tan\left(\dfrac{\pi}{4}\right)} = \dfrac{\left(-\dfrac{1}{2\sqrt{2}} + 1\right)}{\left(1 - \left(-\dfrac{1}{2\sqrt{2}}\right)\cdot 1\right)} = \dfrac{\left(\dfrac{-1 + 2\sqrt{2}}{2\sqrt{2}}\right)}{\left(\dfrac{2\sqrt{2} + 1}{2\sqrt{2}}\right)}$

$$= \left(\dfrac{2\sqrt{2} - 1}{2\sqrt{2} + 1}\right)\cdot\left(\dfrac{2\sqrt{2} - 1}{2\sqrt{2} - 1}\right) = \dfrac{9 - 4\sqrt{2}}{7}$$

31. $\sin\left(\dfrac{\pi}{2} + \theta\right) = \sin\left(\dfrac{\pi}{2}\right)\cos\theta + \cos\left(\dfrac{\pi}{2}\right)\sin\theta = 1\cdot\cos\theta + 0\cdot\sin\theta = \cos\theta$

33. $\sin(\pi - \theta) = \sin(\pi)\cos\theta - \cos(\pi)\sin\theta = 0 \cdot \cos\theta - (-1)\sin\theta = \sin\theta$

35. $\sin(\pi + \theta) = \sin(\pi)\cos\theta + \cos(\pi)\sin\theta = 0 \cdot \cos\theta + (-1)\sin\theta = -\sin\theta$

37. $\tan(\pi - \theta) = \dfrac{\tan(\pi) - \tan\theta}{1 + \tan(\pi)\tan\theta} = \dfrac{0 - \tan\theta}{1 + 0 \cdot \tan\theta} = \dfrac{-\tan\theta}{1} = -\tan\theta$

39. $\sin\left(\dfrac{3\pi}{2} + \theta\right) = \sin\left(\dfrac{3\pi}{2}\right)\cos\theta + \cos\left(\dfrac{3\pi}{2}\right)\sin\theta = -1 \cdot \cos\theta + 0 \cdot \sin\theta = -\cos\theta$

41. $\sin(\alpha + \beta) + \sin(\alpha - \beta) = \sin\alpha\cos\beta + \cos\alpha\sin\beta + \sin\alpha\cos\beta - \cos\alpha\sin\beta$
$$= 2\sin\alpha\cos\beta$$

43. $\dfrac{\sin(\alpha + \beta)}{\sin\alpha\cos\beta} = \dfrac{\sin\alpha\cos\beta + \cos\alpha\sin\beta}{\sin\alpha\cos\beta} = \dfrac{\sin\alpha\cos\beta}{\sin\alpha\cos\beta} + \dfrac{\cos\alpha\sin\beta}{\sin\alpha\cos\beta} = 1 + \cot\alpha\tan\beta$

45. $\dfrac{\cos(\alpha + \beta)}{\cos\alpha\cos\beta} = \dfrac{\cos\alpha\cos\beta - \sin\alpha\sin\beta}{\cos\alpha\cos\beta} = \dfrac{\cos\alpha\cos\beta}{\cos\alpha\cos\beta} - \dfrac{\sin\alpha\sin\beta}{\cos\alpha\cos\beta} = 1 - \tan\alpha\tan\beta$

47. $\dfrac{\sin(\alpha + \beta)}{\sin(\alpha - \beta)} = \dfrac{\sin\alpha\cos\beta + \cos\alpha\sin\beta}{\sin\alpha\cos\beta - \cos\alpha\sin\beta} = \dfrac{\left(\dfrac{\sin\alpha\cos\beta}{\cos\alpha\cos\beta} + \dfrac{\cos\alpha\sin\beta}{\cos\alpha\cos\beta}\right)}{\left(\dfrac{\sin\alpha\cos\beta}{\cos\alpha\cos\beta} - \dfrac{\cos\alpha\sin\beta}{\cos\alpha\cos\beta}\right)} = \dfrac{\tan\alpha + \tan\beta}{\tan\alpha - \tan\beta}$

49. $\cot(\alpha + \beta) = \dfrac{\cos(\alpha + \beta)}{\sin(\alpha + \beta)} = \dfrac{\cos\alpha\cos\beta - \sin\alpha\sin\beta}{\sin\alpha\cos\beta + \cos\alpha\sin\beta}$

$$= \dfrac{\left(\dfrac{\cos\alpha\cos\beta}{\sin\alpha\sin\beta} - \dfrac{\sin\alpha\sin\beta}{\sin\alpha\sin\beta}\right)}{\left(\dfrac{\sin\alpha\cos\beta}{\sin\alpha\sin\beta} + \dfrac{\cos\alpha\sin\beta}{\sin\alpha\sin\beta}\right)} = \dfrac{\cot\alpha\cot\beta - 1}{\cot\beta + \cot\alpha}$$

51. $\sec(\alpha + \beta) = \dfrac{1}{\cos(\alpha + \beta)} = \dfrac{1}{\cos\alpha\cos\beta - \sin\alpha\sin\beta}$

$$= \dfrac{\left(\dfrac{1}{\sin\alpha\sin\beta}\right)}{\left(\dfrac{\cos\alpha\cos\beta}{\sin\alpha\sin\beta} - \dfrac{\sin\alpha\sin\beta}{\sin\alpha\sin\beta}\right)} = \dfrac{\csc\alpha\csc\beta}{\cot\alpha\cot\beta - 1}$$

53. $\sin(\alpha - \beta)\sin(\alpha + \beta) = (\sin\alpha\cos\beta - \cos\alpha\sin\beta)(\sin\alpha\cos\beta + \cos\alpha\sin\beta)$
$$= \sin^2\alpha\cos^2\beta - \cos^2\alpha\sin^2\beta = \sin^2\alpha(1 - \sin^2\beta) - (1 - \sin^2\alpha)\sin^2\beta$$
$$\sin^2\alpha - \sin^2\alpha\sin^2\beta - \sin^2\beta + \sin^2\alpha\sin^2\beta = \sin^2\alpha - \sin^2\beta$$

55. $\sin(\theta + k\pi) = \sin\theta\cos(k\pi) + \cos\theta\sin(k\pi) = \sin\theta(-1)^k + \cos\theta \cdot 0$

          $= (-1)^k \sin\theta, \; k$ any integer

57. $\sin\left(\sin^{-1}\left(\dfrac{1}{2}\right) + \cos^{-1}(0)\right) = \sin\left(\dfrac{\pi}{6} + \dfrac{\pi}{2}\right) = \sin\left(\dfrac{\pi}{6}\right)\cos\left(\dfrac{\pi}{2}\right) + \cos\left(\dfrac{\pi}{6}\right)\sin\left(\dfrac{\pi}{2}\right) = \dfrac{1}{2}\cdot 0 + \dfrac{\sqrt{3}}{2}\cdot 1 = \dfrac{\sqrt{3}}{2}$

59. $\sin\left[\sin^{-1}\left(\dfrac{3}{5}\right) - \cos^{-1}\left(-\dfrac{4}{5}\right)\right]$

Let $\alpha = \sin^{-1}\left(\dfrac{3}{5}\right)$ and $\beta = \cos^{-1}\left(-\dfrac{4}{5}\right)$. α is in quadrant I; β is in quadrant II.

Then $\sin\alpha = \dfrac{3}{5}, \; -\dfrac{\pi}{2} \leq \alpha \leq \dfrac{\pi}{2}$, and $\cos\beta = -\dfrac{4}{5}, 0 \leq \beta \leq \pi$.

$$\cos\alpha = \sqrt{1 - \sin^2\alpha} = \sqrt{1 - \left(\dfrac{3}{5}\right)^2} = \sqrt{1 - \dfrac{9}{25}} = \sqrt{\dfrac{16}{25}} = \dfrac{4}{5}$$

$$\sin\beta = \sqrt{1 - \cos^2\beta} = \sqrt{1 - \left(-\dfrac{4}{5}\right)^2} = \sqrt{1 - \dfrac{16}{25}} = \sqrt{\dfrac{9}{25}} = \dfrac{3}{5}$$

$$\sin\left[\sin^{-1}\left(\dfrac{3}{5}\right) - \cos^{-1}\left(-\dfrac{4}{5}\right)\right] = \sin(\alpha - \beta) = \sin\alpha\cos\beta - \cos\alpha\sin\beta$$

$$= \left(\dfrac{3}{5}\right)\cdot\left(-\dfrac{4}{5}\right) - \left(\dfrac{4}{5}\right)\cdot\left(\dfrac{3}{5}\right) = -\dfrac{12}{25} - \dfrac{12}{25} = -\dfrac{24}{25}$$

61. $\cos\left(\tan^{-1}\left(\dfrac{4}{3}\right) + \cos^{-1}\left(\dfrac{5}{13}\right)\right)$

Let $\alpha = \tan^{-1}\left(\dfrac{4}{3}\right)$ and $\beta = \cos^{-1}\left(\dfrac{5}{13}\right)$. α is in quadrant I; β is in quadrant I.

Then $\tan\alpha = \dfrac{4}{3}, \; -\dfrac{\pi}{2} < \alpha < \dfrac{\pi}{2}$, and $\cos\beta = \dfrac{5}{13}, 0 \leq \beta \leq \pi$.

$$\sec\alpha = \sqrt{1 + \tan^2\alpha} = \sqrt{1 + \left(\dfrac{4}{3}\right)^2} = \sqrt{1 + \dfrac{16}{9}} = \sqrt{\dfrac{25}{9}} = \dfrac{5}{3}; \;\; \cos\alpha = \dfrac{3}{5}$$

$$\sin\alpha = \sqrt{1 - \cos^2\alpha} = \sqrt{1 - \left(\dfrac{3}{5}\right)^2} = \sqrt{1 - \dfrac{9}{25}} = \sqrt{\dfrac{16}{25}} = \dfrac{4}{5}$$

$$\sin\beta = \sqrt{1 - \cos^2\beta} = \sqrt{1 - \left(\dfrac{5}{13}\right)^2} = \sqrt{1 - \dfrac{25}{169}} = \sqrt{\dfrac{144}{169}} = \dfrac{12}{13}$$

$$\cos\left(\tan^{-1}\left(\dfrac{4}{3}\right) + \cos^{-1}\left(\dfrac{5}{13}\right)\right) = \cos(\alpha + \beta) = \cos\alpha\cos\beta - \sin\alpha\sin\beta$$

$$= \left(\dfrac{3}{5}\right)\cdot\left(\dfrac{5}{13}\right) - \left(\dfrac{4}{5}\right)\cdot\left(\dfrac{12}{13}\right) = \dfrac{15}{65} - \dfrac{48}{65} = -\dfrac{33}{65}$$

63. $\cos\left(\sin^{-1}\left(\dfrac{5}{13}\right)-\tan^{-1}\left(\dfrac{3}{4}\right)\right)$

Let $\alpha=\sin^{-1}\left(\dfrac{5}{13}\right)$ and $\beta=\tan^{-1}\left(\dfrac{3}{4}\right)$. α is in quadrant I; β is in quadrant I.

Then $\sin\alpha=\dfrac{5}{13}$, $-\dfrac{\pi}{2}\le\alpha\le\dfrac{\pi}{2}$, and $\tan\beta=\dfrac{3}{4}$, $-\dfrac{\pi}{2}<\beta<\dfrac{\pi}{2}$.

$$\cos\alpha=\sqrt{1-\sin^2\alpha}=\sqrt{1-\left(\dfrac{5}{13}\right)^2}=\sqrt{1-\dfrac{25}{169}}=\sqrt{\dfrac{144}{169}}=\dfrac{12}{13}$$

$$\sec\beta=\sqrt{1+\tan^2\beta}=\sqrt{1+\left(\dfrac{3}{4}\right)^2}=\sqrt{1+\dfrac{9}{16}}=\sqrt{\dfrac{25}{16}}=\dfrac{5}{4};\ \ \cos\beta=\dfrac{4}{5}$$

$$\sin\beta=\sqrt{1-\cos^2\beta}=\sqrt{1-\left(\dfrac{4}{5}\right)^2}=\sqrt{1-\dfrac{16}{25}}=\sqrt{\dfrac{9}{25}}=\dfrac{3}{5}$$

$$\cos\left[\sin^{-1}\left(\dfrac{5}{13}\right)-\tan^{-1}\left(\dfrac{3}{4}\right)\right]=\cos(\alpha-\beta)$$

$$=\cos\alpha\cos\beta+\sin\alpha\sin\beta=\dfrac{12}{13}\cdot\dfrac{4}{5}+\dfrac{5}{13}\cdot\dfrac{3}{5}=\dfrac{48}{65}+\dfrac{15}{65}=\dfrac{63}{65}$$

65. $\tan\left(\sin^{-1}\left(\dfrac{3}{5}\right)+\dfrac{\pi}{6}\right)$

Let $\alpha=\sin^{-1}\left(\dfrac{3}{5}\right)$. α is in quadrant I.

Then $\sin\alpha=\dfrac{3}{5}$, $0\le\alpha\le\pi$.

$$\cos\alpha=\sqrt{1-\sin^2\alpha}=\sqrt{1-\left(\dfrac{3}{5}\right)^2}=\sqrt{1-\dfrac{9}{25}}=\sqrt{\dfrac{16}{25}}=\dfrac{4}{5}$$

$$\tan\alpha=\dfrac{\sin\alpha}{\cos\alpha}=\dfrac{\left(\dfrac{3}{5}\right)}{\left(\dfrac{4}{5}\right)}=\dfrac{3}{5}\cdot\dfrac{5}{4}=\dfrac{3}{4}$$

$$\tan\left(\sin^{-1}\left(\dfrac{3}{5}\right)+\dfrac{\pi}{6}\right)==\dfrac{\tan\left(\sin^{-1}\left(\dfrac{3}{5}\right)\right)+\tan\left(\dfrac{\pi}{6}\right)}{1-\tan\left(\sin^{-1}\left(\dfrac{3}{5}\right)\right)\tan\left(\dfrac{\pi}{6}\right)}=\dfrac{\left(\dfrac{3}{4}+\dfrac{\sqrt{3}}{3}\right)}{\left(1-\dfrac{3}{4}\cdot\dfrac{\sqrt{3}}{3}\right)}$$

$$=\dfrac{\left(\dfrac{9+4\sqrt{3}}{12}\right)}{\left(\dfrac{12-3\sqrt{3}}{12}\right)}=\left(\dfrac{9+4\sqrt{3}}{12}\right)\left(\dfrac{12}{12-3\sqrt{3}}\right)=\dfrac{9+4\sqrt{3}}{12-3\sqrt{3}}$$

67. $\tan\left(\sin^{-1}\left(\dfrac{4}{5}\right)+\cos^{-1}(1)\right)$

Let $\alpha=\sin^{-1}\left(\dfrac{4}{5}\right)$ and $\beta=\cos^{-1}(1)$; α is in quadrant I.

Then $\sin\alpha=\dfrac{4}{5},\ -\dfrac{\pi}{2}\le\alpha\le\dfrac{\pi}{2}$ and $\cos\beta=1,\ 0\le\beta\le\pi..$

$\cos\beta=1,\ 0\le\beta\le\pi\Rightarrow\beta=0\ \therefore\cos^{-1}1=0$

$\cos\alpha=\sqrt{1-\sin^2\alpha}=\sqrt{1-\left(\dfrac{4}{5}\right)^2}=\sqrt{1-\dfrac{16}{25}}=\sqrt{\dfrac{9}{25}}=\dfrac{3}{5}$

$\tan\alpha=\dfrac{\sin\alpha}{\cos\alpha}=\dfrac{\left(\dfrac{4}{5}\right)}{\left(\dfrac{3}{5}\right)}=\dfrac{4}{5}\cdot\dfrac{5}{3}=\dfrac{4}{3}$

$\tan\left(\sin^{-1}\left(\dfrac{4}{5}\right)+\cos^{-1}(1)\right)=\dfrac{\tan\left(\sin^{-1}\left(\dfrac{4}{5}\right)\right)+\tan\left(\cos^{-1}(1)\right)}{1-\tan\left(\sin^{-1}\left(\dfrac{4}{5}\right)\right)\tan\left(\cos^{-1}(1)\right)}=\dfrac{\left(\dfrac{4}{3}+0\right)}{\left(1-\dfrac{4}{3}\cdot0\right)}=\dfrac{\left(\dfrac{4}{3}\right)}{(1)}=\dfrac{4}{3}$

69. $\cos\left(\cos^{-1}u+\sin^{-1}v\right)$

Let $\alpha=\cos^{-1}u$ and $\beta=\sin^{-1}v$.

Then $\cos\alpha=u,\ 0\le\alpha\le\pi$, and $\sin\beta=v,\ -\dfrac{\pi}{2}\le\beta\le\dfrac{\pi}{2}$

$\sin\alpha=\sqrt{1-\cos^2\alpha}=\sqrt{1-u^2}$

$\cos\beta=\sqrt{1-\sin^2\beta}=\sqrt{1-v^2}$

$\cos\left(\cos^{-1}u+\sin^{-1}v\right)=\cos(\alpha+\beta)=\cos\alpha\cos\beta-\sin\alpha\sin\beta=u\sqrt{1-v^2}-v\sqrt{1-u^2}$

71. $\sin\left(\tan^{-1}u-\sin^{-1}v\right)$

Let $\alpha=\tan^{-1}u$ and $\beta=\sin^{-1}v$.

Then $\tan\alpha=u,\ -\dfrac{\pi}{2}<\alpha<\dfrac{\pi}{2}$, and $\sin\beta=v,\ -\dfrac{\pi}{2}\le\beta\le\dfrac{\pi}{2}$

$\sec\alpha=\sqrt{\tan^2\alpha+1}=\sqrt{u^2+1};\quad\cos\alpha=\dfrac{1}{\sqrt{u^2+1}}$

$\sin\alpha=\sqrt{1-\cos^2\alpha}=\sqrt{1-\dfrac{1}{u^2+1}}=\sqrt{\dfrac{u^2+1-1}{u^2+1}}=\sqrt{\dfrac{u^2}{u^2+1}}=\dfrac{u}{\sqrt{u^2+1}}$

$\cos\beta=\sqrt{1-\sin^2\beta}=\sqrt{1-v^2}$

$\sin\left(\tan^{-1}u-\sin^{-1}v\right)=\sin(\alpha-\beta)=\sin\alpha\cos\beta-\cos\alpha\sin\beta$

$=\dfrac{u}{\sqrt{u^2+1}}\cdot\sqrt{1-v^2}-\dfrac{1}{\sqrt{u^2+1}}\cdot v=\dfrac{u\sqrt{1-v^2}-v}{\sqrt{u^2+1}}$

73. $\tan\left(\sin^{-1}u - \cos^{-1}v\right)$

Let $\alpha = \sin^{-1}u$ and $\beta = \cos^{-1}v$.

Then $\sin\alpha = u$, $-\dfrac{\pi}{2} \le \alpha \le \dfrac{\pi}{2}$, and $\cos\beta = v$, $0 \le \beta \le \pi$

$$\cos\alpha = \sqrt{1 - \sin^2\alpha} = \sqrt{1 - u^2}\,; \qquad \tan\alpha = \frac{\sin\alpha}{\cos\alpha} = \frac{u}{\sqrt{1 - u^2}}$$

$$\sin\beta = \sqrt{1 - \cos^2\beta} = \sqrt{1 - v^2}\,; \qquad \tan\beta = \frac{\sin\beta}{\cos\beta} = \frac{\sqrt{1 - v^2}}{v}$$

$$\tan\left(\sin^{-1}u - \cos^{-1}v\right) = \tan(\alpha - \beta) = \frac{\tan\alpha - \tan\beta}{1 + \tan\alpha\tan\beta} = \frac{\left(\dfrac{u}{\sqrt{1 - u^2}} - \dfrac{\sqrt{1 - v^2}}{v}\right)}{\left(1 + \dfrac{u}{\sqrt{1 - u^2}} \cdot \dfrac{\sqrt{1 - v^2}}{v}\right)}$$

$$= \frac{\left(\dfrac{uv - \sqrt{1 - u^2}\sqrt{1 - v^2}}{v\sqrt{1 - u^2}}\right)}{\left(\dfrac{v\sqrt{1 - u^2} + u\sqrt{1 - v^2}}{v\sqrt{1 - u^2}}\right)} = \frac{uv - \sqrt{1 - u^2}\sqrt{1 - v^2}}{v\sqrt{1 - u^2} + u\sqrt{1 - v^2}}$$

75.   Show that $\sin^{-1}v + \cos^{-1}v = \dfrac{\pi}{2}$.

Let $\alpha = \sin^{-1}v$ and $\beta = \cos^{-1}v$.

Then $\sin\alpha = v = \cos\beta$, and since $\sin\alpha = \cos\left(\dfrac{\pi}{2} - \alpha\right)$, $\cos\left(\dfrac{\pi}{2} - \alpha\right) = \cos\beta$.

If $v \ge 0$, then $0 \le \alpha \le \dfrac{\pi}{2}$, so that $\left(\dfrac{\pi}{2} - \alpha\right)$ and β both lie in the interval $\left[0, \dfrac{\pi}{2}\right]$.

If $v < 0$, then $-\dfrac{\pi}{2} \le \alpha < 0$, so that $\left(\dfrac{\pi}{2} - \alpha\right)$ and β both lie in the interval $\left[\dfrac{\pi}{2}, \pi\right]$. Either

way, $\cos\left(\dfrac{\pi}{2} - \alpha\right) = \cos\beta$ implies $\dfrac{\pi}{2} - \alpha = \beta$, or $\alpha + \beta = \dfrac{\pi}{2}$. Thus, $\sin^{-1}v + \cos^{-1}v = \dfrac{\pi}{2}$.

77.   Show that $\tan^{-1}\left(\dfrac{1}{v}\right) = \dfrac{\pi}{2} - \tan^{-1}v$, if $v > 0$.

Let $\alpha = \tan^{-1}\left(\dfrac{1}{v}\right)$ and $\beta = \tan^{-1}v$. Because $\dfrac{1}{v}$ must be defined, $v \ne 0$ and so $\alpha, \beta \ne 0$.

Then $\tan\alpha = \dfrac{1}{v} = \dfrac{1}{\tan\beta} = \cot\beta$, and since $\tan\alpha = \cot\left(\dfrac{\pi}{2} - \alpha\right)$, $\cot\left(\dfrac{\pi}{2} - \alpha\right) = \cot\beta$.

Because $v > 0$, $0 < \alpha < \dfrac{\pi}{2}$ and so $\dfrac{\pi}{2} - \alpha$ and β both lie in the interval

$\left(0, \dfrac{\pi}{2}\right)$. Then, $\cot\left(\dfrac{\pi}{2} - \alpha\right) = \cot\beta$ implies $\dfrac{\pi}{2} - \alpha = \beta$ or $\alpha = \dfrac{\pi}{2} - \beta$.

Thus, $\tan^{-1}\left(\dfrac{1}{v}\right) = \dfrac{\pi}{2} - \tan^{-1}v$, if $v > 0$.

79. $\sin\left(\sin^{-1}v + \cos^{-1}v\right) = \sin\left(\sin^{-1}v\right)\cos\left(\cos^{-1}v\right) + \cos\left(\sin^{-1}v\right)\sin\left(\cos^{-1}v\right)$

$$= v \cdot v + \sqrt{1-v^2}\,\sqrt{1-v^2} = v^2 + 1 - v^2 = 1$$

81. $\dfrac{\sin(x+h) - \sin x}{h} = \dfrac{\sin x \cos h + \cos x \sin h - \sin x}{h} = \dfrac{\cos x \sin h - \sin x + \sin x \cos h}{h}$

$$= \cos x \cdot \left(\frac{\sin h}{h}\right) - \sin x \cdot \left(\frac{1 - \cos h}{h}\right)$$

83. $\tan\left(\dfrac{\pi}{2} - \theta\right) = \dfrac{\tan\left(\dfrac{\pi}{2}\right) - \tan\theta}{1 + \tan\left(\dfrac{\pi}{2}\right)\tan\theta}$; This is impossible because $\tan\left(\dfrac{\pi}{2}\right)$ is undefined.

$$\tan\left(\frac{\pi}{2} - \theta\right) = \frac{\sin\left(\dfrac{\pi}{2} - \theta\right)}{\cos\left(\dfrac{\pi}{2} - \theta\right)} = \frac{\cos\theta}{\sin\theta} = \cot\theta$$

85. $\tan\theta = \tan\left(\theta_2 - \theta_1\right) = \dfrac{\tan\theta_2 - \tan\theta_1}{1 + \tan\theta_2 \tan\theta_1} = \dfrac{m_2 - m_1}{1 + m_2 m_1}$

87.   The first step in the derivation

$$\tan\left(\theta + \frac{\pi}{2}\right) = \frac{\tan\theta + \tan\left(\dfrac{\pi}{2}\right)}{1 - \tan\theta \tan\left(\dfrac{\pi}{2}\right)}$$

is impossible because $\tan\left(\dfrac{\pi}{2}\right)$ is undefined.

Analytic Trigonometry

6.5 Double-Angle and Half-Angle Formulas

1. $\sin\theta = \dfrac{3}{5}, \quad 0 < \theta < \dfrac{\pi}{2}; \qquad$ thus, $\quad 0 < \dfrac{\theta}{2} < \dfrac{\pi}{4} \implies \dfrac{\theta}{2}$ is in quadrant I.

$y = 3, \; r = 5$

$x^2 + 3^2 = 5^2, \; x > 0 \implies x^2 = 25 - 9 = 16, \; x > 0 \implies x = 4$

$\cos\theta = \dfrac{4}{5}$

(a) $\sin(2\theta) = 2\sin\theta\cos\theta = 2 \cdot \dfrac{3}{5} \cdot \dfrac{4}{5} = \dfrac{24}{25}$

(b) $\cos(2\theta) = \cos^2\theta - \sin^2\theta = \left(\dfrac{4}{5}\right)^2 - \left(\dfrac{3}{5}\right)^2 = \dfrac{16}{25} - \dfrac{9}{25} = \dfrac{7}{25}$

(c) $\sin\left(\dfrac{\theta}{2}\right) = \sqrt{\dfrac{1-\cos\theta}{2}} = \sqrt{\dfrac{1-\dfrac{4}{5}}{2}} = \sqrt{\dfrac{\dfrac{1}{5}}{2}} = \sqrt{\dfrac{1}{10}} = \dfrac{1}{\sqrt{10}} = \dfrac{\sqrt{10}}{10}$

(d) $\cos\left(\dfrac{\theta}{2}\right) = \sqrt{\dfrac{1+\cos\theta}{2}} = \sqrt{\dfrac{1+\dfrac{4}{5}}{2}} = \sqrt{\dfrac{\dfrac{9}{5}}{2}} = \sqrt{\dfrac{9}{10}} = \dfrac{3}{\sqrt{10}} = \dfrac{3\sqrt{10}}{10}$

3. $\tan\theta = \dfrac{4}{3}, \quad \pi < \theta < \dfrac{3\pi}{2}; \qquad$ thus, $\quad \dfrac{\pi}{2} < \dfrac{\theta}{2} < \dfrac{3\pi}{4} \implies \dfrac{\theta}{2}$ is in quadrant II.

$x = -3, \; y = -4$

$r^2 = (-3)^2 + (-4)^2 = 9 + 16 = 25 \implies r = 5$

$\sin\theta = -\dfrac{4}{5}, \quad \cos\theta = -\dfrac{3}{5}$

(a) $\sin(2\theta) = 2\sin\theta\cos\theta = 2 \cdot \left(-\dfrac{4}{5}\right) \cdot \left(-\dfrac{3}{5}\right) = \dfrac{24}{25}$

(b) $\cos(2\theta) = \cos^2\theta - \sin^2\theta = \left(-\dfrac{3}{5}\right)^2 - \left(-\dfrac{4}{5}\right)^2 = \dfrac{9}{25} - \dfrac{16}{25} = -\dfrac{7}{25}$

(c) $\sin\left(\dfrac{\theta}{2}\right) = \sqrt{\dfrac{1-\cos\theta}{2}} = \sqrt{\dfrac{1-\left(-\dfrac{3}{5}\right)}{2}} = \sqrt{\dfrac{\left(\dfrac{8}{5}\right)}{2}} = \sqrt{\dfrac{4}{5}} = \dfrac{2}{\sqrt{5}} = \dfrac{2\sqrt{5}}{5}$

(d) $\cos\left(\dfrac{\theta}{2}\right) = -\sqrt{\dfrac{1+\cos\theta}{2}} = -\sqrt{\dfrac{1+\left(-\dfrac{3}{5}\right)}{2}} = -\sqrt{\dfrac{\left(\dfrac{2}{5}\right)}{2}} = -\sqrt{\dfrac{1}{5}} = -\dfrac{1}{\sqrt{5}} = -\dfrac{\sqrt{5}}{5}$

5. $\cos\theta = -\dfrac{\sqrt{6}}{3}, \quad \dfrac{\pi}{2} < \theta < \pi;$ thus, $\dfrac{\pi}{4} < \dfrac{\theta}{2} < \dfrac{\pi}{2} \Rightarrow \dfrac{\theta}{2}$ is in quadrant I.

$x = -\sqrt{6}, \ r = 3$

$(-\sqrt{6})^2 + y^2 = 3^2 \ \Rightarrow \ y^2 = 9 - 6 = 3 \ \Rightarrow \ y = \sqrt{3}$

$\sin\theta = \dfrac{\sqrt{3}}{3}$

(a) $\sin(2\theta) = 2\sin\theta\cos\theta = 2\cdot\left(\dfrac{\sqrt{3}}{3}\right)\cdot\left(-\dfrac{\sqrt{6}}{3}\right) = -\dfrac{2\sqrt{18}}{9} = -\dfrac{6\sqrt{2}}{9} = -\dfrac{2\sqrt{2}}{3}$

(b) $\cos(2\theta) = \cos^2\theta - \sin^2\theta = \left(-\dfrac{\sqrt{6}}{3}\right)^2 - \left(\dfrac{\sqrt{3}}{3}\right)^2 = \dfrac{6}{9} - \dfrac{3}{9} = \dfrac{3}{9} = \dfrac{1}{3}$

(c) $\sin\left(\dfrac{\theta}{2}\right) = \sqrt{\dfrac{1-\cos\theta}{2}} = \sqrt{\dfrac{\left(1-\left(-\dfrac{\sqrt{6}}{3}\right)\right)}{2}} = \sqrt{\dfrac{\left(\dfrac{3+\sqrt{6}}{3}\right)}{2}} = \sqrt{\dfrac{3+\sqrt{6}}{6}}$

(d) $\cos\left(\dfrac{\theta}{2}\right) = \sqrt{\dfrac{1+\cos\theta}{2}} = \sqrt{\dfrac{\left(1+\left(-\dfrac{\sqrt{6}}{3}\right)\right)}{2}} = \sqrt{\dfrac{\left(\dfrac{3-\sqrt{6}}{3}\right)}{2}} = \sqrt{\dfrac{3-\sqrt{6}}{6}}$

7. $\sec\theta = 3, \ \sin\theta > 0; \quad 0 < \theta < \dfrac{\pi}{2};$ thus, $0 < \dfrac{\theta}{2} < \dfrac{\pi}{4} \Rightarrow \dfrac{\theta}{2}$ is in quadrant I.

$x = 1, \ r = 3$

$1^2 + y^2 = 3^2 \ \Rightarrow \ y^2 = 9 - 1 = 8 \ \Rightarrow \ y = 2\sqrt{2}$

$\sin\theta = \dfrac{2\sqrt{2}}{3}, \quad \cos\theta = \dfrac{1}{3}$

(a) $\sin(2\theta) = 2\sin\theta\cos\theta = 2\cdot\dfrac{2\sqrt{2}}{3}\cdot\dfrac{1}{3} = \dfrac{4\sqrt{2}}{9}$

(b) $\cos(2\theta) = \cos^2\theta - \sin^2\theta = \left(\dfrac{1}{3}\right)^2 - \left(\dfrac{2\sqrt{2}}{3}\right)^2 = \dfrac{1}{9} - \dfrac{8}{9} = -\dfrac{7}{9}$

(c) $\sin\left(\dfrac{\theta}{2}\right) = \sqrt{\dfrac{1-\cos\theta}{2}} = \sqrt{\dfrac{\left(1-\dfrac{1}{3}\right)}{2}} = \sqrt{\dfrac{\left(\dfrac{2}{3}\right)}{2}} = \sqrt{\dfrac{1}{3}} = \dfrac{1}{\sqrt{3}} = \dfrac{\sqrt{3}}{3}$

(d) $\cos\left(\dfrac{\theta}{2}\right) = \sqrt{\dfrac{1+\cos\theta}{2}} = \sqrt{\dfrac{\left(1+\dfrac{1}{3}\right)}{2}} = \sqrt{\dfrac{\left(\dfrac{4}{3}\right)}{2}} = \sqrt{\dfrac{2}{3}} = \dfrac{\sqrt{6}}{3}$

9. $\cot\theta = -2,\ \ \sec\theta < 0;\ \ \dfrac{\pi}{2} < \theta < \pi;\ $ thus, $\ \dfrac{\pi}{4} < \dfrac{\theta}{2} < \dfrac{\pi}{2}\ \Rightarrow\ \dfrac{\theta}{2}$ is in quadrant I.

   $x = -2,\ \ y = 1$

   $r^2 = (-2)^2 + 1^2 = 4 + 1 = 5\ \ \Rightarrow\ \ r = \sqrt{5}$

   $\sin\theta = \dfrac{1}{\sqrt{5}} = \dfrac{\sqrt{5}}{5},\ \ \cos\theta = -\dfrac{2}{\sqrt{5}} = -\dfrac{2\sqrt{5}}{5}$

   (a) $\sin(2\theta) = 2\sin\theta\cos\theta = 2\cdot\left(\dfrac{\sqrt{5}}{5}\right)\cdot\left(-\dfrac{2\sqrt{5}}{5}\right) = -\dfrac{20}{25} = -\dfrac{4}{5}$

   (b) $\cos(2\theta) = \cos^2\theta - \sin^2\theta = \left(-\dfrac{2\sqrt{5}}{5}\right)^2 - \left(\dfrac{\sqrt{5}}{5}\right)^2 = \dfrac{20}{25} - \dfrac{5}{25} = \dfrac{15}{25} = \dfrac{3}{5}$

   (c) $\sin\left(\dfrac{\theta}{2}\right) = \sqrt{\dfrac{1-\cos\theta}{2}} = \sqrt{\dfrac{\left(1-\left(-\dfrac{2\sqrt{5}}{5}\right)\right)}{2}} = \sqrt{\dfrac{\left(\dfrac{5+2\sqrt{5}}{5}\right)}{2}} = \sqrt{\dfrac{5+2\sqrt{5}}{10}}$

   (d) $\cos\left(\dfrac{\theta}{2}\right) = \sqrt{\dfrac{1+\cos\theta}{2}} = \sqrt{\dfrac{\left(1+\left(-\dfrac{2\sqrt{5}}{5}\right)\right)}{2}} = \sqrt{\dfrac{\left(\dfrac{5-2\sqrt{5}}{5}\right)}{2}} = \sqrt{\dfrac{5-2\sqrt{5}}{10}}$

11. $\tan\theta = -3,\ \ \sin\theta < 0;\ \ \dfrac{3\pi}{2} < \theta < 2\pi;\ $ thus, $\ \dfrac{3\pi}{4} < \dfrac{\theta}{2} < \pi\ $ or $\ \dfrac{\theta}{2}$ is in quadrant II.

   $x = 1,\ \ y = -3$

   $r^2 = 1^2 + (-3)^2 = 1 + 9 = 10\ \ \Rightarrow\ \ r = \sqrt{10}$

   $\sin\theta = \dfrac{-3}{\sqrt{10}} = -\dfrac{3\sqrt{10}}{10},\ \ \cos\theta = \dfrac{1}{\sqrt{10}} = \dfrac{\sqrt{10}}{10}$

   (a) $\sin(2\theta) = 2\sin\theta\cos\theta = 2\cdot\left(-\dfrac{3\sqrt{10}}{10}\right)\cdot\left(\dfrac{\sqrt{10}}{10}\right) = -\dfrac{60}{100} = -\dfrac{3}{5}$

   (b) $\cos(2\theta) = \cos^2\theta - \sin^2\theta = \left(\dfrac{\sqrt{10}}{10}\right)^2 - \left(-\dfrac{3\sqrt{10}}{10}\right)^2 = \dfrac{10}{100} - \dfrac{90}{100} = -\dfrac{80}{100} = -\dfrac{4}{5}$

   (c) $\sin\left(\dfrac{\theta}{2}\right) = \sqrt{\dfrac{1-\cos\theta}{2}} = \sqrt{\dfrac{\left(1-\dfrac{\sqrt{10}}{10}\right)}{2}} = \sqrt{\dfrac{\left(\dfrac{10-\sqrt{10}}{10}\right)}{2}} = \sqrt{\dfrac{10-\sqrt{10}}{20}} = \dfrac{1}{2}\sqrt{\dfrac{10-\sqrt{10}}{5}}$

   (d) $\cos\left(\dfrac{\theta}{2}\right) = -\sqrt{\dfrac{1+\cos\theta}{2}} = -\sqrt{\dfrac{\left(1+\dfrac{\sqrt{10}}{10}\right)}{2}} = -\sqrt{\dfrac{\left(\dfrac{10+\sqrt{10}}{10}\right)}{2}} = -\sqrt{\dfrac{10+\sqrt{10}}{20}}$

   $= -\dfrac{1}{2}\sqrt{\dfrac{10+\sqrt{10}}{5}}$

13.　$\sin(22.5°) = \sin\left(\dfrac{45°}{2}\right) = \sqrt{\dfrac{1-\cos(45°)}{2}} = \sqrt{\dfrac{\left(1-\dfrac{\sqrt{2}}{2}\right)}{2}} = \sqrt{\dfrac{2-\sqrt{2}}{4}} = \dfrac{\sqrt{2-\sqrt{2}}}{2}$

15.　$\tan\left(\dfrac{7\pi}{8}\right) = \tan\left(\dfrac{\left(\dfrac{7\pi}{4}\right)}{2}\right) = -\sqrt{\dfrac{1-\cos\left(\dfrac{7\pi}{4}\right)}{1+\cos\left(\dfrac{7\pi}{4}\right)}} = -\sqrt{\dfrac{1-\dfrac{\sqrt{2}}{2}}{1+\dfrac{\sqrt{2}}{2}}} = -\sqrt{\left(\dfrac{2-\sqrt{2}}{2+\sqrt{2}}\right)\cdot\left(\dfrac{2-\sqrt{2}}{2-\sqrt{2}}\right)}$

$= -\sqrt{\dfrac{6-4\sqrt{2}}{2}} = -\sqrt{3-2\sqrt{2}}$

17.　$\cos(165°) = \cos\left(\dfrac{330°}{2}\right) = -\sqrt{\dfrac{1+\cos(330°)}{2}} = -\sqrt{\dfrac{\left(1+\dfrac{\sqrt{3}}{2}\right)}{2}} = -\sqrt{\dfrac{2+\sqrt{3}}{4}} = -\dfrac{\sqrt{2+\sqrt{3}}}{2}$

19.　$\sec\left(\dfrac{15\pi}{8}\right) = \dfrac{1}{\cos\left(\dfrac{15\pi}{8}\right)} = \dfrac{1}{\cos\left(\dfrac{\left(\dfrac{15\pi}{4}\right)}{2}\right)} = \dfrac{1}{\sqrt{\dfrac{\left(1+\cos\left(\dfrac{15\pi}{4}\right)\right)}{2}}} = \dfrac{1}{\sqrt{\dfrac{\left(1+\dfrac{\sqrt{2}}{2}\right)}{2}}} = \dfrac{1}{\sqrt{\dfrac{2+\sqrt{2}}{4}}}$

$= \left(\dfrac{2}{\sqrt{2+\sqrt{2}}}\right)\cdot\left(\dfrac{\sqrt{2+\sqrt{2}}}{\sqrt{2+\sqrt{2}}}\right) = \left(\dfrac{2\sqrt{2+\sqrt{2}}}{2+\sqrt{2}}\right)\cdot\left(\dfrac{2-\sqrt{2}}{2-\sqrt{2}}\right)$

$= \dfrac{2\left(2-\sqrt{2}\right)\sqrt{2+\sqrt{2}}}{2} = \left(2-\sqrt{2}\right)\sqrt{2+\sqrt{2}}$

21.　$\sin\left(-\dfrac{\pi}{8}\right) = \sin\left(\dfrac{\left(-\dfrac{\pi}{4}\right)}{2}\right) = -\sqrt{\dfrac{1-\cos\left(-\dfrac{\pi}{4}\right)}{2}} = -\sqrt{\dfrac{\left(1-\dfrac{\sqrt{2}}{2}\right)}{2}} = -\sqrt{\dfrac{2-\sqrt{2}}{4}} = -\dfrac{\sqrt{2-\sqrt{2}}}{2}$

23.　$\sin^4\theta = \left(\sin^2\theta\right)^2 = \left(\dfrac{1-\cos(2\theta)}{2}\right)^2 = \dfrac{1}{4}\left(1-2\cos(2\theta)+\cos^2(2\theta)\right)$

$= \dfrac{1}{4} - \dfrac{1}{2}\cos(2\theta) + \dfrac{1}{4}\cos^2(2\theta) = \dfrac{1}{4} - \dfrac{1}{2}\cos(2\theta) + \dfrac{1}{4}\left(\dfrac{1+\cos(4\theta)}{2}\right)$

$= \dfrac{1}{4} - \dfrac{1}{2}\cos(2\theta) + \dfrac{1}{8} + \dfrac{1}{8}\cos(4\theta) = \dfrac{3}{8} - \dfrac{1}{2}\cos(2\theta) + \dfrac{1}{8}\cos(4\theta)$

25.　$\sin(4\theta) = \sin(2(2\theta)) = 2\sin(2\theta)\cos(2\theta) = 2\left(2\sin\theta\cos\theta\right)\left(1-2\sin^2\theta\right)$

$= \cos\theta\left(4\sin\theta - 8\sin^3\theta\right)$

27. Use the result of problem 25 to help solve the problem:

$$\sin(5\theta) = \sin(4\theta + \theta) = \sin(4\theta)\cos\theta + \cos(4\theta)\sin\theta$$

$$= \cos\theta\left(4\sin\theta - 8\sin^3\theta\right)\cos\theta + \cos(2(2\theta))\sin\theta$$

$$= \cos^2\theta\left(4\sin\theta - 8\sin^3\theta\right) + \left(1 - 2\sin^2(2\theta)\right)\sin\theta$$

$$= \left(1 - \sin^2\theta\right)\left(4\sin\theta - 8\sin^3\theta\right) + \sin\theta\left(1 - 2(2\sin\theta\cos\theta)^2\right)$$

$$= 4\sin\theta - 12\sin^3\theta + 8\sin^5\theta + \sin\theta\left(1 - 8\sin^2\theta\cos^2\theta\right)$$

$$= 4\sin\theta - 12\sin^3\theta + 8\sin^5\theta + \sin\theta - 8\sin^3\theta\left(1 - \sin^2\theta\right)$$

$$= 5\sin\theta - 12\sin^3\theta + 8\sin^5\theta - 8\sin^3\theta + 8\sin^5\theta$$

$$= 5\sin\theta - 20\sin^3\theta + 16\sin^5\theta$$

29. $\cos^4\theta - \sin^4\theta = \left(\cos^2\theta + \sin^2\theta\right)\left(\cos^2\theta - \sin^2\theta\right) = 1\cdot\cos(2\theta) = \cos(2\theta)$

31. $\cot(2\theta) = \dfrac{1}{\tan(2\theta)} = \dfrac{1}{\left(\dfrac{2\tan\theta}{1-\tan^2\theta}\right)} = \dfrac{1-\tan^2\theta}{2\tan\theta} = \dfrac{\left(1-\dfrac{1}{\cot^2\theta}\right)}{\left(\dfrac{2}{\cot\theta}\right)} = \dfrac{\left(\dfrac{\cot^2\theta-1}{\cot^2\theta}\right)}{\left(\dfrac{2}{\cot\theta}\right)} = \dfrac{\cot^2\theta-1}{2\cot\theta}$

33. $\sec(2\theta) = \dfrac{1}{\cos(2\theta)} = \dfrac{1}{2\cos^2\theta-1} = \dfrac{1}{\left(\dfrac{2}{\sec^2\theta}-1\right)} = \dfrac{1}{\left(\dfrac{2-\sec^2\theta}{\sec^2\theta}\right)} = \dfrac{\sec^2\theta}{2-\sec^2\theta}$

35. $\cos^2(2\theta) - \sin^2(2\theta) = \cos(2(2\theta)) = \cos(4\theta)$

37. $\dfrac{\cos(2\theta)}{1+\sin(2\theta)} = \dfrac{\cos^2\theta-\sin^2\theta}{1+2\sin\theta\cos\theta} = \dfrac{(\cos\theta-\sin\theta)(\cos\theta+\sin\theta)}{\cos^2\theta+\sin^2\theta+2\sin\theta\cos\theta}$

$$= \dfrac{(\cos\theta-\sin\theta)(\cos\theta+\sin\theta)}{(\cos\theta+\sin\theta)(\cos\theta+\sin\theta)} = \dfrac{\cos\theta-\sin\theta}{\cos\theta+\sin\theta} = \dfrac{\left(\dfrac{\cos\theta}{\sin\theta}-\dfrac{\sin\theta}{\sin\theta}\right)}{\left(\dfrac{\cos\theta}{\sin\theta}+\dfrac{\sin\theta}{\sin\theta}\right)} = \dfrac{\cot\theta-1}{\cot\theta+1}$$

39. $\sec^2\left(\dfrac{\theta}{2}\right) = \dfrac{1}{\cos^2\left(\dfrac{\theta}{2}\right)} = \dfrac{1}{\left(\dfrac{1+\cos\theta}{2}\right)} = \dfrac{2}{1+\cos\theta}$

41. $\cot^2\left(\dfrac{\theta}{2}\right) = \dfrac{1}{\tan^2\left(\dfrac{\theta}{2}\right)} = \dfrac{1}{\left(\dfrac{1-\cos\theta}{1+\cos\theta}\right)} = \dfrac{1+\cos\theta}{1-\cos\theta} = \dfrac{\left(1+\dfrac{1}{\sec\theta}\right)}{\left(1-\dfrac{1}{\sec\theta}\right)} = \dfrac{\left(\dfrac{\sec\theta+1}{\sec\theta}\right)}{\left(\dfrac{\sec\theta-1}{\sec\theta}\right)} = \dfrac{\sec\theta+1}{\sec\theta-1}$

43. $\dfrac{\left(1 - \tan^2\left(\dfrac{\theta}{2}\right)\right)}{\left(1 + \tan^2\left(\dfrac{\theta}{2}\right)\right)} = \dfrac{\left(1 - \dfrac{1-\cos\theta}{1+\cos\theta}\right)}{\left(1 + \dfrac{1-\cos\theta}{1+\cos\theta}\right)} = \dfrac{\left(\dfrac{1+\cos\theta - (1-\cos\theta)}{1+\cos\theta}\right)}{\left(\dfrac{1+\cos\theta + 1-\cos\theta}{1+\cos\theta}\right)} = \dfrac{2\cos\theta}{2} = \cos\theta$

45. $\dfrac{\sin(3\theta)}{\sin\theta} - \dfrac{\cos(3\theta)}{\cos\theta} = \dfrac{\sin(3\theta)\cos\theta - \cos(3\theta)\sin\theta}{\sin\theta\cos\theta} = \dfrac{\sin(3\theta - \theta)}{\sin\theta\cos\theta}$

$\qquad\qquad = \dfrac{\sin 2\theta}{\sin\theta\cos\theta} = \dfrac{2\sin\theta\cos\theta}{\sin\theta\cos\theta} = 2$

47. $\tan(3\theta) = \tan(2\theta + \theta) = \dfrac{\tan(2\theta) + \tan\theta}{1 - \tan(2\theta)\tan\theta} = \dfrac{\left(\dfrac{2\tan\theta}{1-\tan^2\theta} + \tan\theta\right)}{\left(1 - \dfrac{2\tan\theta}{1-\tan^2\theta}\cdot\tan\theta\right)}$

$\qquad\qquad = \dfrac{\left(\dfrac{2\tan\theta + \tan\theta - \tan^3\theta}{1-\tan^2\theta}\right)}{\left(\dfrac{1-\tan^2\theta - 2\tan^2\theta}{1-\tan^2\theta}\right)} = \dfrac{3\tan\theta - \tan^3\theta}{1 - 3\tan^2\theta}$

49. $\dfrac{1}{2}\cdot\left(\ln\left|1-\cos(2\theta)\right| - \ln 2\right) = \dfrac{1}{2}\cdot\ln\left|\dfrac{1-\cos 2\theta}{2}\right| = \ln\left(\left|\dfrac{1-\cos(2\theta)}{2}\right|^{1/2}\right) = \ln\left(\left|\sin^2\theta\right|^{1/2}\right) = \ln\left|\sin\theta\right|$

51. $\sin\left(2\sin^{-1}\left(\dfrac{1}{2}\right)\right) = \sin\left(2\left(\dfrac{\pi}{6}\right)\right) = \sin\left(\dfrac{\pi}{3}\right) = \dfrac{\sqrt{3}}{2}$

53. $\cos\left(2\sin^{-1}\left(\dfrac{3}{5}\right)\right) = 1 - 2\sin^2\left(\sin^{-1}\left(\dfrac{3}{5}\right)\right) = 1 - 2\left(\dfrac{3}{5}\right)^2 = 1 - 2\left(\dfrac{9}{25}\right) = 1 - \dfrac{18}{25} = \dfrac{7}{25}$

55. $\tan\left[2\cos^{-1}\left(-\dfrac{3}{5}\right)\right]$

Let $\alpha = \cos^{-1}\left(-\dfrac{3}{5}\right)$. α is in quadrant II.

Then $\cos\alpha = -\dfrac{3}{5}$, $0 \le \alpha \le \pi$.

$\sec\alpha = -\dfrac{5}{3}$; $\tan\alpha = -\sqrt{\sec^2\alpha - 1} = -\sqrt{\left(-\dfrac{5}{3}\right)^2 - 1} = -\sqrt{\dfrac{25}{9} - 1} = -\sqrt{\dfrac{16}{9}} = -\dfrac{4}{3}$

$\tan\left[2\cos^{-1}\left(-\dfrac{3}{5}\right)\right] = \tan 2\alpha = \dfrac{2\tan\alpha}{1-\tan^2\alpha} = \dfrac{2\left(-\dfrac{4}{3}\right)}{\left(1 - \left(-\dfrac{4}{3}\right)^2\right)} = \dfrac{\left(-\dfrac{8}{3}\right)}{\left(1 - \dfrac{16}{9}\right)} = \dfrac{\left(-\dfrac{8}{3}\right)}{\left(-\dfrac{7}{9}\right)} = \left(-\dfrac{8}{3}\right)\cdot\left(-\dfrac{9}{7}\right) = \dfrac{24}{7}$

57. $\sin\left(2\cos^{-1}\left(\dfrac{4}{5}\right)\right)$

Let $\alpha = \cos^{-1}\left(\dfrac{4}{5}\right)$. α is in quadrant I.

Then $\cos\alpha = \dfrac{4}{5}$, $0 \le \alpha \le \pi$.

$$\sin\alpha = \sqrt{1-\cos^2\alpha} = \sqrt{1-\left(\dfrac{4}{5}\right)^2} = \sqrt{1-\dfrac{16}{25}} = \sqrt{\dfrac{9}{25}} = \dfrac{3}{5}$$

$$\sin\left[2\cos^{-1}\left(\dfrac{4}{5}\right)\right] = \sin 2\alpha = 2\sin\alpha\cos\alpha = 2\cdot\dfrac{3}{5}\cdot\dfrac{4}{5} = \dfrac{24}{25}$$

59. $\sin^2\left[\dfrac{1}{2}\cdot\cos^{-1}\left(\dfrac{3}{5}\right)\right] = \dfrac{1-\cos\left(\cos^{-1}\left(\dfrac{3}{5}\right)\right)}{2} = \dfrac{\left(1-\dfrac{3}{5}\right)}{2} = \dfrac{\left(\dfrac{2}{5}\right)}{2} = \dfrac{1}{5}$

61. $\sec\left(2\tan^{-1}\left(\dfrac{3}{4}\right)\right)$

Let $\alpha = \tan^{-1}\left(\dfrac{3}{4}\right)$. α is in quadrant I.

Then $\tan\alpha = \dfrac{3}{4}$, $-\dfrac{\pi}{2} < \alpha < \dfrac{\pi}{2}$.

$$\sec\alpha = \sqrt{\tan^2\alpha+1} = \sqrt{\left(\dfrac{3}{4}\right)^2+1} = \sqrt{\dfrac{9}{16}+1} = \sqrt{\dfrac{25}{16}} = \dfrac{5}{4};\ \ \cos\alpha = \dfrac{4}{5}$$

$$\sec\left[2\tan^{-1}\left(\dfrac{3}{4}\right)\right] = \sec(2\alpha) = \dfrac{1}{\cos(2\alpha)} = \dfrac{1}{2\cos^2\alpha-1} = \dfrac{1}{\left(2\left(\dfrac{4}{5}\right)^2-1\right)} = \dfrac{1}{\left(2\cdot\left(\dfrac{16}{25}\right)-1\right)} = \dfrac{1}{\left(\dfrac{7}{25}\right)} = \dfrac{25}{7}$$

63. If $x = 2\tan\theta$, then:

$$\sin(2\theta) = 2\sin\theta\cos\theta = \dfrac{2\sin\theta}{\cos\theta}\cdot\dfrac{\cos^2\theta}{1} = \dfrac{\left(2\cdot\dfrac{\sin\theta}{\cos\theta}\right)}{\left(\dfrac{1}{\cos^2\theta}\right)} = \dfrac{2\tan\theta}{\sec^2\theta} = \dfrac{2\tan\theta}{1+\tan^2\theta}\cdot\dfrac{4}{4}$$

$$= \dfrac{4(2\tan\theta)}{4+(2\tan\theta)^2} = \dfrac{4x}{4+x^2}$$

65. Solve for C:

$$\frac{1}{2} \cdot \sin^2 x + C = -\frac{1}{4} \cdot \cos(2x)$$

$$C = -\frac{1}{4} \cdot \cos(2x) - \frac{1}{2} \cdot \sin^2 x = -\frac{1}{4} \cdot \left(\cos(2x) + 2\sin^2 x\right)$$

$$= -\frac{1}{4} \cdot \left(1 - 2\sin^2 x + 2\sin^2 x\right) = -\frac{1}{4} \cdot (1) = -\frac{1}{4}$$

67. $z = \tan\left(\dfrac{\alpha}{2}\right) = \dfrac{1 - \cos\alpha}{\sin\alpha}$

$$z\sin\alpha = 1 - \cos\alpha$$

$$z\sin\alpha = 1 - \sqrt{1 - \sin^2 \alpha}$$

$$z\sin\alpha - 1 = -\sqrt{1 - \sin^2 \alpha}$$

$$z^2 \sin^2 \alpha - 2z\sin\alpha + 1 = 1 - \sin^2 \alpha$$

$$z^2 \sin^2 \alpha + \sin^2 \alpha = 2z\sin\alpha$$

$$\sin^2 \alpha(z^2 + 1) = 2z\sin\alpha$$

$$\sin\alpha(z^2 + 1) = 2z$$

$$\sin\alpha = \frac{2z}{z^2 + 1}$$

69. Let b represent the base of the triangle.

$$\cos\left(\frac{\theta}{2}\right) = \frac{h}{s} \implies h = s\cos\left(\frac{\theta}{2}\right) \quad \text{and} \quad \sin\left(\frac{\theta}{2}\right) = \frac{\left(\frac{1}{2}b\right)}{s} \implies b = 2s\sin\left(\frac{\theta}{2}\right)$$

$$A = \frac{1}{2}b \cdot h = \frac{1}{2} \cdot \left(2s\sin\left(\frac{\theta}{2}\right)\right)\left(s\cos\left(\frac{\theta}{2}\right)\right) = s^2 \sin\left(\frac{\theta}{2}\right)\cos\left(\frac{\theta}{2}\right) = \frac{1}{2} \cdot s^2 \sin\theta$$

71. $f(x) = \sin^2 x = \dfrac{1 - \cos(2x)}{2}$

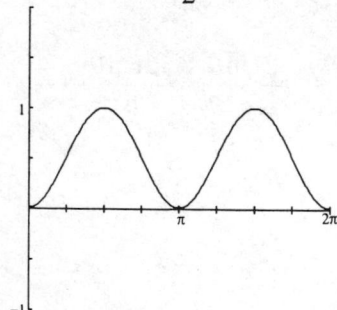

73. $\sin\left(\dfrac{\pi}{24}\right) = \sin\left(\dfrac{\left(\dfrac{\pi}{12}\right)}{2}\right) = \sqrt{\dfrac{1-\cos\left(\dfrac{\pi}{12}\right)}{2}} = \sqrt{\dfrac{1-\left(\dfrac{1}{4}\left(\sqrt{6}+\sqrt{2}\right)\right)}{2}} = \sqrt{\dfrac{1}{2}-\dfrac{1}{8}\left(\sqrt{6}+\sqrt{2}\right)}$

$= \sqrt{\dfrac{8-2\left(\sqrt{6}+\sqrt{2}\right)}{16}} = \dfrac{\sqrt{8-2\left(\sqrt{6}+\sqrt{2}\right)}}{4}$

$\cos\left(\dfrac{\pi}{24}\right) = \cos\left(\dfrac{\left(\dfrac{\pi}{12}\right)}{2}\right) = \sqrt{\dfrac{1+\cos\left(\dfrac{\pi}{12}\right)}{2}} = \sqrt{\dfrac{1+\left(\dfrac{1}{4}\left(\sqrt{6}+\sqrt{2}\right)\right)}{2}} = \sqrt{\dfrac{1}{2}+\dfrac{1}{8}\left(\sqrt{6}+\sqrt{2}\right)}$

$= \sqrt{\dfrac{8+2\left(\sqrt{6}+\sqrt{2}\right)}{16}} = \dfrac{\sqrt{8+2\left(\sqrt{6}+\sqrt{2}\right)}}{4}$

75. $\sin^3\theta + \sin^3(\theta+120°) + \sin^3(\theta+240°)$

$= \sin^3\theta + \left(\sin\theta\cos(120°)+\cos\theta\sin(120°)\right)^3 + \left(\sin\theta\cos(240°)+\cos\theta\sin(240°)\right)^3$

$= \sin^3\theta + \left(-\dfrac{1}{2}\cdot\sin\theta + \dfrac{\sqrt{3}}{2}\cdot\cos\theta\right)^3 + \left(-\dfrac{1}{2}\cdot\sin\theta - \dfrac{\sqrt{3}}{2}\cdot\cos\theta\right)^3$

$= \sin^3\theta + \dfrac{1}{8}\cdot\left(-\sin^3\theta + 3\sqrt{3}\sin^2\theta\cos\theta - 9\sin\theta\cos^2\theta + 3\sqrt{3}\cos^3\theta\right)$

$\qquad -\dfrac{1}{8}\left(\sin^3\theta + 3\sqrt{3}\sin^2\theta\cos\theta + 9\sin\theta\cos^2\theta + 3\sqrt{3}\cos^3\theta\right)$

$= \sin^3\theta - \dfrac{1}{8}\cdot\sin^3\theta + \dfrac{3\sqrt{3}}{8}\cdot\sin^2\theta\cos\theta - \dfrac{9}{8}\cdot\sin\theta\cos^2\theta + \dfrac{3\sqrt{3}}{8}\cdot\cos^3\theta$

$\qquad -\dfrac{1}{8}\cdot\sin^3\theta - \dfrac{3\sqrt{3}}{8}\cdot\sin^2\theta\cos\theta - \dfrac{9}{8}\cdot\sin\theta\cos^2\theta - \dfrac{3\sqrt{3}}{8}\cdot\cos^3\theta$

$= \dfrac{3}{4}\cdot\sin^3\theta - \dfrac{9}{4}\cdot\sin\theta\cos^2\theta = \dfrac{3}{4}\cdot\left(\sin^3\theta - 3\sin\theta\left(1-\sin^2\theta\right)\right)$

$= \dfrac{3}{4}\cdot\left(\sin^3\theta - 3\sin\theta + 3\sin^3\theta\right) = \dfrac{3}{4}\cdot\left(4\sin^3\theta - 3\sin\theta\right) = -\dfrac{3}{4}\cdot\sin(3\theta)$

(See the formula for $\sin(3\theta)$ on page 495 of the text.)

77. (a) $R(\theta) = \dfrac{v_0^2\sqrt{2}}{16}\cos\theta(\sin\theta - \cos\theta) = \dfrac{v_0^2\sqrt{2}}{16}(\cos\theta\sin\theta - \cos^2\theta)$

$= \dfrac{v_0^2\sqrt{2}}{16}\cdot\dfrac{1}{2}(2\cos\theta\sin\theta - 2\cos^2\theta) = \dfrac{v_0^2\sqrt{2}}{32}\left(\sin2\theta - 2\left(\dfrac{1+\cos2\theta}{2}\right)\right)$

$= \dfrac{v_0^2\sqrt{2}}{32}\left(\sin(2\theta) - 1 - \cos(2\theta)\right) = \dfrac{v_0^2\sqrt{2}}{32}\left(\sin(2\theta) - \cos(2\theta) - 1\right)$

(b)

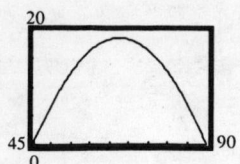

(c)   Using the MAXIMUM feature on the calculator:
        R has the largest value when $\theta \approx 67.5°$.

Analytic Trigonometry

6.6 Product-to-Sum and Sum-to-Product Formulas

For Problems 1-9, use the formulas:

$$\sin\alpha\sin\beta = \frac{1}{2}\cdot[\cos(\alpha-\beta)-\cos(\alpha+\beta)] \qquad \cos\alpha\cos\beta = \frac{1}{2}\cdot[\cos(\alpha-\beta)+\cos(\alpha+\beta)]$$

$$\sin\alpha\cos\beta = \frac{1}{2}\cdot[\sin(\alpha+\beta)+\sin(\alpha-\beta)]$$

1. $\sin(4\theta)\sin(2\theta) = \frac{1}{2}\cdot[\cos(4\theta-2\theta)-\cos(4\theta+2\theta)] = \frac{1}{2}\cdot[\cos(2\theta)-\cos(6\theta)]$

3. $\sin(4\theta)\cos(2\theta) = \frac{1}{2}\cdot[\sin(4\theta+2\theta)+\sin(4\theta-2\theta)] = \frac{1}{2}\cdot[\sin(6\theta)+\sin(2\theta)]$

5. $\cos(3\theta)\cos(5\theta) = \frac{1}{2}\cdot[\cos(3\theta-5\theta)+\cos(3\theta+5\theta)] = \frac{1}{2}\cdot[\cos(-2\theta)+\cos(8\theta)]$

$$= \frac{1}{2}\cdot[\cos(2\theta)+\cos(8\theta)]$$

7. $\sin\theta\sin(2\theta) = \frac{1}{2}\cdot[\cos(\theta-2\theta)-\cos(\theta+2\theta)] = \frac{1}{2}\cdot[\cos(-\theta)-\cos(3\theta)]$

$$= \frac{1}{2}\cdot[\cos\theta-\cos(3\theta)]$$

9. $\sin\left(\frac{3\theta}{2}\right)\cos\left(\frac{\theta}{2}\right) = \frac{1}{2}\cdot\left[\sin\left(\frac{3\theta}{2}+\frac{\theta}{2}\right)+\sin\left(\frac{3\theta}{2}-\frac{\theta}{2}\right)\right] = \frac{1}{2}\cdot[\sin(2\theta)+\sin\theta]$

For Problems 11-17, use the formulas:

$$\sin\alpha+\sin\beta = 2\sin\left(\frac{\alpha+\beta}{2}\right)\cos\left(\frac{\alpha-\beta}{2}\right) \qquad \sin\alpha-\sin\beta = 2\sin\left(\frac{\alpha-\beta}{2}\right)\cos\left(\frac{\alpha+\beta}{2}\right)$$

$$\cos\alpha+\cos\beta = 2\cos\left(\frac{\alpha+\beta}{2}\right)\cos\left(\frac{\alpha-\beta}{2}\right) \qquad \cos\alpha-\cos\beta = -2\sin\left(\frac{\alpha+\beta}{2}\right)\sin\left(\frac{\alpha-\beta}{2}\right)$$

11. $\sin(4\theta)-\sin(2\theta) = 2\sin\left(\frac{4\theta-2\theta}{2}\right)\cos\left(\frac{4\theta+2\theta}{2}\right) = 2\sin\theta\cos(3\theta)$

13. $\cos(2\theta) + \cos(4\theta) = 2\cos\left(\dfrac{2\theta+4\theta}{2}\right)\cos\left(\dfrac{2\theta-4\theta}{2}\right) = 2\cos(3\theta)\cos(-\theta) = 2\cos(3\theta)\cos\theta$

15. $\sin\theta + \sin(3\theta) = 2\sin\left(\dfrac{\theta+3\theta}{2}\right)\cos\left(\dfrac{\theta-3\theta}{2}\right) = 2\sin(2\theta)\cos(-\theta) = 2\sin(2\theta)\cos\theta$

17. $\cos\left(\dfrac{\theta}{2}\right) - \cos\left(\dfrac{3\theta}{2}\right)$

$= -2\sin\left(\dfrac{\left(\dfrac{\theta}{2}+\dfrac{3\theta}{2}\right)}{2}\right)\sin\left(\dfrac{\left(\dfrac{\theta}{2}-\dfrac{3\theta}{2}\right)}{2}\right) = -2\sin\theta\sin\left(-\dfrac{\theta}{2}\right) = -2\sin\theta\left(-\sin\left(\dfrac{\theta}{2}\right)\right) = 2\sin\theta\sin\left(\dfrac{\theta}{2}\right)$

19. $\dfrac{\sin\theta + \sin(3\theta)}{2\sin(2\theta)} = \dfrac{2\sin(2\theta)\cos(-\theta)}{2\sin(2\theta)} = \cos(-\theta) = \cos\theta$

21. $\dfrac{\sin(4\theta) + \sin(2\theta)}{\cos(4\theta) + \cos(2\theta)} = \dfrac{2\sin(3\theta)\cos\theta}{2\cos(3\theta)\cos\theta} = \dfrac{\sin(3\theta)}{\cos(3\theta)} = \tan(3\theta)$

23. $\dfrac{\cos\theta - \cos(3\theta)}{\sin\theta + \sin(3\theta)} = \dfrac{-2\sin(2\theta)\sin(-\theta)}{2\sin(2\theta)\cos(-\theta)} = \dfrac{-(-\sin\theta)}{\cos\theta} = \tan\theta$

25. $\sin\theta[\sin\theta + \sin(3\theta)] = \sin\theta[2\sin(2\theta)\cos(-\theta)] = \cos\theta[2\sin(2\theta)\sin\theta]$

$= \cos\theta\left[2\cdot\dfrac{1}{2}(\cos\theta - \cos(3\theta))\right] = \cos\theta(\cos\theta - \cos(3\theta))$

27. $\dfrac{\sin(4\theta) + \sin(8\theta)}{\cos(4\theta) + \cos(8\theta)} = \dfrac{2\sin(6\theta)\cos(-2\theta)}{2\cos(6\theta)\cos(-2\theta)} = \dfrac{\sin(6\theta)}{\cos(6\theta)} = \tan(6\theta)$

29. $\dfrac{\sin(4\theta) + \sin(8\theta)}{\sin(4\theta) - \sin(8\theta)} = \dfrac{2\sin(6\theta)\cos(-2\theta)}{2\sin(-2\theta)\cos(6\theta)} = \dfrac{\sin(6\theta)\cos(2\theta)}{-\sin(2\theta)\cos(6\theta)}$

$= -\tan(6\theta)\cot(2\theta) = -\dfrac{\tan(6\theta)}{\tan(2\theta)}$

31. $\dfrac{\sin\alpha + \sin\beta}{\sin\alpha - \sin\beta} = \dfrac{2\sin\left(\dfrac{\alpha+\beta}{2}\right)\cos\left(\dfrac{\alpha-\beta}{2}\right)}{2\sin\left(\dfrac{\alpha-\beta}{2}\right)\cos\left(\dfrac{\alpha+\beta}{2}\right)} = \tan\left(\dfrac{\alpha+\beta}{2}\right)\cot\left(\dfrac{\alpha-\beta}{2}\right)$

33. $\dfrac{\sin\alpha + \sin\beta}{\cos\alpha + \cos\beta} = \dfrac{2\sin\left(\dfrac{\alpha+\beta}{2}\right)\cos\left(\dfrac{\alpha-\beta}{2}\right)}{2\cos\left(\dfrac{\alpha+\beta}{2}\right)\cos\left(\dfrac{\alpha-\beta}{2}\right)} = \tan\left(\dfrac{\alpha+\beta}{2}\right)$

35. $1 + \cos(2\theta) + \cos(4\theta) + \cos(6\theta) = \cos 0 + \cos(6\theta) + \cos(2\theta) + \cos(4\theta)$
$$= 2\cos(3\theta)\cos(-3\theta) + 2\cos(3\theta)\cos(-\theta) = 2\cos^2(3\theta) + 2\cos(3\theta)\cos\theta$$
$$= 2\cos(3\theta)(\cos(3\theta) + \cos\theta) = 2\cos(3\theta)2\cos(2\theta)\cos\theta$$
$$= 4\cos\theta\cos(2\theta)\cos(3\theta)$$

37. (a) $y = \sin[2\pi(852)t] + \sin[2\pi(1209)t]$
$$= 2\sin\left(\frac{2\pi(852)t + 2\pi(1209)t}{2}\right)\cos\left(\frac{2\pi(852)t - 2\pi(1209)t}{2}\right)$$
$$= 2\sin(2061\pi t)\cos(357\pi t)$$
(b)   The maximum value of y is 2.
(c)

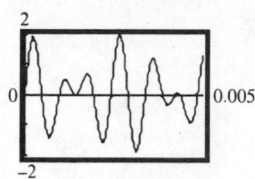

39. $\sin(2\alpha) + \sin(2\beta) + \sin(2\gamma)$
$$= 2\sin\left(\frac{2\alpha + 2\beta}{2}\right)\cos\left(\frac{2\alpha - 2\beta}{2}\right) + \sin(2\gamma)$$
$$= 2\sin(\alpha + \beta)\cos(\alpha - \beta) + 2\sin\gamma\cos\gamma$$
$$= 2\sin(\pi - \gamma)\cos(\alpha - \beta) + 2\sin\gamma\cos\gamma$$
$$= 2\sin\gamma\cos(\alpha - \beta) + 2\sin\gamma\cos\gamma = 2\sin\gamma[\cos(\alpha - \beta) + \cos\gamma]$$
$$= 2\sin\gamma\left(2\cos\left(\frac{\alpha - \beta + \gamma}{2}\right)\cos\left(\frac{\alpha - \beta - \gamma}{2}\right)\right)$$
$$= 4\sin\gamma\cos\left(\frac{\pi}{2} - \beta\right)\cos\left(\alpha - \frac{\pi}{2}\right) = 4\sin\gamma\sin\beta\sin\alpha = 4\sin\alpha\sin\beta\sin\gamma$$

41. Add the two sum formulas for $\sin(\alpha + \beta)$ and $\sin(\alpha - \beta)$ and solve:
$$\sin(\alpha + \beta) = \sin\alpha\cos\beta + \cos\alpha\sin\beta$$
$$\sin(\alpha - \beta) = \sin\alpha\cos\beta - \cos\alpha\sin\beta$$
$$\sin(\alpha + \beta) + \sin(\alpha - \beta) = 2\sin\alpha\cos\beta$$
$$\sin\alpha\cos\beta = \frac{1}{2}\cdot[\sin(\alpha + \beta) + \sin(\alpha - \beta)]$$

43. $2\cos\left(\dfrac{\alpha + \beta}{2}\right)\cos\left(\dfrac{\alpha - \beta}{2}\right) = 2\cdot\dfrac{1}{2}\left[\cos\left(\dfrac{\alpha + \beta}{2} - \dfrac{\alpha - \beta}{2}\right) + \cos\left(\dfrac{\alpha + \beta}{2} + \dfrac{\alpha - \beta}{2}\right)\right]$
$$= \cos\left(\frac{2\beta}{2}\right) + \cos\left(\frac{2\alpha}{2}\right) = \cos\beta + \cos\alpha$$
Therefore, $\cos\alpha + \cos\beta = 2\cos\left(\dfrac{\alpha + \beta}{2}\right)\cos\left(\dfrac{\alpha - \beta}{2}\right)$

Chapter 6

Analytic Trigonometry

6.7   Trigonometric Equations (I)

1. $\sin \theta = \dfrac{1}{2}$

$\theta = \dfrac{\pi}{6} + 2k\pi$ or $\theta = \dfrac{5\pi}{6} + 2k\pi$, where k is any integer

Six solutions are $\theta = \dfrac{\pi}{6}, \dfrac{5\pi}{6}, \dfrac{13\pi}{6}, \dfrac{17\pi}{6}, \dfrac{25\pi}{6}, \dfrac{29\pi}{6}$

3. $\tan \theta = -\dfrac{\sqrt{3}}{3}$

$\theta = \dfrac{5\pi}{6} + k\pi$, where k is any integer

Six solutions are $\theta = \dfrac{5\pi}{6}, \dfrac{11\pi}{6}, \dfrac{17\pi}{6}, \dfrac{23\pi}{6}, \dfrac{29\pi}{6}, \dfrac{35\pi}{6}$

5. $\cos \theta = 0$

$\theta = \dfrac{\pi}{2} + 2k\pi$ or $\theta = \dfrac{3\pi}{2} + 2k\pi$, where k is any integer

Six solutions are $\theta = \dfrac{\pi}{2}, \dfrac{3\pi}{2}, \dfrac{5\pi}{2}, \dfrac{7\pi}{2}, \dfrac{9\pi}{2}, \dfrac{11\pi}{2}$

7. $\cos(2\theta) = -\dfrac{1}{2}$

$2\theta = \dfrac{2\pi}{3} + 2k\pi$ $\Rightarrow$ $\theta = \dfrac{\pi}{3} + k\pi$, where k is any integer

$2\theta = \dfrac{4\pi}{3} + 2k\pi$ $\Rightarrow$ $\theta = \dfrac{2\pi}{3} + k\pi$, where k is any integer

Six solutions are $\theta = \dfrac{\pi}{3}, \dfrac{2\pi}{3}, \dfrac{4\pi}{3}, \dfrac{5\pi}{3}, \dfrac{7\pi}{3}, \dfrac{8\pi}{3}$

9. $\sin\left(\dfrac{\theta}{2}\right) = -\dfrac{\sqrt{3}}{2}$

$\dfrac{\theta}{2} = \dfrac{4\pi}{3} + 2k\pi$ $\Rightarrow$ $\theta = \dfrac{8\pi}{3} + 4k\pi$, where k is any integer

$\dfrac{\theta}{2} = \dfrac{5\pi}{3} + 2k\pi$ $\Rightarrow$ $\theta = \dfrac{10\pi}{3} + 4k\pi$, where k is any integer

Six solutions are $\theta = \dfrac{8\pi}{3}, \dfrac{10\pi}{3}, \dfrac{20\pi}{3}, \dfrac{22\pi}{3}, \dfrac{32\pi}{3}, \dfrac{34\pi}{3}$

11. $2\sin\theta + 3 = 2$

$$2\sin\theta = -1 \Rightarrow \sin\theta = -\frac{1}{2}$$

$$\theta = \frac{7\pi}{6} + 2k\pi \quad \text{or} \quad \theta = \frac{11\pi}{6} + 2k\pi, \ k \text{ is any integer}$$

The solutions on the interval $[0, 2\pi)$ are $\theta = \frac{7\pi}{6}, \frac{11\pi}{6}$.

13. $4\cos^2\theta = 1$

$$\cos^2\theta = \frac{1}{4} \Rightarrow \cos\theta = \pm\frac{1}{2}$$

$$\theta = \frac{\pi}{3} + k\pi \quad \text{or} \quad \theta = \frac{2\pi}{3} + k\pi, \ k \text{ is any integer}$$

The solutions on the interval $[0, 2\pi)$ are $\theta = \frac{\pi}{3}, \frac{2\pi}{3}, \frac{4\pi}{3}, \frac{5\pi}{3}$.

15. $2\sin^2\theta - 1 = 0$

$$2\sin^2\theta = 1 \Rightarrow \sin^2\theta = \frac{1}{2} \Rightarrow \sin\theta = \pm\sqrt{\frac{1}{2}} = \pm\frac{\sqrt{2}}{2}$$

$$\theta = \frac{\pi}{4} + k\pi \quad \text{or} \quad \theta = \frac{3\pi}{4} + k\pi, \ k \text{ is any integer}$$

The solutions on the interval $[0, 2\pi)$ are $\theta = \frac{\pi}{4}, \frac{3\pi}{4}, \frac{5\pi}{4}, \frac{7\pi}{4}$.

17. $\sin(3\theta) = -1$

$$3\theta = \frac{3\pi}{2} + 2k\pi \quad \Rightarrow \quad \theta = \frac{\pi}{2} + \frac{2k\pi}{3}, \text{ where } k \text{ is any integer}$$

The solutions on the interval $[0, 2\pi)$ are $\theta = \frac{\pi}{2}, \frac{7\pi}{6}, \frac{11\pi}{6}$

19. $\cos(2\theta) = -\frac{1}{2}$

$$2\theta = \frac{2\pi}{3} + 2k\pi \quad \Rightarrow \quad \theta = \frac{\pi}{3} + k\pi, \text{ where } k \text{ is any integer}$$

$$2\theta = \frac{4\pi}{3} + 2k\pi \quad \Rightarrow \quad \theta = \frac{2\pi}{3} + k\pi, \text{ where } k \text{ is any integer}$$

The solutions on the interval $[0, 2\pi)$ are $\theta = \frac{\pi}{3}, \frac{2\pi}{3}, \frac{4\pi}{3}, \frac{5\pi}{3}$

21. $\sec\left(\frac{3\theta}{2}\right) = -2$

$$\frac{3\theta}{2} = \frac{2\pi}{3} + 2k\pi \quad \Rightarrow \quad \theta = \frac{4\pi}{9} + \frac{4k\pi}{3}, \text{ where } k \text{ is any integer}$$

$$\frac{3\theta}{2} = \frac{4\pi}{3} + 2k\pi \quad \Rightarrow \quad \theta = \frac{8\pi}{9} + \frac{4k\pi}{3}, \text{ where } k \text{ is any integer}$$

The solutions on the interval $[0, 2\pi)$ are $\theta = \frac{4\pi}{9}, \frac{8\pi}{9}, \frac{16\pi}{9}$

23. $\cos\left(2\theta - \dfrac{\pi}{2}\right) = -1$

    $2\theta - \dfrac{\pi}{2} = \pi + 2k\pi \;\Rightarrow\; 2\theta = \dfrac{3\pi}{2} + 2k\pi \;\Rightarrow\; \theta = \dfrac{3\pi}{4} + k\pi,\; k \text{ is any integer}$

    The solutions on the interval $[0, 2\pi)$ are $\theta = \dfrac{3\pi}{4}, \dfrac{7\pi}{4}$.

25. $\tan\left(\dfrac{\theta}{2} + \dfrac{\pi}{3}\right) = 1$

    $\dfrac{\theta}{2} + \dfrac{\pi}{3} = \dfrac{\pi}{4} + k\pi \;\Rightarrow\; \dfrac{\theta}{2} = -\dfrac{\pi}{12} + k\pi \;\Rightarrow\; \theta = -\dfrac{\pi}{6} + 2k\pi,\; k \text{ is any integer}$

    The solutions on the interval $[0, 2\pi)$ are $\theta = \dfrac{11\pi}{6}$.

27. $2\sin\theta + 1 = 0 \;\Rightarrow\; 2\sin\theta = -1 \;\Rightarrow\; \sin\theta = -\dfrac{1}{2}$

    $\theta = \dfrac{7\pi}{6} + 2k\pi \quad \text{or} \quad \theta = \dfrac{11\pi}{6} + 2k\pi,\; k \text{ is any integer}$

    The solutions on the interval $[0, 2\pi)$ are $\theta = \dfrac{7\pi}{6}, \dfrac{11\pi}{6}$.

29. $\tan\theta + 1 = 0 \;\Rightarrow\; \tan\theta = -1$

    $\theta = \dfrac{3\pi}{4} + k\pi,\; k \text{ is any integer}$

    The solutions on the interval $[0, 2\pi)$ are $\theta = \dfrac{3\pi}{4}, \dfrac{7\pi}{4}$.

31. $4\sec\theta + 6 = -2 \;\Rightarrow\; 4\sec\theta = -8 \;\Rightarrow\; \sec\theta = -2$

    $\theta = \dfrac{2\pi}{3} + 2k\pi \quad \text{or} \quad \theta = \dfrac{4\pi}{3} + 2k\pi,\; k \text{ is any integer}$

    The solutions on the interval $[0, 2\pi)$ are $\theta = \dfrac{2\pi}{3}, \dfrac{4\pi}{3}$.

33. $3\sqrt{2}\cos\theta + 2 = -1 \;\Rightarrow\; 3\sqrt{2}\cos\theta = -3 \;\Rightarrow\; \cos\theta = -\dfrac{1}{\sqrt{2}} = -\dfrac{\sqrt{2}}{2}$

    $\theta = \dfrac{3\pi}{4} + 2k\pi \quad \text{or} \quad \theta = \dfrac{5\pi}{4} + 2k\pi,\; k \text{ is any integer}$

    The solutions on the interval $[0, 2\pi)$ are $\theta = \dfrac{3\pi}{4}, \dfrac{5\pi}{4}$.

35. $\sin\theta = 0.4$

    $\theta \approx 0.4115168 \quad \text{or} \quad \theta \approx \pi - 0.4115168 \approx 2.7300758$

    $\theta \approx 0.41, 2.73$

37. $\tan\theta = 5$

    $\theta \approx 1.3734008 \quad \text{or} \quad \theta \approx \pi + 1.3734008 \approx 4.5149934$

    $\theta \approx 1.37, 4.51$

39. $\cos\theta = -0.9$

   $\theta \approx 2.6905658 \quad \text{or} \quad \theta \approx 2\pi - 2.6905658 \approx 3.5926195$
   $\theta \approx 2.69, \, 3.59$

41. $\sec\theta = -4 \;\Rightarrow\; \cos\theta = -\dfrac{1}{4}$

   $\theta \approx 1.8234766 \quad \text{or} \quad \theta \approx 2\pi - 1.8234766 \approx 4.4597087$
   $\theta \approx 1.82, \, 4.46$

43. Use Snell's Law to solve:

   $$\frac{\sin(40^\circ)}{\sin\theta_2} = 1.33 \Rightarrow \sin(40^\circ) = 1.33\sin\theta_2 \Rightarrow \sin\theta_2 = \frac{\sin(40^\circ)}{1.33} \approx 0.4833$$

   $$\theta_2 = \sin^{-1}(0.4833) \approx 28.9^\circ$$

45. Calculate the index of refraction for each:

   $\theta_1 = 10^\circ, \; \theta_2 = 7^\circ 45' = 7.75^\circ \qquad \dfrac{\sin\theta_1}{\sin\theta_2} = \dfrac{\sin(10^\circ)}{\sin(7.75^\circ)} \approx 1.2877$

   $\theta_1 = 20^\circ, \; \theta_2 = 15^\circ 30' = 15.5^\circ \qquad \dfrac{\sin\theta_1}{\sin\theta_2} = \dfrac{\sin(20^\circ)}{\sin(15.5^\circ)} \approx 1.2798$

   $\theta_1 = 30^\circ, \; \theta_2 = 22^\circ 30' = 22.5^\circ \qquad \dfrac{\sin\theta_1}{\sin\theta_2} = \dfrac{\sin(30^\circ)}{\sin(22.5^\circ)} \approx 1.3066$

   $\theta_1 = 40^\circ, \; \theta_2 = 29^\circ 0' = 29^\circ \qquad \dfrac{\sin\theta_1}{\sin\theta_2} = \dfrac{\sin(40^\circ)}{\sin(29^\circ)} \approx 1.3259$

   $\theta_1 = 50^\circ, \; \theta_2 = 35^\circ 0' = 35^\circ \qquad \dfrac{\sin\theta_1}{\sin\theta_2} = \dfrac{\sin(50^\circ)}{\sin(35^\circ)} \approx 1.3356$

   $\theta_1 = 60^\circ, \; \theta_2 = 40^\circ 30' = 40.5^\circ \qquad \dfrac{\sin\theta_1}{\sin\theta_2} = \dfrac{\sin(60^\circ)}{\sin(40.5^\circ)} \approx 1.3335$

   $\theta_1 = 70^\circ, \; \theta_2 = 45^\circ 30' = 45.5^\circ \qquad \dfrac{\sin\theta_1}{\sin\theta_2} = \dfrac{\sin(70^\circ)}{\sin(45.5^\circ)} \approx 1.3175$

   $\theta_1 = 80^\circ, \; \theta_2 = 50^\circ 0' = 50^\circ \qquad \dfrac{\sin\theta_1}{\sin\theta_2} = \dfrac{\sin(80)^\circ}{\sin(50^\circ)} \approx 1.2856$

   The results range from 1.28 to 1.34 and are surprisingly close to Snell's Law.

47. Calculate the index of refraction:

   $\theta_1 = 40^\circ, \; \theta_2 = 26^\circ \qquad \dfrac{\sin\theta_1}{\sin\theta_2} = \dfrac{\sin(40^\circ)}{\sin(26^\circ)} \approx 1.47$

49. If θ is the original angle of incidence and ϕ is the angle of refraction, then $\dfrac{\sin\theta}{\sin\phi} = n_2$. The angle of incidence of the emerging beam is also ϕ, and the index of refraction is $\dfrac{1}{n_2}$. Thus, θ is the angle of refraction of the emerging beam. The two beams are parallel since the original angle of incidence and the angle of refraction of the emerging beam are equal.

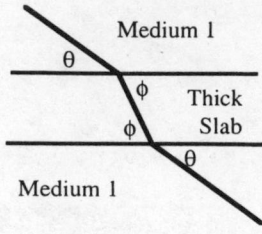

Chapter 6

Analytic Trigonometry

6.8   Trigonometric Equations (II)

1. $2\cos^2\theta + \cos\theta = 0 \Rightarrow \cos\theta(2\cos\theta + 1) = 0$

$\quad\quad \cos\theta = 0 \quad \Rightarrow \theta = \dfrac{\pi}{2}, \dfrac{3\pi}{2}$

$\quad$ or $\quad 2\cos\theta + 1 = 0$

$\quad\quad 2\cos\theta = -1 \Rightarrow \cos\theta = -\dfrac{1}{2} \Rightarrow \theta = \dfrac{2\pi}{3}, \dfrac{4\pi}{3}$

3. $2\sin^2\theta - \sin\theta - 1 = 0 \Rightarrow (2\sin\theta + 1)(\sin\theta - 1) = 0$

$\quad 2\sin\theta + 1 = 0 \quad \Rightarrow 2\sin\theta = -1 \quad \Rightarrow \sin\theta = -\dfrac{1}{2} \Rightarrow \theta = \dfrac{7\pi}{6}, \dfrac{11\pi}{6}$

$\quad$ or $\quad \sin\theta - 1 = 0 \Rightarrow \sin\theta = 1 \Rightarrow \theta = \dfrac{\pi}{2}$

5. $(\tan\theta - 1)(\sec\theta - 1) = 0$

$\quad \tan\theta - 1 = 0 \quad \Rightarrow \tan\theta = 1 \Rightarrow \theta = \dfrac{\pi}{4}, \dfrac{5\pi}{4}$

or $\sec\theta - 1 = 0 \Rightarrow \sec\theta = 1 \Rightarrow \theta = 0$

7. $\sin^2\theta - \cos^2\theta = 1 + \cos\theta$

$\quad \left(1 - \cos^2\theta\right) - \cos^2\theta = 1 + \cos\theta \Rightarrow 1 - 2\cos^2\theta = 1 + \cos\theta$

$\quad 2\cos^2\theta + \cos\theta = 0 \Rightarrow (\cos\theta)(2\cos\theta + 1) = 0$

$\quad\quad\quad \cos\theta = 0 \Rightarrow \theta = \dfrac{\pi}{2}, \dfrac{3\pi}{2}$

$\quad\quad\quad$ or $2\cos\theta + 1 = 0 \Rightarrow \cos\theta = -\dfrac{1}{2} \Rightarrow \theta = \dfrac{2\pi}{3}, \dfrac{4\pi}{3}$

9. $\sin^2\theta = 6(\cos\theta + 1)$

$\quad\quad 1 - \cos^2\theta = 6(\cos\theta + 1) \Rightarrow 1 - \cos^2\theta = 6\cos\theta + 6$

$\quad\quad \cos^2\theta + 6\cos\theta + 5 = 0 \Rightarrow (\cos\theta + 5)(\cos\theta + 1) = 0$

$\quad\quad\quad\quad \cos\theta + 5 = 0 \Rightarrow \cos\theta = -5,$ which is impossible

$\quad\quad\quad\quad$ or $\cos\theta + 1 \Rightarrow \cos\theta = -1 \Rightarrow \theta = \pi$

$\quad$ the solution is $\theta = \pi$.

11. $\cos(2\theta) + 6\sin^2\theta = 4$

$$1 - 2\sin^2\theta + 6\sin^2\theta = 4 \Rightarrow 4\sin^2\theta = 3 \Rightarrow \sin^2\theta = \frac{3}{4} \Rightarrow \sin\theta = \pm\frac{\sqrt{3}}{2}$$

$$\theta = \frac{\pi}{3}, \frac{2\pi}{3}, \frac{4\pi}{3}, \frac{5\pi}{3}$$

13. $\cos\theta = \sin\theta$

$\dfrac{\sin\theta}{\cos\theta} = 1 \Rightarrow \tan\theta = 1$

$$\theta = \frac{\pi}{4}, \frac{5\pi}{4}$$

15. $\tan\theta = 2\sin\theta$

$\dfrac{\sin\theta}{\cos\theta} = 2\sin\theta$

$\sin\theta = 2\sin\theta\cos\theta$

$0 = 2\sin\theta\cos\theta - \sin\theta$

$0 = \sin\theta(2\cos\theta - 1)$

$$2\cos\theta - 1 = 0 \Rightarrow \cos\theta = \frac{1}{2} \Rightarrow \theta = \frac{\pi}{3}, \frac{5\pi}{3}$$

$$\text{or} \ \ \sin\theta = 0 \Rightarrow \theta = 0, \pi$$

17. $\sin\theta = \csc\theta$

$\sin\theta = \dfrac{1}{\sin\theta}$

$\sin^2\theta = 1 \Rightarrow \sin\theta = \pm 1$

$$\theta = \frac{\pi}{2}, \frac{3\pi}{2}$$

19. $\cos(2\theta) = \cos\theta$

$2\cos^2\theta - 1 = \cos\theta \Rightarrow 2\cos^2\theta - \cos\theta - 1 = 0 \Rightarrow (2\cos\theta + 1)(\cos\theta - 1) = 0$

$$2\cos\theta + 1 = 0 \Rightarrow \cos\theta = -\frac{1}{2} \Rightarrow \theta = \frac{2\pi}{3}, \frac{4\pi}{3}$$

$\text{or} \ \ \cos\theta - 1 = 0 \Rightarrow \cos\theta = 1 \Rightarrow \theta = 0$

21. $\sin(2\theta) + \sin(4\theta) = 0$

$\sin(2\theta) + 2\sin(2\theta)\cos(2\theta) = 0$

$\sin(2\theta)(1 + 2\cos(2\theta)) = 0$

$$1 + 2\cos(2\theta) = 0 \Rightarrow \cos(2\theta) = -\frac{1}{2} \Rightarrow 2\theta = \frac{2\pi}{3} + 2k\pi \Rightarrow \theta = \frac{\pi}{3} + k\pi$$

$$2\theta = \frac{4\pi}{3} + 2k\pi \Rightarrow \theta = \frac{2\pi}{3} + k\pi$$

$$\text{or} \ \ \sin(2\theta) = 0 \Rightarrow 2\theta = 0 + 2k\pi \Rightarrow \theta = k\pi$$

$$2\theta = \pi + 2k\pi \Rightarrow \theta = \frac{\pi}{2} + k\pi$$

$$\theta = 0, \frac{\pi}{3}, \frac{\pi}{2}, \frac{2\pi}{3}, \pi, \frac{4\pi}{3}, \frac{3\pi}{2}, \frac{5\pi}{3}$$

23. $$\cos(4\theta) - \cos(6\theta) = 0$$
$$\cos(5\theta - \theta) - \cos(5\theta + \theta) = 0$$
$$-2\sin(5\theta)\sin(-\theta) = 0$$
$$2\sin(5\theta)\sin\theta = 0$$

$$\sin(5\theta) = 0 \Rightarrow 5\theta = 0 + 2k\pi \quad \Rightarrow \quad \theta = \frac{2k\pi}{5}$$

$$5\theta = \pi + 2k\pi \quad \Rightarrow \quad \theta = \frac{\pi}{5} + \frac{2k\pi}{5}$$

$$\text{or} \quad \sin\theta = 0 \Rightarrow \theta = 0 + 2k\pi$$
$$\theta = \pi + 2k\pi$$

$$\theta = 0, \frac{\pi}{5}, \frac{2\pi}{5}, \frac{3\pi}{5}, \frac{4\pi}{5}, \pi, \frac{6\pi}{5}, \frac{7\pi}{5}, \frac{8\pi}{5}, \frac{9\pi}{5}$$

25. $$1 + \sin\theta = 2\cos^2\theta$$
$$1 + \sin\theta = 2(1 - \sin^2\theta) \Rightarrow 1 + \sin\theta = 2 - 2\sin^2\theta$$
$$2\sin^2\theta + \sin\theta - 1 = 0 \Rightarrow (2\sin\theta - 1)(\sin\theta + 1) = 0$$

$$2\sin\theta - 1 = 0 \Rightarrow \sin\theta = \frac{1}{2} \Rightarrow \theta = \frac{\pi}{6}, \frac{5\pi}{6}$$

$$\text{or} \quad \sin\theta + 1 = 0 \Rightarrow \sin\theta = -1 \Rightarrow \theta = \frac{3\pi}{2}$$

27. $$\tan^2\theta = \frac{3}{2}\sec\theta$$

$$\sec^2\theta - 1 = \frac{3}{2}\sec\theta \Rightarrow 2\sec^2\theta - 2 = 3\sec\theta$$

$$2\sec^2\theta - 3\sec\theta - 2 = 0 \Rightarrow (2\sec\theta + 1)(\sec\theta - 2) = 0$$

$$2\sec\theta + 1 = 0 \quad \Rightarrow \sec\theta = -\frac{1}{2}, \text{ which is impossible}$$

$$\text{or} \quad \sec\theta - 2 = 0 \Rightarrow \sec\theta = 2 \Rightarrow \theta = \frac{\pi}{3}, \frac{5\pi}{3}$$

the solutions are $\theta = \dfrac{\pi}{3}, \dfrac{5\pi}{3}$

29. $$3 - \sin\theta = \cos(2\theta)$$
$$3 - \sin\theta = 1 - 2\sin^2\theta$$
$$2\sin^2\theta - \sin\theta + 2 = 0$$
This is a quadratic equation in $\sin\theta$. The discriminant is $b^2 - 4ac = 1 - 16 = -15 < 0$.
The equation has no real solutions.

31. $$\sec^2\theta + \tan\theta = 0$$
$$\tan^2\theta + 1 + \tan\theta = 0$$
This is a quadratic equation in $\tan\theta$. The discriminant is $b^2 - 4ac = 1 - 4 = -3 < 0$.
The equation has no real solutions.

33. $\sin\theta - \sqrt{3}\cos\theta = 1$

Divide each side by 2:

$$\frac{1}{2}\sin\theta - \frac{\sqrt{3}}{2}\cos\theta = \frac{1}{2}$$

Rewrite in the difference of two angles form where

$$\cos\phi = \frac{1}{2} \text{ and } \sin\phi = \frac{\sqrt{3}}{2} \text{ and } \phi = \frac{\pi}{3}:$$

$$\sin\theta\cos\phi - \cos\theta\sin\phi = \frac{1}{2} \Rightarrow \sin(\theta - \phi) = \frac{1}{2}$$

$$\theta - \phi = \frac{\pi}{6} \quad \text{or} \quad \theta - \phi = \frac{5\pi}{6}$$

$$\theta - \frac{\pi}{3} = \frac{\pi}{6} \quad \text{or} \quad \theta - \frac{\pi}{3} = \frac{5\pi}{6}$$

$$\theta = \frac{\pi}{2} \quad \text{or} \quad \theta = \frac{7\pi}{6}$$

35. $\tan(2\theta) + 2\sin\theta = 0$

$$\frac{\sin(2\theta)}{\cos(2\theta)} + 2\sin\theta = 0$$

$$\frac{\sin 2\theta + 2\sin\theta\cos 2\theta}{\cos 2\theta} = 0 \Rightarrow 2\sin\theta\cos\theta + 2\sin\theta(2\cos^2\theta - 1) = 0$$

$$2\sin\theta\left(\cos\theta + 2\cos^2\theta - 1\right) = 0 \Rightarrow 2\sin\theta\left(2\cos^2\theta + \cos\theta - 1\right) = 0$$

$$2\sin\theta(2\cos\theta - 1)(\cos\theta + 1) = 0$$

$$2\cos\theta - 1 = 0 \Rightarrow \cos\theta = \frac{1}{2} \Rightarrow \theta = \frac{\pi}{3}, \frac{5\pi}{3}$$

$$\text{or } 2\sin\theta = 0 \Rightarrow \sin\theta = 0 \Rightarrow \theta = 0, \pi$$

$$\text{or } \cos\theta + 1 = 0 \Rightarrow \cos\theta = -1 \Rightarrow \theta = \pi$$

the solutions are $\theta = 0, \dfrac{\pi}{3}, \pi, \dfrac{5\pi}{3}$

37. $\sin\theta + \cos\theta = \sqrt{2}$

Divide each side by $\sqrt{2}$: $\dfrac{1}{\sqrt{2}}\sin\theta + \dfrac{1}{\sqrt{2}}\cos\theta = 1$

Rewrite in the sum of two angles form where $\cos\phi = \dfrac{1}{\sqrt{2}}$ and $\sin\phi = \dfrac{1}{\sqrt{2}}$ and $\phi = \dfrac{\pi}{4}:$

$$\sin\theta\cos\phi + \cos\theta\sin\phi = 1 \Rightarrow \sin(\theta + \phi) = 1$$

$$\theta + \phi = \frac{\pi}{2}$$

$$\theta + \frac{\pi}{4} = \frac{\pi}{2} \Rightarrow \theta = \frac{\pi}{4}$$

39.   Use INTERSECT to solve by graphing $y_1 = \cos x$, $y_2 = e^x$.

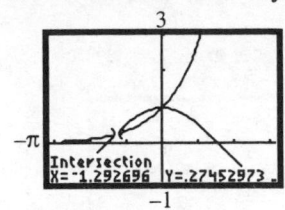 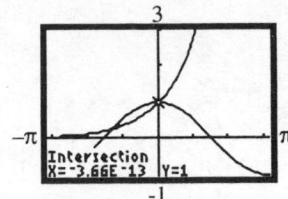

$x \approx -1.29, 0$

41.   Use INTERSECT to solve by graphing $y_1 = 2\sin x$, $y_2 = 0.7x$.

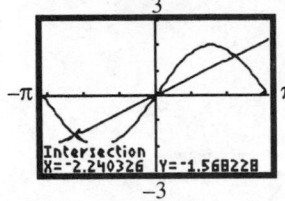 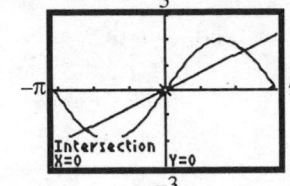 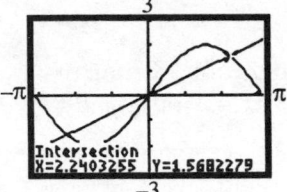

$x \approx -2.24, 0, 2.24$

43.   Use INTERSECT to solve $y_1 = \cos x$, $y_2 = x^2$.

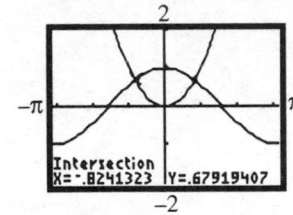 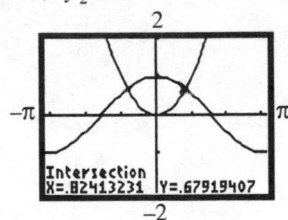

$x \approx -0.82, 0.82$

45. $x + 5\cos x = 0$
       Find the intersection of
       $y_1 = x + 5\cos x$ and $y_2 = 0$:

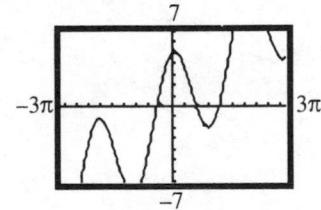

$x \approx -1.31, 1.98, 3.84$

47. $22x - 17\sin x = 3$
       Find the intersection of
       $y_1 = 22x - 17\sin x$ and $y_2 = 3$:

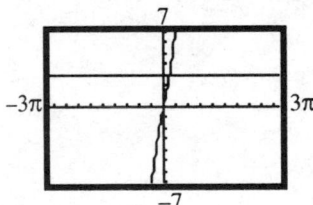

$x \approx 0.52$

387

49. $\sin x + \cos x = x$
     Find the intersection of
     $y_1 = \sin x + \cos x$ and $y_2 = x$:

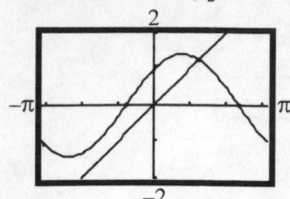

     $x \approx 1.26$

51. $x^2 - 2\cos x = 0$
     Find the intersection of
     $y_1 = x^2 - 2\cos x$ and $y_2 = 0$:

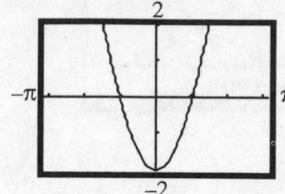

     $x \approx -1.02, 1.02$

53. $x^2 - 2\sin 2x = 3x$
     Find the intersection of
     $y_1 = x^2 - 2\sin 2x$ and $y_2 = 3x$:

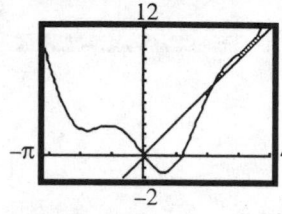

     $x \approx 0, 2.15$

55. $6\sin x - e^x = 2, \ x > 0$
     Find the intersection of
     $y_1 = 6\sin x - e^x$ and $y_2 = 2$:

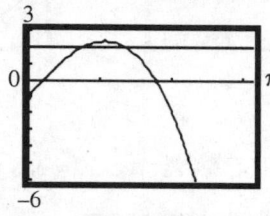

     $x \approx 0.76, 1.35$

57. (a) Solve: $\cos(2\theta) + \cos\theta = 0, \ 0° < \theta < 90°$

     $$2\cos^2\theta - 1 + \cos\theta = 0 \Rightarrow 2\cos^2\theta + \cos\theta - 1 = 0$$

     $$(2\cos\theta - 1)(\cos\theta + 1) = 0$$

     $$2\cos\theta - 1 = 0 \ \Rightarrow \cos\theta = \frac{1}{2} \Rightarrow \theta = 60°, 300°$$

     $$\text{or} \ \cos\theta + 1 = 0 \Rightarrow \cos\theta = -1 \Rightarrow \theta = 180°$$

     The solution is 60°.

     (b) Solve: $\cos(2\theta) + \cos\theta = 0, \ 0° < \theta < 90°$

     $$2\cos\left(\frac{3\theta}{2}\right)\cos\left(\frac{\theta}{2}\right) = 0 \Rightarrow \cos\left(\frac{3\theta}{2}\right) = 0 \ \text{ or } \ \cos\left(\frac{\theta}{2}\right) = 0$$

     $$\frac{3\theta}{2} = 90° \ \Rightarrow \ \theta = 60°$$

     $$\frac{3\theta}{2} = 270° \ \Rightarrow \ \theta = 180°$$

     $$\frac{\theta}{2} = 90° \ \Rightarrow \ \theta = 180°$$

     $$\frac{\theta}{2} = 270° \ \Rightarrow \ \theta = 540°$$

     The solution is 60°.

     (c) $A(60°) = 16\sin(60°)(\cos(60°) + 1) = 16 \cdot \dfrac{\sqrt{3}}{2}\left(\dfrac{1}{2} + 1\right) = 12\sqrt{3} \ \text{in}^2 \approx 20.78 \ \text{in}^2$

(d)   Graph and use the MAXIMUM feature:

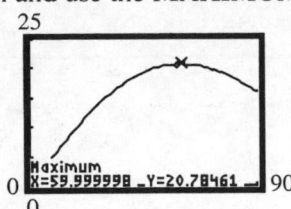

The maximum area is approximately 20.78 in^2 when the angle is 60°.

59.   Graph:

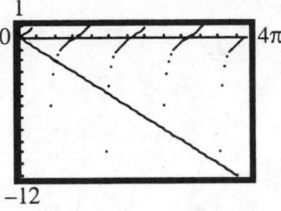

The first two positive solutions are 2.03 and 4.91.

61.   (a) $107 = \dfrac{(34.8)^2 \sin(2\theta)}{9.8}$

$\sin(2\theta) = \dfrac{107(9.8)}{(34.8)^2} \approx 0.8659$

$2\theta = \sin^{-1}(0.8659) \approx 59.98°$ or $120.02°$

$\therefore\quad \theta \approx 29.99°$ or $60.01°$

(b)   Graph and use the MAXIMUM feature:
    The maximum distance is 123.58 meters when the angle is 45°.

(c)   Graph:

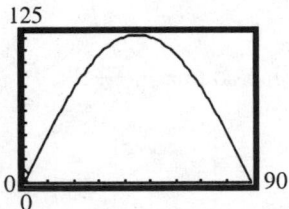

Analytic Trigonometry

6.R   Chapter Review

1. $\sin^{-1}(1)$

We are finding the angle θ, $-\dfrac{\pi}{2} \le \theta \le \dfrac{\pi}{2}$, whose sine equals 1.

$$\sin\theta = 1 \qquad -\frac{\pi}{2} \le \theta \le \frac{\pi}{2}$$

$$\theta = \frac{\pi}{2} \Rightarrow \sin^{-1}(1) = \frac{\pi}{2}$$

3. $\tan^{-1}(1)$

We are finding the angle θ, $-\dfrac{\pi}{2} < \theta < \dfrac{\pi}{2}$, whose tangent equals 1.

$$\tan\theta = 1 \qquad -\frac{\pi}{2} < \theta < \frac{\pi}{2}$$

$$\theta = \frac{\pi}{4} \Rightarrow \tan^{-1}(1) = \frac{\pi}{4}$$

5. $\cos^{-1}\left(-\dfrac{\sqrt{3}}{2}\right)$

We are finding the angle θ, $0 \le \theta \le \pi$, whose cosine equals $-\dfrac{\sqrt{3}}{2}$.

$$\cos\theta = -\frac{\sqrt{3}}{2} \qquad 0 \le \theta \le \pi$$

$$\theta = \frac{5\pi}{6} \Rightarrow \cos^{-1}\left(-\frac{\sqrt{3}}{2}\right) = \frac{5\pi}{6}$$

7. $\sec^{-1}\left(\sqrt{2}\right)$

Find the angle θ, $0 \le \theta \le \pi$, whose secant equals $\sqrt{2}$.

$$\sec\theta = \sqrt{2} \qquad 0 \le \theta \le \pi$$

$$\theta = \frac{\pi}{4}$$

9. $\tan\left(\sin^{-1}\left(-\dfrac{\sqrt{3}}{2}\right)\right)$

Find the angle θ, $-\dfrac{\pi}{2} \le \theta \le \dfrac{\pi}{2}$, whose sine equals $-\dfrac{\sqrt{3}}{2}$.

$$\sin\theta = -\dfrac{\sqrt{3}}{2} \qquad -\dfrac{\pi}{2} \le \theta \le \dfrac{\pi}{2}$$

$$\theta = -\dfrac{\pi}{3} \Rightarrow \tan\left(\sin^{-1}\left(-\dfrac{\sqrt{3}}{2}\right)\right) = \tan\left(-\dfrac{\pi}{3}\right) = -\sqrt{3}$$

11. $\sec\left(\tan^{-1}\left(\dfrac{\sqrt{3}}{3}\right)\right)$

Find the angle θ, $-\dfrac{\pi}{2} < \theta < \dfrac{\pi}{2}$, whose tangent is $\dfrac{\sqrt{3}}{3}$

$$\tan\theta = \dfrac{\sqrt{3}}{3}, \qquad -\dfrac{\pi}{2} < \theta < \dfrac{\pi}{2}$$

$$\theta = \dfrac{\pi}{6} \Rightarrow \sec\left(\tan^{-1}\left(\dfrac{\sqrt{3}}{3}\right)\right) = \sec\left(\dfrac{\pi}{6}\right) = \dfrac{2\sqrt{3}}{3}$$

13. $\sin\left(\tan^{-1}\left(\dfrac{3}{4}\right)\right)$

Since $\tan\theta = \dfrac{3}{4}$, $-\dfrac{\pi}{2} < \theta < \dfrac{\pi}{2}$, let $x = 4$ and $y = 3$. Solve for r:

$$16 + 9 = r^2 \Rightarrow r^2 = 25 \Rightarrow r = 5$$

θ is in quadrant I.

$$\sin\left(\tan^{-1}\left(\dfrac{3}{4}\right)\right) = \sin\theta = \dfrac{y}{r} = \dfrac{3}{5}$$

15. $\tan\left(\sin^{-1}\left(-\dfrac{4}{5}\right)\right)$

Since $\sin\theta = -\dfrac{4}{5}$, $-\dfrac{\pi}{2} \le \theta \le \dfrac{\pi}{2}$, let $y = -4$ and $r = 5$. Solve for x:

$$x^2 + 16 = 25 \Rightarrow x^2 = 9 \Rightarrow x = \pm 3$$

Since θ is in quadrant IV, $x = 3$.

$$\tan\left(\sin^{-1}\left(-\dfrac{4}{5}\right)\right) = \tan\theta = \dfrac{y}{x} = \dfrac{-4}{3} = -\dfrac{4}{3}$$

17. $\sin^{-1}\left(\cos\left(\dfrac{2\pi}{3}\right)\right) = \sin^{-1}\left(-\dfrac{1}{2}\right) = -\dfrac{\pi}{6}$

19. $\tan^{-1}\left(\tan\left(\dfrac{7\pi}{4}\right)\right) = \tan^{-1}(-1) = -\dfrac{\pi}{4}$

21. $\tan\theta\cot\theta - \sin^2\theta = \tan\theta \cdot \dfrac{1}{\tan\theta} - \sin^2\theta = 1 - \sin^2\theta = \cos^2\theta$

23. $\cos^2\theta(1 + \tan^2\theta) = \cos^2\theta \cdot \sec^2\theta = \cos^2\theta \cdot \dfrac{1}{\cos^2\theta} = 1$

25. $4\cos^2\theta + 3\sin^2\theta = \cos^2\theta + 3\cos^2\theta + 3\sin^2\theta = \cos^2\theta + 3(\cos^2\theta + \sin^2\theta)$
$$= \cos^2\theta + 3 \cdot 1 = \cos^2\theta + 3 = 3 + \cos^2\theta$$

27. $\dfrac{1-\cos\theta}{\sin\theta} + \dfrac{\sin\theta}{1-\cos\theta} = \dfrac{(1-\cos\theta)^2 + \sin^2\theta}{\sin\theta(1-\cos\theta)} = \dfrac{1 - 2\cos\theta + \cos^2\theta + \sin^2\theta}{\sin\theta(1-\cos\theta)}$
$$= \dfrac{1 - 2\cos\theta + 1}{\sin\theta(1-\cos\theta)} = \dfrac{2 - 2\cos\theta}{\sin\theta(1-\cos\theta)} = \dfrac{2(1-\cos\theta)}{\sin\theta(1-\cos\theta)} = \dfrac{2}{\sin\theta} = 2\csc\theta$$

29. $\dfrac{\cos\theta}{\cos\theta - \sin\theta} = \dfrac{\cos\theta}{\cos\theta - \sin\theta} \cdot \dfrac{\left(\dfrac{1}{\cos\theta}\right)}{\left(\dfrac{1}{\cos\theta}\right)} = \dfrac{1}{\left(1 - \dfrac{\sin\theta}{\cos\theta}\right)} = \dfrac{1}{1 - \tan\theta}$

31. $\dfrac{\csc\theta}{1 + \csc\theta} = \dfrac{\left(\dfrac{1}{\sin\theta}\right)}{\left(1 + \dfrac{1}{\sin\theta}\right)} = \dfrac{\left(\dfrac{1}{\sin\theta}\right)}{\left(\dfrac{\sin\theta + 1}{\sin\theta}\right)} = \left(\dfrac{1}{1 + \sin\theta}\right) \cdot \left(\dfrac{1 - \sin\theta}{1 - \sin\theta}\right) = \dfrac{1 - \sin\theta}{1 - \sin^2\theta} = \dfrac{1 - \sin\theta}{\cos^2\theta}$

33. $\csc\theta - \sin\theta = \dfrac{1}{\sin\theta} - \sin\theta = \dfrac{1 - \sin^2\theta}{\sin\theta} = \dfrac{\cos^2\theta}{\sin\theta} = \cos\theta \cdot \dfrac{\cos\theta}{\sin\theta} = \cos\theta\cot\theta$

35. $\dfrac{1 - \sin\theta}{\sec\theta} = \cos\theta(1 - \sin\theta) = \cos\theta(1 - \sin\theta) \cdot \dfrac{1 + \sin\theta}{1 + \sin\theta} = \dfrac{\cos\theta(1 - \sin^2\theta)}{1 + \sin\theta}$
$$= \dfrac{\cos\theta(\cos^2\theta)}{1 + \sin\theta} = \dfrac{\cos^3\theta}{1 + \sin\theta}$$

37. $\cot\theta - \tan\theta = \dfrac{\cos\theta}{\sin\theta} - \dfrac{\sin\theta}{\cos\theta} = \dfrac{\cos^2\theta - \sin^2\theta}{\sin\theta\cos\theta} = \dfrac{1 - \sin^2\theta - \sin^2\theta}{\sin\theta\cos\theta} = \dfrac{1 - 2\sin^2\theta}{\sin\theta\cos\theta}$

39. $\dfrac{\cos(\alpha + \beta)}{\cos\alpha\sin\beta} = \dfrac{\cos\alpha\cos\beta - \sin\alpha\sin\beta}{\cos\alpha\sin\beta} = \dfrac{\cos\alpha\cos\beta}{\cos\alpha\sin\beta} - \dfrac{\sin\alpha\sin\beta}{\cos\alpha\sin\beta} = \cot\beta - \tan\alpha$

41. $\dfrac{\cos(\alpha - \beta)}{\cos\alpha\cos\beta} = \dfrac{\cos\alpha\cos\beta + \sin\alpha\sin\beta}{\cos\alpha\cos\beta} = \dfrac{\cos\alpha\cos\beta}{\cos\alpha\cos\beta} + \dfrac{\sin\alpha\sin\beta}{\cos\alpha\cos\beta} = 1 + \tan\alpha\tan\beta$

43. $(1 + \cos\theta)\left(\tan\left(\dfrac{\theta}{2}\right)\right) = (1 + \cos\theta) \cdot \dfrac{\sin\theta}{1 + \cos\theta} = \sin\theta$

45. $2\cot\theta\cot(2\theta) = 2 \cdot \dfrac{\cos\theta}{\sin\theta} \cdot \dfrac{\cos2\theta}{\sin2\theta} = \dfrac{2\cos\theta(\cos^2\theta - \sin^2\theta)}{\sin\theta2\sin\theta\cos\theta} = \dfrac{\cos^2\theta - \sin^2\theta}{\sin^2\theta}$
$$= \cot^2\theta - 1$$

47. $1 - 8\sin^2\theta\cos^2\theta = 1 - 2(2\sin\theta\cos\theta)^2 = 1 - 2\sin^2(2\theta) = \cos(4\theta)$

49. $\dfrac{\sin(2\theta) + \sin(4\theta)}{\cos(2\theta) + \cos(4\theta)} = \dfrac{2\sin(3\theta)\cos(-\theta)}{2\cos(3\theta)\cos(-\theta)} = \dfrac{\sin(3\theta)}{\cos(3\theta)} = \tan(3\theta)$

51. $\dfrac{\cos(2\theta) - \cos(4\theta)}{\cos(2\theta) + \cos(4\theta)} - \tan\theta\tan(3\theta) = \dfrac{-2\sin(3\theta)\sin(-\theta)}{2\cos(3\theta)\cos(-\theta)} - \tan\theta\tan(3\theta)$

$\qquad = \dfrac{2\sin(3\theta)\sin\theta}{2\cos(3\theta)\cos\theta} - \tan\theta\tan(3\theta) = \tan(3\theta)\tan\theta - \tan\theta\tan(3\theta) = 0$

53. $\sin(165°) = \sin(120° + 45°) = \sin(120°)\cos(45°) + \cos(120°)\sin(45°)$

$\qquad = \left(\dfrac{\sqrt{3}}{2}\right)\cdot\left(\dfrac{\sqrt{2}}{2}\right) + \left(-\dfrac{1}{2}\right)\cdot\left(\dfrac{\sqrt{2}}{2}\right) = \dfrac{1}{4}\left(\sqrt{6} - \sqrt{2}\right)$

55. $\cos\left(\dfrac{5\pi}{12}\right) = \cos\left(\dfrac{3\pi}{12} + \dfrac{2\pi}{12}\right) = \cos\left(\dfrac{\pi}{4}\right)\cos\left(\dfrac{\pi}{6}\right) - \sin\left(\dfrac{\pi}{4}\right)\sin\left(\dfrac{\pi}{6}\right) = \dfrac{\sqrt{2}}{2}\cdot\dfrac{\sqrt{3}}{2} - \dfrac{\sqrt{2}}{2}\cdot\dfrac{1}{2}$

$\qquad = \dfrac{1}{4}\left(\sqrt{6} - \sqrt{2}\right)$

57. $\cos(80°)\cos(20°) + \sin(80°)\sin(20°) = \cos(80° - 20°) = \cos(60°) = \dfrac{1}{2}$

59. $\tan\left(\dfrac{\pi}{8}\right) = \tan\left(\dfrac{\left(\dfrac{\pi}{4}\right)}{2}\right) = \sqrt{\dfrac{1 - \cos\left(\dfrac{\pi}{4}\right)}{1 + \cos\left(\dfrac{\pi}{4}\right)}} = \sqrt{\dfrac{1 - \dfrac{\sqrt{2}}{2}}{1 + \dfrac{\sqrt{2}}{2}}} = \sqrt{\left(\dfrac{2 - \sqrt{2}}{2 + \sqrt{2}}\right)\cdot\left(\dfrac{2 - \sqrt{2}}{2 - \sqrt{2}}\right)}$

$\qquad = \sqrt{\dfrac{6 - 4\sqrt{2}}{2}} = \sqrt{3 - 2\sqrt{2}}$

61. $\sin\alpha = \dfrac{4}{5},\ 0 < \alpha < \dfrac{\pi}{2};\qquad \sin\beta = \dfrac{5}{13},\ \dfrac{\pi}{2} < \beta < \pi$

$\cos\alpha = \dfrac{3}{5},\ \tan\alpha = \dfrac{4}{3},\ \cos\beta = -\dfrac{12}{13},\ \tan\beta = -\dfrac{5}{12},\ 0 < \dfrac{\alpha}{2} < \dfrac{\pi}{4},\ \dfrac{\pi}{4} < \dfrac{\beta}{2} < \dfrac{\pi}{2}$

(a) $\sin(\alpha + \beta) = \sin\alpha\cos\beta + \cos\alpha\sin\beta = \left(\dfrac{4}{5}\right)\cdot\left(-\dfrac{12}{13}\right) + \left(\dfrac{3}{5}\right)\cdot\left(\dfrac{5}{13}\right) = \dfrac{-48 + 15}{65} = -\dfrac{33}{65}$

(b) $\cos(\alpha + \beta) = \cos\alpha\cos\beta - \sin\alpha\sin\beta = \left(\dfrac{3}{5}\right)\cdot\left(-\dfrac{12}{13}\right) - \left(\dfrac{4}{5}\right)\cdot\left(\dfrac{5}{13}\right) = \dfrac{-36 - 20}{65} = -\dfrac{56}{65}$

(c) $\sin(\alpha - \beta) = \sin\alpha\cos\beta - \cos\alpha\sin\beta = \left(\dfrac{4}{5}\right)\cdot\left(-\dfrac{12}{13}\right) - \left(\dfrac{3}{5}\right)\cdot\left(\dfrac{5}{13}\right) = \dfrac{-48 - 15}{65} = -\dfrac{63}{65}$

(d) $\tan(\alpha + \beta) = \dfrac{\tan\alpha + \tan\beta}{1 - \tan\alpha\tan\beta} = \dfrac{\left(\dfrac{4}{3} + \left(-\dfrac{5}{12}\right)\right)}{\left(1 - \left(\dfrac{4}{3}\right)\cdot\left(-\dfrac{5}{12}\right)\right)} = \dfrac{\left(\dfrac{11}{12}\right)}{\left(\dfrac{14}{9}\right)} = \dfrac{11}{12}\cdot\dfrac{9}{14} = \dfrac{33}{56}$

(e) $\sin(2\alpha) = 2\sin\alpha\cos\alpha = 2\cdot\dfrac{4}{5}\cdot\dfrac{3}{5} = \dfrac{24}{25}$

(f) $\cos(2\beta) = \cos^2\beta - \sin^2\beta = \left(-\dfrac{12}{13}\right)^2 - \left(\dfrac{5}{13}\right)^2 = \dfrac{144}{169} - \dfrac{25}{169} = \dfrac{119}{169}$

(g) $\sin\left(\dfrac{\beta}{2}\right) = \sqrt{\dfrac{1-\cos\beta}{2}} = \sqrt{\dfrac{1-\left(-\dfrac{12}{13}\right)}{2}} = \sqrt{\dfrac{\left(\dfrac{25}{13}\right)}{2}} = \sqrt{\dfrac{25}{26}} = \dfrac{5}{\sqrt{26}} = \dfrac{5\sqrt{26}}{26}$

(h) $\cos\left(\dfrac{\alpha}{2}\right) = \sqrt{\dfrac{1+\cos\alpha}{2}} = \sqrt{\dfrac{1+\dfrac{3}{5}}{2}} = \sqrt{\dfrac{\left(\dfrac{8}{5}\right)}{2}} = \sqrt{\dfrac{4}{5}} = \dfrac{2}{\sqrt{5}} = \dfrac{2\sqrt{5}}{5}$

63. $\sin\alpha = -\dfrac{3}{5}, \ \pi < \alpha < \dfrac{3\pi}{2};\qquad \cos\beta = \dfrac{12}{13}, \ \dfrac{3\pi}{2} < \beta < 2\pi$

$\cos\alpha = -\dfrac{4}{5}, \tan\alpha = \dfrac{3}{4}, \sin\beta = -\dfrac{5}{13}, \tan\beta = -\dfrac{5}{12}, \ \dfrac{\pi}{2} < \dfrac{\alpha}{2} < \dfrac{3\pi}{4}, \ \dfrac{3\pi}{4} < \dfrac{\beta}{2} < \pi$

(a) $\sin(\alpha+\beta) = \sin\alpha\cos\beta + \cos\alpha\sin\beta = \left(-\dfrac{3}{5}\right)\cdot\left(\dfrac{12}{13}\right) + \left(-\dfrac{4}{5}\right)\cdot\left(-\dfrac{5}{13}\right) = \dfrac{-36+20}{65} = -\dfrac{16}{65}$

(b) $\cos(\alpha+\beta) = \cos\alpha\cos\beta - \sin\alpha\sin\beta = \left(-\dfrac{4}{5}\right)\cdot\left(\dfrac{12}{13}\right) - \left(-\dfrac{3}{5}\right)\cdot\left(-\dfrac{5}{13}\right) = \dfrac{-48-15}{65} = -\dfrac{63}{65}$

(c) $\sin(\alpha-\beta) = \sin\alpha\cos\beta - \cos\alpha\sin\beta = \left(-\dfrac{3}{5}\right)\cdot\left(\dfrac{12}{13}\right) - \left(-\dfrac{4}{5}\right)\cdot\left(-\dfrac{5}{13}\right) = \dfrac{-36-20}{65} = -\dfrac{56}{65}$

(d) $\tan(\alpha+\beta) = \dfrac{\tan\alpha + \tan\beta}{1 - \tan\alpha\tan\beta} = \dfrac{\left(\dfrac{3}{4} + \left(-\dfrac{5}{12}\right)\right)}{\left(1 - \left(\dfrac{3}{4}\right)\cdot\left(-\dfrac{5}{12}\right)\right)} = \dfrac{\left(\dfrac{1}{3}\right)}{\left(\dfrac{21}{16}\right)} = \dfrac{1}{3}\cdot\dfrac{16}{21} = \dfrac{16}{63}$

(e) $\sin(2\alpha) = 2\sin\alpha\cos\alpha = 2\cdot\left(-\dfrac{3}{5}\right)\cdot\left(-\dfrac{4}{5}\right) = \dfrac{24}{25}$

(f) $\cos(2\beta) = \cos^2\beta - \sin^2\beta = \left(\dfrac{12}{13}\right)^2 - \left(-\dfrac{5}{13}\right)^2 = \dfrac{144}{169} - \dfrac{25}{169} = \dfrac{119}{169}$

(g) $\sin\left(\dfrac{\beta}{2}\right) = \sqrt{\dfrac{1-\cos\beta}{2}} = \sqrt{\dfrac{\left(1-\left(\dfrac{12}{13}\right)\right)}{2}} = \sqrt{\dfrac{\left(\dfrac{1}{13}\right)}{2}} = \sqrt{\dfrac{1}{26}} = \dfrac{1}{\sqrt{26}} = \dfrac{\sqrt{26}}{26}$

(h) $\cos\left(\dfrac{\alpha}{2}\right) = -\sqrt{\dfrac{1+\cos\alpha}{2}} = -\sqrt{\dfrac{\left(1+\left(-\dfrac{4}{5}\right)\right)}{2}} = -\sqrt{\dfrac{\left(\dfrac{1}{5}\right)}{2}} = -\sqrt{\dfrac{1}{10}} = -\dfrac{1}{\sqrt{10}} = -\dfrac{\sqrt{10}}{10}$

65. $\tan\alpha = \dfrac{3}{4}, \ \pi < \alpha < \dfrac{3\pi}{2};\qquad \tan\beta = \dfrac{12}{5}, \ 0 < \beta < \dfrac{\pi}{2}$

$\sin\alpha = -\dfrac{3}{5}, \cos\alpha = -\dfrac{4}{5}, \sin\beta = \dfrac{12}{13}, \cos\beta = \dfrac{5}{13}, \ \dfrac{\pi}{2} < \dfrac{\alpha}{2} < \dfrac{3\pi}{4}, \ 0 < \dfrac{\beta}{2} < \dfrac{\pi}{4}$

(a) $\sin(\alpha+\beta) = \sin\alpha\cos\beta + \cos\alpha\sin\beta = \left(-\dfrac{3}{5}\right)\cdot\left(\dfrac{5}{13}\right) + \left(-\dfrac{4}{5}\right)\cdot\left(\dfrac{12}{13}\right) = \dfrac{-15-48}{65} = -\dfrac{63}{65}$

(b) $\cos(\alpha+\beta) = \cos\alpha\cos\beta - \sin\alpha\sin\beta = \left(-\dfrac{4}{5}\right)\cdot\left(\dfrac{5}{13}\right) - \left(-\dfrac{3}{5}\right)\cdot\left(\dfrac{12}{13}\right) = \dfrac{-20+36}{65} = \dfrac{16}{65}$

(c) $\sin(\alpha - \beta) = \sin\alpha\cos\beta - \cos\alpha\sin\beta = \left(-\dfrac{3}{5}\right)\cdot\left(\dfrac{5}{13}\right) - \left(-\dfrac{4}{5}\right)\cdot\left(\dfrac{12}{13}\right) = \dfrac{-15 + 48}{65} = \dfrac{33}{65}$

(d) $\tan(\alpha + \beta) = \dfrac{\tan\alpha + \tan\beta}{1 - \tan\alpha\tan\beta} = \dfrac{\left(\dfrac{3}{4} + \left(\dfrac{12}{5}\right)\right)}{\left(1 - \left(\dfrac{3}{4}\right)\cdot\left(\dfrac{12}{5}\right)\right)} = \dfrac{\left(\dfrac{15 + 48}{20}\right)}{\left(-\dfrac{4}{5}\right)} = \left(\dfrac{63}{20}\right)\cdot\left(-\dfrac{5}{4}\right) = -\dfrac{63}{16}$

(e) $\sin(2\alpha) = 2\sin\alpha\cos\alpha = 2\cdot\left(-\dfrac{3}{5}\right)\cdot\left(-\dfrac{4}{5}\right) = \dfrac{24}{25}$

(f) $\cos(2\beta) = \cos^2\beta - \sin^2\beta = \left(\dfrac{5}{13}\right)^2 - \left(\dfrac{12}{13}\right)^2 = \dfrac{25}{169} - \dfrac{144}{169} = -\dfrac{119}{169}$

(g) $\sin\left(\dfrac{\beta}{2}\right) = \sqrt{\dfrac{1 - \cos\beta}{2}} = \sqrt{\dfrac{\left(1 - \left(\dfrac{5}{13}\right)\right)}{2}} = \sqrt{\dfrac{\left(\dfrac{8}{13}\right)}{2}} = \sqrt{\dfrac{4}{13}} = \dfrac{2}{\sqrt{13}} = \dfrac{2\sqrt{13}}{13}$

(h) $\cos\left(\dfrac{\alpha}{2}\right) = -\sqrt{\dfrac{1 + \cos\alpha}{2}} = -\sqrt{\dfrac{\left(1 + \left(-\dfrac{4}{5}\right)\right)}{2}} = -\sqrt{\dfrac{\left(\dfrac{1}{5}\right)}{2}} = -\sqrt{\dfrac{1}{10}} = -\dfrac{1}{\sqrt{10}} = -\dfrac{\sqrt{10}}{10}$

67. $\sec\alpha = 2,\ -\dfrac{\pi}{2} < \alpha < 0;\qquad \sec\beta = 3,\ \dfrac{3\pi}{2} < \beta < 2\pi$

$\sin\alpha = -\dfrac{\sqrt{3}}{2},\ \cos\alpha = \dfrac{1}{2},\ \tan\alpha = -\sqrt{3},\ \sin\beta = -\dfrac{2\sqrt{2}}{3},\ \cos\beta = \dfrac{1}{3},\ \tan\beta = -2\sqrt{2},$

$-\dfrac{\pi}{4} < \dfrac{\alpha}{2} < 0,\quad \dfrac{3\pi}{4} < \dfrac{\beta}{2} < \pi$

(a) $\sin(\alpha + \beta) = \sin\alpha\cos\beta + \cos\alpha\sin\beta = \left(-\dfrac{\sqrt{3}}{2}\right)\cdot\left(\dfrac{1}{3}\right) + \left(\dfrac{1}{2}\right)\cdot\left(-\dfrac{2\sqrt{2}}{3}\right) = \dfrac{-\sqrt{3} - 2\sqrt{2}}{6}$

(b) $\cos(\alpha + \beta) = \cos\alpha\cos\beta - \sin\alpha\sin\beta = \left(\dfrac{1}{2}\right)\cdot\left(\dfrac{1}{3}\right) - \left(-\dfrac{\sqrt{3}}{2}\right)\cdot\left(-\dfrac{2\sqrt{2}}{3}\right) = \dfrac{1 - 2\sqrt{6}}{6}$

(c) $\sin(\alpha - \beta) = \sin\alpha\cos\beta - \cos\alpha\sin\beta = \left(-\dfrac{\sqrt{3}}{2}\right)\cdot\left(\dfrac{1}{3}\right) - \left(\dfrac{1}{2}\right)\cdot\left(-\dfrac{2\sqrt{2}}{3}\right) = \dfrac{-\sqrt{3} + 2\sqrt{2}}{6}$

(d) $\tan(\alpha + \beta) = \dfrac{\tan\alpha + \tan\beta}{1 - \tan\alpha\tan\beta} = \dfrac{\left(-\sqrt{3} + \left(-2\sqrt{2}\right)\right)}{\left(1 - \left(-\sqrt{3}\right)\left(-2\sqrt{2}\right)\right)} = \left(\dfrac{-\sqrt{3} - 2\sqrt{2}}{1 - 2\sqrt{6}}\right)\cdot\left(\dfrac{1 + 2\sqrt{6}}{1 + 2\sqrt{6}}\right)$

$= \dfrac{-9\sqrt{3} - 8\sqrt{2}}{-23} = \dfrac{9\sqrt{3} + 8\sqrt{2}}{23}$

(e) $\sin(2\alpha) = 2\sin\alpha\cos\alpha = 2\cdot\left(-\dfrac{\sqrt{3}}{2}\right)\cdot\left(\dfrac{1}{2}\right) = -\dfrac{\sqrt{3}}{2}$

(f) $\cos(2\beta) = \cos^2\beta - \sin^2\beta = \left(\dfrac{1}{3}\right)^2 - \left(-\dfrac{2\sqrt{2}}{3}\right)^2 = \dfrac{1}{9} - \dfrac{8}{9} = -\dfrac{7}{9}$

(g) $\sin\left(\dfrac{\beta}{2}\right) = \sqrt{\dfrac{1 - \cos\beta}{2}} = \sqrt{\dfrac{\left(1 - \left(\dfrac{1}{3}\right)\right)}{2}} = \sqrt{\dfrac{\left(\dfrac{2}{3}\right)}{2}} = \sqrt{\dfrac{1}{3}} = \dfrac{1}{\sqrt{3}} = \dfrac{\sqrt{3}}{3}$

(h) $\quad \cos\left(\dfrac{\alpha}{2}\right) = \sqrt{\dfrac{1+\cos\alpha}{2}} = \sqrt{\dfrac{\left(1+\dfrac{1}{2}\right)}{2}} = \sqrt{\dfrac{\left(\dfrac{3}{2}\right)}{2}} = \sqrt{\dfrac{3}{4}} = \dfrac{\sqrt{3}}{2}$

69. $\quad \sin\alpha = -\dfrac{2}{3},\ \pi < \alpha < \dfrac{3\pi}{2};\qquad \cos\beta = -\dfrac{2}{3},\ \pi < \beta < \dfrac{3\pi}{2}$

$\cos\alpha = -\dfrac{\sqrt{5}}{3},\ \tan\alpha = \dfrac{2\sqrt{5}}{5},\ \sin\beta = -\dfrac{\sqrt{5}}{3},\ \tan\beta = \dfrac{\sqrt{5}}{2},\ \dfrac{\pi}{2} < \dfrac{\alpha}{2} < \dfrac{3\pi}{4},\ \dfrac{\pi}{2} < \dfrac{\beta}{2} < \dfrac{3\pi}{4}$

(a) $\quad \sin(\alpha+\beta) = \sin\alpha\cos\beta + \cos\alpha\sin\beta = \left(-\dfrac{2}{3}\right)\cdot\left(-\dfrac{2}{3}\right) + \left(-\dfrac{\sqrt{5}}{3}\right)\cdot\left(-\dfrac{\sqrt{5}}{3}\right) = \dfrac{4+5}{9} = 1$

(b) $\quad \cos(\alpha+\beta) = \cos\alpha\cos\beta - \sin\alpha\sin\beta = \left(-\dfrac{\sqrt{5}}{3}\right)\cdot\left(-\dfrac{2}{3}\right) - \left(-\dfrac{2}{3}\right)\cdot\left(-\dfrac{\sqrt{5}}{3}\right) = \dfrac{2\sqrt{5}-2\sqrt{5}}{9} = 0$

(c) $\quad \sin(\alpha-\beta) = \sin\alpha\cos\beta - \cos\alpha\sin\beta = \left(-\dfrac{2}{3}\right)\cdot\left(-\dfrac{2}{3}\right) - \left(-\dfrac{\sqrt{5}}{3}\right)\cdot\left(-\dfrac{\sqrt{5}}{3}\right) = \dfrac{4-5}{9} = -\dfrac{1}{9}$

(d) $\quad \tan(\alpha+\beta) = \dfrac{\tan\alpha+\tan\beta}{1-\tan\alpha\tan\beta} = \dfrac{\left(\dfrac{2\sqrt{5}}{5}+\dfrac{\sqrt{5}}{2}\right)}{\left(1-\left(\dfrac{2\sqrt{5}}{5}\right)\cdot\left(\dfrac{\sqrt{5}}{2}\right)\right)} = \dfrac{\left(\dfrac{4\sqrt{5}+5\sqrt{5}}{10}\right)}{\left(\dfrac{10-10}{10}\right)} = \dfrac{\left(\dfrac{9\sqrt{5}}{10}\right)}{0}$ Undefined

(e) $\quad \sin(2\alpha) = 2\sin\alpha\cos\alpha = 2\cdot\left(-\dfrac{2}{3}\right)\cdot\left(-\dfrac{\sqrt{5}}{3}\right) = \dfrac{4\sqrt{5}}{9}$

(f) $\quad \cos(2\beta) = \cos^2\beta - \sin^2\beta = \left(-\dfrac{2}{3}\right)^2 - \left(-\dfrac{\sqrt{5}}{3}\right)^2 = \dfrac{4}{9} - \dfrac{5}{9} = -\dfrac{1}{9}$

(g) $\quad \sin\left(\dfrac{\beta}{2}\right) = \sqrt{\dfrac{1-\cos\beta}{2}} = \sqrt{\dfrac{1-\left(-\dfrac{2}{3}\right)}{2}} = \sqrt{\dfrac{\left(\dfrac{5}{3}\right)}{2}} = \sqrt{\dfrac{5}{6}} = \dfrac{\sqrt{30}}{6}$

(h) $\quad \cos\left(\dfrac{\alpha}{2}\right) = -\sqrt{\dfrac{1+\cos\alpha}{2}} = -\sqrt{\dfrac{\left(1+\left(-\dfrac{\sqrt{5}}{3}\right)\right)}{2}} = -\sqrt{\dfrac{\left(\dfrac{3-\sqrt{5}}{3}\right)}{2}} = -\sqrt{\dfrac{3-\sqrt{5}}{6}} = -\dfrac{\sqrt{18-6\sqrt{5}}}{6}$

71. $\quad \cos\left(\sin^{-1}\left(\dfrac{3}{5}\right) - \cos^{-1}\left(\dfrac{1}{2}\right)\right)$

Let $\alpha = \sin^{-1}\left(\dfrac{3}{5}\right)$ and $\beta = \cos^{-1}\left(\dfrac{1}{2}\right)$. α is in quadrant I; β is in quadrant I.

Then $\sin\alpha = \dfrac{3}{5},\ -\dfrac{\pi}{2} \le \alpha \le \dfrac{\pi}{2}$, and $\cos\beta = \dfrac{1}{2}, 0 \le \beta \le \pi$.

$\cos\alpha = \sqrt{1-\sin^2\alpha} = \sqrt{1-\left(\dfrac{3}{5}\right)^2} = \sqrt{1-\dfrac{9}{25}} = \sqrt{\dfrac{16}{25}} = \dfrac{4}{5}$

$\sin\beta = \sqrt{1-\cos^2\beta} = \sqrt{1-\left(\dfrac{1}{2}\right)^2} = \sqrt{1-\dfrac{1}{4}} = \sqrt{\dfrac{3}{4}} = \dfrac{\sqrt{3}}{2}$

$$\cos\left(\sin^{-1}\left(\frac{3}{5}\right)-\cos^{-1}\left(\frac{1}{2}\right)\right)=\cos(\alpha-\beta)=\cos\alpha\cos\beta+\sin\alpha\sin\beta$$

$$=\left(\frac{4}{5}\right)\cdot\left(\frac{1}{2}\right)+\left(\frac{3}{5}\right)\cdot\left(\frac{\sqrt{3}}{2}\right)=\frac{4+3\sqrt{3}}{10}$$

73. $\tan\left[\sin^{-1}\left(-\frac{1}{2}\right)-\tan^{-1}\left(\frac{3}{4}\right)\right]$

Let $\alpha=\sin^{-1}\left(-\frac{1}{2}\right)$ and $\beta=\tan^{-1}\left(\frac{3}{4}\right)$. α is in quadrant IV; β is in quadrant I.

Then $\sin\alpha=-\frac{1}{2},\ -\frac{\pi}{2}\le\alpha\le\frac{\pi}{2}$, and $\tan\beta=\frac{3}{4},-\frac{\pi}{2}<\beta<\frac{\pi}{2}$.

$$\cos\alpha=\sqrt{1-\sin^2\alpha}=\sqrt{1-\left(-\frac{1}{2}\right)^2}=\sqrt{1-\frac{1}{4}}=\sqrt{\frac{3}{4}}=\frac{\sqrt{3}}{2};\quad\tan\alpha=-\frac{1}{\sqrt{3}}=-\frac{\sqrt{3}}{3}$$

$$\tan\left[\sin^{-1}\left(-\frac{1}{2}\right)-\tan^{-1}\left(\frac{3}{4}\right)\right]=\tan(\alpha-\beta)=\frac{\tan\alpha-\tan\beta}{1+\tan\alpha\tan\beta}=\frac{\left(-\frac{\sqrt{3}}{3}-\frac{3}{4}\right)}{\left(1+\left(-\frac{\sqrt{3}}{3}\right)\cdot\left(\frac{3}{4}\right)\right)}$$

$$=\frac{\left(\frac{-4\sqrt{3}-9}{12}\right)}{\left(1-\frac{3\sqrt{3}}{12}\right)}=\left(\frac{-9-4\sqrt{3}}{12-3\sqrt{3}}\right)\left(\frac{12+3\sqrt{3}}{12+3\sqrt{3}}\right)$$

$$=\frac{-144-75\sqrt{3}}{117}=\frac{-48-25\sqrt{3}}{39}$$

75. $\sin\left[2\cos^{-1}\left(-\frac{3}{5}\right)\right]$

Let $\alpha=\cos^{-1}\left(-\frac{3}{5}\right)$. α is in quadrant II.

Then $\cos\alpha=-\frac{3}{5},\ 0\le\alpha\le\pi$.

$$\sin\alpha=\sqrt{1-\cos^2\alpha}=\sqrt{1-\left(-\frac{3}{5}\right)^2}=\sqrt{1-\frac{9}{25}}=\sqrt{\frac{16}{25}}=\frac{4}{5}$$

$$\sin\left[2\cos^{-1}\left(-\frac{3}{5}\right)\right]=\sin 2\alpha=2\sin\alpha\cos\alpha=2\cdot\left(\frac{4}{5}\right)\cdot\left(-\frac{3}{5}\right)=-\frac{24}{25}$$

77. $\cos\theta=\frac{1}{2}$

$\theta=\frac{\pi}{3}+2k\pi$ or $\theta=\frac{5\pi}{3}+2k\pi$, where k is any integer

The solutions on the interval $[0,2\pi)$ are $\theta=\frac{\pi}{3},\frac{5\pi}{3}$

79. $2\cos\theta + \sqrt{2} = 0 \;\Rightarrow\; 2\cos\theta = -\sqrt{2} \;\Rightarrow\; \cos\theta = -\dfrac{\sqrt{2}}{2}$

$\theta = \dfrac{3\pi}{4} + 2k\pi$ or $\theta = \dfrac{5\pi}{4} + 2k\pi,\; k$ is any integer

The solutions on the interval $[0,\, 2\pi)$ are $\theta = \dfrac{3\pi}{4},\, \dfrac{5\pi}{4}$.

81. $\sin(2\theta) + 1 = 0 \;\Rightarrow\; \sin(2\theta) = -1$

$2\theta = \dfrac{3\pi}{2} + 2k\pi \;\Rightarrow\; \theta = \dfrac{3\pi}{4} + k\pi,$ where k is any integer

The solutions on the interval $[0,\, 2\pi)$ are $\theta = \dfrac{3\pi}{4},\, \dfrac{7\pi}{4}$.

83. $\tan(2\theta) = 0$

$2\theta = 0 + k\pi \;\Rightarrow\; \theta = \dfrac{k\pi}{2},$ where k is any integer

The solutions on the interval $[0,\, 2\pi)$ are $\theta = 0,\, \dfrac{\pi}{2},\, \pi,\, \dfrac{3\pi}{2}$.

85. $\sec^2\theta = 4$

$\sec\theta = \pm 2 \Rightarrow \cos\theta = \pm\dfrac{1}{2}$

$\theta = \dfrac{\pi}{3} + k\pi,$ where k is any integer

$\theta = \dfrac{2\pi}{3} + k\pi,$ where k is any integer

The solutions on the interval $[0,\, 2\pi)$ are $\theta = \dfrac{\pi}{3},\dfrac{2\pi}{3},\dfrac{4\pi}{3},\dfrac{5\pi}{3}$

87. $\sin\theta = \tan\theta$

$\sin\theta = \dfrac{\sin\theta}{\cos\theta} \Rightarrow \sin\theta\cos\theta = \sin\theta$

$\sin\theta\cos\theta - \sin\theta = 0 \Rightarrow \sin\theta(\cos\theta - 1) = 0$

$\cos\theta - 1 = 0 \Rightarrow \cos\theta = 1 \to \theta = \pi$

or $\sin\theta = 0 \Rightarrow \theta = 0$

The solutions on the interval $[0,\, 2\pi)$ are $\theta = 0, \pi$.

89. $\sin\theta + \sin(2\theta) = 0$

$\sin\theta + 2\sin\theta\cos\theta = 0 \Rightarrow \sin\theta(1 + 2\cos\theta) = 0$

$1 + 2\cos\theta = 0 \Rightarrow \cos\theta = -\dfrac{1}{2} \;\Rightarrow\; \theta = \dfrac{2\pi}{3}, \dfrac{4\pi}{3}$

or $\sin\theta = 0 \Rightarrow \theta = 0, \pi$

The solutions on the interval $[0,\, 2\pi)$ are $\theta = 0, \dfrac{2\pi}{3}, \pi, \dfrac{4\pi}{3}$.

91. $\sin(2\theta) - \cos\theta - 2\sin\theta + 1 = 0$

$2\sin\theta\cos\theta - \cos\theta - 2\sin\theta + 1 = 0 \Rightarrow \cos\theta(2\sin\theta - 1) - 1(2\sin\theta - 1) = 0$

$(2\sin\theta - 1)(\cos\theta - 1) = 0$

$$\sin\theta = \frac{1}{2} \Rightarrow \theta = \frac{\pi}{6}, \frac{5\pi}{6}$$

or $\cos\theta = 1 \Rightarrow \theta = 0$

The solutions on the interval $[0, 2\pi)$ are $\theta = 0, \dfrac{\pi}{6}, \dfrac{5\pi}{6}$.

93. $2\sin^2\theta - 3\sin\theta + 1 = 0$

$(2\sin\theta - 1)(\sin\theta - 1) = 0$

$$2\sin\theta - 1 = 0 \Rightarrow \sin\theta = \frac{1}{2} \Rightarrow \theta = \frac{\pi}{6}, \frac{5\pi}{6}$$

or $\sin\theta - 1 = 0 \Rightarrow \sin\theta = 1 \Rightarrow \theta = \dfrac{\pi}{2}$

The solutions on the interval $[0, 2\pi)$ are $\theta = \dfrac{\pi}{6}, \dfrac{\pi}{2}, \dfrac{5\pi}{6}$.

95. $4\sin^2\theta = 1 + 4\cos\theta$

$4(1 - \cos^2\theta) = 1 + 4\cos\theta \Rightarrow 4 - 4\cos^2\theta = 1 + 4\cos\theta$

$4\cos^2\theta + 4\cos\theta - 3 = 0 \Rightarrow (2\cos\theta - 1)(2\cos\theta + 3) = 0$

$$2\cos\theta - 1 = 0 \Rightarrow \cos\theta = \frac{1}{2} \Rightarrow \theta = \frac{\pi}{3}, \frac{5\pi}{3}$$

or $2\cos\theta + 3 = 0 \Rightarrow \cos\theta = -\dfrac{3}{2}$, which is impossible

The solutions on the interval $[0, 2\pi)$ are $\theta = \dfrac{\pi}{3}, \dfrac{5\pi}{3}$.

97. $\sin(2\theta) = \sqrt{2}\cos\theta$

$2\sin\theta\cos\theta = \sqrt{2}\cos\theta \Rightarrow 2\sin\theta\cos\theta - \sqrt{2}\cos\theta = 0 \Rightarrow \cos\theta(2\sin\theta - \sqrt{2}) = 0$

$$\cos\theta = 0 \Rightarrow \theta = \frac{\pi}{2}, \frac{3\pi}{2} \quad \text{or} \quad 2\sin\theta - \sqrt{2} = 0 \Rightarrow \sin\theta = \frac{\sqrt{2}}{2} \Rightarrow \theta = \frac{\pi}{4}, \frac{3\pi}{4}$$

The solutions on the interval $[0, 2\pi)$ are $\theta = \dfrac{\pi}{4}, \dfrac{\pi}{2}, \dfrac{3\pi}{4}, \dfrac{3\pi}{2}$.

99. $\sin\theta - \cos\theta = 1$

Divide each side by $\sqrt{2}$:

$$\frac{1}{\sqrt{2}}\sin\theta - \frac{1}{\sqrt{2}}\cos\theta = \frac{1}{\sqrt{2}}$$

Rewrite in the difference of two angles form where

$$\cos\phi = \frac{1}{\sqrt{2}} \text{ and } \sin\phi = \frac{1}{\sqrt{2}} \text{ and } \phi = \frac{\pi}{4}:$$

$$\sin\theta\cos\phi - \cos\theta\sin\phi = \frac{1}{\sqrt{2}}$$

$$\sin(\theta - \phi) = \frac{\sqrt{2}}{2}$$

$$\theta - \phi = \frac{\pi}{4} \quad \text{or} \quad \theta - \phi = \frac{3\pi}{4}$$

$$\theta - \frac{\pi}{4} = \frac{\pi}{4} \quad \text{or} \quad \theta - \frac{\pi}{4} = \frac{3\pi}{4}$$

$$\theta = \frac{\pi}{2} \quad \text{or} \quad \theta = \pi$$

The solutions on the interval $[0, 2\pi)$ are $\theta = \dfrac{\pi}{2}, \pi$.

101. $\sin^{-1}(0.7) \approx 0.78$

103. $\tan^{-1}(-2) \approx -1.11$

105. $\sec^{-1}(3) = \cos^{-1}\left(\dfrac{1}{3}\right) \approx 1.23$

107. $2x = 5\cos x$
     Find the intersection of
     $y_1 = 2x$ and $y_2 = 5\cos x$:

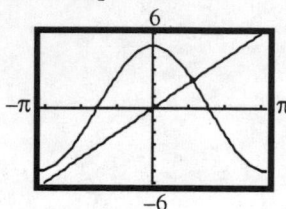

$x \approx 1.11$

109. $2\sin x + 3\cos x = 4x$
     Find the intersection of
     $y_1 = 2\sin x + 3\cos x$ and $y_2 = 4x$:

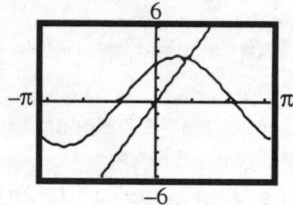

$x \approx 0.87$

111. $\sin x = \ln x$
     Find the intersection of
     $y_1 = \sin x$ and $y_2 = \ln x$:

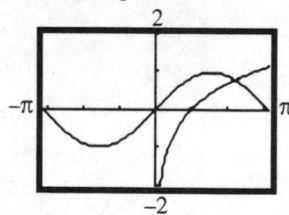

$x \approx 2.22$

Analytic Trigonometry

6.CR Cumulative Review

1. $3x^2 + x - 1 = 0$

$$x = \frac{-b \pm \sqrt{b^2 - 4ac}}{2a} = \frac{-1 \pm \sqrt{1^2 - 4(3)(-1)}}{2(3)} = \frac{-1 \pm \sqrt{1 + 12}}{6} = \frac{-1 \pm \sqrt{13}}{6}$$

   The solution set is $\left\{ \dfrac{-1 - \sqrt{13}}{6}, \dfrac{-1 + \sqrt{13}}{6} \right\}$.

3. $3x + y^2 = 9$

   x-intercepts: $3x + 0^2 = 9 \Rightarrow 3x = 9 \Rightarrow x = 3;\ (3,0)$

   y-intercepts: $3(0) + y^2 = 9 \Rightarrow y^2 = 9 \Rightarrow y = \pm 3;\ (0,-3), (0,3)$

   Test for symmetry:

   x-axis: Replace y by $-y$ so $3x + (-y)^2 = 9 \Rightarrow 3x + y^2 = 9$

   which is equivalent to $3x + y^2 = 9$.

   y-axis: Replace x by $-x$ so $3(-x) + y^2 = 9 \Rightarrow -3x + y^2 = 9$,

   which is not equivalent to $3x + y^2 = 9$.

   Origin: Replace x by $-x$ and y by $-y$ so $3(-x) + (-y)^2 = 9 \Rightarrow -3x + y^2 = 9$,

   which is equivalent to $3x + y^2 = 9$.

   Therefore, the graph is symmetric with respect to the x-axis.

5. $y = 3e^x - 2$

   Using the graph of $y = e^x$, stretch
   vertically by a factor of 3, and vertically
   shift down 2 units.

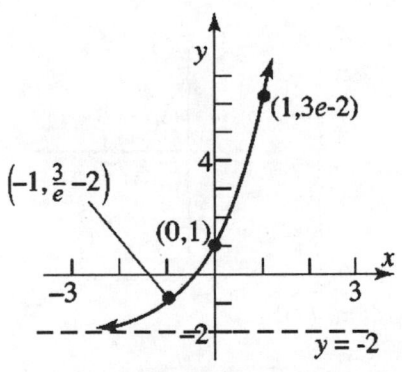

7. (a) $y = x^3$ Inverse function: $y = \sqrt[3]{x}$

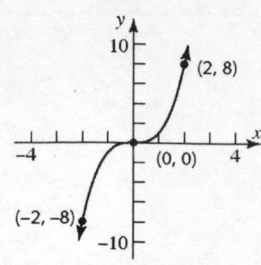   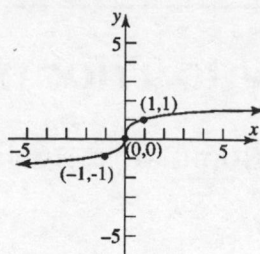

(b) $y = e^x$ Inverse function: $y = \ln x$

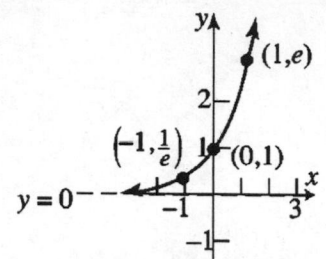   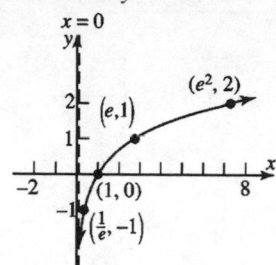

(c) $y = \sin x, \; -\dfrac{\pi}{2} \le x \le \dfrac{\pi}{2}$ Inverse function: $y = \sin^{-1} x$,

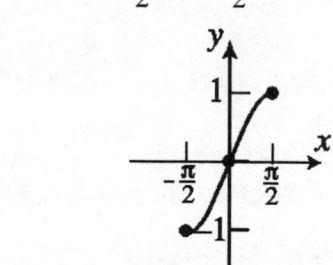   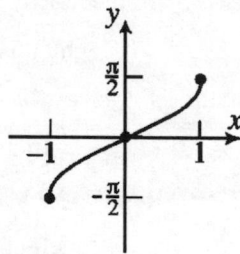

(d) $y = \cos x, \; 0 \le x \le \pi$ Inverse function: $y = \cos^{-1} x$,

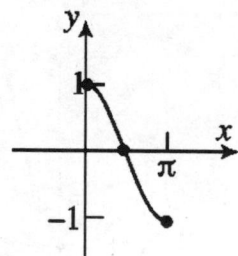   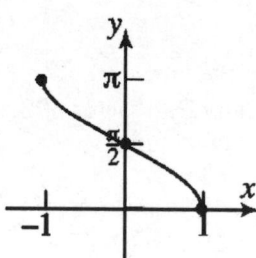

9. $\cos\left(\tan^{-1}(2)\right)$

Find the angle θ, $-\dfrac{\pi}{2} < \theta < \dfrac{\pi}{2}$, whose tangent equals 2.

$\tan\theta = 2, \quad -\dfrac{\pi}{2} < \theta < \dfrac{\pi}{2} \Rightarrow \theta$ is in Quadrant I

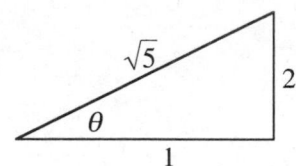

$$\cos\theta = \frac{1}{\sqrt{5}} \cdot \frac{\sqrt{5}}{\sqrt{5}} = \frac{\sqrt{5}}{5}, \text{ Therefore, } \cos\left(\tan^{-1}(2)\right) = \frac{\sqrt{5}}{5}.$$

11. $f(x) = 2x^5 - x^4 - 4x^3 + 2x^2 + 2x - 1$

(a)   Step 1: $f(x)$ has at most 5 real zeros.

Step 2:   Possible rational zeros: $p = \pm 1;\quad q = \pm 1, \pm 2;\quad \dfrac{p}{q} = \pm 1, \pm\dfrac{1}{2}$

Step 3:   Using the Bounds on Zeros Theorem:

$$f(x) = 2\left(x^5 - 0.5x^4 - 2x^3 + x^2 + x - 0.5\right)$$

$$a_4 = -0.5, \quad a_3 = -2, \quad a_2 = 1, \quad a_1 = 1, \quad a_0 = -0.5$$

$$\text{Max }\left\{1, |-0.5| + |\ 1| + |\ 1\ | + |-2| + |-0.5|\ \right\} = \text{Max }\{1, 5\} = 5$$

$$1 + \text{Max }\left\{|-0.5|, |\ 1|, |\ 1\ |, |-2|, |-0.5|\ \right\} = 1 + 2 = 3$$

The smaller of the two numbers is 3. Thus, every zero of f lies
between −3 and 3.
Graphing using the bounds and ZOOM-FIT: (Second graph has a better window.)

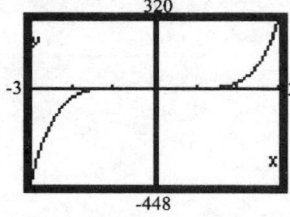 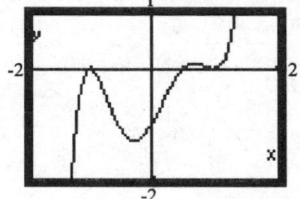

Step 4:   From the graph it appears that there are x-intercepts at −1, 0.5 and 1.
Using synthetic division with −1:

```
-1)2  -1  -4   2   2  -1
        -2   3   1  -3   1
    ────────────────────
     2  -3  -1   3  -1   0
```

Since the remainder is 0, $x - (-1) = x + 1$ is a factor. The other factor is the
quotient: $2x^4 - 3x^3 - x^2 + 3x - 1$.
Using synthetic division with −1 on the quotient:

```
1)2  -3  -1   3  -1
       2  -1  -2   1
   ─────────────────
    2  -1  -2   1   0
```

Since the remainder is 0, $x - 1$ is a factor. The other factor is the
quotient: $2x^3 - x^2 - 2x + 1$.

Using synthetic division with 0.5 on the quotient:

$$0.5 \overline{)\begin{array}{rrrr} 2 & -1 & -2 & 1 \\ & 1 & 0 & -1 \\ \hline 2 & 0 & -2 & 0 \end{array}}$$

Since the remainder is 0, $x - 0.5$ is a factor. The other factor is the quotient: $2x^2 - 2 = 2(x^2 - 1) = 2(x+1)(x-1)$.

Factoring, $f(x) = (x+1)(x-1)(x-0.5)2(x+1)(x-1) = 2(x-0.5)(x+1)^2(x-1)^2$

The real zeros are -1 and 1 (multiplicity 2), and 0.5 (multiplicity 1).

(b)   x-intercepts: $-1, 0.5$ and 1; y-intercept: -1

(c) $y = 2x^5$

(d)   graphing utility:

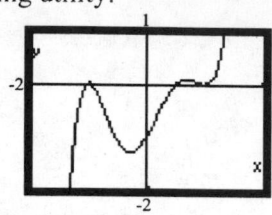

(e)   4 turning points; use MAXIMUM and MINIMUM to locate local maxima at $(-1,0)$, $(0.69, 0.1)$ and local minima at $(1, 0)$, $(-0.290, -1.325)$

(f)   graphing by hand

Interval	$(-\infty, -1)$	$(-1, 0.5)$	$(0.5, 1)$	$(1, \infty)$
Number Chosen	-2	0	0.6	2
Value of f	$f(-2) = -45$	$f(0) = -1$	$f(0.6) \approx 0.8$	$f(2) = 27$
Location of Graph	Below x-axis	Below x-axis	Above x-axis	Above x-axis
Point on Graph	$(-2, -45)$	$(0, -1)$	$(0.6, 0.8)$	$(2, 27)$

f is above the x-axis for $(0.5, 1) \cup (1, \infty)$; f is below the x-axis for $(-\infty, -1) \cup (1, 0.5)$

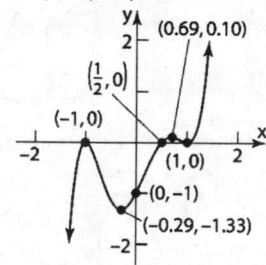

Applications of Trigonometric Functions

7.1 Right Triangle Trigonometry

1. opposite = 5; adjacent = 12
   Find the hypotenuse:
   $$5^2 + 12^2 = (\text{hypotenuse})^2$$
   $$(\text{hypotenuse})^2 = 25 + 144 = 169 \Rightarrow \text{hypotenuse} = 13$$
   $$\sin\theta = \frac{\text{opp}}{\text{hyp}} = \frac{5}{13} \qquad \cos\theta = \frac{\text{adj}}{\text{hyp}} = \frac{12}{13} \qquad \tan\theta = \frac{\text{opp}}{\text{adj}} = \frac{5}{12}$$
   $$\csc\theta = \frac{\text{hyp}}{\text{opp}} = \frac{13}{5} \qquad \sec\theta = \frac{\text{hyp}}{\text{adj}} = \frac{13}{12} \qquad \cot\theta = \frac{\text{adj}}{\text{opp}} = \frac{12}{5}$$

3. opposite = 2; adjacent = 3
   Find the hypotenuse:
   $$2^2 + 3^2 = (\text{hypotenuse})^2$$
   $$(\text{hypotenuse})^2 = 4 + 9 = 13 \Rightarrow \text{hypotenuse} = \sqrt{13}$$
   $$\sin\theta = \frac{\text{opp}}{\text{hyp}} = \frac{2}{\sqrt{13}} = \frac{2\sqrt{13}}{13} \qquad \cos\theta = \frac{\text{adj}}{\text{hyp}} = \frac{3}{\sqrt{13}} = \frac{3\sqrt{13}}{13} \qquad \tan\theta = \frac{\text{opp}}{\text{adj}} = \frac{2}{3}$$
   $$\csc\theta = \frac{\text{hyp}}{\text{opp}} = \frac{\sqrt{13}}{2} \qquad \sec\theta = \frac{\text{hyp}}{\text{adj}} = \frac{\sqrt{13}}{3} \qquad \cot\theta = \frac{\text{adj}}{\text{opp}} = \frac{3}{2}$$

5. adjacent = 2; hypotenuse = 4
   Find the opposite side:
   $$(\text{opposite})^2 + 2^2 = 4^2$$
   $$(\text{opposite})^2 = 16 - 4 = 12 \Rightarrow \text{opposite} = \sqrt{12} = 2\sqrt{3}$$
   $$\sin\theta = \frac{\text{opp}}{\text{hyp}} = \frac{2\sqrt{3}}{4} = \frac{\sqrt{3}}{2} \qquad \cos\theta = \frac{\text{adj}}{\text{hyp}} = \frac{2}{4} = \frac{1}{2} \qquad \tan\theta = \frac{\text{opp}}{\text{adj}} = \frac{2\sqrt{3}}{2} = \sqrt{3}$$
   $$\csc\theta = \frac{\text{hyp}}{\text{opp}} = \frac{4}{2\sqrt{3}} = \frac{2\sqrt{3}}{3} \qquad \sec\theta = \frac{\text{hyp}}{\text{adj}} = \frac{4}{2} = 2 \qquad \cot\theta = \frac{\text{adj}}{\text{opp}} = \frac{2}{2\sqrt{3}} = \frac{\sqrt{3}}{3}$$

7. opposite = $\sqrt{2}$; adjacent = 1
   Find the hypotenuse:
   $$\left(\sqrt{2}\right)^2 + 1^2 = (\text{hypotenuse})^2$$
   $$(\text{hypotenuse})^2 = 2 + 1 = 3 \Rightarrow \text{hypotenuse} = \sqrt{3}$$

$$\sin\theta = \frac{\text{opp}}{\text{hyp}} = \frac{\sqrt{2}}{\sqrt{3}} = \frac{\sqrt{6}}{3} \qquad \cos\theta = \frac{\text{adj}}{\text{hyp}} = \frac{1}{\sqrt{3}} = \frac{\sqrt{3}}{3} \qquad \tan\theta = \frac{\text{opp}}{\text{adj}} = \frac{\sqrt{2}}{1} = \sqrt{2}$$

$$\csc\theta = \frac{\text{hyp}}{\text{opp}} = \frac{\sqrt{3}}{\sqrt{2}} = \frac{\sqrt{6}}{2} \qquad \sec\theta = \frac{\text{hyp}}{\text{adj}} = \frac{\sqrt{3}}{1} = \sqrt{3} \qquad \cot\theta = \frac{\text{adj}}{\text{opp}} = \frac{1}{\sqrt{2}} = \frac{\sqrt{2}}{2}$$

9. opposite = 1; hypotenuse = $\sqrt{5}$
    Find the adjacent side:
    $$1^2 + (\text{adjacent})^2 = \left(\sqrt{5}\right)^2$$

    $$(\text{adjacent})^2 = 5 - 1 = 4 \Rightarrow \text{adjacent} = 2$$

    $$\sin\theta = \frac{\text{opp}}{\text{hyp}} = \frac{1}{\sqrt{5}} = \frac{\sqrt{5}}{5} \qquad \cos\theta = \frac{\text{adj}}{\text{hyp}} = \frac{2}{\sqrt{5}} = \frac{2\sqrt{5}}{5} \qquad \tan\theta = \frac{\text{opp}}{\text{adj}} = \frac{1}{2}$$

    $$\csc\theta = \frac{\text{hyp}}{\text{opp}} = \frac{\sqrt{5}}{1} = \sqrt{5} \qquad \sec\theta = \frac{\text{hyp}}{\text{adj}} = \frac{\sqrt{5}}{2} \qquad \cot\theta = \frac{\text{adj}}{\text{opp}} = \frac{2}{1} = 2$$

11. $\sin(38°) - \cos(52°) = \sin(38°) - \sin(90° - 52°) = \sin(38°) - \sin(38°) = 0$

13. $\dfrac{\cos(10°)}{\sin(80°)} = \dfrac{\sin(90° - 10°)}{\sin(80°)} = \dfrac{\sin(80°)}{\sin(80°)} = 1$

15. $1 - \cos^2(20°) - \cos^2(70°) = \sin^2(20°) - \sin^2(90° - 70°) = \sin^2(20°) - \sin^2(20°) = 0$

17. $\tan(20°) - \dfrac{\cos(70°)}{\cos(20°)} = \tan 20° - \dfrac{\sin(90° - 70°)}{\cos(20°)} = \tan(20°) - \dfrac{\sin(20°)}{\cos(20°)} = \tan(20°) - \tan(20°) = 0$

19. $\cos(35°)\sin(55°) + \sin(35°)\cos(55°) = \sin(55° + 35°) = \sin(90°) = 1$

21. $b = 5, \ \beta = 20°$
    $$\sin\beta = \frac{b}{c} \ \rightarrow \ \sin(20°) = \frac{5}{c} \ \rightarrow \ c = \frac{5}{\sin(20°)} \approx \frac{5}{0.3420} \approx 14.62$$

    $$\tan\beta = \frac{b}{a} \ \rightarrow \ \tan(20°) = \frac{5}{a} \ \rightarrow \ a = \frac{5}{\tan(20°)} \approx \frac{5}{0.3640} \approx 13.74$$

    $$\alpha = 90° - \beta = 90° - 20° = 70°$$

23. $a = 6, \ \beta = 40°$
    $$\cos\beta = \frac{a}{c} \ \rightarrow \ \cos(40°) = \frac{6}{c} \ \rightarrow \ c = \frac{6}{\cos(40°)} \approx \frac{6}{0.7660} \approx 7.83$$

    $$\tan\beta = \frac{b}{a} \ \rightarrow \ \tan(40°) = \frac{b}{6} \ \rightarrow \ b = 6\tan(40°) \approx 6 \cdot (0.8391) \approx 5.03$$

    $$\alpha = 90° - \beta = 90° - 40° = 50°$$

25. $b = 4, \ \alpha = 10°$

$\tan\alpha = \dfrac{a}{b} \ \rightarrow \ \tan(10°) = \dfrac{a}{4} \ \rightarrow \ a = 4\tan(10°) \approx 4 \cdot (0.1763) \approx 0.71$

$\cos\alpha = \dfrac{b}{c} \ \rightarrow \ \cos(10°) = \dfrac{4}{c} \ \rightarrow \ c = \dfrac{4}{\cos(10°)} \approx \dfrac{4}{0.9848} \approx 4.06$

$\beta = 90° - \alpha = 90° - 10° = 80°$

27. $a = 5, \ \alpha = 25°$

$\cot\alpha = \dfrac{b}{a} \ \rightarrow \ \cot(25°) = \dfrac{b}{5} \ \rightarrow \ b = 5\cot(25°) \approx 5 \cdot (2.1445) \approx 10.72$

$\csc\alpha = \dfrac{c}{a} \ \rightarrow \ \csc(25°) = \dfrac{c}{5} \ \rightarrow \ c = 5\csc(25°) \approx 5 \cdot (2.3662) \approx 11.83$

$\beta = 90° - \alpha = 90° - 25° = 65°$

29. $c = 9, \ \beta = 20°$

$\sin\beta = \dfrac{b}{c} \ \rightarrow \ \sin(20°) = \dfrac{b}{9} \ \rightarrow \ b = 9\sin(20°) \approx 9 \cdot (0.3420) \approx 3.08$

$\cos\beta = \dfrac{a}{c} \ \rightarrow \ \cos(20°) = \dfrac{a}{9} \ \rightarrow \ a = 9\cos(20°) \approx 9 \cdot (0.9397) \approx 8.46$

$\alpha = 90° - \alpha = 90° - 20° = 70°$

31. $a = 5, \ b = 3$

$c^2 = a^2 + b^2 = 5^2 + 3^2 = 25 + 9 = 34 \ \rightarrow \ c = \sqrt{34} \approx 5.83$

$\tan\alpha = \dfrac{a}{b} = \dfrac{5}{3} \approx 1.6667 \ \rightarrow \ \alpha \approx 59.0°$

$\beta = 90° - \alpha = 90° - 59.0° = 31.0°$

33. $a = 2, \ c = 5$

$c^2 = a^2 + b^2 \ \rightarrow \ b^2 = c^2 - a^2 = 5^2 - 2^2 = 25 - 4 = 21 \ \rightarrow \ b = \sqrt{21} \approx 4.58$

$\sin\alpha = \dfrac{a}{c} = \dfrac{2}{5} = 0.4000 \ \rightarrow \ \alpha \approx 23.6°$

$\beta = 90° - \alpha = 90° - 23.6° = 66.4°$

35. $c = 8, \ \alpha = 35°$

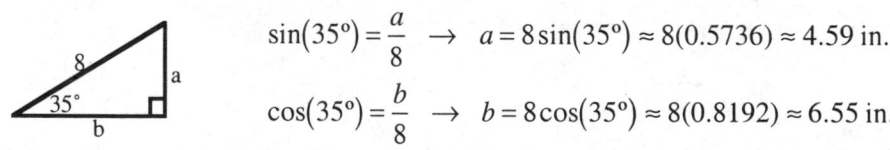

$\sin(35°) = \dfrac{a}{8} \ \rightarrow \ a = 8\sin(35°) \approx 8(0.5736) \approx 4.59 \text{ in.}$

$\cos(35°) = \dfrac{b}{8} \ \rightarrow \ b = 8\cos(35°) \approx 8(0.8192) \approx 6.55 \text{ in.}$

37. $\alpha = 25°, \ a = 5$

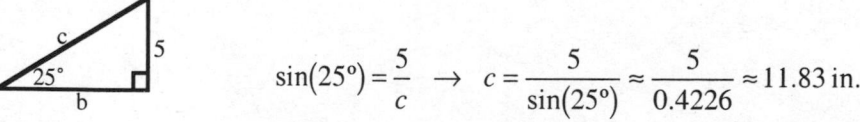

$\sin(25°) = \dfrac{5}{c} \ \rightarrow \ c = \dfrac{5}{\sin(25°)} \approx \dfrac{5}{0.4226} \approx 11.83 \text{ in.}$

$\alpha = 25°, \quad b = 5$

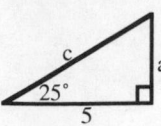

$$\cos(25°) = \frac{5}{c} \quad \rightarrow \quad c = \frac{5}{\cos(25°)} \approx \frac{5}{0.9063} \approx 5.52 \text{ in.}$$

39. $c = 5, \quad a = 2$

$$\sin \alpha = \frac{2}{5} = 0.4000 \quad \rightarrow \quad \alpha \approx 23.6 \quad \rightarrow \quad \beta = 90° - \alpha = 90° - 23.6° \approx 66.4°$$

41. $\tan(35°) = \dfrac{b}{100} \quad \rightarrow \quad b = 100\tan(35°) \approx 100(0.7002) \approx 70.02 \text{ feet}$

43. $\tan(85.361°) = \dfrac{a}{80} \quad \rightarrow \quad a = 80\tan(85.361°) \approx 80(12.3239) \approx 985.91 \text{ feet}$

45.

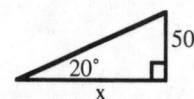

$$\tan(20°) = \frac{50}{x} \quad \rightarrow \quad x = \frac{50}{\tan(20°)} \approx \frac{50}{0.3640} \approx 137.37 \text{ meters}$$

47.

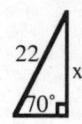

$$\sin(70°) = \frac{x}{22} \quad \rightarrow \quad x = 22\sin(70°) \approx 22(0.9397) \approx 20.67 \text{ feet}$$

49.   opposite side = 10 feet, adjacent side = 35 feet

$$\tan\theta = \frac{10}{35} \approx 0.2857 \quad \rightarrow \quad \theta = \tan^{-1}\left(\frac{10}{35}\right) \approx 15.9°$$

51.   Let h represent the height of Lincoln's face.

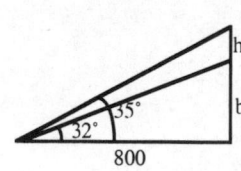

$$\tan(32°) = \frac{b}{800} \quad \rightarrow \quad b = 800\tan(32°) \approx 800(0.6249) \approx 499.9$$

$$\tan(35°) = \frac{b+h}{800} \quad \rightarrow \quad b+h = 800\tan(35°) \approx 800(0.7002) \approx 560.2$$

$$h = (b+h) - b = 560.2 - 499.9 \approx 60.27 \text{ feet}$$

53.

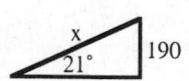

$$\sin(21°) = \frac{190}{x} \quad \rightarrow \quad x = \frac{190}{\sin(21°)} \approx \frac{190}{0.3584} \approx 530.18 \text{ ft.}$$

55.

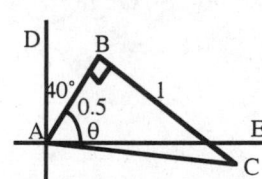

$$\tan(35.1°) = \frac{x}{789} \quad \rightarrow \quad x = 789\tan(35.1°) \approx 789(0.7028) \approx 554.52 \text{ ft}$$

57. (a) $\tan(15°) = \frac{30}{x} \quad \rightarrow \quad x = \frac{30}{\tan(15°)} \approx \frac{30}{0.2679} \approx 111.96 \text{ feet}$

The truck is traveling at 111.96 ft/sec.

$$\frac{111.96 \text{ ft}}{\text{sec}} \cdot \frac{1 \text{ mile}}{5280 \text{ ft}} \cdot \frac{3600 \text{ sec}}{\text{hr}} \approx 76.34 \text{ mi / hr}$$

(b) $\tan(20°) = \frac{30}{x} \quad \rightarrow \quad x = \frac{30}{\tan(20°)} \approx \frac{30}{0.3640} \approx 82.42 \text{ feet}$

The truck is traveling at 82.42 ft/sec.

$$\frac{82.42 \text{ ft}}{\text{sec}} \cdot \frac{1 \text{ mile}}{5280 \text{ ft}} \cdot \frac{3600 \text{ sec}}{\text{hr}} \approx 56.20 \text{ mi / hr}$$

(c) A ticket is issued for traveling at a speed of 60 mi/hr or more.

$$\frac{60 \text{ mi}}{\text{hr}} \cdot \frac{5280 \text{ ft}}{\text{mi}} \cdot \frac{1\text{hr}}{3600 \text{ sec}} = 88 \text{ ft / sec.}$$

If $\tan \theta < \frac{30}{88}$, the trooper should issue a ticket.

A ticket is issued if $\theta < 18.8°$.

59. Find angle θ: (see the figure)

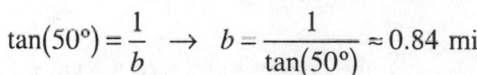

$\tan\theta = \frac{1}{0.5} = 2 \quad \rightarrow \quad \theta \approx 63.4°$

$\angle DAC = 40° + 63.4° = 103.4°$

$\angle EAC = 103.4° - 90° = 13.4°$

The bearing the control tower should use is S76.6°E.

61. $\tan \alpha = \frac{10-6}{15} = \frac{4}{15} \quad \rightarrow \quad \alpha = 14.9°$

63. The length of the highway $= x + y + z$

$\sin(40°) = \frac{1}{x} \quad \rightarrow \quad x = \frac{1}{\sin(40°)} \approx 1.56 \text{ mi}$

$\sin(50°) = \frac{1}{z} \quad \rightarrow \quad z = \frac{1}{\sin(50°)} \approx 1.31 \text{ mi}$

$\tan(40°) = \frac{1}{a} \quad \rightarrow \quad a = \frac{1}{\tan(40°)} \approx 1.19 \text{ mi}$

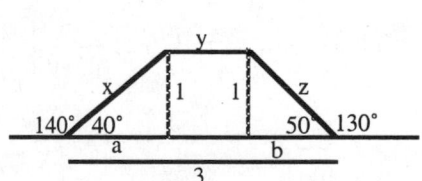

$\tan(50°) = \frac{1}{b} \quad \rightarrow \quad b = \frac{1}{\tan(50°)} \approx 0.84 \text{ mi}$

$a + y + b = 3 \quad \rightarrow \quad y = 3 - a - b = 3 - 1.19 - 0.84 = 0.97 \text{ mi}$

The length of the highway is: $1.56 + 0.97 + 1.31 = 3.83$ miles.

65.   In order to see George's head and feet the camera must be x feet from George.
      Solve:

$$\tan(20°) = \frac{4}{x} \quad \rightarrow \quad x = \frac{4}{\tan(20°)} \approx 10.99 \text{ feet}$$

      The camera will need to be moved back 1 foot to see George's feet.

67. $\sin\theta = \frac{y}{1} = y; \qquad \cos\theta = \frac{x}{1} = x$

      (a) $A = 2xy = 2\cos\theta\sin\theta$
      (b) $2\cos\theta\sin\theta = 2\sin\theta\cos\theta = \sin(2\theta)$
      (c)   The largest value of the sine function is 1. Solve:
            $$\sin(2\theta) = 1$$

            $$2\theta = \frac{\pi}{2}$$

            $$\theta = \frac{\pi}{4}$$

      (d) $x = \cos\left(\frac{\pi}{4}\right) = \frac{\sqrt{2}}{2} \qquad y = \sin\left(\frac{\pi}{4}\right) = \frac{\sqrt{2}}{2}$

            The dimensions are $\sqrt{2}$ by $\frac{\sqrt{2}}{2}$.

69.   Consider the diagram showing the lighthouse at point L, relative to the center of Earth,
      using the radius of Earth as 3960 miles. Let P refer to the furthest point on the horizon from
      which the light is visible. Note also that $362 \text{ feet} = \frac{362}{5280}$ miles.

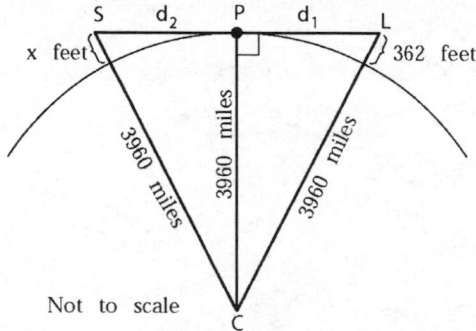

      Apply the Pythagorean Theorem to the right triangle, $\triangle CPL$,

      $$(3960)^2 + (d_1)^2 = \left(3960 + \frac{362}{5280}\right)^2$$

      $$(d_1)^2 = \left(3960 + \frac{362}{5280}\right)^2 - (3960)^2 \rightarrow d_1 = \sqrt{\left(3960 + \frac{362}{5280}\right)^2 - (3960)^2} \approx 23.30 \text{ miles.}$$

      Verify the ship information:
            Let S refer to the ship's location, and let x equal the height, in feet, of the ship.
            We need $d_1 + d_2 \geq 40$. Since $d_1 \approx 23.30$ miles, we need
                $d_2 \geq 40 - 23.30 = 16.70$ miles.

Apply the Pythagorean Theorem to the right triangle, ΔCPS ,

$$\left(3960\right)^2 + \left(16.7\right)^2 = \left(3960 + x\right)^2 \rightarrow \sqrt{\left(3960\right)^2 + \left(16.7\right)^2} = 3960 + x$$

$$x = \sqrt{\left(3960\right)^2 + \left(16.7\right)^2} - 3960 \approx 0.035 \ \text{miles} \ \approx \ 185.93 \ \text{feet.}$$

The ship would have to be at least 186 feet tall to see the lighthouse from 40 miles away.

Verify the airplane information:

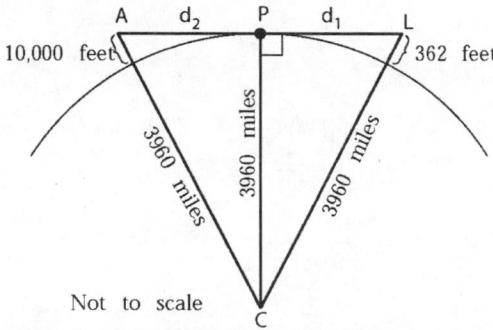

Apply the Pythagorean Theorem to the right triangle, ΔCPL ,

$$\left(3960\right)^2 + \left(d_1\right)^2 = \left(3960 + \frac{362}{5280}\right)^2$$

$$\left(d_1\right)^2 = \left(3960 + \frac{362}{5280}\right)^2 - \left(3960\right)^2 \rightarrow d_1 = \sqrt{\left(3960 + \frac{362}{5280}\right)^2 - \left(3960\right)^2} \approx 23.30 \ \text{miles.}$$

Let A refer to the airplane's location. The distance from the plane to point P is d_2. We want to show that $d_1 + d_2 \geq 120$.

Assume the altitude of the airplane is 10,000 feet $= \dfrac{10000}{5280} \approx 1.89$ miles..

Apply the Pythagorean Theorem to the right triangle, ΔCPA ,

$$\left(3960\right)^2 + \left(d_2\right)^2 = \left(3960 + \frac{10000}{5280}\right)^2$$

$$\left(d_2\right)^2 = \left(3960 + \frac{10000}{5280}\right)^2 - \left(3960\right)^2$$

$$d_2 = \sqrt{\left(3960 + \frac{10000}{5280}\right)^2 - \left(3960\right)^2} \approx 122.49 \ \text{miles.}$$

Therefore, $d_1 + d_2 \approx 23.30 + 122.49 = 145.79 \geq 120$.

A plane at an altitude of 6233 feet could see the lighthouse from 120 miles away.

Chapter 7

Applications of Trigonometric Functions

7.2   The Law of Sines

1. $c = 5$, $\beta = 45°$, $\gamma = 95°$
     $\alpha = 180° - \beta - \gamma = 180° - 45° - 95° = 40°$

$$\frac{\sin\alpha}{a} = \frac{\sin\gamma}{c} \quad \rightarrow \quad \frac{\sin(40°)}{a} = \frac{\sin(95°)}{5} \quad \rightarrow \quad a = \frac{5\sin(40°)}{\sin(95°)} \approx 3.23$$

$$\frac{\sin\beta}{b} = \frac{\sin\gamma}{c} \quad \rightarrow \quad \frac{\sin(45°)}{b} = \frac{\sin(95°)}{5} \quad \rightarrow \quad b = \frac{5\sin(45°)}{\sin(95°)} \approx 3.55$$

3. $b = 3$, $\alpha = 50°$, $\gamma = 85°$
     $\beta = 180° - \alpha - \gamma = 180° - 50° - 85° = 45°$

$$\frac{\sin\alpha}{a} = \frac{\sin\beta}{b} \quad \rightarrow \quad \frac{\sin(50°)}{a} = \frac{\sin(45°)}{3} \quad \rightarrow \quad a = \frac{3\sin(50°)}{\sin(45°)} \approx 3.25$$

$$\frac{\sin\gamma}{c} = \frac{\sin\beta}{b} \quad \rightarrow \quad \frac{\sin(85°)}{c} = \frac{\sin(45°)}{3} \quad \rightarrow \quad c = \frac{3\sin(85°)}{\sin(45°)} \approx 4.23$$

5. $b = 7$, $\alpha = 40°$, $\beta = 45°$
     $\gamma = 180° - \alpha - \beta = 180° - 40° - 45° = 95°$

$$\frac{\sin\alpha}{a} = \frac{\sin\beta}{b} \quad \rightarrow \quad \frac{\sin(40°)}{a} = \frac{\sin(45°)}{7} \quad \rightarrow \quad a = \frac{7\sin(40°)}{\sin(45°)} \approx 6.36$$

$$\frac{\sin\gamma}{c} = \frac{\sin\beta}{b} \quad \rightarrow \quad \frac{\sin(95°)}{c} = \frac{\sin(45°)}{7} \quad \rightarrow \quad c = \frac{7\sin(95°)}{\sin(45°)} \approx 9.86$$

7. $b = 2$, $\beta = 40°$, $\gamma = 100°$
     $\alpha = 180° - \beta - \gamma = 180° - 40° - 100° = 40°$

$$\frac{\sin\alpha}{a} = \frac{\sin\beta}{b} \quad \rightarrow \quad \frac{\sin(40°)}{a} = \frac{\sin(40°)}{2} \quad \rightarrow \quad a = \frac{2\sin(40°)}{\sin(40°)} = 2$$

$$\frac{\sin\gamma}{c} = \frac{\sin\beta}{b} \quad \rightarrow \quad \frac{\sin(100°)}{c} = \frac{\sin(40°)}{2} \quad \rightarrow \quad c = \frac{2\sin(100°)}{\sin(40°)} \approx 3.06$$

9. $\alpha = 40°$, $\beta = 20°$, $a = 2$
    $\gamma = 180° - \alpha - \beta = 180° - 40° - 20° = 120°$

    $\dfrac{\sin\alpha}{a} = \dfrac{\sin\beta}{b}$ $\rightarrow$ $\dfrac{\sin(40°)}{2} = \dfrac{\sin(20°)}{b}$ $\rightarrow$ $b = \dfrac{2\sin(20°)}{\sin(40°)} \approx 1.06$

    $\dfrac{\sin\gamma}{c} = \dfrac{\sin\alpha}{a}$ $\rightarrow$ $\dfrac{\sin(120°)}{c} = \dfrac{\sin(40°)}{2}$ $\rightarrow$ $c = \dfrac{2\sin(120°)}{\sin(40°)} \approx 2.69$

11. $\beta = 70°$, $\gamma = 10°$, $b = 5$
    $\alpha = 180° - \beta - \gamma = 180° - 70° - 10° = 100°$

    $\dfrac{\sin\alpha}{a} = \dfrac{\sin\beta}{b}$ $\rightarrow$ $\dfrac{\sin(100°)}{a} = \dfrac{\sin(70°)}{5}$ $\rightarrow$ $a = \dfrac{5\sin(100°)}{\sin(70°)} \approx 5.24$

    $\dfrac{\sin\gamma}{c} = \dfrac{\sin\beta}{b}$ $\rightarrow$ $\dfrac{\sin(10°)}{c} = \dfrac{\sin(70°)}{5}$ $\rightarrow$ $c = \dfrac{5\sin(10°)}{\sin(70°)} \approx 0.92$

13. $\alpha = 110°$, $\gamma = 30°$, $c = 3$
    $\beta = 180° - \alpha - \gamma = 180° - 110° - 30° = 40°$

    $\dfrac{\sin\alpha}{a} = \dfrac{\sin\gamma}{c}$ $\rightarrow$ $\dfrac{\sin(110°)}{a} = \dfrac{\sin(30°)}{3}$ $\rightarrow$ $a = \dfrac{3\sin(110°)}{\sin(30°)} \approx 5.64$

    $\dfrac{\sin\gamma}{c} = \dfrac{\sin\beta}{b}$ $\rightarrow$ $\dfrac{\sin(30°)}{3} = \dfrac{\sin(40°)}{b}$ $\rightarrow$ $b = \dfrac{3\sin(40°)}{\sin(30°)} \approx 3.86$

15. $\alpha = 40°$, $\beta = 40°$, $c = 2$
    $\gamma = 180° - \alpha - \beta = 180° - 40° - 40° = 100°$

    $\dfrac{\sin\alpha}{a} = \dfrac{\sin\gamma}{c}$ $\rightarrow$ $\dfrac{\sin(40°)}{a} = \dfrac{\sin(100°)}{2}$ $\rightarrow$ $a = \dfrac{2\sin(40°)}{\sin(100°)} \approx 1.31$

    $\dfrac{\sin\beta}{b} = \dfrac{\sin\gamma}{c}$ $\rightarrow$ $\dfrac{\sin(40°)}{b} = \dfrac{\sin(100°)}{2}$ $\rightarrow$ $b = \dfrac{2\sin(40°)}{\sin(100°)} \approx 1.31$

17. $a = 3$, $b = 2$, $\alpha = 50°$

    $\dfrac{\sin\beta}{b} = \dfrac{\sin\alpha}{a}$ $\rightarrow$ $\dfrac{\sin\beta}{2} = \dfrac{\sin(50°)}{3}$ $\rightarrow$ $\sin\beta = \dfrac{2\sin(50°)}{3} \approx 0.5107$

    $\rightarrow$ $\beta = 30.7°$ or $\beta = 149.3°$

    The second value is discarded because $\alpha + \beta > 180°$.
    $\gamma = 180° - \alpha - \beta = 180° - 50° - 30.7° = 99.3°$

    $\dfrac{\sin\gamma}{c} = \dfrac{\sin\alpha}{a}$ $\rightarrow$ $\dfrac{\sin(99.3°)}{c} = \dfrac{\sin(50°)}{3}$ $\rightarrow$ $c = \dfrac{3\sin(99.3°)}{\sin(50°)} \approx 3.86$

    One triangle: $\beta \approx 30.7°$, $\gamma \approx 99.3°$, $c \approx 3.86$

19. $b = 5$, $c = 3$, $\beta = 100°$

$$\frac{\sin\beta}{b} = \frac{\sin\gamma}{c} \rightarrow \frac{\sin(100°)}{5} = \frac{\sin\gamma}{3} \rightarrow \sin\gamma = \frac{3\sin(100°)}{5} = 0.5909$$

$$\rightarrow \gamma = 36.2° \text{ or } \gamma = 143.8°$$

The second value is discarded because $\beta + \gamma > 180°$.

$\alpha = 180° - \beta - \gamma = 180° - 100° - 36.2° = 43.8°$

$$\frac{\sin\beta}{b} = \frac{\sin\alpha}{a} \rightarrow \frac{\sin(100°)}{5} = \frac{\sin(43.8°)}{a} \rightarrow a = \frac{5\sin(43.8°)}{\sin(100°)} \approx 3.51$$

One triangle: $\alpha \approx 43.8°$, $\gamma \approx 36.2°$, $a \approx 3.51$

21. $a = 4$, $b = 5$, $\alpha = 60°$

$$\frac{\sin\beta}{b} = \frac{\sin\alpha}{a} \rightarrow \frac{\sin\beta}{5} = \frac{\sin(60°)}{4} \rightarrow \sin\beta = \frac{5\sin(60°)}{4} = 1.0825$$

There is no angle β for which $\sin\beta > 1$. Therefore, there is no triangle with the given measurements.

23. $b = 4$, $c = 6$, $\beta = 20°$

$$\frac{\sin\beta}{b} = \frac{\sin\gamma}{c} \rightarrow \frac{\sin(20°)}{4} = \frac{\sin\gamma}{6} \rightarrow \sin\gamma = \frac{6\sin(20°)}{4} \approx 0.5130$$

$$\rightarrow \gamma_1 = 30.9° \text{ or } \gamma_2 = 149.1°$$

For both values, $\beta + \gamma < 180°$. Therefore, there are two triangles.

$\alpha_1 = 180° - \beta - \gamma_1 = 180° - 20° - 30.9° = 129.1°$

$$\frac{\sin\beta}{b} = \frac{\sin\alpha_1}{a_1} \rightarrow \frac{\sin(20°)}{4} = \frac{\sin(129.1°)}{a_1} \rightarrow a_1 = \frac{4\sin(129.1°)}{\sin(20°)} \approx 9.07$$

$\alpha_2 = 180° - \beta - \gamma_2 = 180° - 20° - 149.1° = 10.9°$

$$\frac{\sin\beta}{b} = \frac{\sin\alpha_2}{a_2} \rightarrow \frac{\sin(20°)}{4} = \frac{\sin(10.9°)}{a_2} \rightarrow a_2 = \frac{4\sin(10.9°)}{\sin(20°)} \approx 2.20$$

Two triangles: $\alpha_1 \approx 129.1°$, $\gamma_1 \approx 30.9°$, $a_1 \approx 9.08$
    or $\alpha_2 \approx 10.9°$, $\gamma_2 \approx 149.1°$, $a_2 \approx 2.21$

25. $a = 2$, $c = 1$, $\gamma = 100°$

$$\frac{\sin\gamma}{c} = \frac{\sin\alpha}{a} \rightarrow \frac{\sin(100°)}{1} = \frac{\sin\alpha}{2} \rightarrow \sin\alpha = \frac{2\sin(100°)}{1} \approx 1.9696$$

There is no angle α for which $\sin\alpha > 1$. Therefore, there is no triangle with the given measurements.

27. $a = 2$, $c = 1$, $\gamma = 25°$

$$\frac{\sin\alpha}{a} = \frac{\sin\gamma}{c} \rightarrow \frac{\sin\alpha}{2} = \frac{\sin(25°)}{1} \rightarrow \sin\alpha = \frac{2\sin(25°)}{1} \approx 0.8452$$

$$\rightarrow \alpha_1 = 57.7° \text{ or } \alpha_2 = 122.3°$$

For both values, $\alpha + \gamma < 180°$. Therefore, there are two triangles.

$$\beta_1 = 180° - \alpha_1 - \gamma = 180° - 57.7° - 25° = 97.3°$$

$$\frac{\sin\beta_1}{b_1} = \frac{\sin\gamma}{c} \quad \rightarrow \quad \frac{\sin(97.3°)}{b_1} = \frac{\sin(25°)}{1} \quad \rightarrow \quad b_1 = \frac{1\sin(97.3°)}{\sin(25°)} \approx 2.35$$

$$\beta_2 = 180° - \alpha_2 - \gamma = 180° - 122.3° - 25° = 32.7°$$

$$\frac{\sin\beta_2}{b_2} = \frac{\sin\gamma}{c} \quad \rightarrow \quad \frac{\sin(32.7°)}{b_2} = \frac{\sin(25°)}{1} \quad \rightarrow \quad b_2 = \frac{1\sin(32.7°)}{\sin(25°)} \approx 1.28$$

Two triangles: $\alpha_1 \approx 57.7°$, $\beta_1 \approx 97.3°$, $b_1 \approx 2.35$

or $\alpha_2 \approx 122.3°$, $\beta_2 \approx 32.7°$, $b_2 \approx 1.28$

29. (a) Find γ; then use the Law of Sines:

$$\gamma = 180° - 60° - 55° = 65°$$

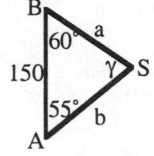

$$\frac{\sin(55°)}{a} = \frac{\sin(65°)}{150} \quad \rightarrow \quad a = \frac{150\sin(55°)}{\sin(65°)} \approx 135.58 \text{ miles}$$

$$\frac{\sin(60°)}{b} = \frac{\sin(65°)}{150} \quad \rightarrow \quad b = \frac{150\sin(60°)}{\sin(65°)} \approx 143.3 \text{ miles}$$

(b) $t = \dfrac{a}{r} = \dfrac{135.6}{200} \approx 0.68$ hours or ≈ 41 minutes

31. $\angle CAB = 180° - 25° = 155°$ $\angle ABC = 180° - 155° - 15° = 10°$

Let c represent the distance from A to B.

$$\frac{\sin(15°)}{c} = \frac{\sin(10°)}{1000} \quad \rightarrow \quad c = \frac{1000\sin(15°)}{\sin(10°)} \approx 1490.48 \text{ feet}$$

The length of the proposed ski lift is approximately 1490 feet.

33. Find the distance from B to the plane:

$$\gamma = 180° - 40° - 35° = 105° \qquad (\gamma = \angle APB)$$

$$\frac{\sin(40°)}{x} = \frac{\sin(105°)}{1000} \quad \rightarrow \quad x = \frac{1000\sin(40°)}{\sin(105°)} \approx 665.5 \text{ feet}$$

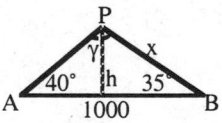

Find the height:

$$\sin(35°) = \frac{h}{x} = \frac{h}{665.5} \quad \rightarrow \quad h = (665.5)\sin(35°) \approx 381.69 \text{ feet}$$

The plane is 381.69 feet high.

35. (a) $\angle ABC = 180° - 40° = 140°$

Find the angle at city C:

$$\frac{\sin C}{150} = \frac{\sin(140°)}{300} \quad \rightarrow \quad \sin C = \frac{150\sin(140°)}{300} \approx 0.3214 \quad \rightarrow \quad C \approx 18.7°$$

Find the angle at city A:

$$A = 180° - 140° - 18.7° = 21.3°$$

$$\frac{\sin(21.3°)}{y} = \frac{\sin(140°)}{300} \quad \rightarrow \quad y = \frac{300\sin(21.3°)}{\sin(140°)} \approx 169.18 \text{ miles}$$

The distance from city B to city C is approximately 169.18 miles.

(b)   To find the angle to turn, subtract angle C from 180°:

$$180° - 18.7° = 161.3°$$

The pilot needs to turn through an angle of 161.3° to return to city A.

37.   Find angle β ($\angle ACB$):

$$\frac{\sin\beta}{123} = \frac{\sin(60°)}{184.5} \rightarrow \sin\beta = \frac{123\sin(60°)}{184.5} \approx 0.5774$$

$$\beta \approx 35.3°$$

$$\angle CAB = 180° - 60° - 35.3° \approx 84.7°$$

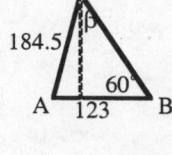

Find the perpendicular distance:

$$\sin(84.7°) = \frac{h}{184.5} \rightarrow h = 184.5\sin(84.7°) = 183.72 \text{ feet}$$

39. $\alpha = 180° - 140° = 40°$ $\beta = 180° - 135° = 45°$

$\gamma = 180° - 40° - 45° = 95°$

$$\frac{\sin(40°)}{a} = \frac{\sin(95°)}{2} \rightarrow a = \frac{2\sin(40°)}{\sin 95°} \approx 1.290 \text{ mi}$$

$$\frac{\sin(45°)}{b} = \frac{\sin(95°)}{2} \rightarrow b = \frac{2\sin(45°)}{\sin(95°)} \approx 1.420 \text{ mi}$$

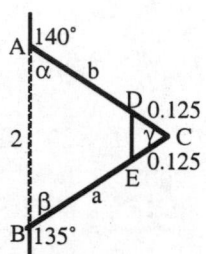

$$\overline{BE} = 1.290 - 0.125 = 1.165 \text{ mi}$$

$$\overline{AD} = 1.420 - 0.125 = 1.295 \text{ mi}$$

$$\angle CDE = \angle CED = \frac{180° - 95°}{2} = 42.5°$$

For the isosceles triangle, $\dfrac{\sin(95°)}{DE} = \dfrac{\sin(42.5°)}{0.125} \rightarrow DE = \dfrac{0.125\sin(95°)}{\sin(42.5°)} \approx 0.184 \text{ miles}$

The length of the highway is $1.165 + 1.295 + 0.184 = 2.64$ miles.

41. $\angle ABD = 180° - 30° = 150°$ $\gamma = 180° - 150° - 20° = 10°$

$$\frac{\sin(150°)}{y} = \frac{\sin(10°)}{1} \rightarrow y = \frac{1\sin(150°)}{\sin(10°)} \approx 2.88 \text{ mi}$$

$$\frac{\sin\beta}{2.88} = \frac{\sin(20°)}{1} \rightarrow \sin\beta = \frac{2.88\sin(20°)}{1} \approx 0.9850$$

$$\beta \approx 80° \text{ or } \beta \approx 100°$$

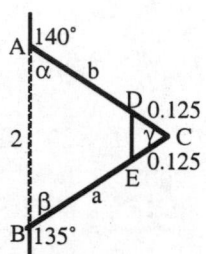

$$\alpha = 180° - 80° - 30° = 70° \text{ or } \alpha = 180° - 100° - 30° = 50°$$

$$\frac{\sin(70°)}{x} = \frac{\sin(30°)}{1} \Rightarrow x = \frac{\sin(70°)}{\sin(30°)} \approx 1.88 \text{ mi}$$

$$\frac{\sin(50°)}{x} = \frac{\sin(30°)}{1} \Rightarrow x = \frac{\sin(50°)}{\sin(30°)} \approx 1.53 \text{ mi}$$

The ship is about 1.88 miles from the harbor, or the ship is about 1.53 miles from the harbor.

43. Using the Law of Sines:

$$\frac{\sin(46.27°)}{x} = \frac{\sin(90° - 46.27°)}{y + 100}$$

$$(y + 100)\sin(46.27°) = x\sin(43.73°)$$

$$y\sin(46.27°) + 100\sin(46.27°) = x\sin(43.73°)$$

$$y = \frac{x\sin(43.73°) - 100\sin(46.27°)}{\sin(46.27°)}$$

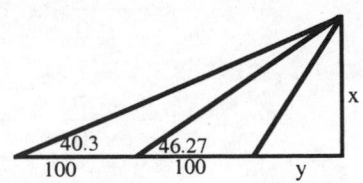

$$\frac{\sin(40.3°)}{x} = \frac{\sin(90° - 40.3°)}{y + 200}$$

$$(y + 200)\sin(40.3°) = x\sin(49.7°)$$

$$y\sin(40.3°) + 200\sin(40.3°) = x\sin(49.7°)$$

$$y = \frac{x\sin(49.7°) - 200\sin(40.3°)}{\sin(40.3°)}$$

Set the two equations equal to each other and solve:

$$\frac{x\sin(43.73°) - 100\sin(46.27°)}{\sin(46.27°)} = \frac{x\sin(49.7°) - 200\sin(40.3°)}{\sin(40.3°)}$$

$$x\sin(43.73°) \cdot \sin(40.3°) - 100\sin(46.27°) \cdot \sin(40.3°)$$
$$= x\sin(49.7°) \cdot \sin(46.27°) - 200\sin(40.3°) \cdot \sin(46.27°)$$

$$x\sin(43.73°) \cdot \sin(40.3°) - x\sin(49.7°) \cdot \sin(46.27°)$$
$$= 100\sin(46.27°) \cdot \sin(40.3°) - 200\sin(40.3°) \cdot \sin(46.27°)$$

$$x = \frac{100\sin(46.27°) \cdot \sin(40.3°) - 200\sin(40.3°) \cdot \sin(46.27°)}{\sin(43.73°) \cdot \sin(40.3°) - \sin(49.7°) \cdot \sin(46.27°)} \approx 449.36 \text{ feet}$$

45. Using the Law of Sines:

$$\frac{\sin(30°)}{h} = \frac{\sin(60°)}{x} \quad \rightarrow \quad x = \frac{h\sin(60°)}{\sin(30°)}$$

$$\frac{\sin(20°)}{h} = \frac{\sin(70°)}{x + 40} \quad \rightarrow \quad x = \frac{h\sin(70°)}{\sin(20°)} - 40$$

$$\frac{h\sin(60°)}{\sin(30°)} = \frac{h\sin(70°)}{\sin(20°)} - 40$$

$$h\left(\frac{\sin(60°)}{\sin(30°)} - \frac{\sin(70°)}{\sin(20°)}\right) = -40 \rightarrow h = \frac{-40}{\left(\dfrac{\sin(60°)}{\sin(30°)} - \dfrac{\sin(70°)}{\sin(20°)}\right)} \approx 39.39 \text{ feet}$$

47. Find the distance from B to the helicopter:
$$\gamma = 180° - 40° - 25° = 115° \qquad (\gamma = \angle APB)$$

$$\frac{\sin(40°)}{x} = \frac{\sin(115°)}{100} \;\rightarrow\; x = \frac{100\sin(40°)}{\sin(115°)} \approx 70.9 \text{ feet}$$

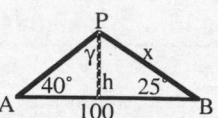

Find the height:
$$\sin(25°) = \frac{h}{x} \approx \frac{h}{70.9} \;\rightarrow\; h \approx 70.9(\sin 25°) \approx 29.97 \text{ feet}$$

The helicopter is about 30 feet high.

49. $$\frac{a-b}{c} = \frac{a}{c} - \frac{b}{c} = \frac{\sin\alpha}{\sin\gamma} - \frac{\sin\beta}{\sin\gamma} = \frac{\sin\alpha - \sin\beta}{\sin\gamma} = \frac{2\sin\left(\dfrac{\alpha-\beta}{2}\right)\cos\left(\dfrac{\alpha+\beta}{2}\right)}{\sin\left(2\cdot\dfrac{\gamma}{2}\right)}$$

$$= \frac{2\sin\left(\dfrac{\alpha-\beta}{2}\right)\cos\left(\dfrac{\alpha+\beta}{2}\right)}{2\sin\left(\dfrac{\gamma}{2}\right)\cos\left(\dfrac{\gamma}{2}\right)} = \frac{\sin\left(\dfrac{\alpha-\beta}{2}\right)\cos\left(\dfrac{\pi}{2}-\dfrac{\gamma}{2}\right)}{\sin\left(\dfrac{\gamma}{2}\right)\cos\left(\dfrac{\gamma}{2}\right)} = \frac{\sin\left(\dfrac{\alpha-\beta}{2}\right)\sin\left(\dfrac{\gamma}{2}\right)}{\sin\left(\dfrac{\gamma}{2}\right)\cos\left(\dfrac{\gamma}{2}\right)}$$

$$= \frac{\sin\left(\dfrac{\alpha-\beta}{2}\right)}{\cos\left(\dfrac{\gamma}{2}\right)} = \frac{\sin\left(\dfrac{1}{2}(\alpha-\beta)\right)}{\cos\left(\dfrac{1}{2}\gamma\right)}$$

51. Derive the Law of Tangents:

$$\frac{a-b}{a+b} = \frac{\left(\dfrac{a-b}{c}\right)}{\left(\dfrac{a+b}{c}\right)} = \frac{\left(\dfrac{\sin\left(\dfrac{1}{2}(\alpha-\beta)\right)}{\cos\left(\dfrac{1}{2}\gamma\right)}\right)}{\left(\dfrac{\cos\left(\dfrac{1}{2}(\alpha-\beta)\right)}{\sin\left(\dfrac{1}{2}\gamma\right)}\right)} = \frac{\sin\left(\dfrac{1}{2}(\alpha-\beta)\right)}{\cos\left(\dfrac{1}{2}\gamma\right)} \cdot \frac{\sin\left(\dfrac{1}{2}\gamma\right)}{\cos\left(\dfrac{1}{2}(\alpha-\beta)\right)} = \tan\left(\dfrac{1}{2}(\alpha-\beta)\right)\tan\left(\dfrac{1}{2}\gamma\right)$$

$$= \tan\left(\dfrac{1}{2}(\alpha-\beta)\right)\tan\left(\dfrac{1}{2}(\pi-(\alpha+\beta))\right) = \tan\left(\dfrac{1}{2}(\alpha-\beta)\right)\tan\left(\dfrac{\pi}{2}-\left(\dfrac{\alpha+\beta}{2}\right)\right)$$

$$= \tan\left(\dfrac{1}{2}(\alpha-\beta)\right)\cot\left(\dfrac{\alpha+\beta}{2}\right) = \frac{\tan\left(\dfrac{1}{2}(\alpha-\beta)\right)}{\tan\left(\dfrac{1}{2}(\alpha+\beta)\right)}$$

Applications of Trigonometric Functions

7.3   The Law of Cosines

1. $a = 2, \; c = 4, \; \beta = 45° \qquad b^2 = a^2 + c^2 - 2ac\cos\beta$

$b^2 = 2^2 + 4^2 - 2 \cdot 2 \cdot 4\cos(45°) = 20 - 16 \cdot \dfrac{\sqrt{2}}{2} = 20 - 8\sqrt{2} \approx 8.6863$

$b \approx 2.95$

$a^2 = b^2 + c^2 - 2bc\cos\alpha \;\; \rightarrow \;\; 2bc\cos\alpha = b^2 + c^2 - a^2 \;\; \rightarrow \;\; \cos\alpha = \dfrac{b^2 + c^2 - a^2}{2bc}$

$\cos\alpha = \dfrac{2.95^2 + 4^2 - 2^2}{2(2.95)(4)} = \dfrac{20.6863}{23.6} \approx 0.8765 \rightarrow \alpha \approx 28.7°$

$c^2 = a^2 + b^2 - 2ab\cos\gamma \;\; \rightarrow \;\; \cos\gamma = \dfrac{a^2 + b^2 - c^2}{2ab} = \dfrac{2^2 + 2.95^2 - 4^2}{2(2)(2.95)} \approx -0.2794 \rightarrow \gamma \approx 106.3°$

3. $a = 2, \; b = 3, \; \gamma = 95° \qquad c^2 = a^2 + b^2 - 2ab\cos\gamma$
$c^2 = 2^2 + 3^2 - 2 \cdot 2 \cdot 3\cos(95°) = 13 - 12 \cdot (-0.0872) \approx 14.0459$

$c \approx 3.75$

$a^2 = b^2 + c^2 - 2bc\cos\alpha \;\; \rightarrow \;\; \cos\alpha = \dfrac{b^2 + c^2 - a^2}{2bc} = \dfrac{3^2 + 3.75^2 - 2^2}{2(3)(3.75)} \approx 0.8472 \rightarrow \alpha \approx 32.1°$

$b^2 = a^2 + c^2 - 2ac\cos\beta \;\; \rightarrow \;\; \cos\beta = \dfrac{a^2 + c^2 - b^2}{2ac} = \dfrac{2^2 + 3.75^2 - 3^2}{2(2)(3.75)} \approx 0.6042 \rightarrow \beta \approx 52.9°$

5. $a = 6, \; b = 5, \; c = 8$

$a^2 = b^2 + c^2 - 2bc\cos\alpha \;\; \rightarrow \;\; \cos\alpha = \dfrac{b^2 + c^2 - a^2}{2bc} = \dfrac{5^2 + 8^2 - 6^2}{2(5)(8)} \approx 0.6625 \rightarrow \alpha \approx 48.5°$

$b^2 = a^2 + c^2 - 2ac\cos\beta \;\; \rightarrow \;\; \cos\beta = \dfrac{a^2 + c^2 - b^2}{2ac} = \dfrac{6^2 + 8^2 - 5^2}{2(6)(8)} \approx 0.7813 \rightarrow \beta = 38.6°$

$c^2 = a^2 + b^2 - 2ab\cos\gamma \;\; \rightarrow \;\; \cos\gamma = \dfrac{a^2 + b^2 - c^2}{2ab} = \dfrac{6^2 + 5^2 - 8^2}{2(6)(5)} \approx -0.0500 \rightarrow \gamma = 92.9°$

7.　　$a = 9, \ b = 6, \ c = 4$

$$a^2 = b^2 + c^2 - 2bc\cos\alpha \ \rightarrow \ \cos\alpha = \frac{b^2 + c^2 - a^2}{2bc} = \frac{6^2 + 4^2 - 9^2}{2(6)(4)} \approx -0.6042 \rightarrow \alpha \approx 127.2°$$

$$b^2 = a^2 + c^2 - 2ac\cos\beta \ \rightarrow \ \cos\beta = \frac{a^2 + c^2 - b^2}{2ac} = \frac{9^2 + 4^2 - 6^2}{2(9)(4)} \approx 0.8472 \rightarrow \beta = 32.1°$$

$$c^2 = a^2 + b^2 - 2ab\cos\gamma \ \rightarrow \ \cos\gamma = \frac{a^2 + b^2 - c^2}{2ab} = \frac{9^2 + 6^2 - 4^2}{2(9)(6)} \approx 0.9352 \rightarrow \gamma = 20.7°$$

9.　　$a = 3, \ b = 4, \ \gamma = 40°$

$$c^2 = a^2 + b^2 - 2ab\cos\gamma$$

$$c^2 = 3^2 + 4^2 - 2 \cdot 3 \cdot 4\cos(40°) \approx 6.6149$$

$$c \approx 2.57$$

$$a^2 = b^2 + c^2 - 2bc\cos\alpha \ \rightarrow \ \cos\alpha = \frac{b^2 + c^2 - a^2}{2bc} = \frac{4^2 + 2.57^2 - 3^2}{2(4)(2.57)} \approx 0.6617 \rightarrow \alpha \approx 48.6°$$

$$b^2 = a^2 + c^2 - 2ac\cos\beta \ \rightarrow \ \cos\beta = \frac{a^2 + c^2 - b^2}{2ac} = \frac{3^2 + 2.57^2 - 4^2}{2(3)(2.57)} \approx -0.0256 \rightarrow \beta \approx 91.4°$$

11.　　$b = 1, \ c = 3, \ \alpha = 80°$

$$a^2 = b^2 + c^2 - 2bc\cos\alpha$$

$$a^2 = 1^2 + 3^2 - 2 \cdot 1 \cdot 3\cos(80°) \approx 8.9581$$

$$a \approx 2.99$$

$$c^2 = a^2 + b^2 - 2ab\cos\gamma \ \rightarrow \ \cos\gamma = \frac{a^2 + b^2 - c^2}{2ab} = \frac{2.99^2 + 1^2 - 3^2}{2(2.99)(1)} \approx 0.1572 \rightarrow \gamma \approx 80.8°$$

$$b^2 = a^2 + c^2 - 2ac\cos\beta \ \rightarrow \ \cos\beta = \frac{a^2 + c^2 - b^2}{2ac} = \frac{2.99^2 + 3^2 - 1^2}{2(2.99)(3)} \approx 0.9443 \rightarrow \beta = 19.2°$$

13.　　$a = 3, \ c = 2, \ \beta = 110°$

$$b^2 = a^2 + c^2 - 2ac\cos\beta$$

$$b^2 = 3^2 + 2^2 - 2 \cdot 3 \cdot 2\cos(110°) \approx 17.1042$$

$$b \approx 4.14$$

$$a^2 = b^2 + c^2 - 2bc\cos\alpha \ \rightarrow \ \cos\alpha = \frac{b^2 + c^2 - a^2}{2bc} = \frac{4.14^2 + 2^2 - 3^2}{2(4.14)(2)} \approx 0.7331 \rightarrow \alpha \approx 43.0°$$

$$c^2 = a^2 + b^2 - 2ab\cos\gamma \ \rightarrow \ \cos\gamma = \frac{a^2 + b^2 - c^2}{2ab} = \frac{3^2 + 4.14^2 - 2^2}{2(3)(4.14)} \approx 0.8913 \rightarrow \gamma \approx 27.0°$$

15.　　$a = 2, \ b = 2, \ \gamma = 50°$

$$c^2 = a^2 + b^2 - 2ab\cos\gamma$$

$$c^2 = 2^2 + 2^2 - 2 \cdot 2 \cdot 2\cos(50°) \approx 2.8577$$

$$c \approx 1.69$$

$$a^2 = b^2 + c^2 - 2bc\cos\alpha \;\rightarrow\; \cos\alpha = \frac{b^2 + c^2 - a^2}{2bc} = \frac{2^2 + 1.69^2 - 2^2}{2(2)(1.69)} \approx 0.4225 \rightarrow \alpha \approx 65.0°$$

$$b^2 = a^2 + c^2 - 2ac\cos\beta \;\rightarrow\; \cos\beta = \frac{a^2 + c^2 - b^2}{2ac} = \frac{2^2 + 1.69^2 - 2^2}{2(2)(1.69)} \approx 0.4225 \rightarrow \beta \approx 65.0°$$

17. $a = 12, \; b = 13, \; c = 5$

$$a^2 = b^2 + c^2 - 2bc\cos\alpha \;\rightarrow\; \cos\alpha = \frac{b^2 + c^2 - a^2}{2bc} = \frac{13^2 + 5^2 - 12^2}{2(13)(5)} \approx 0.3846 \rightarrow \alpha \approx 67.4°$$

$$b^2 = a^2 + c^2 - 2ac\cos\beta \;\rightarrow\; \cos\beta = \frac{a^2 + c^2 - b^2}{2ac} = \frac{12^2 + 5^2 - 13^2}{2(12)(5)} = 0 \rightarrow \beta = 90°$$

$$c^2 = a^2 + b^2 - 2ab\cos\gamma \;\rightarrow\; \cos\gamma = \frac{a^2 + b^2 - c^2}{2ab} = \frac{12^2 + 13^2 - 5^2}{2(12)(13)} \approx 0.9231 \rightarrow \gamma \approx 22.6°$$

19. $a = 2, \; b = 2, \; c = 2$

$$a^2 = b^2 + c^2 - 2bc\cos\alpha \;\rightarrow\; \cos\alpha = \frac{b^2 + c^2 - a^2}{2bc} = \frac{2^2 + 2^2 - 2^2}{2(2)(2)} = 0.5 \rightarrow \alpha = 60°$$

$$b^2 = a^2 + c^2 - 2ac\cos\beta \;\rightarrow\; \cos\beta = \frac{a^2 + c^2 - b^2}{2ac} = \frac{2^2 + 2^2 - 2^2}{2(2)(2)} = 0.5 \rightarrow \beta = 60°$$

$$c^2 = a^2 + b^2 - 2ab\cos\gamma \;\rightarrow\; \cos\gamma = \frac{a^2 + b^2 - c^2}{2ab} = \frac{2^2 + 2^2 - 2^2}{2(2)(2)} = 0.5 \rightarrow \gamma = 60°$$

21. $a = 5, \; b = 8, \; c = 9$

$$a^2 = b^2 + c^2 - 2bc\cos\alpha \;\rightarrow\; \cos\alpha = \frac{b^2 + c^2 - a^2}{2bc} = \frac{8^2 + 9^2 - 5^2}{2(8)(9)} \approx 0.8333 \rightarrow \alpha \approx 33.6°$$

$$b^2 = a^2 + c^2 - 2ac\cos\beta \;\rightarrow\; \cos\beta = \frac{a^2 + c^2 - b^2}{2ac} = \frac{5^2 + 9^2 - 8^2}{2 \cdot 5 \cdot 9} \approx 0.4667 \rightarrow \beta \approx 62.2°$$

$$c^2 = a^2 + b^2 - 2ab\cos\gamma \;\rightarrow\; \cos\gamma = \frac{a^2 + b^2 - c^2}{2ab} = \frac{5^2 + 8^2 - 9^2}{2 \cdot 5 \cdot 8} = 0.1000 \rightarrow \gamma \approx 84.3°$$

23. $a = 10, \; b = 8, \; c = 5$

$$a^2 = b^2 + c^2 - 2bc\cos\alpha \;\rightarrow\; \cos\alpha = \frac{b^2 + c^2 - a^2}{2bc} = \frac{8^2 + 5^2 - 10^2}{2(8)(5)} \approx -0.1375 \rightarrow \alpha \approx 97.9°$$

$$b^2 = a^2 + c^2 - 2ac\cos\beta \;\rightarrow\; \cos\beta = \frac{a^2 + c^2 - b^2}{2ac} = \frac{10^2 + 5^2 - 8^2}{2(10)(5)} \approx 0.6100 \rightarrow \beta \approx 52.4°$$

$$c^2 = a^2 + b^2 - 2ab\cos\gamma \;\rightarrow\; \cos\gamma = \frac{a^2 + b^2 - c^2}{2ab} = \frac{10^2 + 8^2 - 5^2}{2(10)(8)} \approx 0.8688 \rightarrow \gamma \approx 29.7°$$

25. Find the third side of the triangle using the Law of Cosines:
$$a = 50, \; b = 70, \; \gamma = 70°$$
$$c^2 = a^2 + b^2 - 2ab\cos\gamma = 50^2 + 70^2 - 2 \cdot 50 \cdot 70\cos(70°) \approx 5005.86 \rightarrow c \approx 70.75$$
The houses are approximately 70.75 feet apart.

27.   (a)   After 15 minutes, the plane would have flown 220(0.25) = 55 miles.
            Find the third side of the triangle:
            $a = 55$, $b = 330$, $\gamma = 10°$
            $c^2 = a^2 + b^2 - 2ab\cos\gamma = 55^2 + 330^2 - 2\cdot 55\cdot 330\cos(10°) \approx 76176.48 \rightarrow c \approx 276$
            Find the measure of the angle opposite the 330 side:
$$\cos\beta = \frac{a^2 + c^2 - b^2}{2ac} = \frac{55^2 + 276^2 - 330^2}{2(55)(276)} \approx -0.9782 \rightarrow \beta \approx 168°$$
            The pilot should turn through an angle of $180° - 168° = 12°$.
      (b)   If the total trip is to be done in 90 minutes, and 15 minutes were used already, then
            there are 75 minutes or 1.25 hours to complete the trip. The plane must travel 276
            miles in 1.25 hours.
$$r = \frac{276}{1.25} = 220.8 \text{ miles / hour}$$
            The pilot must maintain a speed of 220.8 mi/hr to complete the whole trip in 90
            minutes.

29.   (a)   Find x in the figure:
$$x^2 = 60.5^2 + 90^2 - 2(60.5)90\cos(45°) \approx 4059.86$$
            $x \approx 63.7$ feet

            It is about 63.7 feet from the pitching rubber

            to first base.

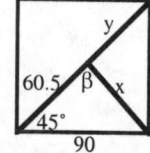

      (b)   Use the Pythagorean Theorem to find y in the figure:
            $90^2 + 90^2 = (60.5 + y)^2 \rightarrow 8100 + 8100 = (60.5 + y)^2$

            $16200 = (60.5 + y)^2 \rightarrow 60.5 + y \approx 127.3 \rightarrow y \approx 66.8$ feet
            It is about 66.8 feet from the pitching rubber to second base.
      (c)   Find β in the figure by using the Law of Cosines:
$$\cos\beta = \frac{60.5^2 + 63.7^2 - 90^2}{2(60.5)(63.7)} \approx -0.0496 \rightarrow \beta \approx 92.8°$$
            The pitcher needs to turn through an angle of 92.8° to face first base.

31.   (a)   Find x by using the Law of Cosines:

            $x^2 = 500^2 + 100^2 - 2(500)100\cos(80°) \approx 242,635$

            $x \approx 492.6$ feet

            The guy wire needs to be about 492.6 feet long.

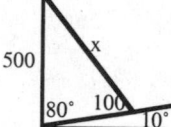

      (b)   Use the Pythagorean Theorem to find the value of y:

            $y^2 = 100^2 + 250^2 = 72500$

            $y = 269.3$ feet

            The guy wire needs to be about 269.3 feet long.

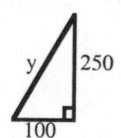

33. Find x by using the Law of Cosines:

$$x^2 = 400^2 + 90^2 - 2(400)90\cos(45°) \approx 117,188.3$$

$x \approx 342.3$ feet

It is approximately 342.3 feet from dead center

to third base.

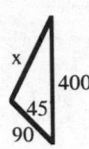

35. Use the Law of Cosines:

$$L^2 = x^2 + r^2 - 2xr\cos\theta$$

$$x^2 - 2xr\cos\theta + r^2 - L^2 = 0$$

$$x = \frac{2r\cos\theta + \sqrt{(2r\cos\theta)^2 - 4(1)(r^2 - L^2)}}{2(1)}$$

$$x = \frac{2r\cos\theta + \sqrt{4r^2\cos^2\theta - 4(r^2 - L^2)}}{2}$$

$$x = r\cos\theta + \sqrt{r^2\cos^2\theta + L^2 - r^2}$$

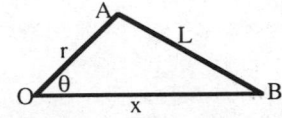

37. $$\cos\left(\frac{\gamma}{2}\right) = \sqrt{\frac{1 + \cos\gamma}{2}} = \sqrt{\frac{\left(1 + \dfrac{a^2 + b^2 - c^2}{2ab}\right)}{2}} = \sqrt{\frac{2ab + a^2 + b^2 - c^2}{4ab}} = \sqrt{\frac{(a+b)^2 - c^2}{4ab}}$$

$$= \sqrt{\frac{(a+b+c)(a+b-c)}{4ab}} = \sqrt{\frac{2s(2s - c - c)}{4ab}} = \sqrt{\frac{4s(s-c)}{4ab}} = \sqrt{\frac{s(s-c)}{ab}}$$

39. $$\frac{\cos\alpha}{a} + \frac{\cos\beta}{b} + \frac{\cos\gamma}{c} = \frac{b^2 + c^2 - a^2}{2bca} + \frac{a^2 + c^2 - b^2}{2acb} + \frac{a^2 + b^2 - c^2}{2abc}$$

$$= \frac{b^2 + c^2 - a^2 + a^2 + c^2 - b^2 + a^2 + b^2 - c^2}{2abc} = \frac{a^2 + b^2 + c^2}{2abc}$$

Chapter 7

Applications of Trigonometric Functions

7.4 The Area of a Triangle

1. $a = 2, \ c = 4, \ \beta = 45°$

$A = \dfrac{1}{2} ac \sin \beta = \dfrac{1}{2}(2)(4) \sin(45°) \approx 2.83$

3. $a = 2, \ b = 3, \ \gamma = 95°$

$A = \dfrac{1}{2} ab \sin \gamma = \dfrac{1}{2}(2)(3) \sin(95°) \approx 2.99$

5. $a = 6, \ b = 5, \ c = 8$

$s = \dfrac{1}{2}(a+b+c) = \dfrac{1}{2}(6+5+8) = \dfrac{19}{2}$

$A = \sqrt{s(s-a)(s-b)(s-c)} = \sqrt{\dfrac{19}{2}\left(\dfrac{7}{2}\right)\left(\dfrac{9}{2}\right)\left(\dfrac{3}{2}\right)} = \sqrt{\dfrac{3591}{16}} \approx 14.98$

7. $a = 9, \ b = 6, \ c = 4$

$s = \dfrac{1}{2}(a+b+c) = \dfrac{1}{2}(9+6+4) = \dfrac{19}{2}$

$A = \sqrt{s(s-a)(s-b)(s-c)} = \sqrt{\dfrac{19}{2}\left(\dfrac{1}{2}\right)\left(\dfrac{7}{2}\right)\left(\dfrac{11}{2}\right)} = \sqrt{\dfrac{1463}{16}} \approx 9.56$

9. $a = 3, \ b = 4, \ \gamma = 40°$

$A = \dfrac{1}{2} ab \sin \gamma = \dfrac{1}{2}(3)(4) \sin(40°) \approx 3.86$

11. $b = 1, \ c = 3, \ \alpha = 80°$

$A = \dfrac{1}{2} bc \sin \alpha = \dfrac{1}{2}(1)(3) \sin(80°) \approx 1.48$

13. $a = 3, \ c = 2, \ \beta = 110°$

$A = \dfrac{1}{2} ac \sin \beta = \dfrac{1}{2}(3)(2) \sin(110°) \approx 2.82$

15. $a = 2, \ b = 2, \ \gamma = 50°$

$A = \dfrac{1}{2} ab \sin \gamma = \dfrac{1}{2}(2)(2) \sin(50°) \approx 1.53$

17. $a = 12, \ b = 13, \ c = 5$

$s = \dfrac{1}{2}(a+b+c) = \dfrac{1}{2}(12+13+5) = 15$

$A = \sqrt{s(s-a)(s-b)(s-c)} = \sqrt{15(3)(2)(10)} = \sqrt{900} = 30$

19. $a = 2, \ b = 2, \ c = 2$

$s = \dfrac{1}{2}(a+b+c) = \dfrac{1}{2}(2+2+2) = 3$

$A = \sqrt{s(s-a)(s-b)(s-c)} = \sqrt{3(1)(1)(1)} = \sqrt{3} \approx 1.73$

21. $a = 5, \ b = 8, \ c = 9$

$s = \dfrac{1}{2}(a+b+c) = \dfrac{1}{2}(5+8+9) = 11$

$A = \sqrt{s(s-a)(s-b)(s-c)} = \sqrt{11(6)(3)(2)} = \sqrt{396} \approx 19.90$

23. $a = 10, \ b = 8, \ c = 5$

$s = \dfrac{1}{2}(a+b+c) = \dfrac{1}{2}(10+8+5) = \dfrac{23}{2}$

$A = \sqrt{s(s-a)(s-b)(s-c)} = \sqrt{\dfrac{23}{2}\left(\dfrac{3}{2}\right)\left(\dfrac{7}{2}\right)\left(\dfrac{13}{2}\right)} = \sqrt{\dfrac{6279}{16}} \approx 19.81$

25. Area of a sector $= \dfrac{1}{2}r^2\theta$ where θ is in radians.

$\theta = 70° \cdot \dfrac{\pi}{180} = \dfrac{7\pi}{18}$

Area of the sector $= \dfrac{1}{2} \cdot 8^2 \cdot \dfrac{7\pi}{18} = \dfrac{112\pi}{9} \approx 39.10$ square feet

Area of the triangle $= \dfrac{1}{2} \cdot 8 \cdot 8\sin(70°) = 32\sin(70°) \approx 30.07$ square feet

Area of the segment $= 39.10 - 30.07 = 9.03$ square feet

27. Find the area of the lot using Heron's Formula:

$a = 100, \ b = 50, \ c = 75$

$s = \dfrac{1}{2}(a+b+c) = \dfrac{1}{2}(100+50+75) = \dfrac{225}{2}$

$A = \sqrt{s(s-a)(s-b)(s-c)} = \sqrt{\dfrac{225}{2}\left(\dfrac{25}{2}\right)\left(\dfrac{125}{2}\right)\left(\dfrac{75}{2}\right)} = \sqrt{\dfrac{52,734,375}{16}} \approx 1815.46$

The cost is \$3 times the area: Cost $= \$3(1815.46) = \$5446.38$

29. The area of the shaded region = the area of the semicircle − the area of the triangle.

Area of the semicircle $= \dfrac{1}{2}\pi r^2 = \dfrac{1}{2}\pi(4)^2 = 8\pi$ square centimeters

The triangle is a right triangle. Find the other leg:

$6^2 + b^2 = 8^2 \ \rightarrow \ b^2 = 64 - 36 = 28 \ \rightarrow \ b = \sqrt{28} = 2\sqrt{7}$

Area of the triangle $= \dfrac{1}{2} \cdot 6 \cdot 2\sqrt{7} = 6\sqrt{7}$ square centimeters

Area of the shaded region $= 8\pi - 6\sqrt{7} \approx 9.26$ square centimeters

31. Use the Law of Sines in the area of the triangle formula:

$$A = \frac{1}{2}ab\sin\gamma = \frac{1}{2}a\sin\gamma\left(\frac{a\sin\beta}{\sin\alpha}\right) = \frac{a^2\sin\beta\sin\gamma}{2\sin\alpha}$$

33. $\alpha = 40°$, $\beta = 20°$, $a = 2$ $\gamma = 180° - \alpha - \beta = 180° - 40° - 20° = 120°$

$$A = \frac{a^2\sin\beta\sin\gamma}{2\sin\alpha} = \frac{2^2\sin(20°)\sin(120°)}{2\sin(40°)} \approx \frac{4(0.3420)(0.8660)}{2(0.6428)} \approx 0.92$$

35. $\beta = 70°$, $\gamma = 10°$, $b = 5$ $\alpha = 180° - \beta - \gamma = 180° - 70° - 10° = 100°$

$$A = \frac{b^2\sin\alpha\sin\gamma}{2\sin\beta} = \frac{5^2\sin(100°)\sin(10°)}{2\sin(70°)} \approx \frac{25(0.9848)(0.1736)}{2(0.9397)} \approx 2.27$$

37. $\alpha = 110°$, $\gamma = 30°$, $c = 3$ $\beta = 180° - \alpha - \gamma = 180° - 110° - 30° = 40°$

$$A = \frac{c^2\sin\alpha\sin\beta}{2\sin\gamma} = \frac{3^2\sin(110°)\sin(40°)}{2\sin(30°)} \approx \frac{9(0.9397)(0.6428)}{2(0.5000)} \approx 5.44$$

39. The area is the sum of the area of a triangle and a sector.

Area of the triangle $= \frac{1}{2}r \cdot r\sin(\pi - \theta) = \frac{1}{2}r^2\sin(\pi - \theta)$

Area of the sector $= \frac{1}{2}r^2\theta$

$$A = \frac{1}{2}r^2\sin(\pi - \theta) + \frac{1}{2}r^2\theta = \frac{1}{2}r^2\left(\sin(\pi - \theta) + \theta\right)$$

$$= \frac{1}{2}r^2\left(\sin\pi\cos\theta - \cos\pi\sin\theta + \theta\right) = \frac{1}{2}r^2\left(0 + \sin\theta + \theta\right) = \frac{1}{2}r^2\left(\theta + \sin\theta\right)$$

41. The grazing area must be considered in sections. A_1 represents $\frac{3}{4}$ of a circle:

$$A_1 = \frac{3}{4}\pi(100)^2 = 7500\pi \approx 23{,}562 \text{ square feet}$$

Angles are needed to find A_2 and A_3: (see the figure)

In $\triangle ABC$, $\angle CBA = 45°$, $AB = 10$, $AC = 90$

Find $\angle BCA$:

$$\frac{\sin\angle CBA}{90} = \frac{\sin\angle BCA}{10} \quad \rightarrow \quad \frac{\sin(45°)}{90} = \frac{\sin\angle BCA}{10}$$

$$\sin\angle BCA = \frac{10\sin(45°)}{90} \approx 0.0786$$

$$\angle BCA \approx 4.5°$$

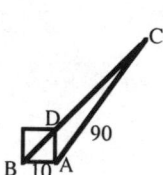

$m\angle BAC = 180° - 45° - 4.5° = 130.5°$

$m\angle DAC = 130.5° - 90° = 40.5°$

Area of $A_3 = \dfrac{1}{2}(10)(90)\sin(40.5°) \approx 292$ square feet

Area of sector $A_2 = \dfrac{1}{2}(90)^2\left(49.5° \cdot \dfrac{\pi}{180}\right) \approx 3499$ square feet

Since the cow can go in either direction around the barn, A_2 and A_3 must be doubled. Total grazing area is: $23,562 + 2(3499) + 2(292) = 31,144$ square feet

43. $h_1 = \dfrac{2K}{a}, \quad h_2 = \dfrac{2K}{b}, \quad h_3 = \dfrac{2K}{c}$ where K is the area of the triangle.

$\dfrac{1}{h_1} + \dfrac{1}{h_2} + \dfrac{1}{h_3} = \dfrac{a}{2K} + \dfrac{b}{2K} + \dfrac{c}{2K} = \dfrac{a+b+c}{2K} = \dfrac{2s}{2K} = \dfrac{s}{K}$

45. $h = \dfrac{a\sin\beta\sin\gamma}{\sin\alpha}$ where h is the altitude to side a.

In $\triangle OAB$, c is opposite angle AOB. The two adjacent angles are $\dfrac{\alpha}{2}$ and $\dfrac{\beta}{2}$.

Then $r = \dfrac{c \cdot \sin\left(\dfrac{\alpha}{2}\right)\sin\left(\dfrac{\beta}{2}\right)}{\sin(\angle AOB)}$

$\angle AOB = \pi - \left(\dfrac{\alpha}{2} + \dfrac{\beta}{2}\right)$

$\sin(\angle AOB) = \sin\left(\pi - \left(\dfrac{\alpha}{2} + \dfrac{\beta}{2}\right)\right) = \sin\left(\dfrac{\alpha}{2} + \dfrac{\beta}{2}\right) = \sin\left(\dfrac{\alpha+\beta}{2}\right) = \cos\left(\dfrac{\pi}{2} - \left(\dfrac{\alpha+\beta}{2}\right)\right)$

$= \cos\left(\dfrac{\pi - (\alpha+\beta)}{2}\right) = \cos\left(\dfrac{\gamma}{2}\right)$

Thus, $r = \dfrac{c \cdot \sin\left(\dfrac{\alpha}{2}\right)\sin\left(\dfrac{\beta}{2}\right)}{\cos\left(\dfrac{\gamma}{2}\right)}$

47. Use the result of Problem 46:

$\cot\left(\dfrac{\alpha}{2}\right) + \cot\left(\dfrac{\beta}{2}\right) + \cot\left(\dfrac{\gamma}{2}\right) = \dfrac{s-a}{r} + \dfrac{s-b}{r} + \dfrac{s-c}{r} = \dfrac{s-a+s-b+s-c}{r}$

$= \dfrac{3s - (a+b+c)}{r} = \dfrac{3s - 2s}{r} = \dfrac{s}{r}$

Chapter **7**

Applications of Trigonometric Functions

7.5   Simple Harmonic Motion; Damped Motion

1. $d = -5\cos(\pi t)$

3. $d = -6\cos(2t)$

5. $d = -5\sin(\pi t)$

7. $d = -6\sin(2t)$

9. $d = 5\sin(3t)$
   (a)   Simple harmonic
   (b)   5 meters
   (c) $\dfrac{2\pi}{3}$ seconds
   (d) $\dfrac{3}{2\pi}$ oscillation/second

11. $d = 6\cos(\pi t)$
   (a)   Simple harmonic
   (b)   6 meters
   (c)   2 seconds
   (d) $\dfrac{1}{2}$ oscillation/second

13. $d = -3\sin\left(\dfrac{1}{2}t\right)$
   (a)   Simple harmonic
   (b)   3 meters
   (c) 4π seconds
   (d) $\dfrac{1}{4\pi}$ oscillation/second

15. $d = 6 + 2\cos(2\pi t)$
   (a)   Simple harmonic
   (b)   2 meters
   (c)   1 second
   (d)   1 oscillation/second

17.   (a)

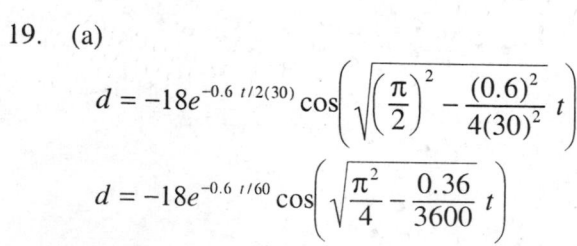

(b)

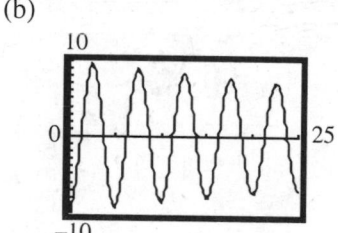

19.   (a)

$$d = -18e^{-0.6\,t/2(30)}\cos\left(\sqrt{\left(\dfrac{\pi}{2}\right)^2 - \dfrac{(0.6)^2}{4(30)^2}}\;t\right)$$

$$d = -18e^{-0.6\,t/60}\cos\left(\sqrt{\dfrac{\pi^2}{4} - \dfrac{0.36}{3600}}\;t\right)$$

(b)

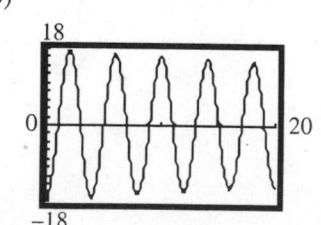

21. (a)
$$d = -5e^{-0.8\ t/2(10)} \cos\left(\sqrt{\left(\frac{2\pi}{3}\right)^2 - \frac{(0.8)^2}{4(10)^2}}\ t \right)$$

$$d = -5e^{-0.8\ t/20} \cos\left(\sqrt{\frac{4\pi^2}{9} - \frac{0.64}{400}}\ t \right)$$

(b)

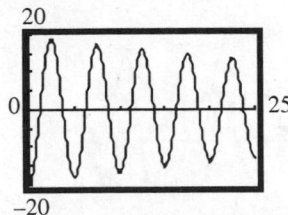

23. (a) Damped motion with a bob of mass 20 kg and a damping factor of 0.7.
(b) 20 meters downward
(c) Graph:

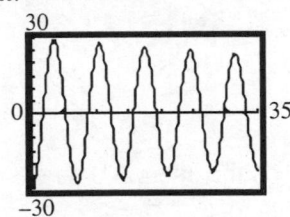

(d) The maximum displacement after one oscillation is 18.32 meters.
(e) It approaches zero, since $e^{-0.7\ t/40} \to 0$ as $t \to \infty$.

25. (a) Damped motion with a bob of mass 40 kg and a damping factor of 0.6.
(b) 30 meters downward
(c) Graph:

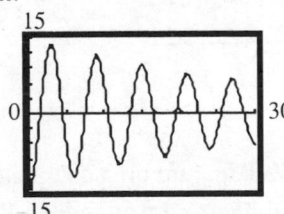

(d) The maximum displacement after one oscillation is 28.46 meters.
(e) It approaches zero, since $e^{-0.6\ t/80} \to 0$ as $t \to \infty$.

27. (a) Damped motion with a bob of mass 15 kg and a damping factor of 0.9.
(b) 15 meters downward
(c) Graph:

(d) The maximum displacement after one oscillation is 12.53 meters.
(e) It approaches zero, since $e^{-0.9\ t/30} \to 0$ as $t \to \infty$.

29. $f(x) = x + \cos x$

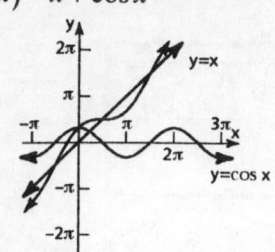

31. $f(x) = x - \sin x$

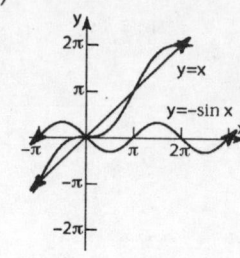

33. $f(x) = \sin x + \cos x$

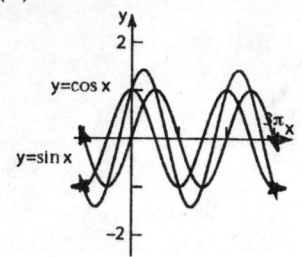

35. $f(x) = \sin x + \sin(2x)$

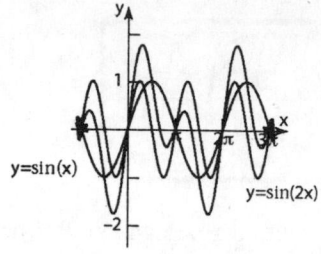

37. (a)   Graph:

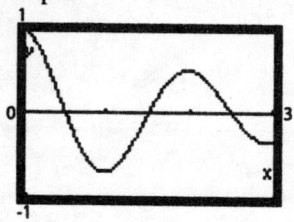

(b)   The graph of V touches the graph of
$y = e^{-1.9\,t}$ when $t = 0, 2$.
The graph of V touches the graph of
$y = -e^{-1.9\,t}$ when $t = 1, 3$.

(c)   To solve the inequality $-0.4 < e^{-t/3} \cdot \cos(\pi t) < 0.4$ on the interval $0 \le t \le 3$, we consider
the graphs of $y = -0.4$; $y = e^{-t/3} \cdot \cos(\pi t)$; and $y = 0.4$.

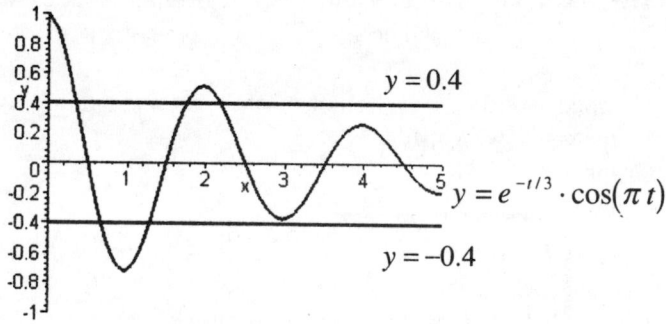

On the interval $0 \le t \le 3$, we can use the INTERSECT feature on a calculator to determine
that $y = e^{-t/3} \cdot \cos(\pi t)$ intersects $y = 0.4$ when $t \approx 0.35$, $t \approx 1.75$ and $t \approx 2.19$,
$y = e^{-t/3} \cdot \cos(\pi t)$ intersects $y = -0.4$ when $t \approx 0.67$ and $t \approx 1.28$ and the graph shows that
$-0.4 < e^{-t/3} \cdot \cos(\pi t) < 0.4$ when $t = 3$. Therefore, the voltage V is between -0.4 and 0.4 on
the intervals $0.35 < t < 0.67$, $1.28 < t < 1.75$ and $2.19 < t \le 3$.

39. $y = \sin(2\pi(852)t) + \sin(2\pi(1209)t)$

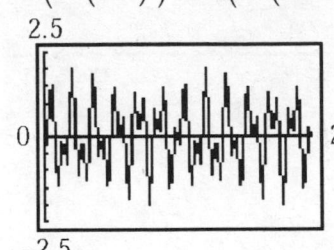

41.   Graph $f(x) = \dfrac{\sin x}{x}$:

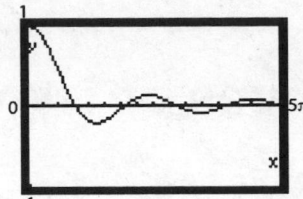

As x approaches 0, $\dfrac{\sin x}{x}$ approaches 1.

43.   Graphing:

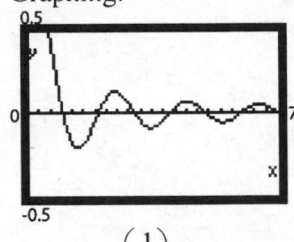

$$y = \left(\frac{1}{x}\right)\sin x \qquad\qquad y = \left(\frac{1}{x^2}\right)\sin x \qquad\qquad y = \left(\frac{1}{x^3}\right)\sin x$$

As x gets larger, the graph of $y = \left(\dfrac{1}{x^n}\right)\sin x$ gets closer to $y = 0$.

Applications of Trigonometric Functions

7.R   Chapter Review

1. opposite = 3; hypotenuse = 8
    Find the adjacent side:
    $$(\text{adjacent})^2 + 3^2 = 8^2$$
    $$(\text{adjacent})^2 = 64 - 9 = 55 \Rightarrow \text{adjacent} = \sqrt{55}$$

    $\sin\theta = \dfrac{\text{opp}}{\text{hyp}} = \dfrac{3}{8}$ $\cos\theta = \dfrac{\text{adj}}{\text{hyp}} = \dfrac{\sqrt{55}}{8}$ $\tan\theta = \dfrac{\text{opp}}{\text{adj}} = \dfrac{3}{\sqrt{55}} \cdot \dfrac{\sqrt{55}}{\sqrt{55}} = \dfrac{3\sqrt{55}}{55}$

    $\csc\theta = \dfrac{\text{hyp}}{\text{opp}} = \dfrac{8}{3}$ $\sec\theta = \dfrac{\text{hyp}}{\text{adj}} = \dfrac{8}{\sqrt{55}} \cdot \dfrac{\sqrt{55}}{\sqrt{55}} = \dfrac{8\sqrt{55}}{55}$ $\cot\theta = \dfrac{\text{adj}}{\text{opp}} = \dfrac{\sqrt{55}}{3}$

3. adjacent = 5; opposite = 6
    Find the hypotenuse:
    $$5^2 + 6^2 = (\text{hypotenuse})^2$$
    $$25 + 36 = (\text{hypotenuse})^2 \Rightarrow \text{hypotenuse} = \sqrt{61}$$

    $\sin\theta = \dfrac{\text{opp}}{\text{hyp}} = \dfrac{6}{\sqrt{61}} \cdot \dfrac{\sqrt{61}}{\sqrt{61}} = \dfrac{6\sqrt{61}}{61}$ $\cos\theta = \dfrac{\text{adj}}{\text{hyp}} = \dfrac{5}{\sqrt{61}} \cdot \dfrac{\sqrt{61}}{\sqrt{61}} = \dfrac{5\sqrt{61}}{61}$ $\tan\theta = \dfrac{\text{opp}}{\text{adj}} = \dfrac{6}{5}$

    $\csc\theta = \dfrac{\text{hyp}}{\text{opp}} = \dfrac{\sqrt{61}}{6}$ $\sec\theta = \dfrac{\text{hyp}}{\text{adj}} = \dfrac{\sqrt{61}}{5}$ $\cot\theta = \dfrac{\text{adj}}{\text{opp}} = \dfrac{5}{6}$

5. $\sec(25°) - \csc(65°) = \sec(25°) - \sec(25°) - 0$

7. $\sec\left(\dfrac{\pi}{8}\right) - \dfrac{1}{\sin\left(\dfrac{3\pi}{8}\right)} = \sec\left(\dfrac{\pi}{8}\right) - \csc\left(\dfrac{3\pi}{8}\right) = \sec\left(\dfrac{\pi}{8}\right) - \sec\left(\dfrac{\pi}{8}\right) = 0$

9. $c = 10,\ \beta = 20°$

    $\sin\beta = \dfrac{b}{c} \ \rightarrow\ \sin(20°) = \dfrac{b}{10} \ \rightarrow\ b = 10\sin(20°) \approx 10(0.3420) \approx 3.42$

    $\cos\beta = \dfrac{a}{c} \ \rightarrow\ \cos(20°) = \dfrac{a}{10} \ \rightarrow\ a = 10\cos(20°) \approx 10(0.9397) \approx 9.40$

    $\alpha = 90° - \beta = 90° - 20° = 70°$

11. $b = 2, \ c = 5$

$c^2 = a^2 + b^2 \ \rightarrow \ a^2 = c^2 - b^2 = 5^2 - 2^2 = 25 - 4 = 21 \ \rightarrow \ a = \sqrt{21} \approx 4.58$

$\sin\beta = \dfrac{b}{c} = \dfrac{2}{5} = 0.4000 \ \rightarrow \ \beta \approx 23.6°$

$\alpha = 90° - \beta = 90° - 23.6° = 66.4°$

13. $\alpha = 50°, \ \beta = 30°, \ a = 1$

$\gamma = 180° - \alpha - \beta = 180° - 50° - 30° = 100°$

$\dfrac{\sin\alpha}{a} = \dfrac{\sin\beta}{b} \ \rightarrow \ \dfrac{\sin(50°)}{1} = \dfrac{\sin(30°)}{b} \ \rightarrow \ b = \dfrac{1\sin(30°)}{\sin(50°)} \approx 0.65$

$\dfrac{\sin\gamma}{c} = \dfrac{\sin\alpha}{a} \ \rightarrow \ \dfrac{\sin(100°)}{c} = \dfrac{\sin(50°)}{1} \ \rightarrow \ c = \dfrac{1\sin(100°)}{\sin(50°)} \approx 1.29$

15. $a = 5, \ c = 2, \ \alpha = 100°$

$\dfrac{\sin\gamma}{c} = \dfrac{\sin\alpha}{a} \ \rightarrow \ \dfrac{\sin\gamma}{2} = \dfrac{\sin(100°)}{5} \ \rightarrow \ \sin\gamma = \dfrac{2\sin(100°)}{5} \approx 0.3939$

$\rightarrow \ \gamma = 23.2° \ \text{ or } \ \gamma = 156.8°$

The second value is discarded because $\alpha + \gamma > 180°$.

$\beta = 180° - \alpha - \gamma = 180° - 100° - 23.2° = 56.8°$

$\dfrac{\sin\beta}{b} = \dfrac{\sin\alpha}{a} \ \rightarrow \ \dfrac{\sin(56.8°)}{b} = \dfrac{\sin(100°)}{5} \ \rightarrow \ b = \dfrac{5\sin(56.8°)}{\sin(100°)} \approx 4.25$

17. $a = 3, \ c = 1, \ \gamma = 110°$

$\dfrac{\sin\gamma}{c} = \dfrac{\sin\alpha}{a} \ \rightarrow \ \dfrac{\sin(110°)}{1} = \dfrac{\sin\alpha}{3} \ \rightarrow \ \sin\alpha = \dfrac{3\sin(110°)}{1} \approx 2.8191$

There is no angle α for which $\sin\alpha > 1$. Therefore, there is no triangle with the given measurements.

19. $a = 3, \ c = 1, \ \beta = 100°$

$b^2 = a^2 + c^2 - 2ac\cos\beta = 3^2 + 1^2 - 2 \cdot 3 \cdot 1\cos(100°) \approx 11.0419 \rightarrow b \approx 3.32$

$a^2 = b^2 + c^2 - 2bc\cos\alpha \ \rightarrow \ \cos\alpha = \dfrac{b^2 + c^2 - a^2}{2bc} = \dfrac{3.32^2 + 1^2 - 3^2}{2(3.32)(1)} \approx 0.4552 \rightarrow \alpha \approx 62.8°$

$c^2 = a^2 + b^2 - 2ab\cos\gamma \rightarrow \cos\gamma = \dfrac{a^2 + b^2 - c^2}{2ab} = \dfrac{3^2 + 3.32^2 - 1^2}{2(3)(3.32)} \approx 0.9549 \rightarrow \gamma \approx 17.2°$

21. $a = 2, \ b = 3, \ c = 1$

$a^2 = b^2 + c^2 - 2bc\cos\alpha \ \rightarrow \ \cos\alpha = \dfrac{b^2 + c^2 - a^2}{2bc} = \dfrac{3^2 + 1^2 - 2^2}{2(3)(1)} = 1.000 \rightarrow \alpha \approx 0°$

No triangle exists with an angle of $0°$.

23. $a = 1$, $b = 3$, $\gamma = 40°$

$c^2 = a^2 + b^2 - 2ab\cos\gamma$

$c^2 = 1^2 + 3^2 - 2 \cdot 1 \cdot 3\cos(40°) \approx 5.4037$

$c \approx 2.32$

$a^2 = b^2 + c^2 - 2bc\cos\alpha \;\;\rightarrow\;\; \cos\alpha = \dfrac{b^2 + c^2 - a^2}{2bc} = \dfrac{3^2 + 2.32^2 - 1^2}{2(3)(2.32)} \approx 0.9614 \rightarrow \alpha \approx 16.1°$

$b^2 = a^2 + c^2 - 2ac\cos\beta \;\;\rightarrow\;\; \cos\beta = \dfrac{a^2 + c^2 - b^2}{2ac} = \dfrac{1^2 + 2.32^2 - 3^2}{2(1)(2.32)} \approx -0.5641 \rightarrow \beta \approx 123.9°$

25. $a = 5$, $b = 3$, $\alpha = 80°$

$\dfrac{\sin\beta}{b} = \dfrac{\sin\alpha}{a} \;\;\rightarrow\;\; \dfrac{\sin\beta}{3} = \dfrac{\sin(80°)}{5} \;\;\rightarrow\;\; \sin\beta = \dfrac{3\sin(80°)}{5} \approx 0.5909$

$\rightarrow\;\; \beta = 36.2°$ or $\beta = 143.8°$

The second value is discarded because $\alpha + \beta > 180°$.

$\gamma = 180° - \alpha - \beta = 180° - 80° - 36.2° = 63.8°$

$\dfrac{\sin\gamma}{c} = \dfrac{\sin\alpha}{a} \;\;\rightarrow\;\; \dfrac{\sin(63.8°)}{c} = \dfrac{\sin(80°)}{5} \;\;\rightarrow\;\; c = \dfrac{5\sin(63.8°)}{\sin(80°)} \approx 4.55$

27. $a = 1$, $b = \dfrac{1}{2}$, $c = \dfrac{4}{3}$

$a^2 = b^2 + c^2 - 2bc\cos\alpha \rightarrow \cos\alpha = \dfrac{b^2 + c^2 - a^2}{2bc}$

$= \dfrac{\left(\left(\dfrac{1}{2}\right)^2 + \left(\dfrac{4}{3}\right)^2 - 1^2\right)}{2\left(\dfrac{1}{2}\right)\left(\dfrac{4}{3}\right)} \approx 0.7708 \rightarrow \alpha \approx 39.6°$

$b^2 = a^2 + c^2 - 2ac\cos\beta \rightarrow \cos\beta = \dfrac{a^2 + c^2 - b^2}{2ac}$

$= \dfrac{\left(1^2 + \left(\dfrac{4}{3}\right)^2 - \left(\dfrac{1}{2}\right)^2\right)}{2(1)\left(\dfrac{4}{3}\right)} \approx 0.9479 \rightarrow \beta \approx 18.6°$

$\gamma = 180° - \alpha - \beta \approx 180° - 39.6° - 18.6° \approx 121.9°$

29. $a = 3$, $b = 4$, $\alpha = 10°$

$\dfrac{\sin\beta}{b} = \dfrac{\sin\alpha}{a} \;\;\rightarrow\;\; \dfrac{\sin\beta}{4} = \dfrac{\sin(10°)}{3} \;\;\rightarrow\;\; \sin\beta = \dfrac{4\sin(10°)}{3} \approx 0.2315$

$\rightarrow\;\; \beta_1 \approx 13.4°$ or $\beta_2 \approx 166.6°$

For both values, $\alpha + \beta < 180°$. Therefore, there are two triangles.

$$\gamma_1 = 180° - \alpha - \beta_1 = 180° - 10° - 13.4° \approx 156.6°$$

$$\frac{\sin\alpha}{a} = \frac{\sin\gamma_1}{c_1} \quad \rightarrow \quad \frac{\sin(10°)}{3} = \frac{\sin(156.6°)}{c_1} \quad \rightarrow \quad c_1 = \frac{3\sin(156.6°)}{\sin(10°)} \approx 6.86$$

$$\gamma_2 = 180° - \alpha - \beta_2 = 180° - 10° - 166.6° \approx 3.4°$$

$$\frac{\sin\alpha}{a} = \frac{\sin\gamma_2}{c_2} \quad \rightarrow \quad \frac{\sin(10°)}{3} = \frac{\sin(3.4°)}{c_2} \quad \rightarrow \quad c_2 = \frac{3\sin(3.4°)}{\sin(10°)} \approx 1.02$$

Two triangles: $\beta_1 \approx 13.4°$, $\gamma_1 \approx 156.6°$, $c_1 \approx 6.86$
or $\beta_2 \approx 166.6°$, $\gamma_2 \approx 3.4°$, $c_2 \approx 1.02$

31. $b = 4$, $c = 5$, $\alpha = 70°$

$$a^2 = b^2 + c^2 - 2bc\cos\alpha$$

$$a^2 = 4^2 + 5^2 - 2 \cdot 4 \cdot 5\cos(70°) \approx 27.3192$$

$$a \approx 5.23$$

$$c^2 = a^2 + b^2 - 2ab\cos\gamma \quad \rightarrow \quad \cos\gamma = \frac{a^2 + b^2 - c^2}{2ab} = \frac{5.23^2 + 4^2 - 5^2}{2(5.23)(4)} \approx 0.4386 \rightarrow \gamma \approx 64.0°$$

$$\beta = 180° - \alpha - \gamma = 180° - 70° - 64° \approx 46.0°$$

33. $a = 2$, $b = 3$, $\gamma = 40°$

$$A = \frac{1}{2}ab\sin\gamma = \frac{1}{2}(2)(3)\sin(40°) \approx 1.93$$

35. $b = 4$, $c = 10$, $\alpha = 70°$

$$A = \frac{1}{2}bc\sin\alpha = \frac{1}{2}(4)(10)\sin(70°) \approx 18.79$$

37. $a = 4$, $b = 3$, $c = 5$

$$s = \frac{1}{2}(a + b + c) = \frac{1}{2}(4 + 3 + 5) = 6$$

$$A = \sqrt{s(s-a)(s-b)(s-c)} = \sqrt{6(2)(3)(1)} = \sqrt{36} = 6$$

39. $a = 4$, $b = 2$, $c = 5$

$$s = \frac{1}{2}(a + b + c) = \frac{1}{2}(4 + 2 + 5) = \frac{11}{2}$$

$$A = \sqrt{s(s-a)(s-b)(s-c)} = \sqrt{\frac{11}{2}\left(\frac{3}{2}\right)\left(\frac{7}{2}\right)\left(\frac{1}{2}\right)} = \sqrt{\frac{231}{16}} \approx 3.80$$

41. $\alpha = 50°$, $\beta = 30°$, $a = 1$ $\qquad \gamma = 180° - \alpha - \beta = 180° - 50° - 30° = 100°$

$$A = \frac{a^2\sin\beta\sin\gamma}{2\sin\alpha} = \frac{1^2\sin(30°)\sin(100°)}{2\sin(50°)} \approx \frac{1(0.5000)(0.9848)}{2(0.7660)} \approx 0.32$$

43. Use right triangle methods:

$$\tan(65°) = \frac{500}{b} \quad \rightarrow \quad b = \frac{500}{\tan(65°)} \approx 233.15$$

$$\tan(25°) = \frac{500}{a+b} \quad \rightarrow \quad a+b = \frac{500}{\tan(25°)} \approx 1072.25$$

$$a = 1072.25 - 233.15 = 839.1 \text{ feet}$$

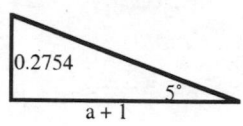

The lake is approximately 839 feet long.

45. $\tan(25°) = \dfrac{b}{50} \quad \rightarrow \quad b = 50\tan(25°) \approx 50(0.4663) \approx 23.32 \text{ feet}$

47. 1454 ft ≈ 0.2754 miles

$$\tan(5°) \approx \frac{0.2754}{a+1}$$

$$a+1 = \frac{0.2754}{\tan(5°)} \approx 3.15 \quad \rightarrow \quad a \approx 2.15 \text{ miles}$$

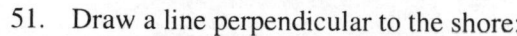

The boat is about 2.15 miles offshore.

49. $\angle ABC = 180° - 20° = 160°$

Find the angle at city C:

$$\frac{\sin C}{100} = \frac{\sin(160°)}{300} \quad \rightarrow \quad \sin C = \frac{100\sin(160°)}{300} \approx 0.1140 \quad \rightarrow \quad C \approx 6.55°$$

Find the angle at city A:

$$A = 180° - 160° - 6.55° = 13.45°$$

$$\frac{\sin(13.45°)}{y} = \frac{\sin(160°)}{300} \quad \rightarrow \quad y = \frac{300\sin(13.45°)}{\sin(160°)} \approx 204.07 \text{ miles}$$

The distance from city B to city C is approximately 204 miles.

51. Draw a line perpendicular to the shore:

(a) $\angle ACB = 12° + 30° = 42°$

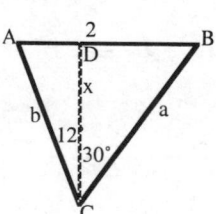

$\angle ABC = 90° - 30° = 60°$

$\angle CAB = 90° - 12° = 78°$

$$\frac{\sin(60°)}{b} = \frac{\sin(42°)}{2} \quad \rightarrow \quad b = \frac{2\sin(60°)}{\sin(42°)} \approx 2.59 \text{ miles}$$

(b) $\dfrac{\sin(78°)}{a} = \dfrac{\sin(42°)}{2} \quad \rightarrow \quad a = \dfrac{2\sin(78°)}{\sin(42°)} \approx 2.92 \text{ miles}$

(c) $\cos(12°) = \dfrac{x}{2.59} \quad \rightarrow \quad x = 2.59\cos(12°) \approx 2.53 \text{ miles}$

53. (a) After 4 hours, the yacht would have sailed 18(4) = 72 miles.
         Find the third side of the triangle determines the distance from the island:
         $a = 72, \ b = 200, \ \gamma = 15°$

         $c^2 = a^2 + b^2 - 2ab\cos\gamma$

         $c^2 = 72^2 + 200^2 - 2 \cdot 72 \cdot 200\cos(15°) \approx 17365.34$

         $c \approx 131.8$ miles
         The yacht is about 131.8 miles from the island.

    (b) Find the measure of the angle opposite the 200 side:

         $$\cos\beta = \frac{a^2 + c^2 - b^2}{2ac} \to \cos\beta = \frac{72^2 + 131.8^2 - 200^2}{2(72)(131.8)} \approx -0.9192 \to \beta \approx 156.8°$$

         The yacht should turn through an angle of $180° - 156.8° = 23.1°$ to correct its course.

    (c) The original trip would have taken: $t = \dfrac{200}{18} \approx 11.1$ hours.

         The actual trip takes: $t = 4 + \dfrac{131.8}{18} \approx 4 + 7.3 \approx 11.3$ hours.

         The trip takes about 0.21 hour or 12 minutes longer.

55. Find the lengths of the two unknown sides of the middle triangle:

     $$x^2 = 100^2 + 125^2 - 2(100)(125)\cos(50°) \approx 9555.31 \to x \approx 97.75 \text{ feet}$$

     $$y^2 = 70^2 + 50^2 - 2(70)(50)\cos(100°) \approx 8615.54 \to y \approx 92.82 \text{ feet}$$

     Find the areas of the three triangles:

     $$A_1 = \frac{1}{2}(100)(125)\sin(50°) \approx 4787.78 \text{ ft}^2; \quad A_2 = \frac{1}{2}(50)(70)\sin(100°) \approx 1723.41 \text{ ft}^2$$

     $$s = \frac{1}{2}(50 + 97.75 + 92.82) = 120.285$$

     $$A_3 = \sqrt{120.285(70.285)(22.535)(27.465)} \approx 2287.47$$

     The approximate area of the lake is $4787.78 + 1723.41 + 2287.47 = 8798.67$ sq.ft.

57. Area of the segment = area of the sector - area of the triangle.

     $$\text{Area of sector} = \frac{1}{2}r^2\theta = \frac{1}{2} \cdot 6^2\left(50 \cdot \frac{\pi}{180}\right) \approx 15.708 \text{ in}^2$$

     $$\text{Area of triangle} = \frac{1}{2}ab\sin\theta = \frac{1}{2} \cdot 6 \cdot 6\sin(50°) \approx 13.789 \text{ in}^2$$

     $$\text{Area of segment} \approx 15.708 - 13.789 = 1.92 \text{ in}^2$$

59. Extend the tangent line until it meets a line extended through the centers of the pulleys.
     Label these extensions x and y. The distance between the points of tangency is z. Two
     similar triangles are formed.
     Therefore:

     $$\frac{24 + y}{y} = \frac{6.5}{2.5}$$

     where $24 + y$ is the hypotenuse of the larger
     triangle and y is the hypotenuse of the
     smaller triangle.

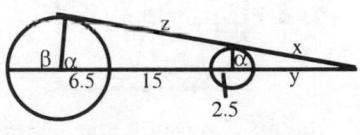

Solve for y:
$$6.5y = 2.5(24 + y) \rightarrow 6.5y = 60 + 2.5y \rightarrow 4y = 60 \rightarrow y = 15$$
Use the Pythagorean Theorem to find x:

$$x^2 + 2.5^2 = 15^2 \rightarrow x^2 = 225 - 6.25 = 218.75 \rightarrow x \approx 14.79$$
Use the Pythagorean Theorem to find z:

$$(z + 14.79)^2 + 6.5^2 = (24 + 15)^2 \rightarrow (z + 14.79)^2 = 1521 - 42.25 = 1478.75$$

$$z + 14.79 = 38.45 \rightarrow z = 23.66$$
Find α:

$$\cos\alpha = \frac{2.5}{15} \approx 0.1667 \quad \rightarrow \quad \alpha \approx 1.4033 \text{ radians}$$

$$\beta = \pi - 1.4033 \approx 1.7383 \text{ radians}$$
The arc length on the larger pulley is: $6.5(1.7383) = 11.30$ inches.
The arc length on the smaller pulley is $2.5(1.4033) = 3.51$ inches.
The distance between the points of tangency is 23.66 inches.
The length of the belt is: $2(11.30 + 3.51 + 23.66) = 76.94$ inches.

61. $d = 6\sin(2t)$
   (a) Simple harmonic
   (b) 6 feet
   (c) π seconds
   (d) $\dfrac{1}{\pi}$ oscillation/second

63. $d = -2\cos(\pi t)$
   (a) Simple harmonic
   (b) 2 feet
   (c) 2 seconds
   (d) $\dfrac{1}{2}$ oscillation/second

65. $d = -4\cos\left(\dfrac{2\pi t}{3}\right)$

67. (a)

$$d = -15e^{-0.75t/2(40)} \cos\left(\sqrt{\left(\frac{2\pi}{5}\right)^2 - \frac{(0.75)^2}{4(40)^2}}\; t\right)$$

$$d = -15e^{-0.75t/80} \cos\left(\sqrt{\frac{4\pi^2}{25} - \frac{0.5625}{6400}}\; t\right)$$

(b)

69. (a) Damped motion with a bob of mass 20 kg and a damping factor of 0.6.
   (b) 15 meters downward
   (c) Graph:

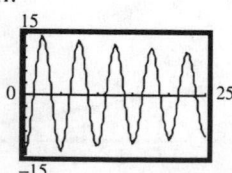

   (d) The maximum displacement after one oscillation is 13.92 meters.

(e)   It approaches zero, since $e^{-0.6\ t/40} \to 0$ as $t \to \infty$.

71. $y = 2\sin(x) + \cos(2x), \qquad 0 \le x \le 2\pi$

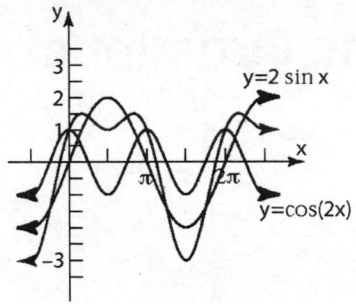

Applications of Trigonometric Functions

7.CR   Cumulative Review

1. $3x^2 + 1 = 4x \Rightarrow 3x^2 - 4x + 1 = 0$

    $(3x-1)(x-1) = 0$

    $x = \dfrac{1}{3}$ or $x = 1$

    The solution set is $\left\{\dfrac{1}{3}, 1\right\}$.

3. $f(x) = \sqrt{x^2 - 3x - 4}$

    f will be defined provided $g(x) = x^2 - 3x - 4 \geq 0$.

    $x^2 - 3x - 4 \geq 0$

    $(x-4)(x+1) \geq 0$

    $x = 4$, $x = -1$ are the zeros.

Interval	Test Number	$g(x)$	Positive/Negative
$-\infty < x < -1$	-2	6	Positive
$-1 < x < 4$	0	-4	Negative
$4 < x < \infty$	5	6	Positive

    The domain of $f(x) = \sqrt{x^2 - 3x - 4}$ is $\left\{x \mid -\infty < x \leq -1 \text{ or } 4 \leq x < \infty\right\}$.

5. $y = -2\cos(2x - \pi) = -2\cos\left(2\left(x - \dfrac{\pi}{2}\right)\right)$

    Amplitude : $|A| = |-3| = 2$

    Period : $T = \dfrac{2\pi}{2} = \pi$

    Phase Shift : $\dfrac{\phi}{\omega} = \dfrac{\pi}{2}$

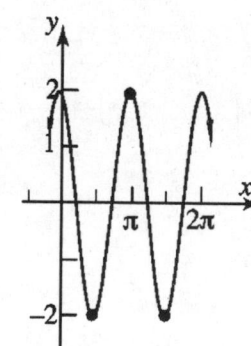

7. (a) $y = e^x, \ 0 \le x \le 4$

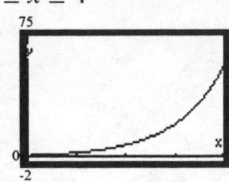

(b) $y = \sin x, \ 0 \le x \le 4$

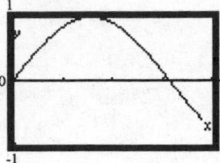

(c) $y = e^x \sin x, \ 0 \le x \le 4$

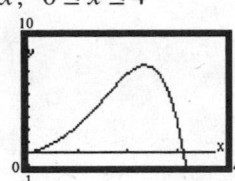

(d) $y = 2x + \sin x, \ 0 \le x \le 4$

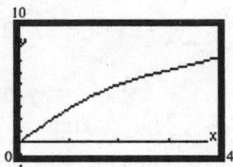

9.   Given the triangle

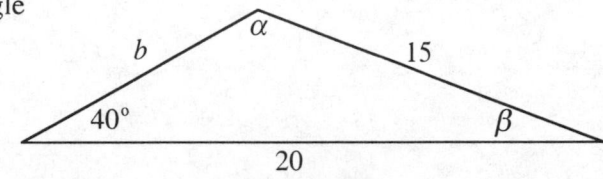

$$\frac{\sin(40^\circ)}{15} = \frac{\sin(\alpha)}{20} \Rightarrow \sin(\alpha) = (20)\left(\frac{\sin(40^\circ)}{15}\right) \approx 0.8570$$

$$\sin(\alpha) \approx 0.8570 \Rightarrow \alpha \approx \sin^{-1}(0.8570)$$

$$\alpha \approx 58.99^\circ \ \text{ or } \ \alpha \approx 121.02^\circ$$

Case 1: $\alpha \approx 58.99^\circ \Rightarrow \beta = 180^\circ - \left(40^\circ + 58.99^\circ\right) = 81.01^\circ$

$$\frac{\sin(40^\circ)}{15} = \frac{\sin(\beta)}{b} \Rightarrow \frac{\sin(40^\circ)}{15} = \frac{\sin(81.01^\circ)}{b}$$

$$\Rightarrow b\sin(40^\circ) = 15\sin(81.01^\circ)$$

$$b = \frac{15\sin(81.01^\circ)}{\sin(40^\circ)} \approx 23.05$$

Therefore, $\alpha \approx 58.99^\circ, \ \beta \approx 81.01^\circ, \ b \approx 23.05$.

Case 2: $\alpha \approx 121.02^\circ \Rightarrow \beta = 180^\circ - \left(40^\circ + 121.02^\circ\right) = 18.98^\circ$

$$\frac{\sin(40^\circ)}{15} = \frac{\sin(\beta)}{b} \Rightarrow \frac{\sin(40^\circ)}{15} = \frac{\sin(18.98^\circ)}{b}$$

$$\Rightarrow b\sin(40^\circ) = 15\sin(18.98^\circ)$$

$$b = \frac{15\sin(18.98^\circ)}{\sin(40^\circ)} \approx 7.59$$

Therefore, $\alpha \approx 121.02^\circ, \ \beta \approx 18.98^\circ, \ b \approx 7.59$.

11. $R(x) = \dfrac{2x^2 - 7x - 4}{x^2 + 2x - 15} = \dfrac{(2x+1)(x-4)}{(x-3)(x+5)}$ $p(x) = 2x^2 - 7x - 4;\ q(x) = x^2 + 2x - 15;$
$$n = 2;\ m = 2$$

Step 1:   Domain: $\{x \mid x \neq -5,\ x \neq 3\}$

Step 2: R is in lowest terms.

Step 3:   (a)   The x-intercepts are the zeros of $p(x)$: -0.5 and 4

(b)   The y-intercept is $R(0) = \dfrac{2 \cdot 0^2 - 7 \cdot 0 - 4}{0^2 + 2 \cdot 0 - 15} = \dfrac{-4}{-15} = \dfrac{4}{15}$.

Step 4: $R(-x) = \dfrac{2(-x)^2 - 7(-x) - 4}{(-x)^2 + 2(-x) - 15} = \dfrac{2x^2 + 7x - 4}{x^2 - 2x - 15}$; this is neither $R(x)$ nor $-R(x)$, so there is no symmetry.

Step 5:   The vertical asymptotes are the zeros of $q(x)$: $x = -5$ and $x = 3$

Step 6:   Since $n = m$, the line $y = 2$ is the horizontal asymptote.

$R(x)$ intersects $y = 2$ at $\left(\dfrac{26}{11}, 2\right)$, since:

$$\frac{2x^2 - 7x - 4}{x^2 + 2x - 15} = 2 \Rightarrow 2x^2 - 7x - 4 = 2x^2 + 4x - 30$$

$$-11x = -26 \Rightarrow x = \frac{26}{11}$$

Step 7:   Graphing utility:

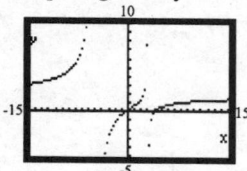

Step 8:   Graphing by hand:

Interval	$(-\infty, -5)$	$(-5, -0.5)$	$(-0.5, 3)$	$(3, 4)$	$(4, \infty)$
Number Chosen	-6	-1	0	3.5	5
Value of f	$f(-6) \approx 12.22$	$f(-1) = -0.3125$	$f(0) \approx 0.27$	$f(3.5) \approx -0.94$	$f(5) = 0.55$
Location of Graph	Above x-axis	Below x-axis	Above x-axis	Below x-axis	Above x-axis
Point on Graph	$(-6, 12.22)$	$(-1, -0.3125)$	$(0, 0.27)$	$(3.5, -0.94)$	$(5, 0.55)$

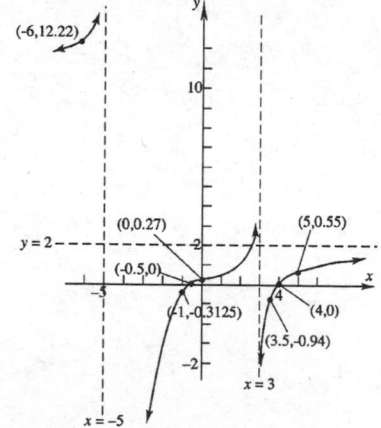

13. $\log_3(x+8) + \log_3(x) = 2$

$\log_3\big((x+8)(x)\big) = 2$

$(x+8)(x) = 3^2$

$x^2 + 8x = 9$

$x^2 + 8x - 9 = 0$

$(x+9)(x-1) = 0$

$x = -9$ or $x = 1$

Since $x = -9$ makes the original logarithms undefined, the only solution is $x = 1$.

Chapter 8

Polar Coordinates; Vectors

8.1 Polar Coordinates

1. A   3. C   5. B   7. A

9. $(3, 90°)$

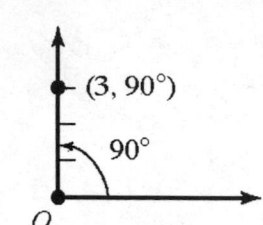

11. $(-2, 0)$

13. $\left(6, \dfrac{\pi}{6}\right)$

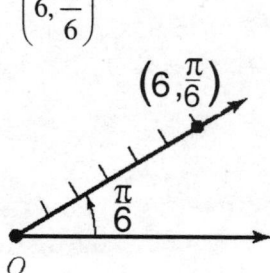

15. $(-2, 135°)$

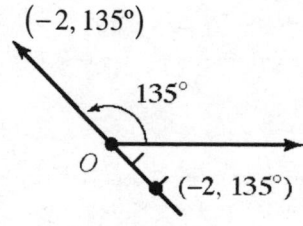

17. $\left(-1, -\dfrac{\pi}{3}\right)$

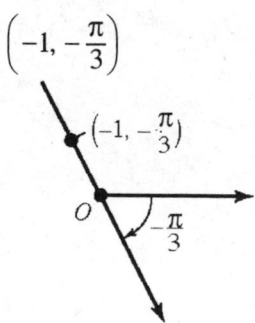

19. $(-2, -\pi)$

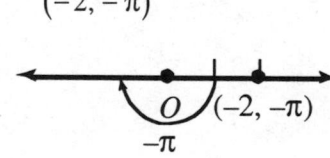

21. $\left(5, \dfrac{2\pi}{3}\right)$

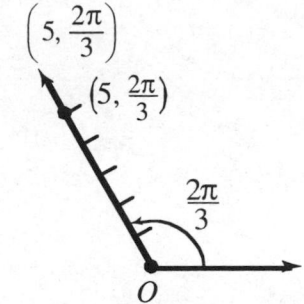

        (a) $r > 0,\ -2\pi \le \theta < 0$ $\left(5, -\dfrac{4\pi}{3}\right)$

        (b) $r < 0,\ 0 \le \theta < 2\pi$ $\left(-5, \dfrac{5\pi}{3}\right)$

        (c) $r > 0,\ 2\pi \le \theta < 4\pi$ $\left(5, \dfrac{8\pi}{3}\right)$

23. $(-2, 3\pi)$

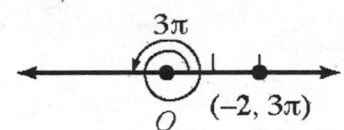

        (a) $r > 0,\ -2\pi \le \theta < 0$ $(2, -2\pi)$

        (b) $r < 0,\ 0 \le \theta < 2\pi$ $(-2, \pi)$

        (c) $r > 0,\ 2\pi \le \theta < 4\pi$ $(2, 2\pi)$

25. $\left(1, \dfrac{\pi}{2}\right)$

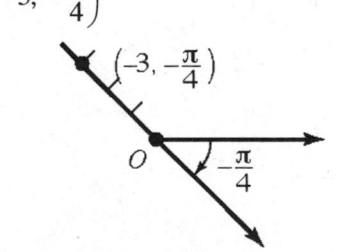

        (a) $r > 0,\ -2\pi \le \theta < 0$ $\left(1, -\dfrac{3\pi}{2}\right)$

        (b) $r < 0,\ 0 \le \theta < 2\pi$ $\left(-1, \dfrac{3\pi}{2}\right)$

        (c) $r > 0,\ 2\pi \le \theta < 4\pi$ $\left(1, \dfrac{5\pi}{2}\right)$

27. $\left(-3, -\dfrac{\pi}{4}\right)$

        (a) $r > 0,\ -2\pi \le \theta < 0$ $\left(3, -\dfrac{5\pi}{4}\right)$

        (b) $r < 0,\ 0 \le \theta < 2\pi$ $\left(-3, \dfrac{7\pi}{4}\right)$

        (c) $r > 0,\ 2\pi \le \theta < 4\pi$ $\left(3, \dfrac{11\pi}{4}\right)$

29. $x = r\cos\theta = 3\cos\left(\dfrac{\pi}{2}\right) = 3 \cdot 0 = 0$

     $y = r\sin\theta = 3\sin\left(\dfrac{\pi}{2}\right) = 3 \cdot 1 = 3$

The rectangular coordinates of the point $\left(3, \dfrac{\pi}{2}\right)$ are $(0, 3)$.

31. $x = r\cos\theta = -2\cos(0) = -2 \cdot 1 = -2$
     $y = r\sin\theta = -2\sin(0) = -2 \cdot 0 = 0$

The rectangular coordinates of the point $(-2, 0)$ are $(-2, 0)$.

33. $x = r\cos\theta = 6\cos(150°) = 6 \cdot \left(-\dfrac{\sqrt{3}}{2}\right) = -3\sqrt{3}$

      $y = r\sin\theta = 6\sin(150°) = 6 \cdot \dfrac{1}{2} = 3$

The rectangular coordinates of the point $(6, 150°)$ are $\left(-3\sqrt{3}, 3\right)$.

35. $x = r\cos\theta = -2\cos\left(\dfrac{3\pi}{4}\right) = -2 \cdot \left(-\dfrac{\sqrt{2}}{2}\right) = \sqrt{2}$

      $y = r\sin\theta = -2\sin\left(\dfrac{3\pi}{4}\right) = -2 \cdot \dfrac{\sqrt{2}}{2} = -\sqrt{2}$

The rectangular coordinates of the point $\left(-2, \dfrac{3\pi}{4}\right)$ are $\left(\sqrt{2}, -\sqrt{2}\right)$.

37. $x = r\cos\theta = -1\cos\left(-\dfrac{\pi}{3}\right) = -1 \cdot \dfrac{1}{2} = -\dfrac{1}{2}$

      $y = r\sin\theta = -1\sin\left(-\dfrac{\pi}{3}\right) = -1 \cdot \left(-\dfrac{\sqrt{3}}{2}\right) = \dfrac{\sqrt{3}}{2}$

The rectangular coordinates of the point $\left(-1, -\dfrac{\pi}{3}\right)$ are $\left(-\dfrac{1}{2}, \dfrac{\sqrt{3}}{2}\right)$.

39. $x = r\cos\theta = -2\cos(-180°) = -2 \cdot -1 = 2$
      $y = r\sin\theta = -2\sin(-180°) = -2 \cdot 0 = 0$
The rectangular coordinates of the point $(-2, -180°)$ are $(2, 0)$.

41. $x = r\cos\theta = 7.5\cos(110°) \approx 7.5(-0.3420) \approx -2.57$
      $y = r\sin\theta = 7.5\sin(110°) \approx 7.5(0.9397) \approx 7.05$

The rectangular coordinates of the point $(7.5, 110°)$ are $(-2.57, 7.05)$.

43. $x = r\cos\theta = 6.3\cos(3.8) \approx 6.3(-0.7910) \approx -4.98$
      $y = r\sin\theta = 6.3\sin(3.8) \approx 6.3(-0.6119) \approx -3.85$

The rectangular coordinates of the point $(6.3, 3.8)$ are $(-4.98, -3.85)$.

45. $r = \sqrt{x^2 + y^2} = \sqrt{3^2 + 0^2} = \sqrt{9} = 3 \qquad \theta = \tan^{-1}\left(\dfrac{y}{x}\right) = \tan^{-1}\left(\dfrac{0}{3}\right) = \tan^{-1}(0) = 0$

Polar coordinates of the point $(3, 0)$ are $(3, 0)$.

47. $r = \sqrt{x^2 + y^2} = \sqrt{(-1)^2 + 0^2} = \sqrt{1} = 1 \qquad \theta = \tan^{-1}\left(\dfrac{y}{x}\right) = \tan^{-1}\left(\dfrac{0}{-1}\right) = \tan^{-1}(0) = 0$

The point lies on the negative x-axis thus $\theta = \pi$.
Polar coordinates of the point $(-1, 0)$ are $(1, \pi)$.

49. The point $(1, -1)$ lies in quadrant IV.

$r = \sqrt{x^2 + y^2} = \sqrt{1^2 + (-1)^2} = \sqrt{2}$ $\theta = \tan^{-1}\left(\dfrac{y}{x}\right) = \tan^{-1}\left(\dfrac{-1}{1}\right) = \tan^{-1}(-1) = -\dfrac{\pi}{4}$

Polar coordinates of the point $(1, -1)$ are $\left(\sqrt{2}, -\dfrac{\pi}{4}\right)$.

51. The point $\left(\sqrt{3}, 1\right)$ lies in quadrant I.

$r = \sqrt{x^2 + y^2} = \sqrt{\left(\sqrt{3}\right)^2 + 1^2} = \sqrt{4} = 2$ $\theta = \tan^{-1}\left(\dfrac{y}{x}\right) = \tan^{-1}\left(\dfrac{1}{\sqrt{3}}\right) = \dfrac{\pi}{6}$

Polar coordinates of the point $\left(\sqrt{3}, 1\right)$ are $\left(2, \dfrac{\pi}{6}\right)$.

53. The point $(1.3, -2.1)$ lies in quadrant IV.

$r = \sqrt{x^2 + y^2} = \sqrt{1.3^2 + (-2.1)^2} = \sqrt{6.1} \approx 2.47$

$\theta = \tan^{-1}\left(\dfrac{y}{x}\right) = \tan^{-1}\left(\dfrac{-2.1}{1.3}\right) \approx \tan^{-1}(-1.6154) \approx -1.02$

Polar coordinates of the point $(1.3, -2.1)$ are $(2.47, -1.02)$.

55. The point $(8.3, 4.2)$ lies in quadrant I.

$r = \sqrt{x^2 + y^2} = \sqrt{8.3^2 + 4.2^2} = \sqrt{86.53} \approx 9.30$

$\theta = \tan^{-1}\left(\dfrac{y}{x}\right) = \tan^{-1}\left(\dfrac{4.2}{8.3}\right) \approx \tan^{-1}(0.5060) \approx 0.47$

Polar coordinates of the point $(8.3, 4.2)$ are $(9.30, 0.47)$.

57. $2x^2 + 2y^2 = 3$
$2\left(x^2 + y^2\right) = 3$

$2r^2 = 3 \Rightarrow r^2 = \dfrac{3}{2}$

59. $x^2 = 4y$
$(r\cos\theta)^2 = 4r\sin\theta$
$r^2\cos^2\theta - 4r\sin\theta = 0$

61. $2xy = 1$
$2(r\cos\theta)(r\sin\theta) = 1$
$2r^2\sin\theta\cos\theta = 1 \Rightarrow r^2\sin 2\theta = 1$

63. $x = 4$
$r\cos\theta = 4$

65. $r = \cos\theta$
$r^2 = r\cos\theta$
$x^2 + y^2 = x \Rightarrow x^2 - x + y^2 = 0$

67. $r^2 = \cos\theta$
$r^3 = r\cos\theta$
$\left(x^2 + y^2\right)^{3/2} = x \Rightarrow \left(x^2 + y^2\right)^{3/2} - x = 0$

69. $r = 2$
$\sqrt{x^2 + y^2} = 2 \Rightarrow x^2 + y^2 = 4$

71. $r = \dfrac{4}{1 - \cos\theta}$

$r(1 - \cos\theta) = 4 \Rightarrow r - r\cos\theta = 4$

$\sqrt{x^2 + y^2} - x = 4 \Rightarrow \sqrt{x^2 + y^2} = x + 4$

$x^2 + y^2 = x^2 + 8x + 16$

$y^2 = 8(x + 2)$

73.   Rewrite the polar coordinates in rectangular form:

$P_1 = (r_1, \theta_1) \quad \rightarrow \quad P_1 = (r_1\cos\theta_1, r_1\sin\theta_1)$

$P_2 = (r_2, \theta_2) \quad \rightarrow \quad P_2 = (r_2\cos\theta_2, r_2\sin\theta_2)$

$d = \sqrt{\left(r_2\cos\theta_2 - r_1\cos\theta_1\right)^2 + \left(r_2\sin\theta_2 - r_1\sin\theta_1\right)^2}$

$= \sqrt{r_2^2\cos^2\theta_2 - 2r_1r_2\cos\theta_2\cos\theta_1 + r_1^2\cos^2\theta_1 + r_2^2\sin^2\theta_2 - 2r_1r_2\sin\theta_2\sin\theta_1 + r_1^2\sin^2\theta_1}$

$= \sqrt{r_2^2\left(\cos^2\theta_2 + \sin^2\theta_2\right) + r_1^2\left(\cos^2\theta_1 + \sin^2\theta_1\right) - 2r_1r_2\left(\cos\theta_2\cos\theta_1 + \sin\theta_2\sin\theta_1\right)}$

$= \sqrt{r_2^2 + r_1^2 - 2r_1r_2\cos\left(\theta_2 - \theta_1\right)}$

Chapter 8

Polar Coordinates; Vectors

8.2 Polar Equations and Graphs

1. $r = 4$
   The equation is of the form $r = a$, $a > 0$.
   It is a circle, center at the pole and radius 4.
   Transform to rectangular form:
   $$r = 4$$
   $$r^2 = 16$$
   $$x^2 + y^2 = 16$$

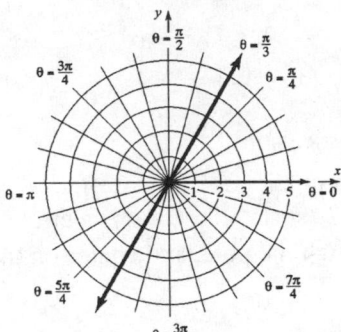

3. $\theta = \dfrac{\pi}{3}$
   The equation is of the form $\theta = \alpha$. It is a
   line, passing through the pole at an angle
   of $\dfrac{\pi}{3}$.
   Transform to rectangular form:
   $$\theta = \frac{\pi}{3} \Rightarrow \tan\theta = \tan\left(\frac{\pi}{3}\right)$$
   $$\frac{y}{x} = \sqrt{3}$$
   $$y = \sqrt{3}x$$

5. $r\sin\theta = 4$
   The equation is of the form $r\sin\theta = b$. It is
   a horizontal line, 4 units above the pole.
   Transform to rectangular form:
   $$r\sin\theta = 4$$
   $$y = 4$$

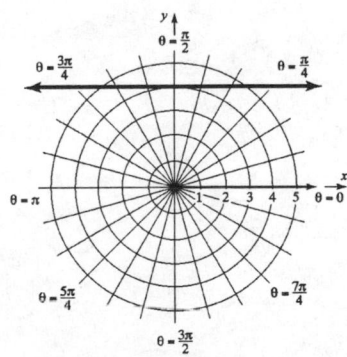

7. $r\cos\theta = -2$

     The equation is of the form $r\cos\theta = a$. It is
     a vertical line, 2 units to the left of the pole.
     Transform to rectangular form:

$$r\cos\theta = -2$$
$$x = -2$$

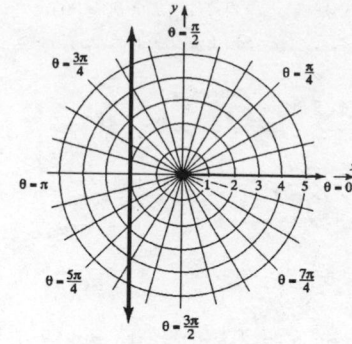

9. $r = 2\cos\theta$

     The equation is of the form
     $r = 2a\cos\theta, \ a > 0$. It is a circle, passing
     through the pole, and center on the polar
     axis.

     Transform to rectangular form:

$$r = 2\cos\theta$$
$$r^2 = 2r\cos\theta$$
$$x^2 + y^2 = 2x$$
$$x^2 - 2x + y^2 = 0$$
$$(x-1)^2 + y^2 = 1$$

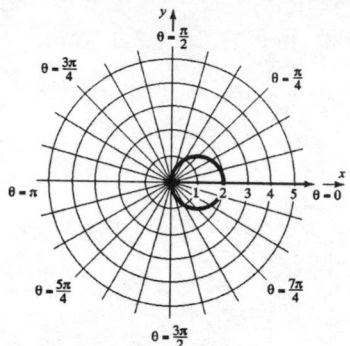

11. $r = -4\sin\theta$

     The equation is of the form
     $r = 2a\sin\theta, \ a > 0$. It is a circle, passing
     through the pole, and center on the line
     $\theta = \dfrac{\pi}{2}$.

     Transform to rectangular form:

$$r = -4\sin\theta$$
$$r^2 = -4r\sin\theta$$
$$x^2 + y^2 = -4y$$
$$x^2 + y^2 + 4y = 0$$
$$x^2 + (y+2)^2 = 4$$

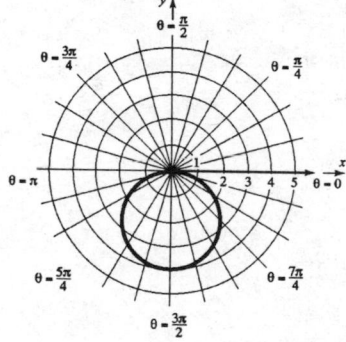

13. $r \sec \theta = 4$

Transform to rectangular form:
$$r \sec \theta = 4$$
$$r \cdot \frac{1}{\cos \theta} = 4$$
$$r = 4 \cos \theta$$
$$r^2 = 4r \cos \theta$$
$$x^2 + y^2 = 4x$$
$$x^2 - 4x + y^2 = 0$$
$$(x - 2)^2 + y^2 = 4$$

The equation is a circle, passing through the pole, center on the polar axis and radius 2.

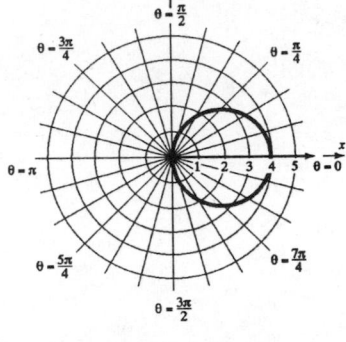

15. $r \csc \theta = -2$

Transform to rectangular form:
$$r \csc \theta = -2$$
$$r \cdot \frac{1}{\sin \theta} = -2$$
$$r = -2 \sin \theta$$
$$r^2 = -2r \sin \theta$$
$$x^2 + y^2 = -2y$$
$$x^2 + y^2 + 2y = 0$$
$$x^2 + (y + 1)^2 = 1$$

The equation is a circle, passing through the pole, center on the line $\theta = \frac{\pi}{2}$ and radius 1.

17. E   19. F   21. H   23. D

25. D   27. F   29. A

31. $r = 2 + 2\cos \theta$ The graph will be a cardioid. Check for symmetry:
Polar axis: Replace θ by $-\theta$. The result is $r = 2 + 2\cos(-\theta) = 2 + 2\cos \theta$.
The graph is symmetric with respect to the polar axis.

The line $\theta = \frac{\pi}{2}$: Replace θ by $\pi - \theta$.
$$r = 2 + 2\cos(\pi - \theta) = 2 + 2(\cos(\pi)\cos \theta + \sin(\pi)\sin \theta)$$
$$= 2 + 2(-\cos \theta + 0) = 2 - 2\cos \theta$$
The test fails.

The pole: Replace r by $-r$. $-r = 2 + 2\cos \theta$. The test fails.

Due to symmetry to the polar axis, assign values to θ from 0 to π.

θ	0	$\frac{\pi}{6}$	$\frac{\pi}{3}$	$\frac{\pi}{2}$	$\frac{2\pi}{3}$	$\frac{5\pi}{6}$	π
$r = 2 + 2\cos \theta$	4	$2 + \sqrt{3} \approx 3.7$	3	2	1	$2 - \sqrt{3} \approx 0.3$	0

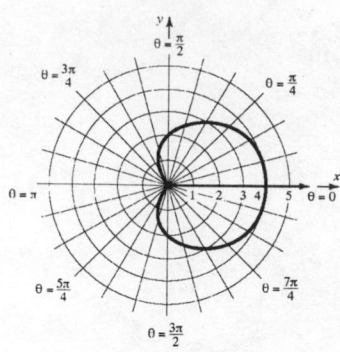

33. $r = 3 - 3\sin\theta$ The graph will be a cardioid. Check for symmetry:
Polar axis: Replace θ by $-\theta$. The result is $r = 3 - 3\sin(-\theta) = 3 + 3\sin\theta$.
          The test fails.

The line $\theta = \dfrac{\pi}{2}$: Replace θ by $\pi - \theta$.

$$r = 3 - 3\sin(\pi - \theta) = 3 - 3(\sin(\pi)\cos\theta - \cos(\pi)\sin\theta)$$
$$= 3 - 3(0 + \sin\theta) = 3 - 3\sin\theta$$

The graph is symmetric with respect to the line $\theta = \dfrac{\pi}{2}$.

The pole: Replace r by $-r$. $-r = 3 - 3\sin\theta$. The test fails.

Due to symmetry to the line $\theta = \dfrac{\pi}{2}$, assign values to θ from $-\dfrac{\pi}{2}$ to $\dfrac{\pi}{2}$.

θ	$-\dfrac{\pi}{2}$	$-\dfrac{\pi}{3}$	$-\dfrac{\pi}{6}$	0	$\dfrac{\pi}{6}$	$\dfrac{\pi}{3}$	$\dfrac{\pi}{2}$
$r = 3 - 3\sin\theta$	6	$3 + \dfrac{3\sqrt{3}}{2} \approx 5.6$	$\dfrac{9}{2}$	3	$\dfrac{3}{2}$	$3 - \dfrac{3\sqrt{3}}{2} \approx 0.4$	0

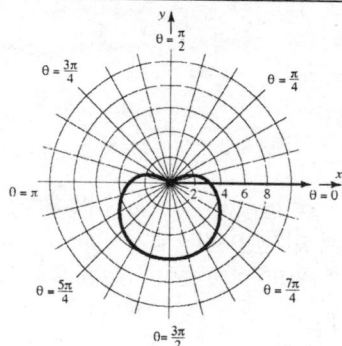

35. $r = 2 + \sin\theta$ The graph will be a limacon without an inner loop.
Check for symmetry:
Polar axis: Replace θ by $-\theta$. The result is $r = 2 + \sin(-\theta) = 2 - \sin\theta$.
          The test fails.

The line $\theta = \dfrac{\pi}{2}$: Replace θ by $\pi - \theta$.

$$r = 2 + \sin(\pi - \theta) = 2 + (\sin(\pi)\cos\theta - \cos(\pi)\sin\theta)$$
$$= 2 + (0 + \sin\theta) = 2 + \sin\theta$$

The graph is symmetric with respect to the line $\theta = \dfrac{\pi}{2}$.

The pole: Replace r by $-r$. $-r = 2 + \sin\theta$. The test fails.

Due to symmetry to the line $\theta = \dfrac{\pi}{2}$, assign values to θ from $-\dfrac{\pi}{2}$ to $\dfrac{\pi}{2}$.

θ	$-\dfrac{\pi}{2}$	$-\dfrac{\pi}{3}$	$-\dfrac{\pi}{6}$	0	$\dfrac{\pi}{6}$	$\dfrac{\pi}{3}$	$\dfrac{\pi}{2}$
$r = 2 + \sin\theta$	1	$2 - \dfrac{\sqrt{3}}{2} \approx 1.1$	$\dfrac{3}{2}$	2	$\dfrac{5}{2}$	$2 + \dfrac{\sqrt{3}}{2} \approx 2.9$	3

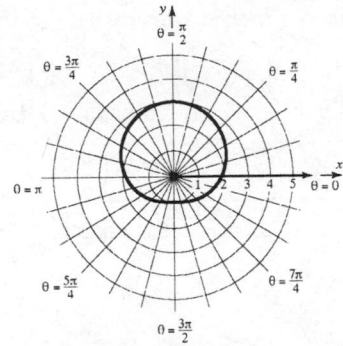

37. $r = 4 - 2\cos\theta$ The graph will be a limacon without an inner loop.
   Check for symmetry:
   Polar axis: Replace θ by $-\theta$. The result is $r = 4 - 2\cos(-\theta) = 4 - 2\cos\theta$.
            The graph is symmetric with respect to the polar axis.

   The line $\theta = \dfrac{\pi}{2}$: Replace θ by $\pi - \theta$.

   $$r = 4 - 2\cos(\pi - \theta) = 4 - 2(\cos(\pi)\cos\theta + \sin(\pi)\sin\theta)$$

   $$= 4 - 2(-\cos\theta + 0) = 4 + 2\cos\theta$$

            The test fails.
   The pole: Replace r by $-r$. $-r = 4 - 2\cos\theta$. The test fails.
   Due to symmetry to the polar axis, assign values to θ from 0 to π.

θ	0	$\dfrac{\pi}{6}$	$\dfrac{\pi}{3}$	$\dfrac{\pi}{2}$	$\dfrac{2\pi}{3}$	$\dfrac{5\pi}{6}$	π
$r = 4 - 2\cos\theta$	2	$4 - \sqrt{3} \approx 2.3$	3	4	5	$4 + \sqrt{3} \approx 5.7$	6

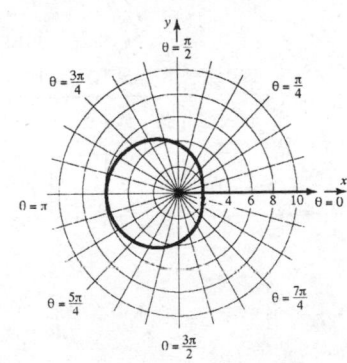

39. $r = 1 + 2\sin\theta$ The graph will be a limacon with an inner loop.
Check for symmetry:
Polar axis: Replace θ by $-\theta$. The result is $r = 1 + 2\sin(-\theta) = 1 - 2\sin\theta$.
    The test fails.
The line $\theta = \dfrac{\pi}{2}$: Replace θ by $\pi - \theta$.

$$r = 1 + 2\sin(\pi - \theta) = 1 + 2(\sin(\pi)\cos\theta - \cos(\pi)\sin\theta)$$
$$= 1 + 2(0 + \sin\theta) = 1 + 2\sin\theta$$

The graph is symmetric with respect to the line $\theta = \dfrac{\pi}{2}$.

The pole: Replace r by $-r$. $-r = 1 + 2\sin\theta$. The test fails.

Due to symmetry to the line $\theta = \dfrac{\pi}{2}$, assign values to θ from $-\dfrac{\pi}{2}$ to $\dfrac{\pi}{2}$.

θ	$-\dfrac{\pi}{2}$	$-\dfrac{\pi}{3}$	$-\dfrac{\pi}{6}$	0	$\dfrac{\pi}{6}$	$\dfrac{\pi}{3}$	$\dfrac{\pi}{2}$
$r = 1 + 2\sin\theta$	-1	$1 - \sqrt{3} \approx -0.7$	0	1	2	$1 + \sqrt{3} \approx 2.7$	3

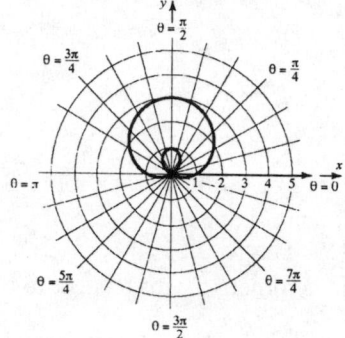

41. $r = 2 - 3\cos\theta$ The graph will be a limacon with an inner loop.
Check for symmetry:
Polar axis: Replace θ by $-\theta$. The result is $r = 2 - 3\cos(-\theta) = 2 - 3\cos\theta$.
        The graph is symmetric with respect to the polar axis.
The line $\theta = \dfrac{\pi}{2}$: Replace θ by $\pi - \theta$.

$$r = 2 - 3\cos(\pi - \theta) = 2 - 3(\cos(\pi)\cos\theta + \sin(\pi)\sin\theta)$$
$$= 2 - 3(-\cos\theta + 0) = 2 + 3\cos\theta$$

        The test fails.
The pole: Replace r by $-r$. $-r = 2 - 3\cos\theta$. The test fails.
Due to symmetry to the polar axis, assign values to θ from 0 to π.

θ	0	$\dfrac{\pi}{6}$	$\dfrac{\pi}{3}$	$\dfrac{\pi}{2}$	$\dfrac{2\pi}{3}$	$\dfrac{5\pi}{6}$	π
$r = 2 - 3\cos\theta$	-1	$2 - \dfrac{3\sqrt{3}}{2} \approx -0.6$	$\dfrac{1}{2}$	2	$\dfrac{7}{2}$	$2 + \dfrac{3\sqrt{3}}{2} \approx 4.6$	5

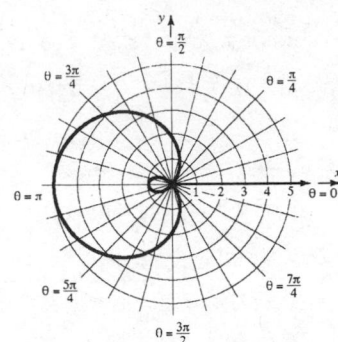

43. $r = 3\cos(2\theta)$ The graph will be a rose with four petals. Check for symmetry:
Polar axis: Replace θ by $-\theta$. $r = 3\cos(2(-\theta)) = 3\cos(-2\theta) = 3\cos(2\theta)$.
            The graph is symmetric with respect to the polar axis.

The line $\theta = \dfrac{\pi}{2}$: Replace θ by $\pi - \theta$.

$$r = 3\cos(2(\pi - \theta)) = 3\cos(2\pi - 2\theta)$$
$$= 3(\cos(2\pi)\cos(2\theta) + \sin(2\pi)\sin(2\theta)) = 3(\cos 2\theta + 0) = 3\cos(2\theta)$$

The graph is symmetric with respect to the line $\theta = \dfrac{\pi}{2}$.

The pole: Since the graph is symmetric to both the polar axis and the line
                $\theta = \dfrac{\pi}{2}$, it is also symmetric to the pole.

Due to symmetry, assign values to θ from 0 to $\dfrac{\pi}{2}$.

θ	0	$\dfrac{\pi}{6}$	$\dfrac{\pi}{4}$	$\dfrac{\pi}{3}$	$\dfrac{\pi}{2}$
$r = 3\cos 2\theta$	3	$\dfrac{3}{2}$	0	$-\dfrac{3}{2}$	-3

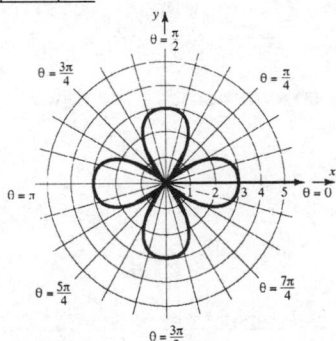

45. $r = 4\sin(5\theta)$ The graph will be a rose with five petals. Check for symmetry:
Polar axis: Replace θ by $-\theta$. $r = 4\sin(5(-\theta)) = 4\sin(-5\theta) = -4\sin(5\theta)$.
                The test fails.

The line $\theta = \dfrac{\pi}{2}$: Replace θ by $\pi - \theta$.

$$r = 4\sin(5(\pi - \theta)) = 4\sin(5\pi - 5\theta)$$
$$= 4(\sin(5\pi)\cos(5\theta) - \cos(5\pi)\sin(5\theta)) = 4(0 + \sin(5\theta)) = 4\sin(5\theta)$$

The graph is symmetric with respect to the line $\theta = \dfrac{\pi}{2}$.

The pole: Replace r by $-r$. $-r = 4\sin(5\theta)$. The test fails.

Due to symmetry to the line $\theta = \dfrac{\pi}{2}$, assign values to θ from $-\dfrac{\pi}{2}$ to $\dfrac{\pi}{2}$.

θ	$-\dfrac{\pi}{2}$	$-\dfrac{\pi}{3}$	$-\dfrac{\pi}{4}$	$-\dfrac{\pi}{6}$	0	$\dfrac{\pi}{6}$	$\dfrac{\pi}{4}$	$\dfrac{\pi}{3}$	$\dfrac{\pi}{2}$
$r = 4\sin(5\theta)$	-4	$2\sqrt{3} \approx 3.5$	$-2\sqrt{2} \approx -2.8$	-2	0	2	$-2\sqrt{2} \approx -2.8$	$-2\sqrt{3} \approx -3.5$	4

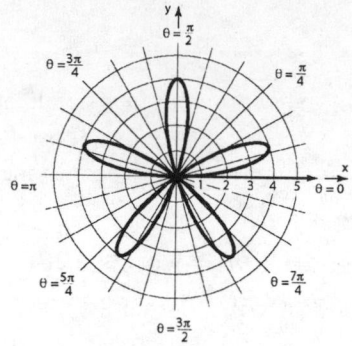

47. $r^2 = 9\cos(2\theta)$ The graph will be a lemniscate. Check for symmetry:

Polar axis: Replace θ by $-\theta$. $r^2 = 9\cos(2(-\theta)) = 9\cos(-2\theta) = 9\cos(2\theta)$.

The graph is symmetric with respect to the polar axis.

The line $\theta = \dfrac{\pi}{2}$: Replace θ by $\pi - \theta$.

$$r^2 = 9\cos(2(\pi - \theta)) = 9\cos(2\pi - 2\theta)$$

$$= 9(\cos(2\pi)\cos 2\theta + \sin(2\pi)\sin 2\theta) = 9(\cos 2\theta + 0) = 9\cos(2\theta)$$

The graph is symmetric with respect to the line $\theta = \dfrac{\pi}{2}$.

The pole: Since the graph is symmetric to both the polar axis and the line

$\theta = \dfrac{\pi}{2}$, it is also symmetric to the pole.

Due to symmetry, assign values to θ from 0 to $\dfrac{\pi}{2}$.

θ	0	$\dfrac{\pi}{6}$	$\dfrac{\pi}{4}$	$\dfrac{\pi}{3}$	$\dfrac{\pi}{2}$
$r = \pm\sqrt{9\cos(2\theta)}$	± 3	$\pm\dfrac{3\sqrt{2}}{2} \approx \pm 2.1$	0	not defined	not defined

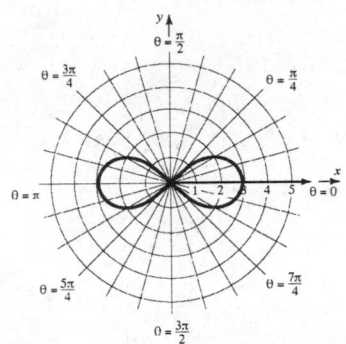

49. $r = 2^{\theta}$ The graph will be a spiral. Check for symmetry:
Polar axis: Replace θ by $-\theta$. $r = 2^{-\theta}$. The test fails.

The line $\theta = \dfrac{\pi}{2}$: Replace θ by $\pi - \theta$. $r = 2^{\pi - \theta}$. The test fails.

The pole: Replace r by $-r$. $-r = 2^{\theta}$. The test fails.

θ	$-\pi$	$-\dfrac{\pi}{2}$	$-\dfrac{\pi}{4}$	0	$\dfrac{\pi}{4}$	$\dfrac{\pi}{2}$	π	$\dfrac{3\pi}{2}$	2π
$r = 2^{\theta}$	0.1	0.3	0.6	1	1.7	3.0	8.8	26.2	77.9

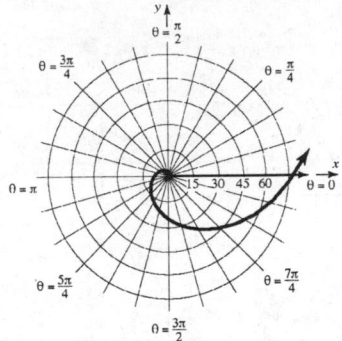

51. $r = 1 - \cos\theta$ The graph will be a cardioid. Check for symmetry:
Polar axis: Replace θ by $-\theta$. The result is $r = 1 - \cos(-\theta) = 1 - \cos\theta$.
                The graph is symmetric with respect to the polar axis.

The line $\theta = \dfrac{\pi}{2}$: Replace θ by $\pi - \theta$.

$$r = 1 - \cos(\pi - \theta) = 1 - (\cos(\pi)\cos\theta + \sin(\pi)\sin\theta)$$

$$= 1 - (-\cos\theta + 0) = 1 + \cos\theta. \quad \text{The test fails}$$

The pole: Replace r by $-r$. $-r = 1 - \cos\theta$. The test fails.
Due to symmetry to the polar axis, assign values to θ from 0 to π.

θ	0	$\dfrac{\pi}{6}$	$\dfrac{\pi}{3}$	$\dfrac{\pi}{2}$	$\dfrac{2\pi}{3}$	$\dfrac{5\pi}{6}$	π
$r = 1 - \cos\theta$	0	$1 - \dfrac{\sqrt{3}}{2} \approx 0.1$	$\dfrac{1}{2}$	1	$\dfrac{3}{2}$	$1 + \dfrac{\sqrt{3}}{2} \approx 1.9$	2

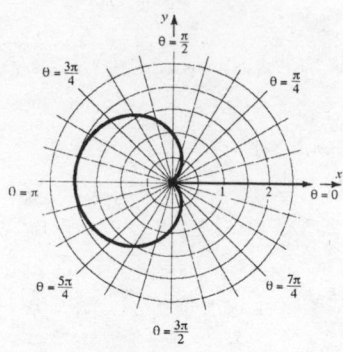

53. $r = 1 - 3\cos\theta$ The graph will be a limacon with an inner loop.
Check for symmetry:
Polar axis: Replace θ by $-\theta$. The result is $r = 1 - 3\cos(-\theta) = 1 - 3\cos\theta$.
             The graph is symmetric with respect to the polar axis.
The line $\theta = \dfrac{\pi}{2}$: Replace θ by $\pi - \theta$.

$$r = 1 - 3\cos(\pi - \theta) = 1 - 3(\cos(\pi)\cos\theta + \sin(\pi)\sin\theta)$$

$$= 1 - 3(-\cos\theta + 0) = 1 + 3\cos\theta$$

The test fails.
The pole: Replace r by $-r$. $-r = 1 - 3\cos\theta$. The test fails.
Due to symmetry to the polar axis, assign values to θ from 0 to π.

θ	0	$\dfrac{\pi}{6}$	$\dfrac{\pi}{3}$	$\dfrac{\pi}{2}$	$\dfrac{2\pi}{3}$	$\dfrac{5\pi}{6}$	π
$r = 1 - 3\cos\theta$	-2	$1 - \dfrac{3\sqrt{3}}{2} \approx -1.6$	$-\dfrac{1}{2}$	1	$\dfrac{5}{2}$	$1 + \dfrac{3\sqrt{3}}{2} \approx 3.6$	4

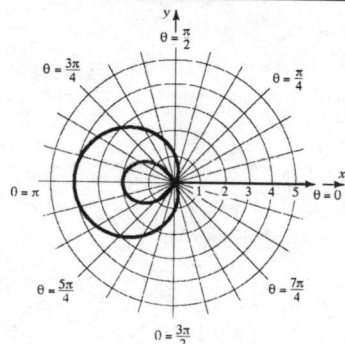

55. $r = \dfrac{2}{1 - \cos\theta}$ Check for symmetry:

Polar axis: Replace θ by $-\theta$. The result is $r = \dfrac{2}{1 - \cos(-\theta)} = \dfrac{2}{1 - \cos\theta}$.
             The graph is symmetric with respect to the polar axis.
The line $\theta = \dfrac{\pi}{2}$: Replace θ by $\pi - \theta$.

$$r = \frac{2}{1-\cos(\pi-\theta)} = \frac{2}{1-(\cos(\pi)\cos\theta + \sin(\pi)\sin\theta)}$$

$$= \frac{2}{1-(-\cos\theta + 0)} = \frac{2}{1+\cos\theta}$$

The test fails.

The pole: Replace r by $-r$. $-r = \dfrac{2}{1-\cos\theta}$. The test fails.

Due to symmetry to the polar axis, assign values to θ from 0 to π.

θ	0	$\dfrac{\pi}{6}$	$\dfrac{\pi}{3}$	$\dfrac{\pi}{2}$	$\dfrac{2\pi}{3}$	$\dfrac{5\pi}{6}$	π
$r = \dfrac{2}{1-\cos\theta}$	undefined	$\dfrac{2}{\left(1-\dfrac{\sqrt{3}}{2}\right)} \approx 14.9$	4	2	$\dfrac{4}{3}$	$\dfrac{2}{\left(1+\dfrac{\sqrt{3}}{2}\right)} \approx 1.1$	1

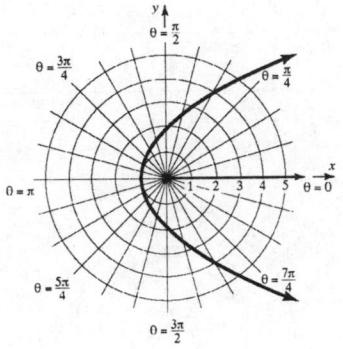

57. $r = \dfrac{1}{3-2\cos\theta}$ Check for symmetry:

Polar axis: Replace θ by $-\theta$. The result is $r = \dfrac{1}{3-2\cos(-\theta)} = \dfrac{1}{3-2\cos\theta}$.

The graph is symmetric with respect to the polar axis.

The line $\theta = \dfrac{\pi}{2}$: Replace θ by $\pi - \theta$.

$$r = \frac{1}{3-2\cos(\pi-\theta)} = \frac{1}{3-2(\cos(\pi)\cos\theta + \sin(\pi)\sin\theta)}$$

$$= \frac{1}{3-2(-\cos\theta + 0)} = \frac{1}{3+2\cos\theta}$$

The test fails.

The pole: Replace r by $-r$. $-r = \dfrac{1}{3-2\cos\theta}$. The test fails.

Due to symmetry to the polar axis, assign values to θ from 0 to π.

θ	0	$\dfrac{\pi}{6}$	$\dfrac{\pi}{3}$	$\dfrac{\pi}{2}$	$\dfrac{2\pi}{3}$	$\dfrac{5\pi}{6}$	π
$r = \dfrac{1}{3-2\cos\theta}$	1	$\dfrac{1}{3-\sqrt{3}} \approx 0.8$	$\dfrac{1}{2}$	$\dfrac{1}{3}$	$\dfrac{1}{4}$	$\dfrac{1}{3+\sqrt{3}} \approx 0.2$	$\dfrac{1}{5}$

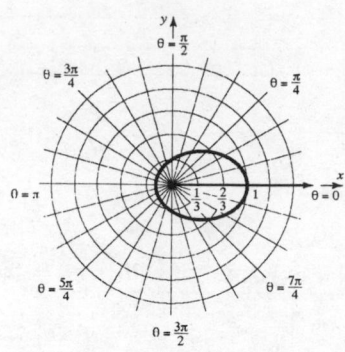

59. $r = \theta,\ \theta \geq 0$ Check for symmetry:

Polar axis: Replace θ by $-\theta$. $r = -\theta$. The test fails.

The line $\theta = \dfrac{\pi}{2}$: Replace θ by $\pi - \theta$. $r = \pi - \theta$. The test fails.

The pole: Replace r by $-r$. $-r = \theta$. The test fails.

θ	0	$\dfrac{\pi}{6}$	$\dfrac{\pi}{3}$	$\dfrac{\pi}{2}$	π	$\dfrac{3\pi}{2}$	2π
$r = \theta$	0	$\dfrac{\pi}{6} \approx 0.5$	$\dfrac{\pi}{3} \approx 1.0$	$\dfrac{\pi}{2} \approx 1.6$	$\pi \approx 3.1$	$\dfrac{3\pi}{2} \approx 4.7$	$2\pi \approx 6.3$

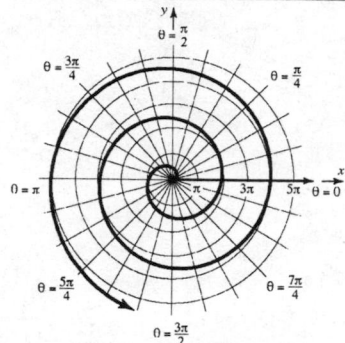

61. $r = \csc\theta - 2 = \dfrac{1}{\sin\theta} - 2,\ \ 0 < \theta < \pi$ Check for symmetry:

Polar axis: Replace θ by $-\theta$. $r = \csc(-\theta) - 2 = -\csc\theta - 2$.

          The test fails.

The line $\theta = \dfrac{\pi}{2}$: Replace θ by $\pi - \theta$.

$$r = \csc(\pi - \theta) - 2 = \frac{1}{\sin(\pi - \theta)} - 2$$

$$= \frac{1}{\sin(\pi)\cos\theta - \cos(\pi)\sin\theta} - 2 = \frac{1}{\sin\theta} - 2 = \csc\theta - 2$$

The graph is symmetric with respect to the line $\theta = \dfrac{\pi}{2}$.

The pole: Replace r by $-r$. $-r = \csc\theta - 2$. The test fails.

Due to symmetry, assign values to θ from 0 to $\dfrac{\pi}{2}$.

θ	0	$\dfrac{\pi}{6}$	$\dfrac{\pi}{4}$	$\dfrac{\pi}{3}$	$\dfrac{\pi}{2}$
$r = \csc\theta - 2$	not defined	0	$\sqrt{2} - 2 \approx -0.6$	$\dfrac{2\sqrt{3}}{3} - 2 \approx -0.8$	-1

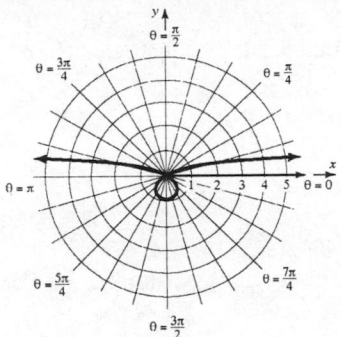

63. $r = \tan\theta, \quad -\dfrac{\pi}{2} < \theta < \dfrac{\pi}{2}$ Check for symmetry:

Polar axis: Replace θ by $-\theta$. $r = \tan(-\theta) = -\tan\theta$. The test fails.

The line $\theta = \dfrac{\pi}{2}$: Replace θ by $\pi - \theta$.

$$r = \tan(\pi - \theta) = \frac{\tan(\pi) - \tan\theta}{1 + \tan(\pi)\tan\theta} = \frac{-\tan\theta}{1} = -\tan\theta$$

The test fails.

The pole: Replace r by $-r$. $-r = \tan\theta$. The test fails.

θ	$-\dfrac{\pi}{3}$	$-\dfrac{\pi}{4}$	$-\dfrac{\pi}{6}$	0	$\dfrac{\pi}{6}$	$\dfrac{\pi}{4}$	$\dfrac{\pi}{3}$
$r = \tan\theta$	$-\sqrt{3} \approx -1.7$	-1	$-\dfrac{\sqrt{3}}{3} \approx -0.6$	0	$\dfrac{\sqrt{3}}{3} \approx 0.6$	1	$\sqrt{3} \approx 1.7$

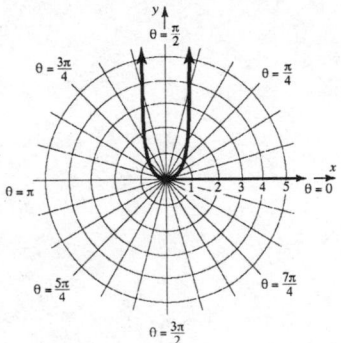

65. Convert the equation to rectangular form:
       $r\sin\theta = a \Rightarrow y = a$

The graph of $r\sin\theta = a$ is a horizontal line a units above the pole if $a > 0$, and $|a|$ units below the pole, if $a < 0$.

67. Convert the equation to rectangular form:

$$r = 2a\sin\theta, \, a > 0$$

$$r^2 = 2ar\sin\theta$$

$$x^2 + y^2 = 2ay \Rightarrow x^2 + y^2 - 2ay = 0 \Rightarrow x^2 + (y-a)^2 = a^2$$

Circle: radius a, center at rectangular coordinates $(0, a)$.

69. Convert the equation to rectangular form:

$$r = 2a\cos\theta, \, a > 0$$

$$r^2 = 2ar\cos\theta$$

$$x^2 + y^2 = 2ax \Rightarrow x^2 - 2ax + y^2 = 0 \Rightarrow (x-a)^2 + y^2 = a^2$$

Circle: radius a, center at rectangular coordinates $(a, 0)$.

71. (a) $r^2 = \cos\theta$: $r^2 = \cos(\pi - \theta) \Rightarrow r^2 = -\cos\theta$ Test fails.

$(-r)^2 = \cos(-\theta) \Rightarrow r^2 = \cos\theta$ New test works.

(b) $r^2 = \sin\theta$: $r^2 = \sin(\pi - \theta) \Rightarrow r^2 = \sin\theta$ Test works.

$(-r)^2 = \sin(-\theta) \Rightarrow r^2 = -\sin\theta$ New test fails.

Chapter 8

Polar Coordinates; Vectors

8.3   The Complex Plane; De Moivre's Theorem

1. $r = \sqrt{x^2 + y^2} = \sqrt{1^2 + 1^2} = \sqrt{2}$

$\tan\theta = \dfrac{y}{x} = 1 \quad \rightarrow \quad \theta = 45°$

The polar form of $z = 1 + i$ is
$z = r(\cos\theta + i\sin\theta)$

$= \sqrt{2}\left(\cos(45°) + i\sin(45°)\right)$

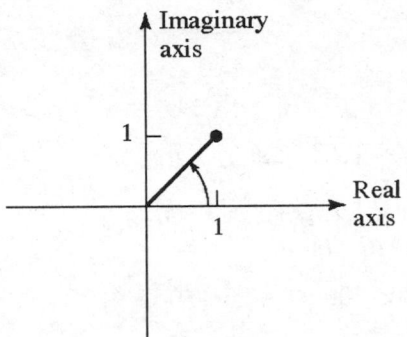

3. $r = \sqrt{x^2 + y^2} = \sqrt{\left(\sqrt{3}\right)^2 + (-1)^2} = \sqrt{4} = 2$

$\tan\theta = \dfrac{y}{x} = \dfrac{-1}{\sqrt{3}} = -\dfrac{\sqrt{3}}{3} \quad \rightarrow \quad \theta = 330°$

The polar form of $z = \sqrt{3} - i$ is
$z = r(\cos\theta + i\sin\theta)$

$= 2\left(\cos(330°) + i\sin(330°)\right)$

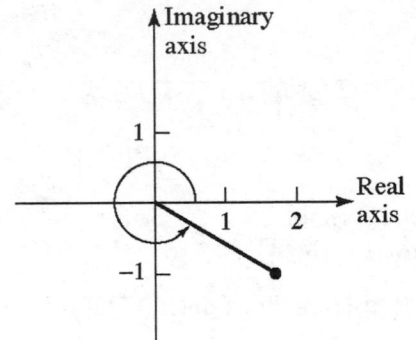

5. $r = \sqrt{x^2 + y^2} = \sqrt{0^2 + (-3)^2} = \sqrt{9} = 3$

$\tan\theta = \dfrac{y}{x} = \dfrac{-3}{0} = \text{undefined} \rightarrow \theta = 270°$

The polar form of $z = -3i$ is
$z = r(\cos\theta + i\sin\theta)$

$= 3\left(\cos(270°) + i\sin(270°)\right)$

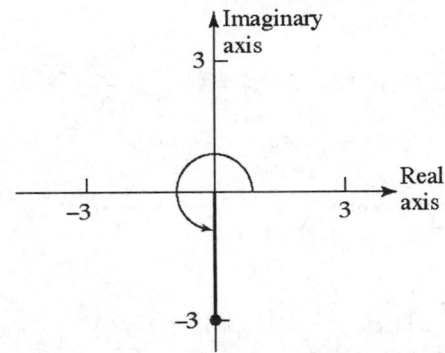

7. $r = \sqrt{x^2 + y^2} = \sqrt{4^2 + (-4)^2} = \sqrt{32} = 4\sqrt{2}$

$\tan\theta = \dfrac{y}{x} = \dfrac{-4}{4} = -1 \rightarrow \theta = 315°$

The polar form of $z = 4 - 4i$ is

$z = r(\cos\theta + i\sin\theta)$

$\quad = 4\sqrt{2}(\cos(315°) + i\sin(315°))$

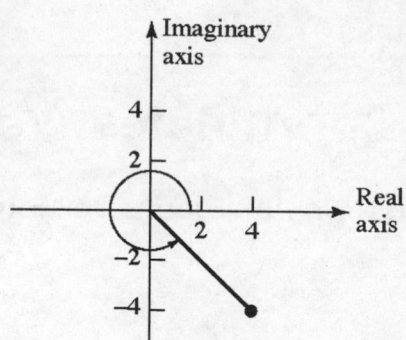

9. $r = \sqrt{x^2 + y^2} = \sqrt{3^2 + (-4)^2} = \sqrt{25} = 5$

$\tan\theta = \dfrac{y}{x} = \dfrac{-4}{3} \rightarrow \theta \approx 306.9°$

The polar form of $z = 3 - 4i$ is

$z = r(\cos\theta + i\sin\theta)$

$\quad = 5(\cos(306.9°) + i\sin(306.9°))$

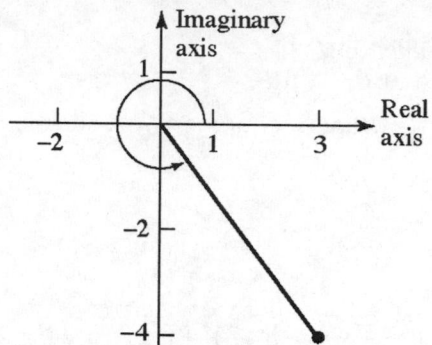

11. $r = \sqrt{x^2 + y^2} = \sqrt{(-2)^2 + 3^2} = \sqrt{13}$

$\tan\theta = \dfrac{y}{x} = \dfrac{3}{-2} = -\dfrac{3}{2} \rightarrow \theta \approx 123.7°$

The polar form of $z = -2 + 3i$ is

$z = r(\cos\theta + i\sin\theta)$

$\quad = \sqrt{13}(\cos(123.7°) + i\sin(123.7°))$

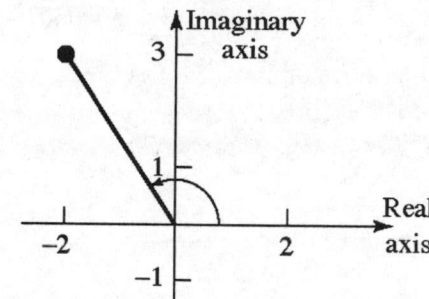

13. $2(\cos(120°) + i\sin(120°)) = 2\left(-\dfrac{1}{2} + \dfrac{\sqrt{3}}{2}i\right) = -1 + \sqrt{3}\,i$

15. $4\left(\cos\left(\dfrac{7\pi}{4}\right) + i\sin\left(\dfrac{7\pi}{4}\right)\right) = 4\left(\dfrac{\sqrt{2}}{2} - \dfrac{\sqrt{2}}{2}i\right) = 2\sqrt{2} - 2\sqrt{2}\,i$

17. $3\left(\cos\left(\dfrac{3\pi}{2}\right) + i\sin\left(\dfrac{3\pi}{2}\right)\right) = 3(0 - 1i) = -3i$

19. $0.2(\cos(100°) + i\sin(100°)) \approx 0.2(-0.1736 + 0.9848i) = -0.0347 + 0.1970\,i$

21. $2\left(\cos\left(\dfrac{\pi}{18}\right)+i\sin\left(\dfrac{\pi}{18}\right)\right)\approx 2(0.9848+0.1736i)=1.9696+0.3472i$

23. $z\cdot w=2(\cos(40°)+i\sin(40°))\cdot 4(\cos(20°)+i\sin(20°))$

   $=2\cdot 4(\cos(40°+20°)+i\sin(40°+20°))=8(\cos(60°)+i\sin(60°))$

   $\dfrac{z}{w}=\dfrac{2(\cos(40°)+i\sin(40°))}{4(\cos(20°)+i\sin(20°))}=\dfrac{2}{4}(\cos(40°-20°)+i\sin(40°-20°))$

   $=\dfrac{1}{2}(\cos(20°)+i\sin(20°))$

25. $z\cdot w=3(\cos(130°)+i\sin(130°))\cdot 4(\cos(270°)+i\sin(270°))$

   $=3\cdot 4(\cos(130°+270°)+i\sin(130°+270°))=12(\cos(400°)+i\sin(400°))$

   $=12(\cos(400°-360°)+i\sin(400°-360°))=12(\cos(40°)+i\sin(40°))$

   $\dfrac{z}{w}=\dfrac{3(\cos(130°)+i\sin(130°))}{4(\cos(270°)+i\sin(270°))}=\dfrac{3}{4}(\cos(130°-270°)+i\sin(130°-270°))$

   $=\dfrac{3}{4}(\cos(-140°)+i\sin(-140°))=\dfrac{3}{4}(\cos(220°)+i\sin(220°))$

27. $z\cdot w=2\left(\cos\left(\dfrac{\pi}{8}\right)+i\sin\left(\dfrac{\pi}{8}\right)\right)\cdot 2\left(\cos\left(\dfrac{\pi}{10}\right)+i\sin\left(\dfrac{\pi}{10}\right)\right)=2\cdot 2\left(\cos\left(\dfrac{\pi}{8}+\dfrac{\pi}{10}\right)+i\sin\left(\dfrac{\pi}{8}+\dfrac{\pi}{10}\right)\right)$

   $=4\left(\cos\left(\dfrac{9\pi}{40}\right)+i\sin\left(\dfrac{9\pi}{40}\right)\right)$

   $\dfrac{z}{w}=\dfrac{2\left(\cos\left(\dfrac{\pi}{8}\right)+i\sin\left(\dfrac{\pi}{8}\right)\right)}{2\left(\cos\left(\dfrac{\pi}{10}\right)+i\sin\left(\dfrac{\pi}{10}\right)\right)}=\dfrac{2}{2}\left(\cos\left(\dfrac{\pi}{8}-\dfrac{\pi}{10}\right)+i\sin\left(\dfrac{\pi}{8}-\dfrac{\pi}{10}\right)\right)=\cos\left(\dfrac{\pi}{40}\right)+i\sin\left(\dfrac{\pi}{40}\right)$

29. $z=2+2i\qquad r=\sqrt{2^2+2^2}=\sqrt{8}=2\sqrt{2}\qquad \tan\theta=\dfrac{2}{2}=1\qquad \theta=45°$

   $z=2\sqrt{2}(\cos(45°)+i\sin(45°))$

   $w=\sqrt{3}-i\qquad r=\sqrt{\left(\sqrt{3}\right)^2+(-1)^2}=\sqrt{4}=2\qquad \tan\theta=\dfrac{-1}{\sqrt{3}}=-\dfrac{\sqrt{3}}{3}\qquad \theta=330°$

   $w=2(\cos(330°)+i\sin(330°))$

$$z \cdot w = 2\sqrt{2}\left(\cos(45°) + i\sin(45°)\right) \cdot 2\left(\cos(330°) + i\sin(330°)\right)$$

$$= 2\sqrt{2} \cdot 2\left(\cos(45° + 330°) + i\sin(45° + 330°)\right) = 4\sqrt{2}\left(\cos(375°) + i\sin(375°)\right)$$

$$= 4\sqrt{2}\left(\cos(375° - 360°) + i\sin(375° - 360°)\right) = 4\sqrt{2}\left(\cos(15°) + i\sin(15°)\right)$$

$$\frac{z}{w} = \frac{2\sqrt{2}\left(\cos(45°) + i\sin(45°)\right)}{2\left(\cos(330°) + i\sin(330°)\right)} = \frac{2\sqrt{2}}{2}\left(\cos(45° - 330°) + i\sin(45° - 330°)\right)$$

$$= \sqrt{2}\left(\cos(-285°) + i\sin(-285°)\right) = \sqrt{2}\left(\cos(75°) + i\sin(75°)\right)$$

31. $\left[4\left(\cos(40°) + i\sin(40°)\right)\right]^3 = 4^3\left(\cos(3 \cdot 40°) + i\sin(3 \cdot 40°)\right) = 64\left(\cos(120°) + i\sin(120°)\right)$

$$= 64\left(-\frac{1}{2} + \frac{\sqrt{3}}{2}i\right) = -32 + 32\sqrt{3}\, i$$

33. $\left[2\left(\cos\left(\dfrac{\pi}{10}\right) + i\sin\left(\dfrac{\pi}{10}\right)\right)\right]^5 = 2^5\left(\cos\left(5 \cdot \dfrac{\pi}{10}\right) + i\sin\left(5 \cdot \dfrac{\pi}{10}\right)\right) = 32\left(\cos\left(\dfrac{\pi}{2}\right) + i\sin\left(\dfrac{\pi}{2}\right)\right)$

$$= 32(0 + 1\,i) = 0 + 32\,i$$

35. $\left[\sqrt{3}\left(\cos(10°) + i\sin(10°)\right)\right]^6 = \left(\sqrt{3}\right)^6\left(\cos(6 \cdot 10°) + i\sin(6 \cdot 10°)\right) = 27\left(\cos(60°) + i\sin(60°)\right)$

$$= 27\left(\frac{1}{2} + \frac{\sqrt{3}}{2}i\right) = \frac{27}{2} + \frac{27\sqrt{3}}{2}i$$

37. $\left[\sqrt{5}\left(\cos\left(\dfrac{3\pi}{16}\right) + i\sin\left(\dfrac{3\pi}{16}\right)\right)\right]^4 = \left(\sqrt{5}\right)^4\left(\cos\left(4 \cdot \dfrac{3\pi}{16}\right) + i\sin\left(4 \cdot \dfrac{3\pi}{16}\right)\right)$

$$= 25\left(\cos\left(\frac{3\pi}{4}\right) + i\sin\left(\frac{3\pi}{4}\right)\right) = 25\left(-\frac{\sqrt{2}}{2} + \frac{\sqrt{2}}{2}i\right) = -\frac{25\sqrt{2}}{2} + \frac{25\sqrt{2}}{2}i$$

39. $1 - i \qquad r = \sqrt{1^2 + (-1)^2} = \sqrt{2} \qquad \tan\theta = \dfrac{-1}{1} = -1 \qquad \theta = \dfrac{7\pi}{4}$

$$1 - i = \sqrt{2}\left(\cos\left(\frac{7\pi}{4}\right) + i\sin\left(\frac{7\pi}{4}\right)\right)$$

$$(1 - i)^5 = \left[\sqrt{2}\left(\cos\left(\frac{7\pi}{4}\right) + i\sin\left(\frac{7\pi}{4}\right)\right)\right]^5 = \left(\sqrt{2}\right)^5\left(\cos\left(5 \cdot \frac{7\pi}{4}\right) + i\sin\left(5 \cdot \frac{7\pi}{4}\right)\right)$$

$$= 4\sqrt{2}\left(\cos\left(\frac{35\pi}{4}\right) + i\sin\left(\frac{35\pi}{4}\right)\right) = 4\sqrt{2}\left(-\frac{\sqrt{2}}{2} + \frac{\sqrt{2}}{2}i\right) = -4 + 4\,i$$

41. $\sqrt{2} - i$ $r = \sqrt{\left(\sqrt{2}\right)^2 + (-1)^2} = \sqrt{3}$ $\tan\theta = \dfrac{-1}{\sqrt{2}} = -\dfrac{\sqrt{2}}{2}$ $\theta \approx 324.7°$

$\sqrt{2} - i \approx \sqrt{3}\left(\cos(324.7°) + i\sin(324.7°)\right)$

$\left(\sqrt{2} - i\right)^6 \approx \left[\sqrt{3}\left(\cos(324.7°) + i\sin(324.7°)\right)\right]^6 = \left(\sqrt{3}\right)^6\left(\cos(6 \cdot 324.7°) + i\sin(6 \cdot 324.7°)\right)$

$= 27\left(\cos(1948.2°) + i\sin(1948.2°)\right) \approx 27(-0.8499 + 0.5270\,i)$

$= -22.95 + 14.23\,i$

43. $1 + i$ $r = \sqrt{1^2 + 1^2} = \sqrt{2}$ $\tan\theta = \dfrac{1}{1} = 1$ $\theta = 45°$

$1 + i = \sqrt{2}\left(\cos 45° + i\sin 45°\right)$

The three complex cube roots of $1 + i = \sqrt{2}\left(\cos(45°) + i\sin(45°)\right)$ are:

$z_k = \sqrt[3]{\sqrt{2}}\left[\cos\left(\dfrac{45°}{3} + \dfrac{360°k}{3}\right) + i\sin\left(\dfrac{45°}{3} + \dfrac{360°k}{3}\right)\right]$

$= \sqrt[6]{2}\left[\cos(15° + 120°k) + i\sin(15° + 120°k)\right]$

$z_0 = \sqrt[6]{2}\left[\cos(15° + 120° \cdot 0) + i\sin(15° + 120° \cdot 0)\right] = \sqrt[6]{2}\left(\cos(15°) + i\sin(15°)\right)$

$z_1 = \sqrt[6]{2}\left[\cos(15° + 120° \cdot 1) + i\sin(15° + 120° \cdot 1)\right] = \sqrt[6]{2}\left(\cos(135°) + i\sin(135°)\right)$

$z_2 = \sqrt[6]{2}\left[\cos(15° + 120° \cdot 2) + i\sin(15° + 120° \cdot 2)\right] = \sqrt[6]{2}\left(\cos(255°) + i\sin(255°)\right)$

45. $4 - 4\sqrt{3}\,i$ $r = \sqrt{4^2 + \left(-4\sqrt{3}\right)^2} = \sqrt{64} = 8$ $\tan\theta = \dfrac{-4\sqrt{3}}{4} = -\sqrt{3}$ $\theta = 300°$

$4 - 4\sqrt{3}\,i = 8\left(\cos(300°) + i\sin(300°)\right)$

The four complex fourth roots of $4 - 4\sqrt{3}\,i = 8\left(\cos(300°) + i\sin(300°)\right)$ are:

$z_k = \sqrt[4]{8}\left[\cos\left(\dfrac{300°}{4} + \dfrac{360°k}{4}\right) + i\sin\left(\dfrac{300°}{4} + \dfrac{360°k}{4}\right)\right]$

$= \sqrt[4]{8}\left[\cos(75° + 90°k) + i\sin(75° + 90°k)\right]$

$z_0 = \sqrt[4]{8}\left[\cos(75° + 90° \cdot 0) + i\sin(75° + 90° \cdot 0)\right] = \sqrt[4]{8}\left(\cos(75°) + i\sin(75°)\right)$

$z_1 = \sqrt[4]{8}\left[\cos(75° + 90° \cdot 1) + i\sin(75° + 90° \cdot 1)\right] = \sqrt[4]{8}\left(\cos(165°) + i\sin(165°)\right)$

$z_2 = \sqrt[4]{8}\left[\cos(75° + 90° \cdot 2) + i\sin(75° + 90° \cdot 2)\right] = \sqrt[4]{8}\left(\cos(255°) + i\sin(255°)\right)$

$z_3 = \sqrt[4]{8}\left[\cos(75° + 90° \cdot 3) + i\sin(75° + 90° \cdot 3)\right] = \sqrt[4]{8}\left(\cos(345°) + i\sin(345°)\right)$

47. $-16i$ $r = \sqrt{0^2 + (-16)^2} = \sqrt{256} = 16$ $\tan\theta = \dfrac{-16}{0} = \text{undefined}$ $\theta = 270°$

$-16i = 16\left(\cos(270°) + i\sin(270°)\right)$

The four complex fourth roots of $-16i = 16\left(\cos 270° + i\sin 270°\right)$ are:

$z_k = \sqrt[4]{16}\left[\cos\left(\dfrac{270°}{4} + \dfrac{360°k}{4}\right) + i\sin\left(\dfrac{270°}{4} + \dfrac{360°k}{4}\right)\right]$

$= 2\left[\cos(67.5° + 90°k) + i\sin(67.5° + 90°k)\right]$

$$z_0 = 2\left[\cos(67.5° + 90°\cdot 0) + i\sin(67.5° + 90°\cdot 0)\right] = 2\left(\cos(67.5°) + i\sin(67.5°)\right)$$

$$z_1 = 2\left[\cos(67.5° + 90°\cdot 1) + i\sin(67.5° + 90°\cdot 1)\right] = 2\left(\cos(157.5°) + i\sin(157.5°)\right)$$

$$z_2 = 2\left[\cos(67.5° + 90°\cdot 2) + i\sin(67.5° + 90°\cdot 2)\right] = 2\left(\cos(247.5°) + i\sin(247.5°)\right)$$

$$z_3 = 2\left[\cos(67.5° + 90°\cdot 3) + i\sin(67.5° + 90°\cdot 3)\right] = 2\left(\cos(337.5°) + i\sin(337.5°)\right)$$

49. i $r = \sqrt{0^2 + 1^2} = \sqrt{1} = 1$ $\tan\theta = \dfrac{1}{0} = $ undefined $\theta = 90°$

    $i = 1\left(\cos(90°) + i\sin(90°)\right)$

The five complex fifth roots of $\ i = 1\left(\cos(90°) + i\sin(90°)\right)\ $ are:

$$z_k = \sqrt[5]{1}\left[\cos\left(\frac{90°}{5} + \frac{360°k}{5}\right) + i\sin\left(\frac{90°}{5} + \frac{360°k}{5}\right)\right]$$

$$= 1\left[\cos(18° + 72°k) + i\sin(18° + 72°k)\right]$$

$$z_0 = 1\left[\cos(18° + 72°\cdot 0) + i\sin(18° + 72°\cdot 0)\right] = \cos(18°) + i\sin(18°)$$

$$z_1 = 1\left[\cos(18° + 72°\cdot 1) + i\sin(18° + 72°\cdot 1)\right] = \cos(90°) + i\sin(90°)$$

$$z_2 = 1\left[\cos(18° + 72°\cdot 2) + i\sin(18° + 72°\cdot 2)\right] = \cos(162°) + i\sin(162°)$$

$$z_3 = 1\left[\cos(18° + 72°\cdot 3) + i\sin(18° + 72°\cdot 3)\right] = \cos(234°) + i\sin(234°)$$

$$z_4 = 1\left[\cos(18° + 72°\cdot 4) + i\sin(18° + 72°\cdot 4)\right] = \cos(306°) + i\sin(306°)$$

51. $1 = 1 + 0i$ $r = \sqrt{1^2 + 0^2} = \sqrt{1} = 1$ $\tan\theta = \dfrac{0}{1} = 0$ $\theta = 0°$

    $1 + 0i = 1\left(\cos(0°) + i\sin(0°)\right)$

The four complex fourth roots of $\ 1 + 0i = 1\left(\cos(0°) + i\sin(0°)\right)\ $ are:

$$z_k = \sqrt[4]{1}\left[\cos\left(\frac{0°}{4} + \frac{360°k}{4}\right) + i\sin\left(\frac{0°}{4} + \frac{360°k}{4}\right)\right]$$

$$= 1\left[\cos(90°k) + i\sin(90°k)\right]$$

$$z_0 = \cos(90°\cdot 0) + i\sin(90°\cdot 0) = \cos(0°) + i\sin(0°) = 1 + 0i = 1$$

$$z_1 = \cos(90°\cdot 1) + i\sin(90°\cdot 1) = \cos(90°) + i\sin(90°) = 0 + 1i = i$$

$$z_2 = \cos(90°\cdot 2) + i\sin(90°\cdot 2) = \cos(180°) + i\sin(180°) = -1 + 0i = -1$$

$$z_3 = \cos(90°\cdot 3) + i\sin(90°\cdot 3) = \cos(270°) + i\sin(270°) = 0 - 1i = -i$$

The complex roots are: $1, i, -1, -i$.

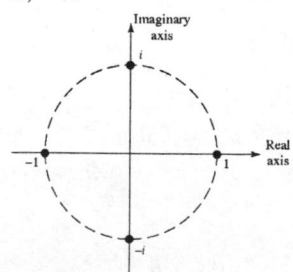

53.   Let $w = r(\cos\theta + i\sin\theta)$ be a complex number. If $w \neq 0$, there are n distinct nth roots of w, given by the formula:
$$z_k = \sqrt[n]{r}\left(\cos\left(\frac{\theta}{n} + \frac{2k\pi}{n}\right) + i\sin\left(\frac{\theta}{n} + \frac{2k\pi}{n}\right)\right), \text{ where } k = 0, 1, 2, \dots, n-1$$
$|z_k| = \sqrt[n]{r}$ for all k

55.   Examining the formula for the distinct complex nth roots of the complex number $w = r(\cos\theta + i\sin\theta)$,
$$z_k = \sqrt[n]{r}\left(\cos\left(\frac{\theta}{n} + \frac{2k\pi}{n}\right) + i\sin\left(\frac{\theta}{n} + \frac{2k\pi}{n}\right)\right), \text{ where } k = 0, 1, 2, \dots, n-1$$

we see that the z_k are spaced apart by an angle of $\dfrac{2\pi}{n}$.

Polar Coordinates; Vectors

8.4   Vectors

1.   **v + w**

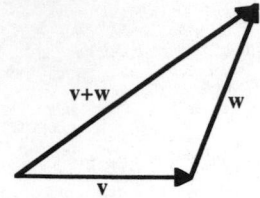

3.   **3v**

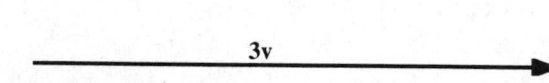

5.   **v − w**

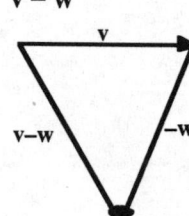

7.   **3v + u − 2w**

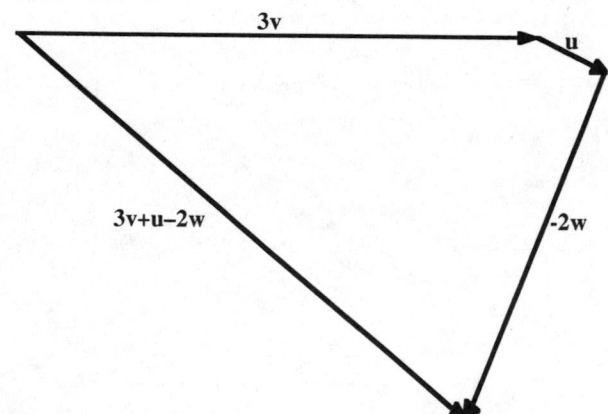

9.   True

11.   False   **C = −F + E − D**

13.   False   **D − E = H + G**

15.   True

17.   If $\|\mathbf{v}\| = 4$, then $\|3\mathbf{v}\| = |3|\|\mathbf{v}\| = 3(4) = 12.$

19. $P = (0, 0), Q = (3, 4)$ $\mathbf{v} = (3-0)\mathbf{i} + (4-0)\mathbf{j} = 3\mathbf{i} + 4\mathbf{j}$

21. $P = (3, 2), Q = (5, 6)$ $\mathbf{v} = (5-3)\mathbf{i} + (6-2)\mathbf{j} = 2\mathbf{i} + 4\mathbf{j}$

23. $P = (-2, -1), Q = (6, -2)$ $\mathbf{v} = (6-(-2))\mathbf{i} + (-2-(-1))\mathbf{j} = 8\mathbf{i} - \mathbf{j}$

25. $P = (1, 0), Q = (0, 1)$ $\mathbf{v} = (0-1)\mathbf{i} + (1-0)\mathbf{j} = -\mathbf{i} + \mathbf{j}$

27. For $\mathbf{v} = 3\mathbf{i} - 4\mathbf{j}$, $\|\mathbf{v}\| = \sqrt{3^2 + (-4)^2} = \sqrt{25} = 5$

29. For $\mathbf{v} = \mathbf{i} - \mathbf{j}$, $\|\mathbf{v}\| = \sqrt{1^2 + (-1)^2} = \sqrt{2}$

31. For $\mathbf{v} = -2\mathbf{i} + 3\mathbf{j}$, $\|\mathbf{v}\| = \sqrt{(-2)^2 + 3^2} = \sqrt{13}$

33. $\mathbf{v} = 3\mathbf{i} - 5\mathbf{j}, \;\; \mathbf{w} = -2\mathbf{i} + 3\mathbf{j}$
     $2\mathbf{v} + 3\mathbf{w} = 2(3\mathbf{i} - 5\mathbf{j}) + 3(-2\mathbf{i} + 3\mathbf{j}) = 6\mathbf{i} - 10\mathbf{j} - 6\mathbf{i} + 9\mathbf{j} = -\mathbf{j}$

35. $\mathbf{v} = 3\mathbf{i} - 5\mathbf{j}, \;\; \mathbf{w} = -2\mathbf{i} + 3\mathbf{j}$
     $\|\mathbf{v} - \mathbf{w}\| = \|(3\mathbf{i} - 5\mathbf{j}) - (-2\mathbf{i} + 3\mathbf{j})\| = \|5\mathbf{i} - 8\mathbf{j}\| = \sqrt{5^2 + (-8)^2} = \sqrt{89}$

37. $\mathbf{v} = 3\mathbf{i} - 5\mathbf{j}, \;\; \mathbf{w} = -2\mathbf{i} + 3\mathbf{j}$
     $\|\mathbf{v}\| - \|\mathbf{w}\| = \|3\mathbf{i} - 5\mathbf{j}\| - \|-2\mathbf{i} + 3\mathbf{j}\| = \sqrt{3^2 + (-5)^2} - \sqrt{(-2)^2 + 3^2} = \sqrt{34} - \sqrt{13}$

39. $\mathbf{u} = \dfrac{\mathbf{v}}{\|\mathbf{v}\|} = \dfrac{5\mathbf{i}}{\|5\mathbf{i}\|} = \dfrac{5\mathbf{i}}{\sqrt{25+0}} = \dfrac{5\mathbf{i}}{5} = \mathbf{i}$

41. $\mathbf{u} = \dfrac{\mathbf{v}}{\|\mathbf{v}\|} = \dfrac{3\mathbf{i} - 4\mathbf{j}}{\|3\mathbf{i} - 4\mathbf{j}\|} = \dfrac{3\mathbf{i} - 4\mathbf{j}}{\sqrt{3^2 + (-4)^2}} = \dfrac{3\mathbf{i} - 4\mathbf{j}}{\sqrt{25}} = \dfrac{3\mathbf{i} - 4\mathbf{j}}{5} = \dfrac{3}{5}\mathbf{i} - \dfrac{4}{5}\mathbf{j}$

43. $\mathbf{u} = \dfrac{\mathbf{v}}{\|\mathbf{v}\|} = \dfrac{\mathbf{i} - \mathbf{j}}{\|\mathbf{i} - \mathbf{j}\|} = \dfrac{\mathbf{i} - \mathbf{j}}{\sqrt{1^2 + (-1)^2}} = \dfrac{\mathbf{i} - \mathbf{j}}{\sqrt{2}} = \dfrac{1}{\sqrt{2}}\mathbf{i} - \dfrac{1}{\sqrt{2}}\mathbf{j} = \dfrac{\sqrt{2}}{2}\mathbf{i} - \dfrac{\sqrt{2}}{2}\mathbf{j}$

45. Let $\mathbf{v} = a\mathbf{i} + b\mathbf{j}$. We want $\|\mathbf{v}\| = 4$ and $a = 2b$.
    $\|\mathbf{v}\| = \sqrt{a^2 + b^2} = \sqrt{(2b)^2 + b^2} = \sqrt{5b^2}$

    $\sqrt{5b^2} = 4 \;\to\; 5b^2 = 16 \;\to\; b^2 = \dfrac{16}{5} \;\to\; b = \pm\sqrt{\dfrac{16}{5}} = \pm\dfrac{4}{\sqrt{5}} = \pm\dfrac{4\sqrt{5}}{5}$

    $a = 2b = \pm\dfrac{8\sqrt{5}}{5}$ therefore, $\mathbf{v} = \dfrac{8\sqrt{5}}{5}\mathbf{i} + \dfrac{4\sqrt{5}}{5}\mathbf{j}$ or $\mathbf{v} = -\dfrac{8\sqrt{5}}{5}\mathbf{i} - \dfrac{4\sqrt{5}}{5}\mathbf{j}$

47. $\mathbf{v} = 2\mathbf{i} - \mathbf{j}, \;\; \mathbf{w} = x\mathbf{i} + 3\mathbf{j}$ $\|\mathbf{v} + \mathbf{w}\| = 5$
     $\|\mathbf{v} + \mathbf{w}\| = \|2\mathbf{i} - \mathbf{j} + x\mathbf{i} + 3\mathbf{j}\| = \|(2+x)\mathbf{i} + 2\mathbf{j}\| = \sqrt{(2+x)^2 + 2^2}$
        $= \sqrt{x^2 + 4x + 4 + 4} = \sqrt{x^2 + 4x + 8}$

Solve for x:
$$\sqrt{x^2 + 4x + 8} = 5 \Rightarrow x^2 + 4x + 8 = 25 \Rightarrow x^2 + 4x - 17 = 0$$
$$x = \frac{-4 \pm \sqrt{16 - 4(1)(-17)}}{2(1)} = \frac{-4 \pm \sqrt{84}}{2} = \frac{-4 \pm 2\sqrt{21}}{2} = -2 \pm \sqrt{21}$$
$$x = -2 + \sqrt{21} \approx 2.58 \text{ or } x = -2 - \sqrt{21} \approx -6.58$$

49. $\|\mathbf{v}\| = 5, \quad \alpha = 60°$

$$\mathbf{v} = \|\mathbf{v}\|(\cos\alpha \mathbf{i} + \sin\alpha \mathbf{j}) = 5(\cos(60°)\mathbf{i} + \sin(60°)\mathbf{j}) = 5\left(\frac{1}{2}\mathbf{i} + \frac{\sqrt{3}}{2}\mathbf{j}\right) = \frac{5}{2}\mathbf{i} + \frac{5\sqrt{3}}{2}\mathbf{j}$$

51. $\|\mathbf{v}\| = 14, \quad \alpha = 120°$

$$\mathbf{v} = \|\mathbf{v}\|(\cos\alpha \mathbf{i} + \sin\alpha \mathbf{j}) = 14(\cos(120°)\mathbf{i} + \sin(120°)\mathbf{j}) = 14\left(-\frac{1}{2}\mathbf{i} + \frac{\sqrt{3}}{2}\mathbf{j}\right) = -7\mathbf{i} + 7\sqrt{3}\mathbf{j}$$

53. $\|\mathbf{v}\| = 25, \quad \alpha = 330°$

$$\mathbf{v} = \|\mathbf{v}\|(\cos\alpha \mathbf{i} + \sin\alpha \mathbf{j}) = 25(\cos(330°)\mathbf{i} + \sin(330°)\mathbf{j}) = 25\left(\frac{\sqrt{3}}{2}\mathbf{i} - \frac{1}{2}\mathbf{j}\right) = \frac{25\sqrt{3}}{2}\mathbf{i} - \frac{25}{2}\mathbf{j}$$

55. $\mathbf{F} = 40(\cos(30°)\mathbf{i} + \sin(30°)\mathbf{j}) = 40\left(\frac{\sqrt{3}}{2}\mathbf{i} + \frac{1}{2}\mathbf{j}\right) = 20\sqrt{3}\mathbf{i} + 20\mathbf{j}$

57. $\mathbf{F}_1 = 40(\cos(30°)\mathbf{i} + \sin(30°)\mathbf{j}) = 40\left(\frac{\sqrt{3}}{2}\mathbf{i} + \frac{1}{2}\mathbf{j}\right) = 20\sqrt{3}\mathbf{i} + 20\mathbf{j}$

$$\mathbf{F}_2 = 60(\cos(-45°)\mathbf{i} + \sin(-45°)\mathbf{j}) = 60\left(\frac{\sqrt{2}}{2}\mathbf{i} - \frac{\sqrt{2}}{2}\mathbf{j}\right) = 30\sqrt{2}\mathbf{i} - 30\sqrt{2}\mathbf{j}$$

$$\mathbf{F}_1 + \mathbf{F}_2 = 20\sqrt{3}\mathbf{i} + 20\mathbf{j} + 30\sqrt{2}\mathbf{i} - 30\sqrt{2}\mathbf{j} = \left(20\sqrt{3} + 30\sqrt{2}\right)\mathbf{i} + \left(20 - 30\sqrt{2}\right)\mathbf{j}$$

magnitude of $\mathbf{F}_1 + \mathbf{F}_2 = \sqrt{\left(20\sqrt{3} + 30\sqrt{2}\right)^2 + \left(20 - 30\sqrt{2}\right)^2} \approx 80.26$ Newtons

direction of $\mathbf{F}_1 + \mathbf{F}_2 = \tan^{-1}\left(\dfrac{20 - 30\sqrt{2}}{20\sqrt{3} + 30\sqrt{2}}\right) \approx -16.22°$

59. Let $\mathbf{F}_1$ be the tension on the left cable and $\mathbf{F}_2$ be the tension on the right cable.
Let $\mathbf{F}_3$ represent the force of the weight of the box.

$\mathbf{F}_1 = \|\mathbf{F}_1\|(\cos(155°)\mathbf{i} + \sin(155°)\mathbf{j}) \approx \|\mathbf{F}_1\|(-0.9063\mathbf{i} + 0.4226\mathbf{j})$

$\mathbf{F}_2 = \|\mathbf{F}_2\|(\cos(40°)\mathbf{i} + \sin(40°)\mathbf{j}) \approx \|\mathbf{F}_2\|(0.7660\mathbf{i} + 0.6428\mathbf{j})$

$\mathbf{F}_3 = -1000\mathbf{j}$

For equilibrium, the sum of the force vectors must be zero.

$\mathbf{F}_1 + \mathbf{F}_2 + \mathbf{F}_3 = -0.9063\|\mathbf{F}_1\|\mathbf{i} + 0.4226\|\mathbf{F}_1\|\mathbf{j} + 0.7660\|\mathbf{F}_2\|\mathbf{i} + 0.6428\|\mathbf{F}_2\|\mathbf{j} - 1000\mathbf{j}$

$\quad = \left(-0.9063\|\mathbf{F}_1\| + 0.7660\|\mathbf{F}_2\|\right)\mathbf{i} + \left(0.4226\|\mathbf{F}_1\| + 0.6428\|\mathbf{F}_2\| - 1000\right)\mathbf{j} = 0$

Set the **i** and **j** components equal to zero and solve:

$$\begin{cases} -0.9063\|\mathbf{F}_1\| + 0.7660\|\mathbf{F}_2\| = 0 \;\Rightarrow\; \|\mathbf{F}_2\| = \dfrac{0.9063}{0.7660}\|\mathbf{F}_1\| = 1.1832\|\mathbf{F}_1\| \\[2mm] 0.4226\|\mathbf{F}_1\| + 0.6428\|\mathbf{F}_2\| - 1000 = 0 \end{cases}$$

$$0.4226\|\mathbf{F}_1\| + 0.6428\big(1.1832\|\mathbf{F}_1\|\big) - 1000 = 0 \Rightarrow 1.1832\|\mathbf{F}_1\| = 1000$$

$$\|\mathbf{F}_1\| = 845.2 \text{ pounds}; \quad \|\mathbf{F}_2\| = 1.1832(845.2) = 1000 \text{ pounds}$$

The tension in the left cable is about 845.2 pounds and the tension in the right cable is about 1000 pounds.

61. Let $\mathbf{F}_1$ be the tension on the left end of the rope and $\mathbf{F}_2$ be the tension on the right end of the rope. Let $\mathbf{F}_3$ represent the force of the weight of the tightrope walker.

$$\mathbf{F}_1 = \|\mathbf{F}_1\|\big(\cos(175.8°)\mathbf{i} + \sin(175.8°)\mathbf{j}\big) \approx \|\mathbf{F}_1\|(-0.9973\mathbf{i} + 0.0732\mathbf{j})$$

$$\mathbf{F}_2 = \|\mathbf{F}_2\|\big(\cos(3.7°)\mathbf{i} + \sin(3.7°)\mathbf{j}\big) \approx \|\mathbf{F}_2\|(0.9979\mathbf{i} + 0.0645\mathbf{j})$$

$$\mathbf{F}_3 = -150\mathbf{j}$$

For equilibrium, the sum of the force vectors must be zero.

$$\mathbf{F}_1 + \mathbf{F}_2 + \mathbf{F}_3 = -0.9973\|\mathbf{F}_1\|\mathbf{i} + 0.0732\|\mathbf{F}_1\|\mathbf{j} + 0.9979\|\mathbf{F}_2\|\mathbf{i} + 0.0645\|\mathbf{F}_2\|\mathbf{j} - 150\mathbf{j}$$

$$= \big(-0.9973\|\mathbf{F}_1\| + 0.9979\|\mathbf{F}_2\|\big)\mathbf{i} + \big(0.0732\|\mathbf{F}_1\| + 0.0645\|\mathbf{F}_2\| - 150\big)\mathbf{j} = 0$$

Set the **i** and **j** components equal to zero and solve:

$$\begin{cases} -0.9973\|\mathbf{F}_1\| + 0.9979\|\mathbf{F}_2\| = 0 \;\Rightarrow\; \|\mathbf{F}_2\| = \dfrac{0.9973}{0.9979}\|\mathbf{F}_1\| = 0.9994\|\mathbf{F}_1\| \\[2mm] 0.0732\|\mathbf{F}_1\| + 0.0645\|\mathbf{F}_2\| - 150 = 0 \end{cases}$$

$$0.0732\|\mathbf{F}_1\| + 0.0645\big(0.9994\|\mathbf{F}_1\|\big) - 150 = 0 \Rightarrow 0.1377\|\mathbf{F}_1\| = 150$$

$$\|\mathbf{F}_1\| = 1089.3 \text{ pounds}; \quad \|\mathbf{F}_2\| = 0.9994(1089.3) = 1088.6 \text{ pounds}$$

The tension in the left end of the rope is about 1089.3 pounds and the tension in the right end of the rope is about 1088.6 pounds.

63. The given forces are: $\mathbf{F}_1 = -3\mathbf{i}; \quad \mathbf{F}_2 = -\mathbf{i} + 4\mathbf{j}; \quad \mathbf{F}_3 = 4\mathbf{i} - 2\mathbf{j}; \quad \mathbf{F}_4 = -4\mathbf{j}$
A vector $\mathbf{x} = a\mathbf{i} + b\mathbf{j}$ needs to be added for equilibrium. Find vector $\mathbf{x} = a\mathbf{i} + b\mathbf{j}$:

$$\mathbf{F}_1 + \mathbf{F}_2 + \mathbf{F}_3 + \mathbf{F}_4 + \mathbf{x} = 0$$

$$-3\mathbf{i} + (-\mathbf{i} + 4\mathbf{j}) + (4\mathbf{i} - 2\mathbf{j}) + (-4\mathbf{j}) + (a\mathbf{i} + b\mathbf{j}) = 0 \Rightarrow 0\mathbf{i} - 2\mathbf{j} + (a\mathbf{i} + b\mathbf{j}) = 0$$

$$a\mathbf{i} + (-2 + b)\mathbf{j} = 0 \Rightarrow a = 0$$

$$-2 + b = 0 \;\Rightarrow\; b = 2$$

Therefore, $\mathbf{x} = 2\mathbf{j}$.

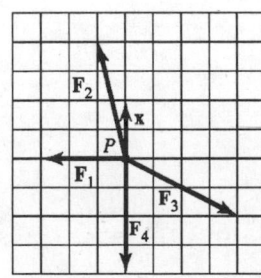

Polar Coordinates; Vectors

8.5   The Dot Product

1. $\mathbf{v} = \mathbf{i} - \mathbf{j}, \quad \mathbf{w} = \mathbf{i} + \mathbf{j}$

(a) $\mathbf{v} \bullet \mathbf{w} = 1(1) + (-1)(1) = 1 - 1 = 0$

(b) $\cos\theta = \dfrac{\mathbf{v} \bullet \mathbf{w}}{\|\mathbf{v}\|\|\mathbf{w}\|} = \dfrac{0}{\sqrt{1^2 + (-1)^2}\sqrt{1^2 + 1^2}} = \dfrac{0}{\sqrt{2}\sqrt{2}} = \dfrac{0}{2} = 0 \quad \rightarrow \quad \theta = 90°$

(c)   The vectors are orthogonal.

3. $\mathbf{v} = 2\mathbf{i} + \mathbf{j}, \quad \mathbf{w} = \mathbf{i} + 2\mathbf{j}$

(a) $\mathbf{v} \bullet \mathbf{w} = 2(1) + 1(2) = 2 + 2 = 4$

(b) $\cos\theta = \dfrac{\mathbf{v} \bullet \mathbf{w}}{\|\mathbf{v}\|\|\mathbf{w}\|} = \dfrac{4}{\sqrt{2^2 + 1^2}\sqrt{1^2 + 2^2}} = \dfrac{4}{\sqrt{5}\sqrt{5}} = \dfrac{4}{5} = 0.8 \quad \rightarrow \quad \theta \approx 36.87°$

(c)   The vectors are neither parallel nor orthogonal.

5. $\mathbf{v} = \sqrt{3}\,\mathbf{i} - \mathbf{j}, \quad \mathbf{w} = \mathbf{i} + \mathbf{j}$

(a) $\mathbf{v} \bullet \mathbf{w} = \sqrt{3}\,(1) + (-1)(1) = \sqrt{3} - 1$

(b) $\cos\theta = \dfrac{\mathbf{v} \bullet \mathbf{w}}{\|\mathbf{v}\|\|\mathbf{w}\|} = \dfrac{\sqrt{3} - 1}{\sqrt{\left(\sqrt{3}\right)^2 + (-1)^2}\sqrt{1^2 + 1^2}} = \dfrac{\sqrt{3} - 1}{\sqrt{4}\sqrt{2}} = \dfrac{\sqrt{3} - 1}{2\sqrt{2}} = \dfrac{\sqrt{6} - \sqrt{2}}{4}$

$\theta \approx 75°$

(c)   The vectors are neither parallel nor orthogonal.

7. $\mathbf{v} = 3\mathbf{i} + 4\mathbf{j}, \quad \mathbf{w} = 4\mathbf{i} + 3\mathbf{j}$

(a) $\mathbf{v} \bullet \mathbf{w} = 3(4) + 4(3) = 12 + 12 = 24$

(b) $\cos\theta = \dfrac{\mathbf{v} \bullet \mathbf{w}}{\|\mathbf{v}\|\|\mathbf{w}\|} = \dfrac{24}{\sqrt{3^2 + 4^2}\sqrt{4^2 + 3^2}} = \dfrac{24}{\sqrt{25}\sqrt{25}} = \dfrac{24}{25} = 0.96 \quad \rightarrow \quad \theta \approx 16.26°$

(c)   The vectors are neither parallel nor orthogonal.

9. $\mathbf{v} = 4\mathbf{i}, \quad \mathbf{w} = \mathbf{j}$

(a) $\mathbf{v} \bullet \mathbf{w} = 4(0) + 0(1) = 0 + 0 = 0$

(b) $\cos\theta = \dfrac{\mathbf{v} \bullet \mathbf{w}}{\|\mathbf{v}\|\|\mathbf{w}\|} = \dfrac{0}{\sqrt{4^2 + 0^2}\sqrt{0^2 + 1^2}} = \dfrac{0}{4 \cdot 1} = \dfrac{0}{4} = 0 \quad \rightarrow \quad \theta = 90°$

(c)   The vectors are orthogonal.

11. $\mathbf{v} = \mathbf{i} - a\mathbf{j}, \quad \mathbf{w} = 2\mathbf{i} + 3\mathbf{j}$

Two vectors are orthogonal if the dot product is zero. Solve for a:

$$\mathbf{v} \bullet \mathbf{w} = 1(2) + (-a)(3) = 2 - 3a$$

$$2 - 3a = 0 \rightarrow 3a = 2 \rightarrow a = \frac{2}{3}$$

13. $\mathbf{v} = 2\mathbf{i} - 3\mathbf{j}, \quad \mathbf{w} = \mathbf{i} - \mathbf{j}$

$$\mathbf{v}_1 = \text{proj}_{\mathbf{w}}\mathbf{v} = \frac{\mathbf{v} \bullet \mathbf{w}}{\left(\|\mathbf{w}\|\right)^2}\mathbf{w} = \frac{2(1) + (-3)(-1)}{\left(\sqrt{1^2 + (-1)^2}\right)^2}(\mathbf{i} - \mathbf{j}) = \frac{5}{2}(\mathbf{i} - \mathbf{j}) = \frac{5}{2}\mathbf{i} - \frac{5}{2}\mathbf{j}$$

$$\mathbf{v}_2 = \mathbf{v} - \mathbf{v}_1 = (2\mathbf{i} - 3\mathbf{j}) - \left(\frac{5}{2}\mathbf{i} - \frac{5}{2}\mathbf{j}\right) = -\frac{1}{2}\mathbf{i} - \frac{1}{2}\mathbf{j}$$

15. $\mathbf{v} = \mathbf{i} - \mathbf{j}, \quad \mathbf{w} = \mathbf{i} + 2\mathbf{j}$

$$\mathbf{v}_1 = \text{proj}_{\mathbf{w}}\mathbf{v} = \frac{\mathbf{v} \bullet \mathbf{w}}{\left(\|\mathbf{w}\|\right)^2}\mathbf{w} = \frac{1(1) + (-1)(2)}{\left(\sqrt{1^2 + 2^2}\right)^2}(\mathbf{i} + 2\mathbf{j}) = -\frac{1}{5}(\mathbf{i} + 2\mathbf{j}) = -\frac{1}{5}\mathbf{i} - \frac{2}{5}\mathbf{j}$$

$$\mathbf{v}_2 = \mathbf{v} - \mathbf{v}_1 = (\mathbf{i} - \mathbf{j}) - \left(-\frac{1}{5}\mathbf{i} - \frac{2}{5}\mathbf{j}\right) = \frac{6}{5}\mathbf{i} - \frac{3}{5}\mathbf{j}$$

17. $\mathbf{v} = 3\mathbf{i} + \mathbf{j}, \quad \mathbf{w} = -2\mathbf{i} - \mathbf{j}$

$$\mathbf{v}_1 = \text{proj}_{\mathbf{w}}\mathbf{v} = \frac{\mathbf{v} \bullet \mathbf{w}}{\left(\|\mathbf{w}\|\right)^2}\mathbf{w} = \frac{3(-2) + 1(-1)}{\left(\sqrt{(-2)^2 + (-1)^2}\right)^2}(-2\mathbf{i} - \mathbf{j}) = -\frac{7}{5}(-2\mathbf{i} - \mathbf{j}) = \frac{14}{5}\mathbf{i} + \frac{7}{5}\mathbf{j}$$

$$\mathbf{v}_2 = \mathbf{v} - \mathbf{v}_1 = (3\mathbf{i} + \mathbf{j}) - \left(\frac{14}{5}\mathbf{i} + \frac{7}{5}\mathbf{j}\right) = \frac{1}{5}\mathbf{i} - \frac{2}{5}\mathbf{j}$$

19.   Let $\mathbf{v}_a$ = the velocity of the plane in still air.

$\mathbf{v}_w$ = the velocity of the wind.

$\mathbf{v}_g$ = the velocity of the plane relative to the ground.

$\mathbf{v}_g = \mathbf{v}_a + \mathbf{v}_w$

$$\mathbf{v}_a = 550(\cos(225°)\mathbf{i} + \sin(225°)\mathbf{j}) = 550\left(-\frac{\sqrt{2}}{2}\mathbf{i} - \frac{\sqrt{2}}{2}\mathbf{j}\right) = -275\sqrt{2}\,\mathbf{i} - 275\sqrt{2}\,\mathbf{j}$$

$\mathbf{v}_w = 80\mathbf{i}$

$$\mathbf{v}_g = \mathbf{v}_a + \mathbf{v}_w = -275\sqrt{2}\,\mathbf{i} - 275\sqrt{2}\,\mathbf{j} + 80\mathbf{i} = \left(80 - 275\sqrt{2}\right)\mathbf{i} - 275\sqrt{2}\,\mathbf{j}$$

The speed of the plane relative to the ground is:

$$\|\mathbf{v}_g\| = \sqrt{\left(80 - 275\sqrt{2}\right)^2 + \left(-275\sqrt{2}\right)^2} = \sqrt{6400 - 44000\sqrt{2} + 151250 + 151250}$$

$$= \sqrt{246674.6} \approx 496.7 \text{ miles per hour}$$

To find the direction, find the angle between $\mathbf{v}_g$ and a convenient vector such as due south, $-\mathbf{j}$.

$$\cos\theta = \frac{\mathbf{v}_g \bullet -\mathbf{j}}{\|\mathbf{v}_g\|\|-\mathbf{j}\|} = \frac{\left(80 - 275\sqrt{2}\right)\cdot 0 + \left(-275\sqrt{2}\right)(-1)}{496.7\sqrt{0^2 + (-1)^2}} = \frac{275\sqrt{2}}{496.7} \approx 0.7829$$

$$\theta \approx 38.5°$$

The plane is traveling with a ground speed of about 496.7 miles per hour in a direction of 38.5° west of south.

21. Let the positive x-axis point downstream, so that the velocity of the current is $\mathbf{v}_c = 3\mathbf{i}$.
Let $\mathbf{v}_w$ = the velocity of the boat in the water.
Let $\mathbf{v}_g$ = the velocity of the boat relative to the land.
Then $\mathbf{v}_g = \mathbf{v}_w + \mathbf{v}_c$
The speed of the boat is $\|\mathbf{v}_w\| = 20$; we need to find the direction.

Let $\mathbf{v}_w = a\mathbf{i} + b\mathbf{j}$ so $\|\mathbf{v}_w\| = \sqrt{a^2 + b^2} = 20 \ \rightarrow \ a^2 + b^2 = 400$.
Let $\mathbf{v}_g = k\mathbf{j}$.
Since $\mathbf{v}_g = \mathbf{v}_w + \mathbf{v}_c, \ \ k\mathbf{j} = a\mathbf{i} + b\mathbf{j} + 3\mathbf{i} \ \rightarrow \ k\mathbf{j} = (a+3)\mathbf{i} + b\mathbf{j}$
$a + 3 = 0$ and $k = b \ \rightarrow \ a = -3$
$a^2 + b^2 = 400 \ \rightarrow \ 9 + b^2 = 400 \ \rightarrow \ b^2 = 391 \ \rightarrow \ k = b \approx 19.77$
$\mathbf{v}_w = -3\mathbf{i} + 19.77\mathbf{j}$ and $\mathbf{v}_g = 19.77\mathbf{j}$

Find the angle between $\mathbf{v}_w$ and $\mathbf{j}$:

$$\cos\theta = \frac{\mathbf{v}_w \bullet \mathbf{j}}{\|\mathbf{v}_w\|\|\mathbf{j}\|} = \frac{-3 \cdot 0 + 19.77(1)}{20\sqrt{0^2 + 1^2}} = \frac{19.77}{20} \approx 0.9885$$

$$\theta \approx 8.7°$$

The heading of the boat needs to be 8.7° upstream.
The velocity of the boat directly across the river is 19.77 kilometers per hour. The time to cross the river is: $t = \dfrac{0.5}{19.77} \approx 0.025$ hours or $t \approx 1.5$ minutes.

23. Split the force into the components going down the hill and perpendicular to the hill.

$\mathbf{F_d} = \mathbf{F}\sin(8°) = 5300\sin(8°)$

$\quad = 5300(0.1392) \approx 738$ pounds

$\mathbf{F_p} = \mathbf{F}\cos(8°) = 5300\cos(8°)$

$\quad = 5300(0.9903) \approx 5249$ pounds

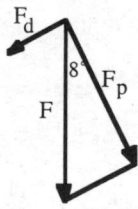

The force required to keep the car from rolling down the hill is about 738 pounds.
The force perpendicular to the hill is approximately 5249 pounds.

25. Let $\mathbf{v}_a$ = the velocity of the plane in still air.
$\mathbf{v}_w$ = the velocity of the wind.
$\mathbf{v}_g$ = the velocity of the plane relative to the ground.
$\mathbf{v}_g = \mathbf{v}_a + \mathbf{v}_w$

$$v_a = 500\big(\cos(45°)i + \sin(45°)j\big) = 500\left(\frac{\sqrt{2}}{2}i + \frac{\sqrt{2}}{2}j\right) = 250\sqrt{2}\,i + 250\sqrt{2}\,j$$

$$v_w = 60\big(\cos(120°)i + \sin(120°)j\big) = 60\left(-\frac{1}{2}i + \frac{\sqrt{3}}{2}j\right) = -30i + 30\sqrt{3}\,j$$

$$v_g = v_a + v_w = 250\sqrt{2}\,i + 250\sqrt{2}\,j - 30i + 30\sqrt{3}\,j$$
$$= \big(-30 + 250\sqrt{2}\big)i + \big(250\sqrt{2} + 30\sqrt{3}\big)j$$

The speed of the plane relative to the ground is:

$$\|v_g\| = \sqrt{\big(-30 + 250\sqrt{2}\big)^2 + \big(250\sqrt{2} + 30\sqrt{3}\big)^2}$$
$$= \sqrt{269129.1} \approx 518.8 \text{ kilometers per hour}$$

To find the direction, find the angle between v_g and a convenient vector such as due north, j.

$$\cos\theta = \frac{v_g \bullet j}{\|v_g\|\|j\|} = \frac{\big(-30 + 250\sqrt{2}\big)\cdot 0 + \big(250\sqrt{2} + 30\sqrt{3}\big)(1)}{518.8\sqrt{0^2 + 1^2}} = \frac{250\sqrt{2} + 30\sqrt{3}}{518.8}$$

$$= \frac{405.5}{518.8} \approx 0.7816 \quad \rightarrow \quad \theta \approx 38.6°$$

The plane is traveling with a ground speed of about 518.8 kilometers per hour in a direction of 38.6° east of north.

27.   Let the positive x-axis point downstream, so that the velocity of the current is $v_c = 3i$.

Let v_w = the velocity of the boat in the water.

Let v_g = the velocity of the boat relative to the land.

Then $v_g = v_w + v_c$

The speed of the boat is $\|v_w\| = 20$; its direction is directly across the river, so

Let $v_w = 20j$.

$$v_g = v_w + v_c = 20j + 3i = 3i + 20j$$

Let $$\|v_g\| = \sqrt{3^2 + 20^2} = \sqrt{409} \approx 20.2 \text{ miles per hour}^.$$

Find the angle between v_g and j:

$$\cos\theta = \frac{v_g \bullet j}{\|v_g\|\|j\|} = \frac{3\cdot 0 + 20(1)}{20.2\sqrt{0^2 + 1^2}} = \frac{20}{20.2} \approx 0.9901$$
$$\theta \approx 8.1°$$

The heading of the boat will be 8.1° downstream.

29. $$F = 3\big(\cos(60°)i + \sin(60°)j\big) = 3\left(\frac{1}{2}i + \frac{\sqrt{3}}{2}j\right) = \frac{3}{2}i + \frac{3\sqrt{3}}{2}j$$

$$W = F \bullet AB = \left(\frac{3}{2}i + \frac{3\sqrt{3}}{2}j\right)\bullet 2i = \frac{3}{2}(2) + \frac{3\sqrt{3}}{2}\cdot 0 = 3 \text{ foot - pounds}$$

31. $\mathbf{F} = 20\left(\cos(30^\circ)\mathbf{i} + \sin(30^\circ)\mathbf{j}\right) = 20\left(\dfrac{\sqrt{3}}{2}\mathbf{i} + \dfrac{1}{2}\mathbf{j}\right) = 10\sqrt{3}\mathbf{i} + 10\mathbf{j}$

$W = \mathbf{F} \bullet AB = \left(10\sqrt{3}\mathbf{i} + 10\mathbf{j}\right) \bullet 100\mathbf{i} = 10\sqrt{3}(100) + 10 \cdot 0 = 1732 \text{ foot - pounds}$

33. Let $\mathbf{u} = a_1\mathbf{i} + b_1\mathbf{j}, \quad \mathbf{v} = a_2\mathbf{i} + b_2\mathbf{j}, \quad \mathbf{w} = a_3\mathbf{i} + b_3\mathbf{j}$

$\mathbf{u} \bullet (\mathbf{v} + \mathbf{w}) = \left(a_1\mathbf{i} + b_1\mathbf{j}\right) \bullet \left(a_2\mathbf{i} + b_2\mathbf{j} + a_3\mathbf{i} + b_3\mathbf{j}\right) = \left(a_1\mathbf{i} + b_1\mathbf{j}\right) \bullet \left(a_2\mathbf{i} + a_3\mathbf{i} + b_2\mathbf{j} + b_3\mathbf{j}\right)$

$= \left(a_1\mathbf{i} + b_1\mathbf{j}\right) \bullet \left((a_2 + a_3)\mathbf{i} + (b_2 + b_3)\mathbf{j}\right) = a_1(a_2 + a_3) + b_1(b_2 + b_3)$

$= a_1 a_2 + a_1 a_3 + b_1 b_2 + b_1 b_3 = a_1 a_2 + b_1 b_2 + a_1 a_3 + b_1 b_3$

$= \left(a_1\mathbf{i} + b_1\mathbf{j}\right) \bullet \left(a_2\mathbf{i} + b_2\mathbf{j}\right) + \left(a_1\mathbf{i} + b_1\mathbf{j}\right) \bullet \left(a_3\mathbf{i} + b_3\mathbf{j}\right) = \mathbf{u} \bullet \mathbf{v} + \mathbf{u} \bullet \mathbf{w}$

35. Let $\mathbf{v} = a\mathbf{i} + b\mathbf{j}$.

Since $\mathbf{v}$ is a unit vector, $\|\mathbf{v}\| = \sqrt{a^2 + b^2} = 1$ or $a^2 + b^2 = 1$

If α is the angle between $\mathbf{v}$ and $\mathbf{i}$, then $\cos\alpha = \dfrac{\mathbf{v} \bullet \mathbf{i}}{\|\mathbf{v}\|\|\mathbf{i}\|}$ or $\cos\alpha = \dfrac{(a\mathbf{i} + b\mathbf{j}) \bullet \mathbf{i}}{1 \cdot 1} = a$.

$\quad a^2 + b^2 = 1$

$\quad \cos^2\alpha + b^2 = 1 \rightarrow b^2 = 1 - \cos^2\alpha \rightarrow b^2 = \sin^2\alpha \rightarrow b = \sin\alpha$

Thus, $\mathbf{v} = \cos\alpha\, \mathbf{i} + \sin\alpha\, \mathbf{j}$

37. Let $\mathbf{v} = a\mathbf{i} + b\mathbf{j}$.

$\text{proj}_i\, \mathbf{v} = \dfrac{\mathbf{v} \bullet \mathbf{i}}{\left(\|\mathbf{i}\|\right)^2}\mathbf{i} = \dfrac{(a\mathbf{i} + b\mathbf{j}) \bullet \mathbf{i}}{\left(\sqrt{1^2 + 0^2}\right)^2}\mathbf{i} = \dfrac{a(1) + b(0)}{1^2}\mathbf{i} = a\mathbf{i}$

$\mathbf{v} \bullet \mathbf{i} = a, \quad \mathbf{v} \bullet \mathbf{j} = b,$

$\mathbf{v} = (\mathbf{v} \bullet \mathbf{i})\mathbf{i} + (\mathbf{v} \bullet \mathbf{j})\mathbf{j}$

39. $(\mathbf{v} - \alpha\mathbf{w}) \bullet \mathbf{w} = \mathbf{v} \bullet \mathbf{w} - \alpha\mathbf{w} \bullet \mathbf{w} = \mathbf{v} \bullet \mathbf{w} - \alpha\left(\|\mathbf{w}\|\right)^2 = \mathbf{v} \bullet \mathbf{w} - \dfrac{\mathbf{v} \bullet \mathbf{w}}{\left(\|\mathbf{w}\|\right)^2}\left(\|\mathbf{w}\|\right)^2 = 0$

Therefore the vectors are orthogonal.

41. If $\mathbf{F}$ is orthogonal to $\mathbf{AB}$, then $W = \mathbf{F} \bullet \mathbf{AB} = 0$.

Chapter 8

Polar Coordinates; Vectors

8.6 Vectors in Space

1. $y = 0$ is the set of all points in the xz-plane.

3. $z = 2$ is the set of all points of the form $(x, y, 2)$; the plane two units above the xy-plane.

5. $x = -4$ is the set of all points of the form $(-4, y, z)$; the plane four units to the left of yz-plane.

7. $x = 1$ and $y = 2$ is the set of all points of the form $(1, 2, z)$; a line parallel to the z-axis.

9. $d = \sqrt{(4-0)^2 + (1-0)^2 + (2-0)^2} = \sqrt{16+1+4} = \sqrt{21}$

11. $d = \sqrt{(0-(-1))^2 + (-2-2)^2 + (1-(-3))^2} = \sqrt{1+16+16} = \sqrt{33}$

13. $d = \sqrt{(3-4)^2 + (2-(-2))^2 + (1-(-2))^2} = \sqrt{1+16+9} = \sqrt{26}$

15. The bottom of the box is formed by the vertices $(0, 0, 0)$, $(2, 0, 0)$, $(0, 1, 0)$, and $(2, 1, 0)$. The top of the box is formed by the vertices $(0, 0, 3)$, $(2, 0, 3)$, $(0, 1, 3)$, and $(2, 1, 3)$.

17. The bottom of the box is formed by the vertices $(1, 2, 3)$, $(3, 2, 3)$, $(3, 4, 3)$, and $(1, 4, 3)$. The top of the box is formed by the vertices $(3, 4, 5)$, $(1, 2, 5)$, $(3, 2, 5)$, and $(1, 4, 5)$.

19. The bottom of the box is formed by the vertices $(-1, 0, 2)$, $(4, 0, 2)$, $(-1, 2, 2)$, and $(4, 2, 2)$. The top of the box is formed by the vertices $(4, 2, 5)$, $(-1, 0, 5)$, $(4, 0, 5)$, and $(-1, 2, 5)$.

21. $\mathbf{v} = (3-0)\mathbf{i} + (4-0)\mathbf{j} + (-1-0)\mathbf{k} = 3\mathbf{i} + 4\mathbf{j} - 1\mathbf{k}$

23. $\mathbf{v} = (5-3)\mathbf{i} + (6-2)\mathbf{j} + (0-(-1))\mathbf{k} = 2\mathbf{i} + 4\mathbf{j} + 1\mathbf{k}$

25. $\mathbf{v} = (6-(-2))\mathbf{i} + (-2-(-1))\mathbf{j} + (4-4)\mathbf{k} = 8\mathbf{i} - 1\mathbf{j}$

27. $\|\mathbf{v}\| = \sqrt{3^2 + (-6)^2 + (-2)^2} = \sqrt{9+36+4} = \sqrt{49} = 7$

29. $\|\mathbf{v}\| = \sqrt{1^2 + (-1)^2 + 1^2} = \sqrt{1+1+1} = \sqrt{3}$

31. $\|\mathbf{v}\| = \sqrt{(-2)^2 + 3^2 + (-3)^2} = \sqrt{4 + 9 + 9} = \sqrt{22}$

33. $2\mathbf{v} + 3\mathbf{w} = 2(3\mathbf{i} - 5\mathbf{j} + 2\mathbf{k}) + 3(-2\mathbf{i} + 3\mathbf{j} - 2\mathbf{k}) = 6\mathbf{i} - 10\mathbf{j} + 4\mathbf{k} - 6\mathbf{i} + 9\mathbf{j} - 6\mathbf{k}$
$\qquad = 0\mathbf{i} - 1\mathbf{j} - 2\mathbf{k}$

35. $\|\mathbf{v} - \mathbf{w}\| = \|(3\mathbf{i} - 5\mathbf{j} + 2\mathbf{k}) - (-2\mathbf{i} + 3\mathbf{j} - 2\mathbf{k})\| = \|3\mathbf{i} - 5\mathbf{j} + 2\mathbf{k} + 2\mathbf{i} - 3\mathbf{j} + 2\mathbf{k}\|$
$\qquad = \|5\mathbf{i} - 8\mathbf{j} + 4\mathbf{k}\| = \sqrt{5^2 + (-8)^2 + 4^2} = \sqrt{25 + 64 + 16} = \sqrt{105}$

37. $\|\mathbf{v}\| - \|\mathbf{w}\| = \|3\mathbf{i} - 5\mathbf{j} + 2\mathbf{k}\| - \|-2\mathbf{i} + 3\mathbf{j} - 2\mathbf{k}\|$
$\qquad = \sqrt{3^2 + (-5)^2 + 2^2} - \sqrt{(-2)^2 + 3^2 + (-2)^2} = \sqrt{38} - \sqrt{17}$

39. $\mathbf{u} = \dfrac{\mathbf{v}}{\|\mathbf{v}\|} = \dfrac{5\mathbf{i}}{\sqrt{5^2 + 0^2 + 0^2}} = \dfrac{5\mathbf{i}}{5} = \mathbf{i}$

41. $\mathbf{u} = \dfrac{\mathbf{v}}{\|\mathbf{v}\|} = \dfrac{3\mathbf{i} - 6\mathbf{j} - 2\mathbf{k}}{\sqrt{3^2 + (-6)^2 + (-2)^2}} = \dfrac{3\mathbf{i} - 6\mathbf{j} - 2\mathbf{k}}{7} = \dfrac{3}{7}\mathbf{i} - \dfrac{6}{7}\mathbf{j} - \dfrac{2}{7}\mathbf{k}$

43. $\mathbf{u} = \dfrac{\mathbf{v}}{\|\mathbf{v}\|} = \dfrac{\mathbf{i} + \mathbf{j} + \mathbf{k}}{\sqrt{1^2 + 1^2 + 1^2}} = \dfrac{\mathbf{i} + \mathbf{j} + \mathbf{k}}{\sqrt{3}} = \dfrac{1}{\sqrt{3}}\mathbf{i} + \dfrac{1}{\sqrt{3}}\mathbf{j} + \dfrac{1}{\sqrt{3}}\mathbf{k} = \dfrac{\sqrt{3}}{3}\mathbf{i} + \dfrac{\sqrt{3}}{3}\mathbf{j} + \dfrac{\sqrt{3}}{3}\mathbf{k}$

45. $\mathbf{v} \bullet \mathbf{w} = (\mathbf{i} - \mathbf{j}) \bullet (\mathbf{i} + \mathbf{j} + \mathbf{k}) = 1 \cdot 1 + (-1)(1) + 0 \cdot 1 = 1 - 1 + 0 = 0$
$\quad \cos\theta = \dfrac{\mathbf{v} \bullet \mathbf{w}}{\|\mathbf{v}\|\|\mathbf{w}\|} = \dfrac{0}{\sqrt{1^2 + (-1)^2 + 0^2}\sqrt{1^2 + 1^2 + 1^2}} = \dfrac{0}{\sqrt{2}\sqrt{3}} = \dfrac{0}{\sqrt{6}} = 0$
$\qquad \theta = \dfrac{\pi}{2}$ radians $= 90°$

47. $\mathbf{v} \bullet \mathbf{w} = (2\mathbf{i} + \mathbf{j} - 3\mathbf{k}) \bullet (\mathbf{i} + 2\mathbf{j} + 2\mathbf{k}) = 2 \cdot 1 + 1(2) + (-3)(2) = 2 + 2 - 6 = -2$
$\quad \cos\theta = \dfrac{\mathbf{v} \bullet \mathbf{w}}{\|\mathbf{v}\|\|\mathbf{w}\|} = \dfrac{-2}{\sqrt{2^2 + 1^2 + (-3)^2}\sqrt{1^2 + 2^2 + 2^2}} = \dfrac{-2}{\sqrt{14}\sqrt{9}} = \dfrac{-2}{3\sqrt{14}} \approx -0.1782$
$\qquad \theta \approx 1.75$ radians $\approx 100.3°$

49. $\mathbf{v} \bullet \mathbf{w} = (3\mathbf{i} - \mathbf{j} + 2\mathbf{k}) \bullet (\mathbf{i} + \mathbf{j} - \mathbf{k}) = 3 \cdot 1 + (-1)(1) + 2(-1) = 3 - 1 - 2 = 0$
$\quad \cos\theta = \dfrac{\mathbf{v} \bullet \mathbf{w}}{\|\mathbf{v}\|\|\mathbf{w}\|} = \dfrac{0}{\sqrt{3^2 + (-1)^2 + 2^2}\sqrt{1^2 + 1^2 + (-1)^2}} = \dfrac{0}{\sqrt{14}\sqrt{3}} = 0$
$\qquad \theta = \dfrac{\pi}{2}$ radians $= 90°$

51. $\mathbf{v} \bullet \mathbf{w} = (3\mathbf{i} + 4\mathbf{j} + \mathbf{k}) \bullet (6\mathbf{i} + 8\mathbf{j} + 2\mathbf{k}) = 3 \cdot 6 + 4 \cdot 8 + 1 \cdot 2 = 18 + 32 + 2 = 52$
$\quad \cos\theta = \dfrac{\mathbf{v} \bullet \mathbf{w}}{\|\mathbf{v}\|\|\mathbf{w}\|} = \dfrac{52}{\sqrt{3^2 + 4^2 + 1^2}\sqrt{6^2 + 8^2 + 2^2}} = \dfrac{52}{\sqrt{26}\sqrt{104}} = \dfrac{52}{52} = 1$
$\qquad \theta = 0$ radians $= 0°$

53. $\cos\alpha = \dfrac{a}{\|\mathbf{v}\|} = \dfrac{3}{\sqrt{3^2 + (-6)^2 + (-2)^2}} = \dfrac{3}{\sqrt{49}} = \dfrac{3}{7} \;\rightarrow\; \alpha \approx 64.6°$

$\cos\beta = \dfrac{b}{\|\mathbf{v}\|} = \dfrac{-6}{\sqrt{3^2 + (-6)^2 + (-2)^2}} = \dfrac{-6}{\sqrt{49}} = -\dfrac{6}{7} \;\rightarrow\; \beta \approx 149.0°$

$\cos\gamma = \dfrac{c}{\|\mathbf{v}\|} = \dfrac{-2}{\sqrt{3^2 + (-6)^2 + (-2)^2}} = \dfrac{-2}{\sqrt{49}} = -\dfrac{2}{7} \;\rightarrow\; \gamma \approx 106.6°$

$\mathbf{v} = 7\big(\cos(64.6°)\mathbf{i} + \cos(149.0°)\mathbf{j} + \cos(106.6°)\mathbf{k}\big)$

55. $\cos\alpha = \dfrac{a}{\|\mathbf{v}\|} = \dfrac{1}{\sqrt{1^2 + 1^2 + 1^2}} = \dfrac{1}{\sqrt{3}} = \dfrac{\sqrt{3}}{3} \;\rightarrow\; \alpha \approx 54.7°$

$\cos\beta = \dfrac{b}{\|\mathbf{v}\|} = \dfrac{1}{\sqrt{1^2 + 1^2 + 1^2}} = \dfrac{1}{\sqrt{3}} = \dfrac{\sqrt{3}}{3} \;\rightarrow\; \beta \approx 54.7°$

$\cos\gamma = \dfrac{c}{\|\mathbf{v}\|} = \dfrac{1}{\sqrt{1^2 + 1^2 + 1^2}} = \dfrac{1}{\sqrt{3}} = \dfrac{\sqrt{3}}{3} \;\rightarrow\; \gamma \approx 54.7°$

$\mathbf{v} = \sqrt{3}\big(\cos(54.7°)\mathbf{i} + \cos(54.7°)\mathbf{j} + \cos(54.7°)\mathbf{k}\big)$

57. $\cos\alpha = \dfrac{a}{\|\mathbf{v}\|} = \dfrac{1}{\sqrt{1^2 + 1^2 + 0^2}} = \dfrac{1}{\sqrt{2}} = \dfrac{\sqrt{2}}{2} \;\rightarrow\; \alpha = 45°$

$\cos\beta = \dfrac{b}{\|\mathbf{v}\|} = \dfrac{1}{\sqrt{1^2 + 1^2 + 0^2}} = \dfrac{1}{\sqrt{2}} = \dfrac{\sqrt{2}}{2} \;\rightarrow\; \beta = 45°$

$\cos\gamma = \dfrac{c}{\|\mathbf{v}\|} = \dfrac{0}{\sqrt{1^2 + 1^2 + 0^2}} = \dfrac{0}{\sqrt{2}} = 0 \;\rightarrow\; \gamma = 90°$

$\mathbf{v} = \sqrt{2}\big(\cos(45°)\mathbf{i} + \cos(45°)\mathbf{j} + \cos(90°)\mathbf{k}\big)$

59. $\cos\alpha = \dfrac{a}{\|\mathbf{v}\|} = \dfrac{3}{\sqrt{3^2 + (-5)^2 + 2^2}} = \dfrac{3}{\sqrt{38}} \;\rightarrow\; \alpha \approx 60.9°$

$\cos\beta = \dfrac{b}{\|\mathbf{v}\|} = \dfrac{-5}{\sqrt{3^2 + (-5)^2 + 2^2}} = -\dfrac{5}{\sqrt{38}} \;\rightarrow\; \beta \approx 144.2°$

$\cos\gamma = \dfrac{c}{\|\mathbf{v}\|} = \dfrac{2}{\sqrt{3^2 + (-5)^2 + 2^2}} = \dfrac{2}{\sqrt{38}} \;\rightarrow\; \gamma \approx 71.1°$

$\mathbf{v} = \sqrt{38}\big(\cos(60.9°)\mathbf{i} + \cos(144.2°)\mathbf{j} + \cos(71.1°)\mathbf{k}\big)$

61. $d(P_0, P) = \sqrt{(x - x_0)^2 + (y - y_0)^2 + (z - z_0)^2} = r$

$\qquad (x - x_0)^2 + (y - y_0)^2 + (z - z_0)^2 = r^2$

63. $(x - 1)^2 + (y - 2)^2 + (z - 2)^2 = 4$

65. $x^2 + y^2 + z^2 + 2x - 2y = 2$

$\qquad x^2 + 2x + y^2 - 2y + z^2 = 2$

$\qquad x^2 + 2x + 1 + y^2 - 2y + 1 + z^2 = 2 + 1 + 1$

$\qquad (x+1)^2 + (y-1)^2 + (z-0)^2 = 4$

$\qquad$ Center: $(-1, 1, 0)$; Radius: 2

67. $x^2 + y^2 + z^2 - 4x + 4y + 2z = 0$

$\qquad x^2 - 4x + y^2 + 4y + z^2 + 2z = 0$

$\qquad x^2 - 4x + 4 + y^2 + 4y + 4 + z^2 + 2z + 1 = 4 + 4 + 1$

$\qquad (x-2)^2 + (y+2)^2 + (z+1)^2 = 9$

$\qquad$ Center: $(2, -2, -1)$; Radius: 3

69. $2x^2 + 2y^2 + 2z^2 - 8x + 4z = -1$

$\qquad x^2 - 4x + y^2 + z^2 + 2z = \dfrac{-1}{2}$

$\qquad x^2 - 4x + 4 + y^2 + z^2 + 2z + 1 = \dfrac{-1}{2} + 4 + 1$

$\qquad (x-2)^2 + (y-0)^2 + (z+1)^2 = \dfrac{9}{2}$

$\qquad$ Center: $(2, 0, -1)$; Radius: $\dfrac{3\sqrt{2}}{2}$

71.   Write the force as a vector:

$\qquad \cos\alpha = \dfrac{2}{\sqrt{2^2 + 1^2 + 2^2}} = \dfrac{2}{\sqrt{9}} = \dfrac{2}{3}; \quad \cos\beta = \dfrac{1}{3}; \quad \cos\gamma = \dfrac{2}{3}$

$\qquad \mathbf{F} = 3\left(\dfrac{2}{3}\mathbf{i} + \dfrac{1}{3}\mathbf{j} + \dfrac{2}{3}\mathbf{k}\right)$

$\qquad W = 3\left(\dfrac{2}{3}\mathbf{i} + \dfrac{1}{3}\mathbf{j} + \dfrac{2}{3}\mathbf{k}\right) \bullet 2\mathbf{j} = 3\left(\dfrac{1}{3} \cdot 2\right) = 2$ joules

73. $W = \mathbf{F} \bullet \mathbf{AB} = (2\mathbf{i} - \mathbf{j} - \mathbf{k}) \bullet (3\mathbf{i} + 2\mathbf{j} - 5\mathbf{k}) = 2 \cdot 3 + (-1)(2) + (-1)(-5) = 9$

Polar Coordinates; Vectors

8.7   The Cross Product

1. $\begin{vmatrix} 3 & 4 \\ 1 & 2 \end{vmatrix} = 3 \cdot 2 - 1 \cdot 4 = 6 - 4 = 2$

3. $\begin{vmatrix} 6 & 5 \\ -2 & -1 \end{vmatrix} = 6(-1) - (-2)(5) = -6 + 10 = 4$

5. $\begin{vmatrix} A & B & C \\ 2 & 1 & 4 \\ 1 & 3 & 1 \end{vmatrix} = \begin{vmatrix} 1 & 4 \\ 3 & 1 \end{vmatrix} A - \begin{vmatrix} 2 & 4 \\ 1 & 1 \end{vmatrix} B + \begin{vmatrix} 2 & 1 \\ 1 & 3 \end{vmatrix} C = (1 - 12)A - (2 - 4)B + (6 - 1)C$

$= -11A + 2B + 5C$

7. $\begin{vmatrix} A & B & C \\ -1 & 3 & 5 \\ 5 & 0 & -2 \end{vmatrix} = \begin{vmatrix} 3 & 5 \\ 0 & -2 \end{vmatrix} A - \begin{vmatrix} -1 & 5 \\ 5 & -2 \end{vmatrix} B + \begin{vmatrix} -1 & 3 \\ 5 & 0 \end{vmatrix} C = (-6 - 0)A - (2 - 25)B + (0 - 15)C$

$= -6A + 23B - 15C$

9. (a) $\mathbf{v} \times \mathbf{w} = \begin{vmatrix} \mathbf{i} & \mathbf{j} & \mathbf{k} \\ 2 & -3 & 1 \\ 3 & -2 & -1 \end{vmatrix} = \begin{vmatrix} -3 & 1 \\ -2 & -1 \end{vmatrix} \mathbf{i} - \begin{vmatrix} 2 & 1 \\ 3 & -1 \end{vmatrix} \mathbf{j} + \begin{vmatrix} 2 & -3 \\ 3 & -2 \end{vmatrix} \mathbf{k} = 5\mathbf{i} + 5\mathbf{j} + 5\mathbf{k}$

(b) $\mathbf{w} \times \mathbf{v} = \begin{vmatrix} \mathbf{i} & \mathbf{j} & \mathbf{k} \\ 3 & -2 & -1 \\ 2 & -3 & 1 \end{vmatrix} = \begin{vmatrix} -2 & -1 \\ -3 & 1 \end{vmatrix} \mathbf{i} - \begin{vmatrix} 3 & -1 \\ 2 & 1 \end{vmatrix} \mathbf{j} + \begin{vmatrix} 3 & -2 \\ 2 & -3 \end{vmatrix} \mathbf{k} = -5\mathbf{i} - 5\mathbf{j} - 5\mathbf{k}$

(c) $\mathbf{w} \times \mathbf{w} = \begin{vmatrix} \mathbf{i} & \mathbf{j} & \mathbf{k} \\ 3 & -2 & -1 \\ 3 & -2 & -1 \end{vmatrix} = \begin{vmatrix} -2 & -1 \\ -2 & -1 \end{vmatrix} \mathbf{i} - \begin{vmatrix} 3 & -1 \\ 3 & -1 \end{vmatrix} \mathbf{j} + \begin{vmatrix} 3 & -2 \\ 3 & -2 \end{vmatrix} \mathbf{k} = 0\mathbf{i} + 0\mathbf{j} + 0\mathbf{k} = \mathbf{0}$

(d) $\mathbf{v} \times \mathbf{v} = \begin{vmatrix} \mathbf{i} & \mathbf{j} & \mathbf{k} \\ 2 & -3 & 1 \\ 2 & -3 & 1 \end{vmatrix} = \begin{vmatrix} -3 & 1 \\ -3 & 1 \end{vmatrix} \mathbf{i} - \begin{vmatrix} 2 & 1 \\ 2 & 1 \end{vmatrix} \mathbf{j} + \begin{vmatrix} 2 & -3 \\ 2 & -3 \end{vmatrix} \mathbf{k} = 0\mathbf{i} + 0\mathbf{j} + 0\mathbf{k} = \mathbf{0}$

11. (a) $\mathbf{v} \times \mathbf{w} = \begin{vmatrix} \mathbf{i} & \mathbf{j} & \mathbf{k} \\ 1 & 1 & 0 \\ 2 & 1 & 1 \end{vmatrix} = \begin{vmatrix} 1 & 0 \\ 1 & 1 \end{vmatrix} \mathbf{i} - \begin{vmatrix} 1 & 0 \\ 2 & 1 \end{vmatrix} \mathbf{j} + \begin{vmatrix} 1 & 1 \\ 2 & 1 \end{vmatrix} \mathbf{k} = 1\mathbf{i} - 1\mathbf{j} - 1\mathbf{k}$

(b) $\mathbf{w} \times \mathbf{v} = \begin{vmatrix} \mathbf{i} & \mathbf{j} & \mathbf{k} \\ 2 & 1 & 1 \\ 1 & 1 & 0 \end{vmatrix} = \begin{vmatrix} 1 & 1 \\ 1 & 0 \end{vmatrix} \mathbf{i} - \begin{vmatrix} 2 & 1 \\ 1 & 0 \end{vmatrix} \mathbf{j} + \begin{vmatrix} 2 & 1 \\ 1 & 1 \end{vmatrix} \mathbf{k} = -1\mathbf{i} + 1\mathbf{j} + 1\mathbf{k}$

(c) $\mathbf{w} \times \mathbf{w} = \begin{vmatrix} \mathbf{i} & \mathbf{j} & \mathbf{k} \\ 2 & 1 & 1 \\ 2 & 1 & 1 \end{vmatrix} = \begin{vmatrix} 1 & 1 \\ 1 & 1 \end{vmatrix} \mathbf{i} - \begin{vmatrix} 2 & 1 \\ 2 & 1 \end{vmatrix} \mathbf{j} + \begin{vmatrix} 2 & 1 \\ 2 & 1 \end{vmatrix} \mathbf{k} = 0\mathbf{i} + 0\mathbf{j} + 0\mathbf{k} = \mathbf{0}$

(d) $\mathbf{v} \times \mathbf{v} = \begin{vmatrix} \mathbf{i} & \mathbf{j} & \mathbf{k} \\ 1 & 1 & 0 \\ 1 & 1 & 0 \end{vmatrix} = \begin{vmatrix} 1 & 0 \\ 1 & 0 \end{vmatrix} \mathbf{i} - \begin{vmatrix} 1 & 0 \\ 1 & 0 \end{vmatrix} \mathbf{j} + \begin{vmatrix} 1 & 1 \\ 1 & 1 \end{vmatrix} \mathbf{k} = 0\mathbf{i} + 0\mathbf{j} + 0\mathbf{k} = \mathbf{0}$

13. (a) $\mathbf{v} \times \mathbf{w} = \begin{vmatrix} \mathbf{i} & \mathbf{j} & \mathbf{k} \\ 2 & -1 & 2 \\ 0 & 1 & -1 \end{vmatrix} = \begin{vmatrix} -1 & 2 \\ 1 & -1 \end{vmatrix} \mathbf{i} - \begin{vmatrix} 2 & 2 \\ 0 & -1 \end{vmatrix} \mathbf{j} + \begin{vmatrix} 2 & -1 \\ 0 & 1 \end{vmatrix} \mathbf{k} = -1\mathbf{i} + 2\mathbf{j} + 2\mathbf{k}$

(b) $\mathbf{w} \times \mathbf{v} = \begin{vmatrix} \mathbf{i} & \mathbf{j} & \mathbf{k} \\ 0 & 1 & -1 \\ 2 & -1 & 2 \end{vmatrix} = \begin{vmatrix} 1 & -1 \\ -1 & 2 \end{vmatrix} \mathbf{i} - \begin{vmatrix} 0 & -1 \\ 2 & 2 \end{vmatrix} \mathbf{j} + \begin{vmatrix} 0 & 1 \\ 2 & -1 \end{vmatrix} \mathbf{k} = 1\mathbf{i} - 2\mathbf{j} - 2\mathbf{k}$

(c) $\mathbf{w} \times \mathbf{w} = \begin{vmatrix} \mathbf{i} & \mathbf{j} & \mathbf{k} \\ 0 & 1 & -1 \\ 0 & 1 & -1 \end{vmatrix} = \begin{vmatrix} 1 & -1 \\ 1 & -1 \end{vmatrix} \mathbf{i} - \begin{vmatrix} 0 & -1 \\ 0 & -1 \end{vmatrix} \mathbf{j} + \begin{vmatrix} 0 & 1 \\ 0 & 1 \end{vmatrix} \mathbf{k} = 0\mathbf{i} + 0\mathbf{j} + 0\mathbf{k} = \mathbf{0}$

(d) $\mathbf{v} \times \mathbf{v} = \begin{vmatrix} \mathbf{i} & \mathbf{j} & \mathbf{k} \\ 2 & -1 & 2 \\ 2 & -1 & 2 \end{vmatrix} = \begin{vmatrix} -1 & 2 \\ -1 & 2 \end{vmatrix} \mathbf{i} - \begin{vmatrix} 2 & 2 \\ 2 & 2 \end{vmatrix} \mathbf{j} + \begin{vmatrix} 2 & -1 \\ 2 & -1 \end{vmatrix} \mathbf{k} = 0\mathbf{i} + 0\mathbf{j} + 0\mathbf{k} = \mathbf{0}$

15. (a) $\mathbf{v} \times \mathbf{w} = \begin{vmatrix} \mathbf{i} & \mathbf{j} & \mathbf{k} \\ 1 & -1 & -1 \\ 4 & 0 & -3 \end{vmatrix} = \begin{vmatrix} -1 & -1 \\ 0 & -3 \end{vmatrix} \mathbf{i} - \begin{vmatrix} 1 & -1 \\ 4 & -3 \end{vmatrix} \mathbf{j} + \begin{vmatrix} 1 & -1 \\ 4 & 0 \end{vmatrix} \mathbf{k} = 3\mathbf{i} - 1\mathbf{j} + 4\mathbf{k}$

(b) $\mathbf{w} \times \mathbf{v} = \begin{vmatrix} \mathbf{i} & \mathbf{j} & \mathbf{k} \\ 4 & 0 & -3 \\ 1 & -1 & -1 \end{vmatrix} = \begin{vmatrix} 0 & -3 \\ -1 & -1 \end{vmatrix} \mathbf{i} - \begin{vmatrix} 4 & -3 \\ 1 & -1 \end{vmatrix} \mathbf{j} + \begin{vmatrix} 4 & 0 \\ 1 & -1 \end{vmatrix} \mathbf{k} = -3\mathbf{i} + 1\mathbf{j} - 4\mathbf{k}$

(c) $\mathbf{w} \times \mathbf{w} = \begin{vmatrix} \mathbf{i} & \mathbf{j} & \mathbf{k} \\ 4 & 0 & -3 \\ 4 & 0 & -3 \end{vmatrix} = \begin{vmatrix} 0 & -3 \\ 0 & -3 \end{vmatrix} \mathbf{i} - \begin{vmatrix} 4 & -3 \\ 4 & -3 \end{vmatrix} \mathbf{j} + \begin{vmatrix} 4 & 0 \\ 4 & 0 \end{vmatrix} \mathbf{k} = 0\mathbf{i} + 0\mathbf{j} + 0\mathbf{k} = \mathbf{0}$

(d) $\mathbf{v} \times \mathbf{v} = \begin{vmatrix} \mathbf{i} & \mathbf{j} & \mathbf{k} \\ 1 & -1 & -1 \\ 1 & -1 & -1 \end{vmatrix} = \begin{vmatrix} -1 & -1 \\ -1 & -1 \end{vmatrix} \mathbf{i} - \begin{vmatrix} 1 & -1 \\ 1 & -1 \end{vmatrix} \mathbf{j} + \begin{vmatrix} 1 & -1 \\ 1 & -1 \end{vmatrix} \mathbf{k} = 0\mathbf{i} + 0\mathbf{j} + 0\mathbf{k} = \mathbf{0}$

17. $\mathbf{u} \times \mathbf{v} = \begin{vmatrix} \mathbf{i} & \mathbf{j} & \mathbf{k} \\ 2 & -3 & 1 \\ -3 & 3 & 2 \end{vmatrix} = \begin{vmatrix} -3 & 1 \\ 3 & 2 \end{vmatrix} \mathbf{i} - \begin{vmatrix} 2 & 1 \\ -3 & 2 \end{vmatrix} \mathbf{j} + \begin{vmatrix} 2 & -3 \\ -3 & 3 \end{vmatrix} \mathbf{k} = -9\mathbf{i} - 7\mathbf{j} - 3\mathbf{k}$

19. $\mathbf{v} \times \mathbf{u} = \begin{vmatrix} \mathbf{i} & \mathbf{j} & \mathbf{k} \\ -3 & 3 & 2 \\ 2 & -3 & 1 \end{vmatrix} = \begin{vmatrix} 3 & 2 \\ -3 & 1 \end{vmatrix} \mathbf{i} - \begin{vmatrix} -3 & 2 \\ 2 & 1 \end{vmatrix} \mathbf{j} + \begin{vmatrix} -3 & 3 \\ 2 & -3 \end{vmatrix} \mathbf{k} = 9\mathbf{i} + 7\mathbf{j} + 3\mathbf{k}$

21. $\mathbf{v} \times \mathbf{v} = \begin{vmatrix} \mathbf{i} & \mathbf{j} & \mathbf{k} \\ -3 & 3 & 2 \\ -3 & 3 & 2 \end{vmatrix} = \begin{vmatrix} 3 & 2 \\ 3 & 2 \end{vmatrix}\mathbf{i} - \begin{vmatrix} -3 & 2 \\ -3 & 2 \end{vmatrix}\mathbf{j} + \begin{vmatrix} -3 & 3 \\ -3 & 3 \end{vmatrix}\mathbf{k} = 0\mathbf{i} + 0\mathbf{j} + 0\mathbf{k} = \mathbf{0}$

23. $(3\mathbf{u}) \times \mathbf{v} = \begin{vmatrix} \mathbf{i} & \mathbf{j} & \mathbf{k} \\ 6 & -9 & 3 \\ -3 & 3 & 2 \end{vmatrix} = \begin{vmatrix} -9 & 3 \\ 3 & 2 \end{vmatrix}\mathbf{i} - \begin{vmatrix} 6 & 3 \\ -3 & 2 \end{vmatrix}\mathbf{j} + \begin{vmatrix} 6 & -9 \\ -3 & 3 \end{vmatrix}\mathbf{k} = -27\mathbf{i} - 21\mathbf{j} - 9\mathbf{k}$

25. $\mathbf{u} \times (2\mathbf{v}) = \begin{vmatrix} \mathbf{i} & \mathbf{j} & \mathbf{k} \\ 2 & -3 & 1 \\ -6 & 6 & 4 \end{vmatrix} = \begin{vmatrix} -3 & 1 \\ 6 & 4 \end{vmatrix}\mathbf{i} - \begin{vmatrix} 2 & 1 \\ -6 & 4 \end{vmatrix}\mathbf{j} + \begin{vmatrix} 2 & -3 \\ -6 & 6 \end{vmatrix}\mathbf{k} = -18\mathbf{i} - 14\mathbf{j} - 6\mathbf{k}$

27. $\mathbf{u} \bullet (\mathbf{u} \times \mathbf{v}) = \mathbf{u} \bullet \begin{vmatrix} \mathbf{i} & \mathbf{j} & \mathbf{k} \\ 2 & -3 & 1 \\ -3 & 3 & 2 \end{vmatrix} = \mathbf{u} \bullet \left(\begin{vmatrix} -3 & 1 \\ 3 & 2 \end{vmatrix}\mathbf{i} - \begin{vmatrix} 2 & 1 \\ -3 & 2 \end{vmatrix}\mathbf{j} + \begin{vmatrix} 2 & -3 \\ -3 & 3 \end{vmatrix}\mathbf{k} \right)$

$= (2\mathbf{i} - 3\mathbf{j} + \mathbf{k}) \bullet (-9\mathbf{i} - 7\mathbf{j} - 3\mathbf{k}) = 2(-9) + (-3)(-7) + 1(-3) = -18 + 21 - 3 = 0$

29. $\mathbf{u} \bullet (\mathbf{v} \times \mathbf{w}) = \mathbf{u} \bullet \begin{vmatrix} \mathbf{i} & \mathbf{j} & \mathbf{k} \\ -3 & 3 & 2 \\ 1 & 1 & 3 \end{vmatrix} = \mathbf{u} \bullet \left(\begin{vmatrix} 3 & 2 \\ 1 & 3 \end{vmatrix}\mathbf{i} - \begin{vmatrix} -3 & 2 \\ 1 & 3 \end{vmatrix}\mathbf{j} + \begin{vmatrix} -3 & 3 \\ 1 & 1 \end{vmatrix}\mathbf{k} \right)$

$= (2\mathbf{i} - 3\mathbf{j} + \mathbf{k}) \bullet (7\mathbf{i} + 11\mathbf{j} - 6\mathbf{k}) = 2 \cdot 7 + (-3)(11) + 1(-6) = 14 - 33 - 6 = -25$

31. $\mathbf{v} \bullet (\mathbf{u} \times \mathbf{w}) = \mathbf{v} \bullet \begin{vmatrix} \mathbf{i} & \mathbf{j} & \mathbf{k} \\ 2 & -3 & 1 \\ 1 & 1 & 3 \end{vmatrix} = \mathbf{v} \bullet \left(\begin{vmatrix} -3 & 1 \\ 1 & 3 \end{vmatrix}\mathbf{i} - \begin{vmatrix} 2 & 1 \\ 1 & 3 \end{vmatrix}\mathbf{j} + \begin{vmatrix} 2 & -3 \\ 1 & 1 \end{vmatrix}\mathbf{k} \right)$

$= (-3\mathbf{i} + 3\mathbf{j} + 2\mathbf{k}) \bullet (-10\mathbf{i} - 5\mathbf{j} + 5\mathbf{k}) = -3(-10) + 3(-5) + 2 \cdot 5 = 30 - 15 + 10 = 25$

33. $\mathbf{u} \times (\mathbf{v} \times \mathbf{v}) = \mathbf{u} \times \begin{vmatrix} \mathbf{i} & \mathbf{j} & \mathbf{k} \\ -3 & 3 & 2 \\ -3 & 3 & 2 \end{vmatrix} = \mathbf{u} \times \left(\begin{vmatrix} 3 & 2 \\ 3 & 2 \end{vmatrix}\mathbf{i} - \begin{vmatrix} -3 & 2 \\ -3 & 2 \end{vmatrix}\mathbf{j} + \begin{vmatrix} -3 & 3 \\ -3 & 3 \end{vmatrix}\mathbf{k} \right)$

$= (2\mathbf{i} - 3\mathbf{j} + \mathbf{k}) \times (0\mathbf{i} + 0\mathbf{j} + 0\mathbf{k}) = \begin{vmatrix} \mathbf{i} & \mathbf{j} & \mathbf{k} \\ 2 & -3 & 1 \\ 0 & 0 & 0 \end{vmatrix}$

$= \begin{vmatrix} -3 & 1 \\ 0 & 0 \end{vmatrix}\mathbf{i} - \begin{vmatrix} 2 & 1 \\ 0 & 0 \end{vmatrix}\mathbf{j} + \begin{vmatrix} 2 & -3 \\ 0 & 0 \end{vmatrix}\mathbf{k} = 0\mathbf{i} + 0\mathbf{j} + 0\mathbf{k} = \mathbf{0}$

35. $\mathbf{u} \times \mathbf{v} = \begin{vmatrix} \mathbf{i} & \mathbf{j} & \mathbf{k} \\ 2 & -3 & 1 \\ -3 & 3 & 2 \end{vmatrix} = \begin{vmatrix} -3 & 1 \\ 3 & 2 \end{vmatrix}\mathbf{i} - \begin{vmatrix} 2 & 1 \\ -3 & 2 \end{vmatrix}\mathbf{j} + \begin{vmatrix} 2 & -3 \\ -3 & 3 \end{vmatrix}\mathbf{k} = -9\mathbf{i} - 7\mathbf{j} - 3\mathbf{k}$ is orthogonal to both

$\mathbf{u}$ and $\mathbf{v}$.

37. A vector that is orthogonal to both $\mathbf{u}$ and $\mathbf{i} + \mathbf{j}$ is $\mathbf{u} \times (\mathbf{i} + \mathbf{j})$.

$\mathbf{u} \times (\mathbf{i} + \mathbf{j}) = \begin{vmatrix} \mathbf{i} & \mathbf{j} & \mathbf{k} \\ 2 & -3 & 1 \\ 1 & 1 & 0 \end{vmatrix} = \begin{vmatrix} -3 & 1 \\ 1 & 0 \end{vmatrix}\mathbf{i} - \begin{vmatrix} 2 & 1 \\ 1 & 0 \end{vmatrix}\mathbf{j} + \begin{vmatrix} 2 & -3 \\ 1 & 1 \end{vmatrix}\mathbf{k} = -1\mathbf{i} + 1\mathbf{j} + 5\mathbf{k}$

39. $\mathbf{u} = P_1P_2 = 1\mathbf{i} + 2\mathbf{j} + 3\mathbf{k}$ $\mathbf{v} = P_1P_3 = -2\mathbf{i} + 3\mathbf{j} + 0\mathbf{k}$

$$\mathbf{u} \times \mathbf{v} = \begin{vmatrix} \mathbf{i} & \mathbf{j} & \mathbf{k} \\ 1 & 2 & 3 \\ -2 & 3 & 0 \end{vmatrix} = \begin{vmatrix} 2 & 3 \\ 3 & 0 \end{vmatrix}\mathbf{i} - \begin{vmatrix} 1 & 3 \\ -2 & 0 \end{vmatrix}\mathbf{j} + \begin{vmatrix} 1 & 2 \\ -2 & 3 \end{vmatrix}\mathbf{k} = -9\mathbf{i} - 6\mathbf{j} + 7\mathbf{k}$$

$\text{Area} = \|\mathbf{u} \times \mathbf{v}\| = \sqrt{(-9)^2 + (-6)^2 + 7^2} = \sqrt{166} \approx 12.9$

41. $\mathbf{u} = P_1P_2 = -3\mathbf{i} + 1\mathbf{j} + 4\mathbf{k}$ $\mathbf{v} = P_1P_3 = -1\mathbf{i} - 4\mathbf{j} + 3\mathbf{k}$

$$\mathbf{u} \times \mathbf{v} = \begin{vmatrix} \mathbf{i} & \mathbf{j} & \mathbf{k} \\ -3 & 1 & 4 \\ -1 & -4 & 3 \end{vmatrix} = \begin{vmatrix} 1 & 4 \\ -4 & 3 \end{vmatrix}\mathbf{i} - \begin{vmatrix} -3 & 4 \\ -1 & 3 \end{vmatrix}\mathbf{j} + \begin{vmatrix} -3 & 1 \\ -1 & -4 \end{vmatrix}\mathbf{k} = 19\mathbf{i} + 5\mathbf{j} + 13\mathbf{k}$$

$\text{Area} = \|\mathbf{u} \times \mathbf{v}\| = \sqrt{19^2 + 5^2 + 13^2} = \sqrt{555} \approx 23.6$

43. $\mathbf{u} = P_1P_2 = 0\mathbf{i} + 1\mathbf{j} + 1\mathbf{k}$ $\mathbf{v} = P_1P_3 = -3\mathbf{i} + 2\mathbf{j} - 2\mathbf{k}$

$$\mathbf{u} \times \mathbf{v} = \begin{vmatrix} \mathbf{i} & \mathbf{j} & \mathbf{k} \\ 0 & 1 & 1 \\ -3 & 2 & -2 \end{vmatrix} = \begin{vmatrix} 1 & 1 \\ 2 & -2 \end{vmatrix}\mathbf{i} - \begin{vmatrix} 0 & 1 \\ -3 & -2 \end{vmatrix}\mathbf{j} + \begin{vmatrix} 0 & 1 \\ -3 & 2 \end{vmatrix}\mathbf{k} = -4\mathbf{i} - 3\mathbf{j} + 3\mathbf{k}$$

$\text{Area} = \|\mathbf{u} \times \mathbf{v}\| = \sqrt{(-4)^2 + (-3)^2 + 3^2} = \sqrt{34} \approx 5.8$

45. $\mathbf{u} = P_1P_2 = 3\mathbf{i} + 0\mathbf{j} - 2\mathbf{k}$ $\mathbf{v} = P_1P_3 = 5\mathbf{i} - 7\mathbf{j} + 3\mathbf{k}$

$$\mathbf{u} \times \mathbf{v} = \begin{vmatrix} \mathbf{i} & \mathbf{j} & \mathbf{k} \\ 3 & 0 & -2 \\ 5 & -7 & 3 \end{vmatrix} = \begin{vmatrix} 0 & -2 \\ -7 & 3 \end{vmatrix}\mathbf{i} - \begin{vmatrix} 3 & -2 \\ 5 & 3 \end{vmatrix}\mathbf{j} + \begin{vmatrix} 3 & 0 \\ 5 & -7 \end{vmatrix}\mathbf{k} = -14\mathbf{i} - 19\mathbf{j} - 21\mathbf{k}$$

$\text{Area} = \|\mathbf{u} \times \mathbf{v}\| = \sqrt{(-14)^2 + (-19)^2 + (-21)^2} = \sqrt{998} \approx 31.6$

47. $$\mathbf{v} \times \mathbf{w} = \begin{vmatrix} \mathbf{i} & \mathbf{j} & \mathbf{k} \\ 1 & 3 & -2 \\ -2 & 1 & 3 \end{vmatrix} = \begin{vmatrix} 3 & -2 \\ 1 & 3 \end{vmatrix}\mathbf{i} - \begin{vmatrix} 1 & -2 \\ -2 & 3 \end{vmatrix}\mathbf{j} + \begin{vmatrix} 1 & 3 \\ -2 & 1 \end{vmatrix}\mathbf{k} = 11\mathbf{i} + 1\mathbf{j} + 7\mathbf{k}$$

$\|\mathbf{v} \times \mathbf{w}\| = \sqrt{11^2 + 1^2 + 7^2} = \sqrt{171}$

$\mathbf{u} = \dfrac{\mathbf{v} \times \mathbf{w}}{\|\mathbf{v} \times \mathbf{w}\|} = \dfrac{11\mathbf{i} + 1\mathbf{j} + 7\mathbf{k}}{\sqrt{171}} = \dfrac{11}{\sqrt{171}}\mathbf{i} + \dfrac{1}{\sqrt{171}}\mathbf{j} + \dfrac{7}{\sqrt{171}}\mathbf{k}$

49. Prove: $\mathbf{u} \times \mathbf{v} = -(\mathbf{v} \times \mathbf{u})$
Let $\mathbf{u} = a_1\mathbf{i} + b_1\mathbf{j} + c_1\mathbf{k}$ and $\mathbf{v} = a_2\mathbf{i} + b_2\mathbf{j} + c_2\mathbf{k}$

$$\mathbf{u} \times \mathbf{v} = \begin{vmatrix} \mathbf{i} & \mathbf{j} & \mathbf{k} \\ a_1 & b_1 & c_1 \\ a_2 & b_2 & c_2 \end{vmatrix} = \begin{vmatrix} b_1 & c_1 \\ b_2 & c_2 \end{vmatrix}\mathbf{i} - \begin{vmatrix} a_1 & c_1 \\ a_2 & c_2 \end{vmatrix}\mathbf{j} + \begin{vmatrix} a_1 & b_1 \\ a_2 & b_2 \end{vmatrix}\mathbf{k}$$

$$= (b_1c_2 - b_2c_1)\mathbf{i} - (a_1c_2 - a_2c_1)\mathbf{j} + (a_1b_2 - a_2b_1)\mathbf{k}$$

$$= -(b_2c_1 - b_1c_2)\mathbf{i} + (a_2c_1 - a_1c_2)\mathbf{j} - (a_2b_1 - a_1b_2)\mathbf{k}$$

$$= -\left((b_2c_1 - b_1c_2)\mathbf{i} - (a_2c_1 - a_1c_2)\mathbf{j} + (a_2b_1 - a_1b_2)\mathbf{k}\right)$$

$$= -\left(\begin{vmatrix} b_2 & c_2 \\ b_1 & c_1 \end{vmatrix}\mathbf{i} - \begin{vmatrix} a_2 & c_2 \\ a_1 & c_1 \end{vmatrix}\mathbf{j} + \begin{vmatrix} a_2 & b_2 \\ a_1 & b_1 \end{vmatrix}\mathbf{k}\right) = -\begin{vmatrix} \mathbf{i} & \mathbf{j} & \mathbf{k} \\ a_2 & b_2 & c_2 \\ a_1 & b_1 & c_1 \end{vmatrix} = -(\mathbf{v} \times \mathbf{u})$$

51.   Prove: $\left\| \mathbf{u} \times \mathbf{v} \right\|^2 = \left\| \mathbf{u} \right\|^2 \left\| \mathbf{v} \right\|^2 - \left(\mathbf{u} \bullet \mathbf{v} \right)^2$

Let $\mathbf{u} = a_1\mathbf{i} + b_1\mathbf{j} + c_1\mathbf{k}$ and $\mathbf{v} = a_2\mathbf{i} + b_2\mathbf{j} + c_2\mathbf{k}$

$$\mathbf{u} \times \mathbf{v} = \begin{vmatrix} \mathbf{i} & \mathbf{j} & \mathbf{k} \\ a_1 & b_1 & c_1 \\ a_2 & b_2 & c_2 \end{vmatrix} = \begin{vmatrix} b_1 & c_1 \\ b_2 & c_2 \end{vmatrix}\mathbf{i} - \begin{vmatrix} a_1 & c_1 \\ a_2 & c_2 \end{vmatrix}\mathbf{j} + \begin{vmatrix} a_1 & b_1 \\ a_2 & b_2 \end{vmatrix}\mathbf{k}$$

$$= (b_1c_2 - b_2c_1)\mathbf{i} - (a_1c_2 - a_2c_1)\mathbf{j} + (a_1b_2 - a_2b_1)\mathbf{k}$$

$$\left\| \mathbf{u} \times \mathbf{v} \right\|^2 = (b_1c_2 - b_2c_1)^2 + (a_1c_2 - a_2c_1)^2 + (a_1b_2 - a_2b_1)^2$$

$$= b_1^2c_2^2 - 2b_1b_2c_1c_2 + b_2^2c_1^2 + a_1^2c_2^2 - 2a_1a_2c_1c_2 + a_2^2c_1^2$$
$$+ a_1^2b_2^2 - 2a_1a_2b_1b_2 + a_2^2b_1^2$$

$$= a_1^2b_2^2 + a_1^2c_2^2 + a_2^2b_1^2 + a_2^2c_1^2 + b_1^2c_2^2 + b_2^2c_1^2 - 2a_1a_2b_1b_2$$
$$- 2a_1a_2c_1c_2 - 2b_1b_2c_1c_2$$

$$\left\| \mathbf{u} \right\|^2 = a_1^2 + b_1^2 + c_1^2$$

$$\left\| \mathbf{v} \right\|^2 = a_2^2 + b_2^2 + c_2^2$$

$$\left(\mathbf{u} \bullet \mathbf{v} \right)^2 = \left(a_1a_2 + b_1b_2 + c_1c_2 \right)^2$$

$$\left\| \mathbf{u} \right\|^2 \left\| \mathbf{v} \right\|^2 - \left(\mathbf{u} \bullet \mathbf{v} \right)^2 = \left(a_1^2 + b_1^2 + c_1^2 \right)\left(a_2^2 + b_2^2 + c_2^2 \right) - \left(a_1a_2 + b_1b_2 + c_1c_2 \right)^2$$

$$= a_1^2a_2^2 + a_1^2b_2^2 + a_1^2c_2^2 + b_1^2a_2^2 + b_1^2b_2^2 + b_1^2c_2^2 + c_1^2a_2^2 + c_1^2b_2^2 + c_1^2c_2^2$$
$$- \left(\begin{aligned} & a_1^2a_2^2 + a_1a_2b_1b_2 + a_1a_2c_1c_2 + a_1a_2b_1b_2 + b_1^2b_2^2 + b_1b_2c_1c_2 \\ & \qquad + a_1a_2c_1c_2 + b_1b_2c_1c_2 + c_1^2c_2^2 \end{aligned} \right)$$

$$= a_1^2b_2^2 + a_1^2c_2^2 + a_2^2b_1^2 + a_2^2c_1^2 + b_1^2c_2^2 + b_2^2c_1^2 - 2a_1a_2b_1b_2$$
$$- 2a_1a_2c_1c_2 - 2b_1b_2c_1c_2$$

53.   If $\mathbf{u}$ and $\mathbf{v}$ are orthogonal, then $\mathbf{u} \bullet \mathbf{v} = 0$. From problem 51, then:

$$\left\| \mathbf{u} \times \mathbf{v} \right\|^2 = \left\| \mathbf{u} \right\|^2 \left\| \mathbf{v} \right\|^2 - \left(\mathbf{u} \bullet \mathbf{v} \right)^2 = \left\| \mathbf{u} \right\|^2 \left\| \mathbf{v} \right\|^2 - (0)^2 = \left\| \mathbf{u} \right\|^2 \left\| \mathbf{v} \right\|^2$$

$$\left\| \mathbf{u} \times \mathbf{v} \right\| = \left\| \mathbf{u} \right\| \left\| \mathbf{v} \right\|$$

55. $\mathbf{u} \bullet \mathbf{v} = 0 \implies \mathbf{u}$ and $\mathbf{v}$ are orthogonal.

$\mathbf{u} \times \mathbf{v} = 0 \implies \mathbf{u}$ and $\mathbf{v}$ are parallel.

Therefore, if $\mathbf{u} \bullet \mathbf{v} = 0$ and $\mathbf{u} \times \mathbf{v} = 0$, then either $\mathbf{u} = \mathbf{0}$ or $\mathbf{v} = \mathbf{0}$.

Polar Coordinates; Vectors

8.R Chapter Review

1. $\left(3, \dfrac{\pi}{6}\right)$

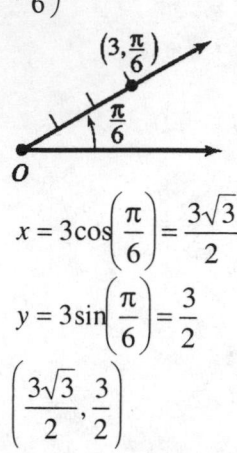

$x = 3\cos\left(\dfrac{\pi}{6}\right) = \dfrac{3\sqrt{3}}{2}$

$y = 3\sin\left(\dfrac{\pi}{6}\right) = \dfrac{3}{2}$

$\left(\dfrac{3\sqrt{3}}{2}, \dfrac{3}{2}\right)$

3. $\left(-2, \dfrac{4\pi}{3}\right)$

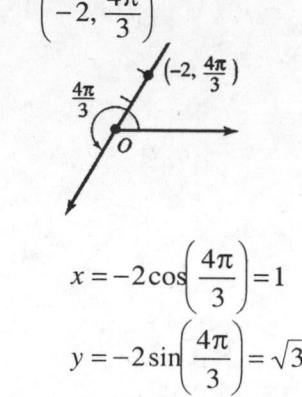

$x = -2\cos\left(\dfrac{4\pi}{3}\right) = 1$

$y = -2\sin\left(\dfrac{4\pi}{3}\right) = \sqrt{3}$

$\left(1, \sqrt{3}\right)$

5. $\left(-3, -\dfrac{\pi}{2}\right)$

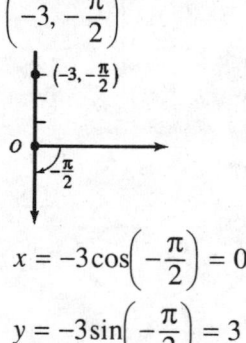

$x = -3\cos\left(-\dfrac{\pi}{2}\right) = 0$

$y = -3\sin\left(-\dfrac{\pi}{2}\right) = 3$

$(0, 3)$

7. The point $(-3, 3)$ lies in quadrant II.

$r = \sqrt{x^2 + y^2} = \sqrt{(-3)^2 + 3^2} = 3\sqrt{2}$ $\theta = \tan^{-1}\left(\dfrac{y}{x}\right) = \tan^{-1}\left(\dfrac{3}{-3}\right) = \tan^{-1}(-1) = -\dfrac{\pi}{4}$

Polar coordinates of the point $(-3, 3)$ are $\left(-3\sqrt{2}, -\dfrac{\pi}{4}\right)$ or $\left(3\sqrt{2}, \dfrac{3\pi}{4}\right)$.

9. The point (0, –2) lies on the negative y-axis.

$r=\sqrt{x^2+y^2}=\sqrt{0^2+(-2)^2}=2 \qquad \theta=\tan^{-1}\left(\frac{y}{x}\right)=\tan^{-1}\left(\frac{-2}{0}\right)=-\frac{\pi}{2}$

Polar coordinates of the point $(0,-2)$ are $\left(2,-\frac{\pi}{2}\right)$ or $\left(-2,\frac{\pi}{2}\right)$.

11. The point (3, 4) lies in quadrant I.

$r=\sqrt{x^2+y^2}=\sqrt{3^2+4^2}=5 \qquad \theta=\tan^{-1}\left(\frac{y}{x}\right)=\tan^{-1}\left(\frac{4}{3}\right)\approx 0.93$

Polar coordinates of the point $(3,4)$ are $(5,0.93)$ or $(-5,4.07)$.

13. $r=2\sin\theta$

$$r^2=2r\sin\theta$$
$$x^2+y^2=2y$$
$$x^2+y^2-2y=0$$
$$x^2+y^2-2y+1=1$$
$$x^2+(y-1)^2=1^2$$

circle with center $(0,1)$, radius = 1.

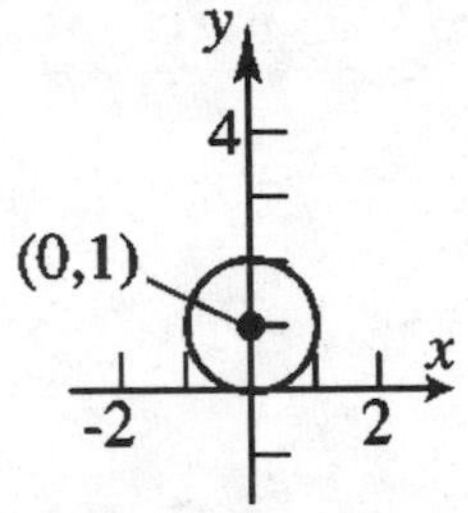

15. $r=5$

$$r^2=25$$
$$x^2+y^2=5^2$$

circle with center $(0,0)$, radius = 5.

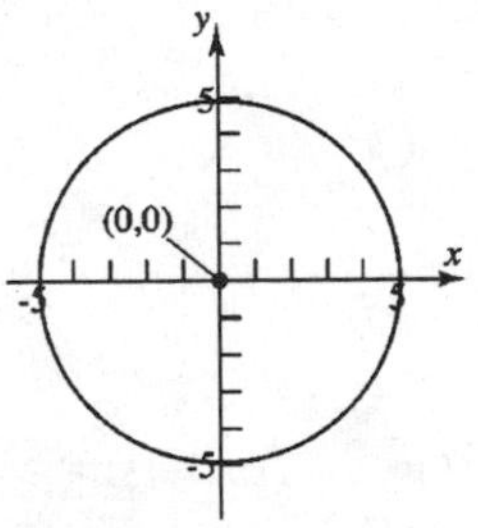

17. $r\cos\theta+3r\sin\theta=6$

$$x+3y=6$$
$$3y=-x+6$$
$$y=-\frac{1}{3}x+2$$

line through $(0,2)$, slope = $-\frac{1}{3}$.

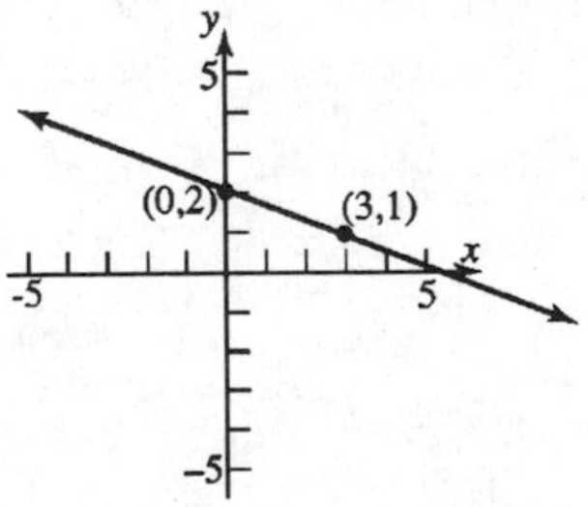

19. $r_1=2+\sin\theta,\ 0\le\theta\le 2\pi$

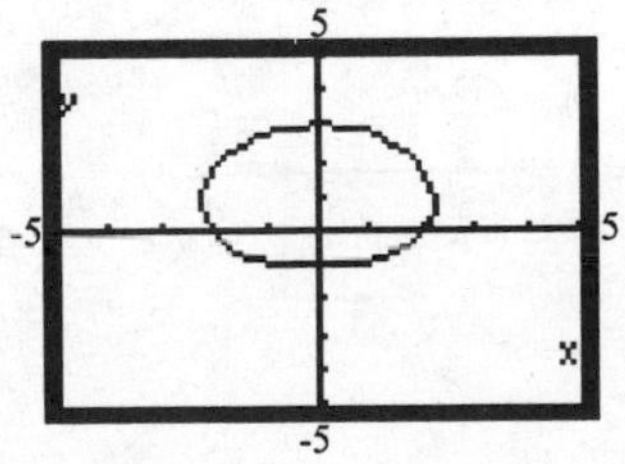

21. $r\tan\theta=2\Rightarrow r_1=\frac{2}{\tan\theta},\quad 0\le\theta\le 2\pi$

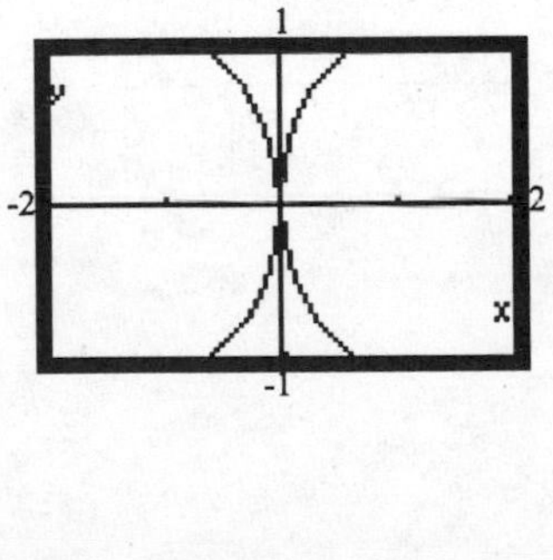

23. $r = 4\cos\theta$ The graph will be a circle. Check for symmetry:
Polar axis: Replace θ by $-\theta$. The result is $r = 4\cos(-\theta) = 4\cos\theta$.
The graph is symmetric with respect to the polar axis.

The line $\theta = \dfrac{\pi}{2}$: Replace θ by $\pi - \theta$.

$$r = 4\cos(\pi - \theta) = 4(\cos(\pi)\cos\theta + \sin(\pi)\sin\theta)$$

$$= 4(-\cos\theta + 0) = -4\cos\theta$$

The test fails.
The pole: Replace r by $-r$. $-r = 4\cos\theta$. The test fails.
Due to symmetry to the polar axis, assign values to θ from 0 to π.

θ	0	$\dfrac{\pi}{6}$	$\dfrac{\pi}{3}$	$\dfrac{\pi}{2}$	$\dfrac{2\pi}{3}$	$\dfrac{5\pi}{6}$	π
$r = 4\cos\theta$	4	$2\sqrt{3} \approx 3.5$	2	0	-2	$-2\sqrt{3} \approx -3.5$	-4

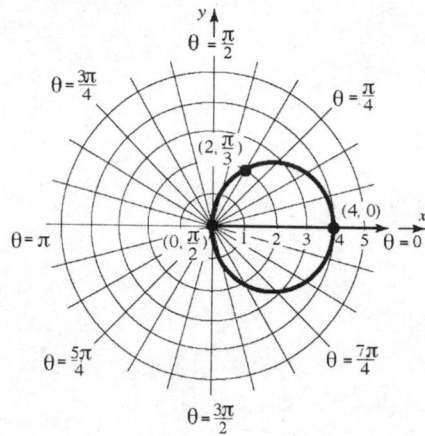

25. $r = 3 - 3\sin\theta$ The graph will be a cardioid. Check for symmetry:
Polar axis: Replace θ by $-\theta$. The result is $r = 3 - 3\sin(-\theta) = 3 + 3\sin\theta$.
                    The test fails.

The line $\theta = \dfrac{\pi}{2}$: Replace θ by $\pi - \theta$.

$$r = 3 - 3\sin(\pi - \theta) = 3 - 3(\sin(\pi)\cos\theta - \cos(\pi)\sin\theta)$$

$$= 3 - 3(0 + \sin\theta) = 3 - 3\sin\theta$$

The graph is symmetric with respect to the line $\theta = \dfrac{\pi}{2}$.
The pole: Replace r by $-r$. $-r = 3 - 3\sin\theta$. The test fails.
Due to symmetry to the line $\theta = \dfrac{\pi}{2}$, assign values to θ from $-\dfrac{\pi}{2}$ to $\dfrac{\pi}{2}$.

θ	$-\dfrac{\pi}{2}$	$-\dfrac{\pi}{3}$	$-\dfrac{\pi}{6}$	0	$\dfrac{\pi}{6}$	$\dfrac{\pi}{3}$	$\dfrac{\pi}{2}$
$r = 3 - 3\sin\theta$	6	$3 + \dfrac{3\sqrt{3}}{2} \approx 5.6$	$\dfrac{9}{2}$	3	$\dfrac{3}{2}$	$3 - \dfrac{3\sqrt{3}}{2} \approx 0.4$	0

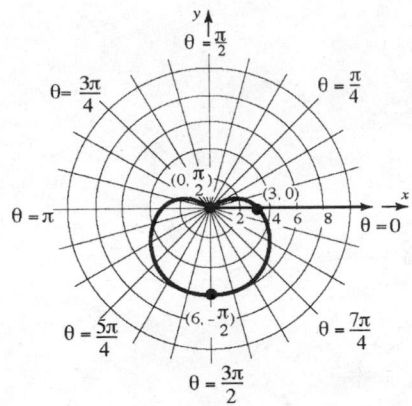

27. $r = 4 - \cos\theta$ The graph will be a limacon without an inner loop.
       Check for symmetry:
       Polar axis: Replace θ by $-\theta$. The result is $r = 4 - \cos(-\theta) = 4 - \cos\theta$.
                   The graph is symmetric with respect to the polar axis.

       The line $\theta = \dfrac{\pi}{2}$: Replace θ by $\pi - \theta$.

$$r = 4 - \cos(\pi - \theta) = 4 - (\cos(\pi)\cos\theta + \sin(\pi)\sin\theta)$$

$$= 4 - (-\cos\theta + 0) = 4 + \cos\theta$$

       The test fails.

       The pole: Replace r by $-r$. $-r = 4 - \cos\theta$. The test fails.
       Due to symmetry to the polar axis, assign values to θ from 0 to π.

θ	0	$\dfrac{\pi}{6}$	$\dfrac{\pi}{3}$	$\dfrac{\pi}{2}$	$\dfrac{2\pi}{3}$	$\dfrac{5\pi}{6}$	π
$r = 4 - \cos\theta$	3	$4 - \dfrac{\sqrt{3}}{2} \approx 3.1$	$\dfrac{7}{2}$	4	$\dfrac{9}{2}$	$4 + \dfrac{\sqrt{3}}{2} \approx 4.9$	5

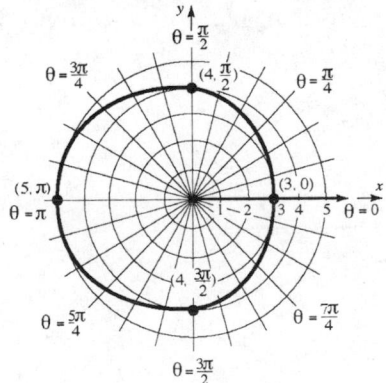

29. $r = \sqrt{x^2 + y^2} = \sqrt{(-1)^2 + (-1)^2} = \sqrt{2}$

       $\tan\theta = \dfrac{y}{x} = \dfrac{-1}{-1} = 1 \implies \theta = 225°$

       The polar form of $z = -1 - i$ is $z = r(\cos\theta + i\sin\theta) = \sqrt{2}\left(\cos(225°) + i\sin(225°)\right)$.

31. $r = \sqrt{x^2 + y^2} = \sqrt{4^2 + (-3)^2} = \sqrt{25} = 5$

$\tan\theta = \dfrac{y}{x} = -\dfrac{3}{4} \implies \theta \approx 323.1°$

The polar form of $z = 4 - 3i$ is $z = r(\cos\theta + i\sin\theta) = 5(\cos(323.1°) + i\sin(323.1°))$.

33. $2(\cos(150°) + i\sin(150°)) = 2\left(-\dfrac{\sqrt{3}}{2} + \dfrac{1}{2}i\right) = -\sqrt{3} + i$

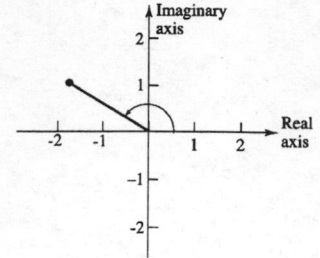

35. $3\left(\cos\left(\dfrac{2\pi}{3}\right) + i\sin\left(\dfrac{2\pi}{3}\right)\right) = 3\left(-\dfrac{1}{2} + \dfrac{\sqrt{3}}{2}i\right) = -\dfrac{3}{2} + \dfrac{3\sqrt{3}}{2}i$

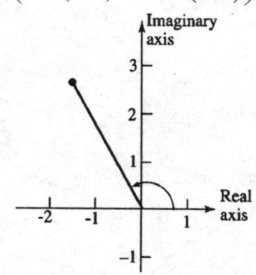

37. $0.1(\cos(350°) + i\sin(350°)) \approx 0.1(0.9848 - 0.1736i) = 0.0985 - 0.0174i$

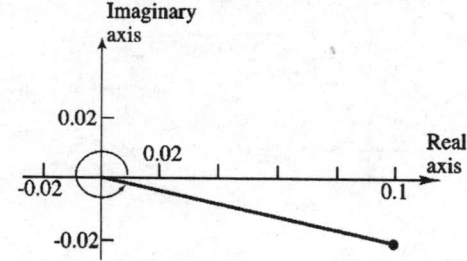

39. $z \cdot w = (\cos(80°) + i\sin(80°)) \cdot (\cos(50°) + i\sin(50°))$

$= 1 \cdot 1(\cos(80° + 50°) + i\sin(80° + 50°)) = \cos(130°) + i\sin(130°)$

$\dfrac{z}{w} = \dfrac{(\cos(80°) + i\sin(80°))}{(\cos(50°) + i\sin(50°))} = \dfrac{1}{1}(\cos(80° - 50°) + i\sin(80° - 50°))$

$= \cos(30°) + i\sin(30°)$

41. $z \cdot w = 3\left(\cos\left(\dfrac{9\pi}{5}\right) + i\sin\left(\dfrac{9\pi}{5}\right)\right) \cdot 2\left(\cos\dfrac{\pi}{5} + i\sin\dfrac{\pi}{5}\right) = 3 \cdot 2\left(\cos\left(\dfrac{9\pi}{5} + \dfrac{\pi}{5}\right) + i\sin\left(\dfrac{9\pi}{5} + \dfrac{\pi}{5}\right)\right)$

$= 6 \cdot (\cos(2\pi) + i\sin(2\pi)) = 6(\cos(0) + i\sin(0))$

$\dfrac{z}{w} = \dfrac{3 \cdot \left(\cos\left(\dfrac{9\pi}{5}\right) + i\sin\left(\dfrac{9\pi}{5}\right)\right)}{2 \cdot \left(\cos\left(\dfrac{\pi}{5}\right) + i\sin\left(\dfrac{\pi}{5}\right)\right)} = \dfrac{3}{2}\left(\cos\left(\dfrac{9\pi}{5} - \dfrac{\pi}{5}\right) + i\sin\left(\dfrac{9\pi}{5} - \dfrac{\pi}{5}\right)\right) = \dfrac{3}{2}\left(\cos\left(\dfrac{8\pi}{5}\right) + i\sin\left(\dfrac{8\pi}{5}\right)\right)$

43. $z \cdot w = 5(\cos(10°) + i\sin(10°)) \cdot (\cos(355°) + i\sin(355°))$

$= 5 \cdot 1(\cos(10° + 355°) + i\sin(10° + 355°)) = 5 \cdot (\cos(365°) + i\sin(365°))$

$= 5 \cdot (\cos(5°) + i\sin(5°))$

$\dfrac{z}{w} = \dfrac{5(\cos(10°) + i\sin(10°))}{(\cos(355°) + i\sin(355°))} = \dfrac{5}{1}(\cos(10° - 355°) + i\sin(10° - 355°))$

$= 5 \cdot (\cos(-345°) + i\sin(-345°)) = 5 \cdot (\cos(15°) + i\sin(15°))$

45. $[3(\cos(20°) + i\sin(20°))]^3 = 3^3(\cos(3 \cdot 20°) + i\sin(3 \cdot 20°)) = 27(\cos(60°) + i\sin(60°))$

$= 27\left(\dfrac{1}{2} + \dfrac{\sqrt{3}}{2}i\right) = \dfrac{27}{2} + \dfrac{27\sqrt{3}}{2}i$

47. $\left[\sqrt{2} \cdot \left(\cos\left(\dfrac{5\pi}{8}\right) + i\sin\left(\dfrac{5\pi}{8}\right)\right)\right]^4 = (\sqrt{2})^4\left(\cos\left(4 \cdot \dfrac{5\pi}{8}\right) + i\sin\left(4 \cdot \dfrac{5\pi}{8}\right)\right)$

$= 4 \cdot \left(\cos\left(\dfrac{5\pi}{2}\right) + i\sin\left(\dfrac{5\pi}{2}\right)\right) = 4(0 + 1\,i) = 4i$

49. $1 - \sqrt{3}\,i \qquad r = \sqrt{1^2 + \left(-\sqrt{3}\right)^2} = 2 \qquad \tan\theta = \dfrac{-\sqrt{3}}{1} = -\sqrt{3} \qquad \theta = 300°$

$1 - \sqrt{3}\,i = 2(\cos(300°) + i\sin(300°))$

$\left(1 - \sqrt{3}\,i\right)^6 = [2(\cos(300°) + i\sin(300°))]^6 = 2^6 \cdot (\cos(6 \cdot 300°) + i\sin(6 \cdot 300°))$

$= 64 \cdot (\cos(1800°) + i\sin(1800°)) = 64 \cdot (\cos(0°) + i\sin(0°)) = 64 + 0\,i = 64$

51. $3 + 4i \qquad r = \sqrt{3^2 + 4^2} = 5 \qquad \tan\theta = \dfrac{4}{3} \qquad \theta \approx 53.1°$

$3 + 4i = 5 \cdot (\cos(53.1°) + i\sin(53.1°))$

$(3 + 4i)^4 = [5^4 \cdot (\cos(4 \cdot 53.1°) + i\sin(4 \cdot 53.1°))]^4 = 625 \cdot (\cos(212.4°) + i\sin(212.4°))$

$\approx 625 \cdot (-0.8443 + i(-0.5358)) = -527.7 - 334.9i$

53. $27 + 0i$ $r = \sqrt{27^2 + 0^2} = 27$ $\tan\theta = \dfrac{0}{27} = 0$ $\theta = 0°$

$27 + 0i = 27\left(\cos(0°) + i\sin(0°)\right)$

The three complex cube roots of $27 = 27(\cos 0° + i\sin 0°)$ are:

$$z_k = \sqrt[3]{27} \cdot \left[\cos\left(\frac{0°}{3} + \frac{360° \cdot k}{3}\right) + i\sin\left(\frac{0°}{3} + \frac{360° \cdot k}{3}\right)\right]$$

$$= 3\left[\cos(120° k) + i\sin(120° \cdot k)\right]$$

$$z_0 = 3\left[\cos(120° \cdot 0) + i\sin(120° \cdot 0)\right] = 3 \cdot \left(\cos(0°) + i\sin(0°)\right) = 3$$

$$z_1 = 3\left[\cos(120° \cdot 1) + i\sin(120° \cdot 1)\right] = 3 \cdot \left(\cos(120°) + i\sin(120°)\right) = -\frac{3}{2} + \frac{3\sqrt{3}}{2}i$$

$$z_2 = 3\left[\cos(120° \cdot 2) + i\sin(120° \cdot 2)\right] = 3 \cdot \left(\cos(240°) + i\sin(240°)\right) = -\frac{3}{2} - \frac{3\sqrt{3}}{2}i$$

55.

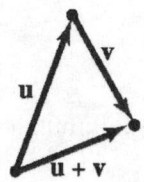

57.

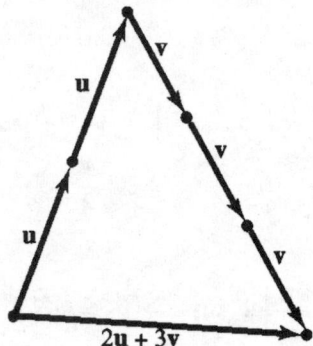

59. $P = (1, -2), Q = (3, -6)$ $\mathbf{v} = (3-1)\mathbf{i} + (-6-(-2))\mathbf{j} = 2\mathbf{i} - 4\mathbf{j}$

$\|\mathbf{v}\| = \sqrt{2^2 + (-4)^2} = \sqrt{20} = 2\sqrt{5}$

61. $P = (0, -2), Q = (-1, 1)$ $\mathbf{v} = (-1-0)\mathbf{i} + (1-(-2))\mathbf{j} = -1\mathbf{i} + 3\mathbf{j}$

$\|\mathbf{v}\| = \sqrt{(-1)^2 + 3^2} = \sqrt{10}$

63. $\mathbf{v} = -2\mathbf{i} + \mathbf{j}, \ \ \mathbf{w} = 4\mathbf{i} - 3\mathbf{j}$

$\mathbf{v} + \mathbf{w} = \left(-2\mathbf{i} + \mathbf{j}\right) + \left(4\mathbf{i} - 3\mathbf{j}\right) = 2\mathbf{i} - 2\mathbf{j}$

65. $\mathbf{v} = -2\mathbf{i} + \mathbf{j}, \ \ \mathbf{w} = 4\mathbf{i} - 3\mathbf{j}$

$4\mathbf{v} - 3\mathbf{w} = 4\left(-2\mathbf{i} + \mathbf{j}\right) - 3\left(4\mathbf{i} - 3\mathbf{j}\right) = -8\mathbf{i} + 4\mathbf{j} - 12\mathbf{i} + 9\mathbf{j} = -20\mathbf{i} + 13\mathbf{j}$

67. $\mathbf{v} = -2\mathbf{i} + \mathbf{j}$

$\|\mathbf{v}\| = \|-2\mathbf{i} + \mathbf{j}\| = \sqrt{(-2)^2 + 1^2} = \sqrt{5}$

69. $\mathbf{v} = -2\mathbf{i} + \mathbf{j}, \ \ \mathbf{w} = 4\mathbf{i} - 3\mathbf{j}$

$\|\mathbf{v}\| + \|\mathbf{w}\| = \|-2\mathbf{i} + \mathbf{j}\| + \|4\mathbf{i} - 3\mathbf{j}\| = \sqrt{(-2)^2 + 1^2} + \sqrt{4^2 + (-3)^2} = \sqrt{5} + 5$

71. $\quad \mathbf{u} = \dfrac{\mathbf{v}}{\|\mathbf{v}\|} = \dfrac{-2\mathbf{i} + \mathbf{j}}{\|-2\mathbf{i} + \mathbf{j}\|} = \dfrac{-2\mathbf{i} + \mathbf{j}}{\sqrt{(-2)^2 + 1^2}} = \dfrac{-2\mathbf{i} + \mathbf{j}}{\sqrt{5}} = -\dfrac{2\sqrt{5}}{5}\mathbf{i} + \dfrac{\sqrt{5}}{5}\mathbf{j}$

73. Let $\mathbf{v} = x \cdot \mathbf{i} + y \cdot \mathbf{j}, \ \|\mathbf{v}\| = 3$, with the angle between $\mathbf{v}$ and $\mathbf{i}$ equal to $60°$.

$\|\mathbf{v}\| = 3 \Rightarrow \sqrt{x^2 + y^2} = 3 \Rightarrow x^2 + y^2 = 9$

The angle between $\mathbf{v}$ and $\mathbf{i}$ equals $60° \Rightarrow \cos(60°) = \dfrac{\mathbf{v} \bullet \mathbf{i}}{\|\mathbf{v}\| \|\mathbf{i}\|} = \dfrac{x}{3 \cdot 1} = \dfrac{x}{3}$.

We also conclude that $\mathbf{v}$ lies in Quadrant I.

$\cos(60°) = \dfrac{x}{3}$

$\dfrac{1}{2} = \dfrac{x}{3} \Rightarrow x = \dfrac{3}{2}$

$x^2 + y^2 = 9 \Rightarrow \left(\dfrac{3}{2}\right)^2 + y^2 = 9$

$y^2 = 9 - \left(\dfrac{3}{2}\right)^2 \Rightarrow y = \pm\sqrt{9 - \left(\dfrac{3}{2}\right)^2} = \pm\sqrt{9 - \dfrac{9}{4}} = \pm\sqrt{\dfrac{36-9}{4}} = \pm\sqrt{\dfrac{27}{4}} = \pm\dfrac{3\sqrt{3}}{2}$

Since $\mathbf{v}$ lies in Quadrant I, $y = \dfrac{3\sqrt{3}}{2}$

So $\mathbf{v} = x \cdot \mathbf{i} + y \cdot \mathbf{j} = \dfrac{3}{2} \cdot \mathbf{i} + \dfrac{3\sqrt{3}}{2} \cdot \mathbf{j}$.

75. $\quad d(P_1, P_2) = \sqrt{(4-1)^2 + (-2-3)^2 + (1-(-2))^2} = \sqrt{9 + 25 + 9} = \sqrt{43}$

77. $\quad \mathbf{v} = (4-1)\mathbf{i} + (-2-3)\mathbf{j} + (1-(-2))\mathbf{k} = 3\mathbf{i} - 5\mathbf{j} + 3\mathbf{k}$

79. $\quad 4\mathbf{v} - 3\mathbf{w} = 4(3\mathbf{i} + \mathbf{j} - 2\mathbf{k}) - 3(-3\mathbf{i} + 2\mathbf{j} - \mathbf{k}) = 12\mathbf{i} + 4\mathbf{j} - 8\mathbf{k} + 9\mathbf{i} - 6\mathbf{j} + 3\mathbf{k}$
$\quad\quad\quad\quad = 21\mathbf{i} - 2\mathbf{j} - 5\mathbf{k}$

81. $\quad \|\mathbf{v} - \mathbf{w}\| = \|(3\mathbf{i} + \mathbf{j} - 2\mathbf{k}) - (-3\mathbf{i} + 2\mathbf{j} - \mathbf{k})\| = \|3\mathbf{i} + \mathbf{j} - 2\mathbf{k} + 3\mathbf{i} - 2\mathbf{j} + \mathbf{k}\|$
$\quad\quad\quad\quad = \|6\mathbf{i} - \mathbf{j} - \mathbf{k}\| = \sqrt{6^2 + (-1)^2 + (-1)^2} = \sqrt{36 + 1 + 1} = \sqrt{38}$

83. $\quad \|\mathbf{v}\| - \|\mathbf{w}\| = \|3\mathbf{i} + \mathbf{j} - 2\mathbf{k}\| - \|-3\mathbf{i} + 2\mathbf{j} - \mathbf{k}\|$
$\quad\quad\quad\quad = \sqrt{3^2 + 1^2 + (-2)^2} - \sqrt{(-3)^2 + 2^2 + (-1)^2} = \sqrt{14} - \sqrt{14} = 0$

85. $\quad \mathbf{v} \times \mathbf{w} = \begin{vmatrix} \mathbf{i} & \mathbf{j} & \mathbf{k} \\ 3 & 1 & -2 \\ -3 & 2 & -1 \end{vmatrix} = \begin{vmatrix} 1 & -2 \\ 2 & -1 \end{vmatrix}\mathbf{i} - \begin{vmatrix} 3 & -2 \\ -3 & -1 \end{vmatrix}\mathbf{j} + \begin{vmatrix} 3 & 1 \\ -3 & 2 \end{vmatrix}\mathbf{k} = 3\mathbf{i} + 9\mathbf{j} + 9\mathbf{k}$

87. Same direction:

$\quad \dfrac{\mathbf{v}}{\|\mathbf{v}\|} = \dfrac{3\mathbf{i} + \mathbf{j} - 2\mathbf{k}}{\sqrt{3^2 + 1^2 + (-2)^2}} = \dfrac{3\mathbf{i} + \mathbf{j} - 2\mathbf{k}}{\sqrt{14}} = \dfrac{3\sqrt{14}}{14}\mathbf{i} + \dfrac{\sqrt{14}}{14}\mathbf{j} - \dfrac{\sqrt{14}}{7}\mathbf{k}$

Opposite direction:
$$\frac{-\mathbf{v}}{\|\mathbf{v}\|} = -\frac{3\sqrt{14}}{14}\mathbf{i} - \frac{\sqrt{14}}{14}\mathbf{j} + \frac{\sqrt{14}}{7}\mathbf{k}$$

89. $\mathbf{v} = -2\mathbf{i} + \mathbf{j}$, $\mathbf{w} = 4\mathbf{i} - 3\mathbf{j}$
$$\mathbf{v} \bullet \mathbf{w} = -2(4) + 1(-3) = -8 - 3 = -11$$
$$\cos\theta = \frac{\mathbf{v} \bullet \mathbf{w}}{\|\mathbf{v}\|\|\mathbf{w}\|} = \frac{-11}{\sqrt{(-2)^2 + 1^2}\sqrt{4^2 + (-3)^2}} = \frac{-11}{\sqrt{5} \cdot 5} = \frac{-11}{5\sqrt{5}} \approx -0.9839$$
$$\theta \approx 169.7°$$

91. $\mathbf{v} = \mathbf{i} - 3\mathbf{j}$, $\mathbf{w} = -\mathbf{i} + \mathbf{j}$
$$\mathbf{v} \bullet \mathbf{w} = 1(-1) + (-3)(1) = -1 - 3 = -4$$
$$\cos\theta = \frac{\mathbf{v} \bullet \mathbf{w}}{\|\mathbf{v}\|\|\mathbf{w}\|} = \frac{-4}{\sqrt{1^2 + (-3)^2}\sqrt{(-1)^2 + 1^2}} = \frac{-4}{\sqrt{10}\sqrt{2}} = \frac{-2}{\sqrt{5}} \approx -0.8944$$
$$\theta \approx 153.4°$$

93. $\mathbf{v} \bullet \mathbf{w} = (\mathbf{i} + \mathbf{j} + \mathbf{k}) \bullet (\mathbf{i} - \mathbf{j} + \mathbf{k}) = 1 \cdot 1 + 1(-1) + 1 \cdot 1 = 1 - 1 + 1 = 1$
$$\cos\theta = \frac{\mathbf{v} \bullet \mathbf{w}}{\|\mathbf{v}\|\|\mathbf{w}\|} = \frac{1}{\sqrt{1^2 + 1^2 + 1^2}\sqrt{1^2 + (-1)^2 + 1^2}} = \frac{1}{\sqrt{3}\sqrt{3}} = \frac{1}{3}$$
$$\theta \approx 70.5°$$

95. $\mathbf{v} \bullet \mathbf{w} = (4\mathbf{i} - \mathbf{j} + 2\mathbf{k}) \bullet (\mathbf{i} - 2\mathbf{j} - 3\mathbf{k}) = 4 \cdot 1 + (-1)(-2) + 2(-3) = 4 + 2 - 6 = 0$
$$\cos\theta = \frac{\mathbf{v} \bullet \mathbf{w}}{\|\mathbf{v}\|\|\mathbf{w}\|} = \frac{0}{\sqrt{4^2 + (-1)^2 + 2^2}\sqrt{1^2 + (-2)^2 + (-3)^2}} = 0$$
$$\theta = 90°$$

97. $\mathbf{v} = 2\mathbf{i} + 3\mathbf{j}$, $\mathbf{w} = -4\mathbf{i} - 6\mathbf{j}$
$$\mathbf{v} \bullet \mathbf{w} = (2)(-4) + (3)(-6) = -8 - 18 = -26$$
$$\cos\theta = \frac{\mathbf{v} \bullet \mathbf{w}}{\|\mathbf{v}\|\|\mathbf{w}\|} = \frac{-26}{\sqrt{2^2 + 3^2}\sqrt{(-4)^2 + (-6)^2}} = \frac{-26}{\sqrt{13} \cdot \sqrt{52}} = \frac{-26}{\sqrt{676}} = -1$$
$$\theta = \cos^{-1}(-1) = 180°$$
Therefore, the vectors are parallel.

99. $\mathbf{v} = 3\mathbf{i} - 4\mathbf{j}$, $\mathbf{w} = -3\mathbf{i} + 4\mathbf{j}$
$$\mathbf{v} \bullet \mathbf{w} = (3)(-3) + (-4)(4) = -9 - 16 = -25$$
$$\cos\theta = \frac{\mathbf{v} \bullet \mathbf{w}}{\|\mathbf{v}\|\|\mathbf{w}\|} = \frac{-25}{\sqrt{3^2 + (-4)^2}\sqrt{(-3)^2 + 4^2}} = \frac{-25}{\sqrt{25} \cdot \sqrt{25}} = \frac{-25}{25} = -1$$
$$\theta = \cos^{-1}(-1) = 180°$$
Therefore, the vectors are parallel.

101. $\mathbf{v}=3\mathbf{i}-2\mathbf{j},\quad \mathbf{w}=4\mathbf{i}+6\mathbf{j}$
$\mathbf{v}\bullet\mathbf{w}=(3)(4)+(-2)(6)=12-12=0$
Therefore, the vectors are perpendicular.

103. $\mathbf{v}=2\mathbf{i}+\mathbf{j},\quad \mathbf{w}=-4\mathbf{i}+3\mathbf{j}$
The decomposition of $\mathbf{v}$ into 2 vectors $\mathbf{v}_1$ and $\mathbf{v}_2$ so that $\mathbf{v}_1$ is parallel to $\mathbf{w}$ and $\mathbf{v}_2$ is perpendicular to $\mathbf{w}$ is given by:
$$\mathbf{v}_1=\text{proj}_\mathbf{w}\mathbf{v}=\frac{\mathbf{v}\bullet\mathbf{w}}{\|\mathbf{w}\|^2}\mathbf{w}\quad\text{and}\quad \mathbf{v}_2=\mathbf{v}-\mathbf{v}_1$$
$$\mathbf{v}_1=\text{proj}_\mathbf{w}\mathbf{v}=\frac{\mathbf{v}\bullet\mathbf{w}}{\|\mathbf{w}\|^2}\mathbf{w}=\frac{(2\mathbf{i}+\mathbf{j})\bullet(-4\mathbf{i}+3\mathbf{j})}{\left(\sqrt{(-4)^2+3^2}\right)^2}(-4\mathbf{i}+3\mathbf{j})=\frac{(2)(-4)+(1)(3)}{25}(-4\mathbf{i}+3\mathbf{j})$$
$$=\left(-\frac{1}{5}\right)(-4\mathbf{i}+3\mathbf{j})=\frac{4}{5}\mathbf{i}-\frac{3}{5}\mathbf{j}$$
$$\mathbf{v}_2=\mathbf{v}-\mathbf{v}_1=2\mathbf{i}+\mathbf{j}-\left(\frac{4}{5}\mathbf{i}-\frac{3}{5}\mathbf{j}\right)=\frac{6}{5}\mathbf{i}+\frac{8}{5}\mathbf{j}$$

105. $\mathbf{v}=2\mathbf{i}+3\mathbf{j},\quad \mathbf{w}=3\mathbf{i}+\mathbf{j}$
The projection of $\mathbf{v}$ onto $\mathbf{w}$ is given by:
$$\text{proj}_\mathbf{w}\mathbf{v}=\frac{\mathbf{v}\bullet\mathbf{w}}{\|\mathbf{w}\|^2}\mathbf{w}=\frac{(2\mathbf{i}+3\mathbf{j})\bullet(3\mathbf{i}+\mathbf{j})}{\left(\sqrt{3^2+1^2}\right)^2}(3\mathbf{i}+\mathbf{j})=\frac{(2)(3)+(3)(1)}{10}(3\mathbf{i}+\mathbf{j})=\left(\frac{9}{10}\right)(3\mathbf{i}+\mathbf{j})$$

107. $\cos\alpha=\dfrac{a}{\|\mathbf{v}\|}=\dfrac{3}{\sqrt{3^2+(-4)^2+2^2}}=\dfrac{3}{\sqrt{29}}\ \rightarrow\ \alpha\approx56.1°$
$\cos\beta=\dfrac{b}{\|\mathbf{v}\|}=\dfrac{-4}{\sqrt{3^2+(-4)^2+2^2}}=-\dfrac{4}{\sqrt{29}}\ \rightarrow\ \beta\approx138.0°$
$\cos\gamma=\dfrac{c}{\|\mathbf{v}\|}=\dfrac{2}{\sqrt{3^2+(-4)^2+2^2}}=\dfrac{2}{\sqrt{29}}\ \rightarrow\ \gamma\approx68.2°$

109. $\mathbf{u}=P_1P_2=1\mathbf{i}+2\mathbf{j}+3\mathbf{k}\qquad \mathbf{v}=P_1P_3=5\mathbf{i}+4\mathbf{j}+1\mathbf{k}$
$$\mathbf{u}\times\mathbf{v}=\begin{vmatrix}\mathbf{i}&\mathbf{j}&\mathbf{k}\\1&2&3\\5&4&1\end{vmatrix}=\begin{vmatrix}2&3\\4&1\end{vmatrix}\mathbf{i}-\begin{vmatrix}1&3\\5&1\end{vmatrix}\mathbf{j}+\begin{vmatrix}1&2\\5&4\end{vmatrix}\mathbf{k}=-10\mathbf{i}+14\mathbf{j}-6\mathbf{k}$$
Area $=\|\mathbf{u}\times\mathbf{v}\|=\sqrt{(-10)^2+14^2+(-6)^2}=\sqrt{332}\approx18.2$

111. $\mathbf{v}\times\mathbf{u}=-(\mathbf{u}\times\mathbf{v})=-(2\mathbf{i}-3\mathbf{j}+\mathbf{k})=-2\mathbf{i}+3\mathbf{j}-\mathbf{k}$

113. Let the positive x-axis point downstream, so that the velocity of the current is $\mathbf{v}_c=2\mathbf{i}$.
Let $\mathbf{v}_w=$ the velocity of the swimmer in the water.
Let $\mathbf{v}_g=$ the velocity of the swimmer relative to the land.
Then $\mathbf{v}_g=\mathbf{v}_w+\mathbf{v}_c$

The speed of the swimmer is $\|\mathbf{v}_w\| = 5$; its direction is directly across the river, so

Let $\mathbf{v}_w = 5\mathbf{j}$.

Let $\mathbf{v}_g = \mathbf{v}_w + \mathbf{v}_c = 5\mathbf{j} + 2\mathbf{i} = 2\mathbf{i} + 5\mathbf{j}$

$$\Rightarrow \|\mathbf{v}_g\| = \sqrt{2^2 + 5^2} = \sqrt{29} \approx 5.4 \text{ miles per hour}$$

Since the river is 1 mile wide, it takes the swimmer 0.2 hours to cross the river. The swimmer will end up $(0.2)(2) = 0.4$ miles downstream.

115. Let $\mathbf{F}_1$ be the tension on the left cable and $\mathbf{F}_2$ be the tension on the right cable.
Let $\mathbf{F}_3$ represent the force of the weight of the box.

$$\mathbf{F}_1 = \|\mathbf{F}_1\|(\cos(140°)\mathbf{i} + \sin(140°)\mathbf{j}) \approx \|\mathbf{F}_1\|(-0.7660\mathbf{i} + 0.6428\mathbf{j})$$

$$\mathbf{F}_2 = \|\mathbf{F}_2\|(\cos(30°)\mathbf{i} + \sin(30°)\mathbf{j}) \approx \|\mathbf{F}_2\|(0.8660\mathbf{i} + 0.5000\mathbf{j})$$

$$\mathbf{F}_3 = -2000\mathbf{j}$$

For equilibrium, the sum of the force vectors must be zero.

$$\mathbf{F}_1 + \mathbf{F}_2 + \mathbf{F}_3 = -0.7660\|\mathbf{F}_1\|\mathbf{i} + 0.6428\|\mathbf{F}_1\|\mathbf{j} + 0.8660\|\mathbf{F}_2\|\mathbf{i} + 0.5000\|\mathbf{F}_2\|\mathbf{j} - 2000\mathbf{j}$$

$$= \left(-0.7660\|\mathbf{F}_1\| + 0.8660\|\mathbf{F}_2\|\right)\mathbf{i} + \left(0.6428\|\mathbf{F}_1\| + 0.5000\|\mathbf{F}_2\| - 2000\right)\mathbf{j}$$

$$= 0$$

Set the $\mathbf{i}$ and $\mathbf{j}$ components equal to zero and solve:

$$\begin{cases} -0.7660\|\mathbf{F}_1\| + 0.8660\|\mathbf{F}_2\| = 0 \quad \rightarrow \quad \|\mathbf{F}_2\| = \dfrac{0.7660}{0.8660}\|\mathbf{F}_1\| = 0.8845\|\mathbf{F}_1\| \\ 0.6428\|\mathbf{F}_1\| + 0.5000\|\mathbf{F}_2\| - 2000 = 0 \end{cases}$$

$$0.6428\|\mathbf{F}_1\| + 0.5000\left(0.8845\|\mathbf{F}_1\|\right) - 2000 = 0$$

$$1.0851\|\mathbf{F}_1\| = 2000 \rightarrow \|\mathbf{F}_1\| \approx 1843 \text{ pounds}; \quad \|\mathbf{F}_2\| = 0.8845(1843) = 1630 \text{ pounds}$$

The tension in the left cable is about 1843 pounds and the tension in the right cable is about 1630 pounds.

117. $\mathbf{F} = 5(\cos(60°)\mathbf{i} + \sin(60°)\mathbf{j}) = 5\left(\dfrac{1}{2}\mathbf{i} + \dfrac{\sqrt{3}}{2}\mathbf{j}\right) = \dfrac{5}{2}\mathbf{i} + \dfrac{5\sqrt{3}}{2}\mathbf{j}$

$\overrightarrow{AB} = 20\mathbf{i}$

$W = \mathbf{F} \bullet \overrightarrow{AB} = \left(\dfrac{5}{2}\mathbf{i} + \dfrac{5\sqrt{3}}{2}\mathbf{j}\right) \bullet 20\mathbf{i} = \left(\dfrac{5}{2}\right)(20) + \left(\dfrac{5\sqrt{3}}{2}\right)(0) = 50 \text{ foot - pounds}$

Polar Coordinates; Vectors

8.CR   Cumulative Review

1. $e^{x^2-9} = 1$

$\ln\left(e^{x^2-9}\right) = \ln(1)$

$x^2 - 9 = 0$

$(x+3)(x-3) = 0$

$x = -3$ or $x = 3$

The solution set is $\{-3,3\}$.

3. center point $(0,1)$; radius $= 3$

$(x-h)^2 + (y-k)^2 = r^2$

$(x-0)^2 + (y-1)^2 = 3^2$

$x^2 + y^2 - 2y + 1 = 9$

$x^2 + y^2 - 2y - 8 = 0$

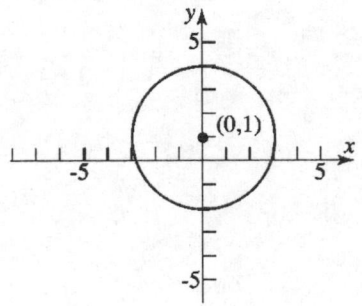

5. $x^2 + y^3 = 2x^4$

Test for symmetry:

x-axis :   Replace y by $-y$ so $x^2 + (-y)^3 = 2x^4 \Rightarrow x^2 - y^3 = 2x^4$

which is not equivalent to $x^2 + y^3 = 2x^4$.

y-axis :   Replace x by $-x$ so $(-x)^2 + y^3 = 2(-x)^4 \Rightarrow x^2 + y^3 = 2x^4$,

which is equivalent to $x^2 + y^3 = 2x^4$.

Origin :   Replace x by $-x$ and y by $-y$ so $(-x)^2 + (-y)^3 = 2(-x)^4 \Rightarrow x^2 - y^3 = 2x^4$,

which is not equivalent to $x^2 + y^3 = 2x^4$.

Therefore, the graph is symmetric with respect to the y-axis.

7. $y = \left| \sin x \right| = \begin{cases} \sin x, & \text{when } \sin x \geq 0 \\ -\sin x, & \text{when } \sin x < 0 \end{cases}$

$= \begin{cases} \sin x, & \text{when } 0 \leq x \leq \pi \\ -\sin x, & \text{when } \pi < x < 2\pi \end{cases}$

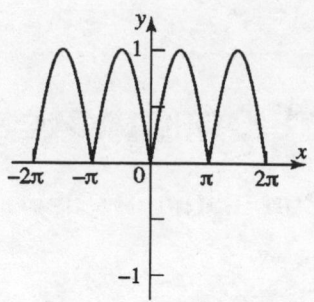

9. $\sin^{-1}\left(-\dfrac{1}{2}\right)$

We are finding the angle θ, $-\dfrac{\pi}{2} \leq \theta \leq \dfrac{\pi}{2}$, whose sine equals $-\dfrac{1}{2}$.

$\sin\theta = -\dfrac{1}{2} \qquad -\dfrac{\pi}{2} \leq \theta \leq \dfrac{\pi}{2}$

$\theta = -\dfrac{\pi}{6} \rightarrow \sin^{-1}\left(-\dfrac{1}{2}\right) = -\dfrac{\pi}{6}$

11. graphing $r = 2$ and $\theta = \dfrac{\pi}{3}$ using polar coordinates:

$r = 2$ yields a circle, centered at $(0,0)$, with radius $= 2$.

$\theta = \dfrac{\pi}{3}$ yields a line passing through the point $(0,0)$, forming an angle of $\theta = \dfrac{\pi}{3}$ with the positive x-axis.

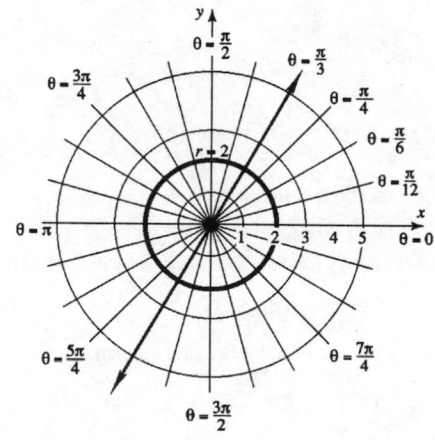

Analytic Geometry
9.2 The Parabola

1. B   3. E   5. H   7. C

9. E   11. D   13. C

15. The focus is (4, 0) and the vertex is (0, 0). Both lie
     on the horizontal line $y = 0$. $a = 4$ and since (4, 0)
     is to the right of (0, 0), the parabola opens to the
     right. The equation of the parabola is:
     $$y^2 = 4ax$$
     $$y^2 = 4 \cdot 4 \cdot x$$
     $$y^2 = 16x$$
     Letting $x = 4$, we find $y^2 = 64$ or $y = \pm 8$.
     The points (4, 8) and (4, –8) define the latus rectum.

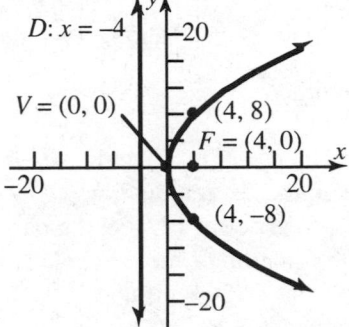

17. The focus is (0, –3) and the vertex is (0, 0). Both lie
     on the vertical line $x = 0$. $a = 3$ and since (0, –3) is
     below (0, 0), the parabola opens down. The
     equation of the parabola is:
     $$x^2 = -4ay$$
     $$x^2 = -4 \cdot 3 \cdot y$$
     $$x^2 = -12y$$
     Letting $y = -3$, we find $x^2 = 36$ or $x = \pm 6$.
     The points (6, 3) and (6, –3) define the latus rectum.

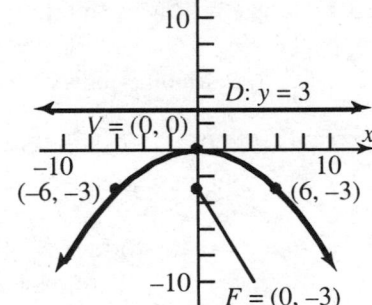

19. The focus is (–2, 0) and the directrix is $x = 2$. The
     vertex is (0, 0). $a = 2$ and since (–2, 0) is to the left
     of (0, 0), the parabola opens to the left. The
     equation of the parabola is:
     $$y^2 = -4ax$$
     $$y^2 = -4 \cdot 2 \cdot x$$
     $$y^2 = -8x$$
     Letting $x = -2$, we find $y^2 = 16$ or $y = \pm 4$. The
     points (–2, 4) and (–2, –4) define the latus rectum.

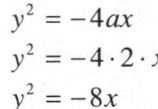

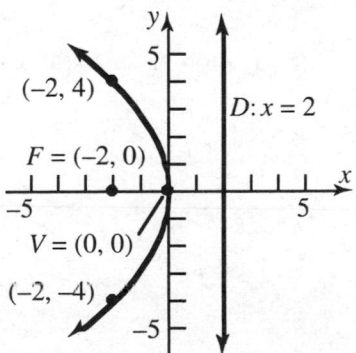

21. The directrix is $y = -\dfrac{1}{2}$ and the vertex is $(0, 0)$. The focus is $\left(0, \dfrac{1}{2}\right)$. $a = \dfrac{1}{2}$ and since $\left(0, \dfrac{1}{2}\right)$ is above $(0, 0)$, the parabola opens up. The equation of the parabola is:

$$x^2 = 4ay$$

$$x^2 = 4 \cdot \frac{1}{2} \cdot y \Rightarrow x^2 = 2y$$

Letting $y = \dfrac{1}{2}$, we find $x^2 = 1$ or $x = \pm 1$.

The points $\left(1, \dfrac{1}{2}\right)$ and $\left(-1, \dfrac{1}{2}\right)$ define the latus rectum.

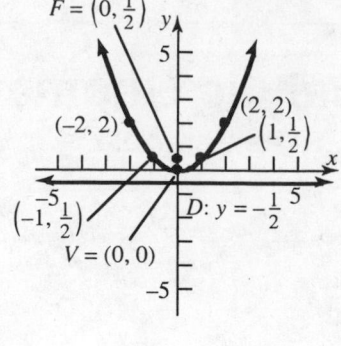

23. The focus is $(2, -5)$ and the vertex is $(2, -3)$. Both lie on the vertical line $x = 2$. $a = 2$ and since $(2, -5)$ is below $(2, -3)$, the parabola opens down. The equation of the parabola is:

$$(x - h)^2 = -4a(y - k)$$
$$(x - 2)^2 = -4 \cdot 2 \cdot (y - (-3))$$
$$(x - 2)^2 = -8(y + 3)$$

Letting $y = -5$, we find $(x - 2)^2 = 16$ or $x - 2 = \pm 4$. So, $x = 6$ or $x = -2$. The points $(6, -5)$ and $(-2, -5)$ define the latus rectum.

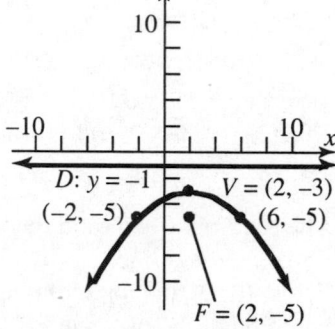

25. Vertex: $(0,0)$. Since the axis of symmetry is vertical, the parabola opens up or down. Since $(2, 3)$ is above $(0, 0)$, the parabola opens up. The equation has the form $x^2 = 4ay$. Substitute the coordinates of $(2, 3)$ into the equation to find a:

$$2^2 = 4a \cdot 3 \Rightarrow 4 = 12a \Rightarrow a = \frac{1}{3}$$

The equation of the parabola is: $x^2 = \dfrac{4}{3}y$. The focus is $\left(0, \dfrac{1}{3}\right)$. Letting $y = \dfrac{1}{3}$,

we find $x^2 = \dfrac{4}{9}$ or $x = \pm\dfrac{2}{3}$. The points $\left(\dfrac{2}{3}, \dfrac{1}{3}\right)$ and $\left(-\dfrac{2}{3}, \dfrac{1}{3}\right)$ define the latus rectum.

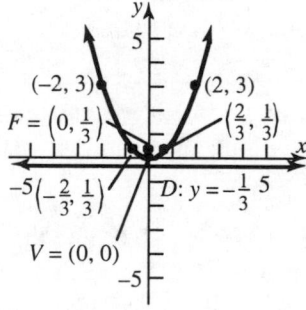

27. The directrix is $y = 2$ and the focus is (–3, 4). This is a vertical case, so the vertex is (–3, 3). $a = 1$ and since (–3, 4) is above $y = 2$, the parabola opens up. The equation of the parabola is: $(x-h)^2 = 4a(y-k)$

$$(x-(-3))^2 = 4\cdot 1\cdot (y-3)$$

$$(x+3)^2 = 4(y-3)$$

Letting $y = 4$, we find $(x+3)^2 = 4$ or $x+3 = \pm 2$. So, $x = -1$ or $x = -5$. The points (–1, 4) and (–5, 4) define the latus rectum.

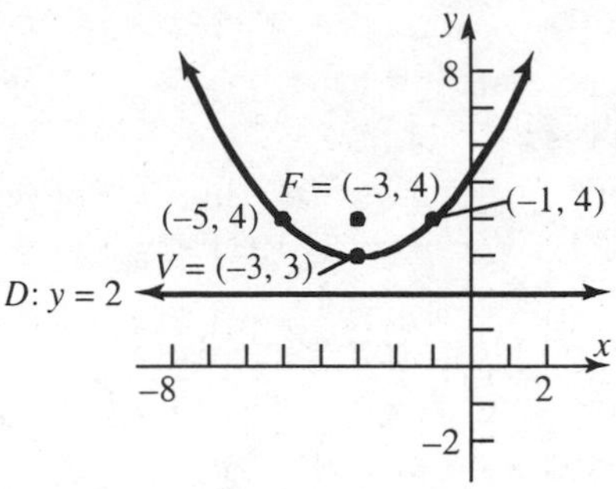

29. The directrix is $x = 1$ and the focus is (–3, –2). This is a horizontal case, so the vertex is (–1, –2). $a = 2$ and since (–3, –2) is to the left of $x = 1$, the parabola opens to the left. The equation of the parabola is: $(y-k)^2 = -4a(x-h)$

$$(y-(-2))^2 = -4\cdot 2\cdot (x-(-1))$$

$$(y+2)^2 = -8(x+1)$$

Letting $x = -3$, we find $(y+2)^2 = 16$ or $y+2 = \pm 4$. So, $y = 2$ or $y = -6$. The points (–3, 2) and (–3, –6) define the latus rectum.

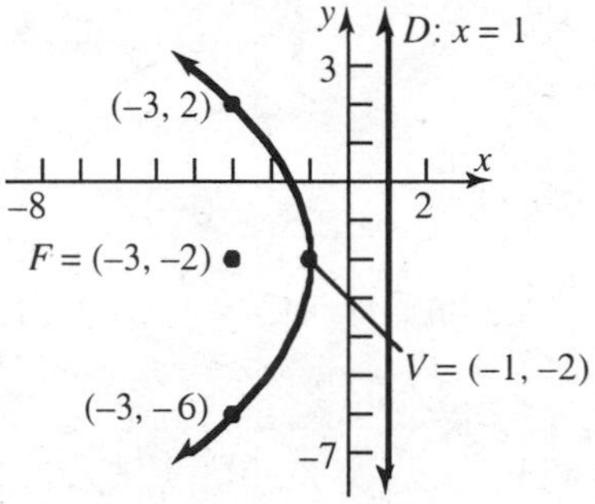

31. The equation $x^2 = 4y$ is in the form $x^2 = 4ay$ where $4a = 4$ or $a = 1$. Thus, we have:

Vertex: (0, 0)
Focus: (0, 1)
Directrix: $y = -1$

To graph, enter: $y_1 = x^2 / 4$

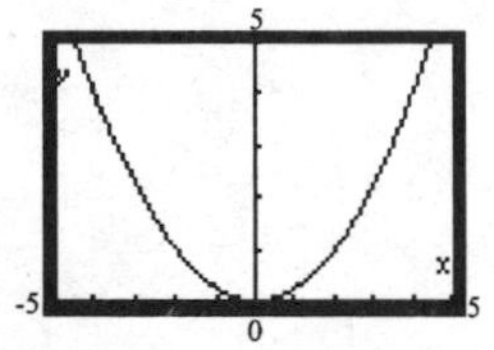

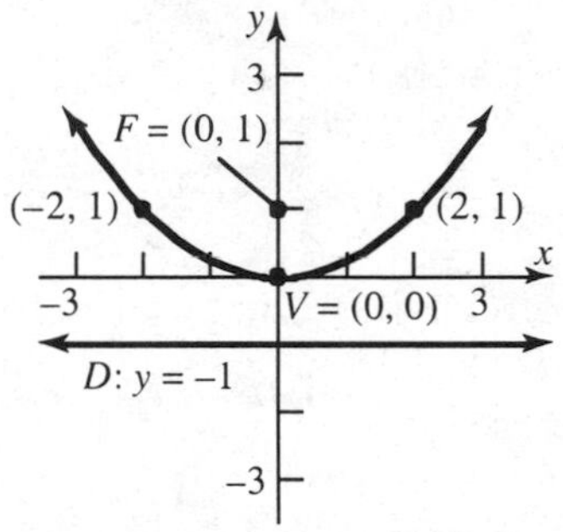

33. The equation $y^2 = -16x$ is in the form $y^2 = -4ax$ where $-4a = -16$ or $a = 4$. Thus, we have:

Vertex: (0, 0)
Focus: (–4, 0)
Directrix: $x = 4$

To graph, enter: $y_1 = \sqrt{-16x}$; $y_2 = -\sqrt{-16x}$

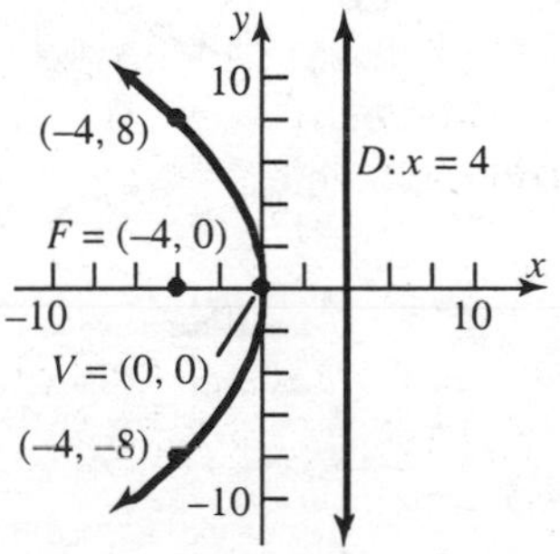

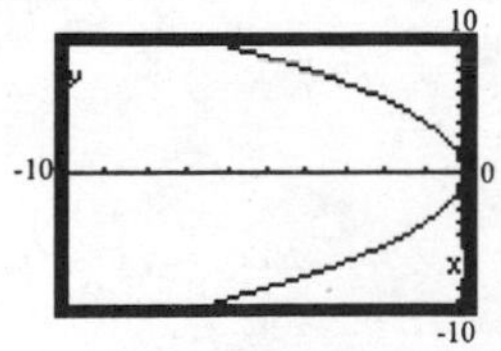

35. The equation $(y-2)^2 = 8(x+1)$ is in the form
$(y-k)^2 = 4a(x-h)$ where $4a = 8$ or $a = 2$,
$h = -1$, and $k = 2$. Thus, we have:

       Vertex: $(-1, 2)$

       Focus: $(1, 2)$

       Directrix: $x = -3$

To graph, enter: $y_1 = 2 + \sqrt{8(x+1)}$;

      $y_2 = 2 - \sqrt{8(x+1)}$

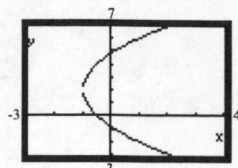

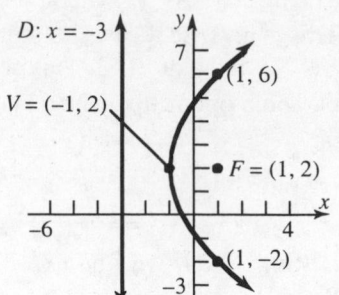

37. The equation $(x-3)^2 = -(y+1)$ is in the form

$(x-h)^2 = -4a(y-k)$ where $-4a = -1$ or $a = \dfrac{1}{4}$,

$h = 3$, and $k = -1$. Thus, we have:

       Vertex: $(3, -1)$

       Focus: $\left(3, -\dfrac{5}{4}\right)$

       Directrix: $y = -\dfrac{3}{4}$

To graph, enter: $y_1 = -1 - (x-3)^2$

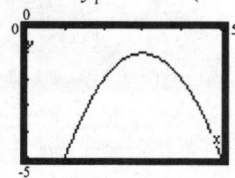

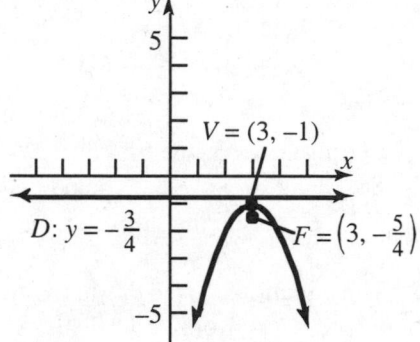

39. The equation $(y+3)^2 = 8(x-2)$ is in the form
$(y-k)^2 = 4a(x-h)$ where $4a = 8$ or $a = 2$,
$h = 2$, and $k = -3$. Thus, we have:

       Vertex: $(2, -3)$

       Focus: $(4, -3)$

       Directrix: $x = 0$

To graph, enter:

    $y_1 = -3 + \sqrt{8(x-2)}$; $y_2 = -3 - \sqrt{8(x-2)}$

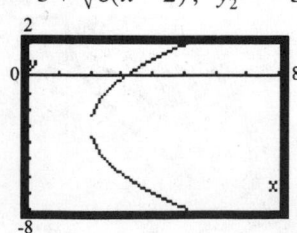

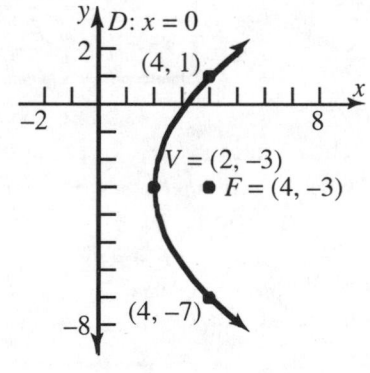

41. Complete the square to put in standard form:
$$y^2 - 4y + 4x + 4 = 0$$
$$y^2 - 4y + 4 = -4x$$
$$(y - 2)^2 = -4x$$

The equation is in the form $(y - k)^2 = -4a(x - h)$ where $-4a = -4$ or $a = 1$, $h = 0$, and $k = 2$.

Thus, we have:

   Vertex: $(0, 2)$
   Focus: $(-1, 2)$
   Directrix: $x = 1$

To graph, enter: $y_1 = 2 + \sqrt{-4x}$; $y_2 = 2 - \sqrt{-4x}$

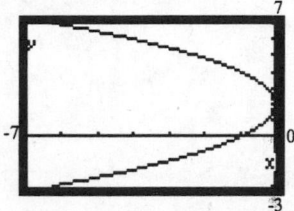

43. Complete the square to put in standard form:
$$x^2 + 8x = 4y - 8$$
$$x^2 + 8x + 16 = 4y - 8 + 16$$
$$(x + 4)^2 = 4(y + 2)$$

The equation is in the form $(x - h)^2 = 4a(y - k)$ where $4a = 4$ or $a = 1$, $h = -4$, and $k = -2$.

Thus, we have:

   Vertex: $(-4, -2)$
   Focus: $(-4, -1)$
   Directrix: $y = -3$

To graph, enter: $y_1 = -2 + (x + 4)^2 / 4$

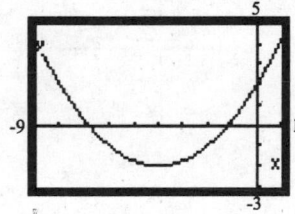

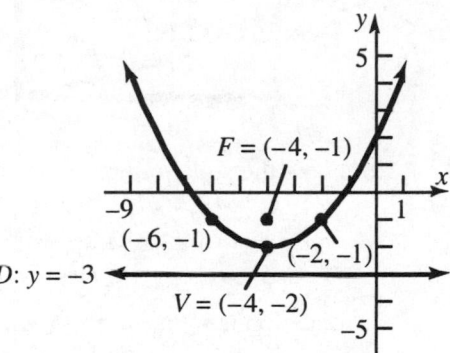

45.   Complete the square to put in standard form:

$$y^2 + 2y - x = 0$$
$$y^2 + 2y + 1 = x + 1$$
$$(y + 1)^2 = x + 1$$

The equation is in the form

$(y - k)^2 = 4a(x - h)$ where $4a = 1$ or $a = \dfrac{1}{4}$,

$h = -1$, and $k = -1$.

Thus, we have:

    Vertex: $(-1, -1)$

    Focus: $\left(-\dfrac{3}{4}, -1\right)$

    Directrix: $x = -\dfrac{5}{4}$

To graph, enter:

$$y_1 = -1 + \sqrt{x + 1}; \quad y_2 = -1 - \sqrt{x + 1}$$

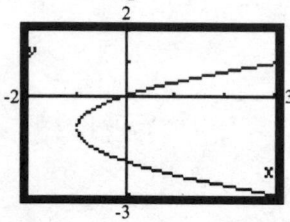

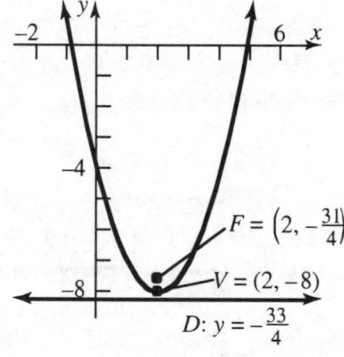

47.   Complete the square to put in standard form:

$$x^2 - 4x = y + 4$$
$$x^2 - 4x + 4 = y + 4 + 4$$
$$(x - 2)^2 = y + 8$$

The equation is in the form $(x - h)^2 = 4a(y - k)$

where $4a = 1$ or $a = \dfrac{1}{4}$, $h = 2$, and $k = -8$. Thus,

we have:

    Vertex: $(2, -8)$

    Focus: $\left(2, -\dfrac{31}{4}\right)$

    Directrix: $y = -\dfrac{33}{4}$

To graph, enter: $y_1 = -8 + (x - 2)^2$

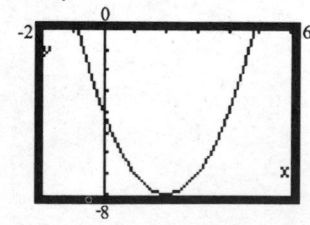

49. $(y-1)^2 = c(x-0)$
     $(y-1)^2 = cx$
     $(2-1)^2 = c(1) \Rightarrow 1 = c$
     $(y-1)^2 = x$

51. $(y-1)^2 = c(x-2)$
     $(0-1)^2 = c(1-2)$
     $1 = -c \Rightarrow c = -1$
     $(y-1)^2 = -(x-2)$

53. $(x-0)^2 = c(y-1)$
     $(2-0)^2 = c(2-1) \Rightarrow 4 = c$
     $x^2 = 4(y-1)$

55. $(y-0)^2 = c(x-(-2))$
     $y^2 = c(x+2)$
     $1^2 = c(0+2) \Rightarrow 1 = 2c \Rightarrow c = \dfrac{1}{2}$
     $y^2 = \dfrac{1}{2}(x+2)$

57. Set up the problem so that the vertex of the parabola is at (0, 0) and it opens up. Then the equation of the parabola has the form: $x^2 = 4ay$. Since the parabola is 10 feet across and 4 feet deep, the points (5, 4) and (–5, 4) are on the parabola.
     Substitute and solve for a: $5^2 = 4a(4) \Rightarrow 25 = 16a \Rightarrow a = \dfrac{25}{16}$
     a is the distance from the vertex to the focus. Thus, the receiver (located at the focus) is $\dfrac{25}{16} = 1.5625$ feet, or 18.75 inches from the base of the dish, along the axis of the parabola.

59. Set up the problem so that the vertex of the parabola is at (0, 0) and it opens up. Then the equation of the parabola has the form: $x^2 = 4ay$. Since the parabola is 4 inches across and 1 inch deep, the points (2, 1) and (–2, 1) are on the parabola.
     Substitute and solve for a: $2^2 = 4a(1) \Rightarrow 4 = 4a \Rightarrow a = 1$, a is the distance from the vertex to the focus. Thus, the bulb (located at the focus) should be 1 inch, from the vertex.

61. Set up the problem so that the vertex of the parabola is at (0, 0) and it opens up. Then the equation of the parabola has the form: $x^2 = cy$.
     The point (300, 80) is a point on the parabola.
     Solve for c and find the equation:
     $300^2 = c(80) \to c = 1125$
     $x^2 = 1125y$

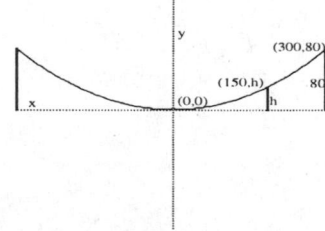

     Since the height of the cable, 150 feet from the center, is to be found, the point (150, h) is a point on the parabola. Solve for h: $150^2 = 1125h \Rightarrow 22500 = 1125h \Rightarrow h = 20$
     The height of the cable, 150 feet from the center, is 20 feet.

63. Set up the problem so that the vertex of the parabola is at (0, 0) and it opens up. Then the equation of the parabola has the form: $x^2 = 4ay$. a is the distance from the vertex to the focus (where the source is located), so $a = 2$. Since the opening is 5 feet across, there is a point (2.5, y) on the parabola.
     Solve for y: $x^2 = 8y \Rightarrow 2.5^2 = 8y \Rightarrow 6.25 = 8y \Rightarrow y = 0.78125$ feet
     The depth of the searchlight should be 0.78125 feet.

65. Set up the problem so that the vertex of the parabola is at (0, 0) and it opens up. Then the equation of the parabola has the form: $x^2 = 4ay$. Since the parabola is 20 feet across and 6 feet deep, the points (10, 6) and (–10, 6) are on the parabola.
Substitute and solve for a: $10^2 = 4a(6) \Rightarrow 100 = 24a \Rightarrow a \approx 4.17$ feet
The heat source will be concentrated 4.17 feet from the base, along the axis of symmetry.

67. Set up the problem so that the vertex of the parabola is at (0, 0) and it opens down. Then the equation of the parabola has the form: $x^2 = cy$.
The point (60, –25) is a point on the parabola.
Solve for c and find the equation:

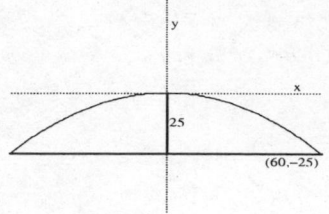

$$60^2 = c(-25) \Rightarrow c = -144$$
$$x^2 = -144y$$

To find the height of the bridge, 10 feet from the center, the point (10, y) is a point on the parabola. Solve for y: $10^2 = -144y \Rightarrow 100 = -144y \Rightarrow y = -0.69$
The height of the bridge, 10 feet from the center, is 25 – 0.69 = 24.31 feet.
To find the height of the bridge, 30 feet from the center, the point (30, y) is a point on the parabola. Solve for y: $30^2 = -144y \Rightarrow 900 = -144y \Rightarrow y = -6.25$
The height of the bridge, 30 feet from the center, is 25 – 6.25 = 18.75 feet.
To find the height of the bridge, 50 feet from the center, the point (50, y) is a point on the parabola. Solve for y: $50^2 = -144y \Rightarrow 2500 = -144y \Rightarrow y = -17.36$
The height of the bridge, 50 feet from the center, is 25 – 17.36 = 7.64 feet.

69. $Ax^2 + Ey = 0 \quad A \neq 0, \ E \neq 0$
$$Ax^2 = -Ey \Rightarrow x^2 = \frac{-E}{A}y$$

This is the equation of a parabola with vertex at (0, 0) and axis of symmetry being the y-axis. The focus is $\left(0, \frac{-E}{4A}\right)$. The directrix is $y = \frac{E}{4A}$.

71. $Ax^2 + Dx + Ey + F = 0 \quad A \neq 0$
(a) If $E \neq 0$, then:
$$Ax^2 + Dx = -Ey - F \Rightarrow A\left(x^2 + \frac{D}{A}x + \frac{D^2}{4A^2}\right) = -Ey - F + \frac{D^2}{4A}$$
$$\left(x + \frac{D}{2A}\right)^2 = \frac{1}{A}\left(-Ey - F + \frac{D^2}{4A}\right) \Rightarrow \left(x + \frac{D}{2A}\right)^2 = \frac{-E}{A}\left(y + \frac{F}{E} - \frac{D^2}{4AE}\right)$$
$$\left(x + \frac{D}{2A}\right)^2 = \frac{-E}{A}\left(y - \frac{D^2 - 4AF}{4AE}\right)$$

This is the equation of a parabola whose vertex is $\left(\frac{-D}{2A}, \frac{D^2 - 4AF}{4AE}\right)$.

(b)   If $E = 0$, then

$$Ax^2 + Dx + F = 0 \Rightarrow x = \frac{-D \pm \sqrt{D^2 - 4AF}}{2A}$$

If $D^2 - 4AF = 0$, then $x = \dfrac{-D}{2A}$ is a vertical line.

(c)   If $E = 0$, then

$$Ax^2 + Dx + F = 0 \Rightarrow x = \frac{-D \pm \sqrt{D^2 - 4AF}}{2A}$$

If $D^2 - 4AF > 0$,

then $x = \dfrac{-D + \sqrt{D^2 - 4AF}}{2A}$ or $x = \dfrac{-D - \sqrt{D^2 - 4AF}}{2A}$ are two vertical lines.

(d)   If $E = 0$, then

$$Ax^2 + Dx + F = 0 \Rightarrow x = \frac{-D \pm \sqrt{D^2 - 4AF}}{2A}$$

If $D^2 - 4AF < 0$, there is no real solution. The graph contains no points.

Analytic Geometry
9.3   The Ellipse

1.   C   3.   B   5.   C   7.   D

9. $\dfrac{x^2}{25}+\dfrac{y^2}{4}=1$

The center of the ellipse is at the origin.
$a=5$, $b=2$. The vertices are $(5,0)$ and
$(-5,0)$. Find the value of c:
$$c^2=a^2-b^2=25-4=21\Rightarrow c=\sqrt{21}$$
The foci are $\left(\sqrt{21},0\right)$ and $\left(-\sqrt{21},0\right)$.
To graph, enter:
$$y_1=2\sqrt{(1-x^2/25)};\quad y_1=-2\sqrt{(1-x^2/25)}$$

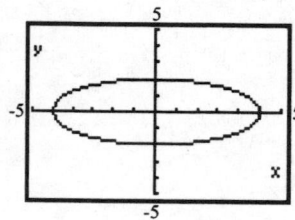

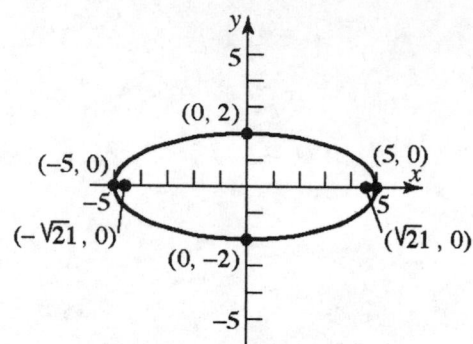

11. $\dfrac{x^2}{9}+\dfrac{y^2}{25}=1$

The center of the ellipse is at the origin.
$a=5$, $b=3$. The vertices are $(0,5)$ and $(0,-5)$.
Find the value of c:
$$c^2=a^2-b^2=25-9=16$$
$$c=4$$
The foci are $(0,4)$ and $(0,-4)$.
To graph, enter:
$$y_1=5\sqrt{(1-x^2/9)};\quad y_1=-5\sqrt{(1-x^2/9)}$$

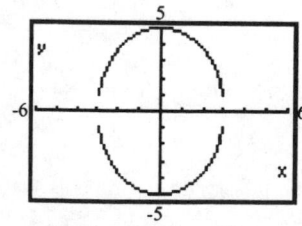

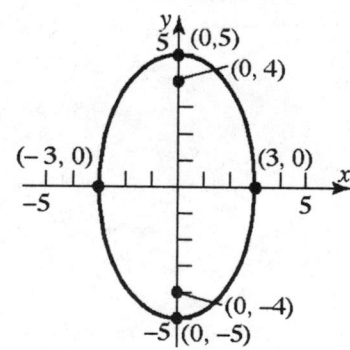

13. $4x^2 + y^2 = 16$

Divide by 16 to put in standard form:

$$\frac{4x^2}{16} + \frac{y^2}{16} = \frac{16}{16} \quad \Rightarrow \quad \frac{x^2}{4} + \frac{y^2}{16} = 1$$

The center of the ellipse is at the origin. $a = 4$, $b = 2$.

The vertices are $(0, 4)$ and $(0, -4)$. Find the value of c:

$$c^2 = a^2 - b^2 = 16 - 4 = 12$$

$$c = \sqrt{12} = 2\sqrt{3}$$

The foci are $\left(0, 2\sqrt{3}\right)$ and $\left(0, -2\sqrt{3}\right)$.

To graph, enter:

$$y_1 = \sqrt{(16 - 4x^2)}; \quad y_1 = -\sqrt{(16 - 4x^2)}$$

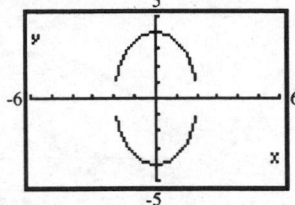

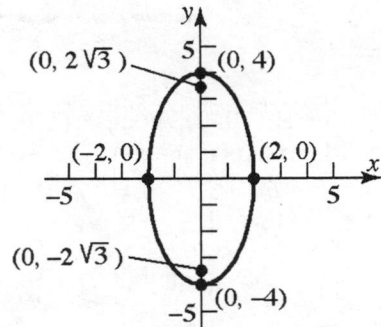

15. $4y^2 + x^2 = 8$

Divide by 8 to put in standard form:

$$\frac{4y^2}{8} + \frac{x^2}{8} = \frac{8}{8} \quad \Rightarrow \quad \frac{x^2}{8} + \frac{y^2}{2} = 1$$

The center of the ellipse is at the origin. $a = \sqrt{8} = 2\sqrt{2}$, $b = \sqrt{2}$.

The vertices are $\left(2\sqrt{2}, 0\right)$ and $\left(-2\sqrt{2}, 0\right)$. Find the value of c:

$$c^2 = a^2 - b^2 = 8 - 2 = 6$$

$$c = \sqrt{6}$$

The foci are $\left(\sqrt{6}, 0\right)$ and $\left(-\sqrt{6}, 0\right)$.

To graph, enter:

$$y_1 = \sqrt{(2 - x^2/4)}; \quad y_1 = -\sqrt{(2 - x^2/4)}$$

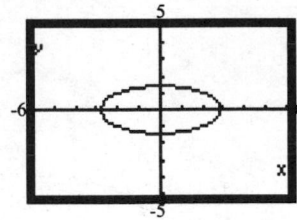

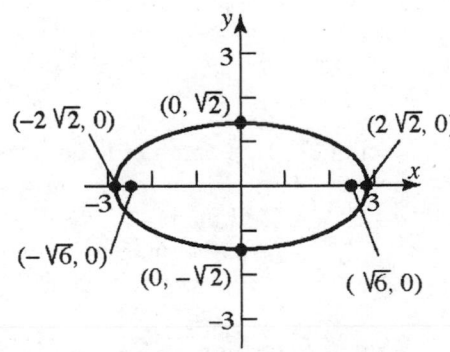

17. $x^2 + y^2 = 16$

This is the equation of a circle whose center is at $(0, 0)$ and radius = 4.

To graph, enter: $y_1 = \sqrt{(16 - x^2)};\ \ y_1 = -\sqrt{(16 - x^2)}$

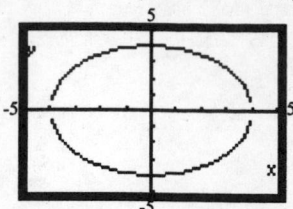

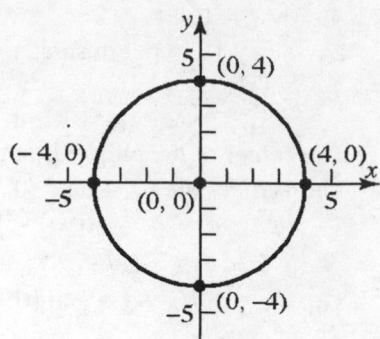

19. Center: $(0, 0)$; Focus: $(3, 0)$; Vertex: $(5, 0)$;
Major axis is the x-axis; $a = 5$; $c = 3$. Find b:
$$b^2 = a^2 - c^2 = 25 - 9 = 16$$
$$b = 4$$

Write the equation: $\dfrac{x^2}{25} + \dfrac{y^2}{16} = 1$

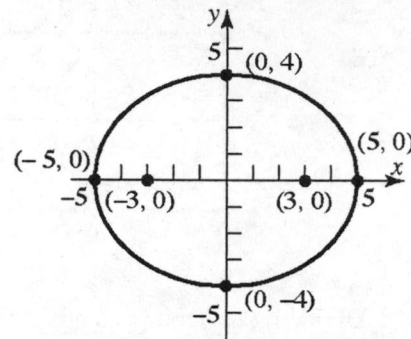

21. Center: $(0, 0)$; Focus: $(0, -4)$; Vertex: $(0, 5)$;
Major axis is the y-axis; $a = 5$; $c = 4$. Find b:
$$b^2 = a^2 - c^2 = 25 - 16 = 9$$
$$b = 3$$

Write the equation: $\dfrac{x^2}{9} + \dfrac{y^2}{25} = 1$

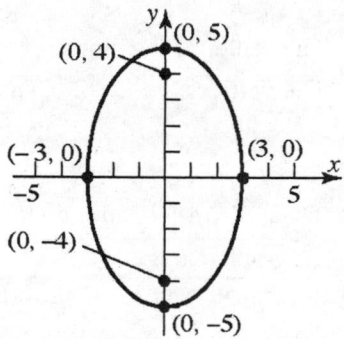

23. Foci: $(\pm 2, 0)$; Length of major axis is 6.
Center: $(0, 0)$; Major axis is the x-axis;
$a = 3$; $c = 2$. Find b:
$$b^2 = a^2 - c^2 = 9 - 4 = 5$$
$$b = \sqrt{5}$$

Write the equation: $\dfrac{x^2}{9} + \dfrac{y^2}{5} = 1$

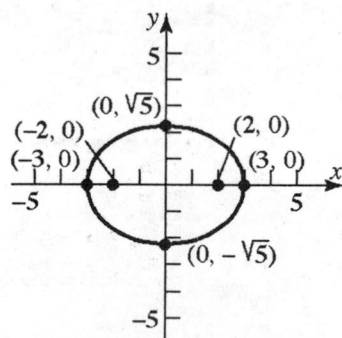

25. Foci: $(0, \pm 3)$; x-intercepts are ± 2. Center: $(0, 0)$;
Major axis is the y-axis; $c = 3$; $b = 2$. Find a:
$$a^2 = b^2 + c^2 = 4 + 9 = 13$$
$$a = \sqrt{13}$$
Write the equation: $\dfrac{x^2}{4} + \dfrac{y^2}{13} = 1$

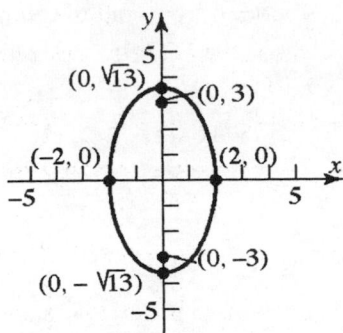

27. Center: $(0, 0)$; Vertex: $(0, 4)$; $b = 1$; Major axis is
the y-axis; $a = 4$; $b = 1$.

Write the equation: $\dfrac{x^2}{1} + \dfrac{y^2}{16} = 1$

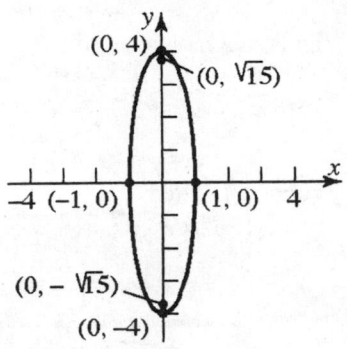

29. $\dfrac{(x+1)^2}{4} + \dfrac{(y-1)^2}{1} = 1$ 31. $\dfrac{(x-1)^2}{1} + \dfrac{y^2}{4} = 1$

33. The equation $\dfrac{(x-3)^2}{4} + \dfrac{(y+1)^2}{9} = 1$ is in the form $\dfrac{(x-h)^2}{b^2} + \dfrac{(y-k)^2}{a^2} = 1$
(major axis parallel to the y-axis) where $a = 3$, $b = 2$, $h = 3$, and $k = -1$.
Solving for c:
$$c^2 = a^2 - b^2 = 9 - 4 = 5 \Rightarrow c = \sqrt{5}$$
Thus, we have:
    Center: $(3, -1)$
    Foci: $\left(3, -1 + \sqrt{5}\right)$, $\left(3, -1 - \sqrt{5}\right)$
    Vertices: $(3, 2)$, $(3, -4)$
To graph, enter: $y_1 = -1 + 3\sqrt{1 - (x-3)^2 / 4}$;
        $y_2 = -1 - 3\sqrt{1 - (x-3)^2 / 4}$

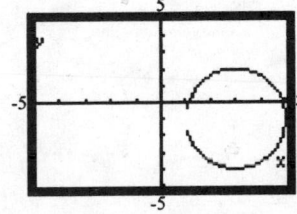

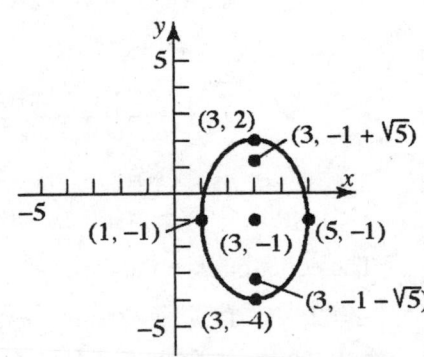

35. Divide by 16 to put the equation in standard form:

$$(x+5)^2 + 4(y-4)^2 = 16$$

$$\frac{(x+5)^2}{16} + \frac{4(y-4)^2}{16} = \frac{16}{16}$$

$$\frac{(x+5)^2}{16} + \frac{(y-4)^2}{4} = 1$$

The equation is in the form $\dfrac{(x-h)^2}{a^2} + \dfrac{(y-k)^2}{b^2} = 1$ (major axis parallel to the x-axis) where $a = 4$, $b = 2$, $h = -5$, and $k = 4$.

Solving for c:

$$c^2 = a^2 - b^2 = 16 - 4 = 12 \Rightarrow c = \sqrt{12} = 2\sqrt{3}$$

Thus, we have:

Center: $(-5, 4)$

Foci: $\left(-5-2\sqrt{3}, 4\right),\ \left(-5+2\sqrt{3}, 4\right)$

Vertices: $(-9, 4),\ (-1, 4)$

To graph, enter:

$$y_1 = 4 + 2\sqrt{1 - (x+5)^2 / 16}\,;$$

$$y_2 = 4 - 2\sqrt{1 - (x+5)^2 / 16}$$

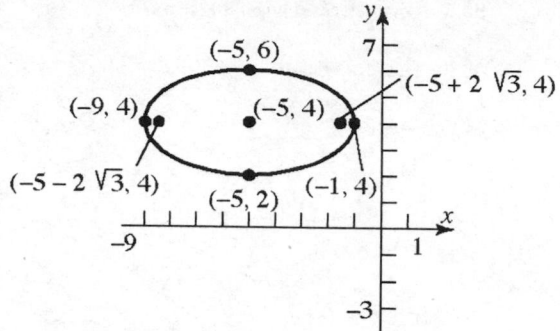

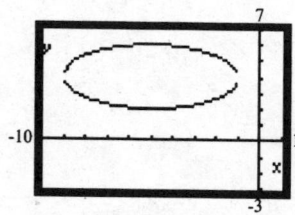

37. Complete the square to put the equation in standard form:

$$x^2 + 4x + 4y^2 - 8y + 4 = 0$$

$$(x^2 + 4x + 4) + 4(y^2 - 2y + 1) = -4 + 4 + 4$$

$$(x+2)^2 + 4(y-1)^2 = 4$$

$$\frac{(x+2)^2}{4} + \frac{4(y-1)^2}{4} = \frac{4}{4}$$

$$\frac{(x+2)^2}{4} + \frac{(y-1)^2}{1} = 1$$

The equation is in the form $\dfrac{(x-h)^2}{a^2} + \dfrac{(y-k)^2}{b^2} = 1$ (major axis parallel to the x-axis) where $a = 2$, $b = 1$, $h = -2$, and $k = 1$.

Solving for c: $c^2 = a^2 - b^2 = 4 - 1 = 3 \Rightarrow c = \sqrt{3}$

Thus, we have:

    Center: $(-2, 1)$

    Foci: $\left(-2-\sqrt{3},1\right),\ \left(-2+\sqrt{3},1\right)$

    Vertices: $(-4, 1),\ (0, 1)$

To graph, enter: $y_1 = 1 + \sqrt{1 - (x+2)^2 / 4}$;

        $y_2 = 1 - \sqrt{1 - (x+2)^2 / 4}$

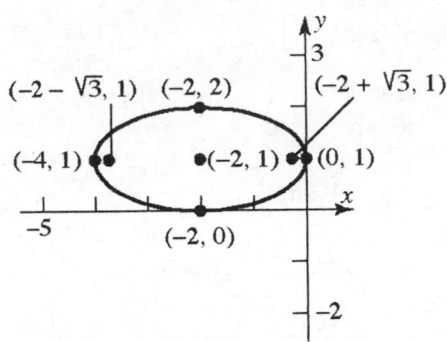

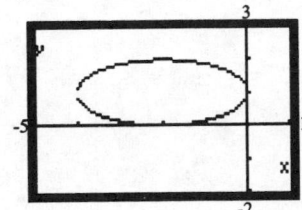

39. Complete the square to put the equation in standard form:

$$2x^2 + 3y^2 - 8x + 6y + 5 = 0 \Rightarrow 2(x^2 - 4x) + 3(y^2 + 2y) = -5$$

$$2(x^2 - 4x + 4) + 3(y^2 + 2y + 1) = -5 + 8 + 3 \Rightarrow 2(x - 2)^2 + 3(y + 1)^2 = 6$$

$$\frac{2(x-2)^2}{6} + \frac{3(y+1)^2}{6} = \frac{6}{6} \Rightarrow \frac{(x-2)^2}{3} + \frac{(y+1)^2}{2} = 1$$

The equation is in the form $\dfrac{(x-h)^2}{a^2} + \dfrac{(y-k)^2}{b^2} = 1$ (major axis parallel to the x-axis) where

$a = \sqrt{3},\ b = \sqrt{2},\ h = 2,$ and $k = -1.$

Solving for c: $c^2 = a^2 - b^2 = 3 - 2 = 1 \Rightarrow c = 1$

Thus, we have:

    Center: $(2, -1)$

    Foci: $(1, -1),\ (3, -1)$

    Vertices: $\left(2 - \sqrt{3}, -1\right),\ \left(2 + \sqrt{3}, -1\right)$

To graph, enter: $y_1 = -1 + \sqrt{2 - 2(x-2)^2 / 3}$;

        $y_2 = -1 - \sqrt{2 - 2(x-2)^2 / 3}$

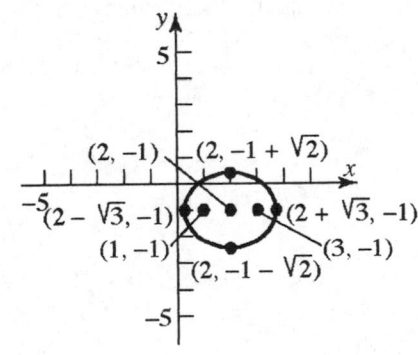

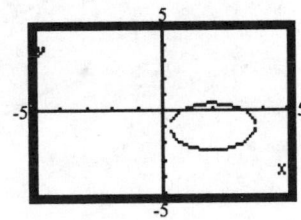

41. Complete the square to put the equation in standard form:

$$9x^2 + 4y^2 - 18x + 16y - 11 = 0 \Rightarrow 9(x^2 - 2x) + 4(y^2 + 4y) = 11$$

$$9(x^2 - 2x + 1) + 4(y^2 + 4y + 4) = 11 + 9 + 16 \Rightarrow 9(x - 1)^2 + 4(y + 2)^2 = 36$$

$$\frac{9(x-1)^2}{36} + \frac{4(y+2)^2}{36} = \frac{36}{36} \Rightarrow \frac{(x-1)^2}{4} + \frac{(y+2)^2}{9} = 1$$

The equation is in the form $\dfrac{(x-h)^2}{b^2} + \dfrac{(y-k)^2}{a^2} = 1$ (major axis parallel to the y-axis) where $a = 3$, $b = 2$, $h = 1$, and $k = -2$.

Solving for c:

$$c^2 = a^2 - b^2 = 9 - 4 = 5 \Rightarrow c = \sqrt{5}$$

Thus, we have:

Center: $(1,-2)$

Foci: $\left(1, -2+\sqrt{5}\right),\ \left(1, -2-\sqrt{5}\right)$

Vertices: $(1, 1),\ (1, -5)$

To graph, enter: $y_1 = -2 + 3\sqrt{1-(x-1)^2/4}$;

$$y_2 = -2 - 3\sqrt{1-(x-1)^2/4}$$

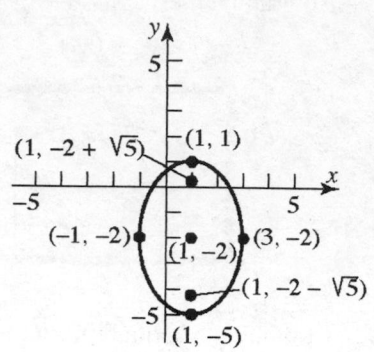

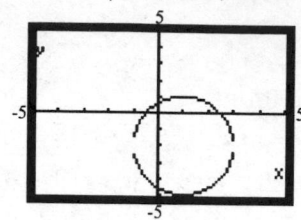

43.   Complete the square to put the equation in standard form:

$$4x^2 + y^2 + 4y = 0 \Rightarrow 4x^2 + y^2 + 4y + 4 = 4$$

$$4x^2 + (y+2)^2 = 4 \Rightarrow \frac{4x^2}{4} + \frac{(y+2)^2}{4} = \frac{4}{4} \Rightarrow \frac{x^2}{1} + \frac{(y+2)^2}{4} = 1$$

The equation is in the form $\dfrac{(x-h)^2}{b^2} + \dfrac{(y-k)^2}{a^2} = 1$ (major axis parallel to the y-axis) where $a = 2$, $b = 1$, $h = 0$, and $k = -2$.

Solving for c:

$$c^2 = a^2 - b^2 = 4 - 1 = 3 \quad \Rightarrow \quad c = \sqrt{3}$$

Thus, we have:

Center: $(0,-2)$

Foci: $\left(0, -2+\sqrt{3}\right),\ \left(0, -2-\sqrt{3}\right)$

Vertices: $(0, 0),\ (0, -4)$

To graph, enter: $y_1 = -2 + 2\sqrt{1-x^2}$;

$$y_2 = -2 - 2\sqrt{1-x^2}$$

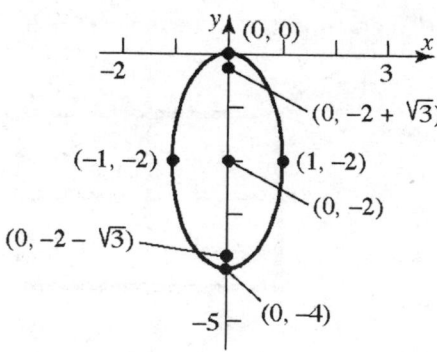

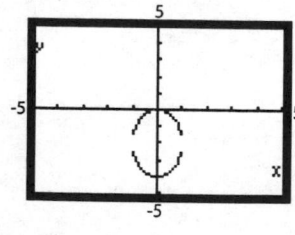

45. Center: $(2, -2)$; Vertex: $(7, -2)$;
Focus: $(4, -2)$; Major axis parallel
to the x-axis; $a = 5$; $c = 2$.
Find b:
$$b^2 = a^2 - c^2 = 25 - 4 = 21$$
$$b = \sqrt{21}$$
Write the equation:
$$\frac{(x-2)^2}{25} + \frac{(y+2)^2}{21} = 1$$

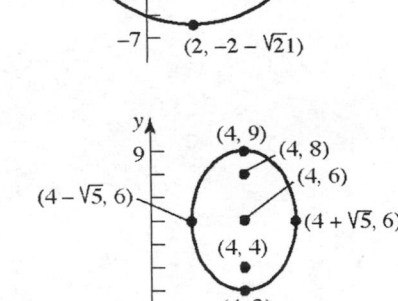

47. Vertices: $(4, 3), (4, 9)$; Focus: $(4, 8)$;
Center: $(4, 6)$; Major axis parallel to the
y-axis; $a = 3$; $c = 2$.
Find b:
$$b^2 = a^2 - c^2 = 9 - 4 = 5$$
$$b = \sqrt{5}$$
Write the equation: $\dfrac{(x-4)^2}{5} + \dfrac{(y-6)^2}{9} = 1$

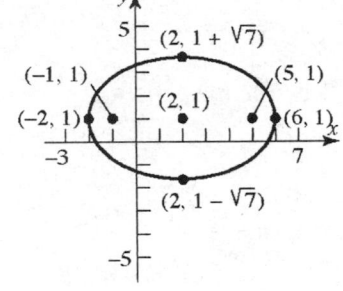

49. Foci: $(5, 1), (-1, 1)$; length of the major axis $= 8$;
Center: $(2, 1)$; Major axis parallel to the x-axis;
$a = 4$; $c = 3$.   Find b:
$$b^2 = a^2 - c^2 = 16 - 9 = 7$$
$$b = \sqrt{7}$$
Write the equation: $\dfrac{(x-2)^2}{16} + \dfrac{(y-1)^2}{7} = 1$

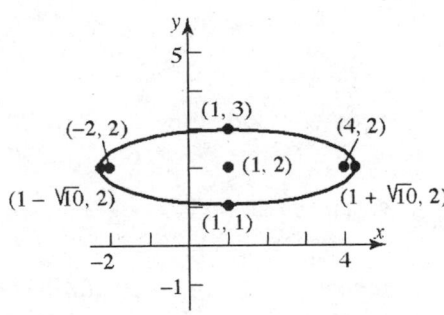

51. Center: $(1, 2)$; Focus: $(4, 2)$; contains the
point $(1, 3)$; Major axis parallel to the
x-axis; $c = 3$.
The equation has the form:
$$\frac{(x-1)^2}{a^2} + \frac{(y-2)^2}{b^2} = 1$$
Since the point $(1, 3)$ is on the curve:
$$\frac{0}{a^2} + \frac{1}{b^2} = 1$$
$$\frac{1}{b^2} = 1 \Rightarrow b^2 = 1 \Rightarrow b = 1$$
Find a:
$$a^2 = b^2 + c^2 = 1 + 9 = 10 \Rightarrow a = \sqrt{10}$$
Write the equation: $\dfrac{(x-1)^2}{10} + \dfrac{(y-2)^2}{1} = 1$

53. Center: $(1, 2)$; Vertex: $(4, 2)$; contains the point $(1, 3)$; Major axis parallel to the x-axis; $a = 3$.

The equation has the form:

$$\frac{(x-1)^2}{a^2} + \frac{(y-2)^2}{b^2} = 1$$

Since the point $(1, 3)$ is on the curve:

$$\frac{0}{9} + \frac{1}{b^2} = 1$$

$$\frac{1}{b^2} = 1 \implies b^2 = 1 \implies b = 1$$

Write the equation: $\dfrac{(x-1)^2}{9} + \dfrac{(y-2)^2}{1} = 1$

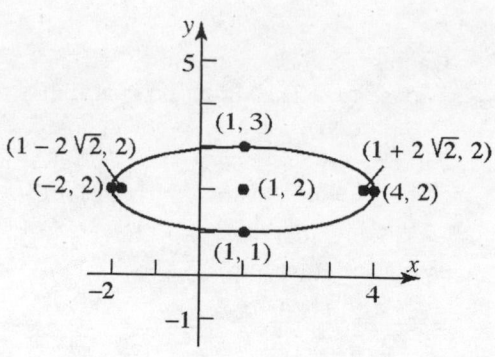

55. Rewrite the equation:

$$y = \sqrt{16 - 4x^2}$$
$$y^2 = 16 - 4x^2, \quad y \geq 0$$
$$4x^2 + y^2 = 16, \quad y \geq 0$$
$$\frac{x^2}{4} + \frac{y^2}{16} = 1, \quad y \geq 0$$

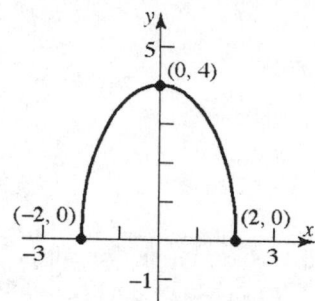

57. Rewrite the equation:

$$y = -\sqrt{64 - 16x^2}$$
$$y^2 = 64 - 16x^2, \quad y \leq 0$$
$$16x^2 + y^2 = 64, \quad y \leq 0$$
$$\frac{x^2}{4} + \frac{y^2}{64} = 1, \quad y \leq 0$$

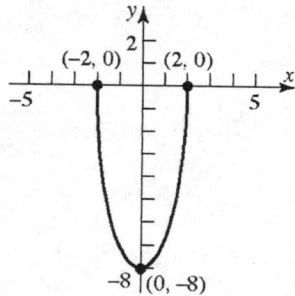

59. The center of the ellipse is $(0, 0)$. The length of the major axis is 20, so $a = 10$. The length of half the minor axis is 6, so $b = 6$. The ellipse is situated with its major axis on the x-axis. The equation is: $\dfrac{x^2}{100} + \dfrac{y^2}{36} = 1$.

61. Assume that the half ellipse formed by the gallery is centered at $(0, 0)$. Since the hall is 100 feet long, $2a = 100$ or $a = 50$. The distance from the center to the foci is 25 feet, so $c = 25$. Find the height of the gallery which is b:

$$b^2 = a^2 - c^2 = 2500 - 625 = 1875 \implies b = \sqrt{1875} \approx 43.3$$

The ceiling will be 43.3 feet high in the center.

63. Place the semielliptical arch so that the x-axis coincides with the water and the y-axis passes through the center of the arch. Since the bridge has a span of 120 feet, the length of the major axis is 120, or $2a = 120$ or $a = 60$. The maximum height of the bridge is 25 feet, so $b = 25$. The equation is: $\dfrac{x^2}{3600} + \dfrac{y^2}{625} = 1$.

The height 10 feet from the center:
$$\frac{10^2}{3600} + \frac{y^2}{625} = 1 \Rightarrow \frac{y^2}{625} = 1 - \frac{100}{3600} \Rightarrow y^2 = 625 \cdot \frac{3500}{3600} \Rightarrow y \approx 24.65 \text{ feet}$$

The height 30 feet from the center:
$$\frac{30^2}{3600} + \frac{y^2}{625} = 1 \Rightarrow \frac{y^2}{625} = 1 - \frac{900}{3600} \Rightarrow y^2 = 625 \cdot \frac{2700}{3600} \Rightarrow y \approx 21.65 \text{ feet}$$

The height 50 feet from the center:
$$\frac{50^2}{3600} + \frac{y^2}{625} = 1 \Rightarrow \frac{y^2}{625} = 1 - \frac{2500}{3600} \Rightarrow y^2 = 625 \cdot \frac{1100}{3600} \Rightarrow y \approx 13.82 \text{ feet}$$

65. Place the semielliptical arch so that the x-axis coincides with the major axis and the y-axis passes through the center of the arch. Since the ellipse is 40 feet wide, the length of the major axis is 40, or $2a = 40$ or $a = 20$. The height is 15 feet at the center, so $b = 15$. The equation is: $\dfrac{x^2}{400} + \dfrac{y^2}{225} = 1$.

The height 10 feet either side of the center:
$$\frac{10^2}{400} + \frac{y^2}{225} = 1 \Rightarrow \frac{y^2}{225} = 1 - \frac{100}{400} \Rightarrow y^2 = 225 \cdot \frac{3}{4} \Rightarrow y \approx 12.99 \text{ feet}$$

The height 20 feet either side of the center:
$$\frac{20^2}{400} + \frac{y^2}{225} = 1 \Rightarrow \frac{y^2}{225} = 1 - \frac{400}{400} \Rightarrow y^2 = 225 \cdot 0 \Rightarrow y \approx 0 \text{ feet}$$

67. Since the mean distance is 93 million miles, $a = 93$ million. The length of the major axis is 186 million. The perihelion is 186 million – 94.5 million = 91.5 million miles. The distance from the center of the ellipse to the sun (focus) is 93 million – 91.5 million = 1.5 million miles; therefore, $c = 1.5$ million. Find b:
$$b^2 = a^2 - c^2 = \left(93 \times 10^6\right)^2 - \left(1.5 \times 10^6\right)^2 = 8.64675 \times 10^{15} \Rightarrow b = 92.99 \times 10^6$$

The equation of the orbit is: $\dfrac{x^2}{\left(93 \times 10^6\right)^2} + \dfrac{y^2}{\left(92.99 \times 10^6\right)^2} = 1$.

69. The mean distance is 507 million – 23.2 million = 483.8 million miles.
The perihelion is 483.8 million – 23.2 million = 460.6 million miles.
Since $a = 483.8 \times 10^6$ and $c = 23.2 \times 10^6$, we can find b:
$$b^2 = a^2 - c^2 = \left(483.8 \times 10^6\right)^2 - \left(23.2 \times 10^6\right)^2 = 2.335242 \times 10^{17} \Rightarrow b = 483.2 \times 10^6$$

The equation of the orbit of Jupiter is: $\dfrac{x^2}{\left(483.8 \times 10^6\right)^2} + \dfrac{y^2}{\left(483.2 \times 10^6\right)^2} = 1$.

71. If the x-axis is placed along the 100 foot length and the y-axis is placed along the 50 foot length, the equation for the ellipse is: $\dfrac{x^2}{50^2} + \dfrac{y^2}{25^2} = 1$.

Find y when x = 40:

$$\frac{40^2}{50^2} + \frac{y^2}{25^2} = 1 \Rightarrow \frac{y^2}{625} = 1 - \frac{1600}{2500} \Rightarrow y^2 = 625 \cdot \frac{9}{25} \Rightarrow y \approx 15 \text{ feet}$$

The width 10 feet from the side is 30 feet.

73. (a) Put the equation in standard ellipse form:

$$Ax^2 + Cy^2 + F = 0 \qquad A \neq 0, \, C \neq 0, \, F \neq 0$$

$$Ax^2 + Cy^2 = -F$$

$$\frac{Ax^2}{-F} + \frac{Cy^2}{-F} = 1$$

$$\frac{x^2}{(-F/A)} + \frac{y^2}{(-F/C)} = 1 \qquad \text{where } -F/A \text{ and } -F/C \text{ are positive}$$

This is the equation of an ellipse with center at (0, 0).

(b) If $A = C$, the equation becomes:

$$Ax^2 + Ay^2 = -F \Rightarrow x^2 + y^2 = \frac{-F}{A}$$

This is the equation of a circle with center at (0, 0) and radius of $\sqrt{\dfrac{-F}{A}}$.

75. Answers will vary.

Analytic Geometry

9.4   The Hyperbola

1. B   3. A   5. B   7. C

9. Center: $(0, 0)$; Focus: $(3, 0)$; Vertex: $(1, 0)$;
    Transverse axis is the x-axis; $a = 1$; $c = 3$.
    Find b:
    $$b^2 = c^2 - a^2 = 9 - 1 = 8$$
    $$b = \sqrt{8} = 2\sqrt{2}$$

    Write the equation: $\dfrac{x^2}{1} - \dfrac{y^2}{8} = 1$

    To graph, enter:
    $$y_1 = \sqrt{8(x^2 - 1)}\,;\ y_2 = -\sqrt{8(x^2 - 1)}$$

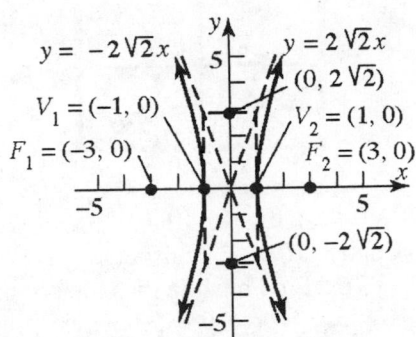

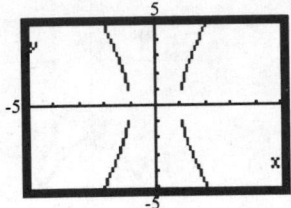

11. Center: $(0, 0)$; Focus: $(0, -6)$; Vertex: $(0, 4)$;
    Transverse axis is the y-axis; $a = 4$; $c = 6$.
    Find b:
    $$b^2 = c^2 - a^2 = 36 - 16 = 20$$
    $$b = \sqrt{20} = 2\sqrt{5}$$

    Write the equation: $\dfrac{y^2}{16} - \dfrac{x^2}{20} = 1$

    To graph, enter:
    $$y_1 = 4\sqrt{1 + x^2 / 20}\,;\ y_2 = -4\sqrt{1 + x^2 / 20}$$

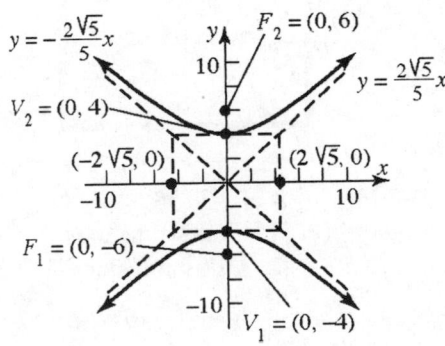

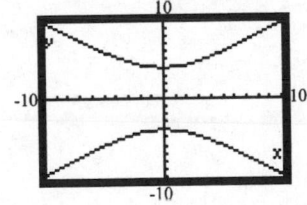

13. Foci: $(-5, 0)$, $(5, 0)$; Vertex: $(3, 0)$
     Center: $(0, 0)$; Transverse axis is the
     x-axis; $a = 3$; $c = 5$. Find b:
     $b^2 = c^2 - a^2 = 25 - 9 = 16 \rightarrow b = 4$

     Write the equation: $\dfrac{x^2}{9} - \dfrac{y^2}{16} = 1$

     To graph, enter:
     $y_1 = 4\sqrt{x^2 / 9 - 1}$; $y_2 = -4\sqrt{x^2 / 9 - 1}$

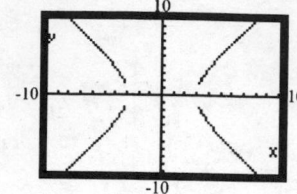

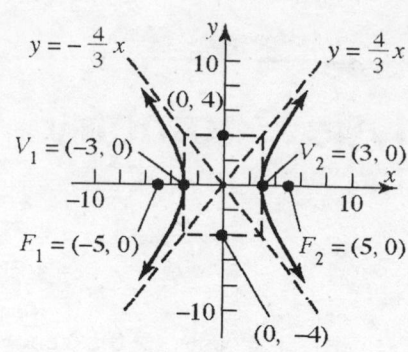

15. Vertices: $(0, -6)$, $(0, 6)$; Asymptote: $y = 2x$;
     Center: $(0, 0)$; Transverse axis is the y-axis;
     $a = 6$. Find b using the slope of the

     asymptote: $\dfrac{a}{b} = \dfrac{6}{b} = 2 \;\rightarrow\; 2b = 6 \;\rightarrow\; b = 3$

     Write the equation: $\dfrac{y^2}{36} - \dfrac{x^2}{9} = 1$

     To graph, enter:
     $y_1 = 6\sqrt{1 + x^2 / 9}$; $y_2 = -6\sqrt{1 + x^2 / 9}$

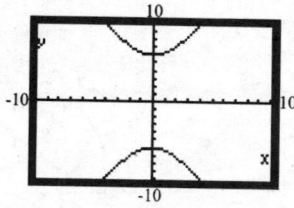

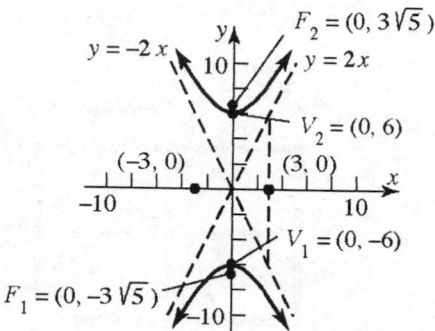

17. Foci: $(-4, 0)$, $(4, 0)$; Asymptote: $y = -x$;
     Center: $(0, 0)$; Transverse axis is the
     x-axis; $c = 4$. Using the slope of the

     asymptote: $-\dfrac{b}{a} = -1 \rightarrow -b = -a \rightarrow b = a$

Find b:
$$b^2 = c^2 - a^2 \rightarrow a^2 + b^2 = c^2 \quad (c = 4)$$
$$b^2 + b^2 = 16 \rightarrow 2b^2 = 16 \rightarrow b^2 = 8$$
$$b = \sqrt{8} = 2\sqrt{2}$$
$$a = \sqrt{8} = 2\sqrt{2} \quad (a = b)$$

Write the equation: $\dfrac{x^2}{8} - \dfrac{y^2}{8} = 1$

To graph, enter: $y_1 = \sqrt{x^2 - 8}$; $y_2 = -\sqrt{x^2 - 8}$

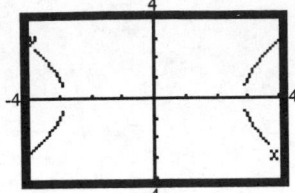

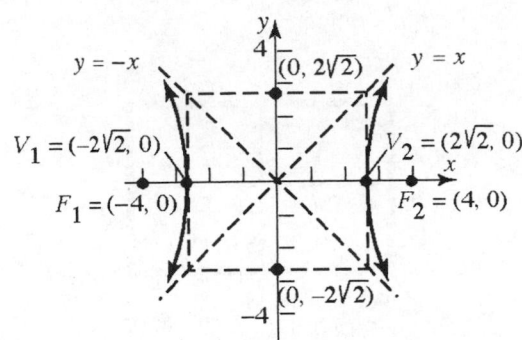

19. $\dfrac{x^2}{25} - \dfrac{y^2}{9} = 1$

The center of the hyperbola is at $(0, 0)$.
$a = 5$, $b = 3$. The vertices are $(5, 0)$ and
$(-5, 0)$. Find the value of c:
$c^2 = a^2 + b^2 = 25 + 9 = 34 \rightarrow c = \sqrt{34}$
The foci are $\left(\sqrt{34}, 0\right)$ and $\left(-\sqrt{34}, 0\right)$.

The transverse axis is $y = 0$.

The asymptotes are $y = \dfrac{3}{5}x$; $y = -\dfrac{3}{5}x$.

To graph, enter:
$y_1 = 3\sqrt{(x^2/25 - 1)}$; $y_2 = -3\sqrt{(x^2/25 - 1)}$

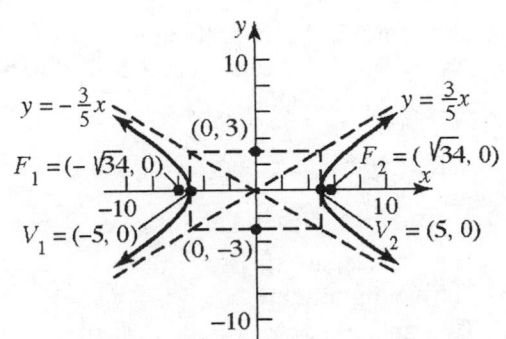

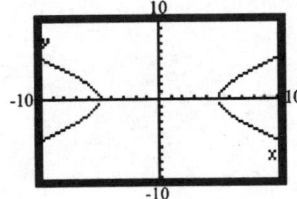

21. $4x^2 - y^2 = 16$

Divide both sides by 16 to put in standard form: $\dfrac{4x^2}{16} - \dfrac{y^2}{16} = \dfrac{16}{16} \rightarrow \dfrac{x^2}{4} - \dfrac{y^2}{16} = 1$

The center of the hyperbola is at $(0, 0)$.
$a = 2$, $b = 4$. The vertices are $(2, 0)$ and
$(-2, 0)$. Find the value of c:
$c^2 = a^2 + b^2 = 4 + 16 = 20$
$c = \sqrt{20} = 2\sqrt{5}$
The foci are $\left(2\sqrt{5}, 0\right)$ and $\left(-2\sqrt{5}, 0\right)$.
The transverse axis is $y = 0$.
The asymptotes are $y = 2x$; $y = -2x$.

To graph, enter:

$$y_1 = 4\sqrt{(x^2/4-1)}; \quad y_2 = -4\sqrt{(x^2/4-1)}$$

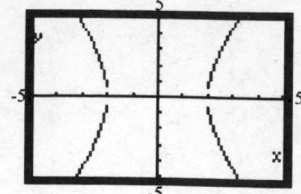

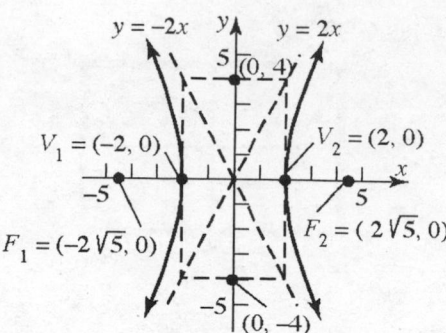

23. $y^2 - 9x^2 = 9$

Divide both sides by 9 to put in standard form: $\dfrac{y^2}{9} - \dfrac{9x^2}{9} = \dfrac{9}{9} \rightarrow \dfrac{y^2}{9} - \dfrac{x^2}{1} = 1$

The center of the hyperbola is at (0, 0).
$a = 3, \ b = 1$. The vertices are (0, 3) and (0, –3). Find the value of c:

$$c^2 = a^2 + b^2 = 9 + 1 = 10$$
$$c = \sqrt{10}$$

The foci are $\left(0, \sqrt{10}\right)$ and $\left(0, -\sqrt{10}\right)$.

The transverse axis is $x = 0$.

The asymptotes are $y = 3x; \ y = -3x$.

To graph, enter:

$$y_1 = \sqrt{(9x^2 + 9)}; \quad y_2 = -\sqrt{(9x^2 + 9)}$$

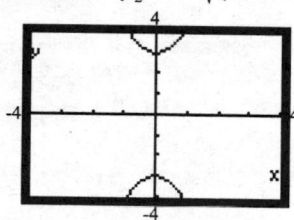

25. $y^2 - x^2 = 25$

Divide both sides by 25 to put in standard form: $\dfrac{y^2}{25} - \dfrac{x^2}{25} = 1$

The center of the hyperbola is at (0, 0).
$a = 5, \ b = 5$. The vertices are (0, 5) and (0, –5).

Find the value of c:

$$c^2 = a^2 + b^2 = 25 + 25 = 50$$
$$c = \sqrt{50} = 5\sqrt{2}$$

The foci are $\left(0, 5\sqrt{2}\right)$ and $\left(0, -5\sqrt{2}\right)$.

The transverse axis is $x = 0$.

The asymptotes are $y = x; \ y = -x$.

To graph, enter:
$$y_1 = \sqrt{(x^2 + 25)}; \quad y_2 = -\sqrt{(x^2 + 25)}$$

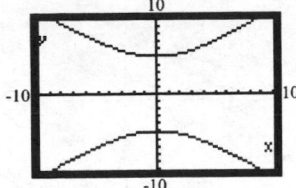

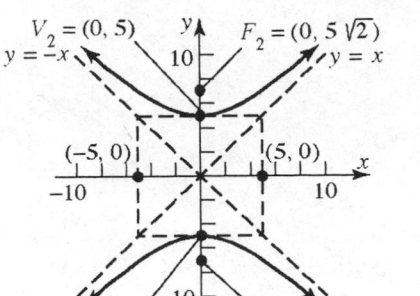

27. $x^2 - y^2 = 1$

29. $\dfrac{y^2}{36} - \dfrac{x^2}{9} = 1$

31. Center: $(4, -1)$; Focus: $(7, -1)$;
Vertex: $(6, -1)$; Transverse axis is
parallel to the x-axis; $a = 2$; $c = 3$.
Find b:
$$b^2 = c^2 - a^2 = 9 - 4 = 5 \rightarrow b = \sqrt{5}$$
Write the equation:
$$\frac{(x - 4)^2}{4} - \frac{(y + 1)^2}{5} = 1$$

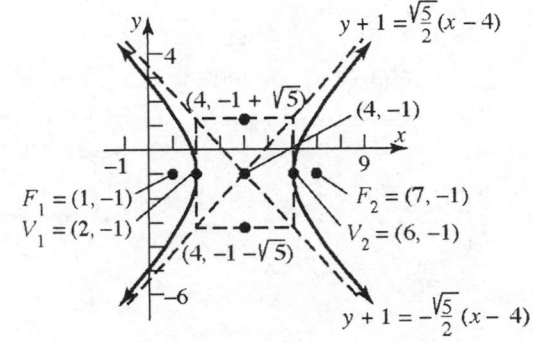

33. Center: $(-3, -4)$; Focus:$(-3, -8)$;
Vertex: $(-3, -2)$; Transverse axis
is parallel to the y-axis;
$a = 2$; $c = 4$. Find b:
$$b^2 = c^2 - a^2 = 16 - 4 = 12$$
$$b = \sqrt{12} = 2\sqrt{3}$$
Write the equation:
$$\frac{(y + 4)^2}{4} - \frac{(x + 3)^2}{12} = 1$$

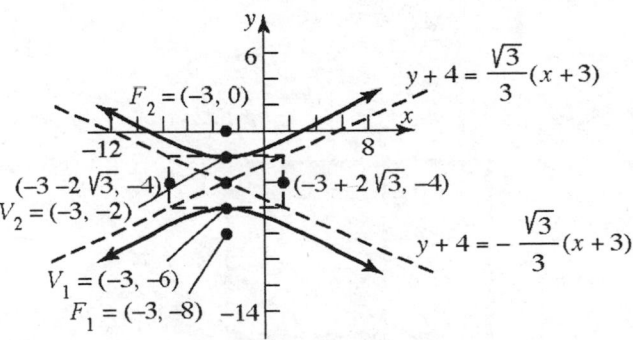

35. Foci: $(3, 7), (7, 7)$; Vertex: $(6, 7)$;
Center: $(5, 7)$; Transverse axis is parallel to
the x-axis; $a = 1$; $c = 2$.
Find b:
$$b^2 = c^2 - a^2 = 4 - 1 = 3$$
$$b = \sqrt{3}$$
Write the equation: $\dfrac{(x - 5)^2}{1} - \dfrac{(y - 7)^2}{3} = 1$

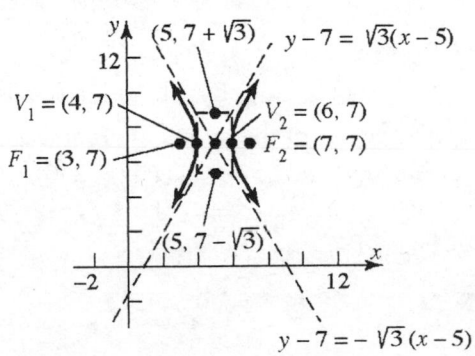

37. Vertices: $(-1, -1)$, $(3, -1)$;
Center: $(1, -1)$; Transverse
axis is parallel to the x-axis;
$a = 2$.

Asymptote: $\dfrac{x-1}{2} = \dfrac{y+1}{3}$

Using the slope of the
asymptote:

$\dfrac{b}{a} = \dfrac{b}{2} = \dfrac{3}{2} \;\rightarrow\; b = 3$

Write the equation:

$\dfrac{(x-1)^2}{4} - \dfrac{(y+1)^2}{9} = 1$

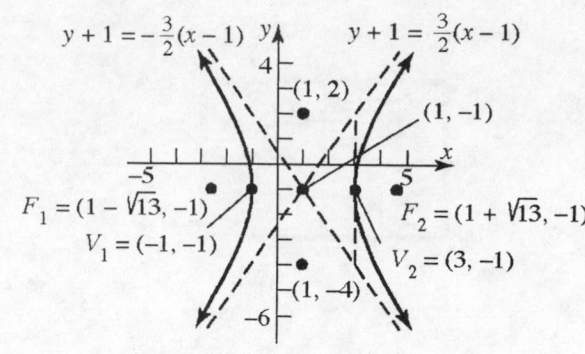

39. $\dfrac{(x-2)^2}{4} - \dfrac{(y+3)^2}{9} = 1$

The center of the hyperbola is at $(2, -3)$. $a = 2$, $b = 3$.
The vertices are $(0, -3)$ and $(4, -3)$. Find the value of c:

$$c^2 = a^2 + b^2 = 4 + 9 = 13 \rightarrow c = \sqrt{13}$$

Foci: $\left(2 - \sqrt{13}, -3\right)$ and $\left(2 + \sqrt{13}, -3\right)$.

Transverse axis: $y = -3$.

Asymptotes: $y + 3 = \dfrac{3}{2}(x-2)$;

$y + 3 = -\dfrac{3}{2}(x-2)$.

To graph, enter:

$y_1 = -3 + 3\sqrt{((x-2)^2/4 - 1)}$;

$y_2 = -3 - 3\sqrt{((x-2)^2/4 - 1)}$

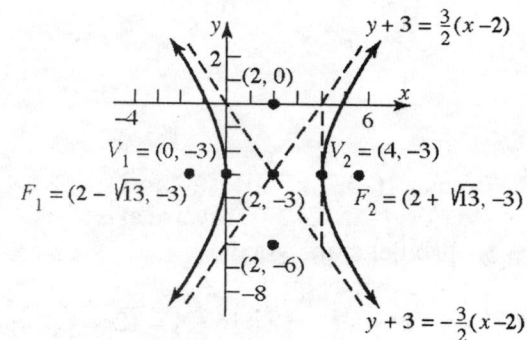

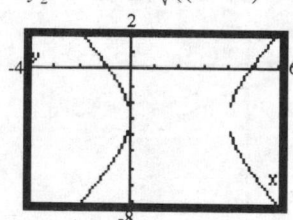

41. $(y-2)^2 - 4(x+2)^2 = 4$

Divide both sides by 4 to put in standard form: $\dfrac{(y-2)^2}{4} - \dfrac{(x+2)^2}{1} = 1$

The center of the hyperbola is at $(-2, 2)$. $a = 2$, $b = 1$.
The vertices are $(-2, 4)$ and $(-2, 0)$. Find the value of c:

$$c^2 = a^2 + b^2 = 4 + 1 = 5 \rightarrow c = \sqrt{5}$$

Foci: $\left(-2, 2 - \sqrt{5}\right)$ and $\left(-2, 2 + \sqrt{5}\right)$.

Transverse axis: $x = -2$.

Asymptotes:

$y - 2 = 2(x + 2)$; $y - 2 = -2(x + 2)$.

To graph, enter: $y_1 = 2 + 2\sqrt{((x + 2)^2 + 1)}$;

$$y_2 = 2 - 2\sqrt{((x + 2)^2 + 1)}$$

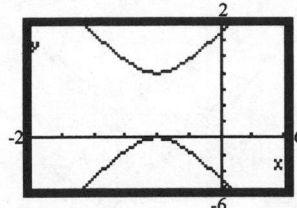

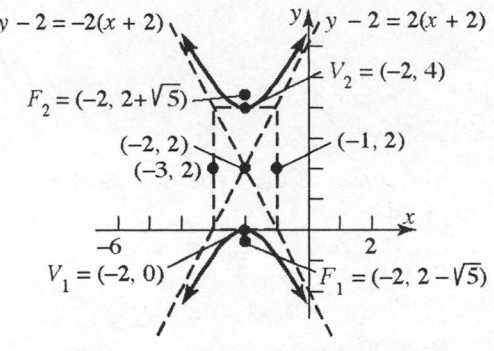

43. $(x + 1)^2 - (y + 2)^2 = 4$

Divide both sides by 4 to put in standard form: $\dfrac{(x + 1)^2}{4} - \dfrac{(y + 2)^2}{4} = 1$

The center of the hyperbola is at $(-1, -2)$. $a = 2$, $b = 2$.

The vertices are $(-3, -2)$ and $(1, -2)$.

Find the value of c:

$$c^2 = a^2 + b^2 = 4 + 4 = 8$$

$$c = \sqrt{8} = 2\sqrt{2}$$

Foci: $\left(-1 - 2\sqrt{2}, -2\right)$

and $\left(-1 + 2\sqrt{2}, -2\right)$.

Transverse axis: $y = -2$.

Asymptotes: $y + 2 = x + 1$;

$$y + 2 = -(x + 1).$$

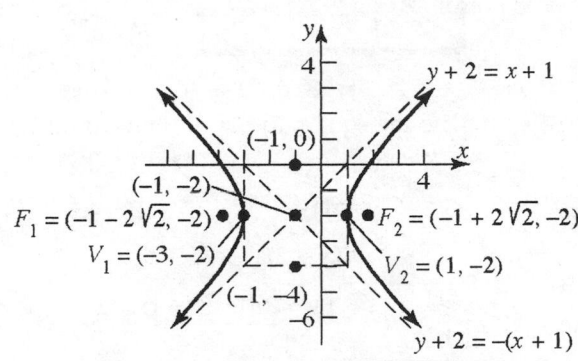

To graph, enter: $y_1 = -2 + 2\sqrt{((x + 1)^2 / 4 - 1)}$;

$$y_2 = -2 - 2\sqrt{((x + 1)^2 / 4 - 1)}$$

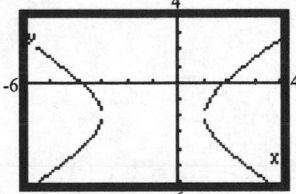

527

45. Complete the square to put in standard form:
$$x^2 - y^2 - 2x - 2y - 1 = 0$$

$$(x^2 - 2x + 1) - (y^2 + 2y + 1) = 1 + 1 - 1$$

$$(x-1)^2 - (y+1)^2 = 1$$

The center of the hyperbola is at
$(1, -1)$. $a = 1$, $b = 1$.
The vertices are $(0, -1)$ and $(2, -1)$.
Find the value of c:
$$c^2 = a^2 + b^2 = 1 + 1 = 2 \Rightarrow c = \sqrt{2}$$
Foci: $\left(1 - \sqrt{2}, -1\right)$ and $\left(1 + \sqrt{2}, -1\right)$.

Transverse axis: $y = -1$.
Asymptotes:
$y + 1 = x - 1$; $y + 1 = -(x - 1)$.
To graph, enter:

$$y_1 = -1 + \sqrt{((x-1)^2 - 1)};$$

$$y_2 = -1 - \sqrt{((x-1)^2 - 1)}$$

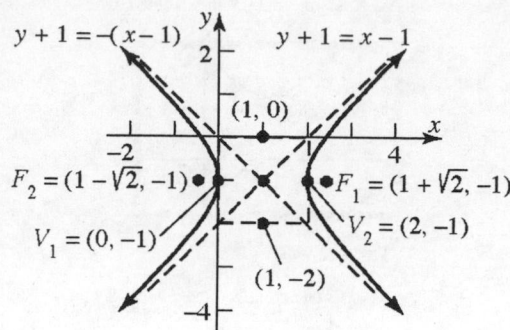

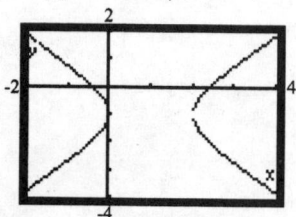

47. Complete the square to put in standard form:
$$y^2 - 4x^2 - 4y - 8x - 4 = 0$$

$$(y^2 - 4y + 4) - 4(x^2 + 2x + 1) = 4 + 4 - 4$$

$$(y-2)^2 - 4(x+1)^2 = 4$$

$$\frac{(y-2)^2}{4} - \frac{(x+1)^2}{1} = 1$$

The center of the hyperbola is at $(-1, 2)$.
$a = 2$, $b = 1$.
The vertices are $(-1, 4)$ and $(-1, 0)$. Find the
value of c:
$$c^2 = a^2 + b^2 = 4 + 1 = 5 \Rightarrow c = \sqrt{5}$$
Foci: $\left(-1, 2 - \sqrt{5}\right)$ and $\left(-1, 2 + \sqrt{5}\right)$.

Transverse axis: $x = -1$.
Asymptotes: $y - 2 = 2(x + 1)$; $y - 2 = -2(x + 1)$

To graph, enter: $y_1 = 2 + 2\sqrt{((x+1)^2 + 1)}$;
$$y_2 = 2 - 2\sqrt{((x+1)^2 + 1)}$$

49. Complete the square to put in standard form:
$$4x^2 - y^2 - 24x - 4y + 16 = 0$$
$$4(x^2 - 6x + 9) - (y^2 + 4y + 4) = -16 + 36 - 4$$
$$4(x-3)^2 - (y+2)^2 = 16$$
$$\frac{(x-3)^2}{4} - \frac{(y+2)^2}{16} = 1$$

The center of the hyperbola is at
$(3, -2)$. $a = 2$, $b = 4$.
The vertices are $(1, -2)$
and $(5, -2)$. Find the value of c:
$$c^2 = a^2 + b^2 = 4 + 16 = 20$$
$$c = \sqrt{20} = 2\sqrt{5}$$
Foci:
$\left(3 - 2\sqrt{5}, -2\right)$ and $\left(3 + 2\sqrt{5}, -2\right)$.
Transverse axis: $y = -2$.
Asymptotes: $y + 2 = 2(x - 3)$;
$$y + 2 = -2(x - 3)$$

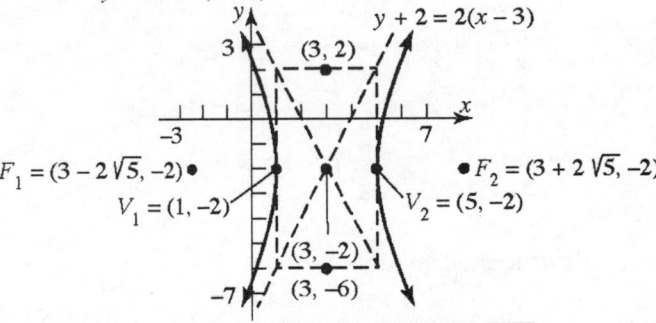

To graph, enter: $y_1 = -2 + 4\sqrt{((x-3)^2 / 4 - 1)}$;
$$y_2 = -2 - 4\sqrt{((x-3)^2 / 4 - 1)}$$

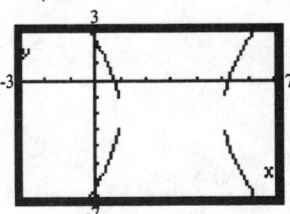

51. Complete the square to put in standard form:
$$y^2 - 4x^2 - 16x - 2y - 19 = 0$$
$$(y^2 - 2y + 1) - 4(x^2 + 4x + 4) = 19 + 1 - 16$$
$$(y-1)^2 - 4(x+2)^2 = 4$$
$$\frac{(y-1)^2}{4} - \frac{(x+2)^2}{1} = 1$$
The center of the hyperbola is at $(-2, 1)$. $a = 2$, $b = 1$.

The vertices are $(-2, 3)$ and $(-2, -1)$. Find the value of c:

$$c^2 = a^2 + b^2 = 4 + 1 = 5 \Rightarrow c = \sqrt{5}$$

Foci: $\left(-2, 1 - \sqrt{5}\right)$ and $\left(-2, 1 + \sqrt{5}\right)$.

Transverse axis: $x = -2$.

Asymptotes: $y - 1 = 2(x + 2); y - 1 = -2(x + 2)$.

To graph, enter:

$$y_1 = 1 + 2\sqrt{((x + 2)^2 + 1)};$$

$$y_2 = 1 - 2\sqrt{((x + 2)^2 + 1)}$$

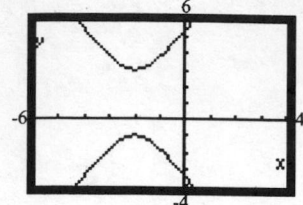

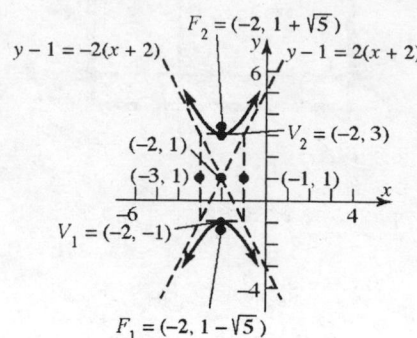

53. Rewrite the equation:

$$y = \sqrt{16 + 4x^2}$$

$$y^2 = 16 + 4x^2, \qquad y \geq 0$$

$$y^2 - 4x^2 = 16, \qquad\quad y \geq 0$$

$$\frac{y^2}{16} - \frac{x^2}{4} = 1, \qquad\quad y \geq 0$$

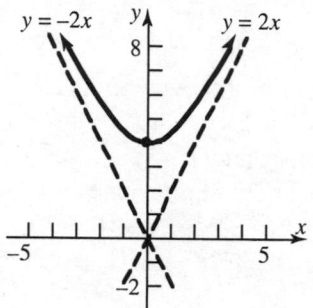

55. Rewrite the equation:

$$y = -\sqrt{-25 + x^2}$$

$$y^2 = -25 + x^2, \qquad y \leq 0$$

$$x^2 - y^2 = 25, \qquad\quad y \leq 0$$

$$\frac{x^2}{25} - \frac{y^2}{25} = 1, \qquad\quad y \leq 0$$

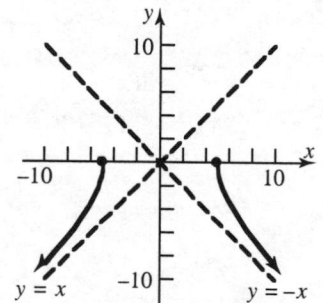

57. (a) Set up a coordinate system so that the two stations lie on the x-axis and the origin is midway between them. The ship lies on a hyperbola whose foci are the locations of the two stations. Since the time difference is 0.00038 seconds and the speed of the signal is 186,000 miles per second, the difference in the distances of the ships from each station is: $(186,000)(0.00038) \approx 70.68$ miles

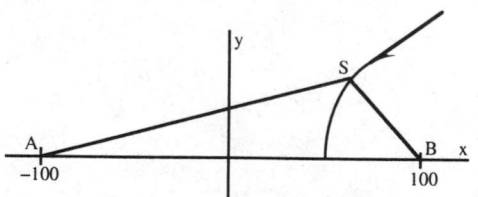

The difference of the distances from the ship to each station, 70.68, equals $2a$, so $a = 35.34$ and the vertex of the corresponding hyperbola is at $(35.34, 0)$. Since the focus is at $(100, 0)$, following this hyperbola, the ship would reach shore 64.66 miles from the master station.

(b) The ship should follow a hyperbola with a vertex at $(80, 0)$. For this hyperbola, $a = 80$, so the constant difference of the distances from the ship to each station is 160. The time difference the ship should look for is:

$$\text{time} = \frac{160}{186,000} \approx 0.00086 \text{ seconds}$$

(c) Find the equation of the hyperbola with vertex at $(80, 0)$ and a focus at $(100, 0)$. The form of the equation of the hyperbola is:

$$\frac{x^2}{a^2} - \frac{y^2}{b^2} = 1 \quad \text{where } a = 80.$$

Since $c = 100$ and $b^2 = c^2 - a^2 \rightarrow b^2 = 100^2 - 80^2 = 3600$.

The equation of the hyperbola is: $\dfrac{x^2}{6400} - \dfrac{y^2}{3600} = 1$.

Since the ship is 50 miles off shore, we have $y = 50$. Solve the equation for x:

$$\frac{x^2}{6400} - \frac{50^2}{3600} = 1 \rightarrow \frac{x^2}{6400} = 1 + \frac{2500}{3600} = \frac{61}{36}$$

$$x^2 = 6400 \cdot \frac{61}{36} \Rightarrow x \approx 104 \text{ miles}$$

The ship's location is $(104, 50)$.

59. (a) Set up a rectangular coordinate system so that the two devices lie on the x-axis and the origin is midway between them. The devices serve as foci to the hyperbola so $c = \dfrac{2000}{2} = 1000$. Since the explosion occurs 200 feet from point B, the vertex of the hyperbola is $(800, 0)$; therefore, $a = 800$. Finding b:

$$b^2 = c^2 - a^2 \rightarrow b^2 = 1000^2 - 800^2 = 360000 \rightarrow b = 600$$

The equation of the hyperbola is:

$$\frac{x^2}{800^2} - \frac{y^2}{600^2} = 1$$

If $x = 1000$, find y:

$$\frac{1000^2}{800^2} - \frac{y^2}{600^2} = 1 \rightarrow \frac{y^2}{600^2} = \frac{1000^2}{800^2} - 1 = \frac{600^2}{800^2}$$

$$y^2 = 600^2 \cdot \frac{600^2}{800^2} \Rightarrow y = 450 \text{ feet}$$

The second detonation should take place 450 feet north of point B.

61. If the eccentricity is close to 1, then $c \approx a$ and $b \approx 0$. When b is close to 0, the hyperbola is very narrow, because the slopes of the asymptotes are close to 0.

If the eccentricity is very large, then c is much larger than a and b is very large. The result is a hyperbola that is very wide.

63. $\dfrac{x^2}{4} - y^2 = 1$ $(a = 2,\ b = 1)$

is a hyperbola with horizontal
transverse axis, centered at $(0, 0)$ and
has asymptotes: $y = \pm\dfrac{1}{2}x$

$y^2 - \dfrac{x^2}{4} = 1$ $(a = 1,\ b = 2)$

is a hyperbola with vertical transverse
axis, centered at $(0, 0)$ and has

asymptotes: $y = \pm\dfrac{1}{2}x$

Since the two hyperbolas have the
same asymptotes, they are conjugate.

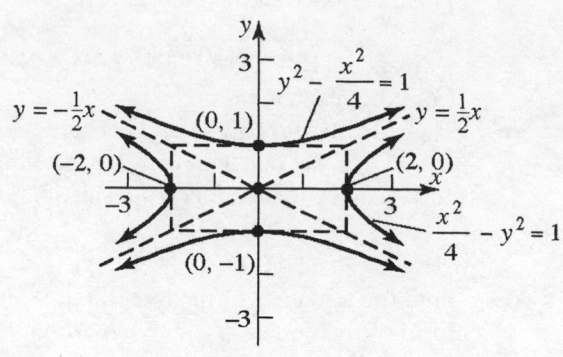

65. Put the equation in standard hyperbola form:
$$Ax^2 + Cy^2 + F = 0 \qquad A \neq 0,\ C \neq 0,\ F \neq 0$$

$$Ax^2 + Cy^2 = -F \rightarrow \frac{Ax^2}{-F} + \frac{Cy^2}{-F} = 1$$

$$\frac{x^2}{-F/A} + \frac{y^2}{-F/C} = 1$$

Since $-F/A$ and $-F/C$ have opposite signs, this is a hyperbola with center at $(0, 0)$.

Chapter 9

Analytic Geometry

9.5   Rotation of Axes; General Form of a Conic

1. $x^2 + 4x + y + 3 = 0$
$A = 1$ and $C = 0$; $AC = (1)(0) = 0$. Since $AC = 0$, the equation defines a parabola.

3. $6x^2 + 3y^2 - 12x + 6y = 0$
$A = 6$ and $C = 3$; $AC = (6)(3) = 18$. Since $AC > 0$ and $A \neq C$, the equation defines an ellipse.

5. $3x^2 - 2y^2 + 6x + 4 = 0$
$A = 3$ and $C = -2$; $AC = (3)(-2) = -6$. Since $AC < 0$, the equation defines a hyperbola.

7. $2y^2 - x^2 - y + x = 0$
$A = -1$ and $C = 2$; $AC = (-1)(2) = -2$. Since $AC < 0$, the equation defines a hyperbola.

9. $x^2 + y^2 - 8x + 4y = 0$
$A = 1$ and $C = 1$; $AC = (1)(1) = 1$. Since $AC > 0$ and $A = C$, the equation defines a circle.

11. $x^2 + 4xy + y^2 - 3 = 0$
$A = 1, B = 4,$ and $C = 1$; $\cot(2\theta) = \dfrac{A-C}{B} = \dfrac{1-1}{4} = \dfrac{0}{4} = 0 \;\rightarrow\; 2\theta = \dfrac{\pi}{2} \;\rightarrow\; \theta = \dfrac{\pi}{4}$

$x = x'\cos\left(\dfrac{\pi}{4}\right) - y'\sin\left(\dfrac{\pi}{4}\right) = \dfrac{\sqrt{2}}{2}x' - \dfrac{\sqrt{2}}{2}y' = \dfrac{\sqrt{2}}{2}(x' - y')$

$y = x'\sin\left(\dfrac{\pi}{4}\right) + y'\cos\left(\dfrac{\pi}{4}\right) = \dfrac{\sqrt{2}}{2}x' + \dfrac{\sqrt{2}}{2}y' = \dfrac{\sqrt{2}}{2}(x' + y')$

13. $5x^2 + 6xy + 5y^2 - 8 = 0$
$A = 5, B = 6,$ and $C = 5$; $\cot(2\theta) = \dfrac{A-C}{B} = \dfrac{5-5}{6} = \dfrac{0}{6} = 0 \;\rightarrow\; 2\theta = \dfrac{\pi}{2} \;\rightarrow\; \theta = \dfrac{\pi}{4}$

$x = x'\cos\left(\dfrac{\pi}{4}\right) - y'\sin\left(\dfrac{\pi}{4}\right) = \dfrac{\sqrt{2}}{2}x' - \dfrac{\sqrt{2}}{2}y' = \dfrac{\sqrt{2}}{2}(x' - y')$

$y = x'\sin\left(\dfrac{\pi}{4}\right) + y'\cos\left(\dfrac{\pi}{4}\right) = \dfrac{\sqrt{2}}{2}x' + \dfrac{\sqrt{2}}{2}y' = \dfrac{\sqrt{2}}{2}(x' + y')$

15. $13x^2 - 6\sqrt{3}xy + 7y^2 - 16 = 0$

$A = 13$, $B = -6\sqrt{3}$, and $C = 7$; $\cot(2\theta) = \dfrac{A-C}{B} = \dfrac{13-7}{-6\sqrt{3}} = \dfrac{6}{-6\sqrt{3}} = -\dfrac{\sqrt{3}}{3}$

$$2\theta = \frac{2\pi}{3} \;\rightarrow\; \theta = \frac{\pi}{3}$$

$$x = x'\cos\left(\frac{\pi}{3}\right) - y'\sin\left(\frac{\pi}{3}\right) = \frac{1}{2}x' - \frac{\sqrt{3}}{2}y' = \frac{1}{2}\left(x' - \sqrt{3}y'\right)$$

$$y = x'\sin\left(\frac{\pi}{3}\right) + y'\cos\left(\frac{\pi}{3}\right) = \frac{\sqrt{3}}{2}x' + \frac{1}{2}y' = \frac{1}{2}\left(\sqrt{3}x' + y'\right)$$

17. $4x^2 - 4xy + y^2 - 8\sqrt{5}\,x - 16\sqrt{5}\,y = 0$

$A = 4$, $B = -4$, and $C = 1$; $\cot(2\theta) = \dfrac{A-C}{B} = \dfrac{4-1}{-4} = -\dfrac{3}{4}$; $\cos 2\theta = -\dfrac{3}{5}$

$$\sin\theta = \sqrt{\frac{1-\left(-\dfrac{3}{5}\right)}{2}} = \sqrt{\frac{4}{5}} = \frac{2}{\sqrt{5}} = \frac{2\sqrt{5}}{5}$$

$$\cos\theta = \sqrt{\frac{1+\left(-\dfrac{3}{5}\right)}{2}} = \sqrt{\frac{1}{5}} = \frac{1}{\sqrt{5}} = \frac{\sqrt{5}}{5}$$

$$x = x'\cos\theta - y'\sin\theta = \frac{\sqrt{5}}{5}x' - \frac{2\sqrt{5}}{5}y' = \frac{\sqrt{5}}{5}\left(x' - 2y'\right)$$

$$y = x'\sin\theta + y'\cos\theta = \frac{2\sqrt{5}}{5}x' + \frac{\sqrt{5}}{5}y' = \frac{\sqrt{5}}{5}\left(2x' + y'\right)$$

19. $25x^2 - 36xy + 40y^2 - 12\sqrt{13}\,x - 8\sqrt{13}\,y = 0$

$A = 25$, $B = -36$, and $C = 40$; $\cot(2\theta) = \dfrac{A-C}{B} = \dfrac{25-40}{-36} = \dfrac{5}{12}$; $\cos 2\theta = \dfrac{5}{13}$

$$\sin\theta = \sqrt{\frac{\left(1-\dfrac{5}{13}\right)}{2}} = \sqrt{\frac{4}{13}} = \frac{2}{\sqrt{13}} = \frac{2\sqrt{13}}{13};$$

$$\cos\theta = \sqrt{\frac{\left(1+\dfrac{5}{13}\right)}{2}} = \sqrt{\frac{9}{13}} = \frac{3}{\sqrt{13}} = \frac{3\sqrt{13}}{13}$$

$$x = x'\cos\theta - y'\sin\theta = \frac{3\sqrt{13}}{13}x' - \frac{2\sqrt{13}}{13}y' = \frac{\sqrt{13}}{13}\left(3x' - 2y'\right)$$

$$y = x'\sin\theta + y'\cos\theta = \frac{2\sqrt{13}}{13}x' + \frac{3\sqrt{13}}{13}y' = \frac{\sqrt{13}}{13}\left(2x' + 3y'\right)$$

21. $x^2 + 4xy + y^2 - 3 = 0; \quad \theta = 45°$ (see Problem 11)

$$\left(\frac{\sqrt{2}}{2}(x'-y')\right)^2 + 4\left(\frac{\sqrt{2}}{2}(x'-y')\right)\left(\frac{\sqrt{2}}{2}(x'+y')\right) + \left(\frac{\sqrt{2}}{2}(x'+y')\right)^2 - 3 = 0$$

$$\frac{1}{2}\left(x'^2 - 2x'y' + y'^2\right) + 2\left(x'^2 - y'^2\right) + \frac{1}{2}\left(x'^2 + 2x'y' + y'^2\right) - 3 = 0$$

$$\frac{1}{2}x'^2 - x'y' + \frac{1}{2}y'^2 + 2x'^2 - 2y'^2 + \frac{1}{2}x'^2 + x'y' + \frac{1}{2}y'^2 = 3$$

$$3x'^2 - y'^2 = 3$$

$$\frac{x'^2}{1} - \frac{y'^2}{3} = 1$$

Hyperbola; center at the origin,
transverse axis is the x'-axis, vertices at (±1, 0).

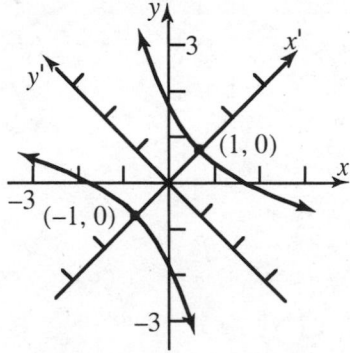

23. $5x^2 + 6xy + 5y^2 - 8 = 0; \quad \theta = 45°$ (see Problem 13)

$$5\left(\frac{\sqrt{2}}{2}(x'-y')\right)^2 + 6\left(\frac{\sqrt{2}}{2}(x'-y')\right)\left(\frac{\sqrt{2}}{2}(x'+y')\right) + 5\left(\frac{\sqrt{2}}{2}(x'+y')\right)^2 - 8 = 0$$

$$\frac{5}{2}\left(x'^2 - 2x'y' + y'^2\right) + 3\left(x'^2 - y'^2\right) + \frac{5}{2}\left(x'^2 + 2x'y' + y'^2\right) - 8 = 0$$

$$\frac{5}{2}x'^2 - 5x'y' + \frac{5}{2}y'^2 + 3x'^2 - 3y'^2 + \frac{5}{2}x'^2 + 5x'y' + \frac{5}{2}y'^2 = 8$$

$$8x'^2 + 2y'^2 = 8$$

$$\frac{x'^2}{1} + \frac{y'^2}{4} = 1$$

Ellipse; center at the origin,
major axis is the y'-axis, vertices at (0, ±2).

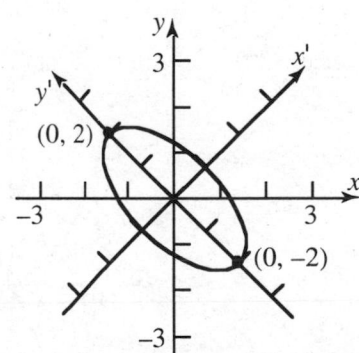

25. $13x^2 - 6\sqrt{3}\,xy + 7y^2 - 16 = 0;\quad \theta = 60°$ (see Problem 15)

$13\left(\frac{1}{2}\left(x' - \sqrt{3}y'\right)\right)^2 - 6\sqrt{3}\left(\frac{1}{2}\left(x' - \sqrt{3}y'\right)\right)\left(\frac{1}{2}\left(\sqrt{3}\,x' + y'\right)\right) + 7\left(\frac{1}{2}\left(\sqrt{3}\,x' + y'\right)\right)^2 - 16 = 0$

$\frac{13}{4}\left(x'^2 - 2\sqrt{3}x'\,y' + 3y'^2\right) - \frac{3\sqrt{3}}{2}\left(\sqrt{3}x'^2 - 2x'\,y' - \sqrt{3}y'^2\right) + \frac{7}{4}\left(3x'^2 + 2\sqrt{3}x'\,y' + y'^2\right) = 16$

$\frac{13}{4}x'^2 - \frac{13\sqrt{3}}{2}x'\,y' + \frac{39}{4}y'^2 - \frac{9}{2}x'^2 + 3\sqrt{3}x'\,y' + \frac{9}{2}y'^2 + \frac{21}{4}x'^2 + \frac{7\sqrt{3}}{2}x'\,y' + \frac{7}{4}y'^2 = 16$

$4x'^2 + 16y'^2 = 16$

$\frac{x'^2}{4} + \frac{y'^2}{1} = 1$

Ellipse; center at the origin,
major axis is the x'-axis, vertices at $(\pm 2, 0)$.

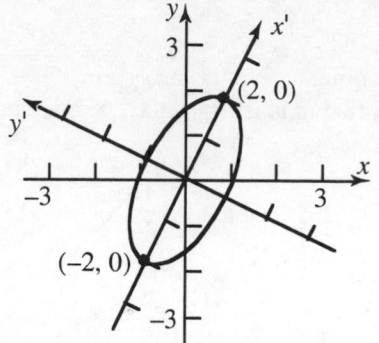

27. $4x^2 - 4xy + y^2 - 8\sqrt{5}\,x - 16\sqrt{5}\,y = 0;\quad \theta = 63.4°$ (see Problem 17)

$4\left(\frac{\sqrt{5}}{5}\left(x' - 2y'\right)\right)^2 - 4\left(\frac{\sqrt{5}}{5}\left(x' - 2y'\right)\right)\left(\frac{\sqrt{5}}{5}\left(2x' + y'\right)\right) + \left(\frac{\sqrt{5}}{5}\left(2x' + y'\right)\right)^2$

$- 8\sqrt{5}\left(\frac{\sqrt{5}}{5}\left(x' - 2y'\right)\right) - 16\sqrt{5}\left(\frac{\sqrt{5}}{5}\left(2x' + y'\right)\right) = 0$

$\frac{4}{5}\left(x'^2 - 4x'\,y' + 4y'^2\right) - \frac{4}{5}\left(2x'^2 - 3x'\,y' - 2y'^2\right) + \frac{1}{5}\left(4x'^2 + 4x'\,y' + y'^2\right)$

$- 8x' + 16y' - 32x' - 16y' = 0$

$\frac{4}{5}x'^2 - \frac{16}{5}x'\,y' + \frac{16}{5}y'^2 - \frac{8}{5}x'^2 + \frac{12}{5}x'\,y' + \frac{8}{5}y'^2 + \frac{4}{5}x'^2 + \frac{4}{5}x'\,y' + \frac{1}{5}y'^2 - 40x' = 0$

$5y'^2 - 40x' = 0$

$y'^2 = 8x'$

Parabola; vertex at the origin, focus at $(2, 0)$.

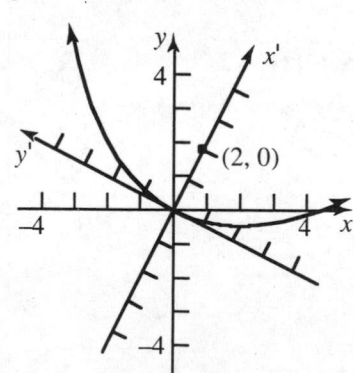

29. $25x^2 - 36xy + 40y^2 - 12\sqrt{13}x - 8\sqrt{13}y = 0;\quad \theta \approx 33.7°$ (see Problem 19)

$$25\left(\frac{\sqrt{13}}{13}(3x'-2y')\right)^2 - 36\left(\frac{\sqrt{13}}{13}(3x'-2y')\right)\left(\frac{\sqrt{13}}{13}(2x'+3y')\right) + 40\left(\frac{\sqrt{13}}{13}(2x'+3y')\right)^2$$

$$-12\sqrt{13}\left(\frac{\sqrt{13}}{13}(3x'-2y')\right) - 8\sqrt{13}\left(\frac{\sqrt{13}}{13}(2x'+3y')\right) = 0$$

$$\frac{25}{13}\left(9x'^2 - 12x'y' + 4y'^2\right) - \frac{36}{13}\left(6x'^2 + 5x'y' - 6y'^2\right) + \frac{40}{13}\left(4x'^2 + 12x'y' + 9y'^2\right)$$

$$- 36x' + 24y' - 16x' - 24y' = 0$$

$$\frac{225}{13}x'^2 - \frac{300}{13}x'y' + \frac{100}{13}y'^2 - \frac{216}{13}x'^2 - \frac{180}{13}x'y' + \frac{216}{13}y'^2$$

$$+ \frac{160}{13}x'^2 + \frac{480}{13}x'y' + \frac{360}{13}y'^2 - 52x' = 0$$

$13x'^2 + 52y'^2 - 52x' = 0$

$x'^2 - 4x' + 4y'^2 = 0$

$(x' - 2)^2 + 4y'^2 = 4$

$\dfrac{(x' - 2)^2}{4} + \dfrac{y'^2}{1} = 1$

Ellipse; center at (2, 0),
major axis is the x'-axis, vertices at (4, 0) and (0, 0).

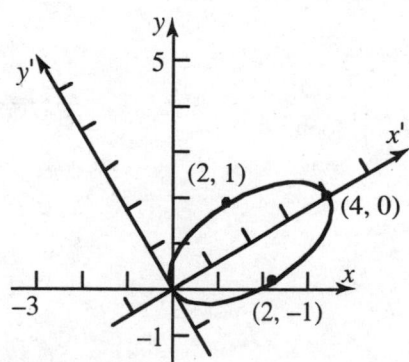

31. $16x^2 + 24xy + 9y^2 - 130x + 90y = 0$

$A = 16, B = 24$, and $C = 9$; $\cot(2\theta) = \dfrac{A - C}{B} = \dfrac{16 - 9}{24} = \dfrac{7}{24}$ $\rightarrow$ $\cos(2\theta) = \dfrac{7}{25}$

$\sin\theta = \sqrt{\dfrac{\left(1 - \dfrac{7}{25}\right)}{2}} = \sqrt{\dfrac{9}{25}} = \dfrac{3}{5}$; $\cos\theta = \sqrt{\dfrac{\left(1 + \dfrac{7}{25}\right)}{2}} = \sqrt{\dfrac{16}{25}} = \dfrac{4}{5}$ $\rightarrow$ $\theta \approx 36.9°$

$x = x'\cos\theta - y'\sin\theta = \dfrac{4}{5}x' - \dfrac{3}{5}y' = \dfrac{1}{5}(4x' - 3y')$

$y = x'\sin\theta + y'\cos\theta = \dfrac{3}{5}x' + \dfrac{4}{5}y' = \dfrac{1}{5}(3x' + 4y')$

$16\left(\dfrac{1}{5}(4x' - 3y')\right)^2 + 24\left(\dfrac{1}{5}(4x' - 3y')\right)\left(\dfrac{1}{5}(3x' + 4y')\right) + 9\left(\dfrac{1}{5}(3x' + 4y')\right)^2$

$$-130\left(\dfrac{1}{5}(4x' - 3y')\right) + 90\left(\dfrac{1}{5}(3x' + 4y')\right) = 0$$

$\dfrac{16}{25}(16x'^2 - 24x'y' + 9y'^2) + \dfrac{24}{25}(12x'^2 + 7x'y' - 12y'^2) + \dfrac{9}{25}(9x'^2 + 24x'y' + 16y'^2)$

$$-104x' + 78y' + 54x' + 72y' = 0$$

$\dfrac{256}{25}x'^2 - \dfrac{384}{25}x'y' + \dfrac{144}{25}y'^2 + \dfrac{288}{25}x'^2 + \dfrac{168}{25}x'y' - \dfrac{288}{25}y'^2$

$$+\dfrac{81}{25}x'^2 + \dfrac{216}{25}x'y' + \dfrac{144}{25}y'^2 - 50x' + 150y' = 0$$

$25x'^2 - 50x' + 150y' = 0$

$x'^2 - 2x' = -6y'$

$(x' - 1)^2 = -6y' + 1$

$(x' - 1)^2 = -6\left(y' - \dfrac{1}{6}\right)$

Parabola; vertex at $\left(1, \dfrac{1}{6}\right)$, focus at $\left(1, -\dfrac{4}{3}\right)$.

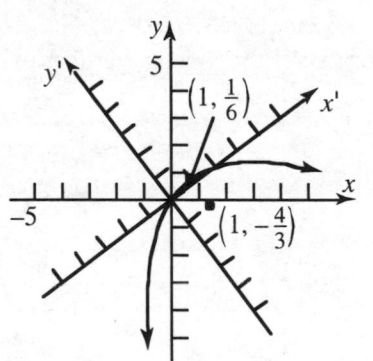

33. $A = 1.\ B = 3,\ C = -2$ $B^2 - 4AC = 3^2 - 4(1)(-2) = 17 > 0$; hyperbola

35. $A = 1.\ B = -7,\ C = 3$ $B^2 - 4AC = (-7)^2 - 4(1)(3) = 37 > 0$; hyperbola

37. $A = 9.\ B = 12,\ C = 4$ $B^2 - 4AC = 12^2 - 4(9)(4) = 0$; parabola

39. $A = 10.\ B = -12,\ C = 4$ $B^2 - 4AC = (-12)^2 - 4(10)(4) = -16 < 0$; ellipse

41. $A = 3.\ B = -2,\ C = 1$ $B^2 - 4AC = (-2)^2 - 4(3)(1) = -8 < 0$; ellipse

43.   See equation 6 on page 770.

$A' = A\cos^2\theta + B\sin\theta\cos\theta + C\sin^2\theta$

$B' = B(\cos^2\theta - \sin^2\theta) + 2(C-A)(\sin\theta\cos\theta)$

$C' = A\sin^2\theta - B\sin\theta\cos\theta + C\cos^2\theta$

$D' = D\cos\theta + E\sin\theta$

$E' = -D\sin\theta + E\cos\theta$

$F' = F$

45. $B'^2 - 4A'C'$

$= \left[B(\cos^2\theta - \sin^2\theta) + 2(C-A)\sin\theta\cos\theta\right]^2$

$\quad - 4\left(A\cos^2\theta + B\sin\theta\cos\theta + C\sin^2\theta\right)\left(A\sin^2\theta - B\sin\theta\cos\theta + C\cos^2\theta\right)$

$= B^2\left(\cos^4\theta - 2\cos^2\theta\sin^2\theta + \sin^4\theta\right) + 4B(C-A)\sin\theta\cos\theta(\cos^2\theta - \sin^2\theta)$

$\quad + 4(C-A)^2\sin^2\theta\cos^2\theta - 4\left[A^2\sin^2\theta\cos^2\theta - AB\sin\theta\cos^3\theta + AC\cos^4\theta\right.$

$\quad + AB\sin^3\theta\cos\theta - B^2\sin^2\theta\cos^2\theta + BC\sin\theta\cos^3\theta + AC\sin^4\theta$

$\quad \left. - BC\sin^3\theta\cos\theta + C^2\sin^2\theta\cos^2\theta\right]$

$= B^2\left(\cos^4\theta - 2\cos^2\theta\sin^2\theta + \sin^4\theta + 4\sin^2\theta\cos^2\theta\right)$

$\quad + BC\left(4\sin\theta\cos\theta(\cos^2\theta - \sin^2\theta) - 4\sin\theta\cos^3\theta + 4\sin^3\theta\cos\theta\right)$

$\quad - AB\left(4\sin\theta\cos\theta(\cos^2\theta - \sin^2\theta) - 4\sin\theta\cos^3\theta + 4\sin^3\theta\cos\theta\right)$

$\quad + 4C^2\left(\sin^2\theta\cos^2\theta - \sin^2\theta\cos^2\theta\right) - 4AC\left(2\sin^2\theta\cos^2\theta + \cos^4\theta + \sin^4\theta\right)$

$\quad + 4A^2\left(\sin^2\theta\cos^2\theta - \sin^2\theta\cos^2\theta\right)$

$= B^2\left(\cos^4\theta + 2\sin^2\theta\cos^2\theta + \sin^4\theta\right) - 4AC\left(\cos^4\theta + 2\sin^2\theta\cos^2\theta + \sin^4\theta\right)$

$= B^2\left(\cos^2\theta + \sin^2\theta\right)^2 - 4AC\left(\cos^2\theta + \sin^2\theta\right)^2 = B^2 - 4AC$

47. $d^2 = (y_2 - y_1)^2 + (x_2 - x_1)^2$

$= \left(x_2'\sin\theta + y_2'\cos\theta - x_1'\sin\theta - y_1'\cos\theta\right)^2$

$\qquad\qquad + \left(x_2'\cos\theta - y_2'\sin\theta - x_1'\cos\theta + y_1'\sin\theta\right)^2$

$= \left((x_2'-x_1')\sin\theta + (y_2'-y_1')\cos\theta\right)^2 + \left((x_2'-x_1')\cos\theta - (y_2'-y_1')\sin\theta\right)^2$

$= (x_2'-x_1')^2\sin^2\theta + 2(x_2'-x_1')(y_2'-y_1')\sin\theta\cos\theta + (y_2'-y_1')^2\cos^2\theta$

$\qquad + (x_2'-x_1')^2\cos^2\theta - 2(x_2'-x_1')(y_2'-y_1')\sin\theta\cos\theta + (y_2'-y_1')^2\sin^2\theta$

$= (x_2'-x_1')^2\sin^2\theta + (x_2'-x_1')^2\cos^2\theta + (y_2'-y_1')^2\cos^2\theta + (y_2'-y_1')^2\sin^2\theta$

$= (x_2'-x_1')^2 + (y_2'-y_1')^2$

49.   Answers will vary.

Chapter 9

Analytic Geometry

9.6   Polar Equations of Conics

1. $e = 1;$ $p = 1;$ parabola; directrix is perpendicular to the polar axis and 1 unit to the right of the pole.

3. $r = \dfrac{4}{2 - 3\sin\theta} = \dfrac{4}{2\left(1 - \dfrac{3}{2}\sin\theta\right)} = \dfrac{2}{\left(1 - \dfrac{3}{2}\sin\theta\right)};$ $ep = 2,\ e = \dfrac{3}{2};\ p = \dfrac{4}{3}$

Hyperbola; directrix is parallel to the polar axis and $\dfrac{4}{3}$ units below the pole.

5. $r = \dfrac{3}{4 - 2\cos\theta} = \dfrac{3}{4\left(1 - \dfrac{1}{2}\cos\theta\right)} = \dfrac{\left(\dfrac{3}{4}\right)}{\left(1 - \dfrac{1}{2}\cos\theta\right)};$ $ep = \dfrac{3}{4},\ e = \dfrac{1}{2};\ p = \dfrac{3}{2}$

Ellipse; directrix is perpendicular to the polar axis and $\dfrac{3}{2}$ units to the left of the pole.

7. $r = \dfrac{1}{1 + \cos\theta}$

$ep = 1,\ e = 1,\ p = 1$
Parabola; directrix is perpendicular to the
polar axis 1 unit to the right of the pole;
vertex is $\left(\dfrac{1}{2}, 0\right)$.

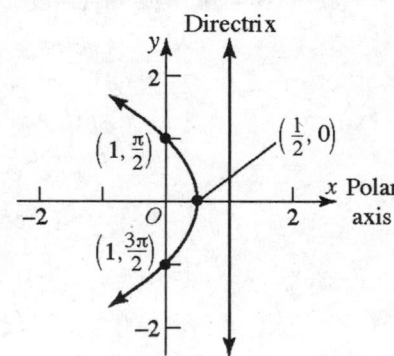

9. $\quad r = \dfrac{8}{4 + 3\sin\theta}$

$$r = \dfrac{8}{4\left(1 + \dfrac{3}{4}\sin\theta\right)} = \dfrac{2}{\left(1 + \dfrac{3}{4}\sin\theta\right)}$$

$ep = 2, \; e = \dfrac{3}{4}, \; p = \dfrac{8}{3}$

Ellipse; directrix is parallel to the polar axis
$\dfrac{8}{3}$ units above the pole; vertices are
$\left(\dfrac{8}{7}, \dfrac{\pi}{2}\right)$ and $\left(8, \dfrac{3\pi}{2}\right)$.

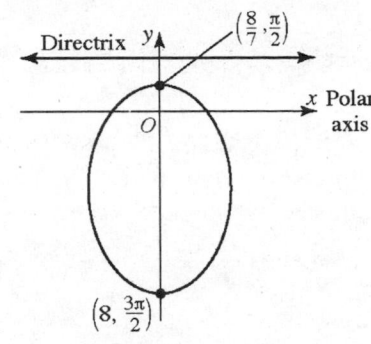

11. $\quad r = \dfrac{9}{3 - 6\cos\theta}$

$$r = \dfrac{9}{3(1 - 2\cos\theta)} = \dfrac{3}{1 - 2\cos\theta}$$

$ep = 3, \; e = 2, \; p = \dfrac{3}{2}$

Hyperbola; directrix is perpendicular to the
polar axis $\dfrac{3}{2}$ units to the left of the pole;
vertices are $(-3, 0)$ and $(1, \pi)$.

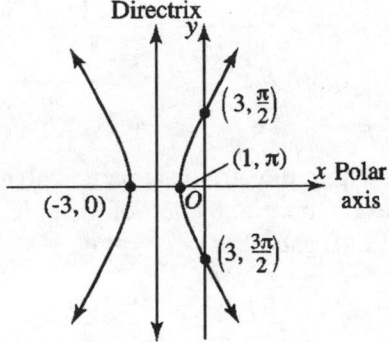

13. $\quad r = \dfrac{8}{2 - \sin\theta}$

$$r = \dfrac{8}{2\left(1 - \dfrac{1}{2}\sin\theta\right)} = \dfrac{4}{1 - \dfrac{1}{2}\sin\theta}$$

$ep = 4, \; e = \dfrac{1}{2}, \; p = 8$

Ellipse; directrix is parallel to the polar axis
8 units below the pole; vertices are
$\left(8, \dfrac{\pi}{2}\right)$ and $\left(\dfrac{8}{3}, \dfrac{3\pi}{2}\right)$.

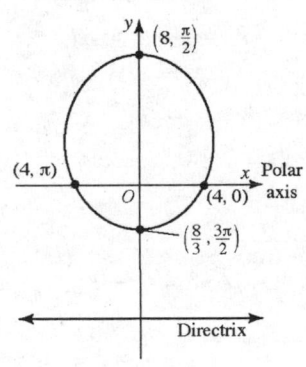

15. $r(3 - 2\sin\theta) = 6 \quad \rightarrow \quad r = \dfrac{6}{3 - 2\sin\theta}$

$r = \dfrac{6}{3\left(1 - \dfrac{2}{3}\sin\theta\right)} = \dfrac{2}{1 - \dfrac{2}{3}\sin\theta}$

$ep = 2, \; e = \dfrac{2}{3}, \; p = 3$

Ellipse; directrix is parallel to the polar axis
3 units below the pole; vertices are
$\left(6, \dfrac{\pi}{2}\right)$ and $\left(\dfrac{6}{5}, \dfrac{3\pi}{2}\right)$.

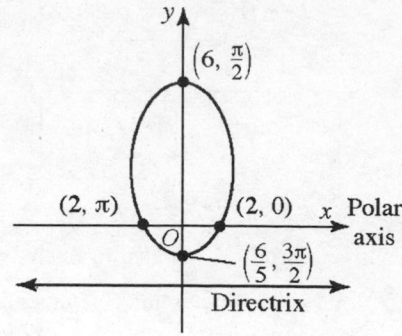

17. $r = \dfrac{6\sec\theta}{2\sec\theta - 1} = \dfrac{6}{2 - \cos\theta}$

$r = \dfrac{6}{2\left(1 - \dfrac{1}{2}\cos\theta\right)} = \dfrac{3}{1 - \dfrac{1}{2}\cos\theta}$

$ep = 3, \; e = \dfrac{1}{2}, \; p = 6$

Ellipse; directrix is perpendicular to the polar
axis 6 units to the left of the pole; vertices are
$(6, 0)$ and $(2, \pi)$.

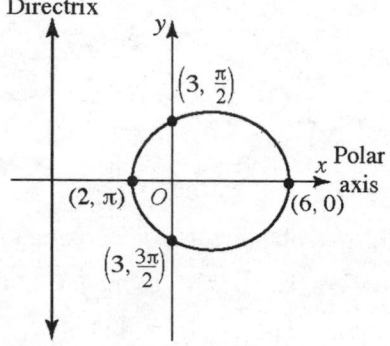

19. $r = \dfrac{1}{1 + \cos\theta}$

$r + r\cos\theta = 1$

$r = 1 - r\cos\theta$

$r^2 = (1 - r\cos\theta)^2$

$x^2 + y^2 = (1 - x)^2$

$x^2 + y^2 = 1 - 2x + x^2$

$y^2 + 2x - 1 = 0$

21. $r = \dfrac{8}{4 + 3\sin\theta}$

$4r + 3r\sin\theta = 8$

$4r = 8 - 3r\sin\theta$

$16r^2 = (8 - 3r\sin\theta)^2$

$16(x^2 + y^2) = (8 - 3y)^2$

$16x^2 + 16y^2 = 64 - 48y + 9y^2$

$16x^2 + 7y^2 + 48y - 64 = 0$

23. $r = \dfrac{9}{3 - 6\cos\theta}$

$3r - 6r\cos\theta = 9$

$3r = 9 + 6r\cos\theta$

$r = 3 + 2r\cos\theta$

$r^2 = (3 + 2r\cos\theta)^2$

$x^2 + y^2 = (3 + 2x)^2$

$x^2 + y^2 = 9 + 12x + 4x^2$

$3x^2 - y^2 + 12x + 9 = 0$

25. $r = \dfrac{8}{2 - \sin\theta}$

$2r - r\sin\theta = 8$

$2r = 8 + r\sin\theta$

$4r^2 = (8 + r\sin\theta)^2$

$4(x^2 + y^2) = (8 + y)^2$

$4x^2 + 4y^2 = 64 + 16y + y^2$

$4x^2 + 3y^2 - 16y - 64 = 0$

27. $r(3 - 2\sin\theta) = 6$

$$3r - 2r\sin\theta = 6$$
$$3r = 6 + 2r\sin\theta$$
$$9r^2 = (6 + 2r\sin\theta)^2$$
$$9(x^2 + y^2) = (6 + 2y)^2$$
$$9x^2 + 9y^2 = 36 + 24y + 4y^2$$
$$9x^2 + 5y^2 - 24y - 36 = 0$$

29. $r = \dfrac{6\sec\theta}{2\sec\theta - 1}$

$$r = \dfrac{6}{2 - \cos\theta}$$
$$2r - r\cos\theta = 6$$
$$2r = 6 + r\cos\theta$$
$$4r^2 = (6 + r\cos\theta)^2$$
$$4(x^2 + y^2) = (6 + x)^2$$
$$4x^2 + 4y^2 = 36 + 12x + x^2$$
$$3x^2 + 4y^2 - 12x - 36 = 0$$

31. $r = \dfrac{ep}{1 + e\sin\theta}$

$e = 1; \quad p = 1$

$$r = \dfrac{1}{1 + \sin\theta}$$

33. $r = \dfrac{ep}{1 - e\cos\theta}$

$e = \dfrac{4}{5}; \quad p = 3$

$$r = \dfrac{\left(\dfrac{12}{5}\right)}{\left(1 - \dfrac{4}{5}\cos\theta\right)} = \dfrac{12}{5 - 4\cos\theta}$$

35. $r = \dfrac{ep}{1 - e\sin\theta}$

$e = 6; \quad p = 2$

$$r = \dfrac{12}{1 - 6\sin\theta}$$

37. $d(F, P) = e \cdot d(D, P)$ $d(D, P) = p - r\cos\theta$

$$r = e(p - r\cos\theta)$$
$$r = ep - er\cos\theta$$

$$r + er\cos\theta = ep \rightarrow r(1 + e\cos\theta) = ep \rightarrow r = \dfrac{ep}{1 + e\cos\theta}$$

39. $d(F, P) = e \cdot d(D, P)$ $d(D, P) = p + r\sin\theta$

$$r = e(p + r\sin\theta)$$
$$r = ep + er\sin\theta$$

$$r - er\sin\theta = ep \rightarrow r(1 - e\sin\theta) = ep \rightarrow r = \dfrac{ep}{1 - e\sin\theta}$$

Chapter 9

Analytic Geometry

9.7   Plane Curves and Parametric Equations

1. $x = 3t + 2,\ \ y = t + 1,\ \ 0 \le t \le 4$

$x = 3(y - 1) + 2$
$x = 3y - 3 + 2$
$x = 3y - 1$
$x - 3y + 1 = 0$

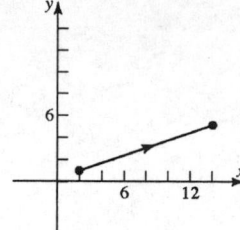

3. $x = t + 2,\ \ y = \sqrt{t},\ \ t \ge 0$

$y = \sqrt{x - 2}$

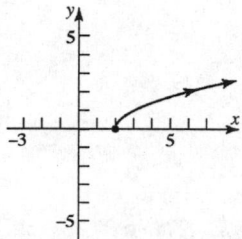

5. $x = t^2 + 4,\ \ y = t^2 - 4,\ \ -\infty < t < \infty$

$y = (x - 4) - 4$
$y = x - 8$
For $-\infty < t < 0$ the movement is to
the left. For $0 < t < \infty$ the
movement is to the right.

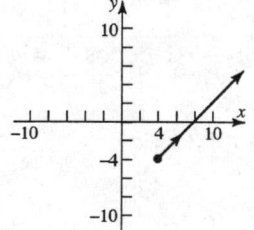

7. $x = 3t^2,\ \ y = t + 1,\ \ -\infty < t < \infty$
$x = 3(y - 1)^2$

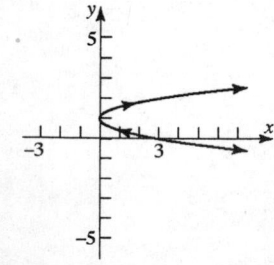

9. $x = 2e^t, \ y = 1 + e^t, \ t \geq 0$

$y = 1 + \dfrac{x}{2}$

$2y = 2 + x$

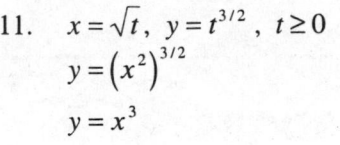

11. $x = \sqrt{t}, \ y = t^{3/2}, \ t \geq 0$

$y = \left(x^2\right)^{3/2}$

$y = x^3$

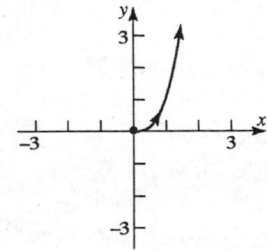

13. $x = 2\cos t, \ y = 3\sin t, \ 0 \leq t \leq 2\pi$

$\dfrac{x}{2} = \cos t \qquad \dfrac{y}{3} = \sin t$

$\left(\dfrac{x}{2}\right)^2 + \left(\dfrac{y}{3}\right)^2 = \cos^2 t + \sin^2 t = 1$

$\dfrac{x^2}{4} + \dfrac{y^2}{9} = 1$

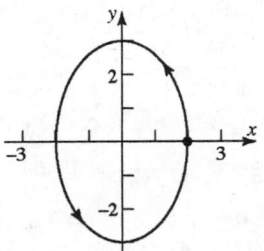

15. $x = 2\cos t, \ y = 3\sin t, \ -\pi \leq t \leq 0$

$\dfrac{x}{2} = \cos t \qquad \dfrac{y}{3} = \sin t$

$\left(\dfrac{x}{2}\right)^2 + \left(\dfrac{y}{3}\right)^2 = \cos^2 t + \sin^2 t = 1$

$\dfrac{x^2}{4} + \dfrac{y^2}{9} = 1 \qquad y \leq 0$

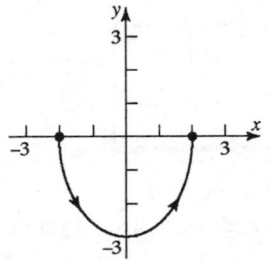

17. $x = \sec t, \ y = \tan t, \ 0 \leq t \leq \dfrac{\pi}{4}$

$\sec^2 t = 1 + \tan^2 t$

$x^2 = 1 + y^2$

$x^2 - y^2 = 1$

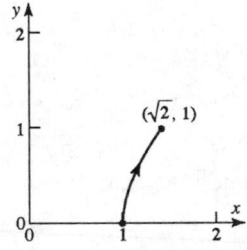

19. $x = \sin^2 t, \; y = \cos^2 t, \; 0 \le t \le 2\pi$

$$\sin^2 t + \cos^2 t = 1$$
$$x + y = 1$$

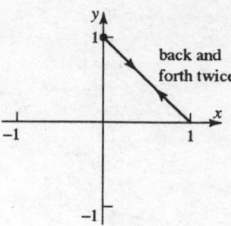

back and
forth twice

21.   (a)   Use equation (2):
$$x = 50\cos(90°)t = 0$$

$$y = -\frac{1}{2}(32)t^2 + \left(50\sin(90°)\right)t + 6 = -16t^2 + 50t + 6$$

     (b)   The ball is in the air until $y = 0$. Solve: $-16t^2 + 50t + 6 = 0$

$$t = \frac{-50 \pm \sqrt{50^2 - 4(-16)(6)}}{2(-16)} = \frac{-50 \pm \sqrt{2884}}{-32} \approx -0.12 \text{ or } 3.24$$

         The ball is in the air for about 3.24 seconds. (The negative solution is extraneous.)

     (c)   The maximum height occurs at the vertex of the quadratic function.
$$t = \frac{-b}{2a} = \frac{-50}{2(-16)} = 1.5625 \text{ seconds}$$

         Evaluate the function to find the maximum height:
$$-16(1.5625)^2 + 50(1.5625) + 6 = 45.0625$$
         The maximum height is 45.0625 feet.

     (d)   For plotting convenience, we use $x = 3$ so that the line is not on top of the y-axis.

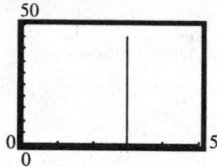

23.   (a)   Train: Use equation (2) with $g = 2, \; v_0 = 0, \; h = 0$

$$x_1 = \frac{1}{2}(2)t^2 + 0 \cdot t + 0 = t^2$$

         Let $y_1 = 1$ for plotting convenience.
         Bill:
$$x_2 = 5(t - 5)$$

         Let $y_2 = 3$ for plotting convenience.

     (b)   Bill will catch the train if $x_1 = x_2$.

$$t^2 = 5(t - 5) \rightarrow t^2 = 5t - 25 \rightarrow t^2 - 5t + 25 = 0$$

         Since $b^2 - 4ac = (-5)^2 - 4(1)(25) = 25 - 100 = -75 < 0$, the equation has no real
solution. Thus, Bill will not catch the train.

(c)

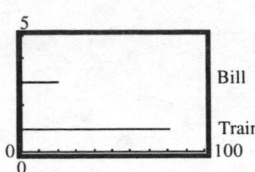

25. (a) Use equation (2):
$$x = (145\cos(20°))t$$

$$y = -\frac{1}{2}(32)t^2 + (145\sin(20°))t + 5$$

(b) The ball is in the air until $y = 0$. Solve: $-16t^2 + (145\sin 20°)t + 5 = 0$

$$t = \frac{-145\sin(20°) \pm \sqrt{(145\sin(20°))^2 - 4(-16)(5)}}{2(-16)}$$

$$\approx \frac{-49.59 \pm \sqrt{2779.46}}{-32} \approx -0.10 \text{ or } 3.20$$

The ball is in the air for about 3.20 seconds. (The negative solution is extraneous.)

(c) The maximum height occurs at the vertex of the quadratic function.

$$t = \frac{-b}{2a} = \frac{-145\sin(20°)}{2(-16)} \approx 1.55 \text{ seconds}$$

Evaluate the function to find the maximum height:

$$-16(1.55)^2 + 145\sin(20°)(1.55) + 5 = 43.43$$

The maximum height is about 43.43 feet.

(d) Find the horizontal displacement:
$$x = (145\cos(20°))(3.20) \approx 436 \text{ feet}$$

(e)

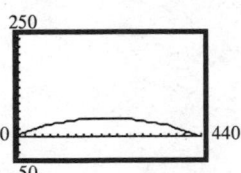

27. (a) Use equation (2):
$$x = (40\cos(45°))t$$

$$y = -\frac{1}{2}(9.8)t^2 + (40\sin(45°))t + 300$$

(b) The ball is in the air until $y = 0$. Solve: $-4.9t^2 + (40\sin(45°))t + 300 = 0$

$$t = \frac{-20\sqrt{2} \pm \sqrt{(20\sqrt{2})^2 - 4(-4.9)(300)}}{2(-4.9)} = \frac{-20\sqrt{2} \pm \sqrt{6680}}{-9.8} \approx -5.45 \text{ or } 11.23$$

The ball is in the air for about 11.23 seconds. (The negative solution is extraneous.)

(c) The maximum height occurs at the vertex of the quadratic function.

$$t = \frac{-b}{2a} = \frac{-20\sqrt{2}}{2(-4.9)} \approx 2.89 \text{ seconds}$$

Evaluate the function to find the maximum height:
$$-4.9(2.89)^2 + 20\sqrt{2}(2.89) + 300 = 340.8 \text{ meters}$$

(d)   Find the horizontal displacement:
$$x = (40\cos(45°))(11.23) \approx 317.6 \text{ meters}$$

(e)

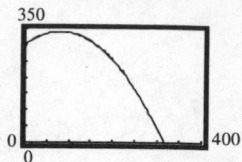

29.   (a)   At $t = 0$, the Paseo is 5 miles from the intersection (at (0, 0)) traveling east (along the x-axis) at 40 mph. Thus, $x = 40t - 5$ describes the position of the Paseo as a function of time. The Bonneville, at $t = 0$, is 4 miles from the intersection traveling north (along the y-axis) at 30 mph. Thus, $y = 30t - 4$ describes the position of the Bonneville as a function of time.

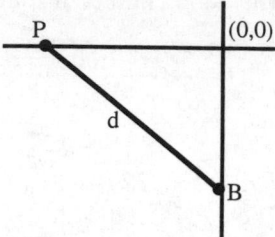

Let d represent the distance between the cars. Use the Pythagorean Theorem to find the distance: $d = \sqrt{(40t - 5)^2 + (30t - 4)^2}$

(b)   From part (a):

Paseo: $x = 40t - 5$ Bonneville: $x = 0$
            $y = 0$ $y = 30t - 4$

(c)   Note this is a function graph not a parametric graph.

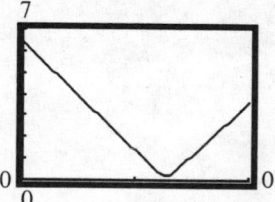

(d)   The minimum distance between the cars is 0.2 miles and occurs at 0.128 hours = 7.68 minutes

(e)

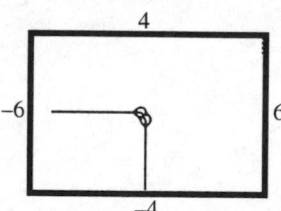

31. $x = t,\ y = 4t - 1$ $x = t + 1,\ y = 4t + 3$

33. $x = t,\ y = t^2 + 1$ $x = t - 1,\ y = t^2 - 2t + 2$

35. $x = t,\ y = t^3$ $x = \sqrt[3]{t},\ y = t$

37. $x = t^{3/2},\ y = t$ $x = t,\ y = t^{2/3}$

39. $x = t + 2,\ y = t;\ \ 0 \le t \le 5$

41. $x = 3\cos t,\ y = 2\sin t;\ \ 0 \le t \le 2\pi$

43. $x = 2\cos(\omega t),\ y = -3\sin(\omega t)$

$\dfrac{2\pi}{\omega} = 2\ \rightarrow\ \omega = \pi$

$x = 2\cos(\pi t),\ y = -3\sin(\pi t),\ \ 0 \le t \le 2$

45. $x = 2\sin(\omega)t,\ y = 3\cos(\omega t)$

$\dfrac{2\pi}{\omega} = 1\ \rightarrow\ \omega = 2\pi$

$x = 2\sin(2\pi t),\ y = 3\cos(2\pi t),\ \ 0 \le t \le 1$

47. C_1 C_2

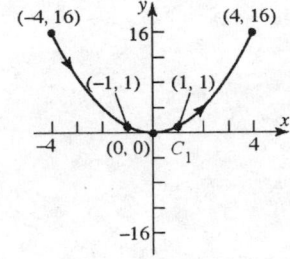

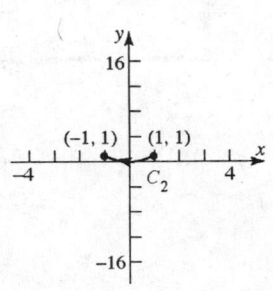

C_3 C_4

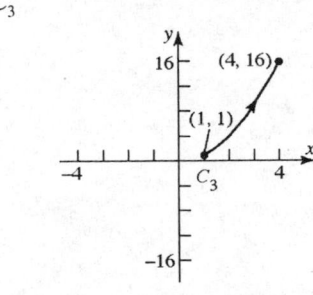

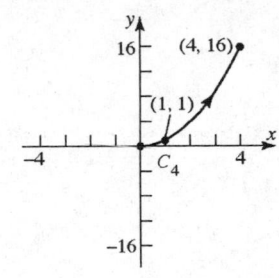

49. $x = (x_2 - x_1)t + x_1,\quad y = (y_2 - y_1)t + y_1,\ \ -\infty < t < \infty$

$\dfrac{x - x_1}{x_2 - x_1} = t$

$y = (y_2 - y_1)\left(\dfrac{x - x_1}{x_2 - x_1}\right) + y_1 \Rightarrow y - y_1 = \left(\dfrac{y_2 - y_1}{x_2 - x_1}\right)(x - x_1)$

This is the two point form for the equation of a line.

Its orientation is from (x_1, y_1) to (x_2, y_2).

51. $x = t \sin t$, $y = t \cos t$

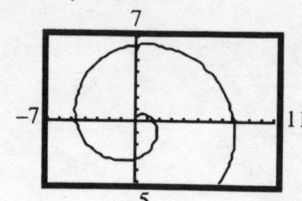

53. $x = 4 \sin t - 2 \sin(2t)$
$y = 4 \cos t - 2 \cos(2t)$

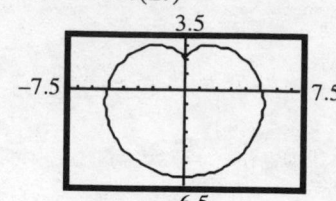

55.   (a) $x(t) = \cos^3 t$, $y(t) = \sin^3 t$, $0 \le t \le 2\pi$

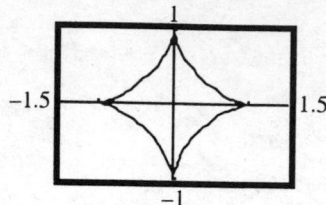

(b) $\cos t = x^{2/3}$, $\sin t = y^{2/3}$

$\cos^2 t + \sin^2 t = \left(x^{1/3}\right)^2 + \left(y^{1/3}\right)^2$

$x^{2/3} + y^{2/3} = 1$

Analytic Geometry

9.R Chapter Review

1. $y^2 = -16x$
    This is a parabola.
    $a = 4$
    Vertex: $(0, 0)$
    Focus: $(-4, 0)$
    Directrix: $x = 4$

3. $\dfrac{x^2}{25} - y^2 = 1$
    This is a hyperbola.
    $a = 5, \; b = 1$.

    Find the value of c:
    $$c^2 = a^2 + b^2 = 25 + 1 = 26 \Rightarrow c = \sqrt{26}$$
    Center: $(0, 0)$
    Vertices: $(5, 0), (-5, 0)$
    Foci: $\left(\sqrt{26}, 0\right), \left(-\sqrt{26}, 0\right)$
    Asymptotes: $y = \dfrac{1}{5}x; \;\; y = -\dfrac{1}{5}x$

5. $\dfrac{y^2}{25} + \dfrac{x^2}{16} = 1$
    This is an ellipse.
    $a = 5, \; b = 4$.

    Find the value of c:
    $$c^2 = a^2 - b^2 = 25 - 16 = 9$$
    $$c = 3$$
    Center: $(0, 0)$
    Vertices: $(0, 5), (0, -5)$
    Foci: $(0, 3), (0, -3)$

7. $x^2 + 4y = 4$
    This is a parabola.
    Write in standard form:
    $$x^2 = -4y + 4$$
    $$x^2 = -4(y - 1)$$

    $a = 1$
    Vertex: $(0, 1)$
    Focus: $(0, 0)$
    Directrix: $y = 2$

9. $4x^2 - y^2 = 8$
    This is a hyperbola.
    Write in standard form:
    $$\dfrac{x^2}{2} - \dfrac{y^2}{8} = 1$$
    $a = \sqrt{2}, \; b = \sqrt{8} = 2\sqrt{2}$.

    Find the value of c:
    $$c^2 = a^2 + b^2 = 2 + 8 = 10$$
    $$c = \sqrt{10}$$
    Center: $(0, 0)$
    Vertices: $\left(-\sqrt{2}, 0\right), \left(\sqrt{2}, 0\right)$
    Foci: $\left(-\sqrt{10}, 0\right), \left(\sqrt{10}, 0\right)$
    Asymptotes: $y = 2x; \;\; y = -2x$

11. $x^2 - 4x = 2y$
This is a parabola.
Write in standard form:
$$x^2 - 4x + 4 = 2y + 4$$
$$(x-2)^2 = 2(y+2)$$

$a = \dfrac{1}{2}$

Vertex: $(2, -2)$

Focus: $\left(2, -\dfrac{3}{2}\right)$

Directrix: $y = -\dfrac{5}{2}$

13. $y^2 - 4y - 4x^2 + 8x = 4$
This is a hyperbola.
Write in standard form:
$$(y^2 - 4y + 4) - 4(x^2 - 2x + 1) = 4 + 4 - 4$$
$$(y-2)^2 - 4(x-1)^2 = 4$$
$$\dfrac{(y-2)^2}{4} - \dfrac{(x-1)^2}{1} = 1$$
$a = 2, \ b = 1.$

Find the value of c :
$$c^2 = a^2 + b^2 = 4 + 1 = 5$$
$$c = \sqrt{5}$$
Center: $(1, 2)$
Vertices: $(1, 0), (1, 4)$
Foci: $\left(1, 2 - \sqrt{5}\right), \left(1, 2 + \sqrt{5}\right)$
Asymptotes:
$$y - 2 = 2(x - 1); \quad y - 2 = -2(x - 1)$$

15. $4x^2 + 9y^2 - 16x - 18y = 11$
This is an ellipse.
Write in standard form:
$$4x^2 + 9y^2 - 16x - 18y = 11$$
$$4(x^2 - 4x + 4) + 9(y^2 - 2y + 1) = 11 + 16 + 9$$
$$4(x-2)^2 + 9(y-1)^2 = 36$$
$$\dfrac{(x-2)^2}{9} + \dfrac{(y-1)^2}{4} = 1$$
$a = 3, \ b = 2.$

Find the value of c:
$$c^2 = a^2 - b^2 = 9 - 4 = 5$$
$$c = \sqrt{5}$$
Center: $(2, 1)$
Vertices: $(-1, 1), (5, 1)$
Foci: $\left(2 - \sqrt{5}, 1\right), \left(2 + \sqrt{5}, 1\right)$

17. $4x^2 - 16x + 16y + 32 = 0$
This is a parabola.
Write in standard form:
$$4(x^2 - 4x + 4) = -16y - 32 + 16$$
$$4(x-2)^2 = -16(y+1)$$
$$(x-2)^2 = -4(y+1)$$

$a = -1$
Vertex: $(2, -1)$
Focus: $(2, -2)$
Directrix: $y = 0$

19. $9x^2 + 4y^2 - 18x + 8y = 23$
This is an ellipse.
Write in standard form:
$$9(x^2 - 2x + 1) + 4(y^2 + 2y + 1) = 23 + 9 + 4$$
$$9(x-1)^2 + 4(y+1)^2 = 36$$
$$\dfrac{(x-1)^2}{4} + \dfrac{(y+1)^2}{9} = 1$$
$a = 3, \ b = 2.$

Find the value of c:
$$c^2 = a^2 - b^2 = 9 - 4 = 5$$
$$c = \sqrt{5}$$
Center: $(1, -1)$
Vertices: $(1, -4), (1, 2)$
Foci: $\left(1, -1 - \sqrt{5}\right), \left(1, -1 + \sqrt{5}\right)$

21. Parabola: The focus is $(-2, 0)$ and the directrix is
    $x = 2$. The vertex is $(0, 0)$. $a = 2$ and since $(-2, 0)$
    is to the left of $(0, 0)$, the parabola opens to the left.
    The equation of the parabola is:
    $$y^2 = -4ax$$
    $$y^2 = -4 \cdot 2 \cdot x$$
    $$y^2 = -8x$$

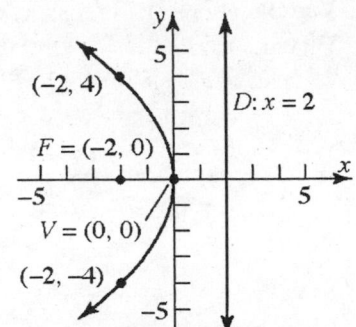

23. Hyperbola: Center: $(0, 0)$;
    Focus: $(0, 4)$;   Vertex: $(0, -2)$;
    Transverse axis is the y-axis;
    $a = 2$; $c = 4$.
    Find b:
    $$b^2 = c^2 - a^2 = 16 - 4 = 12$$
    $$b = \sqrt{12} = 2\sqrt{3}$$
    Write the equation: $\dfrac{y^2}{4} - \dfrac{x^2}{12} = 1$

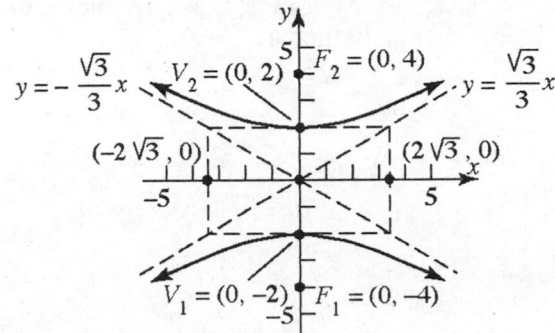

25. Ellipse: Foci: $(-3, 0)$, $(3, 0)$; Vertex: $(4, 0)$;
    Center: $(0, 0)$; Major axis is the x-axis;
    $a = 4$; $c = 3$. Find b:
    $$b^2 = a^2 - c^2 = 16 - 9 = 7$$
    $$b = \sqrt{7}$$
    Write the equation: $\dfrac{x^2}{16} + \dfrac{y^2}{7} = 1$

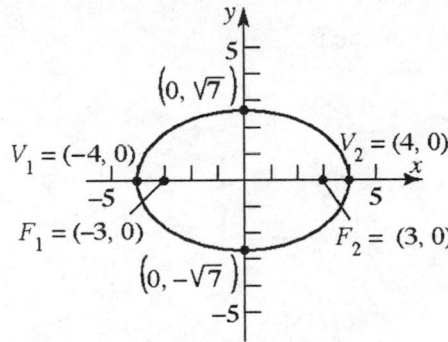

27. Parabola: The focus is $(2, -4)$ and the vertex is
    $(2, -3)$. Both lie on the vertical line $x = 2$. $a = 1$
    and since $(2, -4)$ is below $(2, -3)$, the parabola
    opens down. The equation of the parabola is:
    $$(x - h)^2 = -4a(y - k)$$
    $$(x - 2)^2 = -4 \cdot 1 \cdot (y - (-3))$$
    $$(x - 2)^2 = -4(y + 3)$$

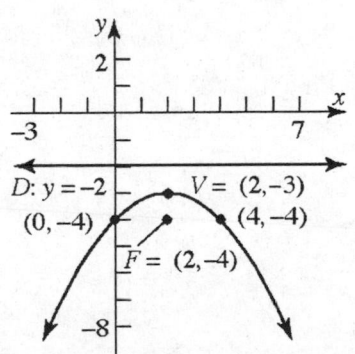

553

29. Hyperbola: Center: $(-2, -3)$; Focus: $(-4, -3)$;
    Vertex: $(-3, -3)$; Transverse axis is parallel to the
    x-axis; $a = 1$; $c = 2$. Find b:

    $$b^2 = c^2 - a^2 = 4 - 1 = 3$$

    $$b = \sqrt{3}$$

    Write the equation: $\dfrac{(x+2)^2}{1} - \dfrac{(y+3)^2}{3} = 1$

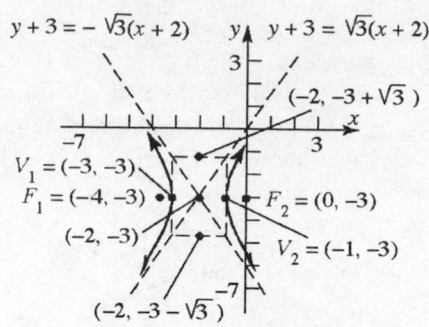

31. Ellipse: Foci: $(-4, 2)$, $(-4, 8)$; Vertex: $(-4, 10)$;
    Center: $(-4, 5)$; Major axis is parallel to the y-axis;
    $a = 5$; $c = 3$. Find b:

    $$b^2 = a^2 - c^2 = 25 - 9 = 16$$

    $$b = 4$$

    Write the equation: $\dfrac{(x+4)^2}{16} + \dfrac{(y-5)^2}{25} = 1$

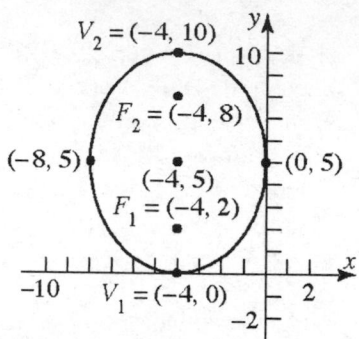

33. Hyperbola: Center: $(-1, 2)$;
    $a = 3$; $c = 4$; Transverse axis parallel to
    the x-axis;
    Find b:

    $$b^2 = c^2 - a^2 = 16 - 9 = 7$$

    $$b = \sqrt{7}$$

    Write the equation:

    $$\dfrac{(x+1)^2}{9} - \dfrac{(y-2)^2}{7} = 1$$

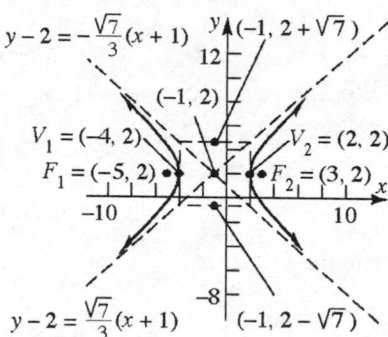

35. Hyperbola: Vertices: $(0, 1)$, $(6, 1)$; Asymptote: $3y + 2x - 9 = 0$; Center: $(3, 1)$;

    Transverse axis is parallel to the x-axis; $a = 3$; The slope of the asymptote is $-\dfrac{2}{3}$;

    Find b:

    $$\dfrac{-b}{a} = \dfrac{-b}{3} = \dfrac{-2}{3}$$

    $$-3b = -6$$

    $$b = 2$$

Write the equation: $\dfrac{(x-3)^2}{9} - \dfrac{(y-1)^2}{4} = 1$

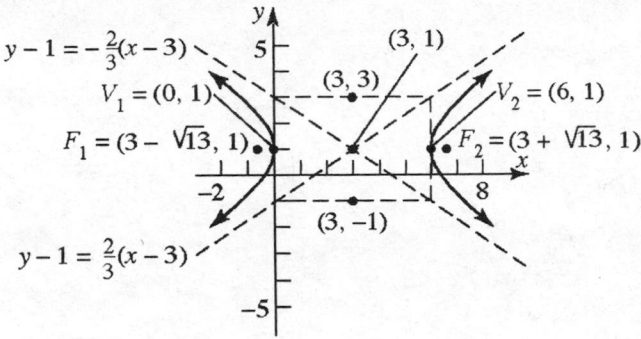

$y - 1 = -\dfrac{2}{3}(x-3)$

$V_1 = (0, 1)$

$F_1 = (3 - \sqrt{13}, 1)$

$(3, 3)$

$(3, 1)$

$V_2 = (6, 1)$

$F_2 = (3 + \sqrt{13}, 1)$

$(3, -1)$

$y - 1 = \dfrac{2}{3}(x-3)$

37. $y^2 + 4x + 3y - 8 = 0$

$A = 0$ and $C = 1$; $AC = (0)(1) = 0$. Since $AC = 0$, the equation defines a parabola.

39. $x^2 + 2y^2 + 4x - 8y + 2 = 0$

$A = 1$ and $C = 2$; $AC = (1)(2) = 2$. Since $AC > 0$ and $A \neq C$, the equation defines an ellipse.

41. $9x^2 - 12xy + 4y^2 + 8x + 12y = 0$

$A = 9$, $B = -12$, $C = 4$ $B^2 - 4AC = (-12)^2 - 4(9)(4) = 0$; parabola

43. $4x^2 + 10xy + 4y^2 - 9 = 0$

$A = 4$, $B = 10$, $C = 4$ $B^2 - 4AC = 10^2 - 4(4)(4) = 36 > 0$; hyperbola

45. $x^2 - 2xy + 3y^2 + 2x + 4y - 1 = 0$

$A = 1$, $B = -2$, $C = 3$ $B^2 - 4AC = (-2)^2 - 4(1)(3) = -8 < 0$; ellipse

47. $2x^2 + 5xy + 2y^2 - \dfrac{9}{2} = 0$

$A = 2, B = 5,$ and $C = 2$; $\cot(2\theta) = \dfrac{A-C}{B} = \dfrac{2-2}{5} = 0 \;\rightarrow\; 2\theta = \dfrac{\pi}{2} \;\rightarrow\; \theta = \dfrac{\pi}{4}$

$x = x'\cos\theta - y'\sin\theta = \dfrac{\sqrt{2}}{2}x' - \dfrac{\sqrt{2}}{2}y' = \dfrac{\sqrt{2}}{2}(x' - y')$

$y = x'\sin\theta + y'\cos\theta = \dfrac{\sqrt{2}}{2}x' + \dfrac{\sqrt{2}}{2}y' = \dfrac{\sqrt{2}}{2}(x' + y')$

$2\left(\dfrac{\sqrt{2}}{2}(x' - y')\right)^2 + 5\left(\dfrac{\sqrt{2}}{2}(x' - y')\right)\left(\dfrac{\sqrt{2}}{2}(x' + y')\right) + 2\left(\dfrac{\sqrt{2}}{2}(x' + y')\right)^2 - \dfrac{9}{2} = 0$

$\left(x'^2 - 2x'y' + y'^2\right) + \dfrac{5}{2}\left(x'^2 - y'^2\right) + \left(x'^2 + 2x'y' + y'^2\right) - \dfrac{9}{2} = 0$

$\dfrac{9}{2}x'^2 - \dfrac{1}{2}y'^2 = \dfrac{9}{2} \rightarrow 9x'^2 - y'^2 = 9 \rightarrow \dfrac{x'^2}{1} - \dfrac{y'^2}{9} = 1$

Hyperbola; center at $(0, 0)$, transverse axis is the x'-axis, vertices at $(\pm 1, 0)$.

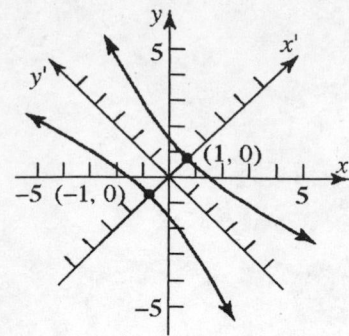

49. $6x^2 + 4xy + 9y^2 - 20 = 0$

$A = 6, B = 4,$ and $C = 9;$ $\cot(2\theta) = \dfrac{A - C}{B} = \dfrac{6 - 9}{4} = -\dfrac{3}{4} \rightarrow \cos(2\theta) = -\dfrac{3}{5}$

$\sin\theta = \sqrt{\dfrac{1 - \left(-\dfrac{3}{5}\right)}{2}} = \sqrt{\dfrac{4}{5}} = \dfrac{2\sqrt{5}}{5};$ $\cos\theta = \sqrt{\dfrac{1 + \left(-\dfrac{3}{5}\right)}{2}} = \sqrt{\dfrac{1}{5}} = \dfrac{\sqrt{5}}{5} \rightarrow \theta \approx 63.4°$

$x = x'\cos\theta - y'\sin\theta = \dfrac{\sqrt{5}}{5}x' - \dfrac{2\sqrt{5}}{5}y' = \dfrac{\sqrt{5}}{5}(x' - 2y')$

$y = x'\sin\theta + y'\cos\theta = \dfrac{2\sqrt{5}}{5}x' + \dfrac{\sqrt{5}}{5}y' = \dfrac{\sqrt{5}}{5}(2x' + y')$

$6\left(\dfrac{\sqrt{5}}{5}(x' - 2y')\right)^2 + 4\left(\dfrac{\sqrt{5}}{5}(x' - 2y')\right)\left(\dfrac{\sqrt{5}}{5}(2x' + y')\right) + 9\left(\dfrac{\sqrt{5}}{5}(2x' + y')\right)^2 - 20 = 0$

$\dfrac{6}{5}\left(x'^2 - 4x'y' + 4y'^2\right) + \dfrac{4}{5}\left(2x'^2 - 3x'y' - 2y'^2\right) + \dfrac{9}{5}\left(4x'^2 + 4x'y' + y'^2\right) - 20 = 0$

$\dfrac{6}{5}x'^2 - \dfrac{24}{5}x'y' + \dfrac{24}{5}y'^2 + \dfrac{8}{5}x'^2 - \dfrac{12}{5}x'y' - \dfrac{8}{5}y'^2 + \dfrac{36}{5}x'^2 + \dfrac{36}{5}x'y' + \dfrac{9}{5}y'^2 = 20$

$10x'^2 + 5y'^2 = 20 \rightarrow \dfrac{x'^2}{2} + \dfrac{y'^2}{4} = 1$

Ellipse; center at the origin, major axis is the y'-axis, vertices at $(0, \pm 2)$.

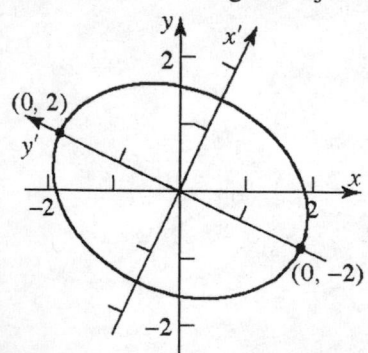

51. $4x^2 - 12xy + 9y^2 + 12x + 8y = 0$

$A = 4, B = -12,$ and $C = 9$; $\cot(2\theta) = \dfrac{A - C}{B} = \dfrac{4 - 9}{-12} = \dfrac{5}{12} \;\rightarrow\; \cos(2\theta) = \dfrac{5}{13}$

$\sin\theta = \sqrt{\dfrac{1 - \dfrac{5}{13}}{2}} = \sqrt{\dfrac{4}{13}} = \dfrac{2\sqrt{13}}{13};\quad \cos\theta = \sqrt{\dfrac{1 + \dfrac{5}{13}}{2}} = \sqrt{\dfrac{9}{13}} = \dfrac{3\sqrt{13}}{13} \;\rightarrow\; \theta \approx 33.7°$

$x = x'\cos\theta - y'\sin\theta = \dfrac{3\sqrt{13}}{13}x' - \dfrac{2\sqrt{13}}{13}y' = \dfrac{\sqrt{13}}{13}(3x' - 2y')$

$y = x'\sin\theta + y'\cos\theta = \dfrac{2\sqrt{13}}{13}x' + \dfrac{3\sqrt{13}}{13}y' = \dfrac{\sqrt{13}}{13}(2x' + 3y')$

$4\left(\dfrac{\sqrt{13}}{13}(3x' - 2y')\right)^2 - 12\left(\dfrac{\sqrt{13}}{13}(3x' - 2y')\right)\left(\dfrac{\sqrt{13}}{13}(2x' + 3y')\right) + 9\left(\dfrac{\sqrt{13}}{13}(2x' + 3y')\right)^2$

$\qquad\qquad\qquad + 12\left(\dfrac{\sqrt{13}}{13}(3x' - 2y')\right) + 8\left(\dfrac{\sqrt{13}}{13}(2x' + 3y')\right) = 0$

$\dfrac{4}{13}(9x'^2 - 12x'y' + 4y'^2) - \dfrac{12}{13}(6x'^2 + 5x'y' - 6y'^2) + \dfrac{9}{13}(4x'^2 + 12x'y' + 9y'^2)$

$\qquad\qquad\qquad + \dfrac{36\sqrt{13}}{13}x' - \dfrac{24\sqrt{13}}{13}y' + \dfrac{16\sqrt{13}}{13}x' + \dfrac{24\sqrt{13}}{13}y' = 0$

$\dfrac{36}{13}x'^2 - \dfrac{48}{13}x'y' + \dfrac{16}{13}y'^2 - \dfrac{72}{13}x'^2 - \dfrac{60}{13}x'y' + \dfrac{72}{13}y'^2$

$\qquad\qquad\qquad + \dfrac{36}{13}x'^2 + \dfrac{108}{13}x'y' + \dfrac{81}{13}y'^2 + 4\sqrt{13}x' = 0$

$13y'^2 + 4\sqrt{13}x' = 0$

$y'^2 = -\dfrac{4\sqrt{13}}{13}x'$

Parabola; vertex at the origin,

focus at $\left(-\dfrac{\sqrt{13}}{13}, 0\right)$.

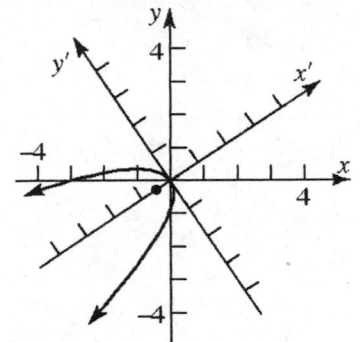

53. $r = \dfrac{4}{1 - \cos\theta}$

$ep = 4,\; e = 1,\; p = 4$

Parabola; directrix is perpendicular to the polar axis 4 units to the left of the pole; vertex is $(2, \pi)$.

55. $r = \dfrac{6}{2 - \sin\theta} = \dfrac{3}{\left(1 - \dfrac{1}{2}\sin\theta\right)}$

$ep = 3, \quad e = \dfrac{1}{2}, \quad p = 6$

Ellipse; directrix is parallel to the polar axis 6 units below the pole; vertices are $\left(6, \dfrac{\pi}{2}\right)$ and $\left(2, \dfrac{3\pi}{2}\right)$.

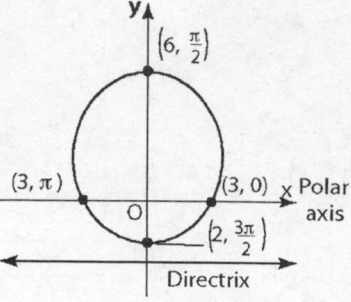

57. $r = \dfrac{8}{4 + 8\cos\theta} = \dfrac{2}{1 + 2\cos\theta}$

$ep = 2, \quad e = 2, \quad p = 1$

Hyperbola; directrix is perpendicular to the polar axis 1 unit to the right of the pole; vertices are $\left(\dfrac{2}{3}, 0\right)$ and $(-2, \pi)$.

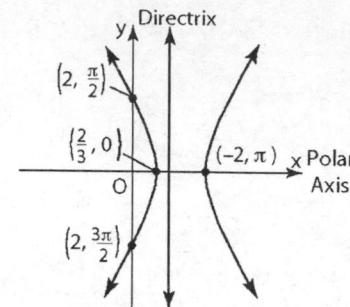

59. $r = \dfrac{4}{1 - \cos\theta}$

$r - r\cos\theta = 4$

$\qquad r = 4 + r\cos\theta$

$\qquad r^2 = (4 + r\cos\theta)^2$

$\quad x^2 + y^2 = (4 + x)^2$

$\quad x^2 + y^2 = 16 + 8x + x^2$

$y^2 - 8x - 16 = 0$

61. $r = \dfrac{8}{4 + 8\cos\theta}$

$4r + 8r\cos\theta = 8$

$\qquad 4r = 8 - 8r\cos\theta$

$\qquad r = 2 - 2r\cos\theta$

$\qquad r^2 = (2 - 2r\cos\theta)^2$

$\quad x^2 + y^2 = (2 - 2x)^2$

$\quad x^2 + y^2 = 4 - 8x + 4x^2$

$3x^2 - y^2 - 8x + 4 = 0$

63. $x = 4t - 2, \quad y = 1 - t, \quad -\infty < t < \infty$

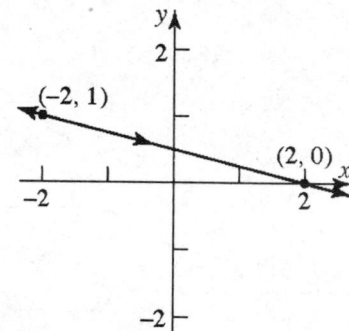

$x = 4(1 - y) - 2$

$x = 4 - 4y - 2$

$x + 4y = 2$

65. $x = 3\sin t, \quad y = 4\cos t + 2, \quad 0 \le t \le 2\pi$

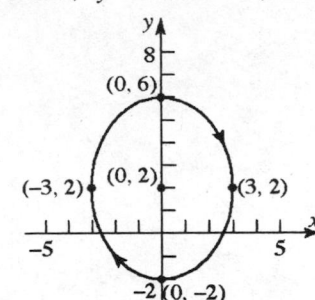

$$\frac{x}{3} = \sin t, \quad \frac{y-2}{4} = \cos t$$

$$\sin^2 t + \cos^2 t = 1$$

$$\left(\frac{x}{3}\right)^2 + \left(\frac{y-2}{4}\right)^2 = 1$$

$$\frac{x^2}{9} + \frac{(y-2)^2}{16} = 1$$

67. $x = \sec^2 t, \quad y = \tan^2 t, \quad 0 \le t \le \dfrac{\pi}{4}$

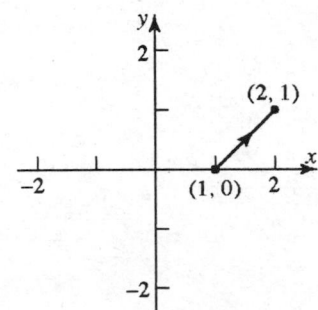

69. $y = -2x + 4$

$\quad x = t, \quad y = -2t + 4$

$\quad x = \dfrac{4 - t}{2}, \quad y = t$

$\tan^2 t + 1 = \sec^2 t \rightarrow y + 1 = x$

71. $\dfrac{x}{4} = \cos(\omega t); \quad \dfrac{y}{3} = \sin(\omega t)$

$\dfrac{2\pi}{\omega} = 4 \Rightarrow 4\omega = 2\pi \Rightarrow \omega = \dfrac{2\pi}{4} = \dfrac{\pi}{2}$

$\therefore \quad \dfrac{x}{4} = \cos\left(\dfrac{\pi}{2}t\right); \quad \dfrac{y}{3} = \sin\left(\dfrac{\pi}{2}t\right) \Rightarrow x = 4\cos\left(\dfrac{\pi}{2}t\right); \quad y = 3\sin\left(\dfrac{\pi}{2}t\right), \quad 0 \le t \le 4$

73.   Write the equation in standard form:

$$4x^2 + 9y^2 = 36 \rightarrow \frac{x^2}{9} + \frac{y^2}{4} = 1$$

The center of the ellipse is (0, 0). The major axis is the x-axis.

$a = 3; \quad b = 2; \quad c^2 = a^2 - b^2 = 9 - 4 = 5 \rightarrow c = \sqrt{5}.$

For the ellipse:

Vertices: (–3, 0), (3, 0);   Foci: $\left(-\sqrt{5}, 0\right), \left(\sqrt{5}, 0\right)$

For the hyperbola:

Foci: (–3, 0), (3, 0);   Vertices: $\left(-\sqrt{5}, 0\right), \left(\sqrt{5}, 0\right)$;   Center: (0, 0)

$a = \sqrt{5}; \quad c = 3; \quad b^2 = c^2 - a^2 = 9 - 5 = 4 \rightarrow b = 2$

The equation of the hyperbola is: $\dfrac{x^2}{5} - \dfrac{y^2}{4} = 1$

75. Let (x, y) be any point in the collection of points.

The distance from (x, y) to $(3, 0) = \sqrt{(x-3)^2 + y^2}$.

The distance from (x, y) to the line $x = \frac{16}{3}$ is $\left| x - \frac{16}{3} \right|$.

Relating the distances, we have:

$$\sqrt{(x-3)^2 + y^2} = \frac{3}{4}\left| x - \frac{16}{3} \right| \rightarrow (x-3)^2 + y^2 = \frac{9}{16}\left(x - \frac{16}{3} \right)^2$$

$$x^2 - 6x + 9 + y^2 = \frac{9}{16}\left(x^2 - \frac{32}{3}x + \frac{256}{9} \right)$$

$$16x^2 - 96x + 144 + 16y^2 = 9x^2 - 96x + 256$$

$$7x^2 + 16y^2 = 112 \rightarrow \frac{7x^2}{112} + \frac{16y^2}{112} = 1 \rightarrow \frac{x^2}{16} + \frac{y^2}{7} = 1$$

The set of points is an ellipse.

77. Locate the parabola so that the vertex is at $(0, 0)$ and opens up. It then has the equation: $x^2 = 4ay$. Since the light source is located at the focus and is 1 foot from the base, $a = 1$. The diameter is 2, so the point $(1, y)$ is located on the parabola. Solve for y:

$$1^2 = 4(1)y \rightarrow 1 = 4y \rightarrow y = 0.25 \text{ feet}$$

The mirror should be 0.25 feet deep or 3 inches deep.

79. Place the semielliptical arch so that the x-axis coincides with the water and the y-axis passes through the center of the arch. Since the bridge has a span of 60 feet, the length of the major axis is 60, or $2a = 60$ or $a = 30$. The maximum height of the bridge is 20 feet, so $b = 20$. The equation is: $\frac{x^2}{900} + \frac{y^2}{400} = 1$.

The height 5 feet from the center:

$$\frac{5^2}{900} + \frac{y^2}{400} = 1 \rightarrow \frac{y^2}{400} = 1 - \frac{25}{900} \rightarrow y^2 = 400 \cdot \frac{875}{900} \rightarrow y \approx 19.72 \text{ feet}$$

The height 10 feet from the center:

$$\frac{10^2}{900} + \frac{y^2}{400} = 1 \rightarrow \frac{y^2}{400} = 1 - \frac{100}{900} \rightarrow y^2 = 400 \cdot \frac{800}{900} \rightarrow y \approx 18.86 \text{ feet}$$

The height 20 feet from the center:

$$\frac{20^2}{900} + \frac{y^2}{400} = 1 \rightarrow \frac{y^2}{400} = 1 - \frac{400}{900} \rightarrow y^2 = 400 \cdot \frac{500}{900} \rightarrow y \approx 14.91 \text{ feet}$$

81. (a) Set up a coordinate system so that the
         two stations lie on the x-axis and the
         origin is midway between them. The
         ship lies on a hyperbola whose foci are
         the locations of the two stations. Since
         the time difference is 0.00032 seconds

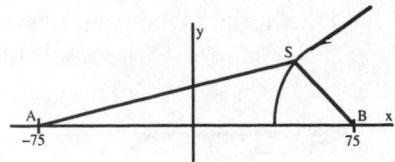

         and the speed of the signal is 186,000
         miles per second, the difference in the
         distances of the ships from each station
         is: (186,000)(0.00032) = 59.52 miles
         The difference of the distances from the ship to each station, 59.52, equals
         $a = 29.76$ and the vertex of the corresponding hyperbola is at (29.76, 0). Since the
         focus is at (75, 0), following this hyperbola, the ship would reach shore 45.24 miles
         from the master station.

    (b) The ship should follow a hyperbola with a vertex at (60, 0). For this hyperbola,
         $a = 60$, so the constant difference of the distances from the ship to each station is
         120. The time difference the ship should look for is:
         $$\text{time} = \frac{120}{186,000} = 0.000645 \text{ seconds}$$

    (c) Find the equation of the hyperbola with vertex at (60, 0) and a focus at (75, 0). The
         form of the equation of the hyperbola is:
         $$\frac{x^2}{a^2} - \frac{y^2}{b^2} = 1 \quad \text{where } a = 60.$$
         Since $c = 75$ and $b^2 = c^2 - a^2 \;\rightarrow\; b^2 = 75^2 - 60^2 = 2025.$

         The equation of the hyperbola is: $\dfrac{x^2}{3600} - \dfrac{y^2}{2025} = 1.$
         Since the ship is 20 miles off shore, we have $y = 20$.
         Solve the equation for x:
         $$\frac{x^2}{3600} - \frac{20^2}{2025} = 1 \;\rightarrow\; \frac{x^2}{3600} = 1 + \frac{400}{2025} = \frac{97}{81} \;\rightarrow\; x^2 = 3600 \cdot \frac{97}{81}$$
         $$x \approx 66 \text{ miles}$$
         The ship's location is (66, 20).

83. (a) Use equation (2):
         $$x = \left(100\cos(35^\circ)\right)t; \qquad y = -\frac{1}{2}(32)t^2 + \left(100\sin(35^\circ)\right)t + 6$$

    (b) The ball is in the air until $y = 0$. Solve: $-16t^2 + \left(100\sin(35^\circ)\right)t + 6 = 0$
         $$t = \frac{-100\sin(35^\circ) \pm \sqrt{\left(100\sin(35^\circ)\right)^2 - 4(-16)(6)}}{2(-16)} \approx \frac{-57.36 \pm \sqrt{3673.9}}{-32} \approx -0.10 \text{ or } 3.69$$
         The ball is in the air for about 3.69 seconds. (The negative solution is extraneous.)

    (c) The maximum height occurs at the vertex of the quadratic function.
         $$t = \frac{-b}{2a} = \frac{-100\sin(35^\circ)}{2(-16)} \approx 1.79 \text{ seconds}$$
         Evaluate the function to find the maximum height:
         $$-16(1.79)^2 + \left(100\sin(35^\circ)\right)(1.79) + 6 \approx 57.4 \text{ feet}$$

(d) Find the horizontal displacement:

$$x = \left(100\cos(35°)\right)(3.69) \approx 302 \text{ feet}$$

(e)

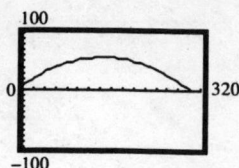

Analytic Geometry

9.CR Cumulative Review

1. $\sin(2\theta) = 0.5$

   $2\theta = \dfrac{\pi}{6} + 2k\pi \implies \theta = \dfrac{\pi}{12} + k\pi$

   or $2\theta = \dfrac{5\pi}{6} + 2k\pi \implies \theta = \dfrac{5\pi}{12} + k\pi$,

   where k is any integer.

3. center point $(0,4)$; radius $= 4$

   $(x - h)^2 + (y - k)^2 = r^2$

   $(x - 0)^2 + (y - 4)^2 = 4^2$

   $x^2 + y^2 - 8y + 16 = 16$

   $x^2 + y^2 - 8y = 0$

   converting to polar coordinates:

   $r^2 - 8r\sin\theta = 0$

   $r^2 = 8r\sin\theta$

   $r = 8\sin\theta$

   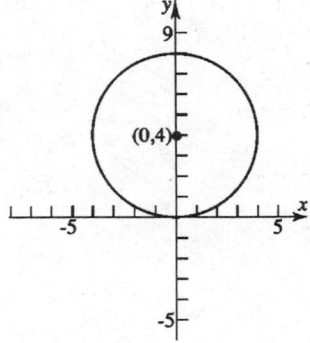

5. Multiply each side of the second equation by -2 and add the equations:

   $\begin{cases} 2x - 3y = 1 \\ x + 4y = 6 \end{cases}$ $\begin{array}{l} \xrightarrow{} \quad 2x - 3y = 1 \\ \xrightarrow{-2} \quad \underline{-2x - 8y = -12} \end{array}$

   $\qquad\qquad\qquad\qquad -11y = -11 \implies y = 1$

   Substitute and solve for x:

   $2x - 3(1) = 1 \implies 2x - 3 = 1 \implies 2x = 4 \implies x = 2$

   The solution of the system is $x = 2$, $y = 1$.

7. $x + y \leq 6$

Graph the line $x + y = 6$. Use a solid line since the inequality uses $\leq$.

Choose a test point not on the line, such as $(0, 0)$.
Since $0 + 0 \leq 6$ is true, shade the same side of the line containing $(0, 0)$.

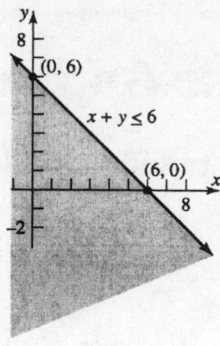

9. $\cot(2\theta) = 1$, where $0° < \theta < 90°$

$2\theta = \dfrac{\pi}{4} + k\pi \implies \theta = \dfrac{\pi}{8} + \dfrac{k\pi}{2}$, where k is any integer.

On the interval $0° < \theta < 90°$, the solution is $\theta = \dfrac{\pi}{8} = 22.5°$.

Chapter 10

Systems of Equations and Inequalities

10.1 Systems of Linear Equations: Two Equations Containing Two Variables

1. Substituting the values of the variables:
$$\begin{cases} 2x - y = 5 & \rightarrow \quad 2(2) - (-1) = 4 + 1 = 5 \\ 5x + 2y = 8 & \rightarrow \quad 5(2) + 2(-1) = 10 - 2 = 8 \end{cases}$$
Each equation is satisfied, so $x = 2$, $y = -1$ is a solution to the system of equations.

3. Substituting the values of the variables:
$$\begin{cases} 3x - 4y = 4 & \rightarrow \quad 3(2) - 4\left(\dfrac{1}{2}\right) = 6 - 2 = 4 \\ \dfrac{1}{2}x - 3y = -\dfrac{1}{2} & \rightarrow \quad \dfrac{1}{2}(2) - 3\left(\dfrac{1}{2}\right) = 1 - \dfrac{3}{2} = -\dfrac{1}{2} \end{cases}$$
Each equation is satisfied, so $x = 2$, $y = \dfrac{1}{2}$ is a solution to the system of equations.

5. Substituting the values of the variables:
$$\begin{cases} x - y = 3 & \rightarrow \quad 4 - 1 = 3 \\ \dfrac{1}{2}x + y = 3 & \rightarrow \quad \dfrac{1}{2}(4) + 1 = 2 + 1 = 3 \end{cases}$$
Each equation is satisfied, so $x = 4$, $y = 1$ is a solution to the system of equations.

7. Substituting the values of the variables:
$$\begin{cases} 3x + 3y + 2z = 4 & \rightarrow 3(1) + 3(-1) + 2(2) = 3 - 3 + 4 = 4 \\ x - y - z = 0 & \rightarrow 1 - (-1) - 2 = 1 + 1 - 2 = 0 \\ 2y - 3z = -8 & \rightarrow 2(-1) - 3(2) = -2 - 6 = -8 \end{cases}$$
Each equation is satisfied, so $x = 1$, $y = -1, z = 2$ is a solution to the system.

9. Solve the first equation for y, substitute into the second equation and solve:
$$\begin{cases} x + y = 8 & \rightarrow \quad y = 8 - x \\ x - y = 4 \end{cases}$$
$$x - (8 - x) = 4 \rightarrow x - 8 + x = 4$$
$$2x = 12$$
$$x = 6$$
Since $x = 6$, $y = 8 - 6 = 2$
The solution of the system is $x = 6$, $y = 2$.

11. Multiply each side of the first equation by 3 and add the equations:

$$\begin{cases} 5x - y = 13 & \xrightarrow{\;3\;} & 15x - 3y = 39 \\ 2x + 3y = 12 & \longrightarrow & 2x + 3y = 12 \end{cases}$$

$$17x = 51 \Rightarrow x = 3$$

Substitute and solve for y:
$$5(3) - y = 13 \rightarrow 15 - y = 13 \rightarrow -y = -2 \rightarrow y = 2$$
The solution of the system is $x = 3,\ y = 2$.

13. Solve the first equation for x and substitute into the second equation:

$$\begin{cases} 3x = 24 & \rightarrow & x = 8 \\ x + 2y = 0 \end{cases}$$

$$8 + 2y = 0 \rightarrow 2y = -8 \rightarrow y = -4$$
The solution of the system is $x = 8,\ y = -4$.

15. Multiply each side of the first equation by 2 and each side of the second equation by 3 to eliminate y:

$$\begin{cases} 3x - 6y = 2 & \xrightarrow{\;2\;} & 6x - 12y = 4 \\ 5x + 4y = 1 & \xrightarrow{\;3\;} & 15x + 12y = 3 \end{cases}$$

$$21x = 7 \Rightarrow x = \frac{1}{3}$$

Substitute and solve for y:
$$3\left(\frac{1}{3}\right) - 6y = 2 \rightarrow 1 - 6y = 2 \rightarrow -6y = 1 \rightarrow y = -\frac{1}{6}$$

The solution of the system is $x = \frac{1}{3},\ y = -\frac{1}{6}$.

17. Solve the first equation for y, substitute into the second equation and solve:

$$\begin{cases} 2x + y = 1 & \rightarrow & y = 1 - 2x \\ 4x + 2y = 3 \end{cases}$$

$$4x + 2(1 - 2x) = 3 \rightarrow 4x + 2 - 4x = 3 \rightarrow 0x = 1$$
This has no solution, so the system is inconsistent.

19. Solve the first equation for y, substitute into the second equation and solve:

$$\begin{cases} 2x - y = 0 & \rightarrow & 2x = y \\ 3x + 2y = 7 \end{cases}$$
$$3x + 2(2x) = 7 \rightarrow 3x + 4x = 7 \rightarrow 7x = 7 \rightarrow x = 1$$
Since $x = 1,\ y = 2(1) = 2$ The solution of the system is $x = 1,\ y = 2$.

21. Solve the first equation for x, substitute into the second equation and solve:

$$\begin{cases} x + 2y = 4 & \rightarrow & x = 4 - 2y \\ 2x + 4y = 8 \end{cases}$$
$$2(4 - 2y) + 4y = 8$$
$$8 - 4y + 4y = 8 \Rightarrow 0y = 0$$

These equations are dependent. Any real number is a solution for y.
The solution of the system is $x = 4 - 2y$, where y is any real number.

23. Multiply each side of the first equation by –5, and add the equations to eliminate x:

$$\begin{cases} 2x - 3y = -1 \\ 10x + \ y = 11 \end{cases} \quad \begin{array}{l} \xrightarrow{-5} \\ \longrightarrow \end{array} \quad \begin{array}{l} -10x + 15y = \ 5 \\ \underline{10x + \ \ y = 11} \\ 16y = 16 \\ y = 1 \end{array}$$

Substitute and solve for x:

$$2x - 3(1) = -1 \rightarrow 2x - 3 = -1 \rightarrow 2x = 2 \rightarrow x = 1$$

The solution of the system is $x = 1, \ y = 1$.

25. Solve the second equation for x, substitute into the first equation and solve:

$$\begin{cases} 2x + 3y = 6 \\ x - \ y = \dfrac{1}{2} \end{cases} \quad \rightarrow \quad x = y + \dfrac{1}{2}$$

$$2\left(y + \dfrac{1}{2}\right) + 3y = 6 \rightarrow 2y + 1 + 3y = 6 \rightarrow 5y = 5 \rightarrow y = 1$$

Since $y = 1, \ x = 1 + \dfrac{1}{2} = \dfrac{3}{2}$, the solution of the system is $x = \dfrac{3}{2}, \ y = 1$.

27. Multiply each side of the first equation by –6 and each side of the second equation by 12 to eliminate x:

$$\begin{cases} \dfrac{1}{2}x + \dfrac{1}{3}y = \ 3 \\ \dfrac{1}{4}x - \dfrac{2}{3}y = -1 \end{cases} \quad \begin{array}{l} \xrightarrow{-6} \\ \xrightarrow{12} \end{array} \quad \begin{array}{l} -3x - 2y = -18 \\ \underline{3x - 8y = -12} \\ {-10y} = -30 \\ y = 3 \end{array}$$

Substitute and solve for x:

$$\dfrac{1}{2}x + \dfrac{1}{3}(3) = 3 \rightarrow \dfrac{1}{2}x + 1 = 3 \rightarrow \dfrac{1}{2}x = 2 \rightarrow x = 4$$

The solution of the system is $x = 4, \ y = 3$.

29. Add the equations to eliminate y and solve for x:

$$\begin{cases} 3x - 5y = \ 3 \\ 15x + 5y = 21 \end{cases}$$

$$\overline{18x = 24} \Rightarrow x = \dfrac{4}{3}$$

Substitute and solve for y: $3\left(\dfrac{4}{3}\right) - 5y = 3 \rightarrow 4 - 5y = 3 \rightarrow -5y = -1 \rightarrow y = \dfrac{1}{5}$

The solution of the system is $x = \dfrac{4}{3}, \ y = \dfrac{1}{5}$.

31.   Rewrite letting $a = \dfrac{1}{x}$, $b = \dfrac{1}{y}$:

$$\begin{cases} \dfrac{1}{x} + \dfrac{1}{y} = 8 \quad \longrightarrow \quad a + b = 8 \\ \dfrac{3}{x} - \dfrac{5}{y} = 0 \quad \longrightarrow \quad 3a - 5b = 0 \end{cases}$$

Solve the first equation for a, substitute into the second equation and solve:

$$\begin{cases} a + b = 8 \quad \to \quad a = 8 - b \\ 3a - 5b = 0 \end{cases}$$

$$3(8 - b) - 5b = 0 \to 24 - 3b - 5b = 0 \to -8b = -24 \to b = 3$$

Since $b = 3$, $a = 8 - 3 = 5$

Thus, $x = \dfrac{1}{a} = \dfrac{1}{5}$, $y = \dfrac{1}{b} = \dfrac{1}{3}$. The solution of the system is $x = \dfrac{1}{5}$, $y = \dfrac{1}{3}$.

33.   Graph the two equations as y_1 and y_2, and use INTERSECT to solve:

$$\begin{cases} y_1 = \sqrt{2}x - 20\sqrt{7} \\ y_2 = -0.1x + 20 \end{cases}$$

The solution of the system is
$x = 48.15$, $y = 15.18$.

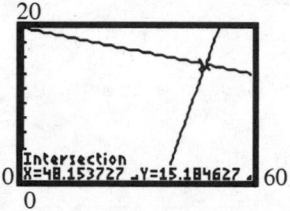

35.   Solve for y in each equation, graph the two equations as y_1 and y_2, and use INTERSECT to solve:

$$\begin{cases} \sqrt{2}x + \sqrt{3}y + \sqrt{6} = 0 \\ \sqrt{3}x - \sqrt{2}y + 60 = 0 \end{cases}$$

$$\begin{cases} y_1 = \dfrac{-\sqrt{2}x - \sqrt{6}}{\sqrt{3}} \\ y_2 = \dfrac{\sqrt{3}x + 60}{\sqrt{2}} \end{cases}$$

The solution of the system is $x = -21.48$, $y = 16.12$.

36. *(see graph)*

37.   Solve for y in each equation, graph the two equations as y_1 and y_2, and use INTERSECT to solve:

$$\begin{cases} \sqrt{3}x + \sqrt{2}y = \sqrt{0.3} \\ 100x - 95y = 20 \end{cases}$$

$$\begin{cases} y_1 = \dfrac{-\sqrt{3}x + \sqrt{0.3}}{\sqrt{2}} \\ y_2 = \dfrac{100x - 20}{95} \end{cases}$$

The solution of the system is $x = 0.26$, $y = 0.07$.

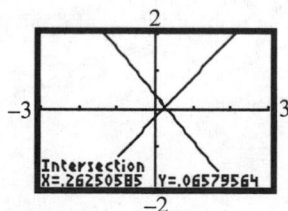

39. Solve the system by substitution:
$$Q_s = Q_d$$
$$-200 + 50p = 1000 - 25p \rightarrow 75p = 1200 \rightarrow p = 16$$
Therefore, $Q_s = -200 + 50(16) = -200 + 800 = 600$
The equilibrium price is $16 and the equilibrium quantity is 600 T-shirts.

41. Let l be the length of the rectangle and w be the width of the rectangle. Then:
$$2l + 2w = 90 \Rightarrow l = 2w$$
Solve by substitution:
$$2(2w) + 2w = 90 \Rightarrow 4w + 2w = 90 \Rightarrow 6w = 90 \Rightarrow w = 15 \text{ feet}$$
$$l = 2(15) = 30 \text{ feet}$$
The dimensions of the floor are 15 feet by 30 feet.

43. Let x = the cost of one cheeseburger and y = the cost of one shake. Then:
$$4x + 2y = 790 \Rightarrow 2y = x + 15$$
Solve by substitution:
$$4x + x + 15 = 790 \rightarrow 5x = 775 \rightarrow x = 155$$
$$2y = 155 + 15 \rightarrow 2y = 170 \rightarrow y = 85$$
A cheeseburger cost $1.55 and a shake costs $0.85.

45. Let x = the number of pounds of cashews.
Then $x + 30$ is the number of pounds in the mixture.
The value of the cashews is $5x$.
The value of the peanuts is $1.50(30) = 45$.
The value of the mixture is $3(x + 30)$.
Setting up a value equation:
$$5x + 45 = 3(x + 30)$$
$$5x + 45 = 3x + 90$$
$$2x = 45$$
$$x = 22.5$$
22.5 pounds of cashews should be used in the mixture.

47. Let x = the plane's air speed and y = the wind speed.

	Rate	Time	Distance
With Wind	$x + y$	3	600
Against	$x - y$	4	600

$$(x + y)(3) = 600 \quad \rightarrow \quad x + y = 200$$
$$(x - y)(4) = 600 \quad \rightarrow \quad x - y = 150$$
Solving by elimination:
$$2x = 350$$
$$x = 175$$
$$y = 200 - x = 200 - 175 = 25$$
The airspeed of the plane is 175 mph, and the wind speed is 25 mph.

49. Let x = the number of one design.
Let y = the number of the second design.
Then $x + y$ = the total number of sets of dishes.
$25x + 45y$ = the cost of the dishes.
Setting up the equations and solving by substitution:
$$\begin{cases} x + y = 200 \\ 25x + 45y = 7400 \end{cases} \rightarrow y = 200 - x$$
$$25x + 45(200 - x) = 7400$$
$$25x + 9000 - 45x = 7400$$
$$-20x = -1600 \rightarrow x = 80$$
$$y = 200 - 80 = 120$$
80 sets of the $25 dishes and 120 sets of the $45 dishes should be ordered.

51. Let x = the cost per package of bacon.
Let y = the cost of a carton of eggs.
Set up a system of equations for the problem:
$$\begin{cases} 3x + 2y = 7.45 \\ 2x + 3y = 6.45 \end{cases}$$
Multiply each side of the first equation by 3 and each side of the second equation by –2 and solve by elimination:
$$\begin{cases} 3x + 2y = 7.45 & \xrightarrow{\;3\;} & 9x + 6y = 22.35 \\ 2x + 3y = 6.45 & \xrightarrow{\;-2\;} & -4x - 6y = -12.90 \end{cases}$$
$$5x = 9.45 \Rightarrow x = 1.89$$
Substitute and solve for y:
$$3(1.89) + 2y = 7.45$$
$$5.67 + 2y = 7.45 \Rightarrow 2y = 1.78 \Rightarrow y = 0.89$$
A package of bacon costs $1.89 and a carton of eggs cost $0.89.
The refund for 2 packages of bacon and 2 cartons of eggs will be $5.56.

53. Let x = the # of mg of liquid 1.
Let y = the # of mg of liquid 2.
Setting up the equations and solving by substitution:
$$\begin{cases} 0.2x + 0.4y = 40 & \text{vitamin C} \\ 0.3x + 0.2y = 30 & \text{vitamin D} \end{cases}$$
multiplying each equation by 10 yields
$$\begin{cases} 2x + 4y = 400 & \xrightarrow{} & 2x + 4y = 400 \\ 3x + 2y = 300 & \xrightarrow{\;2\;} & 6x + 4y = 600 \end{cases}$$
subtracting the bottom equation from the top equation yields
$$2x + 4y - (6x + 4y) = -200$$
$$2x - 6x = -200 \Rightarrow -4x = -200 \Rightarrow x = 50$$

$$\Rightarrow 2(50) + 4y = 400 \Rightarrow 100 + 4y = 400$$

$$\Rightarrow 4y = 300 \Rightarrow y = \frac{300}{4} = 75$$

So 50 mg of liquid 1 should be mixed with 75 mg of liquid 2.

55. Solve the system by substitution:
$$R = C$$

$$8x = 4.5x + 17500 \rightarrow 3.5x = 17500 \rightarrow x = 5000$$
5000 units must be produced and sold for the firm to break-even.

57. Answers will vary.

Chapter 10

Systems of Equations and Inequalities

10.2 Systems of Linear Equations: Three Equations Containing Three Variables

1. Substituting the values of the variables:

$$\begin{cases} 3x + 3y + 2z = 4 & \longrightarrow \quad 3(1) + 3(-1) + 2(2) = 3 - 3 + 4 = 4 \\ x - y - z = 0 & \longrightarrow \quad 1 - (-1) - 2 = 1 + 1 - 2 = 0 \\ 2y - 3z = -8 & \longrightarrow \quad 2(-1) - 3(2) = -2 - 6 = -8 \end{cases}$$

Each equation is satisfied, so $x = 1$, $y = -1$, $z = 2$ is a solution to the system of equations.

3. Multiply each side of the first equation by -2 and add to the second equation to eliminate x:

$$\begin{cases} x - y = 6 & \xrightarrow{-2} & -2x + 2y & = -12 \\ 2x - 3z = 16 & \longrightarrow & 2x \quad\quad - 3z = 16 \\ 2y + z = 4 & & \overline{\quad\quad 2y - 3z = 4} \end{cases}$$

Multiply each side of the result by -1 and add to the original third equation to eliminate y:

$$\begin{aligned} 2y - 3z = 4 & \xrightarrow{-1} & -2y + 3z = -4 \\ 2y + z = 4 & \longrightarrow & \underline{2y + z = 4} \\ & & 4z = 0 \Rightarrow z = 0 \end{aligned}$$

Substituting and solving for the other variables:

$$\begin{aligned} 2y + 0 &= 4 & 2x - 3(0) &= 16 \\ 2y &= 4 & 2x &= 16 \\ y &= 2 & x &= 8 \end{aligned}$$

The solution is $x = 8$, $y = 2$, $z = 0$.

5. Multiply each side of the first equation by -2 and add to the second equation to eliminate x; and multiply each side of the first equation by 3 and add to the third equation to eliminate x:

$$\begin{cases} x - 2y + 3z = 7 & \xrightarrow{-2} & -2x + 4y - 6z = -14 \\ 2x + y + z = 4 & \longrightarrow & \underline{2x + y + z = 4} \\ -3x + 2y - 2z = -10 & & 5y - 5z = -10 & \xrightarrow{1/5} & y - z = -2 \\ & \xrightarrow{3} & 3x - 6y + 9z = 21 \\ & \longrightarrow & \underline{-3x + 2y - 2z = -10} \\ & & -4y + 7z = 11 \end{cases}$$

Multiply each side of the first result by 4 and add to the second result to eliminate y:

$$y - z = -2 \xrightarrow{\ 4\ } 4y - 4z = -8$$

$$-4y + 7z = 11 \longrightarrow -4y + 7z = 11$$

$$3z = 3 \rightarrow z = 1$$

Substituting and solving for the other variables:

$$y - 1 = -2$$
$$y = -1 \qquad x - 2(-1) + 3(1) = 7$$
$$x + 2 + 3 = 7 \rightarrow x = 2$$

The solution is $x = 2,\ y = -1,\ z = 1$.

7.   Add the first and second equations to eliminate z:

$$\begin{cases} x - y - z = 1 \longrightarrow x - y - z = 1 \\ 2x + 3y + z = 2 \longrightarrow 2x + 3y + z = 2 \\ 3x + 2y = 0 \qquad 3x + 2y = 3 \end{cases}$$

Multiply each side of the result by -1 and add to the original third equation to eliminate y:

$$3x + 2y = 3 \xrightarrow{\ -1\ } -3x - 2y = -3$$

$$3x + 2y = 0 \longrightarrow 3x + 2y = 0$$

$$0 = -3$$

This result has no solution, so the system is inconsistent.

9.   Add the first and second equations to eliminate x; and multiply the first equation by -3 and add to the third equation to eliminate x:

$$\begin{cases} x - y - z = 1 \longrightarrow x - y - z = 1 \\ -x + 2y - 3z = -4 \longrightarrow -x + 2y - 3z = -4 \\ 3x - 2y - 7z = 0 \qquad y - 4z = -3 \end{cases}$$

$$\xrightarrow{\ -3\ } -3x + 3y + 3z = -3$$

$$\longrightarrow 3x - 2y - 7z = 0$$

$$y - 4z = -3$$

Multiply each side of the first result by -1 and add to the second result to eliminate y:

$$y - 4z = -3 \xrightarrow{\ -1\ } -y + 4z = 3$$

$$y - 4z = -3 \longrightarrow y - 4z = -3$$

$$0 = 0$$

The system is dependent. If z is any real number, then $y = 4z - 3$.

Solving for x in terms of z in the first equation:

$$x - (4z - 3) - z = 1$$
$$x - 4z + 3 - z = 1$$
$$x - 5z + 3 = 1$$
$$x = 5z - 2$$

The solution is $x = 5z - 2,\ y = 4z - 3,\ z$ is any real number.

11. Multiply the first equation by –2 and add to the second equation to eliminate x; and add the first and third equations to eliminate x:

$$\begin{cases} 2x - 2y + 3z = 6 \\ 4x - 3y + 2z = 0 \\ -2x + 3y - 7z = 1 \end{cases} \xrightarrow{-2} \begin{array}{r} -4x + 4y - 6z = -12 \\ 4x - 3y + 2z = 0 \\ \hline y - 4z = -12 \end{array}$$

$$\xrightarrow{} \begin{array}{r} 2x - 2y + 3z = 6 \\ -2x + 3y - 7z = 1 \\ \hline y - 4z = 7 \end{array}$$

Multiply each side of the first result by –1 and add to the second result to eliminate y:

$$\begin{array}{r} y - 4z = -12 \\ y - 4z = 7 \end{array} \xrightarrow{-1} \begin{array}{r} -y + 4z = 12 \\ y - 4z = 7 \\ \hline 0 = 19 \end{array}$$

This result has no solution, so the system is inconsistent.

13. Add the first and second equations to eliminate z; and multiply the second equation by 2 and add to the third equation to eliminate z:

$$\begin{cases} x + y - z = 6 \\ 3x - 2y + z = -5 \\ x + 3y - 2z = 14 \end{cases} \xrightarrow{} \begin{array}{r} x + y - z = 6 \\ 3x - 2y + z = -5 \\ \hline 4x - y = 1 \end{array}$$

$$\xrightarrow{2} \begin{array}{r} 6x - 4y + 2z = -10 \\ x + 3y - 2z = 14 \\ \hline 7x - y = 4 \end{array}$$

Multiply each side of the first result by –1 and add to the second result to eliminate y:

$$\begin{array}{r} 4x - y = 1 \\ 7x - y = 4 \end{array} \xrightarrow{-1} \begin{array}{r} -4x + y = -1 \\ 7x - y = 4 \\ \hline 3x = 3 \\ x = 1 \end{array}$$

Substituting and solving for the other variables:

$$\begin{array}{ll} 4(1) - y = 1 & 3(1) - 2(3) + z = -5 \\ -y = -3 & 3 - 6 + z = -5 \\ y = 3 & z = -2 \end{array}$$

The solution is $x = 1$, $y = 3$, $z = -2$.

15. Add the first and second equations to eliminate z; and multiply the second equation by 3 and add to the third equation to eliminate z:

$$\begin{cases} x + 2y - z = -3 \\ 2x - 4y + z = -7 \\ -2x + 2y - 3z = 4 \end{cases} \xrightarrow{} \begin{array}{r} x + 2y - z = -3 \\ 2x - 4y + z = -7 \\ \hline 3x - 2y = -10 \end{array}$$

$$\xrightarrow{\;3\;}\quad 6x - 12y + 3z = -21$$
$$\xrightarrow{}\quad \underline{-2x + 2y - 3z = 4}$$
$$4x - 10y = -17$$

Multiply each side of the first result by –5 and add to the second result to eliminate y:

$$3x - 2y = -10 \quad \xrightarrow{\;-5\;}\quad -15x + 10y = 50$$
$$4x - 10y = -17 \quad \xrightarrow{}\quad \underline{4x - 10y = -17}$$
$$-11x = 33$$
$$x = -3$$

Substituting and solving for the other variables:

$$3(-3) - 2y = -10$$
$$-9 - 2y = -10$$
$$-2y = -1$$
$$y = \frac{1}{2}$$

$$-3 + 2\left(\frac{1}{2}\right) - z = -3$$
$$-3 + 1 - z = -3$$
$$-z = -1$$
$$z = 1$$

The solution is $x = -3$, $y = \dfrac{1}{2}$, $z = 1$.

17. $y = ax^2 + bx + c$

At (–1, 4) the equation becomes:

$$4 = a(-1)^2 + b(-1) + c$$
$$4 = a - b + c \Rightarrow a - b + c = 4$$

At (2, 3) the equation becomes:

$$3 = a(2)^2 + b(2) + c$$
$$3 = 4a + 2b + c \Rightarrow 4a + 2b + c = 3$$

At (0, 1) the equation becomes:

$$1 = a(0)^2 + b(0) + c \Rightarrow c = 1$$

The system of equations is:

$$\begin{cases} a - b + c = 4 \\ 4a + 2b + c = 3 \\ \phantom{4a + 2b + {}}c = 1 \end{cases}$$

Substitute $c = 1$ into the first and second equations and simplify:

$$\begin{cases} a - b + 1 = 4 & \rightarrow & a - b = 3 & \rightarrow & a = b + 3 \\ 4a + 2b + 1 = 3 & \rightarrow & 4a + 2b = 2 \end{cases}$$

Solve the first equation for a, substitute into the second equation and solve:

$$4(b + 3) + 2b = 2$$
$$4b + 12 + 2b = 2$$
$$6b = -10 \Rightarrow b = -\frac{5}{3} \Rightarrow a = -\frac{5}{3} + 3 = \frac{4}{3}$$

The solution is $a = \dfrac{4}{3}$, $b = -\dfrac{5}{3}$, $c = 1$. So the equation is $y = \dfrac{4}{3}x^2 - \dfrac{5}{3}x + 1$.

19. Substitute the expression for I_2 into the second and third equations and simplify:

$$\begin{cases} I_2 = I_1 + I_3 \\ 5 - 3I_1 - 5I_2 = 0 \quad \rightarrow \quad 5 - 3I_1 - 5(I_1 + I_3) = 0 \quad \rightarrow \quad -8I_1 - 5I_3 = -5 \\ 10 - 5I_2 - 7I_3 = 0 \quad \rightarrow \quad 10 - 5(I_1 + I_3) - 7I_3 = 0 \quad \rightarrow \quad -5I_1 - 12I_3 = -10 \end{cases}$$

Multiply both sides of the second equation by 5 and multiply both sides of the third equation by –8 to eliminate I_1:

$$\begin{array}{rl} -8I_1 - 5I_3 = -5 & \xrightarrow{\;5\;} \quad -40I_1 - 25I_3 = -25 \\ -5I_1 - 12I_3 = -10 & \xrightarrow{\;-8\;} \quad \underline{40I_1 + 96I_3 = 80} \\ & \qquad\qquad\qquad\quad 71I_3 = 55 \\ & \qquad\qquad\qquad\quad\;\; I_3 = \dfrac{55}{71} \end{array}$$

Substituting and solving for the other variables:

$$-8I_1 - 5\left(\frac{55}{71}\right) = -5 \qquad\qquad I_2 = \frac{10}{71} + \frac{55}{71}$$

$$-8I_1 - \frac{275}{71} = -5 \qquad\qquad\quad I_2 = \frac{65}{71}$$

$$-8I_1 = -\frac{80}{71}$$

$$I_1 = \frac{10}{71}$$

The solution is $I_1 = \dfrac{10}{71},\ I_2 = \dfrac{65}{71},\ I_3 = \dfrac{55}{71}$.

21. Let x = the number of orchestra seats.
Let y = the number of main seats.
Let z = the number of balcony seats.
Since the total number of seats is 500, $x + y + z = 500$.
Since the total revenue is \$17,100 if all seats are sold, $50x + 35y + 25z = 17{,}100$.
If only half of the orchestra seats are sold, the revenue is \$14,600. So,

$$50\left(\frac{1}{2}x\right) + 35y + 25z = 14{,}600$$

Multiply each side of the first equation by –25 and add to the second equation to eliminate z; and multiply each side of the third equation by –1 and add to the second equation to eliminate z:

$$\begin{cases} x + y + z = 500 \\ 50x + 35y + 25z = 17100 \\ 25x + 35y + 25z = 14600 \end{cases} \quad \begin{array}{l} \xrightarrow{-25} \quad -25x - 25y - 25z = -12500 \\ \xrightarrow{} \quad \underline{50x + 35y + 25z = 17100} \\ \qquad\qquad\quad 25x + 10y = 4600 \end{array}$$

$$\begin{array}{l} \xrightarrow{} \quad 50x + 35y + 25z = 17100 \\ \xrightarrow{-1} \quad \underline{-25x - 35y - 25z = -14600} \\ \qquad\quad 25x = 2500 \\ \qquad\qquad\;\; x = 100 \end{array}$$

Substituting and solving for the other variables:

$$25(100) + 10y = 4600 \qquad 100 + 210 + z = 500$$
$$2500 + 10y = 4600 \qquad 310 + z = 500$$
$$10y = 2100 \qquad z = 190$$
$$y = 210$$

There are 100 orchestra seats, 210 main seats, and 190 balcony seats.

23. Let x = the number of servings of chicken.
Let y = the number of servings of corn.
Let z = the number of servings of 2% milk.
Protein equation: $30x + 3y + 9z = 66$
Carbohydrate equation: $35x + 16y + 13z = 94.5$
Calcium equation: $200x + 10y + 300z = 910$
Multiply each side of the first equation by –16 and multiply each side of the second equation by 3 and add them to eliminate y; and multiply each side of the second equation by –5 and multiply each side of the third equation by 8 and add to eliminate y:

$$\begin{cases} 30x + 3y + 9z = 66 \\ 35x + 16y + 13z = 94.5 \\ 200x + 10y + 300z = 910 \end{cases}$$

$$\xrightarrow{-16} \quad -480x - 48y - 144z = -1056$$
$$\xrightarrow{3} \quad \underline{105x + 48y + 39z = 283.5}$$
$$-375x \qquad -105z = -772.5$$

$$\xrightarrow{-5} \quad -175x - 80y - 65z = -472.5$$
$$\xrightarrow{8} \quad \underline{1600x + 80y + 2400z = 7280}$$
$$1425x \qquad + 2335z = 6807.5$$

Multiply each side of the first result by 19 and multiply each side of the second result by 5 to eliminate x:

$$-375x - 105z = -772.5 \quad \xrightarrow{19} \quad -7125x - 1995z = -14677.5$$
$$1425x + 2335z = 6807.5 \quad \xrightarrow{5} \quad \underline{7125x + 11675z = 34037.5}$$
$$9680z = 19360$$
$$z = 2$$

Substituting and solving for the other variables:

$$-375x - 105(2) = -772.5 \qquad 30(1.5) + 3y + 9(2) = 66$$
$$-375x - 210 = -772.5 \qquad 45 + 3y + 18 = 66$$
$$-375x = -562.5 \qquad 3y = 3$$
$$x = 1.5 \qquad y = 1$$

The dietitian should serve 1.5 servings of chicken, 1 serving of corn, and 2 servings of 2% milk.

25. Let x = the price of 1 hamburger.
Let y = the price of 1 order of fries.
Let z = the price of 1 drink.
We can construct the system

$$\begin{cases} 8x + 6y + 6z = 26.10 \\ 10x + 6y + 8z = 31.60 \end{cases}$$

A system involving only 2 equations that contain 3 or more unknowns cannot be solved uniquely. In other words, we can create as many solutions as we want by choosing a specific value for one of the variables and then solving the resulting 2 x 2 system.

For example, suppose we know that

$1.75 < $ hamburger price $< \$2.25$

$0.75 < $ fries price $< \$1.00$

$0.60 < $ fries price $< \$0.90$

Pick a specific value for x, y or z $x = \$2.00$

2x2 system

$$\begin{cases} 8(2) + 6y + 6z = 26.10 \\ 10(2) + 6y + 8z = 31.60 \end{cases} \Rightarrow \begin{cases} 16 + 6y + 6z = 26.10 \\ 20 + 6y + 8z = 31.60 \end{cases}$$

$$\Rightarrow \begin{cases} 6y + 6z = 10.10 \\ 6y + 8z = 11.60 \end{cases}$$

Solution $x = \$2.00,\ y = \$0.93,\ z = \$0.75$

Pick a specific value for x, y or z $y = \$0.90$

2x2 system

$$\begin{cases} 8x + 6(0.9) + 6z = 26.10 \\ 10x + 6(0.9) + 8z = 31.60 \end{cases} \Rightarrow \begin{cases} 8x + 5.4 + 6z = 26.10 \\ 10x + 5.4 + 8z = 31.60 \end{cases}$$

$$\Rightarrow \begin{cases} 8x + 6z = 20.7 \\ 10x + 8z = 26.2 \end{cases}$$

Solution $x = \$2.00,\ y = \$0.90,\ z = \$0.65$

Pick a specific value for x, y or z $z = \$0.80$

2x2 system

$$\begin{cases} 8x + 6y + 6(.8) = 26.10 \\ 10x + 6y + 8(.8) = 31.60 \end{cases} \Rightarrow \begin{cases} 8x + 6y + 4.8 = 26.10 \\ 10x + 6y + 6.4 = 31.60 \end{cases}$$

$$\Rightarrow \begin{cases} 8x + 6y = 21.3 \\ 10x + 6y = 25.2 \end{cases}$$

Solution $x = \$1.95,\ \ y = \$0.95,\ \ z = \$0.80$

27.   Let x = Beth's time working alone.
Let y = Bill's time working alone.
Let z = Edie's time working alone.
We can use the following tables to organize our work:

	Beth	Bill	Edie	Together
Hours to do job	x	y	z	10
Part of job done in 1 hour	$\dfrac{1}{x}$	$\dfrac{1}{y}$	$\dfrac{1}{z}$	$\dfrac{1}{10}$

Equation: $\dfrac{1}{x}+\dfrac{1}{y}+\dfrac{1}{z}=\dfrac{1}{10}$

	Bill	Edie	Together
Hours to do job	y	z	15
Part of job done in 1 hour	$\dfrac{1}{y}$	$\dfrac{1}{z}$	$\dfrac{1}{15}$

Equation: $\dfrac{1}{y}+\dfrac{1}{z}=\dfrac{1}{15}$

	Beth	Bill	Edie	All three	Beth and Bill
Hours to do job	x	y	z	4	8
Part of job done in 1 hour	$\dfrac{1}{x}$	$\dfrac{1}{y}$	$\dfrac{1}{z}$	$\dfrac{1}{4}$	$\dfrac{1}{8}$

Equation: $4\left(\dfrac{1}{x}+\dfrac{1}{y}+\dfrac{1}{z}\right)+8\left(\dfrac{1}{x}+\dfrac{1}{y}\right)=1 \rightarrow \dfrac{12}{x}+\dfrac{12}{y}+\dfrac{4}{z}=1$

We can construct the system

$$\begin{cases} \dfrac{1}{x}+\dfrac{1}{y}+\dfrac{1}{z}=\dfrac{1}{10} \\[2mm] \dfrac{1}{y}+\dfrac{1}{z}=\dfrac{1}{15} \\[2mm] \dfrac{12}{x}+\dfrac{12}{y}+\dfrac{4}{z}=1 \end{cases} \longrightarrow$$

subtracting the second equation from the first equation yields

$$\dfrac{1}{x}+\dfrac{1}{y}+\dfrac{1}{z}=\dfrac{1}{10}$$
$$\dfrac{1}{y}+\dfrac{1}{z}=\dfrac{1}{15}$$
$$\overline{\dfrac{1}{x}=\dfrac{1}{10}-\dfrac{1}{15}}$$

$\rightarrow \dfrac{1}{x}=\dfrac{1}{30} \rightarrow x=30$

Plugging $x = 30$ into the original system yields

$$\begin{cases} \dfrac{1}{30}+\dfrac{1}{y}+\dfrac{1}{z}=\dfrac{1}{10} \longrightarrow \dfrac{1}{y}+\dfrac{1}{z}=\dfrac{1}{10}-\dfrac{1}{30} \longrightarrow \dfrac{1}{y}+\dfrac{1}{z}=\dfrac{1}{15} \\[2mm] \dfrac{1}{y}+\dfrac{1}{z}=\dfrac{1}{15} \\[2mm] \dfrac{12}{30}+\dfrac{12}{y}+\dfrac{4}{z}=1 \longrightarrow \dfrac{12}{y}+\dfrac{4}{z}=1-\dfrac{12}{30} \longrightarrow \dfrac{12}{y}+\dfrac{4}{z}=\dfrac{3}{5} \end{cases}$$

Now consider the system

$$\begin{cases} \dfrac{1}{y}+\dfrac{1}{z}=\dfrac{1}{15} \xrightarrow{\;-12\;} \dfrac{-12}{y}+\dfrac{-12}{z}=\dfrac{-12}{15} \\[2ex] \dfrac{12}{y}+\dfrac{4}{z}=\dfrac{3}{5} \longrightarrow \quad \dfrac{12}{y}+\dfrac{4}{z}=\dfrac{3}{5} \end{cases}$$

adding these 2 equations yields

$$\dfrac{-12}{y}+\dfrac{-12}{z}=\dfrac{-12}{15}$$

$$\dfrac{\dfrac{12}{y}+\dfrac{4}{z}=\dfrac{3}{5}}{\dfrac{-12}{z}+\dfrac{4}{z}=\dfrac{-12}{15}+\dfrac{3}{5}} \quad \rightarrow \dfrac{-8}{z}=\dfrac{-3}{15} \rightarrow \dfrac{8}{z}=\dfrac{1}{5} \rightarrow z=40$$

plugging $z = 40$ into the equation

$$\dfrac{12}{y}+\dfrac{4}{z}=\dfrac{3}{5} \Rightarrow \dfrac{12}{y}+\dfrac{4}{40}=\dfrac{3}{5}$$

$$\dfrac{12}{y}+\dfrac{1}{10}=\dfrac{3}{5} \Rightarrow \dfrac{12}{y}=\dfrac{3}{5}-\dfrac{1}{10} \Rightarrow \dfrac{12}{y}=\dfrac{1}{2} \Rightarrow y=24$$

So, working alone, it would take Beth 30 hours, Bill 24 hours and Edie 40 hours to finish the job.

Systems of Equations and Inequalities

10.3 Systems of Linear Equations: Matrices

1. Writing the augmented matrix for the system of equations:
$$\begin{cases} x - 5y = 5 \\ 4x + 3y = 6 \end{cases} \rightarrow \begin{bmatrix} 1 & -5 & | & 5 \\ 4 & 3 & | & 6 \end{bmatrix}$$

3. Writing the augmented matrix for the system of equations:
$$\begin{cases} 2x + 3y - 6 = 0 \\ 4x - 6y + 2 = 0 \end{cases} \rightarrow \begin{cases} 2x + 3y = 6 \\ 4x - 6y = -2 \end{cases} \rightarrow \begin{bmatrix} 2 & 3 & | & 6 \\ 4 & -6 & | & -2 \end{bmatrix}$$

5. Writing the augmented matrix for the system of equations:
$$\begin{cases} 0.01x - 0.03y = 0.06 \\ 0.13x + 0.10y = 0.20 \end{cases} \rightarrow \begin{bmatrix} 0.01 & -0.03 & | & 0.06 \\ 0.13 & 0.10 & | & 0.20 \end{bmatrix}$$

7. Writing the augmented matrix for the system of equations:
$$\begin{cases} x - y + z = 10 \\ 3x + 3y \quad = 5 \\ x + y + 2z = 2 \end{cases} \rightarrow \begin{bmatrix} 1 & -1 & 1 & | & 10 \\ 3 & 3 & 0 & | & 5 \\ 1 & 1 & 2 & | & 2 \end{bmatrix}$$

9. Writing the augmented matrix for the system of equations:
$$\begin{cases} x + y - z = 2 \\ 3x - 2y \quad = 2 \\ 5x + 3y - z = 1 \end{cases} \rightarrow \begin{bmatrix} 1 & 1 & -1 & | & 2 \\ 3 & -2 & 0 & | & 2 \\ 5 & 3 & -1 & | & 1 \end{bmatrix}$$

11. Writing the augmented matrix for the system of equations:
$$\begin{cases} x - y - z = 10 \\ 2x + y + 2z = -1 \\ -3x + 4y = 5 \\ 4x - 5y + z = 0 \end{cases} \rightarrow \begin{bmatrix} 1 & -1 & -1 & | & 10 \\ 2 & 1 & 2 & | & -1 \\ -3 & 4 & 0 & | & 5 \\ 4 & -5 & 1 & | & 0 \end{bmatrix}$$

13. $$\begin{bmatrix} 1 & -3 & | & -2 \\ 2 & -5 & | & 5 \end{bmatrix} \rightarrow \begin{bmatrix} 1 & -3 & | & -2 \\ 0 & 1 & | & 9 \end{bmatrix}$$
$$R_2 = -2r_1 + r_2$$

15. $\begin{vmatrix} 1 & -3 & 4 & | & 3 \\ 2 & -5 & 6 & | & 6 \\ -3 & 3 & 4 & | & 6 \end{vmatrix} \rightarrow \begin{vmatrix} 1 & -3 & 4 & | & 3 \\ 0 & 1 & -2 & | & 0 \\ -3 & 3 & 4 & | & 6 \end{vmatrix} \rightarrow \begin{vmatrix} 1 & -3 & 4 & | & 3 \\ 0 & 1 & -2 & | & 0 \\ 0 & -6 & 16 & | & 15 \end{vmatrix}$

(a) $R_2 = -2r_1 + r_2$ (b) $R_3 = 3r_1 + r_3$

17. $\begin{vmatrix} 1 & -3 & 2 & | & -6 \\ 2 & -5 & 3 & | & -4 \\ -3 & -6 & 4 & | & 6 \end{vmatrix} \rightarrow \begin{vmatrix} 1 & -3 & 2 & | & -6 \\ 0 & 1 & -1 & | & 8 \\ -3 & -6 & 4 & | & 6 \end{vmatrix} \rightarrow \begin{vmatrix} 1 & -3 & 2 & | & -6 \\ 0 & 1 & -1 & | & 8 \\ 0 & -15 & 10 & | & -12 \end{vmatrix}$

(a) $R_2 = -2r_1 + r_2$ (b) $R_3 = 3r_1 + r_3$

19. $\begin{vmatrix} 1 & -3 & 1 & | & -2 \\ 2 & -5 & 6 & | & -2 \\ -3 & 1 & 4 & | & 6 \end{vmatrix} \rightarrow \begin{vmatrix} 1 & -3 & 1 & | & -2 \\ 0 & 1 & 4 & | & 2 \\ -3 & 1 & 4 & | & 6 \end{vmatrix} \rightarrow \begin{vmatrix} 1 & -3 & 1 & | & -2 \\ 0 & 1 & 4 & | & 2 \\ 0 & -8 & 7 & | & 0 \end{vmatrix}$

(a) $R_2 = -2r_1 + r_2$ (b) $R_3 = 3r_1 + r_3$

21. $\begin{cases} x = 5 \\ y = -1 \end{cases}$ consistent $x = 5,\ y = -1$

23. $\begin{cases} x = 1 \\ y = 2 \\ 0 = 3 \end{cases}$ inconsistent

25. $\begin{cases} x + 2z = -1 \\ y - 4z = -2 \\ \quad 0 = 0 \end{cases}$ consistent $x = -1 - 2z,\ y = -2 + 4z,\ z$ is any real number

27. $\begin{cases} x_1 = 1 \\ x_2 + x_4 = 2 \\ x_3 + 2x_4 = 3 \end{cases}$ consistent $x_1 = 1, x_2 = 2 - x_4, x_3 = 3 - 2x_4, x_4$ is any real number

29. $\begin{cases} x_1 + 4x_4 = 2 \\ x_2 + + x_3 + 3x_4 = 3 \\ \qquad 0 = 0 \end{cases}$ consistent $x_1 = 2 - 4x_4, x_2 = 3 - x_3 - 3x_4,$
x_3, x_4 are any real numbers

31.

$$\begin{cases} x_1 + x_4 = -2 \\ x_2 + 2x_4 = 2 \\ x_3 - x_4 = 0 \\ \quad 0 = 0 \end{cases}$$ consistent $x_1 = -2 - x_4, x_2 = 2 - 2x_4, x_3 = x_4,$

x_4 is any real number

33. $\begin{cases} x + y = 8 \\ x - y = 4 \end{cases}$ can be written as: $\begin{bmatrix} 1 & 1 & | & 8 \\ 1 & -1 & | & 4 \end{bmatrix}$

$\rightarrow \begin{bmatrix} 1 & 1 & | & 8 \\ 0 & -2 & | & -4 \end{bmatrix} \rightarrow \begin{bmatrix} 1 & 1 & | & 8 \\ 0 & 1 & | & 2 \end{bmatrix} \rightarrow \begin{bmatrix} 1 & 0 & | & 6 \\ 0 & 1 & | & 2 \end{bmatrix}$

$R_2 = -r_1 + r_2 \quad R_2 = -\frac{1}{2}r_2 \quad R_1 = -r_2 + r_1$

The solution is $x = 6$, $y = 2$.

35. $\begin{cases} 2x - 4y = -2 \\ 3x + 2y = 3 \end{cases}$ can be written as: $\begin{bmatrix} 2 & -4 & | & -2 \\ 3 & 2 & | & 3 \end{bmatrix}$

$\rightarrow \begin{bmatrix} 1 & -2 & | & -1 \\ 3 & 2 & | & 3 \end{bmatrix} \rightarrow \begin{bmatrix} 1 & -2 & | & -1 \\ 0 & 8 & | & 6 \end{bmatrix} \rightarrow \begin{bmatrix} 1 & -2 & | & -1 \\ 0 & 1 & | & \frac{3}{4} \end{bmatrix} \rightarrow \begin{bmatrix} 1 & 0 & | & \frac{1}{2} \\ 0 & 1 & | & \frac{3}{4} \end{bmatrix}$

$R_1 = \frac{1}{2}r_1 \qquad R_2 = -3r_1 + r_2 \quad R_2 = \frac{1}{8}r_2 \qquad R_1 = 2r_2 + r_1$

The solution is $x = \dfrac{1}{2}$, $y = \dfrac{3}{4}$.

37. $\begin{cases} x + 2y = 4 \\ 2x + 4y = 8 \end{cases}$ can be written as: $\begin{bmatrix} 1 & 2 & | & 4 \\ 2 & 4 & | & 8 \end{bmatrix}$

$\rightarrow \begin{bmatrix} 1 & 2 & | & 4 \\ 0 & 0 & | & 0 \end{bmatrix}$

$R_2 = -2r_1 + r_2$

This is a dependent system and the solution is $x = -2y + 4$, y is any real number.

39. $\begin{cases} 2x + 3y = 6 \\ x - y = \dfrac{1}{2} \end{cases}$ can be written as: $\begin{bmatrix} 2 & 3 & | & 6 \\ 1 & -1 & | & \frac{1}{2} \end{bmatrix}$

$\rightarrow \begin{bmatrix} 1 & \frac{3}{2} & | & 3 \\ 1 & -1 & | & \frac{1}{2} \end{bmatrix} \rightarrow \begin{bmatrix} 1 & \frac{3}{2} & | & 3 \\ 0 & -\frac{5}{2} & | & -\frac{5}{2} \end{bmatrix} \rightarrow \begin{bmatrix} 1 & \frac{3}{2} & | & 3 \\ 0 & 1 & | & 1 \end{bmatrix} \rightarrow \begin{bmatrix} 1 & 0 & | & \frac{3}{2} \\ 0 & 1 & | & 1 \end{bmatrix}$

$R_1 = \frac{1}{2}r_1 \qquad R_2 = -r_1 + r_2 \quad R_2 = -\frac{2}{5}r_2 \quad R_1 = -\frac{3}{2}r_2 + r_1$

The solution is $x = \dfrac{3}{2}$, $y = 1$.

41. $\begin{cases} 3x - 5y = 3 \\ 15x + 5y = 21 \end{cases}$ can be written as: $\begin{bmatrix} 3 & -5 & | & 3 \\ 15 & 5 & | & 21 \end{bmatrix}$

$$\rightarrow \begin{bmatrix} 1 & -\frac{5}{3} & | & 1 \\ 15 & 5 & | & 21 \end{bmatrix} \rightarrow \begin{bmatrix} 1 & -\frac{5}{3} & | & 1 \\ 0 & 30 & | & 6 \end{bmatrix} \rightarrow \begin{bmatrix} 1 & -\frac{5}{3} & | & 1 \\ 0 & 1 & | & \frac{1}{5} \end{bmatrix} \rightarrow \begin{bmatrix} 1 & 0 & | & \frac{4}{3} \\ 0 & 1 & | & \frac{1}{5} \end{bmatrix}$$

$R_1 = \frac{1}{3} r_1$ $R_2 = -15 r_1 + r_2$ $R_2 = \frac{1}{30} r_2$ $R_1 = \frac{5}{3} r_2 + r_1$

The solution is $x = \dfrac{4}{3}, y = \dfrac{1}{5}$.

43. $\begin{cases} x - y = 6 \\ 2x - 3z = 16 \\ 2y + z = 4 \end{cases}$ can be written as: $\begin{bmatrix} 1 & -1 & 0 & | & 6 \\ 2 & 0 & -3 & | & 16 \\ 0 & 2 & 1 & | & 4 \end{bmatrix}$

$$\rightarrow \begin{bmatrix} 1 & -1 & 0 & | & 6 \\ 0 & 2 & -3 & | & 4 \\ 0 & 2 & 1 & | & 4 \end{bmatrix} \rightarrow \begin{bmatrix} 1 & -1 & 0 & | & 6 \\ 0 & 1 & -\frac{3}{2} & | & 2 \\ 0 & 2 & 1 & | & 4 \end{bmatrix} \rightarrow \begin{bmatrix} 1 & 0 & -\frac{3}{2} & | & 8 \\ 0 & 1 & -\frac{3}{2} & | & 2 \\ 0 & 0 & 4 & | & 0 \end{bmatrix} \rightarrow \begin{bmatrix} 1 & 0 & -\frac{3}{2} & | & 8 \\ 0 & 1 & -\frac{3}{2} & | & 2 \\ 0 & 0 & 1 & | & 0 \end{bmatrix}$$

$R_2 = -2 r_1 + r_2$ $R_2 = \frac{1}{2} r_2$ $R_1 = r_2 + r_1$ $R_3 = \frac{1}{4} r_3$

                                 $R_3 = -2 r_2 + r_3$

$$\rightarrow \begin{bmatrix} 1 & 0 & 0 & | & 8 \\ 0 & 1 & 0 & | & 2 \\ 0 & 0 & 1 & | & 0 \end{bmatrix}$$

$R_1 = \frac{3}{2} r_3 + r_1$

$R_2 = \frac{3}{2} r_3 + r_2$

The solution is $x = 8, y = 2, z = 0$.

45. $\begin{cases} x - 2y + 3z = 7 \\ 2x + y + z = 4 \\ -3x + 2y - 2z = -10 \end{cases}$ can be written as: $\begin{bmatrix} 1 & -2 & 3 & | & 7 \\ 2 & 1 & 1 & | & 4 \\ -3 & 2 & -2 & | & -10 \end{bmatrix}$

$$\rightarrow \begin{bmatrix} 1 & -2 & 3 & | & 7 \\ 0 & 5 & -5 & | & -10 \\ 0 & -4 & 7 & | & 11 \end{bmatrix} \rightarrow \begin{bmatrix} 1 & -2 & 3 & | & 7 \\ 0 & 1 & -1 & | & -2 \\ 0 & -4 & 7 & | & 11 \end{bmatrix} \rightarrow \begin{bmatrix} 1 & 0 & 1 & | & 3 \\ 0 & 1 & -1 & | & -2 \\ 0 & 0 & 3 & | & 3 \end{bmatrix}$$

$R_2 = -2 r_1 + r_2$ $R_2 = \frac{1}{5} r_2$ $R_1 = 2 r_2 + r_1$

$R_3 = 3 r_1 + r_3$ $R_3 = 4 r_2 + r_3$

$$\rightarrow \begin{bmatrix} 1 & 0 & 1 & | & 3 \\ 0 & 1 & -1 & | & -2 \\ 0 & 0 & 1 & | & 1 \end{bmatrix} \rightarrow \begin{bmatrix} 1 & 0 & 0 & | & 2 \\ 0 & 1 & 0 & | & -1 \\ 0 & 0 & 1 & | & 1 \end{bmatrix}$$

$R_3 = \frac{1}{3} r_3$ $R_1 = -r_3 + r_1$

                  $R_2 = r_3 + r_2$

The solution is $x = 2, y = -1, z = 1$.

47. $\begin{cases} 2x - 2y - 2z = 2 \\ 2x + 3y + z = 2 \\ 3x + 2y = 0 \end{cases}$ can be written as: $\begin{bmatrix} 2 & -2 & -2 & | & 2 \\ 2 & 3 & 1 & | & 2 \\ 3 & 2 & 0 & | & 0 \end{bmatrix}$

$$\rightarrow \begin{bmatrix} 1 & -1 & -1 & | & 1 \\ 2 & 3 & 1 & | & 2 \\ 3 & 2 & 0 & | & 0 \end{bmatrix} \rightarrow \begin{bmatrix} 1 & -1 & -1 & | & 1 \\ 0 & 5 & 3 & | & 0 \\ 0 & 5 & 3 & | & -3 \end{bmatrix} \rightarrow \begin{bmatrix} 1 & -1 & -1 & | & 1 \\ 0 & 5 & 3 & | & 0 \\ 0 & 0 & 0 & | & -3 \end{bmatrix}$$

$\quad R_1 = \frac{1}{2} r_1 \qquad\qquad R_2 = -2r_1 + r_2 \qquad R_3 = -r_2 + r_3$
$$R_3 = -3r_1 + r_3$$

There is no solution. The system is inconsistent.

49. $\begin{cases} -x + y + z = -1 \\ -x + 2y - 3z = -4 \\ 3x - 2y - 7z = 0 \end{cases}$ can be written as: $\begin{bmatrix} -1 & 1 & 1 & | & -1 \\ -1 & 2 & -3 & | & -4 \\ 3 & -2 & -7 & | & 0 \end{bmatrix}$

$$\rightarrow \begin{bmatrix} 1 & -1 & -1 & | & 1 \\ -1 & 2 & -3 & | & -4 \\ 3 & -2 & -7 & | & 0 \end{bmatrix} \rightarrow \begin{bmatrix} 1 & -1 & -1 & | & 1 \\ 0 & 1 & -4 & | & -3 \\ 0 & 1 & -4 & | & -3 \end{bmatrix} \rightarrow \begin{bmatrix} 1 & 0 & -5 & | & -2 \\ 0 & 1 & -4 & | & -3 \\ 0 & 0 & 0 & | & 0 \end{bmatrix} \rightarrow \begin{matrix} x - 5z = -2 \\ y - 4z = -3 \end{matrix}$$

$\quad R_1 = -r_1 \qquad\qquad R_2 = r_1 + r_2 \qquad\qquad R_1 = r_2 + r_1$
$$R_3 = -3r_1 + r_3 \qquad\quad R_3 = -r_2 + r_3$$

The solution is $x = 5z - 2, y = 4z - 3, z$ is any real number.

51. $\begin{cases} 2x - 2y + 3z = 6 \\ 4x - 3y + 2z = 0 \\ -2x + 3y - 7z = 1 \end{cases}$ can be written as: $\begin{bmatrix} 2 & -2 & 3 & | & 6 \\ 4 & -3 & 2 & | & 0 \\ -2 & 3 & -7 & | & 1 \end{bmatrix}$

$$\rightarrow \begin{bmatrix} 1 & -1 & \frac{3}{2} & | & 3 \\ 4 & -3 & 2 & | & 0 \\ -2 & 3 & -7 & | & 1 \end{bmatrix} \rightarrow \begin{bmatrix} 1 & -1 & \frac{3}{2} & | & 3 \\ 0 & 1 & -4 & | & -12 \\ 0 & 1 & -4 & | & 7 \end{bmatrix} \rightarrow \begin{bmatrix} 1 & 0 & -\frac{5}{2} & | & -9 \\ 0 & 1 & -4 & | & -12 \\ 0 & 0 & 0 & | & 19 \end{bmatrix}$$

$\quad R_1 = \frac{1}{2} r_1 \qquad\qquad R_2 = -4r_1 + r_2 \qquad\qquad R_1 = r_2 + r_1$
$$R_3 = 2r_1 + r_3 \qquad\qquad R_3 = -r_2 + r_3$$

There is no solution. The system is inconsistent.

53. $\begin{cases} x + y - z = 6 \\ 3x - 2y + z = -5 \\ x + 3y - 2z = 14 \end{cases}$ can be written as: $\begin{bmatrix} 1 & 1 & -1 & | & 6 \\ 3 & -2 & 1 & | & -5 \\ 1 & 3 & -2 & | & 14 \end{bmatrix}$

$$\rightarrow \begin{bmatrix} 1 & 1 & -1 & | & 6 \\ 0 & -5 & 4 & | & -23 \\ 0 & 2 & -1 & | & 8 \end{bmatrix} \rightarrow \begin{bmatrix} 1 & 1 & -1 & | & 6 \\ 0 & 1 & -\frac{4}{5} & | & \frac{23}{5} \\ 0 & 2 & -1 & | & 8 \end{bmatrix} \rightarrow \begin{bmatrix} 1 & 0 & -\frac{1}{5} & | & \frac{7}{5} \\ 0 & 1 & -\frac{4}{5} & | & \frac{23}{5} \\ 0 & 0 & \frac{3}{5} & | & -\frac{6}{5} \end{bmatrix}$$

$R_2 = -3r_1 + r_2 \qquad R_2 = -\frac{1}{5} r_2 \qquad R_1 = -r_2 + r_1$
$R_3 = -r_1 + r_3 \qquad\qquad\qquad\qquad R_3 = -2r_2 + r_3$

$$\rightarrow \begin{vmatrix} 1 & 0 & -\frac{1}{5} & \frac{7}{5} \\ 0 & 1 & -\frac{4}{5} & \frac{23}{5} \\ 0 & 0 & 1 & -2 \end{vmatrix} \rightarrow \begin{bmatrix} 1 & 0 & 0 & 1 \\ 0 & 1 & 0 & 3 \\ 0 & 0 & 1 & -2 \end{bmatrix}$$

$$R_3 = \tfrac{5}{3} r_3 \qquad\qquad R_1 = \tfrac{1}{5} r_3 + r_1$$
$$R_2 = \tfrac{4}{5} r_3 + r_2$$

The solution is $x = 1$, $y = 3$, $z = -2$.

55. $\begin{cases} x + 2y - z = -3 \\ 2x - 4y + z = -7 \\ -2x + 2y - 3z = 4 \end{cases}$ can be written as: $\begin{bmatrix} 1 & 2 & -1 & -3 \\ 2 & -4 & 1 & -7 \\ -2 & 2 & -3 & 4 \end{bmatrix}$

$$\rightarrow \begin{bmatrix} 1 & 2 & -1 & -3 \\ 0 & -8 & 3 & -1 \\ 0 & 6 & -5 & -2 \end{bmatrix} \rightarrow \begin{bmatrix} 1 & 2 & -1 & -3 \\ 0 & 1 & -\frac{3}{8} & \frac{1}{8} \\ 0 & 6 & -5 & -2 \end{bmatrix} \rightarrow \begin{bmatrix} 1 & 0 & -\frac{1}{4} & -\frac{13}{4} \\ 0 & 1 & -\frac{3}{8} & \frac{1}{8} \\ 0 & 0 & -\frac{11}{4} & -\frac{11}{4} \end{bmatrix}$$

$$R_2 = -2r_1 + r_2 \qquad R_2 = -\tfrac{1}{8}r_2 \qquad R_1 = -2r_2 + r_1$$
$$R_3 = 2r_1 + r_3 \qquad\qquad\qquad\qquad R_3 = -6r_2 + r_3$$

$$\rightarrow \begin{bmatrix} 1 & 0 & -\frac{1}{4} & -\frac{13}{4} \\ 0 & 1 & -\frac{3}{8} & \frac{1}{8} \\ 0 & 0 & 1 & 1 \end{bmatrix} \rightarrow \begin{bmatrix} 1 & 0 & 0 & -3 \\ 0 & 1 & 0 & \frac{1}{2} \\ 0 & 0 & 1 & 1 \end{bmatrix}$$

$$R_3 = -\tfrac{4}{11}r_3 \qquad\qquad R_1 = \tfrac{1}{4}r_3 + r_1$$
$$R_2 = \tfrac{3}{8}r_3 + r_2$$

The solution is $x = -3$, $y = \dfrac{1}{2}$, $z = 1$.

57. $\begin{cases} 3x + y - z = \dfrac{2}{3} \\ 2x - y + z = 1 \\ 4x + 2y = \dfrac{8}{3} \end{cases}$ can be written as: $\begin{bmatrix} 3 & 1 & -1 & \frac{2}{3} \\ 2 & -1 & 1 & 1 \\ 4 & 2 & 0 & \frac{8}{3} \end{bmatrix}$

$$\rightarrow \begin{bmatrix} 1 & \frac{1}{3} & -\frac{1}{3} & \frac{2}{9} \\ 2 & -1 & 1 & 1 \\ 4 & 2 & 0 & \frac{8}{3} \end{bmatrix} \rightarrow \begin{bmatrix} 1 & \frac{1}{3} & -\frac{1}{3} & \frac{2}{9} \\ 0 & -\frac{5}{3} & \frac{5}{3} & \frac{5}{9} \\ 0 & \frac{2}{3} & \frac{4}{3} & \frac{16}{9} \end{bmatrix} \rightarrow \begin{bmatrix} 1 & \frac{1}{3} & -\frac{1}{3} & \frac{2}{9} \\ 0 & 1 & -1 & -\frac{1}{3} \\ 0 & \frac{2}{3} & \frac{4}{3} & \frac{16}{9} \end{bmatrix} \rightarrow \begin{bmatrix} 1 & 0 & 0 & \frac{1}{3} \\ 0 & 1 & -1 & -\frac{1}{3} \\ 0 & 0 & 2 & 2 \end{bmatrix}$$

$$R_1 = \tfrac{1}{3}r_1 \qquad R_2 = -2r_1 + r_2 \qquad R_2 = -\tfrac{3}{5}r_2 \qquad R_1 = -\tfrac{1}{3}r_2 + r_1$$
$$R_3 = -4r_1 + r_3 \qquad\qquad\qquad R_3 = -\tfrac{2}{3}r_2 + r_3$$

$$\rightarrow \begin{vmatrix} 1 & 0 & 0 & | & \frac{1}{3} \\ 0 & 1 & -1 & | & -\frac{1}{3} \\ 0 & 0 & 1 & | & 1 \end{vmatrix} \rightarrow \begin{bmatrix} 1 & 0 & 0 & | & \frac{1}{3} \\ 0 & 1 & 0 & | & \frac{2}{3} \\ 0 & 0 & 1 & | & 1 \end{bmatrix}$$

$$R_3 = \tfrac{1}{2}r_3 \qquad\qquad R_2 = r_3 + r_2$$

The solution is $x = \dfrac{1}{3}, y = \dfrac{2}{3}, z = 1$.

59. $\begin{cases} x + y + z + w = 4 \\ 2x - y + z \quad\;\; = 0 \\ 3x + 2y + z - w = 6 \\ x - 2y - 2z + 2w = -1 \end{cases}$ can be written as: $\begin{bmatrix} 1 & 1 & 1 & 1 & | & 4 \\ 2 & -1 & 1 & 0 & | & 0 \\ 3 & 2 & 1 & -1 & | & 6 \\ 1 & -2 & -2 & 2 & | & -1 \end{bmatrix}$

$$\rightarrow \begin{vmatrix} 1 & 1 & 1 & 1 & | & 4 \\ 0 & -3 & -1 & -2 & | & -8 \\ 0 & -1 & -2 & -4 & | & -6 \\ 0 & -3 & -3 & 1 & | & -5 \end{vmatrix} \rightarrow \begin{vmatrix} 1 & 1 & 1 & 1 & | & 4 \\ 0 & -1 & -2 & -4 & | & -6 \\ 0 & -3 & -1 & -2 & | & -8 \\ 0 & -3 & -3 & 1 & | & -5 \end{vmatrix} \rightarrow \begin{vmatrix} 1 & 1 & 1 & 1 & | & 4 \\ 0 & 1 & 2 & 4 & | & 6 \\ 0 & -3 & -1 & -2 & | & -8 \\ 0 & -3 & -3 & 1 & | & -5 \end{vmatrix}$$

$$\begin{array}{lll} R_2 = -2r_1 + r_2 & \text{Interchange } r_2 \text{ and } r_3 & R_2 = -r_2 \\ R_3 = -3r_1 + r_3 & & \\ R_4 = -r_1 + r_4 & & \end{array}$$

$$\rightarrow \begin{vmatrix} 1 & 0 & -1 & -3 & | & -2 \\ 0 & 1 & 2 & 4 & | & 6 \\ 0 & 0 & 5 & 10 & | & 10 \\ 0 & 0 & 3 & 13 & | & 13 \end{vmatrix} \rightarrow \begin{vmatrix} 1 & 0 & -1 & -3 & | & -2 \\ 0 & 1 & 2 & 4 & | & 6 \\ 0 & 0 & 1 & 2 & | & 2 \\ 0 & 0 & 3 & 13 & | & 13 \end{vmatrix} \rightarrow \begin{bmatrix} 1 & 0 & 0 & -1 & | & 0 \\ 0 & 1 & 0 & 0 & | & 2 \\ 0 & 0 & 1 & 2 & | & 2 \\ 0 & 0 & 0 & 7 & | & 7 \end{bmatrix}$$

$$\begin{array}{lll} R_1 = -r_2 + r_1 & R_3 = \tfrac{1}{5}r_3 & R_1 = r_3 + r_1 \\ R_3 = 3r_2 + r_3 & & R_2 = -2r_3 + r_2 \\ R_4 = 3r_2 + r_4 & & R_4 = -3r_3 + r_4 \end{array}$$

$$\rightarrow \begin{vmatrix} 1 & 0 & 0 & -1 & | & 0 \\ 0 & 1 & 0 & 0 & | & 2 \\ 0 & 0 & 1 & 2 & | & 2 \\ 0 & 0 & 0 & 1 & | & 1 \end{vmatrix} \rightarrow \begin{vmatrix} 1 & 0 & 0 & 0 & | & 1 \\ 0 & 1 & 0 & 0 & | & 2 \\ 0 & 0 & 1 & 0 & | & 0 \\ 0 & 0 & 0 & 1 & | & 1 \end{vmatrix}$$

$$\begin{array}{ll} R_4 = \tfrac{1}{7}r_4 & R_1 = r_4 + r_1 \\ & R_3 = -2r_4 + r_3 \end{array}$$

The solution is $x = 1, y = 2, z = 0, w = 1$.

61. $\begin{cases} x + 2y + z = 1 \\ 2x - y + 2z = 2 \\ 3x + y + 3z = 3 \end{cases}$ can be written as: $\begin{bmatrix} 1 & 2 & 1 & | & 1 \\ 2 & -1 & 2 & | & 2 \\ 3 & 1 & 3 & | & 3 \end{bmatrix}$

$$\rightarrow \begin{bmatrix} 1 & 2 & 1 & | & 1 \\ 0 & -5 & 0 & | & 0 \\ 0 & -5 & 0 & | & 0 \end{bmatrix} \rightarrow \begin{bmatrix} 1 & 2 & 1 & | & 1 \\ 0 & -5 & 0 & | & 0 \\ 0 & 0 & 0 & | & 0 \end{bmatrix} \rightarrow \begin{array}{l} x + 2y + z = 1 \\ -5y = 0 \end{array}$$

$R_2 = -2r_1 + r_2 \quad R_3 = -r_2 + r_3$
$R_3 = -3r_1 + r_3$

Substitute and solve:
$$y = 0$$

$$x + 2(0) + z = 1 \Rightarrow x + z = 1 \Rightarrow x = 1 - z$$

The solution is $y = 0$, $x = 1 - z$, z is any real number.

63. $\begin{cases} x - y + z = 5 \\ 3x + 2y - 2z = 0 \end{cases}$ can be written as: $\begin{bmatrix} 1 & -1 & 1 & | & 5 \\ 3 & 2 & -2 & | & 0 \end{bmatrix}$

$$\rightarrow \begin{bmatrix} 1 & -1 & 1 & | & 5 \\ 0 & 5 & -5 & | & -15 \end{bmatrix} \rightarrow \begin{bmatrix} 1 & -1 & 1 & | & 5 \\ 0 & 1 & -1 & | & -3 \end{bmatrix} \rightarrow \begin{bmatrix} 1 & 0 & 0 & | & 2 \\ 0 & 1 & -1 & | & -3 \end{bmatrix}$$

$R_2 = -3r_1 + r_2 \qquad\qquad R_2 = \frac{1}{5}r_2 \qquad\qquad R_1 = r_2 + r_1$

The matrix in the third step represents the system $\begin{cases} x = 2 \\ y - z = -3 \end{cases}$

Therefore the solution is $x = 2; y = -3 + z; \ z$ is any real number

or $x = 2; z = y + 3; \ y$ is any real number

65. $\begin{cases} 2x + 3y - z = 3 \\ x - y - z = 0 \\ -x + y + z = 0 \\ x + y + 3z = 5 \end{cases}$ can be written as: $\begin{bmatrix} 2 & 3 & -1 & | & 3 \\ 1 & -1 & -1 & | & 0 \\ -1 & 1 & 1 & | & 0 \\ 1 & 1 & 3 & | & 5 \end{bmatrix}$

$$\rightarrow \begin{bmatrix} 1 & -1 & -1 & | & 0 \\ 2 & 3 & -1 & | & 3 \\ -1 & 1 & 1 & | & 0 \\ 1 & 1 & 3 & | & 5 \end{bmatrix} \longrightarrow \begin{bmatrix} 1 & -1 & -1 & | & 0 \\ 0 & 5 & 1 & | & 3 \\ 0 & 0 & 0 & | & 0 \\ 0 & 2 & 4 & | & 5 \end{bmatrix} \rightarrow \begin{bmatrix} 1 & -1 & -1 & | & 0 \\ 0 & 5 & 1 & | & 3 \\ 0 & 2 & 4 & | & 5 \\ 0 & 0 & 0 & | & 0 \end{bmatrix}$$

interchange r_1 and r_2 $R_2 = -2r_1 + r_2$ interchange r_3 and r_4

$R_3 = r_1 + r_3$

$R_4 = -r_1 + r_4$

$$\rightarrow \left| \begin{array}{ccc|c} 1 & -1 & -1 & 0 \\ 0 & 1 & -7 & -7 \\ 0 & 2 & 4 & 5 \\ 0 & 0 & 0 & 0 \end{array} \right| \longrightarrow \left| \begin{array}{ccc|c} 1 & 0 & -8 & -7 \\ 0 & 1 & -7 & -7 \\ 0 & 1 & 18 & 19 \\ 0 & 0 & 0 & 0 \end{array} \right| \rightarrow \left[\begin{array}{ccc|c} 1 & 0 & -8 & -7 \\ 0 & 1 & -7 & -7 \\ 0 & 0 & 1 & \frac{19}{18} \\ 0 & 0 & 0 & 0 \end{array} \right]$$

$$R_2 = -2r_3 + r_2 \qquad\qquad R_1 = r_2 + r_1 \qquad R_3 = \frac{1}{18}r_3$$
$$R_3 = -2r_2 + r_3$$

The matrix in the last step represents the system $\begin{cases} x - 8z = -7 \\ y - 7z = -7 \\ \qquad z = \dfrac{19}{18} \end{cases}$

Therefore the solution is

$$z = \frac{19}{18}$$

$$x = -7 + 8z = -7 + 8\left(\frac{19}{18}\right) = \frac{13}{9}; \quad y = -7 + 7z = -7 + 7\left(\frac{19}{18}\right) = \frac{7}{18}$$

67. $\begin{cases} 4x + y + z - w = 4 \\ x - y + 2z + 3w = 3 \end{cases}$ can be written as: $\left[\begin{array}{cccc|c} 4 & 1 & 1 & -1 & 4 \\ 1 & -1 & 2 & 3 & 3 \end{array} \right]$

$$\rightarrow \left[\begin{array}{cccc|c} 1 & -1 & 2 & 3 & 3 \\ 4 & 1 & 1 & -1 & 4 \end{array} \right] \longrightarrow \left[\begin{array}{cccc|c} 1 & -1 & 2 & 3 & 3 \\ 0 & 5 & -7 & -13 & -8 \end{array} \right]$$

interchange r_1 and r_2 $\qquad\qquad R_2 = -4r_1 + r_2$

The matrix in the last step represents the system $\begin{cases} x - y + 2z + 3w = 3 \\ 5y - 7z - 13w = -8 \end{cases}$

The second equation yields

$$5y - 7z - 13w = -8 \rightarrow 5y = -8 + 7z + 13w \rightarrow y = -\frac{8}{5} + \frac{7}{5}z + \frac{13}{5}w$$

The first equation yields

$$x - y + 2z + 3w = 3 \rightarrow x = 3 + y - 2z - 3w$$

substituting for y

$$x = 3 + \left(-\frac{8}{5} + \frac{7}{5}z + \frac{13}{5}w\right) - 2z - 3w$$

$$x = -\frac{3}{5}z - \frac{2}{5}w + \frac{7}{5}$$

Therefore the solution is

$$x = -\frac{3}{5}z - \frac{2}{5}w + \frac{7}{5}; \quad y = -\frac{8}{5} + \frac{7}{5}z + \frac{13}{5}w$$

z and w are any real numbers

69. Each of the points must satisfy the equation $y = ax^2 + bx + c$.

    $(1,2)$: $2 = a + b + c$

    $(-2,-7)$: $-7 = 4a - 2b + c$

    $(2,-3)$: $-3 = 4a + 2b + c$

Set up a matrix and solve:

$$\begin{bmatrix} 1 & 1 & 1 & | & 2 \\ 4 & -2 & 1 & | & -7 \\ 4 & 2 & 1 & | & -3 \end{bmatrix} \rightarrow \begin{bmatrix} 1 & 1 & 1 & | & 2 \\ 0 & -6 & -3 & | & -15 \\ 0 & -2 & -3 & | & -11 \end{bmatrix} \rightarrow \begin{bmatrix} 1 & 1 & 1 & | & 2 \\ 0 & 1 & \frac{1}{2} & | & \frac{5}{2} \\ 0 & -2 & -3 & | & -11 \end{bmatrix} \rightarrow \begin{bmatrix} 1 & 0 & \frac{1}{2} & | & -\frac{1}{2} \\ 0 & 1 & \frac{1}{2} & | & \frac{5}{2} \\ 0 & 0 & -2 & | & -6 \end{bmatrix}$$

$$\begin{array}{llll} R_2 = -4r_1 + r_2 & R_2 = -\frac{1}{6}r_2 & R_1 = -r_2 + r_1 \\ R_3 = -4r_1 + r_3 & & R_3 = 2r_2 + r_3 \end{array}$$

$$\rightarrow \begin{bmatrix} 1 & 0 & \frac{1}{2} & | & -\frac{1}{2} \\ 0 & 1 & \frac{1}{2} & | & \frac{5}{2} \\ 0 & 0 & 1 & | & 3 \end{bmatrix} \rightarrow \begin{bmatrix} 1 & 0 & 0 & | & -2 \\ 0 & 1 & 0 & | & 1 \\ 0 & 0 & 1 & | & 3 \end{bmatrix}$$

$$\begin{array}{ll} R_3 = -\frac{1}{2}r_3 & R_1 = -\frac{1}{2}r_3 + r_1 \\ & R_2 = -\frac{1}{2}r_3 + r_2 \end{array}$$

The solution is $a = -2$, $b = 1$, $c = 3$; so the equation is $y = -2x^2 + x + 3$.

71. Each of the points must satisfy the equation $f(x) = ax^3 + bx^2 + cx + d$.

    $f(-3) = -112$: $-27a + 9b - 3c + d = -112$

    $f(-1) = -2$: $-a + b - c + d = -2$

    $f(1) = 4$: $a + b + c + d = 4$

    $f(2) = 13$: $8a + 4b + 2c + d = 13$

Set up a matrix and solve:

$$\begin{bmatrix} -27 & 9 & -3 & 1 & | & -112 \\ -1 & 1 & -1 & 1 & | & -2 \\ 1 & 1 & 1 & 1 & | & 4 \\ 8 & 4 & 2 & 1 & | & 13 \end{bmatrix} \rightarrow \begin{bmatrix} 1 & 1 & 1 & 1 & | & 4 \\ -1 & 1 & -1 & 1 & | & -2 \\ -27 & 9 & -3 & 1 & | & -112 \\ 8 & 4 & 2 & 1 & | & 13 \end{bmatrix} \rightarrow \begin{bmatrix} 1 & 1 & 1 & 1 & | & 4 \\ 0 & 2 & 0 & 2 & | & 2 \\ 0 & 36 & 24 & 28 & | & -4 \\ 0 & -4 & -6 & -7 & | & -19 \end{bmatrix}$$

Interchange r_3 and r_1 $R_2 = r_1 + r_2$

$$R_3 = 27r_1 + r_3$$

$$R_4 = -8r_1 + r_4$$

$$\rightarrow \begin{bmatrix} 1 & 1 & 1 & 1 & | & 4 \\ 0 & 1 & 0 & 1 & | & 1 \\ 0 & 36 & 24 & 28 & | & -4 \\ 0 & -4 & -6 & -7 & | & -19 \end{bmatrix} \rightarrow \begin{bmatrix} 1 & 0 & 1 & 0 & | & 3 \\ 0 & 1 & 0 & 1 & | & 1 \\ 0 & 0 & 24 & -8 & | & -40 \\ 0 & 0 & -6 & -3 & | & -15 \end{bmatrix} \rightarrow \begin{bmatrix} 1 & 0 & 1 & 0 & | & 3 \\ 0 & 1 & 0 & 1 & | & 1 \\ 0 & 0 & 1 & -\frac{1}{3} & | & -\frac{5}{3} \\ 0 & 0 & -6 & -3 & | & -15 \end{bmatrix}$$

$$\begin{array}{lll} R_2 = \frac{1}{2}r_2 & R_1 = -r_2 + r_1 & R_3 = \frac{1}{24}r_3 \\ & R_3 = -36r_2 + r_3 & \\ & R_4 = 4r_2 + r_4 & \end{array}$$

$$\rightarrow \begin{bmatrix} 1 & 0 & 0 & \frac{1}{3} & \frac{14}{3} \\ 0 & 1 & 0 & 1 & 1 \\ 0 & 0 & 1 & -\frac{1}{3} & -\frac{5}{3} \\ 0 & 0 & 0 & -5 & -25 \end{bmatrix} \rightarrow \begin{bmatrix} 1 & 0 & 0 & \frac{1}{3} & \frac{14}{3} \\ 0 & 1 & 0 & 1 & 1 \\ 0 & 0 & 1 & -\frac{1}{3} & -\frac{5}{3} \\ 0 & 0 & 0 & 1 & 5 \end{bmatrix} \rightarrow \begin{bmatrix} 1 & 0 & 0 & 0 & 3 \\ 0 & 1 & 0 & 0 & -4 \\ 0 & 0 & 1 & 0 & 0 \\ 0 & 0 & 0 & 1 & 5 \end{bmatrix}$$

$$R_1 = -r_3 + r_1 \qquad\qquad R_4 = -\tfrac{1}{5} r_4 \qquad\qquad R_1 = -\tfrac{1}{3} r_4 + r_1$$
$$R_4 = 6 r_3 + r_4 \qquad\qquad\qquad\qquad\qquad\qquad R_2 = -r_4 + r_2$$
$$\qquad\qquad\qquad\qquad\qquad\qquad\qquad\qquad\qquad R_3 = \tfrac{1}{3} r_4 + r_3$$

The solution is $a = 3$, $b = -4$, $c = 0$, $d = 5$; so the equation is $f(x) = 3x^3 - 4x^2 + 5$.

73. Let x = the number of servings of salmon steak.
Let y = the number of servings of baked eggs.
Let z = the number of servings of acorn squash.
Protein equation: $30x + 15y + 3z = 78$
Carbohydrate equation: $20x + 2y + 25z = 59$
Vitamin A equation: $2x + 20y + 32z = 75$
Set up a matrix and solve:

$$\begin{bmatrix} 30 & 15 & 3 & 78 \\ 20 & 2 & 25 & 59 \\ 2 & 20 & 32 & 75 \end{bmatrix} \rightarrow \begin{bmatrix} 2 & 20 & 32 & 75 \\ 20 & 2 & 25 & 59 \\ 30 & 15 & 3 & 78 \end{bmatrix} \rightarrow \begin{bmatrix} 1 & 10 & 16 & 37.5 \\ 20 & 2 & 25 & 59 \\ 30 & 15 & 3 & 78 \end{bmatrix}$$

Interchange r_3 and r_1 $R_1 = \tfrac{1}{2} r_1$

$$\rightarrow \begin{bmatrix} 1 & 10 & 16 & 37.5 \\ 0 & -198 & -295 & -691 \\ 0 & -285 & -477 & -1047 \end{bmatrix} \rightarrow \begin{bmatrix} 1 & 10 & 16 & 37.5 \\ 0 & -198 & -295 & -691 \\ 0 & 0 & -\frac{3457}{66} & -\frac{3457}{66} \end{bmatrix}$$

$$R_2 = -20 r_1 + r_2 \qquad\qquad\qquad R_3 = -\tfrac{95}{66} r_2 + r_3$$
$$R_3 = -30 r_1 + r_3$$

$$\rightarrow \begin{bmatrix} 1 & 10 & 16 & 37.5 \\ 0 & -198 & -295 & -691 \\ 0 & 0 & 1 & 1 \end{bmatrix}$$

$$R_3 = -\tfrac{66}{3457} r_3$$

Substitute $z = 1$ and solve:
$$-198y - 295(1) = -691 \qquad\qquad x + 10(2) + 16(1) = 37.5$$
$$-198y = -396 \qquad\qquad\qquad x + 36 = 37.5$$
$$y = 2 \qquad\qquad\qquad\qquad x = 1.5$$

The dietitian should serve 1.5 servings of salmon steak, 2 servings of baked eggs, and 1 serving of acorn squash.

75. Let x = the amount invested in Treasury bills.
Let y = the amount invested in Treasury bonds.
Let z = the amount invested in corporate bonds.
Total investment equation: $x + y + z = 10000$
Annual income equation: $0.06x + 0.07y + 0.08z = 680$

Condition on investment equation: $z = 0.5x$

Set up a matrix and solve:

$$\begin{vmatrix} 1 & 1 & 1 & 10000 \\ 0.06 & 0.07 & 0.08 & 680 \\ 1 & 0 & -2 & 0 \end{vmatrix} \rightarrow \begin{vmatrix} 1 & 1 & 1 & 10000 \\ 0 & 0.01 & 0.02 & 80 \\ 0 & -1 & -3 & -10000 \end{vmatrix} \rightarrow \begin{vmatrix} 1 & 1 & 1 & 10000 \\ 0 & 1 & 2 & 8000 \\ 0 & -1 & -3 & -10000 \end{vmatrix}$$

$$R_2 = -0.06\,r_1 + r_2 \qquad\qquad R_2 = 100\,r_2$$

$$R_3 = -r_1 + r_3$$

$$\rightarrow \begin{vmatrix} 1 & 0 & -1 & 2000 \\ 0 & 1 & 2 & 8000 \\ 0 & 0 & -1 & -2000 \end{vmatrix} \rightarrow \begin{vmatrix} 1 & 0 & -1 & 2000 \\ 0 & 1 & 2 & 8000 \\ 0 & 0 & 1 & 2000 \end{vmatrix} \rightarrow \begin{vmatrix} 1 & 0 & 0 & 4000 \\ 0 & 1 & 0 & 4000 \\ 0 & 0 & 1 & 2000 \end{vmatrix}$$

$$R_1 = -r_2 + r_1 \qquad R_3 = -r_3 \qquad R_1 = r_3 + r_1$$

$$R_3 = r_2 + r_3 \qquad\qquad\qquad R_2 = -2r_3 + r_2$$

Carletta should invest \$4000 in Treasury bills, \$4000 in Treasury bonds, and \$2000 in corporate bonds.

77. Let x = the number of Deltas produced.
Let y = the number of Betas produced.
Let z = the number of Sigmas produced.
Painting equation: $10x + 16y + 8z = 240$
Drying equation: $3x + 5y + 2z = 69$
Polishing equation: $2x + 3y + z = 41$
Set up a matrix and solve:

$$\begin{bmatrix} 10 & 16 & 8 & 240 \\ 3 & 5 & 2 & 69 \\ 2 & 3 & 1 & 41 \end{bmatrix} \rightarrow \begin{bmatrix} 1 & 1 & 2 & 33 \\ 3 & 5 & 2 & 69 \\ 2 & 3 & 1 & 41 \end{bmatrix} \rightarrow \begin{bmatrix} 1 & 1 & 2 & 33 \\ 0 & 2 & -4 & -30 \\ 0 & 1 & -3 & -25 \end{bmatrix} \rightarrow \begin{bmatrix} 1 & 1 & 2 & 33 \\ 0 & 1 & -2 & -15 \\ 0 & 1 & -3 & -25 \end{bmatrix}$$

$$R_1 = -3r_2 + r_1 \quad R_2 = -3r_1 + r_2 \qquad R_2 = \tfrac{1}{2}r_2$$

$$R_3 = -2r_1 + r_3$$

$$\rightarrow \begin{bmatrix} 1 & 0 & 4 & 48 \\ 0 & 1 & -2 & -15 \\ 0 & 0 & -1 & -10 \end{bmatrix} \rightarrow \begin{bmatrix} 1 & 0 & 4 & 48 \\ 0 & 1 & -2 & -15 \\ 0 & 0 & 1 & 10 \end{bmatrix} \rightarrow \begin{bmatrix} 1 & 0 & 0 & 8 \\ 0 & 1 & 0 & 5 \\ 0 & 0 & 1 & 10 \end{bmatrix}$$

$$R_1 = -r_2 + r_1 \qquad R_3 = -r_3 \qquad R_1 = -4r_3 + r_1$$

$$R_3 = -r_2 + r_3 \qquad\qquad\qquad R_2 = 2r_3 + r_2$$

The company should produce 8 Deltas, 5 Betas, and 10 Sigmas.

79. Rewrite the system as set up and solve the matrix:

$$\begin{cases} -4 + 8 - 2I_2 = 0 \\ 8 = 5I_4 + I_1 \\ 4 = 3I_3 + I_1 \\ I_3 + I_4 = I_1 \end{cases} \rightarrow \begin{cases} 2I_2 = 4 \\ I_1 + 5I_4 = 8 \\ I_1 + 3I_3 = 4 \\ I_1 - I_3 - I_4 = 0 \end{cases}$$

$$\begin{vmatrix} 0 & 2 & 0 & 0 & 4 \\ 1 & 0 & 0 & 5 & 8 \\ 1 & 0 & 3 & 0 & 4 \\ 1 & 0 & -1 & -1 & 0 \end{vmatrix} \rightarrow \begin{vmatrix} 1 & 0 & 0 & 5 & 8 \\ 0 & 2 & 0 & 0 & 4 \\ 1 & 0 & 3 & 0 & 4 \\ 1 & 0 & -1 & -1 & 0 \end{vmatrix} \rightarrow \begin{vmatrix} 1 & 0 & 0 & 5 & 8 \\ 0 & 1 & 0 & 0 & 2 \\ 0 & 0 & 3 & -5 & -4 \\ 0 & 0 & -1 & -6 & -8 \end{vmatrix}$$

Interchange r_2 and r_1 $R_2 = \tfrac{1}{2} r_2$

$R_3 = -r_1 + r_3$

$R_4 = -r_1 + r_4$

$$\rightarrow \begin{vmatrix} 1 & 0 & 0 & 5 & 8 \\ 0 & 1 & 0 & 0 & 2 \\ 0 & 0 & -1 & -6 & -8 \\ 0 & 0 & 3 & -5 & -4 \end{vmatrix} \rightarrow \begin{vmatrix} 1 & 0 & 0 & 5 & 8 \\ 0 & 1 & 0 & 0 & 2 \\ 0 & 0 & 1 & 6 & 8 \\ 0 & 0 & 0 & -23 & -28 \end{vmatrix} \rightarrow \begin{vmatrix} 1 & 0 & 0 & 5 & 8 \\ 0 & 1 & 0 & 0 & 2 \\ 0 & 0 & 1 & 6 & 8 \\ 0 & 0 & 0 & 1 & \frac{28}{23} \end{vmatrix}$$

Interchange r_3 and r_4 $R_3 = -r_3$ $R_4 = -\tfrac{1}{23} r_4$

$R_4 = -3r_3 + r_4$

$$\rightarrow \begin{bmatrix} 1 & 0 & 0 & 0 & \frac{44}{23} \\ 0 & 1 & 0 & 0 & 2 \\ 0 & 0 & 1 & 0 & \frac{16}{23} \\ 0 & 0 & 0 & 1 & \frac{28}{23} \end{bmatrix}$$

$R_1 = -5r_4 + r_1$

$R_3 = -6r_4 + r_3$

The solution is $I_1 = \dfrac{44}{23}$, $I_2 = 2$, $I_3 = \dfrac{16}{23}$, $I_4 = \dfrac{28}{23}$.

81. Let x = the amount invested in Treasury bills.
     Let y = the amount invested in Treasury bonds.
     Let z = the amount invested in corporate bonds.

(a) Total investment equation: $x + y + z = 20000$
     Annual income equation: $0.07x + 0.09y + 0.11z = 2000$
     Set up a matrix and solve:

$$\begin{vmatrix} 1 & 1 & 1 & 20000 \\ .07 & .09 & .11 & 2000 \end{vmatrix} \rightarrow \begin{vmatrix} 1 & 1 & 1 & 20000 \\ 7 & 9 & 11 & 200000 \end{vmatrix} \rightarrow \begin{vmatrix} 1 & 1 & 1 & 20000 \\ 0 & 2 & 4 & 60000 \end{vmatrix}$$

$R_2 = 100r_2$ $R_2 = r_2 - 7r_1$

$$\rightarrow \begin{bmatrix} 1 & 1 & 1 & 20000 \\ 0 & 1 & 2 & 30000 \end{bmatrix} \rightarrow \begin{bmatrix} 1 & 0 & -1 & -10000 \\ 0 & 1 & 2 & 30000 \end{bmatrix}$$

$R_2 = \tfrac{1}{2} r_2$ $R_1 = r_1 - r_2$

The matrix in the last step represents the system $\begin{cases} x - z = -10000 \\ y + 2z = 30000 \end{cases}$

Therefore the solution is

$$x = -10000 + z; \quad y = 30000 - 2z; \quad z \text{ is any real number}$$

Possible investment strategies:

Amount invested at

7%	9%	11%
0	10000	10000
1000	8000	11000
2000	6000	12000
3000	4000	13000
4000	2000	14000
5000	0	15000

(b) Total investment equation: $x + y + z = 25000$

Annual income equation: $0.07x + 0.09y + 0.11z = 2000$

Set up a matrix and solve:

$$\begin{bmatrix} 1 & 1 & 1 & | & 25000 \\ .07 & .09 & .11 & | & 2000 \end{bmatrix} \rightarrow \begin{bmatrix} 1 & 1 & 1 & | & 25000 \\ 7 & 9 & 11 & | & 200000 \end{bmatrix} \rightarrow \begin{bmatrix} 1 & 1 & 1 & | & 25000 \\ 0 & 2 & 4 & | & 25000 \end{bmatrix}$$

$$R_2 = 100r_2 \qquad\qquad R_2 = r_2 - 7r_1$$

$$\rightarrow \begin{bmatrix} 1 & 1 & 1 & | & 25000 \\ 0 & 1 & 2 & | & 12500 \end{bmatrix} \rightarrow \begin{bmatrix} 1 & 0 & -1 & | & 12500 \\ 0 & 1 & 2 & | & 12500 \end{bmatrix}$$

$$R_2 = \tfrac{1}{2}r_2 \qquad\qquad R_1 = r_1 - r_2$$

The matrix in the last step represents the system $\begin{cases} x - z = 12500 \\ y + 2z = 12500 \end{cases}$

Therefore the solution is

$$x = 12500 + z; \quad y = 12500 - 2z; \quad z \text{ is any real number}$$

Possible investment strategies:

Amount invested at

7%	9%	11%
12500	12500	0
14500	8500	2000
16500	4500	4000
18750	0	6250

(c) Total investment equation: $x + y + z = 30000$

Annual income equation: $0.07x + 0.09y + 0.11z = 2000$

Set up a matrix and solve:

$$\begin{bmatrix} 1 & 1 & 1 & | & 30000 \\ .07 & .09 & .11 & | & 2000 \end{bmatrix} \rightarrow \begin{bmatrix} 1 & 1 & 1 & | & 30000 \\ 7 & 9 & 11 & | & 200000 \end{bmatrix} \rightarrow \begin{bmatrix} 1 & 1 & 1 & | & 30000 \\ 0 & 2 & 4 & | & -10000 \end{bmatrix}$$

$$R_2 = 100r_2 \qquad\qquad R_1 = r_2 - 7r_1$$

$$\rightarrow \begin{bmatrix} 1 & 1 & 1 & | & 30000 \\ 0 & 1 & 2 & | & -5000 \end{bmatrix} \rightarrow \begin{bmatrix} 1 & 0 & -1 & | & 35000 \\ 0 & 1 & 2 & | & -5000 \end{bmatrix}$$

$$R_2 = \tfrac{1}{2}r_2 \qquad\qquad R_1 = r_1 - r_2$$

The matrix in the last step represents the system $\begin{cases} x - z = 35000 \\ y + 2z = -5000 \end{cases}$

Therefore the solution is

$$x = 35000 + z; \quad y = -5000 - 2z; \quad z \text{ is any real number}$$

One possible investment strategy

Amount invested at

7%	9%	11%
30000	0	0

This will yield ($30000)(.07) = $2100, which is more than the required income.

83.   Let x = the amount of liquid 1.
      Let y = the amount of liquid 2.
      Let z = the amount of liquid 3.
      $.20x + .40y + .30z = 40$ Vitamin C
      $.30x + .20y + .50z = 30$ Vitamin D
      multiplying each equation by 10 yields
      $2x + 4y + 3z = 400$
      $3x + 2y + 5z = 300$

Set up a matrix and solve: $\begin{bmatrix} 2 & 4 & 3 & | & 400 \\ 3 & 2 & 5 & | & 300 \end{bmatrix} \rightarrow \begin{bmatrix} 1 & 2 & \frac{3}{2} & | & 200 \\ 3 & 2 & 5 & | & 300 \end{bmatrix} \rightarrow \begin{bmatrix} 1 & 2 & \frac{3}{2} & | & 200 \\ 0 & -4 & \frac{1}{2} & | & -300 \end{bmatrix}$

$\qquad\qquad R_1 = \frac{1}{2}r_1 \qquad\qquad R_2 = r_2 - 3r_1$

$\rightarrow \begin{bmatrix} 1 & 2 & \frac{3}{2} & | & 200 \\ 0 & 1 & -\frac{1}{8} & | & 75 \end{bmatrix} \rightarrow \begin{bmatrix} 1 & 0 & \frac{7}{4} & | & 50 \\ 0 & 1 & -\frac{1}{8} & | & 75 \end{bmatrix}$

$R_2 = -\frac{1}{4}r_2 \qquad\qquad R_1 = r_1 - 2r_2$

The matrix in the last step represents the system $\begin{cases} x + \frac{7}{4}z = 50 \\ y - \frac{1}{8}z = 75 \end{cases}$

Therefore the solution is $x = 50 - \dfrac{7}{4}z; \quad y = 75 + \dfrac{1}{8}z; \quad z$ is any real number

Possible combinations:

Liquid 1	Liquid 2	Liquid 3
50mg	75mg	0mg
36mg	76mg	8mg
22mg	77mg	16mg
8mg	78mg	24mg

85 – 87.   Answers will vary.

Systems of Equations and Inequalities

10.4 Systems of Linear Equations: Determinants

1. (a)   Evaluating the determinant: $\begin{vmatrix} 3 & 1 \\ 4 & 2 \end{vmatrix} = 3(2) - 4(1) = 6 - 4 = 2$

    (b) Use MATRIX EDIT to create the matrix   Then compute det([A]):

$$A = \begin{bmatrix} 3 & 1 \\ 4 & 2 \end{bmatrix}$$

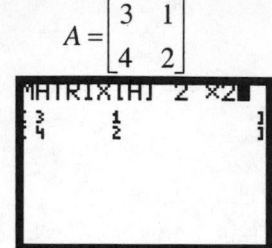

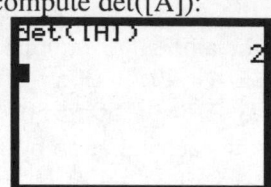

3. (a) Evaluating the determinant: $\begin{vmatrix} 6 & 4 \\ -1 & 3 \end{vmatrix} = 6(3) - (-1)(4) = 18 + 4 = 22$

(b) Use MATRIX EDIT to create the matrix   Then compute det([A]):

$$A = \begin{bmatrix} 6 & 4 \\ -1 & 3 \end{bmatrix}$$

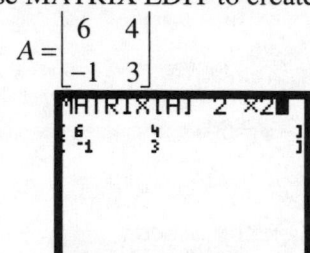

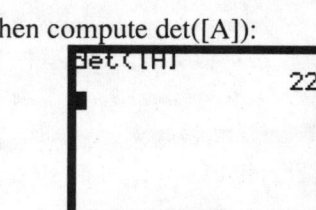

5. (a) Evaluating the determinant: $\begin{vmatrix} -3 & -1 \\ 4 & 2 \end{vmatrix} = -3(2) - 4(-1) = -6 + 4 = -2$

    (b) Use MATRIX EDIT to create the matrix   Then compute det([A]):

$$A = \begin{bmatrix} -3 & -1 \\ 4 & 2 \end{bmatrix}$$

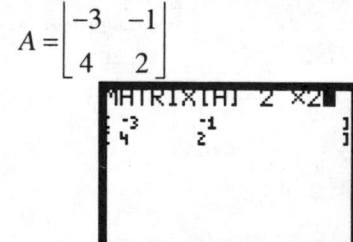

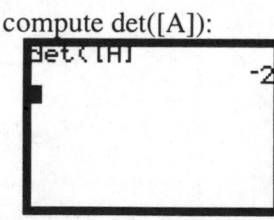

7. (a) Evaluating the determinant:

$$\begin{vmatrix} 3 & 4 & 2 \\ 1 & -1 & 5 \\ 1 & 2 & -2 \end{vmatrix} = 3\begin{vmatrix} -1 & 5 \\ 2 & -2 \end{vmatrix} - 4\begin{vmatrix} 1 & 5 \\ 1 & -2 \end{vmatrix} + 2\begin{vmatrix} 1 & -1 \\ 1 & 2 \end{vmatrix}$$

$$= 3\big[(-1)(-2) - 2(5)\big] - 4\big[1(-2) - 1(5)\big] + 2\big[1(2) - 1(-1)\big]$$

$$= 3(2 - 10) - 4(-2 - 5) + 2(2 + 1) = 3(-8) - 4(-7) + 2(3) = -24 + 28 + 6 = 10$$

(b) Use MATRIX EDIT to create the matrix

$$A = \begin{bmatrix} 3 & 4 & 2 \\ 1 & -1 & 5 \\ 1 & 2 & -2 \end{bmatrix}$$

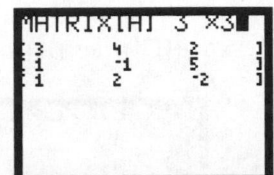

Then compute det([A]):

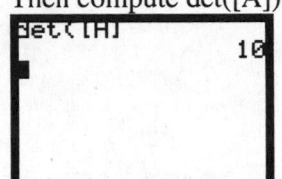

9. (a) Evaluating the determinant:

$$\begin{vmatrix} 4 & -1 & 2 \\ 6 & -1 & 0 \\ 1 & -3 & 4 \end{vmatrix} = 4\begin{vmatrix} -1 & 0 \\ -3 & 4 \end{vmatrix} - (-1)\begin{vmatrix} 6 & 0 \\ 1 & 4 \end{vmatrix} + 2\begin{vmatrix} 6 & -1 \\ 1 & -3 \end{vmatrix}$$

$$= 4\big[-1(4) - 0(-3)\big] + 1\big[6(4) - 1(0)\big] + 2\big[6(-3) - 1(-1)\big]$$

$$= 4(-4) + 1(24) + 2(-17)$$

$$= -16 + 24 - 34$$

$$= -26$$

(b) Use MATRIX EDIT to create the matrix

$$A = \begin{bmatrix} 4 & -1 & 2 \\ 6 & -1 & 0 \\ 1 & -3 & 4 \end{bmatrix}$$

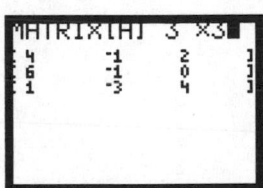

Then compute det([A]):

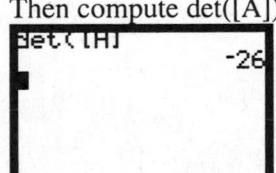

11. (a) Set up and evaluate the determinants to use Cramer's Rule:

$$\begin{cases} x + y = 8 \\ x - y = 4 \end{cases}$$

$$D = \begin{vmatrix} 1 & 1 \\ 1 & -1 \end{vmatrix} = -1 - 1 = -2$$

$$D_x = \begin{vmatrix} 8 & 1 \\ 4 & -1 \end{vmatrix} = -8 - 4 = -12; \quad D_y = \begin{vmatrix} 1 & 8 \\ 1 & 4 \end{vmatrix} = 4 - 8 = -4$$

Find the solutions by Cramer's Rule:

$$x = \frac{D_x}{D} = \frac{-12}{-2} = 6 \qquad y = \frac{D_y}{D} = \frac{-4}{-2} = 2$$

(b) Use MATRIX EDIT to create the matrices

$$A = \begin{bmatrix} 1 & 1 \\ 1 & -1 \end{bmatrix}; \quad B = \begin{bmatrix} 8 & 1 \\ 4 & -1 \end{bmatrix}; \quad C = \begin{bmatrix} 1 & 8 \\ 1 & 4 \end{bmatrix}$$

Then compute

$$x = \det([B])/\det([A]);$$

$$y = \det([C])/\det([A])$$

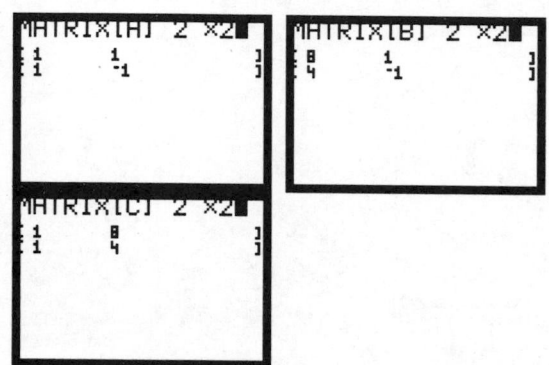

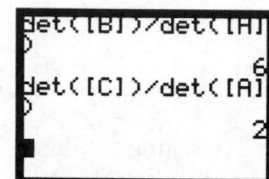

13. (a) Set up and evaluate the determinants to use Cramer's Rule:

$$\begin{cases} 5x - y = 13 \\ 2x + 3y = 12 \end{cases}$$

$$D = \begin{vmatrix} 5 & -1 \\ 2 & 3 \end{vmatrix} = 15 + 2 = 17$$

$$D_x = \begin{vmatrix} 13 & -1 \\ 12 & 3 \end{vmatrix} = 39 + 12 = 51$$

$$D_y = \begin{vmatrix} 5 & 13 \\ 2 & 12 \end{vmatrix} = 60 - 26 = 34$$

Find the solutions by Cramer's Rule:

$$x = \frac{D_x}{D} = \frac{51}{17} = 3 \qquad y = \frac{D_y}{D} = \frac{34}{17} = 2$$

(b) Use MATRIX EDIT to create the matrices

$$A = \begin{bmatrix} 5 & -1 \\ 2 & 3 \end{bmatrix}; \quad B = \begin{bmatrix} 13 & -1 \\ 12 & 3 \end{bmatrix}; \quad C = \begin{bmatrix} 5 & 13 \\ 2 & 12 \end{bmatrix}$$

Then compute

$$x = \det([B])/\det([A]);$$

$$y = \det([C])/\det([A])$$

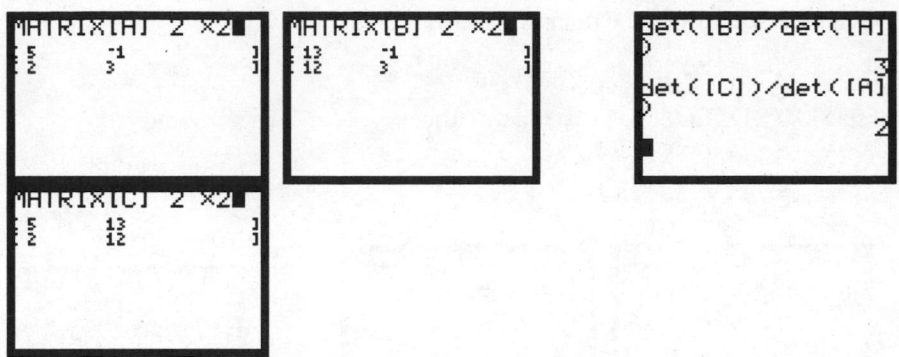

15. (a) Set up and evaluate the determinants to use Cramer's Rule:

$$\begin{cases} 3x \quad\quad = 24 \\ x + 2y = \ \ 0 \end{cases}$$

$$D = \begin{vmatrix} 3 & 0 \\ 1 & 2 \end{vmatrix} = 6 - 0 = 6$$

$$D_x = \begin{vmatrix} 24 & 0 \\ 0 & 2 \end{vmatrix} = 48 - 0 = 48; \quad D_y = \begin{vmatrix} 3 & 24 \\ 1 & 0 \end{vmatrix} = 0 - 24 = -24$$

Find the solutions by Cramer's Rule:

$$x = \frac{D_x}{D} = \frac{48}{6} = 8 \qquad y = \frac{D_y}{D} = \frac{-24}{6} = -4$$

(b) Use **MATRIX EDIT** to create the matrices   Then compute

$$A = \begin{vmatrix} 3 & 0 \\ 1 & 2 \end{vmatrix}; \quad B = \begin{vmatrix} 24 & 0 \\ 0 & 2 \end{vmatrix}; \quad C = \begin{vmatrix} 3 & 24 \\ 1 & 0 \end{vmatrix}$$

$$x = \det([B])/\det([A]);$$

$$y = \det([C])/\det([A])$$

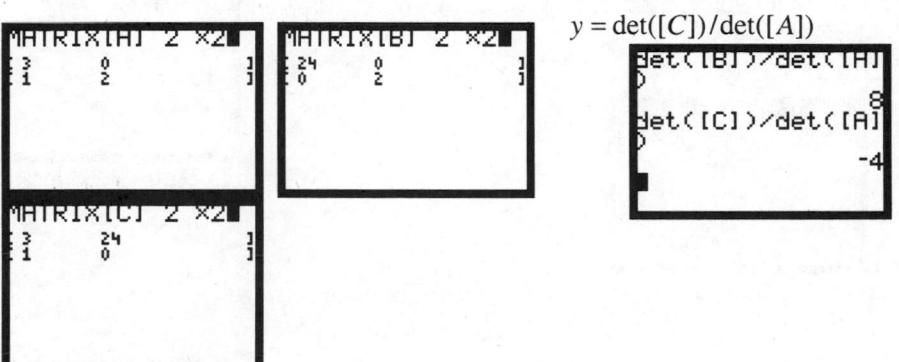

17. (a) Set up and evaluate the determinants to use Cramer's Rule:

$$\begin{cases} 3x - 6y = 24 \\ 5x + 4y = \ 12 \end{cases}$$

$$D = \begin{vmatrix} 3 & -6 \\ 5 & 4 \end{vmatrix} = 12 - (-30) = 42;; D_x = \begin{vmatrix} 24 & -6 \\ 12 & 4 \end{vmatrix} = 96 - (-72) = 168$$

$$D_y = \begin{vmatrix} 3 & 24 \\ 5 & 12 \end{vmatrix} = 36 - 120 = -84$$

Find the solutions by Cramer's Rule:

$$x = \frac{D_x}{D} = \frac{168}{42} = 4 \qquad y = \frac{D_y}{D} = \frac{-84}{42} = -2$$

(b) Use MATRIX EDIT to create the matrices

$$A = \begin{vmatrix} 3 & -6 \\ 5 & 4 \end{vmatrix}; \quad B = \begin{vmatrix} 24 & -6 \\ 12 & 4 \end{vmatrix}; \quad C = \begin{vmatrix} 3 & 24 \\ 5 & 12 \end{vmatrix}$$

Then compute

$$x = \det([B])/\det([A]);$$

$$y = \det([C])/\det([A])$$

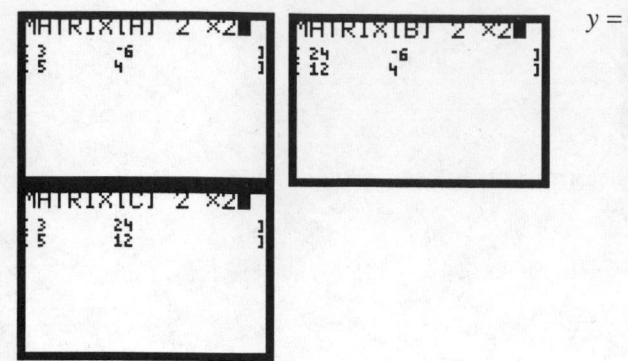

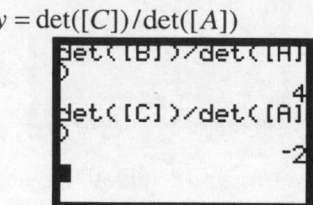

19. (a) Set up and evaluate the determinants to use Cramer's Rule:

$$\begin{cases} 3x - 2y = 4 \\ 6x - 4y = 0 \end{cases}$$

$$D = \begin{vmatrix} 3 & -2 \\ 6 & -4 \end{vmatrix} = -12 - (-12) = 0$$

Since $D = 0$, Cramer's Rule does not apply.

(b) Use MATRIX EDIT to create the matrix

$$A = \begin{vmatrix} 3 & -2 \\ 6 & -4 \end{vmatrix}$$

Then compute det([A]):

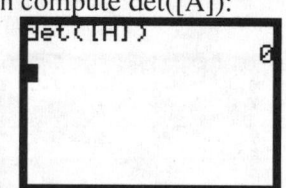

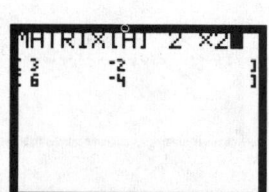

Cramer's Rule does not apply.

21. Set up and evaluate the determinants to use Cramer's Rule:

$$\begin{cases} 2x - 4y = -2 \\ 3x + 2y = 3 \end{cases}$$

$$D = \begin{vmatrix} 2 & -4 \\ 3 & 2 \end{vmatrix} = 4 - (-12) = 16$$

$$D_x = \begin{vmatrix} -2 & -4 \\ 3 & 2 \end{vmatrix} = -4 - (-12) = 8; \quad D_y = \begin{vmatrix} 2 & -2 \\ 3 & 3 \end{vmatrix} = 6 - (-6) = 12$$

Find the solutions by Cramer's Rule:

$$x = \frac{D_x}{D} = \frac{8}{16} = \frac{1}{2} \qquad y = \frac{D_y}{D} = \frac{12}{16} = \frac{3}{4}$$

(b) Use MATRIX EDIT to create the matrices

$$A = \begin{vmatrix} 2 & -4 \\ 3 & 2 \end{vmatrix}; \quad B = \begin{vmatrix} -2 & -4 \\ 3 & 2 \end{vmatrix}; \quad C = \begin{vmatrix} 2 & -2 \\ 3 & 3 \end{vmatrix}$$

Then compute

$x = \det([B])/\det([A]);$

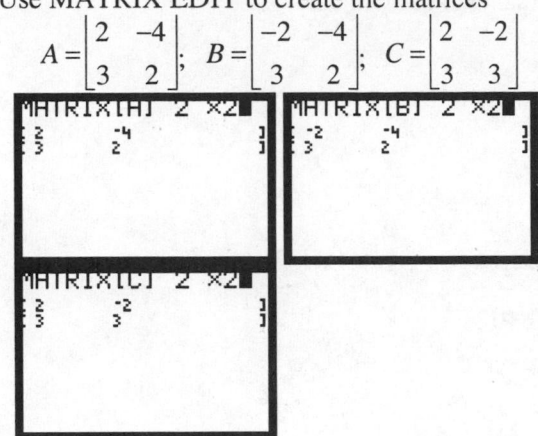

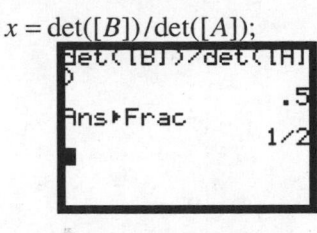

$y = \det([C])/\det([A])$

23. (a) Set up and evaluate the determinants to use Cramer's Rule:

$$\begin{cases} 2x - 3y = -1 \\ 10x + 10y = 5 \end{cases}$$

$$D = \begin{vmatrix} 2 & -3 \\ 10 & 10 \end{vmatrix} = 20 - (-30) = 50$$

$$D_x = \begin{vmatrix} -1 & -3 \\ 5 & 10 \end{vmatrix} = -10 - (-15) = 5$$

$$D_y = \begin{vmatrix} 2 & -1 \\ 10 & 5 \end{vmatrix} = 10 - (-10) = 20$$

Find the solutions by Cramer's Rule:

$$x = \frac{D_x}{D} = \frac{5}{50} = \frac{1}{10} \qquad y = \frac{D_y}{D} = \frac{20}{50} = \frac{2}{5}$$

(b) Use MATRIX EDIT to create the matrices

$$A = \begin{vmatrix} 2 & -3 \\ 10 & 10 \end{vmatrix}; \quad B = \begin{vmatrix} -1 & -3 \\ 5 & 10 \end{vmatrix}; \quad C = \begin{vmatrix} 2 & -1 \\ 10 & 5 \end{vmatrix}$$

Then compute

$x = \det([B])/\det([A]);$

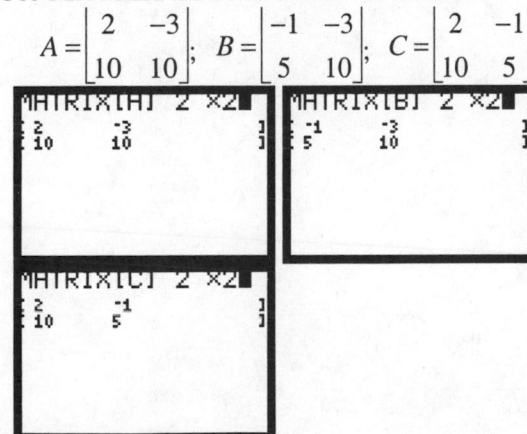

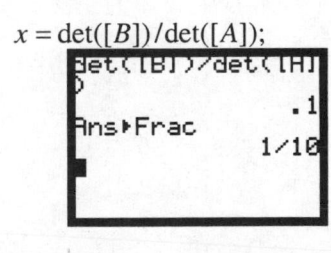

$y = \det([C])/\det([A])$

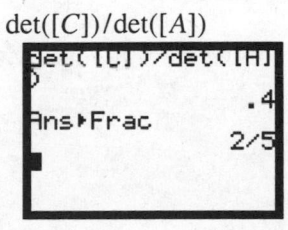

25. (a) Set up and evaluate the determinants to use Cramer's Rule:

$$\begin{cases} 2x + 3y = 6 \\ x - y = \dfrac{1}{2} \end{cases}$$

$$D = \begin{vmatrix} 2 & 3 \\ 1 & -1 \end{vmatrix} = -2 - 3 = -5$$

$$D_x = \begin{vmatrix} 6 & 3 \\ \frac{1}{2} & -1 \end{vmatrix} = -6 - \frac{3}{2} = -\frac{15}{2}; \quad D_y = \begin{vmatrix} 2 & 6 \\ 1 & \frac{1}{2} \end{vmatrix} = 1 - 6 = -5$$

Find the solutions by Cramer's Rule:

$$x = \frac{D_x}{D} = \frac{\left(-\dfrac{15}{2}\right)}{-5} = \frac{3}{2} \qquad y = \frac{D_y}{D} = \frac{-5}{-5} = 1$$

(b) Use **MATRIX EDIT** to create the matrices   Then compute

$$A = \begin{vmatrix} 2 & 3 \\ 1 & -1 \end{vmatrix}; \quad B = \begin{vmatrix} 6 & 3 \\ \frac{1}{2} & -1 \end{vmatrix}; \quad C = \begin{vmatrix} 2 & 6 \\ 1 & \frac{1}{2} \end{vmatrix}$$

$$x = \det([B])/\det([A]);$$

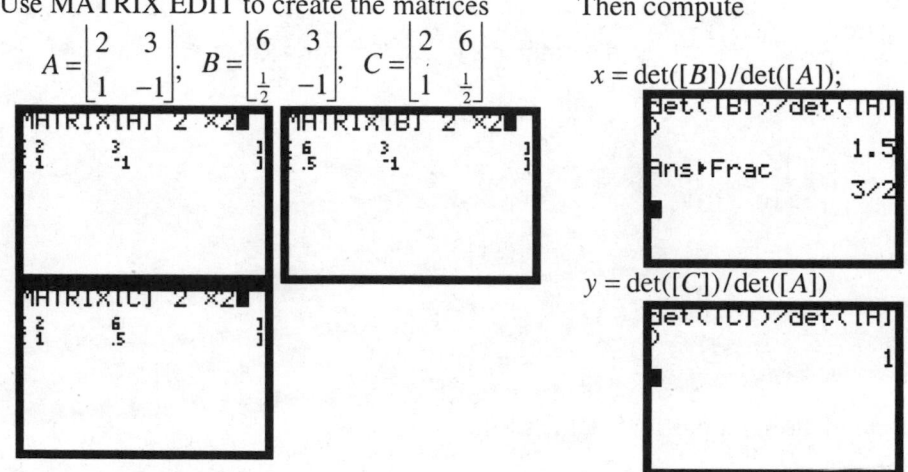

$$y = \det([C])/\det([A])$$

27. Set up and evaluate the determinants to use Cramer's Rule:

$$\begin{cases} 3x - 5y = 3 \\ 15x + 5y = 21 \end{cases}$$

$$D = \begin{vmatrix} 3 & -5 \\ 15 & 5 \end{vmatrix} = 15 - (-75) = 90$$

$$D_x = \begin{vmatrix} 3 & -5 \\ 21 & 5 \end{vmatrix} = 15 - (-105) = 120$$

$$D_y = \begin{vmatrix} 3 & 3 \\ 15 & 21 \end{vmatrix} = 63 - 45 = 18$$

Find the solutions by Cramer's Rule:

$$x = \frac{D_x}{D} = \frac{120}{90} = \frac{4}{3} \qquad y = \frac{D_y}{D} = \frac{18}{90} = \frac{1}{5}$$

(b) Use MATRIX EDIT to create the matrices   Then compute

$$A = \begin{vmatrix} 3 & -5 \\ 15 & 5 \end{vmatrix}; \quad B = \begin{vmatrix} 3 & -5 \\ 21 & 5 \end{vmatrix}; \quad C = \begin{vmatrix} 3 & 3 \\ 15 & 21 \end{vmatrix}$$

$x = \det([B])/\det([A]);$

$y = \det([C])/\det([A])$

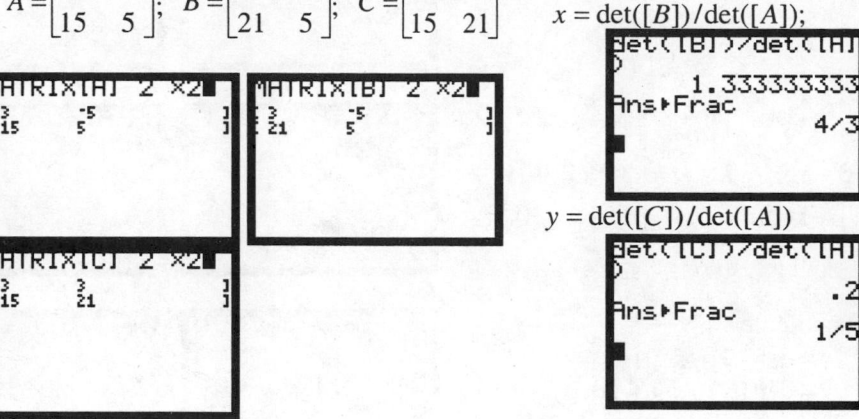

29. Set up and evaluate the determinants to use Cramer's Rule:

$$\begin{cases} x + y - z = 6 \\ 3x - 2y + z = -5 \\ x + 3y - 2z = 14 \end{cases}$$

$$D = \begin{vmatrix} 1 & 1 & -1 \\ 3 & -2 & 1 \\ 1 & 3 & -2 \end{vmatrix} = 1 \begin{vmatrix} -2 & 1 \\ 3 & -2 \end{vmatrix} - 1 \begin{vmatrix} 3 & 1 \\ 1 & -2 \end{vmatrix} + (-1) \begin{vmatrix} 3 & -2 \\ 1 & 3 \end{vmatrix}$$

$$= 1(4-3) - 1(-6-1) - 1(9+2) = 1 + 7 - 11 = -3$$

$$D_x = \begin{vmatrix} 6 & 1 & -1 \\ -5 & -2 & 1 \\ 14 & 3 & -2 \end{vmatrix} = 6 \begin{vmatrix} -2 & 1 \\ 3 & -2 \end{vmatrix} - 1 \begin{vmatrix} -5 & 1 \\ 14 & -2 \end{vmatrix} + (-1) \begin{vmatrix} -5 & -2 \\ 14 & 3 \end{vmatrix}$$

$$= 6(4-3) - 1(10-14) - 1(-15+28) = 6 + 4 - 13 = -3$$

$$D_y = \begin{vmatrix} 1 & 6 & -1 \\ 3 & -5 & 1 \\ 1 & 14 & -2 \end{vmatrix} = 1 \begin{vmatrix} -5 & 1 \\ 14 & -2 \end{vmatrix} - 6 \begin{vmatrix} 3 & 1 \\ 1 & -2 \end{vmatrix} + (-1) \begin{vmatrix} 3 & -5 \\ 1 & 14 \end{vmatrix}$$

$$= 1(10-14) - 6(-6-1) - 1(42+5) = -4 + 42 - 47 = -9$$

$$D_z = \begin{vmatrix} 1 & 1 & 6 \\ 3 & -2 & -5 \\ 1 & 3 & 14 \end{vmatrix} = 1 \begin{vmatrix} -2 & -5 \\ 3 & 14 \end{vmatrix} - 1 \begin{vmatrix} 3 & -5 \\ 1 & 14 \end{vmatrix} + 6 \begin{vmatrix} 3 & -2 \\ 1 & 3 \end{vmatrix}$$

$$= 1(-28+15) - 1(42+5) + 6(9+2) = -13 - 47 + 66 = 6$$

Find the solutions by Cramer's Rule:

$$x = \frac{D_x}{D} = \frac{-3}{-3} = 1 \qquad y = \frac{D_y}{D} = \frac{-9}{-3} = 3 \qquad z = \frac{D_z}{D} = \frac{6}{-3} = -2$$

(b) Use MATRIX EDIT to create the matrices

$$A = \begin{vmatrix} 1 & 1 & -1 \\ 3 & -2 & 1 \\ 1 & 3 & -2 \end{vmatrix}; \quad B = \begin{vmatrix} 6 & 1 & -1 \\ -5 & -2 & 1 \\ 14 & 3 & -2 \end{vmatrix};$$

$$C = \begin{vmatrix} 1 & 6 & -1 \\ 3 & -5 & 1 \\ 1 & 14 & -2 \end{vmatrix}; \quad D = \begin{vmatrix} 1 & 1 & 6 \\ 3 & -2 & -5 \\ 1 & 3 & 14 \end{vmatrix}$$

Then compute

$x = \det([B])/\det([A]);$
$y = \det([C])/\det([A]);$
$z = \det([D])/\det([A])$

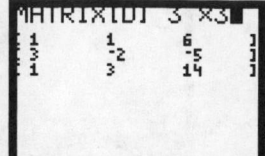

31. (a) Set up and evaluate the determinants to use Cramer's Rule:

$$\begin{cases} x + 2y - z = -3 \\ 2x - 4y + z = -7 \\ -2x + 2y - 3z = 4 \end{cases}$$

$$D = \begin{vmatrix} 1 & 2 & -1 \\ 2 & -4 & 1 \\ -2 & 2 & -3 \end{vmatrix} = 1\begin{vmatrix} -4 & 1 \\ 2 & -3 \end{vmatrix} - 2\begin{vmatrix} 2 & 1 \\ -2 & -3 \end{vmatrix} + (-1)\begin{vmatrix} 2 & -4 \\ -2 & 2 \end{vmatrix}$$

$$= 1(12-2) - 2(-6+2) - 1(4-8) = 10 + 8 + 4 = 22$$

$$D_x = \begin{vmatrix} -3 & 2 & -1 \\ -7 & -4 & 1 \\ 4 & 2 & -3 \end{vmatrix} = -3\begin{vmatrix} -4 & 1 \\ 2 & -3 \end{vmatrix} - 2\begin{vmatrix} -7 & 1 \\ 4 & -3 \end{vmatrix} + (-1)\begin{vmatrix} -7 & -4 \\ 4 & 2 \end{vmatrix}$$

$$= -3(12-2) - 2(21-4) - 1(-14+16) = -30 - 34 - 2 = -66$$

$$D_y = \begin{vmatrix} 1 & -3 & -1 \\ 2 & -7 & 1 \\ -2 & 4 & -3 \end{vmatrix} = 1\begin{vmatrix} -7 & 1 \\ 4 & -3 \end{vmatrix} - (-3)\begin{vmatrix} 2 & 1 \\ -2 & -3 \end{vmatrix} + (-1)\begin{vmatrix} 2 & -7 \\ -2 & 4 \end{vmatrix}$$

$$= 1(21-4) + 3(-6+2) - 1(8-14) = 17 - 12 + 6 = 11$$

$$D_z = \begin{vmatrix} 1 & 2 & -3 \\ 2 & -4 & -7 \\ -2 & 2 & 4 \end{vmatrix} = 1\begin{vmatrix} -4 & -7 \\ 2 & 4 \end{vmatrix} - 2\begin{vmatrix} 2 & -7 \\ -2 & 4 \end{vmatrix} + (-3)\begin{vmatrix} 2 & -4 \\ -2 & 2 \end{vmatrix}$$

$$= 1(-16+14) - 2(8-14) - 3(4-8) = -2 + 12 + 12 = 22$$

Find the solutions by Cramer's Rule:

$$x = \frac{D_x}{D} = \frac{-66}{22} = -3 \qquad y = \frac{D_y}{D} = \frac{11}{22} = \frac{1}{2} \qquad z = \frac{D_z}{D} = \frac{22}{22} = 1$$

(b) Use MATRIX EDIT to create the matrices

$$A = \begin{vmatrix} 1 & 2 & -1 \\ 2 & -4 & 1 \\ -2 & 2 & -3 \end{vmatrix}; \quad B = \begin{vmatrix} -3 & 2 & -1 \\ -7 & -4 & 1 \\ 4 & 2 & -3 \end{vmatrix};$$

$$C = \begin{vmatrix} 1 & -3 & -1 \\ 2 & -7 & 1 \\ -2 & 4 & -3 \end{vmatrix}; \quad D = \begin{vmatrix} 1 & 2 & -3 \\ 2 & -4 & -7 \\ -2 & 2 & 4 \end{vmatrix}$$

Then compute
$$x = \det([B])/\det([A]);$$
$$y = \det([C])/\det([A]);$$
$$z = \det([D])/\det([A])$$

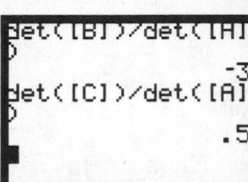
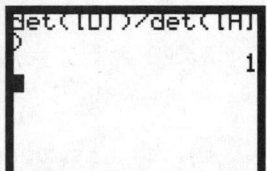

33. (a) Set up and evaluate the determinants to use Cramer's Rule:

$$\begin{cases} x - 2y + 3z = 1 \\ 3x + y - 2z = 0 \\ 2x - 4y + 6z = 2 \end{cases}$$

$$D = \begin{vmatrix} 1 & -2 & 3 \\ 3 & 1 & -2 \\ 2 & -4 & 6 \end{vmatrix} = 1\begin{vmatrix} 1 & -2 \\ -4 & 6 \end{vmatrix} - (-2)\begin{vmatrix} 3 & -2 \\ 2 & 6 \end{vmatrix} + 3\begin{vmatrix} 3 & 1 \\ 2 & -4 \end{vmatrix}$$

$$= 1(6 - 8) + 2(18 + 4) + 3(-12 - 2) = -2 + 44 - 42 = 0$$

Since $D = 0$, Cramer's Rule does not apply.

(b) Use MATRIX EDIT to create the matrices

$$A = \begin{vmatrix} 1 & -2 & 3 \\ 3 & 1 & -2 \\ 2 & -4 & 6 \end{vmatrix}$$

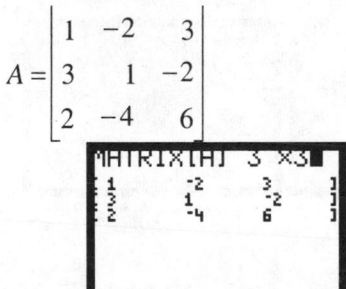

Then compute det([A]):

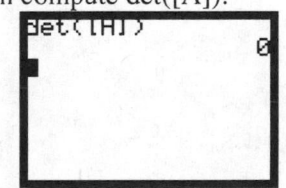

Cramer's Rule does not apply.

35. (a) Set up and evaluate the determinants to use Cramer's Rule:

$$\begin{cases} x + 2y - z = 0 \\ 2x - 4y + z = 0 \\ -2x + 2y - 3z = 0 \end{cases}$$

$$D = \begin{vmatrix} 1 & 2 & -1 \\ 2 & -4 & 1 \\ -2 & 2 & -3 \end{vmatrix} = 1\begin{vmatrix} -4 & 1 \\ 2 & -3 \end{vmatrix} - 2\begin{vmatrix} 2 & 1 \\ -2 & -3 \end{vmatrix} + (-1)\begin{vmatrix} 2 & -4 \\ -2 & 2 \end{vmatrix}$$

$$= 1(12 - 2) - 2(-6 + 2) - 1(4 - 8) = 10 + 8 + 4 = 22$$

$$D_x = \begin{vmatrix} 0 & 2 & -1 \\ 0 & -4 & 1 \\ 0 & 2 & -3 \end{vmatrix} = 0 \quad \text{(By Theorem 12)}$$

$$D_y = \begin{vmatrix} 1 & 0 & -1 \\ 2 & 0 & 1 \\ -2 & 0 & -3 \end{vmatrix} = 0 \quad \text{(By Theorem 12)}; \quad D_z = \begin{vmatrix} 1 & 2 & 0 \\ 2 & -4 & 0 \\ -2 & 2 & 0 \end{vmatrix} = 0 \quad \text{(By Theorem 12)}$$

Find the solutions by Cramer's Rule:

$$x = \frac{D_x}{D} = \frac{0}{22} = 0 \qquad y = \frac{D_y}{D} = \frac{0}{22} = 0 \qquad z = \frac{D_z}{D} = \frac{0}{22} = 0$$

(b) Use MATRIX EDIT to create the matrices

$$A = \begin{vmatrix} 1 & 2 & -1 \\ 2 & -4 & 1 \\ -2 & 2 & -3 \end{vmatrix}; \quad B = \begin{vmatrix} 0 & 2 & -1 \\ 0 & -4 & 1 \\ 0 & 2 & -3 \end{vmatrix};$$

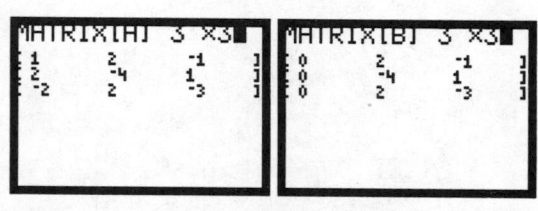

$$C = \begin{vmatrix} 1 & 0 & -1 \\ 2 & 0 & 1 \\ -2 & 0 & -3 \end{vmatrix}; \quad D = \begin{vmatrix} 1 & 2 & 0 \\ 2 & -4 & 0 \\ -2 & 2 & 0 \end{vmatrix}$$

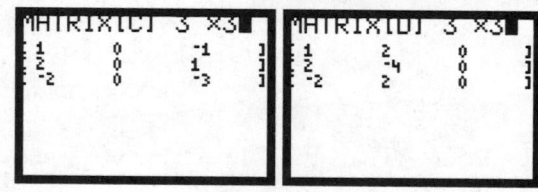

Then compute
$$x = \det([B])/\det([A]);$$
$$y = \det([C])/\det([A]);$$
$$z = \det([D])/\det([A])$$

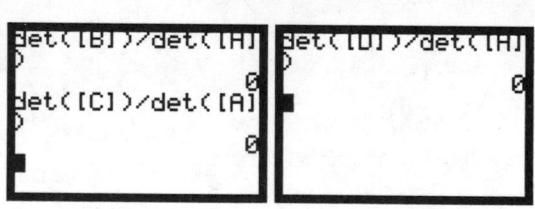

37. (a) Set up and evaluate the determinants to use Cramer's Rule:

$$\begin{cases} x - 2y + 3z = 0 \\ 3x + y - 2z = 0 \\ 2x - 4y + 6z = 0 \end{cases}$$

$$D = \begin{vmatrix} 1 & -2 & 3 \\ 3 & 1 & -2 \\ 2 & -4 & 6 \end{vmatrix} = 1 \begin{vmatrix} 1 & -2 \\ -4 & 6 \end{vmatrix} - (-2) \begin{vmatrix} 3 & -2 \\ 2 & 6 \end{vmatrix} + 3 \begin{vmatrix} 3 & 1 \\ 2 & -4 \end{vmatrix}$$

$$= 1(6 - 8) + 2(18 + 4) + 3(-12 - 2) = -2 + 44 - 42 = 0$$

Since $D = 0$, Cramer's Rule does not apply.

(b) Use MATRIX EDIT to create the matrices

$$A = \begin{vmatrix} 1 & -2 & 3 \\ 3 & 1 & -2 \\ 2 & -4 & 6 \end{vmatrix}$$

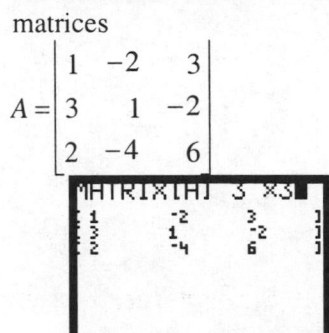

Then compute det([A]):

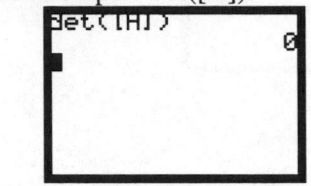

Cramer's Rule does not apply.

39. (a) Rewrite the system letting $u = \dfrac{1}{x}$ and $v = \dfrac{1}{y}$:

$$\begin{cases} \dfrac{1}{x} + \dfrac{1}{y} = 8 \\ \dfrac{3}{x} - \dfrac{5}{y} = 0 \end{cases} \quad \rightarrow \quad \begin{cases} u + v = 8 \\ 3u - 5v = 0 \end{cases}$$

Set up and evaluate the determinants to use Cramer's Rule:

$$D = \begin{vmatrix} 1 & 1 \\ 3 & -5 \end{vmatrix} = -5 - 3 = -8$$

$$D_u = \begin{vmatrix} 8 & 1 \\ 0 & -5 \end{vmatrix} = -40 - 0 = -40; \quad D_v = \begin{vmatrix} 1 & 8 \\ 3 & 0 \end{vmatrix} = 0 - 24 = -24$$

Find the solutions by Cramer's Rule:

$$u = \frac{D_u}{D} = \frac{-40}{-8} = 5 \Rightarrow \frac{1}{x} = 5 \Rightarrow x = \frac{1}{5}$$

$$v = \frac{D_v}{D} = \frac{-24}{-8} = 3 \Rightarrow \frac{1}{y} = 3 \Rightarrow y = \frac{1}{3}$$

The solutions are $x = \dfrac{1}{5}$, $y = \dfrac{1}{3}$.

(b) Use MATRIX EDIT to create the matrices   Then compute

$$A = \begin{vmatrix} 1 & 1 \\ 3 & -5 \end{vmatrix}; \quad B = \begin{vmatrix} 8 & 1 \\ 0 & -5 \end{vmatrix}; \quad C = \begin{vmatrix} 1 & 8 \\ 3 & 0 \end{vmatrix}$$

$u = \det([B])/\det([A]);$
$v = \det([C])/\det([A])$

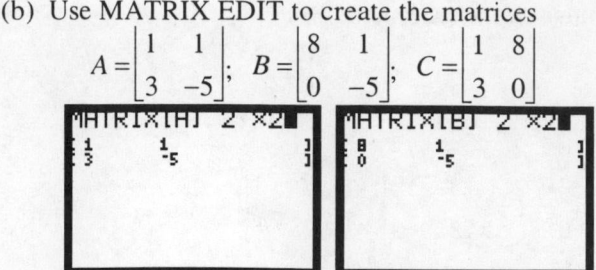

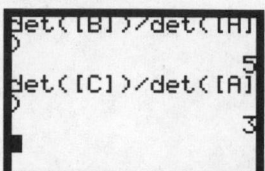

41. Solve for x:

$$\begin{vmatrix} x & x \\ 4 & 3 \end{vmatrix} = 3x - 4x = -x \Rightarrow -x = 5 \Rightarrow x = -5$$

43. Solve for x:

$$\begin{vmatrix} x & 1 & 1 \\ 4 & 3 & 2 \\ -1 & 2 & 5 \end{vmatrix} = x\begin{vmatrix} 3 & 2 \\ 2 & 5 \end{vmatrix} - 1\begin{vmatrix} 4 & 2 \\ -1 & 5 \end{vmatrix} + 1\begin{vmatrix} 4 & 3 \\ -1 & 2 \end{vmatrix}$$

$$= x(15 - 4) - (20 + 2) + (8 + 3) = 11x - 22 + 11 = 11x - 11$$

So, $11x - 11 = 2 \Rightarrow 11x = 13 \Rightarrow x = \dfrac{13}{11}$

45. Solve for x:

$$\begin{vmatrix} x & 2 & 3 \\ 1 & x & 0 \\ 6 & 1 & -2 \end{vmatrix} = x\begin{vmatrix} x & 0 \\ 1 & -2 \end{vmatrix} - 2\begin{vmatrix} 1 & 0 \\ 6 & -2 \end{vmatrix} + 3\begin{vmatrix} 1 & x \\ 6 & 1 \end{vmatrix}$$

$$= x(-2x - 0) - 2(-2 - 0) + 3(1 - 6x)$$

$$= -2x^2 + 4 + 3 - 18x = -2x^2 - 18x + 7$$

So, $-2x^2 - 18x + 7 = 7$

$$-2x^2 - 18x = 0 \Rightarrow -2x(x + 9) = 0 \Rightarrow x = 0 \text{ or } x = -9$$

47. Let $\begin{vmatrix} x & y & z \\ u & v & w \\ 1 & 2 & 3 \end{vmatrix} = 4$

Then $\begin{vmatrix} 1 & 2 & 3 \\ u & v & w \\ x & y & z \end{vmatrix} = -4$ by Theorem 11

The value of the determinant changes sign when two rows are interchanged.

Problems 49 – 53 use the Laws for Determinants in reverse order.

49.   Let $\begin{vmatrix} x & y & z \\ u & v & w \\ 1 & 2 & 3 \end{vmatrix} = 4$

$$\begin{vmatrix} x & y & z \\ -3 & -6 & -9 \\ u & v & w \end{vmatrix} = -3\begin{vmatrix} x & y & z \\ 1 & 2 & 3 \\ u & v & w \end{vmatrix} = -3(-1)\begin{vmatrix} x & y & z \\ u & v & w \\ 1 & 2 & 3 \end{vmatrix} = 3(4) = 12$$

$\qquad\qquad$ Theorem 14 $\qquad\qquad$ Theorem 11

51.   Let $\begin{vmatrix} x & y & z \\ u & v & w \\ 1 & 2 & 3 \end{vmatrix} = 4$

$$\begin{vmatrix} 1 & 2 & 3 \\ x-3 & y-6 & z-9 \\ 2u & 2v & 2w \end{vmatrix} = 2\begin{vmatrix} 1 & 2 & 3 \\ x-3 & y-6 & z-9 \\ u & v & w \end{vmatrix} = 2(-1)\begin{vmatrix} x-3 & y-6 & z-9 \\ 1 & 2 & 3 \\ u & v & w \end{vmatrix}$$

$\qquad\qquad\qquad$ Theorem 14 $\qquad\qquad\qquad$ Theorem 11

$$= 2(-1)(-1)\begin{vmatrix} x-3 & y-6 & z-9 \\ u & v & w \\ 1 & 2 & 3 \end{vmatrix} = 2(-1)(-1)\begin{vmatrix} x & y & z \\ u & v & w \\ 1 & 2 & 3 \end{vmatrix} = 2(-1)(-1)(4) = 8$$

$\qquad\qquad$ Theorem 11 $\qquad\qquad\qquad\qquad$ Theorem 15 $\ (R_1 = -3r_3 + r_1)$

53.   Let $\begin{vmatrix} x & y & z \\ u & v & w \\ 1 & 2 & 3 \end{vmatrix} = 4$

$$\begin{vmatrix} 1 & 2 & 3 \\ 2x & 2y & 2z \\ u-1 & v-2 & w-3 \end{vmatrix} = 2\begin{vmatrix} 1 & 2 & 3 \\ x & y & z \\ u-1 & v-2 & w-3 \end{vmatrix} = 2(-1)\begin{vmatrix} x & y & z \\ 1 & 2 & 3 \\ u-1 & v-2 & w-3 \end{vmatrix}$$

$\qquad\qquad\qquad$ Theorem 14 $\qquad\qquad\qquad$ Theorem 11

$$= 2(-1)(-1)\begin{vmatrix} x & y & z \\ u-1 & v-2 & w-3 \\ 1 & 2 & 3 \end{vmatrix} = 2(-1)(-1)\begin{vmatrix} x & y & z \\ u & v & w \\ 1 & 2 & 3 \end{vmatrix} = 2(-1)(-1)(4) = 8$$

$\qquad\qquad$ Theorem 11 $\qquad\qquad\qquad\qquad$ Theorem 15 $\ (R_2 = -r_3 + r_2)$

55.   Expanding the determinant:

$$\begin{vmatrix} x & y & 1 \\ x_1 & y_1 & 1 \\ x_2 & y_2 & 1 \end{vmatrix} = x\begin{vmatrix} y_1 & 1 \\ y_2 & 1 \end{vmatrix} - y\begin{vmatrix} x_1 & 1 \\ x_2 & 1 \end{vmatrix} + 1\begin{vmatrix} x_1 & y_1 \\ x_2 & y_2 \end{vmatrix}$$

$$= x(y_1 - y_2) - y(x_1 - x_2) + (x_1y_2 - x_2y_1) = 0$$

$$x(y_1 - y_2) + y(x_2 - x_1) = x_2 y_1 - x_1 y_2$$
$$y(x_2 - x_1) = x_2 y_1 - x_1 y_2 + x(y_2 - y_1)$$
$$y(x_2 - x_1) - y_1(x_2 - x_1) = x_2 y_1 - x_1 y_2 + x(y_2 - y_1) - y_1(x_2 - x_1)$$
$$(x_2 - x_1)(y - y_1) = x(y_2 - y_1) + x_2 y_1 - x_1 y_2 - y_1 x_2 + y_1 x_1$$
$$(x_2 - x_1)(y - y_1) = (y_2 - y_1)x - (y_2 - y_1)x_1$$
$$(x_2 - x_1)(y - y_1) = (y_2 - y_1)(x - x_1)$$
$$(y - y_1) = \frac{(y_2 - y_1)}{(x_2 - x_1)}(x - x_1)$$

57. Expanding the determinant:

$$\begin{vmatrix} x^2 & x & 1 \\ y^2 & y & 1 \\ z^2 & z & 1 \end{vmatrix} = x^2 \begin{vmatrix} y & 1 \\ z & 1 \end{vmatrix} - x \begin{vmatrix} y^2 & 1 \\ z^2 & 1 \end{vmatrix} + 1 \begin{vmatrix} y^2 & y \\ z^2 & z \end{vmatrix}$$

$$= x^2(y - z) - x(y^2 - z^2) + 1(y^2 z - z^2 y)$$
$$= x^2(y - z) - x(y - z)(y + z) + yz(y - z)$$
$$= (y - z)\left[x^2 - xy - xz + yz \right]$$
$$= (y - z)\left[x(x - y) - z(x - y) \right] = (y - z)(x - y)(x - z)$$

59. Evaluating the determinant to show the relationship:

$$\begin{vmatrix} a_{13} & a_{12} & a_{11} \\ a_{23} & a_{22} & a_{21} \\ a_{33} & a_{32} & a_{31} \end{vmatrix} = a_{13} \begin{vmatrix} a_{22} & a_{21} \\ a_{32} & a_{31} \end{vmatrix} - a_{12} \begin{vmatrix} a_{23} & a_{21} \\ a_{33} & a_{31} \end{vmatrix} + a_{11} \begin{vmatrix} a_{23} & a_{22} \\ a_{33} & a_{32} \end{vmatrix}$$

$$= a_{13}(a_{22}a_{31} - a_{21}a_{32}) - a_{12}(a_{23}a_{31} - a_{21}a_{33}) + a_{11}(a_{23}a_{32} - a_{22}a_{33})$$
$$= a_{13}a_{22}a_{31} - a_{13}a_{21}a_{32} - a_{12}a_{23}a_{31} + a_{12}a_{21}a_{33} + a_{11}a_{23}a_{32} - a_{11}a_{22}a_{33}$$
$$= -a_{11}a_{22}a_{33} + a_{11}a_{23}a_{32} + a_{12}a_{21}a_{33} - a_{12}a_{23}a_{31} - a_{13}a_{21}a_{32} + a_{13}a_{22}a_{31}$$
$$= -a_{11}(a_{22}a_{33} - a_{23}a_{32}) + a_{12}(a_{21}a_{33} - a_{23}a_{31}) - a_{13}(a_{21}a_{32} - a_{22}a_{31})$$
$$= -a_{11} \begin{vmatrix} a_{22} & a_{23} \\ a_{32} & a_{33} \end{vmatrix} + a_{12} \begin{vmatrix} a_{21} & a_{23} \\ a_{31} & a_{33} \end{vmatrix} - a_{13} \begin{vmatrix} a_{21} & a_{22} \\ a_{31} & a_{32} \end{vmatrix}$$
$$= -\left[a_{11} \begin{vmatrix} a_{22} & a_{23} \\ a_{32} & a_{33} \end{vmatrix} - a_{12} \begin{vmatrix} a_{21} & a_{23} \\ a_{31} & a_{33} \end{vmatrix} + a_{13} \begin{vmatrix} a_{21} & a_{22} \\ a_{31} & a_{32} \end{vmatrix} \right]$$
$$= -\begin{vmatrix} a_{11} & a_{12} & a_{13} \\ a_{21} & a_{22} & a_{23} \\ a_{31} & a_{32} & a_{33} \end{vmatrix}$$

61. Set up a 3 by 3 determinant in which the first column and third column are the same and evaluate:

$$\begin{vmatrix} a & b & a \\ c & d & c \\ e & f & e \end{vmatrix} = -b \begin{vmatrix} c & c \\ e & e \end{vmatrix} + d \begin{vmatrix} a & a \\ e & e \end{vmatrix} - f \begin{vmatrix} a & a \\ c & c \end{vmatrix}$$

$$= -b(ce - ce) + d(ae - ae) - f(ac - ac)$$

$$= -b(0) + d(0) - f(0) = 0$$

Systems of Equations and Inequalities

10.5 Matrix Algebra

1. (a) $A + B = \begin{vmatrix} 0 & 3 & -5 \\ 1 & 2 & 6 \end{vmatrix} + \begin{vmatrix} 4 & 1 & 0 \\ -2 & 3 & -2 \end{vmatrix} = \begin{vmatrix} 0+4 & 3+1 & -5+0 \\ 1+(-2) & 2+3 & 6+(-2) \end{vmatrix} = \begin{vmatrix} 4 & 4 & -5 \\ -1 & 5 & 4 \end{vmatrix}$

(b) Use MATRIX EDIT to create the matrices   Then compute $[A]+[B]$:

$$A = \begin{vmatrix} 0 & 3 & -5 \\ 1 & 2 & 6 \end{vmatrix}; \quad B = \begin{vmatrix} 4 & 1 & 0 \\ -2 & 3 & -2 \end{vmatrix}$$

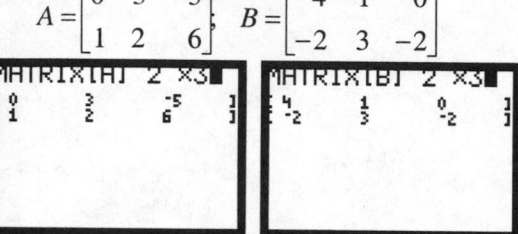
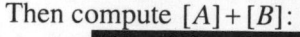
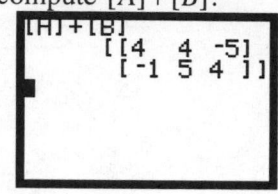

3. (a) $4A = 4\begin{bmatrix} 0 & 3 & -5 \\ 1 & 2 & 6 \end{bmatrix} = \begin{bmatrix} 4\cdot0 & 4\cdot3 & 4(-5) \\ 4\cdot1 & 4\cdot2 & 4\cdot6 \end{bmatrix} = \begin{bmatrix} 0 & 12 & -20 \\ 4 & 8 & 24 \end{bmatrix}$

(b) Use MATRIX EDIT to create the matrix   Then compute $4*[A]$:

$$A = \begin{vmatrix} 0 & 3 & -5 \\ 1 & 2 & 6 \end{vmatrix}$$

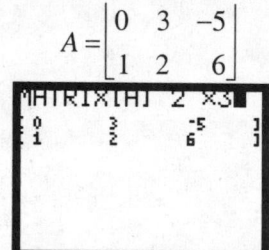
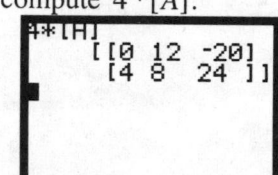

5. (a) $3A - 2B = 3\begin{bmatrix} 0 & 3 & -5 \\ 1 & 2 & 6 \end{bmatrix} - 2\begin{bmatrix} 4 & 1 & 0 \\ -2 & 3 & -2 \end{bmatrix}$

$$= \begin{bmatrix} 0 & 9 & -15 \\ 3 & 6 & 18 \end{bmatrix} - \begin{bmatrix} 8 & 2 & 0 \\ -4 & 6 & -4 \end{bmatrix} = \begin{bmatrix} -8 & 7 & -15 \\ 7 & 0 & 22 \end{bmatrix}$$

(b) Use MATRIX EDIT to create the matrices   Then compute $3*[A]-2*[B]$:

$$A = \begin{vmatrix} 0 & 3 & -5 \\ 1 & 2 & 6 \end{vmatrix}; \quad B = \begin{vmatrix} 4 & 1 & 0 \\ -2 & 3 & -2 \end{vmatrix}$$

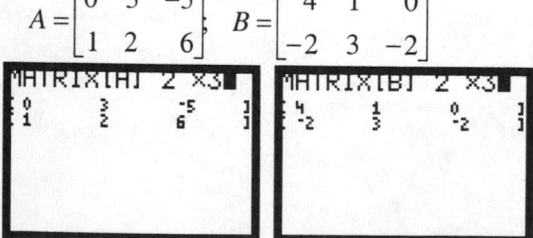
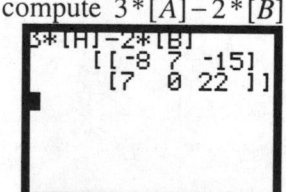

7.   (a) $AC = \begin{bmatrix} 0 & 3 & -5 \\ 1 & 2 & 6 \end{bmatrix} \cdot \begin{bmatrix} 4 & 1 \\ 6 & 2 \\ -2 & 3 \end{bmatrix} = \begin{bmatrix} 0(4) + 3(6) + (-5)(-2) & 0(1) + 3(2) + (-5)(3) \\ 1(4) + 2(6) + 6(-2) & 1(1) + 2(2) + 6(3) \end{bmatrix} = \begin{bmatrix} 28 & -9 \\ 4 & 23 \end{bmatrix}$

(b) Use MATRIX EDIT to create the matrices   Then compute [A]*[C]:

$A = \begin{bmatrix} 0 & 3 & -5 \\ 1 & 2 & 6 \end{bmatrix}; \quad C = \begin{bmatrix} 4 & 1 \\ 6 & 2 \\ -2 & 3 \end{bmatrix}$

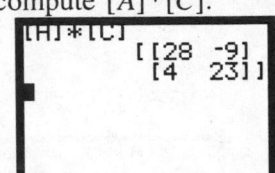

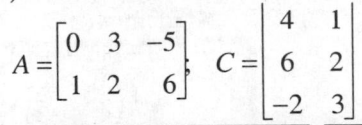

9.   (a) $CA = \begin{bmatrix} 4 & 1 \\ 6 & 2 \\ -2 & 3 \end{bmatrix} \cdot \begin{bmatrix} 0 & 3 & -5 \\ 1 & 2 & 6 \end{bmatrix} = \begin{bmatrix} 4(0) + 1(1) & 4(3) + 1(2) & 4(-5) + 1(6) \\ 6(0) + 2(1) & 6(3) + 2(2) & 6(-5) + 2(6) \\ -2(0) + 3(1) & -2(3) + 3(2) & -2(-5) + 3(6) \end{bmatrix} = \begin{bmatrix} 1 & 14 & -14 \\ 2 & 22 & -18 \\ 3 & 0 & 28 \end{bmatrix}$

(b) Use MATRIX EDIT to create the matrices   Then compute [C]*[A]:

$A = \begin{bmatrix} 0 & 3 & -5 \\ 1 & 2 & 6 \end{bmatrix}; \quad C = \begin{bmatrix} 4 & 1 \\ 6 & 2 \\ -2 & 3 \end{bmatrix}$

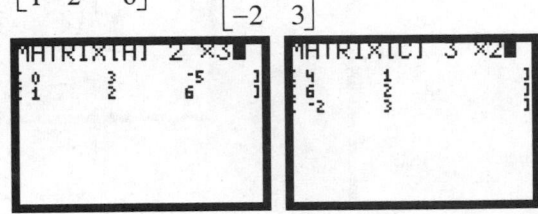

11.   (a) $C(A + B) = \begin{bmatrix} 4 & 1 \\ 6 & 2 \\ -2 & 3 \end{bmatrix} \left(\begin{bmatrix} 0 & 3 & -5 \\ 1 & 2 & 6 \end{bmatrix} + \begin{bmatrix} 4 & 1 & 0 \\ -2 & 3 & -2 \end{bmatrix} \right) = \begin{bmatrix} 4 & 1 \\ 6 & 2 \\ -2 & 3 \end{bmatrix} \cdot \begin{bmatrix} 4 & 4 & -5 \\ -1 & 5 & 4 \end{bmatrix} = \begin{bmatrix} 15 & 21 & -16 \\ 22 & 34 & -22 \\ -11 & 7 & 22 \end{bmatrix}$

(b) Use MATRIX EDIT to create the matrices   Then compute $[C]*([A] + [B])$:

$A = \begin{bmatrix} 0 & 3 & -5 \\ 1 & 2 & 6 \end{bmatrix}; \quad B = \begin{bmatrix} 4 & 1 & 0 \\ -2 & 3 & -2 \end{bmatrix}; \quad C = \begin{bmatrix} 4 & 1 \\ 6 & 2 \\ -2 & 3 \end{bmatrix}$

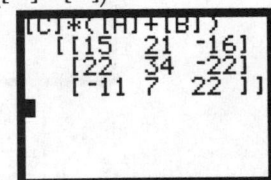

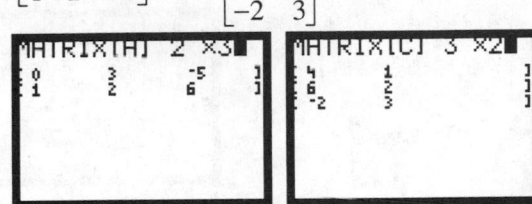

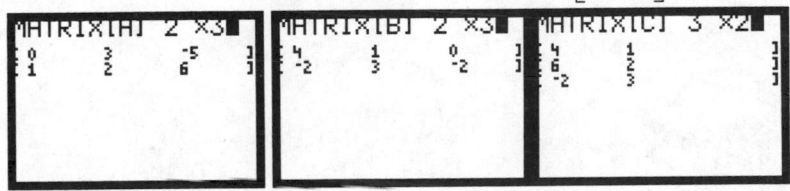

13. (a) $AC - 3I_2 = \begin{bmatrix} 0 & 3 & -5 \\ 1 & 2 & 6 \end{bmatrix} \cdot \begin{bmatrix} 4 & 1 \\ 6 & 2 \\ -2 & 3 \end{bmatrix} - 3 \begin{bmatrix} 1 & 0 \\ 0 & 1 \end{bmatrix} = \begin{bmatrix} 28 & -9 \\ 4 & 23 \end{bmatrix} - \begin{bmatrix} 3 & 0 \\ 0 & 3 \end{bmatrix} = \begin{bmatrix} 25 & -9 \\ 4 & 20 \end{bmatrix}$

(b) Use MATRIX EDIT to create the matrices

$$A = \begin{bmatrix} 0 & 3 & -5 \\ 1 & 2 & 6 \end{bmatrix}; \quad C = \begin{bmatrix} 4 & 1 \\ 6 & 2 \\ -2 & 3 \end{bmatrix}$$

Then compute
$[A]*[C] - 3*\text{identity}(2)$:

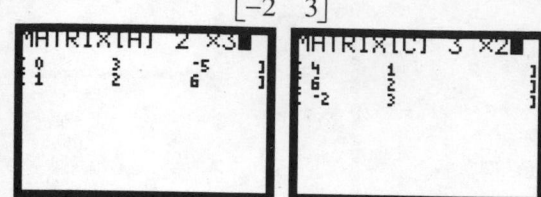

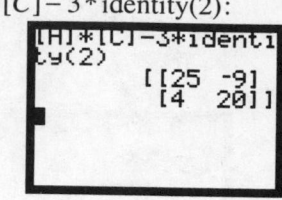

15. (a) $CA - CB = \begin{bmatrix} 4 & 1 \\ 6 & 2 \\ -2 & 3 \end{bmatrix} \cdot \begin{bmatrix} 0 & 3 & -5 \\ 1 & 2 & 6 \end{bmatrix} - \begin{bmatrix} 4 & 1 \\ 6 & 2 \\ -2 & 3 \end{bmatrix} \cdot \begin{bmatrix} 4 & 1 & 0 \\ -2 & 3 & -2 \end{bmatrix}$

$$= \begin{bmatrix} 1 & 14 & -14 \\ 2 & 22 & -18 \\ 3 & 0 & 28 \end{bmatrix} - \begin{bmatrix} 14 & 7 & -2 \\ 20 & 12 & -4 \\ -14 & 7 & -6 \end{bmatrix} = \begin{bmatrix} -13 & 7 & -12 \\ -18 & 10 & -14 \\ 17 & -7 & 34 \end{bmatrix}$$

(b) Use MATRIX EDIT to create the matrices

$$A = \begin{bmatrix} 0 & 3 & -5 \\ 1 & 2 & 6 \end{bmatrix}; \quad B = \begin{bmatrix} 4 & 1 & 0 \\ -2 & 3 & -2 \end{bmatrix}; \quad C = \begin{bmatrix} 4 & 1 \\ 6 & 2 \\ -2 & 3 \end{bmatrix}$$

Then compute $[C]*[A] - [C]*[B]$:

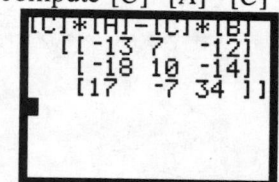

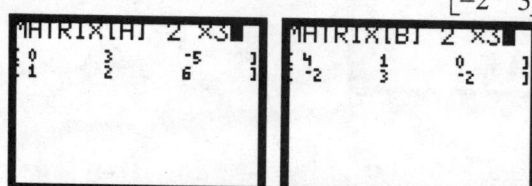

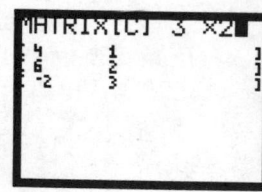

17. (a) $\begin{bmatrix} 2 & -2 \\ 1 & 0 \end{bmatrix} \begin{bmatrix} 2 & 1 & 4 & 6 \\ 3 & -1 & 3 & 2 \end{bmatrix}$

$$= \begin{bmatrix} 2(2) + (-2)(3) & 2(1) + (-2)(-1) & 2(4) + (-2)(3) & 2(6) + (-2)(2) \\ 1(2) + 0(3) & 1(1) + 0(-1) & 1(4) + 0(3) & 1(6) + 0(2) \end{bmatrix}$$

$$= \begin{bmatrix} -2 & 4 & 2 & 8 \\ 2 & 1 & 4 & 6 \end{bmatrix}$$

(b) Use MATRIX EDIT to create the matrices   Then compute $[A]*[B]$:

$$A = \begin{vmatrix} 2 & -2 \\ 1 & 0 \end{vmatrix}; \quad B = \begin{vmatrix} 2 & 1 & 4 & 6 \\ 3 & -1 & 3 & 2 \end{vmatrix}$$

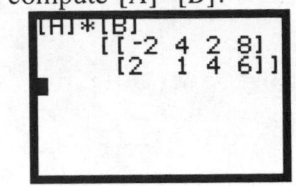

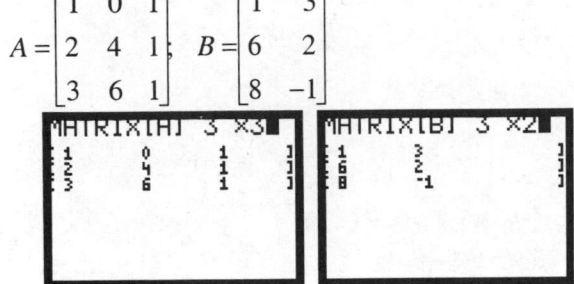

19. (a) $\begin{vmatrix} 1 & 0 & 1 \\ 2 & 4 & 1 \\ 3 & 6 & 1 \end{vmatrix}\begin{vmatrix} 1 & 3 \\ 6 & 2 \\ 8 & -1 \end{vmatrix} = \begin{vmatrix} 1(1)+0(6)+1(8) & 1(3)+0(2)+1(-1) \\ 2(1)+4(6)+1(8) & 2(3)+4(2)+1(-1) \\ 3(1)+6(6)+1(8) & 3(3)+6(2)+1(-1) \end{vmatrix} = \begin{vmatrix} 9 & 2 \\ 34 & 13 \\ 47 & 20 \end{vmatrix}$

(b) Use MATRIX EDIT to create the matrices   Then compute $[A]*[B]$:

$$A = \begin{vmatrix} 1 & 0 & 1 \\ 2 & 4 & 1 \\ 3 & 6 & 1 \end{vmatrix}; \quad B = \begin{vmatrix} 1 & 3 \\ 6 & 2 \\ 8 & -1 \end{vmatrix}$$

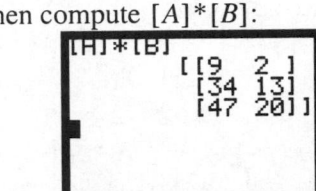

21. Augment the matrix with the identity and use row operations to find the inverse:

$$A = \begin{bmatrix} 2 & 1 \\ 1 & 1 \end{bmatrix} \rightarrow \begin{bmatrix} 2 & 1 & | & 1 & 0 \\ 1 & 1 & | & 0 & 1 \end{bmatrix}$$

$$\rightarrow \begin{bmatrix} 1 & 1 & | & 0 & 1 \\ 2 & 1 & | & 1 & 0 \end{bmatrix} \rightarrow \begin{bmatrix} 1 & 1 & | & 0 & 1 \\ 0 & -1 & | & 1 & -2 \end{bmatrix} \rightarrow \begin{bmatrix} 1 & 1 & | & 0 & 1 \\ 0 & 1 & | & -1 & 2 \end{bmatrix} \rightarrow \begin{bmatrix} 1 & 0 & | & 1 & -1 \\ 0 & 1 & | & -1 & 2 \end{bmatrix}$$

Interchange $R_2 = -2r_1 + r_2$ $R_2 = -r_2$ $R_1 = -r_2 + r_1$

r_1 and r_2

$$A^{-1} = \begin{bmatrix} 1 & -1 \\ -1 & 2 \end{bmatrix}$$

23. Augment the matrix with the identity and use row operations to find the inverse:

$$A = \begin{bmatrix} 6 & 5 \\ 2 & 2 \end{bmatrix} \rightarrow \begin{bmatrix} 6 & 5 & | & 1 & 0 \\ 2 & 2 & | & 0 & 1 \end{bmatrix}$$

$$\rightarrow \begin{bmatrix} 2 & 2 & | & 0 & 1 \\ 6 & 5 & | & 1 & 0 \end{bmatrix} \rightarrow \begin{bmatrix} 2 & 2 & | & 0 & 1 \\ 0 & -1 & | & 1 & -3 \end{bmatrix} \rightarrow \begin{bmatrix} 1 & 1 & | & 0 & \frac{1}{2} \\ 0 & 1 & | & -1 & 3 \end{bmatrix} \rightarrow \begin{bmatrix} 1 & 0 & | & 1 & -\frac{5}{2} \\ 0 & 1 & | & -1 & 3 \end{bmatrix}$$

Interchange $R_2 = -3r_1 + r_2$ $R_1 = \frac{1}{2}r_1$ $R_1 = -r_2 + r_1$

r_1 and r_2 $R_2 = -r_2$

$$A^{-1} = \begin{bmatrix} 1 & -\frac{5}{2} \\ -1 & 3 \end{bmatrix}$$

25. Augment the matrix with the identity and use row operations to find the inverse:

$$A = \begin{bmatrix} 2 & 1 \\ a & a \end{bmatrix} \rightarrow \begin{bmatrix} 2 & 1 & | & 1 & 0 \\ a & a & | & 0 & 1 \end{bmatrix} \text{ where } a \neq 0.$$

$$\rightarrow \begin{bmatrix} 1 & \frac{1}{2} & | & \frac{1}{2} & 0 \\ a & a & | & 0 & 1 \end{bmatrix} \rightarrow \begin{bmatrix} 1 & \frac{1}{2} & | & \frac{1}{2} & 0 \\ 0 & \frac{1}{2}a & | & -\frac{1}{2}a & 1 \end{bmatrix} \rightarrow \begin{bmatrix} 1 & \frac{1}{2} & | & \frac{1}{2} & 0 \\ 0 & 1 & | & -1 & \frac{2}{a} \end{bmatrix} \rightarrow \begin{bmatrix} 1 & 0 & | & 1 & -\frac{1}{a} \\ 0 & 1 & | & -1 & \frac{2}{a} \end{bmatrix}$$

$$R_1 = \tfrac{1}{2}r_1 \qquad R_2 = -ar_1 + r_2 \qquad R_2 = \left(\tfrac{2}{a}\right)r_2 \qquad R_1 = -\tfrac{1}{2}r_2 + r_1$$

$$A^{-1} = \begin{bmatrix} 1 & -\frac{1}{a} \\ -1 & \frac{2}{a} \end{bmatrix}$$

27. Augment the matrix with the identity and use row operations to find the inverse:

$$A = \begin{bmatrix} 1 & -1 & 1 \\ 0 & -2 & 1 \\ -2 & -3 & 0 \end{bmatrix} \rightarrow \begin{bmatrix} 1 & -1 & 1 & | & 1 & 0 & 0 \\ 0 & -2 & 1 & | & 0 & 1 & 0 \\ -2 & -3 & 0 & | & 0 & 0 & 1 \end{bmatrix}$$

$$\rightarrow \begin{bmatrix} 1 & -1 & 1 & | & 1 & 0 & 0 \\ 0 & -2 & 1 & | & 0 & 1 & 0 \\ 0 & -5 & 2 & | & 2 & 0 & 1 \end{bmatrix} \rightarrow \begin{bmatrix} 1 & -1 & 1 & | & 1 & 0 & 0 \\ 0 & 1 & -\frac{1}{2} & | & 0 & -\frac{1}{2} & 0 \\ 0 & -5 & 2 & | & 2 & 0 & 1 \end{bmatrix} \rightarrow \begin{bmatrix} 1 & 0 & \frac{1}{2} & | & 1 & -\frac{1}{2} & 0 \\ 0 & 1 & -\frac{1}{2} & | & 0 & -\frac{1}{2} & 0 \\ 0 & 0 & -\frac{1}{2} & | & 2 & -\frac{5}{2} & 1 \end{bmatrix}$$

$$R_3 = 2r_1 + r_3 \qquad\qquad R_2 = -\tfrac{1}{2}r_2 \qquad\qquad\qquad R_1 = r_2 + r_1$$
$$R_3 = 5r_2 + r_3$$

$$\rightarrow \begin{bmatrix} 1 & 0 & \frac{1}{2} & | & 1 & -\frac{1}{2} & 0 \\ 0 & 1 & -\frac{1}{2} & | & 0 & -\frac{1}{2} & 0 \\ 0 & 0 & 1 & | & -4 & 5 & -2 \end{bmatrix} \rightarrow \begin{bmatrix} 1 & 0 & 0 & | & 3 & -3 & 1 \\ 0 & 1 & 0 & | & -2 & 2 & -1 \\ 0 & 0 & 1 & | & -4 & 5 & -2 \end{bmatrix}$$

$$R_3 = -2r_3 \qquad\qquad R_1 = -\tfrac{1}{2}r_3 + r_1$$
$$R_2 = \tfrac{1}{2}r_3 + r_2$$

$$A^{-1} = \begin{bmatrix} 3 & -3 & 1 \\ -2 & 2 & -1 \\ -4 & 5 & -2 \end{bmatrix}$$

29. Augment the matrix with the identity and use row operations to find the inverse:

$$A = \begin{bmatrix} 1 & 1 & 1 \\ 3 & 2 & -1 \\ 3 & 1 & 2 \end{bmatrix} \rightarrow \begin{bmatrix} 1 & 1 & 1 & | & 1 & 0 & 0 \\ 3 & 2 & -1 & | & 0 & 1 & 0 \\ 3 & 1 & 2 & | & 0 & 0 & 1 \end{bmatrix}$$

$$\rightarrow \begin{bmatrix} 1 & 1 & 1 & | & 1 & 0 & 0 \\ 0 & -1 & -4 & | & -3 & 1 & 0 \\ 0 & -2 & -1 & | & -3 & 0 & 1 \end{bmatrix} \rightarrow \begin{bmatrix} 1 & 1 & 1 & | & 1 & 0 & 0 \\ 0 & 1 & 4 & | & 3 & -1 & 0 \\ 0 & -2 & -1 & | & -3 & 0 & 1 \end{bmatrix} \rightarrow \begin{bmatrix} 1 & 0 & -3 & | & -2 & 1 & 0 \\ 0 & 1 & 4 & | & 3 & -1 & 0 \\ 0 & 0 & 7 & | & 3 & -2 & 1 \end{bmatrix}$$

$$R_2 = -3r_1 + r_2 \qquad\qquad R_2 = -r_2 \qquad\qquad\qquad R_1 = -r_2 + r_1$$
$$R_3 = -3r_1 + r_3 \qquad\qquad\qquad\qquad\qquad\qquad R_3 = 2r_2 + r_3$$

$$\rightarrow \begin{vmatrix} 1 & 0 & -3 & | & -2 & 1 & 0 \\ 0 & 1 & 4 & | & 3 & -1 & 0 \\ 0 & 0 & 1 & | & \frac{3}{7} & -\frac{2}{7} & \frac{1}{7} \end{vmatrix} \rightarrow \begin{bmatrix} 1 & 0 & 0 & | & -\frac{5}{7} & \frac{1}{7} & \frac{3}{7} \\ 0 & 1 & 0 & | & \frac{9}{7} & \frac{1}{7} & -\frac{4}{7} \\ 0 & 0 & 1 & | & \frac{3}{7} & -\frac{2}{7} & \frac{1}{7} \end{bmatrix}$$

$R_3 = \frac{1}{7} r_3$ $\qquad\qquad R_1 = 3r_3 + r_1$

$\qquad\qquad\qquad\qquad R_2 = -4r_3 + r_2$

$$A^{-1} = \begin{bmatrix} -\frac{5}{7} & \frac{1}{7} & \frac{3}{7} \\ \frac{9}{7} & \frac{1}{7} & -\frac{4}{7} \\ \frac{3}{7} & -\frac{2}{7} & \frac{1}{7} \end{bmatrix}$$

31. Rewrite the system of equations in matrix form:

$\begin{cases} 2x + y = 8 \\ x + y = 5 \end{cases}$ $A = \begin{bmatrix} 2 & 1 \\ 1 & 1 \end{bmatrix}$, $X = \begin{bmatrix} x \\ y \end{bmatrix}$, $B = \begin{bmatrix} 8 \\ 5 \end{bmatrix}$

Find the inverse of A and solve $X = A^{-1}B$:

From Problem 21, $A^{-1} = \begin{bmatrix} 1 & -1 \\ -1 & 2 \end{bmatrix}$ and $X = A^{-1}B = \begin{bmatrix} 1 & -1 \\ -1 & 2 \end{bmatrix}\begin{bmatrix} 8 \\ 5 \end{bmatrix} = \begin{bmatrix} 3 \\ 2 \end{bmatrix}$.

The solution is $x = 3$, $y = 2$.

33. Rewrite the system of equations in matrix form:

$\begin{cases} 2x + y = 0 \\ x + y = 5 \end{cases}$ $A = \begin{bmatrix} 2 & 1 \\ 1 & 1 \end{bmatrix}$, $X = \begin{bmatrix} x \\ y \end{bmatrix}$, $B = \begin{bmatrix} 0 \\ 5 \end{bmatrix}$

Find the inverse of A and solve $X = A^{-1}B$:

From Problem 21, $A^{-1} = \begin{bmatrix} 1 & -1 \\ -1 & 2 \end{bmatrix}$ and $X = A^{-1}B = \begin{bmatrix} 1 & -1 \\ -1 & 2 \end{bmatrix}\begin{bmatrix} 0 \\ 5 \end{bmatrix} = \begin{bmatrix} -5 \\ 10 \end{bmatrix}$.

The solution is $x = -5$, $y = 10$.

35. Rewrite the system of equations in matrix form:

$\begin{cases} 6x + 5y = 7 \\ 2x + 2y = 2 \end{cases}$ $A = \begin{bmatrix} 6 & 5 \\ 2 & 2 \end{bmatrix}$, $X = \begin{bmatrix} x \\ y \end{bmatrix}$, $B = \begin{bmatrix} 7 \\ 2 \end{bmatrix}$

Find the inverse of A and solve $X = A^{-1}B$:

From Problem 23, $A^{-1} = \begin{bmatrix} 1 & -\frac{5}{2} \\ -1 & 3 \end{bmatrix}$ and $X = A^{-1}B = \begin{bmatrix} 1 & -\frac{5}{2} \\ -1 & 3 \end{bmatrix}\begin{bmatrix} 7 \\ 2 \end{bmatrix} = \begin{bmatrix} 2 \\ -1 \end{bmatrix}$.

The solution is $x = 2$, $y = -1$.

37. Rewrite the system of equations in matrix form:

$\begin{cases} 6x + 5y = 13 \\ 2x + 2y = 5 \end{cases}$ $A = \begin{bmatrix} 6 & 5 \\ 2 & 2 \end{bmatrix}$, $X = \begin{bmatrix} x \\ y \end{bmatrix}$, $B = \begin{bmatrix} 13 \\ 5 \end{bmatrix}$

Find the inverse of A and solve $X = A^{-1}B$:

From Problem 23, $A^{-1} = \begin{bmatrix} 1 & -\frac{5}{2} \\ -1 & 3 \end{bmatrix}$ and $X = A^{-1}B = \begin{bmatrix} 1 & -\frac{5}{2} \\ -1 & 3 \end{bmatrix}\begin{bmatrix} 13 \\ 5 \end{bmatrix} = \begin{bmatrix} \frac{1}{2} \\ 2 \end{bmatrix}$.

The solution is $x = \frac{1}{2}$, $y = 2$.

39. Rewrite the system of equations in matrix form:

$\begin{cases} 2x + y = -3 \\ ax + ay = -a \end{cases}$ $\quad a \ne 0$ $\qquad A = \begin{bmatrix} 2 & 1 \\ a & a \end{bmatrix}, \quad X = \begin{bmatrix} x \\ y \end{bmatrix}, \quad B = \begin{bmatrix} -3 \\ -a \end{bmatrix}$

Find the inverse of A and solve $X = A^{-1}B$:

From Problem 25, $A^{-1} = \begin{bmatrix} 1 & -\frac{1}{a} \\ -1 & \frac{2}{a} \end{bmatrix}$ and $X = A^{-1}B = \begin{bmatrix} 1 & -\frac{1}{a} \\ -1 & \frac{2}{a} \end{bmatrix}\begin{bmatrix} -3 \\ -a \end{bmatrix} = \begin{bmatrix} -2 \\ 1 \end{bmatrix}$.

The solution is $x = -2,\ y = 1$.

41. Rewrite the system of equations in matrix form:

$\begin{cases} 2x + y = \dfrac{7}{a} \\ ax + ay = 5 \end{cases}$ $\quad a \ne 0$ $\qquad A = \begin{bmatrix} 2 & 1 \\ a & a \end{bmatrix}, \quad X = \begin{bmatrix} x \\ y \end{bmatrix}, \quad B = \begin{bmatrix} \frac{7}{a} \\ 5 \end{bmatrix}$

Find the inverse of A and solve $X = A^{-1}B$:

From Problem 25, $A^{-1} = \begin{bmatrix} 1 & -\frac{1}{a} \\ -1 & \frac{2}{a} \end{bmatrix}$ and $X = A^{-1}B = \begin{bmatrix} 1 & -\frac{1}{a} \\ -1 & \frac{2}{a} \end{bmatrix}\begin{bmatrix} \frac{7}{a} \\ 5 \end{bmatrix} = \begin{bmatrix} \frac{2}{a} \\ \frac{3}{a} \end{bmatrix}$.

The solution is $x = \dfrac{2}{a},\ y = \dfrac{3}{a}$.

43. Rewrite the system of equations in matrix form:

$\begin{cases} x - y + z = 0 \\ -2y + z = -1 \\ -2x - 3y = -5 \end{cases}$ $\qquad A = \begin{bmatrix} 1 & -1 & 1 \\ 0 & -2 & 1 \\ -2 & -3 & 0 \end{bmatrix}, \quad X = \begin{bmatrix} x \\ y \\ z \end{bmatrix}, \quad B = \begin{bmatrix} 0 \\ -1 \\ -5 \end{bmatrix}$

Find the inverse of A and solve $X = A^{-1}B$:

From Problem 27,

$$A^{-1} = \begin{bmatrix} 3 & -3 & 1 \\ -2 & 2 & -1 \\ -4 & 5 & -2 \end{bmatrix} \text{ and } X = A^{-1}B = \begin{bmatrix} 3 & -3 & 1 \\ -2 & 2 & -1 \\ -4 & 5 & -2 \end{bmatrix}\begin{bmatrix} 0 \\ -1 \\ -5 \end{bmatrix} = \begin{bmatrix} -2 \\ 3 \\ 5 \end{bmatrix}.$$

The solution is $x = -2,\ y = 3,\ z = 5$.

45. Rewrite the system of equations in matrix form:

$\begin{cases} x - y + z = 2 \\ -2y + z = 2 \\ -2x - 3y = \dfrac{1}{2} \end{cases}$ $\qquad A = \begin{bmatrix} 1 & -1 & 1 \\ 0 & -2 & 1 \\ -2 & -3 & 0 \end{bmatrix}, \quad X = \begin{bmatrix} x \\ y \\ z \end{bmatrix}, \quad B = \begin{bmatrix} 2 \\ 2 \\ \frac{1}{2} \end{bmatrix}$

Find the inverse of A and solve $X = A^{-1}B$:

From Problem 27,

$$A^{-1} = \begin{bmatrix} 3 & -3 & 1 \\ -2 & 2 & -1 \\ -4 & 5 & -2 \end{bmatrix} \text{ and } X = A^{-1}B = \begin{bmatrix} 3 & -3 & 1 \\ -2 & 2 & -1 \\ -4 & 5 & -2 \end{bmatrix}\begin{bmatrix} 2 \\ 2 \\ \frac{1}{2} \end{bmatrix} = \begin{bmatrix} \frac{1}{2} \\ -\frac{1}{2} \\ 1 \end{bmatrix}.$$

The solution is $x = \dfrac{1}{2},\ y = -\dfrac{1}{2},\ z = 1$.

47. Rewrite the system of equations in matrix form:

$$\begin{cases} x + y + z = 9 \\ 3x + 2y - z = 8 \\ 3x + y + 2z = 1 \end{cases} \quad A = \begin{bmatrix} 1 & 1 & 1 \\ 3 & 2 & -1 \\ 3 & 1 & 2 \end{bmatrix}, \quad X = \begin{bmatrix} x \\ y \\ z \end{bmatrix}, \quad B = \begin{bmatrix} 9 \\ 8 \\ 1 \end{bmatrix}$$

Find the inverse of A and solve $X = A^{-1}B$:

From Problem 29,

$$A^{-1} = \begin{bmatrix} -\frac{5}{7} & \frac{1}{7} & \frac{3}{7} \\ \frac{9}{7} & \frac{1}{7} & -\frac{4}{7} \\ \frac{3}{7} & -\frac{2}{7} & \frac{1}{7} \end{bmatrix} \text{ and } X = A^{-1}B = \begin{bmatrix} -\frac{5}{7} & \frac{1}{7} & \frac{3}{7} \\ \frac{9}{7} & \frac{1}{7} & -\frac{4}{7} \\ \frac{3}{7} & -\frac{2}{7} & \frac{1}{7} \end{bmatrix}\begin{bmatrix} 9 \\ 8 \\ 1 \end{bmatrix} = \begin{bmatrix} -\frac{34}{7} \\ \frac{85}{7} \\ \frac{12}{7} \end{bmatrix}.$$

The solution is $x = -\dfrac{34}{7}, y = \dfrac{85}{7}, z = \dfrac{12}{7}$.

49. Rewrite the system of equations in matrix form:

$$\begin{cases} x + y + z = 2 \\ 3x + 2y - z = \dfrac{7}{3} \\ 3x + y + 2z = \dfrac{10}{3} \end{cases} \quad A = \begin{bmatrix} 1 & 1 & 1 \\ 3 & 2 & -1 \\ 3 & 1 & 2 \end{bmatrix}, \quad X = \begin{bmatrix} x \\ y \\ z \end{bmatrix}, \quad B = \begin{bmatrix} 2 \\ \frac{7}{3} \\ \frac{10}{3} \end{bmatrix}$$

Find the inverse of A and solve $X = A^{-1}B$:

From Problem 29,

$$A^{-1} = \begin{bmatrix} -\frac{5}{7} & \frac{1}{7} & \frac{3}{7} \\ \frac{9}{7} & \frac{1}{7} & -\frac{4}{7} \\ \frac{3}{7} & -\frac{2}{7} & \frac{1}{7} \end{bmatrix} \text{ and } X = A^{-1}B = \begin{bmatrix} -\frac{5}{7} & \frac{1}{7} & \frac{3}{7} \\ \frac{9}{7} & \frac{1}{7} & -\frac{4}{7} \\ \frac{3}{7} & -\frac{2}{7} & \frac{1}{7} \end{bmatrix}\begin{bmatrix} 2 \\ \frac{7}{3} \\ \frac{10}{3} \end{bmatrix} = \begin{bmatrix} \frac{1}{3} \\ 1 \\ \frac{2}{3} \end{bmatrix}.$$

The solution is $x = \dfrac{1}{3}, y = 1, z = \dfrac{2}{3}$.

51. Augment the matrix with the identity and use row operations to find the inverse:

$$A = \begin{bmatrix} 4 & 2 \\ 2 & 1 \end{bmatrix} \rightarrow \begin{bmatrix} 4 & 2 & | & 1 & 0 \\ 2 & 1 & | & 0 & 1 \end{bmatrix}$$

$$\rightarrow \begin{bmatrix} 4 & 2 & | & 1 & 0 \\ 0 & 0 & | & -\frac{1}{2} & 1 \end{bmatrix} \rightarrow \begin{bmatrix} 1 & \frac{1}{2} & | & \frac{1}{4} & 0 \\ 0 & 0 & | & -\frac{1}{2} & 1 \end{bmatrix}$$

$$R_2 = -\tfrac{1}{2}r_1 + r_2 \qquad R_1 = \tfrac{1}{4}r_1$$

There is no way to obtain the identity matrix on the left; thus, there is no inverse.

53. Augment the matrix with the identity and use row operations to find the inverse:

$$A = \begin{bmatrix} 15 & 3 \\ 10 & 2 \end{bmatrix} \rightarrow \begin{bmatrix} 15 & 3 & | & 1 & 0 \\ 10 & 2 & | & 0 & 1 \end{bmatrix} \rightarrow \begin{bmatrix} 15 & 3 & | & 1 & 0 \\ 0 & 0 & | & -\frac{2}{3} & 1 \end{bmatrix} \rightarrow \begin{bmatrix} 1 & \frac{1}{5} & | & \frac{1}{15} & 0 \\ 0 & 0 & | & -\frac{2}{3} & 1 \end{bmatrix}$$

$$R_2 = -\tfrac{2}{3}r_1 + r_2 \qquad R_1 = \tfrac{1}{15}r_1$$

There is no way to obtain the identity matrix on the left; thus, there is no inverse.

55. Augment the matrix with the identity and use row operations to find the inverse:

$$A = \begin{bmatrix} -3 & 1 & -1 \\ 1 & -4 & -7 \\ 1 & 2 & 5 \end{bmatrix} \rightarrow \left[\begin{array}{ccc|ccc} -3 & 1 & -1 & 1 & 0 & 0 \\ 1 & -4 & -7 & 0 & 1 & 0 \\ 1 & 2 & 5 & 0 & 0 & 1 \end{array} \right]$$

$$\rightarrow \left[\begin{array}{ccc|ccc} 1 & 2 & 5 & 0 & 0 & 1 \\ 1 & -4 & -7 & 0 & 1 & 0 \\ -3 & 1 & -1 & 1 & 0 & 0 \end{array} \right] \rightarrow \left[\begin{array}{ccc|ccc} 1 & 2 & 5 & 0 & 0 & 1 \\ 0 & -6 & -12 & 0 & 1 & -1 \\ 0 & 7 & 14 & 1 & 0 & 3 \end{array} \right] \rightarrow \left[\begin{array}{ccc|ccc} 1 & 2 & 5 & 0 & 0 & 1 \\ 0 & 1 & 2 & 0 & -\frac{1}{6} & \frac{1}{6} \\ 0 & 7 & 14 & 1 & 0 & 3 \end{array} \right]$$

Interchange r_1 and r_3 　　 $R_2 = -r_1 + r_2$ 　　　　　 $R_2 = -\frac{1}{6}r_2$

$R_3 = 3r_1 + r_3$

$$\rightarrow \left[\begin{array}{ccc|ccc} 1 & 0 & 1 & 0 & \frac{1}{3} & \frac{2}{3} \\ 0 & 1 & 2 & 0 & -\frac{1}{6} & \frac{1}{6} \\ 0 & 0 & 0 & 1 & \frac{7}{6} & \frac{11}{6} \end{array} \right]$$

$R_1 = -2r_2 + r_1$

$R_3 = -7r_2 + r_3$

There is no way to obtain the identity matrix on the left; thus, there is no inverse.

57. Use **MATRIX EDIT** to create the matrix

$$A = \begin{vmatrix} 25 & 61 & -12 \\ 18 & -2 & 4 \\ 8 & 35 & 21 \end{vmatrix}$$

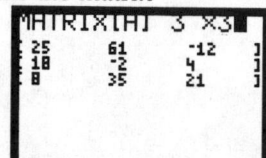

Then compute $[A]^\wedge(-1)$:

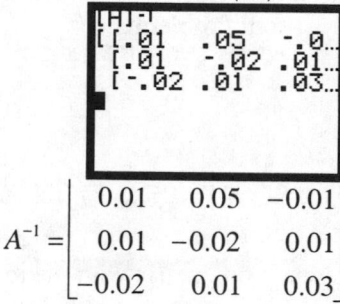

$$A^{-1} = \begin{bmatrix} 0.01 & 0.05 & -0.01 \\ 0.01 & -0.02 & 0.01 \\ -0.02 & 0.01 & 0.03 \end{bmatrix}$$

59. Use **MATRIX EDIT** to create the matrix

$$A = \begin{vmatrix} 44 & 21 & 18 & 6 \\ -2 & 10 & 15 & 5 \\ 21 & 12 & -12 & 4 \\ -8 & -16 & 4 & 9 \end{vmatrix}$$

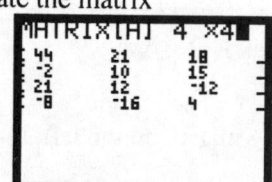

Then compute $[A]^\wedge(-1)$:

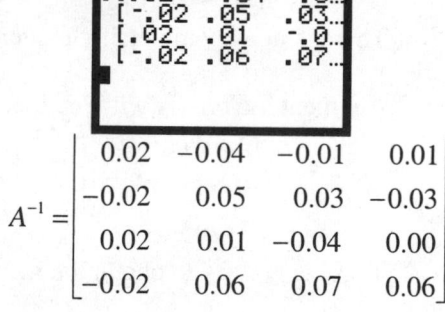

$$A^{-1} = \begin{bmatrix} 0.02 & -0.04 & -0.01 & 0.01 \\ -0.02 & 0.05 & 0.03 & -0.03 \\ 0.02 & 0.01 & -0.04 & 0.00 \\ -0.02 & 0.06 & 0.07 & 0.06 \end{bmatrix}$$

61. Use MATRIX EDIT to create the matrices

$$A = \begin{vmatrix} 25 & 61 & -12 \\ 18 & -12 & 7 \\ 3 & 4 & -1 \end{vmatrix} ; B = \begin{vmatrix} 10 \\ -9 \\ 12 \end{vmatrix}$$

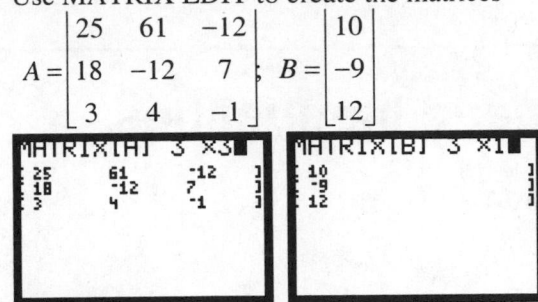

Then compute $([A]^{\wedge}(-1)) * [B]$:

$x = 4.57, \ y = -6.44, \ z = -24.07$

63. Use MATRIX EDIT to create the matrices

$$A = \begin{vmatrix} 25 & 61 & -12 \\ 18 & -12 & 7 \\ 3 & 4 & -1 \end{vmatrix} ; B = \begin{vmatrix} 21 \\ 7 \\ -2 \end{vmatrix}$$

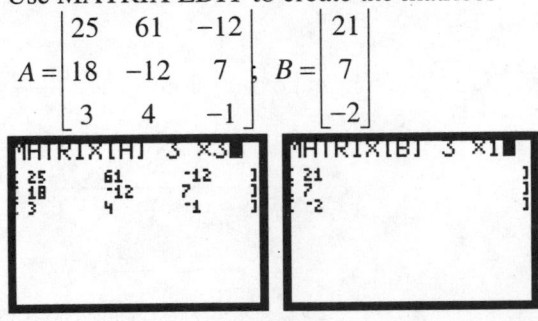

Then compute $([A]^{\wedge}(-1)) * [B]$:

$x = -1.19, \ y = 2.46, \ z = 8.27$

65. (a) The rows of the 2 by 3 matrix represent stainless steel and aluminum. The columns represent 10-gallon, 5-gallon, and 1-gallon.

The 2 by 3 matrix is: The 3 by 2 matrix is:

$$\begin{bmatrix} 500 & 350 & 400 \\ 700 & 500 & 850 \end{bmatrix} \qquad \begin{bmatrix} 500 & 700 \\ 350 & 500 \\ 400 & 850 \end{bmatrix}$$

(b) The 3 by 1 matrix representing the amount of material is:

$$\begin{bmatrix} 15 \\ 8 \\ 3 \end{bmatrix}$$

(c) The days usage of materials is:

$$\begin{bmatrix} 500 & 350 & 400 \\ 700 & 500 & 850 \end{bmatrix} \cdot \begin{bmatrix} 15 \\ 8 \\ 3 \end{bmatrix} = \begin{bmatrix} 11,500 \\ 17,050 \end{bmatrix}$$

11,500 pounds of stainless steel and 17,050 pounds of aluminum are used each day.

(d) The 1 by 2 matrix representing cost is:

$$\begin{bmatrix} 0.10 & 0.05 \end{bmatrix}$$

(e) The total cost of the days production was:

$$\begin{bmatrix} 0.10 & 0.05 \end{bmatrix} \cdot \begin{bmatrix} 11,500 \\ 17,050 \end{bmatrix} = \begin{bmatrix} 2002.50 \end{bmatrix}$$

The total cost of the days production was $2,002.50.

67. Answers will vary.

Chapter 10

Systems of Equations and Inequalities

10.6 Partial Fraction Decomposition

1. The rational expression $\dfrac{x}{x^2-1}$ is proper, since the degree of the numerator is less than the degree of the denominator.

3. The rational expression $\dfrac{x^2+5}{x^2-4}$ is improper, so perform the division:

$$
\begin{array}{r}
1 \\
x^2-4\overline{)x^2+5} \\
\underline{x^2-4} \\
9
\end{array}
$$

The proper rational expression is:
$$\frac{x^2+5}{x^2-4}=1+\frac{9}{x^2-4}$$

5. The rational expression $\dfrac{5x^3+2x-1}{x^2-4}$ is improper, so perform the division:

$$
\begin{array}{r}
5x \\
x^2-4\overline{)5x^3+0x^2+\ 2x-1} \\
\underline{5x^3\qquad\ -20x} \\
22x-1
\end{array}
$$

The proper rational expression is:
$$\frac{5x^3+2x-1}{x^2-4}=5x+\frac{22x-1}{x^2-4}$$

7. The rational expression $\dfrac{x(x-1)}{(x+4)(x-3)}=\dfrac{x^2-x}{x^2+x-12}$ is improper, so perform the division:

$$
\begin{array}{r}
1 \\
x^2+x-12\overline{)x^2-x+\ 0} \\
\underline{x^2+x-12} \\
-2x+12
\end{array}
$$

The proper rational expression is: $\dfrac{x(x-1)}{(x+4)(x-3)}=1+\dfrac{-2x+12}{x^2+x-12}$

9. Find the partial fraction decomposition:

$$\frac{4}{x(x-1)}=\frac{A}{x}+\frac{B}{x-1}$$
$$4=A(x-1)+Bx \quad \text{(Multiply both sides by } x(x-1)\text{.)}$$
Let $x=1$: then $4=A(0)+B \rightarrow B=4$

Let $x=0$: then $4=A(-1)+B(0) \rightarrow A=-4$

$$\frac{4}{x(x-1)}=\frac{-4}{x}+\frac{4}{x-1}$$

11. Find the partial fraction decomposition:

$$\frac{1}{x(x^2+1)} = \frac{A}{x} + \frac{Bx+C}{x^2+1}$$

$$1 = A(x^2+1) + (Bx+C)x \quad \text{(Multiply both sides by } x(x^2+1).)$$

Let $x = 0$:   then $1 = A(1) + (B(0)+C)(0) \rightarrow A = 1$

Let $x = 1$:   then $1 = A(1+1) + (B(1)+C)(1) \rightarrow 1 = 2A + B + C$

so $1 = 2(1) + B + C \rightarrow -1 = B + C$

Let $x = -1$: then $1 = A(1+1) + (B(-1)+C)(-1) \rightarrow 1 = 2A + B - C$

so $1 = 2(1) + B - C \rightarrow -1 = B - C$

Solve the system of equations:

$$B + C = -1$$
$$\underline{B - C = -1}$$
$$2B \quad\quad = -2 \quad\quad\quad -1 + C = -1$$
$$B \quad\quad = -1 \quad\quad\quad\quad C = 0$$

$$\frac{1}{x(x^2+1)} = \frac{1}{x} + \frac{-x}{x^2+1}$$

13. Find the partial fraction decomposition:

$$\frac{x}{(x-1)(x-2)} = \frac{A}{x-1} + \frac{B}{x-2}$$

$$x = A(x-2) + B(x-1) \quad \text{(Multiply both sides by } (x-1)(x-2).)$$

Let $x = 1$:   then $1 = A(1-2) + B(1-1) \rightarrow 1 = -A \rightarrow A = -1$

Let $x = 2$:   then $2 = A(2-2) + B(2-1) \rightarrow 2 = B \rightarrow B = 2$

$$\frac{x}{(x-1)(x-2)} = \frac{-1}{x-1} + \frac{2}{x-2}$$

15. Find the partial fraction decomposition:

$$\frac{x^2}{(x-1)^2(x+1)} = \frac{A}{x-1} + \frac{B}{(x-1)^2} + \frac{C}{x+1} \quad \text{(Multiply both sides by } (x-1)^2(x+1).)$$

$$x^2 = A(x-1)(x+1) + B(x+1) + C(x-1)^2$$

Let $x = 1$:   then $1^2 = A(1-1)(1+1) + B(1+1) + C(1-1)^2$

$$\rightarrow 1 = 2B \rightarrow B = \frac{1}{2}$$

Let $x = -1$: then $(-1)^2 = A(-1-1)(-1+1) + B(-1+1) + C(-1-1)^2$

$$\rightarrow 1 = 4C \rightarrow C = \frac{1}{4}$$

Let $x = 0$:   then $0^2 = A(0-1)(0+1) + B(0+1) + C(0-1)^2$

$$\rightarrow 0 = -A + B + C \rightarrow A = \frac{1}{2} + \frac{1}{4} = \frac{3}{4}$$

$$\frac{x^2}{(x-1)^2(x+1)} = \frac{\left(\dfrac{3}{4}\right)}{x-1} + \frac{\left(\dfrac{1}{2}\right)}{(x-1)^2} + \frac{\left(\dfrac{1}{4}\right)}{x+1}$$

17. Find the partial fraction decomposition:

$$\frac{1}{x^3 - 8} = \frac{1}{(x-2)(x^2 + 2x + 4)} = \frac{A}{x-2} + \frac{Bx + C}{x^2 + 2x + 4}$$

(Multiply both sides by $(x-2)(x^2 + 2x + 4)$.)

$$1 = A(x^2 + 2x + 4) + (Bx + C)(x - 2)$$

Let $x = 2$:   then $1 = A(2^2 + 2(2) + 4) + (B(2) + C)(2 - 2)$

$$\rightarrow 1 = 12A \;\rightarrow\; A = \frac{1}{12}$$

Let $x = 0$:   then $1 = A(0^2 + 2(0) + 4) + (B(0) + C)(0 - 2)$

$$\rightarrow 1 = 4A - 2C \rightarrow 1 = 4\left(\frac{1}{12}\right) - 2C$$

$$\rightarrow -2C = \frac{2}{3} \;\rightarrow\; C = -\frac{1}{3}$$

Let $x = 1$:   then $1 = A(1^2 + 2(1) + 4) + (B(1) + C)(1 - 2)$

$$\rightarrow 1 = 7A - B - C \;\rightarrow\; 1 = 7\left(\frac{1}{12}\right) - B + \frac{1}{3} \;\rightarrow\; B = -\frac{1}{12}$$

$$\frac{1}{x^3 - 8} = \frac{\left(\frac{1}{12}\right)}{x - 2} + \frac{-\left(\frac{1}{12}\right)x - \frac{1}{3}}{x^2 + 2x + 4}$$

19. Find the partial fraction decomposition:

$$\frac{x^2}{(x-1)^2(x+1)^2} = \frac{A}{x-1} + \frac{B}{(x-1)^2} + \frac{C}{x+1} + \frac{D}{(x+1)^2}$$

(Multiply both sides by $(x-1)^2(x+1)^2$.)

$$x^2 = A(x-1)(x+1)^2 + B(x+1)^2 + C(x-1)^2(x+1) + D(x-1)^2$$

Let $x = 1$:   then $1^2 = A(1-1)(1+1)^2 + B(1+1)^2 + C(1-1)^2(1+1) + D(1-1)^2$

$$\rightarrow 1 = 4B \;\rightarrow\; B = \frac{1}{4}$$

Let $x = -1$:   then

$$(-1)^2 = A(-1-1)(-1+1)^2 + B(-1+1)^2 + C(-1-1)^2(-1+1) + D(-1-1)^2$$

$$\rightarrow 1 = 4D \;\rightarrow\; D = \frac{1}{4}$$

Let $x = 0$:   then

$$0^2 = A(0-1)(0+1)^2 + B(0+1)^2 + C(0-1)^2(0+1) + D(0-1)^2$$

$$\rightarrow 0 = -A + B + C + D \;\rightarrow\; A - C = \frac{1}{4} + \frac{1}{4} = \frac{1}{2}$$

Let $x = 2$:   then

$$2^2 = A(2-1)(2+1)^2 + B(2+1)^2 + C(2-1)^2(2+1) + D(2-1)^2$$

$$\rightarrow\ 4 = 9A + 9B + 3C + D\ \rightarrow\ 9A + 3C = 4 - \frac{9}{4} - \frac{1}{4} = \frac{3}{2}$$

$$\rightarrow\ 3A + C = \frac{1}{2}$$

Solve the system of equations:

$$A - C = \frac{1}{2}$$

$$\underline{3A + C = \frac{1}{2}}$$

$$4A\quad = 1$$

$$A\quad = \frac{1}{4}\ \rightarrow\ \frac{3}{4} + C = \frac{1}{2}\ \rightarrow\ C = -\frac{1}{4}$$

$$\frac{x^2}{(x-1)^2(x+1)^2} = \frac{\left(\frac{1}{4}\right)}{x-1} + \frac{\left(\frac{1}{4}\right)}{(x-1)^2} + \frac{\left(-\frac{1}{4}\right)}{x+1} + \frac{\left(\frac{1}{4}\right)}{(x+1)^2}$$

21. Find the partial fraction decomposition:

$$\frac{x-3}{(x+2)(x+1)^2} = \frac{A}{x+2} + \frac{B}{x+1} + \frac{C}{(x+1)^2}$$

(Multiply both sides by $(x+2)(x+1)^2$.)

$$x - 3 = A(x+1)^2 + B(x+2)(x+1) + C(x+2)$$

Let $x = -2$:   then $-2 - 3 = A(-2+1)^2 + B(-2+2)(-2+1) + C(-2+2)$

$$\rightarrow\ -5 = A\ \rightarrow\ A = -5$$

Let $x = -1$:   then $-1 - 3 = A(-1+1)^2 + B(-1+2)(-1+1) + C(-1+2)$

$$\rightarrow\ -4 = C\ \rightarrow\ C = -4$$

Let $x = 0$:   then

$$0 - 3 = A(0+1)^2 + B(0+2)(0+1) + C(0+2)\ \rightarrow\ -3 = A + 2B + 2C$$

$$\rightarrow\ -3 = -5 + 2B + 2(-4)\ \rightarrow\ 2B = 10\ \rightarrow\ B = 5$$

$$\frac{x-3}{(x+2)(x+1)^2} = \frac{-5}{x+2} + \frac{5}{x+1} + \frac{-4}{(x+1)^2}$$

23. Find the partial fraction decomposition:

$$\frac{x+4}{x^2(x^2+4)} = \frac{A}{x} + \frac{B}{x^2} + \frac{Cx+D}{x^2+4}\ \text{(Multiply both sides by } x^2(x^2+4).)$$

$$x + 4 = Ax(x^2+4) + B(x^2+4) + (Cx+D)x^2$$

Let $x = 0$:   then $0 + 4 = A(0)(0^2+4) + B(0^2+4) + (C0+D)(0)^2$

$$\rightarrow\ 4 = 4B\ \rightarrow\ B = 1$$

Let $x = 1$:   then $1 + 4 = A(1)(1^2+4) + B(1^2+4) + (C(1)+D)(1)^2$

$$\rightarrow\ 5 = 5A + 5B + C + D\ \rightarrow\ 5 = 5A + 5 + C + D$$

$$\rightarrow\ 5A + C + D = 0$$

Let $x = -1$: then

$$-1 + 4 = A(-1)((-1)^2 + 4) + B((-1)^2 + 4) + (C(-1) + D)(-1)^2$$

$$\rightarrow \; 3 = -5A + 5B - C + D \;\rightarrow\; 3 = -5A + 5 - C + D$$

$$\rightarrow \; -5A - C + D = -2$$

Let $x = 2$:   then $2 + 4 = A(2)(2^2 + 4) + B(2^2 + 4) + (C(2) + D)(2)^2$

$$\rightarrow \; 6 = 16A + 8B + 8C + 4D \;\rightarrow\; 6 = 16A + 8 + 8C + 4D$$

$$\rightarrow \; 16A + 8C + 4D = -2$$

Solve the system of equations:

$$\begin{array}{r} 5A + C + D = 0 \\ -5A - C + D = -2 \\ \hline 2D = -2 \end{array} \qquad 5A + C - 1 = 0$$

$$D = -1 \qquad\qquad C = 1 - 5A$$

$$16A + 8(1 - 5A) + 4(-1) = -2$$

$$16A + 8 - 40A - 4 = -2 \qquad\qquad C = 1 - 5\left(\frac{1}{4}\right)$$

$$-24A = -6$$

$$A = \frac{1}{4} \qquad\qquad C = 1 - \frac{5}{4} = -\frac{1}{4}$$

$$\frac{x + 4}{x^2(x^2 + 4)} = \frac{\left(\dfrac{1}{4}\right)}{x} + \frac{1}{x^2} + \frac{\left(-\dfrac{1}{4}x - 1\right)}{x^2 + 4}$$

25. Find the partial fraction decomposition:

$$\frac{x^2 + 2x + 3}{(x + 1)(x^2 + 2x + 4)} = \frac{A}{x + 1} + \frac{Bx + C}{x^2 + 2x + 4} \quad \text{(Multiply both sides by } (x + 1)(x^2 + 2x + 4).)$$

$$x^2 + 2x + 3 = A(x^2 + 2x + 4) + (Bx + C)(x + 1)$$

Let $x = -1$: then

$$(-1)^2 + 2(-1) + 3 = A((-1)^2 + 2(-1) + 4) + (B(-1) + C)(-1 + 1)$$

$$\rightarrow \; 2 = 3A \;\rightarrow\; A = \frac{2}{3}$$

Let $x = 0$:   then $0^2 + 2(0) + 3 = A(0^2 + 2(0) + 4) + (B(0) + C)(0 + 1)$

$$\rightarrow \; 3 = 4A + C \;\rightarrow\; 3 = 4\left(\frac{2}{3}\right) + C \;\rightarrow\; C = \frac{1}{3}$$

Let $x = 1$: then

$$1^2 + 2(1) + 3 = A(1^2 + 2(1) + 4) + (B(1) + C)(1 + 1)$$

$$\rightarrow \; 6 = 7A + 2B + 2C \;\rightarrow\; 6 = 7\left(\frac{2}{3}\right) + 2B + 2\left(\frac{1}{3}\right)$$

$$\rightarrow \; 2B = 6 - \frac{14}{3} - \frac{2}{3} \;\rightarrow\; 2B = \frac{2}{3} \;\rightarrow\; B = \frac{1}{3}$$

$$\frac{x^2 + 2x + 3}{(x + 1)(x^2 + 2x + 4)} = \frac{\left(\dfrac{2}{3}\right)}{x + 1} + \frac{\left(\dfrac{1}{3}x + \dfrac{1}{3}\right)}{x^2 + 2x + 4}$$

27. Find the partial fraction decomposition:

$$\frac{x}{(3x-2)(2x+1)} = \frac{A}{3x-2} + \frac{B}{2x+1} \quad \text{(Multiply both sides by } (3x-2)(2x+1).)$$

$$x = A(2x+1) + B(3x-2)$$

Let $x = -\frac{1}{2}$: then $-\frac{1}{2} = A\left(2\left(-\frac{1}{2}\right)+1\right) + B\left(3\left(-\frac{1}{2}\right)-2\right) \rightarrow -\frac{1}{2} = -\frac{7}{2}B \rightarrow B = \frac{1}{7}$

Let $x = \frac{2}{3}$: then $\frac{2}{3} = A\left(2\left(\frac{2}{3}\right)+1\right) + B\left(3\left(\frac{2}{3}\right)-2\right) \rightarrow \frac{2}{3} = \frac{7}{3}A \rightarrow A = \frac{2}{7}$

$$\frac{x}{(3x-2)(2x+1)} = \frac{\left(\frac{2}{7}\right)}{3x-2} + \frac{\left(\frac{1}{7}\right)}{2x+1}$$

29. Find the partial fraction decomposition:

$$\frac{x}{x^2+2x-3} = \frac{x}{(x+3)(x-1)} = \frac{A}{x+3} + \frac{B}{x-1}$$

$$\text{(Multiply both sides by } (x+3)(x-1).)$$

$$x = A(x-1) + B(x+3)$$

Let $x = 1$: then $1 = A(1-1) + B(1+3) \rightarrow 1 = 4B \rightarrow B = \frac{1}{4}$

Let $x = -3$: then $-3 = A(-3-1) + B(-3+3) \rightarrow -3 = -4A \rightarrow A = \frac{3}{4}$

$$\frac{x}{x^2+2x-3} = \frac{\left(\frac{3}{4}\right)}{x+3} + \frac{\left(\frac{1}{4}\right)}{x-1}$$

31. Find the partial fraction decomposition:

$$\frac{x^2+2x+3}{(x^2+4)^2} = \frac{Ax+B}{x^2+4} + \frac{Cx+D}{(x^2+4)^2}$$

$$\text{(Multiply both sides by } (x^2+4)^2.)$$

$$x^2+2x+3 = (Ax+B)(x^2+4) + Cx+D$$
$$x^2+2x+3 = Ax^3 + Bx^2 + 4Ax + 4B + Cx + D$$
$$x^2+2x+3 = Ax^3 + Bx^2 + (4A+C)x + 4B+D$$
$$A = 0$$
$$B = 1$$
$$4A+C = 2 \rightarrow 4(0)+C = 2 \rightarrow C = 2$$
$$4B+D = 3 \rightarrow 4(1)+D = 3 \rightarrow D = -1$$
$$\frac{x^2+2x+3}{(x^2+4)^2} = \frac{1}{x^2+4} + \frac{2x-1}{(x^2+4)^2}$$

33. Find the partial fraction decomposition:

$$\frac{7x+3}{x^3-2x^2-3x}=\frac{7x+3}{x(x-3)(x+1)}=\frac{A}{x}+\frac{B}{x-3}+\frac{C}{x+1}$$

(Multiply both sides by $x(x-3)(x+1)$.)

$$7x+3=A(x-3)(x+1)+Bx(x+1)+Cx(x-3)$$

Let $x=0$:   then $7(0)+3=A(0-3)(0+1)+B(0)(0+1)+C(0)(0-3)$

$$\rightarrow\ 3=-3A\ \rightarrow\ A=-1$$

Let $x=3$:   then $7(3)+3=A(3-3)(3+1)+B(3)(3+1)+C(3)(3-3)$

$$\rightarrow\ 24=12B\ \rightarrow\ B=2$$

Let $x=-1$:   then

$$7(-1)+3=A(-1-3)(-1+1)+B(-1)(-1+1)+C(-1)(-1-3)$$

$$\rightarrow\ -4=4C\ \rightarrow\ C=-1$$

$$\frac{7x+3}{x^3-2x^2-3x}=\frac{-1}{x}+\frac{2}{x-3}+\frac{-1}{x+1}$$

35. Perform synthetic division to find a factor:

$$2\overline{)1\quad-4\quad\ \ 5\quad-2}$$
$$\underline{\qquad\quad 2\quad-4\quad\ \ 2}$$
$$1\quad-2\quad\ \ 1\quad\ \ 0$$

$$x^3-4x^2+5x-2=(x-2)(x^2-2x+1)=(x-2)(x-1)^2$$

Find the partial fraction decomposition:

$$\frac{x^2}{x^3-4x^2+5x-2}=\frac{x^2}{(x-2)(x-1)^2}=\frac{A}{x-2}+\frac{B}{x-1}+\frac{C}{(x-1)^2}$$

(Multiply both sides by $(x-2)(x-1)^2$.)

$$x^2=A(x-1)^2+B(x-2)(x-1)+C(x-2)$$

Let $x=2$:   then $2^2=A(2-1)^2+B(2-2)(2-1)+C(2-2)$

$$\rightarrow\ 4=A\ \rightarrow\ A=4$$

Let $x=1$:   then $1^2=A(1-1)^2+B(1-2)(1-1)+C(1-2)$

$$\rightarrow\ 1=-C\ \rightarrow\ C=-1$$

Let $x=0$:   then $0^2=A(0-1)^2+B(0-2)(0-1)+C(0-2)$

$$\rightarrow\ 0=A+2B-2C\ \rightarrow\ 0=4+2B-2(-1)$$

$$\rightarrow\ 2B=-6\ \rightarrow\ B=-3$$

$$\frac{x^2}{x^3-4x^2+5x-2}=\frac{4}{x-2}+\frac{-3}{x-1}+\frac{-1}{(x-1)^2}$$

37. Find the partial fraction decomposition:

$$\frac{x^3}{(x^2+16)^3}=\frac{Ax+B}{x^2+16}+\frac{Cx+D}{(x^2+16)^2}+\frac{Ex+F}{(x^2+16)^3}$$ (Multiply both sides by $(x^2+16)^3$.)

$$x^3=(Ax+B)(x^2+16)^2+(Cx+D)(x^2+16)+Ex+F$$

$$x^3=(Ax+B)(x^4+32x^2+256)+Cx^3+Dx^2+16Cx+16D+Ex+F$$

$$x^3=Ax^5+Bx^4+32Ax^3+32Bx^2+256Ax+256B+Cx^3+Dx^2$$

$$+16Cx+16D+Ex+F$$

$$x^3 = Ax^5 + Bx^4 + (32A + C)x^3 + (32B + D)x^2 + (256A + 16C + E)x$$
$$+ (256B + 16D + F)$$
$$A = 0$$
$$B = 0$$
$$32A + C = 1 \;\rightarrow\; 32(0) + C = 1 \;\rightarrow\; C = 1$$
$$32B + D = 0 \;\rightarrow\; 32(0) + D = 0 \;\rightarrow\; D = 0$$
$$256A + 16C + E = 0 \;\rightarrow\; 256(0) + 16(1) + E = 0 \;\rightarrow\; E = -16$$
$$256B + 16D + F = 0 \;\rightarrow\; 256(0) + 16(0) + F = 0 \;\rightarrow\; F = 0$$
$$\frac{x^3}{(x^2 + 16)^3} = \frac{x}{(x^2 + 16)^2} + \frac{-16x}{(x^2 + 16)^3}$$

39. Find the partial fraction decomposition:

$$\frac{4}{2x^2 - 5x - 3} = \frac{4}{(x-3)(2x+1)} = \frac{A}{x-3} + \frac{B}{2x+1}$$

(Multiply both sides by $(x - 3)(2x + 1)$.)

$$4 = A(2x + 1) + B(x - 3)$$

Let $x = -\dfrac{1}{2}$: then $4 = A\left(2\left(-\dfrac{1}{2}\right) + 1\right) + B\left(-\dfrac{1}{2} - 3\right)$

$$\rightarrow \; 4 = -\frac{7}{2}B \;\rightarrow\; B = -\frac{8}{7}$$

Let $x = 3$: then $4 = A(2(3) + 1) + B(3 - 3) \;\rightarrow\; 4 = 7A \;\rightarrow\; A = \dfrac{4}{7}$

$$\frac{4}{2x^2 - 5x - 3} = \frac{4}{(x-3)(2x+1)} = \frac{\left(\dfrac{4}{7}\right)}{x-3} + \frac{\left(-\dfrac{8}{7}\right)}{2x+1}$$

41. Find the partial fraction decomposition:

$$\frac{2x + 3}{x^4 - 9x^2} = \frac{2x + 3}{x^2(x-3)(x+3)} = \frac{A}{x} + \frac{B}{x^2} + \frac{C}{x-3} + \frac{D}{x+3}$$ (Multiply both sides by $x^2(x - 3)(x + 3)$.)

$$2x + 3 = Ax(x - 3)(x + 3) + B(x - 3)(x + 3) + Cx^2(x + 3) + Dx^2(x - 3)$$

Let $x = 0$: then

$$2 \cdot 0 + 3 = A \cdot 0(0 - 3)(0 + 3) + B(0 - 3)(0 + 3) + C \cdot 0^2(0 + 3) + D \cdot 0^2(0 - 3)$$

$$\rightarrow \; 3 = -9B \;\rightarrow\; B = -\frac{1}{3}$$

Let $x = 3$: then

$$2 \cdot 3 + 3 = A \cdot 3(3 - 3)(3 + 3) + B(3 - 3)(3 + 3) + C \cdot 3^2(3 + 3) + D \cdot 3^2(3 - 3)$$

$$\rightarrow \; 9 = 54C \;\rightarrow\; C = \frac{1}{6}$$

Let $x = -3$:   then

$$2(-3) + 3 = A(-3)(-3-3)(-3+3) + B(-3-3)(-3+3) + C(-3)^2(-3+3)$$
$$+ D(-3)^2(-3-3)$$

$$\rightarrow -3 = -54D \rightarrow D = \frac{1}{18}$$

Let $x = 1$:   then

$$2 \cdot 1 + 3 = A \cdot 1(1-3)(1+3) + B(1-3)(1+3) + C \cdot 1^2(1+3) + D \cdot 1^2(1-3)$$

$$\rightarrow 5 = -8A - 8B + 4C - 2D$$

$$\rightarrow 5 = -8A - 8\left(-\frac{1}{3}\right) + 4\left(\frac{1}{6}\right) - 2\left(\frac{1}{18}\right)$$

$$\rightarrow 5 = -8A + \frac{8}{3} + \frac{2}{3} - \frac{1}{9} \rightarrow -8A = \frac{16}{9} \rightarrow A = -\frac{2}{9}$$

$$\frac{2x+3}{x^4 - 9x^2} = \frac{2x+3}{x^2(x-3)(x+3)} = \frac{\left(-\frac{2}{9}\right)}{x} + \frac{\left(-\frac{1}{3}\right)}{x^2} + \frac{\left(\frac{1}{6}\right)}{x-3} + \frac{\left(\frac{1}{18}\right)}{x+3}$$

Systems of Equations and Inequalities

10.7 Systems of Nonlinear Equations

1. $\begin{cases} y = x^2 + 1 \\ y = x + 1 \end{cases}$

Graph: $y_1 = x^2 + 1;$ $y_2 = x + 1$

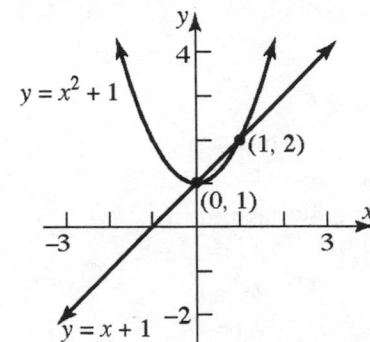

$(0, 1)$ and $(1, 2)$ are the intersection points.

Solve by substitution:
$$x^2 + 1 = x + 1$$
$$x^2 - x = 0$$
$$x(x - 1) = 0$$
$$x = 0 \ \text{ or } \ x = 1$$
$$y = 1 \qquad y = 2$$
Solutions: $(0, 1)$ and $(1, 2)$

3. $\begin{cases} y = \sqrt{36 - x^2} \\ y = 8 - x \end{cases}$

Graph: $y_1 = \sqrt{36 - x^2};$ $y_2 = 8 - x$

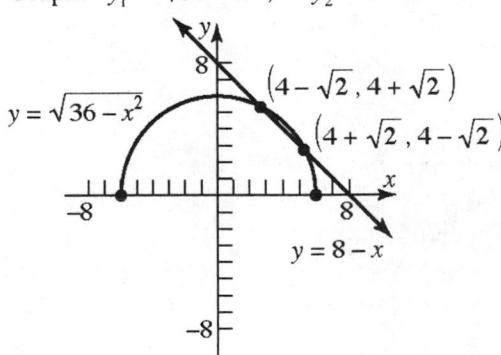

$(2.59, 5.41)$ and $(5.41, 2.59)$ are the intersection
points.

Solve by substitution:
$$\sqrt{36 - x^2} = 8 - x$$
$$36 - x^2 = 64 - 16x + x^2$$
$$2x^2 - 16x + 28 = 0$$
$$x^2 - 8x + 14 = 0$$
$$x = \frac{8 \pm \sqrt{64 - 56}}{2}$$
$$x = \frac{8 \pm 2\sqrt{2}}{2}$$
$$x = 4 \pm \sqrt{2}$$
If $x = 4 + \sqrt{2},\ y = 8 - \left(4 + \sqrt{2}\right) = 4 - \sqrt{2}$

If $x = 4 - \sqrt{2},\ y = 8 - \left(4 - \sqrt{2}\right) = 4 + \sqrt{2}$

Solutions:
$$\left(4 + \sqrt{2}, 4 - \sqrt{2}\right) \text{ and } \left(4 - \sqrt{2}, 4 + \sqrt{2}\right)$$

5. $\begin{cases} y = \sqrt{x} \\ y = 2 - x \end{cases}$

Graph: $y_1 = \sqrt{x}$; $y_2 = 2 - x$

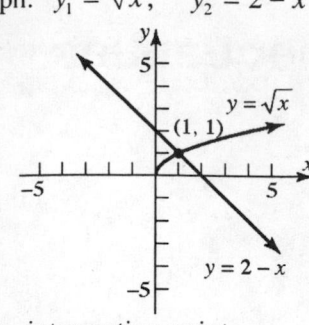

(1, 1) is the intersection point.

Solve by substitution:
$$\sqrt{x} = 2 - x$$
$$x = 4 - 4x + x^2$$
$$x^2 - 5x + 4 = 0$$
$$(x - 4)(x - 1) = 0$$
$$x = 4 \quad \text{or} \quad x = 1$$
$$y = -2 \quad \text{or} \quad y = 1$$
Eliminate $(4, -2)$; it does not check.
Solution: $(1, 1)$

7. $\begin{cases} x = 2y \\ x = y^2 - 2y \end{cases}$

Solve each equation for y in order to enter it into the graphing utility:
$$y^2 - 2y + 1 = x + 1 \rightarrow (y - 1)^2 = x + 1 \rightarrow y - 1 = \pm\sqrt{x + 1} \rightarrow y = 1 \pm \sqrt{x + 1}$$

Graph: $y_1 = \dfrac{x}{2}$; $y_2 = 1 + \sqrt{x + 1}$; $y_3 = 1 - \sqrt{x + 1}$

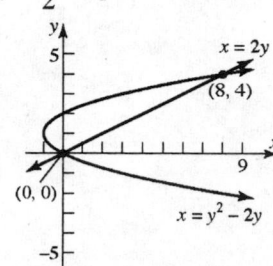

Solve by substitution:
$$2y = y^2 - 2y$$
$$y^2 - 4y = 0$$
$$y(y - 4) = 0$$
$$y = 0 \quad \text{or} \quad y = 4$$
$$x = 0 \quad \text{or} \quad x = 8$$
Solutions: $(0, 0)$ and $(8, 4)$

$(0, 0)$ and $(8, 4)$ are the intersection points.

9. $\begin{cases} x^2 + y^2 = 4 \\ x^2 + 2x + y^2 = 0 \end{cases}$

Graph: $y_1 = \sqrt{4 - x^2}$; $y_2 = -\sqrt{4 - x^2}$; $y_3 = \sqrt{-x^2 - 2x}$; $y_4 = -\sqrt{-x^2 - 2x}$

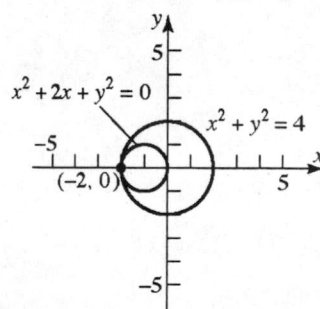

Substitute 4 for $x^2 + y^2$ in the second equation:
$$2x + 4 = 0$$
$$2x = -4$$
$$x = -2$$
$$y = \sqrt{4 - (-2)^2} = 0$$
Solution: $(-2, 0)$

$(-2, 0)$ is the intersection point.   Note: This intersection point is impossible to find on your graphing utility unless you have just the right window and make an excellent guess.

632

11. $\begin{cases} y = 3x - 5 \\ x^2 + y^2 = 5 \end{cases}$

Graph: $y_1 = 3x - 5$; $y_2 = \sqrt{5 - x^2}$;
$$y_3 = -\sqrt{5 - x^2}$$

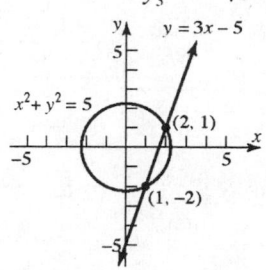

$(1, -2)$ and $(2, 1)$ are the intersection points.

Solve by substitution:
$$x^2 + (3x - 5)^2 = 5$$
$$x^2 + 9x^2 - 30x + 25 = 5$$
$$10x^2 - 30x + 20 = 0$$
$$x^2 - 3x + 2 = 0$$
$$(x - 1)(x - 2) = 0$$
$$x = 1 \qquad \text{or } x = 2$$
$$y = 3(1) - 5 \qquad y = 3(2) - 5$$
$$y = -2 \qquad y = 1$$

Solutions: $(1, -2)$ and $(2, 1)$

13. $\begin{cases} x^2 + y^2 = 4 \\ y^2 - x = 4 \end{cases}$

Graph: $y_1 = \sqrt{4 - x^2}$; $y_2 = -\sqrt{4 - x^2}$; $y_3 = \sqrt{x + 4}$; $y_4 = -\sqrt{x + 4}$

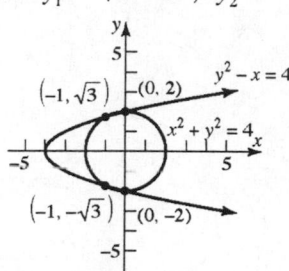

$(-1, 1.73)$, $(-1, -1.73)$, $(0, 2)$, and $(0, -2)$ are the intersection points.

Substitute $x + 4$ for y^2 in the first equation:
$$x^2 + x + 4 = 4$$
$$x^2 + x = 0$$
$$x(x + 1) = 0$$
$$x = 0 \qquad \text{or } x = -1$$
$$y^2 = 4 \qquad y^2 = 3$$
$$y = \pm 2 \qquad y^2 = \pm\sqrt{3}$$

Solutions:
$$(0, -2), (0, 2), \left(-1, \sqrt{3}\right), \left(-1, -\sqrt{3}\right)$$

15. $\begin{cases} xy = 4 \\ x^2 + y^2 = 8 \end{cases}$

Graph: $y_1 = \dfrac{4}{x}$; $y_2 = \sqrt{8 - x^2}$;
$$y_3 = -\sqrt{8 - x^2}$$

$(-2, -2)$ and $(2, 2)$ are the intersection points.

Solve by substitution:
$$x^2 + \left(\frac{4}{x}\right)^2 = 8$$
$$x^2 + \frac{16}{x^2} = 8$$
$$x^4 + 16 = 8x^2$$
$$x^4 - 8x^2 + 16 = 0$$
$$\left(x^2 - 4\right)^2 = 0$$
$$x^2 - 4 = 0$$
$$x^2 = 4$$
$$x = 2 \text{ or } x = -2$$
$$y = 2 \qquad y = -2$$

Solutions: $(-2, -2)$ and $(2, 2)$

17. $\begin{cases} x^2 + y^2 = 4 \\ \quad\ y = x^2 - 9 \end{cases}$

　　　Graph: $y_1 = x^2 - 9$; $y_2 = \sqrt{4 - x^2}$;

　　　　　　　　$y_3 = -\sqrt{4 - x^2}$

No solution; Inconsistent.

Solve by substitution:
$$x^2 + (x^2 - 9)^2 = 4$$
$$x^2 + x^4 - 18x^2 + 81 = 4$$
$$x^4 - 17x^2 + 77 = 0$$
$$x^2 = \frac{17 \pm \sqrt{289 - 4(77)}}{2}$$
$$x^2 = \frac{17 \pm \sqrt{-19}}{2}$$

There are no real solutions to this expression; Inconsistent.

19. $\begin{cases} y = x^2 - 4 \\ y = 6x - 13 \end{cases}$

　　　Graph: $y_1 = x^2 - 4$; $\quad y_2 = 6x - 13$

(3,5) is the intersection point.

Solve by substitution:
$$x^2 - 4 = 6x - 13$$
$$x^2 - 6x + 9 = 0$$
$$(x - 3)^2 = 0$$
$$x - 3 = 0$$
$$x = 3$$
$$y = 6(3) - 13 = 5$$

Solutions: (3,5)

21.　Solve the second equation for y, substitute into the first equation and solve:

$\begin{cases} 2x^2 + y^2 = 18 \\ \quad\ xy = 4 \ \rightarrow\ y = \dfrac{4}{x} \end{cases}$

$$2x^2 + \left(\frac{4}{x}\right)^2 = 18 \Rightarrow 2x^2 + \frac{16}{x^2} = 18$$
$$2x^4 + 16 = 18x^2$$
$$2x^4 - 18x^2 + 16 = 0$$
$$x^4 - 9x^2 + 8 = 0$$
$$\left(x^2 - 8\right)\left(x^2 - 1\right) = 0$$
$$x^2 = 8 \ \Rightarrow x = \pm\sqrt{8} = \pm 2\sqrt{2}$$
$$\text{or}\ \ x^2 = 1 \Rightarrow x = \pm 1$$

If $x = 2\sqrt{2}$: $y = \dfrac{4}{2\sqrt{2}} = \sqrt{2}$

If $x = -2\sqrt{2}$: $y = \dfrac{4}{-2\sqrt{2}} = -\sqrt{2}$

If $x = 1$: $y = \dfrac{4}{1} = 4$

If $x = -1$: $y = \dfrac{4}{-1} = -4$

Solutions: $\left(2\sqrt{2}, \sqrt{2}\right), \left(-2\sqrt{2}, -\sqrt{2}\right), (1, 4), (-1, -4)$

23. Substitute the first equation into the second equation and solve:
$$\begin{cases} y = 2x + 1 \\ 2x^2 + y^2 = 1 \end{cases}$$

$$2x^2 + (2x + 1)^2 = 1 \rightarrow 2x^2 + 4x^2 + 4x + 1 = 1 \rightarrow 6x^2 + 4x = 0 \rightarrow 2x(3x + 2) = 0$$

$$2x = 0 \rightarrow x = 0$$

$$\text{or} \quad 3x + 2 = 0 \rightarrow x = -\frac{2}{3}$$

If $x = 0$: $y = 2(0) + 1 = 1$

If $x = -\dfrac{2}{3}$: $y = 2\left(-\dfrac{2}{3}\right) + 1 = -\dfrac{4}{3} + 1 = -\dfrac{1}{3}$

Solutions: $(0, 1), \left(-\dfrac{2}{3}, -\dfrac{1}{3}\right)$

25. Solve the first equation for y, substitute into the second equation and solve:
$$\begin{cases} x + y + 1 = 0 \quad \rightarrow \quad y = -x - 1 \\ x^2 + y^2 + 6y - x = -5 \end{cases}$$

$$x^2 + (-x - 1)^2 + 6(-x - 1) - x = -5$$

$$x^2 + x^2 + 2x + 1 - 6x - 6 - x = -5$$

$$2x^2 - 5x = 0$$

$$x(2x - 5) = 0$$

$$x = 0 \quad \text{or} \quad x = \frac{5}{2}$$

If $x = 0$: $y = -(0) - 1 = -1$

If $x = \dfrac{5}{2}$: $y = -\dfrac{5}{2} - 1 = -\dfrac{7}{2}$

Solutions: $(0, -1), \left(\dfrac{5}{2}, -\dfrac{7}{2}\right)$

27. Solve the second equation for y, substitute into the first equation and solve:

$$\begin{cases} 4x^2 - 3xy + 9y^2 = 15 \\ \qquad 2x + 3y = 5 \ \rightarrow \ y = -\frac{2}{3}x + \frac{5}{3} \end{cases}$$

$$4x^2 - 3x\left(-\frac{2}{3}x + \frac{5}{3}\right) + 9\left(-\frac{2}{3}x + \frac{5}{3}\right)^2 = 15 \rightarrow 4x^2 + 2x^2 - 5x + 4x^2 - 20x + 25 = 15$$

$$10x^2 - 25x + 10 = 0 \rightarrow 2x^2 - 5x + 2 = 0$$

$$(2x-1)(x-2) = 0 \rightarrow x = \frac{1}{2} \ \text{ or } \ x = 2$$

If $x = \frac{1}{2}$: $y = -\frac{2}{3}\left(\frac{1}{2}\right) + \frac{5}{3} = \frac{4}{3}$

If $x = 2$: $y = -\frac{2}{3}(2) + \frac{5}{3} = \frac{1}{3}$

Solutions: $\left(\frac{1}{2}, \frac{4}{3}\right), \left(2, \frac{1}{3}\right)$

29. Multiply each side of the second equation by 4 and add the equations to eliminate y:

$$\begin{cases} \ x^2 - 4y^2 = -7 \ \longrightarrow \quad x^2 - 4y^2 = -7 \\ 3x^2 + \ y^2 = 31 \ \xrightarrow{\ 4\ } \ \underline{12x^2 + 4y^2 = 124} \end{cases}$$

$$13x^2 \qquad = 117 \rightarrow x^2 = 9 \rightarrow x = \pm 3$$

If $x = 3$: $3(3)^2 + y^2 = 31 \ \rightarrow \ y^2 = 4 \ \rightarrow \ y = \pm 2$

If $x = -3$: $3(-3)^2 + y^2 = 31 \ \rightarrow \ y^2 = 4 \ \rightarrow \ y = \pm 2$

Solutions: $(3, 2), (3, -2), (-3, 2), (-3, -2)$

31. Multiply each side of the first equation by 5 and each side of the second equation by 3 to eliminate y:

$$\begin{cases} 7x^2 - 3y^2 + 5 = 0 \rightarrow 7x^2 - 3y^2 = -5 \ \xrightarrow{\ 5\ } \ 35x^2 - 15y^2 = -25 \\ 3x^2 + 5y^2 = 12 \rightarrow \quad 3x^2 + 5y^2 = 12 \ \xrightarrow{\ 3\ } \ \underline{9x^2 + 15y^2 = \ \ 36} \end{cases}$$

$$44x^2 \qquad = \ 11 \rightarrow x^2 = \frac{1}{4} \rightarrow x = \pm\frac{1}{2}$$

If $x = \frac{1}{2}$: $3\left(\frac{1}{2}\right)^2 + 5y^2 = 12 \ \rightarrow \ 5y^2 = \frac{45}{4} \ \rightarrow \ y^2 = \frac{9}{4} \ \rightarrow \ y = \pm\frac{3}{2}$

If $x = -\frac{1}{2}$: $3\left(-\frac{1}{2}\right)^2 + 5y^2 = 12 \ \rightarrow \ 5y^2 = \frac{45}{4} \ \rightarrow \ y^2 = \frac{9}{4} \ \rightarrow \ y = \pm\frac{3}{2}$

Solutions: $\left(\frac{1}{2}, \frac{3}{2}\right), \left(\frac{1}{2}, -\frac{3}{2}\right), \left(-\frac{1}{2}, \frac{3}{2}\right), \left(-\frac{1}{2}, -\frac{3}{2}\right)$

33. Multiply each side of the second equation by 2 and add to eliminate xy:

$$\begin{cases} x^2 + 2xy = 10 & \longrightarrow & x^2 + 2xy = 10 \\ 3x^2 - xy = 2 & \xrightarrow{2} & 6x^2 - 2xy = 4 \end{cases}$$

$$7x^2 \qquad = 14 \rightarrow x^2 = 2 \rightarrow x = \pm\sqrt{2}$$

If $x = \sqrt{2}$: $3\left(\sqrt{2}\right)^2 - \sqrt{2} \cdot y = 2 \rightarrow -\sqrt{2} \cdot y = -4 \rightarrow y = \dfrac{4}{\sqrt{2}} \rightarrow y = 2\sqrt{2}$

If $x = -\sqrt{2}$: $3\left(-\sqrt{2}\right)^2 - \left(-\sqrt{2}\right)y = 2 \rightarrow \sqrt{2} \cdot y = -4 \rightarrow y = \dfrac{-4}{\sqrt{2}} \rightarrow y = -2\sqrt{2}$

Solutions: $\left(\sqrt{2}, 2\sqrt{2}\right), \left(-\sqrt{2}, -2\sqrt{2}\right)$

35. Multiply each side of the first equation by 2 and add the equations to eliminate y:

$$\begin{cases} 2x^2 + y^2 = 2 \rightarrow 2x^2 + y^2 = 2 & \xrightarrow{2} & 4x^2 + 2y^2 = 4 \\ x^2 - 2y^2 + 8 = 0 \rightarrow x^2 - 2y^2 = -8 & \longrightarrow & x^2 - 2y^2 = -8 \end{cases}$$

$$5x^2 \qquad = -4 \rightarrow x^2 = -\dfrac{4}{5}$$

No solution. The system is inconsistent.

37. Multiply each side of the second equation by 2 and add the equations to eliminate y:

$$\begin{cases} x^2 + 2y^2 = 16 & \longrightarrow & x^2 + 2y^2 = 16 \\ 4x^2 - y^2 = 24 & \xrightarrow{2} & 8x^2 - 2y^2 = 48 \end{cases}$$

$$9x^2 \qquad = 64 \rightarrow x^2 = \dfrac{64}{9} \rightarrow x = \pm\dfrac{8}{3}$$

If $x = \dfrac{8}{3}$: $\left(\dfrac{8}{3}\right)^2 + 2y^2 = 16 \rightarrow 2y^2 = \dfrac{80}{9} \rightarrow y^2 = \dfrac{40}{9} \rightarrow y = \pm\dfrac{2\sqrt{10}}{3}$

If $x = -\dfrac{8}{3}$: $\left(-\dfrac{8}{3}\right)^2 + 2y^2 = 16 \rightarrow 2y^2 = \dfrac{80}{9} \rightarrow y^2 = \dfrac{40}{9} \rightarrow y = \pm\dfrac{2\sqrt{10}}{3}$

Solutions: $\left(\dfrac{8}{3}, \dfrac{2\sqrt{10}}{3}\right), \left(\dfrac{8}{3}, \dfrac{-2\sqrt{10}}{3}\right), \left(-\dfrac{8}{3}, \dfrac{2\sqrt{10}}{3}\right), \left(-\dfrac{8}{3}, \dfrac{-2\sqrt{10}}{3}\right)$

39. Multiply each side of the second equation by 2 and add the equations to eliminate y:

$$\begin{cases} \dfrac{5}{x^2} - \dfrac{2}{y^2} + 3 = 0 \rightarrow \dfrac{5}{x^2} - \dfrac{2}{y^2} = -3 & \longrightarrow & \dfrac{5}{x^2} - \dfrac{2}{y^2} = -3 \\ \dfrac{3}{x^2} + \dfrac{1}{y^2} = 7 \rightarrow \dfrac{3}{x^2} + \dfrac{1}{y^2} = 7 & \xrightarrow{2} & \dfrac{6}{x^2} + \dfrac{2}{y^2} = 14 \end{cases}$$

$$\dfrac{11}{x^2} \qquad = 11$$

$$11 = 11x^2$$

$$x^2 = 1 \rightarrow x = \pm 1$$

If $x = 1$: $\dfrac{3}{(1)^2} + \dfrac{1}{y^2} = 7 \;\rightarrow\; \dfrac{1}{y^2} = 4 \;\rightarrow\; y^2 = \dfrac{1}{4} \;\rightarrow\; y = \pm\dfrac{1}{2}$

If $x = -1$: $\dfrac{3}{(-1)^2} + \dfrac{1}{y^2} = 7 \;\rightarrow\; \dfrac{1}{y^2} = 4 \;\rightarrow\; y^2 = \dfrac{1}{4} \;\rightarrow\; y = \pm\dfrac{1}{2}$

Solutions: $\left(1, \dfrac{1}{2}\right), \left(1, -\dfrac{1}{2}\right), \left(-1, \dfrac{1}{2}\right), \left(-1, -\dfrac{1}{2}\right)$

41. Multiply each side of the first equation by –2 and add the equations to eliminate x:

$$\begin{cases} \dfrac{1}{x^4} + \dfrac{6}{y^4} = 6 & \xrightarrow{\;-2\;} & \dfrac{-2}{x^4} - \dfrac{12}{y^4} = -12 \\[3mm] \dfrac{2}{x^4} - \dfrac{2}{y^4} = 19 & \longrightarrow & \dfrac{2}{x^4} - \dfrac{2}{y^4} = 19 \end{cases}$$

$$\dfrac{-14}{y^4} = 7 \rightarrow -14 = 7y^4 \rightarrow y^4 = -2$$

There are no real solutions. The system is inconsistent.

43. Factor the first equation, solve for x, substitute into the second equation and solve:

$$\begin{cases} x^2 - 3xy + 2y^2 = 0 \;\rightarrow\; (x - 2y)(x - y) = 0 \;\rightarrow\; x = 2y \text{ or } x = y \\ x^2 + xy = 6 \end{cases}$$

Substitute $x = 2y$ and solve: Substitute $x = y$ and solve:

$$\begin{aligned} x^2 + xy &= 6 & x^2 + xy &= 6 \\ (2y)^2 + (2y)y &= 6 & y^2 + y \cdot y &= 6 \\ 4y^2 + 2y^2 &= 6 \rightarrow 6y^2 = 6 & y^2 + y^2 &= 6 \rightarrow 2y^2 = 6 \\ y^2 &= 1 \rightarrow y = \pm 1 & y^2 &= 3 \rightarrow y = \pm\sqrt{3} \end{aligned}$$

If $y = 1$: $x = 2 \cdot 1 = 2$ If $y = \sqrt{3}$: $x = \sqrt{3}$

If $y = -1$: $x = 2(-1) = -2$ If $y = -\sqrt{3}$: $x = -\sqrt{3}$

Solutions: $(2, 1), (-2, -1), \left(\sqrt{3}, \sqrt{3}\right), \left(-\sqrt{3}, -\sqrt{3}\right)$

45. Multiply each side of the second equation by –y and add the equations to eliminate y:

$$\begin{cases} y^2 + y + x^2 - x - 2 = 0 & \longrightarrow & y^2 + y + x^2 - x - 2 = 0 \\[2mm] y + 1 + \dfrac{x-2}{y} = 0 & \xrightarrow{\;-y\;} & -y^2 - y \quad\;\; - x + 2 = 0 \end{cases}$$

$$x^2 - 2x \quad = 0 \rightarrow x(x - 2) = 0$$

$$x = 0 \text{ or } x = 2$$

If $x = 0$: $y^2 + y + 0^2 - 0 - 2 = 0 \;\rightarrow\; y^2 + y - 2 = 0 \;\rightarrow\; (y + 2)(y - 1) = 0$

$\rightarrow y = -2$ or $y = 1$

If $x = 2$: $y^2 + y + 2^2 - 2 - 2 = 0 \;\rightarrow\; y^2 + y = 0 \;\rightarrow\; y(y + 1) = 0$

$\rightarrow y = 0$ or $y = -1$ Note: $y \neq 0$ because of division by zero.

Solutions: $(0, -2), (0, 1), (2, -1)$

47. Rewrite each equation in exponential form:
$$\begin{cases} \log_x y = 3 \;\rightarrow\; y = x^3 \\ \log_x(4y) = 5 \;\rightarrow\; 4y = x^5 \end{cases}$$
Substitute the first equation into the second and solve:
$$4x^3 = x^5$$
$$x^5 - 4x^3 = 0 \rightarrow x^3(x^2 - 4) = 0 \rightarrow x^3 = 0 \;\text{ or }\; x^2 = 4 \rightarrow x = 0 \;\text{ or }\; x = \pm 2$$
The base of a logarithm must be positive, thus $x \neq 0$ and $x \neq -2$.
   If $x = 2$: $y = 2^3 = 8$
Solution: $(2, 8)$

49. Rewrite each equation in exponential form:
$$\begin{cases} \ln x = 4\ln y \;\rightarrow\; x = e^{4\ln y} = e^{\ln y^4} = y^4 \\ \log_3 x = 2 + 2\log_3 y \;\rightarrow\; x = 3^{2 + 2\log_3 y} = 3^2 \cdot 3^{2\log_3 y} = 3^2 \cdot 3^{\log_3 y^2} = 9y^2 \end{cases}$$

So we have the system
$$\begin{cases} x = y^4 \\ x = 9y^2 \end{cases}$$
Therefore we have
$$9y^2 = y^4 \rightarrow 9y^2 - y^4 = 0 \rightarrow y^2(9 - y^2) = 0$$
$$y^2(3 + y)(3 - y) = 0 \rightarrow y = 0 \;\text{ or }\; y = -3 \;\text{ or }\; y = 3$$
Since $\ln y$ is undefined when $y \leq 0$, the only solution is $y = 3$.
   If $y = 3$: $x = y^4 \rightarrow x = 3^4 = 81$
Solution: $(81, 3)$

51. Solve the first equation for x, substitute into the second equation and solve:
$$\begin{cases} x + 2y = 0 \;\rightarrow\; x = -2y \\ (x - 1)^2 + (y - 1)^2 = 5 \end{cases}$$
$$(-2y - 1)^2 + (y - 1)^2 = 5$$
$$4y^2 + 4y + 1 + y^2 - 2y + 1 = 5 \rightarrow 5y^2 + 2y - 3 = 0$$
$$(5y - 3)(y + 1) = 0$$
$$y = \frac{3}{5} = 0.6 \;\text{ or }\; y = -1$$
$$x = -\frac{6}{5} = -1.2 \;\text{ or }\; x = 2$$
The points of intersection are $(-1.2, 0.6), (2, -1)$.

53. Complete the square on the second equation, substitute into the first equation and solve:
$$\begin{cases} (x - 1)^2 + (y + 2)^2 = 4 \\ y^2 + 4y - x + 1 = 0 \;\rightarrow\; y^2 + 4y + 4 = x - 1 + 4 \rightarrow (y + 2)^2 = x + 3 \end{cases}$$

$$(x-1)^2 + x + 3 = 4$$
$$x^2 - 2x + 1 + x + 3 = 4$$
$$x^2 - x = 0$$
$$x(x-1) = 0$$
$$x = 0 \quad \text{or} \quad x = 1$$

If $x = 0$: $(y+2)^2 = 0 + 3$

$$y + 2 = \pm\sqrt{3} \rightarrow y = -2 \pm \sqrt{3}$$

If $x = 1$: $(y+2)^2 = 1 + 3$

$$y + 2 = \pm 2 \rightarrow y = -2 \pm 2$$

The points of intersection are:
$\left(0, -2-\sqrt{3}\right), \left(0, -2+\sqrt{3}\right), (1, -4), (1, 0).$

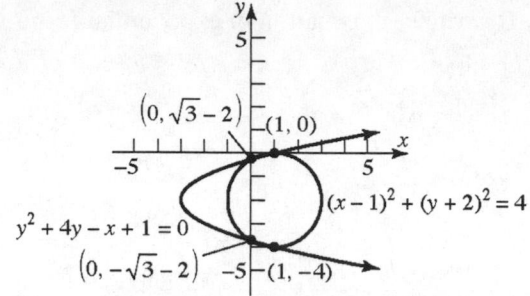

55. Solve the first equation for x, substitute into the second equation and solve:

$$\begin{cases} y = \dfrac{4}{x-3} \rightarrow x - 3 = \dfrac{4}{y} \rightarrow x = \dfrac{4}{y} + 3 \\ x^2 - 6x + y^2 + 1 = 0 \end{cases}$$

$$\left(\frac{4}{y} + 3\right)^2 - 6\left(\frac{4}{y} + 3\right) + y^2 + 1 = 0$$

$$\frac{16}{y^2} + \frac{24}{y} + 9 - \frac{24}{y} - 18 + y^2 + 1 = 0$$

$$\frac{16}{y^2} + y^2 - 8 = 0$$

$$16 + y^4 - 8y^2 = 0$$

$$y^4 - 8y^2 + 16 = 0$$

$$(y^2 - 4)^2 = 0$$

$$y^2 - 4 = 0$$

$$y^2 = 4$$

$$y = \pm 2$$

If $y = 2$: $x = \dfrac{4}{2} + 3 = 5$

If $y = -2$: $x = \dfrac{4}{-2} + 3 = 1$

The points of intersection are: $(1, -2), (5, 2)$.

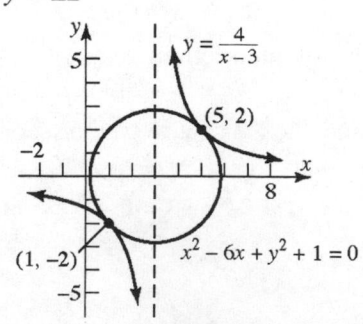

57. Graph: $y_1 = x \wedge (2/3)$; $y_2 = e \wedge (-x)$
Use INTERSECT to solve:

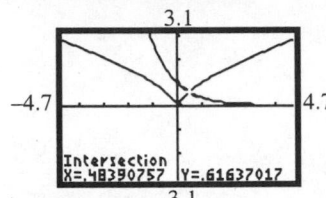

Solution: $(0.48, 0.62)$

59. Graph: $y_1 = \sqrt[3]{(2-x^2)}$; $y_2 = 4/x^3$
Use INTERSECT to solve:

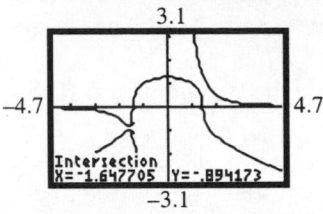

Solution: $(-1.65, -0.89)$

61. Graph:
$$y_1 = \sqrt[4]{(12-x^4)}; y_2 = -\sqrt[4]{(12-x^4)};$$
$$y_3 = \sqrt{2/x}; y_4 = -\sqrt{2/x}$$
Use INTERSECT to solve:

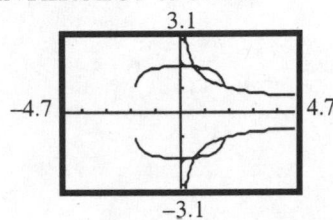

Solutions: $(0.58, 1.86)$, $(1.81, 1.05)$,
$(1.81, -1.05)$, $(0.58, -1.86)$

63. Graph: $y_1 = 2/x$; $y_2 = \ln x$
Use INTERSECT to solve:

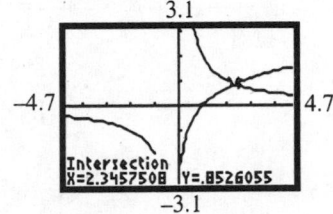

Solution: $(2.35, 0.85)$

65. Let x and y be the two numbers. The system of equations is:
$$\begin{cases} x - y = 2 \\ x^2 + y^2 = 10 \end{cases}$$
Solve the first equation for x, substitute into the second equation and solve:
$$(y+2)^2 + y^2 = 10 \rightarrow y^2 + 4y + 4 + y^2 = 10$$
$$2y^2 + 4y - 6 = 0 \rightarrow y^2 + 2y - 3 = 0 \rightarrow (y+3)(y-1) = 0 \rightarrow y = -3 \text{ or } y = 1$$
If $y = -3$: $x = -3 + 2 = -1$
If $y = 1$: $x = 1 + 2 = 3$
The two numbers are 1 and 3 or -1 and -3.

67. Let x and y be the two numbers. The system of equations is:
$$\begin{cases} xy = 4 \\ x^2 + y^2 = 8 \end{cases}$$
Solve the first equation for x, substitute into the second equation and solve:
$$\left(\frac{4}{y}\right)^2 + y^2 = 8 \rightarrow \frac{16}{y^2} + y^2 = 8 \rightarrow 16 + y^4 = 8y^2$$
$$y^4 - 8y^2 + 16 = 0 \rightarrow (y^2 - 4)^2 = 0 \rightarrow y^2 - 4 = 0 \rightarrow y^2 = 4 \rightarrow y = \pm 2$$

$$\text{If } y = 2: \qquad x = \frac{4}{2} = 2$$

$$\text{If } y = -2: \qquad x = \frac{4}{-2} = -2$$

The two numbers are 2 and 2 or –2 and –2.

69. Let x and y be the two numbers. The system of equations is:

$$\begin{cases} x - y = xy \\ \dfrac{1}{x} + \dfrac{1}{y} = 5 \end{cases}$$

Solve the first equation for x, substitute into the second equation and solve:

$$x - xy = y \rightarrow x(1 - y) = y \rightarrow x = \frac{y}{1 - y}$$

$$\frac{1}{\left(\dfrac{y}{1-y}\right)} + \frac{1}{y} = 5 \rightarrow \frac{1-y}{y} + \frac{1}{y} = 5 \rightarrow \frac{2-y}{y} = 5 \rightarrow 2 - y = 5y \rightarrow 6y = 2 \rightarrow y = \frac{1}{3}$$

$$\text{If } y = \frac{1}{3}: \qquad x = \frac{\left(\dfrac{1}{3}\right)}{\left(1 - \dfrac{1}{3}\right)} = \frac{\left(\dfrac{1}{3}\right)}{\left(\dfrac{2}{3}\right)} = \frac{1}{2} \quad \therefore \text{ The two numbers are } \frac{1}{2} \text{ and } \frac{1}{3}.$$

71. $$\begin{cases} \dfrac{a}{b} = \dfrac{2}{3} \\ a + b = 10 \end{cases}$$

Solve the second equation for a, substitute into the first equation and solve:

$$\frac{10 - b}{b} = \frac{2}{3} \rightarrow 3(10 - b) = 2b \Rightarrow 30 - 3b = 2b \Rightarrow 30 = 5b \Rightarrow b = 6 \Rightarrow a = 4$$

$$a + b = 10; \quad b - a = 2 \quad \therefore \text{ The ratio of } a + b \text{ to } b - a \text{ is } \frac{10}{2} = 5.$$

73. Let x = the width of the rectangle.
Let y = the length of the rectangle.

$$\begin{cases} 2x + 2y = 16 \\ xy = 15 \end{cases}$$

Solve the first equation for y, substitute into the second equation and solve:

$$2x + 2y = 16 \rightarrow 2y = 16 - 2x \rightarrow y = 8 - x$$

$$x(8 - x) = 15 \rightarrow 8x - x^2 = 15 \rightarrow x^2 - 8x + 15 = 0 \rightarrow (x - 5)(x - 3) = 0$$

$$x = 5 \text{ or } x = 3$$

$$y = 3 \qquad y = 5$$

The dimensions of the rectangle are 3 inches by 5 inches.

75. Let x = the radius of the first circle.
Let y = the radius of the second circle.

$$\begin{cases} 2\pi x + 2\pi y = 12\pi \\ \pi x^2 + \pi y^2 = 20\pi \end{cases}$$

Solve the first equation for y, substitute into the second equation and solve:

$$2\pi x + 2\pi y = 12\pi$$
$$x + y = 6$$
$$y = 6 - x$$

$$\pi x^2 + \pi y^2 = 20\pi$$
$$x^2 + y^2 = 20$$
$$x^2 + (6-x)^2 = 20$$
$$x^2 + 36 - 12x + x^2 = 20$$
$$2x^2 - 12x + 16 = 0$$
$$x^2 - 6x + 8 = 0$$
$$(x-4)(x-2) = 0$$
$$x = 4 \text{ or } x = 2$$
$$y = 2 \qquad y = 4$$

The radii of the circles are 2 centimeters and 4 centimeters.

77. The tortoise takes $9 + 3 = 12$ minutes or 0.2 hour longer to complete the race than the hare.
   Let r = the rate of the hare.
   Let t = the time for the hare to complete the race.
   Then $t + 0.2$ = the time for the tortoise and $r - 0.5$ = the rate for the tortoise.
   Since the length of the race is 21 meters, the distance equations are:

$$\begin{cases} rt = 21 \\ (r-0.5)(t+0.2) = 21 \end{cases}$$

   Solve the first equation for r, substitute into the second equation and solve:

$$\left(\frac{21}{t} - 0.5\right)(t + 0.2) = 21 \rightarrow 21 + \frac{4.2}{t} - 0.5t - 0.1 = 21$$

$$10t \cdot \left(21 + \frac{4.2}{t} - 0.5t - 0.1\right) = 10t \cdot (21)$$

$$210t + 42 - 5t^2 - t = 210t \rightarrow 5t^2 + t - 42 = 0 \rightarrow (5t - 14)(t + 3) = 0$$

$$t = \frac{14}{5} = 2.8 \text{ or } t = -3$$

   $t = -3$ makes no sense, since time cannot be negative.
   Solve for r: $r = \dfrac{21}{2.8} = 7.5$

   The average speed of the hare is 7.5 meters per hour, and the average speed for the tortoise is 7 meters per hour.

79. Let x = the width of the cardboard. Let y = the length of the cardboard.
   The width of the box will be $x - 4$, the length of the box will be $y - 4$, and the height is 2.
   The volume is $V = (x-4)(y-4)(2)$.
   Solve the system of equations:

$$\begin{cases} xy = 216 \\ 2(x-4)(y-4) = 224 \end{cases}$$

   Solve the first equation for y, substitute into the second equation and solve:

$$(2x - 8)\left(\frac{216}{x} - 4\right) = 224 \rightarrow 432 - 8x - \frac{1728}{x} + 32 = 224$$

$$432x - 8x^2 - 1728 + 32x = 224x \rightarrow -8x^2 + 240x - 1728 = 0$$

$$x^2 - 30x + 216 = 0 \rightarrow (x - 12)(x - 18) = 0$$

$$x = 12 \ \text{ or } \ x = 18$$

$$y = 18 \qquad y = 12$$

The cardboard should be 12 centimeters by 18 centimeters.

81. Find equations relating area and perimeter:

$$\begin{cases} x^2 + y^2 = 4500 \\ 3x + 3y + (x - y) = 300 \end{cases}$$

Solve the second equation for y, substitute into the first equation and solve:

$$\begin{array}{ll}
4x + 2y = 300 & x^2 + (150 - 2x)^2 = 4500 \\
2y = 300 - 4x & x^2 + 22500 - 600x + 4x^2 = 4500 \\
y = 150 - 2x & 5x^2 - 600x + 18000 = 0 \\
 & x^2 - 120x + 3600 = 0 \\
 & (x - 60)^2 = 0 \\
 & x - 60 = 0 \\
 & x = 60 \\
 & y = 150 - 2(60) = 30
\end{array}$$

The sides of the squares are 30 feet and 60 feet.

83. Solve the system for l and w:

$$\begin{cases} 2l + 2w = P \\ \quad lw = A \end{cases}$$

Solve the first equation for l, substitute into the second equation and solve:

$$2l = P - 2w \ \rightarrow \ l = \frac{P}{2} - w$$

$$\left(\frac{P}{2} - w\right)w = A \rightarrow \frac{P}{2}w - w^2 = A \rightarrow w^2 - \frac{P}{2}w + A = 0$$

$$w = \frac{\left(\dfrac{P}{2} \pm \sqrt{\dfrac{P^2}{4} - 4A}\right)}{2} = \frac{\left(\dfrac{P}{2} \pm \sqrt{\dfrac{P^2 - 16A}{4}}\right)}{2} = \frac{\left(\dfrac{P}{2} \pm \dfrac{\sqrt{P^2 - 16A}}{2}\right)}{2}$$

$$w = \frac{P \pm \sqrt{P^2 - 16A}}{4}$$

If $w = \dfrac{P + \sqrt{P^2 - 16A}}{4}$ then $l = \dfrac{P}{2} - \dfrac{P + \sqrt{P^2 - 16A}}{4} = \dfrac{P - \sqrt{P^2 - 16A}}{4}$

If $w = \dfrac{P - \sqrt{P^2 - 16A}}{4}$ then $l = \dfrac{P}{2} - \dfrac{P - \sqrt{P^2 - 16A}}{4} = \dfrac{P + \sqrt{P^2 - 16A}}{4}$

If it is required that length be greater than width, then the solution is:

$$w = \frac{P - \sqrt{P^2 - 16A}}{4} \ \text{ and } \ l = \frac{P + \sqrt{P^2 - 16A}}{4}$$

85. Solve the equation:
$$m^2 - 4(2m - 4) = 0 \rightarrow m^2 - 8m + 16 = 0 \rightarrow (m - 4)^2 = 0 \rightarrow m - 4 = 0 \rightarrow m = 4$$
Use the point-slope equation with slope 4 and the point (2, 4) to obtain the equation of the tangent line:
$$y - 4 = 4(x - 2) \rightarrow y - 4 = 4x - 8 \rightarrow y = 4x - 4$$

87. Solve the system:
$$\begin{cases} y = x^2 + 2 \\ y = mx + b \end{cases}$$
Solve the system by substitution:
$$x^2 + 2 = mx + b \rightarrow x^2 - mx + 2 - b = 0$$
Note that the tangent line passes through (1, 3).
Find the relation between m and b: $3 = m(1) + b \rightarrow b = 3 - m$
Substitute into the quadratic to eliminate b: $x^2 - mx + 2 - (3 - m) = 0 \rightarrow x^2 - mx + (m - 1) = 0$
Find when the discriminant is 0:
$$(-m)^2 - 4(1)(m - 1) = 0 \rightarrow m^2 - 4m + 4 = 0 \rightarrow (m - 2)^2 = 0$$
$$m - 2 = 0 \rightarrow m = 2 \ \rightarrow \ b = 3 - 2 = 1$$
The equation of the tangent line is $y = 2x + 1$.

89. Solve the system:
$$\begin{cases} 2x^2 + 3y^2 = 14 \\ \qquad y = mx + b \end{cases}$$
Solve the system by substitution:
$$2x^2 + 3(mx + b)^2 = 14 \rightarrow 2x^2 + 3m^2 x^2 + 6mbx + 3b^2 = 14$$
$$(3m^2 + 2)x^2 + 6mbx + 3b^2 - 14 = 0$$
Note that the tangent line passes through (1, 2). Find the relation between m and b:
$$2 = m(1) + b \rightarrow b = 2 - m$$
Substitute into the quadratic to eliminate b:
$$(3m^2 + 2)x^2 + 6m(2 - m)x + 3(2 - m)^2 - 14 = 0$$
$$(3m^2 + 2)x^2 + (12m - 6m^2)x + 12 - 12m + 3m^2 - 14 = 0$$
$$(3m^2 + 2)x^2 + (12m - 6m^2)x + (3m^2 - 12m - 2) = 0$$
Find when the discriminant is 0:
$$(12m - 6m^2)^2 - 4(3m^2 + 2)(3m^2 - 12m - 2) = 0$$
$$144m^2 - 144m^3 + 36m^4 - 4(9m^4 - 36m^3 - 24m - 4) = 0$$
$$144m^2 - 144m^3 + 36m^4 - 36m^4 + 144m^3 + 96m + 16 = 0$$
$$144m^2 + 96m + 16 = 0$$
$$9m^2 + 6m + 1 = 0$$
$$(3m + 1)^2 = 0$$
$$3m + 1 = 0$$
$$m = -\frac{1}{3} \qquad b = 2 - \left(-\frac{1}{3}\right) = \frac{7}{3}$$
The equation of the tangent line is $y = -\frac{1}{3}x + \frac{7}{3}$.

91.   Solve the system:
$$\begin{cases} x^2 - y^2 = 3 \\ \quad y = mx + b \end{cases}$$
Solve the system by substitution:
$$x^2 - (mx+b)^2 = 3 \rightarrow x^2 - m^2x^2 - 2mbx - b^2 = 3 \rightarrow (1-m^2)x^2 - 2mbx - b^2 - 3 = 0$$
Note that the tangent line passes through (2, 1). Find the relation between m and b:
$$1 = m(2) + b \rightarrow b = 1 - 2m$$
Substitute into the quadratic to eliminate b:
$$(1-m^2)x^2 - 2m(1-2m)x - (1-2m)^2 - 3 = 0$$
$$(1-m^2)x^2 + (-2m+4m^2)x - 1 + 4m - 4m^2 - 3 = 0$$
$$(1-m^2)x^2 + (-2m+4m^2)x + (-4m^2 + 4m - 4) = 0$$
Find when the discriminant is 0:
$$(-2m+4m^2)^2 - 4(1-m^2)(-4m^2+4m-4) = 0$$
$$4m^2 - 16m^3 + 16m^4 - 4(4m^4 - 4m^3 + 4m - 4) = 0$$
$$4m^2 - 16m^3 + 16m^4 - 16m^4 + 16m^3 - 16m + 16 = 0$$
$$4m^2 - 16m + 16 = 0 \rightarrow m^2 - 4m + 4 = 0$$
$$(m-2)^2 = 0 \rightarrow m - 2 = 0 \rightarrow m = 2 \;\rightarrow\; b = 1 - 2(2) = -3$$
The equation of the tangent line is $y = 2x - 3$.

93.   Solve for r_1 and r_2:
$$\begin{cases} r_1 + r_2 = -\dfrac{b}{a} \\[2mm] \quad r_1 r_2 = \dfrac{c}{a} \end{cases}$$
Substitute and solve:
$$r_1 = -r_2 - \frac{b}{a} \rightarrow \left(-r_2 - \frac{b}{a}\right)r_2 = \frac{c}{a} \rightarrow -r_2^2 - \frac{b}{a}r_2 - \frac{c}{a} = 0 \rightarrow ar_2^2 + br_2 + c = 0$$
$$r_2 = \frac{-b \pm \sqrt{b^2 - 4ac}}{2a}$$
$$r_1 = -r_2 - \frac{b}{a} = -\left(\frac{-b \pm \sqrt{b^2 - 4ac}}{2a}\right) - \frac{2b}{2a} = \frac{-b \mp \sqrt{b^2 - 4ac}}{2a}$$
The solutions are: $\dfrac{-b + \sqrt{b^2 - 4ac}}{2a}$ and $\dfrac{-b - \sqrt{b^2 - 4ac}}{2a}$.

95.   Since the area of the square piece of sheet metal is 100 square feet, the sheet's dimensions are 10 feet by 10 feet. Let x = the length of the cut.

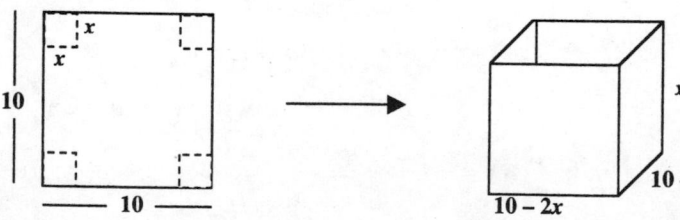

The dimensions of the box are $\text{length} = 10 - 2x;$ $\text{width} = 10 - 2x;$ $\text{height} = x$

Note that each of these expressions must be positive. So we must have

$$x > 0 \text{ and } 10 - 2x > 0 \Rightarrow x < 5, \text{ that is, } 0 < x < 5.$$

So the volume of the box is given by

$$V = (\text{length}) \cdot (\text{width}) \cdot (\text{height}) = (10 - 2x)(10 - 2x)(x) = (10 - 2x)^2 (x)$$

(a) In order to get a volume equal to 9 cubic feet, we solve $(10 - 2x)^2 (x) = 9$.

$$(10 - 2x)^2 (x) = 9 \Rightarrow (100 - 40x + 4x^2)x = 9 \Rightarrow 100x - 40x^2 + 4x^3 = 9$$

So we need to solve the equation $4x^3 - 40x^2 + 100x - 9 = 0$.

Graphing the function $y_1 = 4x^3 - 40x^2 + 100x - 9$ on a calculator yields the graph

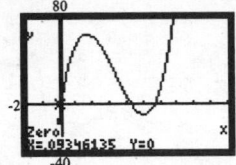 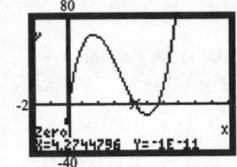 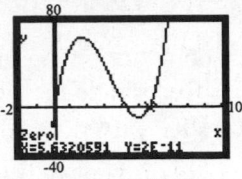

The graph indicates that there three real zeros on the interval $[0, 6]$.

Using the ZERO feature of a graphing calculator, we find that the three roots shown occur at $x \approx 0.09$, $x \approx 4.27$ and $x \approx 5.63$.

But we've already noted that we must have $0 < x < 5$, so the only practical values for the cut are $x \approx 0.09$ feet and $x \approx 4.27$ feet.

(b) If the sheet metal has dimensions k feet by k feet, then the volume equation becomes

$$V = (k - 2x)(k - 2x)(x) = (k - 2x)^2 (x) = 9$$

Solving for k we get the quadratic equation in the variable k.

$$xk^2 - 4x^2 k + 4x^3 - 9 = 0$$

Using the quadratic formula we get:

$$k = \frac{-(-4x^2) \pm \sqrt{(-4x^2)^2 - 4(x)(4x^3 - 9)}}{2x} = \frac{4x^2 \pm \sqrt{16x^4 - 16x^4 + 36x}}{2x}$$

$$= \frac{4x^2 \pm \sqrt{36x}}{2x} = \frac{4x^2 \pm 6\sqrt{x}}{2x}$$

Therefore, we get a real solution for k provided $x > 0$ and $4x^2 \pm 6\sqrt{x} \geq 0$.

$$4x^2 \pm 6\sqrt{x} \geq 0 \Rightarrow 4x^2 \geq 6\sqrt{x} \Rightarrow 16x^4 \geq 36x$$

$$16x^4 - 36x \geq 0 \Rightarrow 4x(4x^3 - 9) \geq 0$$

This last inequality holds provided $x \geq \sqrt[3]{\dfrac{9}{4}}$.

Systems of Equations and Inequalities

10.8 Systems of Inequalities

1. $x \geq 0$

(a) Graph the line $x = 0$. Use a solid line since the inequality uses $\geq$.
Choose a test point not on the line, such as $(2, 0)$.
Since $2 \geq 0$ is true, shade the side of the line containing $(2, 0)$.

(b) Use the VERTICAL function and the SHADE function in the calculator's DRAW menu:

Vertical 0

Shade$(-5,5,0,2)$

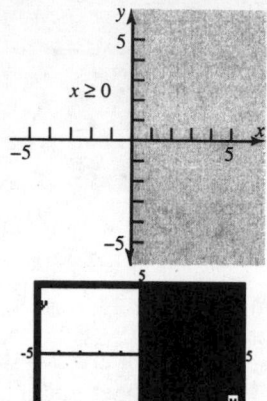

3. $x \geq 4$

(a) Graph the line $x = 4$. Use a solid line since the inequality uses $\geq$.
Choose a test point not on the line, such as $(5, 0)$.
Since $5 \geq 0$ is true, shade the side of the line containing $(5, 0)$.

(b) Use the VERTICAL function and the SHADE function in the calculator's DRAW menu:
Vertical 4

Shade$(-5,5,4,8)$.

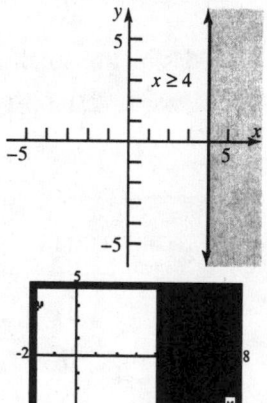

5. $x + y > 1$

(a) Graph the line $x + y = 1$. Use a dotted line since the inequality uses $>$.
Choose a test point not on the line, such as $(0, 0)$.
Since $0 + 0 > 1$ is false, shade the opposite side of the line from $(0, 0)$.

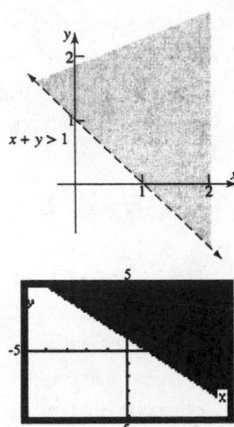

(b) Use the SHADE function in the calculator's DRAW menu: Shade$(-x + 1, 5)$.

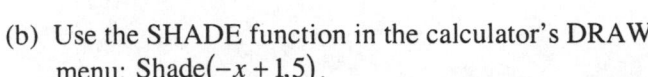

7. $2x + y \geq 6$

(a) Graph the line $2x + y = 6$. Use a solid line since the inequality uses $\geq$.
Choose a test point not on the line, such as $(0, 0)$.
Since $2(0) + 0 \geq 6$ is false, shade the opposite side of the line from $(0, 0)$.

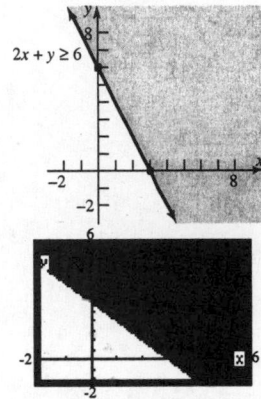

(b) Use the SHADE function in the calculator's DRAW menu: Shade$(-2x + 6, 6)$.

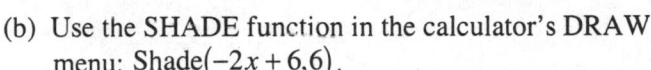

9. $x^2 + y^2 > 1$

Graph the circle $x^2 + y^2 > 1$. Use a dashed line since the inequality uses $>$.
Choose a test point not on the circle, such as $(0, 0)$.
Since $0^2 + 0^2 > 1$ is false, shade the opposite side of the circle from $(0, 0)$.

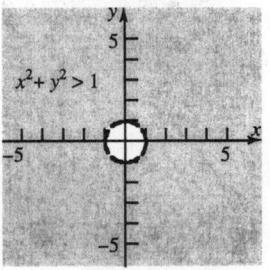

11. $y \leq x^2 - 1$

Graph the parabola $y = x^2 - 1$. Use a solid line since the inequality uses $\leq$.
Choose a test point not on the parabola, such as $(0, 0)$.
Since $0 \leq 0^2 - 1$ is false, shade the opposite side of the parabola from $(0, 0)$.

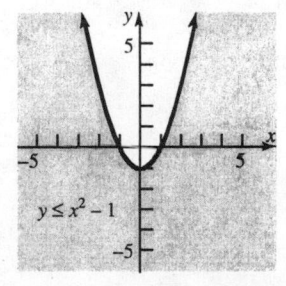

13. $y > |x| - 3$

Graph $y = |x| - 3$. Use a dashed line since the inequality uses >.

Choose a test point not on the graph, such as $(0, 0)$.

Since $0 > 0 - 3$ is true, shade the same side of the graph as $(0, 0)$.

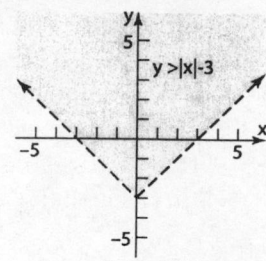

15. $xy \geq 4$

Graph the hyperbola $xy = 4$. Use a solid line since the inequality uses $\geq$.

Choose a test point not on the hyperbola, such as $(0, 0)$.

Since $0 \cdot 0 \geq 4$ is false, shade the opposite side of the hyperbola from $(0, 0)$.

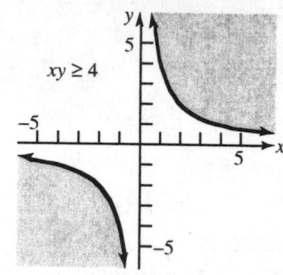

17. $\begin{cases} x + y \leq 2 \\ 2x + y \geq 4 \end{cases}$

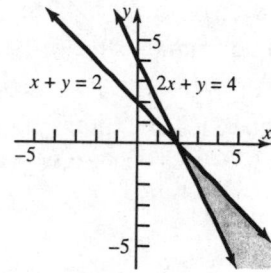

(a) Graph the line $x + y = 2$. Use a solid line since the inequality uses $\leq$.
     Choose a test point not on the line, such as $(0, 0)$. Since $0 + 0 \leq 2$ is true, shade the side of the line containing $(0, 0)$.

(b) Graph the line $2x + y = 4$. Use a solid line since the inequality uses $\geq$.
     Choose a test point not on the line, such as $(0, 0)$. Since $2(0) + 0 \geq 4$ is false, shade the opposite side of the line from $(0, 0)$.

(c) The overlapping region is the solution.

19. $\begin{cases} 2x - y \leq 4 \\ 3x + 2y \geq -6 \end{cases}$

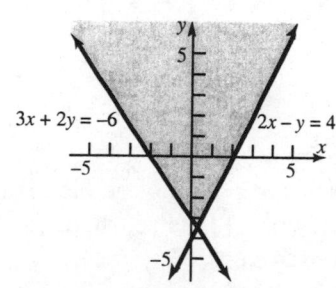

(a)   Graph the line $2x - y = 4$. Use a solid line since the inequality uses $\leq$.
Choose a test point not on the line, such as $(0, 0)$. Since $2(0) - 0 \leq 4$ is true, shade
the side of the line containing $(0, 0)$.

(b)   Graph the line $3x + 2y = -6$. Use a solid line since the inequality uses $\geq$.
Choose a test point not on the line, such as $(0, 0)$. Since $3(0) + 2(0) \geq -6$ is true,
shade the side of the line containing $(0, 0)$.

(c)   The overlapping region is the solution.

21. $\begin{cases} 2x - 3y \leq 0 \\ 3x + 2y \leq 6 \end{cases}$

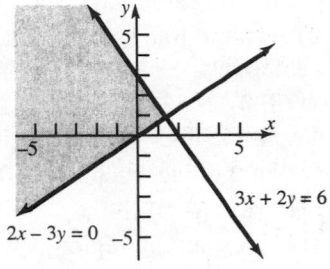

(a)   Graph the line $2x - 3y = 0$. Use a solid line since the inequality uses $\leq$.
Choose a test point not on the line, such as $(0, 3)$. Since $2(0) - 3(3) \leq 0$ is true,
shade the side of the line containing $(0, 3)$.

(b)   Graph the line $3x + 2y = 6$. Use a solid line since the inequality uses $\leq$.
Choose a test point not on the line, such as $(0, 0)$. Since $3(0) + 2(0) \leq 6$ is true,
shade the side of the line containing $(0, 0)$.

(c)   The overlapping region is the solution.

23. $\begin{cases} x^2 + y^2 \leq 9 \\ x + y \geq 3 \end{cases}$

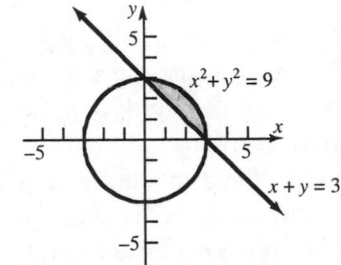

(a)   Graph the circle $x^2 + y^2 = 9$. Use a solid line since the inequality uses $\geq$. Choose a test
point not on the circle, such as $(0, 0)$. Since $0^2 + 0^2 \leq 9$ is true, shade the same side of the
circle as $(0, 0)$.

(b)   Graph the line $x + y = 3$. Use a solid line since the inequality uses $\geq$. Choose a test point
not on the line, such as $(0, 0)$. Since $0 + 0 \geq 3$ is false, shade the opposite side of the line
from $(0, 0)$.

(c)   The overlapping region is the solution.

25. $\begin{cases} y \geq x^2 - 4 \\ y \leq x - 2 \end{cases}$

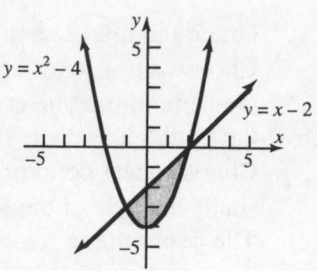

(a) Graph the parabola $y = x^2 - 4$.
Use a solid line since the inequality uses $\geq$. Choose a test point not on the parabola, such as (0, 0). Since $0 \geq 0^2 - 4$ is true, shade the same side of the parabola as (0, 0).

(b) Graph the line $y = x - 2$. Use a solid line since the inequality uses $\leq$. Choose a test point not on the line, such as (0, 0). Since $0 \leq 0 - 2$ is false, shade the opposite side of the line from (0, 0).

(c) The overlapping region is the solution.

27. $\begin{cases} xy \geq 4 \\ y \geq x^2 + 1 \end{cases}$

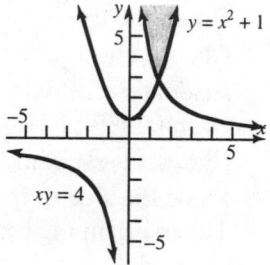

(a) Graph the hyperbola $xy = 4$.
Use a solid line since the inequality uses $\geq$. Choose a test point not on the parabola, such as (0, 0). Since $0 \cdot 0 \geq 4$ is false, shade the opposite side of the hyperbola from (0, 0).

(b) Graph the parabola $y = x^2 + 1$. Use a solid line since the inequality uses $\geq$. Choose a test point not on the parabola, such as (0, 0). Since $0 \geq 0^2 + 1$ is false, shade the opposite side of the parabola from (0, 0).

(c) The overlapping region is the solution.

29. $\begin{cases} x - 2y \leq 6 \\ 2x - 4y \geq 0 \end{cases}$

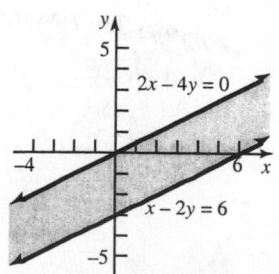

(a)   Graph the line $x - 2y = 6$. Use a solid line since the inequality uses $\leq$.
       Choose a test point not on the line, such as $(0, 0)$. Since $0 - 2(0) \leq 6$ is true, shade
       the side of the line containing $(0, 0)$.

(b)   Graph the line $2x - 4y = 0$. Use a solid line since the inequality uses $\geq$.
       Choose a test point not on the line, such as $(0, 2)$. Since $2(0) - 4(2) \geq 0$ is false,
       shade the opposite side of the line from $(0, 2)$.

(c)   The overlapping region is the solution.

31. $\begin{cases} 2x + y \geq -2 \\ 2x + y \geq 2 \end{cases}$

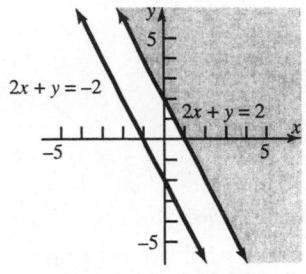

(a)   Graph the line $2x + y = -2$. Use a solid line since the inequality uses $\geq$.
       Choose a test point not on the line, such as $(0, 0)$. Since $2(0) + 0 \geq -2$ is true, shade
       the side of the line containing $(0, 0)$.

(b)   Graph the line $2x + y = 2$. Use a solid line since the inequality uses $\geq$.
       Choose a test point not on the line, such as $(0, 0)$. Since $2(0) + 0 \geq 2$ is false, shade
       the opposite side of the line from $(0, 0)$.

(c)   The overlapping region is the solution.

33. $\begin{cases} 2x + 3y \geq 6 \\ 2x + 3y \leq 0 \end{cases}$

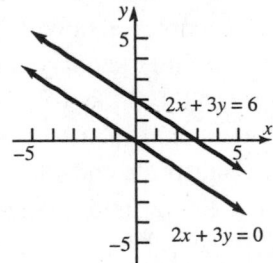

(a)   Graph the line $2x + 3y = 6$. Use a solid line since the inequality uses $\geq$.
       Choose a test point not on the line, such as $(0, 0)$. Since $2(0) + 3(0) \geq 6$ is false,
       shade the opposite side of the line from $(0, 0)$.

(b)   Graph the line $2x + 3y = 0$. Use a solid line since the inequality uses $\leq$.
       Choose a test point not on the line, such as $(0, 2)$. Since $2(0) + 3(2) \leq 0$ is false,
       shade the opposite side of the line from $(0, 2)$.

(c)   Since the regions do not overlap, the solution is an empty set.

35. Graph the system of linear inequalities:

$$\begin{cases} x \geq 0 \\ y \geq 0 \\ 2x + y \leq 6 \\ x + 2y \leq 6 \end{cases}$$

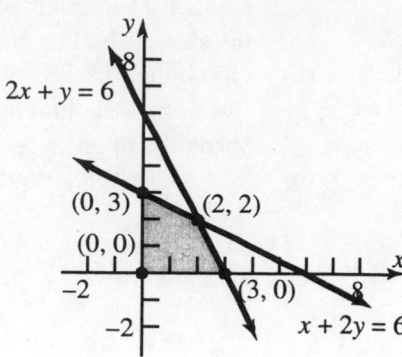

(a) Graph $x \geq 0$; $y \geq 0$. Shaded region is the first quadrant.

(b) Graph the line $2x + y = 6$. Use a solid line since the inequality uses $\leq$.
Choose a test point not on the line, such as $(0, 0)$. Since $2(0) + 0 \leq 6$ is true, shade the side of the line containing $(0, 0)$.

(c) Graph the line $x + 2y = 6$. Use a solid line since the inequality uses $\leq$.
Choose a test point not on the line, such as $(0, 0)$. Since $0 + 2(0) \leq 6$ is true, shade the side of the line containing $(0, 0)$.

(d) The overlapping region is the solution.

(e) The graph is bounded.

(f) Find the vertices:
The x-axis and y-axis intersect at $(0, 0)$.
The intersection of $x + 2y = 6$ and the y-axis is $(0, 3)$.
The intersection of $2x + y = 6$ and the x-axis is $(3, 0)$.
To find the intersection of $x + 2y = 6$ and $2x + y = 6$, solve the system:

$$\begin{cases} x + 2y = 6 & \rightarrow \quad x = 6 - 2y \\ 2x + y = 6 \end{cases}$$

Substitute and solve:

$$2(6 - 2y) + y = 6 \rightarrow 12 - 4y + y = 6 \rightarrow -3y = -6 \rightarrow y = 2$$

$$x = 6 - 2(2) = 6 - 4 = 2$$

The point of intersection is $(2, 2)$.
The four corner points are $(0, 0)$, $(0, 3)$, $(3, 0)$, and $(2, 2$

37. Graph the system of linear inequalities:

$$\begin{cases} x \geq 0 \\ y \geq 0 \\ x + y \geq 2 \\ 2x + y \geq 4 \end{cases}$$

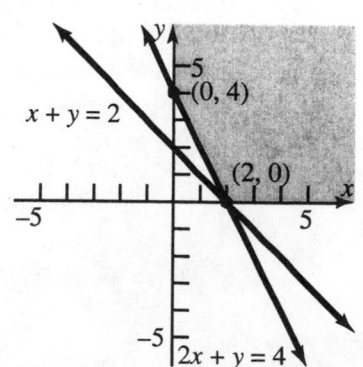

(a)   Graph $x \geq 0$; $y \geq 0$. Shaded region is the first quadrant.

(b)   Graph the line $x + y = 2$. Use a solid line since the inequality uses $\geq$.

   Choose a test point not on the line, such as (0, 0). Since $0 + 0 \geq 2$ is false, shade the opposite side of the line from (0, 0).

(c)   Graph the line $2x + y = 4$. Use a solid line since the inequality uses $\geq$.

   Choose a test point not on the line, such as (0, 0). Since $2(0) + 0 \geq 4$ is false, shade the opposite side of the line from (0, 0).

(d)   The overlapping region is the solution.

(e)   The graph is unbounded.

(f)   Find the vertices:

   The intersection of $x + y = 2$ and the x-axis is (2, 0).

   The intersection of $2x + y = 4$ and the y-axis is (0, 4).

   The two corner points are (2, 0), and (0, 4).

39.   Graph the system of linear inequalities:

$$\begin{cases} x \geq 0 \\ y \geq 0 \\ x + y \geq 2 \\ 2x + 3y \leq 12 \\ 3x + y \leq 12 \end{cases}$$

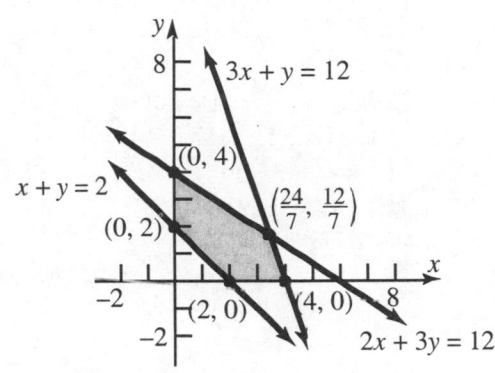

(a)   Graph $x \geq 0$; $y \geq 0$. Shaded region is the first quadrant.

(b)   Graph the line $x + y = 2$. Use a solid line since the inequality uses $\geq$.

   Choose a test point not on the line, such as (0, 0). Since $0 + 0 \geq 2$ is false, shade the opposite side of the line from (0, 0).

(c)   Graph the line $2x + 3y = 12$. Use a solid line since the inequality uses $\leq$.

   Choose a test point not on the line, such as (0, 0). Since $2(0) + 3(0) \leq 12$ is true, shade the side of the line containing (0, 0).

(d)   Graph the line $3x + y = 12$. Use a solid line since the inequality uses $\leq$.

   Choose a test point not on the line, such as (0, 0). Since $3(0) + 0 \leq 12$ is true, shade the side of the line containing (0, 0).

(e)   The overlapping region is the solution.

(f)   The graph is bounded.

(g)   Find the vertices:

   The intersection of $x + y = 2$ and the y-axis is (0, 2).

   The intersection of $x + y = 2$ and the x-axis is (2, 0).

   The intersection of $2x + 3y = 12$ and the y-axis is (0, 4).

   The intersection of $3x + y = 12$ and the x-axis is (4, 0).

To find the intersection of $2x + 3y = 12$ and $3x + y = 12$, solve the system:
$$\begin{cases} 2x + 3y = 12 \\ 3x + y = 12 \end{cases} \rightarrow \quad y = 12 - 3x$$
Substitute and solve:
$$2x + 3(12 - 3x) = 12 \rightarrow 2x + 36 - 9x = 12$$
$$-7x = -24 \rightarrow x = \frac{24}{7}$$
$$y = 12 - 3\left(\frac{24}{7}\right) = 12 - \frac{72}{2} = \frac{12}{7}$$
The point of intersection is $\left(\frac{24}{7}, \frac{12}{7}\right)$.

The five corner points are $(0, 2)$, $(0, 4)$, $(2, 0)$, $(4, 0)$, and $\left(\frac{24}{7}, \frac{12}{7}\right)$.

41.   Graph the system of linear inequalities:

$$\begin{cases} x \geq 0 \\ y \geq 0 \\ x + y \geq 2 \\ x + y \leq 8 \\ 2x + y \leq 10 \end{cases}$$

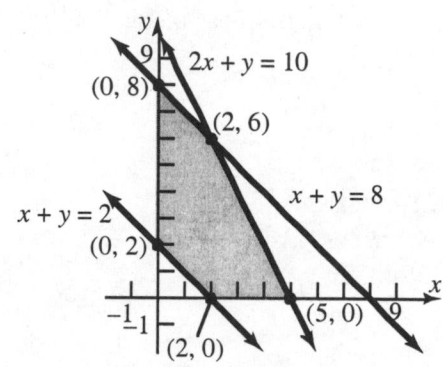

(a)   Graph $x \geq 0$; $y \geq 0$. Shaded region is the first quadrant.
(b)   Graph the line $x + y = 2$. Use a solid line since the inequality uses $\geq$.
       Choose a test point not on the line, such as $(0, 0)$. Since $0 + 0 \geq 2$ is false, shade the opposite side of the line from $(0, 0)$.
(c)   Graph the line $x + y = 8$. Use a solid line since the inequality uses $\leq$.
       Choose a test point not on the line, such as $(0, 0)$. Since $0 + 0 \leq 8$ is true, shade the side of the line containing $(0, 0)$.
(d)   Graph the line $2x + y = 10$. Use a solid line since the inequality uses $\leq$.
       Choose a test point not on the line, such as $(0, 0)$. Since $2(0) + 0 \leq 10$ is true, shade the side of the line containing $(0, 0)$.
(e)   The overlapping region is the solution.
(f)   The graph is bounded.
(g)   Find the vertices:
       The intersection of $x + y = 2$ and the y-axis is $(0, 2)$.
       The intersection of $x + y = 2$ and the x-axis is $(2, 0)$.
       The intersection of $x + y = 8$ and the y-axis is $(0, 8)$.
       The intersection of $2x + y = 10$ and the x-axis is $(5, 0)$.
       To find the intersection of $x + y = 8$ and $2x + y = 10$, solve the system:
       $$\begin{cases} x + y = 8 \\ 2x + y = 10 \end{cases} \rightarrow \quad y = 8 - x$$

Substitute and solve:
$$2x + 8 - x = 10 \rightarrow x = 2$$
$$y = 8 - 2 = 6$$
The point of intersection is (2, 6).
The five corner points are (0, 2), (0, 8), (2, 0), (5, 0), and (2, 6).

43.   Graph the system of linear inequalities:

$$\begin{cases} x \geq 0 \\ y \geq 0 \\ x + 2y \geq 1 \\ x + 2y \leq 10 \end{cases}$$

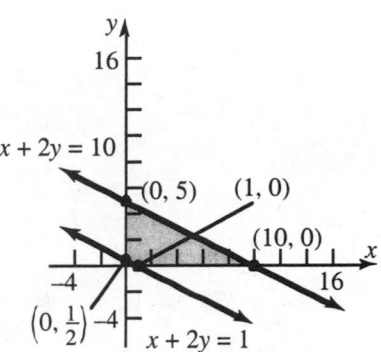

(a)   Graph $x \geq 0$; $y \geq 0$. Shaded region is the first quadrant.

(b)   Graph the line $x + 2y = 1$. Use a solid line since the inequality uses $\geq$.
Choose a test point not on the line, such as (0, 0). Since $0 + 2(0) \geq 1$ is false, shade
the opposite side of the line from (0, 0).

(c)   Graph the line $x + 2y = 10$. Use a solid line since the inequality uses $\leq$.
Choose a test point not on the line, such as (0, 0). Since $0 + 2(0) \leq 10$ is true,
shade the side of the line containing (0, 0).

(d)   The overlapping region is the solution.

(e)   The graph is bounded.

(f)   Find the vertices:
The intersection of $x + 2y = 1$ and the y-axis is (0, 0.5).
The intersection of $x + 2y = 1$ and the x-axis is (1, 0).
The intersection of $x + 2y = 10$ and the y-axis is (0, 5).
The intersection of $x + 2y = 10$ and the x-axis is (10, 0).
The four corner points are (0, 0.5), (0, 5), (1, 0), and (10, 0).

45.   The system of linear inequalities is:

$$\begin{cases} x \geq 0 \\ y \geq 0 \\ x \leq 4 \\ x + y \leq 6 \end{cases}$$

47.   The system of linear inequalities is:

$$\begin{cases} x \geq 0 \\ y \geq 15 \\ x \leq 20 \\ x + y \leq 50 \\ x - y \leq 0 \end{cases}$$

49.   (a)   Let x = the amount invested in Treasury bills.
Let y = the amount invested in corporate bonds.
The constraints are:
$x \geq 0, y \geq 0$ A non-negative amount must be invested.
$x + y \leq 50000$ Total investment cannot exceed \$50,000.

$y \leq 10000$ Amount invested in corporate bonds must not exceed
                      $10,000.

$x \geq 35000$ Amount invested in Treasury bills must be at least $35,000.

$x > y$ Amount invested in Treasury bills must be greater than the amount
                      invested in corporate bonds.

(b)   Graph the system.

$$\begin{cases} x \geq 0 \\ y \geq 0 \\ x + y \leq 50000 \\ y \leq 10000 \\ x \geq 35000 \\ x > y \end{cases}$$

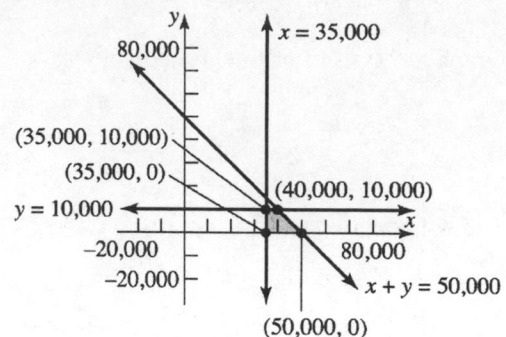

The corner points are (35000, 0), (35000, 10000), (40000, 10000), (50000, 0).

51.   (a)   Let x = the # of packages of the economy blend.
            Let y = the # of packages of the superior blend.
      The constraints are:

      $x \geq 0, y \geq 0$ A non-negative # of packages must be produced.

      $4x + 8y \leq 75 \cdot 16$ Total amount of grade A coffee cannot exceed 75
                                    pounds. (Note: 75 pounds = (75)(16) ounces.)

      $12x + 8y \leq 120 \cdot 16$ Total amount of grade B coffee cannot exceed 120
                                    pounds. (Note: 120 pounds = (120)(16) ounces.)

      We can simplify the equations
            $4x + 8y \leq 75 \cdot 16 \rightarrow x + 2y \leq 75 \cdot 4 \rightarrow x + 2y \leq 300$
            $12x + 8y \leq 120 \cdot 16 \rightarrow 3x + 2y \leq 120 \cdot 4 \rightarrow 3x + 2y \leq 480$

      (b)   Graph the system.

$$\begin{cases} x \geq 0 \\ y \geq 0 \\ x + 2y \leq 300 \\ 3x + 2y \leq 480 \end{cases}$$

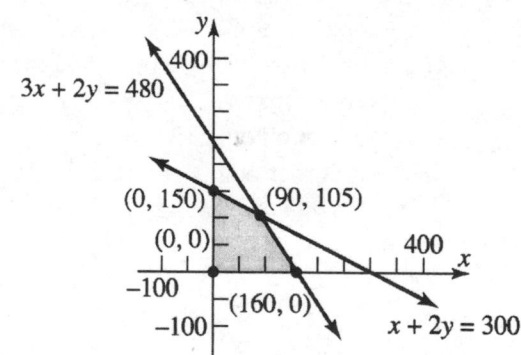

The corner points are (0, 0), (0, 150), (90, 105), (160, 0).

53.   (a)   Let x = the # of microwaves.
            Let y = the # of printers.
      The constraints are:

      $x \geq 0, y \geq 0$ A non-negative # of items must be shipped.

      $30x + 20y \leq 1600$ Total cargo weight cannot exceed 1600 pounds.

      $2x + 3y \leq 150$ Total cargo volume cannot exceed 150 cubic feet.

(b)   Graph the system.

$$\begin{cases} x \geq 0 \\ y \geq 0 \\ 30x + 20y \leq 1600 \\ 2x + 3y \leq 150 \end{cases}$$

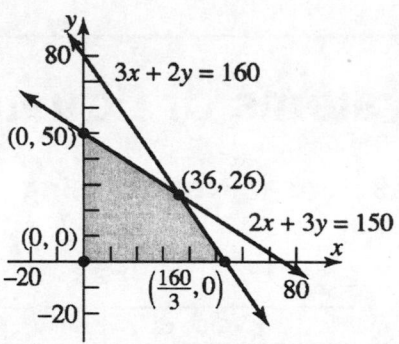

The corner points are $(0, 0)$, $(0, 50)$, $(36, 26)$, $(53.3, 0)$.

Systems of Equations and Inequalities

10.9 Linear Programming

1. $z = x + y$

Vertex	Value of $z = x + y$
(0, 3)	$z = 0 + 3 = 3$
(0, 6)	$z = 0 + 6 = 6$
(5, 6)	$z = 5 + 6 = 11$
(5, 2)	$z = 5 + 2 = 7$
(4, 0)	$z = 4 + 0 = 4$

The maximum value is 11 at (5, 6), and the minimum value is 3 at (0, 3).

3. $z = x + 10y$

Vertex	Value of $z = x + 10y$
(0, 3)	$z = 0 + 10(3) = 30$
(0, 6)	$z = 0 + 10(6) = 60$
(5, 6)	$z = 5 + 10(6) = 65$
(5, 2)	$z = 5 + 10(2) = 25$
(4, 0)	$z = 4 + 10(0) = 4$

The maximum value is 65 at (5, 6), and the minimum value is 4 at (4, 0).

5. $z = 5x + 7y$

Vertex	Value of $z = 5x + 7y$
(0, 3)	$z = 5(0) + 7(3) = 21$
(0, 6)	$z = 5(0) + 7(6) = 42$
(5, 6)	$z = 5(5) + 7(6) = 67$
(5, 2)	$z = 5(5) + 7(2) = 39$
(4, 0)	$z = 5(4) + 7(0) = 20$

The maximum value is 67 at (5, 6), and the minimum value is 20 at (4, 0).

7.   Maximize $z = 2x + y$

Subject to $x \geq 0$, $y \geq 0$, $x + y \leq 6$, $x + y \geq 1$

Graph the constraints.

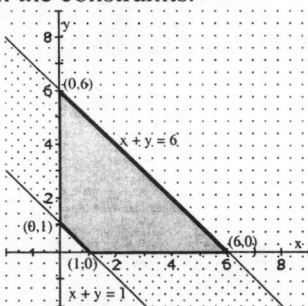

The corner points are $(0, 1)$, $(1, 0)$, $(0, 6)$, $(6, 0)$.

Evaluate the objective function:

Vertex	Value of $z = 2x + y$
$(0, 1)$	$z = 2(0) + 1 = 1$
$(0, 6)$	$z = 2(0) + 6 = 6$
$(1, 0)$	$z = 2(1) + 0 = 2$
$(6, 0)$	$z = 2(6) + 0 = 12$

The maximum value is 12 at $(6, 0)$.

9.   Minimize $z = 2x + 5y$

Subject to $x \geq 0$, $y \geq 0$, $x + y \geq 2$, $x \leq 5$, $y \leq 3$

Graph the constraints.

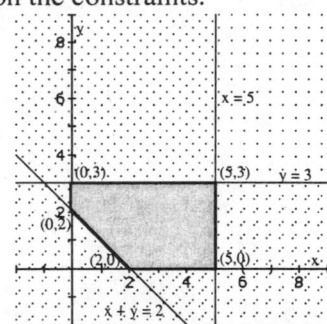

The corner points are $(0, 2)$, $(2, 0)$, $(0, 3)$, $(5, 0)$, $(5, 3)$.

Evaluate the objective function:

Vertex	Value of $z = 2x + 5y$
$(0, 2)$	$z = 2(0) + 5(2) = 10$
$(0, 3)$	$z = 2(0) + 5(3) = 15$
$(2, 0)$	$z = 2(2) + 5(0) = 4$
$(5, 0)$	$z = 2(5) + 5(0) = 10$
$(5, 3)$	$z = 2(5) + 5(3) = 25$

The minimum value is 4 at $(2, 0)$.

11. Maximize $z = 3x + 5y$

Subject to $x \geq 0,\ y \geq 0,\ x + y \geq 2,\ 2x + 3y \leq 12,\ 3x + 2y \leq 12$

Graph the constraints.

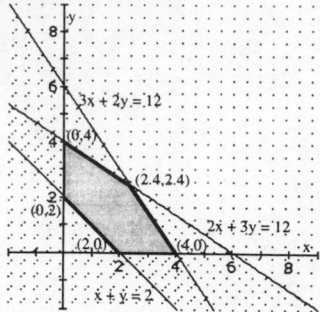

To find the intersection of $2x + 3y = 12$ and $3x + 2y = 12$, solve the system:

$$\begin{cases} 2x + 3y = 12 \\ 3x + 2y = 12 \end{cases} \rightarrow \quad y = 6 - \frac{3}{2}x$$

Substitute and solve:

$$2x + 3\left(6 - \frac{3}{2}x\right) = 12 \rightarrow 2x + 18 - \frac{9}{2}x = 12 \rightarrow -\frac{5}{2}x = -6$$

$$x = \frac{12}{5} \qquad y = 6 - \frac{3}{2}\left(\frac{12}{5}\right) = 6 - \frac{18}{5} = \frac{12}{5}$$

The point of intersection is $(2.4, 2.4)$.

The corner points are (0, 2), (2, 0), (0, 4), (4, 0), (2.4, 2.4).

Evaluate the objective function:

Vertex	Value of $z = 3x + 5y$
(0, 2)	z = 3(0) + 5(2) = 10
(0, 4)	z = 3(0) + 5(4) = 20
(2, 0)	z = 3(2) + 5(0) = 6
(4, 0)	z = 3(4) + 5(0) = 12
(2.4, 2.4)	z = 3(2.4) + 5(2.4) = 19.2

The maximum value is 20 at (0, 4).

13. Minimize $z = 5x + 4y$

Subject to $x \geq 0,\ y \geq 0,\ x + y \geq 2,\ 2x + 3y \leq 12,\ 3x + y \leq 12$

Graph the constraints.

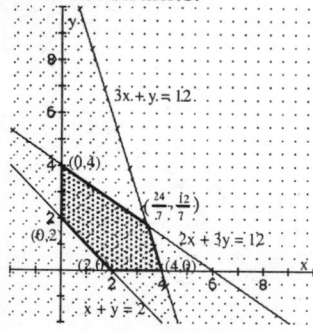

To find the intersection of $2x + 3y = 12$ and $3x + y = 12$, solve the system:

$$\begin{cases} 2x + 3y = 12 \\ 3x + y = 12 \quad \rightarrow \quad y = 12 - 3x \end{cases}$$

Substitute and solve:

$$2x + 3(12 - 3x) = 12 \rightarrow 2x + 36 - 9x = 12 \rightarrow -7x = -24 \rightarrow x = \frac{24}{7}$$

$$y = 12 - 3\left(\frac{24}{7}\right) = 12 - \frac{72}{7} = \frac{12}{7}$$

The point of intersection is $\left(\frac{24}{7}, \frac{12}{7}\right)$.

The corner points are $(0, 2)$, $(2, 0)$, $(0, 4)$, $(4, 0)$, $\left(\frac{24}{7}, \frac{12}{7}\right)$.

Evaluate the objective function:

Vertex	Value of $z = 5x + 4y$
$(0, 2)$	$z = 5(0) + 4(2) = 8$
$(0, 4)$	$z = 5(0) + 4(4) = 16$
$(2, 0)$	$z = 5(2) + 4(0) = 10$
$(4, 0)$	$z = 5(4) + 4(0) = 20$
$\left(\dfrac{24}{7}, \dfrac{12}{7}\right)$	$z = 5\left(\dfrac{24}{7}\right) + 4\left(\dfrac{12}{7}\right) = \dfrac{120}{7} + \dfrac{48}{7} = \dfrac{168}{7} = 24$

The minimum value is 8 at $(0, 2)$.

15. Maximize $z = 5x + 2y$

Subject to $x \geq 0$, $y \geq 0$, $x + y \leq 10$, $2x + y \geq 10$, $x + 2y \geq 10$

Graph the constraints.

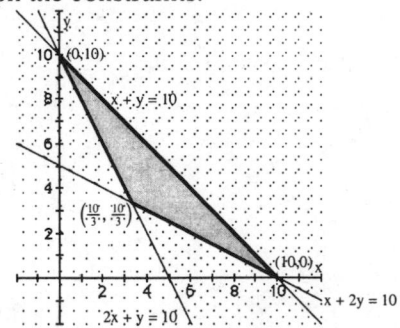

To find the intersection of $2x + y = 10$ and $x + 2y = 10$, solve the system:

$$\begin{cases} 2x + y = 10 \quad \rightarrow \quad y = 10 - 2x \\ x + 2y = 10 \end{cases}$$

Substitute and solve:

$$x + 2(10 - 2x) = 10 \rightarrow x + 20 - 4x = 10 \rightarrow -3x = -10 \rightarrow x = \frac{10}{3}$$

$$y = 10 - 2\left(\frac{10}{3}\right) = 10 - \frac{20}{3} = \frac{10}{3}$$

The point of intersection is $\left(\frac{10}{3}, \frac{10}{3}\right)$.

The corner points are $(0, 10)$, $(10, 0)$, $\left(\dfrac{10}{3}, \dfrac{10}{3} \right)$.

Evaluate the objective function:

Vertex	Value of $z = 5x + 2y$
$(0, 10)$	$z = 5(0) + 2(10) = 20$
$(10, 0)$	$z = 5(10) + 2(0) = 50$
$\left(\dfrac{10}{3}, \dfrac{10}{3} \right)$	$z = 5\left(\dfrac{10}{3}\right) + 2\left(\dfrac{10}{3}\right) = \dfrac{50}{3} + \dfrac{20}{3} = \dfrac{70}{3} = 23\frac{1}{3}$

The maximum value is 50 at $(10, 0)$.

17. Let x = the number of downhill skis produced.
Let y = the number of cross-country skis produced.
The total profit is: $P = 70x + 50y$. Profit is to be maximized; thus, this is the objective function.
The constraints are:

   $x \geq 0$, $y \geq 0$ A positive number of skis must be produced.

   $2x + y \leq 40$ Only 40 hours of manufacturing time is available.

   $x + y \leq 32$ Only 32 hours of finishing time is available.

Graph the constraints.

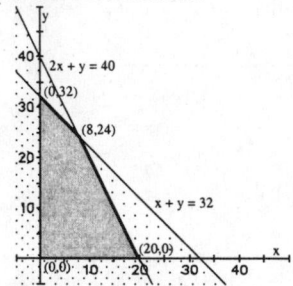

To find the intersection of $x + y = 32$ and $2x + y = 40$, solve the system:

$$\begin{cases} x + y = 32 & \rightarrow \quad y = 32 - x \\ 2x + y = 40 \end{cases}$$

   Substitute and solve:
   $$2x + 32 - x = 40$$

   $$x = 8$$

   $$y = 32 - 8 = 24$$

   The point of intersection is $(8, 24)$.
The corner points are $(0, 0)$, $(0, 32)$, $(20, 0)$, $(8, 24)$.
Evaluate the objective function:

Vertex	Value of $P = 70x + 50y$
$(0, 0)$	$P = 70(0) + 50(0) = 0$
$(0, 32)$	$P = 70(0) + 50(32) = 1600$
$(20, 0)$	$P = 70(20) + 50(0) = 1400$
$(8, 24)$	$P = 70(8) + 50(24) = 1760$

The maximum profit is $1760, when 8 downhill skis and 24 cross-country skis are produced.
With the increase of the manufacturing time to 48 hours, we do the following.

The constraints are:

$x \geq 0, \; y \geq 0$ A positive number of skis must be produced.

$2x + y \leq 48$ Only 48 hours of manufacturing time is available.

$x + y \leq 32$ Only 32 hours of finishing time is available.

Graph the constraints.

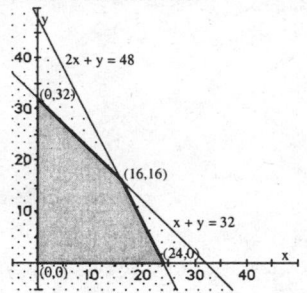

To find the intersection of $x + y = 32$ and $2x + y = 48$, solve the system:

$$\begin{cases} x + y = 32 & \rightarrow \quad y = 32 - x \\ 2x + y = 48 \end{cases}$$

Substitute and solve:

$$2x + 32 - x = 48$$

$$x = 16$$

$$y = 32 - 16 = 16$$

The point of intersection is $(16, 16)$.

The corner points are $(0, 0)$, $(0, 32)$, $(24, 0)$, $(16, 16)$.

Evaluate the objective function:

Vertex	Value of $P = 70x + 50y$
$(0, 0)$	$P = 70(0) + 50(0) = 0$
$(0, 32)$	$P = 70(0) + 50(32) = 1600$
$(24, 0)$	$P = 70(24) + 50(0) = 1680$
$(16, 16)$	$P = 70(16) + 50(16) = 1920$

The maximum profit is \$1920, when 16 downhill skis and 16 cross-country skis are produced.

19.   Let x = the number of acres of corn planted.

Let y = the number of acres of soybeans planted.

The total profit is: $P = 250x + 200y$. Profit is to be maximized; thus, this is the objective function.

The constraints are:

$x \geq 0, \; y \geq 0$ A non-negative number of acres must be planted.

$x + y \leq 100$ Acres available to plant.

$60x + 40y \leq 1800$ Money available for cultivation costs.

$60x + 60y \leq 2400$ Money available for labor costs.

Graph the constraints.

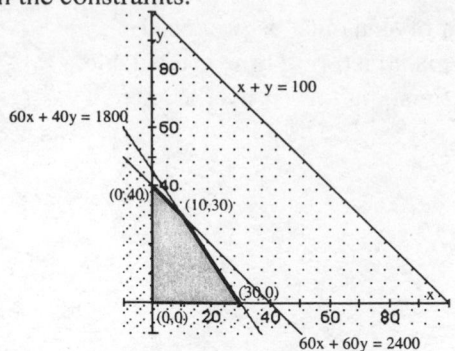

To find the intersection of $60x + 40y = 1800$ and $60x + 60y = 2400$, solve the system:

$$\begin{cases} 60x + 40y = 1800 & \to & 60x = 1800 - 40y \\ 60x + 60y = 2400 \end{cases}$$

Substitute and solve:

$$1800 - 40y + 60y = 2400$$

$$20y = 600 \Rightarrow y = 30$$

$$60x = 1800 - 40(30)$$

$$60x = 600 \Rightarrow x = 10$$

The point of intersection is $(10, 30)$.

The corner points are $(0, 0)$, $(0, 40)$, $(30, 0)$, $(10, 30)$.

Evaluate the objective function:

Vertex	Value of $P = 250x + 200y$
$(0, 0)$	$P = 250(0) + 200(0) = 0$
$(0, 40)$	$P = 250(0) + 200(40) = 8000$
$(30, 0)$	$P = 250(30) + 200(0) = 7500$
$(10, 30)$	$P = 250(10) + 200(30) = 8500$

The maximum profit is $8500, when 10 acres of corn and 30 acres of soybeans are planted.

21. Let $x =$ the number of hours that machine 1 is operated.

Let $y =$ the number of hours that machine 2 is operated.

The total cost is: $C = 50x + 30y$. Cost is to be minimized; thus, this is the objective function.

The constraints are:

$x \geq 0, \quad y \geq 0$ A positive number of hours must be used.

$x \leq 10$ \qquad\qquad 10 hours available on machine 1.

$y \leq 10$ \qquad\qquad 10 hours available on machine 2.

$60x + 40y \geq 240$ \quad At least 240 8-inch pliers must be produced.

$70x + 20y \geq 140$ \quad At least 140 6-inch pliers must be produced.

Graph the constraints.

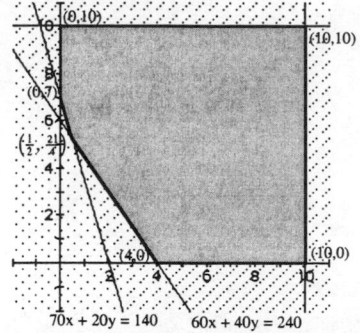

To find the intersection of $60x + 40y = 240$ and $70x + 20y = 140$, solve the system:
$$\begin{cases} 60x + 40y = 240 \\ 70x + 20y = 140 \end{cases} \rightarrow \quad 20y = 140 - 70x$$

Substitute and solve:
$$60x + 2(140 - 70x) = 240$$
$$60x + 280 - 140x = 240$$
$$-80x = -40$$
$$x = 0.5$$
$$20y = 140 - 70(0.5)$$
$$20y = 105$$
$$y = 5.25$$

The point of intersection is $(0.5, 5.25)$.

The corner points are (0, 7), (0, 10), (4, 0), (10, 0), (10, 10), $(0.5, 5.25)$.

Evaluate the objective function:

Vertex	Value of $C = 50x + 30y$
(0, 7)	$C = 50(0) + 30(7) = 210$
(0, 10)	$C = 50(0) + 30(10) = 300$
(4, 0)	$C = 50(4) + 30(0) = 200$
(10, 0)	$C = 50(10) + 30(0) = 500$
(10, 10)	$C = 50(10) + 30(10) = 800$
$(0.5, 5.25)$	$C = 50(0.5) + 30(5.25) = 182.50$

The minimum cost is \$182.50, when machine 1 is used for 0.5 hours and machine 2 is used for 5.25 hours.

23. Let x = the number of pounds of ground beef.
Let y = the number of pounds of ground pork.

The total cost is: $C = 0.75x + 0.45y$. Cost is to be minimized; thus, this is the objective function.

The constraints are:

$x \geq 0, \ y \geq 0$ A positive number of pounds must be used.

$x \leq 200$ Only 200 pounds of ground beef are available.

$y \geq 50$ At least 50 pounds of ground pork must be used.

$0.75x + 0.60y \geq 0.70(x + y) \rightarrow 0.05x \geq 0.10y$ Leanness condition to be met.

Graph the constraints.

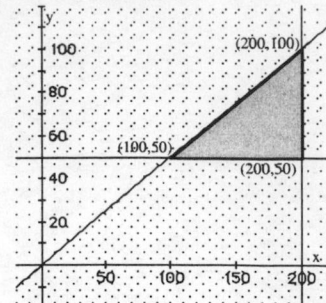

The corner points are (100, 50), (200, 50), (200, 100).
Evaluate the objective function:

Vertex	Value of $C = 0.75x + 0.45y$
(100, 50)	$C = 0.75(100) + 0.45(50) = 97.50$
(200, 50)	$C = 0.75(200) + 0.45(50) = 172.50$
(200,100)	$C = 0.75(200) + 0.45(100) = 195.00$

The minimum cost is $97.50, when 100 pounds of ground beef and 50 pounds of ground pork are used.

25. Let $x =$ the number of racing skates manufactured.
Let $y =$ the number of figure skates manufactured.
The total profit is: $P = 10x + 12y$. Profit is to be maximized; thus, this is the objective function.
The constraints are:

$$x \geq 0, \ y \geq 0 \qquad \text{A positive number of skates must be manufactured.}$$
$$6x + 4y \leq 120 \qquad \text{Only 120 hours are available for fabrication.}$$
$$x + 2y \leq 40 \qquad \text{Only 40 hours are available for finishing.}$$

Graph the constraints.

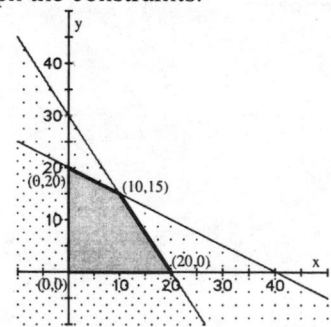

To find the intersection of $6x + 4y = 120$ and $x + 2y = 40$, solve the system:

$$\begin{cases} 6x + 4y = 120 \\ x + 2y = \ 40 \end{cases} \rightarrow \ x = 40 - 2y$$

Substitute and solve:

$$6(40 - 2y) + 4y = 120 \rightarrow 240 - 12y + 4y = 120$$
$$-8y = -120 \rightarrow y = 15 \rightarrow x = 40 - 2(15) = 10$$

The point of intersection is (10, 15).
The corner points are (0, 0), (0, 20), (20, 0), (10, 15).

Evaluate the objective function:

Vertex	Value of $P = 10x + 12y$
(0, 0)	$P = 10(0) + 12(0) = 0$
(0, 20)	$P = 10(0) + 12(20) = 240$
(20, 0)	$P = 10(20) + 12(0) = 200$
(10, 15)	$P = 10(10) + 12(15) = 280$

The maximum profit is $280, when 10 racing skates and 15 figure skates are produced.

27. Let x = the number of metal fasteners.
Let y = the number of plastic fasteners.
The total cost is: $C = 9x + 4y$. Cost is to be minimized; thus, this is the objective function.
The constraints are:

$x \geq 2, \ y \geq 2$ At least 2 of each fastener must be made.

$x + y \geq 6$ At least 6 fasteners are needed.

$4x + 2y \leq 24$ Only 24 hours are available.

Graph the constraints.

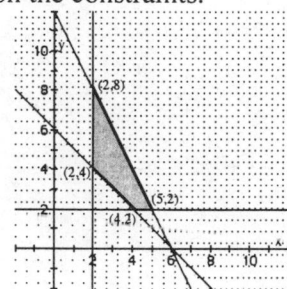

The corner points are (2, 4), (2, 8), (4, 2), (5, 2).

Evaluate the objective function:

Vertex	Value of $C = 9x + 4y$
(2, 4)	$C = 9(2) + 4(4) = 34$
(2, 8)	$C = 9(2) + 4(8) = 50$
(4, 2)	$C = 9(4) + 4(2) = 44$
(5, 2)	$C = 9(5) + 4(2) = 53$

The minimum cost is $34, when 2 metal fasteners and 4 plastic fasteners are ordered.

29. Let x = the number of first-class seats.
Let y = the number of coach seats.
The constraints are:

$8 \leq x \leq 16$ Restriction on first-class seats.

$80 \leq y \leq 120$ Restriction on coach seats.

(a) $\dfrac{x}{y} \leq \dfrac{1}{12}$ Ratio of seats.

If $y = 120$, then $\dfrac{x}{120} \leq \dfrac{1}{12} \rightarrow 12x \leq 120 \rightarrow x \leq 10$

The maximum revenue will be obtained with 120 coach seats and 10 first-class seats. (Note that the first-class seats meet their constraint.)

(b) $\dfrac{x}{y} \le \dfrac{1}{8}$ Ratio of seats.

If $y = 120$, then $\dfrac{x}{120} \le \dfrac{1}{8} \rightarrow 8x \le 120 \rightarrow x \le 15$

The maximum revenue will be obtained with 120 coach seats and 15 first-class seats. (Note that the first-class seats meet their constraint.)

Systems of Equations and Inequalities

10.R Chapter Review

1. Solve the first equation for y, substitute into the second equation and solve:
$$\begin{cases} 2x - y = 5 \\ 5x + 2y = 8 \end{cases} \rightarrow \quad y = 2x - 5$$
$$5x + 2(2x - 5) = 8 \rightarrow 5x + 4x - 10 = 8$$
$$9x = 18 \rightarrow x = 2 \rightarrow y = 2(2) - 5 = 4 - 5 = -1$$
The solution is $x = 2, \ y = -1$.

3. Solve the second equation for x, substitute into the first equation and solve:
$$\begin{cases} 3x - 4y = 4 \\ x - 3y = \dfrac{1}{2} \end{cases} \rightarrow \quad x = 3y + \dfrac{1}{2}$$
$$3\left(3y + \dfrac{1}{2}\right) - 4y = 4 \rightarrow 9y + \dfrac{3}{2} - 4y = 4 \rightarrow 5y = \dfrac{5}{2} \rightarrow y = \dfrac{1}{2} \rightarrow x = 3\left(\dfrac{1}{2}\right) + \dfrac{1}{2} = 2$$
The solution is $x = 2, \ y = \dfrac{1}{2}$.

5. Solve the first equation for x, substitute into the second equation and solve:
$$\begin{cases} x - 2y - 4 = 0 \\ 3x + 2y - 4 = 0 \end{cases} \rightarrow \quad x = 2y + 4$$
$$3(2y + 4) + 2y - 4 = 0 \rightarrow 6y + 12 + 2y - 4 = 0$$
$$8y = -8 \rightarrow y = -1 \rightarrow x = 2(-1) + 4 = 2$$
The solution is $x = 2, \ y = -1$.

7. Substitute the first equation into the second equation and solve:
$$\begin{cases} y = 2x - 5 \\ x = 3y + 4 \end{cases}$$
$$x = 3(2x - 5) + 4 \rightarrow x = 6x - 15 + 4$$
$$-5x = -11 \rightarrow x = \dfrac{11}{5} \rightarrow y = 2\left(\dfrac{11}{5}\right) - 5 = -\dfrac{3}{5}$$
The solution is $x = \dfrac{11}{5}, \ y = -\dfrac{3}{5}$.

9. Multiply each side of the first equation by 3 and each side of the second equation by -6 and add:

$$\begin{cases} x - 3y + 4 = 0 \quad \xrightarrow{\;3\;} \quad 3x - 9y + 12 = 0 \\ \dfrac{1}{2}x - \dfrac{3}{2}y + \dfrac{4}{3} = 0 \quad \xrightarrow{\;-6\;} \quad \dfrac{-3x + 9y - 8 = 0}{4 \neq 0} \end{cases}$$

There is no solution to the system. The system is inconsistent.

11. Multiply each side of the first equation by 2 and each side of the second equation by 3 and add to eliminate y:

$$\begin{cases} 2x + 3y - 13 = 0 \quad \xrightarrow{\;2\;} \quad 4x + 6y - 26 = 0 \\ 3x - 2y \quad\;\; = 0 \quad \xrightarrow{\;3\;} \quad \dfrac{9x - 6y \quad\quad = 0}{13x \quad\quad - 26 = 0} \end{cases}$$

$$13x = 26$$
$$x = 2$$

Substitute and solve for y:
$$3(2) - 2y = 0$$
$$-2y = -6$$
$$y = 3$$

The solution of the system is $x = 2$, $y = 3$.

13. Multiply each side of the second equation by -3 and add to eliminate x:

$$\begin{cases} 3x - 2y = 8 \quad \xrightarrow{\quad\;\;} \quad 3x - 2y = \;\;8 \\ x - \dfrac{2}{3}y = 12 \quad \xrightarrow{\;-3\;} \quad \dfrac{-3x + 2y = -36}{0 \neq -28} \end{cases}$$

The system has no solution, so the system is inconsistent.

15. Multiply each side of the first equation by -2 and add to the second equation to eliminate x; and multiply each side of the first equation by -3 and add to the third equation to eliminate x:

$$\begin{cases} x + 2y - z = \;\;\;6 \quad \xrightarrow{\;-2\;} \quad -2x - 4y + 2z = -12 \\ 2x - \;\;y + 3z = -13 \quad \xrightarrow{\quad\;\;} \quad \dfrac{2x - \;\;y + 3z = -13}{-5y + 5z = -25} \xrightarrow{\;-1/5\;} y - z = 5 \\ 3x - 2y + 3z = -16 \end{cases}$$

$$\xrightarrow{\;-3\;} \quad -3x - 6y + 3z = -18$$
$$\xrightarrow{\quad\;\;} \quad \dfrac{3x - 2y + 3z = -16}{-8y + 6z = -34}$$

Multiply each side of the first result by 8 and add to the second result to eliminate y:

$$\begin{array}{r} y - z = \;\;\;5 \quad \xrightarrow{\;8\;} \quad 8y - 8z = \;\;40 \\ -8y + 6z = -34 \quad \xrightarrow{\quad\;\;} \quad \dfrac{-8y + 6z = -34}{-2z = \;\;\;6} \\ z = -3 \end{array}$$

Substituting and solving for the other variables:

$$y - (-3) = 5 \qquad\qquad x + 2(2) - (-3) = 6$$
$$y = 2 \qquad\qquad\qquad x + 4 + 3 = 6$$
$$\qquad\qquad\qquad x = -1$$

The solution is $x = -1$, $y = 2$, $z = -3$.

17. Multiply the first equation by -1 and the second equation by 2 and then add to eliminate x; then multiply the second equation by -5 and add to the third equation to eliminate x:

$$\begin{cases} 2x - 4y + z = -15 & \xrightarrow{\;-1\;} -2x + 4y - z = 15 \\ x + 2y - 4z = 27 & \xrightarrow{\;2\;} \quad \underline{2x + 4y - 8z = 54} \\ 5x - 6y - 2z = -3 & \qquad\qquad 8y - 9z \;\; = 69 \end{cases}$$

$$x + 2y - 4z = 27 \xrightarrow{\;-5\;} \quad -5x - 10y + 20z = -135$$
$$5x - 6y - 2z = -3 \longrightarrow \quad \underline{5x - 6y - 2z \qquad = -3}$$
$$-16y + 18z \;\; = -138$$

Now eliminate y in the resulting system:

$$8y - 9z = \quad 69 \xrightarrow{\;2\;} 16y - 18z = 138$$
$$-16y + 18z = -138 \longrightarrow \underline{-16y + 18z = -138}$$
$$0 = 0$$

$$-16y + 18z = -138 \Rightarrow 18z + 138 = 16y \Rightarrow y = \frac{9}{8}z + \frac{69}{8}, \; z \text{ is any real number}$$

Substituting into the second equation and solving for the other x:

$$x + 2\left(\frac{9}{8}z + \frac{69}{8}\right) - 4z = 27$$

$$x + \frac{9}{4}z + \frac{69}{4} - 4z = 27 \Rightarrow x = \frac{7}{4}z + \frac{39}{4}, \; z \text{ is any real number}$$

The solution is $x = \frac{7}{4}z + \frac{39}{4}$, $y = \frac{9}{8}z + \frac{69}{8}$, z is any real number.

19. $\begin{cases} 3x + 2y = 8 \\ x + 4y = -1 \end{cases}$

21. $A + C = \begin{bmatrix} 1 & 0 \\ 2 & 4 \\ -1 & 2 \end{bmatrix} + \begin{bmatrix} 3 & -4 \\ 1 & 5 \\ 5 & -2 \end{bmatrix} = \begin{bmatrix} 4 & -4 \\ 3 & 9 \\ 4 & 0 \end{bmatrix}$

23. $6A = 6 \cdot \begin{bmatrix} 1 & 0 \\ 2 & 4 \\ -1 & 2 \end{bmatrix} = \begin{bmatrix} 6 & 0 \\ 12 & 24 \\ -6 & 12 \end{bmatrix}$

25. $AB = \begin{bmatrix} 1 & 0 \\ 2 & 4 \\ -1 & 2 \end{bmatrix} \cdot \begin{bmatrix} 4 & -3 & 0 \\ 1 & 1 & -2 \end{bmatrix} = \begin{bmatrix} 4 & -3 & 0 \\ 12 & -2 & -8 \\ -2 & 5 & -4 \end{bmatrix}$

27. $CB = \begin{bmatrix} 3 & -4 \\ 1 & 5 \\ 5 & -2 \end{bmatrix} \cdot \begin{bmatrix} 4 & -3 & 0 \\ 1 & 1 & -2 \end{bmatrix} = \begin{bmatrix} 8 & -13 & 8 \\ 9 & 2 & -10 \\ 18 & -17 & 4 \end{bmatrix}$

29. Augment the matrix with the identity and use row operations to find the inverse:

$$A = \begin{bmatrix} 4 & 6 \\ 1 & 3 \end{bmatrix} \rightarrow \begin{bmatrix} 4 & 6 & 1 & 0 \\ 1 & 3 & 0 & 1 \end{bmatrix}$$

$$\rightarrow \begin{bmatrix} 1 & 3 & 0 & 1 \\ 4 & 6 & 1 & 0 \end{bmatrix} \rightarrow \begin{bmatrix} 1 & 3 & 0 & 1 \\ 0 & -6 & 1 & -4 \end{bmatrix} \rightarrow \begin{bmatrix} 1 & 3 & 0 & 1 \\ 0 & 1 & -\frac{1}{6} & \frac{2}{3} \end{bmatrix} \rightarrow \begin{bmatrix} 1 & 0 & \frac{1}{2} & -1 \\ 0 & 1 & -\frac{1}{6} & \frac{2}{3} \end{bmatrix}$$

Interchange $R_2 = -4r_1 + r_2$ $R_2 = -\frac{1}{6}r_2$ $R_1 = -3r_2 + r_1$

r_1 and r_2

$$A^{-1} = \begin{bmatrix} \frac{1}{2} & -1 \\ -\frac{1}{6} & \frac{2}{3} \end{bmatrix}$$

31. Augment the matrix with the identity and use row operations to find the inverse:

$$A = \begin{bmatrix} 1 & 3 & 3 \\ 1 & 2 & 1 \\ 1 & -1 & 2 \end{bmatrix} \rightarrow \begin{bmatrix} 1 & 3 & 3 & 1 & 0 & 0 \\ 1 & 2 & 1 & 0 & 1 & 0 \\ 1 & -1 & 2 & 0 & 0 & 1 \end{bmatrix}$$

$$\rightarrow \begin{bmatrix} 1 & 3 & 3 & 1 & 0 & 0 \\ 0 & -1 & -2 & -1 & 1 & 0 \\ 0 & -4 & -1 & -1 & 0 & 1 \end{bmatrix} \rightarrow \begin{bmatrix} 1 & 3 & 3 & 1 & 0 & 0 \\ 0 & 1 & 2 & 1 & -1 & 0 \\ 0 & -4 & -1 & -1 & 0 & 1 \end{bmatrix} \rightarrow \begin{bmatrix} 1 & 0 & -3 & -2 & 3 & 0 \\ 0 & 1 & 2 & 1 & -1 & 0 \\ 0 & 0 & 7 & 3 & -4 & 1 \end{bmatrix}$$

$R_2 = -r_1 + r_2$ $R_2 = -r_2$ $R_1 = -3r_2 + r_1$

$R_3 = -r_1 + r_3$ $R_3 = 4r_2 + r_3$

$$\rightarrow \begin{bmatrix} 1 & 0 & -3 & -2 & 3 & 0 \\ 0 & 1 & 2 & 1 & -1 & 0 \\ 0 & 0 & 1 & \frac{3}{7} & -\frac{4}{7} & \frac{1}{7} \end{bmatrix} \rightarrow \begin{bmatrix} 1 & 0 & 0 & -\frac{5}{7} & \frac{9}{7} & \frac{3}{7} \\ 0 & 1 & 0 & \frac{1}{7} & \frac{1}{7} & -\frac{2}{7} \\ 0 & 0 & 1 & \frac{3}{7} & -\frac{4}{7} & \frac{1}{7} \end{bmatrix} \longrightarrow A^{-1} = \begin{bmatrix} -\frac{5}{7} & \frac{9}{7} & \frac{3}{7} \\ \frac{1}{7} & \frac{1}{7} & -\frac{2}{7} \\ \frac{3}{7} & -\frac{4}{7} & \frac{1}{7} \end{bmatrix}$$

$R_3 = \frac{1}{7}r_3$ $R_1 = 3r_3 + r_1$

$R_2 = -2r_3 + r_2$

33. Augment the matrix with the identity and use row operations to find the inverse:

$$A = \begin{bmatrix} 4 & -8 \\ -1 & 2 \end{bmatrix} \rightarrow \begin{bmatrix} 4 & -8 & 1 & 0 \\ -1 & 2 & 0 & 1 \end{bmatrix}$$

$$\rightarrow \begin{bmatrix} -1 & 2 & 0 & 1 \\ 4 & -8 & 1 & 0 \end{bmatrix} \rightarrow \begin{bmatrix} -1 & 2 & 0 & 1 \\ 0 & 0 & 1 & 4 \end{bmatrix} \rightarrow \begin{bmatrix} 1 & -2 & 0 & -1 \\ 0 & 0 & 1 & 4 \end{bmatrix}$$

Interchange $R_2 = 4r_1 + r_2$ $R_1 = -r_1$

r_1 and r_2

There is no inverse because there is no way to obtain the identity on the left side.
The matrix is singular.

35. $\begin{cases} 3x - 2y = 1 \\ 10x + 10y = 5 \end{cases}$ can be written as: $\begin{bmatrix} 3 & -2 & | & 1 \\ 10 & 10 & | & 5 \end{bmatrix}$

$\rightarrow \begin{bmatrix} 3 & -2 & | & 1 \\ 1 & 16 & | & 2 \end{bmatrix} \rightarrow \begin{bmatrix} 1 & 16 & | & 2 \\ 3 & -2 & | & 1 \end{bmatrix} \rightarrow \begin{bmatrix} 1 & 16 & | & 2 \\ 0 & -50 & | & -5 \end{bmatrix} \rightarrow \begin{bmatrix} 1 & 16 & | & 2 \\ 0 & 1 & | & \frac{1}{10} \end{bmatrix} \rightarrow \begin{bmatrix} 1 & 0 & | & \frac{2}{5} \\ 0 & 1 & | & \frac{1}{10} \end{bmatrix}$

$R_2 = -3r_1 + r_2$ Interchange $R_2 = -3r_1 + r_2$ $R_2 = -\frac{1}{50}r_2$ $R_1 = -16r_2 + r_1$
          r_1 and r_2

The solution is $x = \dfrac{2}{5}$, $y = \dfrac{1}{10}$.

37. $\begin{cases} 5x + 6y - 3z = 6 \\ 4x - 7y - 2z = -3 \\ 3x + y - 7z = 1 \end{cases}$ can be written as: $\begin{bmatrix} 5 & 6 & -3 & | & 6 \\ 4 & -7 & -2 & | & -3 \\ 3 & 1 & -7 & | & 1 \end{bmatrix}$

$\rightarrow \begin{bmatrix} 1 & 13 & -1 & | & 9 \\ 4 & -7 & -2 & | & -3 \\ 3 & 1 & -7 & | & 1 \end{bmatrix} \rightarrow \begin{bmatrix} 1 & 13 & -1 & | & 9 \\ 0 & -59 & 2 & | & -39 \\ 0 & -38 & -4 & | & -26 \end{bmatrix} \rightarrow \begin{bmatrix} 1 & 13 & -1 & | & 9 \\ 0 & 1 & -\frac{2}{59} & | & \frac{39}{59} \\ 0 & -38 & -4 & | & -26 \end{bmatrix}$

$R_1 = -r_2 + r_1$ $R_2 = -4r_1 + r_2$ $R_2 = -\frac{1}{59}r_2$
               $R_3 = -3r_1 + r_3$

$\rightarrow \begin{bmatrix} 1 & 0 & -\frac{33}{59} & | & \frac{24}{59} \\ 0 & 1 & -\frac{2}{59} & | & \frac{39}{59} \\ 0 & 0 & -\frac{312}{59} & | & -\frac{52}{59} \end{bmatrix} \rightarrow \begin{bmatrix} 1 & 0 & -\frac{33}{59} & | & \frac{24}{59} \\ 0 & 1 & -\frac{2}{59} & | & \frac{39}{59} \\ 0 & 0 & 1 & | & \frac{1}{6} \end{bmatrix} \rightarrow \begin{bmatrix} 1 & 0 & 0 & | & \frac{1}{2} \\ 0 & 1 & 0 & | & \frac{2}{3} \\ 0 & 0 & 1 & | & \frac{1}{6} \end{bmatrix}$

$R_1 = -13r_2 + r_1$ $R_3 = -\frac{59}{312}r_3$ $R_1 = \frac{33}{59}r_3 + r_1$
$R_3 = 38r_2 + r_3$ $R_2 = \frac{2}{59}r_3 + r_2$

The solution is $x = \dfrac{1}{2}$, $y = \dfrac{2}{3}$, $z = \dfrac{1}{6}$.

39. $\begin{cases} x - 2z = 1 \\ 2x + 3y = -3 \\ 4x - 3y - 4z = 3 \end{cases}$ can be written as: $\begin{bmatrix} 1 & 0 & -2 & | & 1 \\ 2 & 3 & 0 & | & -3 \\ 4 & -3 & -4 & | & 3 \end{bmatrix}$

$\rightarrow \begin{bmatrix} 1 & 0 & -2 & | & 1 \\ 0 & 3 & 4 & | & -5 \\ 0 & -3 & 4 & | & -1 \end{bmatrix} \rightarrow \begin{bmatrix} 1 & 0 & -2 & | & 1 \\ 0 & 1 & \frac{4}{3} & | & -\frac{5}{3} \\ 0 & -3 & 4 & | & -1 \end{bmatrix} \rightarrow \begin{bmatrix} 1 & 0 & -2 & | & 1 \\ 0 & 1 & \frac{4}{3} & | & -\frac{5}{3} \\ 0 & 0 & 8 & | & -6 \end{bmatrix}$

$R_2 = -2r_1 + r_2$ $R_2 = \frac{1}{3}r_2$ $R_3 = 3r_2 + r_3$
$R_3 = -4r_1 + r_3$

$\rightarrow \begin{bmatrix} 1 & 0 & -2 & | & 1 \\ 0 & 1 & \frac{4}{3} & | & -\frac{5}{3} \\ 0 & 0 & 1 & | & -\frac{3}{4} \end{bmatrix} \rightarrow \begin{bmatrix} 1 & 0 & 0 & | & -\frac{1}{2} \\ 0 & 1 & 0 & | & -\frac{2}{3} \\ 0 & 0 & 1 & | & -\frac{3}{4} \end{bmatrix}$

$R_3 = \frac{1}{8}r_3$ $R_1 = 2r_3 + r_1$
                $R_2 = -\frac{4}{3}r_3 + r_2$

The solution is $x = -\dfrac{1}{2}$, $y = -\dfrac{2}{3}$, $z = -\dfrac{3}{4}$.

41. $\begin{cases} x - y + z = 0 \\ x - y - 5z = 6 \\ 2x - 2y + z = 1 \end{cases}$ can be written as: $\begin{bmatrix} 1 & -1 & 1 & | & 0 \\ 1 & -1 & -5 & | & 6 \\ 2 & -2 & 1 & | & 1 \end{bmatrix}$

$$\rightarrow \begin{bmatrix} 1 & -1 & 1 & | & 0 \\ 0 & 0 & -6 & | & 6 \\ 0 & 0 & -1 & | & 1 \end{bmatrix} \rightarrow \begin{bmatrix} 1 & -1 & 1 & | & 0 \\ 0 & 0 & 1 & | & -1 \\ 0 & 0 & -1 & | & 1 \end{bmatrix} \rightarrow \begin{bmatrix} 1 & -1 & 0 & | & 1 \\ 0 & 0 & 1 & | & -1 \\ 0 & 0 & 0 & | & 0 \end{bmatrix} \rightarrow \begin{cases} x = y + 1 \\ z = -1 \end{cases}$$

$R_2 = -r_1 + r_2$ $\qquad R_2 = -\frac{1}{6}r_2$ $\qquad R_1 = -r_2 + r_1$

$R_3 = -2r_1 + r_3$ $\qquad\qquad\qquad\qquad R_3 = r_2 + r_3$

The solution is $x = y + 1$, $z = -1$, y is any real number..

43. $\begin{vmatrix} x - y - z - t = 1 \\ 2x + y - z + 2t = 3 \\ x - 2y - 2z - 3t = 0 \\ 3x - 4y + z + 5t = -3 \end{vmatrix}$ can be written as : $\begin{vmatrix} 1 & -1 & -1 & -1 & | & 1 \\ 2 & 1 & -1 & 2 & | & 3 \\ 1 & -2 & -2 & -3 & | & 0 \\ 3 & -4 & 1 & 5 & | & -3 \end{vmatrix}$

$$\rightarrow \begin{vmatrix} 1 & -1 & -1 & -1 & | & 1 \\ 0 & 3 & 1 & 4 & | & 1 \\ 0 & -1 & -1 & -2 & | & -1 \\ 0 & -1 & 4 & 8 & | & -6 \end{vmatrix} \rightarrow \begin{vmatrix} 1 & -1 & -1 & -1 & | & 1 \\ 0 & -1 & -1 & -2 & | & -1 \\ 0 & 3 & 1 & 4 & | & 1 \\ 0 & -1 & 4 & 8 & | & -6 \end{vmatrix} \rightarrow \begin{vmatrix} 1 & -1 & -1 & -1 & | & 1 \\ 0 & 1 & 1 & 2 & | & 1 \\ 0 & 3 & 1 & 4 & | & 1 \\ 0 & -1 & 4 & 8 & | & -6 \end{vmatrix}$$

$R_2 = -2r_1 + r_2$ $\qquad$ Interchange r_2 and r_3 $\qquad R_2 = -r_2$

$R_3 = -r_1 + r_3$

$R_4 = -3r_1 + r_4$

$$\rightarrow \begin{vmatrix} 1 & 0 & 0 & 1 & | & 2 \\ 0 & 1 & 1 & 2 & | & 1 \\ 0 & 0 & -2 & -2 & | & -2 \\ 0 & 0 & 5 & 10 & | & -5 \end{vmatrix} \rightarrow \begin{vmatrix} 1 & 0 & 0 & 1 & | & 2 \\ 0 & 1 & 1 & 2 & | & 1 \\ 0 & 0 & 1 & 1 & | & 1 \\ 0 & 0 & 1 & 2 & | & -1 \end{vmatrix} \rightarrow \begin{vmatrix} 1 & 0 & 0 & 1 & | & 2 \\ 0 & 1 & 1 & 2 & | & 1 \\ 0 & 0 & 1 & 1 & | & 1 \\ 0 & 0 & 0 & 1 & | & -2 \end{vmatrix}$$

$R_1 = r_2 + r_1$ $\qquad\qquad R_3 = -\frac{1}{2}r_3$ $\qquad\qquad R_4 = r_4 - r_3$

$R_3 = -3r_2 + r_3$ $\qquad R_4 = \frac{1}{5}r_4$

$R_4 = r_2 + r_4$

$$\rightarrow \begin{vmatrix} 1 & 0 & 0 & 0 & | & 4 \\ 0 & 1 & 0 & 0 & | & 2 \\ 0 & 0 & 1 & 0 & | & 3 \\ 0 & 0 & 0 & 1 & | & -2 \end{vmatrix}$$

$R_1 = r_1 - r_4$

$R_2 = r_2 - r_3 - r_4$

$R_3 = r_3 - r_4$

The solution is $x = 4$, $y = 2$, $z = 3$, $t = -2$.

45. Evaluating the determinant:

$$\begin{vmatrix} 3 & 4 \\ 1 & 3 \end{vmatrix} = 3(3) - 4(1) = 9 - 4 = 5$$

47. Evaluating the determinant:
$$\begin{vmatrix} 1 & 4 & 0 \\ -1 & 2 & 6 \\ 4 & 1 & 3 \end{vmatrix} = 1\begin{vmatrix} 2 & 6 \\ 1 & 3 \end{vmatrix} - 4\begin{vmatrix} -1 & 6 \\ 4 & 3 \end{vmatrix} + 0\begin{vmatrix} -1 & 2 \\ 4 & 1 \end{vmatrix}$$
$$= 1[2(3) - 6(1)] - 4[-1(3) - 6(4)] + 0[-1(1) - 2(4)]$$
$$= 1(6 - 6) - 4(-3 - 24) + 0(-1 - 8)$$
$$= 1(0) - 4(-27) + 0(-9) = 0 + 108 + 0 = 108$$

49. Evaluating the determinant:
$$\begin{vmatrix} 2 & 1 & -3 \\ 5 & 0 & 1 \\ 2 & 6 & 0 \end{vmatrix} = 2\begin{vmatrix} 0 & 1 \\ 6 & 0 \end{vmatrix} - 1\begin{vmatrix} 5 & 1 \\ 2 & 0 \end{vmatrix} + (-3)\begin{vmatrix} 5 & 0 \\ 2 & 6 \end{vmatrix}$$
$$= 2(0 - 6) - 1(0 - 2) + (-3)(30 - 0) = -12 + 2 - 90 = -100$$

51. Set up and evaluate the determinants to use Cramer's Rule:
$$\begin{cases} x - 2y = 4 \\ 3x + 2y = 4 \end{cases}$$
$$D = \begin{vmatrix} 1 & -2 \\ 3 & 2 \end{vmatrix} = 1(2) - 3(-2) = 2 + 6 = 8; \quad D_x = \begin{vmatrix} 4 & -2 \\ 4 & 2 \end{vmatrix} = 4(2) - 4(-2) = 8 + 8 = 16$$
$$D_y = \begin{vmatrix} 1 & 4 \\ 3 & 4 \end{vmatrix} = 1(4) - 4(3) = 4 - 12 = -8$$
Find the solutions by Cramer's Rule: $x = \dfrac{D_x}{D} = \dfrac{16}{8} = 2 \qquad y = \dfrac{D_y}{D} = \dfrac{-8}{8} = -1$

53. Set up and evaluate the determinants to use Cramer's Rule:
$$\begin{cases} 2x + 3y = 13 \\ 3x - 2y = 0 \end{cases}$$
$$D = \begin{vmatrix} 2 & 3 \\ 3 & -2 \end{vmatrix} = -4 - 9 = -13; \quad D_x = \begin{vmatrix} 13 & 3 \\ 0 & -2 \end{vmatrix} = -26 - 0 = -26; \quad D_y = \begin{vmatrix} 2 & 13 \\ 3 & 0 \end{vmatrix} = 0 - 39 = -39$$
Find the solutions by Cramer's Rule: $x = \dfrac{D_x}{D} = \dfrac{-26}{-13} = 2 \qquad y = \dfrac{D_y}{D} = \dfrac{-39}{-13} = 3$

55. Set up and evaluate the determinants to use Cramer's Rule:
$$\begin{cases} x + 2y - z = 6 \\ 2x - y + 3z = -13 \\ 3x - 2y + 3z = -16 \end{cases}$$
$$D = \begin{vmatrix} 1 & 2 & -1 \\ 2 & -1 & 3 \\ 3 & -2 & 3 \end{vmatrix} = 1\begin{vmatrix} -1 & 3 \\ -2 & 3 \end{vmatrix} - 2\begin{vmatrix} 2 & 3 \\ 3 & 3 \end{vmatrix} + (-1)\begin{vmatrix} 2 & -1 \\ 3 & -2 \end{vmatrix}$$
$$= 1(-3 + 6) - 2(6 - 9) - 1(-4 + 3) = 3 + 6 + 1 = 10$$

$$D_x = \begin{vmatrix} 6 & 2 & -1 \\ -13 & -1 & 3 \\ -16 & -2 & 3 \end{vmatrix} = 6\begin{vmatrix} -1 & 3 \\ -2 & 3 \end{vmatrix} - 2\begin{vmatrix} -13 & 3 \\ -16 & 3 \end{vmatrix} + (-1)\begin{vmatrix} -13 & -1 \\ -16 & -2 \end{vmatrix}$$

$$= 6(-3+6) - 2(-39+48) - 1(26-16) = 18 - 18 - 10 = -10$$

$$D_y = \begin{vmatrix} 1 & 6 & -1 \\ 2 & -13 & 3 \\ 3 & -16 & 3 \end{vmatrix} = 1\begin{vmatrix} -13 & 3 \\ -16 & 3 \end{vmatrix} - 6\begin{vmatrix} 2 & 3 \\ 3 & 3 \end{vmatrix} + (-1)\begin{vmatrix} 2 & -13 \\ 3 & -16 \end{vmatrix}$$

$$= 1(-39+48) - 6(6-9) - 1(-32+39) = 9 + 18 - 7 = 20$$

$$D_z = \begin{vmatrix} 1 & 2 & 6 \\ 2 & -1 & -13 \\ 3 & -2 & -16 \end{vmatrix} = 1\begin{vmatrix} -1 & -13 \\ -2 & -16 \end{vmatrix} - 2\begin{vmatrix} 2 & -13 \\ 3 & -16 \end{vmatrix} + 6\begin{vmatrix} 2 & -1 \\ 3 & -2 \end{vmatrix}$$

$$= 1(16-26) - 2(-32+39) + 6(-4+3) = -10 - 14 - 6 = -30$$

Find the solutions by Cramer's Rule:

$$x = \frac{D_x}{D} = \frac{-10}{10} = -1 \qquad y = \frac{D_y}{D} = \frac{20}{10} = 2 \qquad z = \frac{D_z}{D} = \frac{-30}{10} = -3$$

57. Let $\begin{vmatrix} x & y \\ a & b \end{vmatrix} = 8$

Then $\begin{vmatrix} 2x & y \\ 2a & b \end{vmatrix} = 16$ by Theorem 14.

The value of the determinant is multiplied by k when the elements of a column are multiplied by k.

59. Find the partial fraction decomposition:

$$\frac{6}{x(x-4)} = \frac{A}{x} + \frac{B}{x-4} \quad \text{(Multiply both sides by } x(x-4).)$$

$$6 = A(x-4) + Bx$$

Let $x = 4$: then $6 = A(4-4) + B(4) \rightarrow 4B = 6 \rightarrow B = \frac{3}{2}$

Let $x = 0$: then $6 = A(0-4) + B(0) \rightarrow -4A = 6 \rightarrow A = -\frac{3}{2}$

$$\frac{6}{x(x-4)} = \frac{\left(-\frac{3}{2}\right)}{x} + \frac{\left(\frac{3}{2}\right)}{x-4}$$

61. Find the partial fraction decomposition:

$$\frac{x-4}{x^2(x-1)} = \frac{A}{x} + \frac{B}{x^2} + \frac{C}{x-1} \text{(Multiply both sides by } x^2(x-1).)$$

$$x - 4 = Ax(x-1) + B(x-1) + Cx^2$$

Let $x = 1$: then $1 - 4 = A(1)(1-1) + B(1-1) + C(1)^2 \rightarrow -3 = C \rightarrow C = -3$

Let $x = 0$: then $0 - 4 = A(0)(0-1) + B(0-1) + C(0)^2 \rightarrow -4 = -B \rightarrow B = 4$

Let $x = 2$: then $2 - 4 = A(2)(2-1) + B(2-1) + C(2)^2 \rightarrow -2 = 2A + B + 4C$

$\rightarrow 2A = -2 - 4 - 4(-3) \rightarrow 2A = 6 \rightarrow A = 3$

$$\frac{x-4}{x^2(x-1)} = \frac{3}{x} + \frac{4}{x^2} + \frac{-3}{x-1}$$

63. Find the partial fraction decomposition:

$$\frac{x}{(x^2+9)(x+1)} = \frac{A}{x+1} + \frac{Bx+C}{x^2+9} \quad \text{(Multiply both sides by } (x+1)(x^2+9).)$$

$$x = A(x^2+9) + (Bx+C)(x+1)$$

Let $x = -1$: then $-1 = A((-1)^2 + 9) + (B(-1) + C)(-1+1)$

$$\rightarrow -1 = A(10) + (-B+C)(0) \rightarrow -1 = 10A \rightarrow A = -\frac{1}{10}$$

Let $x = 1$: then $1 = A(1^2 + 9) + (B(1) + C)(1+1) \rightarrow 1 = 10A + 2B + 2C$

$$\rightarrow 1 = 10\left(-\frac{1}{10}\right) + 2B + 2C \rightarrow 2 = 2B + 2C \rightarrow B + C = 1$$

Let $x = 0$: then $0 = A(0^2 + 9) + (B(0) + C)(0+1) \rightarrow 0 = 9A + C$

$$\rightarrow 0 = 9\left(-\frac{1}{10}\right) + C \rightarrow C = \frac{9}{10} B = 1 - C \rightarrow B = 1 - \frac{9}{10} \rightarrow B = \frac{1}{10}$$

$$\frac{x}{(x^2+9)(x+1)} = \frac{-\left(\frac{1}{10}\right)}{x+1} + \frac{\left(\frac{1}{10}x + \frac{9}{10}\right)}{x^2+9}$$

65. Find the partial fraction decomposition:

$$\frac{x^3}{(x^2+4)^2} = \frac{Ax+B}{x^2+4} + \frac{Cx+D}{(x^2+4)^2} \quad \text{(Multiply both sides by } (x^2+4)^2.)$$

$$x^3 = (Ax+B)(x^2+4) + Cx + D$$

$$x^3 = Ax^3 + Bx^2 + 4Ax + 4B + Cx + D \Rightarrow x^3 = Ax^3 + Bx^2 + (4A+C)x + 4B + D$$

$$A = 1; \quad B = 0$$

$$4A + C = 0 \rightarrow 4(1) + C = 0 \rightarrow C = -4$$

$$4B + D = 0 \rightarrow 4(0) + D = 0 \rightarrow D = 0$$

$$\frac{x^3}{(x^2+4)^2} = \frac{x}{x^2+4} + \frac{-4x}{(x^2+4)^2}$$

67. Find the partial fraction decomposition:

$$\frac{x^2}{(x^2+1)(x^2-1)} = \frac{x^2}{(x^2+1)(x-1)(x+1)} = \frac{A}{x-1} + \frac{B}{x+1} + \frac{Cx+D}{x^2+1}$$

(Multiply both sides by $(x-1)(x+1)(x^2+1)$.)

$$x^2 = A(x+1)(x^2+1) + B(x-1)(x^2+1) + (Cx+D)(x-1)(x+1)$$

Let $x = 1$: then $1^2 = A(1+1)(1^2+1) + B(1-1)(1^2+1) + (C(1)+D)(1-1)(1+1)$

$$\rightarrow 1 = 4A \rightarrow A = \frac{1}{4}$$

Let $x = -1$: then

$$(-1)^2 = A(-1+1)((-1)^2+1) + B(-1-1)((-1)^2+1) + (C(-1)+D)(-1-1)(-1+1)$$

$$\rightarrow 1 = -4B \rightarrow B = -\frac{1}{4}$$

Let $x = 0$: then

$$0^2 = A(0+1)(0^2+1) + B(0-1)(0^2+1) + (C(0)+D)(0-1)(0+1)$$

$$\rightarrow 0 = A - B - D \rightarrow 0 = \frac{1}{4} - \left(-\frac{1}{4}\right) - D \rightarrow D = \frac{1}{2}$$

Let $x = 2$: then

$$2^2 = A(2+1)(2^2+1) + B(2-1)(2^2+1) + (C(2)+D)(2-1)(2+1)$$

$$\rightarrow 4 = 15A + 5B + 6C + 3D \rightarrow 4 = 15\left(\frac{1}{4}\right) + 5\left(-\frac{1}{4}\right) + 6C + 3\left(\frac{1}{2}\right)$$

$$\rightarrow 6C = 4 - \frac{15}{4} + \frac{5}{4} - \frac{3}{2} \rightarrow 6C = 0 \rightarrow C = 0$$

$$\frac{x^2}{(x^2+1)(x^2-1)} = \frac{x^2}{(x^2+1)(x-1)(x+1)} = \frac{\left(\frac{1}{4}\right)}{x-1} + \frac{-\left(\frac{1}{4}\right)}{x+1} + \frac{\left(\frac{1}{2}\right)}{x^2+1}$$

69. Solve the first equation for y, substitute into the second equation and solve:

$$\begin{cases} 2x + y + 3 = 0 \rightarrow y = -2x - 3 \\ x^2 + y^2 = 5 \end{cases}$$

$$x^2 + (-2x-3)^2 = 5 \Rightarrow x^2 + 4x^2 + 12x + 9 = 5$$

$$5x^2 + 12x + 4 = 0 \Rightarrow (5x+2)(x+2) = 0 \Rightarrow x = -\frac{2}{5} \quad \text{or} \quad x = -2$$

$$y = -\frac{11}{5} \qquad y = 1$$

Solutions: $\left(-\frac{2}{5}, -\frac{11}{5}\right), (-2, 1)$.

71. Multiply each side of the second equation by 2 and add the equations to eliminate xy:

$$\begin{cases} 2xy + y^2 = 10 \longrightarrow \quad 2xy + y^2 = 10 \\ -xy + 3y^2 = 2 \xrightarrow{2} \underline{-2xy + 6y^2 = 4} \end{cases}$$

$$7y^2 = 14 \rightarrow y^2 = 2 \rightarrow y = \pm\sqrt{2}$$

If $y = \sqrt{2}$: $\quad 2x(\sqrt{2}) + (\sqrt{2})^2 = 10 \rightarrow 2\sqrt{2}x = 8 \rightarrow x = \frac{8}{2\sqrt{2}} = 2\sqrt{2}$

If $y = -\sqrt{2}$: $\quad 2x(-\sqrt{2}) + (-\sqrt{2})^2 = 10 \rightarrow -2\sqrt{2}x = 8 \rightarrow x = \frac{8}{-2\sqrt{2}} = -2\sqrt{2}$

Solutions: $(2\sqrt{2}, \sqrt{2}), (-2\sqrt{2}, -\sqrt{2})$

73. Substitute into the second equation into the first equation and solve:
$$\begin{cases} x^2 + y^2 = 6y \\ \quad x^2 = 3y \end{cases}$$
$$3y + y^2 = 6y \rightarrow y^2 - 3y = 0 \rightarrow y(y-3) = 0 \rightarrow y = 0 \text{ or } y = 3$$
If $y = 0$: $x^2 = 3(0) \rightarrow x^2 = 0 \rightarrow x = 0$
If $y = 3$: $x^2 = 3(3) \rightarrow x^2 = 9 \rightarrow x = \pm 3$
Solutions: $(0, 0)$, $(-3, 3)$, $(3, 3)$.

75. Factor the second equation, solve for x, substitute into the first equation and solve:
$$\begin{cases} 3x^2 + 4xy + 5y^2 = 8 \\ \quad x^2 + 3xy + 2y^2 = 0 \rightarrow (x + 2y)(x + y) = 0 \rightarrow x = -2y \text{ or } x = -y \end{cases}$$

Substitute $x = -2y$ and solve:
$$3x^2 + 4xy + 5y^2 = 8$$
$$3(-2y)^2 + 4(-2y)y + 5y^2 = 8$$
$$12y^2 - 8y^2 + 5y^2 = 8$$
$$9y^2 = 8$$
$$y^2 = \frac{8}{9} \Rightarrow y = \pm\frac{2\sqrt{2}}{3}$$

If $y = \frac{2\sqrt{2}}{3}$: $x = -2\left(\frac{2\sqrt{2}}{3}\right) = \frac{-4\sqrt{2}}{3}$

If $y = \frac{-2\sqrt{2}}{3}$: $x = -2\left(\frac{-2\sqrt{2}}{3}\right) = \frac{4\sqrt{2}}{3}$

Substitute $x = -y$ and solve:
$$3x^2 + 4xy + 5y^2 = 8$$
$$3(-y)^2 + 4(-y)y + 5y^2 = 8$$
$$3y^2 - 4y^2 + 5y^2 = 8$$
$$4y^2 = 8$$
$$y^2 = 2 \Rightarrow y = \pm\sqrt{2}$$

If $y = \sqrt{2}$: $x = -\sqrt{2}$
If $y = -\sqrt{2}$: $x = \sqrt{2}$

Solutions: $\left(\frac{-4\sqrt{2}}{3}, \frac{2\sqrt{2}}{3}\right), \left(\frac{4\sqrt{2}}{3}, \frac{-2\sqrt{2}}{3}\right), \left(-\sqrt{2}, \sqrt{2}\right), \left(\sqrt{2}, -\sqrt{2}\right)$

77. Multiply each side of the second equation by $-y$ and add the equations to eliminate y:
$$\begin{cases} x^2 - 3x + y^2 + y = -2 \xrightarrow{\quad} x^2 - 3x + y^2 + y = -2 \\ \dfrac{x^2 - x}{y} + y + 1 = 0 \xrightarrow{-y} -x^2 + x - y^2 - y = 0 \end{cases}$$

$$-2x \qquad\quad = -2 \Rightarrow x = 1$$

If $x = 1$: $1^2 - 3(1) + y^2 + y = -2 \rightarrow y^2 + y = 0 \rightarrow y(y + 1) = 0$
$$\rightarrow y = 0 \text{ or } y = -1$$

Note that $y \neq 0$ because that would cause division by zero in the original system.
Solution: $(1, -1)$

79. $3x + 4y \leq 12$

(a) Graph the line $3x + 4y = 12$. Use a solid line since the inequality uses $\leq$.
Choose a test point not on the line, such as (0, 0).
Since $3(0) + 4(0) \leq 12$ is true, shade the side of the line containing (0, 0).

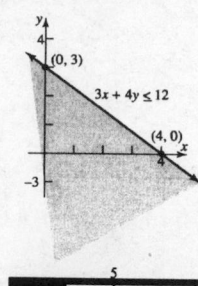

(b) Use the SHADE function in the calculator's DRAW menu: $\text{Shade}\left(-5, -(3/4)x + 3\right)$.

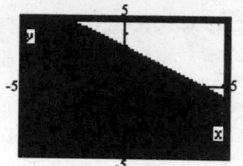

81.   Graph the system of linear inequalities:
$$\begin{cases} -2x + y \leq 2 \\ \quad x + y \geq 2 \end{cases}$$

(a)   Graph the line $-2x + y = 2$. Use a solid line since the inequality uses $\leq$.
Choose a test point not on the line, such as (0, 0). Since $-2(0) + 0 \leq 2$ is true, shade the side of the line containing (0, 0).

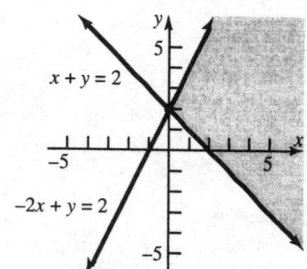

(b)   Graph the line $x + y = 2$. Use a solid line since the inequality uses $\geq$.
Choose a test point not on the line, such as (0, 0). Since $0 + 0 \geq 2$ is false, shade the opposite side of the line from (0, 0).

(c)   The overlapping region is the solution.

(d)   The graph is unbounded.

(e)   Find the vertices:
To find the intersection of $x + y = 2$ and $-2x + y = 2$, solve the system:
$$\begin{cases} \quad x + y = 2 \quad \rightarrow \quad x = 2 - y \\ -2x + y = 2 \end{cases}$$
Substitute and solve:
$$-2(2 - y) + y = 2$$
$$-4 + 2y + y = 2$$
$$3y = 6$$
$$y = 2$$
$$x = 2 - 2 = 0$$
The point of intersection is (0, 2).
The corner point is (0, 2).

83. Graph the system of linear inequalities:

$$\begin{cases} x \geq 0 \\ y \geq 0 \\ x + y \leq 4 \\ 2x + 3y \leq 6 \end{cases}$$

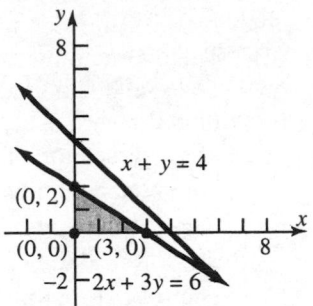

(a) Graph $x \geq 0$; $y \geq 0$. Shaded region is the first quadrant.

(b) Graph the line $x + y = 4$. Use a solid line since the inequality uses $\leq$.
Choose a test point not on the line, such as $(0, 0)$. Since $0 + 0 \leq 4$ is true, shade the side of the line containing $(0, 0)$.

(c) Graph the line $2x + 3y = 6$. Use a solid line since the inequality uses $\leq$.
Choose a test point not on the line, such as $(0, 0)$. Since $2(0) + 3(0) \leq 6$ is true, shade the side of the line containing $(0, 0)$.

(d) The overlapping region is the solution.

(e) The graph is bounded.

(f) Find the vertices:
The x-axis and y-axis intersect at $(0, 0)$.
The intersection of $2x + 3y = 6$ and the y-axis is $(0, 2)$.
The intersection of $2x + 3y = 6$ and the x-axis is $(3, 0)$.
The three corner points are $(0, 0)$, $(0, 2)$, and $(3, 0)$.

85. Graph the system of linear inequalities:

$$\begin{cases} x \geq 0 \\ y \geq 0 \\ 2x + y \leq 8 \\ x + 2y \geq 2 \end{cases}$$

(a) Graph $x \geq 0$; $y \geq 0$. Shaded region is the first quadrant.

(b) Graph the line $2x + y = 8$. Use a solid line since the inequality uses $\leq$.
Choose a test point not on the line, such as $(0, 0)$. Since $2(0) + 0 \leq 8$ is true, shade the side of the line containing $(0, 0)$.

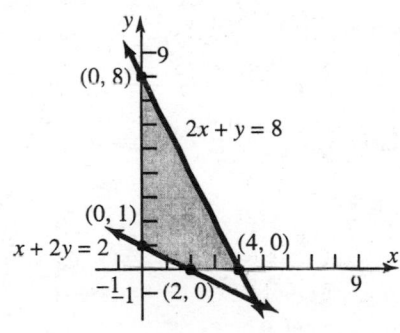

(c)   Graph the line $x + 2y = 2$. Use a solid line since the inequality uses $\geq$.
Choose a test point not on the line, such as (0, 0). Since $0 + 2(0) \geq 2$ is false, shade
the opposite side of the line from (0, 0).

(d)   The overlapping region is the solution.

(e)   The graph is bounded

(f)   Find the vertices:
The intersection of $x + 2y = 2$ and the y-axis is (0, 1).
The intersection of $x + 2y = 2$ and the x-axis is (2, 0).
The intersection of $2x + y = 8$ and the y-axis is (0, 8).
The intersection of $2x + y = 8$ and the x-axis is (4, 0).
The four corner points are (0, 1), (0, 8), (2, 0), and (4, 0).

87.   Graph the system of inequalities:
$$\begin{cases} x^2 + y^2 \leq 16 \\ \quad x + y \geq 2 \end{cases}$$

(a)   Graph the circle $x^2 + y^2 = 16$. Use a solid line since the
inequality uses $\leq$. Choose a test point not on the circle, such
as (0, 0). Since $0^2 + 0^2 \leq 16$ is true, shade the side of the
circle containing (0, 0).

(b)   Graph the line $x + y = 2$. Use a solid line since the
inequality uses $\geq$. Choose a test point not on the line, such
as (0, 0). Since $0 + 0 \geq 2$ is false, shade the opposite side of
the line from (0, 0).

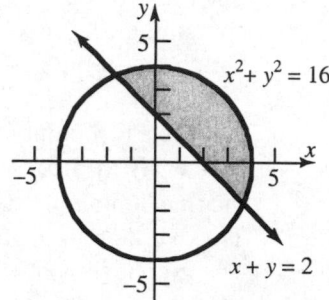

(c)   The overlapping region is the solution.

89.   Graph the system of inequalities:
$$\begin{cases} y \leq x^2 \\ xy \leq 4 \end{cases}$$

(a)   Graph the parabola $y = x^2$. Use a solid line since the
inequality uses $\leq$. Choose a test point not on the parabola,
such as (1, 2). Since $2 \leq 1^2$ is false, shade the opposite side
of the parabola from (1, 2).

(b)   Graph the hyperbola $xy = 4$. Use a solid line since the
inequality uses $\leq$. Choose a test point not on the hyperbola,
such as (1, 2). Since $1 \cdot 2 \leq 4$ is true, shade the same side
of the hyperbola as (1, 2).

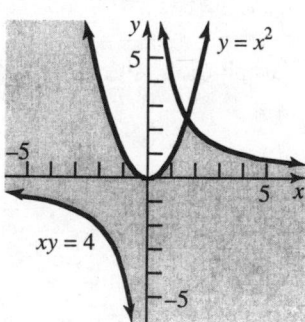

(c)   The overlapping region is the solution.

91. Maximize $z = 3x + 4y$ Subject to $x \geq 0$, $y \geq 0$, $3x + 2y \geq 6$, $x + y \leq 8$
Graph the constraints.

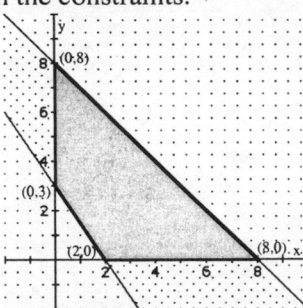

The corner points are $(0, 3)$, $(2, 0)$, $(0, 8)$, $(8, 0)$.

Evaluate the objective function:

Vertex	Value of $z = 3x + 4y$
$(0, 3)$	$z = 3(0) + 4(3) = 12$
$(0, 8)$	$z = 3(0) + 4(8) = 32$
$(2, 0)$	$z = 3(2) + 4(0) = 6$
$(8, 0)$	$z = 3(8) + 4(0) = 24$

The maximum value is 32 at $(0, 8)$.

93. Minimize $z = 3x + 5y$
Subject to $x \geq 0$, $y \geq 0$, $x + y \geq 1$, $3x + 2y \leq 12$, $x + 3y \leq 12$
Graph the constraints.

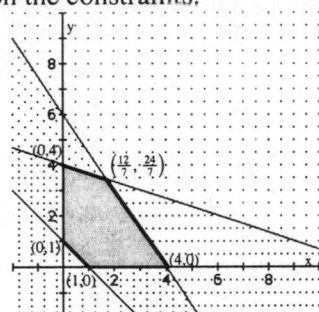

To find the intersection of $3x + 2y = 12$ and $x + 3y = 12$, solve the system:
$$\begin{cases} 3x + 2y = 12 \\ x + 3y = 12 \end{cases} \quad \rightarrow \quad x = 12 - 3y$$

Substitute and solve:

$$3(12 - 3y) + 2y = 12 \rightarrow 36 - 9y + 2y = 12 \rightarrow -7y = -24 \rightarrow y = \frac{24}{7}$$

$$x = 12 - 3\left(\frac{24}{7}\right) = 12 - \frac{72}{7} = \frac{12}{7}$$

The point of intersection is $\left(\dfrac{12}{7}, \dfrac{24}{7}\right)$.

The corner points are $(0, 1)$, $(1, 0)$, $(0, 4)$, $(4, 0)$, $\left(\dfrac{12}{7}, \dfrac{24}{7}\right)$.

Evaluate the objective function:

Vertex	Value of $z = 3x + 5y$
$(0, 1)$	$z = 3(0) + 5(1) = 5$
$(0, 4)$	$z = 3(0) + 5(4) = 20$
$(1, 0)$	$z = 3(1) + 5(0) = 3$
$(4, 0)$	$z = 3(4) + 5(0) = 12$
$\left(\dfrac{12}{7}, \dfrac{24}{7}\right)$	$z = 3\left(\dfrac{12}{7}\right) + 5\left(\dfrac{24}{7}\right) = \dfrac{36}{7} + \dfrac{120}{7} = \dfrac{156}{7} \approx 22.3$

The minimum value is 3 at $(1, 0)$.

95. Multiply each side of the first equation by –2 and eliminate x:
$$\begin{cases} 2x + 5y = 5 \quad \xrightarrow{-2} \quad -4x - 10y = -10 \\ 4x + 10y = A \quad \longrightarrow \quad \underline{4x + 10y = A} \\ \quad 0 = A - 10 \end{cases}$$

If there are to be infinitely many solutions, the sum in elimination should be $0 = 0$.
Therefore, $A - 10 = 0$ or $A = 10$.

97. $y = ax^2 + bx + c$
At $(0, 1)$ the equation becomes:

$1 = a(0)^2 + b(0) + c$

$c = 1$

At $(1, 0)$ the equation becomes:

$0 = a(1)^2 + b(1) + c$

$0 = a + b + c$

$a + b + c = 0$

At $(-2, 1)$ the equation becomes:

$1 = a(-2)^2 + b(-2) + c = 4a - 2b + c \rightarrow 4a - 2b + c = 1$

The system of equations is:
$$\begin{cases} a + b + c = 0 \\ 4a - 2b + c = 1 \\ \phantom{4a - 2b + {}} c = 1 \end{cases}$$

Substitute $c = 1$ into the first and second equations and simplify:
$$\begin{cases} a + b + 1 = 0 \quad \rightarrow \quad a + b = -1 \quad \rightarrow \quad a = -b - 1 \\ 4a - 2b + 1 = 1 \quad \rightarrow \quad 4a - 2b = 0 \end{cases}$$

Solve the first equation for a, substitute into the second equation and solve:

$4(-b - 1) - 2b = 0 \rightarrow -4b - 4 - 2b = 0$

$$-6b = 4 \rightarrow b = -\frac{2}{3} \rightarrow a = \frac{2}{3} - 1 = -\frac{1}{3}$$

The quadratic function is $y = -\dfrac{1}{3}x^2 - \dfrac{2}{3}x + 1$.

99. Let x = the number of pounds of coffee that costs \$3.00 per pound.
Let y = the number of pounds of coffee that costs \$6.00 per pound.
Then $x + y = 100$ represents the total amount of coffee in the blend.
The value of the blend will be represented by the equation: $3x + 6y = 3.90(100)$.
Solve the system of equations:
$$\begin{cases} x + y = 100 \quad \Rightarrow \quad y = 100 - x \\ 3x + 6y = 390 \end{cases}$$

Solve by substitution:
$$3x + 6(100 - x) = 390 \Rightarrow 3x + 600 - 6x = 390$$
$$-3x = -210 \Rightarrow x = 70$$
$$y = 100 - 70 = 30$$
The blend is made up of 70 pounds of the $3 per pound coffee and 30 pounds of the $6 per pound coffee.

101. Let x = the number of small boxes.
Let y = the number of medium boxes.
Let z = the number of large boxes.
Oatmeal raisin equation: $x + 2y + 2z = 15$
Chocolate chip equation: $x + y + 2z = 10$
Shortbread equation: $y + 3z = 11$
Multiply each side of the second equation by -1 and add to the first equation to eliminate x:
$$\begin{cases} x + 2y + 2z = 15 & \longrightarrow & x + 2y + 2z = \ 15 \\ x + \ y + 2z = 10 & \overset{-1}{\longrightarrow} & -x - \ y - 2z = -10 \\ \quad\quad y + 3z = 11 & & y \quad\quad = \ \ 5 \end{cases}$$
Substituting and solving for the other variables:
$$5 + 3z = 11 \qquad\qquad x + 5 + 2(2) = 10$$
$$3z = 6 \qquad\qquad\qquad x + 9 = 10$$
$$z = 2 \qquad\qquad\qquad\quad x = 1$$
1 small box, 5 medium boxes, and 2 large boxes of cookies should be purchased.

103. Let x = the speed of the boat in still water.
Let y = the speed of the river current.
Let d = the distance from Chiritza to the Flotel Orellana (100 kilometers)

	Rate	Time	Distance
trip downstream	$x + y$	$\dfrac{5}{2}$	100
trip downstream	$x - y$	3	100

The system of equations is:
$$\begin{cases} (x + y)\left(\dfrac{5}{2}\right) = 100 & \Rightarrow & 5x + 5y = 200 \\ (x - y)(3) = d & \Rightarrow & 3x - 3y = 100 \end{cases}$$

$$5x + 5y = 200 \xrightarrow{\ 3\ } \quad 15x + 15y = 600$$
$$3x - 3y = 100 \xrightarrow{\ 5\ } \underline{+\ 15x - 15y = 500}$$
$$\qquad\qquad\qquad\qquad 30x = 1100$$
$$\therefore x = \frac{1100}{30} = \frac{110}{3}$$

$$\rightarrow 3\left(\frac{110}{3}\right) - 3y = 100 \Rightarrow 110 - 3y = 100 \Rightarrow 10 = 3y \Rightarrow y = \frac{10}{3}$$

The speed of the boat $= \dfrac{110}{3} \approx 36.67$ km/hr; the speed of the current $= \dfrac{10}{3} \approx 3.33$ km/hr.

105. Let x = the number of hours for Bruce to do the job alone.

Let y = the number of hours for Bryce to do the job alone.

Let z = the number of hours for Marty to do the job alone.

Then $\dfrac{1}{x}$ represents the fraction of the job that Bruce does in one hour.

$\dfrac{1}{y}$ represents the fraction of the job that Bryce does in one hour.

$\dfrac{1}{z}$ represents the fraction of the job that Marty does in one hour.

The equation representing Bruce and Bryce working together is:

$$\frac{1}{x}+\frac{1}{y}=\frac{1}{\left(\dfrac{4}{3}\right)}=\frac{3}{4}=0.75$$

The equation representing Bryce and Marty working together is:

$$\frac{1}{y}+\frac{1}{z}=\frac{1}{\left(\dfrac{8}{5}\right)}=\frac{5}{8}=0.675$$

The equation representing Bruce and Marty working together is:

$$\frac{1}{x}+\frac{1}{z}=\frac{1}{\left(\dfrac{8}{3}\right)}=0.375$$

Solve the system of equations: $\qquad$ Let $u=x^{-1},\ v=y^{-1},\ w=z^{-1}$

$$\begin{cases} x^{-1}+y^{-1}=0.75 \\ y^{-1}+z^{-1}=0.675 \\ x^{-1}+z^{-1}=0.375 \end{cases} \qquad \begin{cases} u+v=0.75 \quad \rightarrow \quad u=0.75-v \\ v+w=0.675 \quad \rightarrow \quad w=0.675-v \\ u+w=0.375 \end{cases}$$

Substitute into the third equation and solve:

$$0.75-v+0.675-v=0.375 \Rightarrow -2v=-1 \Rightarrow v=0.5$$

$$u=0.75-0.5=0.25$$

$$w=0.675-0.5=0.125$$

Solve for x, y, and z: $x=4$, $y=2$, $z=8$ (reciprocals)

Bruce can do the job in 4 hours, Bryce in 2 hours, and Marty in 8 hours.

107. Let x = the number of gasoline engines produced each week.

Let y = the number of diesel engines produced each week.

The total cost is: $C=450x+550y$. Cost is to be minimized; thus, this is the objective

function.

The constraints are:

$\quad 20 \le x \le 60$ $\qquad$ number of gasoline engines needed and capacity each week.

$\quad 15 \le y \le 40$ $\qquad$ number of diesel engines needed and capacity each week.

$\quad x+y \ge 50$ $\qquad$ number of engines produced to prevent layoffs.

Graph the constraints.

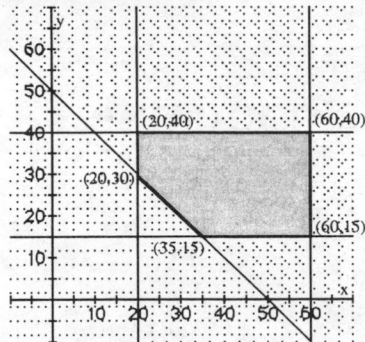

The corner points are (20, 30), (20, 40), (35, 15), (60, 15), (60, 40).
Evaluate the objective function:

Vertex	Value of $C = 450x + 550y$
(20, 30)	C = 450(20) + 550(30) = 25,500
(20, 40)	C = 450(20) + 550(40) = 31,000
(35, 15)	C = 450(35) + 550(15) = 24,000
(60, 15)	C = 450(60) + 550(15) = 35,250
(60, 40)	C = 450(60) + 550(40) = 49,000

The minimum cost is $24,000, when 35 gasoline engines and 15 diesel engines are produced.

The excess capacity is 15 gasoline engines, since only 20 gasoline engines had to be delivered.

Systems of Equations and Inequalities

10.CR **Cumulative Review**

1. $2x^2 - x = -3$
$2x^2 - x + 3 = 0$

$$x = \frac{-(-1) \pm \sqrt{(-1)^2 - 4 \cdot 2 \cdot 3}}{2 \cdot 2} = \frac{1 \pm \sqrt{1 - 24}}{4} = \frac{1 \pm \sqrt{-23}}{4} = \frac{1 \pm \sqrt{23}i}{4}$$

The solution set is $\left\{ \dfrac{1 - \sqrt{23}i}{4}, \dfrac{1 + \sqrt{23}i}{4} \right\}$.

3. $2x^3 - 3x^2 - 8x - 3 = 0$

The graph of $y_1 = 2x^3 - 3x^2 - 8x - 3$
appears to have an x-intercept at $x = 3$.

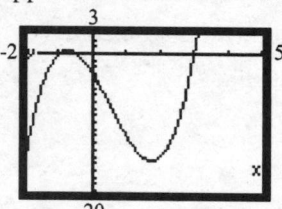

Using synthetic division:

$$3 \overline{)\begin{array}{cccc} 2 & -3 & -8 & -3 \\ & 6 & 9 & 3 \\ \hline 2 & 3 & 1 & 0 \end{array}}$$

Therefore,
$2x^3 - 3x^2 - 8x - 3 = 0$

$(x - 3)(2x^2 + 3x + 1) = 0$

$(x - 3)(2x + 1)(x + 1) = 0$

$x = 3$ or $x = -\dfrac{1}{2}$ or $x = -1$

The solution set is $\left\{ -1, -\dfrac{1}{2}, 3 \right\}$.

5. $\log_3(x - 1) + \log_3(2x + 1) = 2$

$\log_3\big((x - 1)(2x + 1)\big) = 2$

$(x - 1)(2x + 1) = 3^2$

$2x^2 - x - 1 = 9$

$2x^2 - x - 10 = 0$

$(2x - 5)(x + 2) = 0$

$x = \dfrac{5}{2}$ or $x = -2$

Since $x = -2$ makes the original logarithms
undefined, the solution set is $\left\{ \dfrac{5}{2} \right\}$.

7. $f(x) = \dfrac{2x^3}{x^4 + 1}$

$$f(-x) = \frac{2(-x)^3}{(-x)^4 + 1} = \frac{-2x^3}{x^4 + 1} = -f(x),$$

therefore f is an odd function and its graph is
symmetric with respect to the origin.

9. $f(x) = 3^{x-2} + 1$

Using the graph of $y = 3^x$, shift the graph horizontally 2 units to the right, then shift the graph vertically 1 unit upward.

Domain: $(-\infty, \infty)$

Range: $(1, \infty)$

Horizontal Asymptote: $y = 1$

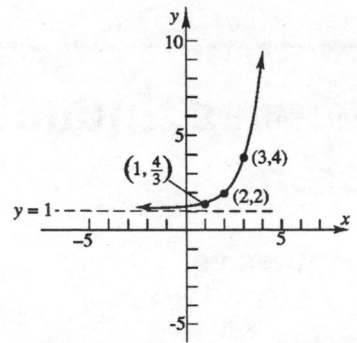

11.   (a)   This relation is not a function since the points $(2, 15900)$ and $(2, 16980)$ cause the graph to fail the Vertical Line Test.

       (b)

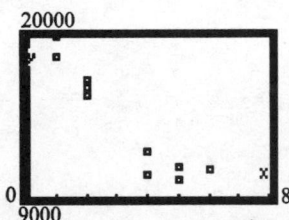

       (c)   Using LINearREGression yields $P = -2054.646a + 20976.213$

       (d)   For each 1 year increase in the age of the car, the price decreases by \$2054.65.

       (e) $P(a) = -2054.646a + 20976.213$

       (f) $P(7) = -2054.646(7) + 20976.213 \approx \$6593.69$

Chapter 11

Sequences; Induction; The Binomial Theorem

11.1 Sequences

1. $a_1 = 1, \ a_2 = 2, \ a_3 = 3, \ a_4 = 4, \ a_5 = 5$

3. $a_1 = \dfrac{1}{1+2} = \dfrac{1}{3}, \ a_2 = \dfrac{2}{2+2} = \dfrac{2}{4} = \dfrac{1}{2}, \ a_3 = \dfrac{3}{3+2} = \dfrac{3}{5}, \ a_4 = \dfrac{4}{4+2} = \dfrac{4}{6} = \dfrac{2}{3},$
    $\quad a_5 = \dfrac{5}{5+2} = \dfrac{5}{7}$

5. $a_1 = (-1)^{1+1}(1^2) = 1, \ a_2 = (-1)^{2+1}(2^2) = -4, \ a_3 = (-1)^{3+1}(3^2) = 9,$
    $\quad a_4 = (-1)^{4+1}(4^2) = -16, \ a_5 = (-1)^{5+1}(5^2) = 25$

7. $a_1 = \dfrac{2^1}{3^1 + 1} = \dfrac{2}{4} = \dfrac{1}{2}, \ a_2 = \dfrac{2^2}{3^2 + 1} = \dfrac{4}{10} = \dfrac{2}{5}, \ a_3 = \dfrac{2^3}{3^3 + 1} = \dfrac{8}{28} = \dfrac{2}{7},$
    $\quad a_4 = \dfrac{2^4}{3^4 + 1} = \dfrac{16}{82} = \dfrac{8}{41}, \ a_5 = \dfrac{2^5}{3^5 + 1} = \dfrac{32}{244} = \dfrac{8}{61}$

9. $a_1 = \dfrac{(-1)^1}{(1+1)(1+2)} = \dfrac{-1}{2 \cdot 3} = -\dfrac{1}{6}, \ a_2 = \dfrac{(-1)^2}{(2+1)(2+2)} = \dfrac{1}{3 \cdot 4} = \dfrac{1}{12},$
    $\quad a_3 = \dfrac{(-1)^3}{(3+1)(3+2)} = \dfrac{-1}{4 \cdot 5} = -\dfrac{1}{20}, \ a_4 = \dfrac{(-1)^4}{(4+1)(4+2)} = \dfrac{1}{5 \cdot 6} = \dfrac{1}{30},$
    $\quad a_5 = \dfrac{(-1)^5}{(5+1)(5+2)} = \dfrac{-1}{6 \cdot 7} = -\dfrac{1}{42}$

11. $a_1 = \dfrac{1}{e^1} = \dfrac{1}{e}, \ a_2 = \dfrac{2}{e^2}, \ a_3 = \dfrac{3}{e^3}, \ a_4 = \dfrac{4}{e^4}, \ a_5 = \dfrac{5}{e^5}$

13. $\dfrac{n}{n+1}$ $\qquad\qquad$ 15. $\dfrac{1}{2^{n-1}}$ $\qquad\qquad$ 17. $(-1)^{n+1}$

19. $(-1)^{n+1} n$

21. $a_1 = 2, \ a_2 = 3 + 2 = 5, \ a_3 = 3 + 5 = 8, \ a_4 = 3 + 8 = 11, \ a_5 = 3 + 11 = 14$

23. $a_1 = -2, \ a_2 = 2 + (-2) = 0, \ a_3 = 3 + 0 = 3, \ a_4 = 4 + 3 = 7, \ a_5 = 5 + 7 = 12$

25. $a_1 = 5$, $a_2 = 2 \cdot 5 = 10$, $a_3 = 2 \cdot 10 = 20$, $a_4 = 2 \cdot 20 = 40$, $a_5 = 2 \cdot 40 = 80$

27. $a_1 = 3$, $a_2 = \dfrac{3}{2}$, $a_3 = \dfrac{\left(\frac{3}{2}\right)}{3} = \dfrac{1}{2}$, $a_4 = \dfrac{\left(\frac{1}{2}\right)}{4} = \dfrac{1}{8}$, $a_5 = \dfrac{\left(\frac{1}{8}\right)}{5} = \dfrac{1}{40}$

29. $a_1 = 1$, $a_2 = 2$, $a_3 = 2 \cdot 1 = 2$, $a_4 = 2 \cdot 2 = 4$, $a_5 = 4 \cdot 2 = 8$

31. $a_1 = A$, $a_2 = A + d$, $a_3 = (A + d) + d = A + 2d$, $a_4 = (A + 2d) + d = A + 3d$,
     $a_5 = (A + 3d) + d = A + 4d$

33. $a_1 = \sqrt{2}$, $a_2 = \sqrt{2 + \sqrt{2}}$, $a_3 = \sqrt{2 + \sqrt{2 + \sqrt{2}}}$, $a_4 = \sqrt{2 + \sqrt{2 + \sqrt{2 + \sqrt{2}}}}$,
     $a_5 = \sqrt{2 + \sqrt{2 + \sqrt{2 + \sqrt{2 + \sqrt{2}}}}}$

35. $\displaystyle\sum_{k=1}^{5}(k + 2) = 3 + 4 + 5 + 6 + 7$

37. $\displaystyle\sum_{k=1}^{8}\dfrac{k^2}{2} = \dfrac{1}{2} + 2 + \dfrac{9}{2} + 8 + \dfrac{25}{2} + 18 + \dfrac{49}{2} + 32$

39. $\displaystyle\sum_{k=0}^{n}\dfrac{1}{3^k} = 1 + \dfrac{1}{3} + \dfrac{1}{9} + \dfrac{1}{27} + \cdots + \dfrac{1}{3^n}$

41. $\displaystyle\sum_{k=0}^{n-1}\dfrac{1}{3^{k+1}} = \dfrac{1}{3} + \dfrac{1}{9} + \dfrac{1}{27} + \cdots + \dfrac{1}{3^n}$

43. $\displaystyle\sum_{k=2}^{n}(-1)^k \ln k = \ln 2 - \ln 3 + \ln 4 - \ln 5 + \cdots + (-1)^n \ln n$

45. $1 + 2 + 3 + \cdots + 20 = \displaystyle\sum_{k=1}^{20} k$ 47. $\dfrac{1}{2} + \dfrac{2}{3} + \dfrac{3}{4} + \cdots + \dfrac{13}{13 + 1} = \displaystyle\sum_{k=1}^{13}\dfrac{k}{k + 1}$

49. $1 - \dfrac{1}{3} + \dfrac{1}{9} - \dfrac{1}{27} + \cdots + (-1)^6\left(\dfrac{1}{3^6}\right) = \displaystyle\sum_{k=0}^{6}(-1)^k\left(\dfrac{1}{3^k}\right)$

51. $3 + \dfrac{3^2}{2} + \dfrac{3^3}{3} + \cdots + \dfrac{3^n}{n} = \displaystyle\sum_{k=1}^{n}\dfrac{3^k}{k}$

53. $a + (a + d) + (a + 2d) + \cdots + (a + nd) = \displaystyle\sum_{k=0}^{n}(a + kd)$

55. (a) $\displaystyle\sum_{k=1}^{10} 5 = \underbrace{5 + 5 + 5 + \ldots + 5}_{10 \text{ times}} = 50$

(b) use the sum(seq) feature:

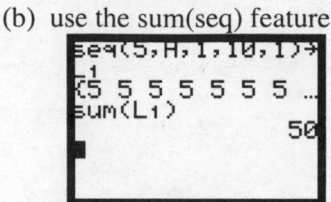

57. (a) $\displaystyle\sum_{k=1}^{6} k = 1 + 2 + 3 + 4 + 5 + 6 = 21$

(b) use the sum(seq) feature

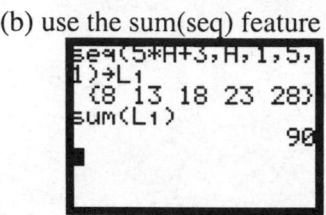

59. (a) $\displaystyle\sum_{k=1}^{5} (5k + 3) = 8 + 13 + 18 + 23 + 28 = 90$

(b) use the sum(seq) feature

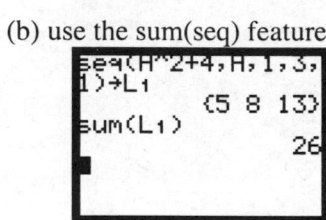

61. (a) $\displaystyle\sum_{k=1}^{3} (k^2 + 4) = 5 + 8 + 13 = 26$

(b) use the sum(seq) feature

63. (a) $\displaystyle\sum_{k=1}^{6} (-1)^k 2^k = (-1)^1 \cdot 2^1 + (-1)^2 \cdot 2^2 + (-1)^3 \cdot 2^3 + (-1)^4 \cdot 2^4 + (-1)^5 \cdot 2^5 + (-1)^6 \cdot 2^6$

$= -2 + 4 - 8 + 16 - 32 + 64 = 42$

(b) use the sum(seq) feature

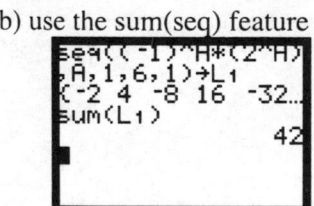

65. (a) $\displaystyle\sum_{k=1}^{4} (k^3 - 1) = 0 + 7 + 26 + 63 = 96$

(b) use the sum(seq) feature

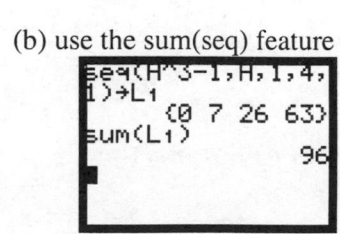

67. (a) $B_2 = 1.01(3000) - 100 = \$2930$

(b) Put the graphing utility in SEQuence mode. Enter Y= as follows, then examine the TABLE:

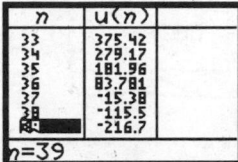

From the table we see that the balance is below $2000 after 14 payments have been made. The balance then is $1953.70.

(c) Scrolling down the table, we find that balance is paid off in the 36th month. The last payment is $83.78. There are 35 payments of $100 and the last payment of $83.78. The total amount paid is: $35(100) + 83.78(1.01) = \$3584.62.$ (we have to add the interest for the last month).

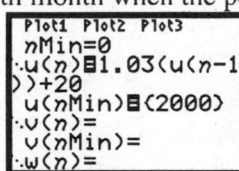

(d) The interest expense is: $3584.62 - 3000.00 = \$584.62$

69. (a) $p_1 = 1.03(2000) + 20 = 2080; \qquad p_2 = 1.03(2080) + 20 = 2162.4$

(b) Scrolling down the table, we find the trout population exceeds 5000 at the beginning of the 26th month when the population is 5084.

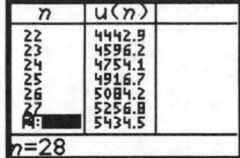

71. (a) Since the fund returns 8% compound annually, this is equivalent to a return of 2% each quarter. Defining a recursive sequence, we have:
$$a_1 = 500, \quad a_n = 1.02a_{n-1} + 500$$

(b) Insert the formulas in your graphing utility and use the table feature to find when the value of the account will exceed $100,000:

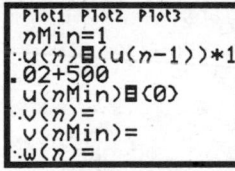
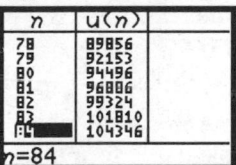

In the 83rd quarter (during the 21st year) the value of the account will exceed $100,000 with a value of $101,810.

(c)   Find the value of the account in 25 years or 100 quarters:

n	u(n)	
96	139042	
97	142323	
98	145670	
99	149083	
100	152565	
101	156116	
102	159738	

$n=102$

The value of the account will be \$156,116.

73.   (a)   Since the interest rate is 6% per annum compounded monthly, this is equivalent to a rate of 0.5% each month. Defining a recursive sequence, we have:
$$a_1 = 150,000, \qquad a_n = 1.005a_{n-1} - 899.33$$

(b) $1.005(150,000) - 899.33 = \$149,850.67$

(c)   Enter the recursive formula in Y= and create the table:

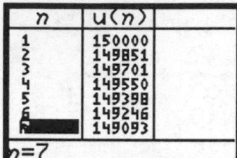

n	u(n)
1	150000
2	149851
3	149701
4	149550
5	149398
6	149246
7	149093

$n=7$

(d)   Scroll through the table:

n	u(n)	
54	140963	
55	140769	
56	140573	
57	140377	
58	140180	
59	139981	
60	139782	

$n=60$

At the beginning of the 59th month, or after 58 payments have been made, the balance is below \$140,000. The balance is \$139,981.

(e)   Scroll through the table:

n	u(n)	
355	5298.7	
356	4425.9	
357	3548.7	
358	2667.1	
359	1781.1	
360	890.65	
361	-4.231	

$n=355$

The loan will be paid off at the end of 360 months or 30 years.
Total amount paid = (359)(\$899.33) + \$890.65(1.005) = \$323,754.57.

(f)   The total interest expense is the difference of the total of the payments and the original loan: $323,754.57 - 150,000 = \$173,754.57$

(g)   (a)   Since the interest rate is 6% per annum compounded monthly, this is equivalent to a rate of 0.5% each month. Defining a recursive sequence, we have:
$$a_1 = 150,000, \qquad a_n = 1.005a_{n-1} - 999.33$$

(b) $1.005(150,000) - 999.33 = \$149,750.67$

(c)   Enter the recursive formula in Y= and create the table:

n	u(n)
1	150000
2	149751
3	149500
4	149248
5	148995
6	148741
7	148485

$n=7$

(d)   Scroll through the table:

n	$u(n)$
34	141079
35	140785
36	140489
37	140192
38	139894
39	139594
40	139293

$n=40$

At the beginning of the 38th month, or after 37 payments have been made, the balance is below $140,000. The balance is $139,894.

(e)   Scroll through the table:

n	$u(n)$
275	4294.6
276	3316.7
277	2333.9
278	1346.3
279	353.69
280	⁻643.9
281	⁻1646

$n=281$

The loan will be paid off at the end of 279 months or 23 years and 3 months. Total amount paid = (279)($999.33) + 353.69(1.005) = $279,168.50

(f)   The total interest expense is the difference of the total of the payments and the original loan: $279,168.50 - 150,000 = $129,168.50

75. $a_1 = 1$, $a_2 = 1$, $a_3 = 2$, $a_4 = 3$, $a_5 = 5$, $a_6 = 8$, $a_7 = 13$, $a_8 = 21$, $a_n = a_{n-1} + a_{n-2}$
$a_8 = a_7 + a_6 = 13 + 8 = 21$

After 7 months there are 21 mature pairs of rabbits.

77.   1, 1, 2, 3, 5, 8, 13   This is the Fibonacci sequence.

79.   To show that $1 + 2 + 3 + ... + (n-1) + n = \dfrac{n(n+1)}{2}$

Let

$S = 1 + 2 + 3 + ... + (n-1) + n$, we can reverse the order to get

$+S = n + (n-1) + (n-2) + ... + 2 + 1$, now add these two lines to get

$2S = [1+n] + [2+(n-1)] + [3+(n-2)] + ... + [(n-1)+2] + [n+1]$

$\underbrace{\qquad\qquad\qquad\qquad\qquad\qquad\qquad\qquad}_{\text{n terms}}$

So we have

$2S = \underbrace{[1+n] + [1+n] + [1+n] + ... + [n+1] + [n+1]}_{\text{n terms}} = n \cdot [n+1]$

$\therefore\ S = \dfrac{n \cdot (n+1)}{2}$

Chapter 11

Sequences; Induction; The Binomial Theorem

11.2 Arithmetic Sequences

1. $d = a_{n+1} - a_n = (n+1+4) - (n+4) = n+5-n-4 = 1$
   $a_1 = 1+4 = 5, \ a_2 = 2+4 = 6, \ a_3 = 3+4 = 7, \ a_4 = 4+4 = 8$

3. $d = a_{n+1} - a_n = (2(n+1)-5) - (2n-5) = 2n+2-5-2n+5 = 2$
   $a_1 = 2\cdot1-5 = -3, \ a_2 = 2\cdot2-5 = -1, \ a_3 = 2\cdot3-5 = 1, \ a_4 = 2\cdot4-5 = 3$

5. $d = a_{n+1} - a_n = (6-2(n+1)) - (6-2n) = 6-2n-2-6+2n = -2$
   $a_1 = 6-2\cdot1 = 4, \ a_2 = 6-2\cdot2 = 2, \ a_3 = 6-2\cdot3 = 0, \ a_4 = 6-2\cdot4 = -2$

7. $d = a_{n+1} - a_n = \left(\dfrac{1}{2} - \dfrac{1}{3}(n+1)\right) - \left(\dfrac{1}{2} - \dfrac{1}{3}n\right) = \dfrac{1}{2} - \dfrac{1}{3}n - \dfrac{1}{3} - \dfrac{1}{2} + \dfrac{1}{3}n = -\dfrac{1}{3}$

   $a_1 = \dfrac{1}{2} - \dfrac{1}{3}\cdot1 = \dfrac{1}{6}, \ a_2 = \dfrac{1}{2} - \dfrac{1}{3}\cdot2 = -\dfrac{1}{6}, \ a_3 = \dfrac{1}{2} - \dfrac{1}{3}\cdot3 = -\dfrac{1}{2}, \ a_4 = \dfrac{1}{2} - \dfrac{1}{3}\cdot4 = -\dfrac{5}{6}$

9. $d = a_{n+1} - a_n = \ln(3^{n+1}) - \ln(3^n) = (n+1)\ln(3) - n\ln(3) = \ln 3(n+1-n) = \ln(3)$
   $a_1 = \ln(3^1) = \ln(3), \ a_2 = \ln(3^2) = 2\ln(3), \ a_3 = \ln(3^3) = 3\ln(3), \ a_4 = \ln(3^4) = 4\ln(3)$

11. $a_n = a + (n-1)d = 2 + (n-1)3 = 2 + 3n - 3 = 3n - 1$
    $a_5 = 3\cdot5 - 1 = 14$

13. $a_n = a + (n-1)d = 5 + (n-1)(-3) = 5 - 3n + 3 = 8 - 3n$
    $a_5 = 8 - 3\cdot5 = -7$

15. $a_n = a + (n-1)d = 0 + (n-1)\dfrac{1}{2} = \dfrac{1}{2}n - \dfrac{1}{2} = \dfrac{1}{2}(n-1)$

    $a_5 = \dfrac{1}{2}\cdot5 - \dfrac{1}{2} = 2$

17. $a_n = a + (n-1)d = \sqrt{2} + (n-1)\sqrt{2} = \sqrt{2} + \sqrt{2}n - \sqrt{2} = \sqrt{2}n$
    $a_5 = 5\sqrt{2}$

19. $a_1 = 2, \ d = 2, \ a_n = a + (n-1)d$
    $a_{12} = 2 + (12-1)2 = 2 + 11(2) = 2 + 22 = 24$

21. $a_1 = 1, \; d = -2 - 1 = -3, \quad a_n = a + (n-1)d$
$a_{10} = 1 + (10-1)(-3) = 1 + 9(-3) = 1 - 27 = -26$

23. $a_1 = a, \; d = (a+b) - a = b, \quad a_n = a + (n-1)d$
$a_8 = a + (8-1)b = a + 7b$

25. $a_8 = a + 7d = 8 \qquad a_{20} = a + 19d = 44$
Solve the system of equations:
$$8 - 7d + 19d = 44$$
$$12d = 36 \rightarrow d = 3 \rightarrow a = 8 - 7(3) = 8 - 21 = -13$$
Recursive formula: $\quad a_1 = -13 \qquad a_n = a_{n-1} + 3$

27. $a_9 = a + 8d = -5 \qquad a_{15} = a + 14d = 31$
Solve the system of equations:
$$-5 - 8d + 14d = 31 \rightarrow 6d = 36 \rightarrow d = 6$$
$$a = -5 - 8(6) = -5 - 48 = -53$$
Recursive formula: $\quad a_1 = -53 \qquad a_n = a_{n-1} + 6$

29. $a_{15} = a + 14d = 0 \qquad a_{40} = a + 39d = -50$
Solve the system of equations:
$$-14d + 39d = -50 \rightarrow 25d = -50 \rightarrow d = -2$$
$$a = -14(-2) = 28$$
Recursive formula: $\quad a_1 = 28 \qquad a_n = a_{n-1} - 2$

31. $a_{14} = a + 13d = -1 \qquad a_{18} = a + 17d = -9$
Solve the system of equations:
$$-1 - 13d + 17d = -9$$
$$4d = -8 \rightarrow d = -2 \rightarrow a = -1 - 13(-2) = -1 + 26 = 25$$
Recursive formula: $\quad a_1 = 25 \qquad a_n = a_{n-1} - 2$

33. $S_n = \dfrac{n}{2}(a + a_n) = \dfrac{n}{2}(1 + (2n-1)) = \dfrac{n}{2}(2n) = n^2$

35. $S_n = \dfrac{n}{2}(a + a_n) = \dfrac{n}{2}(7 + (2 + 5n)) = \dfrac{n}{2}(9 + 5n)$

37. $a_1 = 2, \; d = 4 - 2 = 2, \quad a_n = a + (n-1)d$
$$70 = 2 + (n-1)2 \rightarrow 70 = 2 + 2n - 2 \rightarrow 70 = 2n \rightarrow n = 35$$
$$S_n = \frac{n}{2}(a + a_n) = \frac{35}{2}(2 + 70) = \frac{35}{2}(72) = 35(36) = 1260$$

39. $a_1 = 5, \; d = 9 - 5 = 4, \quad a_n = a + (n-1)d$
$$49 = 5 + (n-1)4 \rightarrow 49 = 5 + 4n - 4 \rightarrow 48 = 4n \rightarrow n = 12$$
$$S_n = \frac{n}{2}(a + a_n) = \frac{12}{2}(5 + 49) = 6(54) = 324$$

41. Using the sum of the sequence feature:

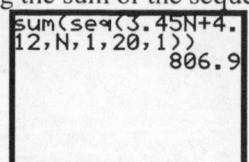

43. $d = 5.2 - 2.8 = 2.4$
        $a = 2.8$
        $36.4 = 2.8 + (n-1)2.4$
        $36.4 = 2.8 + 2.4n - 2.4$
        $36 = 2.4n$
        $n = 15$
        $a_n = 2.8 + (n-1)2.4 = 2.8 + 2.4n - 2.4 = 2.4n + 0.4$

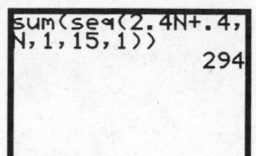

45. $d = 7.48 - 4.9 = 2.58$
        $a = 4.9$
     $66.82 = 4.9 + (n-1)2.58$
     $66.82 = 4.9 + 2.58n - 2.58$
        $64.5 = 2.58n$
        $n = 25$
        $a_n = 4.9 + (n-1)2.58 = 4.9 + 2.58n - 2.58$
        $a_n = 2.58n + 2.32$

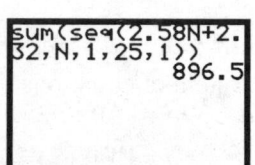

47. Find the common difference of the terms and solve the system of equations:
        $(2x+1) - (x+3) = d \;\rightarrow\; x - 2 = d$
        $(5x+2) - (2x+1) = d \;\rightarrow\; 3x + 1 = d$
        $3x + 1 = x - 2 \rightarrow 2x = -3 \rightarrow x = -1.5$

49. The total number of seats is: $S = 25 + 26 + 27 + \cdots$
     This is the sum of an arithmetic sequence with $d = 1$, $a = 25$, and $n = 30$.
     Find the sum of the sequence: $S_{30} = \dfrac{30}{2}[2(25) + (30-1)(1)] = 15(50 + 29) = 15(79) = 1185$
     There are 1185 seats in the theater.

51. The lighter colored tiles have 20 tiles in the bottom row and 1 tile in the top row. The number
     decreases by 1 as we move up the triangle. This is an arithmetic sequence with
     $a_1 = 20$, $d = -1$, and $n = 20$. Find the sum:
     $$S = \frac{20}{2}[2(20) + (20-1)(-1)] = 10(40 - 19) = 10(21) = 210 \text{ lighter tiles.}$$
     The darker colored tiles have 19 tiles in the bottom row and 1 tile in the top row. The number
     decreases by 1 as we move up the triangle. This is an arithmetic sequence with
     $a_1 = 19$, $d = -1$, and $n = 19$. Find the sum:
     $$S = \frac{19}{2}[2(19) + (19-1)(-1)] = \frac{19}{2}(38 - 18) = \frac{19}{2}(20) = 190 \text{ darker tiles.}$$

53.   Find n in an arithmetic sequence with $a_1 = 10$, $d = 4$, $s_n = 2040$.

$$s_n = \frac{n}{2}\left[2a_1 + (n-1)d\right] \rightarrow 2040 = \frac{n}{2}\left[2(10) + (n-1)4\right]$$

$$4080 = n\left[20 + 4n - 4\right] \rightarrow 4080 = n(4n + 16)$$

$$4080 = 4n^2 + 16n \rightarrow 1020 = n^2 + 4n$$

$$n^2 + 4n - 1020 = 0 \rightarrow (n + 34)(n - 30) = 0 \rightarrow n = -34 \text{ or } n = 30$$

There are 30 rows in the corner section of the stadium.

55.   Answers will vary.

Chapter 11

Sequences; Induction; The Binomial Theorem

11.3 Geometric Sequences; Geometric Series

1. $r = \dfrac{3^{n+1}}{3^n} = 3^{n+1-n} = 3$

   $a_1 = 3^1 = 3, \; a_2 = 3^2 = 9, \; a_3 = 3^3 = 27, \; a_4 = 3^4 = 81$

3. $r = \dfrac{-3\left(\dfrac{1}{2}\right)^{n+1}}{-3\left(\dfrac{1}{2}\right)^n} = \left(\dfrac{1}{2}\right)^{n+1-n} = \dfrac{1}{2}$

   $a_1 = -3\left(\dfrac{1}{2}\right)^1 = -\dfrac{3}{2}, \; a_2 = -3\left(\dfrac{1}{2}\right)^2 = -\dfrac{3}{4}, \; a_3 = -3\left(\dfrac{1}{2}\right)^3 = -\dfrac{3}{8}, \; a_4 = -3\left(\dfrac{1}{2}\right)^4 = -\dfrac{3}{16}$

5. $r = \dfrac{\left(\dfrac{2^{n+1-1}}{4}\right)}{\left(\dfrac{2^{n-1}}{4}\right)} = \dfrac{2^n}{2^{n-1}} = 2^{n-(n-1)} = 2$

   $a_1 = \dfrac{2^{1-1}}{4} = \dfrac{2^0}{2^2} = 2^{-2} = \dfrac{1}{4}, \; a_2 = \dfrac{2^{2-1}}{4} = \dfrac{2^1}{2^2} = 2^{-1} = \dfrac{1}{2}, \; a_3 = \dfrac{2^{3-1}}{4} = \dfrac{2^2}{2^2} = 1,$

   $\qquad a_4 = \dfrac{2^{4-1}}{4} = \dfrac{2^3}{2^2} = 2$

7. $r = \dfrac{2^{\left(\frac{n+1}{3}\right)}}{2^{\left(\frac{n}{3}\right)}} = 2^{\left(\frac{n+1}{3} - \frac{n}{3}\right)} = 2^{1/3}$

   $a_1 = 2^{1/3}, \; a_2 = 2^{2/3}, \; a_3 = 2^{3/3} = 2, \; a_4 = 2^{4/3}$

9. $r = \dfrac{\left(\dfrac{3^{n+1-1}}{2^{n+1}}\right)}{\left(\dfrac{3^{n-1}}{2^n}\right)} = \dfrac{3^n}{3^{n-1}} \cdot \dfrac{2^n}{2^{n+1}} = 3^{n-(n-1)} \cdot 2^{n-(n+1)} = 3 \cdot 2^{-1} = \dfrac{3}{2}$

$a_1 = \dfrac{3^{1-1}}{2^1} = \dfrac{3^0}{2} = \dfrac{1}{2}, \quad a_2 = \dfrac{3^{2-1}}{2^2} = \dfrac{3^1}{2^2} = \dfrac{3}{4}, \quad a_3 = \dfrac{3^{3-1}}{2^3} = \dfrac{3^2}{2^3} = \dfrac{9}{8},$

$a_4 = \dfrac{3^{4-1}}{2^4} = \dfrac{3^3}{2^4} = \dfrac{27}{16}$

11. $\{n + 2\}$ Arithmetic

$d = (n + 1 + 2) - (n + 2) = n + 3 - n - 2 = 1$

13. $\{4n^2\}$ Examine the terms of the sequence: 4, 16, 36, 64, 100, ...

There is no common difference; there is no common ratio; neither.

15. $\left\{3 - \dfrac{2}{3}n\right\}$ Arithmetic

$d = \left(3 - \dfrac{2}{3}(n + 1)\right) - \left(3 - \dfrac{2}{3}n\right) = 3 - \dfrac{2}{3}n - \dfrac{2}{3} - 3 + \dfrac{2}{3}n = -\dfrac{2}{3}$

17. 1, 3, 6, 10, ...   Neither

There is no common difference or common ratio.

19. $\left\{\left(\dfrac{2}{3}\right)^n\right\}$ Geometric $r = \dfrac{\left(\dfrac{2}{3}\right)^{n+1}}{\left(\dfrac{2}{3}\right)^n} = \left(\dfrac{2}{3}\right)^{n+1-n} = \dfrac{2}{3}$

21. $-1, -2, -4, -8, ...$ Geometric $r = \dfrac{-2}{-1} = \dfrac{-4}{-2} = \dfrac{-8}{-4} = 2$

23. $\{3^{n/2}\}$ Geometric $r = \dfrac{3^{\left(\frac{n+1}{2}\right)}}{3^{\left(\frac{n}{2}\right)}} = 3^{\left(\frac{n+1}{2} - \frac{n}{2}\right)} = 3^{1/2}$

25. $a_5 = 2 \cdot 3^{5-1} = 2 \cdot 3^4 = 2 \cdot 81 = 162$ $a_n = 2 \cdot 3^{n-1}$

27. $a_5 = 5(-1)^{5-1} = 5(-1)^4 = 5 \cdot 1 = 5$ $a_n = 5 \cdot (-1)^{n-1}$

29. $a_5 = 0 \cdot \left(\dfrac{1}{2}\right)^{5-1} = 0 \cdot \left(\dfrac{1}{2}\right)^4 = 0$ $a_n = 0 \cdot \left(\dfrac{1}{2}\right)^{n-1} = 0$

31. $a_5 = \sqrt{2} \cdot \left(\sqrt{2}\right)^{5-1} = \sqrt{2} \cdot \left(\sqrt{2}\right)^4 = \sqrt{2} \cdot 4 = 4\sqrt{2}$ $a_n = \sqrt{2} \cdot \left(\sqrt{2}\right)^{n-1} = \left(\sqrt{2}\right)^n$

33. $a = 1$, $r = \dfrac{1}{2}$, $n = 7$ $a_7 = 1 \cdot \left(\dfrac{1}{2}\right)^{7-1} = \left(\dfrac{1}{2}\right)^{6} = \dfrac{1}{64}$

35. $a = 1$, $r = -1$, $n = 9$ $a_9 = 1 \cdot (-1)^{9-1} = (-1)^{8} = 1$

37. $a = 0.4$, $r = 0.1$, $n = 8$ $a_8 = 0.4 \cdot (0.1)^{8-1} = 0.4(0.1)^{7} = 0.00000004$

39. $a = \dfrac{1}{4}$, $r = 2$ $S_n = a\left(\dfrac{1 - r^n}{1 - r}\right) = \dfrac{1}{4}\left(\dfrac{1 - 2^n}{1 - 2}\right) = -\dfrac{1}{4}(1 - 2^n)$

41. $a = \dfrac{2}{3}$, $r = \dfrac{2}{3}$ $S_n = a\left(\dfrac{1 - r^n}{1 - r}\right) = \dfrac{2}{3}\left(\dfrac{1 - \left(\dfrac{2}{3}\right)^n}{1 - \dfrac{2}{3}}\right) = \dfrac{2}{3}\left(\dfrac{1 - \left(\dfrac{2}{3}\right)^n}{\dfrac{1}{3}}\right) = 2\left(1 - \left(\dfrac{2}{3}\right)^n\right)$

43. $a = -1$, $r = 2$ $S_n = a\left(\dfrac{1 - r^n}{1 - r}\right) = -1\left(\dfrac{1 - 2^n}{1 - 2}\right) = 1 - 2^n$

45. Using the sum of the sequence feature:

47. Using the sum of the sequence feature:

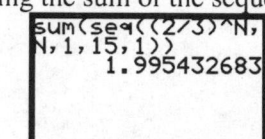

49. Using the sum of the sequence feature:

```
sum(seq(-1*2^N,N
,0,14,1))
           -32767
```

51. $a = 1$, $r = \dfrac{1}{3}$ Since $|r| < 1$, $S_n = \dfrac{a}{1 - r} = \dfrac{1}{\left(1 - \dfrac{1}{3}\right)} = \dfrac{1}{\left(\dfrac{2}{3}\right)} = \dfrac{3}{2}$

53. $a = 8$, $r = \dfrac{1}{2}$ Since $|r| < 1$, $S_n = \dfrac{a}{1 - r} = \dfrac{8}{\left(1 - \dfrac{1}{2}\right)} = \dfrac{8}{\left(\dfrac{1}{2}\right)} = 16$

55. $a = 2$, $r = -\dfrac{1}{4}$ Since $|r| < 1$, $S_n = \dfrac{a}{1 - r} = \dfrac{2}{\left(1 - \left(-\dfrac{1}{4}\right)\right)} = \dfrac{2}{\left(\dfrac{5}{4}\right)} = \dfrac{8}{5}$

57. $a = 5$, $r = \dfrac{1}{4}$ Since $|r| < 1$, $S_n = \dfrac{a}{1-r} = \dfrac{5}{\left(1-\dfrac{1}{4}\right)} = \dfrac{5}{\left(\dfrac{3}{4}\right)} = \dfrac{20}{3}$

59. $a = 6$, $r = -\dfrac{2}{3}$ Since $|r| < 1$, $S_n = \dfrac{a}{1-r} = \dfrac{6}{\left(1-\left(-\dfrac{2}{3}\right)\right)} = \dfrac{6}{\left(\dfrac{5}{3}\right)} = \dfrac{18}{5}$

61. Find the common ratio of the terms and solve the system of equations:

$\dfrac{x+2}{x} = r; \quad \dfrac{x+3}{x+2} = r$

$\dfrac{x+2}{x} = \dfrac{x+3}{x+2} \rightarrow x^2 + 4x + 4 = x^2 + 3x \rightarrow x = -4$

63. This is a geometric series with $a = \$18{,}000$, $r = 1.05$, $n = 5$. Find the 5th term:

$a_5 = 18000(1.05)^{5-1} = 18000(1.05)^4 = \$21{,}879.11$

65. (a) Find the 10th term of the geometric sequence:

$a = 2$, $r = 0.9$, $n = 10$ $a_{10} = 2(0.9)^{10-1} = 2(0.9)^9 = 0.775$ feet

(b) Find n when $a_n < 1$:

$2(0.9)^{n-1} < 1 \Rightarrow (0.9)^{n-1} < 0.5$

$(n-1)\log(0.9) < \log(0.5) \Rightarrow n - 1 > \dfrac{\log(0.5)}{\log(0.9)} \Rightarrow n > \dfrac{\log(0.5)}{\log(0.9)} + 1 \approx 7.58$

On the 8th swing the arc is less than 1 foot.

(c) Find the sum of the first 15 swings:

$S_{15} = 2\left(\dfrac{1-(0.9)^{15}}{1-0.9}\right) = 2\left(\dfrac{1-(0.9)^{15}}{0.1}\right) = 20\left(1-(0.9)^{15}\right) = 15.88$ feet

(d) Find the infinite sum of the geometric series:

$S = \dfrac{2}{1-0.9} = \dfrac{2}{0.1} = 20$ feet

67. Both options are geometric sequences:

Option A: $a = \$20{,}000$; $r = 1.06$; $n = 5$

$a_5 = 20{,}000(1.06)^{5-1} = 20{,}000(1.06)^4 = \$25{,}250$

$S_5 = 20000\left(\dfrac{1-(1.06)^5}{1-1.06}\right) = \$112{,}742$

Option B: $a = \$22{,}000$; $r = 1.03$; $n = 5$

$a_5 = 22{,}000(1.03)^{5-1} = 22{,}000(1.03)^4 = \$24{,}761$

$S_5 = 22000\left(\dfrac{1-(1.03)^5}{1-1.03}\right) = \$116{,}801$

Option A provides more money in the 5th year, while Option B provides the greatest total amount of money over the 5 year period.

69. Option 1: Total Salary $= \$2,000,000(7) + \$100,000(7) = \$14,700,000$

Option 2: Geometric series with: $a = \$2,000,000$, $r = 1.045$, $n = 7$

Find the sum of the geometric series:

$$S = 2,000,000 \left(\frac{1 - (1.045)^7}{1 - 1.045} \right) \approx \$16,038,304$$

Option 3: Arithmetic series with: $a = \$2,000,000$, $d = \$95,000$, $n = 7$

Find the sum of the arithmetic series:

$$S_7 = \frac{7}{2}(2(2,000,000) + (7-1)(95,000)) = \$15,995,000$$

Option 2 provides the most money; Option 1 provides the least money.

71. This is a geometric sequence with $a = 1$, $r = 2$, $n = 64$.
Find the sum of the geometric series:

$$S_{64} = 1\left(\frac{1 - 2^{64}}{1 - 2} \right) = \frac{1 - 2^{64}}{-1} = 2^{64} - 1 \doteq 1.845 \times 10^{19} \text{ grains}$$

73. The common ratio, $r = 0.90 < 1$. The sum is: $S = \dfrac{1}{1 - 0.9} = \dfrac{1}{0.10} = 10$.
The multiplier is 10.

75. This is an infinite geometric series with $a = 4$, and $r = \dfrac{1.03}{1.09}$.

Find the sum: Price $= \dfrac{4}{\left(1 - \dfrac{1.03}{1.09}\right)} \approx \$72.67$.

77. Yes, a sequence can be both arithmetic and geometric. For example, the constant sequence $3, 3, 3, 3, \ldots$ can be viewed as an arithmetic sequence with $a = 3$ and $d = 0$. Alternatively, the same sequence can be viewed as a geometric sequence with $a = 3$ and $r = 1$.

79. Answers will vary.

Chapter **11**

Sequences; Induction; The Binomial Theorem

11.4 Mathematical Induction

1. I: $n = 1$: $2 \cdot 1 = 2$ and $1(1+1) = 2$
   II: If $2 + 4 + 6 + \cdots + 2k = k(k+1)$
          then $2 + 4 + 6 + \cdots + 2k + 2(k+1)$
$$= [2 + 4 + 6 + \cdots + 2k] + 2(k+1) = k(k+1) + 2(k+1)$$
$$= (k+1)(k+2)$$

Conditions I and II are satisfied; the statement is true.

3. I: $n = 1$: $1 + 2 = 3$ and $\dfrac{1}{2} \cdot 1(1+5) = 3$

   II: If $3 + 4 + 5 + \cdots + (k+2) = \dfrac{1}{2} \cdot k(k+5)$
          then $3 + 4 + 5 + \cdots + (k+2) + [(k+1) + 2]$
$$= [3 + 4 + 5 + \cdots + (k+2)] + (k+3) = \dfrac{1}{2} \cdot k(k+5) + (k+3)$$
$$= \dfrac{1}{2}k^2 + \dfrac{5}{2}k + k + 3 = \dfrac{1}{2}k^2 + \dfrac{7}{2}k + 3 = \dfrac{1}{2} \cdot \left(k^2 + 7k + 6\right)$$
$$= \dfrac{1}{2} \cdot (k+1)(k+6)$$

Conditions I and II are satisfied; the statement is true.

5. I: $n = 1$: $3 \cdot 1 - 1 = 2$ and $\dfrac{1}{2} \cdot 1(3 \cdot 1 + 1) = 2$

   II: If $2 + 5 + 8 + \cdots + (3k - 1) = \dfrac{1}{2} \cdot k(3k + 1)$
          then $2 + 5 + 8 + \cdots + (3k - 1) + [3(k+1) - 1]$
$$= [2 + 5 + 8 + \cdots + (3k - 1)] + (3k + 2) = \dfrac{1}{2} \cdot k(3k + 1) + (3k + 2)$$
$$= \dfrac{3}{2}k^2 + \dfrac{1}{2}k + 3k + 2 = \dfrac{3}{2}k^2 + \dfrac{7}{2}k + 2 = \dfrac{1}{2} \cdot \left(3k^2 + 7k + 4\right)$$
$$= \dfrac{1}{2} \cdot (k+1)(3k + 4)$$

Conditions I and II are satisfied; the statement is true.

7.   I: $n = 1$: $2^{1-1} = 1$ and $2^1 - 1 = 1$

    II: If $1 + 2 + 2^2 + \cdots + 2^{k-1} = 2^k - 1$

        then $1 + 2 + 2^2 + \cdots + 2^{k-1} + 2^{k+1-1} = \left[1 + 2 + 2^2 + \cdots + 2^{k-1}\right] + 2^k = 2^k - 1 + 2^k$

$$= 2 \cdot 2^k - 1 = 2^{k+1} - 1$$

Conditions I and II are satisfied; the statement is true.

9.   I: $n = 1$: $4^{1-1} = 1$ and $\dfrac{1}{3} \cdot \left(4^1 - 1\right) = 1$

    II: If $1 + 4 + 4^2 + \cdots + 4^{k-1} = \dfrac{1}{3} \cdot \left(4^k - 1\right)$

        then $1 + 4 + 4^2 + \cdots + 4^{k-1} + 4^{k+1-1} = \left[1 + 4 + 4^2 + \cdots + 4^{k-1}\right] + 4^k = \dfrac{1}{3} \cdot \left(4^k - 1\right) + 4^k$

$$= \frac{1}{3} \cdot 4^k - \frac{1}{3} + 4^k = \frac{4}{3} \cdot 4^k - \frac{1}{3} = \frac{1}{3}\left(4 \cdot 4^k - 1\right) = \frac{1}{3} \cdot \left(4^{k+1} - 1\right)$$

Conditions I and II are satisfied; the statement is true.

11.   I: $n = 1$: $\dfrac{1}{1(1+1)} = \dfrac{1}{2}$ and $\dfrac{1}{1+1} = \dfrac{1}{2}$

    II: If $\dfrac{1}{1 \cdot 2} + \dfrac{1}{2 \cdot 3} + \dfrac{1}{3 \cdot 4} + \cdots + \dfrac{1}{k(k+1)} = \dfrac{k}{k+1}$

        then $\dfrac{1}{1 \cdot 2} + \dfrac{1}{2 \cdot 3} + \dfrac{1}{3 \cdot 4} + \cdots + \dfrac{1}{k(k+1)} + \dfrac{1}{(k+1)(k+1+1)}$

$$= \left[\frac{1}{1 \cdot 2} + \frac{1}{2 \cdot 3} + \frac{1}{3 \cdot 4} + \cdots + \frac{1}{k(k+1)}\right] + \frac{1}{(k+1)(k+2)}$$

$$= \frac{k}{k+1} + \frac{1}{(k+1)(k+2)} = \frac{k}{k+1} \cdot \frac{k+2}{k+2} + \frac{1}{(k+1)(k+2)}$$

$$= \frac{k^2 + 2k + 1}{(k+1)(k+2)} = \frac{(k+1)(k+1)}{(k+1)(k+2)} = \frac{k+1}{k+2}$$

Conditions I and II are satisfied; the statement is true.

13.   I: $n = 1$: $1^2 = 1$ and $\dfrac{1}{6} \cdot 1(1+1)(2 \cdot 1 + 1) = 1$

    II: If $1^2 + 2^2 + 3^2 + \cdots + k^2 = \dfrac{1}{6} \cdot k(k+1)(2k+1)$

        then $1^2 + 2^2 + 3^2 + \cdots + k^2 + (k+1)^2$

$$= \left[1^2 + 2^2 + 3^2 + \cdots + k^2\right] + (k+1)^2 = \frac{1}{6}k(k+1)(2k+1) + (k+1)^2$$

$$= (k+1)\left[\frac{1}{6}k(2k+1) + k + 1\right] = (k+1)\left[\frac{1}{3}k^2 + \frac{1}{6}k + k + 1\right]$$

$$= (k+1)\left[\frac{1}{3}k^2 + \frac{7}{6}k + 1\right] = \frac{1}{6}(k+1)\left[2k^2 + 7k + 6\right]$$

$$= \frac{1}{6} \cdot (k+1)(k+2)(2k+3)$$

Conditions I and II are satisfied; the statement is true.

15.   I: $n = 1$: $5 - 1 = 4$ and $\frac{1}{2} \cdot 1(9 - 1) = 4$

   II: If $4 + 3 + 2 + \cdots + (5 - k) = \frac{1}{2} \cdot k(9 - k)$

      then $4 + 3 + 2 + \cdots + (5 - k) + \left(5 - (k + 1)\right)$

         $= \left[4 + 3 + 2 + \cdots + (5 - k)\right] + (4 - k) = \frac{1}{2} k(9 - k) + (4 - k)$

         $= \frac{9}{2}k - \frac{1}{2}k^2 + 4 - k = -\frac{1}{2}k^2 + \frac{7}{2}k + 4 = -\frac{1}{2} \cdot \left[k^2 - 7k - 8\right]$

         $= -\frac{1}{2} \cdot (k + 1)(k - 8) = \frac{1}{2} \cdot (k + 1)(8 - k) = \frac{1}{2} \cdot (k + 1)[9 - (k + 1)]$

   Conditions I and II are satisfied; the statement is true.

17.   I: $n = 1$: $1(1 + 1) = 2$ and $\frac{1}{3} \cdot 1(1 + 1)(1 + 2) = 2$

   II: If $1 \cdot 2 + 2 \cdot 3 + 3 \cdot 4 + \cdots + k(k + 1) = \frac{1}{3} \cdot k(k + 1)(k + 2)$

      then $1 \cdot 2 + 2 \cdot 3 + 3 \cdot 4 + \cdots + k(k + 1) + (k + 1)(k + 1 + 1)$

         $= \left[1 \cdot 2 + 2 \cdot 3 + 3 \cdot 4 + \cdots + k(k + 1)\right] + (k + 1)(k + 2)$

         $= \frac{1}{3} \cdot k(k + 1)(k + 2) + (k + 1)(k + 2) = (k + 1)(k + 2)\left[\frac{1}{3}k + 1\right]$

         $= \frac{1}{3} \cdot (k + 1)(k + 2)(k + 3)$

   Conditions I and II are satisfied; the statement is true.

19.   I: $n = 1$: $1^2 + 1 = 2$ is divisible by 2
   II: If $k^2 + k$ is divisible by 2
      then $(k + 1)^2 + (k + 1) = k^2 + 2k + 1 + k + 1 = (k^2 + k) + (2k + 2)$

   Since $k^2 + k$ is divisible by 2 and $2k + 2$ is divisible by 2, then $(k + 1)^2 + (k + 1)$

   is divisible by 2. Conditions I and II are satisfied; the statement is true.

21.   I: $n = 1$: $1^2 - 1 + 2 = 2$ is divisible by 2
   II: If $k^2 - k + 2$ is divisible by 2
      then $(k + 1)^2 - (k + 1) + 2 = k^2 + 2k + 1 - k - 1 + 2 = (k^2 - k + 2) + (2k)$

   Since $k^2 - k + 2$ is divisible by 2 and $2k$ is divisible by 2, then

   $(k + 1)^2 - (k + 1) + 2$ is divisible by 2. Conditions I and II are satisfied; the statement is true.

23.   I: $n = 1$: If $x > 1$ then $x^1 = x > 1$.   Show that if $x^k > 1$, then $x^{k+1} > 1$:
   II: Assume, for some natural number k, $x^{k+1} = x^k \cdot x > 1 \cdot x = x > 1$

      that if $x > 1$, then $x^k > 1$. $\uparrow$

                                       $(x^k > 1)$

   Conditions I and II are satisfied; the statement is
   true.

25. I: $n = 1$: $a - b$ is a factor of $a^1 - b^1 = a - b$.
     II: If $a - b$ is a factor of $a^k - b^k$
           Show that $a - b$ is a factor of $a^{k+1} - b^{k+1} = a \cdot a^k - b \cdot b^k$

$$= a \cdot a^k - a \cdot b^k + a \cdot b^k - b \cdot b^k = a\left(a^k - b^k\right) + b^k(a - b)$$

Since $a - b$ is a factor of $a^k - b^k$ and $a - b$ is a factor of $a - b$, then

$a - b$ is a factor of $a^{k+1} - b^{k+1}$.
Conditions I and II are satisfied; the statement is true.

27. $n = 1$: $1^2 - 1 + 41 = 41$ is a prime number.
     $n = 41$: $41^2 - 41 + 41 = 41^2$ is not a prime number.

29. I: $n = 1$: $ar^{1-1} = a$ and $a\left(\dfrac{1 - r^1}{1 - r}\right) = a$

     II: If $a + ar + ar^2 + \cdots + ar^{k-1} = a\left(\dfrac{1 - r^k}{1 - r}\right)$

           then $a + ar + ar^2 + \cdots + ar^{k-1} + ar^{k+1-1}$

$$= \left[a + ar + ar^2 + \cdots + ar^{k-1}\right] + ar^k = a\left(\dfrac{1 - r^k}{1 - r}\right) + ar^k$$

$$= \dfrac{a(1 - r^k) + ar^k(1 - r)}{1 - r} = \dfrac{a - ar^k + ar^k - ar^{k+1}}{1 - r} = a\left(\dfrac{1 - r^{k+1}}{1 - r}\right)$$

Conditions I and II are satisfied; the statement is true.

31. I: $n = 4$: The number of diagonals of a quadrilateral is $\dfrac{1}{2} \cdot 4(4 - 3) = 2$
     II: Assume that for any integer k the number of diagonals of a convex polygon
           with k sides (k vertices) is $\dfrac{1}{2} \cdot k(k - 3)$. A convex polygon with $k + 1$ sides
           ($k + 1$ vertices) consists of a convex polygon with k sides (k vertices) plus
           a triangle for a total of $k + 1$ vertices. The number of diagonals of this
           convex polygon consists of the original ones plus $k - 1$ additional ones,

           namely, $\dfrac{1}{2} \cdot k(k - 3) + (k - 1) = \dfrac{1}{2}k^2 - \dfrac{3}{2}k + k - 1 = \dfrac{1}{2}k^2 - \dfrac{1}{2}k - 1$

$$= \dfrac{1}{2} \cdot \left(k^2 - k - 2\right) = \dfrac{1}{2} \cdot (k + 1)(k - 2)$$

Conditions I and II are satisfied; the statement is true.

33. Answers will vary.

Chapter 11

Sequences; Induction; The Binomial Theorem

11.5 The Binomial Theorem

1. $\dbinom{5}{3} = \dfrac{5!}{3!\,2!} = \dfrac{5 \cdot 4 \cdot 3 \cdot 2 \cdot 1}{3 \cdot 2 \cdot 1 \cdot 2 \cdot 1} = \dfrac{5 \cdot 4}{2 \cdot 1} = 10$

3. $\dbinom{7}{5} = \dfrac{7!}{5!\,2!} = \dfrac{7 \cdot 6 \cdot 5 \cdot 4 \cdot 3 \cdot 2 \cdot 1}{5 \cdot 4 \cdot 3 \cdot 2 \cdot 1 \cdot 2 \cdot 1} = \dfrac{7 \cdot 6}{2 \cdot 1} = 21$

5. $\dbinom{50}{49} = \dfrac{50!}{49!\,1!} = \dfrac{50 \cdot 49!}{49! \cdot 1} = \dfrac{50}{1} = 50$

7. $\dbinom{1000}{1000} = \dfrac{1000!}{1000!\,0!} = \dfrac{1}{1} = 1$

9. $\dbinom{55}{23} = \dfrac{55!}{23!\,32!} = 1.866442159 \times 10^{15}$

11. $\dbinom{47}{25} = \dfrac{47!}{25!\,22!} = 1.483389769 \times 10^{13}$

13. $(x+1)^5 = \dbinom{5}{0}x^5 + \dbinom{5}{1}x^4 + \dbinom{5}{2}x^3 + \dbinom{5}{3}x^2 + \dbinom{5}{4}x^1 + \dbinom{5}{5}x^0$
$$= x^5 + 5x^4 + 10x^3 + 10x^2 + 5x + 1$$

15. $(x-2)^6 = \dbinom{6}{0}x^6 + \dbinom{6}{1}x^5(-2) + \dbinom{6}{2}x^4(-2)^2 + \dbinom{6}{3}x^3(-2)^3 + \dbinom{6}{4}x^2(-2)^4$
$$+ \dbinom{6}{5}x(-2)^5 + \dbinom{6}{6}x^0(-2)^6$$
$$= x^6 + 6x^5(-2) + 15x^4 \cdot 4 + 20x^3(-8) + 15x^2 \cdot 16 + 6x \cdot (-32) + 64$$
$$= x^6 - 12x^5 + 60x^4 - 160x^3 + 240x^2 - 192x + 64$$

17. $(3x+1)^4 = \dbinom{4}{0}(3x)^4 + \dbinom{4}{1}(3x)^3 + \dbinom{4}{2}(3x)^2 + \dbinom{4}{3}(3x) + \dbinom{4}{4}$
$$= 81x^4 + 4 \cdot 27x^3 + 6 \cdot 9x^2 + 4 \cdot 3x + 1 = 81x^4 + 108x^3 + 54x^2 + 12x + 1$$

19. $(x^2 + y^2)^5 = \binom{5}{0}(x^2)^5(y^2)^0 + \binom{5}{1}(x^2)^4(y^2) + \binom{5}{2}(x^2)^3(y^2)^2 + \binom{5}{3}(x^2)^2(y^2)^3$

$$+ \binom{5}{4}x^2(y^2)^4 + \binom{5}{5}(y^2)^5$$

$$= x^{10} + 5x^8 y^2 + 10x^6 y^4 + 10x^4 y^6 + 5x^2 y^8 + y^{10}$$

21. $(\sqrt{x} + \sqrt{2})^6 = \binom{6}{0}(\sqrt{x})^6(\sqrt{2})^0 + \binom{6}{1}(\sqrt{x})^5(\sqrt{2})^1 + \binom{6}{2}(\sqrt{x})^4(\sqrt{2})^2 + \binom{6}{3}(\sqrt{x})^3(\sqrt{2})^3$

$$\binom{6}{4}(\sqrt{x})^2(\sqrt{2})^4 + \binom{6}{5}(\sqrt{x})(\sqrt{2})^5 + \binom{6}{6}(\sqrt{x})^0(\sqrt{2})^6$$

$$= x^3 + 6\sqrt{2}x^{5/2} + 15 \cdot 2x^2 + 20 \cdot 2\sqrt{2}x^{3/2} + 15 \cdot 4x + 6 \cdot 4\sqrt{2}x^{1/2} + 8$$

$$= x^3 + 6\sqrt{2}x^{5/2} + 30x^2 + 40\sqrt{2}x^{3/2} + 60x + 24\sqrt{2}x^{1/2} + 8$$

23. $(ax + by)^5 = \binom{5}{0}(ax)^5 + \binom{5}{1}(ax)^4 \cdot by + \binom{5}{2}(ax)^3(by)^2 + \binom{5}{3}(ax)^2(by)^3$

$$+ \binom{5}{4}ax(by)^4 + \binom{5}{5}(by)^5$$

$$= a^5 x^5 + 5a^4 x^4 by + 10a^3 x^3 b^2 y^2 + 10a^2 x^2 b^3 y^3 + 5axb^4 y^4 + b^5 y^5$$

25. $n = 10, \ j = 4, \ x = x, \ a = 3$

$$\binom{10}{4}x^6 \cdot 3^4 = \frac{10!}{4! \, 6!} \cdot 81x^6 = \frac{10 \cdot 9 \cdot 8 \cdot 7}{4 \cdot 3 \cdot 2 \cdot 1} \cdot 81x^6 = 17{,}010x^6$$

The coefficient of x^6 is 17,010.

27. $n = 12, \ j = 5, \ x = 2x, \ a = -1$

$$\binom{12}{5}(2x)^7 \cdot (-1)^5 = \frac{12!}{5! \, 7!} \cdot 128x^7(-1) = \frac{12 \cdot 11 \cdot 10 \cdot 9 \cdot 8}{5 \cdot 4 \cdot 3 \cdot 2 \cdot 1} \cdot (-128)x^7 = -101{,}376x^7$$

The coefficient of x^7 is $-101{,}376$.

29. $n = 9, \ j = 2, \ x = 2x, \ a = 3$

$$\binom{9}{2}(2x)^7 \cdot 3^2 = \frac{9!}{2! \, 7!} \cdot 128x^7(9) = \frac{9 \cdot 8}{2 \cdot 1} \cdot 128x^7 \cdot 9 = 41{,}472x^7$$

The coefficient of x^7 is $41{,}472$.

31. $n = 7, \ j = 4, \ x = x, \ a = 3$

$$\binom{7}{4}x^3 \cdot 3^4 = \frac{7!}{4! \, 3!} \cdot 81x^3 = \frac{7 \cdot 6 \cdot 5}{3 \cdot 2 \cdot 1} \cdot 81x^3 = 2835x^3$$

33. $n = 9, \ j = 2, \ x = 3x, \ a = -2$

$$\binom{9}{2}(3x)^7 \cdot (-2)^2 = \frac{9!}{2! \, 7!} \cdot 2187x^7 \cdot 4 = \frac{9 \cdot 8}{2 \cdot 1} \cdot 8748x^7 = 314{,}928x^7$$

35.   The constant term in $\binom{12}{j}\left(x^2\right)^{12-j}\left(\dfrac{1}{x}\right)^j$ occurs when:

$$2(12-j)=j \;\rightarrow\; 24-2j=j \;\rightarrow\; 3j=24 \;\rightarrow\; j=8.$$

Evaluate the 9th term:

$$\binom{12}{8}\left(x^2\right)^4 \cdot \left(\frac{1}{x}\right)^8 = \frac{12!}{8!\,4!}x^8 \cdot \frac{1}{x^8} = \frac{12\cdot 11\cdot 10\cdot 9}{4\cdot 3\cdot 2\cdot 1}x^0 = 495$$

37.   The x^4 term in $\binom{10}{j}(x)^{10-j}\left(\dfrac{-2}{\sqrt{x}}\right)^j$ occurs when:

$$10-j-\frac{1}{2}j=4 \;\rightarrow\; -\frac{3}{2}j=-6 \;\rightarrow\; j=4.$$

Evaluate the 5th term:

$$\binom{10}{4}(x)^6 \cdot \left(\frac{-2}{\sqrt{x}}\right)^4 = \frac{10!}{6!\,4!}x^6 \cdot \frac{16}{x^2} = \frac{10\cdot 9\cdot 8\cdot 7}{4\cdot 3\cdot 2\cdot 1}\cdot 16x^4 = 3360x^4$$

The coefficient is 3360.

39. $(1.001)^5 = \left(1+10^{-3}\right)^5 = \binom{5}{0}\cdot 1^5 + \binom{5}{1}\cdot 1^4 \cdot 10^{-3} + \binom{5}{2}\cdot 1^3 \cdot \left(10^{-3}\right)^2 + \binom{5}{3}\cdot 1^2 \cdot \left(10^{-3}\right)^3 + \ldots$

$$= 1 + 5(0.001) + 10(0.000001) + 10(0.000000001) + \ldots$$
$$= 1 + 0.005 + 0.000010 + 0.000000010 + \ldots$$
$$= 1.00501 \quad \text{(correct to 5 decimal places)}$$

41. $\dbinom{n}{n-1} = \dfrac{n!}{(n-1)!\,(n-(n-1))!} = \dfrac{n!}{(n-1)!\,(1)!} = n$

$$\binom{n}{n} = \frac{n!}{n!\,(n-n)!} = \frac{n!}{n!\,0!} = \frac{n!}{n!\cdot 1} = \frac{n!}{n!} = 1$$

43.   Show that $\dbinom{n}{0} + \dbinom{n}{1} + \ldots + \dbinom{n}{n} = 2^n$

$$2^n = (1+1)^n$$

$$= \binom{n}{0}\cdot 1^n + \binom{n}{1}\cdot 1^{n-1}\cdot 1 + \binom{n}{2}\cdot 1^{n-2}\cdot 1^2 + \ldots + \binom{n}{n}\cdot 1^{n-n}\cdot 1^n$$

$$= \binom{n}{0} + \binom{n}{1} + \ldots + \binom{n}{n}$$

45. $\dbinom{5}{0}\left(\dfrac{1}{4}\right)^5 + \dbinom{5}{1}\left(\dfrac{1}{4}\right)^4\left(\dfrac{3}{4}\right) + \dbinom{5}{2}\left(\dfrac{1}{4}\right)^3\left(\dfrac{3}{4}\right)^2 + \dbinom{5}{3}\left(\dfrac{1}{4}\right)^2\left(\dfrac{3}{4}\right)^3$

$$+ \binom{5}{4}\left(\frac{1}{4}\right)\left(\frac{3}{4}\right)^4 + \binom{5}{5}\left(\frac{3}{4}\right)^5 = \left(\frac{1}{4}+\frac{3}{4}\right)^5 = (1)^5 = 1$$

47. We can use Mathematical Induction to prove the Binomial Theorem.

I: $n = 1$: $(x + a)^1 = x + a = \binom{1}{0}x^1 + \binom{1}{1}a^1$. Therefore, the Theorem holds for $n = 1$.

II: Now assume the Theorem holds for some natural number k.
       That is, assume

$$(x + a)^k = \binom{k}{0}x^k + \binom{k}{1}ax^{k-1} + \ldots + \binom{k}{j-1}a^{j-1}x^{k-j+1} + \binom{k}{j}a^j x^{k-j} + \ldots + \binom{k}{k}a^k$$

We now calculate $(x + a)^{k+1}$:

$$(x + a)^{k+1} = (x + a)(x + a)^k$$

$$= (x + a)\left(\binom{k}{0}x^k + \binom{k}{1}ax^{k-1} + \ldots + \binom{k}{j-1}a^{j-1}x^{k-j+1} + \binom{k}{j}a^j x^{k-j} + \ldots + \binom{k}{k}a^k\right)$$

$$= x\left(\binom{k}{0}x^k + \binom{k}{1}ax^{k-1} + \ldots + \binom{k}{j-1}a^{j-1}x^{k-j+1} + \binom{k}{j}a^j x^{k-j} + \ldots + \binom{k}{k}a^k\right)$$

$$+ a\left(\binom{k}{0}x^k + \binom{k}{1}ax^{k-1} + \ldots + \binom{k}{j-1}a^{j-1}x^{k-j+1} + \binom{k}{j}a^j x^{k-j} + \ldots + \binom{k}{k}a^k\right)$$

$$= \left(\binom{k}{0}x^{k+1} + \binom{k}{1}ax^k + \ldots + \binom{k}{j-1}a^{j-1}x^{k-j+2} + \binom{k}{j}a^j x^{k-j+1} + \ldots + \binom{k}{k}a^k x\right)$$

$$+ \left(\binom{k}{0}ax^k + \binom{k}{1}a^2 x^{k-1} + \ldots + \binom{k}{j-1}a^j x^{k-j+1} + \binom{k}{j}a^{j+1}x^{k-j} + \ldots + \binom{k}{k}a^{k+1}\right)$$

$$= \binom{k}{0}x^{k+1} + \left(\binom{k}{1} + \binom{k}{0}\right)ax^k + \ldots + \left(\binom{k}{j} + \binom{k}{j-1}\right)a^j x^{k-j+1} + \ldots + \left(\binom{k}{k} + \binom{k}{k-1}\right)a^k x + \binom{k}{k}a^{k+1}$$

Note that

$$\binom{k}{1} = 1 = \binom{k+1}{0}; \quad \binom{k}{1} + \binom{k}{0} = \binom{k+1}{1}; \quad \binom{k}{2} + \binom{k}{1} = \binom{k+1}{2}; \ldots; \binom{k}{j} + \binom{k}{j-1} = \binom{k+1}{j}$$

and $\binom{k}{k} = 1 = \binom{k+1}{k+1}$.

So we have $(x + a)^{k+1} = \binom{k+1}{0}x^{k+1} + \binom{k+1}{1}ax^k + \ldots + \binom{k+1}{j}a^j x^{k-j+1} + \ldots + \binom{k+1}{k+1}a^{k+1}$

Conditions I and II are satisfied; the Theorem is true.

Chapter 11

Sequences; Induction; The Binomial Theorem

11.R Chapter Review

1. $a_1 = (-1)^1 \dfrac{1+3}{1+2} = -\dfrac{4}{3}, \quad a_2 = (-1)^2 \dfrac{2+3}{2+2} = \dfrac{5}{4}, \quad a_3 = (-1)^3 \dfrac{3+3}{3+2} = -\dfrac{6}{5},$

$\qquad a_4 = (-1)^4 \dfrac{4+3}{4+2} = \dfrac{7}{6}, \quad a_5 = (-1)^5 \dfrac{5+3}{5+2} = -\dfrac{8}{7}$

3. $a_1 = \dfrac{2^1}{1^2} = \dfrac{2}{1} = 2, \quad a_2 = \dfrac{2^2}{2^2} = \dfrac{4}{4} = 1, \quad a_3 = \dfrac{2^3}{3^2} = \dfrac{8}{9}, \quad a_4 = \dfrac{2^4}{4^2} = \dfrac{16}{16} = 1, \quad a_5 = \dfrac{2^5}{5^2} = \dfrac{32}{25}$

5. $a_1 = 3, \quad a_2 = \dfrac{2}{3} \cdot 3 = 2, \quad a_3 = \dfrac{2}{3} \cdot 2 = \dfrac{4}{3}, \quad a_4 = \dfrac{2}{3} \cdot \dfrac{4}{3} = \dfrac{8}{9}, \quad a_5 = \dfrac{2}{3} \cdot \dfrac{8}{9} = \dfrac{16}{27}$

7. $a_1 = 2, \quad a_2 = 2 - 2 = 0, \quad a_3 = 2 - 0 = 2, \quad a_4 = 2 - 2 = 0, \quad a_5 = 2 - 0 = 2$

9. $\displaystyle\sum_{k=1}^{4}(4k+2) = (4 \cdot 1 + 2) + (4 \cdot 2 + 2) + (4 \cdot 3 + 2) + (4 \cdot 4 + 2) = (6) + (10) + (14) + (18) - 48$

11. $1 - \dfrac{1}{2} + \dfrac{1}{3} - \dfrac{1}{4} + \cdots + \dfrac{1}{13} = \displaystyle\sum_{k=1}^{13}(-1)^{k+1}\left(\dfrac{1}{k}\right)$

13. $\{n+5\}$ Arithmetic

$d = (n+1+5) - (n+5) = n + 6 - n - 5 = 1$

$S_n = \dfrac{n}{2}[6 + n + 5] = \dfrac{n}{2}(n+11)$

15. $\{2n^3\}$ Examine the terms of the sequence: 2, 16, 54, 128, 250, ...

There is no common difference; there is no common ratio; neither.

17. $\{2^{3n}\}$ Geometric $\qquad r = \dfrac{2^{3(n+1)}}{2^{3n}} = \dfrac{2^{3n+3}}{2^{3n}} = 2^{3n+3-3n} = 2^3 = 8$

$S_n = 8\left(\dfrac{1-8^n}{1-8}\right) = 8\left(\dfrac{1-8^n}{-7}\right) = \dfrac{8}{7}\left(8^n - 1\right)$

19. 0, 4, 8, 12, ... Arithmetic $\qquad d = 4 - 0 = 4$

$S_n = \dfrac{n}{2}(2(0) + (n-1)4) = \dfrac{n}{2}(4(n-1)) = 2n(n-1)$

715

21. $3, \dfrac{3}{2}, \dfrac{3}{4}, \dfrac{3}{5}, \dfrac{3}{16}, \ldots$ Geometric $r = \dfrac{\left(\dfrac{3}{2}\right)}{3} = \dfrac{3}{2} \cdot \dfrac{1}{3} = \dfrac{1}{2}$

$$S_n = 3\left(\dfrac{1 - \left(\dfrac{1}{2}\right)^n}{1 - \dfrac{1}{2}}\right) = 3\left(\dfrac{1 - \left(\dfrac{1}{2}\right)^n}{\left(\dfrac{1}{2}\right)}\right) = 6\left(1 - \left(\dfrac{1}{2}\right)^n\right)$$

23. Neither.   There is no common difference or common ratio.

25. (a) $\displaystyle\sum_{k=1}^{5}(k^2 + 12) = 13 + 16 + 21 + 28 + 37 = 115$ (b) use the sum(seq) feature:

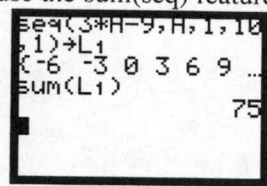

27. (a) $\displaystyle\sum_{k=1}^{10}(3k - 9) = \sum_{k=1}^{10}3k - \sum_{k=1}^{10}9 = 3\sum_{k=1}^{10}k - \sum_{k=1}^{10}9 = 3\left(\dfrac{10(10+1)}{2}\right) - 10(9) = 165 - 90 = 75$

(b) use the sum(seq) feature:

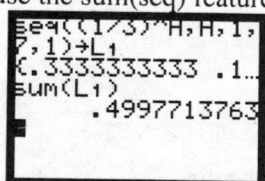

29. (a) $\displaystyle\sum_{k=1}^{7}\left(\dfrac{1}{3}\right)^k = \dfrac{1}{3}\left(\dfrac{1 - \left(\dfrac{1}{3}\right)^7}{1 - \dfrac{1}{3}}\right) = \dfrac{1}{3}\left(\dfrac{1 - \left(\dfrac{1}{3}\right)^7}{\left(\dfrac{2}{3}\right)}\right) = \dfrac{1}{2}\left(1 - \dfrac{1}{2187}\right) = \dfrac{1}{2}\cdot\dfrac{2186}{2187} = \dfrac{1093}{2187} \approx 0.49977$

(b) use the sum(seq) feature:

31. (a) Arithmetic $a = 3,\ d = 4,\ a_n = a + (n-1)d$
$$a_9 = 3 + (9-1)4 = 3 + 8(4) = 3 + 32 = 35$$
(b) scroll to the end of the list:

33. (a) Geometric $a = 1$, $r = \dfrac{1}{10}$, $n = 11$; $a_n = a\, r^{n-1}$

$$a_{11} = 1 \cdot \left(\frac{1}{10}\right)^{11-1} = \left(\frac{1}{10}\right)^{10} = \frac{1}{10{,}000{,}000{,}000}$$

(b) scroll to the end of the list:

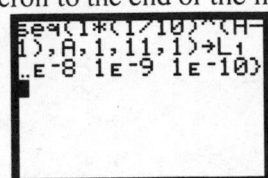

35. (a) Arithmetic $a = \sqrt{2}$, $d = \sqrt{2}$, $n = 9$, $a_n = a + (n-1)d$

$$a_9 = \sqrt{2} + (9-1)\sqrt{2} = \sqrt{2} + 8\sqrt{2} = 9\sqrt{2} \approx 12.7279$$

(b) scroll to the end of the list:

37. $a_7 = a + 6d = 31$ $a_{20} = a + 19d = 96$; $\qquad a_n = a + (n-1)d$

Solve the system of equations:

$$31 - 6d + 19d = 96$$
$$13d = 65$$
$$d = 5$$
$$a = 31 - 6(5) = 31 - 30 = 1$$

General formula: $\{5n - 4\}$

39. $a_{10} = a + 9d = 0$ $a_{18} = a + 17d = 8$; $\qquad a_n = a + (n-1)d$

Solve the system of equations:

$$-9d + 17d = 8$$
$$8d = 8$$
$$d = 1$$
$$a = -9(1) = -9$$

General formula: $\{n - 10\}$

41. $a = 3$, $r = \dfrac{1}{3}$ Since $|r| < 1$, $S_n = \dfrac{a}{1-r} = \dfrac{3}{\left(1 - \dfrac{1}{3}\right)} = \dfrac{3}{\left(\dfrac{2}{3}\right)} = \dfrac{9}{2}$

43. $a = 2$, $r = -\dfrac{1}{2}$ Since $|r| < 1$, $S_n = \dfrac{a}{1-r} = \dfrac{2}{\left(1 - \left(-\dfrac{1}{2}\right)\right)} = \dfrac{2}{\left(\dfrac{3}{2}\right)} = \dfrac{4}{3}$

45. $a = 4$, $r = \dfrac{1}{2}$ Since $|r| < 1$, $S_n = \dfrac{a}{1-r} = \dfrac{4}{\left(1 - \dfrac{1}{2}\right)} = \dfrac{4}{\left(\dfrac{1}{2}\right)} = 8$

47.   I: $n = 1$: $3 \cdot 1 = 3$ and $\dfrac{3 \cdot 1}{2}(1 + 1) = 3$

   II: If $3 + 6 + 9 + \cdots + 3k = \dfrac{3k}{2}(k + 1)$

   then $3 + 6 + 9 + \cdots + 3k + 3(k + 1)$

$$= [3 + 6 + 9 + \cdots + 3k] + 3(k + 1) = \dfrac{3k}{2}(k + 1) + 3(k + 1)$$

$$= (k + 1)\left(\dfrac{3k}{2} + 3\right) = \dfrac{3}{2}(k + 1)(k + 2)$$

Conditions I and II are satisfied; the statement is true.

49.   I: $n = 1$: $2 \cdot 3^{1-1} = 2$ and $3^1 - 1 = 2$
   II: If $2 + 6 + 18 + \cdots + 2 \cdot 3^{k-1} = 3^k - 1$

   then $2 + 6 + 18 + \cdots + 2 \cdot 3^{k-1} + 2 \cdot 3^{k+1-1}$

$$= [2 + 6 + 18 + \cdots + 2 \cdot 3^{k-1}] + 2 \cdot 3^k = 3^k - 1 + 2 \cdot 3^k = 3 \cdot 3^k - 1 = 3^{k+1} - 1$$

Conditions I and II are satisfied; the statement is true.

51.   I: $n = 1$: $(3 \cdot 1 - 2)^2 = 1$ and $\dfrac{1}{2} \cdot 1(6 \cdot 1^2 - 3 \cdot 1 - 1) = 1$

   II: If $1^2 + 4^2 + 7^2 + \cdots + (3k - 2)^2 = \dfrac{1}{2} \cdot k\left(6k^2 - 3k - 1\right)$

   then $1^2 + 4^2 + 7^2 + \cdots + (3k - 2)^2 + \left(3(k + 1) - 2\right)^2$

$$= \left[1^2 + 4^2 + 7^2 + \cdots + (3k - 2)^2\right] + (3k + 1)^2$$

$$= \dfrac{1}{2} \cdot k\left(6k^2 - 3k - 1\right) + (3k + 1)^2$$

$$= \dfrac{1}{2} \cdot \left[6k^3 - 3k^2 - k + 18k^2 + 12k + 2\right] = \dfrac{1}{2} \cdot \left[6k^3 + 15k^2 + 11k + 2\right]$$

$$= \dfrac{1}{2} \cdot (k + 1)\left[6k^2 + 9k + 2\right] = \dfrac{1}{2} \cdot (k + 1)\left[6k^2 + 12k + 6 - 3k - 3 - 1\right]$$

$$= \dfrac{1}{2} \cdot (k + 1)\left[6(k^2 + 2k + 1) - 3(k + 1) - 1\right]$$

$$= \dfrac{1}{2} \cdot (k + 1)\left[6(k + 1)^2 - 3(k + 1) - 1\right]$$

Conditions I and II are satisfied; the statement is true.

53. $\dbinom{5}{2} = \dfrac{5!}{2!\,3!} = \dfrac{5 \cdot 4 \cdot 3 \cdot 2 \cdot 1}{2 \cdot 1 \cdot 3 \cdot 2 \cdot 1} = \dfrac{5 \cdot 4}{2 \cdot 1} = 10$

55. $(x+2)^5 = \binom{5}{0}x^5 + \binom{5}{1}x^4 \cdot 2 + \binom{5}{2}x^3 \cdot 2^2 + \binom{5}{3}x^2 \cdot 2^3 + \binom{5}{4}x^1 \cdot 2^4 + \binom{5}{5} \cdot 2^5$

$\qquad = x^5 + 5 \cdot 2x^4 + 10 \cdot 4x^3 + 10 \cdot 8x^2 + 5 \cdot 16x + 1 \cdot 32$

$\qquad = x^5 + 10x^4 + 40x^3 + 80x^2 + 80x + 32$

57. $(2x+3)^5 = \binom{5}{0}(2x)^5 + \binom{5}{1}(2x)^4 \cdot 3 + \binom{5}{2}(2x)^3 \cdot 3^2 + \binom{5}{3}(2x)^2 \cdot 3^3$

$\qquad\qquad\qquad\qquad + \binom{5}{4}(2x)^1 \cdot 3^4 + \binom{5}{5} \cdot 3^5$

$\qquad = 32x^5 + 5 \cdot 16x^4 \cdot 3 + 10 \cdot 8x^3 \cdot 9 + 10 \cdot 4x^2 \cdot 27 + 5 \cdot 2x \cdot 81 + 1 \cdot 243$

$\qquad = 32x^5 + 240x^4 + 720x^3 + 1080x^2 + 810x + 243$

59. $n = 9, \ j = 2, \ x = x, \ a = 2$

$\qquad \binom{9}{2}x^7 \cdot 2^2 = \dfrac{9!}{2!\,7!} \cdot 4x^7 = \dfrac{9 \cdot 8}{2 \cdot 1} \cdot 4x^7 = 144x^7$

The coefficient of x^7 is 144.

61. $n = 7, \ j = 5, \ x = 2x, \ a = 1$

$\qquad \binom{7}{5}(2x)^2 \cdot 1^5 = \dfrac{7!}{5!\,2!} \cdot 4x^2(1) = \dfrac{7 \cdot 6}{2 \cdot 1} \cdot 4x^2 = 84x^2$

The coefficient of x^2 is 84.

63. This is an arithmetic sequence with $a = 80, \ d = -3, \ n = 25$

(a) $a_{25} = 80 + (25 - 1)(-3) = 80 - 72 = 8$ bricks

(b) $S_{25} = \dfrac{25}{2}(80 + 8) = 25(44) = 1100$ bricks

1100 bricks are needed to build the steps.

65. This is a geometric sequence with $a = 20, \ r = \dfrac{3}{4}$.

(a) After striking the ground the third time, the height is $20\left(\dfrac{3}{4}\right)^3 = \dfrac{135}{16} \approx 8.44$ feet.

(b) After striking the ground the nth time, the height is $20\left(\dfrac{3}{4}\right)^n$ feet.

(c) If the height is less than 6 inches or 0.5 feet, then:

$0.5 = 20\left(\dfrac{3}{4}\right)^n \Rightarrow 0.025 = \left(\dfrac{3}{4}\right)^n \Rightarrow \log(0.025) = n\log\left(\dfrac{3}{4}\right) \Rightarrow n = \dfrac{\log(0.025)}{\log\left(\dfrac{3}{4}\right)} \approx 12.82$

The height is less than 6 inches after the 13th strike.

(d) Since this is a geometric sequence with $|r| < 1$, the distance is the sum of the two infinite geometric series - the distances going down plus the distances going up.

Distance going down: $S_{down} = \dfrac{20}{\left(1 - \dfrac{3}{4}\right)} = \dfrac{20}{\left(\dfrac{1}{4}\right)} = 80$ feet.

Distance going up: $S_{up} = \dfrac{15}{\left(1 - \dfrac{3}{4}\right)} = \dfrac{15}{\left(\dfrac{1}{4}\right)} = 60$ feet.

The total distance traveled is 140 feet.

67.   (a) $b_1 = 5,000, \quad b_n = 1.015b_{n-1} - 100$

$b_2 = 1.015b_{(2-1)} - 100 = 1.015b_1 - 100 = 1.015(5000) - 100 = \$4975$

(b)   Enter the recursive formula in Y= and draw the graph:

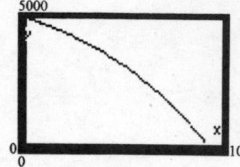

(c)   Scroll through the table:

n	$u(n)$
30	4100
31	4061.5
32	4022.5
33	3982.8
34	3942.5
35	3901.7
36	3860.2

$n = 33$

At the beginning of the 33rd month, or after 32 payments have been made, the balance is below \$4,000. The balance is \$3982.8.

(d)   Scroll through the table:

n	$u(n)$
90	395.82
91	301.75
92	206.28
93	109.37
94	11.014
95	-88.82
96	-190.2

$n = 94$

The balance will be paid off at the end of 94 months or 7years and 10 months.
Total payments $= 100(93) + 11.01 = \$9,311.01$

(e)   The total interest expense is the difference of the total of the payments and the original balance: $100(93) + 11.01 - 5,000 = \$4,311.01$

Chapter 11

Sequences; Induction; The Binomial Theorem

11.CR   Cumulative Review

1. $|x^2| = 9$

   $x^2 = 9$ or $x^2 = -9$

   $x = \pm 3$ or $x = \pm 3i$

3. $2e^x = 5$

   $e^x = \dfrac{5}{2} = 2.5$

   $\ln(e^x) = \ln(2.5)$

   $x = \ln(2.5) \approx 0.916$

5. $\cos^{-1}(-0.5)$

   We are finding the angle θ, $-\pi \le \theta \le \pi$, whose cosine equals -0.5.

       $\cos\theta = -0.5$ $-\pi \le \theta \le \pi$

   $$\theta = \frac{2\pi}{3} \Rightarrow \cos^{-1}(-0.5) = \frac{2\pi}{3}$$

7. Center: $(0, 0)$; Focus: $(0, 3)$; Vertex: $(0, 4)$; Major axis is the y-axis; $a = 4$; $c = 3$.

   Find b: $b^2 = a^2 - c^2 = 16 - 9 = 7 \Rightarrow b = \sqrt{7}$

   Write the equation using rectangular coordinates: $\dfrac{x^2}{7} + \dfrac{y^2}{16} = 1$

   Parametric equations for the ellipse are: $x = \sqrt{7}\cos(\pi t)$; $y = 4\sin(\pi t)$; $0 \le t \le 2$

9. center point $(0,4)$;   passing through the pole $(0,4)$ implies that the radius = 4

   using rectangular coordinates:

   $$(x - h)^2 + (y - k)^2 = r^2$$

   $$(x - 0)^2 + (y - 4)^2 = 4^2$$

   $$x^2 + y^2 - 8y + 16 = 16$$

   $$x^2 + y^2 - 8y = 0$$

   converting to polar coordinates:

   $$r^2 - 8r\sin\theta = 0$$

   $$r^2 = 8r\sin\theta$$

   $$r = 8\sin\theta$$

Chapter 12

Counting and Probability

12.1 Sets and Counting

1. $A \cup B = \{1, 3, 5, 7, 9\} \cup \{1, 5, 6, 7\} = \{1, 3, 5, 6, 7, 9\}$

3. $A \cap B = \{1, 3, 5, 7, 9\} \cap \{1, 5, 6, 7\} = \{1, 5, 7\}$

5. $(A \cup B) \cap C = \left(\{1, 3, 5, 7, 9\} \cup \{1, 5, 6, 7\}\right) \cap \{1, 2, 4, 6, 8, 9\}$
$$= \{1, 3, 5, 6, 7, 9\} \cap \{1, 2, 4, 6, 8, 9\}$$
$$= \{1, 6, 9\}$$

7. $(A \cap B) \cup C = \left(\{1, 3, 5, 7, 9\} \cap \{1, 5, 6, 7\}\right) \cup \{1, 2, 4, 6, 8, 9\}$
$$= \{1, 5, 7\} \cup \{1, 2, 4, 6, 8, 9\}$$
$$= \{1, 2, 4, 5, 6, 7, 8, 9\}$$

9. $(A \cup C) \cap (B \cup C)$
$$= \left(\{1, 3, 5, 7, 9\} \cup \{1, 2, 4, 6, 8, 9\}\right) \cap \left(\{1, 5, 6, 7\} \cup \{1, 2, 4, 6, 8, 9\}\right)$$
$$= \{1, 2, 3, 4, 5, 6, 7, 8, 9\} \cap \{1, 2, 4, 5, 6, 7, 8, 9\}$$
$$= \{1, 2, 4, 5, 6, 7, 8, 9\}$$

11. $\overline{A} = \{0, 2, 6, 7, 8\}$

13. $\overline{A \cap B} = \overline{\{1, 3, 4, 5, 9\} \cap \{2, 4, 6, 7, 8\}} = \overline{\{4\}} = \{0, 1, 2, 3, 5, 6, 7, 8, 9\}$

15. $\overline{A} \cup \overline{B} = \{0, 2, 6, 7, 8\} \cup \{0, 1, 3, 5, 9\} = \{0, 1, 2, 3, 5, 6, 7, 8, 9\}$

17. $\overline{A \cap \overline{C}} = \overline{\{1, 3, 4, 5, 9\} \cap \{0, 2, 5, 7, 8, 9\}} = \overline{\{5, 9\}} = \{0, 1, 2, 3, 4, 6, 7, 8\}$

19. $\overline{A \cup B \cup C} = \overline{\{1, 3, 4, 5, 9\} \cup \{2, 4, 6, 7, 8\} \cup \{1, 3, 4, 6\}}$
$$= \overline{\{1, 2, 3, 4, 5, 6, 7, 8, 9\}} = \{0\}$$

21. $\{a\}, \{b\}, \{c\}, \{d\}, \{a, b\}, \{a, c\}, \{a, d\}, \{b, c\}, \{b, d\}, \{c, d\}, \{a, b, c\}, \{a, b, d\},$
$\{a, c, d\}, \{b, c, d\}, \{a, b, c, d\}, \varnothing$

23. $n(A) = 15, n(B) = 20, n(A \cap B) = 10$
$$n(A \cup B) = n(A) + n(B) - n(A \cap B) = 15 + 20 - 10 = 25$$

25. $n(A \cup B) = 50, n(A \cap B) = 10, n(B) = 20$

$$n(A \cup B) = n(A) + n(B) - n(A \cap B)$$
$$50 = n(A) + 20 - 10$$
$$40 = n(A)$$

27. From the figure:
$$n(A) = 15 + 3 + 5 + 2 = 25$$

29. From the figure:
$$n(A \text{ or } B) = n(A \cup B) = n(A) + n(B) - n(A \cap B) = 25 + 20 - 8 = 37$$

31. From the figure:
$$n(A \text{ but not } C) = n(A) - n(A \cap C) = 25 - 7 = 18$$

33. From the figure:
$$n(A \text{ and } B \text{ and } C) = n(A \cap B \cap C) = 5$$

35. Let $A = \{\text{those who will purchase a major appliance}\}$

$B = \{\text{those who will buy a car}\}$

$n(U) = 500, \ n(A) = 200, \ n(B) = 150, \ n(A \cap B) = 25$

$n(A \cup B) = n(A) + n(B) - n(A \cap B) = 200 + 150 - 25 = 325$

$n(\text{purchase neither}) = 500 - 325 = 175$

$n(\text{purchase only a car}) = 150 - 25 = 125$

37. Construct a Venn diagram:

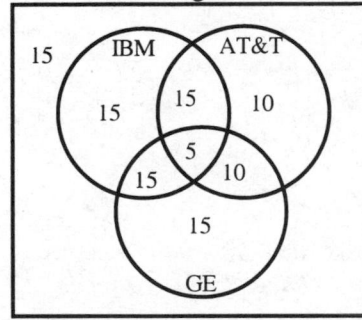

(a)   15
(b)   15
(c)   15
(d)   25
(e)   40

39. (a) $n(\text{married or widowed}) = n(\text{married}) + n(\text{widowed})$
$$= 58{,}986 + 2{,}542 = 61{,}528 \text{ thousand}$$

(b) $n(\text{widowed or divorced}) = n(\text{widowed}) + n(\text{divorced})$
$$= 2{,}542 + 8{,}543 = 11{,}085 \text{ thousand}$$

(c) $n(\text{married, widowed or divorced})$
$$= n(\text{married}) + n(\text{widowed}) + n(\text{divorced})$$
$$= 58{,}986 + 2{,}542 + 8{,}543 = 70{,}071 \text{ thousand}$$

41.   Answers will vary.

Chapter 12

Counting and Probability

12.2 Permutations and Combinations

1. $P(6, 2) = \dfrac{6!}{(6-2)!} = \dfrac{6!}{4!} = \dfrac{6 \cdot 5 \cdot 4!}{4!} = 30$

3. $P(4, 4) = \dfrac{4!}{(4-4)!} = \dfrac{4!}{0!} = \dfrac{4 \cdot 3 \cdot 2 \cdot 1}{1} = 24$

5. $P(7, 0) = \dfrac{7!}{(7-0)!} = \dfrac{7!}{7!} = 1$

7. $P(8, 4) = \dfrac{8!}{(8-4)!} = \dfrac{8!}{4!} = \dfrac{8 \cdot 7 \cdot 6 \cdot 5 \cdot 4!}{4!} = 1680$

9. $C(8, 2) = \dfrac{8!}{(8-2)!\, 2!} = \dfrac{8!}{6!\, 2!} = \dfrac{8 \cdot 7 \cdot 6!}{6! \cdot 2 \cdot 1} = 28$

11. $C(7, 4) = \dfrac{7!}{(7-4)!\, 4!} = \dfrac{7!}{3!\, 4!} = \dfrac{7 \cdot 6 \cdot 5 \cdot 4!}{4! \cdot 3 \cdot 2 \cdot 1} = 35$

13. $C(15, 15) = \dfrac{15!}{(15-15)!\, 15!} = \dfrac{15!}{0!\, 15!} = \dfrac{15!}{15! \cdot 1} = 1$

15. $C(26, 13) = \dfrac{26!}{(26-13)!\, 13!} = \dfrac{26!}{13!\, 13!} = 10,400,600$

17. {*abc, abd, abe, acb, acd, ace, adb, adc, ade, aeb, aec, aed, bac, bad, bae, bca, bcd, bce, bda, bdc, bde, bea, bec, bed, cab, cad, cae, cba, cbd, cbe, cda, cdb, cde, cea, ceb, ced, dab, dac, dae, dba, dbc, dbe, dca, dcb, dce, dea, deb, dec, eab, eac, ead, eba, ebc, ebd, eca, ecb, ecd, eda, edb, edc*}

    $P(5,3) = \dfrac{5!}{(5-3)!} = \dfrac{5!}{2!} = \dfrac{5 \cdot 4 \cdot 3 \cdot 2!}{2!} = 60$

19. {123, 124, 132, 134, 142, 143, 213, 214, 231, 234, 241, 243, 312, 314, 321, 324, 341, 342, 412, 413, 421, 423, 431, 432}

    $P(4,3) = \dfrac{4!}{(4-3)!} = \dfrac{4!}{1!} = \dfrac{4 \cdot 3 \cdot 2 \cdot 1}{1} = 24$

21. {*abc, abd, abe, acd, ace, ade, bcd, bce, bde, cde*}

    $C(5,3) = \dfrac{5!}{(5-3)!\, 3!} = \dfrac{5 \cdot 4 \cdot 3!}{2 \cdot 1 \cdot 3!} = 10$

23.   {123, 124, 134, 234} $C(4,3) = \dfrac{4!}{(4-3)!\,3!} = \dfrac{4 \cdot 3!}{1!\,3!} = 4$

25.   There are 5 choices of shirts and 3 choices of ties; there are (5)(3) = 15 combinations.

27.   There are 4 choices for the first letter in the code and 4 choices for the second letter in the code; there are (4)(4) = 16 possible two-letter codes.

29.   There are two choices for each of three positions; there are (2)(2)(2) = 8 possible three-digit numbers.

31.   To line up the four people, there are 4 choices for the first position, 3 choices for the second position, 2 choices for the third position, and 1 choice for the fourth position. Thus there are (4)(3)(2)(1) = 24 possible ways four people can be lined up.

33.   Since no letter can be repeated, there are 5 choices for the first letter, 4 choices for the second letter, and 3 choices for the third letter. Thus, there are (5)(4)(3) = 60 possible three-letter codes.

35.   There are 26 possible one-letter names. There are (26)(26) = 676 possible two-letter names. There are (26)(26)(26) = 17576 possible three-letter names. Thus, there are 26 + 676 + 17576 = 18,278 possible companies that can be listed on the New York Stock Exchange.

37.   A committee of 4 from a total of 7 students is given by:
$$C(7,4) = \dfrac{7!}{(7-4)!\,4!} = \dfrac{7!}{3!\,4!} = \dfrac{7 \cdot 6 \cdot 5 \cdot 4!}{3 \cdot 2 \cdot 1 \cdot 4!} = 35$$
35 committees are possible.

39.   There are 2 possible answers for each question. Therefore, there are $2^{10} = 1024$ different possible arrangements of the answers.

41.   There are 9 choices for the first digit, and 10 choices for each of the other three digits. Thus, there are (9)(10)(10)(10) = 9000 possible four-digit numbers.

43.   There are 5 choices for the first position, 4 choices for the second position, 3 choices for the third position, 2 choices for the fourth position, and 1 choice for the fifth position. Thus, there are (5)(4)(3)(2)(1) = 120 possible arrangements of the books.

45.   There are 8 choices for the DOW stocks, 15 choices for the NASDAQ stocks, and 4 choices for the global stocks. Thus, there are (8)(15)(4) = 480 different portfolios.

47.   The first person can have any of 365 days, the second person can have any of the remaining 364 days. Thus, there are (365)(364) = 132,860 possible ways two people can have different birthdays.

49. Choosing 2 boys from the 4 boys can be done C(4,2) ways, and choosing 3 girls from the 8 girls can be done in C(8,3) ways. Thus, there are a total of:

$$C(4,2) \cdot C(8,3) = \frac{4!}{(4-2)!\,2!} \cdot \frac{8!}{(8-3)!\,3!} = \frac{4!}{2!\,2!} \cdot \frac{8!}{5!\,3!}$$

$$= \frac{4 \cdot 3!}{2 \cdot 1 \cdot 2 \cdot 1} \cdot \frac{8 \cdot 7 \cdot 6 \cdot 5!}{5!\,3!} = 336$$

51. This is a permutation with repetition. There are $\dfrac{9!}{2!\,2!} = 90,720$ different words.

53. (a) $C(7,2) \cdot C(3,1) = 21 \cdot 3 = 63$
     (b) $C(7,3) = 35$
     (c) $C(3,3) = 1$

55. There are C(100, 22) ways to form the first committee. There are 78 senators left, so there are C(78, 13) ways to form the second committee. There are C(65, 10) ways to form the third committee. There are C(55, 5) ways to form the fourth committee. There are C(50, 16) ways to form the fifth committee. There are C(34, 17) ways to form the sixth committee. There are C(17, 17) ways to form the seventh committee.
The total number of committees =
$$= C(100,22) \cdot C(78,13) \cdot C(65,10) \cdot C(55,5) \cdot C(50,16) \cdot C(34,17) \cdot C(17,17)$$
$$= 1.1568 \times 10^{76}$$

57. There are 9 choices for the first position, 8 choices for the second position, 7 for the third position, etc. There are $9 \cdot 8 \cdot 7 \cdot 6 \cdot 5 \cdot 4 \cdot 3 \cdot 2 \cdot 1 = 9! = 362,880$ possible batting orders.

59. The team must have 1 pitcher and 8 position players (non-pitchers). For pitcher, choose 1 player from a group of 4 players, i.e. C(4, 1). For position players, choose 8 players from a group of 11 players, i.e. C(11, 8). Therefore, the number different teams possible is

$$C(4,1) \cdot C(11,8) = \frac{4!}{(4-1)!\cdot 1!} \cdot \frac{11!}{(11-8)!\cdot 8!} = \frac{4!}{3!} \cdot \frac{11!}{3!\cdot 8!} = \frac{4 \cdot 3!}{3!} \cdot \frac{11 \cdot 10 \cdot 9 \cdot 8!}{3!\cdot 8!}$$

$$= 4 \cdot \left(\frac{11 \cdot 10 \cdot 9}{3!} \right) = 4 \cdot \left(\frac{990}{6} \right) = 4 \cdot 165 = 660$$

61. Choose 2 players from a group of 6 players. Therefore, there are $C(6,2) = 15$ different teams possible.

63. – 65. Answers will vary.

Chapter 12

Counting and Probability

12.3 Probability

1. Probabilities must be between 0 and 1, inclusive. Thus, 0, 0.01, 0.35, and 1 are probabilities.

3. All the probabilities are between 0 and 1.
   The sum of the probabilities is $0.2 + 0.3 + 0.1 + 0.4 = 1$.
   This is a probability model.

5. All the probabilities are between 0 and 1.
   The sum of the probabilities is $0.3 + 0.2 + 0.1 + 0.3 = 0.9$.
   This is not a probability model.

7. The sample space is: $S = \{HH, HT, TH, TT\}$.
   Each outcome is equally likely to occur; so $P(E) = \dfrac{n(E)}{n(S)}$.
   The probabilities are: $P(HH) = \dfrac{1}{4}$, $P(HT) = \dfrac{1}{4}$, $P(TH) = \dfrac{1}{4}$, $P(TT) = \dfrac{1}{4}$.

9. The sample space of tossing two fair coins and a fair die is:
   $S = \{HH1, HH2, HH3, HH4, HH5, HH6, HT1, HT2, HT3, HT4, HT5,$
   $HT6, TH1, TH2, TH3, TH4, TH5, TH6, TT1, TT2, TT3, TT4, TT5, TT6\}$
   There are 24 equally likely outcomes and the probability of each is $\dfrac{1}{24}$.

11. The sample space for tossing three fair coins is:
    $S = \{HHH, HHT, HTH, THH, HTT, THT, TTH, TTT\}$
    There are 8 equally likely outcomes and the probability of each is $\dfrac{1}{8}$.

13. The sample space is:
    S = { 1 Yellow, 1 Red, 1 Green, 2 Yellow, 2 Red, 2 Green, 3 Yellow, 3 Red,
    3 Green, 4 Yellow, 4 Red, 4 Green}
    There are 12 equally likely events and the probability of each is $\dfrac{1}{12}$. The probability of getting a 2 or 4 followed by a Red is $P(2 \text{ Red}) + P(4 \text{ Red}) = \dfrac{1}{12} + \dfrac{1}{12} = \dfrac{1}{6}$.

15. The sample space is:

S = {1 Yellow Forward, 1 Yellow Backward, 1 Red Forward, 1 Red Backward, 1 Green Forward, 1 Green Backward, 2 Yellow Forward, 2 Yellow Backward, 2 Red Forward, 2 Red Backward, 2 Green Forward, 2 Green Backward, 3 Yellow Forward, 3 Yellow Backward, 3 Red Forward, 3 Red Backward, 3 Green Forward, 3 Green Backward, 4 Yellow Forward, 4 Yellow Backward, 4 Red Forward, 4 Red Backward, 4 Green Forward, 4 Green Backward}

There are 24 equally likely events and the probability of each is $\frac{1}{24}$.

The probability of getting a 1, followed by a Red or Green, followed by a Backward is

$$P(1 \text{ Red Backward}) + P(1 \text{ Green Backward}) = \frac{1}{24} + \frac{1}{24} = \frac{1}{12}.$$

17. The sample space is:

S = {1 1 Yellow, 1 1 Red, 1 1 Green, 1 2 Yellow, 1 2 Red, 1 2 Green, 1 3 Yellow, 1 3 Red, 1 3 Green, 1 4 Yellow, 1 4 Red, 1 4 Green, 2 1 Yellow, 2 1 Red, 2 1 Green, 2 2 Yellow, 2 2 Red, 2 2 Green, 2 3 Yellow, 2 3 Red, 2 3 Green, 2 4 Yellow, 2 4 Red, 2 4 Green, 3 1 Yellow, 3 1 Red, 3 1 Green, 3 2 Yellow, 3 2 Red, 3 2 Green, 3 3 Yellow, 3 3 Red, 3 3 Green, 3 4 Yellow, 3 4 Red, 3 4 Green, 4 1 Yellow, 4 1 Red, 4 1 Green, 4 2 Yellow, 4 2 Red, 4 2 Green, 4 3 Yellow, 4 3 Red, 4 3 Green, 4 4 Yellow, 4 4 Red, 4 4 Green}

There are 48 equally likely events and the probability of each is $\frac{1}{48}$.

The probability of getting a 2, followed by a 2 or 4, followed by a Red or Green is

$$P(2 \text{ 2 Red}) + P(2 \text{ 4 Red}) + P(2 \text{ 2 Green}) + P(2 \text{ 4 Green}) = \frac{1}{48} + \frac{1}{48} + \frac{1}{48} + \frac{1}{48} = \frac{1}{12}$$

19. A, B, C, F   21.   B

23. Let $P(\text{tails}) = x$, then $P(\text{heads}) = 4x$

$$x + 4x = 1 \rightarrow 5x = 1 \rightarrow x = \frac{1}{5} \qquad P(\text{tails}) = \frac{1}{5}, \quad P(\text{heads}) = \frac{4}{5}$$

25. $P(2) = P(4) = P(6) = x \qquad P(1) = P(3) = P(5) = 2x$
$P(1) + P(2) + P(3) + P(4) + P(5) + P(6) = 1$

$$2x + x + 2x + x + 2x + x = 1 \rightarrow 9x = 1 \rightarrow x = \frac{1}{9}$$

$$P(2) = P(4) = P(6) = \frac{1}{9} \qquad P(1) = P(3) = P(5) = \frac{2}{9}$$

27. $P(E) = \dfrac{n(E)}{n(S)} = \dfrac{n\{1,2,3\}}{10} = \dfrac{3}{10}$

29. $P(E) = \dfrac{n(E)}{n(S)} = \dfrac{n\{2,4,6,8,10\}}{10} = \dfrac{5}{10} = \dfrac{1}{2}$

31. $P(\text{white}) = \dfrac{n(\text{white})}{n(S)} = \dfrac{5}{5+10+8+7} = \dfrac{5}{30} = \dfrac{1}{6}$

33. The sample space is: S = {BBB, BBG, BGB, GBB, BGG, GBG, GGB, GGG}

$$P(3 \text{ boys}) = \frac{n(3 \text{ boys})}{n(S)} = \frac{1}{8}$$

35. The sample space is:

S = {BBBB, BBBG, BBGB, BGBB, GBBB, BBGG, BGBG, GBBG, BGGB, GBGB, GGBB, BGGG, GBGG, GGBG, GGGB, GGGG}

$$P(1 \text{ girl, } 3 \text{ boys}) = \frac{n(1 \text{ girl, } 3 \text{ boys})}{n(S)} = \frac{4}{16} = \frac{1}{4}$$

37. $$P(\text{sum of two die is } 7) = \frac{n(\text{sum of two die is } 7)}{n(S)}$$

$$= \frac{n\{1,6 \text{ or } 2,5 \text{ or } 3,4 \text{ or } 4,3 \text{ or } 5,2 \text{ or } 6,1\}}{n(S)} = \frac{6}{36} = \frac{1}{6}$$

39. $$P(\text{sum of two die is } 3) = \frac{n(\text{sum of two die is } 3)}{n(S)} = \frac{n\{1,2 \text{ or } 2,1\}}{n(S)} = \frac{2}{36} = \frac{1}{18}$$

41. $$P(A \cup B) = P(A) + P(B) - P(A \cap B) = 0.25 + 0.45 - 0.15 = 0.55$$

43. $$P(A \cup B) = P(A) + P(B) = 0.25 + 0.45 = 0.70$$

45. $$P(A \cup B) = P(A) + P(B) - P(A \cap B)$$
$$0.85 = 0.60 + P(B) - 0.05$$
$$P(B) = 0.85 - 0.60 + 0.05 = 0.30$$

47. $$P(\text{not victim}) = 1 - P(\text{victim}) = 1 - 0.253 = 0.747$$

49. $$P(\text{not in } 70's) = 1 - P(\text{in } 70's) = 1 - 0.3 = 0.7$$

51. $$P(\text{white or green}) = P(\text{white}) + P(\text{green}) = \frac{n(\text{white}) + n(\text{green})}{n(S)} = \frac{9+8}{9+8+3} = \frac{17}{20}$$

53. $$P(\text{not white}) = 1 - P(\text{white}) = 1 - \frac{n(\text{white})}{n(S)} = 1 - \frac{9}{20} = \frac{11}{20}$$

55. $$P(\text{strike or one}) = P(\text{strike}) + P(\text{one}) = \frac{n(\text{strike}) + n(\text{one})}{n(S)} = \frac{3+1}{8} = \frac{4}{8} = \frac{1}{2}$$

57. There are 30 households out of 100 with an income of \$30,000 or more.

$$P(E) = \frac{n(E)}{n(S)} = \frac{n(30,000 \text{ or more})}{n(\text{total households})} = \frac{30}{100} = \frac{3}{10}$$

59. There are 40 households out of 100 with an income of less than \$20,000.

$$P(E) = \frac{n(E)}{n(S)} = \frac{n(\text{less than } \$20,000)}{n(\text{total households})} = \frac{40}{100} = \frac{2}{5}$$

61. (a) $P(1 \text{ or } 2) = P(1) + P(2) = 0.24 + 0.33 = 0.57$

     (b) $P(1 \text{ or more}) = P(1) + P(2) + P(3) + P(4 \text{ or more})$
$$= 0.24 + 0.33 + 0.21 + 0.17 = 0.95$$

     (c) $P(3 \text{ or fewer}) = P(0) + P(1) + P(2) + P(3) = 0.05 + 0.24 + 0.33 + 0.21 = 0.83$

     (d) $P(3 \text{ or more}) = P(3) + P(4 \text{ or more}) = 0.21 + 0.17 = 0.38$

     (e) $P(\text{less than } 2) = P(0) + P(1) = 0.05 + 0.24 = 0.29$

     (f) $P(\text{less than } 1) = P(0) = 0.05$

     (g) $P(1, 2, \text{ or } 3) = P(1) + P(2) + P(3) = 0.24 + 0.33 + 0.21 = 0.78$

     (h) $P(2 \text{ or more}) = P(2) + P(3) + P(4 \text{ or more}) = 0.33 + 0.21 + 0.17 = 0.71$

63. (a) $P(\text{freshman or female}) = P(\text{freshman}) + P(\text{female}) - P(\text{freshman and female})$
$$= \frac{n(\text{freshman}) + n(\text{female}) - n(\text{freshman and female})}{n(S)}$$
$$= \frac{18 + 15 - 8}{33} = \frac{25}{33}$$

     (b) $P(\text{sophomore or male}) = P(\text{sophomore}) + P(\text{male}) - P(\text{sophomore and male})$
$$= \frac{n(\text{sophomore}) + n(\text{male}) - n(\text{sophomore and male})}{n(S)}$$
$$= \frac{15 + 18 - 8}{33} = \frac{25}{33}$$

65. $P(\text{at least 2 with same birthday}) = 1 - P(\text{none with same birthday})$
$$= 1 - \frac{n(\text{different birthdays})}{n(S)}$$
$$= 1 - \frac{365 \cdot 364 \cdot 363 \cdot 362 \cdot 361 \cdot 360 \cdot \ldots \cdot 354}{365^{12}}$$
$$= 1 - 0.833$$
$$= 0.167$$

67. The sample space for picking 5 out of 10 numbers in a particular order contains
$$P(10,5) = \frac{10!}{(10-5)!} = \frac{10!}{5!} = 30,240 \text{ possible outcomes.}$$

One of these is the desired outcome. Thus, the probability of winning is:
$$P(E) = \frac{n(E)}{n(S)} = \frac{n(\text{winning})}{n(\text{total possible outcomes})} = \frac{1}{30240} \approx 0.000033069$$

69. (a) $P(3 \text{ heads}) = \dfrac{C(5,3)}{2^5} = \dfrac{10}{32} = \dfrac{5}{16}$

     (b) $P(0 \text{ heads}) = \dfrac{C(5,0)}{2^5} = \dfrac{1}{32}$

71. (a) $P(\text{sum } = 7 \text{ three times}) = P(\text{sum } = 7) \cdot P(\text{sum } = 7) \cdot P(\text{sum } = 7)$
$$= \frac{1}{6} \cdot \frac{1}{6} \cdot \frac{1}{6} = \frac{1}{216} \approx 0.00463$$

(b) $P(\text{sum} = 7 \text{ or } 11 \text{ at least twice})$
$$= P(\text{sum} = 7 \text{ or } 11) \cdot P(\text{sum} = 7 \text{ or } 11) \cdot P(\text{sum} \neq 7 \text{ or } 11) +$$
$$P(\text{sum} = 7 \text{ or } 11) \cdot P(\text{sum} = 7 \text{ or } 11) \cdot P(\text{sum} = 7 \text{ or } 11)$$
$$= 3\left(\frac{8}{36} \cdot \frac{8}{36} \cdot \frac{28}{36}\right) + \frac{8}{36} \cdot \frac{8}{36} \cdot \frac{8}{36} = 0.049$$

73. $P(\text{all 5 defective}) = \dfrac{n(5 \text{ defective})}{n(S)} = \dfrac{1}{C(30,5)} = 7.02 \times 10^{-6}$

$P(\text{at least 2 defective}) = 1 - (P(\text{none defective}) + P(\text{one defective}))$
$$= 1 - \left(\frac{C(5,0) \cdot C(25,5)}{C(30,5)} + \frac{C(5,1) \cdot C(25,4)}{C(30,5)}\right) = 1 - 0.817 = 0.183$$

75. $P(\text{one of 5 coins is valued at more than \$10,000}) = \dfrac{C(49,4) \cdot C(1,1)}{C(50,5)} = 0.1$

Chapter 12

Counting and Probability

12.4 Obtaining Probabilities From Data

1. (a)

Cause of Death	Probability
Accidents and adverse effects	0.436
Homicide and legal intervention	0.180
Suicide	0.135
Malignant neoplasms	0.055
Diseases of heart	0.035
Human immunodeficiency virus infection	0.006
Congenital anomalies	0.015
Chronic obstructive pulmonary diseases	0.008
Pneumonia and influenza	0.007
Cerebrovascular diseases	0.006
All other causes	0.118

(b) $P(\text{suicide}) = 0.1350 = 13.50\%$

(c) $P(\text{suicide or malignant neoplasms}) = P(\text{suicide}) + P(\text{malignant neoplasms})$
$$= 0.1350 + 0.0555 = 0.1905 = 19.05\%$$

(d) $P(\text{not suicide nor malignant neoplasms}) = 1 - P(\text{suicide or malignant neoplasms})$
$$= 1 - 0.1905 = 0.8095 = 80.95\%$$

3. (a)

Seatbelt Worn	Probability
Never	0.0262
Rarely	0.0678
Sometimes	0.1156
Most of the Time	0.2632
Always	0.5272

(b) $P(\text{Never}) = 0.0262 = 2.62\%$

(c) $P(\text{Never or Rarely}) = P(\text{Never}) + P(\text{Rarely})$
$$= 0.0262 + 0.0678 = 0.0940 = 9.40\%$$

5.   (a)

Weight (in grams)	Probability
Less than 500	0.0015
500 – 999	0.0057
1000 – 1499	0.0073
1500 – 1999	0.0150
2000 – 2499	0.0463
2500 – 2999	0.1650
3000 – 3499	0.3702
3500 – 3999	0.2884
4000 – 4499	0.0851
4500 – 4999	0.0139
5000 or more	0.0016

(b) $P(3500 - 3999) = 0.2884 = 28.84\%$

(c) $P(3500 - 4499) = P(3500 - 3999 \text{ or } 4000 - 4499) = P(3500 - 3999) + P(4000 - 4499)$
$$= 0.2884 + 0.0851 = 0.3735 = 37.35\%$$

(d) $P(\text{not 5000 or more}) = 1 - P(5000 \text{ or more}) = 1 - 0.0016 = 0.9984 = 99.84\%$

7. $P(\text{Titleist}) = \dfrac{n(\text{Titleist})}{n(S)} = \dfrac{4}{10} = \dfrac{2}{5} = 0.40$

9. $P(\text{2boys, 1 girl}) = \dfrac{n(\text{2boys, 1 girl})}{n(S)} = \dfrac{20}{50} = \dfrac{2}{5} = 0.40$

11. $P(\text{tornado, 5 - 6PM}) = \dfrac{n(\text{tornado, 5 - 6PM})}{n(S)} = \dfrac{3262}{28538} = 0.1143 = 11.43\%$

$P(\text{tornado, not 5 - 6PM}) = 1 - P(\text{tornado, 5 - 6PM})$

$$= 1 - \dfrac{3262}{28538} = 1 - 0.1143 = 0.8857 = 88.57\%$$

13. $P(\text{cloudy}) = \dfrac{n(\text{cloudy})}{n(S)} = \dfrac{11.2}{30} = 0.3733 = 37.33\%$

$P(\text{clear}) = \dfrac{n(\text{clear})}{n(S)} = \dfrac{7.3}{30} = 0.2433 = 24.33\%$

$P(\text{not clear}) = 1 - P(\text{clear}) = 1 - 0.2433 = 0.7567 = 75.67\%$

15. $P(\text{thunderstorm day}) = \dfrac{n(\text{thunderstorm day})}{n(S)} = \dfrac{6.4}{30} = 0.2133 = 21.33\%$

$P(\text{not thunderstorm day}) = 1 - P(\text{thunderstorm day})$

$$= 1 - 0.2133 = 0.7867 = 78.67\%$$

17. (a)

Field of Home Run	Probability
Left	0.4194
Left center	0.3387
Center	0.1935
Right center	0.0484
Right	0

(b) P(left field) = 0.4194 = 41.94%

(c) P(center field) = 0.1935 = 19.35%

(d) P(right field) = 0 = 0%

(e)   Answers will vary.

19. Use the randInt function on the calculator as follows (answers will vary):

(a)   randInt(1,6,100)

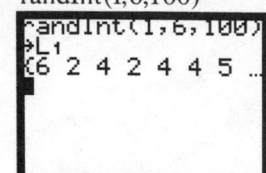   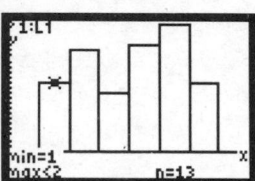

The plot shows the probability of rolling a "1" to be $\dfrac{13}{100} = 0.13$.

(b)   randInt(1,6,100)

   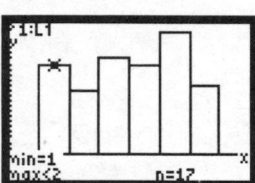

The plot shows the probability of rolling a "1" to be $\dfrac{17}{100} = 0.17$.

(c)   randInt(1,6,500)

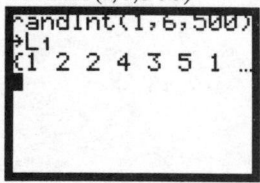   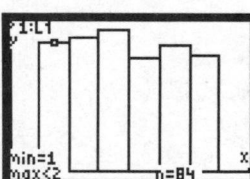

The plot shows the probability of rolling a "1" to be $\dfrac{84}{500} = 0.168$.

(d)   The simulation in part (c) yields the closest estimate to the probability obtained using equally likely outcomes, $\dfrac{1}{6} \approx 0.16\overline{6}$.

Chapter 12

Counting and Probability

12.R Chapter Review

1. {Melody}, {Dave}, {Joanne}, {Erica},
   {Melody, Dave}, {Melody, Joanne}, {Melody, Erica},
   {Dave, Joanne}, {Dave, Erica}, {Joanne, Erica},
   {Melody, Dave, Joanne}, {Melody, Dave, Erica}, {Melody, Joanne, Erica}, {Dave, Joanne, Erica}
   {Melody, Dave, Joanne, Erica}, $\varnothing$

3. $A \cup B = \{1, 3, 5, 7\} \cup \{3, 5, 6, 7, 8\} = \{1, 3, 5, 6, 7, 8\}$

5. $A \cap C = \{1, 3, 5, 7\} \cap \{2, 3, 7, 8, 9\} = \{3, 7\}$

7. $\overline{A} \cup \overline{B} = \overline{\{1, 3, 5, 7\}} \cup \overline{\{3, 5, 6, 7, 8\}} = \{2, 4, 6, 8, 9\} \cup \{1, 2, 4, 9\} = \{1, 2, 4, 6, 8, 9\}$

9. $\overline{B \cap C} = \overline{\{3, 5, 6, 7, 8\} \cap \{2, 3, 7, 8, 9\}} = \overline{\{3, 7, 8\}} = \{1, 2, 4, 5, 6, 9\}$

11. $n(A) = 8, n(B) = 12, n(A \cap B) = 3$
    $n(A \cup B) = n(A) + n(B) - n(A \cap B) = 8 + 12 - 3 = 17$

13. From the figure:
    $n(A) = 20 + 2 + 6 + 1 = 29$

15. From the figure:
    $n(A \text{ and } C) = n(A \cap C) = 1 + 6 = 7$

19. $5! = 5 \cdot 4 \cdot 3 \cdot 2 \cdot 1 = 120$ 21. $P(8,3) = \dfrac{8!}{(8-3)!} = \dfrac{8!}{5!} = \dfrac{8 \cdot 7 \cdot 6 \cdot 5!}{5!} = 336$

23. $C(8,3) = \dfrac{8!}{(8-3)!\,3!} = \dfrac{8!}{5!\,3!} = \dfrac{8 \cdot 7 \cdot 6 \cdot 5!}{5! \cdot 3 \cdot 2 \cdot 1} = 56$

25. There are 2 choices of material, 3 choices of color, and 10 choices of size. The complete assortment would have: $2 \cdot 3 \cdot 10 = 60$ suits.

27. There are two possible outcomes for each game or
    $2 \cdot 2 \cdot 2 \cdot 2 \cdot 2 \cdot 2 \cdot 2 = 2^7 = 128$ outcomes for 7 games.

29. Since order is significant, this is a permutation.

$$P(9,4) = \frac{9!}{(9-4)!} = \frac{9!}{5!} = \frac{9 \cdot 8 \cdot 7 \cdot 6 \cdot 5!}{5!} = 3024 \text{ ways to seat 4 people in 9 seats.}$$

31. Choose 4 runners - order is not significant:

$$C(8,4) = \frac{8!}{(8-4)! \, 4!} = \frac{8!}{4! \, 4!} = \frac{8 \cdot 7 \cdot 6 \cdot 5 \cdot 4!}{4 \cdot 3 \cdot 2 \cdot 1 \cdot 4!} = 70 \text{ ways a squad can be chosen.}$$

33. Choose 14 teams 2 at a time:

$$C(14,2) = \frac{14!}{(14-2)! \, 2!} = \frac{14!}{12! \, 2!} = \frac{14 \cdot 13 \cdot 12!}{12! \cdot 2 \cdot 1} = 91 \text{ ways to pair 14 teams.}$$

35. There are $8 \cdot 10 \cdot 10 \cdot 10 \cdot 10 \cdot 2 = 1{,}600{,}000$ possible phone numbers.

37. There are $24 \cdot 9 \cdot 10 \cdot 10 \cdot 10 = 216{,}000$ possible license plates.

39. Since there are repeated letters:

$$\frac{7!}{2! \cdot 2!} = \frac{7 \cdot 6 \cdot 5 \cdot 4 \cdot 3 \cdot 2 \cdot 1}{2 \cdot 1 \cdot 2 \cdot 1} = 1260 \text{ different words can be formed.}$$

41. (a)

Age	Probability
20 – 24	0.1709
25 – 29	0.1410
30 – 34	0.1344
35 – 39	0.1227
40 – 44	0.0990
45 – 49	0.0816
50 – 54	0.0602
55 – 59	0.0460
60 – 64	0.0389
65 – 69	0.0321
70 – 74	0.0293
75 – 79	0.0248
80 – 84	0.0190

(b) $P(25-29) = 0.1410 = 14.10\%$

(c) $P(20-29) = P(20-24 \text{ or } 25-29) = P(20-24) + P(25-29)$
$$= 0.1709 + 0.1410 = 0.3119 = 31.19\%$$

(d) $P(\text{not } 20-24) = 1 - P(20-24) = 1 - 0.1709 = 0.8291 = 82.91\%$

43. (a) $365 \cdot 364 \cdot 363 \cdot 362 \cdot \ldots \cdot 348 = 8.634628387 \times 10^{45}$

(b) $P(\text{no one has same birthday}) = \dfrac{365 \cdot 364 \cdot 363 \cdot 362 \cdot \ldots \cdot 348}{365^{18}} = 0.6531 = 65.31\%$

(c) $P(\text{at least 2 have same birthday}) = 1 - P(\text{no one has same birthday})$
$$= 1 - 0.6531 = 0.3469 = 34.69\%$$

45. (a) $P(\text{unemployed}) = 0.054 = 5.4\%$

    (b) $P(\text{not unemployed}) = 1 - P(\text{unemployed}) = 1 - 0.054 = 0.946 = 94.6\%$

47. $P(\$1 \text{ bill}) = \dfrac{n(\$1 \text{ bill})}{n(S)} = \dfrac{4}{9}$

49. Let S be all possible selections, let D be a card that is divisible by 5, and let PN be a 1 or a prime number.

$$n(S) = 100$$

$$n(D) = 20 \quad \text{(There are 20 numbers divisible by 5 between 1 and 100.)}$$

$$n(PN) = 26 \quad \text{(There are 25 prime numbers less than or equal to 100.)}$$

$$P(D) = \frac{n(D)}{n(S)} = \frac{20}{100} = \frac{1}{5} = 0.2; \quad P(PN) = \frac{n(PN)}{n(S)} = \frac{26}{100} = \frac{13}{50} = 0.26$$

51. (a) $P(5 \text{ heads}) = \dfrac{n(5 \text{ heads})}{n(S)} = \dfrac{C(10,5)}{2^{10}} = \dfrac{\left(\dfrac{10!}{5!\,5!}\right)}{1024} = \dfrac{252}{1024} \approx 0.2461$

    (b) $P(\text{all heads}) = \dfrac{n(\text{all heads})}{n(S)} = \dfrac{1}{2^{10}} = \dfrac{1}{1024} = 0.00098$

53. (a) $P(\text{all students}) = \dfrac{C(8,5)}{C(18,5)} = \dfrac{56}{8568} \approx 0.0065$

    (b) $P(\text{all faculty}) = \dfrac{C(10,5)}{C(18,5)} = \dfrac{252}{8568} \approx 0.0294$

    (c) $P(2 \text{ students and } 3 \text{ faculty}) = \dfrac{C(8,2) \cdot C(10,3)}{C(18,5)} = \dfrac{28 \cdot 120}{8568} \approx 0.3922$

55. Use the randInt function on the calculator as follows (answers will vary):

    (a) randInt(1,6,100)

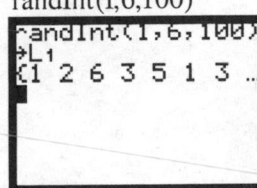 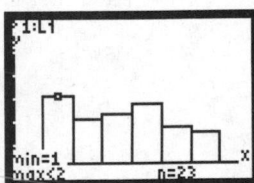

The plot shows the probability of rolling a "1" to be $\dfrac{23}{100} = 0.23$.

    (b) randInt(1,6,100)

 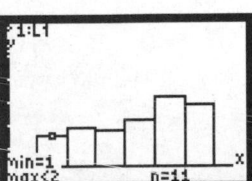

The plot shows the probability of rolling a "1" to be $\dfrac{11}{100} = 0.11$.

(c)   randInt(1,6,500)

The plot shows the probability of rolling a "1" to be $\dfrac{69}{500} = 0.138$.

(d)   The simulation in part (c) yields the closest estimate to the probability obtained using equally likely outcomes, $\dfrac{1}{6} = 0.16\overline{6}$.

Counting and Probability

12.CR   Cumulative Review

1. $3x^2 - 2x = -1 \Rightarrow 3x^2 - 2x + 1 = 0$

$$x = \frac{-b \pm \sqrt{b^2 - 4ac}}{2a} = \frac{-(-2) \pm \sqrt{(-2)^2 - 4(3)(1)}}{2(3)}$$

$$= \frac{4 \pm \sqrt{4-12}}{6} = \frac{4 \pm \sqrt{-8}}{6} = \frac{4 \pm 2\sqrt{2}i}{6} = \frac{2 \pm \sqrt{2}i}{3}$$

The solution set is $\left\{ \dfrac{2 - \sqrt{2}i}{3}, \dfrac{2 + \sqrt{2}i}{3} \right\}$.

3. $y = 2(x+1)^2 - 4$

Using the graph of $y = x^2$, horizontally shift to the left 1 unit, vertically stretch by a factor of 2, and vertically shift down 4 units.

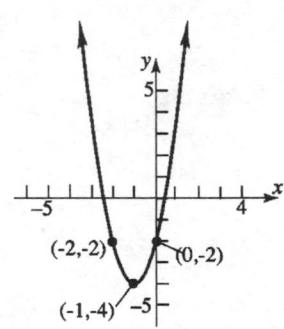

5. $f(x) = 5x^4 - 9x^3 - 7x^2 - 31x - 6$

Step 1: $f(x)$ has at most 4 real zeros.

Step 2:   Possible rational zeros:

$p = \pm 1, \pm 2, \pm 3, \pm 6; \quad q = \pm 1, \pm 5;$

$\dfrac{p}{q} = \pm 1, \pm \dfrac{1}{5}, \pm 2, \pm \dfrac{2}{5}, \pm 3, \pm \dfrac{4}{5}, \pm 6, \pm \dfrac{6}{5}$

Step 3:   Using the Bounds on Zeros Theorem:

$f(x) = 5\left(x^4 - 1.8x^3 - 1.4x^2 - 6.2x - 1.2 \right)$

$a_3 = -1.8, \quad a_2 = -1.4, \quad a_1 = -6.2, \quad a_0 = -1.2$

$\text{Max}\left\{ 1, |-1.2| + |-6.2| + |-1.4| + |-1.8| \right\} = \text{Max}\left\{ 1, 6.2 \right\} = 6.2$

$1 + \text{Max}\left\{ |-1.2|, |-6.2|, |-1.4|, |-1.8| \right\} = 1 + 6.2 = 7.2$

The smaller of the two numbers is 6.2. Thus, every zero of f lies between –6.2 and 6.2.

Graphing using the bounds and ZOOM-FIT: (Second graph has a better window.)

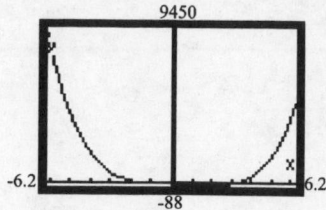
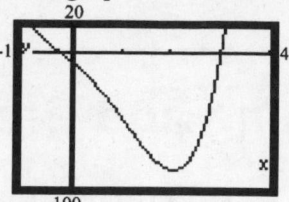

Step 4: From the graph it appears that there are x-intercepts at -0.2 and 3.
Using synthetic division with 3:

$$
\begin{array}{r|rrrrr}
3 & 5 & -9 & -7 & -31 & -6 \\
  &   & 15 & 18 & 33 & 6 \\
\hline
  & 5 & 6 & 11 & 2 & 0
\end{array}
$$

Since the remainder is 0, $x - 3$ is a factor. The other factor is the quotient: $5x^3 + 6x^2 + 11x + 2$.

Using synthetic division with 2 on the quotient:

$$
\begin{array}{r|rrrr}
-0.2 & 5 & 6 & 11 & 2 \\
     &   & -1 & -1 & -2 \\
\hline
     & 5 & 5 & 10 & 0
\end{array}
$$

Since the remainder is 0, $x - (-0.2) = x + 0.2$ is a factor. The other factor is the quotient: $5x^2 + 5x + 10 = 5(x^2 + x + 2)$.

Factoring,

$$f(x) = 5(x^2 + x + 2)(x - 3)(x + 0.2)$$

The real zeros are 3 and $-.02$.

The complex zeros come from solving $x^2 + x + 2 = 0$.

$$x = \frac{-b \pm \sqrt{b^2 - 4ac}}{2a} = \frac{-1 \pm \sqrt{1^2 - 4(1)(2)}}{2(1)} = \frac{-1 \pm \sqrt{1 - 8}}{2} = \frac{-1 \pm \sqrt{-7}}{2} = \frac{-1 \pm \sqrt{7}i}{2}$$

Therefore, the over the set of complex numbers, $f(x) = 5x^4 - 9x^3 - 7x^2 - 31x - 6$ has

zeros $-0.2, 3, \dfrac{-1 - \sqrt{7}i}{2}, \dfrac{-1 + \sqrt{7}i}{2}$.

7. $\log_3\left(9^x\right) = \log_3\left(\left(3^2\right)^x\right) = \log_3\left(3^{2x}\right) = 2x$

9. $y = 3\sin(2x + \pi) = 3\sin\left(2\left(x + \dfrac{\pi}{2}\right)\right)$

Amplitude: $|A| = |3| = 3$

Period: $T = \dfrac{2\pi}{2} = \pi$

Phase Shift: $\dfrac{\phi}{\omega} = \dfrac{-\pi}{2} = -\dfrac{\pi}{2}$

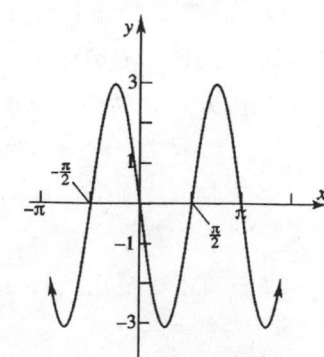

11. Multiply each side of the first equation by –3 and add to the second equation to eliminate x; and multiply each side of the first equation by 2 and add to the third equation to eliminate x:

$$\begin{cases} x - 2y + z = 15 \\ 3x + y - 3z = -8 \\ -2x + 4y - z = -27 \end{cases}$$

$$x - 2y + z = 15 \xrightarrow{\ -3\ } -3x + 6y - 3z = -45$$
$$3x + y - 3z = -8 \longrightarrow \underline{\quad 3x + y - 3z = -8\quad}$$
$$7y - 6z = -53$$

$$x - 2y + z = 15 \xrightarrow{\ 2\ } 2x - 4y + 2z = 30$$
$$-2x + 4y - z = -27 \longrightarrow \underline{\ -2x + 4y - z = -27\ }$$
$$z = 3$$

Substituting and solving for the other variables:
$$z = 3 \Rightarrow 7y - 6(3) = -53$$
$$7y = -35$$
$$y = -5$$

$$z = 3, y = -5 \Rightarrow x - 2(-5) + 3 = 15$$
$$x + 10 + 3 = 15 \Rightarrow x = 2$$

The solution is $x = 2$, $y = -5$, $z = 3$.

A Preview of Calculus:
The Limit, Derivative and Integral of a Function

13.1 Finding Limits Using Tables and Graphs

1. $\lim\limits_{x \to 2} \left(4x^3\right)$

X	Y1
1.99	31.522
1.999	31.952
1.9999	31.995
2.0001	32.005
2.001	32.048
2.01	32.482

Y1◻4X^3

$\lim\limits_{x \to 2} \left(4x^3\right) = 32$

3. $\lim\limits_{x \to 0} \left(\dfrac{x+1}{x^2+1}\right)$

X	Y1
-.01	.9899
-.001	.999
-1E⁻4	.9999
1E⁻4	1.0001
.001	1.001
.01	1.0099

Y1◻(X+1)/(X²+1)

$\lim\limits_{x \to 0} \left(\dfrac{x+1}{x^2+1}\right) = 1$

5. $\lim\limits_{x \to 4} \left(\dfrac{x^2-4x}{x-4}\right)$

X	Y1
3.99	3.99
3.999	3.999
3.9999	3.9999
4.0001	4.0001
4.001	4.001
4.01	4.01

Y1◻(X²-4X)/(X-4)

$\lim\limits_{x \to 4} \left(\dfrac{x^2-4x}{x-4}\right) = 4$

7. $\lim\limits_{x \to 0} \left(e^x + 1\right)$

X	Y1
-.01	1.99
-.001	1.999
-1E⁻4	1.9999
1E⁻4	2.0001
.001	2.001
.01	2.0101

Y1◻e^(X)+1

$\lim\limits_{x \to 0} \left(e^x + 1\right) = 2$

9. $\lim\limits_{x \to 0}\left(\dfrac{\cos x - 1}{x}\right)$

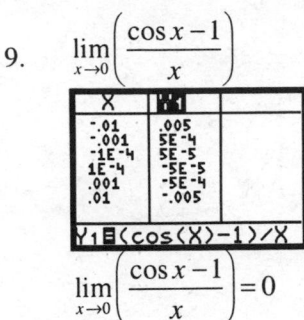

$\lim\limits_{x \to 0}\left(\dfrac{\cos x - 1}{x}\right) = 0$

10. $\lim\limits_{x \to 0}\left(\dfrac{\tan x}{x}\right)$

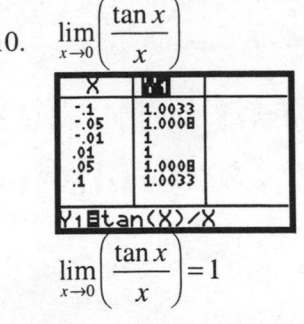

$\lim\limits_{x \to 0}\left(\dfrac{\tan x}{x}\right) = 1$

11. $\lim\limits_{x \to 2} f(x) = 3$

13. $\lim\limits_{x \to 2} f(x) = 4$

15. $\lim\limits_{x \to 3} f(x)$ does not exist because as x gets closer to 3, but is less than 3, $f(x)$ gets closer to 3. However, as x gets closer to 3, but is greater than 3, $f(x)$ gets closer to 6.

17. $f(x) = 3x + 1$

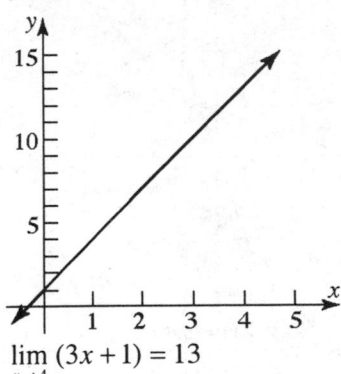

$\lim\limits_{x \to 4}(3x + 1) = 13$

19. $f(x) = 1 - x^2$

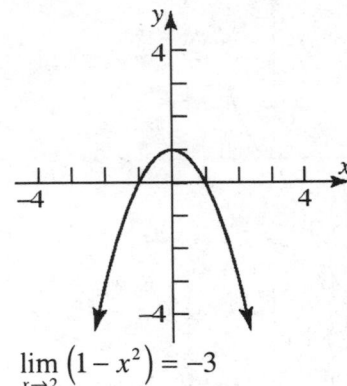

$\lim\limits_{x \to 2}\left(1 - x^2\right) = -3$

21. $f(x) = |2x|$

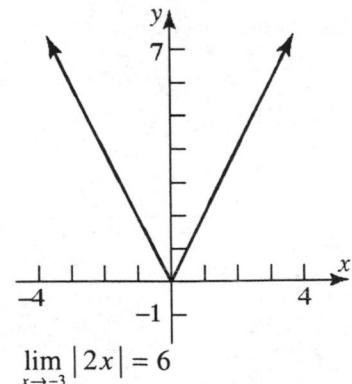

$\lim\limits_{x \to -3}|2x| = 6$

23. $f(x) = \sin x$

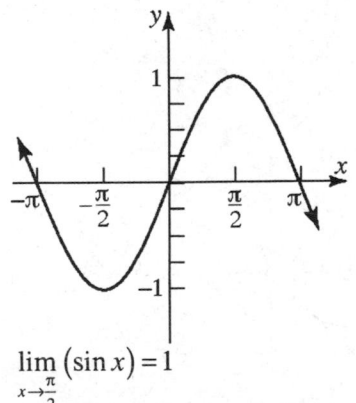

$\lim\limits_{x \to \frac{\pi}{2}}(\sin x) = 1$

25. $f(x) = e^x$

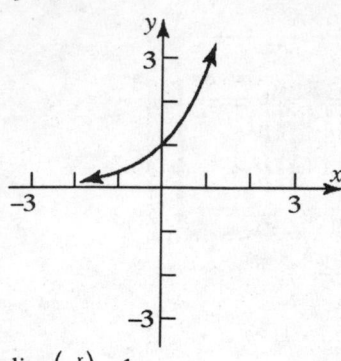

$$\lim_{x \to 0} \left(e^x \right) = 1$$

27. $f(x) = \dfrac{1}{x}$

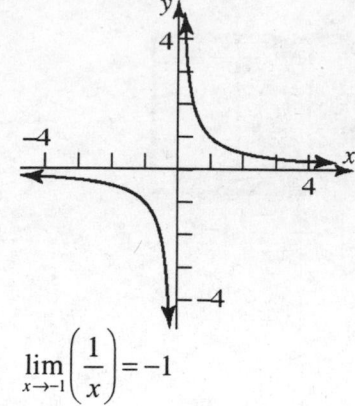

$$\lim_{x \to -1} \left(\frac{1}{x} \right) = -1$$

29. $f(x) = \begin{cases} x^2 & x \ge 0 \\ 2x & x < 0 \end{cases}$

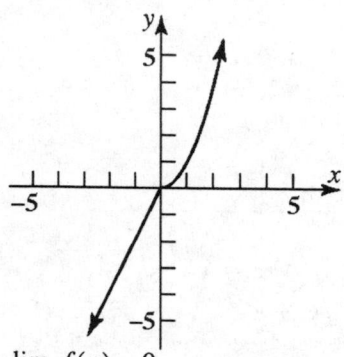

$$\lim_{x \to 0} f(x) = 0$$
since $\displaystyle \lim_{x \to 0^-} f(x) = \lim_{x \to 0^+} f(x) = 0$

31. $f(x) = \begin{cases} 3x & x \le 1 \\ x+1 & x > 1 \end{cases}$

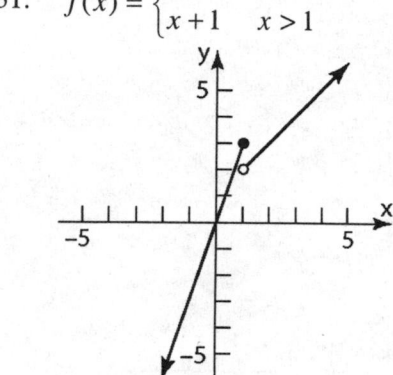

$\displaystyle \lim_{x \to 1} f(x)$ does not exist
since $\displaystyle \lim_{x \to 1^-} f(x) \ne \lim_{x \to 1^+} f(x)$

33. $f(x) = \begin{cases} x & x < 0 \\ 1 & x = 0 \\ 3x & x > 0 \end{cases}$

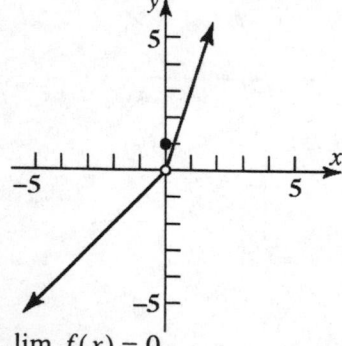

$$\lim_{x \to 0} f(x) = 0$$
since $\displaystyle \lim_{x \to 0^-} f(x) = \lim_{x \to 0^+} f(x) = 0$

35. $f(x) = \begin{cases} \sin x & x \le 0 \\ x^2 & x > 0 \end{cases}$

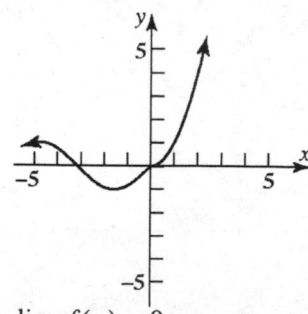

$$\lim_{x \to 0} f(x) = 0$$
since $\displaystyle \lim_{x \to 0^-} f(x) = \lim_{x \to 0^+} f(x) = 0$

37. $\lim\limits_{x \to 1}\left(\dfrac{x^3 - x^2 + x - 1}{x^4 - x^3 + 2x - 2}\right)$

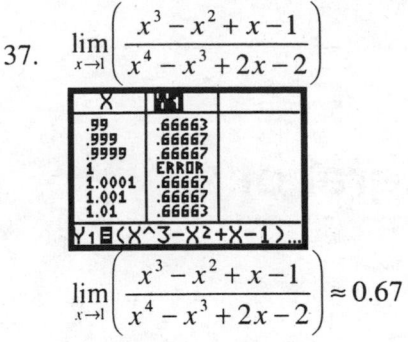

$$\lim\limits_{x \to 1}\left(\dfrac{x^3 - x^2 + x - 1}{x^4 - x^3 + 2x - 2}\right) \approx 0.67$$

39. $\lim\limits_{x \to 2}\left(\dfrac{x^3 - 2x^2 + 4x - 8}{x^2 + x - 6}\right)$

X	Y1
1.99	1.5952
1.999	1.5995
1.9999	1.6
2	ERROR
2.0001	1.6
2.001	1.6005
2.01	1.6048

Y1 ☐ (X^3-2X²+4X-...

$$\lim\limits_{x \to 2}\left(\dfrac{x^3 - 2x^2 + 4x - 8}{x^2 + x - 6}\right) = 1.60$$

41. $\lim\limits_{x \to -1}\left(\dfrac{x^3 + 2x^2 + x}{x^4 + x^3 + 2x + 2}\right)$

X	Y1
-1.01	.01042
-1.001	.001
-1	ERROR
-.9999	-1E-4
-.999	-1E-3
-.99	-.0096

Y1 ☐ (X^3+2X²+X)/...

$$\lim\limits_{x \to -1}\left(\dfrac{x^3 + 2x^2 + x}{x^4 + x^3 + 2x + 2}\right) = 0$$

A Preview of Calculus: The Limit, Derivative and Integral of a Function

13.2 Algebra Techniques for Finding Limits

1. $\lim\limits_{x \to 1} (5) = 5$

3. $\lim\limits_{x \to 4} (x) = 4$

5. $\lim\limits_{x \to 2} (3x + 2) = 3(2) + 2 = 8$

7. $\lim\limits_{x \to -1} (3x^2 - 5x) = 3(-1)^2 - 5(-1) = 8$

9. $\lim\limits_{x \to 1} (5x^4 - 3x^2 + 6x - 9) = 5(1)^4 - 3(1)^2 + 6(1) - 9 = 5 - 3 + 6 - 9 = -1$

11. $\lim\limits_{x \to 1} (x^2 + 1)^3 = \left(\lim\limits_{x \to 1} (x^2 + 1) \right)^3 = \left(1^2 + 1 \right)^3 = 2^3 = 8$

13. $\lim\limits_{x \to 1} \sqrt{5x + 4} = \sqrt{\lim\limits_{x \to 1} (5x + 4)} = \sqrt{5(1) + 4} = \sqrt{9} = 3$

15. $\lim\limits_{x \to 0} \left(\dfrac{x^2 - 4}{x^2 + 4} \right) = \dfrac{\left(\lim\limits_{x \to 0} (x^2 - 4) \right)}{\left(\lim\limits_{x \to 0} (x^2 + 4) \right)} = \dfrac{0^2 - 4}{0^2 + 4} = \dfrac{-4}{4} = -1$

17. $\lim\limits_{x \to 2} \left((3x - 2)^{5/2} \right) = \left(\lim\limits_{x \to 2} (3x - 2) \right)^{5/2} = \left(3(2) - 2 \right)^{5/2} = 4^{5/2} = 32$

19. $\lim\limits_{x \to 2} \left(\dfrac{x^2 - 4}{x^2 - 2x} \right) = \lim\limits_{x \to 2} \left(\dfrac{(x - 2)(x + 2)}{x(x - 2)} \right) = \lim\limits_{x \to 2} \left(\dfrac{x + 2}{x} \right) = \dfrac{2 + 2}{2} = \dfrac{4}{2} = 2$

21. $\lim\limits_{x \to -3} \left(\dfrac{x^2 - x - 12}{x^2 - 9} \right) = \lim\limits_{x \to -3} \left(\dfrac{(x - 4)(x + 3)}{(x - 3)(x + 3)} \right) = \lim\limits_{x \to -3} \left(\dfrac{x - 4}{x - 3} \right) = \dfrac{-3 - 4}{-3 - 3} = \dfrac{-7}{-6} = \dfrac{7}{6}$

23. $\lim\limits_{x \to 1} \left(\dfrac{x^3 - 1}{x - 1} \right) = \lim\limits_{x \to 1} \left(\dfrac{(x - 1)(x^2 + x + 1)}{x - 1} \right) = \lim\limits_{x \to 1} \left(x^2 + x + 1 \right) = 1^2 + 1 + 1 = 3$

25. $\lim\limits_{x \to 1} \left(\dfrac{(x + 1)^2}{x^2 - 1} \right) = \lim\limits_{x \to 1} \left(\dfrac{(x + 1)^2}{(x - 1)(x + 1)} \right) = \lim\limits_{x \to 1} \left(\dfrac{x + 1}{x - 1} \right) = \dfrac{-1 + 1}{-1 - 1} = \dfrac{0}{-2} = 0$

27. $\lim\limits_{x \to 1}\left(\dfrac{x^3 - x^2 + x - 1}{x^4 - x^3 + 2x - 2}\right) = \lim\limits_{x \to 1}\left(\dfrac{x^2(x-1) + 1(x-1)}{x^3(x-1) + 2(x-1)}\right) = \lim\limits_{x \to 1}\left(\dfrac{(x-1)(x^2+1)}{(x-1)(x^3+2)}\right) = \lim\limits_{x \to 1}\left(\dfrac{x^2+1}{x^3+2}\right)$

$$= \dfrac{1^2+1}{1^3+2} = \dfrac{2}{3}$$

29. $\lim\limits_{x \to 2}\left(\dfrac{x^3 - 2x^2 + 4x - 8}{x^2 + x - 6}\right) = \lim\limits_{x \to 2}\left(\dfrac{x^2(x-2) + 4(x-2)}{(x+3)(x-2)}\right) = \lim\limits_{x \to 2}\left(\dfrac{(x-2)(x^2+4)}{(x+3)(x-2)}\right) = \lim\limits_{x \to 2}\left(\dfrac{x^2+4}{x+3}\right)$

$$= \dfrac{2^2+4}{2+3} = \dfrac{8}{5}$$

31. $\lim\limits_{x \to -1}\left(\dfrac{x^3 + 2x^2 + x}{x^4 + x^3 + 2x + 2}\right) = \lim\limits_{x \to -1}\left(\dfrac{x(x^2 + 2x + 1)}{x^3(x+1) + 2(x+1)}\right) = \lim\limits_{x \to -1}\left(\dfrac{x(x+1)^2}{(x+1)(x^3+2)}\right)$

$$= \lim\limits_{x \to -1}\left(\dfrac{x(x+1)}{x^3+2}\right) = \dfrac{-1(-1+1)}{(-1)^3 + 2} = \dfrac{-1(0)}{-1+2} = \dfrac{0}{1} = 0$$

33. $\lim\limits_{x \to 2}\left(\dfrac{f(x) - f(2)}{x - 2}\right) = \lim\limits_{x \to 2}\left(\dfrac{(5x-3) - 7}{x - 2}\right) = \lim\limits_{x \to 2}\left(\dfrac{5x - 10}{x - 2}\right) = \lim\limits_{x \to 2}\left(\dfrac{5(x-2)}{x - 2}\right) = \lim\limits_{x \to 2}(5) = 5$

35. $\lim\limits_{x \to 3}\left(\dfrac{f(x) - f(3)}{x - 3}\right) = \lim\limits_{x \to 3}\left(\dfrac{x^2 - 9}{x - 3}\right) = \lim\limits_{x \to 3}\left(\dfrac{(x-3)(x+3)}{x - 3}\right) = \lim\limits_{x \to 3}(x + 3) = 3 + 3 = 6$

37. $\lim\limits_{x \to -1}\left(\dfrac{f(x) - f(-1)}{x - (-1)}\right) = \lim\limits_{x \to -1}\left(\dfrac{x^2 + 2x - (-1)}{x + 1}\right) = \lim\limits_{x \to -1}\left(\dfrac{x^2 + 2x + 1}{x + 1}\right) = \lim\limits_{x \to -1}\left(\dfrac{(x+1)^2}{x + 1}\right)$

$$= \lim\limits_{x \to -1} x + 1 = -1 + 1 = 0$$

39. $\lim\limits_{x \to 0}\left(\dfrac{f(x) - f(0)}{x - 0}\right) = \lim\limits_{x \to 0}\left(\dfrac{3x^3 - 2x^2 + 4 - 4}{x}\right) = \lim\limits_{x \to 0}\left(\dfrac{3x^3 - 2x^2}{x}\right) = \lim\limits_{x \to 0}(3x^2 - 2x) = 0$

41. $\lim\limits_{x \to 1}\left(\dfrac{f(x) - f(1)}{x - 1}\right) = \lim\limits_{x \to 1}\left(\dfrac{\left(\dfrac{1}{x} - 1\right)}{(x - 1)}\right) = \lim\limits_{x \to 1}\left(\dfrac{\left(\dfrac{1-x}{x}\right)}{(x - 1)}\right) = \lim\limits_{x \to 1}\left(\dfrac{-1(x-1)}{x(x-1)}\right) = \lim\limits_{x \to 1}\left(\dfrac{-1}{x}\right) = \dfrac{-1}{1} = -1$

43. $\lim\limits_{x \to 0}\left(\dfrac{\tan x}{x}\right) = \lim\limits_{x \to 0}\left(\dfrac{\left(\dfrac{\sin x}{\cos x}\right)}{(x)}\right) = \lim\limits_{x \to 0}\left(\dfrac{\sin x}{x} \cdot \dfrac{1}{\cos x}\right) = \left(\lim\limits_{x \to 0}\left(\dfrac{\sin x}{x}\right)\right) \cdot \left(\lim\limits_{x \to 0}\left(\dfrac{1}{\cos x}\right)\right)$

$$= 1 \cdot \left(\dfrac{\lim\limits_{x \to 0} 1}{\lim\limits_{x \to 0}(\cos x)}\right) = 1 \cdot \dfrac{1}{1} = 1$$

45. $\lim\limits_{x \to 0} \left(\dfrac{3\sin x + \cos x - 1}{4x} \right) = \lim\limits_{x \to 0} \left(\dfrac{3\sin x}{4x} + \dfrac{\cos x - 1}{4x} \right) = \lim\limits_{x \to 0} \left(\dfrac{3\sin x}{4x} \right) + \lim\limits_{x \to 0} \left(\dfrac{\cos x - 1}{4x} \right)$

$= \dfrac{3}{4} \lim\limits_{x \to 0} \left(\dfrac{\sin x}{x} \right) + \dfrac{1}{4} \lim\limits_{x \to 0} \left(\dfrac{\cos x - 1}{x} \right) = \dfrac{3}{4} \cdot 1 + \dfrac{1}{4} \cdot 0 = \dfrac{3}{4}$

Chapter 13

A Preview of Calculus: The Limit, Derivative and Integral of a Function

13.3 One-Sided Limits; Continuous Functions

1. Domain: $\{x \mid -8 \leq x < -6 \text{ or } -6 < x < 4 \text{ or } 4 < x \leq 6\}$

3. x-intercepts: $-8, -5, -3$

5. $f(-8) = 0; \ f(-4) = 2$

7. $\lim\limits_{x \to -6^-} f(x) = +\infty$

9. $\lim\limits_{x \to -4^-} f(x) = 2$

11. $\lim\limits_{x \to 2^-} f(x) = 1$

13. $\lim\limits_{x \to 4} f(x)$ does exist. $\lim\limits_{x \to 4} f(x) = 0$ since $\lim\limits_{x \to 4^-} f(x) = \lim\limits_{x \to 4^+} f(x) = 0$

15. f is not continuous at -6 because $f(-6)$ does not exist.

17. f is continuous at 0 because $f(0) = \lim\limits_{x \to 0^-} f(x) = \lim\limits_{x \to 0^+} f(x) = 3$

19. f is not continuous at 4 because $f(4)$ does not exist.

21. $\lim\limits_{x \to 1^+} (2x + 3) = 2(1) + 3 = 5$

23. $\lim\limits_{x \to 1^-} \left(2x^3 + 5x\right) = 2(1)^3 + 5(1) = 2 + 5 = 7$

25. $\lim\limits_{x \to \frac{\pi}{2}^+} (\sin x) = \sin\left(\frac{\pi}{2}\right) = 1$

27. $\lim\limits_{x \to 2^+} \left(\dfrac{x^2 - 4}{x - 2}\right) = \lim\limits_{x \to 2^+} \left(\dfrac{(x + 2)(x - 2)}{x - 2}\right) = \lim\limits_{x \to 2^+} (x + 2) = 2 + 2 = 4$

29. $\lim\limits_{x \to -1^-} \left(\dfrac{x^2 - 1}{x^3 + 1}\right) = \lim\limits_{x \to -1^-} \left(\dfrac{(x + 1)(x - 1)}{(x + 1)(x^2 - x + 1)}\right) = \lim\limits_{x \to -1^-} \left(\dfrac{x - 1}{x^2 - x + 1}\right) = \dfrac{-1 - 1}{(-1)^2 - (-1) + 1} = -\dfrac{2}{3}$

31. $\lim\limits_{x \to -2^{+}} \left(\dfrac{x^{2}+x-2}{x^{2}+2x} \right) = \lim\limits_{x \to -2^{+}} \left(\dfrac{(x+2)(x-1)}{x(x+2)} \right) = \lim\limits_{x \to -2^{+}} \left(\dfrac{x-1}{x} \right) = \dfrac{-2-1}{-2} = \dfrac{-3}{-2} = \dfrac{3}{2}$

33. $f(x) = x^{3} - 3x^{2} + 2x - 6;\ \ c = 2$

    1. $f(2) = 2^{3} - 3 \cdot 2^{2} + 2 \cdot 2 - 6 = -6$

    2. $\lim\limits_{x \to 2^{-}} f(x) = 2^{3} - 3 \cdot 2^{2} + 2 \cdot 2 - 6 = -6$

    3. $\lim\limits_{x \to 2^{+}} f(x) = 2^{3} - 3 \cdot 2^{2} + 2 \cdot 2 - 6 = -6$

Thus, $f(x)$ is continuous at $c = 2$.

35. $f(x) = \dfrac{x^{2}+5}{x-6};\ \ c = 3$

    1. $f(3) = \dfrac{3^{2}+5}{3-6} = \dfrac{14}{-3} = -\dfrac{14}{3}$

    2. $\lim\limits_{x \to 3^{-}} f(x) = \dfrac{3^{2}+5}{3-6} = \dfrac{14}{-3} = -\dfrac{14}{3}$

    3. $\lim\limits_{x \to 3^{+}} f(x) = \dfrac{3^{2}+5}{3-6} = \dfrac{14}{-3} = -\dfrac{14}{3}$

Thus, $f(x)$ is continuous at $c = 3$.

37. $f(x) = \dfrac{x+3}{x-3};\ \ c = 3$

Since $f(x)$ is not defined at $c = 3$, the function is not continuous at $c = 3$.

39. $f(x) = \dfrac{x^{3}+3x}{x^{2}-3x};\ \ c = 0$

Since $f(x)$ is not defined at $c = 0$, the function is not continuous at $c = 0$.

41. $f(x) = \begin{cases} \dfrac{x^{3}+3x}{x^{2}-3x} & \text{if } x \neq 0 \\ 1 & \text{if } x = 0 \end{cases};\ \ c = 0$

    1. $f(0) = 1$

    2. $\lim\limits_{x \to 0^{-}} f(x) = \lim\limits_{x \to 0^{-}} \left(\dfrac{x^{3}+3x}{x^{2}-3x} \right) = \lim\limits_{x \to 0^{-}} \left(\dfrac{x(x^{2}+3)}{x(x-3)} \right) = \lim\limits_{x \to 0^{-}} \left(\dfrac{x^{2}+3}{x-3} \right) = \dfrac{3}{-3} = -1$

Since $\lim\limits_{x \to 0^{-}} f(x) \neq f(c)$, the function is not continuous at $c = 0$.

43. $f(x) = \begin{cases} \dfrac{x^{3}+3x}{x^{2}-3x} & \text{if } x \neq 0 \\ -1 & \text{if } x = 0 \end{cases};\ \ c = 0$

    1. $f(0) = -1$

    2. $\lim\limits_{x \to 0^{-}} f(x) = \lim\limits_{x \to 0^{-}} \left(\dfrac{x^{3}+3x}{x^{2}-3x} \right) = \lim\limits_{x \to 0^{-}} \left(\dfrac{x(x^{2}+3)}{x(x-3)} \right) = \lim\limits_{x \to 0^{-}} \left(\dfrac{x^{2}+3}{x-3} \right) = \dfrac{3}{-3} = -1$

    3. $\lim\limits_{x \to 0^{+}} f(x) = \lim\limits_{x \to 0^{+}} \left(\dfrac{x^{3}+3x}{x^{2}-3x} \right) = \lim\limits_{x \to 0^{+}} \left(\dfrac{x(x^{2}+3)}{x(x-3)} \right) = \lim\limits_{x \to 0^{+}} \left(\dfrac{x^{2}+3}{x-3} \right) = \dfrac{3}{-3} = -1$

The function is continuous at $c = 0$.

45. $f(x) = \begin{cases} \dfrac{x^3 - 1}{x^2 - 1} & \text{if } x < 1 \\ 2 & \text{if } x = 1 \, ; \quad c = 1 \\ \dfrac{3}{x+1} & \text{if } x > 1 \end{cases}$

1. $f(1) = 2$

2. $\lim\limits_{x \to 1^-} f(x) = \lim\limits_{x \to 1^-} \left(\dfrac{x^3 - 1}{x^2 - 1} \right) = \lim\limits_{x \to 1^-} \left(\dfrac{(x-1)(x^2 + x + 1)}{(x-1)(x+1)} \right) = \lim\limits_{x \to 1^-} \left(\dfrac{x^2 + x + 1}{x+1} \right) = \dfrac{3}{2}$

Since $\lim\limits_{x \to 1^-} f(x) \neq f(c)$, the function is not continuous at $c = 1$.

47. $f(x) = \begin{cases} 2e^x & \text{if } x < 0 \\ 2 & \text{if } x = 0 \, ; \quad c = 0 \\ \dfrac{x^3 + 2x^2}{x^2} & \text{if } x > 0 \end{cases}$

1. $f(0) = 2$ 2. $\lim\limits_{x \to 0^-} f(x) = \lim\limits_{x \to 0^-} \left(2e^x \right) = 2e^0 = 2 \cdot 1 = 2$

3. $\lim\limits_{x \to 0^+} f(x) = \lim\limits_{x \to 0^+} \left(\dfrac{x^3 + 2x^2}{x^2} \right) = \lim\limits_{x \to 0^+} \left(\dfrac{x^2(x+2)}{x^2} \right) = \lim\limits_{x \to 0^+} (x + 2) = 0 + 2 = 2$

The function is continuous at $c = 0$.

49. The domain of $f(x) = 2x + 3$ is all real numbers. Therefore, $f(x)$ is continuous everywhere.

51. The domain of $f(x) = 3x^2 + x$ is all real numbers. Therefore, $f(x)$ is continuous everywhere.

53. The domain of $f(x) = 4\sin x$ is all real numbers. Therefore, $f(x)$ is continuous everywhere.

55. The domain of $f(x) = 2\tan x$ is all real numbers except odd integer multiples of $\dfrac{\pi}{2}$.

Therefore, $f(x)$ is continuous everywhere except where $x = \dfrac{k\pi}{2}$ where k is an odd integer.

$f(x)$ is discontinuous at $x = \dfrac{k\pi}{2}$ where k is an odd integer.

57. $f(x) = \dfrac{2x + 5}{x^2 - 4} = \dfrac{2x + 5}{(x-2)(x+2)}$. The domain of $f(x)$ is all real numbers except $x = 2$ and $x = -2$. Therefore, $f(x)$ is continuous everywhere except at $x = 2$ and $x = -2$. $f(x)$ is discontinuous at $x = 2$ and $x = -2$.

59. $f(x) = \dfrac{x - 3}{\ln x}$. The domain of $f(x)$ is $(0, 1)$ or $(1, \infty)$. Thus, $f(x)$ is continuous on the interval $(0, \infty)$ except at $x = 1$. $f(x)$ is discontinuous at $x = 1$.

61. $R(x) = \dfrac{x-1}{x^2-1} = \dfrac{x-1}{(x-1)(x+1)}$. The domain of R is $\{x \,|\, x \neq -1, \, x \neq 1\}$. Thus R is

discontinuous at both -1 and 1. $\displaystyle\lim_{x \to -1^-} R(x) = \lim_{x \to -1^-}\left(\dfrac{x-1}{(x-1)(x+1)}\right) = \lim_{x \to -1^-}\left(\dfrac{1}{x+1}\right) = -\infty$

since when $x < -1$, $\dfrac{1}{x+1} < 0$, and as x approaches -1, $\dfrac{1}{x+1}$ becomes unbounded.

$\displaystyle\lim_{x \to -1^+}\left(\dfrac{1}{x+1}\right) = +\infty$ since when $x > -1$, $\dfrac{1}{x+1} > 0$, and as x approaches -1, $\dfrac{1}{x+1}$ becomes

unbounded. The behavior near $c = 1$ is: $\displaystyle\lim_{x \to 1} R(x) = \lim_{x \to 1}\left(\dfrac{1}{x+1}\right) = \dfrac{1}{2}$. Note there is a hole

in the graph at $\left(1, \dfrac{1}{2}\right)$.

63. $R(x) = \dfrac{x^2+x}{x^2-1} = \dfrac{x(x+1)}{(x-1)(x+1)}$. The domain of R is $\{x \,|\, x \neq -1, \, x \neq 1\}$. Thus R is

discontinuous at both -1 and 1. $\displaystyle\lim_{x \to 1^-} R(x) = \lim_{x \to 1^-}\left(\dfrac{x(x+1)}{(x-1)(x+1)}\right) = \lim_{x \to 1^-}\left(\dfrac{x}{x-1}\right) = -\infty$ since

when $x < 1$, $\dfrac{x}{x-1} < 0$, and as x approaches 1, $\dfrac{x}{x-1}$ becomes unbounded. $\displaystyle\lim_{x \to 1^+}\dfrac{x}{x-1} = +\infty$

since when $x > 1$, $\dfrac{x}{x-1} > 0$, and as x approaches 1, $\dfrac{x}{x-1}$ becomes unbounded. The

behavior near $c = -1$ is: $\displaystyle\lim_{x \to -1} R(x) = \lim_{x \to -1}\left(\dfrac{x}{x-1}\right) = \dfrac{-1}{-2} = \dfrac{1}{2}$. Note there is a hole in the

graph at $\left(-1, \dfrac{1}{2}\right)$.

65. $R(x) = \dfrac{x^3 - x^2 + x - 1}{x^4 - x^3 + 2x - 2} = \dfrac{x^2(x-1) + 1(x-1)}{x^3(x-1) + 2(x-1)} = \dfrac{(x-1)(x^2+1)}{(x-1)(x^3+2)} = \dfrac{x^2+1}{x^3+2}, \ x \neq 1$

There is a vertical asymptote where $x^3 + 2 = 0$. $x = -\sqrt[3]{2}$ is a vertical asymptote. There is a hole in the graph at $x = 1$.

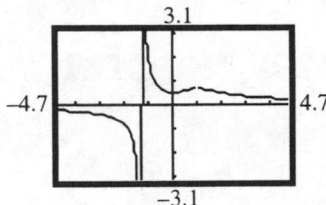

67. $R(x) = \dfrac{x^3 - 2x^2 + 4x - 8}{x^2 + x - 6} = \dfrac{x^2(x-2) + 4(x-2)}{(x+3)(x-2)} = \dfrac{(x-2)(x^2+4)}{(x+3)(x-2)} = \dfrac{x^2+4}{x+3}, \ x \neq 2$

There is a vertical asymptote where $x + 3 = 0$. $x = -3$ is a vertical asymptote. There is a hole in the graph at $x = 2$.

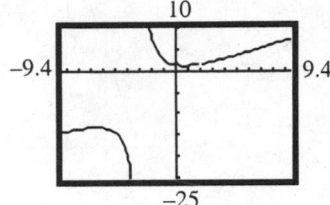

69. $R(x) = \dfrac{x^3 + 2x^2 + x}{x^4 + x^3 + 2x + 2} = \dfrac{x(x^2 + 2x + 1)}{x^3(x+1) + 2(x+1)} = \dfrac{x(x+1)^2}{(x+1)(x^3+2)} = \dfrac{x(x+1)}{x^3+2}, \ x \neq -1$

There is a vertical asymptote where $x^3 + 2 = 0$. $x = -\sqrt[3]{2}$ is a vertical asymptote. There is a hole in the graph at $x = -1$.

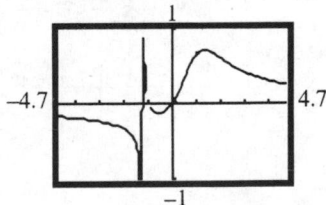

Chapter 13

A Preview of Calculus:
The Limit, Derivative and Integral of a
Function

13.4 The Tangent Problem; The Derivative

1. $f(x) = 3x + 5$ at $(1, 8)$

$$m_{\tan} = \lim_{x \to 1}\left(\frac{f(x) - f(1)}{x - 1}\right) = \lim_{x \to 1}\left(\frac{3x + 5 - 8}{x - 1}\right)$$

$$= \lim_{x \to 1}\left(\frac{3x - 3}{x - 1}\right) = \lim_{x \to 1}\left(\frac{3(x - 1)}{x - 1}\right) = \lim_{x \to 1}(3) = 3$$

Tangent Line: $y - 8 = 3(x - 1)$

$$y - 8 = 3x - 3$$
$$y = 3x + 5$$

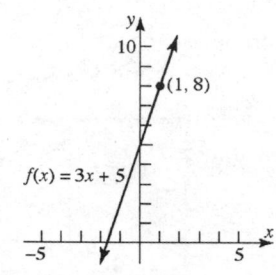

3. $f(x) = x^2 + 2$ at $(-1, 3)$

$$m_{\tan} = \lim_{x \to -1}\left(\frac{f(x) - f(-1)}{x + 1}\right) = \lim_{x \to -1}\left(\frac{x^2 + 2 - 3}{x + 1}\right)$$

$$= \lim_{x \to -1}\left(\frac{x^2 - 1}{x + 1}\right) = \lim_{x \to -1}\left(\frac{(x + 1)(x - 1)}{x + 1}\right)$$

$$= \lim_{x \to -1}(x - 1) = -1 - 1 = -2$$

Tangent Line: $y - 3 = -2(x - (-1))$

$$y - 3 = -2x - 2$$
$$y = -2x + 1$$

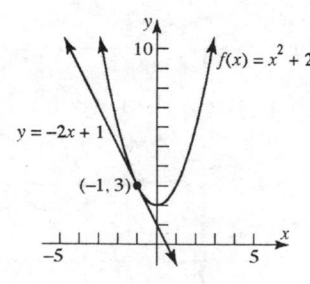

5. $f(x) = 3x^2$ at $(2, 12)$

$$m_{\tan} = \lim_{x \to 2}\left(\frac{f(x) - f(2)}{x - 2}\right) = \lim_{x \to 2}\left(\frac{3x^2 - 12}{x - 2}\right)$$

$$= \lim_{x \to 2}\frac{3\left((x^2) - 4\right)}{x - 2} = \lim_{x \to 2}\left(\frac{3(x + 2)(x - 2)}{x - 2}\right)$$

$$= \lim_{x \to 2}(3(x + 2)) = 3(2 + 2) = 12$$

Tangent Line: $y - 12 = 12(x - 2)$

$$y - 12 = 12x - 24$$
$$y = 12x - 12$$

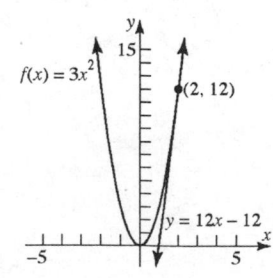

7. $f(x) = 2x^2 + x$ at $(1, 3)$

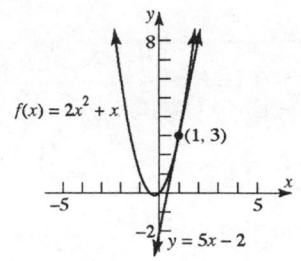

$$m_{\tan} = \lim_{x \to 1}\left(\frac{f(x) - f(1)}{x - 1} \right) = \lim_{x \to 1}\left(\frac{2x^2 + x - 3}{x - 1} \right)$$

$$= \lim_{x \to 1}\left(\frac{(2x + 3)(x - 1)}{x - 1} \right) = \lim_{x \to 1}(2x + 3)$$

$$= 2(1) + 3 = 5$$

Tangent Line: $y - 3 = 5(x - 1)$

$$y - 3 = 5x - 5$$

$$y = 5x - 2$$

9. $f(x) = x^2 - 2x + 3$ at $(-1, 6)$

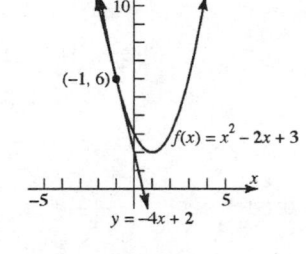

$$m_{\tan} = \lim_{x \to -1} \frac{f(x) - f(-1)}{x + 1} = \lim_{x \to -1} \frac{x^2 - 2x + 3 - 6}{x + 1}$$

$$= \lim_{x \to -1} \frac{x^2 - 2x - 3}{x + 1} = \lim_{x \to -1} \frac{(x + 1)(x - 3)}{x + 1}$$

$$= \lim_{x \to -1}(x - 3) = -1 - 3 = -4$$

Tangent Line: $y - 6 = -4(x - (-1))$

$$y - 6 = -4x - 4$$

$$y = -4x + 2$$

11. $f(x) = x^3 + x$ at $(2, 10)$

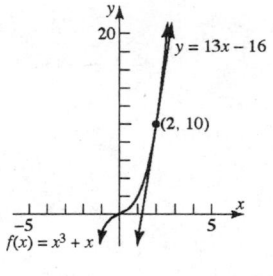

$$m_{\tan} = \lim_{x \to 2}\left(\frac{f(x) - f(2)}{x - 2} \right) = \lim_{x \to 2}\left(\frac{x^3 + x - 10}{x - 2} \right)$$

$$= \lim_{x \to 2}\left(\frac{x^3 - 8 + x - 2}{x - 2} \right)$$

$$= \lim_{x \to 2}\left(\frac{(x - 2)(x^2 + 2x + 4) + x - 2}{x - 2} \right)$$

$$= \lim_{x \to 2}\left(\frac{(x - 2)(x^2 + 2x + 4 + 1)}{x - 2} \right)$$

$$= \lim_{x \to 2}(x^2 + 2x + 5) = 4 + 4 + 5 = 13$$

Tangent Line: $y - 10 = 13(x - 2)$

$$y - 10 = 13x - 26$$

$$y = 13x - 16$$

13. $f(x) = -4x + 5$ at 3

$$f'(3) = \lim_{x \to 3}\left(\frac{f(x) - f(3)}{x - 3} \right) = \lim_{x \to 3}\left(\frac{-4x + 5 - (-7)}{x - 3} \right) = \lim_{x \to 3}\left(\frac{-4x + 12}{x - 3} \right) = \lim_{x \to 3}\left(\frac{-4(x - 3)}{x - 3} \right)$$

$$= \lim_{x \to 3}(-4) = -4$$

15. $f(x) = x^2 - 3$ at 0

$$f'(0) = \lim_{x \to 0}\left(\frac{f(x)-f(0)}{x-0}\right) = \lim_{x \to 0}\left(\frac{x^2-3-(-3)}{x}\right) = \lim_{x \to 0}\left(\frac{x^2}{x}\right) = \lim_{x \to 0}(x) = 0$$

17. $f(x) = 2x^2 + 3x$ at 1

$$f'(1) = \lim_{x \to 1}\left(\frac{f(x)-f(1)}{x-1}\right) = \lim_{x \to 1}\left(\frac{2x^2+3x-5}{x-1}\right) = \lim_{x \to 1}\left(\frac{(2x+5)(x-1)}{x-1}\right) = \lim_{x \to 1}(2x+5) = 7$$

19. $f(x) = x^3 + 4x$ at -1

$$f'(-1) = \lim_{x \to -1}\left(\frac{f(x)-f(-1)}{x-(-1)}\right) = \lim_{x \to -1}\left(\frac{x^3+4x-(-5)}{x+1}\right) = \lim_{x \to -1}\left(\frac{x^3+1+4x+4}{x+1}\right)$$

$$= \lim_{x \to -1}\left(\frac{(x+1)(x^2-x+1)+4(x+1)}{x+1}\right) = \lim_{x \to -1}\left(\frac{(x+1)(x^2-x+1+4)}{x+1}\right)$$

$$= \lim_{x \to -1}(x^2-x+5) = (-1)^2-(-1)+5 = 7$$

21. $f(x) = x^3 + x^2 - 2x$ at 1

$$f'(1) = \lim_{x \to 1}\left(\frac{f(x)-f(1)}{x-1}\right) = \lim_{x \to 1}\left(\frac{x^3+x^2-2x-0}{x-1}\right) = \lim_{x \to 1}\left(\frac{x(x^2+x-2)}{x-1}\right)$$

$$= \lim_{x \to 1}\left(\frac{x(x+2)(x-1)}{x-1}\right) = \lim_{x \to 1}(x(x+2)) = 1(1+2) = 3$$

23. $f(x) = \sin x$ at 0

$$f'(0) = \lim_{x \to 0}\left(\frac{f(x)-f(0)}{x-0}\right) = \lim_{x \to 0}\left(\frac{\sin x-0}{x-0}\right) = \lim_{x \to 0}\left(\frac{\sin x}{x}\right) = 1$$

25. Use NDeriv:

```
nDeriv(3X^3-6X²+
2,X,-2)
            60.000003
```

27. Use NDeriv:

```
nDeriv((-X^3+1)/
(X²+5X+7),X,8)
          -.8587776956
```

29. Use NDeriv:

```
nDeriv(Xsin(X),X
,π/3)
         1.389623659
```

31. Use NDeriv:

```
nDeriv(X²sin(X),
X,π/3)
          2.362110222
```

33. Use NDeriv:

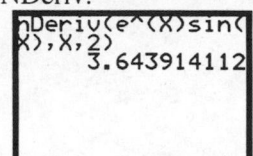

35. $V(r) = 3\pi r^2 \quad$ at $r = 3$

$$V'(3) = \lim_{r \to 3}\left(\frac{V(r) - V(3)}{r - 3}\right) = \lim_{r \to 3}\left(\frac{3\pi r^2 - 27\pi}{r - 3}\right) = \lim_{r \to 3}\left(\frac{3\pi(r^2 - 9)}{r - 3}\right)$$

$$= \lim_{r \to 3}\left(\frac{3\pi(r - 3)(r + 3)}{r - 3}\right) = \lim_{r \to 3}\left(3\pi(r + 3)\right) = 3\pi(3 + 3) = 18\pi$$

At the instant $r = 3$ feet, the volume of the cone is increasing at a rate of 18π cubic feet.

37. $V(r) = \frac{4}{3}\pi r^3 \quad$ at $r = 2$

$$V'(2) = \lim_{r \to 2}\left(\frac{V(r) - V(2)}{r - 2}\right) = \lim_{r \to 2}\left(\frac{\left(\frac{4}{3}\pi r^3 - \frac{32}{3}\pi\right)}{r - 2}\right) = \lim_{r \to 2}\left(\frac{\left(\frac{4}{3}\pi(r^3 - 8)\right)}{r - 2}\right)$$

$$= \lim_{r \to 2}\left(\frac{\left(\frac{4}{3}\pi(r - 2)(r^2 + 2r + 4)\right)}{r - 2}\right) = \lim_{r \to 2}\left(\frac{4}{3}\pi(r^2 + 2r + 4)\right) = \frac{4}{3}\pi(4 + 4 + 4) = 16\pi$$

At the instant $r = 2$ feet, the volume of the sphere is increasing at a rate of 16π cubic feet.

39. (a) $-16t^2 + 96t = 0$
         $-16t(t - 6) = 0$
                 $t = 0$ or $t = 6$
         The ball strikes the ground after 6 seconds.

     (b) $\dfrac{\Delta s}{\Delta t} = \dfrac{s(2) - s(0)}{2 - 0} = \dfrac{-16(2)^2 + 96(2) - 0}{2} = \dfrac{128}{2} = 64$ feet / sec

     (c) $s'(t) = \lim_{t \to t_0}\left(\dfrac{s(t) - s(t_0)}{t - t_0}\right) = \lim_{t \to t_0}\left(\dfrac{-16t^2 + 96t - \left(-16t_0^2 + 96t_0\right)}{t - t_0}\right)$

$$= \lim_{t \to t_0}\left(\frac{-16t^2 + 16t_0^2 + 96t - 96t_0}{t - t_0}\right) = \lim_{t \to t_0}\left(\frac{-16\left(t^2 - t_0^2\right) + 96(t - t_0)}{t - t_0}\right)$$

$$= \lim_{t \to t_0}\left(\frac{-16(t - t_0)(t + t_0) + 96(t - t_0)}{t - t_0}\right) = \lim_{t \to t_0}\left(\frac{(t - t_0)\left(-16(t + t_0) + 96\right)}{t - t_0}\right)$$

$$= \lim_{t \to t_0}\left(-16(t + t_0) + 96\right) = \left(-16(t_0 + t_0) + 96\right) = -32t_0 + 96$$

     (d) $s'(2) = -32(2) + 96 = -64 + 96 = 32$ feet / sec

(e) $s'(t) = 0$

$-32t + 96 = 0$

$-32t = -96$

$t = 3$ seconds

(f) $s(3) = -16(3)^2 + 96(3) = -144 + 288 = 144$ feet

(g) $s'(6) = -32(6) + 96 = -192 + 96 = -96$ feet / sec

41.   (a) $\dfrac{\Delta s}{\Delta t} = \dfrac{s(4) - s(1)}{4 - 1} = \dfrac{917 - 987}{3} = \dfrac{-70}{3} = -23\dfrac{1}{3}$ feet / sec

(b) $\dfrac{\Delta s}{\Delta t} = \dfrac{s(3) - s(1)}{3 - 1} = \dfrac{945 - 987}{2} = \dfrac{-42}{2} = -21$ feet / sec

(c) $\dfrac{\Delta s}{\Delta t} = \dfrac{s(2) - s(1)}{2 - 1} = \dfrac{969 - 987}{1} = \dfrac{-18}{1} = -18$ feet / sec

(d) $s(t) = -2.631t^2 - 10.269t + 999.933$

(e) $s'(1) = \lim\limits_{t \to 1}\left(\dfrac{s(t) - s(1)}{t - 1}\right) = \lim\limits_{t \to 1}\left(\dfrac{-2.631t^2 - 10.269t + 999.933 - 987.033}{t - 1}\right)$

$= \lim\limits_{t \to 1}\left(\dfrac{-2.631t^2 - 10.269t + 12.9}{t - 1}\right)$

$= \lim\limits_{t \to 1}\left(\dfrac{-2.631t^2 + 2.631t - 12.9t + 12.9}{t - 1}\right)$

$= \lim\limits_{t \to 1}\left(\dfrac{-2.631t(t - 1) - 12.9(t - 1)}{t - 1}\right) = \lim\limits_{t \to 1}\left(\dfrac{(-2.631t - 12.9)(t - 1)}{t - 1}\right)$

$= \lim\limits_{t \to 1}(-2.631t - 12.9) = -2.631(1) - 12.9 = -15.531$ feet/sec

The instant $t = 1$, the instantaneous speed of the ball is -15.531 feet / sec.

Chapter 13

A Preview of Calculus: The Limit, Derivative and Integral of a Function

13.5 The Area Problem; The Integral

1. $A \approx f(1) \cdot 1 + f(2) \cdot 1 = 1 \cdot 1 + 2 \cdot 1 = 1 + 2 = 3$

3. $A \approx f(0) \cdot 2 + f(2) \cdot 2 + f(4) \cdot 2 + f(6) \cdot 2 = 10 \cdot 2 + 6 \cdot 2 + 7 \cdot 2 + 5 \cdot 2$
    $= 20 + 12 + 14 + 10 = 56$

5. (a) Graph $f(x) = 3x$:

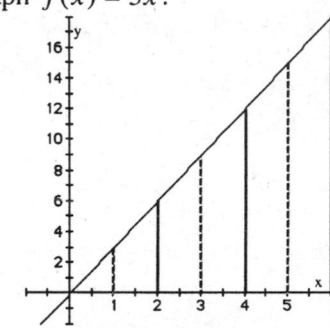

   (b) $A \approx f(0)(2) + f(2)(2) + f(4)(2) = 0(2) + 6(2) + 12(2) = 0 + 12 + 24 = 36$
   (c) $A \approx f(2)(2) + f(4)(2) + f(6)(2) = 6(2) + 12(2) + 18(2) = 12 + 24 + 36 = 72$
   (d) $A \approx f(0)(1) + f(1)(1) + f(2)(1) + f(3)(1) + f(4)(1) + f(5)(1)$
       $= 0(1) + 3(1) + 6(1) + 9(1) + 12(1) + 15(1) = 0 + 3 + 6 + 9 + 12 + 15 = 45$
   (e) $A \approx f(1)(1) + f(2)(1) + f(3)(1) + f(4)(1) + f(5)(1) + f(6)(1)$
       $= 3(1) + 6(1) + 9(1) + 12(1) + 15(1) + 18(1) = 3 + 6 + 9 + 12 + 15 + 18 = 63$
   (f) The actual area is the area of a triangle: $A = \dfrac{1}{2}(6)(18) = 54$

7.   (a)   Graph $f(x) = -3x + 9$:

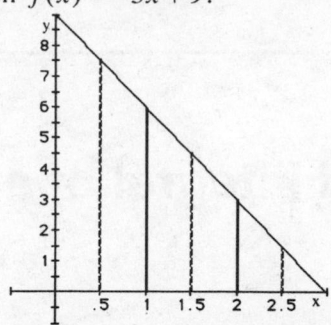

(b) $A \approx f(0)(1) + f(1)(1) + f(2)(1) = 9(1) + 6(1) + 3(1) = 9 + 6 + 3 = 18$

(c) $A \approx f(1)(1) + f(2)(1) + f(3)(1) = 6(1) + 3(1) + 0(1) = 6 + 3 + 0 = 9$

(d) $A \approx f(0)\left(\dfrac{1}{2}\right) + f\left(\dfrac{1}{2}\right)\left(\dfrac{1}{2}\right) + f(1)\left(\dfrac{1}{2}\right) + f\left(\dfrac{3}{2}\right)\left(\dfrac{1}{2}\right) + f(2)\left(\dfrac{1}{2}\right) + f\left(\dfrac{5}{2}\right)\left(\dfrac{1}{2}\right)$

$= 9\left(\dfrac{1}{2}\right) + \dfrac{15}{2}\left(\dfrac{1}{2}\right) + 6\left(\dfrac{1}{2}\right) + \dfrac{9}{2}\left(\dfrac{1}{2}\right) + 3\left(\dfrac{1}{2}\right) + \dfrac{3}{2}\left(\dfrac{1}{2}\right)$

$= \dfrac{9}{2} + \dfrac{15}{4} + 3 + \dfrac{9}{4} + \dfrac{3}{2} + \dfrac{3}{4} = \dfrac{63}{4}$

(e) $A \approx f\left(\dfrac{1}{2}\right)\left(\dfrac{1}{2}\right) + f(1)\left(\dfrac{1}{2}\right) + f\left(\dfrac{3}{2}\right)\left(\dfrac{1}{2}\right) + f(2)\left(\dfrac{1}{2}\right) + f\left(\dfrac{5}{2}\right)\left(\dfrac{1}{2}\right) + f(3)\left(\dfrac{1}{2}\right)$

$= \dfrac{15}{2}\left(\dfrac{1}{2}\right) + 6\left(\dfrac{1}{2}\right) + \dfrac{9}{2}\left(\dfrac{1}{2}\right) + 3\left(\dfrac{1}{2}\right) + \dfrac{3}{2}\left(\dfrac{1}{2}\right) + 0\left(\dfrac{1}{2}\right)$

$= \dfrac{15}{4} + 3 + \dfrac{9}{4} + \dfrac{3}{2} + \dfrac{3}{4} + 0 = \dfrac{45}{4}$

(f)   The actual area is the area of a triangle: $A = \dfrac{1}{2}(3)(9) = \dfrac{27}{2} = 13.5$

9.   (a)   Graph $f(x) = x^2 + 2$, $[0, 4]$:

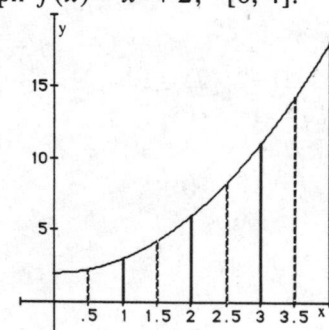

(b) $A \approx f(0)(1) + f(1)(1) + f(2)(1) + f(3)(1) = 2(1) + 3(1) + 6(1) + 11(1)$

$= 2 + 3 + 6 + 11 = 22$

(c) $A \approx f(0)\left(\dfrac{1}{2}\right) + f\left(\dfrac{1}{2}\right)\left(\dfrac{1}{2}\right) + f(1)\left(\dfrac{1}{2}\right) + f\left(\dfrac{3}{2}\right)\left(\dfrac{1}{2}\right) + f(2)\left(\dfrac{1}{2}\right) + f\left(\dfrac{5}{2}\right)\left(\dfrac{1}{2}\right)$

$\qquad\qquad\qquad\qquad\qquad\qquad + f(3)\left(\dfrac{1}{2}\right) + f\left(\dfrac{7}{2}\right)\left(\dfrac{1}{2}\right)$

$= 2\left(\dfrac{1}{2}\right) + \dfrac{9}{4}\left(\dfrac{1}{2}\right) + 3\left(\dfrac{1}{2}\right) + \dfrac{17}{4}\left(\dfrac{1}{2}\right) + 6\left(\dfrac{1}{2}\right) + \dfrac{33}{4}\left(\dfrac{1}{2}\right) + 11\left(\dfrac{1}{2}\right) + \dfrac{57}{4}\left(\dfrac{1}{2}\right)$

$= 1 + \dfrac{9}{8} + \dfrac{3}{2} + \dfrac{17}{8} + 3 + \dfrac{33}{8} + \dfrac{11}{2} + \dfrac{57}{8} = \dfrac{51}{2}$

(d)　$A = \int_0^4 (x^2 + 2)\,dx$

(e)　Use fnInt function:

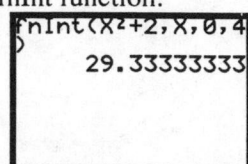

11.　(a)　Graph $f(x) = x^3$,　$[0, 4]$:

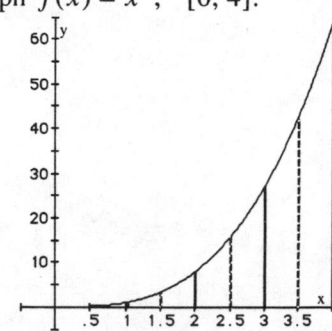

(b)　$A \approx f(0)(1) + f(1)(1) + f(2)(1) + f(3)(1) = 0(1) + 1(1) + 8(1) + 27(1)$
　　　$= 0 + 1 + 8 + 27 = 36$

(c)　$A \approx f(0)\left(\dfrac{1}{2}\right) + f\left(\dfrac{1}{2}\right)\left(\dfrac{1}{2}\right) + f(1)\left(\dfrac{1}{2}\right) + f\left(\dfrac{3}{2}\right)\left(\dfrac{1}{2}\right) + f(2)\left(\dfrac{1}{2}\right) + f\left(\dfrac{5}{2}\right)\left(\dfrac{1}{2}\right)$
$$+ f(3)\left(\dfrac{1}{2}\right) + f\left(\dfrac{7}{2}\right)\left(\dfrac{1}{2}\right)$$

$$= 0\left(\dfrac{1}{2}\right) + \dfrac{1}{8}\left(\dfrac{1}{2}\right) + 1\left(\dfrac{1}{2}\right) + \dfrac{27}{8}\left(\dfrac{1}{2}\right) + 8\left(\dfrac{1}{2}\right) + \dfrac{125}{8}\left(\dfrac{1}{2}\right) + 27\left(\dfrac{1}{2}\right) + \dfrac{343}{8}\left(\dfrac{1}{2}\right)$$

$$= 0 + \dfrac{1}{16} + \dfrac{1}{2} + \dfrac{27}{16} + 4 + \dfrac{125}{16} + \dfrac{27}{2} + \dfrac{343}{16} = 49$$

(d)　$A = \int_0^4 x^3\,dx$

(e)　Use fnInt function:

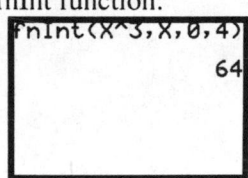

13. (a) Graph $f(x) = \dfrac{1}{x}$, [1, 5]:

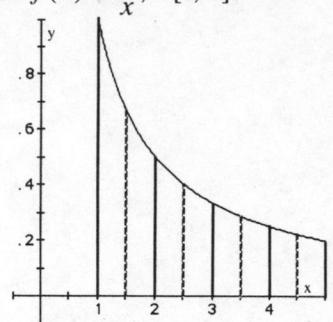

(b) $A \approx f(1)(1) + f(2)(1) + f(3)(1) + f(4)(1) = 1(1) + \dfrac{1}{2}(1) + \dfrac{1}{3}(1) + \dfrac{1}{4}(1)$

$$= 1 + \frac{1}{2} + \frac{1}{3} + \frac{1}{4} = \frac{25}{12}$$

(c) $A \approx f(1)\left(\dfrac{1}{2}\right) + f\left(\dfrac{3}{2}\right)\left(\dfrac{1}{2}\right) + f(2)\left(\dfrac{1}{2}\right) + f\left(\dfrac{5}{2}\right)\left(\dfrac{1}{2}\right) + f(3)\left(\dfrac{1}{2}\right) + f\left(\dfrac{7}{2}\right)\left(\dfrac{1}{2}\right)$

$$+ f(4)\left(\frac{1}{2}\right) + f\left(\frac{9}{2}\right)\left(\frac{1}{2}\right)$$

$$= 1\left(\frac{1}{2}\right) + \frac{2}{3}\left(\frac{1}{2}\right) + \frac{1}{2}\left(\frac{1}{2}\right) + \frac{2}{5}\left(\frac{1}{2}\right) + \frac{1}{3}\left(\frac{1}{2}\right) + \frac{2}{7}\left(\frac{1}{2}\right) + \frac{1}{4}\left(\frac{1}{2}\right) + \frac{2}{9}\left(\frac{1}{2}\right)$$

$$= \frac{1}{2} + \frac{1}{3} + \frac{1}{4} + \frac{1}{5} + \frac{1}{6} + \frac{1}{7} + \frac{1}{8} + \frac{1}{9} = \frac{4690}{2520} \approx 1.829$$

(d) $A = \displaystyle\int_{1}^{5} \frac{1}{x}\,dx$

(e)   Use fnInt function:

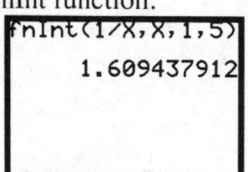

15. (a) Graph $f(x) = e^x$, [−1, 3]:

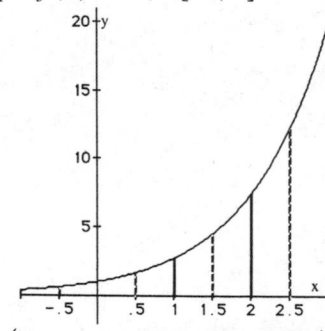

(b) $A \approx \big(f(-1) + f(0) + f(1) + f(2)\big)(1) = (0.3679 + 1 + 2.7183 + 7.3891)(1)$

$$= 11.4753$$

(c) $A \approx \left(f(-1) + f\left(-\dfrac{1}{2}\right) + f(0) + f\left(\dfrac{1}{2}\right) + f(1) + f\left(\dfrac{3}{2}\right) + f(2) + f\left(\dfrac{5}{2}\right)\right) \cdot \left(\dfrac{1}{2}\right)$

$$= (0.3679 + 0.6065 + 1 + 1.6487 + 2.7183 + 4.4817 + 7.3891 + 12.1825)(0.5)$$

$$= 30.3947(0.5) = 15.1974$$

(d) $A = \int_{-1}^{3} e^x \, dx$

(e) Use fnInt function:

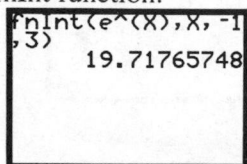

17. (a) Graph $f(x) = \sin x$, $[0, \pi]$:

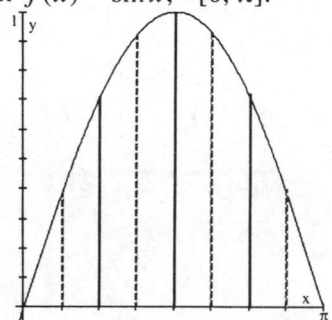

(b) $A \approx \left(f(0) + f\left(\dfrac{\pi}{4}\right) + f\left(\dfrac{\pi}{2}\right) + f\left(\dfrac{3\pi}{4}\right) \right)\left(\dfrac{\pi}{4}\right) = \left(0 + \dfrac{\sqrt{2}}{2} + 1 + \dfrac{\sqrt{2}}{2} \right)\left(\dfrac{\pi}{4}\right)$

$\left(1 + \sqrt{2}\right)\left(\dfrac{\pi}{4}\right) \approx 1.8961$

(c) $A \approx \left(f(0) + f\left(\dfrac{\pi}{8}\right) + f\left(\dfrac{\pi}{4}\right) + f\left(\dfrac{3\pi}{8}\right) + f\left(\dfrac{\pi}{2}\right) + f\left(\dfrac{5\pi}{8}\right) + f\left(\dfrac{3\pi}{4}\right) + f\left(\dfrac{7\pi}{8}\right) \right)\left(\dfrac{\pi}{8}\right)$

$$= (0 + 0.3827 + 0.7071 + 0.9239 + 1 + 0.9239 + 0.7071 + 0.3827)(0.3927)$$

$$= 5.0274(0.3927) = 1.9743$$

(d) $A = \int_{0}^{\pi} \sin x \, dx$

(e) Use fnInt function:

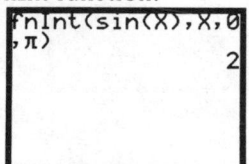

19. (a) The integral represents the area under the graph of $f(x) = 3x + 1$ from 0 to 4.

    (b)   (c)

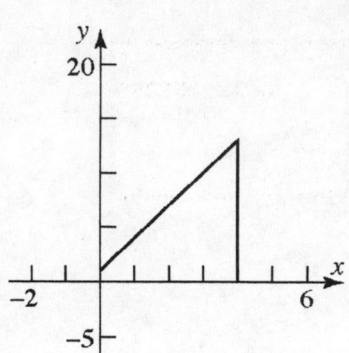

21. (a) The integral represents the area under the graph of $f(x) = x^2 - 1$ from 2 to 5.

    (b)   (c)

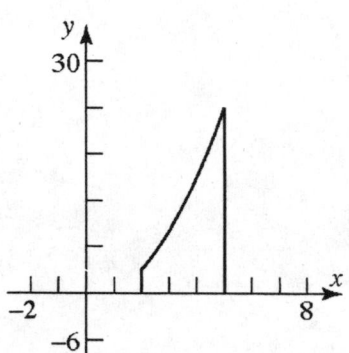

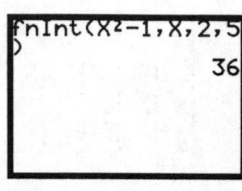

23. (a) The integral represents the area under the graph of $f(x) = \sin x$ from 0 to $\dfrac{\pi}{2}$.

    (b)   (c)

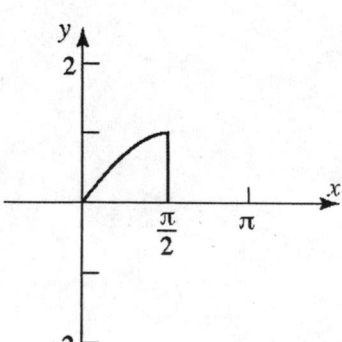

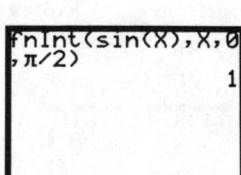

25. (a)   The integral represents the area under the graph of $f(x) = e^x$ from 0 to 2.

     (b)   (c)

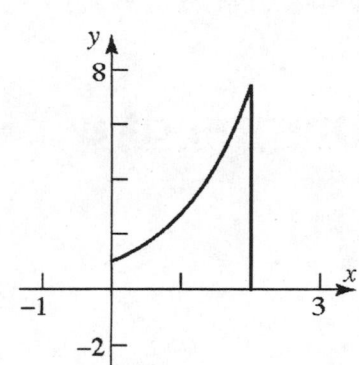

27. Using left endpoints:

$n = 2:$ $0 + 0.5 = 0.5$

$n = 4:$ $0 + 0.125 + 0.25 + 0.375 = 0.75$

$n = 10:$ $0 + 0.02 + 0.04 + 0.06 + \cdots + 0.018 = \dfrac{10}{2}(0 + 0.018) = 0.9$

$n = 100:$ $0 + 0.0002 + 0.0004 + 0.0006 + \cdots + 0.0198 = \dfrac{100}{2}(0 + 0.0198) = 0.99$

Using right endpoints:

$n = 2:$ $0.5 + 1 = 1.5$

$n = 4:$ $0.125 + 0.25 + 0.375 + 0.5 = 1.25$

$n = 10:$ $0.02 + 0.04 + 0.06 + \cdots + 0.20 = \dfrac{10}{2}(0.02 + 0.20) = 1.1$

$n = 100:$ $0.0002 + 0.0004 + 0.0006 + \cdots + 0.02 = \dfrac{100}{2}(0.0002 + 0.02) = 1.01$

A Preview of Calculus:
The Limit, Derivative and Integral of a Function

13.R Chapter Review

1. $\lim\limits_{x\to 2}\left(3x^2 - 2x + 1\right) = 3(2)^2 - 2(2) + 1 = 12 - 4 + 1 = 9$

3. $\lim\limits_{x\to -2}\left(x^2 + 1\right)^2 = \left(\lim\limits_{x\to -2}\left(x^2 + 1\right)\right)^2 = \left((-2)^2 + 1\right)^2 = 5^2 = 25$

5. $\lim\limits_{x\to 3}\sqrt{x^2 + 7} = \sqrt{\lim\limits_{x\to 3}\left(x^2 + 7\right)} = \sqrt{3^2 + 7} = \sqrt{16} = 4$

7. $\lim\limits_{x\to 1^-}\sqrt{1 - x^2} = \sqrt{\lim\limits_{x\to 1^-}\left(1 - x^2\right)} = \sqrt{1 - 1^2} = \sqrt{0} = 0$

9. $\lim\limits_{x\to 2}(5x + 6)^{3/2} = \left(\lim\limits_{x\to 2}(5x + 6)\right)^{3/2} = \left(5(2) + 6\right)^{3/2} = 16^{3/2} = 64$

11. $\lim\limits_{x\to 1}\left(\dfrac{x^2 + x + 2}{x^2 - 9}\right) = \dfrac{\left(\lim\limits_{x\to 1}(x^2 + x + 2)\right)}{\left(\lim\limits_{x\to 1}(x^2 - 9)\right)} = \dfrac{(-1)^2 + (-1) + 2}{(-1)^2 - 9} = \dfrac{2}{-8} = -\dfrac{1}{4}$

13. $\lim\limits_{x\to 1}\left(\dfrac{x - 1}{x^3 - 1}\right) = \lim\limits_{x\to 1}\left(\dfrac{x - 1}{(x - 1)(x^2 + x + 1)}\right) = \lim\limits_{x\to 1}\left(\dfrac{1}{x^2 + x + 1}\right) = \dfrac{1}{1^2 + 1 + 1} = \dfrac{1}{3}$

15. $\lim\limits_{x\to 3}\left(\dfrac{x^2 - 9}{x^2 - x - 12}\right) = \lim\limits_{x\to 3}\left(\dfrac{(x - 3)(x + 3)}{(x - 4)(x + 3)}\right) = \lim\limits_{x\to 3}\left(\dfrac{x - 3}{x - 4}\right) = \dfrac{-3 - 3}{-3 - 4} = \dfrac{-6}{-7} = \dfrac{6}{7}$

17. $\lim\limits_{x\to 1^-}\left(\dfrac{x^2 - 1}{x^3 - 1}\right) = \lim\limits_{x\to 1^-}\left(\dfrac{(x + 1)(x - 1)}{(x - 1)(x^2 + x + 1)}\right) = \lim\limits_{x\to 1^-}\left(\dfrac{x + 1}{x^2 + x + 1}\right) = \dfrac{-1 + 1}{(-1)^2 + (-1) + 1} = \dfrac{0}{1} = 0$

19. $\lim\limits_{x\to 2}\left(\dfrac{x^3 - 8}{x^3 - 2x^2 + 4x - 8}\right) = \lim\limits_{x\to 2}\left(\dfrac{(x - 2)(x^2 + 2x + 4)}{x^2(x - 2) + 4(x - 2)}\right) = \lim\limits_{x\to 2}\left(\dfrac{(x - 2)(x^2 + 2x + 4)}{(x - 2)(x^2 + 4)}\right)$

$= \lim\limits_{x\to 2}\left(\dfrac{x^2 + 2x + 4}{x^2 + 4}\right) = \left(\dfrac{2^2 + 2(2) + 4}{2^2 + 4}\right) = \dfrac{12}{8} = \dfrac{3}{2}$

21. $\lim\limits_{x \to 3}\left(\dfrac{x^4 - 3x^3 + x - 3}{x^3 - 3x^2 + 2x - 6}\right) = \lim\limits_{x \to 3}\left(\dfrac{x^3(x-3) + 1(x-3)}{x^2(x-3) + 2(x-3)}\right) = \lim\limits_{x \to 3}\left(\dfrac{(x-3)(x^3+1)}{(x-3)(x^2+2)}\right) = \lim\limits_{x \to 3}\left(\dfrac{x^3+1}{x^2+2}\right)$

$$= \dfrac{3^3 + 1}{3^2 + 2} = \dfrac{28}{11}$$

23. $f(x) = 3x^4 - x^2 + 2; \ c = 5$
   1. $f(5) = 3(5)^4 - 5^2 + 2 = 1852$
   2. $\lim\limits_{x \to 5^-} f(x) = 3(5)^4 - 5^2 + 2 = 1852$
   3. $\lim\limits_{x \to 5^+} f(x) = 3(5)^4 - 5^2 + 2 = 1852$

   Thus, $f(x)$ is continuous at $c = 5$.

25. $f(x) = \dfrac{x^2 - 4}{x + 2}; \ c = -2$

   Since $f(x)$ is not defined at $c = -2$, the function is not continuous at $c = -2$.

27. $f(x) = \begin{cases} \dfrac{x^2 - 4}{x + 2} & \text{if } x \neq -2 \\ 4 & \text{if } x = -2 \end{cases}; \quad c = -2$

   1. $f(-2) = 4$

   2. $\lim\limits_{x \to -2^-} f(x) = \lim\limits_{x \to -2^-}\left(\dfrac{x^2 - 4}{x + 2}\right) = \lim\limits_{x \to -2^-}\left(\dfrac{(x-2)(x+2)}{x + 2}\right) = \lim\limits_{x \to -2^-}(x - 2) = -4$

   Since $\lim\limits_{x \to -2^-} f(x) \neq f(c)$, the function is not continuous at $c = -2$.

29. $f(x) = \begin{cases} \dfrac{x^2 - 4}{x + 2} & \text{if } x \neq -2 \\ -4 & \text{if } x = -2 \end{cases}; \quad c = -2$

   1. $f(-2) = -4$

   2. $\lim\limits_{x \to -2^-} f(x) = \lim\limits_{x \to -2^-}\left(\dfrac{x^2 - 4}{x + 2}\right) = \lim\limits_{x \to -2^-}\left(\dfrac{(x-2)(x+2)}{x + 2}\right) = \lim\limits_{x \to -2^-}(x - 2) = -4$

   3. $\lim\limits_{x \to -2^+} f(x) = \lim\limits_{x \to -2^+}\left(\dfrac{x^2 - 4}{x + 2}\right) = \lim\limits_{x \to -2^+}\left(\dfrac{(x-2)(x+2)}{x + 2}\right) = \lim\limits_{x \to -2^+}(x - 2) = -4$

   The function is continuous at $c = -2$.

31. Domain: $\{x \mid -6 \leq x < 2 \text{ or } 2 < x < 5 \text{ or } 5 < x \leq 6\}$

33. x-intercepts: $1, 6$ 35. $f(-6) = 2; \ f(-4) = 1$ 37. $\lim\limits_{x \to -4^-} f(x) = 4$

39. $\lim\limits_{x \to -2^-} f(x) = -2$ 41. $\lim\limits_{x \to 2^-} f(x) = -\infty$

43. $\lim\limits_{x \to 0} f(x)$ does not exist. $\lim\limits_{x \to 0^-} f(x) = 4 \neq \lim\limits_{x \to 0^+} f(x) = 1$

45. f is not continuous at -2 because $\lim\limits_{x \to -2^-} f(x) \neq \lim\limits_{x \to -2^+} f(x)$

47. f is not continuous at 0 because $\lim\limits_{x \to 0^-} f(x) \neq \lim\limits_{x \to 0^+} f(x)$

49. f is continuous at 4 because $f(4) = \lim\limits_{x \to 4^-} f(x) = \lim\limits_{x \to 4^+} f(x)$

51. $R(x) = \dfrac{x+4}{x^2 - 16} = \dfrac{x+4}{(x-4)(x+4)}$. The domain of R is $\left\{ x \mid x \neq -4,\, x \neq 4 \right\}$. Thus R is

discontinuous at both -4 and 4. $\lim\limits_{x \to 4^-} R(x) = \lim\limits_{x \to 4^-}\left(\dfrac{x+4}{(x-4)(x+4)} \right) = \lim\limits_{x \to 4^-}\left(\dfrac{1}{x-4} \right) = -\infty$ since

when $x < 4$, $\dfrac{1}{x-4} < 0$, and as x approaches 4, $\dfrac{1}{x-4}$ becomes unbounded.

$\lim\limits_{x \to 4^+}\left(\dfrac{1}{x-4} \right) = +\infty$ since when $x > 4$, $\dfrac{1}{x-4} > 0$, and as x approaches 4, $\dfrac{1}{x-4}$ becomes

unbounded. The behavior near $c = -4$ is: $\lim\limits_{x \to 4} R(x) = \lim\limits_{x \to 4}\left(\dfrac{1}{x-4} \right) = \dfrac{1}{-8}$. Note there is a

hole in the graph at $\left(-4, -\dfrac{1}{8} \right)$.

53. $R(x) = \dfrac{x^3 - 2x^2 + 4x - 8}{x^2 - 11x + 18} = \dfrac{x^2(x-2) + 4(x-2)}{(x-9)(x-2)} = \dfrac{(x-2)(x^2+4)}{(x-9)(x-2)} = \dfrac{x^2+4}{x-9},\ x \neq 2$
      There is a vertical asymptote where $x - 9 = 0$. $x = 9$ is a vertical asymptote. There is a
      hole in the graph at $x = 2$.

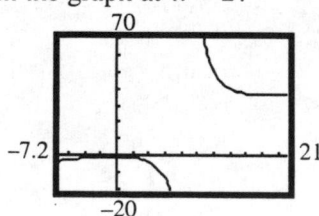

55. $f(x) = 2x^2 + 8x$ at $(1, 10)$

$$m_{\tan} = \lim_{x \to 1}\left(\frac{f(x)-f(1)}{x-1}\right) = \lim_{x \to 1}\left(\frac{2x^2+8x-10}{x-1}\right)$$

$$= \lim_{x \to 1}\left(\frac{2(x+5)(x-1)}{x-1}\right) = \lim_{x \to 1}\left(2(x+5)\right)$$

$$= 2(1+5) = 12$$

Tangent Line: $y - 10 = 12(x - 1)$

$$y - 10 = 12x - 12$$

$$y = 12x - 2$$

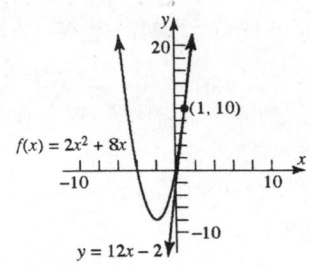

57. $f(x) = x^2 + 2x - 3$ at $(-1, -4)$

$$m_{\tan} = \lim_{x \to 1}\left(\frac{f(x)-f(-1)}{x+1}\right) = \lim_{x \to 1}\left(\frac{x^2+2x-3-(-4)}{x+1}\right)$$

$$= \lim_{x \to 1}\left(\frac{x^2+2x+1}{x+1}\right) = \lim_{x \to 1}\left(\frac{(x+1)^2}{x+1}\right) = \lim_{x \to 1}\left(x+1\right)$$

$$= -1 + 1 = 0$$

Tangent Line: $y - (-4) = 0(x - (-1))$

$$y + 4 = 0$$

$$y = -4$$

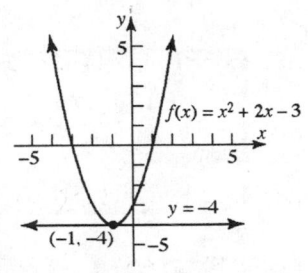

59. $f(x) = x^3 + x^2$ at $(2, 12)$

$$m_{\tan} = \lim_{x \to 2}\left(\frac{f(x)-f(2)}{x-2}\right) = \lim_{x \to 2}\left(\frac{x^3+x^2-12}{x-2}\right)$$

$$= \lim_{x \to 2}\left(\frac{x^3-2x^2+3x^2-12}{x-2}\right)$$

$$= \lim_{x \to 2}\left(\frac{x^2(x-2)+3(x-2)(x+2)}{x-2}\right)$$

$$= \lim_{x \to 2}\left(\frac{(x-2)(x^2+3x+6)}{x-2}\right)$$

$$= \lim_{x \to 2}\left(x^2+3x+6\right) = 4+6+6 = 16$$

Tangent Line : $y - 12 = 16(x - 2)$

$$y - 12 = 16x - 32$$

$$y = 16x - 20$$

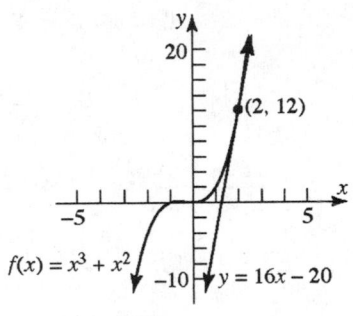

61. $f(x) = -4x^2 + 5$ at 3

$$f'(3) = \lim_{x \to 3}\left(\frac{f(x)-f(3)}{x-3}\right) = \lim_{x \to 3}\left(\frac{-4x^2+5-(-31)}{x-3}\right) = \lim_{x \to 3}\left(\frac{-4x^2+36}{x-3}\right)$$

$$= \lim_{x \to 3}\left(\frac{-4(x^2-9)}{x-3}\right) = \lim_{x \to 3}\left(\frac{-4(x-3)(x+3)}{x-3}\right) = \lim_{x \to 3}\left((-4)(x+3)\right) = -4(6) = -24$$

A Preview of Calculus:
The Limit, Derivative and Integral of a Function

63. $f(x) = x^2 - 3x$ at 0

$$f'(0) = \lim_{x \to 0}\left(\frac{f(x) - f(0)}{x - 0}\right) = \lim_{x \to 0}\left(\frac{x^2 - 3x - 0}{x}\right) = \lim_{x \to 0}\left(\frac{x(x - 3)}{x}\right) = \lim_{x \to 0}(x - 3) = -3$$

65. $f(x) = 2x^2 + 3x + 2$ at 1

$$f'(1) = \lim_{x \to 1}\left(\frac{f(x) - f(1)}{x - 1}\right) = \lim_{x \to 1}\left(\frac{2x^2 + 3x + 2 - 7}{x - 1}\right) = \lim_{x \to 1}\left(\frac{2x^2 + 3x - 5}{x - 1}\right)$$

$$= \lim_{x \to 1}\left(\frac{(2x + 5)(x - 1)}{x - 1}\right) = \lim_{x \to 1}(2x + 5) = 7$$

67. Use NDeriv:

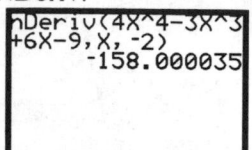

69. Use NDeriv:

71. (a) $-16t^2 + 96t + 112 = 0$

$-16(t^2 - 6t - 7) = 0$

$-16(t + 1)(t - 7) = 0 \Rightarrow t = -1$ or $t = 7$

The ball strikes the ground after 7 seconds.

(b) $-16t^2 + 96t + 112 = 112$

$-16t^2 + 96t = 0$

$-16t(t - 6) = 0 \Rightarrow t = 0$ or $t = 6$

The ball passes the rooftop after 6 seconds.

(c) $\dfrac{\Delta s}{\Delta t} = \dfrac{s(2) - s(0)}{2 - 0} = \dfrac{-16(2)^2 + 96(2) + 112 - 112}{2} = \dfrac{128}{2} = 64$ feet / sec

(d) $s'(t) = \lim_{t \to t_0}\left(\dfrac{s(t) - s(t_0)}{t - t_0}\right) = \lim_{t \to t_0}\left(\dfrac{-16t^2 + 96t + 112 - \left(-16t_0^2 + 96t_0 + 112\right)}{t - t_0}\right)$

$= \lim_{t \to t_0}\left(\dfrac{-16t^2 + 16t_0^2 + 96t - 96t_0}{t - t_0}\right) = \lim_{t \to t_0}\left(\dfrac{-16\left(t^2 - t_0^2\right) + 96(t - t_0)}{t - t_0}\right)$

$= \lim_{t \to t_0}\left(\dfrac{-16(t - t_0)(t + t_0) + 96(t - t_0)}{t - t_0}\right) = \lim_{t \to t_0}\left(\dfrac{(t - t_0)(-16(t + t_0) + 96)}{t - t_0}\right)$

$= \lim_{t \to t_0}\left(-16(t + t_0) + 96\right) = \left(-16(t_0 + t_0) + 96\right) = -32t_0 + 96$

(e) $s'(2) = -32(2) + 96 = -64 + 96 = 32$ feet / sec

(f) $s'(t) = 0$

$-32t + 96 = 0$

$-32t = -96$

$t = 3$ seconds

(g) $s'(6) = -32(6) + 96 = -192 + 96 = -96$ feet / sec

(h) $s'(7) = -32(7) + 96 = -224 + 96 = -128$ feet / sec

73. (a) $\dfrac{\Delta R}{\Delta x} = \dfrac{8775 - 2340}{130 - 25} = \dfrac{6435}{105} = \$61.29$ / watch

(b) $\dfrac{\Delta R}{\Delta x} = \dfrac{6975 - 2340}{90 - 25} = \dfrac{4635}{65} = \$71.31$ / watch

(c) $\dfrac{\Delta R}{\Delta x} = \dfrac{4375 - 2340}{50 - 25} = \dfrac{2035}{25} = \$81.40$ / watch

(d) $R(x) = -0.25x^2 + 100.01x - 1.24$

(e) $R'(25) = \lim\limits_{x \to 25} \left(\dfrac{R(x) - R(25)}{x - 25} \right) = \lim\limits_{x \to 25} \left(\dfrac{-0.25x^2 + 100.01x - 1.24 - 2342.76}{x - 25} \right)$

$= \lim\limits_{x \to 25} \left(\dfrac{-0.25x^2 + 100.01x - 2344}{x - 25} \right)$ (Divide)

$= \lim\limits_{x \to 25} (-0.25x + 93.76) = -0.25(25) + 93.76 = \$87.51$/watch

75. (a) Graph $f(x) = 2x + 3$:

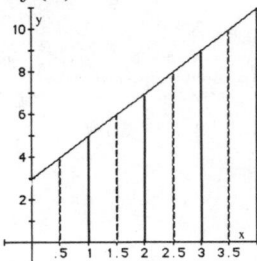

(b) $A \approx \big(f(0) + f(1) + f(2) + f(3) \big)(1) = (3 + 5 + 7 + 9)(1) = 24(1) = 24$

(c) $A \approx \big(f(1) + f(2) + f(3) + f(4) \big)(1) = (5 + 7 + 9 + 11)(1) = 32(1) = 32$

(d) $A \approx \left(f(0) + f\left(\dfrac{1}{2}\right) + f(1) + f\left(\dfrac{3}{2}\right) + f(2) + f\left(\dfrac{5}{2}\right) + f(3) + f\left(\dfrac{7}{2}\right) \right) \cdot \left(\dfrac{1}{2} \right)$

$= (3 + 4 + 5 + 6 + 7 + 8 + 9 + 10)\left(\dfrac{1}{2} \right) = 52\left(\dfrac{1}{2} \right) = 26$

(e) $A \approx \left(f\left(\dfrac{1}{2}\right) + f(1) + f\left(\dfrac{3}{2}\right) + f(2) + f\left(\dfrac{5}{2}\right) + f(3) + f\left(\dfrac{7}{2}\right) + f(4) \right) \cdot \left(\dfrac{1}{2} \right)$

$= (4 + 5 + 6 + 7 + 8 + 9 + 10 + 11)\left(\dfrac{1}{2} \right) = 60\left(\dfrac{1}{2} \right) = 30$

(f) The actual area is the area of a trapezoid: $A = \dfrac{1}{2}(3 + 11)(4) = \dfrac{56}{2} = 28$

77. (a) Graph $f(x) = 4 - x^2$, $[-1, 2]$:

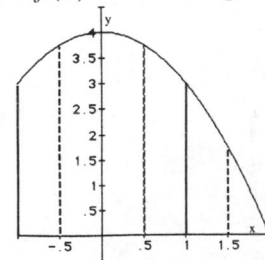

(b) $A \approx \big(f(-1) + f(0) + f(1) \big)(1) = (3 + 4 + 3)(1) = 10(1) = 10$

(c) $A \approx \left(f(-1) + f\left(-\frac{1}{2}\right) + f(0) + f\left(\frac{1}{2}\right) + f(1) + f\left(\frac{3}{2}\right) \right) \cdot \left(\frac{1}{2}\right)$

$= \left(3 + \frac{15}{4} + 4 + \frac{15}{4} + 3 + \frac{7}{4} \right)\left(\frac{1}{2}\right) = \frac{61}{4}\left(\frac{1}{2}\right) = \frac{61}{8} = 7.625$

(d) $A = \int_{-1}^{2} \left(4 - x^2\right) dx$

(e)   Use fnInt function:

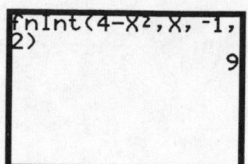

79.   (a)   Graph $f(x) = \dfrac{1}{x^2}$,   [1, 4]:

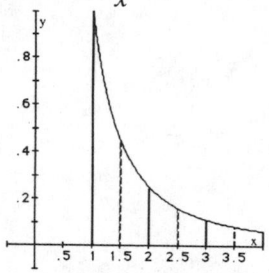

(b) $A \approx (f(1) + f(2) + f(3))(1) = \left(1 + \frac{1}{4} + \frac{1}{9}\right)(1) = \frac{49}{36}(1) = \frac{49}{36} = 1.361$

(c) $A \approx \left(f(1) + f\left(\frac{3}{2}\right) + f(2) + f\left(\frac{5}{2}\right) + f(3) + f\left(\frac{7}{2}\right) \right) \cdot \left(\frac{1}{2}\right)$

$= \left(1 + \frac{4}{9} + \frac{1}{4} + \frac{4}{25} + \frac{1}{9} + \frac{4}{49}\right)\left(\frac{1}{2}\right) = 1.024$

(d) $A = \int_{1}^{4} \left(\frac{1}{x^2}\right) dx$

(e)   Use fnInt function:

81.   (a)   The integral represents the area under the graph of $f(x) = 9 - x^2$ from -1 to 3.
      (b)   Use fnInt function: (c)

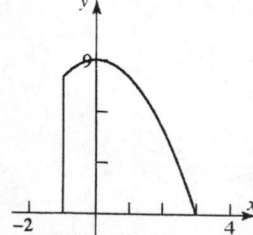

83. (a) The integral represents the area under the graph of $f(x) = e^x$ from -1 to 1.

(b)

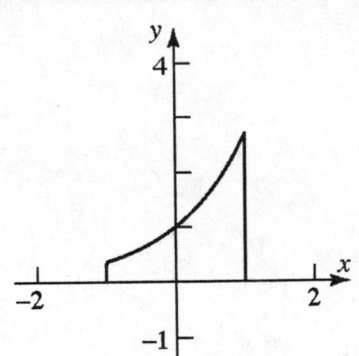

(c)

```
fnInt(e^(X),X,-1
,1)
        2.350402387
```

Appendix Review

A.1 Algebra Review

1. (a) $\{2, 5\}$
    (b) $\{-6, 2, 5\}$
    (c) $\left\{-6, \dfrac{1}{2}, -1.333\ldots, 2, 5\right\}$
    (d) $\{\pi\}$
    (e) $\left\{-6, \dfrac{1}{2}, -1.333\ldots, \pi, 2, 5\right\}$

3. (a) $\{1\}$
    (b) $\{0, 1\}$
    (c) $\left\{0, 1, \dfrac{1}{2}, \dfrac{1}{3}, \dfrac{1}{4}\right\}$
    (d) None
    (e) $\left\{0, 1, \dfrac{1}{2}, \dfrac{1}{3}, \dfrac{1}{4}\right\}$

5. (a) None
    (b) None
    (c) None
    (d) $\left\{\sqrt{2}, \pi, \sqrt{2}+1, \pi+\dfrac{1}{2}\right\}$
    (e) $\left\{\sqrt{2}, \pi, \sqrt{2}+1, \pi+\dfrac{1}{2}\right\}$

7. $x + 2y = -2 + 2\cdot 3 = -2 + 6 = 4$

9. $5xy + 2 = 5(-2)(3) + 2 = -30 + 2 = -28$

11. $\dfrac{2x}{x-y} = \dfrac{2(-2)}{-2-3} = \dfrac{-4}{-5} = \dfrac{4}{5}$

13. $\dfrac{3x + 2y}{2 + y} = \dfrac{3(-2) + 2(3)}{2 + 3} = \dfrac{-6 + 6}{5} = \dfrac{0}{5} = 0$

15. $\dfrac{x^2 - 1}{x}$ Part (c) must be excluded.

The value $x = 0$ must be excluded from the domain because it causes division by 0.

17. $\dfrac{x}{x^2 - 9} = \dfrac{x}{(x-3)(x+3)}$ Part (a) must be excluded.

The values $x = -3$ and $x = 3$ must be excluded from the domain because they cause division by 0.

19. $\dfrac{x^2}{x^2+1}$ None of the given values are excluded.
The domain is all real numbers.

21. $\dfrac{x^2+5x-10}{x^3-x} = \dfrac{x^2+5x-10}{x(x-1)(x+1)}$ Parts (b), (c), and (d) must be excluded.
The values $x = 0$, $x = 1$, and $x = -1$ must be excluded from the domain because they cause division by 0.

23. Since $x = 5$ will cause the denominator of $\dfrac{4}{x-5}$ to equal zero, the domain is the set of all real numbers except 5.

25. Since $x = -4$ will cause the denominator of $\dfrac{x}{x+4}$ to equal zero, the domain is the set of all real numbers except -4.

27.

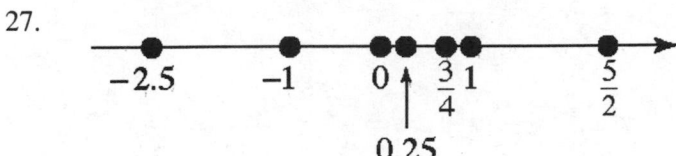

29. $\dfrac{1}{2} > 0$

31. $-1 > -2$

33. $\pi > 3.14$

35. $\dfrac{1}{2} = 0.5$

37. $\dfrac{2}{3} < 0.67$

39. $x > 0$

41. $x < 2$

43. $x \le 1$

45. Graph on the number line: $x \ge -2$

47. Graph on the number line: $x > -1$

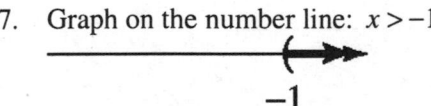

49. $|x+y| = |3+(-2)| = |1| = 1$

51. $|x|+|y| = |3|+|-2| = 3+2 = 5$

53. $\dfrac{|x|}{x} = \dfrac{|3|}{3} = \dfrac{3}{3} = 1$

55. $|4x-5y| = |4(3)-5(-2)| = |12+10| = |22| = 22$

57. $||4x|-|5y|| = ||4(3)|-|5(-2)|| = ||12|-|-10|| = |12-10| = |2| = 2$

59. $d(C,D) = d(0,1) = |1-0| = |1| = 1$

61. $d(D,E) = d(1,3) = |3-1| = |2| = 2$

63. $d(A,E) = d(-3,3) = |3-(-3)| = |6| = 6$

65. $(-4)^2 = (-4)(-4) = 16$

67. $4^{-2} = \dfrac{1}{4^2} = \dfrac{1}{16}$

69. $3^{-6} \cdot 3^4 = 3^{-6+4} = 3^{-2} = \dfrac{1}{3^2} = \dfrac{1}{9}$

71. $\left(3^{-2}\right)^{-1} = 3^{(-2)(-1)} = 3^2 = 9$

73. $\sqrt{25} = \sqrt{5^2} = |5| = 5$

75. $\sqrt{(-4)^2} = |-4| = 4$

77. $\left(8x^3\right)^{-2} = \dfrac{1}{\left(8x^3\right)^2} = \dfrac{1}{8^2 \cdot x^6} = \dfrac{1}{64x^6}$

79. $\left(x^2 y^{-1}\right)^2 = \left(\dfrac{x^2}{y}\right)^2 = \dfrac{x^{2\cdot2}}{y^{1\cdot2}} = \dfrac{x^4}{y^2}$

81. $\dfrac{x^{-2} y^3}{x y^4} = \dfrac{x^{-2}}{x} \cdot \dfrac{y^3}{y^4} = x^{-2-1} y^{3-4} = x^{-3} y^{-1} = \dfrac{1}{x^3} \cdot \dfrac{1}{y} = \dfrac{1}{x^3 y}$

83. $\dfrac{(-2)^3 x^4 (yz)^2}{3^2 x y^3 z^4} = \dfrac{-8 x^4 y^2 z^2}{9 x y^3 z^4} = \dfrac{-8}{9} x^{4-1} y^{2-3} z^{2-4} = \dfrac{-8}{9} x^3 y^{-1} z^{-2} = \dfrac{-8}{9} x^3 \cdot \dfrac{1}{y} \cdot \dfrac{1}{z^2} = -\dfrac{8x^3}{9 y z^2}$

85. $\left(\dfrac{3x^{-1}}{4y^{-1}}\right)^{-2} = \left(\dfrac{3y^1}{4x^1}\right)^{-2} = \dfrac{1}{\left(\dfrac{3y^1}{4x^1}\right)^2} = \dfrac{1}{\left(\dfrac{3^2 y^{1\cdot2}}{4^2 x^{1\cdot2}}\right)} = \dfrac{1}{\left(\dfrac{9y^2}{16x^2}\right)} = \dfrac{16x^2}{9y^2}$

87.

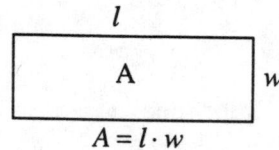

$A = l \cdot w$

89.

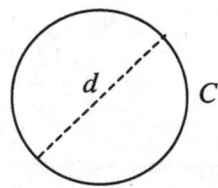

$C = \pi \cdot d$

91.

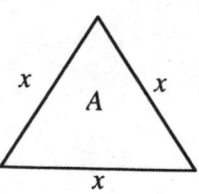

$A = \dfrac{\sqrt{3}}{4} \cdot x^2$

93.

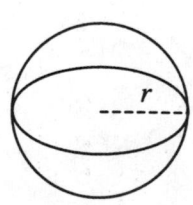

$V = \dfrac{4}{3} \cdot \pi \cdot r^3$

95.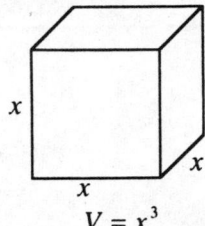

$V = x^3$

97. $|x - 115| \le 5$

   (a) $|x - 115| = |113 - 115| = |-2| = 2 \le 5$ 113 volts is acceptable.

   (b) $|x - 115| = |109 - 115| = |-6| = 6 \nleq 5$ 109 volts is not acceptable.

Appendix Review

A.2 Geometry Review

1. $a = 5, \ b = 12, \quad c^2 = a^2 + b^2 = 5^2 + 12^2 = 25 + 144 = 169 \ \rightarrow \ c = 13$

3. $a = 10, \ b = 24, \quad c^2 = a^2 + b^2 = 10^2 + 24^2 = 100 + 576 = 676 \ \rightarrow \ c = 26$

5. $a = 7, \ b = 24, \quad c^2 = a^2 + b^2 = 7^2 + 24^2 = 49 + 576 = 625 \ \rightarrow \ c = 25$

7. $5^2 = 3^2 + 4^2 \ \rightarrow \ 25 = 9 + 16 \ \rightarrow \ 25 = 25$
The given triangle is a right triangle. The hypotenuse is 5.

9. $6^2 = 4^2 + 5^2 \ \rightarrow \ 36 = 16 + 25 \ \rightarrow \ 36 \neq 41$
The given triangle is not a right triangle.

11. $25^2 = 7^2 + 24^2 \ \rightarrow \ 625 = 49 + 576 \ \rightarrow \ 625 = 625$
The given triangle is a right triangle. The hypotenuse is 25.

13. $6^2 = 3^2 + 4^2 \ \rightarrow \ 36 = 9 + 16 \ \rightarrow \ 36 \neq 25$
The given triangle is not a right triangle.

15. $A = l \cdot w = 4 \cdot 2 = 8 \text{ in}^2$

17. $A = \dfrac{1}{2} b \cdot h = \dfrac{1}{2}(2)(4) = 4 \text{ in}^2$

19. $A = \pi r^2 = \pi (5)^2 = 25\pi \text{ m}^2 \qquad C = 2\pi r = 2\pi(5) = 10\pi \text{ m}$

21. $V = l w h = 8 \cdot 4 \cdot 7 = 224 \text{ ft}^2$

23. $V = \dfrac{4}{3} \pi r^3 = \dfrac{4}{3} \pi \cdot 4^3 = \dfrac{256}{3} \pi \text{ cm}^3 \qquad S = 4\pi r^2 = 4\pi \cdot 4^2 = 64\pi \text{ cm}^2$

25. $V = \pi r^2 h = \pi(9)^2 (8) = 648\pi \text{ in}^3$

27. The diameter of the circle is 2, so its radius is 1. $A = \pi r^2 = \pi (1)^2 = \pi$ square units

29. The diameter of the circle is the length of the diagonal of the square.
$d^2 = 2^2 + 2^2 = 4 + 4 = 8 \ \rightarrow \ d = \sqrt{8} = 2\sqrt{2} \qquad r = \sqrt{2}$
The area of the circle is: $A = \pi r^2 = \pi\left(\sqrt{2}\right)^2 = 2\pi$ square units

31. The total distance traveled is 4 times the circumference of the wheel.
$$\text{Total distance} = 4C = 4(\pi d) = 4\pi \cdot 16 = 64\pi = 201.1 \text{ inches}$$

33. Area of the border = area of EFGH − area of ABCD $= 10^2 - 6^2 = 100 - 36 = 64 \text{ ft}^2$

35. Area of the window = area of the rectangle + area of the semicircle.
$$A = (6)(4) + \frac{1}{2} \cdot \pi \cdot 2^2 = 24 + 2\pi \approx 30.28 \text{ ft}^2$$
Perimeter of the window = 2 heights + width + one-half the circumference.
$$P = 2(6) + 4 + \frac{1}{2} \cdot \pi(4) = 12 + 4 + 2\pi = 16 + 2\pi \approx 22.28 \text{ feet}$$

37. Convert 20 feet to miles, and solve the Pythagorean theorem to find the distance:

$$20 \text{ feet} = 20 \text{ feet} \cdot \frac{1 \text{ mile}}{5280 \text{ feet}} \approx 0.003788 \text{ miles}$$
$$d^2 = (3960 + 0.003788)^2 - 3960^2 = 30$$
$$d \approx 5.477 \text{ miles}$$

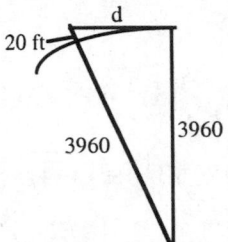

39. Convert 100 feet to miles, and solve the Pythagorean theorem to find the distance:
$$100 \text{ feet} = 100 \text{ feet} \cdot \frac{1 \text{ mile}}{5280 \text{ feet}} = 0.018939 \text{ miles}$$
$$d^2 = (3960 + 0.018939)^2 - 3960^2 = 150$$
$$d \approx 12.247 \text{ miles}$$
Convert 150 feet to miles, and solve the Pythagorean theorem to find the distance:
$$150 \text{ feet} = 150 \text{ feet} \cdot \frac{1 \text{ mile}}{5280 \text{ feet}} \approx 0.028409 \text{ miles}$$
$$d^2 = (3960 + 0.028409)^2 - 3960^2 = 225$$
$$d \approx 15 \text{ miles}$$

41. Given a rectangle with perimeter = 1000 feet, the largest area will be enclosed by a square with dimensions 250 by 250 feet. That is, the area $= 250^2 = 62500$ square feet.

A circular pool with circumference = 1000 feet yields the equation : $2\pi r = 1000 \rightarrow r = \dfrac{500}{\pi}$

The area enclosed by the circular pool is: $A = \pi r^2 = \pi \left(\dfrac{500}{\pi}\right)^2 = \dfrac{500^2}{\pi} \approx 79577.47$ square feet

Therefore, a circular pool will enclose the most area.

Appendix Review

A.3 Polynomials

1. $(x^2 + 4x + 5) + (3x - 3) = x^2 + (4x + 3x) + (5 - 3) = x^2 + 7x + 2$

3. $(x^3 - 2x^2 + 5x + 10) - (2x^2 - 4x + 3) = x^3 - 2x^2 + 5x + 10 - 2x^2 + 4x - 3$
$$= x^3 + (-2x^2 - 2x^2) + (5x + 4x) + (10 - 3)$$
$$= x^3 - 4x^2 + 9x + 7$$

5. $\left(6x^5 + x^3 + x\right) + \left(5x^4 - x^3 + 3x^2\right) = 6x^5 + 5x^4 + 3x^2 + x$

7. $(x^2 - 3x + 1) + 2(3x^2 + x - 4) = x^2 - 3x + 1 + 6x^2 + 2x - 8 = 7x^2 - x - 7$

9. $\left(x^2 - x + 2\right) + \left(2x^2 - 3x + 5\right) - \left(x^2 + 1\right) = x^2 - x + 2 + 2x^2 - 3x + 5 - x^2 - 1$
$$= 2x^2 - 4x + 6$$

11. $(x + 2)(x + 4) = x^2 + 4x + 2x + 8 = x^2 + 6x + 8$

13. $(2x + 5)(x + 2) = 2x^2 + 4x + 5x + 10 = 2x^2 + 9x + 10$

15. $(x - 4)(x + 2) = x^2 + 2x - 4x - 8 = x^2 - 2x - 8$

17. $(x - 3)(x - 2) = x^2 - 2x - 3x + 6 = x^2 - 5x + 6$

19. $(2x + 3)(x - 2) = 2x^2 - 4x + 3x - 6 = 2x^2 - x - 6$

21. $(-x - 2)(-2x - 4) = 2x^2 + 4x + 4x + 8 = 2x^2 + 8x + 8$

23. $(x + 1)(x^2 + 2x - 4) = x^3 + 2x^2 - 4x + x^2 + 2x - 4 = x^3 + 3x^2 - 2x - 4$

25. $(x - 7)(x + 7) = x^2 - 7^2 = x^2 - 49$

26. $(x - 1)(x + 1) = x^2 - 1^2 = x^2 - 1$

27. $(2x + 3)(2x - 3) = (2x)^2 - 3^2 = 4x^2 - 9$

29. $(x + 4)^2 = x^2 + 2 \cdot x \cdot 4 + 4^2 = x^2 + 8x + 16$

31. $(x-4)^2 = x^2 - 2 \cdot x \cdot 4 + 4^2 = x^2 - 8x + 16$

33. $(2x-3)(2x-3) = (2x)^2 + 2(2x)(-3) + (-3)^2 = 4x^2 - 12x + 9$

35. $(x-2)^3 = x^3 - 3 \cdot x^2 \cdot 2 + 3 \cdot x \cdot 2^2 - 2^3 = x^3 - 6x^2 + 12x - 8$

37. $(2x+1)^3 = (2x)^3 + 3(2x)^2(1) + 3(2x) \cdot 1^2 + 1^3 = 8x^3 + 12x^2 + 6x + 1$

39. $x^2 - 36 = (x-6)(x+6)$

41. $2 - 8x^2 = 2(1 - 4x^2) = 2(1 - 2x)(1 + 2x)$

43. $x^2 + 7x + 10 = (x+2)(x+5)$ 45. $x^2 - 10x + 21 = (x-7)(x-3)$

47. $4x^2 - 8x + 32 = 4(x^2 - 2x + 8)$

49. $x^2 + 4x + 16$ is prime because there are no factors of 16 whose sum is 4.

51. $15 + 2x - x^2 = -(x^2 - 2x - 15) = -(x-5)(x+3)$

53. $3x^2 - 12x - 36 = 3(x^2 - 4x - 12) = 3(x-6)(x+2)$

55. $y^4 + 11y^3 + 30y^2 = y^2(y^2 + 11y + 30) = y^2(y+5)(y+6)$

57. $4x^2 + 12x + 9 = (2x+3)^2$

59. $6x^2 + 8x + 2 = 2(3x^2 + 4x + 1) = 2(3x+1)(x+1)$

61. $x^4 - 81 = (x^2 - 9)(x^2 + 9) = (x-3)(x+3)(x^2 + 9)$

63. $x^6 - 2x^3 + 1 = (x^3 - 1)^2 = [(x-1)(x^2 + x + 1)]^2 = (x-1)^2(x^2 + x + 1)^2$

65. $x^7 - x^5 = x^5(x^2 - 1) = x^5(x-1)(x+1)$

66. $x^8 - x^5 = x^5(x^3 - 1) = x^5(x-1)(x^2 + x + 1)$

67. $16x^2 + 24x + 9 = (4x+3)^2$

69. $5 + 16x - 16x^2 = -(16x^2 - 16x - 5) = -(4x-5)(4x+1)$

71. $4y^2 - 16y + 15 = (2y-5)(2y-3)$

73. $1 - 8x^2 - 9x^4 = -(9x^4 + 8x^2 - 1) = -(9x^2 - 1)(x^2 + 1) = -(3x-1)(3x+1)(x^2 + 1)$

75. $x(x+3) - 6(x+3) = (x+3)(x-6)$

77. $(x+2)^2 - 5(x+2) = (x+2)[(x+2)-5] = (x+2)(x-3)$

79. $(3x-2)^3 - 27 = [(3x-2)-3][(3x-2)^2 + 3(3x-2)+9]$
$$= (3x-5)(9x^2 - 12x + 4 + 9x - 6 + 9) = (3x-5)(9x^2 - 3x + 7)$$

81. $3(x^2 + 10x + 25) - 4(x+5) = 3(x+5)^2 - 4(x+5)$
$$= (x+5)[3(x+5)-4] = (x+5)(3x+15-4) = (x+5)(3x+11)$$

83. $x^3 + 2x^2 - x - 2 = x^2(x+2) - (x+2) = (x+2)(x^2 - 1) = (x+2)(x-1)(x+1)$

85. $x^4 - x^3 + x - 1 = x^3(x-1) + (x-1) = (x-1)(x^3 + 1) = (x-1)(x+1)(x^2 - x + 1)$

87. Factors of 4   1, 4   2, 2   −1, −4   −2, −2
     Sum   5   4   −5   −4
     None of the sums of the factors is 0, so $x^2 + 4$ is prime.

89. Answers will vary.

Appendix Review

A.4   Rational Expressions

1. $\dfrac{3x+9}{x^2-9}=\dfrac{3(x+3)}{(x-3)(x+3)}=\dfrac{3}{x-3}$

3. $\dfrac{x^2-2x}{3x-6}=\dfrac{x(x-2)}{3(x-2)}=\dfrac{x}{3}$

5. $\dfrac{24x^2}{12x^2-6x}=\dfrac{24x^2}{6x(2x-1)}=\dfrac{4x}{2x-1}$

7. $\dfrac{y^2-25}{2y^2-8y-10}=\dfrac{(y+5)(y-5)}{2(y^2-4y-5)}=\dfrac{(y+5)(y-5)}{2(y-5)(y+1)}=\dfrac{y+5}{2(y+1)}$

9. $\dfrac{x^2+4x-5}{x^2-2x+1}=\dfrac{(x+5)(x-1)}{(x-1)(x-1)}=\dfrac{x+5}{x-1}$

11. $\dfrac{x^2+5x-14}{2-x}=\dfrac{(x+7)(x-2)}{2-x}=\dfrac{(x+7)(x-2)}{(-1)(-2+x)}=\dfrac{(x+7)(x-2)}{(-1)(x-2)}=-(x+7)$

13. $\dfrac{3x+6}{5x^2}\cdot\dfrac{x}{x^2-4}=\dfrac{3(x+2)}{5x^2}\cdot\dfrac{x}{(x-2)(x+2)}=\dfrac{3}{5x(x-2)}$

15. $\dfrac{4x^2}{x^2-16}\cdot\dfrac{x-4}{2x}=\dfrac{4x^2}{(x-4)(x+4)}\cdot\dfrac{x-4}{2x}=\dfrac{2x}{x+4}$

17. $\dfrac{4x-8}{-3x}\cdot\dfrac{12}{12-6x}=\dfrac{4(x-2)}{-3x}\cdot\dfrac{12}{6(2-x)}=\dfrac{4(x-2)}{-3x}\cdot\dfrac{2}{(-1)(x-2)}=\dfrac{8}{3x}$

19. $\dfrac{x^2-3x-10}{x^2+2x-35}\cdot\dfrac{x^2+4x-21}{x^2+9x+14}=\dfrac{(x-5)(x+2)}{(x+7)(x-5)}\cdot\dfrac{(x+7)(x-3)}{(x+7)(x+2)}=\dfrac{x-3}{x+7}$

21. $\dfrac{\left(\dfrac{6x}{x^2-4}\right)}{\left(\dfrac{3x-9}{2x+4}\right)}=\dfrac{6x}{x^2-4}\cdot\dfrac{2x+4}{3x-9}=\dfrac{6x}{(x-2)(x+2)}\cdot\dfrac{2(x+2)}{3(x-3)}=\dfrac{4x}{(x-2)(x-3)}$

23. $\dfrac{\left(\dfrac{8x}{x^2-1}\right)}{\left(\dfrac{10}{x+1}\right)}=\dfrac{8x}{x^2-1}\cdot\dfrac{x+1}{10}=\dfrac{8x}{(x-1)(x+1)}\cdot\dfrac{x+1}{10}=\dfrac{4x}{5(x-1)}$

Appendix Review

25. $\dfrac{\left(\dfrac{4-x}{4+x}\right)}{\left(\dfrac{4x}{x^2-16}\right)} = \dfrac{4-x}{4+x} \cdot \dfrac{x^2-16}{4x} = \dfrac{4-x}{4+x} \cdot \dfrac{(x+4)(x-4)}{4x} = \dfrac{(4-x)(x-4)}{4x} = -\dfrac{(x-4)^2}{4x}$

27. $\dfrac{\left(\dfrac{x^2+7x+12}{x^2-7x+12}\right)}{\left(\dfrac{x^2+x-12}{x^2-x-12}\right)} = \dfrac{x^2+7x+12}{x^2-7x+12} \cdot \dfrac{x^2-x-12}{x^2+x-12} = \dfrac{(x+3)(x+4)}{(x-3)(x-4)} \cdot \dfrac{(x-4)(x+3)}{(x+4)(x-3)} = \dfrac{(x+3)^2}{(x-3)^2}$

29. $\dfrac{\left(\dfrac{2x^2-x-28}{3x^2-x-2}\right)}{\left(\dfrac{4x^2+16x+7}{3x^2+11x+6}\right)} = \dfrac{2x^2-x-28}{3x^2-x-2} \cdot \dfrac{3x^2+11x+6}{4x^2+16x+7} = \dfrac{(2x+7)(x-4)}{(3x+2)(x-1)} \cdot \dfrac{(3x+2)(x+3)}{(2x+7)(2x+1)}$

$= \dfrac{(x-4)(x+3)}{(x-1)(2x+1)}$

31. $\dfrac{x}{2} + \dfrac{5}{2} = \dfrac{x+5}{2}$

33. $\dfrac{x^2}{2x-3} - \dfrac{4}{2x-3} = \dfrac{x^2-4}{2x-3} = \dfrac{(x+2)(x-2)}{2x-3}$

35. $\dfrac{x+1}{x-3} + \dfrac{2x-3}{x-3} = \dfrac{x+1+2x-3}{x-3} = \dfrac{3x-2}{x-3}$

37. $\dfrac{3x+5}{2x-1} - \dfrac{2x-4}{2x-1} = \dfrac{(3x+5)-(2x-4)}{2x-1} = \dfrac{3x+5-2x+4}{2x-1} = \dfrac{x+9}{2x-1}$

39. $\dfrac{4}{x-2} + \dfrac{x}{2-x} = \dfrac{4}{x-2} - \dfrac{x}{x-2} = \dfrac{4-x}{x-2}$

41. $\dfrac{4}{x-1} - \dfrac{2}{x+2} = \dfrac{4(x+2)}{(x-1)(x+2)} - \dfrac{2(x-1)}{(x+2)(x-1)} = \dfrac{4x+8-2x+2}{(x+2)(x-1)} = \dfrac{2x+10}{(x+2)(x-1)}$

$= \dfrac{2(x+5)}{(x+2)(x-1)}$

43. $\dfrac{x}{x+1} + \dfrac{2x-3}{x-1} = \dfrac{x(x-1)}{(x+1)(x-1)} + \dfrac{(2x-3)(x+1)}{(x-1)(x+1)} = \dfrac{x^2-x+2x^2-x-3}{(x-1)(x+1)}$

$= \dfrac{3x^2-2x-3}{(x-1)(x+1)}$

45. $\dfrac{x-3}{x+2} - \dfrac{x+4}{x-2} = \dfrac{(x-3)(x-2)}{(x+2)(x-2)} - \dfrac{(x+4)(x+2)}{(x-2)(x+2)} = \dfrac{x^2-5x+6-(x^2+6x+8)}{(x+2)(x-2)}$

$= \dfrac{x^2-5x+6-x^2-6x-8}{(x+2)(x-2)} = \dfrac{-11x-2}{(x+2)(x-2)}$

47. $\dfrac{x}{x^2-4}+\dfrac{1}{x}=\dfrac{x^2+x^2-4}{(x)(x^2-4)}=\dfrac{2x^2-4}{(x)(x^2-4)}=\dfrac{2(x^2-2)}{(x)(x-2)(x+2)}$

49. $x^2-4=(x+2)(x-2)$
$x^2-x-2=(x+1)(x-2)$
$\therefore$ LCM is $(x+2)(x-2)(x+1)$

51. $x^3-x=x(x^2-1)=x(x+1)(x-1)$
$x^2-x=x(x-1)$
$\therefore$ LCM is $x(x+1)(x-1)$

53. $4x^3-4x^2+x=x(4x^2-4x+1)=x(2x-1)(2x-1)$
$2x^3-x^2=x^2(2x-1)$
x^3
$\therefore$ LCM is $x^3(2x-1)^2$

55. $x^3-x=x(x^2-1)=x(x+1)(x-1)$
$x^3-2x^2+x=x(x^2-2x+1)=x(x-1)^2$
$x^3-1=(x-1)(x^2+x+1)$
$\therefore$ LCM is $x(x+1)(x-1)^2(x^2+x+1)$

57. $\dfrac{x}{x^2-7x+6}-\dfrac{x}{x^2-2x-24}=\dfrac{x}{(x-6)(x-1)}-\dfrac{x}{(x-6)(x+4)}$
$=\dfrac{x(x+4)}{(x-6)(x-1)(x+4)}-\dfrac{x(x-1)}{(x-6)(x+4)(x-1)}$
$=\dfrac{x^2+4x-x^2+x}{(x-6)(x+4)(x-1)}=\dfrac{5x}{(x-6)(x+4)(x-1)}$

59. $\dfrac{4x}{x^2-4}-\dfrac{2}{x^2+x-6}=\dfrac{4x}{(x-2)(x+2)}-\dfrac{2}{(x+3)(x-2)}$
$=\dfrac{4x(x+3)}{(x-2)(x+2)(x+3)}-\dfrac{2(x+2)}{(x+3)(x-2)(x+2)}$
$=\dfrac{4x^2+12x-2x-4}{(x-2)(x+2)(x+3)}=\dfrac{4x^2+10x-4}{(x-2)(x+2)(x+3)}$
$=\dfrac{2(2x^2+5x-2)}{(x-2)(x+2)(x+3)}$

61. $\dfrac{3}{(x-1)^2(x+1)}+\dfrac{2}{(x-1)(x+1)^2}=\dfrac{3(x+1)+2(x-1)}{(x-1)^2(x+1)^2}=\dfrac{3x+3+2x-2}{(x-1)^2(x+1)^2}$
$=\dfrac{5x+1}{(x-1)^2(x+1)^2}$

63. $\dfrac{x+4}{x^2-x-2}-\dfrac{2x+3}{x^2+2x-8}=\dfrac{x+4}{(x-2)(x+1)}-\dfrac{2x+3}{(x+4)(x-2)}$

$$=\dfrac{(x+4)(x+4)}{(x-2)(x+1)(x+4)}-\dfrac{(2x+3)(x+1)}{(x+4)(x-2)(x+1)}$$

$$=\dfrac{x^2+8x+16-(2x^2+5x+3)}{(x-2)(x+1)(x+4)}=\dfrac{-x^2+3x+13}{(x-2)(x+1)(x+4)}$$

65. $\dfrac{1}{x}-\dfrac{2}{x^2+x}+\dfrac{3}{x^3-x^2}=\dfrac{1}{x}-\dfrac{2}{x(x+1)}+\dfrac{3}{x^2(x-1)}=\dfrac{x(x+1)(x-1)-2x(x-1)+3(x+1)}{x^2(x+1)(x-1)}$

$$=\dfrac{x(x^2-1)-2x^2+2x+3x+3}{x^2(x+1)(x-1)}=\dfrac{x^3-x-2x^2+5x+3}{x^2(x+1)(x-1)}=\dfrac{x^3-2x^2+4x+3}{x^2(x+1)(x-1)}$$

67. $\dfrac{1}{h}\left(\dfrac{1}{x+h}-\dfrac{1}{x}\right)=\dfrac{1}{h}\left(\dfrac{1\cdot x}{(x+h)x}-\dfrac{1(x+h)}{x(x+h)}\right)=\dfrac{1}{h}\left(\dfrac{x-x-h}{x(x+h)}\right)=\dfrac{-h}{hx(x+h)}=\dfrac{-1}{x(x+h)}$

69. $\dfrac{1+\dfrac{1}{x}}{1-\dfrac{1}{x}}=\dfrac{\left(\dfrac{x}{x}+\dfrac{1}{x}\right)}{\left(\dfrac{x}{x}-\dfrac{1}{x}\right)}=\dfrac{\left(\dfrac{x+1}{x}\right)}{\left(\dfrac{x-1}{x}\right)}=\dfrac{x+1}{x}\cdot\dfrac{x}{x-1}=\dfrac{x+1}{x-1}$

71. $\dfrac{x-\dfrac{1}{x}}{x+\dfrac{1}{x}}=\dfrac{\left(\dfrac{x^2}{x}-\dfrac{1}{x}\right)}{\left(\dfrac{x^2}{x}+\dfrac{1}{x}\right)}=\dfrac{\left(\dfrac{x^2-1}{x}\right)}{\left(\dfrac{x^2+1}{x}\right)}=\dfrac{x^2-1}{x}\cdot\dfrac{x}{x^2+1}=\dfrac{(x-1)(x+1)}{x^2+1}$

73. $\dfrac{\left(\dfrac{x+4}{x-2}-\dfrac{x-3}{x+1}\right)}{x+1}=\dfrac{\left(\dfrac{(x+4)(x+1)}{(x-2)(x+1)}-\dfrac{(x-3)(x-2)}{(x+1)(x-2)}\right)}{x+1}=\dfrac{\left(\dfrac{x^2+5x+4-(x^2-5x+6)}{(x-2)(x+1)}\right)}{x+1}$

$$=\dfrac{10x-2}{(x-2)(x+1)}\cdot\dfrac{1}{x+1}=\dfrac{2(5x-1)}{(x-2)(x+1)^2}$$

75. $\dfrac{\left(\dfrac{x-2}{x+2}+\dfrac{x-1}{x+1}\right)}{\left(\dfrac{x}{x+1}-\dfrac{2x-3}{x}\right)}=\dfrac{\left(\dfrac{(x-2)(x+1)}{(x+2)(x+1)}+\dfrac{(x-1)(x+2)}{(x+1)(x+2)}\right)}{\left(\dfrac{x^2}{(x+1)(x)}-\dfrac{(2x-3)(x+1)}{x(x+1)}\right)}=\dfrac{\left(\dfrac{x^2-x-2+x^2+x-2}{(x+2)(x+1)}\right)}{\left(\dfrac{x^2-(2x^2-x-3)}{x(x+1)}\right)}$

$$=\dfrac{\left(\dfrac{2x^2-4}{(x+2)(x+1)}\right)}{\left(\dfrac{-x^2+x+3}{x(x+1)}\right)}=\dfrac{2(x^2-2)}{(x+2)(x+1)}\cdot\dfrac{x(x+1)}{-(x^2-x-3)}=\dfrac{2x(x^2-2)}{-(x+2)(x^2-x-3)}=\dfrac{-2x(x^2-2)}{(x+2)(x^2-x-3)}$$

77. $1 - \dfrac{1}{\left(1 - \dfrac{1}{x}\right)} = 1 - \dfrac{1}{\left(\dfrac{x-1}{x}\right)} = 1 - 1 \cdot \dfrac{x}{x-1} = \dfrac{x-1-x}{x-1} = \dfrac{-1}{x-1}$

79. $\dfrac{1}{f} = (n-1)\left(\dfrac{1}{R_1} + \dfrac{1}{R_2}\right)$

$\dfrac{R_1 \cdot R_2}{f} = (n-1)\left(\dfrac{1}{R_1} + \dfrac{1}{R_2}\right)R_1 \cdot R_2 \rightarrow \dfrac{R_1 \cdot R_2}{f} = (n-1)(R_2 + R_1)$

$\dfrac{f}{R_1 \cdot R_2} = \dfrac{1}{(n-1)(R_2 + R_1)} \rightarrow f = \dfrac{R_1 \cdot R_2}{(n-1)(R_2 + R_1)}$

$f = \dfrac{0.1(0.2)}{(1.5-1)(0.2+0.1)} = \dfrac{0.02}{0.5(0.3)} = \dfrac{0.02}{0.15} = \dfrac{2}{15}$ meters

81. $1 + \dfrac{1}{x} = \dfrac{x+1}{x} \Rightarrow a = 1, b = 1, c = 0$

$1 + \dfrac{1}{1 + \dfrac{1}{x}} = 1 + \dfrac{1}{\left(\dfrac{x+1}{x}\right)} = 1 + \dfrac{x}{x+1} = \dfrac{x+1+x}{x+1} = \dfrac{2x+1}{x+1} \Rightarrow a = 2, b = 1, c = 1$

$1 + \dfrac{1}{1 + \dfrac{1}{1 + \dfrac{1}{x}}} = 1 + \dfrac{1}{\left(\dfrac{2x+1}{x+1}\right)} = 1 + \dfrac{x+1}{2x+1} = \dfrac{2x+1+x+1}{2x+1} = \dfrac{3x+2}{2x+1} \Rightarrow a = 3, b = 2, c = 1$

$1 + \dfrac{1}{1 + \dfrac{1}{1 + \dfrac{1}{1 + \dfrac{1}{x}}}} = 1 + \dfrac{1}{\left(\dfrac{3x+2}{2x+1}\right)} = 1 + \dfrac{2x+1}{3x+2} = \dfrac{3x+2+2x+1}{3x+2} = \dfrac{5x+3}{3x+2} \Rightarrow a = 5, b = 3, c = 2$

If we continue this process, the values of a, b and c produce the following sequences:

$a : 1, 2, 3, 5, 8, 13, 21, ...$

$b : 1, 1, 2, 3, 5, 8, 13, 21, ...$

$c : 0, 1, 1, 2, 3, 5, 8, 13, 21, ...$

In each case we have the *Fibonacci Sequence*, where the next value in the list is obtained from the sum of the previous 2 values in the list.

83. Answers will vary.

Appendix Review

A.5 Polynomial Division; Synthetic Division

1. Divide:

$$\begin{array}{r} 4x^2 - 3x + 1 \\ x\overline{\smash{\big)}4x^3 - 3x^2 + x + 1} \\ \underline{4x^3} \\ -3x^2 + x + 1 \\ \underline{-3x^2} \\ x + 1 \\ \underline{x} \\ 1 \end{array}$$

Check:

$(x)(4x^2 - 3x + 1) + (1) = 4x^3 - 3x^2 + x + 1$

The quotient is $4x^2 - 3x + 1$; the remainder is 1.

3. Divide:

$$\begin{array}{r} 4x^2 + 13x + 53 \\ x - 4\overline{\smash{\big)}4x^3 - 3x^2 + x + 1} \\ \underline{4x^3 - 16x^2} \\ 13x^2 + x \\ \underline{13x^2 - 52x} \\ 53x + 1 \\ \underline{53x - 212} \\ 213 \end{array}$$

Check:

$(x - 4)(4x^2 + 13x + 53) + 213$
$= 4x^3 + 13x^2 + 53x - 16x^2 - 52x - 212 + 213$
$= 4x^3 - 3x^2 + x + 1$

The quotient is $4x^2 + 13x + 53$; the remainder is 213.

5. Divide:

$$\begin{array}{r} 4x - 3 \\ x^2 + 2\overline{\smash{\big)}4x^3 - 3x^2 + x + 1} \\ \underline{4x^3 + 8x} \\ -3x^2 - 7x \\ \underline{-3x^2 - 6} \\ -7x + 7 \end{array}$$

Check:

$(x^2 + 2)(4x - 3) + (-7x + 7)$
$= 4x^3 - 3x^2 + 8x - 6 - 7x + 7$
$= 4x^3 - 3x^2 + x + 1$

The quotient is $4x - 3$; the remainder is $-7x + 7$.

7. Divide:

$$2x^3 - 1\overline{)4x^3 - 3x^2 + x + 1}$$
$$\underline{4x^3 \qquad\qquad -2}$$
$$-3x^2 + x + 3$$

Check :

$$(2x^3 - 1)(2) + (-3x^2 + x + 3)$$
$$= 4x^3 - 2 - 3x^2 + x + 3 = 4x^3 - 3x^2 + x + 1$$

The quotient is 2; the remainder is $-3x^2 + x + 3$.

9. Divide:

$$2x - \frac{5}{2}$$
$$2x^2 + x + 1\overline{)4x^3 - 3x^2 + x + 1}$$
$$\underline{4x^3 + 2x^2 + 2x}$$
$$-5x^2 - x$$
$$\underline{-5x^2 - \frac{5}{3}x - \frac{5}{2}}$$
$$\frac{3}{2}x + \frac{7}{2}$$

Check :

$$\left(2x^2 + x + 12\right)\left(x - \frac{5}{2}\right) + \left(\frac{3}{2}x + \frac{7}{2}\right)$$
$$= 4x^3 - 5x^2 + 2x^2 - \frac{5}{2}x + 2x - \frac{5}{2} + \frac{3}{2}x + \frac{7}{2} = 4x^3 - 3x^2 + x + 1$$

The quotient is $2x - \frac{5}{2}$; the remainder is $\frac{3}{2}x + \frac{7}{2}$.

11. Divide:

$$x - \frac{3}{4}$$
$$4x^2 + 1\overline{)4x^3 - 3x^2 + x + 1}$$
$$\underline{4x^3 \qquad\quad + x}$$
$$-3x^2 \qquad +1$$
$$\underline{-3x^2 \qquad - \frac{3}{4}}$$
$$\frac{7}{4}$$

Check :

$$(4x^2 + 1)\left(x - \frac{3}{4}\right) + \frac{7}{4}$$
$$= 4x^3 - 3x^2 + x - \frac{3}{4} + \frac{7}{4}$$
$$= 4x^3 - 3x^2 + x + 1$$

The quotient is $x - \frac{3}{4}$; the remainder is $\frac{7}{4}$.

13. Divide:

$$x^3 + x^2 + x + 1$$
$$x - 1\overline{)x^4 + 0x^3 + 0x^2 + 0x - 1}$$
$$\underline{x^4 - x^3}$$
$$x^3$$
$$\underline{x^3 - x^2}$$
$$x^2$$
$$\underline{x^2 - x}$$
$$x - 1$$
$$\underline{x - 1}$$
$$0$$

Check:

$$(x - 1)(x^3 + x^2 + x + 1) + 0$$
$$= x^4 + x^3 + x^2 + x - x^3 - x^2 - x - 1$$
$$= x^4 - 1$$

The quotient is $x^3 + x^2 + x + 1$; the remainder is 0.

15. Divide:

$$x^2 - 1 \overline{)\begin{array}{l} x^2 + 1 \\ x^4 + 0x^3 + 0x^2 + 0x - 1 \end{array}}$$

$$\begin{array}{r} x^4 \quad\;\; - \; x^2 \\ \hline x^2 \\ x^2 \quad\;\; - \; 1 \\ \hline 0 \end{array}$$

Check:

$(x^2 - 1)(x^2 + 1) + 0$

$= x^4 + x^2 - x^2 - 1$

$= x^4 - 1$

The quotient is $x^2 + 1$; the remainder is 0.

17. Divide:

$$x - 1 \overline{)\begin{array}{l} -4x^2 - 3x - 3 \\ -4x^3 + \;\; x^2 + 0x - 4 \end{array}}$$

$$\begin{array}{r} -4x^3 + 4x^2 \\ \hline -3x^2 \\ -3x^2 + 3x \\ \hline -3x - 4 \\ -3x + 3 \\ \hline -7 \end{array}$$

Check:

$(x - 1)(-4x^2 - 3x - 3) + (-7)$

$= -4x^3 - 3x^2 - 3x + 4x^2 + 3x + 3 - 7$

$= -4x^3 + x^2 - 4$

The quotient is $-4x^2 - 3x - 3$; the remainder is -7.

19. Divide:

$$x^2 + x + 1 \overline{)\begin{array}{l} x^2 - x - 1 \\ x^4 + 0x^3 - \;\; x^2 + 0x + 1 \end{array}}$$

$$\begin{array}{r} x^4 + \;\; x^3 + \;\; x^2 \\ \hline -x^3 - 2x^2 \\ -x^3 - \;\; x^2 - x \\ \hline -x^2 + x + 1 \\ -x^2 - x - 1 \\ \hline 2x + 2 \end{array}$$

Check:

$(x^2 + x + 1)(x^2 - x - 1) + 2x + 2$

$= x^4 + x^3 + x^2 - x^3 - x^2 - x - x^2 - x - 1 + 2x + 2$

$= x^4 - x^2 + 1$

The quotient is $x^2 - x - 1$; the remainder is $2x + 2$.

21. Divide:

$$-x^2 + 1 \overline{)\begin{array}{l} -x^2 \\ x^4 + 0x^3 - x^2 + 0x + 1 \end{array}}$$

$$\begin{array}{r} x^4 \quad\;\; - \; x^2 \\ \hline 1 \end{array}$$

Check:

$-x^2(-x^2 + 1) + 1$

$= x^4 - x^2 + 1$

The quotient is $-x^2$; the remainder is 1.

23. Divide:

$$\begin{array}{r} x^2 + ax + a^2 \\ x-a\overline{)x^3 + 0x^2 + 0x - a^3} \\ \underline{x^3 - ax^2} \\ ax^2 \\ \underline{ax^2 - a^2x} \\ a^2x - a^3 \\ \underline{a^2x - a^3} \\ 0 \end{array}$$

Check:

$(x-a)(x^2 + ax + a^2) + 0$
$= x^3 + ax^2 + a^2x - ax^2 - a^2x - a^3$
$= x^3 - a^3$

The quotient is $x^2 + ax + a^2$; the remainder is 0.

25. Divide:

$$\begin{array}{r} x^3 + ax^2 + a^2x + a^3 \\ x-a\overline{)x^4 + 0x^3 + 0x^2 + 0x - a^4} \\ \underline{x^4 - ax^3} \\ ax^3 \\ \underline{ax^3 - a^2x^2} \\ a^2x^2 \\ \underline{a^2x^2 - a^3x} \\ a^3x - a^4 \\ \underline{a^3x - a^4} \\ 0 \end{array}$$

Check:

$(x-a)(x^3 + ax^2 + a^2x + a^3) + 0$
$= x^4 + ax^3 + a^2x^2 + a^3x - ax^3 - a^2x^2 - a^3x - a^4$
$= x^4 - a^4$

The quotient is $x^3 + ax^2 + a^2x + a^3$; the remainder is 0.

27. Use synthetic division:

$$\begin{array}{r} 2\overline{)1 \quad -1 \quad 2 \quad 4} \\ \underline{2 \quad 2 \quad 8} \\ 1 \quad 1 \quad 4 \quad 12 \end{array}$$

Quotient: $x^2 + x + 4$ Remainder: 12

29. Use synthetic division:

$$\begin{array}{r} 3\overline{)3 \quad 2 \quad -1 \quad 3} \\ \underline{9 \quad 33 \quad 96} \\ 3 \quad 11 \quad 32 \quad 99 \end{array}$$

Quotient: $3x^2 + 11x + 32$ Remainder: 99

31. Use synthetic division:

$$\begin{array}{r} -3\overline{)1 \quad 0 \quad -4 \quad 0 \quad 1 \quad 0} \\ \underline{-3 \quad 9 \quad -15 \quad 45 \quad -138} \\ 1 \quad -3 \quad 5 \quad -15 \quad 46 \quad -138 \end{array}$$

Quotient: $x^4 - 3x^3 + 5x^2 - 15x + 46$ Remainder: -138

Appendix Review

33. Use synthetic division:

$$\begin{array}{r|rrrrrr} 1) & 4 & 0 & -3 & 0 & 1 & 0 & 5 \\ & & 4 & 4 & 1 & 1 & 2 & 2 \\ \hline & 4 & 4 & 1 & 1 & 2 & 2 & 7 \end{array}$$

Quotient: $4x^5 + 4x^4 + x^3 + x^2 + 2x + 2$ Remainder: 7

35. Use synthetic division:

$$\begin{array}{r|rrrr} -1.1) & 0.1 & 0 & 0.2 & 0 \\ & & -0.11 & 0.121 & -0.3531 \\ \hline & 0.1 & -0.11 & 0.321 & -0.3531 \end{array}$$

Quotient: $0.1x^2 - 0.11x + 0.321$ Remainder: -0.3531

37. Use synthetic division:

$$\begin{array}{r|rrrrr} 1) & 1 & 0 & 0 & 0 & 0 & -1 \\ & & 1 & 1 & 1 & 1 & 1 \\ \hline & 1 & 1 & 1 & 1 & 1 & 0 \end{array}$$

Quotient: $x^4 + x^3 + x^2 + x + 1$ Remainder: 0

39. Use synthetic division:

$$\begin{array}{r|rrrr} 2) & 4 & -3 & -8 & 4 \\ & & 8 & 10 & 4 \\ \hline & 4 & 5 & 2 & 8 \end{array}$$

Remainder $= 8 \neq 0$; therefore $x - 2$ is not a factor of $f(x)$.

41. Use synthetic division:

$$\begin{array}{r|rrrrr} 2) & 3 & -6 & 0 & -5 & 10 \\ & & 6 & 0 & 0 & -10 \\ \hline & 3 & 0 & 0 & -5 & 0 \end{array}$$

Remainder $= 0$; therefore $x - 2$ is a factor of $f(x)$.

43. Use synthetic division:

$$\begin{array}{r|rrrrrrr} -3) & 3 & 0 & 0 & 82 & 0 & 0 & 27 \\ & & -9 & 27 & -81 & -3 & 9 & -27 \\ \hline & 3 & -9 & 27 & 1 & -3 & 9 & 0 \end{array}$$

Remainder $= 0$; therefore $x + 3$ is a factor of $f(x)$.

45. Use synthetic division:

$$\begin{array}{r|rrrrrrr} -4) & 4 & 0 & -64 & 0 & 1 & 0 & -15 \\ & & -16 & 64 & 0 & 0 & -4 & 16 \\ \hline & 4 & -16 & 0 & 0 & 1 & -4 & 1 \end{array}$$

Remainder $= 1 \neq 0$; therefore $x + 3$ is not a factor of $f(x)$.

47. Use synthetic division:

$$\frac{1}{2} \overline{\big)\ 2 \quad -1 \quad 0 \quad 2 \quad -1}$$
$$\phantom{\frac{1}{2}\big)\ 2\ }\ \ 1 \quad 0 \quad 0 \quad 1$$
$$\phantom{\frac{1}{2}\big)}\ \ 2 \quad 0 \quad 0 \quad 2 \quad 0$$

Remainder $= 0$; therefore $x - \dfrac{1}{2}$ is a factor of $f(x)$.

49. Answers will vary.

Appendix Review

A.6 Solving Equations

1. $x + 2 = 10$
$$x + 2 - 2 = 10 - 2$$
$$x = 8$$

3. $2t - 6 = 4$
$$2t - 6 + 6 = 4 + 6$$
$$2t = 10$$
$$\frac{2t}{2} = \frac{10}{2}$$
$$t = 5$$

5. $3 + 2n = 5n + 7$
$$3 + 2n - 2n = 5n + 7 - 2n$$
$$3 = 3n + 7$$
$$3 - 7 = 3n + 7 - 7$$
$$-4 = 3n$$
$$\frac{-4}{3} = \frac{3n}{3}$$
$$-\frac{4}{3} = n$$

7. $x^2 - 7x + 12 = 0$
$$(x - 4)(x - 3) = 0$$
$$x - 4 = 0 \Rightarrow x = 4$$
$$x - 3 = 0 \Rightarrow x = 3$$
Solution set $\{3, 4\}$.

9. $2x^2 + 5x - 3 = 0$
$$(2x - 1)(x + 3) = 0$$
$$2x + 1 = 0 \Rightarrow x = -\frac{1}{2}$$
$$x + 3 = 0 \Rightarrow x = -3$$
Solution set $\left\{-3, -\frac{1}{2}\right\}$.

11. $x^2 - 9 = 0$
$$(x - 3)(x + 3) = 0$$
$$x - 3 = 0 \Rightarrow x = 3$$
$$x + 3 = 0 \Rightarrow x = -3$$
Solution set $\{-3, 3\}$.

13. $x^3 + x^2 - 20x = 0$
$$x(x^2 + x - 20) = 0 \Rightarrow x(x + 5)(x - 4) = 0$$
$$x = 0$$
$$x + 5 = 0 \Rightarrow x = -5$$
$$x - 4 = 0 \Rightarrow x = 4$$
Solution set $\{-5, 0, 4\}$.

15. $4x^2 = x$
$$4x^2 - x = 0 \Rightarrow x(4x - 1) = 0$$
$$x = 0$$
$$4x - 1 = 0 \Rightarrow x = \frac{1}{4}$$
Solution set $\left\{0, \frac{1}{4}\right\}$.

17. $x^3 = 9x$

$$x^3 - 9x = 0 \Rightarrow x(x^2 - 9) = 0$$

$$x(x-3)(x+3) = 0$$

$$x = 0$$

$$x - 3 = 0 \Rightarrow x = 3$$

$$x + 3 = 0 \Rightarrow x = -3$$

Solution set $\{-3, 0, 3\}$.

19. $x^4 = x^2$

$$x^4 - x^2 = 0 \Rightarrow x^2(x^2 - 1) = 0$$

$$x(x-1)(x+1) = 0$$

$$x = 0$$

$$x - 1 = 0 \Rightarrow x = 1$$

$$x + 1 = 0 \Rightarrow x = -1$$

Solution set $\{-1, 0, 1\}$.

21. $x^3 + x^2 + x + 1 = 0$

$$x^2(x+1) + x + 1 = 0 \Rightarrow (x+1)(x^2+1) = 0$$

$$x + 1 = 0 \Rightarrow x = -1$$

$x^2 + 1 = 0$ has no real solution

Solution set $\{-1\}$.

23. $x^3 - 2x^2 - 4x + 8 = 0$

$$x^2(x-2) - 4(x-2) = 0 \Rightarrow (x-2)(x^2-4) = 0$$

$$(x-2)(x-2)(x+2) = 0$$

$$x - 2 = 0 \Rightarrow x = 2$$

$$x + 2 = 0 \Rightarrow x = -2$$

Solution set $\{-2, 2\}$.

25. $x^2 = 25 \Rightarrow x = \pm\sqrt{25} \Rightarrow x = \pm 5$

The solution set is $\{-5, 5\}$.

27. $(x-1)^2 = 4$

$$x - 1 = \pm\sqrt{4} \Rightarrow x - 1 = \pm 2$$

$$x - 1 = 2 \ \text{ or } \ x - 1 = -2$$

$$\Rightarrow x = 3 \ \text{ or } \ x = -1$$

The solution set is $\{-1, 3\}$.

29. $(2x+3)^2 = 9$

$$2x + 3 = \pm\sqrt{9} \Rightarrow 2x + 3 = \pm 3$$

$$2x + 3 = 3 \ \text{ or } \ 2x + 3 = -3$$

$$\Rightarrow x = 0 \ \text{ or } \ x = -3$$

The solution set is $\{-3, 0\}$.

31. $\left(\dfrac{8}{2}\right)^2 = 4^2 = 16$

33. $\left(\dfrac{\left(-\dfrac{1}{2}\right)}{2}\right)^2 = \left(-\dfrac{1}{4}\right)^2 = \dfrac{1}{16}$

35. $x^2 + 4x = 21$

$$x^2 + 4x + 4 = 21 + 4 \Rightarrow (x+2)^2 = 25$$

$$x + 2 = \pm\sqrt{25} \Rightarrow x + 2 = \pm 5$$

$$x = -2 \pm 5 \Rightarrow x = 3 \ \text{ or } \ x = -7$$

The solution set is $\{-7, 3\}$.

37. $x^2 - \dfrac{1}{2}x - \dfrac{3}{16} = 0$

$x^2 - \dfrac{1}{2}x = \dfrac{3}{16}$

$x^2 - \dfrac{1}{2}x + \dfrac{1}{16} = \dfrac{3}{16} + \dfrac{1}{16}$

$\left(x - \dfrac{1}{4}\right)^2 = \dfrac{1}{4}$

$x - \dfrac{1}{4} = \pm\sqrt{\dfrac{1}{4}}$

$x - \dfrac{1}{4} = \pm\dfrac{1}{2}$

$x = \dfrac{1}{4} \pm \dfrac{1}{2} \Rightarrow x = \dfrac{3}{4}$

or $x = -\dfrac{1}{4}$

The solution set is $\left\{-\dfrac{1}{4}, \dfrac{3}{4}\right\}$.

39. $3x^2 + x - \dfrac{1}{2} = 0$

$x^2 + \dfrac{1}{3}x - \dfrac{1}{6} = 0$

$x^2 + \dfrac{1}{3}x = \dfrac{1}{6}$

$x^2 + \dfrac{1}{3}x + \dfrac{1}{36} = \dfrac{1}{6} + \dfrac{1}{36}$

$\left(x + \dfrac{1}{6}\right)^2 = \dfrac{7}{36}$

$x + \dfrac{1}{6} = \pm\sqrt{\dfrac{7}{36}}$

$x + \dfrac{1}{6} = \pm\dfrac{\sqrt{7}}{6}$

$x = -\dfrac{1}{6} \pm \dfrac{\sqrt{7}}{6}$

The solution set is $\left\{-\dfrac{1}{6} + \dfrac{\sqrt{7}}{6}, -\dfrac{1}{6} - \dfrac{\sqrt{7}}{6}\right\}$.

Appendix Review

A.7 Complex Numbers; Quadratic Equations with a Negative Discriminant

1. $(2 - 3i) + (6 + 8i) = (2 + 6) + (-3 + 8)i = 8 + 5i$

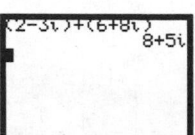

3. $(-3 + 2i) - (4 - 4i) = (-3 - 4) + (2 - (-4))i = -7 + 6i$

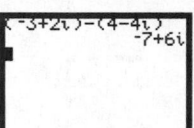

5. $(2 - 5i) - (8 + 6i) = (2 - 8) + (-5 - 6)i = -6 - 11i$

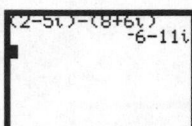

7. $3(2 - 6i) = 6 - 18i$

9. $2i(2 - 3i) = 4i - 6i^2 = 4i - 6(-1) = 6 + 4i$

11. $(3 - 4i)(2 + i) = 6 + 3i - 8i - 4i^2$
$$= 6 - 5i - 4(-1) = 10 - 5i$$

13. $(-6 + i)(-6 - i) = 36 + 6i - 6i - i^2$
$$= 36 - (-1) = 37$$

15. $\dfrac{10}{3 - 4i} = \dfrac{10}{3 - 4i} \cdot \dfrac{3 + 4i}{3 + 4i} = \dfrac{30 + 40i}{9 + 12i - 12i - 16i^2}$
$$= \dfrac{30 + 40i}{9 - 16(-1)} = \dfrac{30 + 40i}{25}$$
$$= \dfrac{30}{25} + \dfrac{40}{25}i = \dfrac{6}{5} + \dfrac{8}{5}i$$

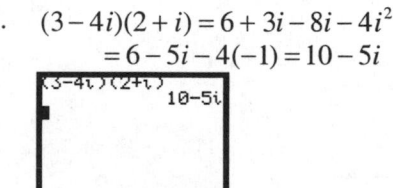

Appendix Review

17. $\dfrac{2+i}{i} = \dfrac{2+i}{i} \cdot \dfrac{-i}{-i} = \dfrac{-2i-i^2}{-i^2}$

 $= \dfrac{-2i-(-1)}{-(-1)} = \dfrac{1-2i}{1} = 1-2i$

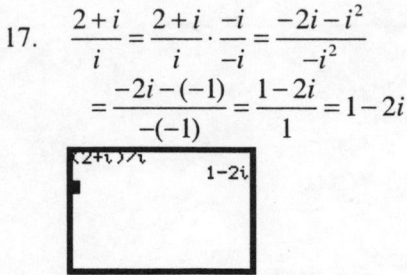

19. $\dfrac{6-i}{1+i} = \dfrac{6-i}{1+i} \cdot \dfrac{1-i}{1-i} = \dfrac{6-6i-i+i^2}{1-i+i-i^2}$

 $= \dfrac{6-7i+(-1)}{1-(-1)} = \dfrac{5-7i}{2} = \dfrac{5}{2} - \dfrac{7}{2}i$

21. $\left(\dfrac{1}{2} + \dfrac{\sqrt{3}}{2}i\right)^2 = \dfrac{1}{4} + 2\left(\dfrac{1}{2}\right)\left(\dfrac{\sqrt{3}}{2}i\right) + \dfrac{3}{4}i^2$

 $= \dfrac{1}{4} + \dfrac{\sqrt{3}}{2}i + \dfrac{3}{4}(-1) = -\dfrac{1}{2} + \dfrac{\sqrt{3}}{2}i$

23. $(1+i)^2 = 1 + 2i + i^2 = 1 + 2i + (-1) = 2i$

25. $i^{23} = i^{22+1} = i^{22} \cdot i = \left(i^2\right)^{11} \cdot i = (-1)^{11}i = -i$

27. $i^{-15} = \dfrac{1}{i^{15}} = \dfrac{1}{i^{14+1}} = \dfrac{1}{i^{14} \cdot i} = \dfrac{1}{\left(i^2\right)^7 \cdot i} = \dfrac{1}{(-1)^7 i} = \dfrac{1}{-i} = \dfrac{1}{-i} \cdot \dfrac{i}{i} = \dfrac{i}{-i^2} = \dfrac{i}{-(-1)} = i$

29. $i^6 - 5 = \left(i^2\right)^3 - 5 = (-1)^3 - 5 = -1 - 5 = -6$

31. $6i^3 - 4i^5 = i^3(6 - 4i^2) = i^2 \cdot i(6 - 4(-1)) = -1 \cdot i(10) = -10i$

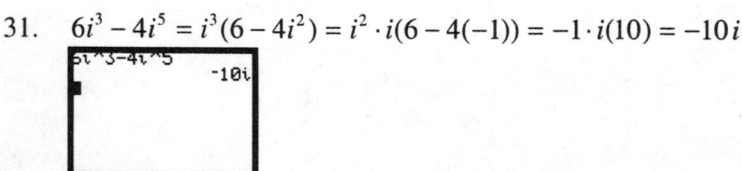

33. $(1+i)^3 = (1+i)(1+i)(1+i) = (1+2i+i^2)(1+i) = (1+2i-1)(1+i) = 2i(1+i)$
$$= 2i + 2i^2 = 2i + 2(-1) = -2 + 2i$$

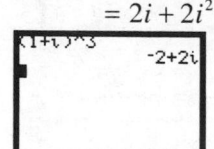

35. $i^7(1+i^2) = i^7(1+(-1)) = i^7(0) = 0$

37. $i^6 + i^4 + i^2 + 1 = \left(i^2\right)^3 + \left(i^2\right)^2 + i^2 + 1 = (-1)^3 + (-1)^2 + (-1) + 1 = -1 + 1 - 1 + 1 = 0$

39. $\sqrt{-4} = 2i$ 41. $\sqrt{-25} = 5i$

43. $\sqrt{(3+4i)(4i-3)} = \sqrt{12i - 9 + 16i^2 - 12i} = \sqrt{-9 + 16(-1)} = \sqrt{-25} = 5i$

45. $x^2 + 4 = 0$
$$a = 1, b = 0, c = 4, \quad b^2 - 4ac = 0^2 - 4(1)(4) = -16$$
$$x = \frac{-0 \pm \sqrt{-16}}{2(1)} = \frac{\pm 4i}{2} = \pm 2i$$
The solution set is $\{\pm 2i\}$.

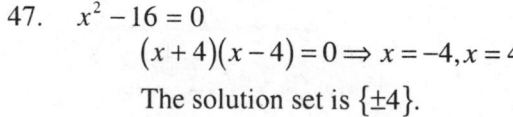

47. $x^2 - 16 = 0$
$$(x+4)(x-4) = 0 \Rightarrow x = -4, x = 4$$
The solution set is $\{\pm 4\}$.

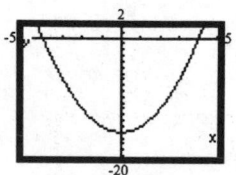

49. $x^2 - 6x + 13 = 0$
$a = 1, b = -6, c = 13,$
$$b^2 - 4ac = (-6)^2 - 4(1)(13) = 36 - 52 = -16$$
$$x = \frac{-(-6) \pm \sqrt{-16}}{2(1)} = \frac{6 \pm 4i}{2} = 3 \pm 2i$$
The solution set is $\{3 - 2i, 3 + 2i\}$.

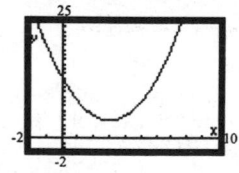

Appendix Review

51. $x^2 - 6x + 10 = 0$

    $a = 1, b = -6, c = 10$

    $b^2 - 4ac = (-6)^2 - 4(1)(10) = 36 - 40 = -4$

    $x = \dfrac{-(-6) \pm \sqrt{-4}}{2(1)} = \dfrac{6 \pm 2i}{2} = 3 \pm i$

    The solution set is $\{3 - i,\ 3 + i\}$.

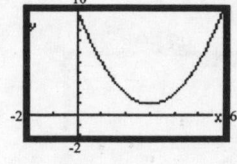

53. $8x^2 - 4x + 1 = 0$

    $a = 8, b = -4, c = 1$

    $b^2 - 4ac = (-4)^2 - 4(8)(1) = 16 - 32 = -16$

    $x = \dfrac{-(-4) \pm \sqrt{-16}}{2(8)} = \dfrac{4 \pm 4i}{16} = \dfrac{1}{4} \pm \dfrac{1}{4}i$

    The solution set is $\left\{ \dfrac{1}{4} - \dfrac{1}{4}i,\ \dfrac{1}{4} + \dfrac{1}{4}i \right\}$.

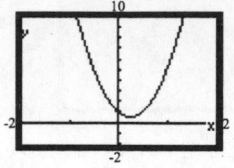

55. $5x^2 - 2x + 1 = 0$

    $a = 5, b = -2, c = 1$

    $b^2 - 4ac = (-2)^2 - 4(5)(1) = 4 - 20 = -16$

    $x = \dfrac{-(-2) \pm \sqrt{-16}}{2(5)} = \dfrac{2 \pm 4i}{10} = \dfrac{1}{5} \pm \dfrac{2}{5}i$

    The solution set is $\left\{ \dfrac{1}{5} - \dfrac{2}{5}i,\ \dfrac{1}{5} + \dfrac{2}{5}i \right\}$.

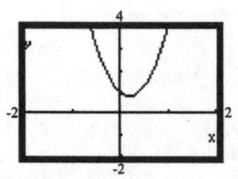

57. $x^2 + x + 1 = 0$

    $a = 1, b = 1, c = 1, \quad b^2 - 4ac = 1^2 - 4(1)(1) = 1 - 4 = -3$

    $x = \dfrac{-1 \pm \sqrt{-3}}{2(1)} = \dfrac{-1 \pm \sqrt{3}\,i}{2} = -\dfrac{1}{2} \pm \dfrac{\sqrt{3}}{2}i$

    The solution set is $\left\{ -\dfrac{1}{2} - \dfrac{\sqrt{3}}{2}i,\ -\dfrac{1}{2} + \dfrac{\sqrt{3}}{2}i \right\}$.

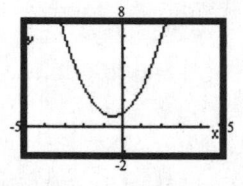

59. $x^3 - 8 = 0$

    $(x - 2)(x^2 + 2x + 4) = 0$

    $x - 2 = 0 \Rightarrow x = 2$

    $x^2 + 2x + 4 = 0$

    $a = 1, b = 2, c = 4$

    $b^2 - 4ac = 2^2 - 4(1)(4) = 4 - 16 = -12$

    $x = \dfrac{-2 \pm \sqrt{-12}}{2(1)} = \dfrac{-2 \pm 2\sqrt{3}\,i}{2} = -1 \pm \sqrt{3}i$

    The solution set is $\left\{ 2,\ -1 - \sqrt{3}i,\ -1 + \sqrt{3}i \right\}$.

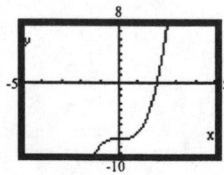

61. $x^4 - 16 = 0$

$(x^2 - 4)(x^2 + 4) = 0 \Rightarrow (x - 2)(x + 2)(x^2 + 4) = 0$

$x - 2 = 0 \Rightarrow x = 2$

$x + 2 = 0 \Rightarrow x = -2$

$x^2 + 4 = 0 \Rightarrow x = \pm 2i$

The solution set is $\{-2,\ 2,\ -2i,\ 2i\}$.

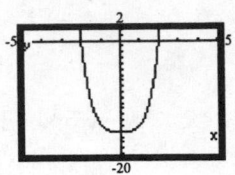

63. $x^4 + 13x^2 + 36 = 0$

$(x^2 + 9)(x^2 + 4) = 0$

$x^2 + 9 = 0 \Rightarrow x = \pm 3i$

$x^2 + 4 = 0 \Rightarrow x = \pm 2i$

The solution set is $\{-3i,\ 3i,\ -2i,\ 2i\}$.

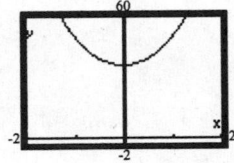

65. $3x^2 - 3x + 4 = 0$

$a = 3, b = -3, c = 4$

$b^2 - 4ac = (-3)^2 - 4(3)(4) = 9 - 48 = -39$

The equation has two complex conjugate solutions.

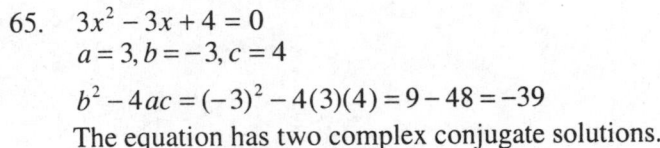
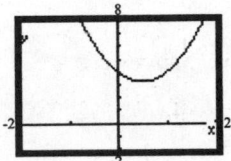

67. $2x^2 + 3x - 4 = 0$

$a = 2, b = 3, c = -4$

$b^2 - 4ac = 3^2 - 4(2)(-4) = 9 + 32 = 41$

The equation has two unequal real solutions.

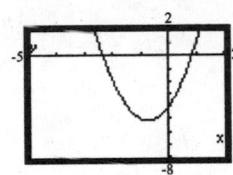

69. $9x^2 - 12x + 4 = 0$

$a = 9, b = -12, c = 4$

$b^2 - 4ac = (-12)^2 - 4(9)(4) = 144 - 144 = 0$

The equation has a repeated real solution.

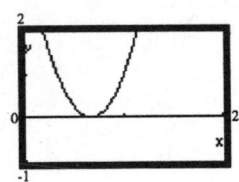

71. The other solution is the conjugate of $2 + 3i$, i.e. $2 - 3i$.

73. $z + \bar{z} = 3 - 4i + \overline{3 - 4i} = 3 - 4i + 3 + 4i = 6$

75. $z \cdot \bar{z} = (3 - 4i)(\overline{3 - 4i}) = (3 - 4i)(3 + 4i) = 9 + 12i - 12i - 16i^2 = 9 - 16(-1) = 25$

77. $z + \bar{z} = a + bi + \overline{a + bi} = a + bi + a - bi = 2a$

$z - \bar{z} = a + bi - (\overline{a + bi}) = a + bi - (a - bi) = a + bi - a + bi = 2bi$

Appendix Review

79. $\overline{z+w} = \overline{(a+bi)+(c+di)} = \overline{(a+c)+(b+d)i} = (a+c)-(b+d)i$

$= (a-bi)+(c-di) = \overline{a+bi} + \overline{c+di} = \overline{z} + \overline{w}$

81. – 83.   Answers will vary.

Appendix Review

A.8   Setting Up Equations; Applications

1.   Let A represent the area of the circle and r the radius.
The area of a circle is the product of π times the square of the radius. $A = \pi r^2$

3.   Let A represent the area of the square and s the length of a side.
The area of the square is the square of the length of a side. $A = s^2$

5.   Let F represent the force, m the mass, and a the acceleration.
Force equals the product of the mass times the acceleration. $F = ma$

7.   Let W represent the work, F the force, and d the distance.
Work equals force times distance. $W = Fd$

9. C = total variable cost in dollars, x = number of dishwashers manufactured.
$C = 150x$

11.

Amount in Bonds	Amount in CD's	Total
x	$x - 2000$	20,000

$x + x - 2000 = 20000$

$2x - 2000 = 20000$

$2x = 22000 \Rightarrow x = 11000$

$11,000 will be invested in bonds. $9,000 will be invested in CD's.

13.

David	Paige	Dan	Total
x	$\dfrac{3}{4}x$	$\dfrac{1}{2}x$	900,000

$x + \dfrac{3}{4}x + \dfrac{1}{2}x = 900{,}000$

$\dfrac{9}{4}x = 900{,}000$

$x = \dfrac{4}{9}(900{,}000) \Rightarrow x = 400{,}000$

David receives $400,000. Paige receives $300,000. Dan receives $200,000.

15.

	Dollars per hour	Number of hours worked	Money earned
Regular wage	x	40	$40x$
Overtime wage	$1.5x$	8	$(1.5x)(8)$

$40x + (1.5x)(8) = 442$ Laura's regular hourly wage is $8.50.

$$40x + 12x = 442 \Rightarrow 52x = 442 \Rightarrow x = \frac{442}{52} = 8.50$$

17. Let x represent the number of touchdowns scored.
Then $x - 2$ represents the number of extra points scored.

	No. scored	Point value	Points earned
touchdown	x	6	$6x$
extra point	$x - 2$	1	$(x-2)(1)$
safety	1	2	$(1)(2)$
field goal	2	3	$(2)(3)$

We know that the total points scored = 41, so we have
$$6x + x - 2 + 2 + 6 = 41$$
$$7x + 6 = 41$$
$$7x = 35$$
$$x = 5$$
The Bears scored 5 touchdowns.

19. l = length, w = width
$$2l + 2w = 60 \qquad \text{Perimeter } = 2l + 2w$$
$$l = w + 8 \qquad \text{The length is 8 more than the width.}$$
$$2(w + 8) + 2w = 60$$
$$2w + 16 + 2w = 60$$
$$4w + 16 = 60 \Rightarrow 4w = 44 \Rightarrow w = 11 \text{ feet}, \quad l = 19 \text{ feet}$$

21.   Let x represent the score on the final exam and construct the table

	Test1	Test2	Test3	Test4	Test5	Final Exam	Final Exam
score	80	83	71	61	95	x	x
weight	1/7	1/7	1/7	1/7	1/7	1/7	1/7

Compute the final average and set equal to 80.
$$\left(\frac{1}{7}\right)(80 + 83 + 71 + 61 + 95 + x + x) = 80$$

Now solve for x:
$$\left(\frac{1}{7}\right)(390 + 2x) = 80 \Rightarrow 390 + 2x = 560 \Rightarrow 2x = 170 \Rightarrow x = 85$$

Brooke needs to score an 85 on the final exam to get an average of 80 in the course.

23.   Let x represent the original price of the house.
Then $0.15x$ represents the reduction in the price of the house.
    original price – reduction = new price
$$x - 0.15x = 125{,}000$$
$$0.85x = 125{,}000$$
$$x = 147{,}058.82$$
The original price of the house was $147,058.82.
The amount of the savings is $0.15(\$147{,}058.82) = \$22{,}058.82$.

25.   Let x represent the price the bookstore pays for the book (publisher price).
Then $0.25x$ represents the mark up on the book.
The selling price of the book is $56.00.
    publisher price + mark up = selling price
$$x + 0.25x = 56.00$$
$$1.25x = 56.00$$
$$x = \frac{56.00}{1.25} = 44.80$$
The bookstore pays $44.80 for the book.

27.   Let x represent the amount of money invested in bonds.
Then $50{,}000 - x$ represents the amount of money invested in CD's.

	Principle	Rate	Time (yrs)	Interest
Bonds	x	0.15	1	$0.15x$
CD's	$50{,}000 - x$	0.07	1	$0.07(50{,}000 - x)$

Since the total interest is to be $6,000, we have:
$$0.15x + 0.07(50{,}000 - x) = 6{,}000$$
$$(100)(0.15x + 0.07(50{,}000 - x)) = (6{,}000)(100)$$
$$15x + 7(50{,}000 - x) = 600{,}000$$
$$15x + 350{,}000 - 7x = 600{,}000$$
$$8x + 350{,}000 = 600{,}000 \Rightarrow 8x = 250{,}000 \Rightarrow x = 31{,}250$$
$31,250 should be invested in bonds at 15% and $18,750 should be invested in CD's at 7%.

29.   Let x represent the amount of money loaned at 8%.
Then $12{,}000 - x$ represents the amount of money loaned at 18%.

	Principle	Rate	Time (yrs)	Interest
Loan at 8%	x	0.08	1	$0.08x$
Loan at 18%	$12{,}000 - x$	0.18	1	$0.18(12{,}000 - x)$

Since the total interest is to be $1,000, we have:
$$0.08x + 0.18(12{,}000 - x) = 1{,}000$$
$$(100)(0.08x + 0.18(12{,}000 - x)) = (1{,}000)(100)$$
$$8x + 18(12{,}000 - x) = 100{,}000$$
$$8x + 216{,}000 - 18x = 100{,}000$$
$$-10x + 216{,}000 = 100{,}000 \Rightarrow -10x = -116{,}000 \Rightarrow x = 11{,}600$$
$11,600 is loaned at 8% and $400 is loaned at 18%.

31. Let x represent the number of pounds of Earl Gray tea.
Then $100 - x$ represents the number of pounds of Orange Pekoe tea.

	No. of pounds	Price per pound	Total Value
Earl Gray	x	$5.00	$5x$
Orange Pekoe	$100 - x$	$3.00	$3(100 - x)$
Blend	100	$4.50	$4.50(100)$

$$5x + 3(100 - x) = 4.50(100)$$
$$5x + 300 - 3x = 450$$
$$2x + 300 = 450$$
$$2x = 150$$
$$x = 75$$

75 pounds of Earl Gray tea must be blended with 25 pounds of Orange Pekoe.

33. Let x represent the number of pounds of cashews.
Then $x + 60$ represents the number of pounds in the mixture.

	No. of pounds	Price per pound	Total Value
cashews	x	$4.00	$4x$
peanuts	60	$1.50	$1.50(60)$
mixture	$x + 60$	$2.50	$2.50(x + 60)$

$$4x + 1.50(60) = 2.50(x + 60)$$
$$4x + 90 = 2.50x + 150$$
$$1.5x = 60$$
$$x = 40$$

40 pounds of cashews must be added to the 60 pounds of peanuts.

35. Let x represent the number of ounces of pure water.

	No. of ounces	Conc. of water	Pure water
Pure water	x	1.00	x
40% solution	20	0.60	$(20)(0.60)$
30% solution	$20 + x$	0.70	$(20 + x)(0.70)$

$$x + (20)(0.60) = (20 + x)(0.70)$$
$$x + 12 = 14 + 0.70x$$
$$0.3x = 2$$
$$x \approx 6.67$$

Approximately 6.67 ounces of pure water should be added.

37. Let r represent the rate of the Metra commuter train.
Then $r + 50$ represents the rate of the Amtrak train.

	Rate	Time	Distance
Metra train	r	3	$3r$
Amtrak train	$r + 50$	1	$r + 50$

$$\text{Amtrak distance} = \text{Metra distance} - 10$$

$$r + 50 = 3r - 10$$

$$60 = 2r \Rightarrow r = 30$$

The Metra commuter train travels at a rate of 30 miles per hour.
The Amtrak train travels at a rate of 80 miles per hour.

39. Let r represent the speed of the current.

	Rate	Time	Distance
Upstream	$16 - r$	$\dfrac{20}{60} = \dfrac{1}{3}$	$\dfrac{16 - r}{3}$
Downstream	$16 + r$	$\dfrac{15}{60} = \dfrac{1}{4}$	$\dfrac{16 + r}{4}$

Since the distance is the same in each direction:

$$\frac{16 - r}{3} = \frac{16 + r}{4}$$

$$4(16 - r) = 3(16 + r)$$

$$64 - 4r = 48 + 3r \Rightarrow 16 = 7r \Rightarrow r = \frac{16}{7} \approx 2.286$$

The speed of the current is approximately 2.286 miles per hour.

41. Let t represent the time it takes to do the job together.

	Time to do job	Part of job done in one minute
Trent	30	$\dfrac{1}{30}$
Lois	20	$\dfrac{1}{20}$
Together	t	$\dfrac{1}{t}$

$$\frac{1}{30} + \frac{1}{20} = \frac{1}{t} \Rightarrow 2t + 3t = 60 \Rightarrow 5t = 60 \Rightarrow t = 12$$

Working together, the job can be done in 12 minutes.

43. Let t represent the time the auxiliary pump needs to run.

	Time to do job alone	Part of job done in one hour	Time on Job	Part of total job done by each pump
Main Pump	4	$\dfrac{1}{4}$	3	$\dfrac{3}{4}$
Auxiliary Pump	9	$\dfrac{1}{9}$	t	$\dfrac{1}{9}t$

Since the two pumps are emptying one tanker, we have:

$$\frac{3}{4} + \frac{1}{9}t = 1 \Rightarrow 27 + 4t = 36 \Rightarrow 4t = 9 \Rightarrow t = \frac{9}{4} = 2.25$$

The auxiliary pump must run for 2.25 hours. It must be started at 9:45 a.m.

45. Let w represent the width of window.
Then $l = w + 2$ represents the length of the window.
Since the area is 143 square feet, we have: $w(w + 2) = 143$

$$w^2 + 2w - 143 = 0 \Rightarrow (w + 13)(w - 11) = 0 \Rightarrow w = -13 \text{ which is not practical}$$

or $w = 11$

The width of the rectangular window is 11 feet and the length is 13 feet.

47. Let l represent the length of the rectangle.
Let w represent the width of the rectangle.
The perimeter is 26 meters and the area is 40 square meters.

$$2l + 2w = 26 \quad \Rightarrow \quad l + w = 13 \quad \Rightarrow \quad w = 13 - l$$

$$lw = 40$$

$$l(13 - l) = 40 \Rightarrow 13l - l^2 = 40 \Rightarrow l^2 - 13l + 40 = 0 \Rightarrow (l - 8)(l - 5) = 0$$

$$l = 8 \text{ or } l = 5$$

$$w = 5 \qquad w = 8$$

The dimensions are 5 meters by 8 meters.

49. Let r represent the speed of the current.

	Rate	Time	Distance
Upstream	$15 - r$	$\dfrac{10}{15 - r}$	10
Downstream	$15 + r$	$\dfrac{10}{15 + r}$	10

Since the total time is 1.5 hours, we have:

$$\frac{10}{15 - r} + \frac{10}{15 + r} = 1.5$$

$$10(15 + r) + 10(15 - r) = 1.5(15 - r)(15 + r)$$

$$150 + 10r + 150 - 10r = 1.5(225 - r^2)$$

$$300 = 1.5(225 - r^2)$$

$$200 = 225 - r^2 \Rightarrow r^2 - 25 = 0$$

$$(r - 5)(r + 5) = 0 \Rightarrow r = 5 \text{ or } r = -5$$

The speed of the current is 5 miles per hour.

51. l = length of the garden
w = width of the garden

(a) The length of the garden is to be twice its width. Thus, $l = 2w$.
The dimensions of the fence are $l + 4$ and $w + 4$.
The perimeter is 46 feet, so:

$$2(l + 4) + 2(w + 4) = 46$$

$$2(2w + 4) + 2(w + 4) = 46$$

$$4w + 8 + 2w + 8 = 46$$

$$6w + 16 = 46$$

$$6w = 30 \Rightarrow w = 5$$

The dimensions of the garden are 5 feet by 10 feet.

(b)   Area $= l \cdot w = 5 \cdot 10 = 50$ square feet

(c)   If the dimensions of the garden are the same, then the length and width of the fence are also the same $(l + 4)$. The perimeter is 46 feet, so:

$$2(l + 4) + 2(l + 4) = 46$$

$$2l + 8 + 2l + 8 = 46$$

$$4l + 16 = 46$$

$$4l = 30$$

$$l = 7.5$$

The dimensions of the garden are 7.5 feet by 7.5 feet.

(d)   Area $= l \cdot w = 7.5(7.5) = 56.25$ square feet.

53.   Let x represent the width of the border measured in feet.

The radius of the pool is 5 feet.

Then $x + 5$ represents the radius of the circle, including both the pool and the border.

The total area of the pool and border is $A_T = \pi(x + 5)^2$.

The area of the pool is $A_P = \pi(5)^2 = 25\pi$.

The area of the border is $A_B = A_T - A_P = \pi(x + 5)^2 - 25\pi$.

Since the concrete is 3 inches or 0.25 feet thick, the volume of the concrete in the border is

$$0.25A_B = 0.25\left(\pi(x + 5)^2 - 25\pi\right)$$

Solving the volume equation:

$$0.25\left(\pi(x + 5)^2 - 25\pi\right) = 27 \Rightarrow \pi\left(x^2 + 10x + 25 - 25\right) = 108 \Rightarrow \pi x^2 + 10\pi x - 108 = 0$$

$$x = \frac{-10\pi \pm \sqrt{(10\pi)^2 - 4(\pi)(-108)}}{2(\pi)}$$

$$= \frac{-31.42 \pm \sqrt{2344.1285}}{6.28} = \frac{-31.42 \pm 48.42}{6.28} = 2.71 \text{ or } -12.71$$

The width of the border is approximately 2.71 feet.

55.   Let x represent the width of the border measured in feet.

The total area is $A_T = (6 + 2x)(10 + 2x)$.

The area of the garden is $A_G = 6 \cdot 10 = 60$.

The area of the border is $A_B = A_T - A_G = (6 + 2x)(10 + 2x) - 60$.

Since the concrete is 3 inches or 0.25 feet thick, the volume of the concrete in the border is

$$0.25A_B = 0.25((6 + 2x)(10 + 2x) - 60)$$

Solving the volume equation:

$$0.25((6 + 2x)(10 + 2x) - 60) = 27$$

$$60 + 32x + 4x^2 - 60 = 108$$

$$4x^2 + 32x - 108 = 0 \Rightarrow x^2 + 8x - 27 = 0$$

$$x = \frac{-8 \pm \sqrt{8^2 - 4(1)(-27)}}{2(1)} = \frac{-8 \pm \sqrt{172}}{2}$$

$$= \frac{-8 \pm 13.11}{2} = 2.56 \text{ or } -10.56 \text{ which is not practical.}$$

The width of the border is approximately 2.56 feet.

57. Let x represent the number of ounces of pure water.
Then $x+1$ represents the number of gallons in the 60% solution.

	No. of gallons	Conc. of Antifreeze	Pure Antifreeze
water	x	0	0
100% antifreeze	1	1.00	1(1)
60% antifreeze	$x+1$	0.60	$0.60(x+1)$

$$0 + 1(1) = 0.60(x+1)$$

$$1 = 0.6x + 0.6 \Rightarrow 0.4 = 0.6x \Rightarrow x = \frac{4}{6} = \frac{2}{3}$$

$\frac{2}{3}$ gallon of pure water should be added.

59. Let x represent the number of pounds of pure cement.
Then $x+20$ represents the number of pounds in the 40% mixture.

	No. of pounds	Conc. of Cement	Pure Cement
Pure Cement	x	1.00	x
25% Cement	20	0.25	0.25(20)
40% Cement	$x+20$	0.40	$0.40(x+20)$

$$x + 0.25(20) = 0.40(x+20)$$

$$x + 5 = 0.4x + 8 \Rightarrow 0.6x = 3 \Rightarrow x = \frac{30}{6} = 5$$

5 pounds of pure cement should be added.

61. Let x represent the number of centimeters the length and width should be reduced.
$12 - x = $ the new length, $7 - x = $ the new width.
The new volume is 90% of the old volume.

$$(12-x)(7-x)(3) = 0.9(12)(7)(3)$$

$$3x^2 - 57x + 252 = 226.8 \Rightarrow 3x^2 - 57x + 25.2 = 0 \Rightarrow x^2 - 19x + 8.4 = 0$$

$$x = \frac{-(-19) \pm \sqrt{(-19)^2 - 4(1)(8.4)}}{2(1)} = \frac{19 \pm \sqrt{327.4}}{2}$$

$$= \frac{19 \pm 18.09}{2} = 0.45 \text{ or } 18.55$$

Since 18.55 exceeds the dimensions, it is discarded.
The dimensions of the new chocolate bar are: 11.55 cm by 6.55 cm by 3 cm.

63. Let x represent the number of grams of pure gold.
Then $60 - x$ represents the number of grams of 12 karat gold to be used.

	No. of grams	Conc. of gold	Pure gold
Pure gold	x	1.00	x
12 karat gold	$60 - x$	$\frac{1}{2}$	$\frac{1}{2}(60-x)$
16 karat gold	60	$\frac{2}{3}$	$\frac{2}{3}(60)$

$$x + \frac{1}{2}(60-x) = \frac{2}{3}(60) \Rightarrow x + 30 - 0.5x = 40 \Rightarrow 0.5x = 10 \Rightarrow x = 20$$

20 grams of pure gold should be mixed with 40 grams of 12 karat gold.

65. Let t represent the time it takes for Mike to catch up with Dan.

	Time to run mile	Time	Part of mile run in one minute	Distance
Mike	6	t	$\dfrac{1}{6}$	$\dfrac{1}{6}t$
Dan	9	$t+1$	$\dfrac{1}{9}$	$\dfrac{1}{9}(t+1)$

Since the distances are the same, we have:
$$\frac{1}{6}t = \frac{1}{9}(t+1) \Rightarrow 3t = 2t+2 \Rightarrow t = 2$$

Mike will pass Dan after 2 minutes, which is a distance of $\dfrac{1}{3}$ mile.

67. Let t represent the time for the tub to fill with the faucets on and the stopper removed.

	Time to do job alone	Part of job done in one minute	Time on Job	Part of total job done by each
Faucets open	15	$\dfrac{1}{15}$	t	$\dfrac{t}{15}$
Stopper removed	20	$-\dfrac{1}{20}$	t	$-\dfrac{t}{20}$

Since one tub is being filled, we have:
$$\frac{t}{15} + \left(-\frac{t}{20}\right) = 1$$

$$4t - 3t = 60 \qquad \therefore \ 60 \text{ minutes is required to fill the tub.}$$

$$t = 60$$

69. Let x represent the amount of money invested in a CD.
Then $100,000 - x$ represents the amount of money invested in the bond.

	Principle	Rate	Time (yrs)	Interest
CD	x	0.09	1	$0.09x$
Bond	$100,000 - x$	0.12	1	$0.12(100,000 - x)$

Since the total interest is to be $0.10(100,000) = \$10,000$, we have:
$$0.09x + 0.12(100,000 - x) = 10,000$$
$$9x + 12(100,000 - x) = 1,000,000$$
$$9x + 1,200,000 - 12x = 1,000,000$$
$$-3x + 1,200,000 = 1,000,000$$
$$-3x = -200,000$$
$$x = 66,667$$

The most money that can be invested in the CD is $66,667 to ensure that the loan payment is made.

71.   Let t_1 and t_2 represent the times for the two segments of the trip.

	Rate	Time	Distance
Chicago to Atlanta	45	t_1	$45t_1$
Atlanta to Miami	55	t_2	$55t_2$

Since Atlanta is halfway between Chicago and Miami, the distances are equal.

$$45t_1 = 55t_2 \implies t_1 = \frac{55}{45}t_2 = \frac{11}{9}t_2$$

Computing the average speed:

$$\text{Avg Speed} = \frac{\text{Distance}}{\text{Time}} = \frac{45t_1 + 55t_2}{t_1 + t_2} = \frac{45\left(\frac{11}{9}t_2\right) + 55t_2}{\frac{11}{9}t_2 + t_2}$$

$$= \frac{55t_2 + 55t_2}{\left(\frac{11t_2 + 9t_2}{9}\right)} = \frac{110t_2}{\left(\frac{20t_2}{9}\right)} = \frac{990t_2}{20t_2} = \frac{99}{2} = 49.5 \text{ miles per hour}$$

The average speed for the trip from Chicago to Miami is 49.5 miles per hour.

73.   Let x be the original selling price of the shirt.

Profit = Revenue − Cost

$$4 = x - 0.40x - 20 \implies 24 = 0.60x \implies x = 40$$

The original price should be $40 to ensure a profit of $4 after the sale.

If the sale is 50% off, the profit is:

$$40 - 0.50(40) - 20 = 40 - 20 - 20 = 0$$

At 50% off there will be no profit.

75.   It is impossible to mix two solutions with a lower concentration and end up with a new solution with a higher concentration.

Appendix Review

A.9 *n*th Roots; Rational Exponents

1. $\sqrt[3]{27} = 3$

3. $\sqrt[3]{-8} = -2$

5. $\sqrt{8} = \sqrt{4 \cdot 2} = 2\sqrt{2}$

7. $\sqrt[3]{-8x^4} = \sqrt[3]{-8 \cdot x^3 \cdot x} = -2x\sqrt[3]{x}$

9. $\sqrt[4]{x^{12}y^8} = \sqrt[4]{\left(x^3\right)^4\left(y^2\right)^4} = x^3 y^2$

11. $\sqrt[4]{\dfrac{x^9 y^7}{x y^3}} = \sqrt[4]{x^8 y^4} = x^2 y$

13. $\sqrt{36x} = 6\sqrt{x}$

15. $\sqrt{3x^2}\sqrt{12x} = \sqrt{36x^2 \cdot x} = 6x\sqrt{x}$

17. $\left(\sqrt{5}\;\sqrt[3]{9}\right)^2 = 5\left(\sqrt[3]{81}\right) = 5\left(\sqrt[3]{27 \cdot 3}\right) = 5 \cdot 3\left(\sqrt[3]{3}\right) = 15\sqrt[3]{3}$

19. $\left(3\sqrt{6}\right)\left(2\sqrt{2}\right) = 6\sqrt{12} = 6\sqrt{4 \cdot 3} = 12\sqrt{3}$

21. $\left(\sqrt{3} + 3\right)\left(\sqrt{3} - 1\right) = \left(\sqrt{3}\right)^2 - \sqrt{3} + 3\sqrt{3} - 3 = 3 + 2\sqrt{3} - 3 = 2\sqrt{3}$

23. $\left(\sqrt{x} - 1\right)^2 = \left(\sqrt{x}\right)^2 - 2\sqrt{x} + 1 = x - 2\sqrt{x} + 1$

25. $3\sqrt{2} - 4\sqrt{8} = 3\sqrt{2} - 4\sqrt{4 \cdot 2} = 3\sqrt{2} - 8\sqrt{2} = -5\sqrt{2}$

27. $\sqrt[3]{16x^4} - \sqrt[3]{2x} = \sqrt[3]{8 \cdot 2 \cdot x^3 \cdot x} - \sqrt[3]{2x} = 2x\sqrt[3]{2x} - \sqrt[3]{2x} = (2x - 1)\sqrt[3]{2x}$

29. $\dfrac{1}{\sqrt{2}} \cdot \dfrac{\sqrt{2}}{\sqrt{2}} = \dfrac{\sqrt{2}}{2}$

31. $\dfrac{-\sqrt{3}}{\sqrt{5}} \cdot \dfrac{\sqrt{5}}{\sqrt{5}} = \dfrac{-\sqrt{15}}{5} = -\dfrac{\sqrt{15}}{5}$

33. $\dfrac{\sqrt{3}}{5 - \sqrt{2}} \cdot \dfrac{5 + \sqrt{2}}{5 + \sqrt{2}} = \dfrac{\sqrt{3}\left(5 + \sqrt{2}\right)}{25 - 2} = \dfrac{\sqrt{3}\left(5 + \sqrt{2}\right)}{23}$

35. $\dfrac{2 - \sqrt{5}}{2 + 3\sqrt{5}} \cdot \dfrac{2 - 3\sqrt{5}}{2 - 3\sqrt{5}} = \dfrac{4 - 6\sqrt{5} - 2\sqrt{5} + 3\left(\sqrt{5}\right)^2}{4 - 9\left(\sqrt{5}\right)^2} = \dfrac{4 - 8\sqrt{5} + 3 \cdot 5}{4 - 9 \cdot 5}$

$$= \dfrac{4 - 8\sqrt{5} + 15}{4 - 45} = \dfrac{19 - 8\sqrt{5}}{-41} = \dfrac{-19 + 8\sqrt{5}}{41}$$

Appendix Review

37. $\dfrac{5}{\sqrt[3]{2}} \cdot \dfrac{\left(\sqrt[3]{2}\right)^2}{\left(\sqrt[3]{2}\right)^2} = \dfrac{5\left(\sqrt[3]{2}\right)^2}{\left(\sqrt[3]{2}\right)^3} = \dfrac{4\left(\sqrt[3]{2^2}\right)}{2} = \dfrac{5\sqrt[3]{4}}{2}$

39. $\dfrac{\sqrt{x+h}-\sqrt{x}}{\sqrt{x+h}+\sqrt{x}} \cdot \dfrac{\sqrt{x+h}-\sqrt{x}}{\sqrt{x+h}-\sqrt{x}} = \dfrac{\left(\sqrt{x+h}\right)^2 - 2\sqrt{x}\cdot\sqrt{x+h} + \left(\sqrt{x}\right)^2}{\left(\sqrt{x+h}\right)^2 - \left(\sqrt{x}\right)^2}$

$= \dfrac{x+h-2\sqrt{x(x+h)}+x}{x+h-x} = \dfrac{2x+h-2\sqrt{x(x+h)}}{h}$

41. $8^{2/3} = \left(2^3\right)^{2/3} = 2^2 = 4$

43. $(-27)^{1/3} = \left((-3)^3\right)^{1/3} = -3$

45. $16^{3/2} = \left(2^4\right)^{3/2} = 2^6 = 64$

47. $9^{-3/2} = \left(3^2\right)^{-3/2} = 3^{-3} = \dfrac{1}{3^3} = \dfrac{1}{27}$

49. $\left(\dfrac{9}{8}\right)^{3/2} = \left(\dfrac{3^2}{2^3}\right)^{3/2} = \dfrac{3^{6/2}}{2^{9/2}} = \dfrac{3^3}{2^{8/2}\cdot 2^{1/2}} = \dfrac{27}{2^4\cdot 2^{1/2}} = \dfrac{27}{16\cdot\sqrt{2}} \cdot \dfrac{\sqrt{2}}{\sqrt{2}} = \dfrac{27\sqrt{2}}{16\cdot 2} = \dfrac{27\sqrt{2}}{32}$

51. $\left(\dfrac{8}{9}\right)^{-3/2} = \left(\dfrac{2^3}{3^2}\right)^{-3/2} = \dfrac{2^{-9/2}}{3^{-6/2}} = \dfrac{2^{-8/2}\cdot 2^{-1/2}}{3^{-3}} = \dfrac{2^{-4}\cdot 2^{-1/2}}{3^{-3}} = \dfrac{27}{16\cdot\sqrt{2}} \cdot \dfrac{\sqrt{2}}{\sqrt{2}} = \dfrac{27\sqrt{2}}{16\cdot 2} = \dfrac{27\sqrt{2}}{32}$

53. $x^{3/4}\cdot x^{1/3}\cdot x^{-1/2} = x^{3/4+1/3-1/2} = x^{(9+4-6)/12} = x^{7/12}$

55. $\left(x^3 y^6\right)^{1/3} = \left(x^3\right)^{1/3}\left(y^6\right)^{1/3} = x\,y^2$

57. $\left(x^2 y\right)^{1/3}\left(x y^2\right)^{2/3} = x^{2/3}y^{1/3}x^{2/3}y^{4/3} = x^{4/3}y^{5/3}$

59. $\left(16x^2 y^{-1/3}\right)^{3/4} = \left(2^4 x^2 y^{-1/3}\right)^{3/4} = 2^3 x^{3/2} y^{-1/4} = \dfrac{8x^{3/2}}{y^{1/4}}$

61. $\dfrac{x}{(1+x)^{1/2}} + 2(1+x)^{1/2} = \dfrac{x + 2(1+x)^{1/2}(1+x)^{1/2}}{(1+x)^{1/2}} = \dfrac{x+2(1+x)}{(1+x)^{1/2}} = \dfrac{x+2+2x}{(1+x)^{1/2}} = \dfrac{3x+2}{(1+x)^{1/2}}$

63. $\dfrac{\left(\sqrt{1+x}-x\cdot\dfrac{1}{2\sqrt{1+x}}\right)}{1+x} = \dfrac{\left(\sqrt{1+x}-\dfrac{x}{2\sqrt{1+x}}\right)}{1+x} = \dfrac{\left(\dfrac{2\sqrt{1+x}\sqrt{1+x}-x}{2\sqrt{1+x}}\right)}{1+x}$

$= \dfrac{2(1+x)-x}{2(1+x)^{1/2}} \cdot \dfrac{1}{1+x} = \dfrac{2+x}{2(1+x)^{3/2}}$

65. $\dfrac{(x+4)^{1/2}-2x(x+4)^{-1/2}}{x+4}=\dfrac{\left((x+4)^{1/2}-\dfrac{2x}{(x+4)^{1/2}}\right)}{x+4}=\dfrac{\left((x+4)^{1/2}\cdot\dfrac{(x+4)^{1/2}}{(x+4)^{1/2}}-\dfrac{2x}{(x+4)^{1/2}}\right)}{x+4}$

$\qquad =\dfrac{\left(\dfrac{x+4-2x}{(x+4)^{1/2}}\right)}{x+4}=\dfrac{-x+4}{(x+4)^{1/2}}\cdot\dfrac{1}{x+4}=\dfrac{-x+4}{(x+4)^{3/2}}$

67. $(x+1)^{3/2}+x\cdot\dfrac{3}{2}(x+1)^{1/2}=(x+1)^{1/2}\left(x+1+\dfrac{3}{2}x\right)=(x+1)^{1/2}\left(\dfrac{5}{2}x+1\right)=\dfrac{1}{2}(x+1)^{1/2}(5x+2)$

69. $6x^{1/2}\left(x^2+x\right)-8x^{3/2}-8x^{1/2}=2x^{1/2}\left(3(x^2+x)-4x-4\right)=2x^{1/2}\left(3x^2-x-4\right)$

$\qquad =2x^{1/2}(3x-4)(x+1)$

71. $x\left(\dfrac{1}{2}\right)\left(8-x^2\right)^{-1/2}(-2x)+\left(8-x^2\right)^{1/2}=-x^2\left(8-x^2\right)^{-1/2}+\left(8-x^2\right)^{1/2}=\left(8-x^2\right)^{-1/2}\left[-x^2+\left(8-x^2\right)\right]$

$\qquad =\left(8-x^2\right)^{-1/2}\left[-x^2+8-x^2\right]=\left(8-x^2\right)^{-1/2}\left[8-2x^2\right]$

$\qquad =\dfrac{8-2x^2}{\left(8-x^2\right)^{1/2}}=\dfrac{2\left(4-x^2\right)}{\left(8-x^2\right)^{1/2}}=\dfrac{2(2-x)(2+x)}{\left(8-x^2\right)^{1/2}}$